AF598622

ENCYCLOPEDIA
OF
GENETICS

INSTRUCTIONS ON GAINING ACCESS TO THE ONLINE VERSION OF THE ENCYCLOPEDIA OF GENETICS:

Access for a limited period to the online version of the Encyclopedia of Genetics is included in the purchase price of the print edition.

This online version has been uniquely and persistently identified by the Digital Object Identifier (DOI)

10.1006/rwgn.2001

By following the link

http://dx.doi.org/10.1006/rwgn.2001

from any Web Browser, buyers of the Encyclopedia of Genetics will find instructions on how to register for access.

If you have any problems with accessing the online version, e-mail:
idealreferenceworks@harcourt.com

ENCYCLOPEDIA OF GENETICS

EDITORS-IN-CHIEF

Sydney Brenner
Jeffrey H. Miller

ASSOCIATE EDITORS

William J. Broughton
Malcolm Ferguson-Smith
Walter Fitch
Nigel D. F. Grindley
Daniel L. Hartl
Jonathan Hodgkin
Charles Kurland
Elizabeth Kutter
Terry H. Rabbitts
Ira Schildkraut
Lee Silver
Gerald R. Smith
Ronald L. Somerville

ACADEMIC PRESS
A Harcourt Science and Technology Company
San Diego San Francisco New York Boston London Sydney Tokyo

This book is printed on acid-free paper.

Academic Press
A division of Harcourt, Inc.
Harcourt Place, 32 Jamestown Road, London NW1 7BY, UK
http://www.academicpress.com

Academic Press
A division of Harcourt, Inc.
525 B Street, Suite 1900, San Diego, California 92101-4495, USA
http://www.academicpress.com

ISBN 0-12-227080-0

Library of Congress Catalog Number: 2001089059
A catalogue record for this book is available from the British Library

Access for a limited period to the online version of the *Encyclopedia of Genetics* is included in the purchase price of the print edition.

This online version has been uniquely and persistently identified by the Digital Object Identifier (DOI)

10.1006/rwgn.2001

By following the link

http://dx.doi.org/10.1006/rwgn.2001

from any Web Browser, buyers of the *Encyclopedia of Genetics* will find instructions on how to register for access.

If you have any problems with accessing the online version, e-mail
idealreferenceworks@harcourt.com

Typeset by Kolam Information Services Pvt Limited, Pondicherry, India
Printed and bound in Spain by Grafos SA Arte Sobre Papel, Barcelona

01 02 03 04 05 06 GF 9 8 7 6 5 4 3 2 1

EDITORS-IN-CHIEF

Sydney Brenner
Molecular Sciences Institute Inc
2168 Shattuck Avenue
Berkeley
CA 94704
USA

Jeffrey H. Miller
University of California, Los Angeles
Department of Microbiology and
Molecular Genetics
405 Hilgard Avenue
Los Angeles
CA 90024
USA

ASSOCIATE EDITORS

William J. Broughton
LBMPS, Université de Genève
Plant Molecular Biology
1 Chemin de l'Imperatrice
CH-1292 Chambésy
Genève
Switzerland

Malcolm Ferguson-Smith
University of Cambridge
Department of Clinical Veterinary Medicine
Madingley Road
Cambridge
CB3 0ES
UK

Walter Fitch
University of California, Irvine
Department of Ecology and Evolutionary Biology
321 Steinhaus Hall
Irvine
CA 92697
USA

Nigel D. F. Grindley
Yale University
Department of Molecular Biophysics and
Biochemistry
PO Box 208144
New Haven
CT 06520-8114
USA

Daniel L. Hartl
Harvard University
Department of Organismic and
Evolutionary Biology
16 Divinity Avenue
Cambridge
MA 02138
USA

Jonathan Hodgkin
University of Oxford
Genetics Unit
Department of Biochemistry
South Parks Road
Oxford
OX1 3QU
UK

Charles Kurland
Uppsala University
Munkarpsv. 21
SE 243 32, Hoor
Sweden

Elizabeth Kutter
The Evergreen State College
Lab 1
Olympia
WA 98505
USA

Terry H. Rabbitts
MCR Laboratory of Molecular Biology
Division of Protein and Nucleic Acid Chemistry
Hills Road
Cambridge
CB2 2QH
UK

Ira Schildkraut
New England Biolabs Inc.
32 Tozer Road
Beverly
MA 01915
USA

Lee Silver
Princeton University
Department of Molecular Biology
Princeton
NJ 08544-1014
USA

Gerald R. Smith
Fred Hutchinson Cancer Research Center
1100 Fairview Avenue North, A1-162
Seattle
WA 98104-1024
USA

Ronald L. Somerville
Purdue University
Department of Biochemistry
West Lafayette
IN 47907-1153
USA

Preface

In the spring of 1953, shortly after Watson and Crick's discovery of the double-helical structure of DNA, I found myself at dinner next to the famous geneticist R.A. Fisher. When I asked him what he thought would be that structure's implication for genetics, he replied firmly "None!" He may have taken this myopic view, because to him genetics was an abstract mathematical subject with laws that were independent of the physical nature of genes. At the time, it was an esoteric subject taught only to a few biologists and regarded as largely irrelevant to medicine. Archibald Garrod's seminal book *Inborn Errors of Metabolism* had made little impact, perhaps because no-one knew what genes were made of or how mutations acted.

The first recognition that mutations act on proteins came in 1949 when Linus Pauling and his collaborators published their paper on "Sickle cell anemia, a molecular disease." They found that patients suffering from this recessively inherited disease had an abnormal hemoglobin that differed from the normal form by the elimination of two negative charges. Eight years later, Vernon Ingram showed that this was due to the replacement of a single glutamic acid residue in each of the identical half-molecules of hemoglobin by a valine. Ingram's discovery was the first specific evidence of the chemical effect of a mutation. It marked the birth of molecular genetics and molecular medicine and it started the transformation of genetics to its central position in molecular biology, biochemistry, and medicine today. Even so, people who have studied genetics as part of their curriculum may not have become familiar with all the hundreds of specialized terms that geneticists have coined. To them and many others at the periphery of genetics this *Encyclopedia* will prove most useful.

Like the *Encyclopedia Britannica* it is a cross between a dictionary and a text book. Hybrids are defined in three lines, while Jonathan Hodgkin's brilliant exposition of past and present research on *Drosophila's* rival, the minute nematode worm *Caenorhabditis elegans*, occupies nearly six pages. This entry is intelligible to non-specialists, but that is not true of some others that were "Greek" to me.

The *Encyclopedia* includes biographies of many of the pioneers, from Gregor Mendel to Ernst Mayr, the great evolutionist who actually contributed several entries. Many of the other contributors are also pioneers, even though they are not yet old enough to figure among the biographies. Sydney Brenner has written the entry on the genetic code which he himself helped to discover 40 years ago; David Weatherall has written on thalassemia to whose exploration he has devoted a lifetime; Malcolm Ferguson-Smith has written on human chromosomal anomalies on which he is the world authority; he and some others made such varied contributions to their fields that they figure among many of the 1650-odd entries. Sydney Brenner, one of the two editors-in-chief, may be the only contributor who does not need the *Encyclopedia*.

What of the future of genetics? Its applications are likely to multiply, and many of the applicants coming from other fields will welcome the *Encyclopedia*. An increasing number of applications will be in medicine. I have heard predictions that in future every newborn child will have its genes screened and the results imprinted on a computer chip that he or she will carry for life. Recorded on it would be all genetic anomalies, susceptibilities to diseases, and intolerance of drugs. In case of illness or accident, that chip would activate an algorithm that automatically prescribes the correct treatment. Unfortunately for such utopias, medicine is more complex. Weatherall has shown elsewhere that the single recessive disease thalassemia is a multiplicity of different diseases and that the same genotype may give rise to widely different phenotypes, depending on environmental and other factors. This complexity arises even when single point mutations do not necessarily lead to disease, but merely to susceptibilities to disease, as in α_1-antitrypsin deficiency or with certain abnormal hemoglobins. The complexity is much greater still in multifactorial diseases like schizophrenia or diabetes. For many reasons good medicine will continue to require wide knowledge, mature judgement, empathy, and wisdom.

There have been glib predictions that the mapping of the human genome will allow most genetic diseases to be cured by either germline or somatic gene therapy, but the former is too risky and the latter is proving extremely difficult and costly. The risks of germline therapy are illustrated by the attempt to create a genetically modified monkey. Scientists injected the gene for the green fluorescent protein from a jellyfish into 222 monkey eggs. After fertilizing them with monkey sperm, they incubated them and implanted a pair each into the wombs of 20 surrogate monkey mothers. Only five of the implants resulted in pregnancies. Only three monkey babies were born and only one of them carried the jellyfish gene, but the monkey does not fluoresce, because the gene, though incorporated into its chromosomes, fails to be expressed. It will be argued that technical improvements may lead to a 100% success rate, but this kind of gene transfer has now been practised for some years in mice and other animals, and it has remained a haphazard affair that would be criminal to apply to humans. Human cloning carries similar risks.

After many failures, somatic gene therapy of a potentially fatal human genetic disease recently succeeded for the first time. A French team cured two baby boys of severe combined immunodeficiency (SCID)-XI disease. They infected cells extracted from the boys' bone marrow with cDNA containing the required gene coupled to a retrovirus-derived vector, and then re-injected them back into the boys' bone marrow. The therapy restored normal immune function that was still intact 10 months later. Another interesting development was the restoration of normal function to the muscle of a dystrophic mouse after injection of fragments of the giant muscular dystrophy gene. On the other hand, it has so far been impossible to cure some of the most common genetic diseases: thalassemia, sickle cell anemia, or cystic fibrosis, because it has proved extremely difficult to express the genes in the correct place in the patients' chromosomes and get them to express the required protein in sufficient quantities in the right tissues. If gene therapy has a bright future, it does not look as though it is just round the corner.

Before the completion of the Human Genome Project, identification of the genes for some human inherited diseases required truly heroic efforts. The search for the Huntington's disease gene occupied up to a hundred people for about 10 years. The same work could now be accomplished by few people in a fraction of the time. This is one of the Human Genome Project's important medical benefits. Others may be the rapid identification of promising new drug targets against diseases ranging from high blood pressure to a variety of cancers, the epidemiology of alleles linked to susceptibility to various diseases and improved basic understanding of human physiology and pathology.

Agriculture offers the greatest scope for applied genetics, but distrust of genetically modified foods has blinded the public to its potential benefits and its vital importance for the avoidance of widespread famines later in this century. Since the early 1960s, food available per head in the developing world has increased by 20% despite a doubling of the population. This outstanding success has been achieved by the introduction of crops improved by crossing and by intensive application of fertilizers, pesticides, and weed-killers. Even so, there are 800 million hungry people and 185 million seriously malnourished pre-school children in the developing world. It is unlikely that the methods that have raised cereal yields hitherto will allow them to be raised again sufficiently to reduce these distressing numbers. Since most fertile land is already intensively cultivated, scientists are trying to introduce genes into crops that would allow them to be grown on poorer soils and in harsher climates, and to make existing crops more nutritious. In the tropics, fungi, bacteria, and viruses still cause huge harvest losses. Scientists are trying to introduce genes that will confer resistance to some of these pests, enabling farmers to use fewer pesticides. Genetically modified plants offer our best hope of feeding a world population that is expected to double in the next 50 years. It will be tragic if the present outcry over genetically modified foods will discourage further research and development in this field. If this *Encyclopedia* helps to promote better public understanding of genetics, this might be the best remedy against irrational fears.

Max F. Perutz
MRC Laboratory of Molecular Biology
Hills Road, Cambridge
CB2 2QH
UK

Introduction

Genetics, the study of inheritance, is fundamental to all of biology. Living organisms are unique among all natural complex systems in that they contain within their genes an internal description encoded in the chemical text of DNA. It is this description and not the organism itself which is handed down from generation to generation and understanding how the genes work to specify the organism constitutes the science of genetics. Furthermore, this constancy is embedded in a vast range of diversity, from bacteria to ourselves, all having arisen by changes in the genes. Understanding evolution is also part of genetics, and an area which will benefit from our increasing ability to determine the complete DNA sequences of genomes. In some sense, these sequences contain a record of genome history and we have now learnt that many genes in our genomes can be found in other organisms, quite unlike us. Indeed some are much the same as those found in bacteria, and can be viewed as molecular fossils, preserved in our genomes.

Although "like begets like" must be one of the oldest observations of mankind, it was only in the 19th century that major scientific advances began. Charles Darwin put forward his theory of the origin of the species by natural selection but he lacked a credible theory of the mechanism of inheritance. He believed in blending inheritance which meant that variation would be continually removed and he was therefore compelled to introduce variation in each generation as an inherited acquired character. Gregor Mendel, working at the same time, discovered the laws of inheritance and showed how the characteristics of the organism could be accounted for by factors which specified them. Mendels' work was rediscovered in 1900 independently by Correns, de Vries and Tchermak and soon after this, Bateson coined the term "gene" for the Mendelian factors and called the science, "genetics."

During the first 50 years of the twentieth century, there was a stream of important discoveries in genetics. We came to understand the relation between genes and chromosomes, and the connection between recombination maps and the physical structure of chromosomes. However, what the genes were made of and what they did remained a mystery until 1953 when Watson and Crick proposed the double helical structure of DNA, which at one blow unified genetics and biochemistry and ushered in the modern era of molecular biology.

Genetics and especially the molecular approach to it is now a pervasive field covering all of biology. In the *Encyclopedia of Genetics*, we have tried to draw together the many strands of what is still a rapidly expanding field, to present a view of all of genetics. This has been a five year effort by over 700 expert authors from all around the world. We have tried to ensure that the breadth of the work has not compromised the depth of the articles, and we hope that readers will be able to find accurate and up-to-date information on all major topics of genetics. We have also included articles on the history of the field as well as the impact of the applications of genetics to medicine and agriculture.

When we began this work, the sequencing of complete genomes was still in its infancy and the sequencing of the human genome was thought to be far in the future. Technological advances and the concentration of resources have brought this to fruition this year, and genetics is a subject very much in the public eye. We hope that at least some of the articles will also be of value to those who are not professional biological or medical scientists, but want to discover more about this field. Many of the articles contain lists for further reading, and the online version of the *Encyclopedia* also includes hypertext links to original articles, abstracts, source items, databases and useful websites, so that readers can seamlessly search other appropriate literature.

We would like to acknowledge the efforts of the Associate Editors, who worked hard in commissioning individual contributors to prepare cutting-edge articles, and who reviewed and edited the manuscripts in a timely manner. Our thanks also go to the Publishers, Academic Press, and the outstanding staff for their commitment, resourcefulness and creative input; and in particular Tessa Picknett, Kate Handyside, Peronel Craddock and the production team, who helped to make this a reality.

Sydney Brenner, Jeffrey H. Miller
Editors-in-Chief

Guide to use

Structure of the *Encyclopedia*

The material in the *Encyclopedia* is arranged as a series of entries in alphabetical order. These range greatly in size, from glossary items, definitions, and short articles, through to major (five pages or more) full articles, as appropriate.

To help you realize the full potential of the material in the *Encyclopedia* we have provided several features to help you find the topic of your choice.

The Contents List

Your first point of reference will probably be the contents list. The complete contents list appearing in all four volumes will provide you with both the volume number and page number of the entry.

Scanning Through the Text

Alternatively you may choose to browse through a volume using the alphabetical order of the entries as your guide. To assist you in identifying your location within the *Encyclopedia*, running headlines indicate the first entry on left-hand pages and the last entry on right-hand pages.

Dummy Entries

You will find 'dummy entries' where obvious synonyms exist, occasional species names/common names are used, or for entries where a topic may be covered within another entry. For example: If you wished to locate the entry which discusses **DNA Code**, the dummy heading would lead you to the entry on **Genetic Code**. The dummy entry for **Fruit Fly** takes you to ***Drosophila melanogaster***.

Cross-References

All of the articles in the *Encyclopedia* have been extensively cross-referenced. To direct the reader to other entries on related topics, the section immediately before the References entitled *See also* in most articles lists related entry titles. For example, the entry **Genetic Code** includes the following cross-references: *See also*: **Adaptor Hypothesis, Codons, Variable Codons, Universal Genetic Code**.

Index

The complete index for the four volumes is provided at the back of the last volume. This will help you to find entries for specific topics that do not appear as such in the list of contents. Index entries differentiate between material that is a whole article, is part of an article, is a figure or contains data presented as a table. Further guidance on the use of the index is supplied on the opening page of the index.

Color Plates

The color figures for each volume have been grouped together in a plate section. The location of this section is cited at the end of the contents list.

Contributors

A full list of contributors and their addresses is given at the beginning of each volume.

Abbreviations

A adenine
aa amino acid
aa-tRNA amino acyl-tRNA
ABC ATP binding cassette
AD Alzheimer's disease
ADA adenosine deaminase
ADC adenocarcinoma
ADH alcohol dehydrogenase
ADP adenosine 5′-diphosphate
AFA acromegaloid facial appearance (syndrome)
AFB aflatoxin B
AFP alpha-fetoprotein
AHA acute hemolytic anemia
AIDS acquired immunodeficiency syndrome
AIL advanced intercross line
Ala alanine
ALL acute lymphoblastic leukemia
AMCA aminomethyl coumarin acetic acid
AMH anti-Müllerian hormone
AML acute myeloid leukemia
Amp ampicillin
AMP adenosine 5′-monophosphate
A_n polyadenylation
AP apyrimidinic (site)
APC adenomatous polyposis coli
APO- apolipoprotein-
APOBEC apo-B mRNA editing cytidine deaminase
APP amyloid precursor protein
ARF ADP-ribosylation factor
Arg arginine
ARS autonomous replication sequence
Asn asparagine
ASO allele specific oligonucleotide
Asp aspartic acid
AT ataxia telangiectasia
ATP adenosine 5′-triphosphate
BAC bacterial artificial chromosome
BER base excision repair
BIC Breast Cancer Information Corp
BIME bacterial interspersed mosaic element
BMD Becker muscular dystrophy
bp base pair
BR Balbiani rings
BS Bloom's syndrome
BSE bovine spongiform encephalopathy
BWS Beckwith–Wiedemann syndrome
C cytosine
C- carboxyl-
CAF chromatin assembly factor
cAMP cyclic AMP
CAP catabolite activator protein
CD campomelic dysplasia
CD circular dichroism
CDC cell division cycle
CDK cyclin-dependent kinase
cDNA complementary DNA
CDS coding sequence
CEN centromere
CF cystic fibrosis
CFTR cystic fibrosis transmembrane conductance regulator
CGH comparative genome hybridisation
CHO Chinese hamster ovary (cells)
CJD Creutzfeldt–Jacob disease
CL cutis laxa
cM centimorgan
CMD congenital muscular dystrophy
CML chronic myeloid leukemia
CMS cytoplasmic male sterility
CoA coenzyme A
Col colicin
COP coat protein
CPE cytoplasmic polyadenylation element
cR centiray
CR conserved region
CRC colorectal adenocarcinoma
Cre cyclization recombination
CREB cAMP response element binding factor
CRP cAMP receptor protein
CT cholera toxin
ctDNA chloroplast DNA
ctf cotransformation frequency
CTP cytidine triphosphate
CV coefficient of variation
CVS chorionic villus sampling
Cys cysteine
Da dalton
DI dentinogenesis imperfecta
DIC differential inference
DMD differentially methylated domain
DMD Duchenne muscular dystrophy
DMEM Dulbecco's Modified Eagle's Medium
DMI deviation from Mendelian inheritance
DMSO dimethylsulfoxide
dN deoxynucleotide
DNA deoxyribonucleic acid
DNP deoxyribonucleoprotein
dNTP deoxyribonucleotide triphosphate
ds double stranded
DSBR double strand break repair
EBN endosperm balanced number
EBV Epstein–Barr virus
ECM extracellular matrix
EDS Ehlers–Danlos syndrome
EF elongation factor
EGF epidermal growth factor
ELISA enzyme linked immunosorbant assay
EMBL European Molecular Biology Laboratories

EN	early nodule	HGMP	Human Genome Mapping Project
EN	endonuclease	His	histidine
ENU	N-ethyl-N-nitrosourea	HIV	human immunodeficiency virus
EP	early promotor	HLA	human leukocyte antigen
ER	endoplasmic reticulum	HLH	helix–loop–helix
eRF	eukaryal release factor	HMC	5′-hydroxymethyl-cytosine
ES	embryonic stem (cells)	HMG	high mobility group
ESI	electrospray ionization	HMW	high molecular weight
ESS	evolutionarily stable strategy	HNPCC	hereditary nonpolyposis colorectal cancer
EST	expressed sequence tagged		
F-factor	fertility-factor	hnRNP	heterogeneous nuclear RNP
FA	fluctuating asymmetry	HPLC	high-performance liquid chromatography
FACS	fluorescence-activated cell sorter		
FAD	flavin-adenine dinucleotide	HPV	human papillomavirus
$FADH_2$	reduced FAD	hsp	heatshock protein
FAK	focal adhesion kinase	HTH	helix–turn–helix
FAP	familial adenoma polyposis	HTLV	human T-cell leukemia virus
FDS	first-division segregation	HV-I, HV-II	hypervariable regions I, II
FFI	familial fatal insomnia	HW equilibrium	Hardy–Weinberg equilibrium
FGF	fibroblast growth factor	IBD	identical by descent
FH	familial hypercholesterolemia	IBS	identical by state
FIGE	field inversion gel electrophoresis	ICM	inner cell mass
FISH	fluorescent *in situ* hybridization	ICSI	intracytoplasmic sperm injection
FITC	fluorescein isothoicyanate	IES	internal eliminated sequences
FRDA	Friedreich's ataxia	IF	initiation factor
FRET	fluorescence energy resonance transfer	Ig	immunoglobulin
FSH	follicle stimulating hormone	IGF	insulin-like growth factor
FSHMD	facioscapulohumeral muscular dystrophy	IHF	integration host factor
		IL	interleukin
G	guanine	ILAR	Institute for Laboratory Animal Research
G-6-P	glucose-6-phosphate		
G-banding	Giemsa-banding	Ile	isoleucine
G-proteins	GTP-binding proteins	IMAC	immobilized metal ion affinity chromatography
GAP	GTPase activating protein		
GDB	genome database	IN	integrase (protein)
GDP	guanosine diphosphate	INR	initiator region
GEF	guanine nucleotide exchange factor	IPTG	isopropylthiogalactoside
GF	growth factor	IS	insertion sequence
GFP	green fluorescent protein	ISH	*in situ* hybridization
GIST	gastrointestinal stromal tumors	ISR	induced systemic resistance
Glu	glutamic acid	IVF	*in vitro* fertilization
Gly	glycine	IVS	intervening sequence
gp	glycoprotein	kb	kilobase
GPCR	G-protein-coupled receptor	KL	kit ligand
gRNA	guide RNA	KO	knockout
GSD	Gerstmann–Straüssler disease	KSS	Kearns–Sayre syndrome
GSS	Gerstmann–Straüssler–Scheinker syndrome	Lac	lactose
		LBC	lampbrush chromosome
GTP	guanosine triphosphate	LCR	locus control region
HA	hemagglutinin	LD	linkage disequilibrium
HAT	histone acetyl transferase	LDL	low-density lipoprotein
Hb	hemoglobin	Leu	leucine
hCG	human chorionic gonadotrophin	LH	luteinizing hormone
HCL	hairy cell leukemia	LHSI/II	light harvesting system I/II
HDAC	histone deacetylase	LINE	long interspersed nuclear element
HDGS	homology-dependent gene silencing	LMC	local mate competition
HDL	high-density lipoprotein	LMW	low molecular weight
Hfr	high frequency recombination	LOD	logarithm of the odds (score)
HGMD	Human Gene Mutation Database	LOH	loss of heterozygosity

Lox	locus of X-over
LPS	lipopolysaccharide
LRC	local resource competition
LRE	local resource enhancement
LRR	leucine-rich repeat
LTR	long terminal repeat
Lys	lysine
m-BCR	minor breakpoint cluster region
M-BCR	major breakpoint cluster region
M-phase	meiosis/mitosis phase
MAP	microtubule associated protein
MAP	mitogen-activated protein
MAPK	mitogen-activated protein kinase
MAR	matrix-attached region
Mb	megabase
MBP	myelin-based protein
MC'F	micro-complement fixation
MCR	mutation cluster region
MCS	multiple cloning site
MDS	myelodysplastic syndrome
Mel	maternal-effect embryonic lethal
MELAS	mitochondrial encephalomyopathy, lactic acidosis, and stroke-like symptoms
MEN	multiple endocrine neoplasia
MERRF	myoclonus epilepsy with ragged red fibers
Met	methionine
MFH	malignant fibrous histiocytoma
MFS	Marfan syndrome
MGD	Mouse Genome Database
MGF	mast cell growth factor
MGI	Mouse Genome Informatics
MHC	major histocompatibility complex
MIC	minimum inhibitory concentration
Mis	Müllerian-inhibiting substance
MLP	major late promotor
MLS	myxoid liposarcoma
mM	millimolar
MMC	maternally inherited myopathy and cardiomyopathy
MMR	mismatch repair
MOI	multiplicity of infection
MPS	mucopolysaccharidosis
MRCA	most recent common ancestor
MRD	minimal residual disease
mRNA	messenger RNA
MS	mass spectroscopy
MSI	microsatellite instability
mtDNA	mitochondrial DNA
Mu element	mutator element
N-	amino-
NAD	nicotinamide adenine dinucleotide
NADH	reduced NAD
NADP	nicotinamide adenine dinucleotide phosphate
NAP	nucleosome assembly protein
NBU	nonreplication Bacteroides units
ncRNA	noncoding RNA
NCS	nonchromosomal striped
NER	nucleotide excision repair
NHL	non-Hodgkin lymphoma
NK	natural killer (cells)
NMD	nonsense-mediated mRNA decay
NMR	nuclear magnetic resonance
NOR	nucleolus organizing region
NOS	nitric oxide synthetase
NPC	nasopharyngeal carcinoma
NPC	nuclear pore complex
NR	nuclear reorganization
nt	nucleotide
OI	osteogenesis imperfecta
OMIM	Online Mendelian Inheritance in Man
OPMD	oculopharyngeal muscular dystrophy
ORC	origin of replication complex
ORF	open reading frame
Ori	origin (of replication)
*Ori*T	origin of transfer
OTU	operational taxonomic unit
PAA	propionic acidemia
PAC	P1 artificial chromosome (vector)
PAC	prostate adenocarcinoma
PAGE	polyacrylamide gel electrophoresis
PAI	pathogenicity islands
PAPP	pregnancy-associated plasma protein
PAR	pseudoautosomal region
P*ax*	paired box-containing genes
PBP	penicillin binding protein
Pc	polycomb
PCO	polycystic ovarian (disease)
PCR	polymerase chain reaction
PCT	plasmacytoma
PDGF	platelet derived growth factor
PE	phosphatidylethanolamine
PEP	phosphoenolpyruvate
PEV	position effect variegation
PFGE	pulsed-field gel electrophoresis
PGPR	plant growth-promoting Rhizobacteria
PH	plekstrin homology
PH	polyhedron
Phe	phenylalanine
Pi	inorganic phosphate
PI	phosphatidylinositol
PIP_2	phosphatidylinositol-4,5-bisphosphate
PKA	protein kinase A
PKU	phenylketonurea
PMF	proton motive force
PMS	postmeiotic segregation
Pol	polymerase
PR	protease
Pro	proline
PS	phosphatidylserine
PS I/II	photosystem I/II
PTC	premature termination codon
PTGS	posttranscriptional gene silencing
Q-(banding)	quinacrine

QTL	quantitative trait loci
R	resistance (eg Amp[R]; ampicillin resistance)
R-plasmids	resistance plasmids
RA	retinoic acid
Ram	ribosomal ambiguity
RB	retinoblastoma
RCC	renal cancer cell
RCL	round cell liposarcoma
RCS	recombinant congenic strains
rDNA	recombinant DNA
RDR	recombination-dependent replication
RED	repeat expansion detection
RER	rough endoplasmic reticulum
REV	reticuloendotheliosis virus
RF	release factor
RF	replicative form
RFLP	restriction fragment length polymorphism
RI	recombinant inbred
Rif	rifampicin
RIM	reproductive isolating mechanism
RIP	repeat induced point (mutation)
RM	restriction modification
RN	recombination nodule
RNA	ribonucleic acid
RNAi	RNA interference
RNP	ribonucleoprotein
ROS	reactive oxygen species
RRF	ribosome recycling factor
rRNA	ribosomal RNA
RSS	recombination signal sequence
RSV	Rous sarcoma virus
RT	reverse transcriptase
RTK	receptor tyrosine kinase
S-phase	synthesis phase
SAR	scaffold-attached region
SAR	systemic acquired resistance
SBT	shifting balance theory
SC	synaptonemal complex
SCE	sister chromatid exchange
SCF	stem cell factor
scRNA	small cytoplasmic RNA
SDP	strain distribution pattern
SDS	second-division segregation
SDS-PAGE	sodium dodecyl-sulfate polyacrylamide gel electrophoresis
SEM	scanning electron microscopy
sen DNA	senescent DNA
Ser	serine
SER	smooth endoplasmic reticulum
SF	steroidogenic factor
SH (domains)	Src homology (domains)
SI	self-incompatibility
SINE	short interspersed nuclear element
SIV	simian immunodeficiency virus
SL	spliced leader
SLF	steel factor
snoRNA	small nucleolar RNA
SNP	single nucleotide polymorphism
snRNA	small nuclear RNA
snRNP	small nuclear ribonucleoprotein
SOD	superoxide dismutase
SRE	sterol response element
SRP	signal recognition particle
SRY	sex-determining region Y
ss	single stranded
SSB	single strand binding (protein)
SSLP	simple sequence length polymorphism
SSR	simple tandem sequence repeats
STR	short tandem repeats
STS	sequence tagged sites
su	subunit
SU	surface (viral)
SV40	simian virus 40
T	thymine
Taq	*Thermus aquatus*
TBP	TATA-box binding protein
TCA	tricarboxylic acid
TCR	T-cell receptor
TDF	testis-determining factor
TE	transposable elements
TEL	telomere
TEM	transmission electron microscopy
Ter	terminator
Tet	tetracyclin
TF	transcription factor
TGF	transforming growth factor
TGS	transcriptional gene silencing
Thr	threonine
TIM	translocase of the inner membrane
TIMPS	tissue inhibitor of metalloproteinases
TIR	terminal inverted repeats
TK	thymidine kinase
Tm	melting temperature
TM	transmembrane
TMV	tobacco mosaic virus
TN	transposon
TNF	tumor necrosis factor
TOM	translocase of the outer membrane
topo	topoisomerase
T_{opt}	optimum temperature
TRD	transmission ratio distortion
TRiC	T-complex polypeptide ring complex
TRITC	tetramethyl rhodamine isothiocyanate
tRNA	transfer RNA
Trp	tryptophan
TSD	Tay–Sachs disease
TSE	transmissible spongiform encephalopathy
TSG	tumor supressor gene
TSS	transcription start site
Tyr	tyrosine
U	uracil
Ub	ubiquitin
UPD	uniparental disomy
URF	unidentified reading frame
USS	uptake signal sequence

UTI upper respiratory tract infection
UTP uridine triphosphate
UTR untranslated region
UV ultraviolet
V gene variable gene
Val valine
VEGF vascular endothelial growth factor
VHL Von Hippel–Lindau disease
VNTR variable number of tandem repeats
VWF Von Willebrand factor
WHO World Health Organization
WS Werner syndrome
WT wild-type
WT1 Wilm's tumor 1
XIC X-inactivation center
Xist X-inactive specific transcript
XP xeroderma pigmentosum
YAC yeast artificial chromosome
ZP zona pellucida

Contributors

S A Aaronson
The Derald H Ruttenberg Cancer Center, Mount Sinai Medical Center, 1425 Madison Avenue, New York, NY 10029, USA

J M Adams
Division of Molecular Genetics of Cancer, The Walter and Eliza Hall Institute of Medical Research, Post Office Royal Melbourne Hospital, Melbourne, VIC 3050, Australia

T H Adams
Monsanto Company, 62 Maritime Drive, Mystic, CT 06355, USA

S Adhya
Laboratory of Molecular Biology, National Cancer Institute, National Institutes of Health, Bethesda, MD 20892, USA

N A Affara
Department of Pathology, University of Cambridge, Cambridge, CB2 1QP, UK

J D Aitchison
Institute for Systems Biology, 4225 Roosevelt Way, Seattle, WA 98105, USA

S-I Aizawa
Department of Biosciences, Teikyo University, 1-1 Toyosatodai, Utsunomiya 320–8551, Japan

Z Alavidze
G Eliava Institute of Bacteriophages, Microbiology and Virology, Tbilisi, 380060, Georgia, USA

D G Albertson
Cancer Research Institute, University of California–San Francisco, San Francisco, CA 94143, USA

K H Albrecht
The Jackson Laboratory, Bar Harbor, ME 04609, USA

R W Alexander
Department of Chemical Biology, Wake Forest University, Winston-Salem, NC 27109, USA

P Alifano
Dipartimento di Biologia e Patologia Cellulare et Moleculare, Università de Napoli Federico, I-80131 Naples, Italy

P Aman
Department of Pathology, Lundberg Laboratory for Cancer Research, Goteborg University, S-41345 Göteborg, Sweden

M Ambrose
Department of Cancer Cell Biology, Harvard School of Public Health, Boston, MA 02215, USA

G F Ames
Division of Biochemistry and Molecular Biology, University of California–Berkeley, Berkeley, CA 94720, USA

P Anderson
Department of Genetics, University of Wisconsin–Madison, Madison, WI 53706, USA

H Antoun
RSVS Pavillian Charles–Eugene Marchand, Université Laval, Québec, GIK 7P4, Canada

C F Aquadro
Department of Molecular Biology and Genetics, Biotechnology Building Cornell University, Ithaca, NY 14853, USA

K Ardlie
Genomics Collaborative, 99 Evie Street, Cambridge, MA 02139, USA

M Arkin
Sunesis Pharmaceuticals, 3696 Haven Avenue, Redwood City, CA 94063, USA

J Arnold
Department of Genetics, University of Georgia, Athens, GA 30602, USA

K Artzt
Section of Molecular Genetics and Microbiology, Institute of Cell and Molecular Biology, University of Texas at Austin, Austin, TX 78712, USA

A Ashworth
Breakthrough Breast Cancer Research Centre, Chester Beatty Laboratories, Institute of Cancer Research, London, SW3 6JB, UK

M A Asmussen
Department of Genetics, University of Georgia, Athens, GA 30602, USA

K J Aufderheide
Department of Biology, Texas A & M University, Austin, TX 77843, USA

J Austin
Department of Molecular Genetics and Cell Biology, University of Chicago, Chicago, IL 60637, USA

J C Avise
Department of Genetics, University of Georgia, Athens, GA 30602, USA

R Baer
Department of Pathology, Institute of Cancer Genetics, College of Physicians and Surgeons of Columbia University, New York, NY 10032, USA

J-L Baert
Institut de Biologie de Lille, F-59021 Lille, France

A Bafico
Mount Sinai Medical Center, The Derald H Ruttenberg Cancer Center, 1425 Madison Avenue, New York, NY 10029, USA

B Bain
Department of Haematology, Imperial College School of Medicine, St Mary's Hospital, London W2 1NY, UK

T T Baird Jr
Department of Pharmaceutical Chemistry, University of California–San Francisco, San Francisco, CA 94143, USA

A Balmain
Cancer Research Institute, University of California–San Francisco, San Francisco, CA 94143, USA

W M Barnes
Department of Biochemistry and Molecular Biophysics, Washington University Medical School, St Louis, MO 63110, USA

M A Barrand
Department of Pharmacology, University of Cambridge, Cambridge, CB2 1QJ, UK

G S Barsh
Department of Pediatrics and Genetics and the Howard Hughes Medical Institute, Stanford University School of Medicine, Beckman Center, Stanford, CA 94305, USA

D P Bartel
The Whitehead Institute for Biochemical Research and Department of Biology, Massachusetts Institute of Technology, Cambridge, MA 02142, USA

R Baumeister
Genzentrum, D-81377 München, Germany

C Beamish
T4 Lab, The Evergreen State College, Olympia, WA 98505, USA

H-A Becker
Max-Planck Institut für Züchtungsforschung, D-50829 Köln, Germany

K M Beckingham
Department of Biochemistry and Cell Biology, Rice University, Houston, TX 77251, USA

C V Beechey
Mammalian Genetics Unit, Medical Research Council, Harwell, Didcot, OX11 0RD, UK

J D Beggs
The Wellcome Centre for Cell Biology, Institute of Cell and Molecular Biology, University of Edinburgh, Edinburgh, EH9 3JR, UK

M Belfort
Wadsworth Center, New York State Department of Health, Albany, NY 12201, USA

P N Benfey
Department of Biology, New York University, New York, NY 10003, USA

J L Bennetzen
Department of Biological Sciences, Purdue University, West Lafayette, IN 47907, USA

D E Bergstrom
The Jackson Laboratory, Bar Harbor, ME 04609, USA

A J Berk
Molecular Biology Institute, University of California–Los Angeles, Los Angeles, CA 90095, USA

M K B Berlyn
E. coli Genetic Stock Center, Department of MCD Biology, Yale University, New Haven, CT 06520, USA

A Bernardi
Laboratoire d'Enzymologie, CNRS, PGE, 1 Avenue de la Terrasse, F-91198 Gif sur Yvette, France

A Berns
The Netherlands Cancer Institute, NL-1066 CX, Amsterdam, The Netherlands

A Bernstein
Program in Molecular Biology and Cancer, Samuel Lunenfeld Research Institute, Mount Sinai Hospital, Toronto, ON, M5G 1X5, Canada

S S Bhattacharya
Department of Molecular Genetics, Institute of Ophthalmology, University College London, London, EC1V 9EL, UK

D L Black
Department of Microbiology and Molecular Genetics, Howard Hughes Medical Institute, University of California–Los Angeles, Los Angeles, CA 90024, USA

R D Blank
University of Wisconsin Medical School, Madison, WI 53792, USA

T Blumenthal
Department of Biochemistry and Molecular Genetics, University of Colorado School of Medicine, Denver, CO 80262, USA

F Bonhomme
Génome, Populations, Interactions UMR 5000, Université de Montpellier, F-34095 Montpellier, France

M I Borges-Walmsley
Division of Infection and Immunity, Institute of Biomedical and Life Sciences, University of Glasgow, Glasgow, G11 6NN, UK

P Boursot
Génome, Populations, Interactions UMR 5000, Université de Montpellier, F-34095 Montpellier, France

Y Boyd
Institute for Animal Health, Compton, RG20 7NN, UK

B Braaten
Department of Molecular, Cellular and Developmental Biology, University of California–Santa Barbara, Santa Barbara, CA 93106, USA

W J Brammar
Department of Biochemistry, University of Leicester, Leicester, LE1 7RH, UK

B Brembs
Department of Neurobiology and Anatomy, The University of Texas–Houston Medical School, Houston, TX 77030, USA

P J Brennan
Department of Pathology and Laboratory of Medicine, University of Pennsylvania School of Medicine, Philadelphia, PA 19104, USA

C Brenner
Faculté de Médecine, Université Libre de Bruxelles, Bruxelles, Belgium

S Brenner
Molecular Sciences Institute, University of California–Berkeley, Berkeley, CA 94704, USA

N J Brewin
Genetics Department, John Innes Centre, Colney, Norwich, NR4 7UH, UK

B A Bridges
MRC Cell Mutation Unit, University of Sussex, Brighton, BN1 9RH, UK

M H Brilliant
Department of Pediatrics, University of Arizona School of Medicine, Tucson, AZ 85724, USA

A Brinker
Max Planck Institut für Biochemie, D-82152 Martinsried, Germany

J Brookfield
Institute of Genetics, University of Nottingham, Greens Medical Centre, NG7 2UH, UK

W J Broughton
Department of Plant Molecular Biology, University of Geneva, CH-1292 Chambésy, Geneva, Switzerland

K E Browman
Bristo-Myers-Squibb Co., Neuroscience and Genitourinary Drug Discovery, Wallingford, CT 06492, USA

M S Brown
Department of Molecular Genetics, University of Texas Southwestern Medical Center, Dallas, TX 75235, USA

T A Brown
Department of Biomolecular Sciences, University of Manchester Institute of Science and Technology, Manchester, M60 1QD, UK

V Brown
Department of Genetics, Emory University School of Medicine, Atlanta, GA 30322, USA

M E Bruce
Neuropathogenesis Unit, Institute for Animal Health, Edinburgh, EH9 3JF, UK

M L Budarf
Division of Human Genetics, Department of Pediatrics, University of Pennsylvania School of Medicine, Philadelphia, PA 19104, USA

L Bülow
Department of Pure and Applied Chemistry, Center for Chemistry and Chemical Engineering, S-22100 Lund, Sweden

A Burchell
Tayside Insititute of Child Health, Ninewells Hospital and Medical School, University of Dundee, Dundee, DD1 9SY, UK

R R Burgess
McArdle Laboratory for Cancer Research, University of Wisconsin–Madison, Madison, WI 53706, USA

T R Bürglin
Department of Cell Biology, University of Basel, CH-4056 Basel, Switzerland

A Bürkle
Department of Gerontology, University of Newcastle upon Tyne, Newcastle upon Tyne, NE4 6BE, UK

J Burn
Department of Human Genetics, University of Newcastle upon Tyne, Newcastle upon Tyne, NE2 4AA, UK

D W Burt
Roslin Institute, Roslin, Midlothian, EH25 9PS, UK

C Caldas
CRC Department of Oncology and The Wellcome Trust Centre for Molecular Mechanisms in Disease, Cambridge Institute for Medical Research, University of Cambridge, Cambridge, CB2 2XY, UK

A Campbell
Department of Biological Sciences, Stanford University, Stanford, CA 94305, USA

S A Camper
Department of Human Genetics, University of Michigan Medical School, Ann Arbor, MI 48109, USA

E Canaani
Department of Molecular Cell Biology, Weizmann Institute of Science, 76100 Rehovot, Israel

E P M Candido
Department of Biochemistry and Molecular Biology, University of British Columbia, Vancouver, British Columbia, V6T1Z3, Canada

J C Carey
Department of Pediatrics, Obstetrics, Gynecology and Nursing, Health Science Center, University of Utah, Salt Lake City, UT 84132, USA

F Carneiro
Faculty of Medicine and IPATIMUP, University of Porto, 4200 Porto, Portugal

A T C Carpenter
Department of Genetics, University of Cambridge, Cambridge, CB2 3EH, UK

D Carr
The Wellcome Trust, 210 Euston Road, London, NW1 2BE, UK

D Carroll
Department of Biochemistry, University of Utah School of Medicine, Salt Lake City, UT 84132, USA

G Caruana
Program in Molecular Biology and Cancer, Samuel Lunenfeld Research Institute, Mount Sinai Hospital, Toronto, Ontario, M5G 1X5, Canada

S M Case
Department of Biology, Salem State College, Salem, MA 01970, USA

M Cashel
Laboratory of Molecular Genetics, NICHD, National Institutes of Health, Bethesda, MD 20892, USA

L A Casselton
Department of Plant Sciences, University of Oxford, Oxford, OX1 3RB, UK

A Castells
Department of Gastroenterology, Barcelona University, Barcelona 8036, Spain

D Catovsky
Academic Department of Haematology and Cytogenetics, Royal Marsden Hospital, London, SW3 6JJ, UK

B M Cattanach
MRC Mammalian Genetics Unit, Harwell, Didcot, OX11 0RD, UK

R Chaganti
Memorial Sloan-Kettering Cancer Center, 1275 York Avenue, New York, NY 10021, USA

R Chakraborty
Human Genetics Center, University of Texas School of Public Health, Houston, TX 77225, USA

M Chalfie
Department of Biological Sciences, Columbia University, New York, NY 10027, USA

M Chandler
Laboratoire de Microbiologie et de Génétique Moléculaire, CNRS, F-31620 Toulouse, France

A C Chandley
20 Comely Bank, Edinburgh, EH4 1AL, UK

F Chedin
Department of Pathology and of Biochemistry, Norris Comprehensive Cancer Center, University of Southern California School of Medicine, Los Angeles 90089, CA, USA

R Chetelat
Department of Vegetable Crops, University of California–Davis, Davis, CA 95616, USA

A D Chisholm
Department of Biology, University of California–Santa Cruz, Santa Cruz, CA 95064, USA

S Chong
New England Biolabs Inc., 32 Tozer Road, Beverly, MA 01915, USA

J Chory
Plant Biology Laboratory, Howard Hughes Medical Institute and The Salk Institute for Biological Studies, La Jolla, CA 92037, USA

G A Churchill
The Jackson Laboratory, Bar Harbor, ME 04609, USA

G Churchward
Department of Microbiology and Immunology, Emory University, Atlanta, GA 30322, USA

A G Clark
Department of Biology, Pennsylvania State University, University Park, PA 16802, USA

S G Clark
Skirball Institute of Biomolecular Medicine, New York University School of Medicine, New York, NY 10016, USA

M A Cleary
Department of Molecular Biology, Princeton University, Princeton, NJ 08544, USA

E H Coe Jr
Plant Genetics Research Unit, United States Department of Agriculture-ARS, University of Missouri, Columbia, MO 65211, USA

N Coleman
Department of Molecular Histopathology, Addenbrooke's Hospital, University of Cambridge, Cambridge, CB2 2QQ, UK

V P Collins
Addenbrooke's Hospital, University of Cambridge, Cambridge, CB2 2QQ, UK

N C Comfort
Center for History of Recent Science, George Washington University, Washington, DC 20052, USA

J M Connor
Department of Medical Genetics, University of Glasgow, Glasgow, G12 8QQ, UK

F Constantini
Department of Genetics and Development, College of Physicians and Surgeons, Columbia University, New York, NY 10032, USA

A Cooke
Department of Pathology, University of Cambridge, Cambridge, CB2 1QP, UK

C S Cooper
Institute of Cancer Research, Haddow Laboratories, Belmont, Sutton, SM2 5NG, UK

V Cormier-Daire
Department of Genetics, Hopitâl Necker Enfants Malades, Paris, France

B C Coughlin
Proceedings of the National Academy of Sciences, Washington, DC 20007, USA

A Coulson
The Sanger Centre, Hinxton, Cambridge, CB10 ISA, UK

D W Cox
Department of Medical Genetics, University of Alberta, Edmonton, Alberta, T6G 2H7, Canada

M M Cox
Department of Biochemistry, University of Wisconsin–Madison, Madison, WI 53706, USA

T M Cox
Department of Medicine, Addenbrooke's Hospital, University of Cambridge, Cambridge, CB2 2QQ, UK

J Coyne
Department of Ecology and Evolution, University of Chicago, Chicago, IL 60637, USA

J C Crabbe
Portland Alcohol Research Center, Department of Veterans Affairs Medical Center, Oregon Health Sciences University, Portland, Oregon 97201, USA

C S Craik
Department of Pharmaceutical Chemistry, University of California–San Francisco, San Francisco, CA 94143, USA

K A Crandall
Department of Zoology, Brigham Young University, Provo, UT 84602, USA

J F Crow
Genetics Laboratory, University of Wisconsin–Madison, Madison, WI 53706, USA

T J Crow
Department of Psychiatry, Warneford Hospital, University of Oxford, Oxford OX3 7JX, UK

J Cruz-Reyes
Department of Biological Chemistry, The Johns Hopkins University School of Medicine, Baltimore, MD 22105, USA

M B Cruzan
Department of Ecology and Evolutionary Biology, University of Tennessee, Knoxville, TN 37996, USA

A K Csink
Department of Biological Sciences, Carnegie Mellon University, Pittsburgh, PA 15213, USA

J Z Dalgaard
Marie Curie Research Institute, The Chase, Oxted, RH8 OTL, UK

A Danchin
Pasteur Research Centre, Hong Kong University, Pokfulam, Hong Kong

G Daniels
Bristol Institute for Transfusion Sciences, Southmead Road, Bristol, BS10 5ND, UK

A Darvasi
Department of Ecology, Systematics and Evolution, Silberman Institute of Life Sciences, The Hebrew University of Jerusalem, Jerusalem 91904, Israel

M T Davisson
The Jackson Laboratory, Bar Harbor, ME 04609, USA

L De Gregorio
Department of Pediatrics and Institute for Molecular Genetics, University of California–San Diego, La Jolla, CA 92093, USA

Y de Launoit
Faculté de Médecine, Université Libre de Bruxelles, Bruxelles, Belgium

D C DeLuca
Department of Biochemistry and Molecular Biology, University of Arkansas for Medical Sciences, Little Rock, AR 72205, USA

P Demant
Division of Molecular Biology, The Netherlands Cancer Institute, NL-1066 CX Amsterdam, The Netherlands

J Dénarié
Laboratorie de Biologie Moléculaire des Relations Plantes-Microorganismes, INRA/CNRS, F-31326 Castanet, France

K M Derbyshire
Division of Infectious Disease, Wadsworth Center, New York State Department of Health, and Department of Biomedical Sciences, School of Public Health, Albany, NY 12201, USA

S A des Etages
Pfizer Inc., Genetic Technologies, Groton, CT 06340, USA

R J Desnick
Department of Human Genetics, Mount Sinai School of Medicine, New York, NY 10029, USA

K M Devos
John Innes Centre, Colney, Norwich, NR4 7UH, UK

J E Donelson
Department of Biochemistry, University of Iowa, Iowa City, IA 52242, USA

D Donnai
University Department of Medical Genetics and Regional Genetics Service, St Mary's Hospital, Manchester, MI3 0JH, UK

R L Dorit
Department of Biology, Yale University, New Haven, CT 06511, USA

K Douglas
Department of Biology, University of Michigan, Ann Arbor, MI 48109, USA

W F Dove
McArdle Laboratory for Cancer Research, University of Wisconsin–Madison, WI 53706, USA

D M Downs
Department of Bacteriology, University of Wisconsin–Madison, Madison, WI 53706, USA

J J Doyle
Department of Plant Biology, L H Bailey Hortorium, Cornell University, Ithaca, NY 14853, USA

J W Drake
National Institute of Environmental Health Sciences, Irvine, NC 27709, USA

M E Dresser
Program in Molecular and Cell Biology, Oklahoma Medical Research Foundation, Oklahoma City, OK 73104, USA

M Driscoll
Department of Molecular Biology and Biochemistry, Nelson Laboratories, The State University of New Jersey, Rutgers, Piscataway, NJ 08854, USA

K Drlica
Public Health Research Institute, New York, NY 10016, USA

K R Dronamraju
Foundation for Genetic Research, Houston, TX 77227, USA

T Dunckley
Department of Molecular and Cellular Biology, Howard Hughes Medical Institute, University of Arizona, Tucson, AZ 85721, USA

J Dvořák
Department of Agronomy and Range Sciences, University of California–Davis, Davis, CA 95616, USA

B D Dyer
Department of Biology, Wheaton College, Norton, MA 02766, USA

M J S Dyer
Department of Haematology, Leicester Royal Infirmary, University of Leicester, Leicester, LE2 7LX, UK

S M Eacker
Departments of Genetics, University of Washington, Seattle, WA 98195, USA

W F Eanes
Department of Ecology and Evolution, State University of New York, Stony Brook, NY 11794, USA

J Eberwine
Department of Pharmacology, University of Pennsylvania Medical Center, Philadelphia, PA 19104, USA

W Eckhart
The Salk Institute for Biological Studies, La Jolla, CA 92037, USA

A W F Edwards
Gonville and Caius College, University of Cambridge, Cambridge, CB2 1TA, UK

R Edwards
Duck End Farm, Park Lane, Dry Drayton, Cambridge CB3 8DB, UK

J C J Eeken
Department of Radiation Genetics and Chemical Mutagenesis, Leiden University Medical Center, 2300 RC Leiden, The Netherlands

R Egel
Department of Genetics, University of Copenhagen, DK-1353 Copenhagen K, Denmark

A L Eggler
Department of Biochemistry, University of Wisconsin–Madison, Madison, WI 53706, USA

L Ehrman
Division of Natural Sciences, State University of New York, Purchase, NY 10577, USA

E M Eicher
The Jackson Laboratory, Bar Harbor, ME 04609, USA

T H Eickbush
Department of Biology, University of Rochester, Rochester, NY 14627, USA

R C Eisensmith
Institute for Gene Therapy and Molecular Medicine, Mount Sinai School of Medicine, New York, NY 10029, USA

W S El-Deiry
Howard Hughes Medical Institute, University of Pennsylvania School of Medicine, Philadelphia, PA 19104, USA

N A Elgerdy
Department of Pathology, South Manchester University Hospital Trust, Manchester, M23 9LT, UK

R P Elinson
Department of Biological Sciences, Bayer School of Natural and Environmental Sciences, Duquesne University, Pittsburgh, PA 15282, USA

A E H Emery
Department of Neurology, Royal Devon and Exeter Hospital, Exeter, EX2 5DW, UK

J A Endicott
Laboratory of Molecular Biophysics, Department of Biochemistry and Oxford Centre for Molecular Sciences, University of Oxford, Oxford, OX1 3QU, UK

W R Engels
Laboratory of Genetics, University of Wisconsin–Madison, Madison, WI 53706, USA

M Estelle
Section of Molecular and Development Biology, Institute of Cellular and Molecular Biology, University of Texas–Austin, Austin TX 78712, USA

E P Evans
Mammalian Genetics Unit, Medical Research Council, Harwell, Didcot, OX11 ORD, UK

T C Evans Jr
New England Biolabs Inc., 32 Tozer Road, Beverly, MA 01915, USA

W J Ewens
Department of Biology, University of Pennsylvania, Philadelphia, PA 19104, USA

J R Fabian
7 Partridgeberry Lane, Keene, NH 03431, USA

B A Fane
Department of Veterinary Science and Microbiology, University of Arizona, Tucson, AZ 85721, USA

C Fankhauser
Department of Molecular Biology, University of Geneva, CH-1211 Genève 4, Switzerland

N Fedoroff
Biotechnology Institute, Pennsylvania State University, University Park, PA 16802, USA

M Feiss
Department of Microbiology, University of Iowa College of Medicine, Iowa City, IA 52242, USA

T V Feldblyum
The Institute for Genomic Research, 9712 Medical Center Drive, Rockville, MD 20850, USA

A C Ferguson-Smith
Department of Anatomy, University of Cambridge, Cambridge, CB2 3DY, UK

M A Ferguson–Smith
Department of Clinical Veterinary Medicine, University of Cambridge, Cambridge, CB3 OES, UK

A R Fersht
Department of Chemistry, University of Cambridge, Cambridge, CB2 1EW, UK

W Filipowicz
Friedrich Miescher Institute, CH-4002 Basel, Switzerland

T M Finan
Department of Biology, McMaster University, Hamilton, Ontario, L8S 4K1, Canada

J R S Fincham
Division of Biology, University of Edinburgh, Edinburgh, EH10 5RY, UK

H Firth
Department of Medical Genetics, Addenbrooke's Hospital, University of Cambridge, Cambridge, CB2 2QQ, UK

W Fitch
Department of Ecology and Evolutionary Biology, University of California–Irvine, Irvine, CA 92697, USA

J Folkman
Department of Surgery, The Children's Hospital and Harvard Medical School, Boston, MA 02115, USA

J Ford
Division of Oncology, Stanford University School of Medicine, Stanford, CA 94305, USA

J Forejt
Institute of Molecular Genetics, Academy of Sciences of the Czech Republic, 14220 Praha 4, Czech Republic

S L Forsburg
Molecular Biology and Virology Laboratory, The Salk Institute for Biological Studies, La Jolla, CA 92037, USA

J A Fossella
Laboratory of Neurobiology and Behavior, Rockefeller University, New York, NY 10021, USA

P L Foster
Department of Biology, Indiana University, Bloomington, IN 47405, USA

M Frame
Beatson Institute for Cancer Research, CRC Beatson Laboratories, Garscube Estate, Switchback Road, Bearsden, Glasgow, G62 1BD, UK

J Frampton
Weatherall Institute of Molecular Medicine, John Radcliffe Hospital, University of Oxford, Oxford, OX3 9DS, UK

W N Frankel
The Jackson Laboratory, Bar Harbor, ME 04609, USA

R Frankham
Key Centre for Biodiversity and Bioresources, Department of Biological Sciences, Macquarie University, Sydney, NSW 2109, Australia

C M Fraser
Institute for Genomic Research, 9712 Medical Center Drive, Rockville, MD 20850, USA

G R Fraser
Department of Clinical Genetics, John Radcliffe Hospital, Oxford, OX3 7LJ, UK

K Fredga
Department of Pediatrics and Institute for Molecular Genetics, University of California, San Diego, La Jolla, CA 92093, USA

S Fredriksson
Department of Pure and Applied Chemistry, Center for Chemistry and Chemical Engineering, Lund University, S-22100 Lund, Sweden

E C Friedberg
Laboratory of Molecular Pathology, Department of Pathology, University of Texas Southwestern Medical Center, Dallas, TX 75235, USA

T Friedmann
Center for Molecular Genetics, University of California–San Diego School of Medicine, La Jolla, CA 92093, USA

D J Futuyma
Department of Ecology and Evolution, State University of New York, Stony Brook, NY 11794, USA

G L Gabor Miklos
GenetixXpress Proprietary Ltd, Palm Beach, Sydney, NSW 2108, Australia

J Gallant
Uppsala University, SE-24332 Hoor, Sweden

J I Garrels
Proteome Division, Incyte Genomics, Beverly, MA 01915, USA

S M Gartler
Departments of Medicine and Genetics, University of Washington, Seattle, WA 98195, USA

A Gavalas
Department of Development Neurobiology, National Institute for Medical Research, Mill Hill, London NW7 IAA, UK

V Gavrias
Monsanto Co., Mystic, CT 06355, USA

A F Gazdar
Hamon Center for Therapeutic Oncology Research, University of Texas Southwestern Medical Center, Dallas, TX 75390, USA

M L Gennaro
Public Health Research Institute, New York, NY 10016, USA

P Gepts
Department of Agronomy and Range Science, University of California–Davis, Davis, CA 95616, USA

J German
Department of Pediatrics, Weill College of Medicine, Cornell University, Ithaca, NY 10021, USA

E Gherardi
Department of Oncology, MRC Centre, Addenbrooke's Hospital, Cambridge, CB2 2QH, UK

F Giannelli
Division of Medical and Molecular Genetics, Guy's, King's, and St Thomas's School of Medicine, King's College London, London, SE1 9RT, UK

G N Gill
University of California–San Diego School of Medicine, La Jolla, CA 92093, USA

M A Gilson
Wadsworth Center, New York State Department of Health, Albany, NY 12201, USA

A C Glasgow
Department of Microbiology, University of Georgia, Athens, GA 30602, USA

M Goldman
Department of Biology, San Francisco State University, San Francisco, CA 94132, USA

M Goldschmidt-Clermont
Department of Molecular Biology, University of Geneva, CH-1211 Genève 4, Switzerland

M Goldsmith
Department of Biological Sciences, University of Rhode Island, Kingston, RI 02881, USA

J L Goldstein
Department of Molecular Genetics, University of Texas Southwestern Medical Center, Dallas, TX 75235, USA

E S Golub
Lykeion Corp., Solana Beach, CA 92075, USA

M F Goodman
Department of Biological Sciences and Chemistry, University of Southern California, Los Angeles, CA 90089, USA

J A Goodrich
Department of Chemistry and Biochemistry, University of Colorado–Boulder, Boulder, CO 80309, USA

J Goodship
Department of Human Genetics, University of Newcastle upon Tyne, Newcastle upon Tyne, NE2 4AA, UK

D M Gordon
Division of Biology and Zoology, Australian National University, Canberra, ACT, Australia

I I Gottesman
245 Terrell Road West, Charlottesville, VA 22901, USA

C S Grant
Department of Surgery, The Mayo Clinic, Rochester, MN, USA

J-P Gratia
Microbial Genetics and Ecology Unit, Brussels University, School of Medicine, B-1070 Brussels, Belgium

J A M Graves
Comparative Genomics Research Group, Research School of Biological Sciences, The Australian National University, Canberra, ACT 2601, Australia

M W Gray
Department of Biochemistry and Molecular Biology, Dalhousie University, Halifax, Nova Scotia, B3H 4H7, Canada

M I Greene
Department of Pathology and Laboratory of Medicine, University of Pennsylvania School of Medicine, Philadelphia, PA 19104, USA

N Gregersen
Research Unit for Molecular Medicine, Aarhus University Hospital and Faculty of Health Sciences, Skejby Hospital, DK-8200 Aarhus N, Denmark

P Gresshoff
Department of Botany, University of Queensland, Brisbane, QLD 4072, Australia

M A Griep
Department of Chemistry, University of Nebraska, Lincoln, NE 68588, USA

T Grigliatti
Department of Biological Sciences, University of British Columbia, Vancouver, British Columbia, V6T 1Z4, Canada

N D F Grindley
Department of Molecular Biophysics and Biochemistry, Yale University, New Haven, CT 06520, USA

A P Grollman
Department of Pharmacological Sciences, State University of New York, Stony Brook, NY 11794, USA

R Grosschedl
Gene Center and Institute of Biochemistry, University of Munich, D-81377 Munich, Germany

P Gruss
Max Planck Institute for Biophysical Chemistry, Department of Molecular Cell Biology, D-37077 Göttingen, Germany

J B Gurdon
Wellcome CRC Institute, Tennis Court Road, Cambridge, CB2 1QR, UK

B S Guttman
Lab 1, The Evergreen State College, Olympia, WA 98505, USA

N Haites
Department of Medical Genetics, University of Aberdeen, Aberdeen, UK

S E Halford
Department of Biochemistry, University of Bristol, Bristol, B58 1TD, UK

R M Hall
Division of Biomolecular Engineering, CSIRO Molecular Science, Sydney Laboratory, North Ryde, NSW 1670, Australia

J L Hamerton
Department of Human Genetics, University of Manitoba, Winnipeg, Manitoba, R3E 0W3, Canada

P Hanawalt
Department of Biological Sciences, Stanford University, Stanford, CA 94305, USA

D B Haniford
Department of Biochemistry, University of Western Ontario, London, Ontario, N6A 5C1, Canada

H Harada
Division of Gastroenterology, University of Pennsylvania, Philadelphia, PA 19104, USA

S C Hardies
Department of Biochemistry, University of Texas Health Science Center, San Antonio, TX 78229, USA

J C Harper
Department of Obstetrics and Gynaecology, University College London, London WC1E 6HX, UK

R Harris
Department of Medical Genetics, University of Manchester, Manchester, M13 9PT, UK

C J Harrison
Department of Haematology, Royal Free Hospital, London, NW3 2QG, UK

D J Harrison
Department of Pathology, University of Edinburgh, Edinburgh, UK

D L Hartl
Department of Organismic and Evolutionary Biology, Harvard University, Cambridge, MA 02138, USA

F U Hartl
Max Planck Institut für Biochemie, D-82152 Martinsried, Germany

M Hasegawa
Institute of Statistical Mathematics, 4-6-7 Minami Avenue, Minato Ku, Tokyo 106-8569, Japan

R Haselkorn
Department of Molecular Genetics and Cell Biology, School of Medical Sciences, University of Chicago, Chicago, IL 60637, USA

P S Hasleton
Department of Pathology, South Manchester University Hospital Trust, Manchester, M23 9LT, UK

N Hastie
MRC Human Genetics Unit, Western General Hospital, Edinburgh, EH4 2XU, UK

P J Hastings
Department of Molecular and Human Genetics, Baylor College of Medicine, Houston, TX 77030, USA

P M Hawkey
Public Health Bacteriology, Public Health Laboratory, Birmingham Heartlands Hospital, Bordesley Green East, Birmingham, B9 5SS, UK

R S Hawley
Section of Molecular and Cellular Biology, University of California–Davis, Davis, CA 95616, USA

T Hazelrigg
Department of Biological Sciences, Columbia University, New York, NY 10027, USA

J K Heath
Department of Biochemistry, University of Birmingham, Birmingham, B15 2TT, UK

J F Heidelberg
Institute for Genomic Research, Rockville, MD 20850, USA

D R Helinski
University of California–San Diego, La Jolla, CA 92093, USA

J G Hengstler
Institut für Toxikologie, Obere Zahlbacher Strasse 67, D-55131 Mainz, Germany

T M Henkin
Department of Microbiology, Ohio State University, Columbus, OH 43210, USA

R K Herman
Department of Genetics and Cell Biology, University of Minnesota, St Paul, MN 55108, USA

I Herskowitz
Department of Biochemistry and Biophysics, University of California–San Francisco, San Francisco, CA 94143, USA

R Hesketh
Department of Biochemistry, University of Cambridge, Cambridge, CB2 1QW, UK

J S Heslop-Harrison
Department of Biology, University of Leicester, Leicester, LE1 7RH, UK

C Heyting
Laboratorium voor Erfelijkheidsleer, NL-6700 HB Wageningen, The Netherlands

P Hieter
Centre for Molecular Medicine and Therapeutics, Department of Medical Genetics, University of British Columbia, Vancouver, British Columbia, V5Z 4H4, Canada

D Higgins
Department of Biochemistry, University College, Cork, Ireland

K L Hill
Department of Microbiology, Immunology and Molecular Genetics, University of California–Los Angeles, Los Angeles, CA 90095, USA

W G Hill
Institute of Cell, Animal and Population Biology, University of Edinburgh, Edinburgh, EH9 3JT, UK

M M Hingorani
Laboratory of DNA Replication, Rockefeller University, New York, NY 10021, USA

M J Hobart
Department of Biological Sciences, De Montfort University, Leicester, LE1 5BH, UK

J Hodgkin
Genetics Unit, Department of Biochemistry, University of Oxford, Oxford, OX1 3QU, England

B Hohn
Friedrich Miescher Institute, CH-4058 Basel, Switzerland

K E Holsinger
Department of Ecology and Evolutionary Biology, University of Connecticut, Storrs, CT 06269, USA

T R Hoover
Department of Microbiology, University of Georgia, Athens, GA 30602, USA

J C Hu
Department of Biochemistry and Biophysics, Texas A & M University, Austin, TX 77843, USA

C Huang
Department of Biological Chemistry, The Johns Hopkins University School of Medicine, Baltimore, MD 21205, USA

R E Huber
Department of Biological Sciences, University of Calgary, Calgary, Alberta, T2N 1N4, Canada

N Hughes-Jones
Formerly member of the Medical Research Council's Mechanisms in Immunopathogy Unit, MRC Centre, Cambridge, CB2 2XY, UK

M Hülskamp
Botanik III, Universität Köln, D-50931 Köln, Germany

M A Hultén
Department of Biological Sciences, University of Warwick, Coventry, CV4 7AL, UK

D M Hunt
Division of Molecular Genetics, Institute of Ophthalmology, University College London, London, EC1V 9EL, UK

S M Huson
Department of Clinical Genetics, John Radcliffe Hospital, Oxford, OX3 7LJ, UK

P G Isaacson
Department of Histopathology, Royal Free and University College School of Medicine, University College London, London, WCIE 6BT, UK

W E Jack
New England Biolabs Inc, 32 Tozer Road, Beverly, MA 01915, USA

I Jackson
MRC Human Genetics Unit, Western General Hospital, Edinburgh, EH4 2XU, UK

P A Jacobs
Wessex Regional Genetics Laboratory, Salisbury District Hospital, Salisbury, SP2 8BJ, UK

E Jansen
Laboratory for Molecular Oncology, Center for Human Genetics, University of Leuven and Flanders Interuniversity for Biotechnology, B-3000, Leuven, Belgium

N G J Jaspers
Department of Cell Biology and Genetics, Erasmus University, NL-3000 DR Rotterdam, The Netherlands

M Jayaram
Section of Molecular Genetics, University of Texas–Austin, Austin, TX 78712, USA

R C Johnson
Department of Biological Chemistry, University of California–Los Angeles, Los Angeles, CA 90095, USA

R Johnson
Department of Biochemistry and Molecular Biology, M D Anderson Cancer Center, University of Texas–Houston, Houston, TX 77030, USA

E M Jorgensen
Department of Biology, University of Utah, Salt Lake City, UT 84112, USA

T H Jukes
Formerly of Department of Integrative Biology, University of California–Berkeley, Berkeley, CA 94720, USA

L W Jurata
Gene Expression Laboratory, The Salk Institute for Biological Studies, La Jolla, CA 92037, USA

M M Kaback
Department of Pediatrics, Children's Hospital and Health Center, San Diego, CA 92123, USA

C I Kado
Department of Plant Pathology, University of California–Davis, Davis, CA 95616, USA

A Kallioniemi
National Human Genome Research Institute, National Institutes of Health, Bethesda, MD 20892, USA

D K Kalousek
Cytogenetics and Embryopathology Laboratory, British Columbia Children's Hospital, Vancouver, British Columbia, V6H 3VH, Canada

C Kane
Department of Molecular and Cell Biology, University of California–Berkeley, Berkeley, CA 94720, USA

D K R Karaolis
Department of Epidemiology and Preventive Medicine, University of Maryland School of Medicine, Baltimore, MD 21201, USA

J Karn
MRC Laboratory of Molecular Biology, Addenbrooke's Hospital, Cambridge, CB2 2QH, UK

R M T Katso
Ludwig Institue for Cancer Research, 91 Riding House Street, London, W1W 8BT, UK

E A Kellogg
Department of Biology, University of Missouri–St Louis, St Louis, MO 63121, USA

T Kelly
Department of Molecular Biology and Genetics, The Johns Hopkins University School of Medicine, Baltimore, MD 21205, USA

K J Kemphues
Department of Molecular Biology and Genetics, Cornell University, Ithaca, NY 14853, USA

M G Kidwell
Department of Ecology and Evolutionary Biology, University of Arizona, Tucson, AZ 85721, USA

M Kimmel
Statistics Department, Rice University, Houston, TX 77251, USA

H Kishino
University of Tokyo, 113-8657 Tokyo, Japan

A J S Klar
Gene Regulation and Chromosome Biology Laboratory, National Cancer Institute at Frederick, Frederick, MD 21702, USA

G Klein
Microbiology and Tumor Biology Center, Karolinska Institute, Stockholm, S-17177, Sweden

L A Klobutcher
Department of Biochemistry, University of Connecticut Health Center, Farmington, CT 06030, USA

J W Kloepper
Department of Entomology and Plant Pathology, Auburn University, AL 36849, USA

M A Koch
Institute of Botany, University of Agricultural Sciences, A-1180 Vienna, Austria

Y Kohara
Center of Genetic Resources Information, National Institute of Genetics, Mishima, 411-8540, Shizuoka-ken, Japan

G S Kopf
Center for Research on Reproduction and Women's Health, University of Pennsylvania School of Medicine, Philadelphia, PA 19104, USA

H C Korswagen
Hubrecht Laboratory, Netherlands Institute for Development Biology, NL-3584 CT Utrecht, The Netherlands

D C Krakauer
Institute for Advanced Study, Princeton University, Princeton, NJ 08540, USA

M Kreitman
Department of Ecology and Evolutionary Biology, University of Chicago, Chicago, IL 60637, USA

K N Kreuzer
Department of Microbiology and Biochemistry, Duke University Medical Center, Durham, NC 27710, USA

C B Krimbas
Department of Philosophy and History of Science, University of Athens, Athens 16771, Greece

H B Krishnan
Department of Plant Pathology, United States Department of Agriculture-ARS, University of Missouri, Columbia, MO 65211, USA

R Krumlauf
Stowers Institute for Medical Research, Kansas City, MO 64110, USA

A Kumar
Department of Molecular Cellular and Developmental Biology, Yale University, New Haven, CT 06520, USA

C Kurland
Uppsala University, S-24332 Hoor, Sweden

M Kutateladze
G Eliava Institute of Bacteriophages, Microbiology and Virology, Tbilisi, 380060, Georgia

E Kutter
Lab1, The Evergreen State College, Olympia, WA 98505, USA

C P Kyriacou
Department of Genetics, University of Leicester, Leicester, LE1 7RH, UK

B N La Du
Department of Pharmacology, University of Michigan Medical School, Ann Arbor, MI 48109, USA

M Labouesse
Institut de Génétique et de Biologie Moléculaire et Cellulaire, CNRS/INSERM/ULP, BP163, F-67404 Illkirch, France

S A Lacks
Department of Biology, Brookhaven National Laboratory, Upton, NY 11973, USA

B C Lamb
Department of Biology, Imperial College, London, SW7 2AZ, UK

A Landy
Department of Biology and Medicine, Brown University, Providence, RI 02912, USA

R A LaRossa
E.I. du Pont de Nemours & Co., Experimental Station, Wilmington, DE 19880, USA

D S Latchman
Windeyer Institute of Medical Sciences, University College London Medical School, London, WIP 6DB, UK

F Latif
Department of Medical Genetics, University of Birmingham, Birmingham, B15 2TT, UK

J Y-K Lau
3098 Holly Hall, Houston, TX 77054, USA

J Laurén
Molecular/Cancer Biology Laboratory, University of Helsinki, SF00014, Helsinki, Finland

J Laval
Institut Gustav Roussy, CNRS, F-94805 Villejuif, France

P Laybourn
Department of Biochemistry and Molecular Biology, Colorado State University, Fort Collins, CO 80523, USA

P Leder
Department of Genetics, Harvard Medical School, Boston, MA 02115, USA

J Y Lee
Whitehead Institute for Biomedical Research and Department of Biology, Massachusetts Institute of Technology, Cambridge, MA 02142, USA

G Lefranc
Laboratoire d'ImmunoGénétique Moléculaire, Institut de Génétique Humaine, Université Montpellier II, F-34396 Montpellier, France

M-P Lefranc
Laboratoire d'ImmunoGénétique Moléculaire, Institut de Génétique Humaine, Université Montpellier II, F-34396 Montpellier Cedex 5, France

J W Lengeler
Fachbereich Biologie/Chermie, AG Genetik, University of Osnabrück, D-49069 Osnabrück, Germany

R E Lenski
Center for Microbial Ecology, Michigan State University, East Lansing, MI 48824, USA

P A Lessard
Department of Biology, Massachusetts Institute of Technology, Cambridge, MA 02138, USA

G Levan
CMB Genetics, University of Göteborg, S-40530 Göteborg, Sweden

J Levinton
Department of Ecology and Evolution, State University of New York, Stony Brook, NY 11794, USA

S B Levy
Center for Adaptation Genetics and Drug Resistance, Tufts University School of Medicine, Boston, MA 02111, USA

D Lew
Duke University Medical Center, Durham, NC 27710, USA

R Lewis
7 Harvest Drive, Scotia, New York, NY 12302, USA

R C Lewontin
Museum of Comparative Zoology, Harvard University, Cambridge, MA 02138, USA

S W L'Hernault
Department of Biology, Emory University, Atlanta, GA 30322, USA

W-H Li
Department of Ecology and Evolution, University of Chicago, Chicago, IL 60637, USA

M R Lieber
Department of Pathology and of Biochemistry, Norris Comprehensive Cancer Center, University of Southern California School of Medicine, Los Angeles, CA 90089, USA

A Liljas
Molecular Biophysics, Center for Chemistry and Chemical Engineering, Lund University, S-22100, Lund, Sweden

D M J Lilley
CRC Nucleic Structure Research Group, Biochemistry Department, University of Dundee, Dundee, DD1 4HN, UK

J Limon
Department of Biology and Genetics, Medical University of Gdansk, 80-210 Gdansk, Poland

G J Lithgow
School of Biological Sciences, University of Manchester, Manchester, M13 9PT, UK

A C Lloyd
MRC Laboratory for Cell Biology, University College London, London, WCIE 6BT, UK

A Long
Department of Ecology and Evolutionary Biology, University of California–Irvine, Irvine, CA 92697, USA

W-E Lönning
Max Planck Institut für Züchtungsforschung, D-50829 Köln, Germany

E J Louis
Department of Biochemistry and Molecular Biology, University of Leicester, Leicester, LE1 7RH, UK

P S Lovett
Department of Biological Sciences, University of Maryland, Catonsville, MD 21228, USA

S T Lovett
Department of Genetics and Molecular Biology, Brandeis University, Waltham, MA 02254, USA

D Low
Department of Molecular, Cellular and Development Biology, University of California–Santa Barbara, Santa Barbara, CA 93106, USA

K B Low
Radiobiology Laboratories, Yale University, New Haven, CT 06520, USA

P Lu
Department of Chemistry, University of Pennsylvania, Philadelphia, PA 19104, USA

S Lusetti
Department of Biochemistry, University of Wisconsin–Madison, Madison, WI 53706, USA

L Luzzatto
Instituto Nazionale per la Ricerca sul Cancro, 16132 Genova, Italy

J Lyndal York
University of Arkansas for Medical Sciences, Little Rock, AR 72205, USA

M F Lyon
MRC Mammalian Genetics Unit, Harwell, Didcot, OX11 0RD, UK

H C Macgregor
Department of Biology, University of Leicester, Leicester, LE1 7RH, UK

J Mager
Department of Genetics, University of North Carolina, Chapel Hill, NC 27599, USA

M E Magnello
Wellcome Trust Centre for the History of Medicine at University College London, London, NW1 1AD, UK

T Magnuson
Department of Genetics, School of Medicine, University of North Carolina, Chapel Hill, NC 27599, USA

E Maher
Department of Medical Genetics, University of Birmingham, Birmingham, B15 2TT, UK

E M Maine
Department of Biology, Syracuse University, Syracuse, NY 13244, USA

A Maitra
Department of Pathology, University of Texas Southwestern Medical Center, Dallas, TX 75390, USA

N Maizels
Department of Immunology, University of Washington Medical School, Seattle, WA 98195, USA

K D Makova
Department of Ecology and Evolution, University of Chicago, Chicago, IL 60637, USA

S Malcolm
Molecular and Clinical Genetics Institute, Institute for Child Health, London, WC1N 1EH, UK

S Maloy
Department of Microbiology, University of Illinois, Urbana, IL 61801, USA

T Maniatis
Department of Molecular and Cellular Biology, Harvard University, Cambridge, MA 02138, USA

A Mansouri
Department of Molecular Cell Biology, Max Planck Institute for Biophysical Chemistry, D-37077 Göttingen, Germany

R Marcus
Department of Haematological Medicine, MRC Centre, Cambridge, CB2 2QH, UK

A Martinez-Arias
Department of Genetics, University of Cambridge, Cambridge, CB2 3EA, UK

P H Masson
Laboratory of Genetics, University of Wisconsin–Madison, Madison, WI 53706, USA

C Mathew
Division of Medical and Molecular Genetics, Guy's, King's and St Thomas's School of Medicine, King's College London, London, SE1 9RT, UK

J Mathur
Botanik III, Universität Köln, D-50931 Köln, Germany

L Maxson
College of Liberal Arts, University of Iowa, Iowa City, IA 52242, USA

E Mayr
Museum of Comparative Zoology, Cambridge, MA 02138, USA

S McKee
Clinical Genetics Unit, Birmingham Women's Hospital, Birmingham, B15 2TG, UK

D W Meinke
Department of Botany, Oklahoma State University, Stillwater, OK 74078, USA

F Meins Jr
Friedrich Miescher Institute, Maulbeerstrasse 66, Basel CH-4058, Switzerland

M Melnick
University of California, Los Angeles, CA 90089, USA

S K Merickel
Department of Biological Chemistry, University of California–Los Angeles, Los Angeles, CA 90095, USA

J Merriam
Department of Biology, University of California–Los Angeles, Los Angeles, CA 90095, USA

J P Métraux
Département de Biologie, Institut de Biologie Végétale, Université de Fribourg, CH-1700 Fribourg, Switzerland

J Michiels
Centre of Microbial and Plant Genetics, Kasteelpark Arenberg 20, B-3001 Heverlee, Belgium

C J Migeon
Department of Pediatrics, The Johns Hopkins Hospital, Baltimore, MD 21287, USA

H I Miller
Hoover Institution, Stanford University, Stanford, CA 94305, USA

J H Miller
Department of Microbiology and Molecular Genetics, University of California–Los Angeles, Los Angeles, CA 90024, USA

O J Miller
Center for Molecular Medicine and Genetics, Wayne State University School of Medicine, Detroit, MI 48201, USA

L Mindich
Public Health Research Institute, New York University, New York, NY 10016, USA

J B Mitton
Department of Environmental, Population and Organismic Biology, University of Colorado–Boulder, Boulder, CO 80309, USA

P B Moens
Department of Biology, York University, Downsview, Ontario, M3J IP3, Canada

G Morata
Centro de Biología Molecular, Consejo Superior de Investigaciones Científicas, Universidad Autónoma de Madrid, 28049 Madrid, Spain

I Mori
Laboratory of Molecular Neurobiology, Division of Biological Science, Graduate School of Science, Nagoya University, 464-8602 Nagoya, Japan

H C Morse
National Institute of Allergy and Infectious Diseases, National Institutes of Health, Bethesda, MD 20892, USA

R K Mortimer
Department of Molecular and Cell Biology, University of California–Berkeley, Berkeley, CA 94720, USA

N E Morton
Genetic Epidemiology Research Group, Human Genetics, Princess Anne Hospital, University of Southampton, Southampton, S016 5YA, UK

R C Moschel
Chemistry of Carcinogenesis Laboratory, National Cancer Institute at Frederick, Frederick, MD 21702, USA

R Mottus
Department of Biological Sciences, University of British Columbia, Vancouver, British Columbia, V6T 1Z4, Canada

L D Mueller
Department of Ecology and Evolutionary Biology, University of California–Irvine, Irvine, CA 92697, USA

B Müller-Hill
Institut für Genetik, Universität Köln, D-50931 Köln, Germany

L M Mulligan
Departments of Pediatrics and Pathology, Queen's University, Kingston, Ontario, K7L 3N6, Canada

E J Murgola
Department of Molecular Genetics, The University of Texas M.D. Anderson Cancer Center, Houston, TX 77030, USA

K Nagai
MRC Laboratory of Molecular Biology, Cambridge, CB2 2QH, UK

M Nei
Department of Biology, Pennsylvania State University, University Park, PA 16802, USA

F C Neidhardt
Department of Microbiology and Immunology, University of Michigan, Ann Arbor, MI 48109, USA

H C M Nelson
Department of Biochemistry and Biophysics, University of Pennsylvania School of Medicine, Philadelphia, PA 19104, USA

K E Nelson
Institute for Genomic Research, 9712 Medical Center Drive, Rockville, MD 20850, USA

C Neuhauser
Department of Ecology, Evolution and Behavior, University of Minnesota, St Paul, MN 55108, USA

G Newton
The Wellcome Trust, 210 Euston Road, London, NW1 2BE, UK

W C Nierman
The Institute for Genomic Research, 9712 Medical Center Drive, Rockville, MD 20850, USA

M E M Noble
Laboratory of Molecular Biophysics, Department of Biochemistry and Oxford Centre for Molecular Sciences, University of Oxford, Oxford, OX1 3QU, UK

K M Noll
Department of Molecular and Cell Biology, University of Connecticut, Storrs, CT 06269, USA

C J Norbury
Imperial Cancer Research Fund, Institute of Molecular Medicine, University of Oxford, Oxford, OX3 7LJ, UK

M A Nowak
Institute for Advanced Study, Princeton University, Princeton, NJ 08540, USA

W L Nyhan
Department of Pediatrics and Institute for Molecular Genetics, University of California–San Diego, La Jolla, CA 92093, USA

S J O'Brien
Laboratory of Genomic Diversity, National Cancer Institute, Frederick Cancer Research and Development Center, Frederick, MD 21701, USA

K O'Connell
Laboratory of Molecular Biology, University of Wisconsin–Madison, Madison, WI 53706, USA

C J O'Kane
Department of Genetics, University of Cambridge, Cambridge, CB2 3EH, UK

S L O'Kane Jr
Department of Biology, University of Northern Iowa, Cedar Falls, IA 50614, USA

O O'Neill
Newnham College, Cambridge, CB3 9DF, UK

J L H O'Riordan
14 Northampton Park, London, N1 2PJ, UK

R Oberbauer
Department of Internal Medicine III, University of Vienna, A-1090 Vienna, Austria

F Oesch
Institut für Toxikologie, Obere Zahlbacher Strasse 67, D-55131 Mainz, Germany

R F Ogle
Department of Obstetrics and Gynecology, University College London, London, WC1E 6HX, UK

T Ohta
National Institute of Genetics, Mishima, 411-8540 Shizuoka-ken, Japan

R Olby
Department of the History and Philosophy of Science, University of Pittsburgh, Pittsburgh, PA 15260, USA

G J Olsen
Department of Chemistry and Life Sciences, University of Illinois, Urbana, IL 61801, USA

T L Orr-Weaver
Whitehead Institute for Biomedical Research and Department of Biology, Massachusetts Institute of Technology, Cambridge, MA 02142, USA

A Orth
Génome, Populations, Interactions UMR 5000, Université de Montpellier, F-34095 Montpellier, France

E A Ostrander
Fred Hutchinson Cancer Research Center, 1100 Fairview Ave N, Seattle, WA 98109, USA

R A Padua
Hematology Department, University of Wales College of Medicine, Cardiff, CF14 4XN, UK

R Palacios
Nitrogen Fixation Research Center, National University of México, Cuernavaca, Mor. CP 62210, Mexico

P Pamilo
Department of Biology, University of Oulu, SF-90014 Oulu, Finland

V E Papaioannou
Department of Genetics and Development, College of Physicians and Surgeons of Columbia University, New York, NY 10032, USA

J Parker
Department of Microbiology, Southern Illinois University Carbondale, Carbondale, IL 62901, USA

R Parker
Department of Molecular and Cellular Biology, Howard Hughes Medical Institute, University of Arizona, Tucson, AZ 85721, USA

H C Passmore
Department of Biological Sciences, Rutgers University, Piscataway, NJ 08855, USA

J Paszkowski
Friedrich Miescher Institute, CH-4002 Basel, Switzerland

M L Pato
Department of Microbiology, University of Colorado Health Sciences Center, Denver, CO 80262, USA

I T Paulsen
Institute for Genomic Research, 9712 Medical Center Drive, Rockville, MD 20850, USA

W J Pavan
Laboratory of Genetic Disease Research, National Human Genome Research Institute, National Institutes of Health, Bothesda, MD 20892, USA

P L Pearson
Division of Medical Genetics, Wilhelmina Children's Hospital, Utrecht University Medical Center, NL-3508 AB, Utrecht, The Netherlands

L Peltonen
Department of Human Genetics, University of California–Los Angeles, School of Medicine, Los Angeles, CA 90095, USA

D Penny
Institute of Molecular Sciences, Massey University, Palmerston North, New Zealand

R R Pera
Departments of Urology, Obstetrics and Gynecology and Reproductive Sciences and Physiology, University of California–San Francisco, San Francisco, CA 94143, USA

J J Perona
Department of Chemistry and Biochemistry, University of California–Santa Barbara, Santa Barbara, CA 93106, USA

X Perret
Laboratoire de Biologie Moléculaire des Plantes Supérieures, Université de Genève, CH-1292 Chambésy, Genève, Switzerland

M M R Petit
Laboratory for Molecular Oncology, Center for Human Genetics, University of Leuven and Flanders Interuniversity Institute for Biotechnology, B-3000 Leuven, Belgium

J F Petrosino
Department of Molecular and Human Genetics, Baylor College of Medicine, Houston, TX 77030, USA

J Phelan
Department of Organismic Biology, University of California, Los Angeles, CA 90095, USA

E Pianka
Section of Integrative Biology, University of Texas–Austin, Austin, TX 78712, USA

S H Pilder
Department of Anatomy and Cell Biology, Temple University School of Medicine, Philadelphia, PA 19104, USA

J M Pipas
Department of Biological Sciences, University of Pittsburgh, Pittsburgh, PA 15260, USA

R H A Plasterk
Hubrecht Laboratory, Netherlands Institute for Development Biology, NL-3584 CT Utrecht, The Netherlands

E Pollak
Department of Statistics, Iowa State University, Ames, IA 50011, USA

J W Pollard
Departments of Developmental and Molecular Biology and Obstetrics and Gynecology and Women's Health, Albert Einstein College of Medicine, NY 10461, USA

D Pomeranz Krummel
Structural Studies Division, Medical Research Council, Laboratory of Molecular Biology, Cambridge, CB2 2QH, UK

F M Pope
Institute of Medical Genetics, University Hospital of Wales, Cardiff, CF4 4XN, UK

M Potter
Laboratory of Genetics, National Cancer Institute, National Institutes of Health, Bethesda, MD 20892, USA

J Poulton
Department of Paediatrics, John Radcliffe Hospital, University of Oxford, Oxford, OX3 9DU, UK

P M Pour
UNMC, Eppley Cancer Center and Department of Pathology and Microbiology, University of Nebraska Medical Center, Omaha, NE 68198, USA

W Powell
Scottish Crop Research Institute, Invergowrie, Dundee, DD2 5DA, UK

D Prangishvili
Department of Microbiology, Universität Regensburg, D-93053 Regensburg, Germany

T Prout
Department of Evolution and Ecology, University of California–Davis, Davis, CA 95616, USA

R E Pyeritz
Department of Human Genetics, University of Pennsylvania School of Medicine, Philadelphia, PA 19104, USA

P Rabbitts
Department of Oncology, Addenbrooke's Hospital, Cambridge, CB2 2QH, UK

T H Rabbitts
Division of Protein and Nucleic Acid Chemistry, MRC Laboratory of Molecular Biology, Cambridge, CB2 2QH, UK

T N K Raju
Department of Pediatrics, University of Illinois–Chicago, Chicago, IL 60612, USA

E A Raleigh
New England Biolabs, 32 Tozer Road, Beverly, MA 01915, USA

J T Reardon
Department of Biochemistry and Biophysics, University of North Carolina School of Medicine, Chapel Hill, NC 27599, USA

G D Recchia
Department of Biochemistry, University of Oxford, Oxford, OX1 3QU, UK

R J Redfield
Department of Zoology, University of British Columbia, Vancouver, British Columbia, V6T 1Z4, Canada

R H Reeves
Department of Physiology, Johns Hopkins University School of Medicine, Baltimore, MD 21205, USA

L J Reha-Krantz
Department of Biological Sciences, University of Alberta, Edmonton, Alberta, T6G 2E9, Canada

S Reichheld
Department of Biology, McMaster University, Hamilton, Ontario, L8S 4K1, Canada

W Reik
Programme in Developmental Genetics, Babraham Institute, Cambridge, CB2 4AT, UK

B J Reinhart
Department of Molecular Biology, Massachusetts General Hospital, Boston, MA 02114, USA

R I Richards
Centre for Medical Genetics, Department of Cytogenetics and Molecular Genetics, Women's and Children's Hospital, North Adelaide, SA 5006, Australia

J P Richardson
Department of Chemistry, Indiana University, Bloomington, IN 47405, USA

N Richardson
Institute of Cancer Research, Chester Beatty Laboratories, London SW3 6JB, UK

R W Ridge
Division of Natural Sciences, International Christian University, Osawa, Mitaka, Tokyo 181-8585, Japan

P Riggs
New England Biolabs, 32 Tozer Road, Beverly, MA 01915, USA

D L Rimoin
Medical Genetics Birth Defects Center, Steven Spielberg Pediatrics Research Center, Cedars–Sinai Medical Center and University of California–Los Angeles School of Medicine, Los Angeles, CA 90048, USA

E M Rinchik
Oak Ridge National Laboratory, Oak Ridge, TN 37831, USA

J L Rinkenberger
Department of Anatomy, University of California–San Francisco, San Francisco, CA 94143, USA

L S Ripley
Department of Microbiology and Molecular Genetics, University of Medicine and Dentistry of New Jersey, New Jersey Medical School, Newark, NJ 07103, USA

F T Robb
Center of Marine Biotechnology, University of Maryland Biotechnology Institute, Baltimore, MD 21202, USA

J-D Rochaix
Department de Biologie Moléculaire, Université de Genève, CH-1211 Genève, Switzerland

C H Rodeck
Department of Obstetrics and Gynecology, University College London, London, WC1E 6HX, UK

T H Roderick
The Jackson Laboratory, Bar Harbor, ME 04609, USA

B A Roe
Department of Chemistry and Biochemistry, University of Oklahoma, Norman, OK 73019, USA

F J Rohlf
Department of Ecology and Evolution, State University of New York, Stony Brook, NY 11794, USA

C Ronson
Department of Microbiology, University of Otago, Dunedin, New Zealand

W A Rosche
Department of Biological Science, University of Tulsa, Tulsa, OK 74104, USA

S M Rosenberg
Department of Molecular and Human Genetics, Biochemistry, and Molecular Virology and Microbiology, Baylor College of Medicine, Houston, TX 77030, USA

J R Roth
Department of Biology, University of Utah, Salt Lake City, UT 84112, USA

N Rougier
Department of Anatomy, University of California–San Francisco, San Francisco, CA 94143, USA

A E Rougvie
Department of Genetics, Cell Biology and Development, University of Minnesota, St Paul, MN 55108, USA

M P Rout
The Rockefeller University, New York, NY 10021, USA

J D Rowley
Section of Hematology and Oncology, University of Chicago Medical School, Chicago, IL 60637, USA

C A Royer
Centre de Biochimie Structurale, Université de Montpellier, F-34090 Montpellier, France

D C Rubinsztein
Department of Medical Genetics, Wellcome Trust Centre for Molecular Mechanisms in Disease, Cambridge Institute for Medical Research, University of Cambridge, Cambridge, CB2 2XY, UK

F Ruddle
Department of Molecular Genetics, Yale University, New Haven, CT 06520, USA

A K Rustgi
Division of Gastroenterology, University of Pennsylvania, Philadelphia, PA 19104, USA

I Ruvinsky
Department of Molecular Biology, Massachusetts General Hospital, Boston, MA 02114, USA

G Ruvkun
Department of Molecular Biology, Massachusetts General Hospital, Boston, MA 02114, USA

H Saedler
Max Planck Institut für Züchtungsforschung, D-50829 Köln, Germany

N Saitou
Laboratory of Evolutionary Genetics, National Institute of Genetics, Mishima, 411-8540 Shizuoka-ken, Japan

M Salas
Centro de Biología Molecular 'Severo Ochoa' (CSIC-UAM), Universidad Autónoma, Canto Blanco, E-28049 Madrid, Spain

L D Samson
Department of Cancer Cell Biology, Harvard School of Public Health, Boston, MA 02215, USA

A Sancar
Department of Biochemistry and Biophysics, University of North Carolina School of Medicine, Chapel Hill, NC 27599, USA

M R Sanderson
Randall Centre for Molecular Mechanisms of Cell Function, Guy's, King's and St Thomas's Hospitals School of Biomedical Sciences, London SE1 1UL, UK

T Sasaki
Rice Genome Research Program, National Institute of Agrobiological Resources, Tsukuba, Ibaraki 305-8602, Japan

B Sauer
Oklahoma Medical Research Foundation, Oklahoma City, OK 73104, USA

R Savarirayan
Victorian Clinical Genetics Service, Royal Children's Hospital, Melbourne, VIC 3052, Australia

T Schedl
Department of Genetics, Washington University, School of Medicine, St Louis, MO 63110, USA

K Schesser
Department of Cell and Molecular Biology, Section of Immunology, Lund University, Lund, S-22184, Sweden

I Schildkraut
New England Biolabs, 32 Tozer Road, Beverly, MA 01915, USA

P Schimmel
Department of Molecular Biology, Skaggs Institute for Chemical Biology, The Scripps Research Institute, La Jolla, CA 92037, USA

T Scholl
Myriad Genetic Laboratories, Salt Lake City, UT 84108, USA

B E Schoner
Lilly Research Laboratories, Lilly Corporate Center, Indianapolis, IN 46285, USA

K Schüler
Institut für Pflanzengenetik und Kulturpflanzenforschung Gatersleben, Genbank Aussenstelle, D-18190 Gross Lüsewitz, Germany

R M Schultz
Division of Biochemistry, Department of Cell Biology, Neurobiology and Anatomy, Stritch School of Medicine, Loyola University Chicago, Maywood, IL 60153, USA

J Scott
Hammersmith Hospital, Imperial College School of Medicine, London, W12 0NN, UK

A G Searle
MRC Mammalian Genetics Unit, Harwell, Didcot, OX11 0RD, UK

D Seemungal
The Wellcome Trust, 210 Euston Road, London NW1 2BE, UK

S Segal
Department of Pediatrics and Medicine, The Children's Hospital of Philadelphia, Philadelphia, PA 19104, USA

A M Segall
Department of Biology and Molecular Biology Institute, San Diego State University, San Diego, CA 92182, USA

M F Seldin
Department of Biological Chemistry and Medicine, University of California–Davis, Davis, CA 95616, USA

M J Seller
Division of Medical and Molecular Genetics, Guy's, King's and St Thoma's Hospitals School of Medicine, London, SE1 9RT, UK

V Sgaramella
Dipartimento di Biologia Cellulare, Università della Calabria, 87030 Arcavacata di Rende CS, Italy

H B Shaffer
Department of Ecology and Evolutionary Biology, University of California, Davis, Davis, CA 95616, USA

P Sham
Institute of Psychiatry, King's College London, London, SE5 8AF, UK

P M Sharp
Institute of Genetics, University of Nottingham, Nottingham, NG7 2UH, UK

D J Sherratt
Department of Biochemistry, University of Oxford, Oxford, OX1 3QU, UK

J Shulman
Department of Molecular and Medical Genetics, University of Toronto, Toronto, Ontario, M5G 1X5, Canada

D O Sillence
Department of Clinical Genetics, The New Children's Hospital, Paramatta, NSW 2124, Australia

L Silver
Department of Molecular Biology, Princeton University, Princeton, NJ 08544, USA

E H Simon
Department of Biological Sciences, Purdue University, West Lafayette, IN 47907, USA

R W Simons
Department of Microbiology, Immunology and Molecular Genetics, University of California–Los Angeles, Los Angeles, CA 90095, USA

A Sinclair
Department of Paediatrics, University of Melbourne and Murdoch Children's Research Institute, Royal Children's Hospital, Melbourne, VIC 3052, Australia

R S Singh
Department of Biology, McMaster University, Hamilton, Ontario, L8S 4K1, Canada

A J Sinskey
Department of Biology, Massachusetts Institute of Technology, Cambridge, MA 02138, USA

T R Skopek
Genetic and Cellular Toxicology Department, Merck Research Laboratories, West Point, PA 19486, USA

B J Smith
Department of Molecular Biology, Princeton University, Princeton, NJ 08544, USA

D W Smith
Department of Biology and Center for Molecular Genetics, University of California–San Diego, San Diego, CA 92093, USA

G R Smith
Fred Hutchinson Cancer Research Center, Seattle, WA 98104, USA

J W Snape
John Innes Centre, Colney, Norwich, NR4 7UH, UK

P H A Sneath
Department of Microbiology and Immunology, University of Leicester, Leicester, LE1 9HN, UK

M Snyder
Department of Molecular Cellular and Developmental Biology, Yale University, New Haven, CT 06520, USA

B Sollner-Webb
Department of Biological Chemistry, The Johns Hopkins University School of Medicine, Baltimore, MD 21205, USA

D Solter
Department of Developmental Biology, Max Planck Institute of Immunology, D-79108 Freiburg, Germany

R L Somerville
Department of Biochemistry, Purdue University, West Lafayette, IN 47907, USA

L C Sowers
Department of Pediatrics and Molecular Medicine, City of Hope National Medical Center, Duart, CA 91010, USA

B T Spear
Department of Microbiology and Immunology, University of Kentucky, Chandler Medical Center, Lexington, KY 40536, USA

T P Speed
Division of Genetics and Bioinformatics, The Walter and Eliza Hall Institute of Medical Research, Melbourne, VIC 3050, Australia

R A Spritz
Human Medical Genetics Program, University of Colorado Health Sciences Center, Denver, CO 80262, USA

N K Spurr
SmithKline Beecham Pharmaceuticals, Harlow, CM19 5AW, UK

G Stacey
Center for Legume Research, University of Tennessee, Knoxville, TN 37996, USA

D Stadler
Department of Genetics, University of Washington, Seattle, WA 98195, USA

C Staehelin
Laboratoire de Biologie Moléculaire des Plantes Supérieures, Université de Genève, CH-1292 Chambésy/Genève, Switzerland

F W Stahl
Institute of Molecular Biology, University of Oregon, Eugene, OR 97403, USA

N Standart
Department of Biochemistry, University of Cambridge, Cambridge, CB2 1QW, UK

G S Stent
Department of Molecular and Cell Biology, University of California–Berkeley, Berkeley, CA 94720, USA

C L Stewart
Laboratory of Cancer and Developmental Biology, National Cancer Institute at Frederick, Frederick, MD 21702, USA

A Stoltzfus
Center for Advanced Research in Biotechnology, 9600 Gudelsky Drive, Rockville, MD 20874, USA

J Stougaard
Laboratory of Gene Expression, Department of Molecular and Structural Biology, University of Aarhus, DK-8000 Aarhus C, Denmark

L D Strausbaugh
Department of Molecular and Cell Biology, University of Connecticut, Storrs, CT 06269, USA

J C Strefford
The Orchid Cancer Appeal, St Bartholomew's Medical College, London, EC1M 6BQ, UK

L Stubbs
Genome Center, Lawrence Livermore National Laboratory, Livermore, CA 94550, USA

Y Sugimoto
Department of Pharmacology, University of Pennsylvania Medical Center, Philadelphia, PA 19104, USA

J Sullivan
Department of Microbiology, University of Otago, Dunedin, New Zealand

K F Sullivan
Department of Cell Biology, The Scripps Research Institute, La Jolla, CA 92037, USA

W C Summers
Department of Therapeutic Radiobiology, Yale University, New Haven, CT 06520, USA

A T Sumner
Saileyknowes Court, North Berwick, EH39 4RG, UK

M Susman
Laboratory of Genetics, University of Wisconsin–Madison, Madison, WI 53706, USA

G R Sutherland
Department of Cytogenetics and Molecular Genetics, Women's and Children's Hospital, Centre for Medical Genetics, North Adelaide, SA 5006, Australia

E Szathmáry
Department of Plant Taxonomy and Ecology, Eötvös University and Collegium Budapest (Institute for Advanced Study), H-1014 Budapest, Hungary

M D Szczelkun
Department of Biochemistry, University of Bristol, Bristol, B58 1TD, UK

T Szczepański
Department of Immunology, Erasmus University, Rotterdam, University Hospital Rotterdam, NL-3000 DR Rotterdam, The Netherlands

D Szymanski
Department of Agronomy, Purdue University, West Lafayette, IN 49707, USA

F Tata
Department of Genetics, University of Leicester, Leicester, LE1 7RH, UK

S Tavaré
Department of Mathematics, University of Southern California, Los Angeles, CA 90089, USA

N Tavernarakis
Department of Molecular Biology and Biochemistry, The State University of New Jersey, Rutgers, Piscataway, NJ 08854, USA

A M R Taylor
CRC Institute for Cancer Studies, University of Birmingham, Birmingham, B15 2TT, UK

C Tease
Department of Biological Sciences, University of Warwick, Coventry, CV4 7AL, UK

G Theissen
Department Molekulare Pflanzengenetik, Max Planck Institut für Züchtungsforschung, D-50829 Köln, Germany

E Thomas
Cold Spring Harbor Laboratory, Cold Spring Harbor, NY 11724, USA

G Thomson
Department of Integrative Biology, University of California–Berkeley, Berkeley, CA 94720, USA

W E Timberlake
Ōcaté Technical Consulting, PO Box 127, Ocate, NM 87734, USA

T D Tlsty
Department of Pathology and University of California–San Francisco Comprehensive Cancer Center, University of California–San Francisco, San Francisco, CA 94143, USA

J L Tolmie
Duncan Guthrie Institute of Medical Genetics, Yorkhill, Glasgow, G3 8SJ, UK

M Tracey
Department of Biological Sciences, Florida International University, Miami, FL 33199, USA

A Travers
MRC Laboratory of Molecular Biology, Cambridge, CB2 2QH, UK

L-C Tsui
Department of Genetics, The Hospital for Sick Children, Toronto, Ontario, M5G 1X8, Canada

P J Turek
Department of Urology, University of California–San Francisco, San Francisco, CA 94143, USA

C Turleau
Service de Cytogénétique, Hôpital Necker Enfants Malades, F-75743 Paris, France

A B Ulrich
Eppley Cancer Center and Department of Pathology and Microbiology, University of Omaha Medical Center, Omaha, NE 68198, USA

F D Urnov
Sangamo Biosciences, Point Richmond Technical Center, Richmond, CA 94804, USA

D W Ussery
Institute of Biotechnology, Technical University of Denmark, DK-2800 Lyngby, Denmark

M K Uyenoyama
Department of Zoology, Duke University, Durham, NC 27708, USA

W J M Van de Ven
Laboratory for Molecular Oncology, Center for Human Genetics, University of Leuven and Flanders Interuniversity Institute for Biotechnology, B-3000 Leuven, Belgium

E van den Berg
Department of Medical Genetics, University of Groningen, NL-9713 AW Groningen, The Netherlands

V H J van der Velden
Department of Immunology, Erasmus University Rotterdam and University Hospital Rotterdam, NL-3000 DR, Rotterdam, The Netherlands

J Van Dommelen
Centre of Microbial and Plant Genetics, Kasteelpark Arenberg 20, B-7001 Heverlee, Belgium

J J M van Dongen
Department of Immunology, Erasmus University Rotterdam, NL-3000 DR Rotterdam, The Netherlands

R A Van Etten
Center for Blood Research, Harvard Medical School, Boston, MA 02115, USA

M Van Montagu
Department of Molecular Genetics, Ghent University, B-9000 Ghent, Belgium

J Vanderleyden
Centre of Microbial and Plant Genetics, Kasteelpark Arenberg 20, B-3001 Heverlee, Belgium

A Varshavsky
Division of Biology, California Institute of Technology, Pasadena, CA 91125, USA

A R Venkitaraman
CRC Department of Oncology and The Wellcome Trust Centre for Molecular Mechanisms in Disease, Cambridge Institute for Medical Research, Cambridge, CB2 2XY, UK

I M Verma
Laboratory of Genetics, The Salk Institute for Biological Studies, La Jolla, CA 92037, USA

M Vidal
Dana-Farber Cancer Institute, Department of Genetics, Harvard Medical School, Boston, MA 02115

T F Vogt
Department of Pharmacology, Merck and Co., Inc., Merck Research Laboratories, West Point, PA 19486, USA

M J Wade
Department of Biology, Indiana University, Bloomington, IN 47405, USA

G P Wagner
Department of Ecology and Evolutionary Biology, Yale University, New Haven, CT 06520, USA

B T Wakimoto
Departments of Genetics, University of Washington, Seattle, WA 98195, USA

A R Walmsley
Division of Infection and Immunity, Institute of Biomedical and Life Sciences, University of Glasgow, Glasgow, G12 8QQ, UK

S T Warren
Department of Biochemistry, Genetics and Pediatrics, University School of Medicine, Atlanta, GA 30322, USA

P M Wassarman
Department of Biochemistry and Molecular Biology, Mount Sinai School of Medicine, New York, NY 10029, USA

M D Waterfield
Ludwig Institute for Cancer Research, 91 Riding House Street, London, W1P 8BT, UK

R K Wayne
Department of Organismic Biology, Ecology and Evolution, University of California–Los Angeles, Los Angeles, CA 90095, USA

D J Weatherall
Institute of Molecular Medicine, John Radcliffe Hospital, University of Oxford, Oxford, OX3 9DU, UK

N F Weeden
Department of Plant Sciences and Plant Pathology, Montana State University, Bozeman, MT 59717, USA

G M Weinstock
Department of Microbiology and Molecular Genetics, University of Texas–Houston Medical School, Houston, TX 77225, USA

A E Weis
Department of Ecology and Evolutionary Biology, University of California–Irvine, Irvine, CA 92697, USA

R D Wells
Institute of Biosciences and Technology, Texas A & M University System Health Science Center, Houston, TX 77030, USA

Z Werb
Department of Anatomy, University of California–San Francisco, San Francisco, CA 94143, USA

T Werner
Institute of Mammalian Genetics, GSF-National Research Center for Environment and Health, D-85764 Neuherberg, Germany

S A West
Institute of Cell, Animal and Population Biology, University of Edinburgh, Edinburgh, EH9 3JT, UK

J A White
MRC Human Biochemical Genetics Unit, The Galton Laboratory, University College London, London, NW1 2HE, UK

D J Wieczorek
Department of Microbiology, University of Iowa College of Medicine, Iowa City, IA 52242, USA

D E Wilcox
Duncan Guthrie Institute of Medical Genetics, University of Glasgow, Glasgow, G3 8SJ, UK

E O Wiley
Department of Ecology and Evolutionary Biology, University of Kansas, Lawrence, KS 66045, USA

A O M Wilkie
Institute of Molecular Medicine, University of Oxford, John Radcliffe Hospital, Oxford, OX3 9DS, UK

L B Willis
Department of Biology, Massachusetts Institute of Technology, Cambridge, MA 02138, USA

H Winking
Institut für Biologie, Medical University of Lübeck, D-2400 Lübeck, Germany

M E Winkler
Infectious Diseases Research, Lilly Research Laboratories, Lilly Corporate Center, Indianapolis, IN 46285, USA

†A P Wolffe
Formerly of Sangamo Biosciences, Point Richmond Technical Center, Richmond, CA 94804, USA

J Wolstenholme
School of Biochemistry and Genetics, University of Newcastle upon Tyne, Newcastle upon Tyne, NE2 4AA, UK

H Y Wong
Museum of Natural History, University of Oxford, Oxford, OX1 3PW, UK

S L C Woo
Institute for Gene Therapy and Molecular Medicine, Mount Sinai School of Medicine, New York, NY 10029, USA

A H Wyllie
Department of Pathology, University of Cambridge, Cambridge, CBZ 1QP, UK

M-Q Xu
New England Biolabs, 32 Tozer Road, Beverly, MA 01915, USA

Y Yamamoto
Department of Genetics, Hyogo College of Medicine, 1-1 Mukogawa-cho, Nishinomiya, 663-8501 Hyogo, Japan

C Yanofsky
Department of Biological Sciences, Stanford University, Stanford, CA 94305, USA

K J Yook
Department of Biology, University of Utah, Salt Lake City, UT 84112, USA

Z-B Zeng
Department of Statistics, North Carolina State University, Raleigh, NC 27695, USA

D O Zharkov
Department of Pharmacological Sciences, State University of New York, Stony Brook, NY 11794, USA

A Zhelonkina
Department of Biological Chemistry, The Johns Hopkins University School of Medicine, Baltimore, MD 21205, USA

W Zillig
Max Planck Institute for Biochemistry, 82152 Martinsried, Germany

F K Zimmermann
Goethestrasse 12, D-64372 Ober-Ramstadt, Germany

G Zubay
Department of Biological Sciences, Columbia University, New York, NY 10027, USA

J W Zyskind
Department of Biology, San Diego State University, San Diego, CA 92182, USA

Contents

A

(A)n tail 1
A DNA 1
Abortive Transduction 1
Acentric Fragment 2
Achondroplasia 2
Acquired Resistance 3
Acridines 3
Acrocentric Chromosome 3
Acrosome 3
Active Site 4
Adaptive Landscapes 4
Adaptor Hypothesis 7
Additive Genetic Variance 7
Adenine 9
Adenocarcinomas 9
Adenoma 12
Adenomatous Polyposis Coli 12
Adenosine Phosphates 14
Adenoviruses 15
Adjacent/Alternate Disjunction 18
Advanced Intercross Lines 19
Advanced/Derived 20
Aflatoxins 20
Aging, Genetics of 21
Agouti 23
Agrobacterium 23
Alanine 25
Albinism 25
Alcoholism 27
Alignment Problem 29
Alkaptonuria 35
Alkyltransferases 36
Allele Frequency 37
Alleles 38
Allelic Exclusion 39
Allopatric 39
Allostery 39
Allotypes 40
Alpha(α_1)-Antitrypsin Deficiency 40
Alpha(α)-Fetoprotein 42
Alternation of Gene Expression 42
Alternative Splicing 47
Altruism 48
Alu Family 48
Alzheimer's Disease 48
Amber Codon 49
Amber Mutation 49
Amber Suppressors 50
Ames, Bruce 50
Ames Test 51
Amino Acids 54
Amino Acid Substitution 57
Amino Terminus 58
Aminoacyl-tRNA 58
Aminoacyl-tRNA Synthetases 59

Aminopterin 60
Amniocentesis 60
Amplicons 61
Anagenesis 62
Analogy 62
Ancestral Inheritance Theory 63
Anchor Locus 64
Anchorage-Independent Growth 65
Androgenone 65
Aneuploid 65
Angiogenesis 66
Annealing 73
Anonymous Locus 74
Antibiotic Resistance 74
Antibiotic-Resistance Mutants 76
Antibody 78
Anticodons 78
Antigen 79
Antigenic Variation 79
Anti-Oncogenes 81
Antiparallel 82
Antisense DNA 82
Antisense RNA 83
Antitermination Factors 84
Antitermination Proteins 85
AP Endonucleases 85
Apert Syndrome 85
Apomorphy 85
Apoptosis 86
Aporepressors 86
Arabidopsis thaliana: Molecular Systematics and Evolution 86
Arabidopsis thaliana: The Premier Model Plant 87
Arabinose 90
Arachnodactyly 90
Arber, Werner 90
Archaea, Genetics of 91
Arginine 95
Artificial Chromosomes, Yeast 95
Artificial Selection 96
Ascobolus 101
Ascobolus immersus 104
Ascus 105
Asilomar Conference 105
ASO (Allele-Specific Oligonucleotide) 105
Asparagine 105
Aspartic Acid 106
Aspergillus nidulans 106
Assortative Mating 111
Assortment 113
Ataxia Telangiectasia 113
ATP (Adenosine Triphosphate) 115
att Sites 116
Attached-X and other Compound Chromosomes 116
Attachment Sites 120
Attenuation 122
Attenuation, Transcriptional 123
AUG Codons 125
Autocrine/Paracrine 126
Autogenous Control 128
Autoimmune Diseases 128

Autonomous Controlling Element 128
Autoradiography 128
Autoregulation 128
Autosomal Inheritance 129
Autosomes 133
Auxotroph 133

B

B6 135
BAC (Bacterial Artificial Chromosome) 135
Bacillus subtilis 135
Backcross 144
Background Selection 145
Bacteria 146
Bacterial Genes 151
Bacterial Genetics 156
Bacterial Transcription Factors 163
Bacterial Transformation 165
Bacteriophage Recombination 168
Bacteriophage Therapy 175
Bacteriophages 179
Baculovirus System 186
Balanced Polymorphism 186
Balanced Translocation 188
BALB/c Mouse 190
Balbiani Rings 190
Barr Body 191
Basal Cell Carcinoma 191
Base Analog Mutagens 191
Base Composition 192
Base Pair (bp) 193
Base Pairing and Base Pair Substitution 193
Base Substitution Mutations 197
Bases 198
Baur, Erwin 199
Bayesian Analysis 203
Bcl-2 Gene Family 205
BCR/ABL Oncogene 207
Bead Theory 208
Beckwith–Wiedemann Syndrome 209
Behavioral Genetics 209
Benzer, Seymour 212
Beta(β)-Galactosidase 212
Bidirectional Replication 214
BIME, ERIC, REP, RIME, and other Short Bacterial Repeated Elements 214
Binomial Distribution 215
Biochemical Genetics 216
Biosynthesis of Small Molecules 218
Biotechnology 224
Birth Defects 227
Blastocyst 227
Blood Group Chimeras 227
Blood Group Systems 228
Bloom's Syndrome 229
Blunt-End Ligation 230
Bombay Blood Group Phenotype 230
Bombyx mori 231
Bootstrapping 233
Bottleneck Effect 233
Boveri, Theodor 235

B-Prolymphocytic Leukemia (B-PLL) 236
Brachydactyly 236
Brachyury Locus 237
Branch Migration 237
Brassicaceae, Molecular Systematics and Evolution of 237
Brassinosteroids 238
BRCA1/BRCA2 239
Breakage and Reunion 241
Break–Copy/Break–Join 241
Breast Cancer 242
Breeding of Animals 243
Brenner, Sydney 244
Buoyant Density 246
Burkitt's Lymphoma 246

C

C Genes 249
C Value 249
C-Value Paradox 249
C. elegans 250
C57BL/6 250
CA Repeats 250
CAAT Box 250
c-*ABL* Gene and Gene Product 250
Caenorhabditis elegans 251
Cairns, John 256
cAMP and Cell Signaling 258
Campbell Model 260
Cancer Susceptibility 262
Candidate Gene 263
Canine Genetics 264
Cap 270
CAP (CRP) 270
Capsid 271
Carcinogens 271
Carcinoid Tumors 272
Carcinoma 274
Carrier 274
Cassette Model 275
Cassette Mutagenesis 278
Castle, William E. 278
Catabolite Gene Activator Protein 278
Catabolite Repression 281
Cats 284
Cattanach's Translocation 284
cDNA 286
Cell Culture 286
Cell Cycle 286
Cell Determination 296
Cell Division Genetics 297
Cell Division in *Caenorhabditis elegans* 298
Cell Lineage 302
Cell Lines 310
Cell Markers: Green Fluorescent Protein (GFP) 311
Cell/Neuron Degeneration 313
Cenancestor 318
Centimorgan (cM) 319
Centric Fusion 319
Centrioles 320
Centromere 320

Chain Initiation, Elongation, and Termination 323
Chaperonins 324
Character 325
Character State 325
Chargaff's Rules 325
Charon Phages 325
Chi Sequences 325
Chiasma 328
Chimera 330
Chimeric Genes, Proteins 333
Chirality 334
Chlamydomonas reinhardtii 334
Chlamydomonas, Historical Model 337
Chloroplasts, Genetics of 337
Christmas Disease 338
Chromatid 338
Chromatid Interference 339
Chromatin 340
Chromomeres 343
Chromosome 344
Chromosome Aberrations 345
Chromosome Banding 348
Chromosome Break 350
Chromosome Bridge 350
Chromosome Dimer Resolution by Site-Specific Recombination 351
Chromosome Mapping 353
Chromosome Movement 368
Chromosome Number 368
Chromosome Painting 368
Chromosome Pairing, Synapsis 369
Chromosome Scaffold 372
Chromosome Structure 373
Chromosome Walking 375
Circular Linkage Map 376
Cis-Acting Locus 377
Cis-Acting Proteins 380
Cis-Dominance 382
Cis–Trans Configurations 383
Cistron 383
Clade 384
Cladistics 384
Cladogenesis 384
Cladograms 384
Class Switching 385
Cleavage 385
Cleft Lip and Cleft Palate 385
Clinical Genetics 386
Clock Mutants 386
Clone 390
Cloned Organisms 390
Cloning Vectors 391
Closed Reading Frame 392
Coalescent 392
Coat Color Mutations, Animals 397
Coding Sequences 401
Coding Strand 402
Codominance 402
Codon Usage Bias 402
Codons 406
Codons, Invariable 406

Coevolution 407
Cognate tRNAs 412
Cohesive Ends 412
Coincidence, Coefficient of 413
Coincidental Evolution 414
Coisogenic Strain 414
Col Factors 415
Cold-Sensitive Mutant 417
Colicins 417
Colinearity 418
Colony Hybridization 420
Color Blindness 420
Colorectal Cancer 422
Commaless Code 423
Commensal 424
Comparative Genomic Hybridization (CGH) 424
Compartmentalization 426
Compatibility Group 427
Complement Loci 427
Complementary DNA (cDNA) 433
Complementation 434
Complementation Map 434
Complementation Test 434
Complete Penetrance 434
Complex Locus 434
Complex Traits 434
Concatemer 435
Concatemer (Genomes) 435
Concatenated Circles 436
Concerted Evolution 436
Concordance 441
Conditional Lethality 441
Congenic Strain 443
Congenital Adrenal Hyperplasia (Adrenogenital Syndrome) 445
Congenital Disorders 448
Conjugation 449
Conjugation, Bacterial 453
Conjugative Transposition 455
Conplastic 456
Consanguinity 456
Consensus Sequence 457
Conservation Genetics 458
Conservative Recombination 462
Conserved Synteny 462
Consomic 462
Constant Regions 462
Constitutive Expression 462
Constitutive Heterochromatin 463
Constitutive Mutations 463
Contig 463
Continuous Variation 463
Contractile Ring 464
Controlling Elements 464
Convergent Evolution 465
Conversion Gradient 465
Coordinate Regulation 468
Copy-Choice Hypothesis 468
Cordycepin 468
Core Particle 468
Corepressor 469

Correlated Response 469
Cosmids 471
Cotransformation 471
Counterselection 472
Covarion Model of Molecular Evolution 473
CpG Islands 477
Craniosynostosis, Genetics of 478
Cre/*lox* – Transgenics 481
Creutzfeldt–Jacob Disease (CJD) 486
Crick, Francis Harry Compton 486
Cri-du-Chat Syndrome 486
Cross 487
Crossing-Over 488
Crossover Suppressor 488
Crouzon Syndrome 489
Crow, James F 489
Crown Gall Tumors 491
Cruciform DNA 493
Cryptic Satellite 495
Cryptic Splice Sites and Cryptic Splicing 495
ct DNA 497
CTP (Cytidine Triphosphate) 497
Cutis laxa 497
Cyclic AMP (cAMP) 499
Cyclin-Dependent Kinases 500
Cysteine 507
Cystic Fibrosis 507
Cytogenetics 509
Cytokinesis 509
Cytoplasm 509
Cytoplasmic Genes 510
Cytoplasmic Inheritance 510
Cytosine 510
Cytoskeleton 510
Cytosol 510

D

D'Herelle, Félix 511
Darlington, Cyril Dean 512
Darwin, Charles 513
Databases, Sequenced Genomes 517
De Vries, Hugo 521
Deficiency 522
Degenerate Code 522
Delbrück, Max 522
Deletion 524
Deletion Mapping 524
Deletion Mapping, Mouse 524
Deletion Mutation 528
Demes 528
Denaturation (Proteins) 529
Derepression 530
Detoxification (SOD) 530
Deuteranopia 530
Developmental Genetics 530
Developmental Genetics of *Caenorhabditis elegans* 531
Developmental Genetics, Mouse 533
Dicentric Chromosome 533
Dictyostelium 533
Dideoxy Sequencing 533

Dideoxynucleotide 533
Differential Segment 533
DiGeorge Syndrome 533
Diploidy 535
Direct Repeats 537
Directed Deletion In Developmental Processes 537
Directed Mutagenesis 537
Directed Mutation 537
Disassortative Mating 537
Discontinuous Replication 537
Discordance 537
Disequilibrium 537
Disjunction 537
Disruptive Selection 538
Distal 538
Divergent Evolution 539
Divergent Transcription 539
D-Loop 539
DNA 540
DNAase 541
DNA-Binding Proteins 542
DNA Cloning 544
DNA Code 550
DNA Damage 550
DNA Denaturation 550
DNA, History of 553
DNA Hybridization 556
DNA Invertases 556
DNA Lesions 556
DNA Ligases 557
DNA Mapping 557
DNA Marker 557
DNA Modification 557
DNA Polymerase η (Eta) 558
DNA Polymerases 558
DNA Recombination 558
DNA Repair 558
DNA Repair Defects and Human Disease 564
DNA Replication 571
DNA Sequencing 572
DNA Structure 572
DNA Supercoiling 575
DNA Synthesis 577
Dobzhansky, Theodosius 577
Dogs 578
Dominance 578
Dosage Compensation 579
Double-Minute Chromosomes 580
Double-Strand Break Repair Model 580
Down Syndrome 583
Down, Up Mutations 584
Downstream 584
Drosophila melanogaster 584
Drug Resistance 585
Duchenne Muscular Dystrophy (or Meryon's Disease) 586
Duffy Blood Groups 588
Dulbecco, Renato 589
Dunn, L.C. 590
Dwarfism 591
Dwarfism, in Mice 591

Dynamic Mutations 593
Dysmorphology 597
Dystrophin 598

E

E.coli 599
Early Genes (in Phage Genomes) 599
EBNA 599
Ectoderm 599
Ectodermal Dysplasias 599
Editing and Proofreading in Translation 601
Effective Population Number 602
Ehlers–Danlos Syndrome 603
Electron Microscopy 605
Electrophoresis 608
Electroporation 609
Elongation 609
Elongation Factors, Translation 610
Embryo Transfer 611
Embryonic Development of the Nematode *Caenorhabditis elegans* 612
Embryonic Development, Mouse 621
Embryonic Stem Cells 623
End Labeling 623
Endoderm 623
Endonucleases 623
End-Product Inhibition 623
Enhancers 624
Enzymes 625
Epidermal Growth Factor (EGF) 626
Epigenetics 628
Episome 638
Epistasis 638
Epstein–Barr Virus (EBV) 641
Equilibrium 644
Equilibrium Population 646
erbA and *erbB* in Human Cancer 648
Error Catastrophe 649
Erythroblastosis Fetalis 650
ES Cells 653
Escherichia coli 653
ESS 659
Established Cell Lines 659
Ethics and Genetics 659
Ets Family 661
Euchromatin 663
Eugenics 663
Eukaryotes 663
Eukaryotic Genes 663
Euploid 663
Evolution 663
Evolution of Gene Families 666
Evolutionarily Stable Strategies 669
Evolutionary Rate 671
Ewing's Tumor 672
Exchange 673
Exchange Pairing 673
Excision Repair 673
Exon 675
Exonucleases 675
Expression Vector 675

Expressivity 675
Extranuclear Genes 676

F

F Factor 677
F1 Generation 680
F1 Hybrid 680
FAB Classification of Leukemia 680
Fabry Disease (α-Galactosidase A Deficiency) 681
1-, 2-, 3-Factor Crosses 681
Facultative Heterochromatin 682
Familial Fatal Insomnia (FFI) 682
Familial Hypercholesterolemia 682
Fanconi's Anemia 683
Fate Map 683
Favism 683
F-Duction 683
Feline Genetics 684
Female Carriers 685
Feral 688
Fertility, Mutations 688
Fertilization 688
Fertilization, Mammalian 690
Filamentous Bacteriophages 695
Filial Generations 698
Filter Hybridization 698
Fingerprinting 698
First and Second Division Segregation 698
FISH (Fluorescent *in situ* Hybridization) 700
Fisher, R.A. 700
Fitness 702
Fitness Landscape 706
Fix Genes 707
Fixation of Alleles 709
Fixation Probability 709
Flagella 711
Flagellar Phase Variation (Biology) 712
FLI1 Oncogene 712
Flower Development, Genetics of 713
Flp Recombinase-Mediated DNA Inversion 717
FMS Oncogene 721
Foldback DNA 722
Follicular Lymphoma 722
Footprinting 722
Ford, Charles 722
Forward Mutations 724
Fosmid 724
Founder Effect 724
Founder Principle 726
Fragile Chromosome Site 726
Fragile X Syndrome 726
Frameshift Mutation 728
Freedom, Degrees of 729
Frequency-Dependent Fitness 730
Frequency-Dependent Selection 730
Frequency-Dependent Selection as Expressed in Rare Male Mating Advantages 731
Fruit Fly 732
Functional Genomics 733
Fundamental Theorem of Natural Selection 733
Fungal Genetics 738

Fungi 738
FUS-CHOP Fusion 738
Fusion Gene 738
Fusion Proteins 739

G

G1 741
G2 741
Galactosemia 741
Galton, Francis 742
Gametes 748
Gametes, Mammalian 750
Gametic Disequilibrium 750
Gametogenesis 752
Gamma Distribution 752
GAP (*RAS* GTPase Activating Protein) 752
Gastrulation 753
Gaucher's Disease 753
G-Banding 757
Gel Electrophoresis 757
Gene 759
Gene Action 761
Gene Amplification 761
Gene Cassettes 771
Gene Conversion 774
Gene Dosage 778
Gene Duplication 778
Gene Expression 780
Gene Family 783
Gene Flow 785
Gene Frequency 790
Gene Insertion 792
Gene Interaction 792
Gene Library 793
Gene Mapping 793
Gene Number 796
Gene Pool 797
Gene Product 797
Gene Rearrangement in Eukaryotic Organisms 798
Gene Rearrangements, Prokaryotic 800
Gene Regulation 803
Gene Replacement 813
Gene Sequencing 813
Gene Silencing 813
Gene Splicing 814
Gene Substitution 814
Gene Targeting 814
Gene Therapy, Human 814
Gene Transfer 819
Gene Trapping 819
Gene Trees 819
Genetic Code 821
Genetic Colonization 822
Genetic Correlation 823
Genetic Counseling 825
Genetic Covariance 826
Genetic Diseases 827
Genetic Distance 828
Genetic Drift 832
Genetic Engineering 834

Genetic Equilibrium 834
Genetic Homeostasis 835
Genetic Load 838
Genetic Mapping 839
Genetic Marker 839
Genetic Material 839
Genetic Migration 839
Genetic Polarity 840
Genetic Ratios 840
Genetic Recombination 841
Genetic Redundancy 845
Genetic Screening 846
Genetic Stock Collections and Centers 846
Genetic Transformation 854
Genetic Translation 854
Genetic Variation 855
Genetics 856
Genome 857
Genome Organization 859
Genome Relationships: Maize and the Grass Model 863
Genome Size 865
Genomic Library 865
Genomics 872
Genotype 873
Genotypic Frequency 873
Germ Cell 873
Giemsa Banding, Mouse Chromosomes 876
Gilbert, Walter 877
Glioma 878
Globin Genes, Human 878
Glucose 6-Phosphate Dehydrogenase (G6PD) Deficiency 881
Glutamic Acid 884
Glutamine 884
Glycine 884
Glycine max (Soybean) 884
Glycolysis 885
Glycosylase Repair 888
Grasses, Synteny, Evolution, and Molecular Systematics 889
Gravitropism in *Arabidopsis thaliana* 890
Group Selection 894
Growth Factors 899
GSD (Gerstmann–Straussler Disease) 900
GT Repeats 900
GT–AG Rule 900
GTP (Guanosine Triphosphate) 900
Guanine 901
Guide RNA 901
Gynogenone 901
Gyrase 902

H

H19 903
H2 Locus 903
Hadulins 903
Hairpin 903
Hairy Cell Leukemia (HCL) 903
Haldane, J.B.S. 904
Haldane–Muller Principle 906
Haldane's Mapping Function 906
Hamilton's Theory 906

Handedness, Left/Right 910
Haploid Number 911
Haploinsufficiency 911
Haplotype 911
Hardy–Weinberg Law 912
Harlequin Chromosomes 914
Heat Shock Proteins 914
Heavy/Light Chains 915
Helicases 915
Helicobacter pylori 916
Helix–Loop–Helix Proteins 917
Helix–Turn–Helix Motif 917
Helper Phage 917
Hemizygote 917
Hemoglobin 917
Hemophilia 917
Hereditary Diseases 920
Hereditary Neoplasia 920
Heritability 921
Hermaphrodite 924
Hershey, Alfred 925
Heteroallele 926
Heterochromatin 926
Heterochronic Mutation 927
Heterochrony 930
Heteroduplexes 930
Heterogenote 932
Heterokaryon 932
Heteropyknosis 932
Heterosis 933
Heterotrimeric G Proteins 933
Heterozygote and Heterozygosis 934
Hfr 936
Hin/Gin-Mediated Site-Specific DNA Inversion 938
Hirschsprung's Disease 942
Histidine 943
Histidine Operon 943
Histocompatibility 947
Histocompatibility Complex Genes 948
Histone Genes 948
Histones 952
Hitchhiking Effect 952
HIV 953
Hodgkin's Disease 953
Hogness Box 954
Holliday Junction 954
Holliday's Model 955
Holocentric Chromosomes 956
Holophyly 958
Homeobox 958
Homeotic Genes 962
Homeotic Mutation 962
Homogeneously Staining Regions 963
Homologous Chromosomes 963
Homologs 964
Homology 964
Homoplasy 969
Homozygosity 970
Hordeum Species 971
Horizontal Transfer 973

Host-Lethal Gene 975
Host-Range Mutant 976
Hot Spot of Recombination 976
Hot Spots 977
Housekeeping Gene 978
Hox Genes 978
Hsp 979
HTLV-1 979
Human Chromosomes 980
Human Genetics 980
Human Genome Project 980
Hunter Syndrome 981
Huntington's Disease 981
Hurler Syndrome 983
Huxley, Thomas Henry 984
Hybrid 984
Hybrid-Arrested Translation 984
Hybrid Dysgenesis 984
Hybrid Sterility, Mouse 985
Hybrid Vigor 987
Hybrid Zone, Mouse 987
Hybridization 989
Hydatidiform Moles 989
Hyperchromicity 991
Hypervariable Region 991

I

Ichthyosis 993
Identity by Descent 995
Idiogram 996
Igf2 Locus 996
Igf 2r Locus 996
Illegitimate Recombination 996
Immunity 998
Immunoglobulin Gene Superfamily 998
Imprinting, Genomic 999
In situ Hybridization 1002
In vitro Evolution 1004
In vitro Fertilization 1008
In vitro Mutagenesis 1010
In vitro Packaging 1014
Inborn Errors of Metabolism 1014
Inbred Strain 1015
129 Inbred Strain 1016
Inbreeding 1016
Inbreeding Depression 1016
Incompatibility 1016
Incomplete Dominance 1016
Incomplete Penetrance 1017
Incross 1017
Indel 1017
Independent Assortment 1017
Independent Segregation 1018
Inducer 1019
Inducible Enzyme, Inducible System 1019
Induction of Prophage 1019
Induction of Transcription 1019
Infertility 1021
Influenza Virus 1026
Inheritance 1026

Inherited Rickets 1027
Initiation Factors 1029
Insertion Sequence 1029
Insertion, Insertional Mutagenesis 1037
Insulinoma 1037
Integrase 1038
Integrase Family of Site-Specific Recombinases 1038
Integration 1041
Integrons 1041
Intelligence and the 'Intelligence Quotient' 1045
Intercross 1046
Interference, Genetic 1046
Interphase 1047
Intersex 1047
Interspecific, Intraspecific Cross 1047
Intervening Sequence 1047
Intron Homing 1047
Introns and Exons 1052
Invariants, Phylogenetic 1053
Inversion 1054
Inverted Repeats 1054
Inverted Terminal Repeats 1054
Isochromosome 1054
Isolation by Distance 1054
Isoleucine 1055
Isomerization (of Holliday Junctions) 1055
Isotype 1057
Isotype Switching 1057

J

J Gene 1059
Jackknifing 1059
Jacob, François 1059
Jukes–Cantor Correction 1059
Jumping Genes 1060

K

Karyotype 1061
kb (Kilobase) 1062
Khorana, Har Gobind 1062
Kimura Correction 1062
Kimura, Motoo 1062
Kin Selection 1063
Kinases (Protein Kinases) 1063
Kinetochore 1064
Klinefelter Syndrome 1065
Knockout 1066
Kornberg, Arthur 1067
Kornberg Enzyme 1068
Kuru 1068

L

lac Mutants 1069
lac Operon 1070
Lactose 1070
Lagging Strand 1071
Lamarck, Jean Baptiste 1071
Lamarckism 1072
Lampbrush Chromosomes 1073
Late Genes 1077

Leader Peptide 1077
Leader Sequence 1078
Leading Strand 1080
Least Squares 1080
Lederberg, Joshua 1080
Leguminosae 1081
Leiomyoma 1085
Lejeune, Jérôme 1085
Lesch–Nyhan Syndrome 1087
Lethal Locus 1088
Lethal Mutation 1088
Leucine 1088
Leukemia 1088
Leukemia, Acute 1091
Leukemia, Chronic 1093
Levan, Albert 1094
Lewis, Edward 1095
Library 1095
Ligation 1096
Light Receptor Kinases 1096
Light, Heavy Chains 1096
LIM Domain Genes 1096
Limb Development 1099
LINE 1103
Linkage 1104
Linkage Disequilibrium 1105
Linkage Group 1106
Linkage Map 1106
Linker DNA 1109
Lipoma and Uterine Leiomyoma 1109
Liposarcoma 1117
Little, Clarence 1117
LMO Family of LIM-Only Genes 1118
Locus 1119
LOD Score 1119
Long-Period Interspersion 1120
Long Terminal Repeats (LTRs) 1120
Loss of Heterozygosity (LOH) 1120
Lotus japonicus 1121
Lung Cancer, Chromosome Studies 1122
Luria, Salvador 1123
Luria–Delbrück Experiment 1124
Lutheran Blood Group 1125
Lycopersicon esculentum (Tomato) 1125
Lyon Hypothesis 1127
Lysenko, T.D./Lysenkoism 1127
Lysine 1130
Lysis 1130
Lysogeny 1130
Lysozyme 1131
Lytic Phage 1132

M

Macronuclear Development, in Ciliates 1133
Macronucleus 1135
Major Histocompatibility Complex (MHC) 1136
Malthus, Thomas 1136
Mammalian Genetics (Mouse Genetics) 1137
Map Distance, Unit 1140
Map Expansion 1140

Mapping Function 1141
Mapping Panel 1144
Marfan Syndrome 1144
Marker 1146
Marker Effect 1146
Marker Rescue 1147
Masked mRNA 1147
Maternal Effect 1148
Maternal Inheritance 1149
Mating Types 1151
Mating-Type Genes and their Switching in Yeasts 1153
Maxam-Gilbert Sequencing 1157
Maximum Likelihood 1157
MC29 Avian Myelocytomatosis Virus 1160
McClintock, Barbara 1161
Medicago truncatula 1162
Medical Genetics 1163
Meiosis 1163
Meiotic Drive, Mouse 1165
Meiotic Product 1167
Melanoma, Cytogenetic Studies 1167
Melting Temperature (Tm) 1168
Mendel, Gregor 1168
Mendel's Laws 1171
Mendelian Genetics 1180
Mendelian Inheritance 1180
Mendelian Population 1181
Mendelian Ratio 1181
Meselson–Radding Model 1181
Meselson–Stahl Experiment 1182
Mesoderm 1184
Messenger RNA (mRNA) 1184
MET 1185
Metabolic Disorders, Mutants 1187
Metacentric Chromosome 1189
Methionine 1189
Metric, Four Point 1189
Metric, Manhattan 1189
Metric, Ultra 1189
Metrics 1190
MIC and MIC-M Classifications of Leukemia 1190
Microarray Technology 1191
Microbial Genetics 1191
Microbial Genomics 1196
Microchromosomes 1203
Micro-complement Fixation (MC'F) 1203
Micronucleus 1204
Microsatellite 1205
Microscopy 1205
Microtubules 1205
Minicells 1205
Minichromosome 1205
Minimal Residual Disease 1206
Minimum Change 1212
Minisatellite 1212
Mismatch Repair (Long/Short Patch) 1213
Mistranslation 1213
Mitochondria 1215
Mitochondria, Genetics of 1217
Mitochondrial DNA (mtDNA) 1219

Mitochondrial Genome 1220
Mitochondrial Inheritance 1222
Mitochondrial Mutants 1222
Mitosis 1224
MLL 1227
Molecular Clock 1229
Molecular Drive 1233
Molecular Genetics 1234
Monochromosomal Somatic Cell Hybrids 1235
Monocistronic mRNA 1235
Monoclonal Antibodies 1235
Monod, Jacques 1237
Monomorphic Locus 1238
Monophyly 1238
Monosomy 1239
Morgan, Thomas Hunt 1239
Morula 1240
MOS 1240
Mosaicism in Humans 1240
Mouse 1242
Mouse, Classical Genetics 1247
Mouse Cleavage 1249
Mouse Leukemia Viruses 1249
Mouse Sex-Reversed Rearrangement 1253
mRNA 1254
mtDNA 1254
Muenke Syndrome 1255
Mule 1255
Muller, Hermann J 1255
Multicopy Plasmids 1257
Multifactorial Inheritance 1257
Multiple Endocrine Neoplasia 1257
Multiplicity of Infection 1258
Multisite Mutation 1258
Mus musculus 1259
Mus musculus castaneus 1261
Mus spretus 1261
Muscular Dystrophies 1262
Mutagenic Specificity 1264
Mutagens 1268
Mutant Allele 1268
Mutation 1268
Mutation, Back 1275
Mutation Frequency 1275
Mutation, Leaky 1276
Mutation Load 1276
Mutation, Missense 1276
Mutation, Null 1277
Mutation Rate 1277
Mutation, Silent 1279
Mutation, Spontaneous 1279
Mutational Analysis 1279
Mutational Site 1285
Mutator Phenotype 1285
Mutators 1286
Myb Oncogene 1287
myc Locus 1287
Myxoid Liposarcoma and *FUS/TLS-CHOP* Fusion Genes 1287

N

N_2, N_3, N_4, etc. 1289
Nasopharyngeal Carcinoma (NPC) 1289
Nathans, Daniel 1290
Natural Selection 1291
Nature–Nurture Controversy 1297
Nearly Neutral Theory 1301
Negative Complementation 1302
Negative Interference 1302
Negative Regulators 1304
Negative Supercoiling 1304
Neoteny 1304
Network 1305
Neu Oncogene 1306
Neurofibromatosis 1307
Neurogenetics in *Caenorhabditis elegans* 1309
Neurogenetics in *Drosophila* 1314
Neuron/Cell Degeneration 1318
Neuronal Guidance 1318
Neuronal Specification 1320
Neurospora crassa 1322
Neutral Drift 1323
Neutral Mutation 1324
Neutral Theory 1326
Nick 1327
Nick Translation 1327
nif Genes 1328
Nirenberg, Marshall Warren 1328
Nod-Box 1329
Nod Factors 1330
Nodulation Genes 1332
Nodulins 1334
Nomenclature of Genetics 1335
Nomenclature of Genetics, Mouse 1336
Nomenclature of Human Genes 1340
Nonautonomous Controlling Elements 1342
Non-Darwinian Evolution 1342
Nondisjunction 1345
Non-Hodgkin's Lymphoma 1347
Non-Mendelian Inheritance 1349
Nonreciprocal Exchange 1350
Nonrepetitive DNA 1350
Nonsense Codon 1350
Nonsense Mutation 1350
Nonsense Suppressor 1350
Nontranscribed Spacer 1350
Northern Blotting 1350
Nuclear Envelope, Transport 1352
Nuclear Import, Nuclear Export 1356
Nuclear Matrix 1356
Nuclear Pore Complex 1356
Nuclear Pores 1356
Nuclear Transfer 1356
Nuclease 1357
Nucleic Acid 1358
Nucleocytoplasmic Transport 1359
Nucleolar Organizer 1359
Nucleolus 1359
Nucleosome 1360

Nucleotide Sequence 1360
Nucleotides and Nucleosides 1360
Nucleus 1361
Nude Mouse 1361
Null Hypothesis 1361
Nullisomy 1362
Nutritional Mutations 1362

O

Ochoa, Severo 1365
Ochre Codon 1366
Ochre Mutation 1366
Ochre Suppressor 1366
Ohno's Law 1367
Okazaki Fragment 1367
Olfaction 1368
Oncogenes 1370
OncoMouse® 1372
Oogenesis in *Caenorhabditis elegans* 1373
Oogenesis, Mouse 1374
Opal Codon 1375
Open Reading Frame 1375
Operators 1376
Operon 1377
Organelles 1377
Organismal Hybridization 1379
Ori Sequences 1381
Origin (*ori*) 1387
Origin of Life, Theories of 1387
Orphan Receptor 1394
Orthology 1394
Oryza sativa (Rice) 1394
Osteogenesis Imperfecta 1395
Outbreeding 1398
Outcross 1400
Overdominance 1400
Overwinding 1401
Ovulation 1401

P

P Elements 1403
P53 Gene 1407
Paedomorphosis 1407
Painter, Theophilus Schickel 1407
Pairing 1410
Palindrome 1410
Panmixis 1410
Paralogy 1411
Paramecia 1411
Paramorphosis 1414
Parapatric 1414
Paraphyly 1414
Parasegment 1415
Parental 1415
Parsimony 1415
Parthenogenesis, Mammalian 1419
Patau Syndrome 1420
Paternal Inheritance 1420
Pathogenicity Islands 1422
Pattern Formation 1424

Pax Genes 1426
pBR322 1428
PCR 1428
Pedigree Analysis 1428
Pedomorphosis 1429
Penetrance 1429
Peptide Bond 1429
Perinuclear Space 1430
Permissive Cells 1430
Petite 1430
P-Glycoprotein 1430
Phage (Bacteriophage) 1434
Phage λ Integration and Excision 1434
Phage Crosses 1437
Phage M13 1438
Phage Mu 1438
Phage Receptor 1441
Phase Variation 1442
Phase Variation (Biology) 1444
Phaseolus vulgaris (Beans) 1444
Phenetics 1446
Phenocopy 1446
Phenogram 1446
Phenotype 1446
Phenotypic Lag 1447
Phenotypic Mixing 1447
Phenylalanine 1447
Phenylketonuria 1447
Philadelphia Chromosome 1449
Phi (ϕ)X174, Genetics of 1450
Photomorphogenesis in Plants, Genetics of 1454
Photoreactivation 1458
Photorepair 1459
Photosynthesis, Genetics of 1459
Phylogenetic Tree 1465
Phylogeny 1465
Phylogeography 1466
Physical Mapping 1467
Piebald Trait 1467
Pigmentation, Mouse 1469
Pim Oncogenes 1470
Pisum sativum (Garden Pea) 1470
Plant Development, Genetics of 1471
Plant Embryogenesis, Genetics of 1474
Plant Growth Promoting Rhizobacteria (PGPR) 1477
Plant Hormones 1480
Plaques 1482
Plasmacytomas in Mice 1482
Plasmids 1485
Pleiotropy 1490
Plesiomorphy 1491
Ploidy 1491
Pneumonia Bacteria 1491
Point Mutations 1494
Polarity 1495
Polaron 1495
Poly(A) Tail 1495
Polyadenylation 1496
Polycistronic mRNA 1496
Polygenes 1496

Polymerase 1497
Polymerase Chain Reaction (PCR) 1499
Polymerase Chain Reaction, Real-Time Quantitative 1503
Polymorphism 1507
Polymorphisms, Tree Reconstruction 1509
Polypeptides 1509
Polyploidy 1509
Polysome (Polyribosome) 1511
Polytene Chromosomes 1511
Population Genetics 1513
Population Substructure 1519
Position Effects 1523
Positive Interference 1530
Positive Regulator Proteins 1530
Positive Supercoiling 1530
Postmeiotic Segregation 1531
Posttranscriptional Modification 1532
Posttranslational Modification 1533
Predator–Prey and Parasite–Host Interactions 1533
Pre-mRNA Splicing 1536
Prenatal Diagnosis 1539
Primary Transcript 1541
Primase 1542
Primer 1546
Primer RNA 1546
Primitive Character 1548
Primosome 1548
Prions 1548
Prisoner's Dilemma 1548
Probe 1548
Procentriole 1548
Processed Pseudogene 1548
Progeny Testing 1548
Prokaryotes 1549
Proline 1549
Promoters 1549
Proofreading 1551
Proofreading Function 1551
Prophage 1551
Prophage, Prophage Induction 1551
Protein Interaction Domains 1551
Protein Secretion Systems 1552
Protein Splicing 1565
Protein Synthesis 1567
Proteins and Protein Structure 1568
Proteolysis 1573
Proteome 1575
Proto-Oncogene 1578
Provirus 1578
Proximal 1579
Pseudoalleles 1579
Pseudoautosomal Linkage, Region 1579
Pseudogene 1579
Pseudoxanthoma Elasticum (PXE) 1580
Psoriasis 1581
Puff 1582
Pulse–Chase 1582
Pulsed Field Gel Electrophoresis (PFGE) 1582
Punnett Square 1585
Purine 1585

Pyrimidine 1586
Pyrimidine Dimers 1586

Q

Q-Banding 1587
QTL Mapping 1587
QTL (Quantitative Trait Locus) 1593
Quantitative Genetics 1595
Quantitative Inheritance 1595
Quantitative Trait 1597

R

R Factor, R Plasmids 1599
R Plasmids 1599
RAD Genes (in Yeast) 1599
Radiation Genetics, Mouse 1599
Random Mating 1601
Ras Gene Family 1602
Rats: Genetics and Cytogenetics 1607
rDNA Amplification 1610
Reading Frame 1612
Rearrangements 1612
Rec Genes 1614
RecA Protein and Homology 1618
RecBCD Enzyme, Pathway 1623
Recessive 1631
Recessive Inheritance 1631
Recessive Lethal 1631
Reciprocal Cross 1631
Reciprocal Recombination 1632
Reciprocal Translocation 1635
Reciprocality 1635
Reckless DNA Degradation 1635
Recombinant 1636
Recombinant Congenic Strains 1636
Recombinant DNA 1637
Recombinant DNA Guidelines 1639
Recombinant DNA Technology 1639
Recombinant Inbred Strains 1639
Recombination Hot Spots, Mouse 1640
Recombination in the Immune System 1640
Recombination, Models of 1642
Recombination Nodules (RNs) 1645
Recombination Pathways 1647
Recombination Suppression 1648
Recombinational Repair 1649
Reed–Sternberg Cells 1649
Reeler Locus 1649
Regulation of DNA Repair 1650
Regulatory Genes 1652
Regulatory RNA 1657
Reiterated Genes 1658
Rel Oncogene 1658
Release (Termination) Factors 1659
Renal Cell Cancer 1659
Renaturation 1661
Repair Mechanisms 1661
Repair of Oxidative DNA Damage 1669
Repetitive (DNA) Sequence 1671
Replication 1675

Replication Errors 1676
Replication Eye 1677
Replication Fork 1677
Replicon 1677
Replisome 1678
Reporter Gene 1678
Repression 1678
Repressor 1678
Reproductive Isolation 1679
Resistance 1686
Resistance Plasmids 1686
Resistance to Antibiotics, Genetics of 1687
Resolvase 1687
Resolvase-Mediated Deletion 1688
Restriction and Modification 1692
Restriction Endonuclease 1693
Restriction Fragment Length Polymorphism (RFLP) 1696
Restriction Map 1696
RET Proto-Oncogene 1696
Reticulation 1696
Retinitis Pigmentosa 1697
Retinoblastoma 1697
Retroposon 1698
Retroregulation 1699
Retrotransposons 1699
Retroviruses 1701
Reverse Genetics 1706
Reverse Mutation 1709
Reverse Transcriptase 1710
Reverse Transcription 1711
Reverse Translation 1711
Reversion 1711
Reversion Tests 1711
Revertants 1713
Rh Blood Group Genes 1713
Rhabdomyosarcoma 1715
Rhizobium 1715
Rho Factor 1716
RI Strain 1719
Ribonucleic Acid 1719
Ribosomal RNA (rRNA) 1719
Ribosome Binding Site 1723
Ribosomes 1723
Ribozymes 1730
Rifamycins 1730
Right/Left Handed DNA 1730
R-Loop 1731
RNA 1732
RNAases 1732
RNA-Binding Domains in Proteins 1733
RNA Editing in Animals 1735
RNA Editing in Plants 1740
RNA Editing in Trypanosomes 1741
RNA Interference 1743
RNA Phages 1744
RNA Polymerase 1746
RNA Turnover 1748
RNA World 1751
Robertsonian Translocation 1752
Rolling Circle Replication 1752

Root Development, Genetics of 1753
RuvAB Enzyme 1756
RuvC Enzyme 1759

S

S Phase 1761
S1 Nuclease 1761
Saccharomyces cerevisiae (Brewer's Yeast) 1761
Saccharomyces Chromosomes 1763
Saethre–Chotzen Syndrome 1765
Salmonella 1766
Sanger, Frederick 1768
Sarcomas 1768
Satellite DNA 1773
Satellited Chromosome 1773
SCE (Sister Chromatid Exchange) 1774
Schizophrenia 1774
Schizosaccharomyces pombe, the Principal Subject of Fission Yeast Genetics 1776
Screening 1779
scRNA 1780
scRNP 1780
SDP (Strain Distribution Pattern) 1780
Second Division Segregation 1780
Secretion 1780
Seed Development, Genetics of 1780
Seed Storage Proteins 1782
Segmental Interchange 1787
Segmentation Genes 1790
Segregation 1790
Segregation Distortion, Mouse 1790
Selection 1791
Selection Coefficient 1791
Selection Differential 1791
Selection Index 1792
Selection Intensity 1793
Selection Limit 1794
Selection Pressure 1795
Selection Techniques 1795
Selective Breeding 1796
Selective Neutrality 1800
Selective Sweep 1803
Self-Fertilization 1805
Selfish DNA 1805
Self-Splicing 1808
Semiconservative Replication 1808
Semidiscontinuous Replication 1808
Semidominance 1809
Sense Codon 1809
Sequence Alignments 1809
Serine 1809
Sex Chromatin 1809
Sex Chromosome Aneuploidy: XYY 1810
Sex Chromosomes 1810
Sex Determination, Human 1811
Sex Determination, Mouse 1816
Sex Linkage 1819
Sex Plasmid 1820
Sex Ratios 1820
Sex Reversal 1822
Sexduction 1823

Sex-Limited Character 1824
Sezary's Syndrome 1824
SH Domains 1824
SH2 Domain 1824
SH3 Domain 1825
Shearing 1825
Shifting Balance Theory of Evolution 1825
Shine–Dalgarno Sequence 1828
Shotgun Cloning 1828
Shuttle Vector 1828
Sickle Cell Anemia 1828
Sigma Factors 1831
Signal Sequence 1834
Signal Transduction 1834
Similarity 1834
SINE 1834
Single-Copy Plasmids 1835
Single-Gene Inheritance 1835
Single Nucleotide Polymorphisms (SNPs) 1838
Single-Strand Annealing 1838
Single-Strand Assimilation 1839
Single-Strand Exchange 1839
Single-Stranded DNA-Binding Proteins (SSBs) 1839
Sister Chromatids 1842
Site-Directed Mutagenesis 1842
Site-Specific Recombination 1842
Skeletal Disorder 1846
Ski Oncoprotein 1846
Smith, Hamilton 1846
Snell, George 1847
SNPs 1848
snRNAs 1848
snRNPs 1848
Solanum tuberosum (Potato) 1848
Soluble RNA 1850
Somatic Mutation 1851
Somatic Pairing 1851
SOS Bypass 1853
SOS Repair 1853
Southern Blotting 1855
Specialized Recombination 1857
Specialized Transduction 1858
Speciation 1860
Species 1864
Species Selection 1869
Species Trees 1869
Specific Locus Test 1870
Specificity 1871
Spermatids 1871
Spermatocytes 1872
Spermatogenesis in *Caenorhabditis elegans* 1873
Spermatogenesis, Mouse 1875
Spermatogonia 1876
Spina Bifida 1876
Spindle 1877
Splicing 1877
Splicing Junctions 1878
Split Genes 1878
Spongiform Encephalopathies (Transmissible), Genetic Aspects of 1879
Spores 1882

Src Family Tyrosine Kinases 1883
SSLP (Simple Sequence Length Polymorphism) 1885
SSR (Simple Sequence Repeat) 1885
Stable Equilibrium 1885
Staggered Cuts 1885
Staphylococcus aureus 1885
Start, Stop Codons 1886
Steel Locus 1887
Stem Cells 1889
Steroids 1889
Sticky Ends 1889
Strain 1889
Strain Distribution Pattern (SDP) 1889
Strand Displacement 1890
Streptomyces 1890
Streptomycin 1890
Stringent Response 1891
Structural Gene 1892
Subcellular RNA Localization 1892
Subcloning 1895
Substitution Mutations 1895
Superinfection Immunity 1895
Superrepressor 1896
Suppression 1897
Suppressor Mutations 1899
Suppressor tRNA 1901
Symbionts, Genetics of 1902
Symbiosis Islands 1902
Symbiosome 1903
Sympatric 1904
Sympatric Speciation 1904
Symplesiomorphy 1904
Synapomorphy 1905
Synapsis, Chromosomes 1906
Synapsis in DNA Transactions 1906
Synaptonemal Complex 1910
Syndactyly 1912
Syngenic 1913
Synovial Sarcoma 1913
Synteny (Syntenic Genes) 1913
Systematics 1913
Systemic Acquired Resistance (SAR) 1913

T

T Cell Receptor Gene Family 1919
t Haplotype 1919
T Phages 1921
TAL Gene Family (*SCL*) 1930
Tandem Repeats 1932
Targeted Mutagenesis, Mouse 1933
TATA Box 1934
Taxonomy, Evolutionary 1934
Taxonomy, Numerical 1937
Tay–Sachs Disease 1941
T-Box Genes 1943
Telomerase 1945
Telomeres 1946
Temperate Phage 1950
Temperature-Sensitive Mutant 1950
Template 1951

Terminal Redundancy 1952
Termination Codon 1952
Termination Factors 1952
Terminator 1952
Test Cross 1952
Testes Determining Locus 1952
Tetrad Analysis 1952
Tetraparental Mouse 1957
Tetratype 1958
Thalassemias 1958
Thermophilic Bacteria 1961
Theta(θ) Replication 1963
Three-Point Cross (Test-Cross) 1964
Threonine 1965
Threshold Characters 1965
Thymine 1965
Thymine Dimer 1966
Ti Plasmids 1966
Tissue Culture 1966
Tjio, Joe-Hin 1969
Tm 1970
TNF 1970
Topoisomerases 1970
Tortoiseshell Coloring 1970
Trans, Cis Configurations 1971
Trans-Acting Factors 1971
Transcribed Spacer 1973
Transcription 1973
Transcription Factor 1983
Transduction 1983
Transfection 1984
Transfer of Genetic Information from *Agrobacterium tumefaciens* to Plants 1984
Transfer RNA (tRNA) 1986
Transformation 1989
Transgenes 1989
Transgenic Animals 1990
Transient Polymorphism 1998
Transition 1999
Translation 1999
Translational Control 2002
Translocation 2003
Transmissible Spongiform Encephalopathy 2007
Transmission Genetics 2008
Transposable Elements 2016
Transposable Elements in Plants 2020
Transposase 2033
Transposon Excision 2033
Transposons as Tools 2034
Trans-Splicing 2040
Transvection 2041
Transversion Mutation 2042
Trees 2042
Trichome Development, Genetics of 2045
Trinucleotide Repeats: Dynamic DNA and Human Disease 2048
Triplet Code, Genetic Evidence 2054
Triploidy 2055
Trisomy 2056
Trisomy 18 2058
Triticum Species (Wheat) 2060
Trivial Equilibrium 2068

tRNA 2070
Trophoblast 2070
Trypsin 2071
Tryptophan 2075
Tryptophan Operon 2076
Tumor Antigens Encoded by Simian Virus 40 2078
Tumor Necrosis Factor (TNF) 2081
Tumor Suppressor Genes 2081
Turner Syndrome 2088
Twisting Number 2089
TY Elements 2090
Tyrosine 2090

U

Ubiquitin 2091
Umber Mutation 2093
Underdominance 2093
Underwinding 2094
Undirectional Replication 2095
Unequal Crossing Over 2095
Uniparental Inheritance 2096
Unique DNA 2099
Universal Genetic Code 2099
Unscheduled DNA Synthesis 2099
Unstable Equilibrium 2100
Up, Down Mutations 2100
Upstream 2100
Upstream, Downstream Site 2100
Uracil 2100
URF 2101
UTP (Uridine Triphosphate) 2101
UVR Genes 2101

V

V Gene 2103
Valine 2103
Variable Codons 2103
Variable Region 2103
Variegation 2103
Vascular Endothelial Growth Factor (VEGF) 2103
Vectors 2104
Velocardiofacial Syndrome 2106
Vertical Transmission 2106
Viroids 2107
Virology 2107
Virulent Phage 2107
Virus 2108
Viruses of the Archaea 2114
Visconti–Delbrück Hypothesis 2116
Vitamins 2117
VNTR (Variable Number of Tandem Repeats) 2120
Von Gierke disease 2120
Von Hippel–Lindau Disease 2122
Von Willebrand Disease 2123

W

W Chromosome 2125
W (White Spotting) Locus 2125
WAF1 2127
Wahlund Effect 2127

Wallace, Alfred Russel 2132
Watson, James Dewey 2133
Watson–Crick Model 2135
WHO Classification of Leukemia 2135
Whole Organism Cloning 2136
Wild-Type (WT) 2138
Wilms' Tumor 2138
Wilson Disease 2139
Wilson, Edmund Beecher 2140
Winding 2140
Wiskott–Aldrich Syndrome 2140
Wobble Hypothesis 2140
Wright, Sewall 2141
WT 2143

X

X Chromosome 2145
X-Chromosome Inactivation 2148
Xenology 2149
Xenopus laevis 2149
Xeroderma Pigmentosum 2150
XIST 2150
X-Ray Crystallography 2154

Y

Y Chromosome (Human) 2155
Y Linkage 2160
YAC (Yeast Artificial Chromosome) 2160
Yanofsky, Charles 2161
Yeast Plasmids 2162
Yeast Two-Hybrid System 2164

Z

Z Chromosome 2165
Z DNA 2165
Zinc Finger Proteins 2165
Zoo Blot 2165
Zygote 2165
Zygotic Lethal Gene 2166

Index 2169

Color Plate Section
Volume 1 between pages 314–315
Volume 2 between pages 840–841
Volume 3 between pages 1446–1447
Volume 4 between pages 1978–1979

S

S Phase

doi: 10.1006/rwgn.2001.2011

S phase is the part of the eukaryotic cell cycle during which DNA synthesis takes place.

***See also:* Cell Cycle**

S1 Nuclease

doi: 10.1006/rwgn.2001.2012

An S1 nuclease is an enzyme that specifically digests single-stranded sequences of DNA.

***See also:* Nuclease**

Saccharomyces cerevisiae (Brewer's Yeast)

R K Mortimer

doi: 10.1006/rwgn.2001.0150

The study of yeast genetics originated in the 1930s with the work of Winge (1935). He pioneered tetrad dissection and discovered the alternation between haploid and diploid phases. He also discovered the homothallism (*HO*) gene (this gene converts haploid spores to diploid cells within two divisions by causing a switch in mating type in two of the four resultant cells) and characterized many genes for the fermentation of sucrose, raffinose, and maltose. Winge used strains that had been isolated by Emil Christian Hansen and Albert Kloecker. Carl Lindegren also was one of the pioneers of yeast genetics and fortuitously had been provided with a strain that had been isolated by Emil Mrak in 1938. This strain was diploid and heterothallic, and Lindegren characterized the bipolar mating system and developed the first genetic map of this organism. He provided strains of both mating types to several laboratories and this helped to get yeast genetics started.

Figure 1 shows the life cycles of homothallic and heterothallic yeasts. There are two principal genes involved, the mating-type locus and the homothallism gene. Both heterothallic and homothallic yeasts have functional mating-type loci. Heterothallic yeasts have a nonfunctional homothallism gene (*ho*), so they divide vegetatively either as haploids or as diploids. The haploids are of two mating types, *a* and α. Haploid cells will mate with cells of the opposite mating type to form diploids. The diploids have both mating-type alleles and will not mate, but they can sporulate. Homothallic strains have functional mating type and homothallism genes, and are diploid and usually homozygous for the homothallism gene.

Nearly all laboratory strains are heterothallic, yet such strains are in the minority among natural yeasts. In a study of 239 natural yeasts that had been isolated from natural (noninoculated) wine fermentations, it was found that 185 were homozygous for *HO*, 26 were homozygous for *ho*, and 28 were heterozygous (*HO*/*ho*) for these two alleles (Mortimer, 2000). It is likely that the laboratory strains had their origins as wine or beer yeasts. Mrak's strain EM93 was found on rotting figs, and it seems likely that this yeast was carried to the figs by insects in a manner similar to that in which they are transported to grapes (Mortimer and Polsinelli, 1999).

The heterothallic yeasts have stable haploid and diploid vegetative phases. By treating the haploid cells with a mutagen such as UV light or ethyl methane sulfonate and plating the survivors on a medium such as yeast extract–peptone–dextrose agar, mutations can be recovered. The mutations can be detected by replica-plating on a selective medium such as minimal medium. Some of the replicas will not grow because they require nutrients not present in the minimal medium. Other selective regimes can detect other classes of mutants. Both dominant and recessive mutants will be detected by this approach.

The next step is to cross those 'mutants' to haploid cells of the opposite mating type which do not

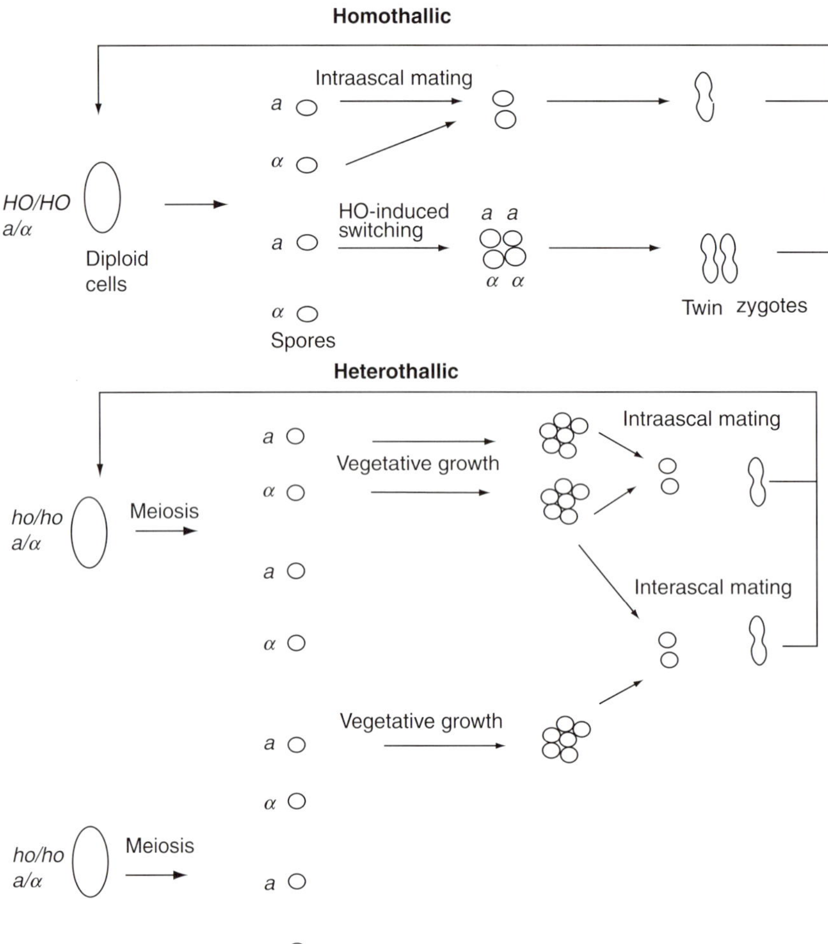

Figure 1 Life cycles of naturally occurring wine yeasts. Homothallic yeast: meiosis of a diploid cell produces four spores. The top two spores shown pair, but only about one in seven such pairings produces a zygote. The third spore shown is left to undergo homothallic switching. In two divisions, two of the four resulting cells switch mating type and mate with the other two to form 'twin zygotes.' Switching of mating type occurs only after a cell has divided at least once. Heterothallic yeast: two diploid cells are considered and they each produce four spores by meiosis. These spores divide by vegetative division and then a cell from each of two spore clones that are of opposite mating type mate to form a zygote. At the bottom, one cell from the upper ascus mates with one cell from the lower ascus. Nearly 100% of cell–cell pairings between heterothallic strains of opposite mating type form zygotes. *HO*, homothallism gene; *ho*, nonfunctional homothallism gene.

carry the mutation. If the diploid is nonmutant, the mutation is recessive; if the diploid has the mutant phenotype, then the mutation is dominant. Nearly all mutations are recessive. If the diploid segregates during meiosis into a ratio of two nonmutant spores to two mutant spores, then the difference between the mutant and wild-type is located in a single gene. This is a direct prediction of Mendel's laws of inheritance. Such crosses also yield segregants that carry the mutation but are of opposite mating type. If, for example, we cross a number of tryptophan-requiring mutants and obtain strains of both mating types for these mutants, then different mutants can be crossed with each other. If the diploid is tryptophan-requiring and both mutations are recessive, then the two mutations are noncomplementing and are in the same gene. If the diploid does not require tryptophan, then the mutations complement and this diploid must be analyzed genetically. If it segregates both tryptophan-requiring and tryptophan-nonrequiring spores, then the mutations are in different genes. However, some of the tryptophan-nonrequiring diploids may segregate only tryptophan-requiring spores. This indicates that the mutations are in the same gene, that is, they are alleles that show interallelic complementation. Such complementation is due to interactions at the level of the protein products of these alleles. If the cross segregates both tryptophan-requiring and tryptophan-nonrequiring spores, then the ratios of these two classes in individual tetrads determines whether these genes are linked to each other or to their respective centromeres. This is determined by the laws of tetrad analysis. For the five tryptophan genes, two are linked to their respective centromeres and the other three are unlinked and are on different chromosomes.

By crossing many different classes of mutants in various combinations, a genetic map was eventually developed. This map describes the locations of over 2000 genes on 16 chromosomes. The total genetic map length is around 4500 cM. The total number of genes determined from the nucleotide sequence of the DNA is about 6200.

Some of the various classes of mutants that have been analyzed and mapped are:

1. Temperature-sensitive mutants that control the cell cycle, protein synthesis, DNA synthesis, and RNA synthesis. These mutants grow at 23 °C but not at 36 °C. The cell-cycle mutations are especially important, because many of these genes have human homologs and are related to cancer.
2. Mutants in genes that control DNA repair. Several of these genes have human homologs and are related to cancer.
3. Suppressor genes and nonsense suppressors. These are mutations in genes that reverse the phenotype of mutations in other genes. A nonsense suppressor reverses the phenotype of any mutation that is caused by a nonsense mutation.
4. Genes controlling fermentation of various sugars. *Saccharomyces cerevisiae* ferments many sugars. The genes for fermentation of sucrose, maltose, and raffinose are dominant and polymeric, that is, cells with any one of several functional genes can ferment the sugar.
5. Mutants in genes that control the synthesis of the cell wall. Yeast has a wall composed mostly of mannan and glucan, and these polysaccharides are in turn synthesized by sets of genes.
6. Mutants in genes that control mitochondrial function. The mitochondria are autonomously replicating, yet they depend on nuclear genes for the synthesis of many of their components.
7. Mutants that affect mating. In addition to the mating-type locus and the homothallism genes are several other genes that affect the expression of the mating-type genes.
8. Mutants that are unable to sporulate (meiotic mutants). Meiosis is a complex process and many genes control this sequence of events.
9. Mutants that affect the uptake of various nutrients. The uptake of nutrients into the cell is complex and involves many genes. There are general amino acid permeases and specific permeases.
10. There are many genes controlling the resistance to various toxic materials such as toxic metals, amino acid analogs, and fungicides.

The nucleotide sequence of this organism was completed in 1996 and this was the first eukaryotic genome to be sequenced. Methods developed in this project have since been applied to the sequencing of other organisms, including humans.

Further Reading

Broach J, Pringle J and Jones E (eds) (1997) *The Molecular and Cellular Biology of the Yeast* Saccharomyces. Plainview, NY: Cold Spring Harbor Laboratory Press.

References

Mortimer RK (2000) Evolution and variation of the yeast genome. *Genome Research* 10: 401–409.

Mortimer R and Polsinelli M (1999) On the origins of wine yeast. *Research in Microbiology* 150: 199–204.

Winge Ø (1935) On haplophase and diplophase in some Saccharomycetes. *Comptes Rendus des Travaux du Laboratoire Carlsberg, Série Physiologique* 21: 77–111.

See also: Mating-Type Genes and Their Switching in Yeasts; *Saccharomyces* Chromosomes; Tetrad Analysis

Saccharomyces Chromosomes

R K Mortimer

doi: 10.1006/rwgn.2001.1401

The physical size of the yeast chromosomes precludes their individual visualization with the light microscope. The development of pulsed-field gel electrophoresis solved this problem. Individual chromosome bands are resolved by this method and these bands can be sized. The method allowed comparisons of different species of yeast, chromosome polymorphisms, translocations, and mapping of genes to chromosomes. The complete nucleotide sequence of the *Saccharomyces cerevisiae* genome revealed the precise sizes of the chromosomes (**Table 1**) and many additional features of them (14 Mb DNA in total). In general this information corroborates earlier genetic data indicating the number of chromosomes (16), the

Table 1 The sizes of the 16 chromosomes of *Saccharomyces cerevisiae*

Chromosome	DNA (bp)
I	230 203
II	813 140
III	316613
IV	1 531 929
V	576 869
VI	270 148
VII	1 090 9365
VIII	562 639
IX	439 885
X	745 440
XI	666 445
XII	1 078 172[a]
XIII	924 430
XIV	784 328
XV	1 091 283
XVI	948 061

[a]plus rDNA. In addition the mitochondrial DNA is 85 779 bp.

order of certain genes on them, and the locations of centromeres.

Structure of Chromosomes

Centromeres

The 16 centromeres of *S. cerevisiae* are approximately 300 bp in length. These sequences bind a protein that is part of the kinetochore assembly. It is to this structure that the spindle fibers attach. There appears to be one spindle fiber per centromere. The centromeres are essential for proper chromosome assortment in both mitosis and meiosis. Genes that are near centromeres tend to segregate in the first meiotic division, and this is how centromeres were identified and mapped.

Telomeres

Telomeres are protein–DNA structures at the termini of linear chromosomes. The telomere appears to act as a repressor of genes in the subtelomeric regions. The genes located in this region include all 25 of the polymeric fermentation genes, the homothallism gene *HO*, and several of the glycolytic genes. The telomeres are special structures whose integrity is maintained by the enzyme telomerase. Shortening of telomeres is proposed as one of the mechanisms that control the life span of cells.

Replication Origins

DNA replication origins are located in intergenic regions and are flanked by open reading frames. Chromosome III has seven such origins. They are referred to as autonomous replication sequences (ARSs).

Nucleosomes

The basic unit of the chromosome fiber is the nucleosome. It is composed of two each of four histone molecules, H2A, H2B, H3, and H4. Histone H1 seems to be involved in wrapping DNA around this octamer. About 200 bp of DNA are associated with each nucleosome, which is approximately 10 nm in diameter.

Chromatin

Chromatin is an assembly of chromosome fibers that are considerably supercoiled. Functional portions of the chromatin are called euchromatin and nonfunctional parts are called heterochromatin.

Spindle Fibers

The spindle fibers form the connections between the centromeres and the centrioles. In yeast there is one fiber per centromere. The fibers pull the sister chromatids apart during anaphase and, because the chromatids are intertwined, this process involves the action of a topoisomerase.

Centrioles

Centrioles are the other attachment points of the spindle fibers. They form the poles of the cell division process.

Artificial Chromosomes

The realization that the essential features of a chromosome were a centromere, two telomeres, a replication origin, and sufficient DNA to form a chromosome, led to the development of yeast artificial chromosomes (YACs). A vector with most of these essential components and a cloning site was developed, and then DNA from any source could be cloned into the cloning site. All that remained was to linearize this molecule and transform it into yeast. YACs are used in many ways. For example, they have been used to clone the DNA of humans, *Drosophila* spp., and other organisms as part of the effort to sequence the genomes of these organisms.

Segregation of Chromosomes

Mitotic Division

Mitosis occurs in haploid, diploid, and higher ploidy yeast cells and produces two cells. There are four steps in mitosis: prophase, metaphase, anaphase, and telophase. Fluorescent *in situ* hybridization was used to study chromosome movement in *S. cerevisiae* mitosis. The features were like those seen in other eukaryotes. Crossing over can occur both within and between genes in diploid and higher-ploidy yeast cells.

Meiotic Division

Meiosis occurs in diploid and higher-ploidy cells. The process involves pairing of homologous chromosomes, recombination between these chromosomes, and two special divisions, meiosis I and meiosis II, which sort the recombined chromatids into four spores. Recombination seems to initiate at specific sites along the chromosomes, and double-strand breaks occur at these sites to initiate recombination.

Recombination Nodules

Recombination nodules are proteinaceous structures, approximately 100 nm in diameter, which are associated with the synaptonemal complexes in early meiotic prophase I. *RAD51* and *DMC1* are yeast genes involved in this process. Their protein products have homologs in mice and lily as well as to the RecA protein of bacteria. *S. cerevisiae* has about 75 recombination nodules in meiotic prophase.

Synaptonemal Complex

Synaptonemal complexes (SCs) are structures formed in the first meiotic prophase when homologous chromosomes pair. Recombination nodules are spaced along these structures where crossing over occurs. Several gene products are needed for proper meiosis, including Zip1. This protein occurs as a dimer with the NH_2 ends in the central region of the SC and the C-terminus in the lateral regions.

Meiotic Recombination

Shorter yeast chromosomes have a higher frequency of crossing-over per kilobase compared with longer chromosomes. In an examination of 10 genetically studied organisms, it was found that the frequencies of crossing-over per unit of physical size (megabase of DNA) was inversely related to the average physical size of the chromosomes (megabase per chromosome; see **Figure 1**). *S. cerevisiae* has the highest frequency of crossing-over per unit of physical size of all these organisms. The frequencies of crossing-over per megabase vary over a 10^5-fold range. Within this range there is a small number of exchanges per chromosome, which must be under evolutionary control in all these organisms.

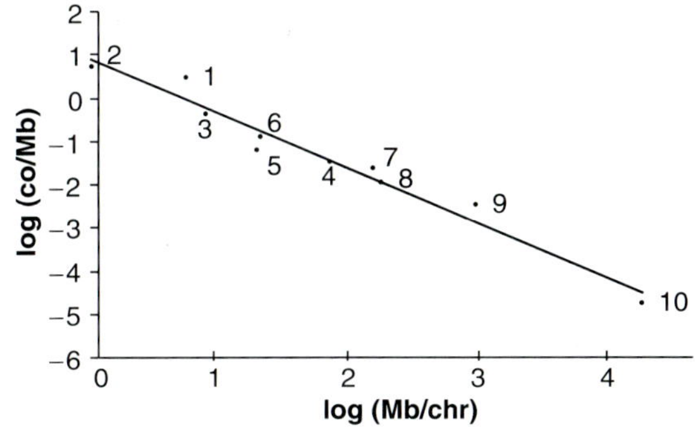

Figure 1 Frequency of crossing-over in relation to chromosome size. (co, crossovers; chr, chromosome.) Points 1–10 represent various organisms.

Organization of Genes on Chromosomes

Functionally related genes in yeast generally are not linked to each other, in contrast to the usual situation in prokaryotes. However, there are some notable exceptions. Three complementation groups of mutations blocking histidine biosynthesis are linked, and these 'genes' *HIS4A*, *-B*, and *-C* control three enzymatic steps in biosynthesis of this nutrient. However, a single protein is encoded by these genes and it has all three enzymatic activities. Similarly, five complementation groups of mutations blocking aromatic amino acid biosynthesis are clustered; the gene *ARO1* encodes a pentafunctional polypeptide. On the other hand, the three separately transcribed genes in galactose fermentation are tightly linked and control three steps in this process. As mentioned earlier, the 25 polymeric fermentation genes controlling the fermentation of sucrose, maltose, melibiose, and α-methyl glucoside are located in the subtelomeric regions of several chromosomes, and many combinations of these genes are linked.

Further Reading

Broach J, Pringle J and Jones E (eds) (1997) *The Molecular and Cellular Biology of the Yeast* Saccharomyces. Plainview, NY: Cold Spring Harbor Laboratory Press.

See also: **Chromosome; Chromosome Structure; *Saccharomyces cerevisiae* (Brewer's Yeast)**

Saethre-Chotzen Syndrome

See: **Craniosynostosis, Genetics of**

Salmonella

J R Roth

doi: 10.1006/rwgn.2001.1148

Salmonella typhimurium (*Salmonella enterica*) – A Genetic System in a Pathogenic Bacterium

Salmonella typhimurium strain LT2 has been the subject of detailed genetic analysis since the discovery of transduction (virus-mediated genetic exchange) in this organism. By applying both transductional and conjugational crosses, a genetic map was constructed and many aspects of physiology have been genetically investigated. The parallel development of genetics in a sister species (*Escherichia coli*, strain K12) has provided a situation for comparison of their shared gene systems and for analysis of the process of bacterial speciation. The *Salmonella* genetic system takes on added importance in that many *Salmonella* isolates are important pathogens and genetics can be used to investigate mechanisms of virulence. *Salmonella typhimurium* causes a typhoid fever in mice; it is a less serious pathogen for humans but is the causative agent of common food-borne enteric infections.

Taxonomy of *Salmonella*

The bacterium *Salmonella typhimurium* strain LT2 belongs to the general group of enteric bacteria, which includes other organisms that have been intensively studied: *E. coli*, *Klebsiella*, and *Citrobacter*. *Salmonella* was recognized long ago as a disease organism (Eberth, 1880) and many isolates, distinguished largely on serological criteria, were given species names within the genus *Salmonella* (e.g., *S. typhimurium, S. typhi, S. dublin*). Subsequent biochemical and genetic characterization revealed that these serovars are very closely related and so the many *Salmonella* species were recently combined into a new, more broadly inclusive group *Salmonella enterica* (Le Minor and Popoff, 1987). According to this nomenclature, the strain used in genetic analysis is designated *Salmonella enterica* serovar *typhimurium* strain LT2. For a discussion of evolutionary genetics of *Salmonella* and its divergence from *E. coli* see Selander *et al.* (1996) Lawrence and Roth (1999).

Phage P22 and Transductional Crosses

The development of *Salmonella* as a genetic system was made possible by discovery of the generalized transducing phage P22 (Zinder, 1992), which permits genetic crosses between *Salmonella* strains. Phage P22, a close relative of the *E. coli* phage lambda, is a temperate phage that replicates its genome and then packages it into protein capsids by a headful-measuring mechanism. This mechanism makes the virus prone to occasional encapsulation of fragments (44 kb) of the host (*Salmonella*) chromosome. Virus particles with such a fragment (transducing particles) can inject bacterial DNA into a new host cell allowing recombination between the injected (transduced) fragment from the donor host and the chromosome of the new recipient bacterium. All regions of the *Salmonella* chromosome are transducible. Most genetic analysis in *Salmonella* is performed using P22-mediated transductional crosses. *Salmonella* is an attractive genetic system largely because its phage (P22) is easy to propagate, stable during storage, and an efficient transducer. Mutational modification of phage P22 has improved its usefulness by increasing its transducing frequency and preventing stable lysogeny so that phage-free recombinants can be obtained for subsequent crosses. The biology of phage P22 has been reviewed by Susskind and Botstein (1978) and the general process of transductional genetic analysis has been surveyed by Masters (1996).

Development of *Salmonella* as a Genetic System

After the discovery of P22 and transduction, Milislav Demerec and his coworkers initiated development of a general genetic system in *Salmonella*. They isolated a variety of mutants, studied the genes involved in several metabolic pathways, and initiated chromosome mapping. Using P22-mediated crosses for fine-structure mapping and conjugational crosses for long-range mapping, a detailed genetic map was developed (Sanderson *et al.*, 1995). The complete genomic DNA base sequences of *S. typhimurium* and of *S. typhi* are nearly complete and several others are in progress.

Among the important findings during early work in *Salmonella* is the observation that genes with related functions are frequently clustered in the bacterial chromosome (Demerec and Hartman, 1959). This later led to discovery of the operon (clusters of genes that are transcribed into mRNA from a single promoter start site). Recent work in *Salmonella* has suggested a mechanism for evolution of bacterial operons based on selection for enhanced horizontal transfer of multiple genes that contribute to a single selectable phenotype (Lawrence and Roth, unpublished data).

Major gene–enzyme systems developed in *Salmonella* are the histidine operon analyzed by Lawrence and Roth (1996) and the leucine operon. These systems contributed to the understanding of operons and their expression and control, and to the biochemistry of the individual synthetic pathways.

One of the genetic methods developed initially in *Salmonella* was the use of transposable genetic elements as a means of making mutations. The transposable element is a DNA sequence that includes a drug resistance determinant and can insert itself into genes of the bacterial chromosome. Mutants made by insertion of such transposons have two phenotypes, a dominant drug-resistance phenotype (conferred by the inserted material) and a recessive null phenotype caused by inactivation of the target gene. These mutations make it possible to transduce mutations selectively into new genetic backgrounds by transductional crosses (Kleckner *et al.*, 1977). In addition, insertion mutations add substantial sequences to various sites in the chromosome; selection for recombination between these sequences makes it possible to create a variety of genomic rearrangements (Roth *et al.*, 1996a). Drug-resistance insertions make it possible to selectively clone many regions of the chromosome. Since the advent of PCR amplification of DNA, insertions have proved useful because they place known sequences at a variety of points in the chromosome that can help in amplification of particular chromosome regions.

Transposon Tn*10*

One of the first transposable elements used in genetic analysis was Tn*10*, a tetracycline-resistance transposon discovered in *Salmonella* (Kleckner *et al.*, 1978). A notable aspect of work in *Salmonella* has been the characterization of this transposon, its structure, and the mechanisms by which it transposes and causes chromosome rearrangements (Kleckner *et al.*, 1991, 1996). The tight regulation of Tn*10*'s transpositional activity made this element an ideal tool for genetic analysis of bacteria.

Analysis of Virulence

The application of genetic methods to the study of pathogenicity revealed that the *Salmonella* chromosome includes several separated blocks of genes (roughly 2% of the total chromosome) encoding functions that are of particular importance for invasion of hosts and evasion of host defense systems. These blocks of genes are known as 'pathogenicity islands' all or some of which are present in various *Salmonella* strains but are absent from *E. coli* (Finlay and Falkow, 1989, 1997; Groisman and Ochman, 1997).

Evolutionary Divergence of *Salmonella enterica* and *Escherichia coli*

Despite the difficulties in the species concept for bacteria, taxonomic criteria make it reasonable to regard *S. enterica* and *E. coli* as independent species that have diverged from a common ancestor living over 100 million years ago. Each of these species has been subjected to intense genetic analysis, including the determination of the complete genome sequence. The divergence of these species may be the first example of an act of speciation for which all genetic changes can be visualized. While the two species show a very low frequency of genetic exchange in nature (Maynard Smith *et al.*, 1993), that low level has been important to their evolution. Early work on DNA hybridization, confirmed by sequence data, suggested that the genomes of *S. enterica* and *E. coli* share about 75% of their sequences. The genes that are present in one species but not the other must encode the species-specific attributes, including those used in taxonomic identification. Some species-specific genes show sequence characteristics suggesting that they entered the chromosome of *S. enterica* or *E. coli* after the divergence of the two organisms. These sequences were acquired after transfer from distantly related (presumably bacterial) species (i.e., by horizontal transfer). Other species-specific genes appear to be ancestral genes that were maintained by one lineage and lost from the other (Lawrence and Roth, 1996; Lawrence and Roth, 1999). *Salmonella*-specific genes include those for synthesis and use of vitamin B_{12}, about 2% of the *Salmonella* genome (Roth *et al.*, 1996b), and the horizontally acquired pathogenicity islands mentioned above. Thus the divergence of *S. enterica* and *E. coli* appears to have occurred by assembly of distinct sets of genes through differential gene loss and acquisition in the two lineages (genomic flux) rather than by sequence divergence of ancestral genes or internal creation of new genes.

References

Demerec M and Hartman P (1959) Complex loci in microorganisms. *Annual Review of Microbiology* 13: 377–406.

Eberth CJ (1880) Die Organismen in den Organen bei *Typhus abdominalis*. *Archiv für Pathologie Anatomie und Physiologie Klinische Medizin* 8: 58–74.

Finlay BB and Falkow S (1989) *Salmonella* as an intracellular parasite. *Molecular Microbiology* 3: 1833–1841.

Finlay BB and Falkow S (1997) Common themes in microbial pathogenicity revisited. *Microbiology and Molecular Biology Reviews* 61: 136–169.

Groisman EA and Ochman H (1997) How *Salmonella* became a pathogen. *Trends in Microbiology* 5: 343–349.

Kleckner N, Roth J and Botstein D (1977) Genetic engineering in vivo using translocatable drug-resistance elements. New methods in bacterial genetics. *Journal of Molecular Biology* 116: 125–159.

Kleckner N, Barker DF, Ross DG and Botstein D (1978) Properties of the translocatable tetracycline-resistance element Tn*10* in *Escherichia coli* and bacteriophage lambda. *Genetics* 90: 427–461.

Kleckner N, Bender J and Gottesman S (1991) Uses of transposons with emphasis on Tn*10*. *Methods in Enzymology* 204: 139–180.

Kleckner N, Chalmers RM, Kwon D, Sakai J and Bolland S (1996) Tn*10* and IS*10* transposition and chromosome rearrangements: mechanism and regulation *in vivo* and *in vitro*. *Current Topics in Microbiology and Immunology* 204: 49–82.

Lawrence JG and Ochman H (1998) Molecular archaeology of the *Escherichia coli* genome. *Proceedings of the National Academy of Sciences, USA* 95: 9413–9417.

Lawrence JG and Roth JR (1996) Selfish operons: horizontal transfer may drive evolution of gene clusters. *Genetics* 143: 1943–1890.

Lawrence JG and Roth JR (1999) Genomic flux: genome evolution by gene loss and acquisition. In: R Charlebois (ed.) *Analysis of Small Genomes*,pp. 263–289. Washington, DC: American Society for Microbiology Press.

Le Minor L and Popoff M (1987) Designation of *Salmonella enterica* sp. nov. nom. rev., as the type and only species of the genus *Salmonella*. *International Journal of Systematic Bacteriology* 37: 465–468.

Masters M (1996) Generalized transduction. In: Neidhardt F, Ingraham J, Low K *et al.* (eds) Escherichia coli *and* Salmonella: *Cellular and Molecular Biology*, 2nd edn, pp. 2421–2441. Washington, DC: American Society for Microbiology Press.

Maynard Smith J, Smith N, O'Rourke M and Spratt B (1993) How clonal are bacteria? *Proceedings of the National Academy of Sciences, USA* 90: 4384–4388.

Roth JR, Galitski T, Haack K, Lawrence J and Miesel L (1996a) Rearrangements of the bacterial chromosome: formation and applications. In: Neidhardt F, Ingraham J, Low K *et al.* (eds) Escherichia coli *and* Salmonella: *Cellular and Molecular Biology*, vol. 2, p. 2256–2276. Washington, DC: American Society for Microbiology Press.

Roth JR, Lawrence JG and Bobik TA (1996b) Cobalamin (coenzyme B_{12}): synthesis and biological significance. *Annual Review of Microbiology* 50: 137–181.

Sanderson KE, Hessel A and Rudd KE (1995) Genetic map of *Salmonella typhimurium*, 8th edn. *Microbiology Reviews* 59: 241–303.

Selander R, Li H and Nelson K (1996) Evolutionary genetics of *Salmonella enterica*. In: Neidhardt F, Ingraham J, Low K *et al.* (eds) Escherichia coli *and* Salmonella: *Cellular and Molecular Biology*, 2nd edn, pp. 2691–2707. Washington, DC: American Society for Microbiology Press.

Susskind MM and Botstein D (1978) Molecular genetics of bacteriophage P22. *Microbiology Reviews* 42: 385–413.

Zinder N (1992) Forty years ago: the discovery of bacterial transduction. *Genetics* 132: 291–294.

See also: **Bacterial Genetics; Transposons as Tools**

Sanger, Frederick

S Brenner

doi: 10.1006/rwgn.2001.1149

Frederick Sanger (1918–) pioneered two areas of biochemistry research. First, he developed methods for the determination of the amino acid sequences of proteins, and when he had completed the structure of insulin, then turned his interest to developing methods for sequencing nucleic acids, first RNA and later DNA. Progressing from the sequence of a small DNA virus, ΦX174, of about 6 kb, to bovine mitochondrial DNA, 17 kb, to the genome of the lambda bacteriophage (50 kb), Sanger pioneered what has now become the genome revolution. He received one Nobel Prize in Chemistry for his insulin work in 1958 and a second one for his work on nucleic acids in 1980.

See also: **DNA Sequencing**

Sarcomas

C S Cooper

doi: 10.1006/rwgn.2001.1617

The term 'sarcoma' is used to describe a heterogeneous group of cancers that exhibit the differential features of various supporting and skeletal tissues of the body such as smooth and striated muscle, fat, fibrous tissue, vessels, bone, and cartilage. This group constitutes a major histogenic class distinct from neoplasms of epithelial origin (carcinomas), blood and lymphorticular origin (leukemias or lymphomas), and of the central nervous system. They are usually named according to the tissue they most resemble, though in some cases such as synovial sarcoma, there is no clear normal tissue homolog (**Table 1**).

Bone and soft tissue sarcomas account for ~2% of human malignancies and 3–4% of cancer deaths with around 8000 cases diagnosed in the USA each year. They include the bone tumors osteosarcoma, and chondrosarcoma (cartilage), as well as the soft tissue sarcomas liposarcoma (fat), leiomyosarcomas (smooth muscle), angiosarcomas (vessels), rhabdomyosarcomas (striated muscle), fibrosarcomas, Kaposi's sarcoma, and malignant fibrous histiocytoma (MFH). Ewing's sarcoma, a highly malignant primary bone tumor, and malignant peripheral nerve sheath tumors are also included. An additional tumor group is fibromatosis, a locally infiltrative but nonmetastasizing fibroblastic proliferation, which is conventionally divided into slow-growing superficial fibromatosis and fast-growing deep fibromatosis (desmoids).

The major sarcoma classes and subclasses are listed in **Table 1**. Sarcomas are relatively more common in children with rhabdomyosarcoma accounting for 6–7% of children's cancers. Benign sarcomas such as lipomas and leiomyosarcomas are up to 100 times more common than their malignant counterparts but there is no evidence supporting the view that these particular benign lesions are the precursors of malignant sarcomas. Considerable differences are observed in the age distribution of occurrence of individual sarcoma classes and subclasses. The embryonal subclass of rhabdomyosarcoma occurs most frequently in children under 15. Ewing's sarcoma, alveolar rhabdomyosarcoma, and synovial sarcoma occur most frequently in adolescents and young adults, while other categories such as liposarcoma, osteosarcomas, and malignant fibrous histocytomas (MFHs) occur predominantly in adults.

Table 1 Major classes and subclasses of sarcomas

Angiosarcoma
Chondrosarcoma
Myxoid
Mesenchymal
Dedifferentiated
Ewing's sarcoma
Fibromatoses
Superficial
Deep (desmoid)
Fibromatoses
Adult
Congenital or infantile
Giant cell tumors (oesteoclastomes)
Kaposi's sarcoma
Leiomyosarcomas
Liposarcomas
Well differentiated
Myxoid
Renal cell
Pleomorphic
Dedifferentiated
Malignant fibrous histiocytoma
Storiform–pleomorphic
Myxoid
Giant cell
Malignant peripheral nerve sheath tumors
Malignant gastrointestinal stromal tumors (GIST)
Osteosarcoma
Rhabdomyosarcomas
Alveolar
Embryonal
Botryoid
Pleomorphic
Synovial sarcomas
Biphasic
Monophasic
Poorly differentiated

Etiology

Several etiological factors have been suggested as possible causes of sarcoma development. Infection with HIV and with herpesvirus type 8 are associated with the development of Kaposi's sarcoma. Trauma and past injury as well as exposure to environmental agents such as dioxin have also been proposed as risk factors. Particularly, the etiology of deep fibromatosis (or desmoids) is thought to involve trauma since they have been reported to arise in surgical scars and bullet wounds. There is good evidence that exposure to radiation is a causal factor with around 0.1% of cancer patients who are treated with radiation and survive 5 years developing bone or soft tissue sarcomas. Immunodeficiency or immunosuppression by drugs such as cyclosporin have been established as risk factors for the development of a variety of sarcoma types. Sarcomas may also arise in other organisms but the histopathological classification is not as sophisticated as that in humans with most lesions being referred to simply as 'sarcomas.' Sarcomas may be induced in many species including rodents, cats, and chickens by type C retroviruses and often arise following treatment of rodents with chemical carcinogens.

Several genetic diseases are associated with the development of soft tissue and bone sarcoma (**Table 2**). The development of deep fibromatosis can occur in patients with familial adenoma polyposis (FAP) in a condition known as Gardner syndrome that often includes benign soft tissue tumors such as lipomas and leiomyomas. Pure FAP and Gardner syndrome both occur as a consequence of germline mutation in the *APC* gene, which encodes a protein that functions in the regulations of cell growth. Mutations in *APC* usually result in a truncated protein that can no longer correctly regulate β-catenin. Detailed analysis of mutations in the *APC* gene in FAP and Gardner patients failed to reveal any correlation between the type of mutations in the *APC* gene and desmoid formation. Indeed it has been found that individuals with a G > T transition mutation in codon 309 of the *APC* gene can either have the range of symptoms associated with Gardner syndrome or have FAP with no evidence of extracolonic manifestation. This observation

Table 2 Inherited predisposition to soft tissue sarcomas

Disorder	Sarcoma type	Gene/location
Gardner's syndrome	Fibromatosis	*APC*
Li–Fraumeni syndrome	Rhabdomyosarcomas plus other types	*p53, CHK2*
Beckwith–Wiedemann syndrome	Rhabdomyosarcoma	
Von Recklinghausen's neurofibromatosis	Malignant peripheral nerve sheath tumours	*NF1*
Paget's disease	Osteosarcoma	18q12
Familial retinoblastoma	Soft tissue sarcomas and osteosarcomas	*RB1*

suggests that other genetic factors may control the conditions specifically associated with the development of desmoids.

The Li–Fraumeni syndrome is an autosomal dominant familial cancer syndrome that is defined by the existence of both a proband with sarcoma and tumors in other first-degree relatives with cancer by the age of 45. The cancers associated with this syndrome include early onset soft tissue sarcomas, osteosarcomas, breast cancers, brain cancers, and leukemias. Germline mutations within the *p53* or *CHK2* genes have been implicated as the cause for many of these families. Hereditary retinoblastoma, which is usually found in children under 5 years of age, is caused by germline mutations in the *RB1* suppressor genes. Of patients with the hereditary form of this disease, 10–20% will develop secondary tumors, usually osteosarcomas and soft tissue tumors, later in life.

Von Recklinghausen's neurofibromatosis (type 1 neurofibromatosis NF1) is an autosomal dominant disorder affecting around 1 in 3500 newborns. Characteristics of this disease are the presence of café-au-lait spots, skin neurofibromas, Lisch nodules, and orthopedic abnormalities. Five percent of individuals develop Schwannoma of the peripheral nerves and 2% of cases develop malignant peripheral nerve sheath tumors. This disorder is caused by germline mutations in the *NF1* gene that encodes a protein called neurofibrin involved in controlling signaling through the RAS pathway.

The Beckwith–Wiedemann syndrome (BWS) is characterized by abdominal wall defects, macroglossia, and gigantism and is associated with an increased risk of developing Wilms' tumor, hepatoblastoma, and rhabdomyosarcoma. The predisposing gene has been localized to 11p15 and analysis of this syndrome strongly suggests that the *BWS* gene is imprinted with insulin growth factor 2 (*IGF2*) and cyclin dependent kinase inhibitor. *CDKN1C* (*p57/kip2*) have been proposed as the candidate genes for this disorder.

Osteosarcomas commonly develop against a background of preexisting bone pathology such as Paget's disease of the bone, multiple enchondromas, multiple osteochondromas, chronic osteomyelitis, fibrous dysplasia, and fractures of bone. In the case of Paget's disease, which accounts for around 10–15% of osteosarcoma in individuals over 30, a gene accounted for the familial form of this disease and has been mapped to chromosome band 18q12.

The detection of inherited mutations that predispose to cancer development is extremely useful in the management of sarcomas through the identification of individuals of high risk of cancer development. For example it is now possible to screen Li–Fraumeni families for mutations in the *p53* and *CHK2* gene, to identify carriers who have high risk of sarcoma development.

Pathology and Prognosis

Despite the definition of major sarcoma categories in AFIP and WHO classification schemes considerable controversy still exists over the definition of some sarcoma types. For example, although a decade ago MFH tumors were one of the most common categories, it has been proposed that many, perhaps most, of these sarcomas should be reclassified as pleomorphic liposarcomas, leiomosarcomas, or rhabdomyosarcomas. Because of this and other problems with the differential diagnosis of sarcoma there is clearly a continued need to search for markers that will improve diagnostic reliability. Histopathological categories and subcategories commonly define disease with distinct clinical behavior and prognosis. Liposarcomas, for example, can be divided into myxoid, round-cell (poorly differentiated), well differentiated, dedifferentiated, and pleomorphic forms. Their clinical behavior closely reflects the state of differentiation with the round-cell tumors having poorer prognosis than the myxoid tumors, while the well-differentiated subclasses have the best outcome. Synovial sarcoma may be divided into monophasic and biphasic variants with calcification, an indicator of a more favorable prognosis, observed in around 20% of cases. Poorly differentiated synovial sarcomas which are associated with a more aggressive and metastatic behavior are also occasionally observed. Several subclasses of rhabdomyosarcoma including embryonal, botyroid embryonal, alveolar, and pleomorphic can be distinguished.

Osteosarcomas can vary widely in appearance and site of occurrence but have in common the presence of an anaplastic mesencymal parenchyma that is punctuated by the formation of osteoid matrix. These tumors are generally aggressive but can vary in behavior depending on the presence of predominant lines of differentiation such as osteoblastic, telangiectatic, or chondroblastic. Pure osteoblastoma, sometimes called giant osteoid osteoma, is in fact usually benign. Similarly pure chondroblastomas are commonly benign in their behavior. A variety of subtypes of chondrosarcoma can be distinguished including myxoid, mesenchymal, and dedifferentiated. Although chondrosarcomas are in general amenable to surgical removal and cure the dedifferentiated variant form represents a particularly aggressive variant.

A variety of factors in addition to histopathological appearance can influence clinical outcome. Grading of sarcomas is based on mitotic activity, extent of recrosis, degree of cellularity, anaplasia, infiltrative growth, matrix formation, calcification, and presence of hemorrhage, although the first two of these factors are usually the most important. A three-grade system (I–III) has proven most useful for predicting survival and likely treatment response with the highest grade faring worst. Histopathogical subtype may be used for establishing grade with Ewing's sarcoma, osteosarcomas, rhabdomyosarcoma, and round-cell and pleomorphic liposarcomas all being considered as high grade. Well-differential and myoxid liposarcomas are considered as low grade.

The American Joint Committee (ASC) system stages sarcomas based on size, infiltration by the primary tumor, involvement of lymph nodes, presence of metastatic tumor, and the sarcoma type and grade. Usually smaller tumors (<5 cm) without lymph node involvement or metastatic tumor are grade I, while large tumors with lymph node involvement and metastases represent the highest stage IV. The difference between these grades is dramatic with most stage I cancers and <10% of Stage IV tumors surviving 5 years.

Clinical Evaluation and Treatment

Once the histopathological diagnosis has been determined a variety of procedures including X-rays, CT scan, and magnetic resonance imaging are used to determine the sarcoma's precise shape and degree of infiltration. Surgical removal is the mainstay of the majority of treatment plans for many sarcoma classes. One of the key problems with surgery is the tendency of the tumor to recur locally. A wide margin around the tumor is usually removed to minimize recurrence, perhaps including muscle groups or even limb amputation. The precise site of the sarcoma can determine the ease with which the sarcoma can be surgically removed and hence overall survival. Sarcomas in the extremities are usually easier to treat, particularly those located distally as they are usually detected when relatively small and are readily accessible. Sarcomas located at sites where removal is technically more difficult, such as in the head and neck or adjacent to vital structures, may by comparison have a poor prognosis. Surgery is often accompanied by radiotherapy, usually postoperatively and particularly for tumors of higher grades and stages. These treatments can achieve good local control in patients, but for intermediate or high-grade sarcomas up to 50% of patients may subsequently develop metastatic disease. Adjuvant chemotherapy may be used in these patients in attempts to improve overall survival. Drugs such as doxorubicin, ifosofamide, vincristine, cisplatin, dactinomycin, dacarbazine, cyclophosphamide, and methotrexate are used in these treatments and administration may occur preoperatively for some sarcomas such as osteosarcomas of the extremities. Although chemotherapy is not curative for most adult sarcomas, certain classes of sarcoma such as childhood rhabdomyosarcoma, osteogenic sarcomas, and Ewing's sarcoma exhibit much better responses. HIV-infected individuals with Kaposi's sarcoma may be treated with a combination antiretroviral therapy consisting of indinavir which targets the viral protease, and one or more viral reverse-transcriptase inhibitors.

Cytogenetic Aberrations

Cytogenetic studies have identified specific chromosome translocations in several classes of sarcomas (**Table 3**). These abnormalities can be found as the only cytogenetic alteration indicating that their formation may be a key event in tumor development and they usually occur in a high proportion of tumors thus offering the prospect of their use in tumor diagnosis. A consistent feature of these translocations that they result in fusion of genes present at two different cytogenetic locations. The new chimeric genes created by the translocation are expressed as chimateric transcripts, which in turn encode novel fusion proteins.

The t(X;18)(p11.2;q11.2) translocation detected in practically all monophasic and biphasic variants of synovial sarcoma was found to result in the fusion of the *SYT* gene on chromosome 18 to either of two closely related genes *SSX1* and *SSX2* located at Xp11.2. The predicted protein encoded by the *SYT–SSX* transcripts most commonly consist of the 379 amino terminal amino acids of SYT (lacking the final eight C-terminal amino acids) fused to the 78

Table 3 Major translocations in human sarcomas

Tumor	Translocation	Genes
Synovial sarcomas	t(X;18)(p11.2;q11.2)	*SYT, SSX1, SSX2*
Myxoid and round-cell liposarcomas	t(12;16)(q13;p11)	*FUS/TLS, CHOP*
Ewing's sarcoma	t(11;22)(q24;q12)	*EWS, FLI1*
Childhood fibrosarcoma	der(15)t(12;15)(p13;q25)	*ETV6/TEL, NTRK3*
Rhabdomyosarcoma	t(2;13)(q35;q14)	*PAX3, FKHR*
	t(1;13)(q36;q14)	*PAX7, FKHR*
Extraskeletal myxoid chondrosarcoma	t(9;22)(q31;q12)	*EWS, CHN/TEC*
Clear-cell sarcomas	t(12;22)(q13;q12)	*EWS, ATF1*
Desmoplastic round-cell tumors	t(11;22)(p13;q12)	*EWS, WT1*
Dermatofibrosarcoma	t(17;22)(q22;q13)	*PDGFβ, COLA1*

C-terminal amino acids of either SSX or SSX2. The use of reverse transcriptase PCR technology for detecting this diagnostic translocation is of particular use in distinguishing the monophasic spindle cell subtype of synovial sarcoma from other types of spindle cell tumor that fall within its differential diagnosis such as fibrosarcoma, leiomyosarcoma, malignant peripheral nerve sheath tumors, MFHs, and hemangiopericytomas. The SYT–SSX1 and SYT–SSX2 fusions have prognostic importance since the metastasis-free survival time is significantly longer in cases involving fusion to SSX2.

The consistent translocation t(12;16)(p18;p11) identified in liposarcomas causes the fusion of N-terminal transcription activation domain of FUS/TLS to the entire open reading frame of CHOP, a protein normally involved in the cellular response to endoplasmic reticulum stress. Interestingly this fusion is found in both myxoid liposarcoma and in a diagnostically and prognostically distinct category designated round-cell liposarcomas. Myxoid liposarcomas are rarely metastatic and associated with favorable survival while round-cell tumors are usually highly metastatic and high-grade.

The translocations t(2;13)(q35;q14) and t(1;13)(p36;q14) are found in the majority of alveolar rhabdomyosarcomas and in a low proportion of tumors diagnosed as embryonal rhabdomyosarcomas. These rearrangements result in the fusion of N-terminal paired box and homeodomains of either PAX3 or PAX7 to the C-terminal domain of the fork head protein FKHR. The aberrant PAX3-FKHR and PAX7-FKHR transcription factors appear to disrupt the normal process of muscle differentiation leading to tumor development.

A t(11;22)(q24;q12) translocations is found in approximately 80% of Ewing's sarcomas, in neuroepithelioma, and in Askin's tumors, indicating a common histiogenesis of these tumors. At the molecular level this translocation results in the fusion of N-terminal EWS transcriptional activation sequence to the FLI1 DNA binding domain. Variants of this fusion have been observed in Ewing's sarcomas in which the *EWS* gene becomes fused to other Ets family members including the *ERG*, *ETV-1*, *EIAF*, and/or *FEV* genes. Interestingly one of the splicing variants of the *EWS-FLI1* (called type 1) appears to define a clinically favourable subset of Ewing's sarcoma. The *EWS* gene also becomes fused to the *ATF2* transcription factor gene in clear-cell sarcoma (also known as malignant melanoma of soft parts) and to the *WT1* gene in desmoplastic small cell tumors, a primitive sarcoma with desmoplastic and multilineage differentiation.

The recurrent translocation t(9;22)(q22;q12) found in extraskeletal myxoid chondrosarcoma also involves fusions of the *EWS* gene in this case to the *CHN/TEC* gene that encodes a novel orphan nuclear receptor containing a zinc finger DNA-binding domain. A variety of other rare fusions have been found including the fusion of the platelet derived growth factor β gene and the collagen *COLA1* gene in dermofibrosarcoma and the fusion of the *ETV6(TEL)* gene, an Ets family member to the *NTRK3* neurotrophin-3 receptor gene in childhood fibrosarcoma.

Oncogene and Suppressor Genes

A variety of somatically acquired alterations in oncogenes and suppressor genes have been observed in sarcoma. Alteration of several genes in the cellular control pathway involved in regulating RB1 status have been observed including mutation and loss of the *RB1* gene itself, loss of the *CDKN6/P16*, $p14^{arf}$, and *p15* genes, and amplification and overexpression of the *CDK4* and *cyclinD1* genes. Alterations have also found in the *p53* suppressor gene pathway. In addition to mutations in the *p53* gene that are found in almost all classes of sarcomas, amplification and overexpression of *MDM2* proto-oncogene which encodes a protein, which can bind to and inhibit the growth-regulating function of p53, is found in up to a third of both bone and soft tissue sarcomas. An interesting feature of these alterations is that mutation of the *RB1* gene and *p53* gene are often found together in individual sarcoma indicating that cooperative may occur between these abnormalities. Activation of members of the *RAS* gene family by point mutations at codons 12, 13, and 6 have been found in some sarcomas such as rhabdomyosarcomas and leiomyosarcomas while amplification and overexpression of a selection of genes including the *SAS* gene and *MYC* gene family members have also been implicated in sarcoma development. In addition to *p53* and *RB1* alterations of other genes implicated in inherited predisposition to sarcoma development have also been observed in spontaneous lesions. For example mutations in the *APC* gene have been observed in sporadic desmoid tumors. Interestingly for this sarcoma a high frequency of alteration of the APC interacting protein β-catenin has also been observed. Molecular cytogenetic studies have identified many other consistent regions of chromosome genes in sarcomas including amplification of 1q21–22 sequences in soft tissue tumours and of 6q and 17p sequence in oesteosarcoma.

Future Perspectives

Many clinical problems remain to be addressed. For some sarcoma categories precise diagnosis remains difficult or controversial. There is an urgent need for new clinical markers that will predict how tumors may behave or whether they will be drug resistant. Survival for many sarcoma classes remain poor and more effective drugs are urgently required. Finally much still remains to be learnt about the molecular mechanisms of development of these tumors. Current technological advances including (1) the completion of the Human Genome Project that will lead to the identification of all human genes, (2) the development of microarray technology for analyzing many thousands of genes simultaneously, (3) the initiation of the Cancer Genome Project that will screen major human tumors including sarcomas for alterations at all genes in the human genome, and (4) the development of new drugs that target key cellular control pathways are likely to have a major impact on sarcoma management over the next decade.

Further Reading

Enzinger FW and Weiss SW (1995) *Soft Tissue Tumors*. St Louis, MO: Mosby.

Parfman HD and Czerniak B (1997) *Bone Tumors*. St Louis, MO: Mosby.

Verweig J, Pinedo HM and Suit H (eds) (1997) *Soft Tissue Sarcomas: Present Achievements and Future Prospects*. Boston, MA: Kluwer Academic Publishers.

***See also:* Adenoma; Beckwith–Wiedemann Syndrome; Cancer Susceptibility; Carcinogens; Ewing's Tumor; Lipoma and Uterine Leiomyoma; Neurofibromatosis; Oncogenes; Retinoblastoma; Retroviruses; Rhabdomyosarcoma; Synovial Sarcoma; Translocation; Wilms' Tumor**

Satellite DNA

***See:* Microsatellite**

Satellited Chromosome

M A Ferguson-Smith

doi: 10.1006/rwgn.2001.1150

'Chromosomal satellite' is the term given to that part of the end of a chromosome which is separated from the rest of the chromosome by a secondary constriction. (The primary constriction refers to the region of the chromosome occupied by the centromere.) It seems that chromosomal satellites were first described by Russian cytologists early in the twentieth century who used the term 'sputnic.' The secondary constriction marks the site of the nucleolar organizer, a region containing multiple copies of the ribosomal genes (see Nucleolus). This region remains attached to the nucleolus during interphase, and nucleolar remnants remaining on the chromosome may lead the chromatin fiber containing the ribosomal genes to persist as a thin thread at metaphase.

In the human karyotype there are five pairs of acrocentric chromosomes (13, 14, 15, 21, and 22) which carry nucleolar organizers on their short arms. Not all are active in any one cell, as revealed by silver staining of the ribosomal nucleoprotein at metaphase. Their activity determines whether or not a satellite is formed. The satellite itself is largely composed of families of repetitive DNA (heterochromatin) usually without the presence of transcribed genes. The satellite is variable in size, and particularly large or duplicated satellites may often be observed as familial traits, inherited through many generations of a pedigree. The satellite region may also be deleted, and these variants as well as the large heteromorphisms are without phenotypic effect on the carrier individual.

During interphase the nucleolar organizers of several satellited chromosomes may together form a common nucleolus. In the following metaphase, with the dissolution (disassembly) of the nucleolus, these chromosomes may be observed still to be attached to one another by their short arms, a phenomenon described as 'satellite association,' and an example of the nonrandom arrangement of chromosomes during the cell cycle. It is believed that this close association may favor recombination between nonhomologous acrocentric chromosomes leading to the formation of Robertsonian translocations. Robertsonian translocations, particularly those between chromosomes 13 and 14, and 13 and 21, are the commonest chromosomal rearrangements found in humans (see Robertsonian Translocation).

See *also:* Human Chromosomes; Nucleolus; Robertsonian Translocation

SCE (Sister Chromatid Exchange)

J Hodgkin

doi: 10.1006/rwgn.2001.1153

Sister chromatid exchange (SCE) is a form of exchange, equivalent to recombination, between the two replicating or replicated chromatids of a chromosome, which can occur at mitosis as well as at meiosis. When chromatids are differentially labeled, as in the harlequin chromosome technique, SCE events can be seen as discontinuities in the uniform labeling of each chromatid.

See *also:* Chromatid; Harlequin Chromosomes

Schizophrenia

T J Crow

doi: 10.1006/rwgn.2001.1154

Origin of the Term

'Schizophrenia' is the term that was introduced by E. Bleuler to refer to a 'group of diseases' that had been identified as 'dementia praecox' by E. Kraepelin at the turn of the nineteenth century. It connotes that group of 'psychoses' (illnesses characterized by the presence of delusions, hallucinations, and thought disorder) that have a poor outcome. Outcome was the criterion that Kraepelin depended upon for his definition. Bleuler argued that there was a fundamental disturbance of thinking – a 'loosening of associations.' Neither criterion is well defined.

Arguably the entity is as well established as the lynchpin of psychiatric thought at the millennium as it was at the end of the nineteenth century. Other conditions, *a fortiori* the affective psychoses (disorders with psychotic symptoms in which mood disturbance predominates), are defined by reference to schizophrenia – these are diagnoses to be considered when schizophrenia has been excluded. The classical Kraepelinian concept that 'schizophrenia' (with psychotic symptoms that cannot be explained in terms of a primary mood change) and 'manic-depressive illness' (in which the mood change is primary) constitute separate disease entities is central to almost all discussions of psychiatric diagnosis and practice. The structure of textbooks and examinations is built around it.

Categories or Dimensions?

But there are serious doubts about whether either Bleuler's or Kraepelin's concept defines a real disease entity. The fundamental flaw was apparent to Kraepelin. It is simply that the concept has no boundaries. In 1920 Kraepelin wrote of:

> the difficulties which prevent us from distinguishing reliably between manic-depressive insanity and dementia praecox. No experienced psychiatrist will deny there is an alarmingly large number of cases in which it seems impossible, in spite of the most careful observation, to make a firm diagnosis ...it is becoming increasingly clear that we cannot distinguish satisfactorily between these two illnesses and this brings home the suspicion that our formulation of the problem may be incorrect.

The failure of the concept was demonstrated by Endicott and coworkers (1982) when they applied seven different sets of operational criteria to a consecutive series of 46 patients meeting any of the criteria for a diagnosis of schizophrenia admitted to the Psychiatric Institute in New York. By the most liberal criterion 44 patients were diagnosed as suffering from schizophrenia; by the most restrictive it was only six. Yet all of these criteria can be traced back through Bleuler to Kraepelin. Something is fundamentally wrong with the concept. What it is, is clear from Endicott *et al.*'s seminal contribution and from the direction of recent research on psychosis. Endicott *et al.* showed that the differences between different sets of criteria are to a large extent related to whether or not patients with an affective component to their illness are included. By the more liberal criteria some who by modal Research

Diagnostic Criteria (RDC) will be diagnosed as manic-depressive will be given a diagnosis of schizophrenia. Much of the recent literature, whether nosological, pathophysiological, or genetic, is consistent with the concept that psychotic disorders represent continua rather than discrete disease entities.

Brain Changes and Syndromes

An important conclusion from imaging and post-mortem studies is that the psychoses are associated with structural changes in the brain. For example there is a degree of enlargement of the cerebral ventricles and there may be a modest reduction in cortical mass. Such changes are not specific to schizophrenic illnesses but occur also, probably to a lesser degree, in the affective psychoses. Thus the presence of structural change must be integrated within the concept of a continuum.

But the nature of the continuum is elusive. Much recent work has focused on dimensions of psychopathology. One concept was that there are separate dimensions of positive (features such as delusions and hallucinations that are pathological by their presence) and negative symptoms (features such as poverty of speech or flattening of affect that represent loss of function) within a disease entity of 'schizophrenia' and that these syndromes had separate underlying pathologies – a neurochemical and a structural component respectively. But it is clear that such syndromes are not limited to illnesses that can be labeled schizophrenic in the original Kraepelinian sense. The existence of such variation raises the question of why if we really had discrete disease entities would we also have dimensions? What could these be other than a single dimension of severity? Alongside this issue must be placed the body of evidence that symptoms apparently characteristic of psychosis are common in the general population, i.e., in people who are never formally considered either by themselves or others to suffer from an illness. Where is the line to be drawn?

E. Kretschmer formulated a challenge to the original Kraepelinian concept:

> We can never do justice to the endogenous psychoses so long as we regard them as isolated unities of disease, having taken them out of their natural heredity environment, and forced them into the limits of a clinical system. Viewed in a large biological framework, however, the endogenous psychoses are nothing other than marked accentuations of normal types of temperament.

This formulation raises the questions of what is 'the natural heredity environment' of psychosis, i.e., its genetic basis, and 'its large biological framework'? These questions impinge upon the origins of humans, and the nature of human diversity. In their solution the epidemiological characteristics of psychosis are crucial.

Nuclear Symptoms, Epidemiology, and Genetic Implications

K. Schneider identified a set of 'nuclear' or 'first rank' symptoms that appear to define a boundary of severity that makes it likely that an individual who experiences them for the first time will be referred to a treatment facility and thus will be enumerated for comparisons across populations. The symptoms include specific types of hallucination (e.g., hearing one's thoughts spoken aloud, and hearing voices discussing one in the third person) and primary experiences concerning one's thoughts (e.g., that thoughts are removed from or inserted into one's head). With the use of these symptoms Jablensky and coworkers (1992) in the World Health Organization (WHO) Ten Country study concluded that

> schizophrenic illnesses are ubiquitous, appear with similar incidence in different cultures and have features that are more remarkable by their similarity across cultures than by their difference.

Thus the predisposition to psychosis is intrinsic to human populations. The biological disadvantage that is associated with such genetic predisposition must be balanced by an advantage. It has been argued that it is related to the speciation characteristic, the capacity for language. Schizophrenia, according to this view, is 'the price that *Homo sapiens* pays for language.' Nuclear symptoms themselves can be conceived as anomalies of the transition from thought to speech. They may represent 'language at the end of its tether.' A response to Kretschmer's challenge therefore is that the 'natural heredity environment' is the genetic change (the 'speciation event') associated with the transition between a precursor hominid and *Homo sapiens*. The genetics of psychosis according to this view is inseparable from the genetics of speciation. The relevant variation is *Homo sapiens*-specific. It may be related to what appears to be the defining feature of the human brain – that it is lateralized. A critical component of language, probably the phonological sequence is confined to one (the 'dominant') hemisphere. This lateralization is associated with subtle anatomical asymmetries and loss of these asymmetries appears to be an accompaniment of the ventricular enlargement and cortical reduction that has been identified in schizophrenic and to a lesser degree in affective psychoses.

A simple view is that the psychoses represent components of the variation that is associated with the development of the human cerebral cortex. This evolved by a change that allowed the two hemispheres to develop with a degree of independence and this genetic change permitted the evolution of language. The nature of the variation that persists within the population even though it is associated with a biological disadvantage (individuals with psychosis are much less likely that the rest of the population to procreate) is a question of theoretical importance. One possibility, consistent with findings in studies of monozygotic twins discordant for psychosis is that it is 'epigenetic' (associated with gene expression) rather than variation in the DNA sequence. The 'larger biological framework' of the genetics of psychosis is the evolution of the capacity for language and the associated revolution in brain function (hemispheric differentiation) that allowed the transition to take place.

Conclusions

The focus on 'schizophrenia' as a disease entity, as widely promoted by the psychiatric textbooks of the past century, does no justice to the nature of psychosis. These phenomena are intrinsic to human populations and tell us about the genetic origin of the characteristic that defines the species, the capacity for language. The nuclear symptoms of schizophrenia (e.g., thoughts inserted into or removed from one's mind, hearing one's thoughts spoken aloud) can be regarded as a window on the transition from thought to speech, i.e., a reflection on the neural basis of language. By the same logic the whole range of psychotic manifestations, including the affective psychoses, tells us about the variation that relates to the core characteristic of the human brain – hemispheric differentiation. Thus the phenomena of psychosis help us to unravel what is distinctive about the function of the brain in *Homo sapiens* and why it is so diverse. They are also relevant to the genetic nature of the transition from a precursor hominid species.

Further Reading

Bruton CJ, Crow TJ, Frith CD *et al.* (1990) Schizophrenia and the brain: a prospective clinico-neuropathological study. *Psychological Medicine* 20: 285–304.

Crow TJ (1990) Molecular pathology of schizophrenia: more than one disease process? *British Medical Journal* 280: 66–68.

Crow TJ (1998) Nuclear schizophrenic symptoms as a window on the relationship between thought and speech. *British Journal of Psychiatry* 173: 103–109.

Crow TJ (1999) Twin studies of psychosis and the genetics of cerebral asymmetry. *British Journal of Psychiatry* 175: 399–401.

Verdoux H, van Os J, Maurice-Tison S *et al.* (1998) Is early adulthood a critical developmental stage for psychosis proneness? A survey of delusional ideation in normal subjects. *Schizophrenia Research* 29: 247–254.

References

Endicott J, Nee J, Fleiss J *et al.* (1982) Diagnostic criteria for schizophrenia: reliabilities and agreement between systems. *Archives of General Psychiatry* 39: 884–889.

Jablensky A, Sartorius N, Ernberg G *et al.* (1992) Schizophrenia: manifestations, incidence and course in different cultures. A World Health Organization Ten Country Study. *Psychological Medicine Supplement* 20: 1–97.

See *also*: Clinical Genetics

Schizosaccharomyces pombe, the Principal Subject of Fission Yeast Genetics

R Egel

doi: 10.1006/rwgn.2001.1649

Schizosaccharomyces pombe is a primitive ascomycetous fungus, also known as fission yeast. It has been extensively used in general and molecular genetics, and its genome has been fully sequenced (August 2000). It is considered a very useful model organism for experimental research on fundamental properties of eukaryotic cells, such as cell cycle control mechanisms, polarized cell growth, signal transduction, and sexual differentiation.

History

The narrow group of fission yeasts harbors only three recognized species, of which *S. pombe* is the best known, by far, experimentally. The other two are *S. japonicus* and *S. octosporus*. These yeasts can be isolated from fermenting saps or juices in tropical and subtropical areas. They are not directly related to the more commonly occurring budding yeasts, which independently have developed a divergent pattern of unicellular growth from another lineage of filamentous ascomycetes. In many respects, the molecular clock of protein evolution seems to have kept a slower pace in the fission yeasts as compared to the more rapidly evolving budding yeasts. Hence, *S. pombe* tends to resemble the common fungal ancestor more faithfully – as well as the common ancestor of fungi and animals. This makes it a particularly useful model organism for eukaryotic cell and molecular biology studies in general. The fission mode of cell division resembles hyphal septation, which commonly occurs in many filamentous fungi, whereas the budding mode may have arisen from more specialized patterns, such as the emergence of microconidia in *Neurospora crassa*.

The first genetic studies of *S. pombe* were done by Urs Leupold in the late 1940s, and essentially all experimental strains in current usage are derived from Leupold's cultures, which had been isolated in 1921 by A. Osterwalder from "an exceedingly over-sulfurized grape juice," originating from southern France. The unique potential of fission yeast for studies of cell division and growth was first recognized by Murdoch Mitchison in the 1950s, and the cell cycle studies of his group were merged with Leupold's genetic approach by Paul Nurse in the 1970s.

General Genetics

Leupold began his genetic analyses by characterizing cross-fertile strains of two mating types (P, plus; or M, minus). These were related to the originally homothallic parent culture (h^{90}, capable of mating-type switching and self-fertile) by various kinds of chromosomal rearrangements in the mating-type region (see Mating-Type Genes and their Switching in Yeasts). This set the stage for doing genetic crosses by mixing compatible haploid strains and harvesting asci or free spores at the end of growth, as meiosis and sporulation directly occur in the zygotes formed upon nutritional limitation. *S. pombe* usually grows as haploid cells from spores, or rather rarely as diploid cells from uncommitted zygotes that failed to sporulate upon transfer to growth medium.

Crosses are routinely analyzed by tetrad dissection or random spore analysis. Useful screening procedures include the staining of spores on agar plates with iodine vapor (reacting with starch-like polyglucans), selective staining of diploid colonies by certain dyes, such as Phloxin B (actively excluded by live cells, but readily adsorbed to the dead cells that occur relatively frequently in diploid cultures), or mutant enrichment after inositol-less death or deoxyglucose-induced cell lysis of growing cells. Diploid growth from uncommitted zygotes is routinely selected for by allelic complementation for a growth requirement. Sterile strains can be crossed by the alternative method of protoplast fusion, after the rigid cell walls have been removed by enzymic digestion.

Leupold's group initiated the systematic coverage of the genetic map by biochemical markers, and later specialized in suppressor tRNA genetics. Other topics pursued in *S. pombe* before the onset of the molecular era include analyses of biosynthetic pathways, such as purine metabolism, intragenic recombination and allelic complementation, mutagenesis and DNA damage repair, conditional cell division cycle mutants, mating-type switching, and mitochondrial genetics.

Reverse Genetics

The modern era of *S. pombe* genetics began in the early 1980s when Paul Nurse and others established gene cloning procedures in fission yeast, soon after the basic protocols had been developed for budding yeast. Since then the number of vector plasmids and other tools has increased considerably. Targeted gene disruption and replacement by linearized DNA constructs work reasonably well, as guided by flanking regions of homology with the target sequence, although the efficiency of insertion at the correct target may vary from place to place in the genome. Random insertion of selectable markers in the absence of sequence homology has been utilized as a novel means for mutagenesis and facile cloning. Systematic screening of green fluorescent protein (GFP)-fusion libraries has proved very useful for the classification of functional proteins as to their subcellular *in vivo* localization. The potential of *S. pombe* as host cells for heterologous expression of selected proteins is actively being investigated.

Genomics

The haploid genome of *S. pombe* amounts to about 14 Mb, distributed among three chromosomes of 5.7, 4.6, and 3.5 Mb. These have become convenient size markers in pulsed-field gel electrophoresis of chromosomal DNA of other organisms. The sequencing of the entire genome, coordinated by the Sanger Centre, was completed in 2001.

There are about 6000 protein-encoding genes. About half of these have introns; half of those, again, have at least two, and so forth, in a uniformly descending series of genes with multiple introns. Most introns are relatively short (40–75 nt), and their spacial distribution is significantly skewed towards the 5′ end of a gene. There is little evidence of differential splicing in general, but some instances of meiosis-specific splicing have been reported.

Ribosomal RNA genes are located at both ends of chromosome III. Other conspicuous features of the chromosomal landscape include centromeres, telomeres, *ars* elements (autonomously replicating sequences), the specifically organized mating-type region (*mat*), and retrotransposon-related repetitive sequences.

Relatively long segments of 60–100 kb are needed to make up functional centromeres. These stretches are characterized by a repetitive, heterochromatin-like organization – highly inaccessible to both transcription and recombination, during vegetative growth as well as during meiosis. A sizable chunk of centromeric-repeat DNA appears to have been transposed to the *mat* region, where it holds a central position in the silenced domain carrying two backup cassettes for mating-type switching.

Chromosomal origins of replication (*ars*) tend to map at 1–4 kb gene-free regions of low complexity. Essential consensus sequences could not be defined, but the number and spacing of multiple binding sites for so-called AT-hook proteins appear to be important. The local configuration of replication origins and pause sites around the functional *mat1* cassette has been implicated in the molecular mechanism of mating-type switching.

Nuclear Division Cycle

The cell cycle is governed by the regular replication and segregation of the chromosomal genome. In fission yeast, the analysis of cell cycle control factors by conditional *cdc* mutants was pioneered by Paul Nurse. The key role was ascribed to the prototype cyclin-dependent protein kinase Cdc2p, which is homologous and functionally equivalent to MPF (maturation promoting factor) activity of developing amphibian oocytes. Its activity is needed independently at two crucial control points, the G_1/S transition to start replication and the G_2/M transition to initiate mitosis. A host of interacting factors, such as cyclins, inhibitors and ancillary protein kinases, have been identified in addition. Furthermore, the metaphase–anaphase transition has been recognized as a specific phase of polyubiquitinylation and proteasome-driven proteolysis to dissolve sister chromatid cohesion.

Modern emphasis has shifted to various 'checkpoint' control systems, which ensure that crucial transitions in the cell cycle can take place only if essential functions of the preceding stage have safely been completed and premutational damage has been repaired. The mitotic cyclins are removed by the proteasome, and cytokinesis commences. Aberrant behavior at this stage was first recognized in Mitsuhiro Yanagida's group as conditional *cut* mutants, when still undivided nuclei were bisected by untimely septation.

Cell Shape and Growth

Morphogenetic processes in fission yeast appear to be relatively simple. Individual cells are cylindrical in shape, and they grow in essentially one dimension at their tips. They arise by cytokinesis, shortly after mitosis, when a centripetally constricting septum bisects the mother cell, and the old cell wall is split apart. The 'new' end of any given cell originates from the most recent septum – the 'old' end being related to an earlier septation event. During mitosis and cytokinesis, cell wall elongation ceases temporarily. Cell growth is then resumed at the old end only for a while, before it becomes bipolar by switching over to the new end as well.

All these transitions are tightly coordinated with the nuclear division cycle, and a host of interacting components has been identified by many groups, focusing on actin localization, organization of microtubules, participation of various motor proteins, signal transduction cascades, and other critical factors. Thereby, the superficially simple processes of linear growth, septation and morphogenesis are being cast into very sophisticated mechanistic networks at the molecular scale.

Sexual Differentiation

In fission yeast, mating occurs only at the end of vegetative growth, upon nutritional starvation. This is mediated by the transcriptional repression of sexually important genes on rich medium, where two vegetative protein kinases are active at a high level. Upon nitrogen depletion, in particular, this repression is relieved and the mating-type genes can take over. These code for two complementary pairs of transcription factors: an early acting pair, specifying the sexual identity of M cells and P cells, respectively, and a later acting pair controlling meiosis.

Mating-competent cells communicate by secreting their own peptide pheromone and responding to the presence of the complementary one, as mediated by specific receptor proteins, spanning the cytoplasmic membrane seven times. The pheromone response leads to loose agglutination *en masse* and tighter pairwise association, before the interactive partners fuse by local dissolution of the separating cell walls. Rapid nuclear movements are likewise induced by the pheromone response, resulting in karyogamy shortly after cytoplasmic fusion. As the two late-acting mating-type genes themselves are induced by the pheromone response, the diploid nucleus of the zygote is primed to undergo meiosis immediately. Finally, upon sporulation, the zygotes are converted into four-spored asci. Azygotic asci can also be produced by rarely occurring diploid cells, if these are heterozygous at the *mat* locus.

Within the differentiating cells, pheromone perception is coupled to both trimeric and Ras-related G proteins, and the signal is further conveyed via a classical MAP kinase cascade, before it results in pheromone-induced transcription in the nucleus. There is important cross-talk to the stress response cascade as well. Various mechanisms contribute to specific desensitization of the pheromone response in the end. Many groups have worked on and added further details to these aspects.

Meiosis and Recombination

Meiosis in fission yeast is remarkable by several unusual features. The number of crossover events, 10–20 per chromosome, is higher than in any other organism analyzed genetically. Yet, this efficient meiotic recombination occurs in the absence of synaptonemal complexes, and crossover interference is not observed either. Most conspicuously, the meiotic prophase nuclei are rapidly pulled back and forth in a series of so-called 'horsetail' movements, led by the spindle pole bodies, to which the partially contracted chromosomes are attached by their clustered telomeres. This mechanical agitation, as well as the organization of chromosomal cores ('linear elements') connecting sister chromatids, is essential for efficient recombination. Systematic mutant screening has identified numerous genes involved in the mechanism of meiotic crossing-over.

Mitochondrial Genetics

The mitochondrial genome of fission yeast (17–22 kb) is one of the smallest of its kind in lower eukaryotes. *S. pombe* is a so-called petite-negative yeast, which cannot form ATP in the absence of mitochondrial DNA, even under nonrespiratory conditions. Mutations affecting this dependency have been obtained, in addition to more commonly occurring respiratory-deficient mutants.

Future Prospects

Even though many scientists have firmly established *S. pombe* as a powerful model system for many basic aspects of eukaryotic cell biology, a lot is still to learn. The systematic evaluation of the information provided by the complete genome sequence has barely begun. In a most significant trend, more and more workers concerned with higher eukaryotes are becoming interested in the fission yeast to look at their favorite ortholog together with potentially interacting proteins so as to learn more about the underlying molecular mechanisms.

Relevant Web Sites

History and Methods: http://www.bio.uva.nl/pombe/
Sequencing program: http:/www.sanger.ac.uk/Projects/S_pombe/
Vectorplasmids: http://pingu.salk.edu/users/forsburg/plasmids.html

Further Reading

Egel R (2000) Fission yeast on the brink of meiosis. *BioEssays* 22: 854–860.

Forsburg SL (1999) The best yeast? *Trends in Genetics* 15: 340–344.

Nasim A, Young P and Johnson BF (eds) (1989) *Molecular Biology of the Fission Yeast*. San Diego, CA: Academic Press.

Verde F (1998) On growth and form: control of cell morphogenesis in fission yeast. *Currents Opinion in Microbiology* 1: 712–718.

Yanagida M (1999) From phage to chromosome biology: a personal account. *Journal of Molecular Biology* 293: 181–185.

***See also:* Cell Division Genetics; Mating-Type Genes and Their Switching in Yeasts; *Saccharomyces* Chromosomes; Yeast Plasmids**

Screening

I Schildkraut

doi: 10.1006/rwgn.2001.1155

Screening is a method of analyzing bacterial isolates for a particular property or phenotype. It is useful for isolating mutant organisms which do not have a selective advantage over the wild-type organism or when the mutation leads to a conditionally lethal phenotype.

Screening for a particular phenotype can be quite straightforward; for example, a distinguishing characteristic can be analyzed based on the color or size of a colony. Differentiating a blue colony from a sea of white colonies is one such example. More involved methodologies include DNA probe analysis, biochemical assay (for enzymatic activity), and antigen/antibody reaction (for the presence of a protein). Screening for a particular phenotype can be quite laborious. For example, analysis of certain enzymatic activities requires each bacterial isolate to be grown in a separate test tube. Then the protein must be extracted and each extract tested biochemically. This can limit the analysis to a small number of isolates.

Screening bacteria for a deleterious characteristic such as antibiotic sensitivity requires growing each of the isolates in the absence of the antibiotic. Each individual isolate must then be tested by picking colonies onto a plate that contains the antibiotic and on a plate that does not contain the antibiotic. Those isolates that can only grow in the absence of the antibiotic and fail to grow in the presence of the antibiotic are the antibiotic-sensitive isolates.

An alternative and simpler method for screening for isolates that do not have the ability to grow under specialized conditions is the method of replica plating. Here, instead of individually picking hundreds of colonies for analysis on different media, a simple device that takes an imprint of all the colonies from a master agar plate is applied to a fresh agar plate, thereby creating a replica of the master plate. The plate can be incubated under unfavorable conditions such as in the presence of an antibiotic. By this method, a relatively large number of colonies can be screened without much effort.

Screening should not be confused with selection. Selection distinguishes among bacteria by their ability to grow under specific conditions. Selections can be set up to isolate a single bacterium with a specific property from a mixture of hundreds of millions of bacteria lacking the property.

***See also:* Antibiotic-Resistance Mutants; Bacterial Genetics; Resistance to Antibiotics, Genetics of; Selection Techniques**

scRNA

doi: 10.1006/rwgn.2001.2013

scRNA is the abbreviation for small cytoplasmic RNAs present in the cytoplasm and (sometimes) nucleus.

***See also:* Cytoplasm; Nucleus**

scRNP

doi: 10.1006/rwgn.2001.2014

scRNP is the abbreviation for small cytoplasmic ribonucleoproteins, i.e. scRNAs associated with proteins.

***See also:* Cytoplasm**

SDP (Strain Distribution Pattern)

***See:* Strain Distribution Pattern (SDP)**

Second Division Segregation

***See:* First and Second Division Segregation**

Secretion

***See:* Protein Secretion Systems**

Seed Development, Genetics of

D W Meinke

doi: 10.1006/rwgn.2001.1676

The seeds of flowering plants have long been viewed as a convenient subject for basic research and an important factor in agriculture and human nutrition. The concept that seeds could also be subjected to genetic analysis dates back to the earliest days of modern genetics. Wrinkled seeds were one of the initial characters studied by Mendel and defective kernel mutants were first described during the formative years of maize genetics. Large-scale analysis of such mutants, however, did not occur until many years later, when the power of genetics and molecular biology were combined. *Arabidopsis* eventually became the model system of choice for the identification of genes with essential functions during seed development while maize has remained the preferred system for genetic and biochemical studies of endosperm maturation. With the completion of the *Arabidopsis* genome, and multinational programs in functional genomics already in progress, it appears likely that every gene required for seed development in this model plant will eventually be identified. The challenge will then be to understand the many regulatory pathways and interesting variations in seed development that are evident throughout the angiosperms.

Seed Development

The angiosperm seed forms through a unique process known as double fertilization, in which one male gamete contributed by the pollen fuses with the egg cell located within the ovule to form the diploid zygote, while the other male gamete fuses with two haploid maternal nuclei in close proximity to the zygote to form a triploid endosperm nucleus.

Following double fertilization, the ovule enlarges and develops into a seed, and the ovary that generated the ovule becomes known as the fruit. The angiosperm zygote develops into an embryo composed of two parts, the embryo proper and the suspensor. The embryo proper ultimately differentiates into the mature embryo, whereas the suspensor degenerates during later stages of development and is not usually present at maturity. The suspensor appears to perform both a structural role in attaching the embryo proper to the surrounding seed coat and an active role in supporting and promoting growth of the embryo. The embryo proper progresses through a series of characteristic morphological stages early in development, establishes the root and shoot apical meristem regions that will produce the vegetative plant following germination, and prepares for seed desiccation at maturity. The endosperm tissue often begins as a free nuclear syncytium that later undergoes cellularization. In dicotyledonous plants such as *Arabidopsis*, the endosperm is absorbed by the developing embryo and is not present in the mature seed. Nutrients required for germination and initial seedling development are stored in embryonic leaves known as cotyledons. In monocots such as maize, a significant amount of starchy endosperm tissue remains at seed maturity to support growth of the young seedling.

Screening for Seed Mutations

Mature seeds are the preferred target for chemical mutagenesis in *Arabidopsis* because they are small, easily suspended in small volumes of mutagen, and contain only a few target cells in the shoot apical meristem. M_1 plants derived from treated seeds are chimeric, with sectors of heterozygous cells adjacent to wild-type cells derived from other parts of the treated meristem. Flowers that form within a mutant sector produce siliques (fruits) with 25% homozygous mutant seeds following self-pollination. Embryo-defective mutations can therefore be identified by screening siliques of M_1 plants for the presence of 25% defective seeds. Mutant seeds often differ from normal seeds in size, color, and embryo morphology. The transparent seed coat facilitates the identification of embryos altered in morphology and pigmentation. Seeds treated with a clearing agent can be viewed under a compound microscope equipped with Nomarski (DIC) optics to reveal striking cellular details within the developing embryo. Although large collections of embryo-defective mutants have been generated by this method of screening immature siliques for abnormal seeds, many additional mutants defective in seed development have been identified by screening germinated seedlings on agar plates for defects in morphology indicative of a disruption of embryo development.

Pollen is the preferred target for chemical mutagenesis in maize because controlled crosses are performed with minimal effort and because mature kernels are much larger than in *Arabidopsis* and contain a more complex shoot meristem. Segregation of mutant kernels can be detected by examining ears at different stages of development. Translocations involving accessory (B) chromosomes of maize, which undergo nondisjunction during pollen development, have in some cases been used to construct discordant kernels in which the embryo and endosperm differ in genotype.

Many of the seed mutations analyzed in recent years have been generated through insertional mutagenesis. The preferred agent for *Arabidopsis* research is a piece of plasmid DNA (T-DNA) from the soil bacterium *Agrobacterium tumefaciens*. More than 200 000 transgenic *Arabidopsis* plants carrying independent T-DNA insertions are available for detailed forward and reverse genetic screens and many additional lines await further characterization. A variety of endogenous transposable elements have been used in place of T-DNA for large-scale gene tagging in maize. Alterations in patterns of kernel pigmentation caused by transposon activity were originally an important factor in the discovery of transposable elements by Barbara McClintock. Since that time, movement of transposable elements during development and their experimental manipulation for the purpose of gene isolation have been characterized in detail. Several large-scale projects have recently been initiated to identify transposon insertions in most of the genes required for endosperm development in maize.

Diversity of Mutant Phenotypes

Thousands of mutants defective in seed development have already been found in maize and *Arabidopsis*. Although the long process of analyzing mutant phenotypes and identifying the disrupted genes will take years, many interesting and informative phenotypes have already been studied in detail. Extensive analysis of maize endosperm mutants altered in storage product accumulation has contributed not only to our understanding of endosperm function but also to the development of plants with improved nutritional qualities. Informative *Arabidopsis* phenotypes include the twin (*twn*) mutants in which the suspensor forms a secondary embryo, viviparous leafy cotyledon (*lec*) mutants characterized by premature germination and partial transformation of cotyledons into leaf-like structures, fertilization-independent (*fis* and *fie*) mutants in which seed development begins in the absence of fertilization, titan (*ttn*) mutants with giant endosperm nuclei and enlarged embryo cells, shoot meristemless (*stm*) mutants, auxotrophic mutants defective in biotin synthesis, and a variety of mutants disrupted in cell division patterns during early stages of development. In contrast to *Drosophila*, where a small number of genes regulate many of the critical events in embryo development, pattern formation during plant embryogenesis appears to be controlled by many genes with a variety of cellular functions. Current topics of interest that arose from mutant analysis include the role of the plant hormone auxin in regulating embryo pattern formation, the relationship between intracellular transport mechanisms and embryo morphogenesis, and the importance of gene silencing in early endosperm development.

Future Prospects

Arabidopsis contains an estimated 500 genes that can mutate to give an embryo-defective phenotype. Identifying knockout mutations in each of these essential genes is a high priority of future efforts in functional genomics. Many additional genes that are expressed during embryo development do not give rise to a mutant phenotype when disrupted. Most of these genes are likely to be duplicated in the genome or encode products that function in redundant cellular or biochemical processes. Understanding the precise roles of these gene products in embryo and endosperm development represents a significant challenge for the future.

Further Reading

Goldberg RB, de Paiva G and Yadegari R (1994) Embryogenesis: zygote to seed. *Science* 266: 605–614.

Laux T and Jürgens G (1997) Embryogenesis: a new start in life. *Plant Cell* 9: 989–1000.

Lopes MA and Larkins BA (1993) Endosperm origin, development, and function. *Plant Cell* 5: 1383–1399.

Meinke DW (1994) Seed development in *Arabidopsis thaliana*. In: Meyerowitz EM and Somerville CR (eds) *Arabidopsis*, pp. 253–295. Plainview, NY: Cold Spring Harbor Laboratory Press.

Meinke DW (1995) Molecular genetics of plant embryogenesis. *Annual Review of Plant Physiology and Plant Molecular Biology* 46: 369–394.

Mordhorst AP, Toonen MAJ and deVries SC (1997) Plant embryogenesis. *Critical Reviews in Plant Science* 16: 535–576.

See also: *Arabidopsis thaliana*: Molecular Systematics and Evolution; Plant Embryogenesis, Genetics of; Root Development, Genetics of; Transfer of Genetic Information from *Agrobacterium tumefaciens* to Plants; Transposable Elements in Plants

Seed Storage Proteins

H B Krishnan and E H Coe, Jr

doi: 10.1006/rwgn.2001.1714

More than 70% of the proteins consumed by humans are derived from storage proteins of legume and cereal seeds. Seed protein content of legumes varies from 20% to 40%, while in cereals it accounts for 7–15% of the dry weight of the seed. Seed storage proteins accumulate in the cotyledon and embryo of dicotyledonous plants and in the endosperm of monocotyledonous plants. These proteins are deposited in specialized membrane-bound organelles called protein bodies. The predominant legume storage proteins are salt-soluble globulins and are grouped under two classes, 7S and 11S, while the major storage proteins of cereals are the alcohol-soluble prolamins. Exceptions are oats and rice, in which the major storage proteins are globulin-like. Because of the abundance of these proteins, they are mainly responsible for the nutritional quality of the human diet. The deficiency of certain essential amino acids in the seed proteins may, however, limit their nutritional quality for monogastric animals. In general, the cereal storage proteins are deficient in lysine, threonine, and tryptophan while the cereal prolamins contain low levels of cysteine and methionine. Most of the seed storage proteins are encoded by multigene families. The cereal prolamins appear to have evolved from a single ancestral gene and similarly the 7S and 11S legume proteins appear to have evolved from a common ancestral protein. The synthesis of seed storage protein is primarily controlled at the level of transcription, but may also be subject to posttranscriptional controls. The transcription of seed protein genes is highly regulated in a spatial and temporal manner.

Properties of Seeds

Starch is just one of the major food value ingredients in seeds. In addition to carbohydrates, seeds contain proteins and lipids. All three are storage constituents of the seed, and they are the source not only of the energy and growth compounds for germination of the seed, but for the consumer of the seed and its products. The proteins confer food value, handling properties, and gustatory qualities. Legume seeds, for example, are widely recognized to be significant sources of protein in the diet. Proteins in wheat flours give them their specific texture, consistency, and baking quality. Indeed, differences among breads and pastas are a reflection of differences among their proteins, in types and in the balance of different classes of proteins. Even the many 'starch' products prepared from seeds carry significant amounts of proteins. For the most part, they are valuable food constituents, but many parents of young children become aware by experience that some cereal grain proteins may cause digestive distress in infants while others do not.

History

Humans have utilized seed crops for centuries as a major source of carbohydrate and protein. For example, domesticated wheat and barley have been found in 7000-year-old Egyptian dwellings. Isolation of the proteins (gluten) of wheat flour from the starch was described in the mid-1700s, soon followed by characterizations of differential solubilities and fractions of the gluten. The protein (prolamin) fraction of maize, zein, was isolated by differential extraction in the early 1800s, and studies on other cereal proteins followed over time. The modern study of seed proteins was stimulated by the classical and innovative work of Thomas Osborne. His work on the classification of seed proteins on the basis of their differential solubility in aqueous and nonaqueous solutions facilitated research on the characterization of seed proteins. The ability to classify the proteins and to measure differences in specific qualities has permitted extensive inheritance studies. In fact, a number of genes that encode seed proteins, or that regulate them, have been cloned.

What Are Seed Storage Proteins?

Seed storage proteins are proteins that accumulate significantly in the developing seed, whose main function is to act as a storage reserve for nitrogen, carbon, and sulfur. These proteins are rapidly mobilized during seed germination and serve as the major source of reduced nitrogen for the growing seedlings. In general, seed storage proteins do not carry out any enzymatic functions. Even though storage proteins from diverse plants are structurally different, they all share some common characteristics. One of the main characteristics of storage proteins is that they accumulate in high levels in specific tissues at a specific stage of development. Seed storage proteins are generally not found in nonseed organs. These proteins accumulate within membrane-bound organelles called protein bodies (**Figure 1**). The sequestration of storage proteins within protein bodies ensures that these proteins are separated from the metabolic compartments of the cell. The expression of storage proteins is also regulated by nutrition. For example, the synthesis of sulfur-rich proteins may be restricted when the plants are grown in soils having low sulfur.

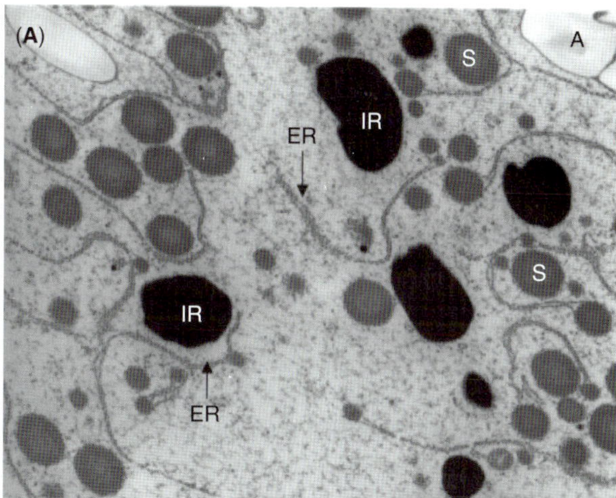

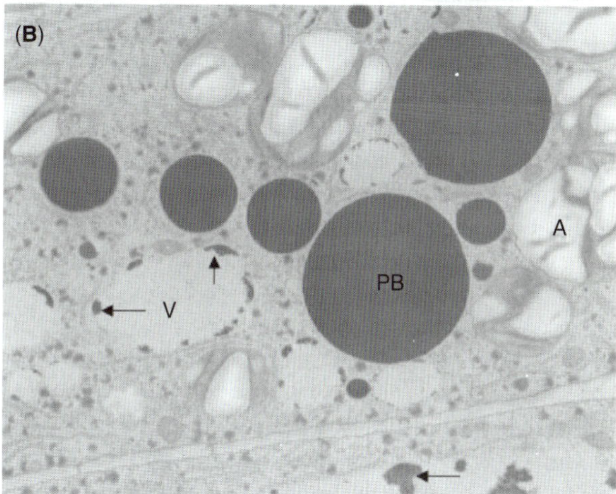

Figure 1 (A) Transmission electron micrograph of rice endosperm showing spherical (S) and irregular-shaped (IR) protein bodies. The spherical protein bodies store prolamins and the irregular-shaped protein bodies accumulate glutelins. Note the direct connection between the rough endoplasmic reticulum (ER) and spherical protein bodies. (B) Protein bodies (PB) in the cotyledons of developing soybean seed. Note the occurrence of protein deposits (arrows) within the vacuoles (V). Large amyloplasts (A) are also seen.

Classification of Seed Proteins

Traditionally seed proteins are classified into four groups based on their solubility properties. Proteins soluble in water are known as albumins, in salt solutions as globulins, in alcoholic solutions as prolamins, and in dilute acid or alkali as glutelins. Even though this classification is not absolute, since one group of proteins can be soluble in more than one solution, it provides a convenient method to group proteins into different classes. Prolamins are found exclusively in the grass family and so far have not been detected in other plant families. The members of the legume family store globulins as the predominant storage reserve. The globulins are broadly classified into 7–8S and 11–12S based on their sedimentation coefficients. In several instances, seed storage proteins are given names that are derived from the Latin generic name of the plant. For example, the storage proteins of maize are called zeins (*Zea mays*), hordein for barley (*Hordeum vulgare*), and glycinin for soybean (*Glycine max*). Currently, the seed proteins are classified on the basis of function and molecular/biochemical relationships. On the basis of their function, seed proteins are classified as storage, structural and metabolic, and protective proteins. The major function of the storage proteins is to serve as the source of nitrogen, carbon, and sulfur. The structural and metabolic proteins are essential for normal growth and development. The protective proteins play a role in providing resistance against pathogens, pests, and desiccation.

Cereal Storage Proteins

Prolamins are the abundant storage proteins of most cereals and consequently were the earliest proteins to be studied. Rice and oats are exceptions in that the major storage proteins are glutelins and globulins, respectively. Prolamins can be subdivided into two categories: those that are soluble in aqueous alcohol and those that are soluble only in the presence of reducing agents such as mercaptoethanol. The prolamins of the second group contain interchain disulfide bonds that render them insoluble in aqueous alcohol. The prolamins of cereals can also be conveniently put into four cereal groups: the Triticeae (wheat, barley, rye, and their relatives), oats, rice, and the Panicoideae (maize, sorghum, and most millets).

Prolamins of the Triticeae

The prolamins of the Triticeae contain high levels of proline and glutamine. They are classified into three families: sulfur-rich (S-rich), sulfur-poor (S-poor), and high-molecular-weight (HMW) prolamins. The S-rich prolamins are the predominant storage proteins representing 80–90% of the total prolamin fractions. The α-gliadins, γ-gliadins, and the low-molecular-weight (LMW) glutenin subunits of wheat, the β and γ-hordeins of barley, and the γ-secalin of rye all belong to this group of prolamins. The amino acid sequence of the S-rich prolamins consists of N-terminal repetitive and C-terminal nonrepetitive domains. The repeats are based on short peptide motifs that are rich in proline and glutamine. The nonrepetitive domain contains most of the cysteine residues that lead to interchain and intrachain disulfide bonds in the polymeric and monomeric S-rich prolamins. A comparison of the nonrepetitive domains of S-rich prolamins reveals three highly conserved regions of 20–30 residues. Three regions, designated A, B, and C, contain the conserved cysteine residue, indicating a common origin from a single ancestral protein. The S-poor prolamins include the ω-gliadins of wheat, ω-secalins of rye, and C hordeins of barley. The S-poor prolamins consist of a repetitive domain near the N-terminus and a nonrepetitive domain at the C-terminus. The HMW prolamins, which contribute to the bread-making quality of wheat, consist of three domains. A nonrepetitive domain is present at the N- and C-termini that flank a central repetitive domain. The repetitive domain contributes to the unusual amino acid composition of these proteins. These proteins contain high levels of glycine and glutamine. Regions related to A, B, and C of S-rich prolamins are also present in HMW prolamins. These similarities indicate an evolutionary relationship among the S-rich, S-poor, and HMW prolamins.

Prolamins of Rice and Oats

The prolamins of oats are called avenins and account for about 10% of the total seed proteins. These proteins also resemble the prolamins of Triticeae in the presence of repeated sequences that occur as two separate blocks. The repeats have close similarity to those present in the S-poor and S-rich prolamins. On the other hand, the rice prolamins, which account for 5–10% of the total seed proteins, do not reveal the repeated sequences that are typical of prolamins of the Triticeae. Rice prolamins fall into four classes. Classes I to III encode the abundant prolamin polypeptides, while class IV represents the minor prolamin component. The class IV prolamin, which is rich in sulfur-containing amino acids (30% methionine and cysteine), has very little sequence homology to the other three classes.

Prolamins of Subfamily Panicoideae

The prolamins of maize (zeins) are divided into four major groups called α, β, γ, and δ zeins. Zeins can be fractionated by sodium dodecyl sulfate-polyacrylamide gel electrophoresis (SDS-PAGE) into polypeptides of 27 000, 22 000, 19 000, 16 000, 15 000, 14 000, and 10 000 Da. The α-zeins, which are made up of polypeptides of 22 000 and 19 000 Da, are the predominant storage proteins of maize, accounting for about 70% of the total protein. The second abundant zein fraction is represented by the γ-zeins made up of 27 000 and 16 000 Da proteins. These proteins are rich in cysteine residues and are soluble in aqueous and alcoholic solutions containing a reducing agent. These proteins may be phylogenetically related to the α, β, γ gliadins of wheat and the β hordein from barley. Proteins with a molecular mass of 14 000 and 15 000 represent the β-zeins, while the δ-zeins are made up of 10 000 and 18 000 Da proteins. Both classes of zeins are rich in methionine. The δ-zeins are structurally unrelated to other zeins, showing some similarity to the methionine-rich Brazil nut 2S storage proteins. In contrast to other classes of zeins, the 18 000 Da δ-zein contains one lysine and two tryptophan residues. Other tropical cereals such as sorghum, Coix, and millets also accumulate prolamins accounting for over 50% of the total seed proteins. The prolamins of sorghum (kafirins), Coix (coixins), pearl millet (pennisetins), and foxtail millet (setarins) have solubility and protein structure similar to zeins.

Legume Storage Proteins

The globulins are the predominant storage proteins in legumes. Unlike the prolamins, the globulins are widely distributed amongst higher plants. They are present not only in dicots but also occur in monocots, gymnosperms, and ferns. These proteins are soluble in dilute salt solutions and have sedimentation coefficients of 7–8S and 11–12S. The 7–8S globulins are generally referred to as vicilin-type globulins, while the 11–12S globulins are called legumin-type globulins. The globulins have been studied in detail from several important legumes including peas, soybean, lupin, peanut, French bean, and broad bean. The amino acid composition of the globulins reveals deficiency in sulfur-containing amino acids with methionine being the most limiting amino acid.

The 11–12S storage proteins are isolated as hexamers. Each of these subunits is made up of an acidic subunit of 40 000 Da and a basic subunit of 20 000 Da. A single disulfide bond holds each of these subunits together. The position of the disulfide bridge is highly conserved amongst 11–12S globulins. Each subunit is synthesized as a precursor protein that undergoes proteolytic cleavage resulting in the basic and acidic subunits. The 7–8S globulins are found as trimers with an apparent molecular mass of 150 000 to 190 000. The 7–8S globulins can be divided into two groups. The first group contains members in which the precursor polypeptides undergo little or no posttranslational modification. The polypeptides belonging to this group are in the range of 76 000 to 40 000 Da. On the other hand, the second group undergoes extensive posttranslational modification resulting in subunits in the molecular mass range of 12 000 to 34 000 Da. Based on the similarity between the C-terminals of the 11–12S and 7–8S globulins, it has been suggested that these two groups of proteins are related to each other and presumably evolved from a common ancestral protein.

Genomic Organization of Seed Storage Protein Genes

Multigene families encode cereal prolamins. The prolamin genes of the Triticeae can be grouped under three multigene families encoding the S-poor prolamins, the S-rich prolamins, and the HMW prolamins. The sequences of these genes vary widely and contain no introns. Classical and molecular genetic studies have shown that most of the prolamin genes of Triticeae are located at complex loci on the homoeologous group 1 chromosomes. The α-zeins are encoded by a multigene family consisting of about 70 to 100 members. Some of these genes contain in-frame stop codons indicating that they are pseudogenes. Several of the α-zein genes have been mapped to the long and short arm of chromosome 4, the short arm of chromosome 7, the long arm of chromosome 10, and near the centromere of chromosome 1. The gene encoding the 27 000 γ-zein has been mapped to the long arm of chromosome 7, while the 10 000 δ-zein has been mapped to the short arm of the same chromosome. Similarly the β-zein is mapped to the short arm, while the 18 000 δ-zein has been mapped to the long arm of chromosome 6. The rice prolamins are also encoded by multigene families consisting of about 80–100 copies per haploid genome. Some of the prolamin genes of rice occur in tandem repeats.

The globulins of legumes are encoded by a small family of genes. For example, the soybean glycinins are encoded by five genes (*Gy1–Gy5*) and are resolved into two groups. These genes are scattered throughout the genome. Most 11S globulins contain three introns while the 7S globulins have five introns. The insertion positions of the introns are highly conserved among the different globulin genes supporting the notion that the 11S and 7S globulin genes share a common precursor. The globulin proteins of maize embryo and endosperm are encoded by two genes, *glb1* and *glb2*. The proteins are quite different between the two. The *glb1* gene has variants found in various strains that produce proteins differing in electrophoretic mobility. In the long-term selection strains Illinois High Protein and Illinois Low Protein, and in some inbred lines, the *glb1* locus produces no protein product. Variants of the *glb2* gene have only been found showing presence or absence of the protein. Derived strains lacking both proteins produce normal kernels, indicating that globulins are not involved in essential functions.

Regulation of Seed Storage Genes

Many different processes regulate seed storage protein gene expression. Differences in the transcription rates, messenger RNA (mRNA) stabilities, translation efficiency rate, and protein degradation rates can all play a role in determining the relative levels of storage proteins. The spatial and temporal regulation of seed storage protein genes is mainly at the transcriptional level. In general, seed protein genes are not transcribed in nonseed tissues. Based on detailed DNA sequence analysis of seed storage protein genes from several cereals and legumes, it is clear that several conserved sequence motifs (regulatory elements) are present in the 5′ promoter regions of these genes. These regulatory elements include the CATGCATG, CACA, and ANCCCA sequences. The CATGCATG motif, which is also called an RY element, is involved in high-level expression in seeds and represses expression in nonseed organs. In addition, an octanucleotide box, GCCAC (c/t)TC, is present in most genes encoding the 7S and 11S globulins. In the case of cereals, a prolamin-specific motif (TGTAAAG) has been identified around −300 in the 5′ regions of prolamin genes. It is believed that the −300 box may have a role in the quantitative level of expression of some prolamin genes. The seed-specific expression is controlled by both positive and negative DNA elements. The positive element will stimulate the transcription, while the negative element will inhibit transcription in nonseed organs. Binding of nuclear proteins (*trans*-acting factors) to conserved elements (*cis*-acting elements) also contributes to transcriptional regulation. In general, the *cis*-acting elements are AT-rich. Sequences that are CA-rich, CATGCATG-like sequences, and G-boxes (CACGTG) are present in genes expressed in seeds and have been shown to bind nuclear factors. In addition to the transcriptional regulation, seed storage protein gene expression may be controlled by post-transcriptional events. Relative RNA stabilities, differential translation, and differential protein stabilities can all influence the accumulation of seed storage proteins.

Seed Storage Protein Mutants

Several high-lysine mutants, both spontaneous and induced, are found in maize, barley, and sorghum. These mutants are easy to identify because they exhibit three common characteristics: reduced accumulation of prolamins, presence of starchy endosperm, and decreased yield. In addition, these high-lysine mutants may be affected in their embryo size. A high-lysine barley line (Hiproly) contains 30% more lysine as compared to normal isolines. Another mutation (*lys* 3a) located in the same chromosome, but not linked to the *lys* gene, also has elevated levels of lysine. Even though the high-lysine gene (*lys*) has been mapped to chromosome 7, the molecular basis of mutation is not understood. Several mutations in maize (*opaque*2, *opaque*7, *opaque*15, *opaque*6, *floury*2, *floury*3, *defective endosperm**B30, and *Mucronate*) cause reduction in the accumulation of maize zeins. The *o*2 mutation affects a regulatory gene and regulates the expression of 22 000 Da α-zein genes. The *o*2 and *fl*2 mutants contain elevated levels of lysine and tryptophan when compared to the wild-type. Through plant breeding programs at The International Maize and Wheat Improvement Center, the soft, starchy phenotype of the *o*2 mutant was converted to hard, translucent endosperm by incorporating genetic modifiers. These high-lysine cultivars are known as Quality Protein Maize (QPM) and are currently being developed for human and livestock consumption.

Targets and Modification

In comparison with legumes, cereals have low protein content. In addition, the amino acid composition of cereal prolamins is not balanced. They are deficient in lysine, an essential amino acid for monogastric animals. As a result, a considerable amount of research is focused on improving the quality and quantity of seed storage proteins both by traditional plant breeding and modern genetic engineering technology. In addition to low protein content, some of the protein components are not easily digestible by monogastric animals. For example, protein digestibility values for cooked sorghum, rice, maize, and wheat are 46%, 63%, 73%, and 81%, respectively. In the case of sorghum, it has been determined that the interior location of α-kafirin within the protein bodies resulted in poor accessibility to digestive enzymes. Recently, sorghum lines with high protein digestibility were discovered and were found to contain morphologically altered protein bodies. Detailed understanding of the molecular basis of altered protein body formation could be used to improve the protein digestibility of other important seed proteins.

In the case of legumes, most studies have been directed at increasing the methionine content. Three types of approaches have been attempted. In the first approach, methionine-rich sequences have been introduced into globulin genes and expressed in transgenic plants. In the second approach, overexpression of endogenous high-sulfur-containing seed proteins has been attempted. In the third approach, heterologous genes encoding high-sulfur proteins have been expressed in transgenic plants.

Genetic engineering technology has enabled scientists to alter the quality and quantity of seed storage proteins. For example, a methionine-rich 2S albumin of Brazil nut has been successfully expressed in tobacco seeds and forage grasses, thereby greatly increasing the methionine content. However, it has been found to have allergenic properties. Similarly, by expressing two key enzymes in the lysine biosynthetic pathway, the overall content of lysine in maize and soybeans has been elevated. By antisense technology, the levels of a 16 000 Da rice allergenic protein have been reduced to one-fifth of the wild-type rice. Furthermore, the lysine content of rice seeds was elevated by expressing the β-phaseolin gene in transgenic rice. All the above examples demonstrate the invaluable contribution of biotechnology in the improvement of seed storage protein quality and quantity.

Further Reading

Habben JE and Larkins BA (1995) Improving protein quality in seeds. In: Kige J and Galili G (eds) *Seed Development and Germination*, pp. 791–810. New York: Marcel Dekker.

Okita TW and Rogers JC (1996) Compartmentation of proteins in the endomembrane system of plant cells. *Annual Review of Plant Physiology and Plant Molecular Biology* 47: 327–350.

Shewry PR and Tatham AS (1999) The characteristics, structures and evolutionary relationships of prolamins. In: Shewry P and Casey R (eds) *Seed Proteins*, pp. 11–33. Norwel, MA: Kluwer Academic Publishers.

Shewry PR, Napier JA and Tatham AS (1995) Seed storage proteins: structures and biosynthesis. *The Plant Cell* 7: 945–956.

Vitale A and Bollini R (1995) Legume storage proteins. In: Kige J and Galili G (eds) *Seed Development and Germination*, pp. 73–102. New York: Marcel Dekker.

***See also:* Grasses, Synteny, Evolution, and Molecular Systematics; Leguminosae; Seed Development, Genetics of**

Segmental Interchange

J R S Fincham

doi: 10.1006/rwgn.2001.1522

When chromosomes are broken, whether spontaneously or as a result of irradiation, the broken ends have a strong tendency to rejoin. Simultaneous breaks in two chromosomes may result in a reciprocal exchange of segments. This process is known as segmental interchange, or reciprocal translocation.

Visualization

A diploid with two chromosomes with interchanged segments and two structurally normal homologs is called an interchange heterozygote. Since, overall, it has a balanced chromosome complement, it will be phenotypically normal unless, as sometimes happens, a dominant mutation has been induced at one of the original breakpoints. But it encounters complications when it undergoes meiosis. The close homologous pairing of chromosome segments at the pachytene stage can occur only through an association of four chromosomes, two normal and two interchanged, with exchange of partners at the loci of the original breaks. Such pachytene configurations can be seen clearly with the light microscope in organisms such as *Zea mays* (maize, corn) in which clear pachytene preparations can be made, and with higher definition with the electron microscope in sexual organisms generally after staining the axial elements of the synaptonemal complex with silver ions. Interchange points can be defined with the highest precision in *Drosophila* species, not in meiotic cells but in the giant nuclei of the salivary gland cells, where the polytene chromosomes display close homologous pairing (**Figure 1**).

Genetic Consequences

The effect of the association of four chromosomes at pachytene depends critically on the positioning of the crossovers (chiasmata) which occur within it. Potentially, meiosis can yield wholly viable products if the chiasmata are confined to regions distal to (i.e., further

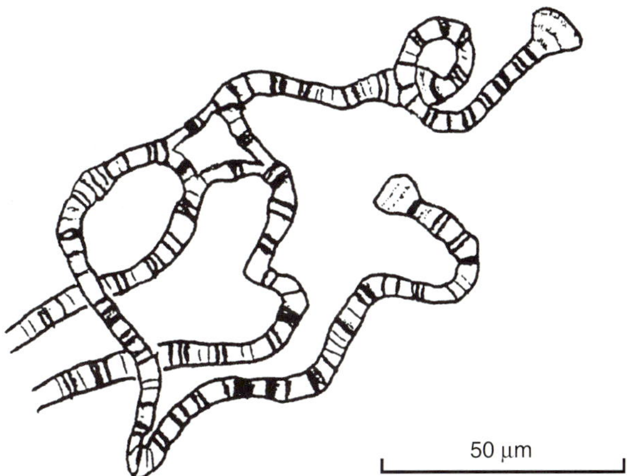

Figure 1 A segmental interchange between centromere-distal segments of chromosomes 2 and 3 in *Drosophila*, seen in the polytene chromosomes of a salivary gland cell nucleus. The centromere-proximal ends of the chromosome arms are not shown. (Drawn from the photograph of Roberts (1970) *Genetics* 65: 429.)

from the centromere than) the exchange points. Then, in the case where all four chromosomes are joined by chiasmata to form a ring or chain of four, alternate orientation of centromeres will result in two structurally normal divided chromosomes (dyads) passing to one pole of the spindle and two interchanged dyads to the other, and all meiotic products will have a balanced chromosome complement (**Figure 2Bi**). If, on the other hand, adjacent centromeres in the ring or chain pass to the same pole, each of the nuclei resulting from the first division will be duplicated in one chromosome segment and deficient in the other, and all products will be inviable. If a chiasma (crossover) forms between an interchange point and the centromere, at least two out of the four products of meiosis will have a duplication and a deficiency whatever the orientation of the centromeres (**Figure 2Biii,iv**). Thus, interchange heterozygotes are expected to suffer some degree of infertility, but can be normally fertile given an appropriate chiasma distribution and metaphase I centromere orientation.

Because viability of the meiotic products of an interchange heterozygote depends on their having a balanced chromosome complement, consisting of either the two structurally normal or the two interchanged chromosomes, allelic differences (markers) close to the interchange points will appear closely linked, even though they are normally located on nonhomologous chromosomes. They will be able to recombine to give viable meiotic products only when there is a crossover between one of them and the exchange point (see **Figure 2B**). Their linkage relationships will be represented by a four-armed linkage map mirroring the pachytene pairing patterns, with loci close to the four-way junction giving little or no effective recombination. In the interchange homozygote, of course, there will be the normal number of independent linear linkage groups with segments of two of them interchanged.

Interchange Complexes in Plants

In some plants, most famously in one section of the genus *Oenothera* (evening primrose), the avoidance of infertility in interchange heterozygotes, by restriction of chiasmata to distal chromosome segments and alternate orientation of centromeres at first metaphase, has become established as a regular system. In these *Oenothera* species there are 14 chromosomes in the diploid set, each with two arms of approximately equal length. The plants are all complex interchange heterozygotes, forming at meiosis a ring of 14 involving the entire genome. At the first anaphase of meiosis the chromosomes are alternately arranged and separate into two sets of seven. The ring is the result of an interchange in each chromosome arm. If one

Figure 2 (Opposite) (A) Diagram showing the pairing and chiasma formation at the first prophase of meiosis in an interchange heterozygote. The two structurally normal chromosomes inherited from one parent are shown as thick and thin lines, respectively, with centromeres labeled 1 and 3. The two chromosomes contributed by the other parent, with centromeres labeled 2 (homologous to 1) and 4 (homologous to 3) have a segmental interchange. The normal and interchanged chromosome pairs are distinguished by markers at three loci, with the normal chromosomes carrying *A*, *B*, and *C*, and the interchanged pair the alleles *a*, *b*, *c*. The *A* and *B* loci are located on the interchanged segments close to the interchange points. The chiasmata (crossovers) are, for convenience in drawing, shown occurring between the two 'inner' chromatids in each case, but this makes no difference to the argument. (B) Quadrivalent associations at metaphase I of meiosis. (i) Chromosomes joined in a ring-of-four following formation of chiasmata in the positions shown by full lines in (A). Viable meiotic products are possible only if centromeres are oriented alternately as shown, with 1 and 3 directed to one spindle pole and 2 and 4 to the other. Separation of the centromeres of adjacent members of the ring to the same pole (1,2/3,4 or 1,4/2,3) will result in all products having segmental duplications and deficiences which will make them inviable. The viable products will always be of the parental constitutions *A B* and *a b* except in the rare event of chiasma formation (crossing-over) in the short space between either locus and the interchange point. Thus, the two loci will appear closely linked even though they are normally on different chromosomes. *C*, far enough from the interchange for frequent chiasma formation in the intervening interval, will give two-out-of-four recombination with the other two markers whenever, as shown in this diagram, such a chiasma occurs. (ii) Constitutions of the four meiotic products resulting from the chiasma distribution and centromere orientation shown in (i). (iii) The corresponding association of four when an additional chiasma (crossover) is formed between a centromere and the interchange point (shown dotted in **Figure 2A**). Viable meiotic products can be formed whether the centromeres separate 1 with 3 and 2 with 4, or (the alternative shown) 1 with 4 and 2 with 3. But in either case two of the four products will have duplications and deficiencies and be inviable. (iv) The two alternative sets of meiotic products resulting from the chiasma distribution shown in (iii). Those duplicated for the *A*/*a*-containing segment and deficient for *B*/*b*, or vice versa, and therefore inviable, are marked X. The viable products are again all *A B* or *a b*.

numbers the exchanged sections of arms 1 to 14, one set will have combinations 1–2 3–4 5–6 7–8 9–10 11–12 13–14, and the other 2–3 4–5 6–7 8–9 10–11 12–13 14–1, the hyphen in each case standing for the centromere and the region around it.

The regular formation of the metaphase I ring of 14 chromosomes, with alternate orientation of centromeres, depends on chiasma localization, such that chiasmata are regularly formed in each chromosome arm but hardly at all in the regions between the centromeres and the exchange points. The two chromosome sets can exchange genes located in their pairing arms, but in the regions between the centromeres and the exchange points they remain distinct and to some extent functionally differentiated, most notably with respect to germ cell production and viability. The meiotic products that form embryo sacs all carry one set and all viable pollen grains carry the other. Thus, a state of 'permanent hybridity' is maintained.

The system depends on the localization of chiasmata and the alternate orientation of centromeres, and it is not, in fact, completely stable. In some stocks a few per cent of plants can show striking differences

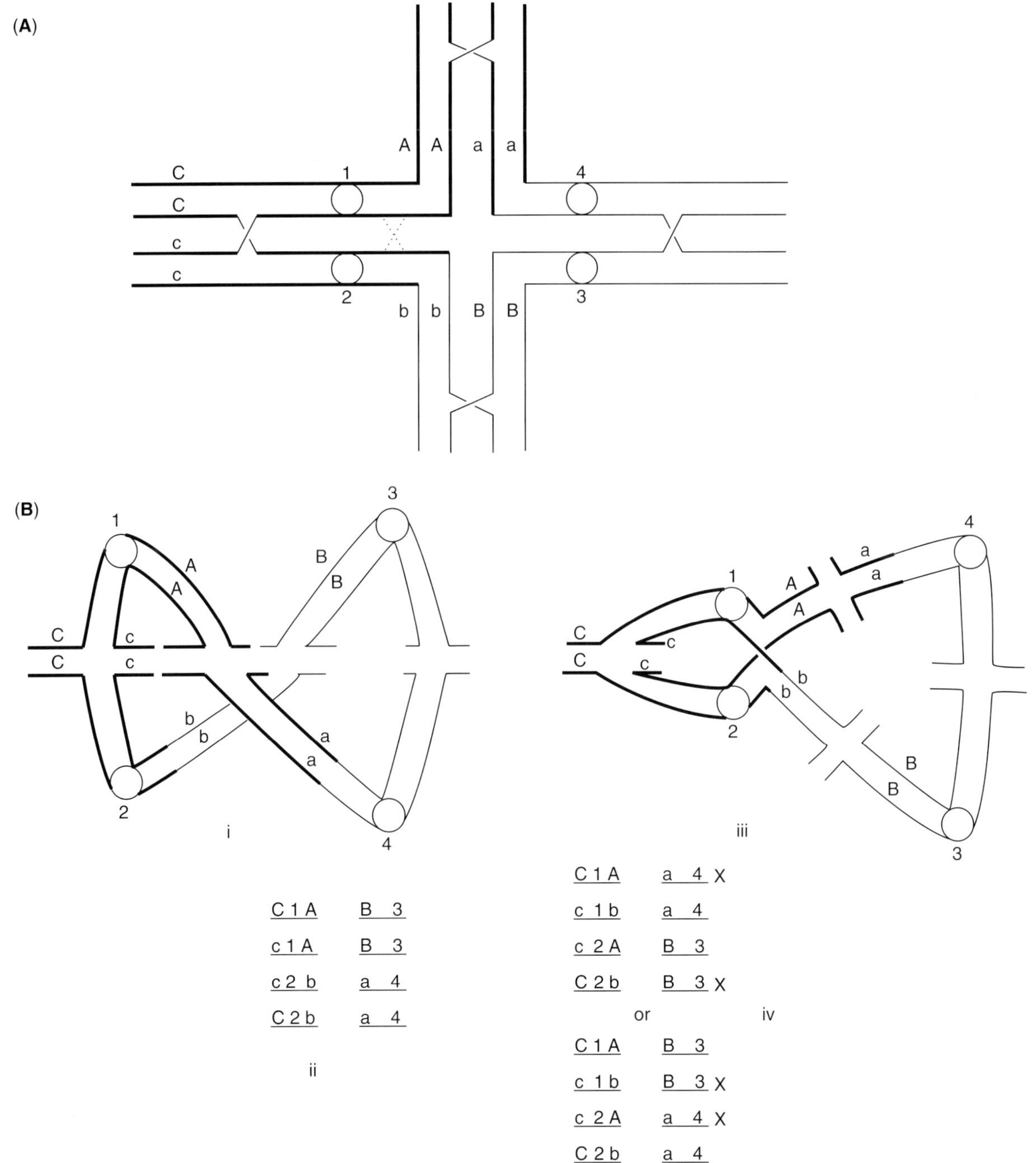

from the standard species type because they have become homozygous for chromosome segments that are normally kept heterozygous. This apparent high mutability of *Oenothera* spp., quite untypical of organisms in general, was a main foundation for Hugo de Vries's mutation theory of evolution, which was very influential in the first two decades of the twentieth century.

Further Reading

Darlington CD (1937) *Recent Advances in Cytology*, 2nd edn. London: Churchill.

See also: Crossing-Over; Crossover Suppressor; De Vries, Hugo; Polytene Chromosomes; Synaptonemal Complex; Translocation

Segmentation Genes

doi: 10.1006/rwgn.2001.2015

Segmentation genes are those required for controlling segmentation in insect embryos.

See also: *Drosophila melanogaster*

Segregation

I Ruvinsky

doi: 10.1006/rwgn.2001.1159

Segregation is a fundamental concept of genetics. It refers to the separation, during gamete formation, of two alleles present at a locus in a diploid individual. This ensures the representation of both alleles in the progeny.

The notion of segregation was introduced by Gregor Mendel in his seminal work 'Experiments in plant hybridization.' He demonstrated that "various kinds of egg and pollen cells were formed in hybrids on the average in equal numbers," thus establishing that heterozygotes produce two types of gametes with regard to each locus. These two types are being produced equally and each represents an allele of the parent. Therefore, Mendel's First Law is also known as 'The Law of Segregation.'

In the early twentieth century it was recognized that segregation of genes during gametogenesis is closely paralleled by segregation of chromosomes during meiosis, the cell division responsible for gamete production. This observation lead to the formulation of the chromosomal theory of heredity.

While in the majority of cases segregation results in equal representation of alleles in offspring, several exceptions exist when the ratio deviates significantly from 1:1. This phenomenon is known as segregation distortion, transmission ratio distortion, or meiotic drive. Among the more spectacular examples are the inheritance of the *t* complex in the mouse, SD locus in *Drosophila*, and *Spore killer* in *Neurospora* where ratios of alleles transmitted to the offspring by a heterozygous parent can be biased as much as 95:5 or even 99:1. The underlying causes of segregation distortion may differ from case to case.

The concept of segregation is also used in a more specific sense in fungal genetics with regard to the arrangement of spores in a linear ascus. Thus, if *A* and *a* are alleles at a particular locus, a situation when spores are arranged in the order *AAaa* (in a tetrad) or *AAAAaaaa* (in an octad) is referred to as a first-division segregation (FDS) pattern. *AaAa* and *AAaaAAaa* (in a tetrad and octad, respectively) is called second-division segregation (SDS). The linear nature of the ascus with respect to the division planes of meiosis allows the inference that in FDS the two alleles were separated in the first meiotic division, whereas in SDS the alleles segregated in the second meiotic division. This information, in turn, can be used to conclude that no recombination (or an even number of events, if the distance is large) has occurred between the locus and the centromere in the first case. In contrast, there was a recombination event (or an odd number of events) in the second case. In this way, analysis of the order of spores in linear asci can be utilized for centromere mapping.

See also: Mendel's Laws; Mendelian Genetics

Segregation Distortion, Mouse

L Silver

doi: 10.1006/rwgn.2001.1160

According to Mendel's First Law, sexually reproducing organisms segregate the two copies they carry of each of their genes equally to their gametes and offspring. Thus, if an organism is heterozygous with an A1 and an A2 allele at its A locus, half of its offspring (on average) will receive the A1 allele, and half will receive the A2 allele. Some unusual genetic entities violate Mendel's First Law. In the mouse, a chromosomal region called the t complex can be present in a mutant form known as a t haplotype. Males that are heterozygous with a wild-type form of the t complex and a t haplotype form can transmit the t haplotype to over 90% of their offspring in a clear violation of Mendel's Law. Since segregation is not equal, as for most genes, this process is referred to as segregation distortion.

***See also:* Meiotic Drive, Mouse**

Selection

***See:* Frequency-Dependent Selection; Frequency-Dependent Selection as Expressed in Rare Male Mating Advantages; Fundamental Theorem of Natural Selection; Natural Selection**

Selection Coefficient

M Tracey

doi: 10.1006/rwgn.2001.1162

Natural selection is differential reproduction and at the single gene level it is modeled mathematically by assigning relative probabilities of reproduction to the various genotypes at a genetic locus. For example, a gene which may mutate to an allele which leads to prereproductive death or to sterility in homozygotes may be assigned a zero probability of reproducing and leaving progeny. In this simple case, we may also assume that the lethal/sterile allele is completely recessive; then the probabilities of reproduction for the heterozygote and alternate homozygote will be one or unity. If we count genotypes at fertilization we will have AA, Aa, and aa individuals before selection acts. The most common model assigns reproductive probabilities of 1 to the AA and Aa genotypes and zero to the aa genotype. These are multiplied by the frequencies of the three genotypes to predict frequencies in the reproductive gene pool. Clearly, only the AA and Aa genotypes will be represented among the individuals reproducing and the fertilization-stage genotypes for the next generation will be determined by the relative frequencies of these two genotypes. The most common formulation of this model uses notation in which the most likely to reproduce genotype(s) are assigned values of 1.0 and the less successful genotype(s) are assigned values of $(1.0 - s)$. In this formulation the variable s is known as the selection coefficient and takes values ranging from zero to 1.0. In some cases selection coefficients may vary with population density or genotype frequency.

***See also:* Frequency-Dependent Selection**

Selection Differential

E Pollak

doi: 10.1006/rwgn.2001.1163

If there is selection aimed at changing the expression of a quantitative trait in a population, individuals that have measurements in a desired range of values are chosen to be parents. So if, for example, high yield is what a plant breeder wants, he or she saves for reproduction plants that have yields in the highest $100p$ percent of the population. The average amount by which chosen individuals exceed the population mean is called the selection differential.

The selection differential does not by itself express the strength of selection. A more revealing way to do this is to divide the selection differential S by the phenotypic standard deviation σ. The ratio, i, that is thus obtained is called the intensity of selection. If, in particular, it is assumed that measurements on a quantitative character are normally distributed and a proportion p of individuals having the highest measurements are selected,

$$i = \frac{z}{p}$$

where z is the ordinate of the standard normal distribution at the point of truncation.

The concept of a selection differential can be generalized. First, one may consider a weighted selection differential. A weight, proportional to the contribution to offspring measured in the next generation, is assigned to a parent (or a pair of parents). Second, if several characters are correlated with the one under selection, the selection differential for any trait may be partitioned into a component that estimates direct selection and a sum of components from indirect selection on all correlated traits. A third generalization applies where selection is based on an index which is a linear combination of measurements on at least two traits. A weight associated with a trait in this expression represents the relative importance assigned to the trait by the breeder. In this case the selection differential in the average amount by which the chosen individuals have index values that exceed the population mean of the index.

Further Reading

Falconer DS and Mackay TFC (1996) *Introduction to Quantitative Genetics*, 4th edn. Harlow, UK: Longman.

Lynch M and Walsh B (1998) *Genetics and Analysis of Quantitative Traits*. Sunderland, MA: Sinauer Associates.

***See also:* QTL (Quantitative Trait Locus); Quantitative Inheritance; Quantitative Trait; Selective Breeding**

Selection Index

W G Hill

doi: 10.1006/rwgn.2001.1443

In programs for the genetic improvement of animals and plants, the aim is usually to improve performance in a number of traits. In pigs, for example, these traits might include traits of the growing animal such as growth rate, leanness, efficiency of conversion of food into growth, and viability; and traits of the parents such as litter size, fertility, and longevity. Furthermore, data are collected on individual candidates for selection and on their relatives, perhaps on different traits and at different times: for example carcass leanness can not be recorded on an animal for breeding, and traits such as litter size can be recorded only on females. The selection index is designed to put together information in an optimal way to enable selection of the individuals likely to have progeny with the highest overall economic performance. The bases for constructing a multitrait selection index were put forward by Fairfield Smith in 1936 and by Hazel in 1943. In 1947 Lush showed how to incorporate information on relatives.

The selection index is essentially a multiple regression predictor of progeny performance or breeding value for an objective, typically economic merit, from a set of observations, y_i, on an individual or its relatives. The index, $I=\sum b_i y_i$, is a weighted sum of these observations and the index weights, b_i, or in matrix notation, $I=\mathbf{b}'\mathbf{y}$. The variances and covariances of the y_i are summarized by **P**, with elements $p_{ij}=cov(y_i, y_j)$. **P** is traditionally called the phenotypic covariance matrix, but the y_i can be family means, or predicted breeding values, or indeed almost any measurement. Economic merit, H, is usually decomposed into a set of breeding values, z_j, for individual traits, so $H=\sum z_j a_j=\mathbf{z}'\mathbf{a}$, where the economic weights, a_j, specify the financial benefit of a unit change in the trait (e.g., the price of an average chicken egg), holding the rest constant. The covariances between the observations and these genetic objectives are summarized by the genetic (co)variance matrix **G**, with $g_{ij}=\mathrm{cov}(y_i, z_j)$.

The weights of the optimal index are given by $\mathbf{b}=\mathbf{P}^{-1}\mathbf{G}\mathbf{a}$; and if the objective is just to improve a single trait, **G** can be represented by a column vector for that trait, with $a_1=1$. These optimal index weights satisfy a number of criteria: they maximize the accuracy of the index (the correlation, r_{IH}, between I and H), the probability that two individuals are correctly ranked, and the selection response; and they minimize the variance of predicted about actual values of H. The index is, of course, optimal only if **P**, **G**, and **a** are known without error. In practice this can never be the case so that accuracy is usually less than predicted, but indices are robust to poor estimates of most of the variables. The index can be expanded in various ways: for example to include data on molecular markers linked to loci influencing the quantitative traits of interest; and to maximize changes in overall performance, but holding other traits constant.

Selection indices are widely used in animal breeding, where information on individual animals is expensive to acquire and progeny group sizes often small, and on trees, but are less used in other plant breeding programs. An example of the use of an index was in pig breeding in Britain in a national program operated in the 1960 and 1970s. Four littermates were taken from each candidate litter to a central testing station and reared to bacon weight, when two were then slaughtered. Indices were computed for selecting the boars, each combining 14 items of information: rate of weight gain, efficiency of conversion of food into gain and backfat depth (recorded ultrasonically) individually on both the boar and his brother, and the average daily gain, food conversion, and six carcass traits, including dissected lean content of a joint, on their two slaughtered sibs.

In order to compare the performance of candidates for selection using their own or relatives' performance, identifiable sources of environmentally caused differences between them have to be eliminated. These include, for example, number of animals in the litter in which they were born, and season, year, or herd of birth. In the classical selection index, these environmental effects are assumed to be negligible, for example if candidates are reared together, or accurately estimated. In many breeding programs using field records, for example on milk production of dairy cows, environmental effects such as herd can not be estimated with sufficient precision. The method of best linear unbiased prediction (BLUP), due to Henderson, deals with this problem. His ideas evolved from 1949, but first came into use two decades later for selection of dairy bulls used in artificial insemination programs and having daughters unevenly distributed over many herds. Now it is common to use an animal model in which a breeding value is computed for each individual, taking account of records on that animal and all its relatives, properly weighted according to the degree of relationship, in a single analysis. In principle, BLUP combines the methods of least squares to estimate identifiable environmental (fixed) effects and selection indices to predict the random effects or breeding values. BLUP is highly computer intensive, but is now standard practice in animal breeding programs, and has largely replaced the traditional selection index.

Further Reading

Cameron ND (1997) *Selection Indices and Prediction of Genetic Merit in Animal Breeding*. Wallingford, UK: CAB International.

Falconer DS and Mackay TFC (1996) *Introduction to Quantitative Genetics*, 4th edn. Harlow, UK: Longman.

See *also*: Artificial Selection; Genetic Correlation; Selective Breeding

Selection Intensity

W G Hill

doi: 10.1006/rwgn.2001.1444

Selection intensity is a measure of the strength of directional selection applied in a selection experiment or breeding program to change a quantitative trait. If selection is practiced on individual performance, the selection applied can be described by the selection differential (S), which is the difference between the mean performance of the selected individuals and that of the population as a whole. The selection differential is measured in units of the trait, for example grams of body weight or number of offspring born in a litter of mice or pigs. The selection intensity, usually denoted i, equals the selection differential measured in phenotypic standard deviations ($\sigma_P = \sqrt{V_P}$), i.e., $i = S/\sigma_P$. Therefore i is a dimensionless quantity, and its magnitude does not depend on the variability of the trait.

The selection intensity is a useful measure because its value can be predicted in an artificial selection program from knowledge of the selection criteria, the proportion of individuals selected and the distribution of the trait. For example with truncation selection, in which the highest performing individuals for the trait are selected, i is a simple function of proportion selected (p) for any distribution. For the normal distribution, tables of i are available (e.g., Falconer and Mackay, 1996), or it can be computed as $i = z/p$, where z is the ordinate of the standardized normal at the truncation point. For example:

p	0.5	0.2	0.1	0.05	0.01	0.001
i	0.798	1.400	1.755	2.063	2.665	3.367

Selection intensity depends somewhat on the size of the population. For a given proportion selected the intensity becomes slightly less as the population size becomes smaller, and values can be computed using order statistics. For example, if N are selected from M recorded, and $N = M/10$, e.g., 2 out of 20:

M	10	20	50	100	200	$\rightarrow \infty$
i	1.539	1.638	1.705	1.730	1.742	1.755

The intensity is further reduced in a predictable way in a small population because the performance of family members is correlated.

If selection is practiced on individual performance, the predicted response to one generation of selection is given by $R = h^2 S = i h^2 \sigma_P$, where i is the mean for males and females if the intensities differ between the sexes. Thus, for example, if the highest-scoring 5% of males and 20% of females of a large population are selected for a trait (say growth rate of pigs) with $\sigma_P = 50$ g per day and $h^2 = 0.3$, the predicted response would be 26 g per day ($= 0.5 \times (2.063 + 1.400) \times 0.3 \times 50$). The selection intensity can be used to measure the amount of selection when selection is not simply on individual performance (mass selection) but on relatives' performance or indeed on any quantitative selection index. It also features in formulas for the change in gene frequency and hence selective value (s) of a gene affecting a quantitative trait under selection: with mass selection, $s = ia/\sigma_P$, where a is the effect of the gene.

Selection intensity depends on reproductive rate, but the breeder can manipulate this, for example, by retaining selected animals for more litters or by using techniques such as artificial insemination. There can be a tradeoff, however: an increase in intensity may be at the expense of increased generation interval and at the cost of increasing inbreeding and reducing variation if few individuals are selected.

Natural Selection

Selection intensity can be computed even where truncation selection is not practiced. In artificial selection this may arise where numbers of offspring are deliberately adjusted according to performance. Under natural selection, individuals contribute to the next generation according to their fitness. Assume that the number of offspring of individual j is X_j and the mean number is μ. Hence the relative fitness of the individual is X_j/μ, and the selection differential in fitness is $\Sigma_j(X_j - \mu)X_j/\mu^2 = V_X/\mu^2$. This is called the index of opportunity for selection. As the standard deviation of fitness is $\sqrt{(V_X/\mu)}$, it follows that $i = \sqrt{(V_X/\mu)}$. This shows that, of course, selection can occur only if there is variability in fitness. Whether selection on fitness, or indeed any other trait, is effective depends then on it having additive genetic variance. Selection intensity is not used to define the magnitude of stabilizing selection, i.e., whereby selection acts mainly to reduce variance in fitness.

Further Reading

Cameron ND (1997) *Selection Indices and Prediction of Genetic Merit in Animal Breeding*. Wallingford, UK: CAB International.

Reference

Falconer DS and Mackay TFC (1996) *Introduction to Quantitative Genetics*, 4th edn. Harlow, UK: Longman.

***See also:* Artificial Selection; Heritability**

Selection Limit

W G Hill

doi: 10.1006/rwgn.2001.1445

In long-term selection experiments for quantitative traits it has often been found that after many generations of selection, a plateau or selection limit has been reached at which there appears to be little or no response despite continued selection. The limit can be explained by fixation of all useful genetic variation or by counteracting effects, for example, natural selection opposing the artificial selection.

It is usually hard to be sure a limit actually has been reached because of sampling due to small numbers of individuals and environmental differences among generations. Even so there are some clear cases (e.g., F. W. Robertson, 1955) where limits have been reached. There are others in which, despite selection for over 50 generations, limits appear not to have been obtained. An example is the Illinois corn experiment for increased oil content, in which response has continued for a century. In the low line, however, little response has occurred in recent generations, but the mean is so near zero there is little opportunity for further change (see graph in the article on Artificial Selection).

A selection limit will occur if a population has run out of useful variation. This is inevitable if there is essentially no phenotypic variation (as for low oil content in the Illinois corn oil experiment). More generally the limit can occur if additive genetic variation is exhausted, i.e., the selected line becomes homozygous for all genes which were segregating in the base and which increase the trait in the desired direction. Residual nonadditive genetic variation can remain at such a limit if the favorable genes are dominant and reach high frequency or if there is overdominance.

The magnitude of the response to the limit in relation to the genetic variation in the base population depends on the numbers of genes affecting the trait and on the distribution of their effects. If very few loci which influenced the trait were segregating in the base population, such that individuals with the extreme genotype were present in it, albeit at low frequency, the limit would not be outside the initial range of the population. Usually, however, it is far outside the initial range, i.e., the total response is many phenotypic standard deviations. An estimate of the number (n) of genes affecting a trait, Wright's effective number, can be obtained by comparing the range (R = high–low divergence) achieved to the additive genetic variation (V_A) in the base population or in an F_2 cross of high and low lines, as $n = R^2/8V_A$.

As a population under selection is necessarily of finite size, desirable genes may be lost by chance, particularly those with a small effect on the trait and particularly if selection is weak and the population size is small. The limit to artificial selection then depends on the probability of fixation of the favorable genes. In a theory of limits to artificial selection, A. Robertson (1966) showed that the fixation probability is proportional to the product of effective population size (N_e), selection intensity (i), the effect of the gene on the trait relative to the phenotypic standard deviation, and its degree of dominance. Prediction of the actual limit to selection is not, however, possible without (usually) unknown information on the distribution of gene effects and frequencies in the base population, but nevertheless there are some practical consequences of the theory. In particular there is a trade-off between short- and long-term response, for the initial response is proportional to the selection intensity, whereas the limit is proportional to $N_e i$, and is maximized if only one-half of the population is selected. Similarly, use of relatives' information in a selection index reduces the limit because relatives are coselected, and so N_e is reduced more than the accuracy of selection is increased.

Fixation is not the only cause of selection limits. In theory, limits can occur if there are overdominant loci or if most of the variance is due to recessive genes, when inbreeding would lead to reduction in performance. More importantly, perhaps, because selected populations become extreme for the trait under selection, but also show correlated responses in other traits, it is to be expected that natural selection opposes artificial selection such that a limit occurs at the balance between these opposing forces. Evidence comes from experiments in which the population mean at the limit falls when either selection in the opposite direction (reversed selection) has been practiced or the population has been maintained without selection (relaxed selection). Natural selection may be a consequence solely of the shift in mean of the correlated traits subject to stabilizing selection, or of increases in frequency of specific genes with effect on the trait under selection but also pleiotropic effects upon fitness. (Extreme examples found are genes that have a large effect on the trait as a heterozygote, but are lethal as a homozygote.)

As new variation in quantitative traits arises by mutation, limits cannot happen as a consequence of running out of variation unless there are so few possible loci and useful alleles at them that all were present at the outset or appeared during the selection process. It is therefore likely that fixation cannot account exclusively for limits, and other factors such as natural selection have to be invoked. All 'limits' may therefore be transient, and renewed responses expected and explained by mutations or, perhaps, by recombination among haplotypes with balanced repulsion for useful genes. Nevertheless, in selection experiments for competitive fitness in bacteria for which responses must derive from mutation, Lenski and colleagues (Lenski and Travisano, 1994) found plateaus in response after thousands of generations of selection.

Further Reading

Falconer DS and Mackay TFC (1996) *Introduction to Quantitative Genetics*, 4th edn. Harlow, UK: Longman.

Hill WG and Caballero A (1992) Artificial selection experiments. *Annual Review of Systematics and Ecology* 23: 287–310.

References

Lenski RE and Travisano M (1994) *Proceedings of the National Academy of Sciences, USA* 91: 6808–6814.

Robertson A (1966) A theory of limits in artificial selection. *Proceedings of the Royal Society of London* B 153: 234–249.

Robertson FW (1955) *Cold Spring Harbor Symposia in Quantitative Biology* 20: 166–177.

***See also:* Additive Genetic Variance; Artificial Selection; Heritability; Selective Breeding**

Selection Pressure

M Tracey

doi: 10.1006/rwgn.2001.1164

Selection, either natural or artificial, involves unequal reproduction among various genetic types. Consider the old example of selection for longer necks in giraffes. In times of scarce browse, the giraffes able to reach the tops of trees would eat more and have more progeny. To the extent that their longer necks and legs were genetically determined we would expect to see taller progeny who would, in turn have taller progeny. There is a point of diminishing returns in this type of selection as the giraffe population becomes taller and taller in response to the hunt for nutrition higher and higher in the trees. Eventually the competition for browse is just as intense at the tops of the trees as it was at other levels. The effectiveness of selection in changing giraffe height genotypes over generations is selection pressure.

Selection pressure depends primarily on the selection differential (see Selection Differential, Selection Intensity) and the amount of genetic variation in the selected population (see Heritability). Consider genetic resistance to pathogens (see Sickle Cell Anemia) in which any differential reproduction among genotypes, the selection differential, takes place only in the presence of the pathogen and magnitude of the reproductive advantage depends on the prevalence of the pathogen. In this example the selection pressure will depend on the prevalence of the pathogen; there are no genotypic differences in the absence of the pathogen and as the disease becomes more common the reproductive differences among the genotypes become more important in altering the reproductive success of different genotypes.

***See also:* Branch Migration; Heritability; Selection Differential; Sickle Cell Anemia**

Selection Techniques

I Schildkraut

doi: 10.1006/rwgn.2001.1165

Selection techniques are used by geneticists to isolate mutations. The techniques involve the process of isolating cells with a mutant phenotype by choosing conditions that favor the survival of the mutant phenotype and disfavor the survival of the parental type. Selective techniques are a powerful tool and are routinely used by microbial geneticists. Selective techniques can be robust when applied to microorganisms because of the numbers of organisms that can be manipulated. Although selective techniques exist for complex organisms, it is much easier to handle a million bacterial cells than a million mice.

Typically a large population of bacterial cells are grown under selective conditions so that the relatively small number of variants that have arisen due to mutation are the only cells that are capable of forming a colony on selective agar plates. Among the most powerful of selective techniques is the selection of antibiotic-resistant cells. In a population of hundreds of millions of antibiotic sensitive cells, a single antibiotic-resistant cell can be isolated with little effort. The selective growth condition in this case contains in addition to all the required nutrients for growth, an antibiotic that would prevent the original population of cells from growing.

Selection techniques vary and can employ the capacity to: (1) grow in the presence of an antibiotic or other inhibitor; (2) utilize a new carbon source; (3) resist bacteriophage infection; (4) convert from auxotrophy to prototrophy; (5) grow at a higher/lower temperature; and so on.

A strong selection usually results in the demanded phenotype, but not necessarily the desired genotype. Usually it is necessary to analyze the new phenotype to verify the genotype. For example, an investigator would like to set up a selection that would increase the catalytic activity (efficiency) of the enzyme responsible for degrading the antibiotic ampicillin and therefore increase the level of resistance. A strain of *Escherichia coli* that is resistant to 20 $\mu g\ ml^{-1}$ of ampicillin but sensitive to 100 $\mu g\ ml^{-1}$ of ampicillin is spread on agar plates that contain 100 $\mu g\ ml^{-1}$ of ampicillin. By increasing the concentration of ampicillin on the agar plates, it is anticipated that mutations would occur that would improve the catalytic efficiency of the enzyme, which degrades the ampicillin. However, after analyzing the colonies that grew on the higher level of ampicillin, it is determined that two different independent classes of mutations can result in resistance to the higher level of ampicillin. One is due to the anticipated improvement of an increased catalytic activity of the enzyme; the other class is due to increased expression of the enzyme (increased number of molecules of the enzyme).

See also: Antibiotic-Resistance Mutants; Screening

Selective Breeding

W G Hill

doi: 10.1006/rwgn.2001.1167

The basis of genetic improvement programs in any organism is selective breeding, where individuals are chosen that are expected to have offspring with desirable properties. This is directed evolution: fitness is defined by the breeder rather than by the individual's ability to survive and reproduce in nature. Selective breeding long predates the discovery of the mechanisms of inheritance. Indeed, Darwin was much influenced by the success of selective breeding in animals and plants in developing the theory of natural selection "There can be no doubt that methodical selection [by man] has effected and will effect wonderful results" (Darwin, 1868). The vast range of phenotypes of dogs (all derived by selection from the wolf) in color, size, conformation, and behavior, is perhaps the clearest example of the power of selective breeding over long periods of time. Improvements in selection of plants and animals for food, illustrated by the greatly increased grain production of modern varieties of plants and of meat or milk by the modern breeds or strains of animals, has enabled an enormous increase in the human population. Selective breeding is used to develop more efficient strains of microorganisms, for example yeast for brewing, and of laboratory stocks with defined properties, such as extreme obesity, for analysis to elucidate the genetic basis of the trait.

Principles of Selective Breeding Programs

For selective breeding to be effective, there must be genetic variation present in the population, a way of identifying individuals for selection that are likely to transmit the desired properties to the descendants, and sufficient spare reproductive capacity so that the population can be bred from only the chosen individuals. For most traits there is considerable variation at the observed or phenotypic level, thus providing plenty of selective opportunity. Indeed, selection has utilized both extreme mutant forms that have arisen or been identified, such as coat color in livestock or dwarfing genes in wheat, and quantitative genetic variation contributed by many unidentified loci. Most traits have a sufficiently high heritability, i.e., proportion of the variation that is genetic (formally additive genetic), for artificial selection to be effective. In other words, individuals that are extreme for the trait(s) of interest are likely to have offspring with somewhat similar properties, albeit less extreme than the selected parents. In practice, selection may not be based just on the individual's own performance, but additionally or exclusively on that of its relatives using a selection index or best-linear unbiased prediction. Spare reproductive capacity to enable selection is the norm in most plants and in most animals, at least among males. Techniques such as artificial insemination can be used to increase the selection intensity.

A breeding program needs a clear set of objectives that is followed over several generations. Typically, this involves the simultaneous improvement of many traits. For example, in wheat these include traits of the product, such as yield and bread-making quality, and agronomic traits such as straw strength, disease resistance, and drought resistance. In dairy cattle, the equivalent objectives include milk yield and protein composition, fertility, longevity, and mastitis resistance.

A further fundamental component of a selective breeding program is the mating system employed by the breeder, which depends on the reproductive system of the organism, for example whether heterosis is important for the traits of commercial interest, and whether homogeneity of product is required.

In most livestock breeding programs, selection is practiced within segregating populations of as large a size as can be managed or is available nationally. In dairy cattle, commercial animals are usually purebred, largely because one breed in particular, the Holstein–Friesian (the black and white), is regarded as superior for milk production. In poultry or pigs that are bred for meat production, selection is typically practiced within three or more segregating populations and crosses made to produce the dam of the commercial animal, and a further three-way cross to produce the commercial offspring. This is mainly to utilize heterosis in reproductive performance of the dam, and to utilize complementary properties of the sire and dam lines. The sire line(s) do not contribute to reproduction except via fertility and can be selected primarily for traits of growth, and the dam lines are selected for both growth and reproductive performance.

In plants, the wide range of reproductive systems leads to a wide range of design of improvement programs. For naturally outcrossing species of plant such as maize (corn), for which there is considerable heterosis in the commercial traits of yield and a need to produce a uniform product, the seed grown commercially is a two-way or higher cross of inbred lines. The selection is practiced both within inbred lines during their formation from crosses of existing commercial or other populations, and among inbred lines on the basis of their own and crossbred performance. It is the selection that leads to the success of a new variety of maize, and not the inbreeding *per se*. In natural selfing species such as wheat, inbred lines are used commercially and are developed by selection within and among selfing inbred lines developed from crosses. In species that are reproduced clonally, such as the potato, selection has to be practiced during a reproductive cycle, but subsequent uniformity of product does not require homozygosity of the variety. Thus, while clonal reproduction offers the opportunity to use a specific genotype widely, it does not offer a route to subsequent improvement for which segregation and recombination are needed.

There are a considerable number of theories and experiments on the design of selective breeding programs in animals and plants, which have been developed over many decades. In livestock, the most influential proponent of the use of genetic principles was J.L. Lush, a disciple of Sewall Wright; and most modern methodology comes from C.R. Henderson whose work, like that of Lush, was mainly motivated by the problems of dairy cattle improvement. In plants, the use of inbreeding and selection was propounded by Mangelsdorf, using the ideas of East and Jones. In general, because they can be grown in large numbers and individual plants of most crops are not valuable, less formal statistical/quantitative genetic methodology has been applied in plant rather than in animal and (more recently) tree improvement programs.

Success in a breeding program does not depend solely on its scientific basis. It requires, as indeed does any business, quality management that actually executes the geneticists' breeding plan, financial strength, foresight, and luck.

Examples of Genetic Changes from Selective Breeding in Practice

The efficacy of selective breeding is obvious from the changes in yield and of the cost of food relative to income. While improvement in yields can be attributed to both genetic selection and to management, they can be separated. This typically requires the comparison of stock of different generations at the same time in the same environment, for example by planting seeds stored for several years alongside seed from modern varieties. Even so, the most spectacular differences are not the products of man's recent efforts, but can be seen among breeds of dog that differ by almost 100-fold in weight. These breeds have been developed over many centuries, and have presumably utilized mutations that have occurred over the long period since domestication. Similarly, much of the improvement during and subsequent to domestication in food plants and animals was not based on knowledge of genetics. Some examples of the changes brought about by more recent selective breeding can be obtained from well-conducted experiments.

Animals

In the past, the same breeds or strains of chickens were used for both commercial egg and meat production, but now they are specialized. Most selection in broiler chickens for meat consumption has been placed on growth rate, as faster growing birds incur less food and housing costs at market weight. Selection has, however, also been placed on many other traits, including meat yield, conformation, leg function, and disease resistance of the broiler and on reproductive performance of the broiler parent. Havenstein *et al.* (1994) estimated genetic progress in broiler chickens by comparing the performance of a population maintained without selection since 1957 and a 1991 commercial strain, in each case fed on a diet typical for 1991 (**Table 1**). Note the threefold increase in growth rate, thereby enabling slaughter at a younger age (e.g., at 6 versus 8 weeks, but consequently with less flavor), with improved feed conversion efficiency and meat yield. There are downsides, however: the birds are fatter, so feed needs to be restricted to broiler hens, and there are increases in mortality and leg abnormality. Birds were also compared on a diet formulated to 1957 rather than 1991 standards (which has about 10% higher energy and protein content). At 6 weeks of age, for example, body weights were 0.51 and 1.77 kg for the 1957 and 1991 strains, respectively, fed on the 1957 diet. Hence most of the change was genetic in nature.

In dairy cattle, as in many other species of livestock, there has been substantial breed substitution (notably toward more specialized dairy types) between countries, particularly in the black and whites. The North American Holstein population was derived from European animals exported during the late nineteenth century. Subsequently, however, because of greater concentration on milk production characteristics, the American population became superior in production to those remaining in Europe. During the last quarter of the twentieth century there has been almost complete replacement of European animals by North American Holsteins. Rates of genetic change have greatly accelerated in recent decades as modern selection and breeding methods have been introduced, and are now in excess of 1% of the mean per year for production traits.

There are some problems and failures, however. Thoroughbred racehorses do not seem to be running much faster now than 50 years ago, judging by winning times recorded in the classic races. Genetic change in some species has left associated fitness problems in their wake; for example, problems with leg weakness in broiler stocks have to be overcome or kept in abeyance by devoting selection effort to such traits in the breeding program. In the developing world, and even in less developed areas of other countries, there has been little input or uptake of genetic change, in line with the lack of change in management practices.

Plants

Plants need to be adapted to their environment, which for field crops can be modified but not controlled, in contrast to animal breeding where housing can be uniform worldwide. Genotype × environment interactions therefore are an important feature of plant breeding: for example, different varieties of maize are used at different latitudes in North America to make best use of the length of the local growing season. Improvements in yield have been well publicized, for example in the 'Green Revolution' where changes in management practices have been accompanied by and have benefited from new varieties. Indeed, for genetic progress to be effective, beneficial changes in management are usually required. In order to isolate the consequences of selective breeding, designed experiments are more suitable. **Table 2** illustrates the extent and basis of the changes in winter-sown wheat (Austin *et al.*, 1980). Varieties that were of major commercial importance in their time were grown in 1978 in England in low and high fertility soils. Note in particular the substantial increases in yield, much of it achieved by shortening of the straw. The latter is associated with an increase in the proportion of the plant mass present in the grain, and enables much heavier

Table 1 Growth in the same trial of broiler chickens bred in 1957 and 1991

Strain	Body weight (kg)		Feed conversion (feed/gain)		Per cent (at 6 weeks)			
	At 6 weeks	At 8 weeks	At 6 weeks	At 8 weeks	Meat	Fat	Deaths	TD[a]
1957	0.63	0.99	2.51	2.65	11.6	8.4	2.2	1.2
1991	2.13	3.11	2.04	2.34	15.6	14.1	9.7	47.5

[a]Tibial dyschondroplasia.

Table 2 Production of wheat in the same trial from varieties bred in different years.

Year variety introduced	Yield(t/ha)		Height(cm)		Harvest index[a]	
	Poor soil	Good soil	Poor soil	Good soil	Poor soil	Good soil
1908	3.30	5.22	112	142	34	36
1953	3.74	5.86	87	110	42	42
1972	4.04	6.54	86	106	46	46
1977	4.63	7.30	64	80	50	48

[a]Grain/(grain + straw)%.

use to be made of nitrogen fertilizer without lodging (i.e., failing to stand). Indeed the 1977 variety was the first to incorporate a dwarfing gene with major effect.

Developments in Selective Breeding Programs

There has been extensive research on optimizing breeding schemes. This includes methods for prediction of breeding value using selection indices and best line unbiased prediction (BLUP) (see Selection Index) for better estimation of genetic parameters such as heritabilities and correlations for use in these predictors, and for balancing selection intensity and effective population size. As decisions among potential breeding animals or plants cannot usually be made until they are mature enough to have records, for example, on milk yield or amount of wood produced, there has been much research into indirect predictors of performance. Yield in first lactation is an excellent indicator of yield in later lactations, but indirect measures such as hormone levels are usually not found to be sufficiently accurate to be useful.

Markers of individual loci associated with performance provide a quite different route. For example, a molecular marker has been used to identify heterozygotes and so eliminate a recessive gene causing stress susceptibility in pigs. With the advent of large numbers of molecular markers and dense linkage maps, the opportunities to use Mendelian variants increases, and studies have been conducted in the major commercial species of plants and animals to identify quantitative trait loci (QTL). This information is intended for use in two ways. One is marker-assisted introgression, where QTL from one population are backcrossed into another (e.g., the dwarfing gene in wheat), using the marker information to bring in the QTL but to exclude as much background as possible. The other is marker-assisted selection, where marker data are used alongside quantitative trait data to increase the accuracy of selection, and to make more accurate early selection; for example, in picking recombinant lines from an F_2 cross, or in selecting among young bulls prior to progeny testing for milk. Such marker-assisted selection is most likely to be effective where there is much linkage disequilibrium, such as in an F_2 cross of inbred lines of plants, rather than in random-mating livestock populations.

Transgenic manipulation is being used commercially in plants, for example, to provide herbicide-resistant soybeans, but is not yet commercially available in animals. Indeed, the consumer response to genetically modified varieties is often highly emotional, even though the actual genetic changes made in effecting the improvement are known, whereas in classical breeding they are not. Selective breeding is increasingly based on a wider range of science and technology and is exposed to greater public interest as new routes to improvement become available.

Further Reading

Cameron ND (1997) *Selection Indices and Prediction of Genetic Merit in Animal Breeding*. Wallingford, UK: CAB International.

Darwin CR (1868) *The Variation of Animals and Plants under Domestication*, 2nd edn. London: John Murray.

Falconer DS and Mackay TFC (1996) *Introduction to Quantitative Genetics*, 4th edn. Harlow, UK: Longman.

Simm G (1998) *Genetic Improvement of Cattle and Sheep*. Ipswich, UK: Farming Press.

Simmonds NW (1979) *Principles of Crop Improvement*. Harlow, UK: Longman.

Van Vleck LD, Pollak EJ and Oltenacu EAB (1987) *Genetics for the Animal Sciences*. New York: WH Freeman.

References

Austin RB, Bingham J, Blackwell RD *et al.* (1980) Genetic improvements in winter wheat yields since 1900 and associated physiological changes. *Journal of Agricultural Science, Cambridge* 94: 675–689.

Havenstein *et al.* (1994) Growth, livability, and feed conversion of 1959 vs. 1991 broilers when fed 'typical' 1957 and 1991 broiler diets. *Poultry Science* 73: 1785–1795.

***See also:* Artificial Selection; Genetic Correlation; Heritability; Selection Index; Selection Intensity**

Selective Neutrality

T Ohta

doi: 10.1006/rwgn.2001.1168

Effect of a Mutant and Neutrality

When a mutant is neither advantageous nor deleterious, and its behavior is determined not by selection but by random genetic drift, the condition is said to be selective neutrality. In the strict sense, it means that a mutant has no effect whatsoever. An example is a nucleotide change of a pseudogene; since the pseudogene has no function, any changes to that gene have no effect. However, it is possible, even in this example, that a nucleotide change has some very small effects on DNA replication and recombination, and hence it is not completely neutral. Such a small effect cannot be recognized by natural selection, and is considered to be practically negligible. Let us define selective neutrality in the narrow sense and that in the broad sense. The former applies to those cases where the effect of a mutant is practically nil in the biological world, whereas the latter includes the cases in which a mutant has some small, but not negligible, effects. For the latter, both random genetic drift and selection become important, and the nearly neutral theory can be applied.

Selective Neutrality in the Narrow Sense

The neutral theory was first put forward by M. Kimura, and later by J.L. King and T.H. Jukes. Under selective neutrality in the narrow sense, which was mainly developed by Kimura and his associates, the behavior of mutant alleles in the population is solely controlled by random genetic drift, and the theory becomes simple. Let us consider the process of accumulation of new mutants within the species in the course of evolution. Suppose that mutant genes are substituted one after another in the finite population of N individuals. Let v be the neutral mutation rate per gene per generation. Since each individual has two homologous genes, there are $2N$ genes in the population, and $2Nv$ new mutants appear in the population in each generation. The rate of mutant substitutions per generation is equal to this number multiplied by the fixation probability. For a neutral mutant, the fixation probability is equal to the initial frequency, $1/(2N)$. Therefore, the rate of substitution, k, becomes:

$$k = v \tag{1}$$

This formula is very simple, i.e., the rate of mutant substitution equals the neutral mutation rate. It should be noticed that the remaining fraction, $1 - 1/(2N)$, of mutants are lost from the population.

In considering the population dynamics of mutant substitutions, one needs to know how long the mutant takes until it fixes in the population. The average time until fixation, $\bar{t}_1$, of a neutral mutant is known to be four times the effective population size, N_e:

$$\bar{t}_1 = 4N_e \tag{2}$$

The effective population size is usually smaller than the actual population size, N. During the process of the substitution, polymorphism of mutant and original alleles appears. Population geneticists often measure polymorphism by heterozygosity, which is the probability that two randomly chosen alleles differ. Under the selective neutrality in the narrow sense, if the population is in equilibrium between mutation and random drift, the heterozygosity, H, is expected to be:

$$H = \frac{4N_e v}{1 + 4N_e v} \tag{3}$$

This formula is again remarkably simple and quite useful. In addition to the heterozygosity, various quantities have been obtained enabling the neutral theory to be tested.

Selective Neutrality in the Broad Sense

Selective neutrality in the broad sense brings many complications to discussion on the subject. It is well known that the rate of molecular evolution is strongly dependent on selective constraints of proteins or nucleic acids, and we know that there are numerous types of mutations, from those with negligible effect to those with large effect. The borderline mutations between the selected and the neutral classes may be important and are called nearly neutral. How do the predictions differ from those under simple neutrality if these mutations are important? Theoretical studies on this problem were mainly carried out by T. Ohta and colleagues. One of the most critical quantities on such mutations is the fixation probability, u. For the simplest case of a semidominant gene with selection coefficient, s, the fixation probability in a finite population of the effective size, N_e, is the function of the product, $N_e s$. It is a continuous monotone function of $N_e s$. Therefore, in discussing the rate of gene substitution, one has to consider all mutants around $N_e s = 0$. The effectiveness of selection is determined by this product, whereas actual species have various population sizes from very small to very large. Hence, the effectiveness of selection differs among species. In addition, physiological conditions may influence weak selection, e.g., a functional constraint on a protein may differ between homoiotherms and poikilotherms.

Considering the importance of negative selection for keeping gene function, it is likely that many nearly neutral mutations are very slightly deleterious, i.e., s is negative. Remember that a slightly deleterious mutation may have a finite probability of fixation depending on the absolute value of the product, $|N_e s|$. If this value increases, the fixation probability becomes smaller, i.e., it decreases as the population size becomes larger, or as the selection intensity becomes stronger. This prediction is related to the molecular clock in an important way. As explained, the fixation probability is higher in a small population than in a large population. On the other hand, the mutation rate would depend on the number of cell generations, and therefore mildly on the generation number. In general, large organisms have a long generation time and small population size and vice versa, and there is a negative correlation between population size and generation time. Then the population-size effect on the fixation probability is expected to be partially canceled by the generation-time effect on mutation rate. This cancellation is likely to be responsible for the molecular clock of protein evolution.

Distribution of Mutants' Effects

Precise formulation of selective neutrality in the broad sense is difficult. Distribution of mutants' effects around strict neutrality is needed for the exact analysis, but it is not known. There are two theoretical approaches to near neutrality, i.e., the shift model and the fixed model. The former is based on the assumption that the distribution of mutants' effects shifts back whenever a mutant fixes in the population. In other words, mutant substitutions are independent of one another. For example, in one study for the shift model, the gamma distribution is assumed for the effects of slightly deleterious mutations, and this distribution remains the same when a mutant fixes, since the population mean shifts back to the original state. On the other hand, in the fixed model, the distribution is fixed irrespective of mutant substitutions, and the population mean moves according to the effect of the fixed mutant. This model is also called the 'house of cards' model. Contrary to the shift model, the effect of each substitution stays and affects subsequent substitutions by changing the mean fitness in the fixed model. Therefore substitutions are interrelated in their effects on fitness. If the evolution of a protein is

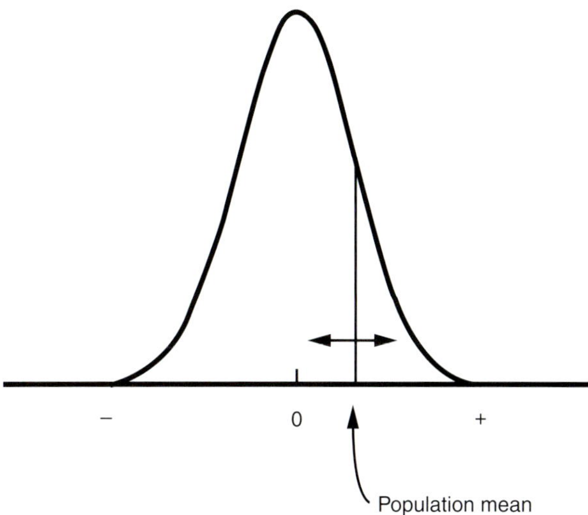

Figure 1 Distribution of selection coefficient of new mutations. (From Ohta, 1992.)

the subject of interest, amino acid substitutions are interrelated, and the fixed model is more realistic than the shift model.

An example of the fixed model where the normal distribution is used for the mutants' effect on fitness is considered here in some detail. **Figure 1** shows the distribution of the selection coefficient of new mutants around the population mean. If selection is strong enough, the mean moves toward the right without fluctuation. For nearly neutral mutations, it moves erratically but tends to increase. When the population mean becomes positive, the average selection coefficient of new mutations becomes negative, i.e., new mutations are slightly deleterious on the average. Then the mutant substitution slows down. The effectiveness of selection is again determined by the product of population size and selection intensity, which is measured by the standard deviation, σ_s, of the normal distribution. According to H. Tachida, the nearly neutral mutations lie in the range $3 \geqq 4N_e\sigma_s \geqq 0.2$, where both random drift and selection affect the population fitness. Although the shift model and the fixed one are different, some patterns of evolution and polymorphisms are similar under both models, i.e., the negative correlation between the evolutionary rate and the population size, and the slow increase of polymorphism with larger population size.

Some Related Observations

What kind of experimental evidence is there for selective neutrality? As compared with the rate of phenotypic evolution, the rate of molecular evolution is remarkably constant, i.e., a molecular clock is often observed. This characteristic of molecular evolution is strong evidence for the neutral theory as it is explained by equation (1). Another characteristic is the fact that the more constrained the protein is, the lower is its rate of evolution. This is also thought to support the neutral theory, since the proportion of neutral mutations is thought to decrease as the functional constraint becomes stronger.

Through comparative studies of DNA sequences, many interesting patterns of evolution and polymorphism have emerged. The difference in average patterns between synonymous and nonsynonymous substitutions of mammalian genes is in accord with selective neutrality in the broad sense, i.e., the generation-time effect is larger for the synonymous substitutions than for the nonsynonymous substitutions.

The next question to ask concerns the variance of the evolutionary rate. One approach is to estimate the index of dispersion, that is the ratio of the variance to the mean number of substitutions. The index becomes 1 under the simple Poisson process of mutant substitutions. By examining sequences of mammalian genes, J.H. Gillespie has shown that the dispersion index, R, is larger than unity, and is often between 1.5 and 10 for nonsynonymous substitutions. His analysis also suggests that synonymous substitutions are less erratic. Is this large index of nonsynonymous substitutions in accord with selective neutrality in the broad sense? Analysis of the fixed model suggests that the dispersion index is only slightly larger than unity. In other words, the interaction effect of mutant substitutions of the fixed model is not big enough to explain the observed value of R. If one incorporates changes in population size, R can be shown to become larger, and to be similar to the observed value.

Real difficulty lies in distinguishing the neutrality in the broad sense from the selection theory. The choice is not one or the other, and there may be cases in which both selection and drift are almost equally important. It would be unwise to decide which of the theories is 'correct' in such cases.

Another aspect of selective neutrality is concerned with polymorphisms. Under strict neutrality, various quantitative predictions can be made and the neutral theory is testable. Many attempts have been made to test the theory. D. Hartl, M. Kreitman, and associates carried out many studies on synonymous and nonsynonymous polymorphisms. Their test compared the relative numbers of synonymous and nonsynonymous substitutions either within a species or among closely related species. Note that the relative numbers should remain the same whether it is measured within species or between species. In some cases, an excess of nonsynonymous differences was found for within-species comparisons, whereas in the other cases, a deficiency of the same differences was observed. Again, it is difficult to tell whether the selection theory or the nearly neutral theory fits better to such cases. The results reflect the short-term effect of large variance of evolutionary rate mentioned above.

Let us turn our attention to synonymous substitutions, which were interpreted to be neutral in the 1970s and 1980s. However, very weak selection has been shown to be operating in relation to the codon usage bias that reflects tRNA abundancy. A large amount of data on codon usage bias are available, and it became clear that the bias is particularly conspicuous for highly expressed genes. Such facts are explained by assuming the presence of the optimum codon usage. The model of very weak selection nicely explains the observed bias of codon usage. In finite populations, most synonymous sites are fixed with biased frequencies. Mutant substitutions occasionally take place by chance. They are either slightly deleterious or slightly advantageous, and mutation–selection–drift equilibrium is expected for codon usage. Thus, most synonymous substitutions are neutral in the broad sense. Average selection intensity is smaller for synonymous than for nonsynonymous substitutions.

Effective Population Size

So far, our discussion is based on the concept of effective population size. In small populations, the range of effectively neutral mutations increases, and vice versa. In actual species, the population size rarely remains constant. For example, speciation often accompanies bottlenecks, and the effective population size may depend on the founding individuals; the range of selective neutrality increases, and more mutations become neutral than after the species expanded. Several examples of rapid molecule evolution in conjunction with bottlenecks have been reported, e.g., ribosomal RNA and protein evolution of Hawaiian *Drosophila*. The effective population size is dependent on linkage to other selected loci. Under strong linkage with many selected loci, the effective size becomes much smaller than the value under free recombination. A structured population brings another complication, since the weak selective force may differ between local colonies, and the effective size may be the local one. Effects of linkage and population structure on selective neutrality are currently being investigated by population geneticists.

Further Reading

Gillespie JH (1991) *The Causes of Molecular Evolution*. Oxford: Oxford University Press.

Hartl DL (1987) *A Primer of Population Genetics*, 2nd edn. Sunderland, MA: Sinauer Associates.

Kimura M (1983) *The Neutral Theory of Molecular Evolution*. Cambridge: Cambridge University Press.

Li W-H (1997) *Molecular Evolution*. Sunderland, MA: Sinauer Associates.

Ohta T (1992) The nearly neutral theory of molecular evolution. *Annual Review of Ecology and Systematics* 23: 263–286.

***See also:* Effective Population Number; Evolutionary Rate; Fixation Probability; Gene Substitution; Genetic Drift; Molecular Clock; Natural Selection; Nearly Neutral Theory; Neutral Theory; Selection Coefficient**

Selective Sweep

M Kreitman

doi: 10.1006/rwgn.2001.1446

Most populations of plants and animals are polymorphic at the nucleotide level. In the fruit fly *Drosophila*, for example, where we have the most comprehensive knowledge of variation, nucleotide diversity, the probability that a randomly chosen location along the DNA is different between two randomly chosen chromosomes, can exceed 1%. For a typical gene 2000 bp in length, nucleotide polymorphisms will be found at several dozen positions along the gene in a population sample of only a dozen randomly chosen chromosomes, and no two gene sequences in the sample will be identical. Many of these polymorphisms will be in locations along the gene that do not have known functionality, such as in introns, and to a first approximation we may assume that this extensive polymorphism is representative of the standing crop of selectively neutral mutations found genome-wide (see Neutral Mutation). Now consider a new mutation that arises in one particular allele of this gene, as defined by the particular combination of these neutral variants along the gene, and further imagine that this is a selectively favorable mutation that is destined to be driven to fixation by positive natural selection. It follows that as this favored allele goes to fixation in the population all of the other alleles in the population will necessarily be driven to extinction, along with the neutal variation that distinguishes these alleles one from another. In other words, neutral variation that is tightly linked (in both a physical and genetic sense) to the site under positive selection will

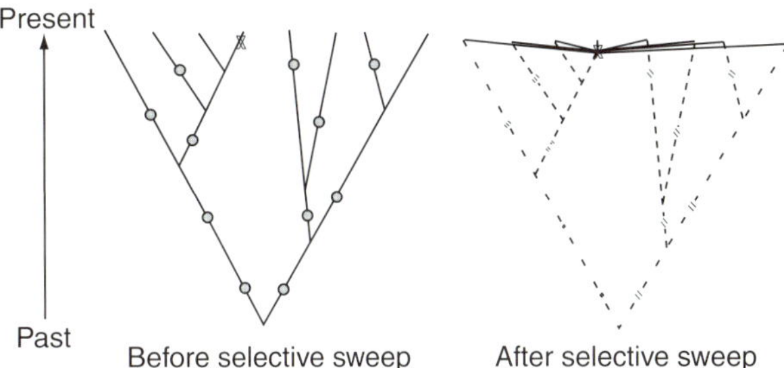

Figure 1 Gene genealogy of gene sequences from eight sample chromosomes before (left) and after (right) a selective sweep.

be swept out of the population as the selectively favored allele sweeps to fixation. Such is the double meaning of a selective 'sweep.'

The process of a selective sweep can be illustrated by considering the "genealogy" of alleles at a locus before and after a selective sweep, as depicted in **Figure 1**. In this example we have sampled eight chromosomes from a population and have traced the ancestry of each of these alleles back in time to each of their most recent common ancestor with another allele. All the alleles 'coalesce' to a single common ancestor of the sample at the bottom of the tree. This tree of relationships is called a gene genealogy. Imposed on this genealogy are the mutations that have occurred along the branches, and these mutations, of course, represent the differences that can be used to distinguish the eight alleles one from another. One of these mutations, demarcated by an X, is an adaptive mutation that has occurred in the recent past. At the instant this allele reaches fixation in the population, every individual will possess this variant, and the genealogy of a random sample of eight chromosomes will look like the one depicted on the right by the solid lines. The dashed lines, representing the old alleles, will have been driven to elimination, and the mutations that distinguish them will also have been lost.

Notice that immediately following a selective sweep, as depicted in this example, not enough time has elapsed for new mutations to have arisen: all alleles are identical in sequence along the entirety of the gene. In addition, the genealogy looks decidedly 'star-like,' with each allele emanating as a spoke from the original allele that incurred the favorable mutation. Both of these features – the loss of linked neutral polymorphism and a star-like genealogy – are two telltale signatures of a recent selective sweep, and each of these signatures of nucleotide polymorphism can be used as a criterion for inferring the existence of a selective sweep in real polymorphism survey data.

In reality the situation is not quite this simple because the fixation of advantageous alleles is not instantaneous, and during the time in which the favored allele increases in frequency it can both incur new mutations and it can also regain some of the old polymorphisms by recombining with another old allele. More generally, a selective sweep will produce a trough of neutral polymorphism that will be restricted to a small interval of tightly linked sites. The length of the trough will be determined by the relationship between the strength of positive selection on the adaptive mutation (how fast the allele increases in frequency) and the rate of recombination between the selected site and a linked site (how fast the associations are broken up). Applying realistic values of the recombination rate across a gene and the strength of positive selection of a favored mutation, a model of selective sweep indicates that the loss of neutral linked polymorphism might extend only several kilobases to either side of the site under selection. In addition, a trough of reduced polymorphism will only be discernible in a survey of nucleotide polymorphism if the selective sweep has occurred in the relatively recent past. For these reasons, it may not be surprising that few convincing examples of selective sweeps have been documented from surveying nucleotide polymorphism in genes.

Perhaps the most convincing example of a selective sweep can be seen in the superoxide dismutase locus (SOD) in *Drosphila melanogaster*. This locus segregates for two different protein variants, and these variants have measurably different enzymatic properties. An extensive survey of nucleotide variation found that all of the sequenced representatives of the more common of the two alleles are completely identical to one another in sequence, suggesting that they all recently derive from a common ancestor. The other less frequent allele, in contrast, carries extensive nucleotide polymorphism among individual copies. The locus was also sequenced in a closely related species, and it carries the same amino acid as the less frequent variant in *D. melanogaster*. A reasonable interpretation of the data is that the more common protein variant is the younger of the two alleles, and that it has been rapidly driven up in frequency in the recent past to become the more common type. This may very well be an example of a selective sweep caught in the act.

Selective sweep is an example of genetic hitchhiking (see Hitchhiking Effect) between a site under selection and a linked site not under selection. This linkage can come about in unexpected ways. The animal mitochondrial genome, for example, a maternally inherited circular genome consisting of 13 genes, is expected to be particularly susceptible to hitchhiking events because it is a nonrecombining genome. In one species of the fruit fly, *D. simulans*, a maternally inherited microorganism, called *Wolbachia*, has a mechanism by which it provides a strong selective advantage to females carrying the infection when they are introduced into a population without the infection. This strong selective advantage and maternal inheritance of both *Wolbachia* and the mitochondrial genome has been shown to cause the mitochondrial variant of the infected female to increase in frequency to near-fixation as it hitchhikes up along with the frequency of *Wolbachia* infection.

If selective sweeps of advantageous variants are common occurrences in genes, then neutral variation levels might be expected to be depressed throughout the genome, but more so in regions of chromosomes characterized by low rates of recombination than in regions having high recombination rates. In *D. melanogaster*, recombination rates vary by one or two orders of magnitude across regions of chromosomes, and true to this prediction, levels of noncoding polymorphism are strongly positively correlated with the recombination rate. Importantly, the rate of divergence between this species and its sibling species *D. simulans* is not correlated with recombination rate, indicating that differences in polymorphism levels are not the result of any difference in the mutation rate. Thus, it is possible that selective sweeps modulate levels of nucleotide variation genome-wide. Unfortunately for this hypothesis, an alternative model of genetic hitchhiking, called background selection (see Background Selection), has also been proposed to explain this correlation between the recombination rate and the level of polymorphism, and at present time both hypotheses remain viable.

Further Reading

Ballard JWO (2000) Comparative genomics of mitochondrial DNA in *Drosophila simulans*. *Journal of Molecular Evolution* 51(1): 64–75.

Hudson RR, Saez AG and Ayala FJ (1997) DNA variation at the Sod locus of *Drosophila melanogaster*: an unfolding story of natural selection. *Proceedings of the National Academy of Sciences, USA* 94(15): 7725–7729.

Kaplan NL, Hudson RR and Langely CH (1989) The 'hitchhiking effect' revisited. *Genetics* 123: 887–899.

Turelli M and Hoffmann AA (1991) Rapid spread of an inherited imcompatibility factor in California *Drosophila*. *Nature* 353: 440–442.

***See also:* Background Selection; Coalescent; *Drosophila melanogaster*; Gene Trees; Hitchhiking Effect; Neutral Mutation**

Self-Fertilization

J Hodgkin

doi: 10.1006/rwgn.2001.1169

Self-fertilization occurs when an individual is capable of generating both male and female gametes, and using the former (sperm) to fertilize the latter (eggs), thereby producing self-progeny. The majority of animal species reproduce by cross-fertilization, either with separate male and female sexes, or with a hermaphrodite sex that is specialized to avoid self-fertilization. Reproduction by cross-fertilization increases genetic diversity by sampling the genomes of two different individuals, and so creating new combinations in their offspring. Self-fertilization can also create diversity in progeny genotypes, because both male and female gametes are usually produced with meiotic recombination, but a purely selfing population will steadily lose heterozygosity at any given locus, and is therefore vulnerable to inbreeding depression. Self-fertility nevertheless occurs widely in the animal kingdom, in a variety of invertebrate groups.

There are several obvious advantages to self-fertility, of which the most important is the avoidance of the twofold 'cost of sex': all individuals in a self-fertile population are capable of producing progeny, in contrast to a population of males and females, in which only the females produce progeny. Self-fertility also has the advantages that a single organism can colonize a new habitat, and that individuals do not need to invest time and resources in finding mating partners.

The same advantages of rapid population growth and efficient colonization apply also to organisms that reproduce amictically (parthenogenetically), but such organisms do not even undergo meiotic recombination, and therefore have no ability to create new genetic combinations from one generation to the next. In general, it is believed that completely amictic populations are more likely to go extinct than those with some degree of genetic exchange. Reproduction by self-fertilization represents a compromise, especially for organisms with the capacity for both self-fertilization and cross-fertilization. Examples of this strategy are provided by species with hermaphrodite sexes that are capable of both selfing and crossing, or species such as the laboratory nematode *Caenorhabditis elegans*, which has populations consisting mostly of self-fertile hermaphrodites with rare males that can cross-fertilize the hermaphrodites.

***See also:* *Caenorhabditis elegans*; Fertilization; Hermaphrodite; Parthenogenesis, Mammalian**

Selfish DNA

H Y Wong

doi: 10.1006/rwgn.2001.1170

Theoretical Concepts

The assertion that organisms are simply DNA's way of producing more DNA has been made so often that it is hard to remember who made it first.

So begins one of two classic papers (Doolittle and Sapienza (1980); Orgel and Crick, 1980), which developed the concept of selfish DNA and sparked off a debate which is still not altogether resolved. Whilst Dawkins (1976, p. 47) also mentioned the idea of selfish DNA, his focus was primarily on 'selfish genes': a phrase he used to describe *all* genes, in order to emphasize that genes are selected solely on the basis of their own propensity to increase in number.

One way in which this increase occurs is when the DNA sequence provides a function that increases the general reproductive output (fitness) of the organism in which it is found, but there are two other ways in which a sequence can increase in number over time. The first is by subverting the process of inheritance such that, at a single locus, an individual heterozygous for the sequence has a greater than 50% chance of passing it to an offspring. This can be accomplished by mechanisms such as meiotic drive which promotes one chromosome to the detriment of the other. The second way is for sequences to replicate across the genome, so that many copies may be found in different locations in the same genome. It should be emphasized that both of these methods only require sequences to be inherited vertically, from parent to offspring. Those replicating sequences that are regularly transmitted horizontally are usually regarded as viruses or virus-like organisms. Sometimes, but not always, the spread of these self-promoting elements reduces the fitness of the bearer. This is only likely to occur in sexually outcrossing species, where the element can be selected relatively independently of the rest of the genome. The result of this collision of interests is genomic conflict, the demonstration of which is probably the clearest evidence of the 'selfish' nature of a sequence.

Definitions

Since the term 'selfish gene' can be used to describe any gene, genetic elements that promote themselves without necessarily benefiting the organism in which they are found, can be referred to as 'ultra-selfish genes,' 'selfish genetic elements,' or 'outlaw genes.' This avoids the terminological confusion witnessed in the use of 'selfish gene' and 'selfish DNA' in the respective titles of the two previously mentioned papers. Nevertheless, it is becoming common to find the name 'selfish gene' being used specifically to refer to these sequences.

'Selfish DNA' is a term often used rather vaguely to refer to a self-promoting element which works at a molecular level. Meiotic drive genes are not usually thought of as selfish DNA, and it seems sensible to restrict the term to self-promoting elements which multiply within a genome (although some spread in both ways). Some authors prefer the less anthropomorphic terminology 'parasitic DNA,' and perhaps more accurate still is the term 'symbiotic DNA.' This emphasizes that the DNA is not distinguished by its physical effects. Indeed, one of the main points emphasized by Doolittle and Sapienza (1980) was that presence or absence of these sequences often has no specific effect on the phenotype. However, gross phenotypic effects can sometimes be seen, such as sterility and other defects in *Drosophila* hybrids. This hybrid dysgenesis is due to the introduction of selfish DNA, in the form of P elements, into genomes which have no mechanism to prevent its excessive duplication. In addition, it is now known that organisms sometimes use selfish elements for particular purposes (see below, 'Coevolution of selfish DNA and host'). It thus seems unwise to define selfish DNA by the absence of a specific phenotype. Nevertheless, one successful strategy for a replicating sequence is to reduce phenotypic effects so as to cause as little disturbance as possible to the organism.

'Junk DNA' is used to refer to DNA which is sequence-independent: the sequence order does not lead to any recognizable function. Selfish DNA or normal genes may lose functionality and become 'junk.' Mutation may then render their origin uncertain.

Genetic Mechanisms

This section presents only a brief summary of some types of selfish DNA. For greater detail, the reader is referred to more specific encyclopedia entries. Generally, selfish DNA sequences provide some of the machinery needed for self-replication, but also rely on cellular DNA replication mechanisms.

Transposable Elements

Transposons are perhaps the most common form of selfish DNA, and eukaryotic genomes contain many nonfunctional remnants of transposons as well as functional elements. They fall into class I and class II elements depending on their method of transposition.

Class I elements (retroelements) use a 'copy-and-paste' mechanism whereby reverse transcriptase makes a DNA copy of the element from transcribed RNA, which is then integrated elsewhere in the genome. One type, the retrotransposons, bear close similarity to retroviruses, and evolution from retrotransposons to retroviruses, and vice versa, is likely. Indeed, the *gypsy* retrotransposon in *Drosophila* can be passed from one individual to another via food, and could be said to be a virus. A few retroelements do not themselves code for reverse transcriptase, presumably requiring it to be provided by other elements. This is the case for *Alu*, an element derived from a normal host gene, which makes up about 5% of human DNA. It could be said that these are hyperparasitic DNA, parasitizing the resources of 'normal' parasitic class I elements, their reduced size giving faster and more accurate replication, as well as making mutational inactivation less likely. Indeed, most transposons have nonautonomous variants which, like *Alu*, 'borrow' some of the genes needed for transposition.

Class II transposons, such as the *Drosophila* P element, use a 'cut-and-paste' method, excising themselves from the genome and reinserting elsewhere. At first sight, this would seem not to increase the number of elements present, but high copy numbers suggest that they can replicate, and it has been suggested that this occurs due to transposition from replicated to unreplicated regions during normal DNA duplication.

Transposons are thought to have an insertion bias for noncoding areas, such as heterochromatin or even preexisting transposons, where their phenotypic effect is minimal. Even so, a large fraction of mutations in most organisms are caused by transposable elements inserting and excising.

Transposition in somatic cells gives no long-term advantage to the sequence, and may endanger the host. This is presumably the reason for transposons such as P elements only being active in germ-line cells. Concentration of transposons on the germ-line is seen in an extreme form in hypotrich ciliates. These unicellular organisms use a separate nucleus for somatic gene transcription, which only contains 5% of the germ-line DNA. A substantial fraction of the excised DNA consists of transposons.

Type I and Type II Introns

These self-splicing introns are unlike other introns: not only do they have a fairly conserved sequence, but in addition to being present in eukaryotic nuclei, multiple copies are also found in prokaryotes and organelles. Their RNA sequence is catalytic, splicing itself out of any length of RNA in which it occurs, hence concealing its presence when inserted in a coding region. Type I introns have the best-studied 'selfish' behavior. They can add themselves to alleles without the intron at the corresponding locus on the homologous chromosome. This is done by coding for a restriction endonuclease which recognizes and chops DNA at the locus, forcing repair to take place using the intron-containing sequence as a template. Alleles containing the intron cannot be cut, as the intron spans the restriction site. This biased gene conversion process is known as homing, and may also be responsible for the transfer of the intron to other loci. Perhaps the most intriguing behavior is endonuclease war, in which different type I introns present in separate bacteriophages, preferentially cut each others splice sites, even when the intron is present. When both phages infect the same host, a selfish DNA battle may ensue, with each intron trying to deplete the available splice sites so as to reduce the risk of being spliced into by the other sequence.

Type II introns code for reverse-transcriptase-like proteins, which suggests mechanisms for autonomous self-replication. Similarities in splicing mechanisms suggest that a type-II-like sequence may well have been ancestral to normal eukaryotic introns.

Supernumerary Chromosomes

Ten to fifteen percent of plant and animal species possess B chromosomes: chromosomes which are generally small and seem to be dispensable, since they are present in some, but not all, individuals. They are often present in multiple copies, and were first suggested as parasitic elements as early as 1945. Their probability of transmission is increased in species that do not use all four meiotic products, by preferential movement into those cells which become true gametes. They may also move preferentially into germline cells in development. These transmission methods can lead to a cumulative increase in the number of B chromosomes.

Plasmids may also be considered as extra chromosomes, and the probable evolution of viruses from plasmids shows their potentially selfish nature.

Tandemly Repeated DNA

This is also known as satellite DNA, and consists of a single sequence repeated many times over. It is caused by slippage, where misalignment during meiotic recombination leaves one chromosome with a higher number of copies and the other with a lower number. A few satellite sequences may have organismal functions, but many do not. Those that do not are often regarded as junk DNA, but are clearly sequence-dependent: although it is never translated into protein, changing a sequence in the array will reduce slippage. They could be said to be 'selfishly' using the mechanism of crossover to replicate, as although their replication is extremely limited and elimination as well as amplification can occur, the world disproportionately contains the results of amplification. Evidence that tandemly repeated DNA can be transferred between loci also suggests the potential for selfish-DNA-like behavior.

Evidence and Controversy

The concept of selfish DNA is now generally accepted: it is indisputable that these sequences multiply in the genome, and that they can cause problems for the organism in which they reside. Debate has instead focused on the role of selfish DNA in evolution.

The C Value Paradox

The C value paradox is that the amount of DNA in a haploid genome (the 1C value) does not seem to correspond strongly to the complexity of an organism, and 1C values can be extremely variable. Some salamanders have more than 30 times the amount of DNA per cell as humans, and within genera such as the sunflowers, *Helianthus*, some species have 1C values four times greater than others. Much DNA in the cell is present as repetitive sequences of varying lengths, often intermediate repeat sequences which are mostly selfish elements. Over 50% of the maize genome probably consists of retroelements.

A strong correlation between the C value and nuclear size, cell size, and cell cycle time has led some to suggest that selection on these factors maintains a C value which is more or less optimal for the organism. According to this hypothesis, the organism requires a certain amount of DNA, which could consist of any sequence. Selfish DNA is particularly good at competing for this 'resource,' hence its presence in the genome. The organism can regulate the C value, for example, by deleting stretches of sequence in heterochromatic regions. The organism thus has the final say in the C value, and selfish DNA does not explain the paradox.

The opposing argument is that selfish DNA can increase the C value to well above that which is best for the organism: conflict between selfish elements and the rest of the genome results in different C values depending on which is winning. Under this view, selfish DNA can explain much of the paradox.

One factor suggesting that organisms have ultimate control over their genome size is the presence of genomes which contain mostly coding DNA, but have no major reason to prevent the build-up of selfish genetic elements. Although bacterial genomes have little surplus DNA, this can be explained by strong selection for rapid replication. Organelle genomes are similarly economical, but this may be due to competition among themselves for representation in the cell or in the gametes. Indeed, *petite* mutants of yeast contain defective mitochondria which are successful due to their increased replication rate, but do not provide a respiratory function (they have been suggested as selfish DNA in their own right). Since the methods by which selection acts on increased genome size are still a topic of debate, it seems likely that the issue will remain controversial.

Intron Evolution

Because prokaryotes have no introns, it is tempting to assume that introns are a late addition to the eukaryotic genome. By contrast, the introns-early hypothesis states that the ancestor of eukaryotes already possessed introns, and that introns were lost in prokaryotes. It has close links with the 'exon theory of genes' which states that exon shuffling of originally small exons, each of which provides a functional domain of a protein, is the origin of the eukaryotic genome we see today. This leads to a selective advantage for possessing introns, which now provide an important organismal function. The introns-late hypothesis considers that introns are primarily the result of more recent, selfish DNA movement.

Generally, introns at different loci have quite different sequences, suggesting that, although they might have been selfish DNA, they have been inactive for a reasonable length of time. More convincingly, a few intron positions are shared between plants and animals. This could be evidence for early introns (although the plant/animal split is later than the prokaryote/eukaryote one), or could conceivably be due to insertional bias.

It seems extremely probable that many genes arose by exon shuffling: an example of gene formation of this sort has been found recently in *Drosophila*. In addition, there is slight evidence that introns do correspond to boundaries between protein domains. Whether this is due to exons evolving to provide this function and relatively recently co-opting introns based on selfish DNA to an organismal function, or whether the system started off in this form, is essentially still unresolved.

Coevolution of Selfish DNA and Host

The possibility that introns may have been selfish DNA, which now provides a function for the rest of the genome, is one of many examples of coevolution which have been suggested by recent research. Another is the telomere structure in *Drosophila* and some ciliates, which consists of repeated retroelements: the ends of the chromosome are extended by retrotransposition. In many plants, the regulatory regions of several genes are encoded by the mobile elements *Tourist* and *Stowaway*, and the use of plasmids to transfer antibiotic resistance between bacteria is well known. Finally, it is often suggested that a function of transposons is to provide new mutations, especially when accidentally transposing parts of the host sequence to new loci. This last suggestion is clearly an important factor in evolution, but is unlikely to provide short-term advantage. It is best interpreted as the host adapting to what is increasingly recognized as a very fluid genome.

Further Reading

Zeyl C and Bell G (1996) Symbiotic DNA in eukaryotic genomes. *Trends in Ecology and Evolution* 11: 10–15.

References

Dawkins R (1976) *The Selfish Gene*. Oxford: Oxford University Press.

Doolittle WF and Sapienza C (1980) Selfish genes, the phenotype paradigm and genome evolution. *Nature* 284: 601–603.

Orgel LE and Crick FHC (1980) Selfish DNA: the ultimate parasite. *Nature* 284: 604–607.

See also: C-Value Paradox; Intron Homing; Introns and Exons; Satellite DNA

Self-Splicing

See: Introns and Exons

Semiconservative Replication

doi: 10.1006/rwgn.2001.2016

Semiconservative replication is the universal system of DNA replication whereby strands of a parental duplex DNA molecule separate, each then acting as a template for the synthesis of a new complementary strand.

See also: Replication

Semidiscontinuous Replication

doi: 10.1006/rwgn.2001.2017

Semidiscontinuous replication is the mode of DNA replication whereby one new strand is synthesized continuously while the other is synthesized discontinuously.

See also: Replication

Semidominance

See: Incomplete Dominance

Sense Codon

J H Miller

doi: 10.1006/rwgn.2001.1173

A codon that specifies an amino acid, as distinct from a nonsense codon that does not specify an amino acid but instead signals chain termination.

See also: Genetic Code

Sequence Alignments

See: Alignment Problem

Serine

E J Murgola

doi: 10.1006/rwgn.2001.1174

Serine is one of the 20 amino acids commonly found in proteins. Its abbreviation is Ser and its single letter designation is S. As one of the nonessential amino acids in humans, it is synthesized by the body and so need not be provided in an individual's diet. It is a precursor of selenocysteine in certain proteins. In bacteria, after a specialized tRNA is aminoacylated with Ser, the amino acid is converted in two steps to selenocysteine.

The chemical structure of serine is given in **Figure 1**.

```
         COOH
          |
  H2N—C—H
          |
     H—C—OH
          |
          H
```

Figure 1 Serine.

See also: Amino Acids; Proteins and Protein Structure

Sex Chromatin

M A Ferguson-Smith

doi: 10.1006/rwgn.2001.1175

In female somatic cells one of the two X chromosomes is genetically inactive and becomes condensed, forming a small mass of dense chromatin which can be seen with the light microscope within the cell nucleus closely applied to the nuclear membrane. This small structure, variously termed the X chromatin, sex chromatin, or Barr body, can be identified in most female somatic tissues in proportion depending on the active state of the tissue and the stage reached in the cell cycle.

The sex chromatin was discovered in 1949 by Murray Barr, a Canadian neurophysiologist, while studying the effects of electrical stimulation of the hypoglossal nerve in a series of cats; he noticed that only half the cats showed this nuclear structure. Fortunately, he had recorded the sex of each experimental subject and quickly recognized that it was only present in female subjects. A similar structure, the nucleolar satellite, can be seen in the early drawings of nerve cells by Ramón y Cojal made in the previous century, although its significance in terms of sexual dimorphism was not recognized at the time.

Barr and his colleagues followed up this observation and soon determined that the sex chromatin body could be recognized in female somatic cells of many mammals including humans, but was absent in male cells. Its association with only one of the two X chromosomes was not recognized until much later (see X-Chromosome Inactivation), but it was put to practical use earlier on as an aid to the investigation of problems of intersex (see Intersex), for example in the diagnosis of female pseudohermaphroditism due to congential adrenal hyperplasia (see Congenital Adrenal Hyperplasia (Adrenogenital Syndrome)).

The most surprising result of nuclear sexing was the discovery that two forms of hypogonadism in humans were associated with paradoxical sex chromatin findings. In 1954, Polani and Lennox found that patients with Turner syndrome (see Turner Syndrome) had 'male' nuclear sex, and were thus presumptively sex-reversed males. This seemed to be confirmed by the incidence of color blindness in cases of the syndrome, which was the same as the incidence in normal males. Then in 1956, Barr and colleagues found that a number of patients with Klinefelter syndrome (see Klinefelter Syndrome) had 'female' nuclear sex, suggesting that they were sex-reversed females. In 1959 Turner patients were shown to have a single X chromosome and no Y, while Klinefelter patients were found to have two Xs and a Y, incidentally demonstrating the dominant sex-determining role of the Y chromosome.

Progress towards the elucidation of the nature of the sex chromatin body came with the discovery of XXXY Klinefelter patients with two Barr bodies and Turner patients with two X chromosomes in which large and small Barr bodies were associated with X chromosome duplications and deletions respectively. These findings pointed to a derivation from a single X chromosome. In 1961 Mary Lyon presented evidence for random inactivation in female mammals based on the patchy distribution of coat colour in mice heterozygous for X-linked coat color genes. It was immediately obvious that X inactivation was associated with the formation of the Barr body.

Sex chromatin is readily studied in smears taken from the buccal mucosa with a spatula and spread onto microscope slides. Once the buccal mucosal cells are fixed and stained by a simple nuclear dye (e.g., cresyl violet), the sex chromatin body can be observed in approximately 30% of cells from normal females, and in no cells from normal males. The procedure has been used to screen various populations of individuals. As a result Klinefelter syndrome has been found to occur in approximately 11% of azoospermic or oligozoospermic males, and 1% of males with learning defects. The same method was introduced in 1960 as a gender verification test for female athletes taking part in the Olympic Games. This resulted in the identification of XY females in approximately 1 in 420 female athletes, and many were unfairly excluded from participation. DNA testing of buccal smears replaced nuclear sexing in the 1980s, but the same discrimination continued until the Olympic Games in Sydney in 2000, when gender verification was finally abandoned.

***See also:* Congenital Adrenal Hyperplasia (Adrenogenital Syndrome); Intersex; Klinefelter Syndrome; Turner Syndrome; X-Chromosome Inactivation**

Sex Chromosome Aneuploidy: XYY

P A Jacobs

doi: 10.1006/rwgn.2001.1397

Normal males have one X and one Y sex chromosome, but individuals with an abnormal number of sex chromosomes are not uncommon in the human population. Approximately 1 in 1000 males has an additional Y chromosome. Such men have a normal appearance, although on average they are considerably taller than XY males. Furthermore, as a group, they have a propensity for aberrant behavior associated with a personality disorder. As a result, XYY men are found with a 30-fold increased frequency among men in maximum security hospitals. The basis for the aberrant behavior of a proportion of XYY males is not understood.

***See also:* Klinefelter Syndrome**

Sex Chromosomes

J A M Graves

doi: 10.1006/rwgn.2001.1176

Sex as a mode of reproduction is widespread in the animal world. The system whereby males make sperm and females make eggs requires a means to differentiate two sexes with distinct anatomy, hormones, and behaviors superimposed on a common body plan. There are several ways to accomplish this. First, sex can be established by different environmental conditions. For instance, egg incubation temperature determines the sex of alligator hatchlings (hotter for males). Second, sex in many animals is determined by different versions (alleles) of a single gene, as in many fish and all amphibians. Last, sex may be determined by a sex chromosome system.

Chromosomal Sex Determination

In all mammals and birds, some reptiles and fish, males and females differ in one pair of chromosomes. Heteromorphic chromosomes also occur in many insects such as the fruit fly *Drosophila*, moths, and butterflies. In fact, sex chromosomes were first spotted in grasshoppers, when it was observed that one chromosome was present in the normal duplicate in females, but was solo in males. This peculiar sex-related chromosome was called the "X" to denote its unknown significance – the name has nothing to do with its shape. In other insects such as the fruit fly, females again had two X chromosomes and males only one, but there was also a small male-specific entity (called a Y). In moths and butterflies, it is the other way around – males have two copies (as per normal) of a sex chromosome (called the Z to avoid confusion), and females have a single Z and a smaller W chromosome. It is the same story in vertebrates; mammals (including humans) have an XX female:XY male system, whereas birds and snakes have a ZW female: ZZ male system.

Sex works by the distribution of the heteromorphic sex chromosomes during spermatogenesis. For instance, in humans and fruit flies, the X and Y chromosomes of an XY male separate into different sperm at meiosis. All eggs carry a single X. An egg fertilized by an X-bearing sperm develops into a female, and an egg fertilized by a Y-bearing sperm develops into a male. In these XX female: XY male species, we call the male the heterogametic sex because he can make two kinds of gametes. In species such as birds and butterflies, the female is the heterogametic sex. She makes two kinds of eggs, Z- and W-bearing, which become female and male when fertilized by Z-bearing sperm.

Sex-Determining Genes

What is it about the X and Y chromosomes in fruit flies and man that determine maleness and femaleness? Appearances are deceptive – it turns out that the two species have completely different ways of doing it. In fruit flies, the Y chromosome is quite irrelevant to sex determination, although it carries genes required for making sperm. It is the different dosages of the X chromosome in females (two) and males (one) that determine the sex of the embryo. We know this because flies that are XO (have a single X but no Y) are male like XY, and flies that are XXY (have a Y as well as two Xs) are female like XX. Several genes were identified in fruit flies because mutants develop into the wrong sex. One of these genes is the key switch that is flipped one way if there is a single X, and the other way if there are two. The ratio of important genes on the X chromosome to genes on other chromosomes determines how the RNA product of this gene is spliced to form alternative products that activate male-specific and female-specific sets of genes.

In mammals (man is typical), the Y chromosome is paramount. XO individuals are females with Turner syndrome, and XXY are male with Klinefelter syndrome. There was a frenzied search of the human Y in the 1980s to identify the gene that triggers testis differentiation, the first step in a hormone-controlled pathway to all the other sex differences. Studies of patients having only parts of a Y chromosome directed the search to a small region of the Y near one end. Candidate genes were isolated from this region and tested by their patterns of expression, and their location and expression in closely and distantly related mammals. The *SRY* gene on Y of humans and other mammals, even kangaroos, was mutated in some XY females, and directed male development when injected into XX mouse eggs. This gene works by regulating other genes in a testis-determining pathway, but it is not yet known exactly which genes, or whether *SRY* turns them on or off. The other genes in the pathway are not on the sex chromosomes in mammals. However, one of them, *DMRT*, turns out to be the sex-determining gene on the Z chromosome in birds that probably works by dosage differences in the male and female.

Sex Chromosome Evolution

Compared to the other chromosomes, the mammalian Y is a genetic wasteland, being small and almost entirely devoid of active genes. Other than *SRY*, the human Y contains only about 20 genes, several concerned with spermatogenesis. The Y is largely genetic junk – dead (pseudo)genes and highly repeated sequences that do not specify proteins. However, we know that the Y was once equivalent to the X. Over the last 200 million years, it lost most of its 2000-odd genes when it became isolated from genetic recombination with the X. Degradation is still continuing, so that in time the Y may disappear entirely and new sex chromosomes may be initiated, as seems to have happened in some unusual rodent species.

See also: **Sex Determination, Human; W Chromosome; X Chromosome; X-Chromosome Inactivation; Z Chromosome**

Sex Determination, Human

A Sinclair

doi: 10.1006/rwgn.2001.1179

Since antiquity people have proposed various mechanisms to account for an individuals sex. Aristotle believed that vigorous intercourse would result in a boy, whereas a gentler approach would yield a girl. Since that time we have developed a more sophisticated understanding of the molecular genetic mechanisms that govern human sex determination. In mammals, sex determination involves the commitment of the embryo to follow either a male or female developmental pathway. The key step in this process is the development of the undifferentiated embryonic gonads into either testes or ovaries. In humans and other mammals, sex is determined genetically at fertilization by the sex chromosome constitution. Two X chromosomes result in female development while the inheritance of an X and a Y chromosome results in male development. It was postulated that the Y chromosome carried a dominant testis-determining factor (TDF, so called because at that time it was an unknown 'factor') which causes the undifferentiated embryonic gonad to develop as a testis. The masculinizing effect of the testis is due to the secretion of the hormones testosterone and anti-Müllerian hormone (AMH; also known as Müllerian inhibitory substance, MIS). AMH causes regression of the embryonic female Müllerian ducts. In the absence of the Y chromosome (and absence of TDF), ovaries will develop. Interestingly, female development will still occur in the absence of ovaries or their hormonal products. Consequently, the decisive event in sex determination is whether or not a testis develops. In mammals, sex determination can be equated with testis determination. The postulated Y-linked testis-determining factor was thought to orchestrate a hierarchy of genes in a pathway leading to testis development. Isolation of the master switch gene *TDF* would allow the stepwise unraveling of the molecular genetic pathway of human sex determination.

SRY: the Master-Switch Testis-Determining Gene

In 1990, Sinclair *et al.* isolated and characterized the *SRY* gene (*S*ex-determining *R*egion on the Y chromosome) from the human Y chromosome and showed it to be the elusive master-switch testis-determining factor (TDF). The *SRY* gene was isolated using DNA from sex-reversed patients who had two X chromosomes but had formed testes and were male. Approximately 80% of these XX males had a small portion of the Y chromosome including *SRY* translocated onto one of their X chromosomes. Consequently, *SRY* was derived from the smallest region on the short arm of the Y chromosome, known to be sex-determining. Another group of sex-reversed patients had XY chromosomes but no testes and were female. In 20% of these XY females there was a loss-of-function mutation in the *SRY* gene. These mutations in XY females confirmed that *SRY* was required for normal testis formation and male sex determination. The other 80% of XY females are thought to have mutations in other genes in the testis pathway. Finally, the *SRY* gene in mouse (*Sry*) was isolated and used to make sex-reversed transgenic mice. These mice carried a 14 kb DNA fragment containing only the *Sry* gene and developed as males with (sterile) testes even though they had two X chromosomes (Koopman *et al.,* 1991). This was final proof that *SRY* is the only Y-linked gene necessary and sufficient to initiate testis development. Consequently, the *SRY* gene is the long-sought after testis-determining factor (TDF).

Current evidence suggests *SRY* is expressed in the pre-Sertoli cells and acts to induce Sertoli cell differentiation. Pre-Sertoli cells promote the formation of testicular cords and Leydig cell formation. Leydig cells in turn produce the key male hormone testosterone and Sertoli cells produce AMH. *SRY* is a single exon gene that encodes a 79 amino acid motif, the HMG box, which is capable of sequence-specific binding and bending of DNA. The SRY protein is thought to act as an architectural transcription factor influencing the expression of other genes by inducing conformational change in the surrounding chromatin. The HMG domain also contains signal motifs for transporting the SRY protein into the nucleus. Outside the HMG box SRY shows little conservation between mammalian species, suggesting that the HMG box is the major component. However, in mice there appear to be other regions outside the HMG box necessary for *Sry* function. *Sry* appears as a brief burst of expression just prior to morphological differentiation of the embryonic testis. This suggests that *SRY* acts as a switch toward Sertoli cell fate but it is not required for the maintenance or function of Sertoli cells. *SRY* must act as switch, activating other genes that are involved with maintaining functioning Sertoli cells. Despite extensive studies we still do not know the genuine *in vivo* target(s) of *SRY*, nor do we know how *SRY* itself is regulated.

SOX9: An Autosomal Testis-Determining Gene

Several autosomal genes are known to be associated with sex-reversal syndromes and presumably play a role downstream of *SRY* in the testis developmental pathway. Unlike the male Y-specific *SRY* gene, these autosomal genes will be present in both males and females but are likely to show differential sex-specific expression. Deletions of loci from chromosomes 9p (short arm), 10q (long arm), and 17q (long arm) can result in XY females with dysgenic ovaries. In the latter case, translocations and deletions of 17q are also associated with campomelic dyplasia (CD), a rare fatal congenital skeletal malformation syndrome, characterized by bowing of the long bones. Three-quarters of the XY individuals with CD develop as phenotypic females or intersexes. Analysis of the 17q translocation breakpoints in CD patients revealed an *SRY*-related HMG-box gene, *SOX9*. CD patients and sex-reversed XY females with CD were both found to have loss-of-function mutations in one allele of *SOX9*. This indicates that haploinsufficiency of *SOX9* is responsible for both the CD skeletal dysplasia and the XY sex reversal. This implies that SOX9 is a key component in the testis-determining pathway.

Duplication of the chromosome 17q region, including *SOX9*, can also result in sex-reversed XX male patients suggesting a gain-of-function mutation in *SOX9* may be responsible. *SOX9* is a classic transcription factor possessing both a DNA-binding HMG box and a transactivation domain. *SOX9* is expressed in a variety of human fetal tissues including brain, testis, and chondrocytes of the hypotrophic zones of developing long bones and ribs. In the mouse, *Sox9* is upregulated specifically in the pre-Sertoli cells suggesting its role is to induce Sertoli cell differentiation. *Sox9* is expressed in the genital ridge, just after the onset of *Sry* expression. This led to speculation that *SRY* may be regulating *SOX9* expression. However, as the human *SOX9* regulatory region is spread over a vast 1Mb region of DNA, it has been difficult to examine interactions between *SRY* and *SOX9*. However, it is clear that *SOX9* is a key downstream component of the sex-determining pathway, inducing Sertoli cell differentiation in the developing testis.

DMRT1: A Conserved Testis-Determining Gene

Sex-determining mechanisms appear to be very different between vertebrates and invertebrates. So it was surprising to find a family of genes related by a DM domain (DNA-binding region) that play a sex-specific role across the different phyla. The *Drosophila dsx* (doublesex) gene and the *Caenorhabditis elegans mab*-3 gene share the DM domain and both genes are involved in the differentiation of sex-specific structures such as the peripheral nervous system and yolk protein development. A search for DM-related genes in humans revealed the *DMRT1* (DM-related transcription factor 1) gene. *DMRT1* mapped to the distal short arm of human chromosome 9, deletions of which are associated with XY female sex-reversal. *DMRT1* is upregulated specifically in the male genital ridge in humans and mice. Birds and reptiles such as the chicken and alligator, respectively, lack the *SRY* gene but do show male specific upregulation of *SOX9* and *DMRT1* in the developing gonads. This suggests an important role for *DMRT1* in vertebrate sex determination. However, there is a complication with *DMRT1* because knockout mice do not develop as sex-reversed XY females. This suggests that *DMRT1* is not a testis-determining gene but it may be that its function is compensated by other genes residing near *DMRT1* on the distal short arm of human chromosome 9. The large deletions of chromosome 9 seen in sex-reversed XY female patients would presumably remove a number of genes from this region. The jury is still out on the role of this new candidate sex-determining gene.

DAX1: Suppression of Testis Development

Some sex-reversal syndromes are known to involve X chromosome rearrangements and result in failure of the testis to develop. Duplications of the Xp21 region have been shown to cause XY female development and this led to the proposal that a dosage-sensitive sex-reversing (DSS) gene must exist. Two active copies of the postulated DSS gene are believed to override the testis-determining signal, resulting in the development of ovaries and XY female sex-reversal. This same region on the X chromosome (Xp21) is also involved in adrenal hypoplasia congenita (AHC), which results in the failure of the adrenals to form properly. A search of the Xp21 region revealed a gene called *DAX1* (DSS-AHC-critical region of the X chromosome, gene l). *DAX1* is likely to be subject to X chromosome inactivation, as Klinefelter syndrome (XXY) individuals are male, not sex-reversed females. Consequently, normal males and females would each be expected to have one functional copy of *DAX1*. The *DAX1* gene encodes an orphan nuclear receptor with a ligand-binding domain but lacking the usual zinc finger DNA-binding motif. Instead the N-terminal domain of *DAX1* has an unusual repeat motif rich in alanine and glycine. Mutations in the *DAX1* gene result in patients with adrenal hypoplasia congenita (AHC). However, it is not clear if *DAX1* also played a role in dosage-sensitive sex reversal. Deletions of *DAX1* in XY patients do not disrupt testis differentiation. However, transgenic mice that over-express *Dax1* can be shown to undergo XY female sex reversal. *Dax1* is expressed at the same time as *Sry*; however, *Dax1* is downregulated in the developing testis but maintains expression levels in the ovary. This suggests that either extra *DAX1* copies in an XY individual can suppress testis development or that *DAX1* may have a role in ovary development. However, knockout *Dax1* XX female mice still developed ovaries and were fertile, suggesting that *DAX1* is not an ovarian-determining gene but acts antagonistically to *SRY* as an anti-testis gene. Lack of *Dax1* in XY male mice resulted in sterility, confirming another, if unexpected, role for *DAX1* in spermatogenesis.

Genes Required for Early Indifferent Gonad Formation

Steroidogenic factor 1 (*SF1*) is an orphan nuclear receptor and a key regulator of steroidogenic enzymes. *SF1* is expressed in all primary steroidogenic tissues, including the adrenal cortex, Leydig cells of the testis, and ovarian follicles. *SF1* encodes a zinc finger DNA-binding domain and a ligand-binding domain. In the mouse embryo, *Sf1* is expressed at the earliest stage of urogenital development in the undifferentiated gonad. *Sf1* shows continued expression throughout testis development but is downregulated during ovarian development. In the testis *SF1* expression is localized to the Sertoli and Leydig cells. Together these data suggest a role for *SF1* in the formation of the urogenital ridge and subsequent sexual differentiation. SF1 is also expressed in the developing hypothalamus and in pituitary gonadotropes suggesting it has a wider role by operating at different levels of the reproductive axis. Null mutant *Sf1* knockout mice (XX and XY) lack both adrenal glands and gonads. These mice develop as phenotypic females but die shortly after birth from adrenal failure. Detailed analysis of these mice showed that in the absence of *Sf1* the genital ridge displayed arrested development and then atrophied. *SF1* clearly acts at several points in the sex-determining hierarchy. One of its known targets is AMH, where it plays a key regulatory function (see 'Regulation of AMH').

The Wilms' tumor 1 (*WT1*) gene is an oncogene associated with a pediatric cancer of the kidney (Wilms' tumor). *WT1* encodes a zinc finger protein and is thought to act as a transcription factor. Heterozygous mutations in the zinc finger domain of *WT1* result in patients with Denys–Drash syndrome, characterized by renal failure and genital abnormalities, including XY female sex reversal. Frasier syndrome, also associated with XY female sex reversal, was shown to be due to a mutation in a splice donor site in *WT1*, causing loss of a specific isoform of *WT1*. The mouse gene, *Wt1* is expressed in the undifferentiated gonads of both males and females at the same time as *Sf1*. Targeted disruption of *Wt1* in mice resulted in the failure of the kidney, ovary, and testis development. This suggests that *Wt1* acts early in urogenital development (possibly in conjunction with *Sf1*) to ensure proper formation of the indifferent embryonic gonad. The mutations in *WT1* that result in XY females in both Denys–Drash and Frasier syndome patients indicate that *WT1* plays an additional role in testis development. The human *WT1* gene is comprised of 10 exons and produces a range of different transcripts using alternative splice sites and translation start sites. It is thought that the different isoforms of *WT1* carry out a variety of different functions. A specific isoform of WT1 (the same isoform abolished by the *WT1* mutations in Frasier syndrome) is thought to interact with SF1 to upregulate AMH expression in the testis (see 'Regulation of AMH' below).

The homeopaired box gene, *Lim1*, appears to play a role in kidney and gonad development as targeted disruption of *Lim1* results in mice lacking these organs. *Lim1* is expressed in early urogenital ridge development in both the mesonephros and metanephros but its role in gonad development is not clear. The mouse gene *M33* has also been implicated in gonad development. *M33* shows similarity to the *Drosophila* polycomb group of genes (*PcG*). Mice carrying a disrupted *M33* gene show retarded gonad development associated with varying degrees of XY sex reversal. In *Drosophila*, PcG proteins regulate the coordinate expression of homeotic genes. So it is possible that *M33* may similarly regulate *Hox* gene expression in the mammalian urogenital ridge. Although *SF1*, *WT1*, *Lim1*, and *M33* all appear to be important in the formation of the indifferent gonad, their precise role is unknown.

Regulation of AMH

Anti-Müllerian hormone (AMH) is synthesized in the Sertoli cells and is one of the first proteins produced by the developing testis. AMH causes regression of the Müllerian ducts within the male embryo, which would otherwise develop as oviducts, uterus, and upper vagina. Within the small regulatory region of the *AMH* gene, binding sites were discovered for *SOX9* and *SF1* (steroidogenic factor 1). SOX9 and SF1 proteins bind to adjacent sites in the regulatory region of *AMH* and cause significant upregulation of AMH expression. In mice, mutation of *Sox9* or *Sf1* binding sites in the *Amh* regulatory region causes abolition or diminution, respectively, of *Amh* expression. SOX9 appears to be essential for initiating AMH expression. SF1 physically interacts with the adjacent SOX9 protein to significantly upregulate AMH transcript levels. A specific isoform of WT1 also appears to physically interact with SF1 to synergistically upregulate AMH expression. As a counter to this, DAX1 has been shown by *in vitro* experiments to repress the synergistic action of SF1 and WT1 on the *AMH* promoter. This suggests that when there is an abnormal double dose of DAX1 and it is present in high levels in the testis it can block the normal upregulation of AMH expression. The normal role of *DAX1* in the ovary may be to prevent the expression of AMH. One member of the *GATA* family of transcription factors, *GATA4*, is expressed in the developing gonads. A *GATA4* binding site is also present in the AMH regulatory region. It is thought that the GATA4 protein may bind to this site and interact with SOX9 to regulate AMH expression. The role of WT1 and GATA4 on AMH regulation has not been confirmed *in vivo*. However, we can clearly state that *SOX9* and *SF1* together upregulate *AMH* expression. While AMH is required for sex-specific differentiation of the reproductive tract, lack of *AMH* does not affect testis development.

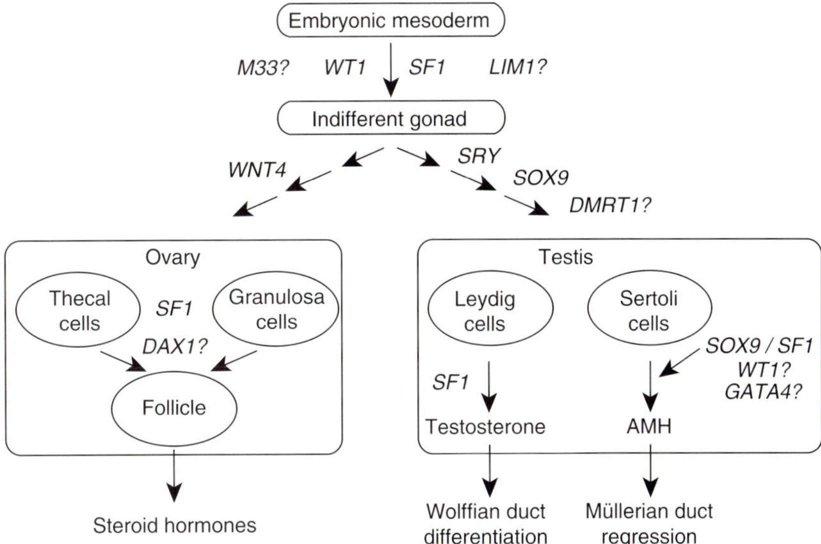

Figure 1 Genetic control of human sex determination. (Adapted from O'Neill M and Sinclair A (1996) The testis-determining gene, SRY. *Advances in Genome Biology* 4: 29–51.)

WNT4: Ovarian Development and Testis Suppression

Wnt4, a member of the *Wnt* gene family of signaling molecules, appears to be required for ovarian development and acts to suppress testis formation. The *Wnt4* gene is initially expressed in the genital ridge and mesonephros of both sexes but as sex-specific gonadal differentiation proceeds it is downregulated in the testis but maintained in the ovary. Male (XY) mice lacking *Wnt4* develop normal testes and Wolffian duct derivatives. By contrast, females (XX) lacking *Wnt4* are masculinized as the Müllerian duct is absent and the Wolffian duct is similar to that of a male. Furthermore, the ovaries of *Wnt4* mutant female mice express genes coding for enzymes normally associated with testosterone production. Finally, the ovaries of these mice display a marked decrease in oocyte development. The implication is that steroid cell precursors of Leydig and theca cells must be present in the indifferent gonad of both males and females. As the testis develops more rapidly than the ovary, the Leydig cells can begin producing testosterone as soon as testis cords form in the embryo. By contrast, in the ovary the theca cells are not steroidogenically active until birth. This data suggests that in normal XX females high levels of *Wnt4* expression act in the indifferent genital ridge to represses testosterone production in the precursors of Leydig cells, allowing theca cells to eventually develop. Presumably in normal (XY) males the downregulation of *Wnt4* expression allows testosterone biosynthesis from Leydig cells to proceed. Consequently, *Wnt4* has three distinct functions in the female pathway: formation of the Müllerian duct, suppression of Leydig cell development in the indifferent genital ridge, and postmeiotic maintenance of oocytes in the ovary.

Future Prospects

Over the past decade there has been an enormous increase in our understanding of the molecular basis of mammalian sex determination. The isolation and characterization of several key genes has begun to elucidate the complexity inherent in the development of a gonad. Unfortunately, we have only a few pieces of this time-dependent three-dimensional jigsaw puzzle. As more of the pieces are found and their interactions in time and space become clear, we will gain a unique insight into the process underlying organogenesis.

Further Reading

Arango NA, Lovell-Badge R and Behringer RR (1999) Targeted mutagenesis of the endogenous mouse *Mis* gene promoter: *in vivo* definition of genetic pathways of vertebrate sexual development. *Cell* 99: 409–419.

Foster JW, Dominguez-Steglich MA, Guioli S *et al.* (1994) Campomelic dysplasia and autosomal sex reversal caused by mutations in an *SRY*-related gene. *Nature* 372: 525–530.

Koopman P, Gubbay J, Vivian N, Goodfellow P and Lovell-Badge R (1991) Male development of chromosomally female mice transgenic for *Sry*. *Nature* 351: 117–121.

Raymond CS, Shamu CE, Shen MM *et al.* (1998) Evidence for evolutionary conservation of sex-determining genes. *Nature* 391: 691–695.

Roberts LM, Shen J and Ingraham HA (1999) New solutions to an ancient riddle: defining the differences between Adam and Eve. *American Journal of Human Genetics* 65: 933–942.

Schafer AJ (1995) Sex determination and its pathology in man. In: Hall JC and Dunlap JC (eds) *Advances in Genetics*, vol. 33, pp. 275–329. San Diego, CA: Academic Press.

Sinclair AH, Berta P, Palmer MS *et al.* (1990) A gene from the human sex determining region encodes a protein with homology to a conserved DNA-binding motif. *Nature* 346: 240–244.

Swain A and Lovell-Badge R (1999) Mammalian sex determination: a molecular drama. *Genes and Development* 13: 755–767.

Swain A, Narvaez V, Burgoyne P, Camerino G and Lovell Badge R (1998) *Daxl* antagonizes *Sry* action in mammalian sex determination. *Nature* 391: 761–767.

***See also:* Sex Determination, Mouse; Sex Reversal; Y Chromosome (Human)**

Sex Determination, Mouse

K H Albrecht and E M Eicher

doi: 10.1006/rwgn.2001.1178

Introduction and Developmental Biology

In mammals, XY fetuses develop testes due to the action of the Y-linked testis determining gene (*Sry*, also known as *Tdy*) and XX fetuses develop ovaries in its absence. As a result of early experiments, primarily by Jost, it is known that sex determination (primary) can be reduced to the genetic choice between the development of an ovary or a testis; subsequent sex differentiation (secondary sex determination) is dependent upon gonadal sex determination. Sex differentiation is regulated by the production of two hormones exported from the developing testis, Müllerian inhibiting substance (MIS, or AMH for anti-Müllerian hormone) and testosterone, which both affect the further development of the reproductive tract. In males, MIS causes the Müllerian ducts to regress and testosterone stimulates the Wolffian ducts to differentiate into the epididymides, vasa deferentia, and seminal vesicles. In females, the absence of MIS allows the Müllerian ducts to differentiate into the oviduct, uterus, and upper part of the vagina, while the absence of testosterone causes the Wolffian ducts to degenerate.

This review will be restricted to a genetic description of the process of sex determination in mice, and will mostly describe testis determination because currently little is known about the genetics of ovary determination.

The genital ridge, the anlagen of both the testis and the ovary, develops on the ventromedial surface of the mesonephros (primitive kidney) and is first visible at approximately 10 days post coitum (dpc) in the mouse. The mesonephros and the genital ridge are derived from the intermediate mesoderm. Mammalian adult gonadal cells can be grouped into four types: germ cells (sperm in the testis, eggs in the ovary), steroid-producing cells (Leydig cells in the testis, theca cells in the ovary), supporting cells (Sertoli cells in the testis, follicle/granulosa cells in the ovary), and connective tissue cells (essentially similar in both organs, except peritubular myoid cells are unique to the testis). Analysis of XX $\leftrightarrow$ XY aggregation chimeras demonstrated that Sertoli cells are almost exclusively XY whereas the other testis cell types can be XX or XY. This result indicates that the cell-autonomous action of *Sry* is needed only in Sertoli cells, and suggests the following working model for testis determination. *Sry* expression in pre-Sertoli cells causes them to differentiate and become organized into testis cords. The formation of testis cords then triggers the remaining cell types to differentiate along the testis development pathway.

Genes

Sry (Sex-determining region, Y Chromosome)

In the early 1950s, it was discovered that the presence or absence of a Y chromosome determined whether a mammal developed as a male or female, respectively. It was hypothesized that a gene on the Y chromosome determined male sex and the human gene was designated *TDF* (testis determining factor) while the mouse gene was designated *Tdy* (testis determining-Y). In 1990, the search for *TDF*/*Tdy* identified the human *SRY* gene in a small Y chromosome region that caused sex reversal when present in XX individuals. (By convention, *Sry* refers to the mouse gene, *SRY* refers to the gene in all other species, and SRY refers to the protein in all species.) Definitive proof equating *Tdy* and *Sry* came from mutation analysis in humans, and gain- and loss-of-function studies in mice. Introduction of a 14.6 kb transgene containing only *Sry* causes XX mice to develop as sterile males. (Sterility is caused by the presence of two X chromosomes and the absence of a Y chromosome.) Conversely, XY^{Tdym1} mice, which have an 11 kb deletion at the *Sry* locus, develop as semifertile females.

SRY contains a 79 amino acid DNA binding motif that was first identified in high mobility group proteins (the HMG domain). SRY binds to the A/TAACAAT/A consensus sequence in linear DNA and induces an $\sim 80^\circ$ bend. These features strongly suggest that SRY is a transcription factor, but it is unknown if *Sry* is a transcription activator or repressor (or both) and no definitive target genes have been identified. SRY also binds in a sequence independent manner to cruciform DNA, the significance of which is obscure. *Sry* can be considered a strange gene for many reasons. It is rapidly evolving and the HMG box is the only obviously conserved portion of the gene (true for *SRY* in general). It is monoexonic and imbedded within large inverted repeats. It generates two very different RNA transcripts, an embryonic linear polyadenylated form and an adult germ-cell circular nonpolyadenylated form. The function of the circular form is unknown; it probably is not translated, it is not present in other species, and it may be an artifact of the inverted repeat structure of the mouse *Sry* locus. In fetal mice, *Sry* reportedly is expressed exclusively in the somatic cells of the developing XY gonad between 10.5 and 12.5 dpc immediately prior to the first visible sign of sex determination, which is the formation of testis cords at 12.5 dpc. However, *SRY* is transcribed in many different tissues in humans and marsupials and throughout testis development in marsupials.

Sox 9 (*Sry*-box containing gene 9)

After the cloning of *Sry*, the Sox gene family was identified by their high sequence homology to the *Sry*–HMG box. *SOX9* is disrupted in patients with campomelic dysplasia (CD), a severe dwarfism syndrome that often is associated with XY sex reversal. Interestingly, *SOX9* is probably haploinsufficient because all CD patients are heterozygous for the mutation and many of the identified mutations are predicted to cause loss of function. In contrast to *Sry*, *Sox9* has two introns and is highly conserved between species both within and outside of the HMG box. In addition to the HMG domain, SOX9 has two domains that are rich in proline, glutamine, or serine, and are necessary for its function as a transcription transactivator. SOX9 binds to and activates transcription from the *Col2a1* (procollagen, type II, alpha I) gene; however, no definitive target genes have been identified in the developing gonad.

Sox9 expression in the developing gonads is sexually dimorphic. Prior to 11.5 dpc, it is expressed in both XX and XY genital ridges at fairly equal levels; but thereafter, expression falls precipitously in XX gonads, and increases in XY gonads. SOX9 expression mirrors this pattern and is largely undetectable in XX gonads after 11.5 dpc. Curiously, SOX9 is localized perinuclearly prior to 11.5 dpc in both sexes, but then becomes localized to the nucleus of Sertoli cells. Subsequently, *Sox9* expression is maintained in fetal and adult testes. In adult humans and mice, *SOX9*/*Sox9* expression is germ-cell independent. In the urogenital system, *Sox9* also is expressed in the Müllerian and Wolffian ducts, the cranial portion of the mesonephros, and in the epididymis. Interestingly, fetal ovaries that transdifferentiate into testes after transplantation under adult kidney capsules activate *Sox9* expression. Gene ablation studies in mouse are currently in progress.

Sf1 (Steroidogenic factor 1, also Known as Adrenal 4-Binding Protein, Ad4BP)

SF1 is encoded by the *Ftzf1* locus (fushi tarazu factor 1 homolog), and was initially isolated as an important regulator of the cytochrome P-450 steroid hydroxylases. It is important to note that the *Ftzf1* locus is genetically complex and produces four related transcripts by alternative promoter usage and splicing: embryonal long terminal repeat binding protein 1 (*ELP1*), *ELP2*, *ELP3*, and *Sf1/Ad4BP*. SF1 is an orphan nuclear receptor and encodes a protein with two zinc fingers, an A or FTZ-F1 box that mediates additional DNA binding, a proline rich domain, and an AF-2 domain both of which may be transactivation domains. SF1 binds DNA as a monomer with specificity for the YCAAGGTCA motif and has been shown to activate the expression of a number of genes, including *Mis*.

Sf1 is expressed in the urogenital ridges of both sexes from 9 dpc, and continues in the genital ridges of both sexes until about 12.5 dpc. After 12.5 dpc expression is extinguished in XX gonads and is undetectable by *in situ* hybridization at 14.5 dpc. Expression in XY gonads continues at high levels in Sertoli cells and in interstitial Leydig cells until about 14.5 dpc when expression becomes restricted to Leydig cells and is extinguished by 17.5 dpc. In adults, SF1 is expressed in testicular Leydig cells, and ovarian theca and granulosa cells. Mice homozygous for a targeted disruption of *Ftzf1* do not develop gonads, adrenal glands, and the ventromedial hypothalamic nucleus, and exhibit impaired gonadotrope function. Fetuses homozygous for the null allele display some mesenchymal cell thickening on the surface of the urogenital ridge where the genital ridge usually develops, but this region does not develop further and begins to degenerate, probably by apoptosis.

Dax 1 (Dosage-Sensitive Sex-Reversal-AHC Critical Region on the X Chromosome, also Known as Adrenal Hypoplasia Congenita Homolog, *ahch*)

DAX1 was cloned from a 160 kb region on human Xp21 that is associated with adrenal hypoplasia congenita and dosage-sensitive sex reversal. Duplication of the chromosomal region including *DAX1* causes sex reversal in XY humans. Like *Sf1*, *Dax1* encodes an orphan nuclear hormone receptor. Unlike *Sf1*, *Dax1* lacks any zinc finger DNA binding motifs and is rapidly evolving in certain regions of the gene. DNA binding activity has been assumed by a region with three and a half repeats of a 67–68 amino acid motif. DAX1 inhibits the transcriptional activity of a number of genes, including *Sf1*. Conversely, in certain cells SF1 activates *Dax1* transcription. However, the significance of these finding to gonadogenesis is unclear because *Dax1* expression persists in the genital ridge of *Ftzf1* null mice.

Dax1 is first expressed in somatic cells of the XX and XY genital ridge at 11.5 dpc. After 12 dpc, *Dax1* expression becomes sexual dimorphic: rapidly decreasing in the XY gonad, but persisting in the XX gonad until 14.5 dpc. However, conflicting data suggests that expression does not become sexually dimorphic. In the adult, *Dax1* is expressed in testicular Sertoli and Leydig cells as well as ovarian stromal cells.

The role of *Dax1* in sex determination is enigmatic. Overexpression seems to inhibit proper testis development. On the other hand, loss of function in knockout mice has no effect on gonadogenesis or sex determination. Adult males homozygous for the null allele are sterile due to loss of stratification of the testicular germinal epithelium, and therefore do not complete spermatogenesis. Adult females are fertile, but display minor abnormalities in oogenesis: some follicles contain multiple oocytes. It is possible that overexpression of DAX1 inhibits the normal functioning of SF1 in testis development and thereby causes the sex-reversal phenotype.

Wt1 (Wilms's Tumor 1)

WT1 mutations in humans are associated with Wilms's tumor (kidney) and urogenital malformation, particularly Denys–Drash and Fraser syndromes. WT1 is a zinc-finger-containing, DNA binding transcription factor that probably also plays a role in mRNA splicing. The protein contains both activation and repression domains, as well as a self-association domain. *Wt1* is expressed in the intermediate mesoderm and coelomic epithelium of 9 dpc mouse embryos, in the mesonephros from 9.5 to 12.5 dpc (excluding the Müllerian and Wolffian ducts), and in the genital ridges and later the developing gonads from 9.5 dpc until at least birth. *Wt1* expression is not sexually dimorphic in the developing gonads, is limited to the sex cords and is excluded from interstitial and germ cells. *Wt1* is expressed in Sertoli cells in adult males.

Fetal mice homozygous for a *Wt1* targeted null allele display a very reduced thickening of the coelomic epithelium at 11 dpc and the gonadal ridge does not develop much further. Clearly, *Wt1* is important for the development of the gonadal ridge from the urogenital ridge, but a role in later gonadogenesis also is possible. In fact, WT1 was recently reported to functionally oppose DAX1 in testis development and modulate SF1-mediated transactivation.

Mis (Müllerian Inhibiting Substance, also Known as *Amh* for Anti-Müllerian Hormone)

Mis is a member of the transforming growth factor-β family that is expressed in developing Sertoli cells from 11.5 dpc until puberty and in ovarian granulosa cells from birth. Clearly, *Mis* is not involved in sex determination because mice homozygous for a targeted null allele of either *Mis*, or its receptor (Amh type 2 receptor, *Amhr2*), develop normal gonads (except for some Leydig cell hyperplasia in adult males) and gonadal phenotypic sex and genetic sex are concordant. However, chronic overexpression partially represses gonadal development and function in both sexes.

Other Genes

Additionally, a number of genes have been implicated as participating in sex determination or gonadogenesis. However, their roles are less well established, and thus are briefly described below.

Lhx1 (LIM homeobox protein 1, also known as Lim1)

Lhx1 is expressed in the developing urogenital ridges (probably the mesonephric portion) from 8.5 dpc, and later becomes restricted to the mesonephric tubules and mesonephric (Wolffian) duct. It is not expressed in the genital ridges or gonads. Most embryos homozygous for a targeted null allele die at ~10 dpc; however, a few surviving neonatal pups lacked kidneys and gonads.

Cbx2 (chromobox homolog 2, also known as M33)

Cbx2, the mouse homolog of the *Polycomb* gene in *Drosophila*, is ubiquitously expressed in 12.5 dpc fetuses. The development of the genital ridge is very retarded in fetuses homozygous for a targeted probable-null allele. Subsequent gonadogenesis is similarly affected particularly testis formation, leading to XY sex reversal. The XY animals range from XY females with two ovaries to hermaphrodites with one ovary and one undescended testis. XX homozygous null females are sterile and occasionally lack one ovary.

Gata4 (GATA-binding protein 4)

GATA4 activates transcription from the *Mis* promoter *in vivo*, and is expressed in gonadal somatic cells in a sexually dimorphic pattern. The GATA4 protein is present in both XX and XY genital ridges at 11.5 dpc, and continues to be expressed in the testis through puberty. However, its expression is extinguished in the developing ovary beginning at 13.5 dpc. Unfortunately, fetuses homozygous for a targeted null *Gata4* allele die by 9.5 dpc thereby precluding an analysis of sex determination in these mice.

See also: **Sex Chromosomes; Sex Determination, Human**

Sex Linkage

J R S Fincham

doi: 10.1006/rwgn.2001.1183

In mammals, many insects and a few flowering plants, separate male and female sexes are determined by a pair of only partly homologous sex chromosomes, called X and Y. The female carries two X chromosomes in each cell nucleus and the male an X and a Y. The male, with the pair of unlike sex chromosomes, is called the heterogametic sex, because it produces two kinds of sperm, X-bearing and Y-bearing in equal numbers. On the female side, all the eggs carry the X chromosome; progeny are female or male depending on whether the egg is fertilized by an X- or a Y-bearing sperm cell.

In birds and reptiles the situation is reversed in that it is the female that is heterogametic. The sex chromosomes are called W and Z; the females are WZ and the males are ZZ.

X and Z chromosomes typically carry as many genes in proportion to their size as the ordinary chromosomes (autosomes), but the Y or W, present in only one sex, is typically nearly inert genetically (though necessary for Y sperm motility). Indeed, in some groups of insects, notably the Orthoptera (grasshoppers and locusts), with XO sex-determining systems, there is no Y chromosome at all, and the sperm cells either have an X chromosome or no sex chromosome. The same applies to the nematode worm *Caenorhabditis elegans*, except that here XO and XX individuals are respectively males and hermaphrodites, not males and females.

The twofold difference between males and females in the dosage of the gene-rich X or Z chromosome, when the great majority of genes are required equally by both sexes, requires some system of dosage compensation. This has been well studied both in mammals and in *Drosophila* (see Dosage). In mammals one or other of the two X chromosomes is largely inactivated in every female somatic cell, resulting in a mosaic phenotype when the two X chromosomes carry distinguishable alleles – as in the case of the tortoiseshell cat.

The key feature of the inheritance of sex-linked genes is that the heterogametic sex inherits them only from the homogametic parent. Consequently the X-chromosome constitutions of human, mouse, or *Drosophila* males are a direct reflection of the segregation and recombination of the X-linked genes of their mothers, without any concealment by dominance from the father. X-linked mutations that are recessive in the female will always show their phenotypic effects in the male because there is no second chromosome to supply a normal dominant allele (**Table 1**). Those few genes (if there are any) present on both X and Y chromosomes show a modified form of sex-linkage called *pseudoautosomal* (see Pseudoautosomal Linkage, Region).

Table 1 Mode of inheritance of an X-linked recessive mutant allele

Cross				*Progeny*		
Female		***Male***		***Female***		***Male***
X^+/X^+ Normal	×	X^m/Y Mutant		X^+/X^m Carrier		X^+/Y Normal
X^m/X^m Mutant	×	X^+/Y Normal		X^m/X^+ Carrier		X^m/Y Mutant
X^+/X^m Carrier	×	X^+/Y Normal	50%	X^+/X^+ Normal	50%	X^+/Y Normal
			50%	X^m/X^+ Carrier	50%	X^m/Y Mutant
X^+/X^m Carrier	×	X^m/Y Mutant	50%	X^+/X^m Carrier	50%	X^+/Y Normal
			50%	X^m/X^m Mutant	50%	X^m/Y Mutant

This Table will apply equally to WZ (bird and reptilian) sex chromosome systems if the X is replaced by Z and Y by W, and the male and female headings are reversed.

Human sex-linked conditions, such as hemophilia, are characteristically transmitted to sons by symptomless mothers. Such conditions are seen in males but hardly ever in females. If a recessive X-linked allele has an overall frequency of p in the whole population of genes, the phenotype will occur at frequency p in males (provided that they are fully viable) and p^2 in females, with heterozygous ('carrier') females at frequency $2p\ (1 - p)$. Deleterious X-linked recessive mutations are expected to be less common in the gene pool than autosomal recessives of comparably severe phenotype because they are always exposed to adverse selection in the male sex. This prediction is borne out by medical statistics.

See also: **Dosage Compensation; Pseudoautosomal Linkage, Region; Sex Linkage; X-Chromosome Inactivation**

Sex Plasmid

doi: 10.1006/rwgn.2001.2018

A sex plasmid (S plasmid) is an episome that is able to initiate the process of conjugation, resulting in the transfer of chromosomal material from one bacterium to another.

***See also:* Conjugation, Bacterial**

Sex Ratios

S A West

doi: 10.1006/rwgn.2001.1184

Research into the sex ratio (proportion of individuals that are male) has been one of the most quantitatively successful areas of evolutionary biology. Relatively simple theory is able to explain why many animal species produce approximately equal numbers of males and females, why certain species have an excess of males or females, and why individuals of some species facultatively shift their offspring sex ratio in response to environmental conditions (Charnov, 1982).

Fisher's Theory of Equal Investment in the Sexes

Fisher (1930) provided an explanation for why most animal species, including humans, produce approximately equal numbers of males and females. If there were an excess of males, they would on average obtain less than one mate, and so the fitness of females would be greater, favoring parents that produced a relative excess of female offspring. In contrast, if there were an excess of females, males would on average obtain more than one mate, and so the fitness of males would be greater, favoring parents that produced a relative excess of male offspring. Consequently, the fitness of males and females is only equal when equal numbers of the two sexes are produced (a sex ratio of 0.5). This argument assumes that equal amounts of resources are put into the production of sons and daughters. If this is not the case then the argument is phrased in terms of investment, and the evolutionarily stable strategy (ESS) is to invest resources equally in male and female offspring.

Biased Sex Ratios

Fisher's principle clearly shows the frequency-dependent nature of selection on the sex ratio, and it provides a null model (equal investment in the sexes) which is the foundation block on which most areas of sex ratio research have been built. However, it assumes that the fitness returns from the production of sons and daughters are identical (or linear). Many different biological mechanisms contradict this assumption and in these cases biased sex ratios are predicted. Two scenarios are reviewed here where there is a rich experimental literature exploring the predictions of many theoretical models: (1) sex-biased interactions between relatives; and (2) differential effects of the environment on male and female fitness.

Sex-Biased Interactions between Relatives

Fisher's principle assumes that the fitness returns from producing sons and daughters do not differ. This will not be the case if there are sex differences in the interactions of offspring with each other, or their parents (Hamilton, 1967).

If production of one sex leads to a greater increase in fitness of the parents or their offspring, then an excess production of that sex is favored by a process called local resource enhancement (LRE). One example of this process is observed in African wild dogs, which are wolf-like social carnivores found in sub-Saharan Africa. These dogs live in packs, and young males help more than young females in the rearing of pups. This favors an excess of males, and, indeed, 60% of offspring are males. Another example is provided by allodapine bees, primitively social bees that communally nest in burrows. Sisters nest together, and increasing nest size leads to fitness benefits through increased survival and reproduction, in part due to the efficiency gained from division of labor. This favors an excess of females and, in this case, less than 20% of offspring are male.

In contrast, if competition between siblings and/or parents is greater for one sex, then an excess production of the other sex is favored by a process called local resource competition (LRC). One example of this comes from African primates in the Galagidae family. In these species female disperse much less than males, and related females compete for resources, especially during the breeding season. This favors an excess of males and, in this case, 70% of offspring are male.

A special case of LRC that has received considerable attention is the competition for mates between brothers in structured populations, which is termed local mate competition (LMC). If brothers compete for mates (including their sisters) before the females disperse, then LMC theory predicts a female-biased sex ratio to reduce this competition. Support for LMC theory has come from a wide range of animals and plants, especially insects (e.g., parasitic wasps, aphids, thrips, and beetles), other arthropods (e.g., mites, spiders), protozoan parasites (e.g., blood parasites such as those causing malaria, and intestinal parasites such as *Toxoplasma*), and flowering plants. In these cases not only do populations show female-biased sex ratios, but in many cases individuals have been shown to adjust the ratio of their offspring in response to variation in the level of LMC.

Both LRE and LRC can occur in the same species. One of the clearest examples of this comes from studies of the Seychelles warbler. This small bird is extremely territorial, and in situations where there are few new territories available, some young will remain at their natal nest and help raise siblings. The majority (80%) of helpers are daughters. Importantly, whether a helper is advantageous or disadvantageous for her parent depends on the quality of the territory occupied, which depends on the availability of insects for food. On high-quality territories, helpers are beneficial from the point of view of their parent, and increase the number of young produced (LRE). On low-quality territories, the increased competition for food with helpers means that their presence is disadvantageous from the point of view of their parent (LRC). As predicted, predominantly (90%) females are laid on high-quality territories where their presence as helpers will be beneficial (LRE), and predominantly (80%) males are laid on low-quality territories, from which they will disperse, and avoid competition with their relatives (LRC).

Differential Effects of Environment on Male and Female Fitness

Fisher's principle assumes that variation in environmental conditions affects the fitness of sons and daughters equally. If this is not the case, then individuals can be selected to adjust the sex of their offspring in response to the environment (Trivers and Willard, 1973).

This idea was first applied to explain sex ratio patterns in mammals caused by variation in maternal condition. For example, in red deer, higher quality (indicated by rank) females are more likely to produce sons, and lower quality females are more likely to produce daughters. This is thought to occur because (1) higher-quality females are able to provide more resources for their offspring, and (2) competition for mates between males is intense, with only the highest quality males being successful, and so sons benefit more from increased resources than daughters. The same concept can explain why in many species of parasitic wasp, where only one individual can develop per host, females lay male eggs on small hosts and female eggs on large hosts. In this case it is presumed to be advantageous because the resultant increase in body size provides a greater benefit for daughters (through effects on fecundity) than sons.

The same concept can also apply in response to variation in mate quality. In several bird species (e.g., zebra finches, collared flycatchers, and blue tits), females have been shown to adjust the sex of their offspring in response to the attractiveness of their mate. This is advantageous when male attractiveness is an indicator of genetic quality and heritable. Consequently, if a female mates with a relatively attractive male there is an advantage to producing sons, who will inherit their father's attractiveness. This pattern is observed, and some bird species show remarkable ability to shift their offspring sex ratio in the predicted way.

Conclusions

Evolutionary biologists have developed an excellent understanding of the selective factors that shape the sex ratio. More generally, studies of the sex ratio have provided some of the best support for the adaptationist approach (West *et al.*, 2000). In particular, they have provided an area in which theory is able to make predictions that can be tested qualitatively, and sometimes quantitatively, with experimental and observational data. Perhaps one of the greatest questions remaining is how do species with chromosomal sex determination, such as mammals and birds, achieve such striking facultative shifts in offspring sex ratios?

References

Charnov EL (1982) *The Theory of Sex Allocation*. Princeton, NJ: Princeton University Press.

Fisher RA (1930) *The Genetical Theory of Natural Selection*. Oxford: Clarendon Press.

Hamilton WD (1967) Extraordinary sex ratios. *Science* 156: 477–488.

Trivers RL and Willard DE (1973) Natural selection of parental ability to vary the sex ratio of offspring. *Science* 179: 90–92.

West SA, Herre EA and Sheldon BC (2000) Recent developments in the study and use of sex allocation. *Science*, 290: 288–290

See *also:* Evolutionarily Stable Strategies; Frequency-Dependent Selection; Frequency-Dependent Selection as Expressed in Rare Male Mating Advantages

Sex Reversal

M A Ferguson-Smith

doi: 10.1006/rwgn.2001.1185

The somatic (or phenotypic) sex of most mammals including man is normally determined at fertilization by the sex chromosome carried by the sperm. X-bearing sperm produce female embryos, while Y-bearing sperm produce male embryos. The *SRY* gene present on the short arm of the Y chromosome is regarded as the primary signal required for inducing the undifferentiated gonad in the early embryo to develop into a testis. Without *SRY* the gonad defaults into ovarian development. However, testis differentiation depends on the action of a number of other genes, some X-linked and others autosomal. It follows that mutations of a number of genes involved in the testis differentiation pathway may lead to a female phenotype in the presence of a normal XY sex chromosome constitution. Less commonly, genetic defects may lead to testis differentiation in the presence of a normal XX sex chromosome complement. These cases of sex reversal are often referred to as XY females and XX males.

XY Females

Individuals with mutations of the *SRY* testis-determining gene develop as immature females of normal to above average height and normal intelligence. The internal genitalia are normal in childhood, with development of a uterus and fallopian tubes and absence of Wolffian derivatives. However, the ovaries are abnormal and devoid of oogonia. In the adult the ovaries are represented by thin strips of ovarian stroma (streak gonads) in the broad ligament, and this results in sexual infantilism due primarily to lack of estrogenic hormones. Breasts and vulva fail to develop, and there is lack of axillary and pubic hair. These features can be corrected by estrogen therapy, which has to continue for life. Pregnancy is possible only by ovum donation. The phenotype of these XY females is usually known as pure gonadal dysgenesis to distinguish it from Turner syndrome, which is also characterized by streak gonads and sexual infantilism but which has in addition short stature and a number of additional features referred to as Turner stigmata. These include webbing of the neck, cubitus valgus, short IVth metacarpals, multiple pigmented naevi, and hypoplastic finger nails (see Turner Syndrome).

Pure gonadal dysgenesis in XY females also occurs in individuals with *SOX9* mutations (campomelic dysplasia), with WT mutations (Denys–Drash syndrome), with duplications of the Xp21 region (*DSS* gene), and with deletions of the short arm of chromosome 9 and the long arm of chromosome 10. The gene locus involved in these last two conditions has not yet been identified precisely. (See Sex Determination, Human for a detailed account of the known genes involved in the sex determination pathway.) In all these genetic aberrations, additional clinical features of varying severity are associated, usually leading to substantial handicap. Investigation of XY females for the underlying genetic defect has been largely responsible for our present understanding of the genetic pathway in mammalian sex differentiation. As the underlying defect has been identified in only a small proportion of XY females, other loci important in sex differentiation remain to be discovered.

Apart from pure gonadal dysgenesis, several other disorders are associated with sex reversal in XY females:

1. Androgen insensitivity (testicular feminization). This condition is the result of mutations of the X-linked androgen receptor gene. The developing testis secretes normal amounts of testosterone but the tissues are unable to respond due to the absence of androgen receptors. Affected individuals have normal female gender, are within the stature range of males, and develop breasts at puberty. However, they fail to menstruate, and pubic and axillary hair is scant. This is often the first indication of the condition. Sometimes patients present with an inguinal hernia in childhood, and this leads to the discovery of testes in the inguinal canals; however, the testes usually remain within the pelvis and are only identified by laparotomy. This reveals a short, blind-ending vagina, the absence of uterus and fallopian tubes, and the failure of development of Wolffian structures. As there is a risk that testicular tumors (dysgerminoma) may develop, the testes are removed and the patient maintained for life on a small daily dose of estrogen.

   The disorder is inherited as an X-linked recessive trait and the carrier state in normal females may sometimes be recognized by the patchy distribution of sex hair. Half the XY offspring of a carrier are at risk of being affected. Incomplete forms of androgen insensitivity are also recognized, due to mutations of the androgen receptor locus distinct from the complete form. Partial masculinization occurs leading to sexual ambiguity at birth and virilization at puberty.

2. 5-α-reductase deficiency (pseudovaginal perineoscrotal hypospadias). Deficiency of 5-α-reductase leads to failure of conversion of testosterone to dihydrotestosterone. XY infants with severe enzyme deficiency have a small hypospadic phallus, a blind vaginal pouch, and absent Müllerian derivatives. Over 50% are raised as XY females. At puberty, the patients virilize, do not develop breasts, and undergo a gender identity change to male gender.
3. Several other rare disorders of steroid biosynthesis may lead to sex reversal in XY females. These include:

   a. testicular unresponsiveness to gonadotrophin
   b. congenital lipoid adrenal hyperplasia
   c. 3-β-hydroxysteroid dehydrogenase deficiency
   d. 17-hydroxylase deficiency
   e. 17 β-hydroxysteroid oxidoreductase deficiency.

   Deficiency of each of these enzymes can be associated with female external genitalia, absence of Müllerian derivatives, and a blind vaginal pouch.

XX Males

Phenotypic males with small testes and, commonly, gynaecomastia associated with an apparently normal female karyotype are usually referred to as XX males. The endocrinological features are identical to Klinefelter syndrome (Klinefelter Syndrome), with the exception that there are no associated learning difficulties and stature is reduced to within the normal female range. In the majority of patients the condition is due to abnormal recombination between the X and Y in paternal meiosis, so that the region containing the SRY gene is transferred to the end of the short arm of the X. This abnormal event is often accompanied by loss of the Xg locus on Xp, revealed by failure of the patient to inherit the paternal Xg(a) allele. Sometimes, the transferred region of the short arm of the Y is large enough to be identified under the microscope.

Some 15% of XX males do not have SRY and the reason for the sex reversal is at present unknown. Hypospadias is more common in these SRY-negative patients and, rarely, several XX males may occur in the same pedigree. The etiology of this group may be similar to that of XX true hermaphroditism and, in fact, pedigrees have been reported in which both conditions occur.

Further Reading

Ferguson-Smith MA and Goodfellow PN (1995) SRY and primary sex reversal syndromes. In: Scriver CR, Beaudet AL, Sly WS and Valle D (eds) *The Metabolic and Molecular Basis of Inherited Disease*,. pp. 739–748. New York: McGraw-Hill.

Grumbach MM and Conte FA (1992) In: Wilson JD and Foster D (eds) *Abnormalities of Sex Differentiation. Williams Textbook of Endocrinology*, 8th edn. Philadelphia, PA: WB Saunders.

***See also:* Congenital Adrenal Hyperplasia (Adrenogenital Syndrome); Klinefelter Syndrome; Sex Determination, Human; Turner Syndrome**

Sexduction

S M Rosenberg and P J Hastings

doi: 10.1006/rwgn.2001.1180

'Sexduction' (also called F-duction) is an old term no longer in general use. Coined by Jacob and Wollman (1961), sexduction uses an analogy with phage-mediated transduction to describe the process of high frequency conjugative transfer to a recipient bacterium of a segment of *Escherichia coli* DNA incorporated in an F′ (F-prime) plasmid.

F′ plasmids are derivatives of the F sex plasmid, a conjugative plasmid of *E. coli* (see F Factor). F′ plasmids have acquired a piece of the *E. coli* chromosomal DNA. (See F Factor, Figure 2 for a description of how F′ plasmids acquire chromosomal DNA.) Because the chromosomal DNA segment is joined covalently into the F plasmid, it can be transferred efficiently into recipient bacteria. The piece of chromosomal DNA transferred is of defined length (the length present in the F′), transfers quickly, and does not require recombination into the recipient bacterium's chromosome for its replication, expression, and heritable transmission. Also, the bacterial recipients of sexduction become male (donors) upon sexduction because they acquire the entire F′ plasmid with its transfer and pilus genes, as well as its chromosomal DNA segment. These properties are unlike the conjugative transfer of chromosomal DNA from Hfr strains of *E. coli*, in which the F plasmid has integrated into the bacterial chromosome (see Hfr). Sexduction is similar to transfer of bacterial DNA in phage-mediated 'transduction' (see Transduction), especially in 'generalized transduction' in which essentially any region of the bacterial chromosome can be transferred to a recipient cell. This is possible because the F can be integrated at many different locations in the chromosome, such that many different F′ plasmids can be formed upon its excision. The key similarity between sexduction and phage-mediated transduction is the finite amount of the chromosomal DNA transferred. With sexduction, unlike transduction, the recipient strain becomes partially diploid (merodiploid) for the segment transferred.

Sexduction was used by bacterial geneticists as an early form of *in vivo* gene cloning. A piece of chromosomal DNA could be selected that was attached to the F plasmid vector, isolated from it, and transferred to other bacteria by conjugation.

Reference

Jacob F and Wollman EF (1961) *Sexuality and the Genetics of Bacteria*. New York: Academic Press.

See *also*: F Factor; F-Duction; Hfr; Transduction

Sex-Limited Character

D E Wilcox

doi: 10.1006/rwgn.2001.1181

A sex-limited character is a trait that is expressed in only one sex. Not all phenotypic differences between the sexes are due to the sex-linked genes that are present in differing amounts in each sex. Autosomal genes, that are equally present in both sexes, can also cause sex-limited characters. An example is the development of horns in only one sex of certain animals. Sex-limited disease can also occur. An example is the development of breast cancer in human females caused by mutations in the autosomal *BRCA1* gene. Males carrying the mutation have a low risk of developing breast cancer but can transmit the disease to their daughters.

Example pedigrees may be seen at http://www.gla.ac.uk/medicalgenetics/encyclopedia.htm

Sezary's Syndrome

M J S Dyer

doi: 10.1006/rwgn.2001.1618

Sezary's syndrome is a form of low-grade cutaneous T-cell non-Hodgkin's lymphoma with involvement of the peripheral blood; the purely cutaneous form of the disease is known as mycosis fungoides. Both forms are rare. In Sezary's syndrome the degree of cutaneous infiltration is often marked, resulting in erythroderma or 'l'homme rouge' appearance. The T cells are clonal, have a characteristic nuclear morphology, and usually express CD4 but neither CD8 nor CD25. The pathogenesis of this disease is not known. There are no consistent cytogenetic or molecular changes. Reports claiming retroviral involvement have not been confirmed.

Further Reading

Siegel RS, Pandolfino T, Guitart J, Rosen S and Kuzel TM (2000) Primary cutaneous T-cell lymphoma: review and current concepts. *Journal of Clinical Oncology* 18: 2908–2925.

See *also*: Genetic Diseases

SH Domains

M Frame

doi: 10.1006/rwgn.2001.1745

The Src homology (SH) domains are protein modules with defined structures and protein–protein interaction function that are found in Src family kinases and also in many other intracellular signal transduction proteins, including, for example, receptor protein tyrosine kinases, phospholipase C-gamma, and the Ras GTP-ase-activating protein. The SH domains exemplify protein–protein interaction domains, which like their cognate recognition motifs are modular in nature. Their widespread occurrence and conserved molecular functions, even in the context of proteins with distinct enzymatic or biological properties, has led to the concept of 'protein recognition codes' (reviewed in Sudol, 1998).

Reference

Sudol (1998) *Oncogene* 17: 1469–1474.

See *also*: SH2 Domain; SH3 Domain; Signal Transduction

SH2 Domain

M Frame

doi: 10.1006/rwgn.2001.1619

Noncatalytic Src homology 2 (SH2) domains recognize tyrosine phosphorylated residues in other proteins, including receptor tyrosine kinase autophosphorylation sites. Phosphorylation of individual tyrosine residues induces the binding of SH2 domain-containing proteins, with specificity being determined by the amino acids flanking the phosphotyrosine, particularly the C-terminal residues. Recruitment of tyrosine phosphorylated peptides by SH2 domain-containing proteins results in heteromeric protein complexes that are temporally and spatially controlled within the cell. This links tyrosine kinase signaling to the formation of protein complexes that are necessary for signal propagation.

***See also:* SH Domains; SH3 Domain; Signal Transduction**

SH3 Domain

M Frame

doi: 10.1006/rwgn.2001.1620

The SH3 domain is a distinct motif that binds target proteins, including proteins associated with the actin cytoskeleton, through sequences containing proline and hydrophobic amino acids. Many proteins that contain SH3 domains also have SH2 domains, and these may act together to modulate specific protein–protein interactions. Some signaling and transforming proteins contain SH3 and/or SH2 domains with no associated catalytic activity, for example, the Grb2 adaptor protein that links receptor tyrosine kinases to the Ras pathway. Thus, adaptor proteins serve to mediate higher order protein complexes at appropriate times and places within the cell and to orchestrate appropriate biological responses. Three-dimensional structures of several individual SH domains have been determined and have provided clues as to the determinants of molecular specificity.

***See also:* SH Domains; SH2 Domain; Signal Transduction**

Shearing

I Schildkraut

doi: 10.1006/rwgn.2001.1188

This is the method of subjecting large DNA molecules to hydrodynamic forces to reduce the size of the molecules. DNA molecules which are hundreds of thousands of base pairs in length can be broken to tens of thousands of base pairs simply by pipetting a solution of DNA. DNA of thousands of base pairs can be reduced to hundreds of base pairs by forcing a solution of DNA through a small-bore needle with a syringe.

Shifting Balance Theory of Evolution

J Coyne

doi: 10.1006/rwgn.2001.1189

The 'shifting balance theory' (SBT) is a theory of evolution that was proposed by Sewall Wright in the early 1930s as an alternative to the view of evolutionary change dominant at the time. That view, based largely on the work of R. A. Ronald Fisher, saw evolution as resulting mainly from directional selection acting on large populations. Wright, on the other hand, believed that such a process could not account for either the nature of adaptations or the pace of evolutionary change, and considered his SBT a more plausible explanation. In contrast to the 'Fisherian' view of evolution requiring only selection on favorable alleles, Wright's SBT assumes that species are subdivided into small populations ('demes'), that these demes exchange migrants based on their degree of adaptation, that there are special forms of epistasis between genes, and that both natural selection and random genetic drift interact to cause evolutionary change.

Adaptive Landscape

The SBT is intimately connected with Wright's notion of the 'adaptive landscape,' which must be understood to fully grasp his theory. In Wright's view, the evolutionary opportunities for a population can be viewed as a type of topographical map. In a three-dimensional analog, the latitude and longitude of a point would correspond respectively to the frequencies of alleles segregating at two loci within a population, and the height of the landscape above any point would represent the mean fitness of a population having the specified allele frequencies at those two loci. 'Peaks' in the landscape would correspond to frequencies of alleles conferring high fitness on a population, and 'valleys' to intervening areas of low fitness. Wright's SBT was designed to explain how an entire species could move from one adaptive peak to successively higher ones – hence becoming more and more adapted – while temporarily undergoing periods of reduced adaptation as it crossed adaptive valleys.

Figure 1 shows a simple adaptive landscape based on two alleles segregating at each of two loci (such landscapes can be envisioned for more than two genes, but are difficult to represent since they require more

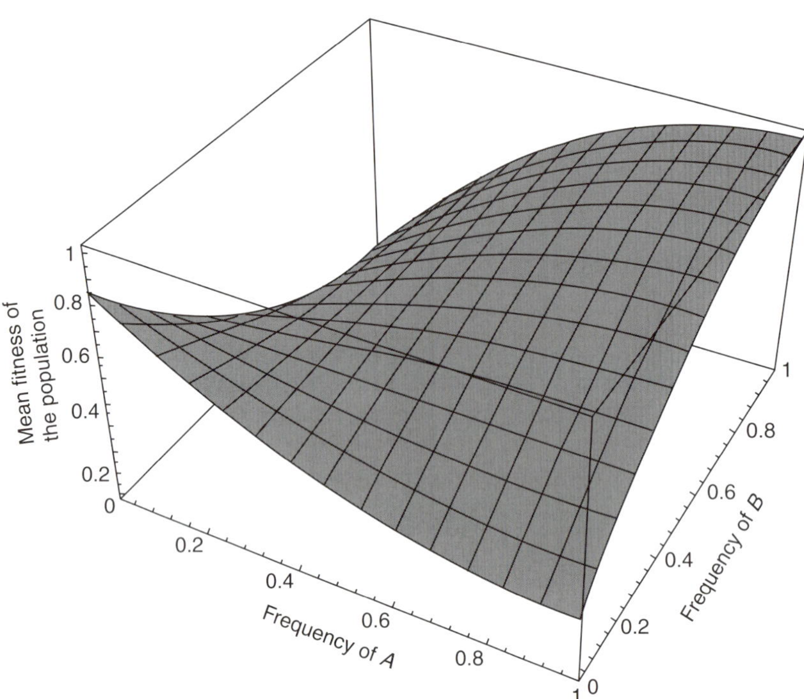

Figure 1 Simple adaptive landscape involving two alleles at each of two loci (*A*, *a* and *B*, *b*). The frequencies of the *A* and *B* alleles are plotted on the horizontal axes, and the 'adaptive landscape' (shaded area) represents the fitness of a population having the given frequencies of the two alleles. The *AABB* and *aabb* genotypes are assumed to be the most well adapted, having relative fitnesses of 1.0 and 0.8, respectively. The fitness of the *A_bb* genotype is assumed to be 0.4, and of the *aaB_* genotype 0.2. These assumptions produce an adaptive landscape having two peaks (corresponding to fixation for the *aabb* or *AABB* genotypes) separated by an adaptive valley. A population cannot move from the *aabb* peak to the *AABB* peak without the help of genetic drift, which may allow a move across the intervening valley.

than three dimensions). This landscape is derived from assuming that a population may have dominant or recessive alleles at two loci (*A*, *a* and *B*, *b*, respectively). Populations consisting entirely of *aabb* or *AABB* genotypes occupy adaptive peaks, while populations consisting either of *AAbb* or *aaBB* genotypes, or which contain more than one genotype, have lower fitness. Plotting the fitness of a population against its genetic constitution yields an adaptive landscape with two peaks: a higher peak (an *AABB* population) and a lower peak (an *aabb* population). As a biological example, the *AABB* and *aabb* genotypes might both produce cryptic white coat colors in arctic mammals, while the *aaB_* or *A_bb* genotypes produced colored coats that are maladaptive; the higher fitness of the *AABB* than of the *aabb* peak might result from other pleiotropic effects of the coat-color genotype on reproduction. In such a case, which involves both epistasis and pleiotropy, adaptive peaks can be separated by adaptive valleys, as seen in **Figure 1**. (While Wright's model assumed pleiotropic effects of identical phenotypes, this is not necessary to produce a multipeaked adaptive landscape, for the peaks can also represent different phenotypes separated by adaptive valleys.)

If one accepts that such fitness landscapes are common in nature, then populations on adaptive peaks are 'stranded,' i.e., unable to reach higher peaks by natural selection alone. A population fixed for genotype *aabb* in **Figure 1**, for example, cannot reach the higher *AABB* peak by selection, for this would require crossing the 'adaptive valley' dividing the peaks, a journey opposed by natural selection. In Wright's view, hilly adaptive topographies are quite common in nature. To become more and more adapted to its environment (or to adapt to a changing environment), species thus require a mechanism for becoming temporarily maladapted so that they may traverse adaptive valleys. This mechanism was supplied by the SBT.

Shifting Balance Process

The SBT posits that the movement of a species from one adaptive peak to a higher one involves three distinct phases of evolution. In phase I, a single population of a species is perched on an adaptive peak, but then undergoes a loss in fitness due to random genetic drift, which counteracts selection and drives the population into an adaptive valley. Once in the valley, the population experiences phase II, coming under the influence of another, higher peak. Natural selection draws this population uphill until it attains a level of adaptation higher than its original state (in **Figure 1**, for example, a population could move from the *aabb* peak to the *AABB* peak). In phase III, the population that has attained the higher peak sends out additional migrants to other populations of the species (the assumption here is that the number of migrants leaving a population is proportional to its size, which is correlated with its degree of adaptation). These migrants would then genetically alter other populations, forcing them off their peaks until the entire species comes to rest on the new and higher peak. If this process occurred repeatedly, it could lead a species to become more and more adapted as it scaled ever higher peaks. Because Wright viewed adaptation in nature as the movement between peaks in an adaptive landscape, he did not believe that mass selection in large populations could lead to greater adaptation, for such populations are largely immune to the genetic drift that enables them to cross adaptive valleys.

Importance and Influence of the Theory

Between 1935 and 1975, the SBT was quite influential, and discussions of the theory and adaptive landscapes appeared frequently in major works of the evolutionary synthesis and still permeate modern textbooks of evolution. The SBT has been attractive for several reasons. First, it introduced the appealing idea of the adaptive landscape, in which evolution is seen as a form of hill climbing. This simple graphical metaphor is able to make complex evolutionary processes visually comprehensible. In addition, unlike the Fisherian view of evolution, which requires only natural selection and large populations, the SBT incorporates a diversity of evolutionary elements, including genetic drift, population structure, selection, and epistasis, and thus may be considered to be more comprehensive.

When evaluating the SBT, one must ask four questions. First, is, as Wright claimed, the Fisherian theory of adaptation insufficient to explain large-scale evolutionary changes? Second, does the SBT work as a theory; that is, does the verbal description given above prove valid when one makes a mathematical model of the entire process? Third, is there evidence from laboratory experiments that evolution can proceed according to the SBT and produce greater adaptation than does mass selection in large populations? Finally, is there evidence from nature that the SBT has been a frequent cause of adaptation?

Wright's main rationale for proposing the SBT, that mass selection would be insufficient to produce the observed diversity of life, has not been substantiated, as there is no evidence that mass selection is too slow to explain either adaptation or biological diversity. Moreover, there are almost no adaptations known whose evolution would involve intermediate steps of lowered fitness, making genetic drift essential to overcome selection during their evolution. Thus one cannot assert the superiority of the SBT over the Fisherian theory on this basis.

Although Wright developed mathematical methods to represent different phases of the SBT, including theories of how allele frequencies in populations respond to the joint pressures of selection, drift, and migration, he never produced a mathematical analysis including more than one phase of the SBT. Subsequent workers have produced such models, and their work has indeed shown that, given Wright's assumptions, the SBT can operate under restricted conditions. Empirical data and some models of evolution do imply the existence of different adaptive peaks, and although the transitions between such peaks need not require genetic drift, they can occur if populations are fairly small and valleys are sufficiently shallow. The primary theoretical problems of the SBT occur in phase III, in which a population arrives on top of a new adaptive peak and, through migration, draws the rest of the species to that same peak. One major obstacle to this process is that, unlike adaptations favored by simple mass selection, adaptations whose fixation requires some genetic drift are often prevented from spreading by physical barriers to gene flow (such as poor habitat). Moreover, the evolution of complex adaptations by the SBT requires that the components of such adaptations arise by peak shifts in different demes, and theory shows it is difficult for the SBT to assemble these components into the whole.

There have been some attempts to test the SBT in the laboratory, which have generally been experiments in which artificial selection is practiced on both large and subdivided populations. In general, these experiments have failed to show a greater response in the structured populations, as might be predicted by the SBT.

It is difficult to test the SBT in nature because of both the complexity of the process and the near-impossibility of establishing whether an existing adaptation involved crossing an adaptive valley via drift. (Movement of a species from one adaptive peak to another need not require genetic drift but can occur through mass selection following a changed environment or a new mutation.) Reviewing specific adaptations in nature, one finds that although there is some evidence for individual phases of the shifting balance theory (i.e., as proposed in phase I, drift can occasionally counteract natural selection), there are almost no empirical observations explained better by the SBT than by simple mass selection.

In view of these theoretical and empirical problems, there is not strong support for the SBT or Wright's assertion that it has been the major engine of evolutionary change. Nevertheless, the theory has left an important legacy, both in the metaphor of the adaptive landscape, which still pervades evolutionary biology, and in the mathematical constituents of the SBT devised by Wright. The most important of these are the general equations for the interaction of diverse evolutionary forces such as drift, selection, migration, and mutation. These equations are still used to solve many problems of theoretical population genetics.

Further Reading

Coyne JA, Barton NH and Turelli M (1997) A critique of Sewall Wright's shifting balance theory of evolution. *Evolution* 51: 643–671.

Wright S (1932) The roles of mutation inbreeding, cross breeding, and selection in evolution, pp. 356–366. *Proceedings of the 6th International Congress on Genetics.*

Wright S (1978) Random drift and the shifting balance theory of evolution. In: Kojima K (ed.) *Mathematical Topics in Population Genetics*, pp. 1–31. New York: Springer-Verlag.

***See also:* Adaptive Landscapes; Wright, Sewall**

Shine–Dalgarno Sequence

doi: 10.1006/rwgn.2001.2019

A Shine–Dalgarno sequence is a polypurine sequence found in bacterial mRNA just before an AUG initiation codon. It is part or all of the sequence 5′-AGGAGG-3′. It is complementary to a highly conserved sequence at the 3′ end of 16S rRNA, and is involved in binding of the ribosome to the mRNA.

***See also:* Messenger RNA (mRNA)**

Shotgun Cloning

doi: 10.1006/rwgn.2001.2101

Shotgun cloning is the cloning of an entire genome in the form of randomly generated fragments.

***See also:* Human Genome Project**

Shuttle Vector

doi: 10.1006/rwgn.2001.2021

A shuttle vector is a cloning vector that is able to replicate in more than one organism, e.g., *Escherichia coli* and *Saccharomyces cerevisiae*. Generally, it is a plasmid constructed to contain the origins of replication of both hosts, and is used to carry foreign genes from one species to another.

***See also:* Cloning Vectors; Vectors**

Sickle Cell Anemia

D J Weatherall

doi: 10.1006/rwgn.2001.1191

A form of anemia associated with elongated and sickle-shaped blood cells was first reported by the American physician James Herrick in 1910. In 1949 Linus Pauling and colleagues found that it results from an inherited structural change in hemoglobin. It was subsequently realized that it is an extremely common disease which occurs principally in sub-Saharan Africa, some of the Mediterranean populations, throughout the Middle East, in localized areas of the Indian subcontinent, and in populations anywhere in the world which originated from these regions (**Figure 1**).

Inheritance

Sickle cell anemia is inherited in a Mendelian recessive fashion. The structure of human hemoglobin changes between fetal and adult life. All the human hemoglobins consist of two different pairs of peptide chains called globin chains, each of which is attached to the oxygen carrying moiety, heme. Fetal hemoglobin consists of a pair of α-globin chains and a pair of γ-globin chains ($\alpha_2\gamma_2$), while adult hemoglobin consists of α and β chains ($\alpha_2\beta_2$). Hemoglobin S differs from hemoglobin A by a single amino acid substitution in the β-globin chain, valine for glutamic acid. The β chains consist of 146 amino acids, numbered from the N-terminal end; this change is at position 6. Individuals with a single sickle cell mutation, that is carriers for the abnormal gene, have one normal β chain gene, β^A, and one abnormal gene, β^S. Thus they have two types of hemoglobin, normal ($\alpha_2\beta_2$) and sickle cell ($\alpha_2\beta_2^S$). This is called the sickle cell trait. Those who inherit a

Figure 1 The world distribution of the sickle cell gene. Hemoglobin E, the second commonest structural hemoglobin variant, is also shown.

β^S gene from both parents can only make sickle cell hemoglobin and hence they have sickle cell disease.

Effect of the Sickle Cell Mutation

The amino acid substitution that causes sickle cell hemoglobin has the remarkable effect of changing the shape of red blood cells from their normal circular, biconcave configuration into an elongated, sickled shape when blood is deoxygenated (**Figure 2**). This occurs because concentrated solutions of hemoglobin S form gels containing fibers of hemoglobin S molecules which are stabilized by interactions of the substituted valine residue. These have the effect of distorting the red cell into a sickle, or sometimes a holly leaf, shape. In addition, the continued distortion of the red cell damages its membrane and causes it to become dehydrated and more rigid than normal. Furthermore, these cells tend to become abnormally adherent to the endothelial lining of blood vessels.

The overall effect of the complex changes which occur to deoxygenated sickle cells is to reduce their survival time in the circulation and also to cause blockage to small blood vessels with subsequent destruction of tissues. Thus sickle cell disease is characterized by a reduction in the number of red cells, or anemia, and a variety of complications due to the sequestration of sickle cells in different organs and the tissue damage that results from this process.

Clinical Features

The sickle cell trait is symptomless. It can be diagnosed by hemoglobin electrophoresis. Patients with sickle cell anemia adapt surprisingly well to their anemia. However, they are prone to a variety of complications. They are particularly susceptible to infection, possibly because sickling of red cells damages the spleen and causes it to shrink and become scarred; the spleen, for reasons that are not well understood, plays a major role in combating infection, particularly early in life. Constant minor blockages of small blood vessels also leads to other chronic complications, including damage to the bones, particularly the heads of the humerus and femur, progressive impairment of kidney function, and chronic leg ulcers.

The most important feature of sickle cell anemia, however, is the occurrence of so-called sickling crises, a name given to a variety of acute complications of the disease. These may begin early in infancy with the 'hand–foot syndrome,' that is painful swelling of the hands and feet due to damage of the growing ends of

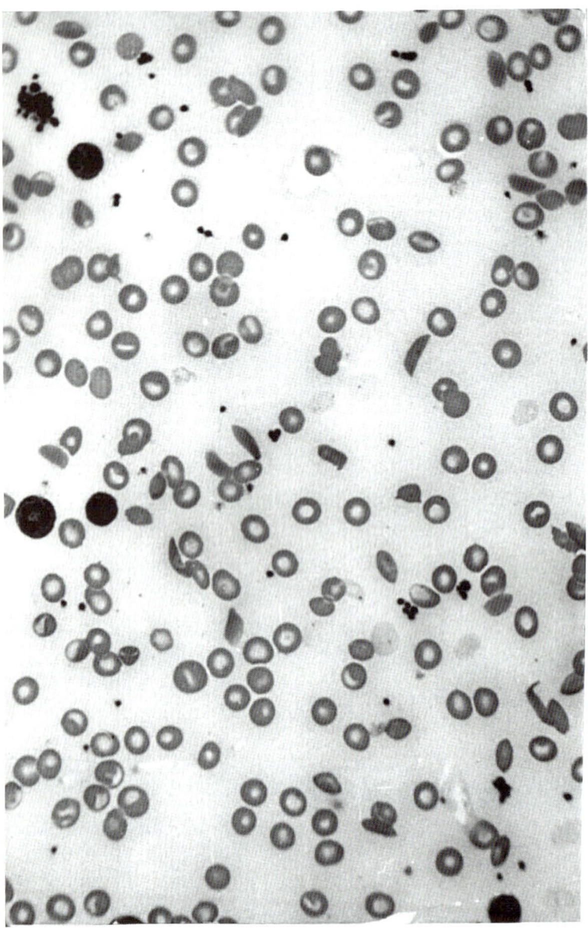

Figure 2 Peripheral blood film of a patient with sickle cell anemia showing sickle cells.

the bones. Painful crises, characterized by widespread bone pain due to local areas of destruction of the marrow and damage to the overlying bone, can occur at any age. Although they are sometimes precipitated by infection or cold, often no cause can be found. In babies there is a form of sickling crisis characterized by rapid progressive anemia and enlargement of the spleen, which probably reflects the sequestration of sickle cells. A similar process occuring in the main blood vessels of the lungs is the basis for the lung syndrome, characterized by increasing breathlessness and anemia. Some children develop curious thickening of the arteries at the base of the brain which may give rise to recurrent strokes. Other complications include prolonged, painful erections, and episodes of profound anemia associated with infections with an agent called parvovirus. Pregnancy may be associated with an increased frequency of painful crises and there is also an increased rate of fetal loss.

The clinical course of sickle cell anemia varies widely; some patients go through life with few complications while others are troubled with frequent crises and a wide variety of pathological sequelae. Although a number of genetic factors have been discovered which are partly responsible for this variability, in many patients it remains unexplained.

Control and Treatment

In between crises no specific treatment is required except for supplements of the vitamin folic acid, which is required for red cell production. As soon as the condition is recognized patients are given prophylactic penicillin, taken in tablet form each day, and they should be immunized to combat infections with several organisms which commonly cause severe infections in this disease. Mild painful crises are managed at home with simple analgesics, but when the pain is more severe hospital admission is required, so that more powerful analgesics can be administered, together with adequate hydration and monitoring of the oxygen levels in the blood. Infections are treated with appropriate antibiotics. The various sequestration crises are medical emergencies which require urgent hospital admission and treatment with transfusion or even exchange transfusion, that is the replacement of most of the patient's blood with that from a normal donor.

Children should be monitored with regular testing of their cerebral circulation and if there is evidence that they are likely to develop a stroke, or they have had an event of this kind, they should be maintained on blood transfusion for an indefinite period. Because of the risks of iron overload following long-term blood transfusion, the decision to embark on a regimen of this type is made only if there is a danger of stroke or the frequency of painful crises is making life intolerable.

Sickle cell disease can be identified prenatally at about 9–12 weeks' gestation. However, in most countries very few pregnancies are terminated for this condition. It can be cured by bone marrow transplantation if suitable donors are available. However, because of the uncertainty of the prognosis, and the risks of this procedure, its place in management is still controversial. Attempts are being made to elevate fetal hemoglobin levels in this disorder because children who make more fetal hemoglobin than usual are protected against some of the effects of the disease. A clinical trial of the agent hydroxyurea, a drug that is used for certain forms of leukemia, has shown that it has this effect and reduces the frequency of painful crises in adults. The long-term safety of this agent has not yet been assessed, however. A great deal of research is being carried out toward discovering other agents that will stimulate fetal hemoglobin synthesis, and into ways of trying to replace the sickle cell gene with a normal β-globin gene, or by other methods of genetic engineering, to correct the defect in sickle cell hemoglobin.

Prognosis

It is now usual for the majority of patients with sickle cell anemia who live in the richer countries, where there is a high quality of medical care, to survive to adult life. This is not the case in sub-Saharan Africa, where the disease is still a major killer in early childhood.

Other Sickling Disorders

The sickle gene may be inherited together with one for another hemoglobin variant, hemoglobin C or D for example, with the production of milder sickling disorders. It may also be inherited together with different forms of β-thalassemia (see Thalassemias).

Further Reading

Bunn HF (1997) Pathogenesis and treatment of sickle cell disease. *New England Journal of Medicine* 337: 762–769.

Charache S, Terrin ML, Moore RD *et al.* (1995) Effect of hydroxyurea on the frequency of painful crises in sickle cell anemia. *New England Journal of Medicine* 332: 1317–1322.

Dover GJ and Platt OS (1998) Sickle cell disease. In: Nathan DG and Orkin SH (eds) *Hematology in Infancy and Childhood*, pp. 62–801. Philadelphia, PA: WB Saunders.

Hebbel RP and Vercellotti GM (1997) The endothelial biology of sickle cell disease. *Journal of Laboratory and Clinical Medicine* 129: 288–293.

Hillery CA (1998) Potential therapeutic approaches for the treatment of vaso-occlusion in sickle cell disease. *Current Opinion in Hematology* 5: 151–155.

Okpala I (1998) The management of crisis in sickle cell disease. *European Journal of Haematology* 60: 1–6.

Serjeant GR (1992) *Sickle Cell Disease*, 2nd edn. New York: Oxford University Press.

Vichinsky EP, Haberkern CM, Neumayr L *et al.* (1995) A comparison of conservative and aggressive transfusion regimens in the perioperative management of sickle cell disease. The Preoperative Transfusion in Sickle Cell Disease Study Group. *New England Journal of Medicine* 333: 206–213.

See also: **Globin Genes, Human; Thalassemias**

Sigma Factors

R R Burgess

doi: 10.1006/rwgn.2001.1192

History

Sigma factors are subunits of all bacterial RNA polymerases. They are responsible for determining the specificity of promoter DNA binding and control how efficiently RNA synthesis (transcription) is initiated. The first sigma factor discovered was the sigma70 (σ^{70}) of the highly studied bacterium *Escherichia coli*. Its discovery in 1968 was an unexpected outcome of attempts to understand the subunit structure of RNA polymerase. It was found that RNA polymerase activity was associated with two protein species. A core polymerase (with subunit structure $\alpha 2\beta\beta'$) can transcribe DNA into RNA inefficiently and nonspecifically. When the sigma subunit, σ^{70}, is added, it can bind to core forming a holoenzyme ($\alpha 2\beta\beta'\sigma$) that is capable of specific engagement with duplex DNA at the beginning of genes (promoters) as well as efficient initiation of transcription. It was hypothesized that multiple sigma factors would be found in *E. coli*, each capable of directing the core polymerase to transcribe a specific set of genes. In this way, by regulating the level of each active sigma factor, the cell could coordinately regulate groups of genes with common functions. During the last 25 years, multiple sigma factors have indeed been found. The seven sigma factors of *E. coli* are listed in **Table 1** along with their gene names, molecular weights, consensus promoter DNA binding sites, and classes of genes they regulate. Sigma factors have also been discovered that are encoded by bacteriophage. By binding to core polymerase these proteins cause preferential transcription of phage genes. In the sporulating bacterium *Bacillus subtilis*, ten sigma factors have been discovered and characterized. These proteins not only regulate classes of genes during vegetative growth, but also orchestrate the development of the spore, in response to nutrient starvation. *E. coli* σ^{70} was essentially the first positive transcription activation factor whose basic mode of action was understood. The concept that groups of genes could be coordinately regulated by transcription initiation factors spurred successful searches for numerous transcription factors in both bacteria and higher organisms.

Basic Role in Transcription: The Sigma Cycle

In the bacterial cell, sigma factors exist in several types of complexes, each of which is important to the transcription process. When a free sigma binds to core to form holoenzyme, it weakens the nonspecific binding of core to DNA and enhances the specific interaction of RNA polymerase with promoter DNA. This positions the polymerase at the beginning of the gene (promoter) to be transcribed. Although most sigmas cannot bind to DNA by themselves, a conformational

Table 1 The seven sigma factors of *Escherichia coli*

Factor[a]	Gene	Number of amino acid residues	Size (kDa)	Consensus binding site[b]	Genes regulated
σ^{70} (σ^{D})	*rpoD*	613	70	TTGACA–N_{17}–TATAAT	Housekeeping
σ^{54} (σ^{N})	*rpoN (ntrA)*	477	54	CTGGCAC–N_5–TTGCA	Nitrogen metabolism
σ^{S}	*rpoS (katF)*	362	38	TTGACA–N_{12}–TGTGCTATACT	Stationary phase
σ^{32} (σ^{H})	*rpoH (htpR)*	284	32	CTTGAA–N_{14}–CCCCATNT	Heat shock
σ^{F} (σ^{28})	*fliA*	239	28	TAAA–N_{15}–GCCGATAA	Flagellar proteins
σ^{E}	*rpoE*	191	24	GAACTT–N_{16}–TCTGA	Extreme heat shock
σ^{fecI}	*fecI*	173	19	GGAAAT–N_{17}–TC	Iron transport

[a]Alternative names are given in parenthesis.
[b]N_x indicates any x number of nucleotides.

change occurs when sigma binds to core that exposes two regions of sigma. The exposed segments recognize two short regions of promoter DNA that lie about 38-32 bases (the –35 region) and 13-8 bases (the –10 region) before the start site of transcription (at position +1). This resulting closed promoter complex then undergoes a conformational change that facilitates a 'melting' process to form an open promoter complex in which the base pairs from positions –10 to +1 are disrupted, allowing the two strands of DNA to separate slightly. The first two nucleoside triphosphates then bind to the open complex, base pairing with the DNA bases in positions +1 and +2 of the template strand. The first two nucleotides become linked by a phosphodiester bond during RNA chain initiation. After initiation, when the RNA chain reaches a length of about 8–10 residues, the sigma factor dissociates from the core, allowing the polymerase to escape from the promoter and traverse the DNA, elongating the RNA chain until the complete gene is transcribed and termination of transcription occurs. The released sigma is free to find another core polymerase and start the sigma cycle anew. It is likely that the conformation of sigma changes significantly both in its interactions with the core polymerase and with the promoter DNA as the above cycle progresses.

It should be noted that not all of the regulation of bacterial transcription is attributable to sigma factors. In addition to the general/global regulation exerted by sigma factors, there are very important effects on transcriptional regulation exerted by promoter-specific negative transcription factors, such as repressors, that bind to operator targets that overlap the promoter, thereby preventing RNA polymerase from binding. There is also regulation by positive transcription factors that bind to specific targets just upstream of promoters. By interacting with the holoenzyme, such factors enhance its ability to bind to the promoter, the efficiency of open promoter complex formation, or the rate of RNA chain initiation.

Sigma70 (σ^{70}) Structure and Function

The gene for σ^{70}, *rpoD*, has been mapped, cloned, and sequenced. By comparing its sequence to those of other sigma factors, regions of strong similarity were identified. This implies that the *E. coli* sigmas (with the exception of σ^{54}) belong to a family of homologous proteins. The most highly conserved segments were designated regions 1, 2, 3, and 4. Predicted helix–loop–helix (HLH) structures, often found in DNA binding proteins, in regions 3.1 and 4.2 suggest that the role of these regions may be to bind DNA. Through a combination of genetic, biochemical, and sequence analysis studies, various structural and functional features have been assigned to segments of the 613 amino acid residue polypeptide chain. These are summarized in **Figure 1**. By analyzing promoter mutations and compensating mutations in σ^{70}, regions 4.2 and 2.4 of σ^{70} were shown to interact with the –35 region and the –10 region of the promoter DNA, respectively, and are thus responsible for the specific DNA binding properties of σ^{70}. Region 2.3 appears to be involved in interacting nonspecifically with single-stranded DNA in the open promoter complex. Region 1.1 seems to interact with region 4.2 to prevent free σ^{70} from binding DNA. Deletion and mutational analysis impliates regions 2.1 and 2.2 of σ^{70} in core binding. The sequence between regions 1.2 and 2.1 is thought to be dispensable because it is found only in σ^{70} and is absent from the other members of the σ^{70} family in *E. coli*. It is also not found in the σ^{70} equivalent in *B. subtilis* (σ^{A}). A highly acidic region, containing 18–22 acidic amino acids, is found around residue 200. Regions 1.2 through 2.4 (amino acids 114–448)

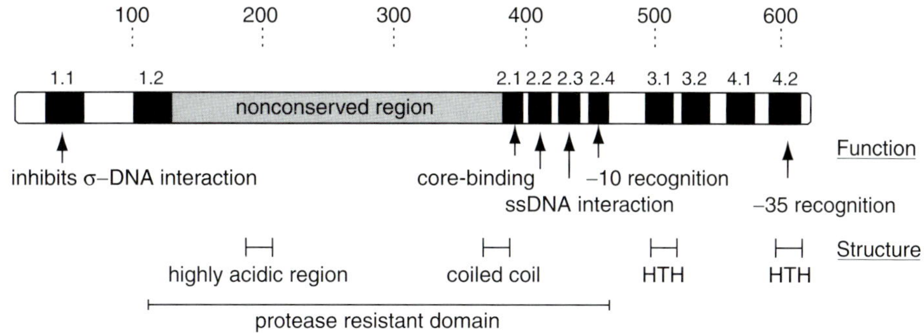

Figure 1 Structural and functional features of σ^{70}.

form a structure that is resistant to digestion by a variety of proteases. This protease-resistant domain was recently crystallized and its three-dimensional structure determined (Malhotra *et al.*, 1996). Regions 1.2 and 2.1 are seen to be α-helices, located close together and forming a coiled-coil structure. The structures of the remaining regions of σ^{70} remain unsolved.

How Sigma Activity is Regulated

It is not understood fully how the regulation of transcription by way of multiple sigma factors is accomplished. The simplest model is one where the amount of each type of holoenzyme is determined by the relative amounts and binding strengths for core polymerase of each sigma. Emerging results for a few of the *E. coli* and *B. subtilis* sigmas suggest that diverse and complex mechanisms are employed to regulate by sigma abundance and activity under different growth conditions. For example, the abundance of σ^{32} appears to be regulated by its stability. During growth at 37 °C, σ^{32} is bound by a heat shock protein, DnaK, resulting in its rapid proteolytic degradation (half-life, 1–2 minutes). At 42 °C, certain, proteins become denatured. DnaK, which binds to partially unfolded proteins, is competed away from σ^{32}. Free σ^{32} rapidly accumulates, binding to core polymerase, thereby stimulating transcription from the promoters of heat shock genes. One of the genes transcribed is DnaK. The consequent elevation in DnaK levels results in the sequestration of σ^{32} and its rapid degradation.

In another example, σ^F is bound to an 'anti-sigma,' FlgM. This makes σ unavailable for the transcription of genes essential for the construction of the flagellum. During nutritional deprivation, the bacterium must make flagella to be able to swim toward new sources of food. The cell first turns on the synthesis of the proteins that make up the base of the flagellum. The first protein to be transported out of the cell through the base pore structure is FlgM. This frees up σ^F to bind to core polymerase, activating the transcription of the remaining flagellar genes. When the flagellum is complete, FlgM levels again build up and σ^F is converted into an inactive σ^F/FlgM sigma/anti-sigma complex.

Future Directions

Extensive work is in progress to determine the precise nature of the interaction between σ^{70} and core polymerase. A major portion of this interaction appears to involve amino acids 260–309 of the β′ subunit of core. This region of β′ is likely to be involved in the interaction of core with most, if not all, of the other sigma factors. Detailed knowledge of this important interaction may allow the design of small molecules that interfere with σ^{70}–β′ interaction. Such compounds could have potential therapeutic use as antibiotics.

Recently the crystal structure of the core RNA polymerase from *Thermus aquaticus* was determined (Zhang *et al.*, 1999), providing us with new insights into sigma binding. Since σ^{70} is likely to undergo major conformational changes during its participation in the sigma cycle, we really need a motion picture rather than a snapshot. More extensive site-directed mutagenesis of key regions of σ^{70} is in progress. Careful study of the effects of these mutations on various σ^{70} functions will provide a more detailed view of the system and a deeper understanding of sigma factors in general carry out their functions.

Work is in progress to measure the cytoplasmic concentration of each sigma factor and the level of the cognate holoenzyme under a variety of physiological states. Such information will provide insight into how the level and activity of each sigma varies with growth condition. The advent of high-density DNA arrays for simultaneously measuring the level of transcription of all 4300 *E. coli* genes will allow us to determine how many of the operons are regulated by each of the seven sigma factors and give us a much more complete picture of how global gene expression is regulated by fluctuations in the amount and activity of sigma factors.

Further Reading

Burgess RR, Travers AA, Dunn JJ and Bautz EKF (1969) Factor stimulating transcription by RNA polymerase. *Nature* 221: 43–46.

Gross CA, Lonetto M and Losick R (1992) Bacterial sigma factors. In: McKnight S and Yamamoto K (eds) *Transcriptional Regulation*, pp. 129–176. Plainview, NY: Cold Spring Harbor Laboratory Press.

Helmann J and Chamberlin M (1988) Structure and function of bacterial sigma factors. *Annual Review of Biochemistry* 57: 839–872.

Lonetto M, Gribskov M and Gross CA (1992) The σ^{70} family: Sequence conservation and evolutionary relationships. *Journal of Bacteriology* 174: 3843–3849.

References

Malhotra A, Severinova E and Darst SA (1996) Crystal structure of a σ^{70} subunit fragment from *E. coli* RNA polymerase. *Cell* 87: 127–136.

Zhang G, Campbell EA, Minakhin L, Richter C, Severinov K and Darst SA (1999) Crystal structure of *Thermus aquaticus* core RNA polymerase at 3.3A resolution. *Cell* 98: 811–824.

See *also:* Bacterial Genetics; Transcription Factor

Signal Sequence

doi: 10.1006/rwgn.2001.2023

A signal sequence is a peptide present on proteins (usually N-terminal) reponsible for cotranslational insertion into membranes of the endoplasmic reticulum. These sequences are usually present on proteins destined to become membrane components or to be secreted. They are highly hydophobic sequences of approximately 20 amino acids, which are normally removed from the growing peptide chain by signal peptidase, a specific protease of the endoplasmic reticulum.

See *also:* Proteins and Protein Structure

Signal Transduction

S Brenner

doi: 10.1006/rwgn.2001.1193

Signals acting on the outside of the cell need to have their effects transmitted across the cell membrane. This signal transduction is mediated by receptors which interact with the external signals and undergo changes in molecular conformation which alters the receptor structure on the inside of the cell. These can be detected by other proteins and passed on through complex chains of protein alterations, usually involving phosphorylation, ultimately terminating in functional changes in the cell.

One system involves the G proteins which bind GTP and which carry information from receptors called seven transmembrane receptors and lead, amongst other things, to stimulation or inhibition of the synthesis of cyclic AMP.

See *also:* cAMP and Cell Signaling

Similarity

P H A Sneath

doi: 10.1006/rwgn.2001.1196

Similarity is the degree to which two entities resemble each other in their properties, and is commonly expressed as the proportion of such resemblances in a defined list of properties. This broad definition may be qualified by restricting the comparison in some way. For example, phenetic similarity refers to similarity in observed properties without considering evolution, cladistic similarity refers to that deduced from phylogenetic principles, and genomic similarity is that from genomic data. Similarity is occasionally called resemblance or relationship. Dissimilarity is the complement of similarity (i.e., identical entities have 100% similarity but zero dissimilarity). This is useful when dissimilarity is expressed as a distance in genetic algorithms.

See *also:* Taxonomy, Evolutionary; Taxonomy, Numerical

SINE

L Silver

doi: 10.1006/rwgn.2001.1197

The general name coined for selfish genetic elements that disperse themselves through the genome by means of an RNA intermediate is 'retroposon'. There are two classes of retroposons. The SINE family is made up of very small DNA elements that require other genetic information to facilitate their dispersion throughout the genome. The LINE family is derived from a full-fledged selfish DNA sequence with a self-encoded reverse transcriptase.

The two major families of highly repetitive elements in the mouse – B1 and B2 – are both of the SINE type with relatively short repeat units of ~140 bp and ~190 bp in length respectively. In humans, the highly repetitive Alu repeat element can also be classified as a SINE type.

The significance of the short repeat length of a SINE element is that it does not provide sufficient capacity for these elements to actually encode their own reverse transcriptase. Nevertheless, SINE elements are able to disperse themselves through the genome, just like LINE elements, by means of an RNA intermediate that undergoes reverse transcription. Clearly, SINEs are dependent on the availability of reverse transcriptase produced elsewhere, perhaps from LINE transcripts or endogenous retroviruses.

All SINE elements, in the mouse genome and elsewhere, appear to have evolved out of small cellular RNA species – most often tRNAs but also (in the case of mice and humans) the 7S cytoplasmic RNA which is one of the components of the signal recognition particle (SRP) essential for protein translocation across the endoplasmic reticulum. Unlike the LINE families, however, SINE families present in the genomes of different organisms appear, for the most part, to have independent origins. The defining event in the evolution of a functional cellular RNA into an altered-function self-replicating SINE element is the accumulation of nucleotide changes in the 3′ region that lead to self-complementarity with the propensity to form hairpin loops. The open end of the hairpin loop can be recognized by reverse transcriptase as a primer for strand elongation. Since hairpin loop formation of this type is likely to be very rare among normal cellular RNAs, the SINE transcripts in a cell will be utilized preferentially as templates for the production of cDNA molecules that are able (somehow) to integrate into the genome at random sites. Like the LINE family, SINE families appear to be evolving by episodic amplification followed by sequence degradation.

***See also:* LINE; Repetitive (DNA) Sequence**

Single-Copy Plasmids

doi: 10.1006/rwgn.2001.2026

Single-copy plasmids are maintained at a level of one plasmid per host chromosome.

***See also:* Multicopy Plasmids; Plasmids**

Single-Gene Inheritance

D E Wilcox

doi: 10.1006/rwgn.2001.1198

Single-gene inheritance occurs when the development of a trait, or phenotype, is largely determined by the presence of mutation in the alleles of a single gene. In pedigrees where the mutation is present in several individuals, the inheritance of the disorder will show a pattern of affected individuals. Characteristic patterns are created depending on whether the mutated gene alleles are dominant or recessive and whether the genes are located on the autosomes or sex chromosomes.

History

The effects of genes were first recognized because mutations caused similar phenotypic differences in several members of a family. The first human disorder to be recognized as a single gene trait was alkaptonuria, which was described by Garrod in 1902. He and Bateson then proposed that affected individuals were homozygous for an underactive recessive gene. The first human gene to be mapped to a chromosome was Wilson's demonstration of the X-linked nature of color blindness in 1911. The first evidence, in any organism, that a mutation in a structural gene could cause an altered amino acid sequence in a protein came in 1956. Ingram, developing the work of Pauling, demonstrated an abnormal hemoglobin polypeptide sequence in sickle cell disease. Since then, many thousand single gene disorders have been identified and characterized. They are catalogued in Online Mendelian Inheritance in Man. At the end of 2000 the database had information on 11 372 autosomal, 674 X-linked, 37 Y-linked, and 60 mitochondrial entries. Although our knowledge of the total number of single genes in the genome is nearing completion, our understanding of genomics, or gene interactions, is still elementary.

Autosomal, Sex-Linked, and Mitochondrial Inheritance Patterns

Genes can be characterized by their location in the genome. Genes on the 22 pairs of autosomes are autosomal, genes on the pair of sex chromosomes are X-linked or Y-linked, and genes on the mitochondrial chromosome are mitochondrial. The nature of transmission of each of these chromosomes, and whether a trait caused by a mutation is dominant or recessive, create distinctive inheritance patterns of affected individuals. Mendel described some of these patterns in the inheritance of pea characteristics and sometimes, single gene inheritance is referred to as Mendelian inheritance. Strictly speaking, Mendelian inheritance does not include those single genes on the mitochondrial chromosome.

In practice, many individual pedigrees with a single gene disorder may not show a typical Mendelian pattern of inheritance. This can occur for a variety of reasons. A single affected individual may be a new mutation. A new dominant mutation may be lethal before reproduction. A healthy parent may have two or more affected offspring as a result of a new dominant mutation occurring in the parent's gonads (gonadal mosaicism). An affected parent may transmit only normal alleles to offspring. Some individuals may carry the mutation but not express the expected phenotype.

Recessive and Dominant Traits

If mutation needs to be present in both alleles (m/m) before altering the phenotype, the trait is recessive. If a mutation in a gene alters the phenotype when it is present only in a single allele in the heterozygote ($+/m$), the trait is dominant.

Molecular Basis of Recessivity

The mutated allele in a heterozygote for a recessive trait ($+/m$) has little or no effect on the phenotype generally because the mutation causes a loss of function with no gain of new function or interference of function of the healthy allele. The normal protein is usually present in the heterozygote at half the level found in the normal homozygote but this is sufficient to maintain the normal phenotype. Examples of recessive disorders include enzyme defects such as alkaptonuria and cell membrane receptor or channel defects such as cystic fibrosis. Some recessive traits such as sickle cell disease, a disorder of hemoglobin, result in two proteins (normal and mutated) being produced and found in the target tissue. These traits may show a change of phenotype in heterozygotes in some environments. In nonmalarial areas the sickle cell heterozygote phenotype is normal but in malarial areas the heterozygotes are fitter than the normal homozygotes because the presence of sickle hemoglobin in the red blood cells interferes with the malarial parasites' life cycle. This is called heterozygote advantage. Other recessive traits, such as Duchenne muscular dystrophy, are caused by mutations that result in little or no mutated protein being found in the target tissue.

Molecular Basis of Dominance

The mutated allele in a heterozygote for a dominant trait ($+/M$) affects the phenotype by either interfering with the function of the normal allele or by gain of new function. Examples of dominant traits include collagen disorders such as osteogenesis imperfecta, a disorder causing brittle bones. In these, the mutated gene codes for a protein that is a subunit of a larger multimeric structural protein. The presence of the mutated subunit protein degrades the function of the final protein. This explains the apparent paradox in which a mutation in a subunit gene, which produces no protein, is recessive and results in a normal phenotype in the heterozygote but an apparently less severe mutation, which produces an abnormal protein, results in a severe dominant trait.

Other dominant traits such as Huntington disease result from a gain of function of the mutated protein. Huntington disease is a progressive neurodegenerative condition that causes dementia and a movement disorder. Affected individuals develop and grow normally until symptoms develop, usually in middle age. The normal and mutated huntingtin protein, which contains an expanded glutamine repeat near the protein N-terminal, are widely expressed but a gain of function, expressed as abnormal metabolism of the mutant protein, leads to aggregation of N-terminal huntingtin fragments. The presence of these aggregates in neurones predisposes to early cell death.

Complete (True) and Partial Dominance

Complete dominance

When a dominant trait is caused by a mutation that causes only a gain, and no loss, of function of the mutant protein, then individuals homozygous for the mutant allele would be expected to have the same phenotype as heterozygotes. Huntington disease is an example of a complete or true dominant condition. The mutant huntingtin proteins in homozygotes have not lost their normal function, allowing normal growth and development until the gain of function causes early neuronal death.

Partial dominance

Many dominant traits show partial dominance. This is demonstrated when individuals who are homozygous for the mutant allele have a more severe phenotype than heterozygotes. In addition to the mutant allele's gain of function, causing dominance of the trait, there is partial loss of normal function that is recessive in the heterozygote. An example is achondroplasia. The phenotype of heterozygotes involves restricted growth but normal life expectancy. In contrast, mutant homozygotes have a very severe skeletal dysplasia and die in infancy.

Other Factors Affecting the Phenotype in Single-Gene Inheritance

Many single-gene disorders, particularly autosomal dominant traits, show considerable variation in the phenotype and individuals with healthy phenotypes may transmit the trait demonstrating the presence of the mutant genotype. Penetrance and expressivity are common factors influencing autosomal dominant phenotypes.

Penetrance

The penetrance of a trait is the proportion of those who have the trait genotype (obligate carriers) who show the trait phenotype. A trait with full penetrance, such as achondroplasia, results in all heterozygotes developing the trait phenotype. Other disorders show reduced penetrance, e.g. breast cancer caused by *BRCA1* mutations shows about 85% penetrance in female heterozygotes. Penetrance may also be age-dependent. For example, achondroplasia is 100% penetrant at birth, neurofibromatosis is near 100% penetrant by the end of the second decade, and Huntington disease is near 100% penetrant if heterozygotes live long enough.

Variable Expressivity

Stable mutations

Variable expressivity refers to variations in the degree of severity of a phenotype. Some single-gene disorders such as adrenoleukodystrophy, a multisystem disorder affecting the adrenal glands and nervous system, show a marked variation in phenotype involving age of onset and extent of involvement of each system. In one family, where each affected member has the same mutation, the phenotype can vary from severely affected children to asymptomatic, nonpenetrant adults. Variable expressivity in disorders with stable mutations may represent the effects of modifying genes and/or environmental factors on the final phenotype.

Unstable mutations

The mechanism that causes variable expressivity in some disorders is related to instability of the causative mutation, even among members of the same family. An example is fragile X syndrome, which causes a variable dysmorphic mental retardation. It is caused by an unstable amplified CGG trinucleotide repeat mutation and the size of an individual's CGG repeat mutation correlates with the severity of the phenotype.

Anticipation

Several autosomal dominant disorders show anticipation where the age of onset is earlier and the phenotype more severe in successive generations. Myotonic dystrophy is an example where the first generation may only develop cataracts in late middle age, the second generation may develop muscular weakness and stiffness in early adult life, and the third generation may have severe congenital onset. Anticipation in myotonic dystrophy is caused by instability of the amplified CTG trinucleotide repeat mutation. The number of repeats tends to increase with each generation, particularly when transmitted by a female. Mildly affected adults in the first generation of an affected family may have only 50 repeats but a congenitally affected infant may have more than 2000.

Mendelian Disorders: Are They Truly Single-Gene?

Some single-gene disorders such as achondroplasia and Duchenne muscular dystrophy show little variation in severity of the phenotype, even in unrelated individuals and can be considered to show true single-gene inheritance. However, disorders with stable mutations showing variable expressivity suggest the possible effects of modifying genes. There are also conditions in which only the susceptibility to the trait is inherited as a single-gene disorder. Examples include autosomal dominant familial cancers such as early onset breast cancer and early onset colon cancer. Cancers are caused by a sequence of genetic changes (which may be triggered by environmental factors) occurring in a clone of somatic cells in the affected tissue. Over time, these somatically inherited mutations, some of which may need to become homozygous, lead to uncontrolled cellular proliferation in the clone. In the familial cancers, the first key step is inherited through the germ line, often as an autosomal dominant. This results in the whole sequence being completed more quickly, giving rise to an earlier age of onset in familial cancers than in sporadic cancers. Some individuals who have inherited the susceptibility mutation may not be exposed to the factors that cause the full subsequent sequence of somatic mutations and so may never develop cancer and be nonpenetrant for the trait. These healthy individuals can pass their susceptibility mutation to their offspring who may not appreciate their own risk of developing cancer.

Finally, mutations in some 'single-gene disorders' may not be in a single-gene at all. In myotonic dystrophy, which is transmitted as an autosomal dominant single-gene disorder, the associated mutation is situated in the 3′-untranslated region of the *DMPK* gene. This is also the promoter region of *SIX5* gene, which is immediately downstream of *DMPK*. The multisystem nature and very variable phenotype of this disorder are not yet explained by knowledge of the function of these two genes, raising the possibility that the myotonic dystrophy mutation affects the expression of additional genes either directly on local genes as a result of disruption of the normal chromatin structure or indirectly through the effects of altered expression of *DMPK* and *SIX5* on the expression of genes elsewhere in the genome.

Further Reading

Connor JM and Ferguson-Smith MA (1997) *Essential Medical Genetics*, 5th edn. Oxford: Blackwell Scientific Publications.

Gelehrter TD, Collins FS and Ginsburg D (1998) *Principles of Medical Genetics*, 2nd edn. Baltimore, MD: Williams & Wilkins.

Online Mendelian Inheritance in Man: http://www.ncbi.nlm.nih.gov/omim/

University of Glasgow, Department of Medical Genetics, Encyclopaedia of Genetics pages contain a number of illustrations and animated diagrams to accompany this article: http://www.gla.ac.uk/medicalgenetics/encyclopedia.htm

Winchester CL, Ferrier RK, Sermoni A, Clark BJ and Johnson KJ (1999) Characterization of the expression of *DMPK* and *SIX5* in the human eye and implications for pathogenesis in myotonic dystrophy. *Human Molecular Genetics* 8(3): 481–492.

See also: **Dominance; Expressivity; Mutation; Penetrance; Recessive Inheritance**

Single Nucleotide Polymorphisms (SNPs)

J Read and S Brenner

doi: 10.1006/rwgn.2001.2094

SNPs (pronounced snips) are single-nucleotide polymorphisms. They are single-base variations in the genetic code that occur about every 1000 bases along the 3 billion bases of the human genome. Researchers believe that knowing the locations of these closely-spaced DNA landmarks will help to discover genes involved in such major human diseases such as asthma, diabetes, heart disease, schizophrenia, and cancer.

See also: **Genetic Diseases; Human Genome Project; Polymorphism**

Single-Strand Annealing

P J Hastings

doi: 10.1006/rwgn.2001.1500

There are normally two mechanisms available for the repair of a DNA double-strand break: homologous recombination and nonhomologous end joining. In special circumstances, there is a third way. When there is a repeated nucleotide sequence in a direct orientation (pointing the same way), exonucleolytic removal of one polynucleotide chain from each of the broken ends can reveal the complementary sequences of the repeated length. Rejoining by complementary base pairing, followed by endonucleolytic removal of the loose ends, filling in gaps, and subsequent ligation, yields a rejoined molecule that has lost one copy of the repeated sequence, and any sequence that was between the repeats. This mechanism can operate to repair double-strand breaks when a homologous duplex molecule is not present, as in a haploid cell in the G_1 phase of the cell cycle.

The single-strand annealing mechanism was first proposed to explain plasmid recombination in mammalian cells and was subsequently studied in detail in *Xenopus* and *Saccharomyces cerevisiae*. **Figure 1** shows a scheme by which single-strand annealing is proposed to occur. The broken ends of the DNA molecule are processed as proposed in other recombination models, by resection of ends of like polarity. When this resection exposes the complementary sequences of the repeated nucleotide sequence, they can anneal with each other by complementary base pairing. The nonhomologous tail requires a flap endonuclease (the products of the *RAD1* and *RAD10* genes in *S. cerevisiae*) for its removal unless the tail is shorter than about 30 nucleotides. The editing function of a DNA polymerase is believed to be able to remove very short tails. As **Figure 1** shows, we expect to need some DNA synthesis to extend the 3′ ends where there has been excessive resection. The final step will be ligation to close the last nicks and restore intact DNA with deletion as described.

In *S. cerevisiae*, the process is highly efficient if the lengths of homology are 400 bp or more. The distance between the repeats can be short, or up to 10 or 20 kb. Time to completion of the repair depends on the distance between the repeats, apparently because of the time required for longer resection. Single-strand annealing differs from crossing-over between the directly repeated sequences in that single-strand annealing is not conservative. The reciprocal product, a circle consisting of the deleted length, is not formed.

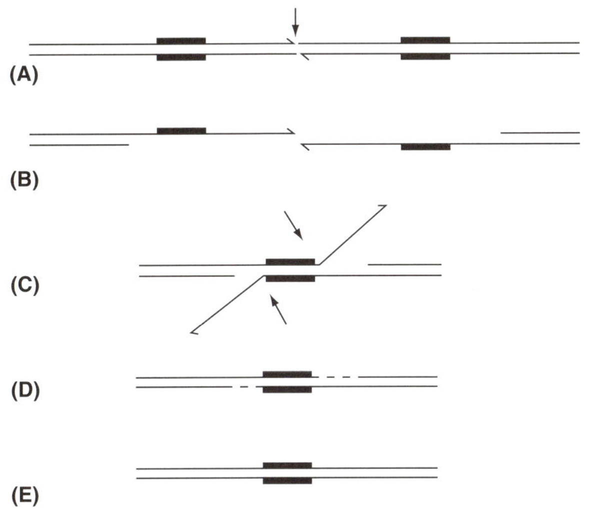

Figure 1 Double-strand break repair and deletion formation by single-strand annealing. Lines represent polynucleotide chains, with broken lines showing new synthesis. Lengths of repeated sequence are shown by a thicker line. Half arrows indicate 3′ ends. The open arrow shows a double-strand break. Solid arrows show places where flap endonuclease activity is required. (A) A double-strand break occurs anywhere within or between the repeated sequence. (B) Resection of 5′ ends by exonuclease reveals complementary sequences in the 3′ tails. (C) Complementary regions become annealed. (D) Nonhomologous tails are removed by a flap endonuclease, and gaps are filled by new synthesis. (E) Ligation yields an intact DNA molecule with one repeat and the sequence between deleted.

Direct repeated sequences, which are common in the human genome, would be subject to removal by this mechanism. These may persist because they have diverged enough to reduce the likelihood of homologous interaction.

Further Reading

Fishman-Lobell J, Rudin N and Haber JE (1992) Two alternative pathways of double-strand break repair that are kinetically separable and independently modulated. *Molecular and Cellular Biology* 12: 1292–1003.

Lin DS, Sperle K and Sternberg N (1984) Model for homologous recombination during transfer of DNA into mouse L-cells: role for DNA ends in the recombination process. *Molecular and Cellular Biology* 4: 1020–1034.

Paque F and Haber JE (1999) Multiple pathways of recombination induced by double-strand breaks in *Saccharomyces cerevisiae*. *Microbiology and Molecular Biology Reviews* 63: 349–404.

See also: **Double-Strand Break Repair Model; Recombinational Repair; Repair Mechanisms**

Single-Strand Assimilation

doi: 10.1006/rwgn.2001.2027

Single-strand assimilation is the ability of RecA protein to cause a DNA strand to displace its homologous strand in a duplex, i.e., the single strand is assimilated into the duplex.

See also: **RecA Protein and Homology**

Single-Strand Exchange

doi: 10.1006/rwgn.2001.2028

Single-strand exchange is a reaction whereby one of the strands of a DNA duplex leaves its former partner and instead pairs with the complementary strand of another molecule, displacing its homolog in the second duplex.

See also: **Homologs**

Single-Stranded DNA-Binding Proteins (SSBs)

A L Eggler

doi: 10.1006/rwgn.2001.1233

The single-stranded DNA-binding protein (SSB) of *Escherichia coli* is a well-studied member of a class of proteins that have essential roles in DNA replication, recombination, and repair. These proteins lack enzymatic activity and contribute to the 'three Rs' of DNA metabolism mainly through their shared ability to bind with high affinity to single-stranded DNA (ssDNA) and with low affinity to double-stranded DNA (dsDNA). Additionally, they organize processes by selectively binding to various participating proteins. Other well-studied SSBs discussed here are gene 32 protein (gp32) from the T4 bacteriophage and replication protein A (RPA) from humans and *Saccharomyces cerevisiae*.

Functional Aspects of Having a High Affinity for ssDNA

Helix Destabilization

In the metabolism of DNA during replication, recombination, and repair, regions of ssDNA are generated. Random ssDNA sequences have a tendency to form internal Watson–Crick bonds, and this secondary structure can interfere with the binding of proteins used in DNA metabolism. SSBs possess a helix destabilization ability, by virtue of their high specificity for ssDNA, which allows them to 'melt' these structures.

Protection of ssDNA

It is generally thought that due to their great abundance in the cell, and their high affinity for ssDNA, SSBs are the first proteins to bind newly formed ssDNA. Not only do they reduce secondary structure formation, but they also prevent unintentional access to the ssDNA by other proteins. Endonucleases are particularly deleterious, as a nick in the single strand generates a double-strand break. If left unrepaired, these breaks can lead to the inability to replicate DNA or gross chromosomal rearrangements during cell division.

Structure of SSBs: How Do They Bind DNA?

SSBs from all organisms are, for the most part, functional homologs. Sequence similarity between various SSBs is limited; however, they share a DNA-binding motif called the OB fold. This fold, consisting of a five-stranded antiparallel β barrel with a terminal α helix, binds rather weakly but specifically to ssDNA. ssDNA binds in a narrow cleft formed by the motif and interacts with various residues through contacts with its backbone, sugar, and base moieties. The relatively weak binding of an OB fold to ssDNA can lead to strong binding of an SSB to ssDNA when more than one motif is present.

The actual sequence, structure, and binding modes differ considerably for SSBs from various species, but most utilize multiple copies of the OB fold to bind ssDNA. SSB from *E. coli*, a monomer of which has a molecular weight of 19 kDa, forms a stable homotetramer and effectively presents four OB folds to ssDNA. Almost the entire sequence of SSB is devoted to forming the OB fold, highlighting the lack of enzymatic activity for this protein.

E. coli SSB can bind to ssDNA in several different modes, depending on the salt concentration. At all salt concentrations, ssDNA is wrapped around SSB, causing the apparent DNA length to be shorter. At relatively low salt concentrations, only two of the tetramer subunits contact the DNA. In this mode, SSB exhibits a high cooperativity in DNA binding. This mode is illustrated in **Figure 1**, in which the apparent length of ssDNA bound with SSB is greatly reduced compared to linear dsDNA. At higher salt concentrations, all four subunits interact with the DNA, and cooperativity is limited to the formation of octamers, leading to a "beads on a string appearance." Intermediate binding modes have also been detected. The level of cooperativity observed in DNA binding is an important consideration when other proteins are competing with SSBs for binding sites.

The other SSBs differ markedly in their structure from *E. coli* SSB. All identified RPA homologs are heterotrimers whose subunits have molecular weights of approximately 70, 30, and 14 kDa. OB folds are found in all of the subunits. Human RPA displays a low cooperativity in its DNA-binding function, while the results for yeast RPA are less clear. T4 gp32 is a stable monomer of 33.5 kDa, and contains an OB fold. Unlike SSB, gp32 does not rely on multiple copies of the OB fold to bind ssDNA tightly. Instead, gp32 binding is aided by the high level of positive cooperativity displayed under all binding conditions.

Roles in Cellular Processes

SSBs play major roles in DNA metabolism in the cell. In reviewing these roles, only a brief outline of replication, recombination, and repair processes are given below. For a more complete review of these processes, see Replication, Genetic Recombination, Recombinational Repair, Mismatch Repair (Long/Short Patch), and Excision Repair.

Replication

Owing to the intrinsic nature of replication, the process involves unwinding dsDNA to generate ssDNA. The capability of SSBs to bind ssDNA and interact functionally and physically with other proteins places them in a crucial position. *E. coli* SSB aids the formation and stabilization of origins of replication, and assembly and modulation of the primosome, allowing for primer synthesis only near the origin of replication used *in vivo*. SSB aids DNA helicases as they unwind DNA. Using ATP, helicases disrupt extensive Watson–Crick base pairing. Without SSB, separated single strands would rapidly reanneal. SSB has severa efects on the DNA polymerase, enhancing both polymerase–template binding and polymerase fidelity, as well as destabilizing secondary structure that would lead to lower polymerase processivity.

RPA serves mainly the same functions as SSB in replication, and protein–protein interactions are

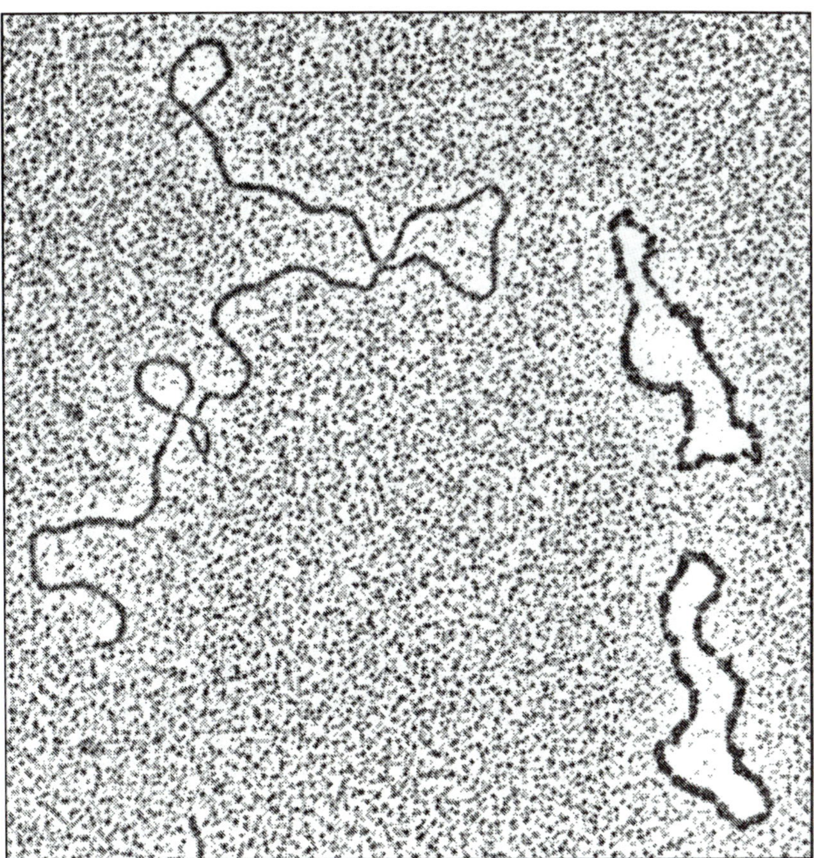

Figure 1 Electron micrograph of circular ssDNA bound with *Escherichia coli* SSB. At the right-hand side are two circular ssDNA molecules bound with SSB protein. At the left-hand side, a linear double-stranded molecule of the same sequence as the circular ssDNA is shown for comparison. SSB causes a large apparent decrease in the contour length of the DNA due to the wrapping of DNA around the SSB homotetramer. (DNA is from ϕX174. Samples were cross-linked and spread using cytochrome *c* grids: courtesy of Ross B. Inman, University of Wisconsin Madison.)

important for some of its functions. For example, formation of the priming complex in SV40 replication requires interaction between human RPA and both T-antigen and DNA polymerase α/primase. Similarly in T4, gp32 interacts with gp43 (DNA polymerase), gp61 (primase), and gp59 (helicase loading factor), ensuring proper assembly of the replication machinery onto ssDNA and successful replication.

Recombination

In recombination, as in replication, a necessary prerequisite to the process is ssDNA. In homologous recombination in *E. coli*, the RecA protein coats the ssDNA, searches for homologous DNA, and facilitates a strand switch, creating a heteroduplex DNA and a displaced single strand. SSB is required for complete binding of RecA to ssDNA, by virtue of its ability to melt out secondary structure as described above. However, SSB and RecA compete for binding sites on ssDNA. While the nucleation of RecA onto ssDNA is inhibited, once a RecA monomer is bound, other monomers bind cooperatively, displace SSB, and coat the ssDNA. The cooperativity of SSB must be low enough to allow RecA to displace it. If the ssDNA is precoated with SSB, RecO, in a complex with RecR, physically interacts with SSB to allow RecA binding.

In addition to aiding the binding of RecA, SSB participates in the formation of certain ssDNA sites for recombination. It modulates the activity of the RecBCD helicase/nuclease, which generates ssDNA regions from double-strand breaks and loads RecA onto the appropriate ssDNA. Finally, after the initiation of DNA strand exchange, SSB aids the reaction by binding to the displaced strand, preventing reinitiation of strand exchange that could lead to extended DNA networks.

RPA and gp32 also serve to facilitate complete binding of the cognate strand exchange proteins Rad51 and UvsX, respectively. As in *E. coli*, mediator proteins aid in binding of Rad51 and UvsX to ssDNA coated with SSBs. Modulation is species specific, with one mediator known in T4, UvsY, and two in yeast, Rad52 and Rad55/57.

Repair

SSBs participate in DNA repair processes, including mismatch repair, nucleotide excision repair, base excision repair, and recombinational repair, as outlined above. Well-studied examples are the involvement of SSB in mismatch repair and RPA in nucleotide excision repair.

Base pair mismatches that arise occasionally during replication are subsequently repaired in a process that takes advantage of the fact that newly synthesized DNA in some organisms is undermethylated. In reconstituted reactions *in vitro*, methyl-directed mismatch repair requires SSB for DNA helicase II-mediated unwinding of DNA, stimulation of exonucleolytic excision of the strand containing the error, and synthesis of a complementary strand.

Human RPA plays several roles in nucleotide excision repair. RPA and the XPA protein initially sense the DNA damage and bind to it. Other factors subsequently bind, and the damaged DNA is cleaved and excised. RPA also stabilizes the ssDNA gap prior to DNA synthesis.

Summary

Although SSBs lack enzymatic activity, they are essential for DNA metabolism in the cell. It is rare for one protein to have key roles in processes with very different mechanisms, as SSBs do; however, these processes are linked by the involvement of ssDNA. Not only do SSBs protect ssDNA and remove secondary structure, they also guide the assembly of the machinery required for DNA metabolism.

See also: **Excision Repair; Genetic Recombination; Mismatch Repair (Long/Short Patch); Recombinational Repair; Replication**

Sister Chromatids

doi: 10.1006/rwgn.2001.2029

A sister chromatid is one of the two chromatids comprising a bivalent. Both chromatids are semiconservative copies produced by replication of the original chromosome.

See also: **Chromatid; Chromosome; Semiconservative Replication**

Site-Directed Mutagenesis

See: ***In vitro*** **Mutagenesis**

Site-Specific Recombination

N D F Grindley

doi: 10.1006/rwgn.2001.1200

Site-specific recombination describes a variety of specialized recombination processes that involve reciprocal exchange between defined DNA sites. In its strictest definition, site specific recombination involves: (1) two DNA partners, (2) a specialized recombinase protein that is responsible for recognizing the sites and breaking and rejoining the DNA, and (3) a mechanism that involves DNA breakage and reunion with conservation of the phosphodiester bond energy (i.e., lacking a requirement for either DNA synthesis or a high-energy nucleotide cofactor). A consequence of these features is that site-specific recombination is not dependent on the cellular machinery for homologous recombination. The prototypes of site-specific recombination (thus defined) are the integration of bacteriophage lambda into the *Escherichia coli* chromosome (see Phage λ Integration and Excision), the resolution of cointegrates derived from transposition of Tn3-related transposons (see Resolvase-Mediated Deletion), and the DNA inversions responsible for flagellar phase variation in *Salmonella* (see Hin/Gin-Mediated Site-Specific DNA Inversion). The strict definition excludes several other specialized recombination processes that have, on occasion, been described as 'site-specific'; these include VDJ joining catalyzed by the RAG1/2 proteins during the development of the immune system (see Integration, T Cell Receptor Gene Family); most DNA transposition events (even when a specific target site is used) including integration of retroviral cDNA (see Retrotransposons, Retroviruses); and the 'homing' of mobile introns (or inteins) (see Intron Homing).

Structural Consequences of Site-Specific Recombination

Recombination sites are naturally polar and recombination respects that polarity, always joining left halves to right (or, as shown in **Figure 1**, arrow heads to tails). Depending on the initial arrangement of the parental recombination sites, site-specific recombination has one of three possible outcomes: integration,

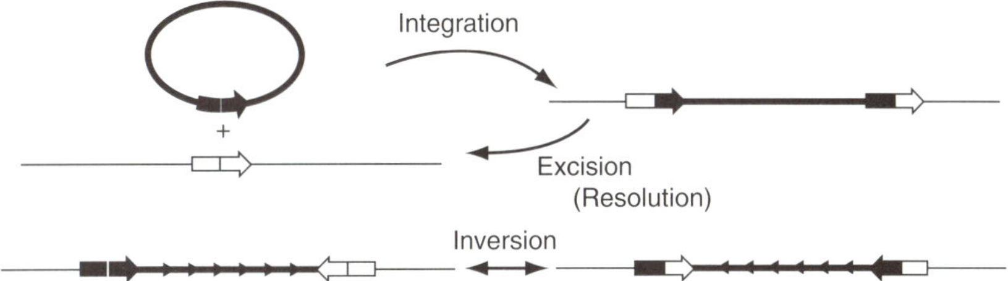

Figure 1 The three consequences of site-specific recombination. The recognition sites for the site-specific recombinase are represented by the broad black-and-white arrows.

excision, or inversion (**Figure 1**). Integration results from recombination between sites on separate DNA molecules (provided at least one of the parental chromosomes is circular) and occurs with a uniquely defined orientation. For sites located on the same chromosome, the outcome is determined by their relative orientation. Thus, excision results from recombination between sites in a head-to-tail orientation, while inversion results from exchange between inverted (head-to-head) sites.

Biological Consequences of Site-Specific Recombination

The three structural outcomes are used for a wide variety of purposes in biological systems and a number of examples are shown in **Table 1**. Most commonly, use of site-specific recombination by an organism or a genetic element is driven by a primary need to physically join or separate DNA segments. However, it is also used as a means of activating or switching gene expression, and generating genetic diversity through the acquisition of advantageous genes or gene segments.

The first category of uses is dominated by three common biological processes. (1) The integration of bacteriophage chromosomes into (and their excision from) the chromosome of their host for the reversible formation of lysogens. This phage strategy is also used by a few classes of transposons, exemplified on the one hand by Tn916 and Tn1545, and on the other by Tn4451 (see Conjugative Transposition) (see **Table 1**). (2) The reduction to a monomeric state of dimers of a variety of circular chromosomes including both plasmids and bacterial chromosomes, to allow separation and correct segregation into daughter cells (see Chromosome Dimer Resolution by Site-Specific Recombination). (3) Cointegrate resolution: the irreversible excision of the transposon-donor vector replicon from the cointegrate intermediate formed during transposition of elements such as Tn3, to regenerate the transposon donor and produce a simple insertion of the Tn in the target DNA (see Resolvase-Mediated Deletion).

In the second category of uses, the primary purpose of the recombination is to juxtapose alternative DNA sequences in ways that affect their expression or coding potential. Inversions provide a relatively common means to achieve these goals in a reversible manner. For example, the inversions mediated by FimE, FimB, and Hin flip the orientation of a transcriptional promoter, switching adjacent genes off and on (see Hin/Gin-Mediated Site-Specific DNA Inversion). The inversions mediated by Gin, Cin, Rci, and Piv bring alternate coding sequences to the downstream segment of an expressed gene, changing the C-terminal portion of the encoded protein in ways that affect its activity or antigenicity (see Alternation of Gene Expression, Hin/Gin-Mediated Site-Specific DNA Inversion). Deletions can also be used to affect gene expression; for example the XisA, XisC, and XisF functions in *Anabaena*, and CisA (SpoIVCA) in *Bacillus subtilis* delete large DNA segments that split specific genes into two inactive portions, to create an active gene fusion (see Gene Rearrangements, Prokaryotic). In these cases, the change is irreversible and is part of a developmentally regulated pathway leading to a terminally differentiated cell type. Finally, a system that combines features of both categories is the acquisition of mobile gene cassettes mediated by the IntI activities of integrons (see Integrons, Gene Cassettes). Integration of a cassette is required both for cassette survival and for its expression, which generally occurs from a promoter adjacent to the insertion site.

The Mechanism of Site-Specific Recombination: An Overview

The process of site-specific recombination can be divided into a series of conceptually simple steps. The recombinase binds to the two recombination sites. The two recombinase-bound sites pair, forming a synaptic complex with crossover sites juxtaposed. The recombinase then catalyzes cleavage and rejoining of the DNA within the synaptic complex. Finally, the synaptic complex breaks down, releasing the recombinant products.

From this description it follows that the minimal components of a site-specific recombination system are a recombinase and a pair of recombination sites. The simplest sites are short duplex DNA segments,

Table 1 Site-specific recombination: a sampling of enzymes and functions

Recombinase	Biological function
λ Integrase family	
λ Int and many phage integrases	Integration and excision of phage genomes
Int of Tn916/Tn1545	Integration and excision: 'transposition' of circular transposons
IntI	Integration and excision of gene cassettes in integrons
Cre	Excision: dimer reduction in phage P1 plasmids
XerCD	Excision: dimer reduction in the *E. coli* and many other bacterial chromosomes, and some plasmids
TnpI of Tn4430	Excision: resolution of cointegrates resulting from transposition of Tn4430
FimB, FimE	Inversion: alternation of gene expression (fimbrial phase variation in *E. coli*)
Rci of R64	Inversion of shufflon segment in plasmid R64 producing various forms of pili
Flp	Inversion: for amplification of yeast 2 μm plasmid
Resolvase family	
TnpR of Tn3/γδ and related transposons	Excision: resolution of cointegrates resulting from transposition
ParA of RP4	Excision: dimer reduction in plasmid RP4
Hin	Inversion: alternation of gene expression (flagellar phase variation) in *Salmonella*
Gin, Cin	Inversion: alternation of gene expression (tail fiber proteins) in phages Mu and P1
Int of ϕC31/Sre of R4[a]	Integration and excision of *Streptomyces* phages ϕC31 and R4
TnpX of Tn4451[a]	Integration and excision of Tn4451 in *Clostridium*
SpoIVCA (CisA)[a]	Excision: for developmentally regulated gene activation in *B. subtilis*
XisF[a]	Excision: for developmentally regulated gene activation in *Anabaena*
Other classes	
Piv	Inversion: alternation of gene expression (pilin phase variation) in *Moraxella*
XisA, XisC	Excision: for developmentally regulated gene activation in *Anabaena*

[a] Unusually large members of the resolvase family.

20 to 30 base pairs in length, that contain an inverted pair of recognition sequences and bind one dimer (or two monomers) of the recombinase. Such sites contain at their center the point of DNA breakage and joining, and are often referred to as the crossover sites. In nature, however, most recombination sites are more complicated containing not only a crossover site, but additional sequences spanning 100 or more base pairs. Such a complex site may operate in combination with a simple crossover site or with another complex partner. The extra DNA contains additional sites of protein recognition and may bind more copies of the recombinase or other protein factors encoded by the host or the genetic element (e.g., phage or transposon) associated with the recombination system. The purpose of these additional DNA-bound proteins may be regulatory, structural, or both. They may initiate or stabilize the pairing of recombination sites, or inhibit inappropriate pairings; they may deliver recombinase catalytic domains to the crossover site; and they may determine the directionality of recombination (for example, promoting deletion but preventing inversion, or vice versa).

As indicated earlier, breakage and rejoining of DNA in site-specific recombination occurs with no loss or gain of nucleotides and with strict conservation of phosphodiester bond energy. To achieve this, a mechanism analogous to that of a topoisomerase (see Topoisomerases) is used; DNA strands are broken not by hydrolysis but rather by direct phosphoryl transfer to a side chain of the recombinase. This side chain, a tyrosine or a serine in all characterized cases, directly attacks the DNA sugar–phosphate backbone at the crossover site in a transesterification reaction, forming a covalent recombinase–DNA intermediate on one side of the break and a free hydroxyl group on the other. Rejoining the DNA strands is accomplished by reversing the process; the free hydroxyls from one recombination partner directly attack the phosphodiester linkage between recombinase and DNA of the other partner, releasing the recombinase and sealing the breaks to produce recombinant products. Intriguingly, the details of the process differ depending on whether the recombinase uses a tyrosine or a serine as the attacking nucleophile (see below).

The Specialized Recombinases and their Mechanisms of Recombination

Despite the many and distinct roles that site-specific recombination plays in biology and the large number of systems that have been identified, comparisons of the recombinase amino acid sequences indicate that nearly all fall into two families. These are the integrase family, named after the prototypical phage lambda integrase, and the resolvase family, named after the cointegrate-resolving recombinase encoded by the transposons Tn3 and $\gamma\delta$. The two families are unrelated in protein sequence or structure and employ different recombinational mechanisms; each family appears to have arisen and evolved separately. Despite the existence of these two distinct families, members of one family are not all associated with a particular set of structural and biological consequences. Thus, although the prototypical integrase is responsible for reversible integration and excision, there is at least one integrase-related cointegrate resolvase, and other members of the family catalyze DNA inversion. Similarly, although the prototypical resolvase catalyzes irreversible excision, related enzymes catalyze inversion, or combined (i.e., reversible) integration and excision (see **Table 1**).

The Integrase Family: Tyrosine Recombinases

Members of the integrase family all possess a tyrosine nucleophile in combination with a totally conserved set of basic amino acid residues, two arginines and a histidine, known as the RHR triad. These residues are essential for full recombinational activity. A particular feature of recombination performed by members of the integrase family is that double-strand breaks are not observed; rather, after each crossover site is nicked by the recombinase, it must be joined to its partner before the second strand can be cut. This produces a cross-strand intermediate called a 'Holliday junction.'

Biochemical and structural analyses have elucidated many of the details of the recombination process (see **Figure 2A**). Within the synaptic complex the two crossover sites are held in antiparallel (head-to-tail) alignment by a tetramer of the recombinase. The initiating catalytic event is attack by a pair of diametrically opposed integrase subunits on one strand of each parental DNA duplex, three or four nucleotides 5′ to the center of the crossover site. The active site tyrosines link to the 3′ phosphates of each nicked strand, liberating a 5′ OH. These free ends melt away from the unbroken complementary strands of the parental duplex, and reach across to the partner duplex, forming an open square with each side composed of a short single-strand segment. The 5′ OHs attack the integrase–DNA phosphotyrosine linkages, releasing the recombinase and forming the first recombinant joint. This religated intermediate, with one pair of recombinant single strands and one pair of parental strands, is the classical Holliday junction: two homologous duplex DNAs connected by a pair of reciprocal single-strand exchanges. The second set of single-strand exchanges, necessary to complete the recombination, occurs in a similar fashion. The other pair of opposed integrase subunits cleaves the unexchanged parental strands, the freed 5′ OH ends again reach across to their partners (forming a heteroduplex with the single strand segment initially exchanged (see **Figure 2A**)) and initiate the fourth and final pair of phosphoryl transfers.

The Resolvase Family: Serine Recombinases

Members of the resolvase family all contain a serine nucleophile in a short, conserved stretch of amino acid

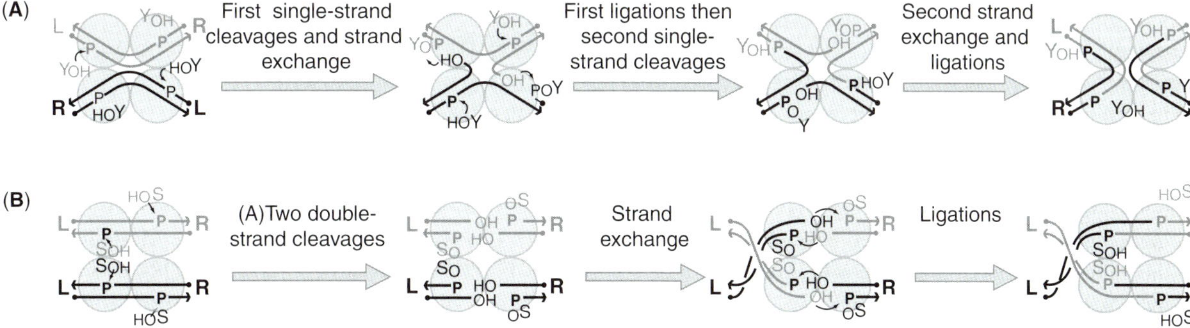

Figure 2 The processes of DNA strand exchange in site-specific recombination. (A) Mechanism of integrase-related recombinases; (B) mechanism of resolvase-related recombinases. Y_{OH} represents each tyrosine nucleophile, S_{OH} represents the serine nucleophile, P is the phosphate at the recombinase cleavage site. DNA 5′ and 3′ ends are represented by the terminal circles and arrowheads, respectively.

residues near to the N-terminus of the protein. The serine plus several other conserved residues, including three arginines, are essential for recombination activity. A defining feature of recombination by members of this family that distinguishes them from integrase-related recombinases is the formation of double-strand breaks at both crossover sites; all strands are broken before any exchange is initiated.

As with the integrase family, strand exchange occurs within a synaptic complex, containing the paired recombinase-bound crossover sites. However, the organization of this complex and, in particular, the movements of DNA ends or the recombinase that effect strand exchange remain a mystery since no structures of the complex have yet been solved. Synapsis triggers the catalytic activity of all four recombinase subunits bound to the crossover sites (**Figure 2B**). The serine nucleophiles attack both strands of the two parental DNAs, at the phosphates positioned one nucleotide 3′ to the center of the crossover sites. This creates staggered breaks with a 3′ single-strand extension of two nucleotides terminated with a 3′ OH, and a recessed 5′ phosphate covalently linked to the recombinase via the active site serine residue. Without dissociation of the complex, the ends somehow reassort from a parental to a recombinant configuration, so that attack by the free 3′ OHs on the phophoserine linkages produces recombinant products (and releases the recombinase).

Other Classes of Recombinase

A few site-specific recombinases, listed at the end of **Table 1**, appear to be unrelated to members of either of the two large families. Moreover, Piv is unrelated to XisA and XisC, suggesting that there are at least two classes of recombinase that await further characterization. Since almost nothing is yet known about these recombinases, they will not be discussed further here.

Further Reading

Grindley NDF (1997) Site-specific recombination: synapsis and strand exchange revealed. *Current Biology* 7: R608–R612.

Mizuuchi K (1997) Polynucleotide transfer reactions in site-specific recombination. *Genes to Cells* 2: 1–12.

Nash HA (1996) Site-specific recombination: integration, excision, resolution, and inversion of defined DNA segments. In: Neidhardt FC, Ingraham J, Low K *et al.* Escherichia coli *and* Salmonella: Cellular and Molecular Biology, 2nd edn, pp. 2363–2376. Washington, DC: ASM Press.

Nunes-Duby, Kwon HJ, Tirumalai RS *et al.* (1998) Similarities and differences among 105 members of the Int family of site-specific recombinases. *Nucleic Acids Research* 26: 391–406.

Skeletal Disorder

***See:* Achondroplasia**

Ski Oncoprotein

M Frame

doi: 10.1006/rwgn.2001.1621

c-Ski is a nuclear protein with transforming and myogenesis-promoting activities. It is the cellular homolog of the v-Ski oncoprotein that is responsible for transformation induced by the SKV avian carcinoma virus. v-Ski, which is produced in infected avian cells as a fusion with N-terminal gag sequences, can induce cell proliferation, morphological transformation, anchorage-independent growth, as well as myoblast differentiation. Ski is a DNA-binding protein that induces expression of muscle-specific genes, and which, under some circumstances such as downstream of TGF-β or nuclear hormone receptor signaling, can act as a transcriptional corepressor by recruiting histone deacetylases to transcription complexes. v-Ski acts in a dominant-negative manner to inhibit transcriptional repression by pRb. Mice lacking the c-*ski* gene have established its role in the expansion of neuroepithelial and skeletal muscle precursors during development. *In vitro*, v-Ski can also induce self-renewal of primary avain hematopoietic progenitors and arrest hematopoietic cell differentiation. c-Ski and the related protein Sno have been detected in human tumor cell lines.

***See also:* Transformation**

Smith, Hamilton

T N K Raju

doi: 10.1006/rwgn.2001.1204

Hamilton O. Smith (1931–), the co-recipient of the 1978 Nobel Prize in Physiology or Medicine, shared the prize with Werner Arber (1929–) and Daniel Nathans (1928–). The Nobel Foundation honored these scientists "for the discovery of restriction enzymes and their applications in molecular biology."

Born in New York City in 1931, Hamilton Smith did much of his schooling in Urbana, a small Midwestern town, where his father was Professor of

Education at the University of Illinois. Although Hamilton majored in mathematics at the University of Illinois, he developed an avid interest in biological subjects, and in 1952 entered the Johns Hopkins Medical School. Four years later, he earned his MD degree, and in 1962 was awarded a postdoctoral fellowship from the National Institutes of Health to pursue genetics research with Myron Levine at the University of Michigan in Ann Arbor.

Smith's initial work was on bacteriophages – viruses that infect bacteria. Specifically, he studied phage P 22, which infects *Salmonella typhimurium*. Smith focused his studies on lysogen, a phenomenon in which viruses inside host cells divide in concert with the host cell, without harming the latter.

By the mid-1960s, Werner Arber, the Swiss scientist, had discovered restriction enzymes – the 'chemical knives' that cut DNA molecules into discrete, smaller segments by acting upon specific chemical sites. In 1966 Smith learned of Arber's remarkable work on restriction enzymes responsible for cutting DNA, and modification enzymes that prevented harm to the host DNA. Smith realized the relevance of these discoveries to his own work, and pursued research on the study of specific restriction enzymes.

In 1970, Hamilton Smith published two classic papers describing the first restriction enzyme from a common bacterium *Haemophilus influenzae*. He also characterized in detail the mechanism of enzyme action. The restriction enzyme from *H. influenzae* degraded foreign DNA to fragments of 1000 bp without affecting the DNA of the host bacterium. Smith showed that all fragments had the same four base pair sequence at each end. The enzyme he had discovered cleaved DNA at specific sequences of 6 bp. Thus, not only was Arber's discovery verified and confirmed, but also the biochemical basis of enzyme action was elucidated. Later work confirmed that restriction enzymes recognize symmetrical base pair sequences and cleave DNA wherever these sequences occur.

Daniel Nathans, a colleague of Hamilton Smith at Johns Hopkins University, pioneered practical applications for restriction enzymes. In his classic 1971 paper, Nathans reported that the restriction enzyme discovered by Smith cleaved the small DNA molecule from a simian virus called SV40 into 11 fragments. Using innovative methods of cleaving and mapping, Nathans later reported the complete genetic map of SV40 DNA – the first DNA mapping obtained by a chemical method.

Nathans' approach was refined by others enabling mapping of increasingly complex DNA structures, including those in human chromosomes. These developments led to the formulation of the basic tenets of genetic engineering in which restriction enzymes began to be used to determine the order of genes in chromosomes and to manufacture 'designer genes.'

The discoveries of Arber, Smith, and Nathans also influenced all of modern molecular genetics and much of the biological sciences. The knowledge and applications so derived are being used to this day in the study of evolutionary biology, the Human Genome Project, the discovery of the biochemical basis of hundreds of human diseases, and gene therapy for many diseases including malignancies.

Even from his childhood, Hamilton Smith's life can be characterized as one filled with an atmosphere of intense intellectualism and scholarly pursuits. In school he studied French, played electronic games, and participated in football and basketball. With his brother, Hamilton Smith collected an assortment of chemistry and electronic paraphernalia in the basement of their house, setting up recreational scientific experiments. Hamilton Smith played the piano, but claims that he was "in no way gifted" at it. However, when he was 13, he heard a recording of Beethoven's Pathetique Sonata performed by Artur Rubinstein – this was to awaken in Smith a lifelong passion for the dramatic beauty of classical music.

***See also:* Arber, Werner; Bacteriophages; Genetic Recombination; Nathans, Daniel; Restriction Endonuclease**

Snell, George

L Silver

doi: 10.1006/rwgn.2001.1205

The conceptual basis for the development of congenic mice was formulated by George Snell at the Jackson Laboratory during the 1940s and it led to the first and only Nobel Prize for work strictly in the field of mouse genetics. Snell was interested in the problem of tissue transplantation. Long before 1944, it was known that tissues could be readily transplanted between individuals of the same inbred strain without immunological rejection, but that mice of different strains would reject tissue transplants from each other. Although these observations were a clear indication of the fact that genetic differences were responsible for tissue rejection, the number and types of genes involved remained entirely unknown. In absentia, these genes were named *histocompatibility* (or *H*) loci. The assumption was that the histocompatibility genes were responsible – directly or indirectly – for the production of tissue (or 'histological') markers that could be distinguished as 'self' or 'nonself' by an animal's immune system. If transplanted tissue and a host recipient carried identical genotypes at all *H* loci, there would be no immunological response and the transplant would 'take.' However, if a single foreign allele at any *H* locus was present in the tissue, it would be recognized as foreign and attacked.

Although the number of histocompatibility loci was unknown, it was assumed to be large because of the rarity with which unrelated individuals – both mice and humans – accept each other's tissues. The logic behind this assumption was the empirical finding that polymorphic loci are most often diallelic and not usually associated with more than three common alleles. If *H* loci showed a similar level of polymorphism, a large number would be required to ensure that there would almost always be at least one allelic difference between any two unrelated individuals. The experimental problem was to identify and characterize each of the histocompatibility loci in isolation from all of the others.

Snell's approach to this problem was to use a novel multigeneration breeding protocol based on repeated backcrossing to trap a single *H* locus from one mouse strain (the donor) in the genetic background of another (the inbred partner). The basic approach caused the newly forming congenic strain to become increasingly similar to the inbred partner at each generation, but only those offspring who remained histoincompatible with the inbred partner were selected to participate in the next round of backcrossing. It was assumed that a difference at any one *H* locus would be sufficient to allow full histoincompatibility. Thus, at the end of the process, Snell expected to find that each independently derived congenic line would have trapped the donor strain allele at a single random *H* locus. With random selection, all *H* loci could be isolated in different congenic strains so long as a large enough number were generated.

With this outcome in mind, Snell began the production of histoincompatible congenic strains (originally called 'congenic resistant' strains) with 125 independent lines of matings. Of these, 27 were carried through to the point at which it was possible to determine which *H* locus had been trapped. Surprisingly, 22 of the 27 lines had trapped the same locus, which was given the name *H-2* (by chance, it was the second one identified). Contrary to expectations, the *H-2* locus (now called the *H2* complex since it is known to be a tightly linked complex of genes) acts, for all effective purposes, as the only strong determinant of histocompatibility. Snell and his predecessors were misled by the false assumption that only a limited number of alleles are possible at any one locus. Instead, a subset of genes within the *H2* complex (known as the class I genes) are the most polymorphic in the genome with hundreds of alleles at each individual locus. The generic term 'major histocompatibility complex' (MHC) is now used to designate this complex locus in mice as well as its homolog in all other mammalian species including humans, where it was historically called HLA.

See also: **Major Histocompatibility Complex (MHC)**

SNPs

See: **Single Nucleotide Polymorphisms (SNPs)**

snRNAs

doi: 10.1006/rwgn.2001.2030

snRNAs (small nuclear RNAs) are an abundant class of RNA found in the nucleus of eukaryotes. Several of the snRNAs are involved in splicing or other RNA processing reactions. They are generally about 100–300 nucleotides long; most are found in complexes with proteins.

See also: **Nucleus**

snRNPs

doi: 10.1006/rwgn.2001.2031

snRNPs (small nuclear ribonucleoproteins) are snRNAs associated with proteins.

See also: **Nucleus**

Solanum tuberosum (Potato)

K Schüler

doi: 10.1006/rwgn.2001.1669

The potato (*Solanum tuberosum* subsp. *tuberosum*) is the fourth most important crop for human nutrition in the world. Potatoes grow under different climatic conditions. The world potato area in 1998 was 17 949 000 hectares, and the amount produced was 295 632 000 t (FAO, 1998). The potato yield, with a world average of 16 t ha^{-1}, ranges from 5–8 t ha^{-1} in some developing countries to 40 t ha^{-1} and more in developed countries. In the last 10 years, potato production has increased at an average of 4.5% per year and the area planted has increased by 2.4% (Zandstra, 1999).

The potato is food for humans and in some regions for animals, and raw material for the food-processing (e.g., potato chips, French fries, dried potatoes) and starch industries. Developing countries recognize more and more the opportunities for potato production. Whereas the potato in developed countries in moderate climates is increasingly used as raw material for the food industry, this crop in developing countries is becoming more important for its original use in human nourishment. Substantial advantages of the potato are its high yield potential in short growth time, the high edible dry-matter content of its tubers, and its high dietary value as staple food. Potato tubers are rich in starch (10–20%), they contain biological high-value protein (2%), ascorbic acid (17 mg per 100 g edible parts), 2.5% roughage, and 1% mineral substances (Ka, Mg, P, Mn).

The potato is native in the southwest of the United States and in the whole of Central and South America, with centers of genetic diversity in the Andean regions of Peru and Bolivia and in Mexico. There are according to Hawkes (1990) 228 wild potato species (tuber-bearing species of the huge genus *Solanum*). The potato species are in cytological respect a polyploid line with the basic number $x = 12$ from $2x$ to $6x$, so diploids are $2n = 24$ (most wild species are diploid), triploids are $2n = 36$, tetraploids are $2n = 48$ (in this group belongs *S. tuberosum* subsp. *tuberosum*), pentaploids are $2n = 60$, and hexaploids are $2n = 72$ chromosomes. In each wild species there are more or less numerous accessions which are varying in important traits (e.g., resistance against diseases).

The different wild potato species are part of the natural plant associations in extraordinary different ecological regions in their native habitats (e.g., high altitudes of 3500–4500 m; hot, dry semidesert conditions; wet mountain rainforests). This results in an adaptability of many species in the most varying environmental conditions with many kinds of abiotical (frost, heat, drought) or biotical (diseases, pests) stress factors. This makes wild potato species immensely useful for potato breeding. The tubers (parts of the stems, not of the roots) of wild species are generally small and grow on long stolons with a distance to the mother plant of 30 cm or even more.

History

Seven thousand to 10 000 years ago in the Central Andean regions in the today's countries Peru and Bolivia native people began to select some species for human use. The evolutionary relations between these seven cultivated species and their probable wild relatives are given by Hawkes (1990). The cultivated species are *S. ajanhuiri* (2*x*), *S. chaucha* (2*x*), *S. curtilobum* (5*x*), *S. juzepczukii* (3*x*), *S. phureja* (2*x*), *S. stenotomum* (2*x*), and *S. tuberosum* (4*x*) with the subspecies *andigena* and *tuberosum*. These species are grown in many different countries in South America. Most important is the tetraploid species *S. tuberosum* subsp. *andigena* which was introduced into Europe in the sixteenth century after the Spanish conquest of America. After adaptation and first simple selection of offspring from selfing berries and crossings, from the middle of the nineteenth century systematic breeding improved disease resistance (especially to *Phytophthora infestans* = late blight) resulting into the present long-day-adapted European and North American cultivars of *S. tuberosum* subsp. *tuberosum*. Evidence for this theory was provided by the "Neotuberosum program" which was carried out in the UK, USA, The Netherlands, and Canada (Bradshaw and Mackay, 1994). A similar process happened thousands year ago in South America. *S. tuberosum* subsp *andigena* migrated from today's Peru and Bolivia into today's Chile; and there, under the same day-length as in Europe, originated also the subspecies *tuberosum*, much earlier than in Europe or North America.

Potato Breeding

The traditional method for clonally propagated potatoes is combination breeding by means of sexual hybridization (crossing) of suitable parents at the tetraploid level. Very important for a successful cross is the combining ability of the parents. Using this method a huge number of valuable cultivars were bred, but this also led sometimes to a limited gene pool and inbreeding, as a result of the relationship between the parents. In the twentieth century, the introduction of wild and other cultivated species and the use of their gene pool became increasingly important because of the need to improve some traits (e.g., resistance to late blight, virus diseases, and nematodes) (Ross, 1986).

Accessions of wild and cultivated species have been collected in numerous expeditions since the 1930s (initiated by Vavilov's *The Theory of the Origin of Cultivated Plants after Darwin*). This has continued to this day, and the specimens are stored, propagated, and evaluated in the potato gene banks (germplasm collections) of the world (Peru, Argentina, Chile, USA, Russia, The Netherlands, Germany, and UK).

The 4*x* cultivated species *S. tuberosum* with its two subspecies is autotetraploid. The tetrasomic inheritance of autotetraploids and the segregation of the very often polygenic traits makes potato breeding difficult.

The ability to make crosses between different species depends on many pre- and postzygotic inhibition mechanisms. The endosperm balanced number (EBN) is responsible for a balanced developing of endosperm and embryo. Dihaploids ($2n = 2x$) of 4*x* breeding lines or cultivars can improve the ability to cross with diploid species. Prebreeding at the diploid level makes the interpretation of segregation and selection of polygenetic traits easier. Besides the classic sexual hybridization there is the possibility of combining genomes asexually by protoplast fusion. This somatic hybridization can be applied to species which are impossible or very difficult to cross sexually (e.g., *S. bulbocastanum* × *S. tuberosum* subsp. *tuberosum*).

Gene mapping with molecular markers and marker, assisted selection is an important tool in making modern potato breeding more efficient. Genes transfer gives new prospects for the future. First steps in this direction are being made, in the improvement of starch quality, and in disease and pest resistance.

The most important goals in breeding potatoes, out of more than 50 traits are: high yield on tubers in different maturity groups; resistance to diseases and pests; resistance to external tuber Damage, quality for food processing; and many others.

The potato is threatened by numerous pathogens, which makes resistance breeding such an important project. Fungal diseases include *Phytophthora infestans, Fusarium* spp., *Synchytrium endobioticum, Phoma foveata, Rhizoctonia solani, Helminthosporium solani, Spongospora subterranea, Colletotrichum coccodes, Verticillium* ssp., and *Sclerotinia sclerotiorum*. Bacterial diseases include *Erwinia* ssp., *Streptomyces scabies, Clavibacter michiganensis*, and *Ralstonia solanacearum*; and viruses causing infections include PVY, PLRV, PVM, PVA, PVX, and PVC. The process of clonal tuber propagation promoted infection and transfer of many diseases, so special extensive phytosanitary treatments are necessary for the production of healthy seed tubers.

In the last decades true potato seeds (TPS) for field cultivation have gained importance especially for countries in hot climate areas. In these regions the production, storage, and transport of seed tubers is difficult and expensive. Here the use of TPS has many advantages: reduction of seed costs (50–250 g TPS per ha instead of 2 t per ha seed tubers), flexibility of TPS in planting time (seed tubers suffer a physiological aging with limited durability), and freedom from tuber-borne or tuber-transmitted diseases (viruses, fungal, and bacterial diseases). Important initiatives in the practical use of TPS in developing countries have been made by the International Potato Center (CIP) in Lima, Peru. Whereas clonal propagation leads to homozygous plants with uniform tubers, TPS progenies are heterozygous and give more uneven produce. The main goal in TPS breeding programs is to improve the progeny uniformity, while maintaining other quality and resistance characteristics. Methods are inbreeding and use of suitable diploid or tetraploid parental lines. TPS production uses natural open pollination, hybrids, synthetic lines, or cytoplasmic male sterility (CMS), this last giving rise to so-called 'cybrids.'

References

Bradshaw JE and Mackay GR (1994) *Potato Genetics*. Wallingford, UK: CAB International.

FAO (1998) *Production Yearbook*, vol. 52: pp. 83–84. Rome: FAO.

Hawkes JG (1990) *The Potato: Evolution, Biodiversity and Genetic Resources*. London: Belhaven Press.

Ross H (1986) *Potato Breeding: Problems and Perspectives*. Berlin: Verlag Paul Parey.

Zandstra HG (1999) Retrospect and future prospects of potato research and development in the world. Keynote address, *Global Conference on Potato*, 6–11 December 1999, New Delhi.

See also: **Genetic Stock Collections and Centers**

Soluble RNA

B S Guttman

doi: 10.1006/rwgn.2001.1206

Soluble RNA is the original term for what is now called transfer RNA (Transfer RNA (tRNA)). The bulk of RNA in a cell (on the order of 80%) is ribosomal, which consists of very large molecules built into particles (ribosomes) that can be collected by high-speed centrifugation. Hoagland *et al.* (1957) demonstrated that cells also contain a large amount of 'soluble' RNA of much lower molecular weight, and they went on to show that these RNAs bind amino acids in the presence of ATP and can transfer these amino acids to microsomal protein (Hoagland *et al.*, 1958). This discovery coincided with Crick's prediction, on theoretical grounds, that such small RNAs should exist. Crick pointed out that amino acids cannot bind specifically to nucleic acids and that if a single amino acid is encoded by a triplet of nucleotides, there is a significant spacing discrepancy between the RNA template and the nascent polypeptide chain. Crick therefore predicted that the protein-synthesizing mechanism ought to include 'adaptor' molecules, probably a type of small RNA, that would recognize a codon at one end and carry an amino acid at the other end. This, of course, is exactly what transfer RNA molecules do.

References

Hoagland MB, Zamecnik PC and Stephenson ML (1957) Intermediate reactions in protein biosynthesis. *Biochimica et Biophysica Acta* 24: 215–216.

Hoagland MB, Stephenson ML, Scott JF, Hecht LI and Zamecnik PC (1958) A soluble ribonucleic acid intermediate in protein synthesis. *Journal of Biological Chemistry* 231: 241–257.

***See also:* Protein Synthesis; Transfer RNA (tRNA)**

Somatic Mutation

doi: 10.1006/rwgn.2001.2032

A somatic mutation is a mutation occurring in a somatic cell. It therefore affects only its descendants and will not be heritable, since it is not present in the germ cells.

***See also:* Mosaicism in Humans**

Somatic Pairing

A K Csink

doi: 10.1006/rwgn.2001.1211

Somatic pairing is the association of the maternal and paternal homologs in nonmeiotic nuclei. When a set of mitotic chromosomes from a typical diploid eukaryotic nucleus is observed under a light microscope, it appears as if all of the pairs of chromosomes were thrown into a bag, shaken, and dumped out into a pile. The chromosomes distribute randomly with respect to each other and the two homologs, one from each parent, are as likely to be found next to each other as they are to any of the other chromosomes. This is expected as, for the most part, homologs find each other only during meiosis I when they align to undergo recombination. Mitotic figures from a dipteran insect, such as the model genetic organism *Drosophila melanogaster*, are organized quite differently (see **Figure 1**). This was first recognized by Stevens (1908). Examining cytological preparations of mitotic nuclei from embryonic cells and somatic gonadal cells, she noted that the homologs in these nuclei were commonly found next to each other. Metz (1916) looked at other insects and tissues and confirmed that homolog pairing was a widespread feature of mitosis in the Diptera.

The extent to which mitotic pairing of homologs reflected the arrangement of the chromosomes in the interphase diploid nucleus was unclear. Thus the term 'mitotic pairing' was often used instead of 'somatic pairing,' as it is now commonly called. One type of interphase nucleus, the polytene nuclei of the dipteran larva, provided an early and incontrovertible link between mitotic pairing and interphase homolog organization. Many larval cells undergo a process of DNA replication without mitosis called endoreduplication which results in a polyploid nucleus that contains as many as 1024 copies of a single chromosome aligned in a rope-like strand. The presence of the two homologs right next to each other gives an appearance of two ropes wrapped around each other. Due to the

Figure 1 Mitotic figures from *Drosophila melanogaster*.

large size and distinctive banding pattern of the polytene chromosomes, it can be quickly recognized that the homologs are closely and precisely apposed. When polytene chromosomes are prepared by squashing for viewing in a microscope, the two homologs only occasionally separate from each other, an event referred to as asynapsis. Synapsis of polytene homologs is maintained even when one of the homologs contains multiple inversions, as in the case of the balancer chromosomes of *D. melanogaster*. Even translocation heterozygotes, where chromosomal rearrangement has moved a substantial fraction of a chromosome arm to another centromere, retain pairing between the unrearranged and the translocated chromosomes.

The diploid interphase nucleus and its chromosomes usually lack visually discernible substructure. As a result, the position of homologous loci is less obvious in diploid interphase nuclei than in either polytene or mitotic nuclei. Development of *in situ* DNA hybridization allowed visualization of specific sequences within the space of the nucleus, which in turn allowed statistical analysis of the distribution of homologs relative to each other. In most diploid organisms, when a probe that is unique to a region of a chromosome is hybridized to a G_1 interphase nucleus, two well-separated spots of hybridization can be seen per nucleus, corresponding to the two homologs. However, in the majority of *D. melanogaster* nuclei, only one spot of hybridization can be resolved. In the diploid tissue of the larva this pairing is seen in 70–100% of the nuclei, with the variability most likely depending on the rapidity of cell division in a given tissue. Notably, somatic pairing is not seen in the very early embryo of *D. melanogaster* and homologous loci do not begin to substantially pair until embryonic nuclear cycle 14, when the cell cycle slows down and acquires a G_1 phase. This agrees with the observation that very close pairing is lost during mitosis and perhaps as early as the onset of DNA replication of the paired locus, although, as seen in mitotic spreads, the homologs are still nonrandomly close to each other.

While the actual observation of somatic pairing in the nuclei of interphase diploid chromosomes of flies required the development of new techniques, genetic observations had long suggested the possibility of cross-talk between homologs during interphase and hence implied their juxtaposition in the nucleus. Remarkably, the potential for such interactions was pointed out in Stevens's first report of somatic pairing in 1908 when she wrote

> One is tempted to suggest that if homologous maternal and paternal chromosomes in the same cell ever exert any influence on each other, such that it is manifest in the heredity of the offspring, there is more opportunity for such influence in these flies than in cases where pairing of homologous chromosomes occurs but once in a generation.

This statement foreshadowed the discovery in *Drosophila* of a variety of genetic phenomena that are dependent on the synapsis of the homologs. The best-described example is the phenomenon of transvection, which is the disruption of allelic complementation by chromosomal rearrangements thought to disrupt somatic pairing. Transvection was first described at the *Ultrabithorax* (*Ubx*) locus. When certain mutant *Ubx* alleles are combined, the phenotype is close to wild-type. However, when chromosomal rearrangements are induced in either one of the two homologs the phenotype becomes more severely mutant, even though the breakpoints do not disrupt the *Ubx* gene. This observation is interpreted to indicate that the two mutant alleles must be close enough to each other in the interphase nucleus to allow cross-talk between the two copies of the gene and rearrangements of the homolog interfere with close somatic pairing. How somatic pairing may promote allelic complementation was further clarified when another example of transvection was described at *yellow*. Here, allelic complementation was described for two alleles, one containing a compromised regulatory region and another containing a mutated coding region. Complementation is not seen when one mutant allele is present on a transgene elsewhere in the genome and the second mutant allele is present at the normal site of *yellow*. Therefore, it seems that the two alleles must be present at their normal location so that somatic pairing and cross-talk can take place. It is thought that the intact regulatory sequences from the first allele are able to promote the transcription of the intact coding region on the homolog to make a functional and properly regulated transcript.

A second example of a genetic phenomenon in *Drosophila* that is dependent on somatic pairing is *trans*-inactivation. Chromosomal rearrangements that bring a gene normally found in euchromatin next to transcriptionally repressive heterochromatin result in the inactivation of that gene in some cells but not in others in a tissue, an effect referred to as position effect variegation (PEV). Most PEV alleles are recessive to the wild-type allele, but in some instances, most notably those involving the *brown* locus, the variegating allele is dominant to the wild-type allele. This dominance can be suppressed by disruption of somatic pairing, implying that the ability of heterochromatin to silence a gene on the opposite chromosome is dependent on the proximity of the two homologs.

The mechanism of somatic pairing is not well understood, which is not surprising since little is known about more general mechanisms organizing the interphase nucleus. The simplest explanation is that chromosomes move about the nucleus in a confined random walk pattern until they find a homologous sequence, at which time pairing interactions are established and maintained. There is some evidence that certain regions of the chromosome pair more quickly than others, but what features could promote such pairing, aside from closer initial position, is unclear. Models based on DNA-pairing or on co-association of DNA-binding proteins have been proposed.

There is no other known group of organisms where somatic pairing is so prevalent as in the Diptera. However, the nonrandom association of homologs in metaphase spreads has been occasionally, if controversially, reported for various plants and animals. More convincing are the reports of premeiotic homologous pairing in germ cell mitoses of organisms other than Diptera. Recent evidence from budding yeast suggests that there is some homolog pairing in both premeiotic and vegetatively growing diploid cells. Intriguingly, one study suggests that the maintenance of imprinting in mammals is correlated with transient association of homologs during certain phases of the cell cycle. Most of these more recent studies used fluorescent *in situ* hybridization to examine directly the position of the homologous loci in interphase nuclei. Undoubtedly, this technique will be applied to many other systems and may begin to tell us how widespread somatic pairing is in groups outside the Diptera.

Further Reading

Ashburner M (1989) *Drosophila: A Laboratory Handbook*. Plainview, NY: Cold Spring Harbor Laboratory Press.

Brown WV (1972) *Textbook of Cytogenetics*. St Louis, MO: Mosby.

Burgess SM, Kleckner N and Weiner BM (1999) Somatic pairing of homologs in budding yeast: existence and modulation. *Genes and Development* 13: 1627–1641.

References

Metz CW (1916) Chromosome studies on the Diptera. II. The paired association of chromosomes in the Diptera, and its significance. *Journal of Experimental Zoology* 21: 213–279.

Stevens NM (1908) A study of the germ cells of certain Diptera, with reference to the heterochromosomes and the phenomena of synapsis. *Journal of Experimental Zoology* 5: 359–383.

***See also:* Chromosome Pairing, Synapsis; Meiosis; Mitosis; Polytene Chromosomes**

SOS Bypass

J H Miller

doi: 10.1006/rwgn.2001.1212

'SOS bypass' is the process of replication past noncoding lesions as a result of induction of the SOS system. Specific DNA polymerases replicate past these lesions, often resulting in mutations.

***See also:* SOS Repair**

SOS Repair

B A Bridges

doi: 10.1006/rwgn.2001.1213

Free-living organisms such as bacteria frequently have a life style in which periods of rapid growth alternate with periods in which growth is inhibited by various stressful conditions. Among the most insidious are those in which their genetic material is subjected to damaging agents such as UV light or chemical mutagens. To maintain the integrity of their genome, organisms have evolved a variety of mechanisms for dealing with such assaults. Many of these can be broadly thought of as DNA repair mechanisms. In many bacteria a mechanism has evolved whereby the genes determining such processes can be expressed when they are actually needed. There are more than 30 genes in *Escherichia coli* whose expression is greatly enhanced when the cellular DNA is damaged: This concept was proposed by Miroslav Radman in 1973 and was termed the SOS response.

SOS-inducible genes are repressed by the product of the *lexA* gene which recognizes and binds to an operator sequence of some 20 nucleotides, known as an SOS box; the *lexA* gene itself has two such boxes. Many types of damage in DNA give rise to single-stranded regions either because of the excision of the damaged region or because the damage causes an interruption in DNA replication with the production of a single-stranded region in one of the daughter chromosomes. The product of another gene under SOS control, *recA*, is normally present in an uninduced bacterium at around 7000 molecules per cell. It binds to single-stranded DNA in the presence of ATP to form a filamentous nucleoprotein in which the RecA protein is said to be in an activated form (RecA*). Lex A protein diffuses to this RecA filament and the

RecA* catalyzes an autoproteolytic reaction in the LexA protein such that it self-cleaves at an Ala–Gly bond near the middle of the protein. The truncated LexA protein is no longer able to bind to SOS boxes and transcription of SOS genes begins. There are *lexA* mutants whose product is uncleavable and which do not therefore show SOS induction (e.g., *lexA3*), and others in which induction is constitutive because cleavage is spontaneous (e.g., *lexA51*). There are also *recA* mutants that cannot be activated to a form that will cleave the LexA repressor (e.g., *recA56*), and others that can catalyze cleavage of the LexA repressor even when not bound to single-stranded DNA (e.g., *recA730* and *recA441* at 42°C). The former are unable to show SOS induction, while the latter show constitutive induction. Strains unable to induce SOS responses are hypersensitive to ultraviolet light and many other mutagens and do not show significant mutagenesis by such agents.

The RecA protein itself has a major role in genetic recombination but, although it is inducible, induced levels are not essential for normal recombination processes. They do, however, seem to be necessary for the recombination processes that are involved in certain types of DNA repair, for example, the repair of double-strand DNA breaks and the recombinational repair of daughter-strand gaps that are formed when certain types of damaged nucleotide encounter the replication fork.

Nucleotide excision repair is a mechanism in which a damaged region of DNA is cut out and replaced by DNA synthesized using the undamaged strand as template. Three very important genes determining this pathway are under SOS control, namely *uvrA*, *uvrB*, and *uvrD*.

Historically, mutagenesis was an important attribute of SOS induction; indeed a crucial experiment that led to the SOS hypothesis was reported by Weigle in 1953 and involved mutagenesis of bacteriophage lambda. It was found that if lambda phage were exposed to UV light and plated on unirradiated bacteria, few if any mutations appeared among the phage progeny. If, however, the bacteria had themselves been independently irradiated then there were many mutants among the progeny of the irradiated phage. Subsequent work showed that the *recA* and *lexA* genes were needed for this 'Weigle mutagenesis' and that the same two genes were required for mutagenesis of the bacterial chromosome by UV light, ionizing radiation, and a wide variety of chemical mutagens. As early as 1967, Evelyn Witkin had noticed that there were similarities between the induction of filamentation and the induction of prophage following UV irradiation and boldly hypothesized that both phenomena reflected the release of repressor action (induction) consequent upon interruption of DNA replication. Radman broadened this concept and argued that there was a whole battery of inducible responses dependent upon this type of induction, and that included among these was one that was needed for mutagenesis to occur both in bacteriophage and the bacteria themselves following exposure to ultraviolet light.

Among the genes found to be induced by UV light was an operon containing two bacterial genes, *umuD* and *umuC*, both of which were needed for SOS mutagenesis. The products of these genes act in a complex consisting of one molecule of UmuC and two molecules of UmuD′, a posttranscriptionally modified form of UmuD. In fact UmuD has the same ability to self-cleave as LexA and does so under the influence of RecA*, so revealing yet another role for RecA protein. It has recently, and rather surprisingly, emerged that the $\mathrm{UmuD'_2}$ UmuC complex constitutes a new DNA polymerase, designated DNA polymerase V, which is able to catalyze synthesis past damaged nucleotides in the template strand. In doing so it inserts incorrect bases that form the induced mutations. Other *E. coli* polymerases can be encouraged to insert bases opposite template damage, but only DNA polymerase V appears to be able to use such a damage/mismatch terminus as a primer for further strand extension. Polymerase V is also prone to error when acting on undamaged template. UmuC contains the polymerase domain in DNA polymerase V and is now known to be representative of a whole family of homologs throughout the evolutionary scale. Homologs of *umuDC*, presumably also determining DNA polymerases, are found on many plasmids, e.g., *mucAB* in pKM101 and *impAB* in TP110. Almost simultaneously with the recognition of polymerase V, it was shown that another SOS-inducible gene, *dinB*, codes for a further DNA polymerase (IV). It was known that *dinB* was required for the phenomenon of indirect mutagenesis which is seen when unirradiated phage are allowed to infect bacteria that have been exposed to UV light. It now appears that polymerase IV, which is, like polymerase V, a low-fidelity enzyme, is induced by the irradiation and makes numerous replication errors while replicating phage DNA, many of them single base deletions. Polymerase IV has little effect on the host DNA and its role in the cell is still unclear. There are hints that it may perform translesion synthesis at certain specific types of damage in template DNA that polymerase V cannot.

The third SOS-inducible DNA polymerase is DNA polymerase II, the product of the *dinA* gene. Its role seems to be to assist in the reassembly of replication forks that have become stalled by encountering particularly difficult types of damage.

As more is revealed about the control of SOS responses it becomes apparent that individual responses are subject to quite sophisticated levels of control. At the crudest level the timing and extent of induction of different genes are controlled by the affinity of LexA protein for their SOS box(es), which is determined by the sequence of the box(es), very few of which are identical. The affinity of LexA repressor for the SOS box of *uvrD*, for example, is 16 times greater than its affinity for the SOS box of *umuDC*.

As mentioned above, the parallels between SOS repair and prophage induction were recognized early on. We can now see that certain lysogenic phages such as lambda have hijacked the activation of RecA protein for their own ends. These phages have evolved a repressor that self-cleaves under the action of activated RecA* protein thus allowing excision of the phage and vegetative reproduction. As far as the phage is concerned it is not the DNA repair responses of the SOS system that are of primary interest but the utilization of RecA*-mediated proteolysis to enable it to bail out from a potentially sinking ship. However, the presence of the SOS system throughout much of the bacterial world and the conservation of many SOS-inducible genes from bacteria to humans testify to the value of SOS repair to the cell. In prokaryotes the primary function of SOS repair is to make available DNA repair and certain other mechanisms when they are needed, with an important secondary function to generate genetic variability when a change in environment may demand it. Thus it is becoming apparent that SOS induction not only occurs when DNA-damaging agents are encountered, but also in other circumstances such as in aging colonies and when there is nutritional stress, although the mechanisms of induction under these conditions need further clarification.

The SOS system was the first paradigm for a global cellular response to DNA damage and it has provided the foundation for subsequent studies in mammalian and other eukaryotic systems including those involving cell cycle control and apoptosis.

See also: **Cell Cycle; DNA Repair; Excision Repair**

Southern Blotting

T A Brown

doi: 10.1006/rwgn.2001.1214

Southern blotting, initially described by Southern (1975), is in essence the transfer of DNA restriction fragments from an electrophoresis gel to a nitrocellulose or nylon membrane in such a way that the DNA banding pattern in the gel is reproduced on the membrane. The DNA fragments become bound to the membrane in a form that is suitable for hybridization analysis with labeled DNA or RNA probes. Southern blotting therefore enables a specific restriction fragment to be detected against a background of many other restriction fragments.

The Methodology for Southern Blotting

The basic methodology for Southern blotting is shown in **Figure 1**. An agarose gel containing an array of DNA fragments is placed on a filter-paper wick which connects with a reservoir of buffer. The membrane is positioned on the gel and a pile of paper towels is placed on top of the membrane. Buffer soaks through the filter-paper wick by capillary action, passes through the gel and membrane, and is absorbed by the paper towels. DNA fragments are carried from the gel to the membrane. The complete transfer of fragments up to 15 kb requires approximately 18 h. This basic methodology can be embellished by alternative forms of transfer, such as electroblotting, which uses electrophoresis rather than capillary action to transfer the DNA, and vacuum blotting, in which buffer is drawn through the gel and membrane under vacuum. Both electroblotting and vacuum blotting are more rapid than the conventional methodology, reducing the transfer time to as little as 30 min.

Nylon membranes are more popular than the nitrocellulose versions because they are tougher and so are unlikely to break during the blotting procedure. In addition nylon membranes can be subjected to multiple rehybridizations. A second advantage is that nylon membranes bind DNA molecules of 50 bp or longer, whereas nitrocellulose membranes do not efficiently bind molecules less than 500 bp. The one major advantage of nitrocellulose is that these membranes give less background hybridization, especially when a nonradioactive label is used.

Prior to blotting, an agarose gel must be pretreated to break the DNA molecules in individual bands into smaller fragments, smaller fragments transferring more efficiently than larger ones. The pretreatment involves soaking the gel in 0.25 mol HCl for 30 min, which cleaves some of the β-N-glycosidic bonds that attach adenine and guanine bases to the sugar components of their nucleotides. This depurination is followed by spontaneous breakage of the polynucleotide chain at the baseless sites that are created. After acid pretreatment, the gel is placed in alkali to disrupt the hydrogen bonds in the double helices, resulting in fragmented single strands at locations corresponding

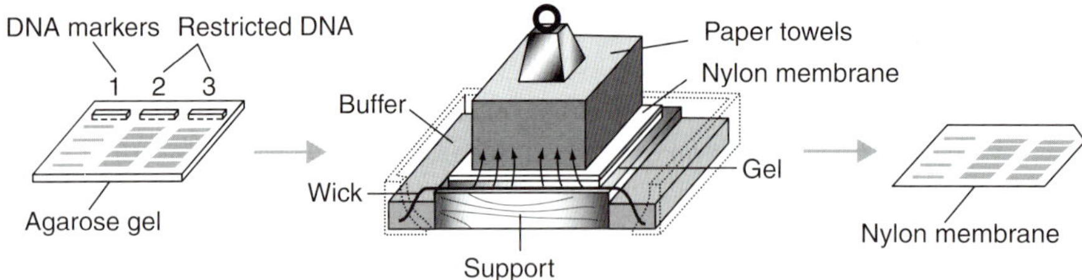

Figure 1 Southern blotting. (Reproduced from Brown TA (1999) *Genomes*, with permission of BIOS Scientific Publishers Ltd, Oxford.)

to the migration positions of the original restriction fragments.

For transfer to a nitrocellulose membrane, the alkali treatment is followed by neutralization of the gel by soaking in a Tris-salt buffer, because DNA does not bind to nitrocellulose above pH 9.0. The blot is then assembled with a high-salt transfer buffer called 20X SSC, which comprises 3.0 mol NaCl + 0.3 mol sodium citrate. This buffer can also be used with a nylon membrane, but if the nylon membrane is positively charged then 0.4 mol NaOH is usually used because this buffer not only transfers the DNA but also binds it covalently to the membrane. With a nitrocellulose or uncharged nylon membrane, the initial attachment of the DNA is reversible and must be made more permanent by a post-treatment: either baking at 80°C for 2 h, which noncovalently attaches DNA to a nitrocellulose membrane, or UV irradiation, which covalently binds DNA to a nylon membrane.

Hybridization Analysis of a Southern Blot

Southern blotting is always a prelude to hybridization analysis, for example with a cloned restriction fragment, a copy DNA (cDNA), or a synthetic oligonucleotide probe. Before hybridization the membrane is placed in a solution containing polymers that attach to any vacant DNA binding sites on the membrane surface, 'blocking' these so that the hybridization probe does not bind nonspecifically to them. Various polymers have been used, including nonbiological ones such as polyvinylpyrrolidone or biological polymers including Ficoll (a carbohydrate), bovine serum albumin (a protein), or dried milk (a complex mixture). DNA can also be used as a blocking agent, providing it is unrelated to the DNA being used as the probe. This prehybridization step takes between 15 min and 3 h at 68 °C, depending on the type of membrane.

Hybridization is performed by placing the membrane in a buffer in a rotating tube containing the hybridization probe, or alternatively in a sealed plastic bag on a shaker. The buffer has a high salt content (e.g., 2X SSC) and a detergent such as 1% sodium dodecyl sulfate is usually included. To increase sensitivity a second polymer might be added at this stage, such as 10% dextran sulphate or 8% polyethylene glycol 6000. These polymers do not block DNA binding sites but instead induce the probe molecules to form networks so that greater amounts attach to the target sites on the membrane.

Specificity is critical during hybridization analysis. The probe DNA must contain a region that is complementary to at least part of the blotted restriction fragment that is being sought. Problems can arise if the probe is partially complementary to other blotted DNA fragments. Hybridization must therefore be carried out under conditions that result in formation of a stable hybrid between the probe and its specific target, but not between the probe and any nonspecific targets. Providing that the probe has been well designed and is more complementary to its specific target than it is to the nonspecific ones, then specificity can be ensured by careful selection of the temperature at which the hybridization is carried out. This is because the highest temperature at which the hybrid between the probe and it specific target is stable (this is called the T_m or melting temperature for the hybrid) will be higher than the highest temperature at which a nonspecific hybrid is stable, because this nonspecific hybrid will be held together by fewer base pairs. If the probe is a restriction fragment or cDNA longer than 100 bp then nonspecific hybridization is usually avoided if the reaction is carried out at 68 °C in a high-salt buffer. With oligonucleotide probes the situation is more complicated because 68 °C might be too high for the formation of any hybrids, including the fully base-paired one. However, the T_m can be estimated from the sequence of the oligonucleotide, using the formula:

$$T_m = (2 \times \text{number of A and T nucleotides}) + (4 \times \text{number of G and C nucleotides})\,^\circ\text{C}$$

This estimation is reasonably accurate for most oligonucleotides whose values of T_m fall between 40 and 90 °C. The initial hybridization is set at a temperature 10 °C or so below the estimated T_m, which allows many hybrids to form, including nonspecific ones. Specificity is subsequently achieved by a series of post-hybridization washes, these being carried out at increasing temperatures so that nonspecific hybrids are disrupted, with the last wash designed to leave just the specific hybrid.

After hybridization, the position of the probe that remains bound to the membrane is determined by autoradiography if a radioactive label has been used, or by an alternative methodology if the probe was nonradioactively labeled. A nylon membrane can be reprobed up to ten times between each hybridization if it is 'stripped' by washing at a high temperature in a buffer containing alkali and detergent to remove the hybridized DNA.

Applications of Southern Blotting

Southern blotting has many applications in molecular biology research. For example, it is used at important stages during gene cloning projects. Genomic DNA, containing the gene to be cloned, is blotted and hybridized to identify one or more restriction fragments containing the desired gene, and Southern blotting is used later in the project when a tentative clone has been isolated, to verify that the clone does indeed contain the desired gene and possibly to identify a smaller subfragment within which the gene lies. A second application of Southern blotting is in restriction fragment length polymorphism (RFLP) analysis, which is important in several contexts including construction of genome maps. An RFLP arises if a restriction site that is present in the genomes of some members of a population is absent, owing to an alteration in the nucleotide sequence, in other individuals. RFLPs are typed by Southern hybridization, using a probe that spans the polymorphic region, the presence or absence of the polymorphic restriction site being determined from the number and sizes of the fragments that are detected.

Reference

Southern EM (1975) Detection of specific sequences among DNA fragments separated by gel electrophoresis. *Journal of Molecular Biology* 98: 503–517.

***See also:* DNA Cloning; Restriction Endonuclease**

Specialized Recombination

N D F Grindley

doi: 10.1006/rwgn.2001.1215

Specialized recombination is a term used to describe recombination events that are distinct from general homologus recombination in that they either are directed to specific DNA sites (or regions), or involve special proteins that are not required for homologous recombination. The term encompasses a variety of recombinational processes of which the best known are transposition and site-specific recombination.

The following are some examples of specialized recombination. For more detailed descriptions see the separate entries below.

DNA Transposition

DNA transposition is the movement of a defined DNA segment (a transposon) from one genomic site to another; the ends of a transposon are specific, but the integration sites generally are relatively random. Movement is catalyzed by a transposon-encoded transposase. Some DNA repair/replication is required to seal the short gaps at the transposon–target junction (and in some cases to duplicate the transposon). (See articles Transposable Elements and Insertion Sequence.)

Retrotransposition

Retrotransposition is the movement of defined DNA segments (retrotransposons) by a process that involves transcription of the element to form an RNA intermediate. In elements such as retroviruses and LTR (long terminal repeat) retrotransposons, the RNA transcript is used as a template to make a double-stranded DNA version of the transposon. Like a conventional transposon, this DNA is then processed and inserted into a target by the element-encoded integrase. In non-LTR retrotransposons, the RNA transcript is copied by reverse transcriptase directly into the target site, using a nick created at the target by the element-encoded endonuclease as the primer for DNA synthesis. (See articles Retrotransposons and Retroviruses.)

Site-Specific Recombination

Site-specific recombination is an exchange between two defined sites resulting in integration, excision, or inversion. Recombination is catalyzed by a site-specific recombinase. DNA cleavage at the recombination site results in an intermediate with the recombinase covalently linked to the ends of the DNA; reversal of this process reseals the DNA to form the recombinant, and releases the recombinase. No replication or repair required. (See Article Site-Specific Recombination.)

V(D)J Joining

This is a process for generating immunoglobulin and T cell receptor (TCR) diversity during mammalian B and T cell development. It involves the precise excision of the DNA segments that separate V and J, or V and D, and D and J coding sequences of the immunoglobulin and TCR loci, coupled to imprecise joining of the V, D, and J coding sequences. The process is catalyzed by Rag1/Rag2 recombinase acting at recombination signal sequences (RSS), and involves DNA synthesis by terminal transferase and cellular DNA double-strand break repair activities (including Ku and DNA-PK). (See articles Immunoglobulin Gene Superfamily and T Cell Receptor Gene Family.)

Immunoglobulin Heavy Chain Class Switch

This is the process for changing the class of an immunoglobulin protein (e.g., from IgM to IgG). It is an imprecise but region-specific form of recombination within the immunoglobulin heavy chain locus, that deletes genomic DNA between the variable (VDJ-encoding) genes and various downstream constant (C_H) genes. Recombination occurs between two 'switch' regions, by an unknown mechanism. (See articles Class Switching and Immunoglobulin Gene Superfamily.)

Intron (and Intein) Homing

Homing is the term for the process for introns from a particular gene to insert into an intronless version of the same gene. There are two distinct mechanisms, a DNA-dependent process and an RNA-dependent process called retrohoming.

1. Group I introns encode a site-specific endonuclease that makes a double-strand break in the intronless (but not the intron-containing) allele. The break is repaired by homology-dependent double-strand break repair, using the uncleaved, intron-containing, allele as the genetic donor; this results in gene conversion of the intronless to the intron-containing form.
2. Group II introns encode a protein with three activities: RNA maturase, DNA endonuclease, and reverse transcriptase (RT); the latter two activities are required for insertion (homing) of the spliced intron RNA. In a complex process, the intron RNA is attached to one strand of the cleaved DNA insertion site, and a cDNA copy is made by the RT primed from the 3′ end of the opposing strand. Host repair/replication (but not recombination) functions are required. (See article Intron Homing.)

Mating-Type Switching in Yeast

Interconversion of yeast haploid cells can occur between the two alternative mating types, a and α, achieved by moving a- or α-specific regulatory genes from silent loci (HMLα and HMRa) to the expression locus, MAT. The genetic identity of MAT is switched by gene conversion, initiated by a double-strand break at the MAT locus catalyzed by the HO site-specific endonuclease. The cleaved MAT locus acts as a target for double-strand break repair using the silent loci HMLα or HMRa as the donors of the genetic information. (See article Mating-Type Genes and their Switching in Yeasts.)

***See also:* Alternation of Gene Expression; Chromosome Dimer Resolution by Site-Specific Recombination; Conjugative Transposition; Flp Recombinase-Mediated DNA Inversion; Gene Cassettes; Gene Rearrangement in Eukaryotic Organisms; Hin/Gin-Mediated Site-Specific DNA Inversion; Integrase Family of Site-Specific Recombinases; Integrons; P Elements; Phage λ Integration and Excision; Resolvase-Mediated Deletion**

Specialized Transduction

W J Brammar

doi: 10.1006/rwgn.2001.1216

Specialized transduction is the virus-mediated transfer of nonviral genetic material to a recipient cell by a process involving the formation of a hybrid genome in which viral genes are substituted by genes derived from the host chromosome. Because it normally applies to a strictly limited set of host genes, the phenomenon has also been called 'restricted' or 'localized transduction.'

Basis of Specialized Transduction

Detailed understanding of the basis of specialized transduction largely derives from studies with coliphage lambda (λ), a temperate phage whose genome in the lysogenic state is integrated at a specific attachment site, *attB*, located between the *gal* and *bio*

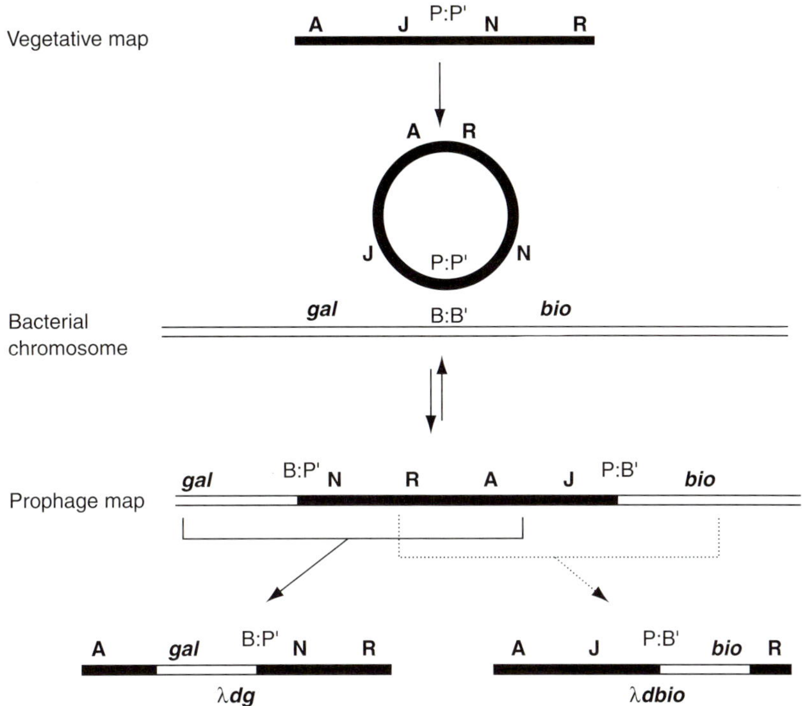

Figure 1 Generation of specialized transducing phages by aberrant excision. The linear genome of bacteriophage λ circularizes on injection into the *Escherichia coli* host cell. The circular genome is integrated into the host chromosome by site-specific recombination via the phage attachment site, P:P′, and the bacterial attachment site, B:B′, between the *gal* and *bio* operons. Normal excision of the prophage is a reversal of integration. Rare, aberrant excision events create transducing phage genomes, in which some phage genes have been replaced by genes from one or other flanking region of the host chromosome. (Reproduced with permission from Campbell, 1962.)

operons on the host chromosome. When λ-lysogens are treated with low doses of UV light or radiomimetic chemicals, the prophages are induced to proliferate, excised from the host chromosome, and eventually give rise to a phage 'lysate.' This lysate is capable of transducing gal^- or bio^- recipient bacteria to the Gal^+ or Bio^+ phenotype at a low frequency (*c.* 1 per 10^6 infecting phage particles). Only genes that are closely linked to *attB* can be transduced, and lytically grown phage lysates are ineffective.

The rare transducing phages are derived by aberrant excision of the prophage from the chromosome of the lysogen. The nonhomologous recombination events that result in the substitution of flanking chromosomal genes for phage genes are independent of λ's normal excision functions encoded by the phage *int* and *xis* genes (**Figure 1**).

The transduced cells are frequently lysogenic for a λ-transducing phage, with the phage genes being flanked by tandem copies of the transduced gene. (The exogenous gene carried by the transducing phage is termed an 'exogenote,' while the endogenous gene is the 'endogenote.' A cell carrying an exogenote is a 'syngenote' or, where the exogenote and endogenote differ in one or more markers, a 'heterogenote.') Excision of the prophage by homologous recombination between the flanking genes readily leads to segregation of the recipient phenotype (1 per 10^3 bacterial divisions).

In most cases the loss of phage genes associated with formation of the transducing derivative leads to a defective phage, incapable of growth in the absence of an associated 'helper' phage. Thus defective gal-transducing phages are referred to as λ*dg* phages. In some cases the genes lost are not essential for phage growth and a plaque-forming transducing phage can be isolated (e.g., *pbio*).

Induction of a host cell lysogenic for λ*dg* does not lead to production of a phage lysate. If the cells are doubly lysogenic, for λ*dg* and λ^+, or if the UV-treated single lysogen is superinfected with λ^+ helper phage, the resulting preparation is a high-frequency transducing (HFT) lysate, with up to half of the phages in the lysate being transducing phages.

Extending the Range of Specialized Transduction

The number of host genes that can be picked up and stably carried by phages such as bacteriophage λ can be greatly extended by the use of donor strains lacking the normal chromosomal attachment site. Lambda integrates its genome and forms stable lysogens about 200-fold less frequently in such hosts, but does so by the normal integration mechanisms at 'secondary' attachment sites. Although about a dozen such sites predominate, where a strong positive selection for gene-inactivation is available, rare λ-integrants can usually be selected. Induction of such lysogens into lytic growth leads to the generation of transducing phages by aberrant excision.

Using *in vitro* recombinant DNA techniques, phages resembling plaque-forming, specialized-transducing phages can readily be generated. When the cloned DNA is homologous with the host cell DNA, phage integration and transduction can be achieved via homologous recombination. Transductants will be syngenotic lysogens so long as the vector phage has a functional immunity system. If the cloned DNA is nonhomologous with the *Escherichia coli* host DNA, integrants can be created by the use of a λ-lysogenic host strain. Using an endogenous prophage and a transducing phage of different immunity specificity allows the double lysogen to be selected by its immunity to the appropriate superinfection.

Other Specialised Transducing Systems

Although most studies have been carried out with λ and its host, *E. coli*, many other phage/host systems supporting specialized transduction have been described and characterized. In most cases, these behave according to the model worked out for λ*gal*- and λ*bio*-transducing phages. Specialized transducing derivatives of generalized transducing phages have been isolated. In some cases these are due to direct transposition of genes from the host chromosome or an episome into the phage genome.

A phenomenon superficially similar to specialized transduction occurs with eukaryotic retroviruses. Transducing retrovirus particles carrying host genes arise by cotranscription of an inserted provirus and an adjacent host gene, followed by splicing, packaging, and recombination into a viral genome.

Uses of Specialized Transducing Phages

Because specialized transduction normally results in a partial diploid, with various lengths of the donor chromosome carried on the exogenote, it has been valuable in the detailed genetic analysis of certain bacterial genes. Complementation tests, dominance tests and deletion mapping are readily carried out using specialized transducing phages. The ease of carrying out electron microscopic heteroduplex mapping with lambdoid phage genomes allows genetic and physical measurements to be correlated.

Transduction with specialized transducing phages involves integration by homologous recombination, a readily reversible event. This integration–excision cycle allows ready exchange of genetic markers between exogenote and endogenote and has proved useful in the manipulation of bacterial genotypes.

Analogs of transducing phages constructed by *in vitro* methods have been used to prepare an ordered array of phages covering the entire *E. coli* chromosome. Such ordered arrays, which facilitate genetic mapping and gene isolation, can readily be constructed for other bacterial genomes.

Further Reading

Campbell AM (1962) Episomes. *Advances in Genetics* 11: 101–145.

Kohara Y, Akiyama K and Isono K (1987) The physical map of the whole *E. coli* chromosome: application of a new strategy for rapid analysis and sorting of a large genomic library. *Cell* 50: 495–508.

Morse ML, Lederberg EM and Lederberg J (1956) Transduction in *Escherichia coli* K-12. *Genetics* 41: 142–156, 758–779.

Weisberg RA (1996) In: Neidhart FC *et al.* (eds) Escherichia coli *and* Salmonella: *Cellular and Molecular Biology,* 2nd edn, pp. 2442–2448. Washington, DC: ASM Press.

See also: **Lysogeny; Transduction**

Speciation

E Mayr

doi: 10.1006/rwgn.2001.1217

One of the most challenging aspects of the diversity of life is that it consists of discrete entities, called species. There was no need to explain this as long as one believed, as did Linnaeus, that "there are as many species as were created at the beginning." But throughout the eighteenth and the first half of the nineteenth centuries, it became ever more obvious that there are processes through which new species originate. The replacement of extinct faunas by new species was one of these phenomena. For Darwin, after his return from the *Beagle*, the origin of species became the foremost research program.

Now, 140 years after the publication of *On the Origin of Species* in 1859, speciation is still an active field of research. This means that there are still unsolved aspects of this process and unresolved controversies. This remaining uncertainty is mainly accounted for by (1) the pluralism of speciation phenomena, and (2) equivocation as to the meaning of the word 'speciation.'

Paleontologists traditionally have referred to the change of phylogenetic lineages as speciation. This process, however, even though resulting in change of the lineage over time, does not produce additional species. To avoid confusion, this process is best referred to as 'phyletic evolution.'

What is usually meant when an author speaks of speciation is the multiplication of species. It is the production of new species by existing ones. Darwin encountered this process and understood its meaning, when he was told that the mockingbirds (*Mimus*) which he had collected on three different islands in the Galapagos were three different species. Because there is only one species on the mainland of South America, opposite the Galapagos Islands, colonists of that species apparently had speciated into three species in the Galapagos archipelago. This led to the question which is still in part controversial today, by what processes does such a multiplication of species take place? Numerous answers to this question have been proposed which require critical analysis.

Type of Speciation

Instantaneous Speciation

When essentialistic thinking was still dominant, speciation could be conceived only by the spontaneous production of a new individual that represented a new kind of organism, a new type. Some of Darwin's contemporaries adopted this solution, such as Lyell and T.H. Huxley. Three leading Mendelians (de Vries, Bateson, and Johannsen) believed, after 1900, that a single mutation could establish a new species. This mode of speciation was defended up to the middle of the century (Goldschmidt, 1940; Willis, 1940; Schindewolf, 1950).

Instantaneous speciation may be defined as the production of a single individual (or the offspring of a single mating) that is reproductively isolated from the species to which the parental stock belongs and that is reproductively and ecologically capable of establishing a new species population.

Even though such instantaneous speciation by a single mutation has been shown not to occur ordinarily in sexually reproducing organisms, it is of course frequent among asexual clones, but the new agamospecies* which are produced by this process are not the equivalent of biological species.

Instantaneous speciation, however, occurs not infrequently in plants and more rarely in animals, as a result of chromosomal restructuring. For instance, the doubling of the chromosome set of a sterile species hybrid in plants may lead to the production of a fully fertile allopolyploid. For reasons not yet fully understood, such a restoration of fertility in species hybrids occurs far less frequently in animals. Instead, animal species hybrids may switch to parthenogenesis and may persist for long periods of time. Chromosomal rearrangements may also lead to the production of new postzygotic isolating mechanisms, if the new population succeeds getting through the first deleterious heterozygous stage. It seems that in animals all cases of seeming instantaneous speciation are accompanied by a shift to parthenogenesis or self-fertilizing hermaphroditism. In plants, however, speciation by polyploidy is common. At least one-third of all plant species are the result of this process.

Geographic Speciation

In this process, "a population which is geographically isolated from its parental species, acquires during this period of isolation genetic differences which promote or guarantee reproductive isolation when the external barriers break down" (Mayr, 1942).

That speciation is a populational process was discovered by several naturalists in the first third of the nineteenth century. Darwin understood it in 1837 when studying the three species of mockingbirds he had discovered on three islands in the Galapagos. Even though Darwin himself later adopted sympatric speciation (see below) and downgraded the importance of geographic speciation, the importance of the latter process continued to be emphasized by leading naturalists such as Moritz Wagner, K. Jordan, D.S. Jordan, Stresemann, Rensch, and Mayr. Under the influence of Darwin and Weismann, sympatric speciation was adop ted s the principal process of speciation until about 1942. It is now acknowledged that geographic speciation is the principal process of speciation in sexually reproducing animals and probably also in plants

Allopatric (geographical) speciation occurs in two forms

1. Dichopatric (splitting) speciation. The range of a more or less widespread species is split into two by a newly arising geographic barrier such as a waterway, a new mountain range, or a vegetational barrier (like a savanna separating two parts of a previously continuous rainforest).

*Agamospecies are sets of asexual clones. Speciation of agamospecies takes place by mutation and by the elimination of less successful clones by natural selection. This creates gaps among sets of clones and if such gaps are wide enough, such sets of clones are considered different species.

2. Peripatric (budding) speciation. A founder population is established beyond the current species border and beyond the gene flow of the parental species. Owing to the normal genetic processes occurring in any population (mutation, errors of sampling, selection, etc.), the isolated population diverges continuously genetically until it has reached the level of distinction that permits it to coexist as a separate species if it establishes contact again with the parental species.

Peripatric speciation differs in a number of ways from dichopatric speciation: (1) the gene pool is very small, answering readily to local selection pressures; (2) the population lives in a new physical and biotic environment and is exposed to new and rather strong selection pressures; (3) the gene pool of the founder population was started by a very small sampling of the genes of the parental population and is apt to lose additional genetic variants through inbreeding. The origin of new epistatic connections is thus favored and this may lead to a rather radical restructuring of the new gene pool. Mayr (1954) has referred to this possibility as a genetic 'revolution'; even though this term is perhaps too strong, there is no doubt about the opportunity for considerable genetic restructuring in founder populations. This is substantiated by students of geographic variation who find that the most distant and most isolated peripatric founder populations are often drastically different from the parental species. By far the majority of the peripatric founder populations are unsuccessful, however, and quickly become extinct.

Evidence for geographic speciation

The evidence for the universality of allopatric speciation is overwhelming (Mayr, 1963). Particularly impressive are the numerous cases of peripatric populations that are 'borderline cases,' that is, populations more or less on the way to become separate species.

Sympatric Speciation

In the 1850s, Darwin switched from allopatric to sympatric speciation as a result of misapplying his Principle of Divergence (Mayr, 1992). He fought for it in his controversy with Moritz Wagner, misled by Wagner's alternative, selection or isolation, as the cause of speciation. Sympatric speciation was the speciation mechanism most widely adopted until the 1940s. It was adopted by nearly all entomologists who worked with host-specific species, in spite of the counterarguments by K. Jordan and E. Poulton. In 1947 Mayr showed the obstacles encountered by sympatric speciation. Maynard Smith showed under what combination of circumstances sympatric speciation nevertheless could occur and Bush (1994) provided considerable evidence for its actual occurrence.

Even though it is still evident that geographic speciation is the most frequent process of speciation both in animals and plants, enough situations have been found in recent years to confirm the occurrence of sympatric speciation. Apparently it proceeds in animals by two very different methods.

By host shift

If an individual of a host-specific insect shifts to another plant host and establishes a population, this founder population might, by sympatric speciation, become a new species, equally host-specific on the new host. The difficulty is reciprocal recolonization. If an insect can switch from plant species A to plant species B, its descendants are most likely able to colonize back to plant species A, for which it was originally particularly adapted. If there is sufficient back and forth of colonization, an insect species might evolve that is successful on both plant species. Selection might well favor such polyphagy as an expansion of the resource base of an insect species. A shift to a new host species might be particularly easy in a peripherally isolated population where the original plant species is rare but a suitable new one is quite abundant.

By acquisition of a new mate preference

It was recently discovered that sympatric speciation occurs quite frequently in certain families of fishes, particularly the Cichlidae. Not all cases of the coexistence of two (or more) very closely related species in the same body of water requires an explanation by sympatric speciation. Sometimes repeated colonization from an outside source is what really had happened. However, when six genetically very similar cichlid species coexist in a crater lake in Cameroon, species that are more similar to each other than to any species outside the lake, sympatric speciation is the only interpretation that makes sense. The mechanism is apparently a switch in mate preference (through sexual selection), but the details of this process have not yet been elucidated.

Acquisition of Isolating Mechanisms

Isolating mechanisms (see Species) are the devices through which species are protected against hybridization with other species. A rigorous definition is: "Isolating mechanisms are biological properties of individuals that prevent the interbreeding of populations that are actually or potentially sympatric." This definition excludes geographical barriers or any other kind of spatial isolation as an isolating mechanism. In sexually reproducing species, isolating mechanisms originate in populations during periods of geographic isolation. Darwin insisted that there cannot be natural selection for the development of isolating mechanisms, rather they are an incidental byproduct of the genetic changes that occur independently in the isolated populations. This is particularly obvious for postzygotic isolating mechanisms (chromosomal changes). As far as prezygotic mechanisms are involved, particularly behavioral barriers in animals, it was long assumed that they were the incidental byproduct of different stochastic processes in the two isolated populations and of the different selection forces to which they are exposed. Evidence has been found recently, however, that 'fashions' in mate selection may play an important role in certain groups of animals. Mate preferences, developed through sexual selection in isolated populations, may (by change of function) become behavioral isolating mechanisms.

Incomplete Speciation and Hybridization

Incipient species often re-establish contact with the parental population owing to range expansion before their isolating mechanisms had been perfected. In the zone of contact, hybridization will now take place and a more or less extensive hybrid zone will develop. Several different outcomes of such an event have been recorded. If the isolating mechanisms were nearly perfect and only a few hybrids occur, the two species will not fuse and natural selection may even lead to an improvement of the isolating mechanisms.

If there is almost indiscriminate hybridization, a more or less permanent hybrid zone will develop due to the continuing elimination of the hybrids and their descendants, since they are of reduced viability. At the same time new hybrids between the two populations continue to be produced. The isolating mechanisms cannot be improved because of this continuous recolonization of the hybrid belt from the two parental populations.

There is some evidence that a highly isolated small hybrid population may, in time, develop its own isolating mechanisms (owing to an absence of recolonization by the two parental populations) and finally become a separate species. This is the most likely explanation for the occurrence of homoploid species coexisting with one or the other parental species, now without hybridizing with them. Such cases have been described in plants and in animals.

Some authors have claimed that two incipient species connected by a hybrid belt might become full species by a process called parapatric speciation. They postulate that the selection pressure against the hybrids would in due time lead to a reduced frequency of hybridization and ultimately to its disappearance. A careful study of all these cases, however, has convinced me that this is unlikely to occur. This is indicated by the high age of some of the hybrid belts between incipient species which had originated in Pleistocene refugia and had re-established contact with each other as much as 8000 to 10 000 years ago.

Cases in which an expanding species begins to overlap the range of a closely related species are different. The most advanced colonists of the expanding species may not be able to find conspecific mates and then mate with individuals of the overlapped species. As the expansion continues, sufficient individuals of their own species become available and the hybridization occurs no longer.

The isolating mechanisms are not always perfect. They are 'leaky' as it is said. The result is occasional hybridization, even between good species. The frequency of such occasional hybridization varies among different kinds of organisms (see Species).

In the prokaryotes, unilateral gene exchange between agamospecies is apparently very frequent, even among such distant groups as the eubacteria and the archaebacteria.

Rates of Speciation

Perhaps the most astonishing aspect of speciation is the enormous difference in the rates of speciation. It may be instantaneous as in the case of allopolyploidy or the shift to parthenogenesis in animal species hybrids. On the other hand, populations that are known to have been isolated from each other millions of years ago may still lack any isolating mechanisms. The American botanist Asa Gray called Darwin's attention to about six or seven eastern North American plants, including the skunk cabbage, which also occur in eastern Asia but have not changed in appearance since their isolation nor acquired cross-sterility. Here speciation had not occurred in a period of six to eight million years of separation.

Rapid speciation may also occur in sexually reproducing organisms. A group of species of cichlid fishes that occur in the southernmost bay of Lake Malawi in east Africa, each endemic to the waters around a rocky island, seem to have evolved in the last 1000 years. The rich fauna of more than four hundred species of cichlid fishes in Lake Victoria evolved since the lakebed was completely dry, perhaps only 25 000 years ago. Even more rapid seems to have been the sympatric speciation of some freshwater fishes, having occurred in less than 1000 years. Mayr infers that in birds an isolation of at least 10 000 years but more likely more than 100 000 years is necessary for the perfecting of isolating mechanisms.

References

Bush GL (1994) Sympatic speciation in animals. *TREE* 9: 285–288.

Goldschmidt R (1940) *The Material Basis of Evolution*. New Haven, CT: Yale University Press.

Mayr E (1942) *Systematics and the Origin of Species*. New York: Columbia University Press.

Mayr E (1947) Ecological factors in speciation. *Evolution* 1: 263–288.

Mayr E (1954) Change of genetic environment and evolution. In: Huxley J, Hardy AC and Fard EB (eds) *Evolution as a Process*, pp. 157–180. London: Allen & Unwin.

Mayr E (1963) *Animal Species and Evolution*. Cambridge, MA: Harvard University Press.

Mayr E (1992) Darwin's principle of divergence. *Journal of the History of Biology* 25: 343–359.

Schindewolf O (1950) *Grundfragen der Paläontologie*. Stuttgart: Schweizerbert.

Willis JC (1940) *The Course of Evolution by Differentiation or Divergent Mutation rather than by Selection*. Cambridge: Cambridge University Press.

See also: **Evolution; Phylogeny; Phylogeography; Species**

Species

E Mayr

doi: 10.1006/rwgn.2001.1218

The word species is used in daily language to refer to different kinds of things. One speaks of different species of metals or species of minerals. This concept of species focuses on degree of difference and is now usually referred to as the typological species concept. This was, on the whole, the species concept of Linnaeus and ruled unchallenged far into the nineteenth century. It fitted well with the essentialistic thinking of that period, but eventually it became obvious that it did not reveal the true nature of species of organisms. This resulted in a search for other, hopefully better, species concepts. Before discussing them, it is, however, necessary to determine what the word 'species' means. In current taxonomy the word species is used for three different concepts.

1. Species as concept. For different biologists, the word 'species' had different meanings. For some it indicated an entity that was different, for others it was a reproductive community. Species definitions reflecting at least seven such different concepts have been proposed and will be discussed below.
2. Species as taxon. A species taxon is a group of natural populations conforming to the definition of a species concept. The species concept serves as the yardstick by which to delimit a taxon against other taxa. A taxon can be described and delimited, but not defined. Such a group of populations has reality, but its rank (species or subspecies) is often difficult to determine. This determination of rank is rarely a problem where two closely related populations are sympatric or are in the process of invading each other's ranges. If the two populations remain reproductively isolated, they are two species; if they freely interbreed, they are only a single species.

   For more than 130 years, naturalists, have agreed that species are not essentialistic classes. But then what are they? Haeckel said, "the species is an individual." However, even though there is an internal cohesion in a species, which makes it a unified system corresponding to an individual, the word 'individual' in the vernacular always applies to a singular object which a species is not. It is definitely counterintuitive to refer to the circa 6 billion human individuals as an individual. It has therefore been suggested to introduce a third ontological category, the biopopulation. A species taxon is a biopopulation. It has the internal cohesion of an individual, but is so to speak a multiple individual. However, a species taxon is definitely not a class.
3. Species as category. A category designates rank or level in a hierarchic classification. The species category is the class whose members are the species taxa. Not only are all biological species placed in this category, but so also are those asexual entities (agamospecies) that are as different from each other as are biological species. Such pluralism in the species category is inevitable since biological species and agamospecies are two rather different natural phenomena. Agamospecies occur in most higher taxa of animals (not in mammals or birds), are very common in plants, and all 'species' of prokaryotes are agamospecies. An agamospecies consists of a set of clones which, in the aggregate, are as different from other agamospecies as are good biological species from each other. Owing to the steady selection against inferior clones, gaps are produced among the clones that correspond to the species borders of biological species.

Species concepts

For a layperson, a species was simply an assemblage of similar entities. But, as the knowledge of nature grew, this vague concept was no longer adequate and a need developed for a more precise definition of the species concept. The first attempt was the typological species concept, but when its weaknesses were discovered, a considerable number of 'better' concepts were proposed by naturalists. Six of these concepts in addition to the typological one will be discussed here.

Typological or Essentialistic Species Concept

This concept developed during and after the Renaissance and particularly in the eighteenth century. It was supported by three sets of observations:

1. The observations of the naturalists of seemingly well defined kinds of species of animals and plants at a given locality.
2. The belief of the Christian naturalists that there are "as many species as there were diverse forms created in the beginning" (Linnaeus).
3. The philosophical view first advanced by the pythagoreans and Plato that the observable variation of nature can be assigned to separate classes characterized by their definition (*eidos*, essence). Each class consists of constant and essentially identical members.

According to this concept, the observed diversity of the universe reflects the existence of a limited number of underlying 'universals' or types. Individuals do not stand in any special relation to each other, being merely manifestations of the same type. Members of a species form a class. Variation is the result of imperfect manifestations of the idea implicit in each species. This concept was the species concept of Linnaeus and his followers. Because this philosophical tradition is also referred to as essentialism, the typological definition is also sometimes called the essentialist species concept. Degree of phenotypic difference is the criterion of species status for the adherent of the typological species concept. For him, a different species is simply that which is different. Morphological evidence is used by all taxonomists, but there is an enormous difference between basing one's species concept entirely on degree of difference and using morphological evidence as an inference in the application of a biological species concept.

The typological species concept was accepted by taxonomists almost unanimously as late as the mid-nineteenth century. It included the acceptance of four postulates:

1. Species consist of similar individuals sharing the same essence.
2. Each species is separated from all others by a sharp discontinuity.
3. Each species is completely constant through time.
4. There are strict limits to the possible variation within any one species.

According to this concept, species are defined as groups of similar individuals that are different from individuals belonging to other species.

Difficulties of the typological species concept

In the last 150 years, more and more exceptions to these criteria were found by working taxonomists. Great phenotypic variation was discovered in many species and the different sexes or different age stages or other intraspeciefic variants were often at first described as different species. When unmasked eventually, they were assigned to their proper species on the basis of biological criteria (life history, etc.). This was in plain conflict with the typological definition.

Equally troublesome was the opposite extreme, the absence of observable phenotypic differences between noninterbreeding coexisting species. On the basis of life history criteria, literally hundreds of morphologically indistinguishable species were described in virtually all higher taxa of animals, from mammals down to protozoans. What was, for instance, traditionally considered *Paramecium aurelia* was finally shown to consist of no less than 14 'sibling species,' as such cryptic species are called. They also occur among flowering plants. Gilbert White, the vicar of Selborne, discovered in 1768 the first sibling species by showing that the leaf warbler *Phylloscopus trochilus* of Linnaeus actually consisted of three different species. It became quite clear in time that the typological species concept was not particularly suitable for sexually reproducing species of organisms. What other concept should instead be adopted?

Biological Species Concept

The shift from a strictly typological to a more biological species concept is already foreshadowed in the writings of Linnaeus. When he discovered that he had described the juvenile goshawk and the female mallard owing to plumage differences as different species, he reduced his names to synonyms, realizing that on the basis of life history criteria they belonged to the same species. He revealed that for him biological criteria had primacy over degree of phenotypic difference. He evidently had asked himself what the real meaning of the word 'species' was and had adopted a new concept.

The meaning of species

Perhaps one should first ask what is meant by concept? The species concept is our view of the role of species in nature. To find this out, we must ask the

Darwinian question, "Why are there species?, what is their meaning in the scheme of things?"

There is no better way of answering these questions than to try to conceive of a world without species. Let us think for instance of a world in which there are only individuals, all belonging to a single mating community. Every individual to varying degrees is different from every other one, and every individual is capable of mating with those others that are most similar to it. In such a world every individual would be, so to speak, the center of a series of concentric circles of increasingly more different individuals. Any two mates would be on the average rather different from each other, and would produce a vast array of genetically different types among their offspring. Now let us assume that one of these recombinations is particularly well adapted for one of the available niches. It is prosperous in this niche, but when the time comes for mating, this superior genetic complex will inevitably be broken up by recombination. There is no mechanism that would prevent such a destruction of genetically superior combinations and there is, therefore, no possibility of the gradual improvement of genetic combinations. The significance of the species now becomes evident. The reproductive isolation of a species is a protective device against the breaking up of its well-integrated coadapted gene system. Through organizing organic diversity into species, a system has been created that permits genetic diversification and the accumulation of favorable genes and gene combinations without any danger of destruction of the basic gene complex. (Mayr, 1963, p. 423)

The meaning of the species is now clear, it is a protected reproductive community. This is expressed in the biological species definition: "A species is a group of interbreeding natural populations that is reproductively isolated from other such groups." Since the interaction of populations is a major aspect of the species concept, the concept is strictly applicable only in the nondimensional situation that is at a given place at a given time. The role of the species concept is to serve as a yardstick in the testing of the species status of populations. A species, thus, is a population (or group of populations), not a type.

The biological species concept is particularly useful for the field naturalist, the ecologist, and the student of behavior. However, in difficult situations it requires an intimate knowledge of natural populations which a museum or herbarium taxonomist may not have.

Isolating mechanisms

Coexisting biological species are prevented from interbreeding by genetic propensities. These are referred to as isolating mechanisms. They are "biological properties of individuals which prevent the interbreeding of populations that are actually or potentially sympatric" (Mayr, 1963, p. 91). Cross-sterility was long considered the exclusive barrier between sympatric species. However, numerous cases have been described in the last 100 years of sympatric species which virtually never interbreed in nature but are shown to be fully fertile in captivity. Their genetic independence is sustained by isolating mechanisms other than sterility.

A large number of different kinds of isolating mechanisms have now been discovered. They consist of premating mechanisms, which prevent the occurrence of copulation, and postmating mechanisms (sterility genes, chromosomal incompatibility, etc.), which diminish or prevent the success of crossing with nonconspecific individuals. Most important among the premating mechanisms, particularly in animals, are behavioral barriers, but important are also seasonal and habitat isolation. Isolating mechanisms occasionally break down and permit sporadic hybridization, particularly in plants. Yet this does not necessarily lead to an eventual fusion of the two species, when the hybrids and their offspring are of sufficiently lowered viability to be eliminated rather quickly from the gene pool.

The 'leaky' nature of isolating mechanisms would seem to invalidate the biological species concept (based on noninterbreeding) or at least its applicability to plants. However, studies of local floras of angiosperms have shown that a very high percentage of plant species have all the characteristics of biological species and are remarkably well isolated reproductively from other sympatric species, even where occasional hybridization occurs.

The perfecting of leaky isolating mechanisms is apparently not easy. Several cases are known in plants (*Quercus*, *Populus*) where the fossil record shows that two species were hybridizing many millions of years ago and still occasionally hybridize, even though, on the whole, they still coexist as two perfectly distinct species. In animals an introgression of genes of one species into another occurs also, perhaps even frequently. Wherever the range of the gray wolf (*Canis lupus*) meets that of the coyote (*Canis latrans*) mitochondrial genes of the coyote are found in the wolf populations. But phenotypically the introgressed individuals look like typical wolves.

The application of the biological species concept to populations may encounter various difficulties. Four of them deserve special discussion.

Agamospecies Asexual organisms do not form biopopulations but produce separate uniparentally reproducing clones. The biological species concept can therefore not be applied. Instead, an assemblage of similar clones is considered an agamospecies (Cain) and ranked in the Linnaean hierarchy as if it were a biological species.

Some clones (agamospecies) are apparently quite isolated from other related clones, because any formerly existing intermediate clones were eliminated by natural selection. As clones they show little variability and are therefore more easily diagnosed than related sexual species. Easy identifiability is helpful in the delimitation of species taxa, but this has nothing to do with the biological meaning of species concepts.

The ranking of geographically isolated populations In many species, there are isolated populations beyond the regular species border, separated by minor geographical barriers. Some of these isolates have somewhat diverged genetically, owing to selection and stochastic processes. Whether or not such populations are still conspecific with the main body of the species cannot be determined directly but must be inferred. The methods on which such inferences are based are described in the taxonomic literature. These methods primarily make use of degrees of morphological difference. It must be emphasized that the rank thus determined is not based on difference but on the probability of interbreeding as inferred from the degree of morphological difference.

Rejection of the applicability of the biological species concept in the delimitation of species taxa has led many botanists to recognize numerous allopatric populations as full species, populations which an ornithologist or lepidopterist would call subspecies. The 'species' of such botanists are of highly unequal biological significance.

Incomplete isolation Two incipient species, after a period of isolation, may spread and come secondarily in contact with each other. If speciation was not yet completed, they will form a parapatric hybrid zone. If there is a complete breakdown of the isolating mechanisms, the two incipient species must be ranked as subspecies. However, if only an occasional hybrid is produced, they are best considered full species. Occasional hybridization between related sympatric species is much more common in plants than in animals, but in spite of its frequency may not lead to a complete breakdown of the barrier between the two species.

Selfing In some sexually reproducing species, self-mating may evolve. This differs from strictly asexual reproduction in that the egg does not develop without fertilization, but is fertilized by a gamete produced by the parent of the egg. An equivalent situation is represented by automixis. Virtually selfing lineages occur in many sexually reproducing species, but nothing is gained by developing a special species concept for such situations. Such selfing lineages are best included with the biological species from which they are derived.

The four mentioned difficulties in the application of the biological species concept are real. However, all endeavors to arrive at a meaningful species concept that is equally applicable to a sexual and a sexual organisms have been a failure. We must accept having two different kinds of species concepts for the two kinds of organisms. The decision to call all peripherally isolated populations full species is biologically unacceptable. It leads to a complete negation of the actual meaning of species and amounts to an unequivocal re-establishment of the typological species concept.

There is no question that there are a number of evolutionary processes (particularly asexuality) to which the biological species concept cannot be applied. In such cases one must adopt plural solutions, as so often in other branches of biology. Alternative mechanisms seem far less acceptable.

Other Species Concepts

The biological and the typological species concepts are the two most widely adopted species concepts. However, in order to correct what some authors considered to be deficiencies of the biological and typological species concepts, a number of other species concepts have been proposed over the years. Some of these cannot properly be considered concepts; they are simply operational instructions for how to delimit species taxa. Many of the controversies which they have raised are still raging.

Nominalistic Species Concept

Nominalists deny the real existence of species. As stated by one of them (Bessey):

> nature produces individuals and nothing more... species have no actual existence in nature. They are mental concepts and nothing more... species have been invented in order that we may refer to great numbers of individuals collectively.

Every naturalist knows from practical experience that this is simply not true. Species of animals are not human constructs, nor are they types in the sense of Plato, rather they are existing entities for which there is no equivalent in the realm of inanimate nature. There is no better refutation of the nominalist claim than the fact that primitive natives refer to the same natural populations as species as do university graduates in the Western world. Even in a local flora, the vast majority of species are exceptionally well demarcated against each other. It is only among asexual organisms that species frequently may have to be delimited rather arbitrarily as stated by the nominalists.

Evolutionary Species Concept

Some paleontologists have advanced a species concept based on evolutionary criteria. In Simpson's definition (1961, p. 153), "an evolutionary species is a lineage (an ancestral-descendent sequence of populations) evolving separately from others and with its own unitary evolutionary role and tendencies." Actually this is the definition of a phyletic lineage, not of a species. It applies equally to almost any isolated population or incipient species; it also fails to explain what a 'unitary role' is and why phyletic lines do not interbreed with each other. What apparently concerned Simpson most was the problem of the delimitation of species taxa in the time dimension, but here his definition is of little help. When we consider a sequence of morphotypes in a single phyletic lineage, how are we to know whether these morphotypes have different unitary evolutionary roles and tendencies and should thus be considered different species or whether all of them have the same unitary evolutionary role and should thus be treated as chronospecies? The principal weakness of the so-called evolutionary species definition is that it fails to account for the causation and maintenance of discontinuities among contemporary species. Furthermore, none of the proponents of an evolutionary species definition has provided a nonarbitrary criterion by which to divide a continuous phyletic lineage into separate species taxa. Hennig (1966) arbitrarily terminated every evolutionary species when a daughter species branched off the parental lineage, ignoring the fact that the parental species may remain unchanged when a new species originates by peripatric speciation. The evolutionary species concept does not account for the selection forces responsible for the existence of species.

Phylogenetic Species Concept

Cladists, in recent years, have proposed one further species concept. It is their endeavor to delimit the branches of the phylogenetic tree and to locate the branching points. To facilitate this they recognize 'phylogenetic species.' "A phylogenetic species is an irreducible (basal) cluster of organisms, diagnosably distinct from other such clusters, and within which there is a parental pattern of ancestry and descent." This definition does not use the word 'population' but indicates by its reference to 'a parental pattern of ancestry and descent' that it refers to a branch. Each branch is initiated by a population designated the stem mother species. A species, for the cladists, is an evolutionary unit characterized by its difference from other populations. Such a population has no other biological significance than its potential to be the starting point of a new phylogenetic lineage. Relying entirely on the amount of difference from other populations, the phylogenetic species is actually an undisguised return to the typological species concept. Almost any isolated subspecies of a traditional classification can be raised to species rank under the phylogenetic definition. The result is a highly uneven value of accepted phylogenetic species. According to the biological species concept, many of them would be subspecies, others incipient species or allospecies, and finally still others well-isolated full species. The phylogenetic species basically has no biological significance and since its criterion, degrees of difference, is purely subjective, it leads to a rather arbitrary determination of species status.

Recognition Species Concept

Paterson proposed a recognition species concept, based on the capacity of members of a species to recognize each other as potential mates. However, it has been shown that Paterson's arguments were based on misunderstandings, and that the recognition species concept is only a version of the biological species concept.

Cohesion Species Concept

The biological species concept describes an ideal situation of a large interactive cohesive gene pool, completely isolated reproductively from other such species. Alas, there are numerous situations which do not quite fit, particularly the agamospecies. But there are also 'leaky' isolating mechanisms, resulting in species hybrids and in the introgression of genetic material into another species, there are selfing lineages and species, and there is the complete abandonment of sexual reproduction and the origin of agamospecies. Templeton proposed to bring all these diverse kinds of species under one hat by accepting a 'cohesion species concept.' It defines "a species as the most inclusive group of organisms having the potential for genetic and/or demographic exchangeability" or "a population of individuals having the potential for phenotypic cohesion through instrinsic cohesion mechanisms." It is difficult to see an underlying Darwinian concept in these definitions. They sound more like instructions for the delimitation of species taxa.

It also seems that many phenomena that should be covered by the cohesion concept do not fit. For instance, where is the cohesion among clones of an agamospecies, ultimately derived from one ancestral species. There is cohesion within each clone, but none whatsoever between clones. Worse, 'the potential for genetic... exchangeability' is shared by many very good species, coyote and wolf being a typical example. Most species of duck cross readily with each other and produce fertile hybrids. And, of course, all prokaryotes freely exchange genes, even the two very different subdivisions of eubacteria and archebacteria. Although the cohesion species concept is not a suitable replacement of the biological species concept, the reading of Templeton's account is recommended as a particularly well-informed discussion of the difficulties of a sound species definition.

References

Hennig W (1996) *Phylogenetic Systematics*. Urbana, IL: University of Illinois Press.

Mayr E (1963) *Animal Species and Evolution*. Cambridge, MA: Harvard University Press.

Simpson GG (1961) *Principles of Animal Taxonomy*. New York: Columbia University Press.

See also: **Population Genetics; Speciation; Species Selection; Species Trees**

Species Selection

J Levinton

doi: 10.1006/rwgn.2001.1219

'Species selection' refers to a process where a variety of factors cause some species to produce more species than others, resulting in a change in proportion of species deriving from different ancestral species.

Species selection is analogous to natural selection, which operates at the population level. The closest natural selection analogy would be to clonal selection, since species are splitting (or becoming extinct) at different rates. The difference in rates explains the differences in relative abundance of species groups, each deriving from a single ancestral species.

Species selection involves selection among species whose long-term result may be a morphological change in the modal morphology or other traits of the whole population of species under consideration. We presume a source of heritable variability, a reproductive mechanism, and differential 'fitness' among units – in this case – species. Differential fitness would include differential rates of speciation of different species morphological types or differential extinction. Speciation events generate among-species morphological variation, while selective mortality, or differential speciation rates, would bias survival in one morphological direction.

Species selection would involve cases where species-level properties bias speciation or extinction rates. For example, if reduced dispersal is fixed in a species it will tend to increase the probability of regional isolation, genetic differentiation, and speciation. One ancestral species bearing the trait of reduced dispersal might therefore produce more species than another species with widespread dispersal. Traits associated with the reduced-dispersal species might hitchhike along and be proliferated in a large group of descendant species, relative to traits hitchhiking with the species that has a low speciation rate.

Species selection is a process that is thought to generate the array of morphological differences among species either to amplify or even supplant the power of natural selection operating within each species. Thus natural selection (within species) is necessary but may be insufficient to explain the breadth of morphological differences within a monophyletic group of species.

Species-level characters such as dispersal type might also influence extinction rates. For example, low dispersal might promote smaller geographic species ranges, which would make such species more vulnerable to extinction. Species with short dispersal might therefore have high speciation rates and high extinction rates, which would suggest that small shifts in the values of these rates could cause very different outcomes in success at the level of species selection.

Species selection is a controversial process that has been suggested to demonstrate how macroevolutionary processes above the level of population could cause large-scale evolutionary trends.

See also: **Natural Selection; Speciation; Species; Species Trees**

Species Trees

N Saitou

doi: 10.1006/rwgn.2001.1480

Any species consists of many individuals, and those are distributed in a certain more or less continuous geographical area. As time goes on, some geological factor, such as mountain formation, shift of river course, creation of channel, may create a geographical barrier to this species. This kind of event prompts to produce two or more geographically isolated populations within the same species, and the evolutionary history (including allele frequency change through random genetic drift, occurrence of new mutation, pattern of natural selection of those populations will be different. Those factors all contribute to differentiation of genetic constitutions of those populations, and they become incipient species. Iteration of this divergence process produces a tree-like structure, as Darwin (1859) showed using a figure (**Figure 1**). This structure is called a species tree.

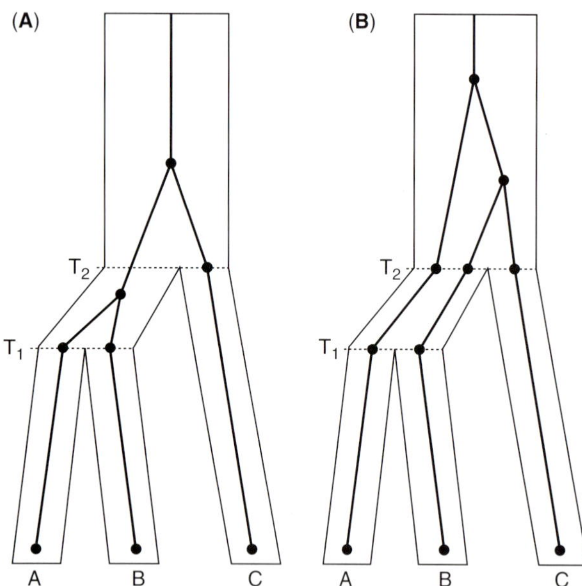

Figure 1 Darwin's diagram of species trees.

Each species is considered to be an abstract point in a species tree. However, one species includes many individuals, and in reality, a species tree is a very rough approximation of the genealogy of genes residing in those individuals. Because species trees are usually estimated from gene tree(s), we should be careful to distinguish two types of trees.

In fact, a gene tree may be different from the corresponding species tree. This difference comes from the existence of gene genealogy in the ancestral species. A simple example is illustrated in **Figure 1A**. A gene sampled from species A has its direct ancestor at the speciation time T_1 generations ago, and so does a gene sampled from species B. Thus the divergence time between the two genes sampled from the different species always overestimates that of species. The amount of overestimation corresponds to the coalescence time in the ancestral species, and its expectation is $2N$ for neutrally evolving nuclear genes of diploid organism, where N is the population size of the ancestral species. Therefore, if the two speciation events (T_1 and T_2) are close enough, the topological relationship of the gene tree may become different from that of the species tree, as shown in **Figure 1B**. Although species A and B are more closely related than to C, genes sampled from species B and C happen to be more closely related with each other than to that sampled from species A. The probability of obtaining an erroneous tree topology is given by Prob(error) $=$ $(2/3)\ e^{-T/2N}$, where $T = T_2 - T_1$ generations. Therefore, a species tree estimated from a single gene may not be correct even if the gene tree was correctly estimated. In this case, we should use more than one gene.

When distantly related species are compared, however, the above problem is not so serious, and topology of a gene tree most often corresponds to that of species tree. This is foundation of 'molecular phylogenetics,' in which molecular data such as nucleotide sequences are used to reconstruct phylogenetic relationships of various species.

Reference

Darwin C (1859) *On the Origin of Species by Means of Natural Selection*. London: John Murray.

See also: **Coalescent; Gene Trees; Taxonomy, Evolutionary; Trees**

Specific Locus Test

L Silver

doi: 10.1006/rwgn.2001.1221

Genetic variation, the existence of at least two forms, is the essential ingredient in all genetic experiments. Phenotypic variation, in particular, is used as a means for uncovering the normal function of a wild-type allele at many loci. It was the availability of many variant phenotypes within the fancy mouse trade that made the house mouse such an ideal organism for studies by early geneticists. In a sense though, the house mouse won by default because, in the absence of domestication and artificial selection, variation in traits visible to the eye is extremely rare, and thus, other small mammals were genetically intractable. Although the fancy mouse variants provided material for a host of early genetic studies, the number of different variants was still limited, and the rate at which new ones arose spontaneously in experimental colonies was exceedingly low: it is now known that, on average, only one gamete in 100 000 is likely to carry a detectable mutation at a particular locus.

During the 1920s, several investigators began investigating the effects of X-rays on reproduction and development. In two laboratories, at least, new mutant alleles were recovered in the offspring of irradiated parents, but the investigators failed to make any connection between irradiation and the induction of these mutations. The connection was finally made by Muller who, in 1927, published his classic paper explaining the induction of heritable mutations by X-rays. Since that time, geneticists who study all of the major experimental organisms – from bacteria to mice – have used both ionizing irradiation and various chemicals as agents of mutagenesis to create novel alleles as tools for understanding gene function.

Large-scale mouse mutagenesis experiments were first begun at two government-based 'atomic energy' laboratories: the Oak Ridge National Laboratory in Oak Ridge, Tennessee, in the US and the MRC Radiobiological Research Unit first at Edinburgh, and then at Harwell, in the UK. Both of these experimental programs were begun after World War II, as a means of quantifying the effects of various forms of radiation on mice and, by extrapolation, humans, to better understand the consequences of detonating nuclear weapons. The US effort was directed by W.L. Russell and the British effort was directed by T.C. Carter. Scientists at both laboratories quickly realized the potential of their newly created resource of mutant animals, and both laboratories have since gone on to generate mutations by chemical agents as well. The studies at Oak Ridge and Harwell were very large; 10 000–60 000 first-generation animals were typically analyzed in an experimental protocol. They have provided most of the empirical data currently available on the mechanisms and rates at which mutations are caused by all well-characterized mutagenic agents in the mouse.

The experiments performed by Russell and Carter, and other colleagues who followed in their footsteps, were designed to obtain discrete values for the mutagenic potential of different radiation protocols. Rather than attempt to examine all animals for all effects of a particular irradiation protocol (as was common in earlier experiments), these mouse geneticists chose instead to look only at the small fraction of animals that were mutated at a small set of well-defined 'specific' loci. The rationale for the 'specific locus test' was that effects on individual loci could be more easily quantitated and that the limited results obtained could still be extrapolated for an estimate of whole genome effects. Russell decided that mutation rates should be followed simultaneously at a sufficient number of loci to distinguish and avoid problems that might be caused by locus-to-locus variations in sensitivity to particular mutagens. He decided further that the same set of loci should be examined in each experiment performed. The seven loci chosen to be followed in the specific locus test were defined by recessive mutations with visible homozygous phenotypes that were easily distinguished in isolation from each other, and had no effect on viability or fertility. The seven loci are agouti (*a* is the recessive nonagouti allele), brown (*b*), albino (*c*), dilute (*d*), short-ear (*se*), pink-eyed dilution (*p*), and piebald (*s*). A special 'marker strain' was constructed that was homozygous for all seven loci.

In its simplest form, the specific locus test is carried out by mating females from the special marker strain to completely wild-type males that have been previously exposed to a potential mutagen. In the absence of any mutations, offspring from this cross will not express any of the seven phenotypes visible in the marker strain mother. However, if the mutagen has induced a mutation at one of the specific loci, the associated mutant phenotype will be uncovered. This test is very efficient because it only requires a single generation of breeding and visual examination is all that is required to score each animal.

Although recessive mutations at all loci other than the specific seven will go undetected in the first generation offspring from this cross, it is possible to detect a dominant mutation at any locus so long as it is viable and produces a gross alteration in heterozygous phenotype such as a skeletal or coat color change. One should realize that the most common effect of any undirected mutagen will be to 'knockout' a gene and, in the vast majority of cases, the resulting null allele will be recessive to the wild-type. There is, however, a very small class of loci at which null alleles will act in a dominant or semidominant fashion to wild-type. These 'haploinsufficient' phenotypes are presumably caused by a developmental sensitivity to gene product dosage. Among the best characterized of the dominant-null mutations are the numerous ones uncovered at the *T* locus (which result in a short tail)- and the *W* locus (which result in white spotting on the coat).

***See also:* Brachyury Locus; Coat Color Mutations, Animals; *W* (White Spotting) Locus**

Specificity

***See:* DNA-Binding Proteins; DNA Hybridization; Restriction Endonuclease**

Spermatids

S W L' Hernault

doi: 10.1006/rwgn.2001.1222

A spermatid is the male animal cell type that results from completion of the second division associated with meiosis. The only location where meiosis occurs in males is within the testis and the cell type from which spermatids form are spermatocytes. Meiosis is the process where the 4*n* nucleus in primary spermatocytes divides to form the 2*n* nucleus of secondary spermatocytes that, in turn, divides to form the 1*n* nucleus of spermatids. While a spermatid is initially a round, immotile cell, it differentiates to becomes the specialized, motile spermatozoon in a process called spermiogenesis. The essential purpose of spermiogenesis is to produce a spermatozoon that is streamlined for swift movement and contains the molecules required for this cell to locate and fertilize an egg. Spermiogenesis does not involve further division of the nucleus but it does involve extensive reorganization and an unusual division of the cytoplasm. This unusual division of the spermatid cytoplasm results in an anucleate cytoplast (essentially a membrane-bounded bag of cytoplasm), that was classically named the cytoplasmic droplet and is now most frequently called the residual body, and a smaller, nucleus-containing spermatid. This smaller spermatid retains mitochondria, centrioles, and certain membrane-bounded organelles. The residual body contains ribosomes and other cellular constituents not required by spermatozoa. There is some variation as to when and how the residual body forms within the different animal phyla. For instance, in the nematodes, the residual body forms at the same point when spermatids are created during the second meiotic division. Nuclei in nematode spermatids apparently do not engage in synthesis of new RNAs and these cells are unable to synthesize any new proteins because ribosomes are discarded into the residual body. Consequently, the dramatic cell shape changes that characterize the transition of a spermatid into a spermatozoon occur solely by rearrangement and modification of preexisting macromolecules. On the other hand, insects and mammals exhibit the more common situation where formation of spermatids during completion of the second meiotic division is temporally and spatially separated from appearance of the residual body. In this case, spermatids (as the spermatogonia and spermatocytes that precede them) remain connected to one another by cytoplasmic passageways and formation of the residual body is the last major cellular event to occur during transition of the spermatid to the spermatozoon. As such, insect and mammalian spermatids retain their ribosomes and the capacity to synthesize new proteins during most of the spermatid stage. In mammals, transcription of new mRNAs also continues during the spermatid stage. These macromolecules can be shared between spermatids through the cytoplasmic passageways that connect adjacent cells, and many are essential for maturation of spermatids into spermatozoa. The spermatid undergoes a number of dramatic changes as it matures, including formation of the flagellum which is required for motility by the spermatozoon. Internal membranes are also reorganized, and the acrosome, which is a specialized structure required for egg penetration during fertilization, forms during this stage. Prior to residual body formation, the spermatid nucleus becomes transcriptionally quiescent as the DNA becomes tightly compacted. This can be associated with a change in the proteins that interact with the DNA in the spermatid nucleus. In mammals, DNA (which is negatively charged) is complexed with histone proteins (that carry a positive charge) in all non-germline cells and all stages of spermatogenesis up to the late spermatid stage. At this point, histones are displaced from the DNA and replaced with protamines, which carry a stronger positive charge than histones. Spermatids in which this protamine replacement has occurred are transcriptionally inactive and the nucleus is physically more compact than the spermatid nucleus that contains histones. In the last stage, the spermatid forms a residual body in which ribosomes and other cellular constituents that are no longer needed are discarded. This more compact configuration of the protamine-containing nucleus permits the head of the spermatozoon to be physically smaller and more streamlined. A great deal remains to be determined about how spermatids form and function. For instance, it is unknown how incomplete cytokinesis that results in cytoplasmic passageways occurs. The mechanism that ensures proper segregation of cytoplasm during separation of the spermatid from the residual body is another poorly understood process.

***See also:* Meiosis**

Spermatocytes

S W L'Hernault

doi: 10.1006/rwgn.2001.1223

Spermatocytes are germ cells in the testes of animals that are engaged in meiosis. The testes contains populations of cells, named spermatogonia, that proliferate by mitotic division to yield new cells. As is usually the case for mitotic division, these cells are equivalent with respect to the DNA content of their nuclei. After one or more mitotic divisions, spermatogonia divide to form spermatocytes. Spermatocytes differ from spermatogonia in that they enter meiosis and engage in the genetic recombination that characterizes this process. They are the only cell type in the male body where meiosis occurs. Genetic recombination shuffles the genome so that new combinations of genetic material are created and the resulting cells have a genetic endowment that differs from the starting spermatogonial cell. After committing to meiosis, two cellular divisions occur where the $4n$ primary spermatocyte divides to form two $2n$ secondary spermatocytes (n refers to one complete copy of the genetic material). In turn, each secondary spermatocyte divides to form two $1n$ spermatids. The genetic recombination associated with meiosis occurs in the primary spermatocyte. In many animals, spermatocytes divide incompletely so that cells remain connected by passageways of cytoplasm. These passageways allow the free exchange of macromolecules, such as proteins and other materials, between connected cells. The significance of these passageways is that spermatocytes are nearly identical with respect to the proteins they contain even though the DNA template in their nucleus can differ due to genetic recombination that had occurred during meiosis. Although there have been some advances in our understanding in recent years, how a spermatocyte exits mitosis and enters meiosis is still poorly understood. It is also unknown how incomplete cell division associated with cytoplasmic passageways occurs during the formation of spermatocytes.

***See also:* Spermatids; Spermatogonia**

Spermatogenesis in *Caenorhabditis elegans*

S W L'Hernault

doi: 10.1006/rwgn.2001.1225

Wild-Type Spermatogenesis

As in most animals, spermatogenesis in the nematode *Caenorhabditis elegans* is a differentiation pathway where spermatogonia ultimately differentiate into spermatozoa. Wild-type *C. elegans* spermatogenesis is summarized in **Figure 1**. The $4n$ primary spermatocyte buds off the rachis (a central syncitial cytoplasmic core) after entering pachytene of meiosis I and undergoes the first meiotic division. Unlike mammalian spermatogenesis where primary spermatocytes remain connected to one another, the *C. elegans* primary spermatocyte is an individualized cell. Subsequent development of the primary spermatocyte can occur *in vitro* in the absence of added hormones, growth factors, or any other supporting cell type(s) such as the Sertoli cell that is required for mammalian spermatogenesis. The two secondary spermatocytes either completely separate or stay linked by a cytoplasmic bridge as they undergo the second meiotic division to yield a total of four haploid spermatids. Asymmetric cytoplasmic partitioning places many cytoplasmic constituents not required for subsequent differentiation into the residual body during formation of haploid spermatids. Resulting nonmotile, apolar spermatids are activated to form motile bipolar spermatozoa in a 5–10-minute differentiation process. Nematode spermatozoa lack the flagellum and acrosome that characterize spermatozoa of many other species, including most vertebrates. Instead, *C. elegans* spermatozoa crawl by directed membrane flow of a single pseudopod that drags the cell body across the substrate. A *C. elegans* spermatogonial cell completes spermatogenesis over several hours rather than days or, in the case of mammals, weeks.

C. elegans Reproductive Biology

Like most animals, *C. elegans* produces both sperm and eggs and the union of these two gametes produces new individuals. Four larval growth stages occur between hatching and formation of the sexually mature adult. While there are *C. elegans* males, this species does not produce females in the conventional sense. In fact, *C. elegans* exists principally as a self-fertile protandrous ('male first') hermaphrodite where both sperm and eggs are produced by the same individual. Anatomical comparison of *C. elegans* to other dioecious *Caenorhabditis* species (which have both a true male and female) reveals that the *C. elegans* hermaphrodite somatic tissues are female. The hermaphrodite germline has been modified so that it functions as a testis during the fourth larval stage and then switches to produce oocytes as an adult. The sperm are stored in a sac named the spermatheca and ovulation places eggs into the spermatheca where they are fertilized. Spermatogenesis in males also begins during the fourth larval stage but continues throughout adult life. When a male copulates with a hermaphrodite, his sperm will preferentially (versus hermaphrodite-produced sperm) fertilize eggs. Spermatogenesis is highly similar in both sexes.

Exploiting *Caenorhabditis elegans* Biology to Obtain Spermatogenesis Mutants

The unusual reproductive biology of *C. elegans* has greatly facilitated genetic analysis of spermatogenesis in this organism. In dioecious organisms, recovery of mutants with specific defects in spermatogenesis is difficult. Such mutants are first detected by the inability of a male to sire progeny, which can have many explanations that are unrelated to spermatogenesis. For instance, males with subtly defective genitalia where (otherwise normal) sperm cannot exit the male would be among the mutants recovered in such a mutant hunt. Compounding this problem,

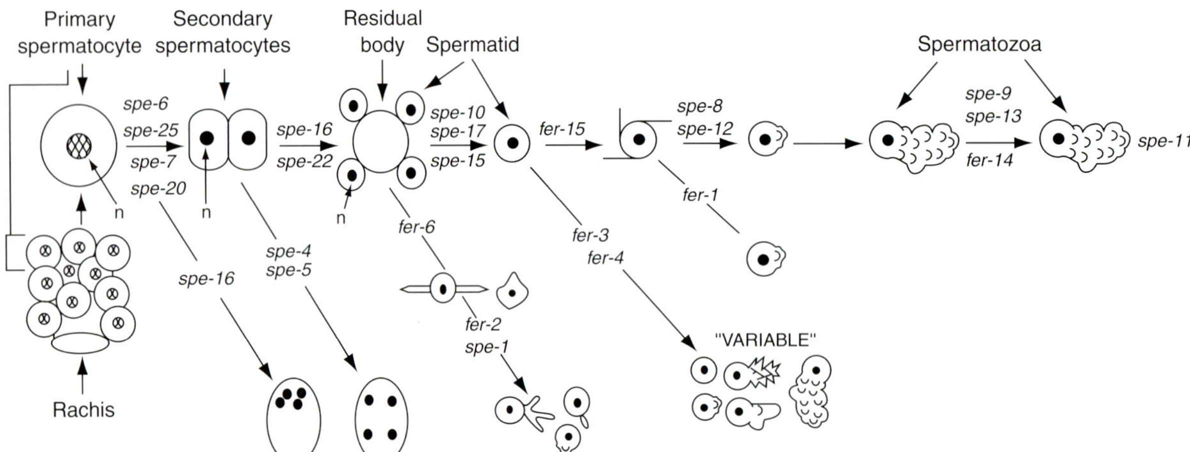

Figure 1 Stages of wild-type spermatogenesis are shown diagramatically as an ordered pathway of morphogenesis (top line). Spermatocytes are initially in syncytium with a central cytoplasmic core named the rachis, and they bud from this structure after initiating meiosis. Spermatocytes subsequently develop as individualized cells and do not require any accessory cells. Along the pathway are listed some of the >40 *spe* and *fer* genes and their approximate point of developmental arrest as determined by light microscopy. For instance, *spe-17* spermatids are abnormal and *spe-9* mutant spermatozoa cannot fertilize oocytes after gamete contact. The last step shows *spe-11* mutant spermatozoa that can fertilize oocytes but cause death of the resulting embryo because they do not provide a component required for embryogenesis. Beneath the pathway are abnormal cells that accumulate when a gene is mutated. n, nucleus. (Modified from and reproduced with permission from L'Hernault SW, Shakes DC and Ward S (1988) *Genetics* 120: 435–452.)

heterozygous siblings must be maintained for all examined mutants so that a mutation responsible for a bona fide spermatogenesis defect can be recovered. The situation in *C. elegans* is much more straightforward because candidate spermatogenesis-defective mutants can be initially identified in hermaphrodites in the absence of mating. Normally, internal self-fertilization in young hermaphrodites is so efficient that virtually every sperm fertilizes an egg, which then start embryogenesis and are subsequently laid on the agar growth plate. A mutation that affects spermatogenesis will abolish self-fertility, and these mutant hermaphrodites will lay numerous oocytes that appear different from embryos under low-power magnification. Self-fertilization in hermaphrodites is not absolutely required for *C. elegans* reproduction and these mutant, self-sterile hermaphrodites can be placed with wild-type males and allowed to mate. If they produce outcross progeny, then inseminated wild-type sperm can correct the sterile phenotype of the mutant hermaphrodite. This indicates the mutant hermaphrodites contained defective sperm and, so far, this technique has allowed identification of >40 genes that appear to affect spermatogenesis. The principal reason it has been possible to identify and recover this large collection is that a mutation affecting spermatogenesis is identified and recovered from the same individual animal.

Figure 1 is a cartoon of light microscopic phenotypes and shows the approximate point where *sp*ermatogenesis or *fer*tilization defective (*spe* or *fer*) mutants affect or arrest further development; about half of the available mutants appear in this figure. In mutants on the lower part of the figure (e.g., *spe-4*), part of the wild-type pathway (upper row in **Figure 1**) of spermatogenesis occurs before abnormal cytology becomes evident. This is a large, diverse set of mutations and many of the processes required for normal spermatogenesis are affected in one or more mutants. Most mutants depicted in **Figure 1** and >20 mutants not shown are incompletely understood. A combination of genetic, molecular, and cell biological tools is being used to analyze these mutants.

Further Reading

L'Hernault SW (1997) Spermatogenesis. In: Riddle DL, Blumenthal T, Meyer BJ and Priess JR (eds) C. elegans, 2nd edn, pp. 274–294. Plainview, NY: Cold Spring Harbor Laboratory Press.

See also: Caenorhabditis elegans

Spermatogenesis, Mouse

G S Kopf

doi: 10.1006/rwgn.2001.1224

Spermatogenesis is the complex developmental process by which haploid spermatozoa, capable of ultimately fertilizing an egg, develop as a consequence of both meiosis and differentiation from diploid stem cells in the testis. Spermatogenesis occurs in the seminiferous tubule compartment of the testis, and this compartment can be clearly influenced by the other compartments of this organ. The seminiferous tubule is composed of a structurally complex epithelium surrounded by interstitial cells (e.g., Leydig cells, muscle-like myoid cells) and fluid. This epithelium is comprised of two basic cell types, the somatic Sertoli cells and the germ cells. The Sertoli cells display a columnar morphology that extends from the base of the epithelium to the lumen of the duct. They surround and nurture the differentiating germ cells, and through Sertoli–Sertoli cell junctions form the blood–testis barrier, which ensures that only spermatogonia and early spermatocytes are exposed to circulating macromolecules in the blood and lymph.

It is known that Sertoli cell products can interact with the germ cells to regulate their differentiation, and there is ample evidence for the reciprocal release of products from the germ cells that can affect Sertoli cell function. Spermatogenesis depends on testosterone that is secreted by the Leydig cells. The control of testosterone production by these cells is regulated at the level of the anterior pituitary through its secretion of luteinizing hormone (LH) in response to hypothalamus-derived gonadotropin-releasing hormone (GnRH). Follicle-stimulating hormone (FSH) is also required for spermatogenesis and is released from the anterior pituitary in response to GnRH released from the hypothalamus. Both testosterone and FSH act directly on the seminiferous tubule epithelium to regulate spermatogenesis.

Spermatogenesis can be divided into three major developmental stages: (1) a period of mitotic cell proliferation; (2) two meiotic cell divisions; and (3) spermiogenesis, a series of morphological changes that give rise to the highly polarized spermatozoon. Spermatogenesis is initiated at the time of puberty with mitotic divisions of the primitive, type A spermatogonia. Subsequent cell divisions of a majority of the daughter cells then ensues, leading to two populations of spermatogonia. One of the populations enters a differentiative pathway and eventually become spermatozoa, whereas the other population undergoes apoptosis, otherwise known as programed cell death. The remaining daughter cells that do not undergo these multiple rounds of cell divisions undergo a different fate, being in some way similar to their progenitors. This population of 'resting' spermatogonia continues to function as stem cells, being required for subsequent rounds of spermatogenesis, thus ensuring the continuous production of spermatozoa. The percentage of type A spermatogonia that adopt these different fates varies from species to species. Those type A spermatogonia that continue to divide eventually differentiate to type B spermatogonia that are committed to enter meiosis. Those germ cells that enter meiosis are termed spermatocytes. Primary spermatocytes undergo the first meiotic division and this division is characterized by a long prophase; as a consequence, these cells can often be seen in histological examinations of the testicular seminiferous epithelium. It is during this first meiotic division that the paired homologous chromosomes of the primary spermatocytes participate in crossing-over, giving rise to genetic recombination. The final result of this first meiotic division is the production of two secondary spermatocytes, each containing the entire set of duplicated autosomal chromosomes and either a duplicated X or duplicated Y chromosome.

The second meiotic division carried out by the secondary spermatocytes is of short duration and thus is more difficult to stage in histological sections. It is during this second meiotic division that each secondary spermatocyte produces two spermatids, each with a haploid number of single chromosomes. These spermatids then enter spermiogenesis, and it is during this time of development that they undergo a series of dramatic differentiation events that eventually conclude with the development of a very highly polarized spermatozoon, the morphology of which is species-specific. Examples of such differentiation events include: (1) the formation of an acrosome, a secretory granule that overlies the nucleus of the spermatozoon and participates in the fertilization process; and (2) the assembly of the spermatozoan flagellum which powers sperm motility. In the mouse spermatogenesis takes approximately 30 days to complete, 14 days of which are devoted to spermiogenesis. Human spermatogenesis, in contrast, takes approximately 64 days to complete, 35 days of which are devoted to spermiogenesis.

An interesting feature of spermatogenesis is that the developing male germ cells fail to complete cytokinesis during mitosis and meiosis; therefore, clones of differentiating daughter cells originating from a single maturing spermatogonium remain connected by cytoplasmic bridges, forming, in effect, a syncytium. These cytoplasmic bridges persist throughout spermiogenesis when the individual spermatozoa are released into the lumen of the seminiferous tubules. Since the haploid spermatozoa undergo a majority of their differentiation after their nuclei have completed meiosis, the presence of cytoplasmic bridges ensures that each developing spermatozoan shares a common cytoplasm with its neighbors, thus supplying them with all of the products of a diploid genome. It must be emphasized that spermatogenesis takes place in clusters that are not necessarily synchronous with one another along the length of the seminiferous tubules, thus ensuring the production of a constant supply of mature spermatozoa. However, owing to the syncytial nature of the development of clonal populations of spermatozoa from a single spermatogonium, development within these syncytia is synchronous.

Further Reading

Handel MA (1998) Meiosis and gametogenesis. In: Pederson RA and Schatten GP (eds) *Current Topics in Developmental Biology*, vol. 37, p. 418. San Diego, CA: Academic Press.

Hecht NB (1998) Molecular mechanisms of male germ cell differentiation. *BioEssays* 20: 555–561.

See also: **Gametogenesis; Spermatids; Spermatocytes**

Spermatogonia

S W L' Hernault

doi: 10.1006/rwgn.2001.1226

Spermatogonia are proliferating germ cells within the testes of animals. They are the only self-renewing adult cell type in males that is capable of contributing to the next generation. Early in the embryonic development of all sexually reproducing animals, a small group of germ cells is formed that are destined to be the exclusive source of gametes within that organism. Germ cells are initially indistinguishable in cellular appearance between males and females but this changes after the germ cells complete migration into the developing gonad. In males, the gonad is named the testis (plural: testes) and the germ cells form spermatogonia within this organ. Spermatogonia are the progenitor cells for all spermatozoa, and spermatogonia applies to all germ cells within the testes that have not entered meiosis. The testis is the only location in a male where meiosis occurs, which is the process where the nucleus is reduced from the typical diploid state (similar to that found in most cells outside the gonad) to the haploid state. For any given organism, the testis is usually one of the most impressive organs in terms of the sheer number of differentiated cells produced; in human males, it is estimated that 10^{12}–10^{13} spermatozoa are produced during his lifetime. This is possible because spermatogonia include a population of self-renewing cells (named stem cells) that divide to produce one cell capable of entering the differentiation pathway that ultimately forms haploid spermatozoa and another stem cell. Generally, spermatogenesis is a polarized, assembly-line-like process where proliferating spermatogonia are spatially separate from cells in meiosis and the spermatids that are near a space or tubule into which they are released. The spermatogonia that function as stem cells are furthest from the position where spermatids are released to become spermatozoa. There are usually several mitotic divisions (e.g., nine in the laboratory mouse) between the time when the spermatogonial stem cell divides and when a cell that enters meiosis and becomes a spermatocyte will form. Future research will reveal how primordial germ cells form, how they differentiate into spermatogonia, and how the spermatogonia retains its stem cell properties.

See also: **Meiosis**

Spina Bifida

M J Seller

doi: 10.1006/rwgn.2001.1227

Spina bifida, literally meaning a cleft in the spine, is a localized congenital malformation and a type of neural tube defect (NTD). It comprises a range of lesions of varying severity along the midline of the back involving any number of vertebrae at any level, most frequently occurring in the lumbosacral region. The minimal expression is spina bifida occulta, which may not even be evident externally or produce any symptoms, but in meningocoele, the meninges herniate through the cleft in the vertebral arches to produce a bulging sac; and in meningomyelocoele, the spinal cord as well as the meninges herniates. The overlying skin is deficient so the nervous tissue is exposed and becomes damaged, leading to motor and sensory deficit of the lower part of the body and often to incontinence. Hydrocephalus may develop too.

Spina bifida arises during early embryogenesis. In weeks 3 and 4 after conception, the future brain and spinal cord originate as a flat plate of cells on the upper surface of the embryo. The lateral edges of this plate elevate, overarch and meet in the midline, fusing from several separate starting points to form the neural tube. This subsequently induces the tissue surrounding it to differentiate, eventually to form the vertebrae. If the neural folds fail to meet and fuse, or if there is incomplete closure, then a gap will occur in the tissue above; this is spina bifida.

Spina bifida, together with the associated and equally common anencephalus where the cephalic part of the neural tube fails to close, has a birth prevalence which varies according to geographical area, ethnicity, socioeconomic status and also temporally. In the United Kingdom during this century, the prevalence has shown a series of peaks and troughs, the peaks occurring roughly every decade, although since 1970 the downward trend has continued and not been reversed. Within the United Kingdom there is a prevalence gradient which is highest in the northwest, decreasing to the southeast. In the 1960s, the highest known prevalence of NTD in the world was in Belfast (8.7/1000 births); today such high rates occur in the Indian Punjab and northeast China. NTD are common in Celts and Sikhs, but uncommon in many Asiatic and black populations. In high prevalence areas there is an association with low socioeconomic status, but NTD are relatively rare amongst Third World peoples who live in abject poverty.

Clues as to possible causes have been sought from these many curious demographic observations. While spina bifida can occasionally be caused by the anticonvulsant sodium valproate, maternal diabetes, and by some specific types of chromosomal imbalance, in the vast majority of cases the cause is unknown. These are regarded as multifactorial, involving both genetic and environmental precipitating factors: environmental, because of the observed association with socioeconomic class and geography; genetic, because of the ethnic specificity and because close relatives of probands have a higher than average risk of being affected themselves.

It is now known that a deficiency of folic acid is involved in the genesis of many spina bifidas, and that maternal therapy with 0.4 mg of folic acid daily prior to, and in the weeks following, conception prevents around 72% of cases. Folate fortification of a staple food item so that all women of childbearing age will be protected has been widely debated, and was recently implemented in the United States. This is primary prevention. Secondary prevention through prenatal diagnosis and termination of affected pregnancies has been practised in the western world since the mid-1970s, and has increasingly influenced the birth prevalence figures. Open lesions in the fetus allow the escape of α-fetoprotein into the amniotic fluid, and elevated levels are also found in the maternal bloodstream in around 75% of cases. This forms the basis of a screening test in the early second trimester. The lesions may also be directly visualized by ultrasound scanning at 16–18 weeks' gestation.

The discovery of prophylactic folate therapy has not yet led to an understanding of the mechanisms involved. But it has stimulated research into metabolic pathways which use folic acid and into searching for mutations in spina bifida subjects in the genes for key enzymes involved. The search also continues for other environmental factors involved in the 25% of cases where folic acid is not a factor. A number of different mouse models of spina bifida exists, and the embryological and molecular mechanisms are being investigated. In addition, experiments involving targeted gene mutations in mice have produced spina bifida, unexpectedly in some cases, so yielding information on the genetic control of normal neural tube formation and closure, which may eventually lead to the identification of the genetic components in the cause of spina bifida.

See also: **Dysmorphology**

Spindle

doi: 10.1006/rwgn.2001.2100

The spindle is the structure formed in a eukaryotic cell at the time of division formed by polymerization of microtubules. The chromosomes attach to the spindle and in metaphase align in a central plate perpendicular to the long axis of the spindle. The spindle plays a role during anaphase in pulling paired chromatids to opposite poles.

See also: **Mitosis**

Splicing

doi: 10.1006/rwgn.2001.2034

Splicing is the process whereby introns are removed from a newly transcribed primary RNA transcript (hnRNA)and exons are joined to produce mature mRNA.

See also: **Introns and Exons; Pre-mRNA Splicing**

Splicing Junctions

doi: 10.1006/rwgn.2001.2035

Splicing junctions are the sequences immediately surrounding the exon–intron boundaries. Right-splicing junctions comprise the boundary between the right end of an intron and the left end of the adjacent exon, whilst left-splicing junctions comprise the boundary between the right end of an exon and the left end of an intron.

See *also:* Introns and Exons

Split Genes

T Maniatis

doi: 10.1006/rwgn.2001.1229

In prokaryotes the gene and messenger RNA (mRNA) are colinear. That is, the sequence of DNA nucleotides in the gene is identical to the sequence of RNA nucleotides in the mRNA encoded by the gene. By contrast, in the vast majority of eukaryotic genes the DNA sequences encoding mRNA are interrupted by noncoding sequences called introns. The sequences that encode the mature mRNA are called exons. These 'split' eukaryotic genes are transcribed by RNA polymerase to produce pre-messenger RNAs (pre-mRNAs) consisting of alternating exons and introns. Many eukaryotic pre-mRNAs are highly complex, containing dozens of exons and introns. The average size of an exon is around 150 nucleotides, whereas introns can be very large, some over 100 000 nucleotides in length.

The process called RNA splicing is required to produce mature mRNAs that encode proteins. RNA splicing involves precise cleavage of the pre-mRNA at the junctions between exons and introns, followed by the covalent joining of adjacent exons. Thus, RNA splicing is similar to splicing an audiotape or film, where the unwanted piece is clipped out with scissors and then taped together to produce the desired sound or images. However, because of the triplet nature of the genetic code, a mistake in cutting the RNA of only one nucleotide will produce an mRNA that has an altered reading frame, thus producing a message that cannot encode the correct protein. Therefore, the splicing process must be very precise.

In the eukaryotic cell, RNA splicing occurs in the nucleus and is carried out by a large RNA–protein complex called the spliceosome, which is capable of recognizing specific RNA sequences located at the exon/intron junctions. A common feature of most eukaryotic introns is a GU dinucleotide at the 5′ end (the 5′ splice site) and an AG dinucleotide at the 3′ end (the 3′ splice site). Once bound to these junctions the spliceosome cuts the RNA, and then joins the adjacent exons. The importance of a precise cleavage at these junctions is dramatically illustrated by the fact that a large number of human genetic diseases are caused by mutations at 5′ or 3′ splice sites, which result in the production of nonfunctional mRNAs during splicing.

Another important consequence of the split gene organization is the generation of multiple proteins from a single pre-mRNA by a process called alternative splicing. In most cases all of the exons present in the pre-mRNA are spliced together to form the mature mRNA. For example, a pre-mRNA containing five exons would give rise to a mature mRNA containing the five exons in the order 12345. However, in some cases an exon may be skipped by the splicing machinery, resulting in an mRNA with the exons in the order 1245, which would encode a different protein. This process can be regulated, so that the transcription of a single gene can lead to the production of different proteins in different cell types. Of course, the number of different exon combinations increases with the number of exons in the gene. In fact, there are examples in which thousands of different proteins can be produced from a single gene by alternative splicing.

Why are eukaryotic genes split? One theory is that exons are the primary unit of protein evolution. According to this theory new genes emerge during evolution by exon shuffling, which occurs by a process called DNA recombination. Thus, two genes with multiple exons could recombine to generate a new gene containing exons from both genes. When the new gene is transcribed and spliced, a novel protein would be produced. Another theory is that the split gene organization evolved because it provides an efficient means of producing multiple proteins from a single gene by alternative splicing.

See *also:* Alternative Splicing; Frameshift Mutation; Introns and Exons

Spongiform Encephalopathies (Transmissible), Genetic Aspects of

M E Bruce

doi: 10.1006/rwgn.2001.1027

The transmissible spongiform encephalopathies (TSEs) or 'prion' diseases are neurodegenerative disorders affecting a range of mammalian species. The devastating epidemic of bovine spongiform encephalopathy (BSE) in the United Kingdom and its subsequent spread to humans and other species have thrown this whole group of diseases into the limelight in recent years. BSE, which was possibly first caused by contamination of animal feed with sheep scrapie offal, has been diagnosed in nearly 180 000 cattle since it was first reported in 1986. Novel TSEs, shown to be related to BSE, have also been recognized in domestic and large cats and in a range of exotic ungulate species in zoological collections. In 1996, a dramatic announcement was made that a new form of human TSE, variant Creutzfeldt–Jakob disease (vCJD), had been identified in 10 patients in the United Kingdom. Laboratory studies soon provided compelling evidence that vCJD was caused by the BSE agent. Since then, the number of vCJD cases has crept slowly upwards, at present (2001) standing at about 100, but it is still too early to predict the eventual size of the outbreak. Given this uncertainty, it is vital to understand the factors governing the transmission of TSEs between and within species. It is well established that genetic information carried by both the host and the agent can have a profound effect on the occurrence and characteristics of TSE disease.

TSE Agents and the Prion Protein

TSEs are caused by unconventional transmissible agents which are unusually resistant to inactivation by heat, chemicals, nucleases and proteases. Extensive studies have so far failed to reveal any infection-specific nucleic acids in highly infective tissue extracts. Rather, infectivity tends to copurify with aggregated forms of the host 'prion protein,' PrP. The conversion of the normal cellular protein, PrP^{c} (an abundant, membrane-bound sialoglycoprotein), into the abnormal form, PrP^{Sc}, involves a conformational change to a predominantly β-pleated structure. PrP^{Sc} accumulates in the brain and other tissues of infected individuals, often forming fibrillar aggregates, and its presence is regarded as a definitive diagnostic feature of these diseases. Moreover, studies of transgenic knockout mice have shown that PrP is required for propagation of TSE infection. These observations have prompted the 'prion' or 'protein-only' hypothesis, originally put forward by Stanley Prusiner in 1982, that TSE agents contain no nucleic acid but consist solely of modified PrP, either PrP^{Sc} itself or an intermediate between PrP^{c} and PrP^{Sc}.

According to the prion hypothesis, when an animal is challenged with a TSE inoculum, conformationally modified PrP from the inoculum interacts with normal host PrP and converts it to PrP^{Sc}. This in turn is suggested to pass on the modification to new PrP molecules, resulting in an accumulation of PrP^{Sc}, which eventually interferes with brain function, leading to death. While the prion hypothesis has gained widespread acceptance, there are still key issues that have not been resolved, namely, the exact modification of the protein that confers infectiousness, the basis of agent strain variation (see below) and the mechanism of propagation of this information in a protein-only structure. Therefore, the existence of an elusive infection-specific informational molecule such as a nucleic acid remains a possibility. If such a molecule exists, then it may be protected and hidden by aggregated PrP, as proposed in the 'virino' hypothesis of Alan Dickinson.

Experimental TSEs in Mice

Host Genetic Effects

Most of our understanding of the biology of the TSEs comes from studies of experimental rodent models. When a mouse is infected with a TSE agent, there is a long asymptomatic incubation period (at least 4 months) before the appearance of neurological disease. If all experimental variables are controlled, the length of the incubation period is remarkably reproducible and depends precisely on interactions between genetic information carried by the host and that carried by the TSE agent. Many years ago, Dickinson established that the major component of the host effect is associated with a single gene, called *Sinc* (scrapie *inc*ubation), although other host genes also have minor effects. Later it became clear from studies in congenic and transgenic mouse lines that the *Sinc* gene (sometimes referred to as *Prn-i*) is located on chromosome 2 and encodes PrP. Only two alleles of the *Sinc*/PrP gene have been found in laboratory mice, encoding PrP proteins that differ by two amino acids (at codons 108 and 189). The effect of the *Sinc*/PrP genotype on the incubation period can be very large (up to hundreds of days with certain TSE isolates), suggesting that PrP is involved in a rate-limiting step in pathogenesis.

Agent Strain Variation

Dickinson also demonstrated that, like conventional microorganisms, TSE agents exhibit strain variation: to date, approximately 20 phenotypically distinct laboratory TSE strains have been identified. The most obvious ways in which TSE strains differ are in the patterns of incubation periods they produce in the three possible *Sinc*/PrP mouse genotypes (the two homozygotes and the heterozygote) and in the type of pathology produced in the brain. TSE strains differ in the length of the incubation period in any single *Sinc* genotype. They also differ in which of the two *Sinc* homozygotes has the shorter incubation period and in the apparent dominance of the two alleles in the heterozygote mouse. It is well established that TSE strains can retain their characteristics over many serial mouse passages, although variant strains may sometimes be selected by changing the passage conditions. It has also been demonstrated that the disease phenotype of TSE strains is independent of the genotype, or even the species, of the animal from which the infective inoculum has been produced. All of these observations indicate that TSE agents carry some form of strain specific information that is independent of the host. It has been suggested that protein-only structures could carry this information in the form of multiple self-perpetuating conformational modifications but it is unclear whether this proposed novel form of genetic information can account for the experimental observations, or whether a separate informational molecule is required.

Although the basis of TSE strain variation is unknown, TSE strain identification based on incubation periods and neuropathology in mice, has been used to unravel the relationships between TSEs occurring naturally in different species. Thus, it has been shown that isolates from BSE, feline spongiform encephalopathy, TSEs of exotic ungulates and, most recently, vCJD produce the same disease phenotype in mice, providing compelling evidence that they are all caused by the same TSE strain. In these cross-species transmissions PrP-independent host genetic effects become much more prominent, but the genes involved have not yet been identified.

Sheep Scrapie

Scrapie has been endemic in sheep in the United Kingdom for at least 250 years. It has long been recognized that there is a strong host genetic influence on the incidence of sheep scrapie, but it is only in recent years that it has been possible to analyze this at the molecular level. As in mice, PrP genotype has a major effect on the incubation period and occurrence of the disease in sheep. This was first described by Nora Hunter, working on a closed flock of Cheviot sheep that had been selectively bred for many years according to their susceptibility or resistance to experimental challenge with scrapie. Later, it was recognized that PrP genotype is also important in natural scrapie. The sheep PrP gene is highly polymorphic, showing amino acid substitutions in at least 12 sites of the coding region. However, only three of these polymorphisms have been shown so far to have a significant influence on the occurrence of scrapie: an alanine(A)/valine (V) polymorphism at codon 136, a histidine(H)/arginine(R) polymorphism at codon 154, and a glutamine(Q)/arginine(R)/histidine(H) polymorphism at codon 171. In terms of these three sites, at least five alleles of the PrP gene are present in sheep (ARQ, ARR, ARH, AHQ, and VRQ).

Given this degree of polymorphism, it is not surprising that PrP genetic effects in natural scrapie are very complicated. The frequency of the above alleles differs markedly between sheep breeds. Also, a high scrapie incidence is associated with different genotypes in different breeds or flocks, possibly because multiple strains of scrapie agent are involved in the natural disease. Indeed, it has been shown experimentally that different TSE isolates target sheep of different PrP genotypes. However, there do appear to be PrP genotypes that confer a high degree of resistance wherever they occur. For example, only one case of scrapie has ever been reported in the ARR/ARR genotype. At the other end of the spectrum, sheep of the rare VRQ/VRQ genotype in the United Kingdom almost always get scrapie and this has led to the suggestion that natural scrapie is a purely genetic disease. However, this and other susceptible genotypes are present in sheep from scrapie-free countries such as Australia and New Zealand, arguing strongly that scrapie is an acquired infection. Furthermore, the occurrence of disease in highly susceptible genotypes within a high-incidence flock can be delayed, or even prevented, by extremely hygienic husbandry during the perinatal period.

BSE

In contrast to the above, the PrP gene in cattle shows very little variation. One major polymorphism, involving a difference the number of an octapeptide repeat in the PrP protein, has been described, but this does not appear to influence the occurrence of BSE.

Human TSEs

TSEs of humans present as sporadic, familial, or acquired disorders. In each category, genetic variation in the PrP gene plays a key role in determining the occurrence and characteristics of the disease. As in sheep, the human PrP gene, which is located on chromosome 20, is very variable. To date, four polymorphisms altering PrP amino acid sequence have been recorded, including three substitutions and a deletion of one of the octapeptide repeats. In addition, at least 23 mutations have been reported: 14 point mutations, 8 insertions of varying numbers of octapeptide repeats, and one stop codon mutation, which results in a truncation of the protein.

Sporadic CJD

Sporadic CJD (sCJD) is a rare condition with a worldwide distribution, occurring at an annual frequency of about one case per million of the population. A common polymorphism, resulting in either methionine or valine at residue 129 of the human PrP protein, influences the occurrence of sCJD. In the general Caucasian population, about 35% of individuals are homozygous for methionine at this site, 15% are homozygous for valine, and 50% are heterozygous. Although sCJD occurs in all three of these genotypes, the heterozygote is substantially underrepresented. It is not known what causes sCJD, as extensive epidemiological surveys have revealed no environmental risk factors. If sCJD is an acquired infection, then the causative agent must be almost ubiquitous. Alternatively, it has been suggested that sCJD occurs as a result of a rare spontaneous conversion of PrP^c to its pathological infectious form.

Acquired Human TSEs

The first clearly acquired TSE to be described in humans was kuru, which was associated with ritualistic cannibalism amongst the Fore people in Papua New Guinea. Since then, it has been recognized that on rare occasions CJD has spread iatrogenically from person to person, by corneal or dura mater grafting, contamination of neurosurgical instruments, or treatment with human pituitary-derived hormones. The largest group of iatrogenic transmissions, to over 100 patients, has involved the administration of CJD-contaminated human growth hormone. Amongst exposed patients, heterozygosity at codon 129 of the PrP gene has resulted in longer incubation periods and perhaps also lower susceptibility to infection.

As described above, there is very strong evidence that vCJD, which is clinically and pathologically distinct from sCJD and occurs in younger patients, is caused by the BSE agent. Although other modes of transmission have not been excluded, it is most likely that BSE infection has been acquired from bovine products in the diet. All vCJD patients so far have been homozygous for methionine at codon 129 of the PrP gene, but it remains to be seen whether valine homozygotes and heterozygotes are also susceptible but with longer incubation periods.

Familial Human TSEs

The reported mutations of the human PrP gene, most of which are rare, are associated with familial neurological diseases showing a range of clinical and neuropathological characteristics. The commonest mutation results in a substitution of lysine for glutamic acid at codon 200, and has been identified in association with large clusters of familial CJD in Jewish people of Libyan origin and in apparently unrelated communities in Slovakia and in Chile. The disease phenotype for this and several other mutations is similar to that of typical sCJD. Another relatively common mutation involves a substitution of asparagine for aspartic acid at codon 178. Interestingly, this mutation is associated with an sCJD-like phenotype when linked to valine at codon 129, but a distinct disease phenotype, fatal familial insomnia, when linked to methionine at codon 129. A quite different clinical and pathological picture, Gerstmann–Straussler–Scheinker disease or GSS, is seen in families carrying several of the other mutations, the most common being a proline-to-leucine mutation at codon 102.

Many of the familial human TSEs, as well as sCJD, have been shown to be transmissible to experimental animals. However, as the penetrance for many of the mutations is high, it is usually assumed that these are purely genetic diseases that occur because the mutant protein readily converts spontaneously to a pathological and transmissible form. Indeed, several mutant human PrPs tend to form aggregates when expressed in cell lines, although these have not, so far, been shown to be infectious. The lesson from sheep scrapie is that inheritance of high susceptibility to infection can sometimes masquerade as a purely genetic disease and the same could be true for at least some of the familial human TSEs.

Concluding Remarks

It is clear that the PrP gene influences the expression of TSE disease in a number of different species, but there are many outstanding questions regarding the mechanisms involved. Currently the effects of PrP variants and mutants are being explored in a range of model systems, including transfected cell lines and transgenic mice. Other studies are exploring the biochemical characteristics of the pathological protein associated with different PrP variants or with different

TSE strains. Hopefully, these studies will clarify the complex interactions between TSE infection and host PrP genotype.

Further Reading

Bruce ME (1993) Scrapie strain variation and mutation. *British Medical Bulletin* 49: 822–838.

Gambetti P, Petersen RB, Parchi P *et al.* (1999) Inherited prion diseases. In: Prusiner SB (ed.) *Prion Biology and Diseases*, pp. 509–583. Plainview, NY: Cold Spring Harbor Laboratory Press.

Hunter N (1997) PrP genetics in sheep and the implications for scrapie and BSE. *Trends in Microbiology* 5: 331–334.

Will RG, Alpers MP, Dormont D, Schonberger LB and Tateishi J (1999) Infectious and sporadic prion diseases. In: Prusiner SB (ed.) *Prion Biology and Diseases*, pp. 465– 507. Plainview, NY: Cold Spring Harbor Laboratory Press.

See *also*: Cell/Neuron Degeneration; Kuru

Spores

J Parker

doi: 10.1006/rwgn.2001.1231

The term 'spore,' derived from the Greek for seed, is generally applied to small, resistant, dormant cells formed by a wide variety of organisms, including bacteria, fungi, protozoans, algae, and plants. Spores are formed as part of sexual or asexual reproductive processes, but when germinated they almost always give rise to new individuals or groups of cells. In this way, most spores can be clearly distinguished from gametes. As one might anticipate from the very wide range of organisms that produce them, there are many very different types of spores. In most organisms, other than the higher plants, spores are much more resistant to environmental agents than the organisms that produce them. Many types of spores can remain dormant for long periods of time.

The endospores produced by certain gram-positive bacteria, such as *Bacillus subtilis*, are typically referred to as spores. Endospores are formed within the bacterial cell, and are themselves highly differentiated, resistant, nongrowing cells. They are formed via the conversion of a vegetative cell by a complicated pathway of gene expression triggered by nutrient exhaustion. The regulation of this pathway serves as a model for the study of differentiation. When conditions favorable to growth return, the endospore can convert rapidly back into a vegetative cell. Endospores have been found to be able to remain dormant for thousands of years under suitable conditions, and there is some evidence for germination after millions of years of dormancy.

The bacterial group Actinomycetes, of which *Streptomyces* is one genus, also produce spores, but these spores are not related to the endospores discussed above. These prokaryotes form mycelia reminiscent of the eukaryotic fungi (see below) and often produce spores on aerial filaments, once again reminiscent of the fungi. Spore production, the morphology of the spores, and spore-producing structures vary widely across the actinomycetes. In *Streptomyces* the multinucleate aerial filaments, called sporophores, form crosswalls which generate single-celled spores, referred to as conidia. Once again this process is reminiscent of that of the fungi, but these prokaryotic organisms are not related to the eukaryotic fungi.

The fungi are a group that includes molds, yeasts, and mushrooms. The molds and mushrooms grow as a mycelium, a mat of cross-branching filaments called hyphae. Fungi reproduce by producing spores, usually unicellular, either sexually or asexually from specialized hyphal compartments. The phylogenetic groupings of the fungi are named according to the mechanism of the production of sexual spores.

The Ascomycetes are fungi that form sexual spores termed ascospores within an enclosed sac (ascus). The Ascomycetes include the yeast *Saccharomyces cerevisiae*. The ordered spores within the ascus of *Saccharomyces* can be dissected and analyzed to yield information on genetic segregation (see Tetrad Analysis). The asexual spores formed by these fungi, called conidia, are often brightly colored and very resistant to drying. Under favorable conditions members of the Zygomycetes, such as *Rhizopus*, form haploid spores asexually in structures called sporangia. When growth conditions are poor, zygosporangia are formed by a sexual process. The Basidiomycetes, which include the mushrooms, puffballs, and rusts, produce sexual spores called basidiospores on the ends of club-shaped structures (basidia). The asexual spores are typically called conidia. Interestingly, some rusts form asexual spores called pycnidiospores or spermatia which act very much like gametes, fusing with another cell before they are able to grow.

There is a large group of protozoans that used to be considered as a single phylum, Sporozoa, but which have been reclassified into four separate phyla: Apicomplexa, Microspora, Acetospora, and Myxospora. All members of these phyla are parasites of animals. As the original taxonomic name implied, many of these organisms form spores, or sporocysts. These protozoan 'spores' are not homologous to spores produced by other organisms. There are differences in nomenclature and life cycles amongst these phyla and the term 'spore' can refer to reproductive, infective, or resistant stages in the often complex life cycles of these organisms.

The fungus-like protists, the phylum Myxomycota (plasmodial slime molds), the phylum Acrasiomycota (cellular slime molds), and the phylum Oomycota (water molds), also produce spores. In the life cycle of a plasmodial slime mold, haploid spores are produced by meiosis from the diploid sporangium, in response to harsh environmental conditions. The spores germinate into active haploid forms which fuse to form the diploid stage. The spores of the cellular slime mold are also haploid, but they are derived by differentiation from existing haploid cells, not by meiosis. The water molds produce encysted, diploid zoospores through an asexual process and diploid spores called oospores through a sexual process. The encysted zoospores are very resistant to environmental conditions, the oospores less so.

All plants, and some algae, have a sexual life cycle that is characterized by alternation of generations. One generation is composed of a haploid multicellular organism/stage called a gametophyte and the other generation is composed of a diploid multicellular organism/stage called a sporophyte. Meiosis in the sporophyte produces haploid cells called spores. In some plants the spores may be of two kinds, a megaspore which forms a female gametophyte and a microspore which forms a male gametophyte. A spore generates the multicellular gametophyte by mitosis; it does not fuse with another haploid cell. In the higher plants the gametophytes do not form independent organisms, but are protected by being retained in the reproductive tissue of the sporophytes.

***See also:* *Bacillus subtilis*; Fungi; *Streptomyces*; Tetrad Analysis**

Src Family Tyrosine Kinases

M Frame

doi: 10.1006/rwgn.2001.1622

Src Family

The Src tyrosine kinases comprise a family of around eight related proteins: Src, Fyn, Yes, Lck, Hck, Lyn, Fgr, and Blk. Of these, Src, Fyn, and Yes are expressed ubiquitously, while the others are mostly found in hematopoietic cells. c-Src is the prototypic family member and is the cellular homolog of v-Src, the transforming protein of Rous sarcoma virus. While the oncogenic and nononcogenic forms of Src both possess intrinsic tyrosine kinase activity, the viral form is deregulated by virtue of a deletion of C-terminal sequences that contain a tyrosine residue (Tyr527 in avian Src), whose phosphorylation confers negative regulation of c-Src's kinase activity.

Structural Organization and Regulation of Src Proteins

In addition to the C-terminal regulatory sequences, each of the Src family kinases is myristylated (and in some cases palmitylated) at its N-terminus and this is required for membrane association. C-terminal to the site of myristylation is a unique domain that is not conserved among family members and is a putative site of serine phosphorylation, a Src homology (SH)3 domain, an SH2 domain, a linker that joins the SH2 domain with the kinase domain, and the conserved regulatory tail region mentioned above.

Over the years, structure/function analysis and crystal structure determinations of Src family members have led to an understanding of the arrangement of Src's structural domains and how this arrangement is altered as Src is regulated. Briefly, the SH3 and SH2 domains of Src interact not only with specific sequences in effector proteins, but also with other regions in the Src protein itself. In particular, when Tyr527 is phosphorylated (carried out in the cell by a tyrosine kinase termed c-Src kinase (CSK)), it binds to its SH2 domain in an intramolecular interaction that has important consequences; first, the SH2 domain can no longer interact with heterologous effector proteins and, second, the intramolecular interaction results in repression of Src's catalytic activity. Conversely, dephosphorylation of Tyr527 results in catalytic activation by releasing the constraints imposed by the SH2 domain–Tyr527 interaction. The SH3 domain of Src also contributes to catalytic repression by forming an intramolecular interaction with sequences in the linker region and kinase domain. Thus, the inactive conformation of Src is maintained by multiple interactions between distinct regions of the protein. Consequently, Src activation requires the disruption of these interactions either via dephosphorylation of Tyr527 or by displacement of the SH3 and/or SH2 domains as a result of high-affinity binding to other proteins. In addition, c-Src activation requires phosphorylation on Tyr416 (believed to be autophosphorylation) that is required to generate the substrate-binding site and ensure correct positioning of the substrate for catalysis.

Thus, full activation of Src kinases requires the release of both SH3- and SH2-mediated intramolecular interactions which allows the catalytic site to adopt an active conformation, as well as phosphorylation of Tyr416 in the kinase domain. The role of other sites of phosphorylation in the Src protein, such as in the unique region, has not been determined.

Src Kinase Activity in Cellular Responses to Environmental Stimuli

The Src family kinases are key components of cellular responses to environmental stimuli. A few examples of Src's role in signaling changes in cell behavior are given below.

Signaling from Receptor Tyrosine Kinases

Src kinases act downstream of the transmembrane receptor protein tyrosine kinases (RPTKs), such as those for platelet-derived growth factor (PDGF) and epidermal growth factor (EGF), and are required to elicit downstream responses after ligand stimulation. Src is activated by both PDGF- and EGF-treatment of cells and becomes associated with the cytoplasmic domains of the activated receptors, via the SH2 domain of Src and specific receptor phosphorylated tyrosine residues that have been mapped (at least in the case of PDGF-R). In addition, Src association is believed to induce further tyrosine phosphorylation of the activated receptors in some cases. Different experimental approaches that interfere with Src's activity have implicated Src in the ability of a variety of growth factors to stimulate DNA synthesis and elicit a mitogenic response, although recent experiments in cells that lack Src, Fyn, and Yes have suggested that there is not an absolute requirement for Src family kinases under all circumstances. It should be noted that the Src/Fyn/Yes triple knockout cells that were used for these studies had been immortalized with SV40 large T antigen, which might account for the apparent discrepancy.

G-Protein-Coupled Receptors

There is now also evidence that ligand-induced activation of G-protein-coupled receptors (GPCRs) activate Src and induce phosphorylation of known Src substrates, such as focal adhesion kinase (FAK) and paxillin. Although the mechanism of Src activation is not known, GPCR-induced transactivation of RPTKS have been implicated in this process.

Integrin Adhesion Receptors

After engagement of integrins as a result of cell adhesion to extracellular matrix (ECM), components of the adhesion sites of integrin clustering (the so-called focal adhesions) are tyrosine phosphorylated. An example of this is FAK, which is phosphorylated on Tyr397 after integrin stimulation, creating a binding site for the Src SH2 domain which results in association of FAK with Src. Further phosphorylation of FAK on additional tyrosine phospho-acceptor sites leads to recruitment of signaling proteins, including the adaptor protein Grb-2 that can link integrins, via FAK, to the Ras-MAP kinase pathway. This type of intracellular signaling pathway, in which Src plays a pivotal role, is likely to contribute to adhesion-dependent cellular responses, although there are probably other ways of inducing adhesion-dependent activation of MAP kinase. Recently, it has also been proposed that FAK mediates the integrin requirement for growth factor-induced MAP kinase activation, and this too might require Src-dependent phosphorylation of FAK.

Oncogenic Transformation by Src

A great deal of information concerning regulation of Src's subcellular localization and the biological consequences of Src activity has been gained by studying cellular transformation induced by oncogenic, deregulated forms of Src protein. Of particular value have been conditional, temperature-dependent mutants of v-Src that have been used to dissect both the intracellular targeting of Src and the ensuing transformation process.

v-Src is targeted to cellular focal adhesions of mesenchymal cells by a process that requires the Src SH3 domain and an intact actin cytoskeletal network maintained by the concerted action of the Rho family of small GTPases and myosin activity. Specifically, inactive v-Src colocalizes with microtubules around the cell nucleus and makes an SH3-and acto-myosin-dependent switch to peripheral focal adhesions at stress fiber termini. Neither myristylation nor the catalytic activity of v-Src is required for translocation to focal adhesions, although these are required for disruption of focal adhesions and the actin cytoskeleton that accompany cell transformation. In addition, Src kinase activity is required for cell motility, mediated by its effects on focal adhesion turnover and actin remodeling. In keeping with this, cells derived from c-Src/Fyn/Yes triple knockout embryos exhibit impaired migration.

In contrast to fibroblasts, c-Src, Fyn, and Yes localize at cadherin-mediated cell–cell adhesions in epithelial cells. However, in a somewhat analogous manner to Src's role in focal adhesion turnover in fibroblasts, Src kinase activity is required for the disassembly of cadherin-mediated epithelial cell–cell adhesions (often termed epithelial cell scattering) that is necessary to free cells from the constraints of their epithelial connections, for example, during wound repair.

As well as v-Src's effects in disturbing the actin/adhesion network, v-Src can also promote cell growth, stimulating both mitogenesis of quiescent cells and rapid transit through the G_1 phase of the cell cycle in growing cells. These effects of v-Src are initiated at the cell periphery and are mediated by intracellular signal transduction pathways, including phosphatidyl inositol (PI) 3-kinase and MAP kinase, that impinge on cell cycle regulators such as cyclin/cyclin-dependent kinases (cdks), the p27cdk inhibitor, and the retinoblastoma protein. In addition, v-Src can also provide a PI 3-kinase-dependent survival signal by suppressing the apoptosis that oncogenically transformed cells are primed to undergo when deprived of serum growth factors or adhesion to ECM. This ability of activated Src to keep cells alive is in keeping with recent reports indicating that Src is critically involved in coupling lymphokine receptor activation with inhibition of apoptosis, and in mediating the VEGF-induced endothelial cell survival that is necessary for angiogenesis. However, at least one Src family member, Lck, has been shown to mediate apoptotic cell death induced by ionizing radiation, indicating that the role of particular Src kinases in regulating life or death decisions might vary and depend on cell context.

Src in Human Cancer

The expression and activity of c-Src is elevated in a variety of human cancers. This has best been documented in colorectal cancer where increases have been reported from normal epithelium through the premalignant stages to invasive and metastatic tumors. Nonetheless, there is a substantial body of evidence indicating that the oncogenic properties of activated Src might contribute to the genesis of human tumors. For this reason, the tyrosine kinase inhibitors that selectively target the Src family will potentially be of value in suppressing Src-dependent aspects of the malignant phenotype in tumor cells, and in further dissecting the molecular mechanisms of Src's biological effects.

More detailed reviews on the structure, regulation and biological activities of the Src family are provided in Further Reading below.

Further Reading

Abram CL and Courtneidge SA (2000) Src family tyrosine kinases and growth factor signalling. *Experimental Cell Research* 254: 1–13.

Brown MT and Cooper JA (1996) Regulation, substrates and functions of src. *Biochimica et Biophysica Acta* 1287: 121–149.

DeMali KA, Godwin SL, Soltoff SP and Kazlauskas A (1999) *Experimental Cell Research* 253: 271–279.

Schlessinger J (2000) New roles for Src Kinases in control of cell survival and angiogenesis. *Cell* 100: 293–296.

Thomas SM and Brugge JS (1997) Cellular functions regulated by Src family kinases *Annual Review of Cell and Developmental Biology* 13: 513–609.

See *also*: Cancer Susceptibility; SH Domains; SH2 Domain; SH3 Domain

SSLP (Simple Sequence Length Polymorphism)

See: Microsatellite

SSR (Simple Sequence Repeat)

See: Microsatellite

Stable Equilibrium

See: Equilibrium

Staggered Cuts

doi: 10.1006/rwgn.2001.2037

Staggered cuts in duplex DNA are made when the two strands are cleaved at different points in close proximity to each other.

See *also*: Restriction Endonuclease

Staphylococcus aureus

J Parker

doi: 10.1006/rwgn.2001.1239

Staphylococcus aureus is a species of gram-positive bacteria which is typically pathogenic, yellow-pigmented, and salt resistant. The organism is often involved in endocarditis, food poisoning, infections of the skin, pneumonia, septic arthritis, and toxic shock syndrome. *S. aureus* is commonly found in the upper respiratory track of healthy individuals and is a notorious hospital-acquired (nosocomial) pathogen.

The chromosome of *S. aureus* is a circular DNA molecule of approximately 2.8 megabase pairs. Genetic analysis has been done with transduction and transformation, as well as by physical methods and sequencing. (*S. aureus* DNA was used as a donor in one of the first published reports of intergeneric gene cloning.) Several plasmids are known and plasmid-borne drug resistance is common.

The emergence of antibiotic-resistant strains of *S. aureus* is a major problem in most hospitals. Virtually all nosocomial strains are resistant to penicillins and there is an increasing number of methicillin-resistant strains that are resistant to multiple antibiotics. Thus far, most strains seem to be susceptible to vancomycin, but vancomycin-tolerant strains have been observed. This means that strains of *S. aureus* may soon appear that lead to infections which cannot be treated by antibiotics.

Pathogenesis involves the production and secretion of cell surface and extracellular proteins that damage the host cells or tissues, or that interfere with the immune system. These proteins can include coagulase, enterotoxins A-E, fibrinolysin, lipase, nuclease, several proteases, α-, β-, and δ-toxins, toxic shock syndrome toxin 1, and over 20 others. The fibrin-clotting enzyme coagulase, also called staphylocoagulase, causes the host protein fibrin to be broken down and deposited on the bacterial cell, possibly helping to protect the bacterium from attack by host cells. The yellow pigment also seems to be protective against killing by phagocytes.

Many of the genes encoding these virulence factors are under the control of a cell-density-dependent, global regulatory system which responds to a peptide produced by the organism itself. The regulatory system involves the *agr* locus. This locus contains two divergent transcription units, RNAII and RNAIII, controlled by promoters P2 and P3, respectively. The RNAIII transcript is an RNA that regulates the genes encoding the cell-surface and extracellular proteins. It acts primarily by regulating transcription of those genes, but in some cases acts as a translational regulator. RNAIII is also the message for δ-toxin, but translation of RNAIII is not involved in regulation. The P2 operon has four genes, *agrA*, *agrB*, *agrC*, and *agrD*. The product of *agrD* is a small protein which is processed to an octapeptide and then excreted from the cell. The processing involves the product of *agrB*. The *agrA* and *agrC* genes encode a two-component, signal-transducing regulatory system, with the *agrA* product (AgrA) being the response regulator and the *agrC* product (AgrC) being the sensor kinase. It is AgrC that binds to the peptide when it reaches a high extracellular concentration near the end of exponential growth. The phosphorylated form of AgrA then presumably activates transcription of both P2 and P3. This leads to very high levels of RNAIII in the cell and the initiation of the virulence response.

***See also:* Drug Resistance; Gene Regulation; Kinases (Protein Kinases)**

Start, Stop Codons

E Thomas

doi: 10.1006/rwgn.2001.1240

Protein-coding genes are transcribed from DNA to messenger RNA (mRNA). The protein is then translated from the mRNA by the ribosome. Only a subsection of the mRNA is translated into protein. A start codon is a codon that signals the ribosome to start translation. Consequently, all translated regions begin with a start codon. A start codon has two functions: as a potential start codon, and as a regular codon for some amino acid. The standard genetic code has a single start codon, AUG, which codes for methionine. Thus, all translated proteins begin with this amino acid. Note that some proteins may undergo posttranslational processing and lose their initial methionine residue. If the start codon appears within the translated region it functions as a regular amino acid codon. Whether a particular AUG is seen as a start codon or as an internal codon depends on the relative location of a ribosome-binding signal on the mRNA. In bacteria, this is generally a so-called Shine–Dalgarno sequence, a subset of the sequence ***GGAGG** that is complementary to a specific sequence near the 3′ end of the 16S ribosomal RNA and lines the mRNA up appropriately to initiate transcription. The first amino acid has to sit unprotected in what then becomes the peptidyl site on the ribosome; to block the N-terminal charge, a formylmethionine version of the methionyl tRNA is used rather than regular met-tRNA to read this initiating codon. Eukaryotes have other ways of distinguishing the start of the actual message.

The other genetic codes used in some organelles and primitive eukaryotes may have multiple start codons, which code for different amino acids. Organisms and organelles that share an otherwise identical genetic code may differ in the start codons they use.

Stop codons are special codons that signal the ribosome to stop translation. Unlike the start codon, the stop codon itself is not translated, and the last amino acid of a protein is the one coded by the codon immediately before the stop codon. All known genetic codes have multiple stop codons, all of which terminate translation. The stop codons in the standard genetic code are UAG, UAA, and UGA. They are referred to as amber, ochre, and opal, respectively.

All translated (coding) regions begin with a start codon and end with a stop codon. They may contain additional start codons (which function as regular amino acid-coding codons), but cannot contain any additional stop codons (since these will terminate translation). In prokaryotes, a single mRNA may contain several coding regions, each of which is bounded by a start codon and a stop codon and usually has its own Shine–Dalgarno sequence.

The trinucleotide sequence of a stop codon signals termination only if it appears within frame, i.e., starting immediately after the previous codon ends. Thus, the trinucleotide sequence of a stop codon will often appear in a translated region off-frame (i.e., the last two nucleotides of one codon and the first nucleotide of the following codon, or the last nucleotide of a codon and the first two nucleotides of the following codon) without affecting translation unless an insertion or deletion mutation shifts the reading frame.

***See also:* Genetic Code; Translation; Translational Control**

Steel Locus

G Caruana and A Bernstein

doi: 10.1006/rwgn.2001.1242

Mice carrying mutations at the *Steel (Sl)* locus located on chromosome 10 display multiple defects in hemapoiesis, gametogenesis, pigmentation, gut motility, and hippocampal-dependent learning. A similar phenotype is also displayed by mice with mutations at the dominant *white spotting (W)* locus. Mosaic analysis involving chimeric embryos and reciprocal bone marrow transplantation experiments in the 1960s and 1970s suggested that the *Sl* and *W* loci, respectively, controlled environmental and intrinsic properties of the stem cells that give rise to the multiple cell lineages affected in these mutants. In the late 1980s and early 1990s this was directly demonstrated with the cloning first of the *W* locus, followed by the identification of the *Sl* gene product, Steel Factor (SLF), as the ligand that binds to the *W* gene product, the Kit receptor tyrosine kinase.

SLF, also referred to as mast cell growth factor (MGF), Kit ligand (KL), and stem cell factor (SCF), is a transmembrane growth factor which is proteolytically cleaved to produce a soluble protein. Two isoforms of SLF exist (SLF^{248} and SLF^{220}) due to alternative splicing around exon 6 which contains the primary proteolytic cleavage site. A second cleavage site, which is used when exon 6 is missing, is present in exon 7 (**Figure 1**). These splice variants are expressed in a tissue-specific manner. Both the transmembrane and soluble forms of SLF are biologically active. SLF stimulation of the Kit receptor results in tyrosine phosphorylation of the receptor and its associated downstream signaling molecules, potentiating the survival, proliferation, and/or differentiation of target cells, such as hematopoietic cells, germ cells, and melanocytes.

There are many independent alleles of the *Sl* locus, with the severity of the phenotype depending on the molecular alteration of the gene. Mice homozygous for lethal *Sl* alleles – *Sl*, Sl^{J}, Sl^{gb}, Sl^{8H}, Sl^{10H}, Sl^{12H}, Sl^{18H} – die *in utero* or shortly after birth of severe macrocytic anemia. These alleles involve deletions within the gene resulting in complete loss of SLF function. In contrast, mice homozygous for the *Steel-Dickie* (Sl^{d}) allele are viable despite displaying all the pleiotropic effects characteristic of disruptions at the *Sl* loci, including anemia, a reduced number of hemapoietic stem cells, a profound mast cell deficiency, a complete lack of pigmentation (white, black-eyed), and sterility in both sexes. The Sl^{d} allele involves a 4-kb intragenic deletion in SLF genomic sequence leading to the loss of the transmembrane and cytoplasmic coding regions of SLF (**Figure 1**). The Sl^{d} mutation is therefore only capable of producing the soluble form of SLF. Thus, the mutant phenotype of Sl^{d} mice suggests that soluble SLF alone cannot provide the normal signal to neighboring cells that express the Kit receptor and that the membrane-bound form of SLF is essential for the normal developmental processes controlled by the *W* and *Sl* loci.

The differential roles of soluble and membrane-bound SLF have been further defined *in vitro*. Soluble SLF produced by fibroblast cells cannot sustain the growth of Kit-expressing hemapoietic cells, primordial germ cells, mast cells, or melanocytes when cocultured. In contrast, membrane-bound SLF expressed by fibroblasts supports the proliferation of these cell types in a contact-dependent manner. The adhesive nature of this latter interaction results in a more sustained phosphorylation of the Kit receptor due to a slower kinetics in the downmodulation of the receptor from the cell surface in comparison to soluble SLF. This differential signaling mediated by the two forms of SLF may determine whether Kit-expressing cells undergo survival, proliferation, and/or differentiation. It has also been proposed that the presentation

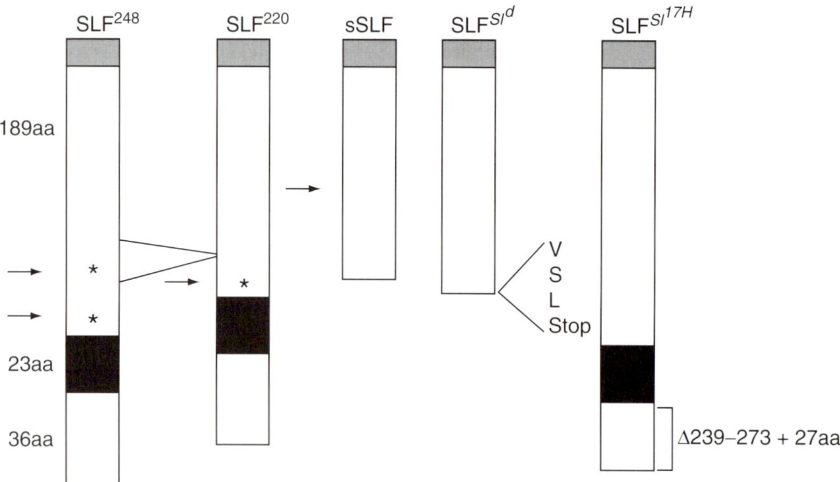

Figure 1 Structure of the alternatively spliced and *Sl* mutant SLF protein products. Diagrammatic representation of the SLF protein SLF^{248} and the alternatively spliced protein product SLF^{220}, lacking 28 amino acids encoded within exon 6. The secretion signal peptide is indicated by a shaded box and the transmembrane domain by a solid black box. Cleavage (denoted by the arrows) of SLF^{248} and SLF^{220} at the proteolytic cleavage sites (represented by asterisks) encoded within exon 6 and exon 7, respectively, result in soluble SLF products. The Sl^{d} mutant SLF product is generated by a 4-kb intragenic deletion in SLF genomic sequence resulting in the loss of the transmembrane and cytoplasmic regions and five amino acids (aa) N-terminal to the transmembrane domain, which are replaced by three additional amino acids and a stop codon. The Sl^{17H} mutant SLF product is the result of a splice donor site mutation that affects splicing of the C-terminal exon that encodes the cytoplasmic tail. This results in the substitution of amino acids 239–273 with 27 additional amino acids.

of membrane-bound SLF along the migratory pathways of hemapoietic progenitor cells, primordial germ cells, and melanoblasts may play a role in guiding these Kit-expressing cells to their final destinations during embryogenesis.

Although soluble SLF is unable to compensate for the loss of membrane-bound SLF, it is apparent from the phenotypes of *Sl* and Sl^{d} mutant mice that soluble SLF is also indispensable *in vivo*. Soluble SLF has extensively been used in the study of hemapoiesis. Alone, soluble SLF acts primarily as a cell survival factor; however, in synergy with a number of cytokines and interleukins, it enhances colony formation by hemapoietic progenitor cells and the proliferation of factor-dependent cell lines.

The Sl^{17H} allele has also provided important insights into the biochemistry of SLF–Kit signaling. The Sl^{17H} allele leads to the substitution of 34 amino acids from the cytoplasmic tail with 27 extraneous amino acids (**Figure 1**). Homozygous Sl^{17H}/Sl^{17H} mice are male sterile due to a block in postnatal spermatogenesis and are white. These results suggest that the cytoplasmic tail of SLF plays a role in either the transport and stability of SLF, its localization within the cell, and/or signaling between SLF and Kit-expressing cells. Sl^{pan} and Sl^{con} are two additional alleles that affect gametogenesis in the adult. Both mutations are the result of DNA rearrangements upstream of the coding sequence of SLF, affecting the levels of SLF mRNA in a tissue-specific manner. In particular, decreased SLF mRNA expression in females causes sterility by affecting ovarian follicle development. These mice display only mild anemia and partial coat pigmentation, demonstrating that these mutations produce only limited impairment of SLF function.

Although the biology of SLF at present is not fully understood, the analysis of mouse mutants disrupted at the *Sl* locus has provided valuable insight into the function of this important growth factor. Further investigation into the distinct roles of membrane-bound versus soluble SLF and the function of the cytoplasmic tail of SLF are still required. Clinical trials involving the use of SLF as a therapeutic agent are also in progress.

Further Reading

Blouin R and Bernstein A (1993) The white spotting and steel hereditary anemias of the mouse. In: Feig SA and Freedman MH (eds) *Clinical Disorders and Experimental Models of Erythropoietic Failure*, p. 157. Boca Raton, FL: CRC Press.

Broudy VC (1997) Stem cell factor and hematopoiesis. *Blood* 90: 1345.

***See also:* W (White Spotting) Locus**

Stem Cells

See: Embryonic Stem Cells

Steroids

R L Somerville

doi: 10.1006/rwgn.1244

Steroids are a diverse group of compounds, mainly though not invariably water-insoluble, that play a major role in physiology as constituents of membranes, as emulsifying agents during the digestive process, and as hormones. From the chemical perspective, the naturally occurring steroids are considered to be derivatives of a hydrocarbon, cyclopentanoperhydrophenanthrene (C17H26). This compound, a product of synthetic organic chemistry, contains three cyclohexane rings and one cyclopentane ring, all fused together to form a puckered structure that is usually referred to as the steroid nucleus. The naturally occurring steroids have various substituents (alkyl, hydroxyl, aldehyde, ketone, carboxylic acid) attached to the four-ring nucleus. Frequently there are one or more double bonds within the steroid nucleus.

Biosynthetically, steroids are formed from five carbon precursors that are chemically related to the hydrocarbon isoprene, in an enzyme-catalyzed sequence of reactions that, in schematic terms, proceeds as follows: C5→C10→C15→C30. The C30 compound, lanosterol, is the precursor to all of the other steroids, including cholesterol (C27), one of the most abundant members of this category of metabolite.

Cholesterol, in addition to its role as a modulator of membrane fluidity, is the precursor of the bile acids (required for efficient digestion of lipids), vitamin D (required for calcium and phosphate metabolism), and the steroid hormones (required for mineral metabolism, blood pressure regulation, reproduction, and manifestation of secondary sexual characteristics). The excessive deposition of cholesterol within the cardiovascular system is one characteristic of atherosclerosis.

***See also:* Familial Hypercholesterolemia**

Sticky Ends

doi: 10.1006/rwgn.2001.2040

Sticky ends (cohesive ends) are short stretches of single strands of DNA that protrude from the ends of duplex DNA, typically generated by staggered cuts in double-stranded DNA, e.g., by restriction endonucleases. Complementary sticky ends can anneal or hybridize to one another and can be joined by DNA ligase, often to create a recombinant molecule.

***See also:* Ligation; Restriction Endonuclease**

Strain

J Parker

doi: 10.1006/rwgn.2001.1246

The term strain is often used by microbiologists to indicate a natural isolate of a particular species. Geneticists, including microbial geneticists, also use the term to indicate a group within a species that has a distinctive genetic trait. For diploid organisms, frequently such a group corresponds to a true-breeding line that is homozygous for genes that contribute to the trait. For instance, experimental geneticists refer to their stocks of true-breeding genotypes, or inbred lines, as strains.

In microbial genetics, a particular strain may have any number of genetic differences compared to other members of the species. However, in animal genetics the term variety is often used to characterize a group within a species that has several distinctive traits.

***See also:* Inbred Strain; Wild-Type (WT)**

Strain Distribution Pattern (SDP)

L Silver

doi: 10.1006/rwgn.2001.1247

Strain distribution pattern is the distribution of two segregating alleles at a single locus across a group of animal samples used for analysis in a linkage study. An SDP is used in the context of backcross data and data obtained from recombinant inbred (RI) strains.

***See also:* Backcross; Recombinant Inbred Strains**

Strand Displacement

doi: 10.1006/rwgn.2001.2041

Strand displacement is a mode of replication of some viruses in which a new DNA strand grows by displacing the previous homologous strand of the duplex.

***See also:* Replication**

Streptomyces

J Parker

doi: 10.1006/rwgn.2001.1248

Streptomyces is a gram-positive genus of filamentous actinomycetes, a large group of spore-forming bacteria that form branching filaments during growth. These filaments form a network called a mycelium. (Both the formation of spores and the pattern of mycelial growth may remind one of fungi, but the actinomycetes are prokaryotes.) The streptomycetes are primarily found in soil.

The streptomycetes have a high G+C base composition (70–74%). The typical prokaryote has a single circular double-stranded DNA molecule as its chromosome, whereas, as a general rule, in *Streptomyces* the chromosome is linear. In most *Streptomyces* species, the chromosomes are about 8 megabase pairs, which is rather large compared with other prokaryotic chromosomes. Many species also contain linear plasmids which range from about 10 to as many as 1000 kilobase pairs, and circular plasmids are also present. The ends of the linear chromosomes and linear plasmids contain an inverted repeat, and each DNA strand has a protein covalently attached to its 5′ end. This protein serves as a primer in DNA replication.

There are a large number of genetic tools available in *Streptomyces*. Several of the plasmids in *Streptomyces* are conjugative and many of these can also mobilize chromosomal genes. Generalized transduction has also been used to map genes. A number of cloning vectors, including shuttle vectors, are also available. Indeed, most of the techniques of molecular genetics, including transposon mutagenesis, can be used.

Many members of the genus *Streptomyces* produce a very large number of secondary metabolites, often as part of the complex pathway leading to sporulation. These secondary metabolites include many useful antibiotics. For instance, *S. aureofaciens* produces tetracycline, *S. clavuligerus* produces cephalosporins and clavulanic acid, *S. erythreus* produces erythromycin, *S. fradiae* produces neomycin, *S. griseus* produces streptomycin, *S. nouseri* produces nystatin, and *S. venezuelae* produces chloramphenicol. Because of the medical importance of many of the antibiotics produced by these organisms, there has been a considerable amount of research on the genetics of antibiotic production.

The genes responsible for antibiotic biosynthesis are located in clusters, which also include genes responsible for resistance to and transport of the antibiotic. Some of these clusters are found on the chromosome, while others are carried by a plasmid. Because of the relatedness of some of the antibiotic-producing genes in fungi and in *Streptomyces* it has been argued that these are examples of horizontal gene transfer which took place between these soil microorganisms.

***See also:* Conjugation, Bacterial; Spores; Streptomycin; Transduction**

Streptomycin

J Parker

doi: 10.1006/rwgn.2001.1249

Streptomycin is an antibiotic of the aminoglycoside group obtained from certain strains of the bacterium *Streptomyces griseus*. The antibiotic binds to the 16S ribosomal RNA (rRNA) of bacterial-type ribosomes and inhibits protein synthesis. Streptomycin increases the frequency of errors in protein synthesis and has been found to be of important use in *in vitro* studies on the accuracy of protein synthesis.

The microbiologist Selman A. Waksman was awarded the Nobel Prize in Physiology or Medicine in 1952 for his discovery of streptomycin, the first antibiotic effective against tuberculosis. (It should be noted that Waksman also introduced the term 'antibiotic.') Although no longer as widely used in the treatment of human infectious disease, streptomycin is still used in combination with other drugs to treat tuberculosis. Reasons for its more restricted clinical use include adverse side effects, such as possible kidney damage and deafness, and streptomycin resistance.

Streptomycin-resistant mutants of bacteria have long been known and intensively studied. The majority of mutants leading to streptomycin resistance in common bacteria are the result of mutations in *rpsL*, the gene encoding the ribosomal protein S12 (a protein in the small subunit of the ribosome), and in *rrs*, the gene encoding 16S ribosomal RNA (rRNA). These mutations are recessive. Since the fast-growing bacteria used as genetic models, such as *Escherichia coli* and *Bacillus subtilis*, have several copies of the genes encoding rRNA, most of the early work on streptomycin resistance involved mutations in *rpsL* (originally named *str*), which exists as a single copy. Mutants in the gene encoding 16S rRNA conferring streptomycin resistance were not uncovered until genetic techniques were available to manipulate these genes *in vitro*.

Ribosomal protein S12 interacts with a highly conserved structure formed by the 16S rRNA, where streptomycin binds. Apparently certain amino acid changes in S12 lead to an alteration or destabilization of this structure. This in turn affects the binding of streptomycin to the ribosome. Some of these mutations lead to streptomycin resistance, but some lead to streptomycin dependence. As mentioned above, streptomycin itself can increase errors in protein synthesis. Interestingly, some streptomycin-resistant mutants restrict the normal level of certain errors, i.e., the ribosomes in these mutants are hyperaccurate. They also have slowed down translation elongation rates. (Certain mutants in ribosomal protein S4, encoded by *rpsD*, are also streptomycin resistant and have hyperaccurate ribosomes.) Like streptomycin, these mutants have proved valuable in investigations of translational accuracy. Several different mutations in *rpsL* are known in enteric bacteria which can lead to streptomycin resistance, but these tend to be clustered at two different regions of the protein: amino acid residues 41 to 45 and 87 to 93. Mutations in similar locations are known in other bacteria.

Mutations in the 16S rRNA encoding gene, *rrs*, which confer resistance to streptomycin have been localized to the region near base 530 and to that near base 915. These regions are part of a putative 'accuracy center' of the ribosome. Mutations in the 915 region not only can lead to streptomycin resistance, but also to changes in translational accuracy. This region seems to be involved with proper selection of tRNA at the ribosomal A site.

Although the causative agent of tuberculosis, *Mycobacterium tuberculosis*, has a reasonably large genome (4.4 million bp) it has only a single copy of each rRNA gene. Therefore, in *M. tuberculosis*, resistance can arise by a mutation in the sole *rrs* gene (or the sole *rpsL* gene). One study found that about 10% of the resistant strains of *M. tuberculosis* isolated from patients have mutations in *rrs*, while 50% have mutations in *rpsL*. However, resistance can also arise by mechanisms other than modification of the target of streptomycin activity, which include uptake and modification of the antibiotic.

Although most antibiotics that act by inhibiting protein synthesis are bacteriostatic, streptomycin is bacteriocidal. It is not completely clear why streptomycin kills bacteria, rather than just stopping growth.

See also:* Antibiotic Resistance; Antibiotic-Resistance Mutants; Resistance to Antibiotics, Genetics of; Ribosomal RNA (rRNA); Ribosomes; *Streptomyces

Stringent Response

M Cashel

doi: 10.1006/rwgn.2001.1250

The bacterial stringent response refers to the many adjustments of gene expression and cell physiology attributable to the accumulation of the (p)ppGpp nucleotides, which are derivatives of GTP (or GDP) bearing pyrophosphoryl substituents on the ribose 3′ hydroxyl. The intracellular level of (p)ppGpp is regulated by mechanisms that sense the availability of different nutrients such as amino acids, carbon sources, nitrogen sources, lipids, and phosphate. The best-understood nutrient limitation condition, mediated by the *relA* gene, involves amino acid deprivation.

The stringent response was first noticed as inhibition of stable RNA accumulation occasioned by amino acid starvation in *Escherichia coli*. The ability of mutants of a single locus to abolish this wild-type 'stringent' RNA control phenotype led to calling the mutant behavior a 'relaxed response' and the mutant gene *relA*. Similar mutant phenotypes are widespread among bacteria distantly related to *E. coli*. In addition to RNA control, many other processes are affected by the stringent response as judged by differential negative or positive mutant effects on regulatory behavior. Negative effects are seen for activities whose functions are presumably superfluous during starvation conditions, such as the synthesis of ribosomes, ribosomal RNA, and transfer RNA. Among functions that can be induced by (p)ppGpp synthesis are synthesis and transport of specific amino acids, accumulation of glycogen and polyphosphate, and induction of the RpoS sigma factor governing stationary phase-specific gene expression. Many of the regulatory outcomes of the stringent response can be viewed as enhancing survival and adaptation to nutritional stress.

Using ATP as a pyrophosphate donor to GTP (or GDP) acceptor substrates, the RelA protein catalyzes (p)ppGpp synthesis on ribosomes. The reaction requires that ribosomes be stalled during translation of mRNA for lack of a bound, codon-specified, charged (aminoacylated) tRNA. Catalysis is activated by uncharged cognate tRNA binding to the otherwise vacant ribosomal acceptor site. Predictions that (p)ppGpp synthesis is activated by increased ratios of uncharged/charged tRNA whenever rates of tRNA aminoacylation fail to keep up with the demands of protein synthesis have been verified with aminoacyl-tRNA synthetase mutants when tRNA levels are artificially varied. A causal role for the (p)ppGpp nucleotides in the stringent response can be demonstrated with engineered gene constructs that allow manipulation of (p)ppGpp abundance atwill in cells that are not nutritionally stressed. Cells with an artificially elevated level of (p)ppGpp mimic many of the major regulatory effects seen during a stringent response provoked by amino acid starvation.

Regulation of (p)ppGpp levels in response to deprivation of nutrients other than amino acids occurs in strains deleted for *relA*. Despite the absence of *relA* function, these starvation protocols elicit responses that share features of the classical stringent response to amino acid limitation. This second source of (p)ppGpp synthesis, in *E. coli*, is a gene called *spoT* that encodes a single bifunctional protein having weak (p)ppGpp synthetic activity as well as a specific (p)ppGpp 3′-pyrophosphoryl hydrolase. Although the SpoT protein sequence shows broad homology with the RelA protein, SpoT is not ribosome associated. The regulation of (p)ppGpp accumulation generally involves inhibition of degradation rather than stimulation of synthesis. The best-studied example (carbon source starvation) leads to (p)ppGpp accumulation through severe inhibition of (p)ppGpp hydrolysis.

Deleting both the *relA* and *spoT* genes of *E. coli* abolishes detectable (p)ppGpp. Such (p)ppGpp^{o} strains appear nearly normal as long as abundant nutrients are provided. However, (p)ppGpp^{o} strains fail to grow on otherwise supportive glucose + salts minimal media unless several amino acids are provided. The corresponding biosynthetic pathways are deduced to be (p)ppGpp-dependent. Survival of (p)ppGpp^{o} strains is also impaired by nutrient starvation, revealing a protective effect of (p)ppGpp during the stringent response. Although extragenic suppressors of these (p)ppGpp^{o} phenotypes map exclusively to genes specifying subunits of the RNA polymerase, the mechanism by which (p)ppGpp inhibits transcription *in vitro* remains elusive.

The stringent response appears to be confined to Eubacteria where specialized roles for (p)ppGpp range from those found in *E. coli* to those contributing to pathogenesis (*Legionella pneumophila*), acid resistance (*Lactococcus lactis*), adaptive catabolism (*Pseudomonas putida*), antibiotic production (*Streptomyces coelicolor*), and quorum sensing for fruiting body development (*Myxococcus xanthus*). In contrast to most Eubacteria, the genomes of some intracellular parasitic bacteria lack genes with Rel/Spo homology; examples are *Rickettsia prowazekii*, *Treponema pallidum*, and *Chlamydia trachomatis*.

Further Reading

Cashel M, Gentry DM, Hernandez VJ and Vinella D (1996) The stringent response. In: Neidhardt FC *et al.* (eds) Escherichia coli *and* Salmonella*: Cellular and Molecular Biology*, 2nd edn, pp. 1458–1496. Washington, DC: ASM Press.

***See also:* Gene Expression; GTP (Guanosine Triphosphate)**

Structural Gene

doi: 10.1006/rwgn.2001.2044

A structural gene is any gene coding for a product (e.g., enzyme, structural protein, tRNA), i.e., any product other than a regulator.

***See also:* Housekeeping Gene**

Subcellular RNA Localization

T Hazelrigg

doi: 10.1006/rwgn.2001.1133

In many types of cells, in diverse species, RNAs are localized to specific cytoplasmic domains. Subcellular RNA localization contributes to the creation of cellular asymmetry by creating spatially unique domains within cells. In some cases, localization of mRNA is coupled to its translational activation, so that only localized transcripts are translated. Together, subcellular RNA localization and localization-dependent translation serve to restrict protein products to specific cellular domains. In recent years considerable advances have been made in understanding the biological functions served by subcellular RNA localization, and the mechanisms behind this localization.

Biological Functions of RNA Localization

Embryonic Patterning

Striking examples of RNA localization occur in oocytes. Many maternally encoded localized RNAs play key roles in embryonic development. In *Xenopus*, several RNAs are localized to the animal or vegetal poles of the oocyte, where their protein products to function in axial embryonic patterning. Examples include *Vg1* mRNA, which encodes a TGFβ-like growth factor, and *VegT* mRNA, which encodes a T-box transcription factor. *Xlsirts*, small noncoding RNAs, are also localized to the vegetal pole of the oocyte, where they are required for correct localization of *Vg1* mRNA. In *Drosophila*, localization of RNAs in the developing oocyte is a key step in setting up both the anterior–posterior and dorsal–ventral axes of the egg and subsequently the embryo. Examples include *bicoid* (*bcd*) mRNA, which is localized to the anterior pole where the Bcd protein, a homeobox-family transcription factor, initiates head and thorax development, and *nanos* (*nos*) mRNA, which is localized to the posterior pole where Nos protein acts as a translational regulator, and is essential for abdomen formation.

Binary Cell Fates

Localization of mRNA is an efficient means of partitioning a factor to one of two daughter cells born from a single cell division. A striking example occurs in the budding yeast *Saccharomyces cerevisae*, where localization of *ASH1* mRNA to the daughter cell, and its exclusion from the mother cell, contributes to the determination of mating type. In the developing *Drosophila* nervous system, asymmetric cell divisions of neuroblasts produce GMCs (ganglion mother cells), the precursors to neurons and glia. Localization of Prospero, a transcription factor required for GMC fate, is enhanced by localization of *prospero* mRNA to the GMC daughter.

Germ Cell Fate

In organisms with distinct germline and soma, the decision to be a germ or somatic cell is usually made early in development. Germ cell fate is often accomplished by sequestering a specialized maternal egg cytoplasm, the germ plasm. In *Drosophila*, the germ plasm is assembled by the action of Oskar (Osk) protein, which is localized at the *Drosophila* posterior pole by localization of *osk* mRNA. Two noncoding RNAs are also localized to the *Drosophila* germ plasm: mitochondrial large ribosomal RNA (mtlrRNA), which is required for the formation of germ cells, and polar granule component (PGC) RNA, required for germ cell development.

Somatic Cells

RNAs are localized in a variety of somatic cells. Examples include β-actin mRNA and myelin basic protein (MBP) mRNA. β-actin mRNA is localized to the leading edges of the lamellipodia of chick fibroblasts. This localization is required for the distinct polarity of this cell type. Localization of MBP mRNA in the processes of oligodendrocytes targets MBP, a protein essential for myelination of the nervous system, to the myelin compartment of these cells.

Extracellular signals stimulate RNA localization. Focal adhesion complexes (FACs) are formed in response to signals that arise when integrin receptors bind to the extracellular matrix (ECM). These signaling events induce the recruitment of mRNA to FACs. In neurons, RNAs are localized by large RNA transport granules, to dendritic domains, to proximal axons, and also to axonal growth cones in developing neurites. Sorting of RNA granules into dendrites is responsive to extracellular signals, including neurotrophic factors, and may contribute to neuronal plasticity. Neural activity modulates the expression and localization of mRNAs in some neurons. For instance, *Arc* (Activity Regulated Cytoskeletal protein) mRNA is sorted to dendrites soon after its expression in response to electrical stimulation. Synaptic activation of specific regions of the brain results in accumulation of ARC transcripts in the synaptically activated dendrites.

Mechanisms of RNA Localization

RNA localization pathways require the transport of RNA through cells and its stable attachment to structures at a final cellular destination. These events utilize *cis*-acting RNA elements, proteins that bind these elements, and cellular structures and regulatory proteins associated with these structures.

Cis-Acting RNA Localization Signals

Within localized RNAs, *cis*-acting signals direct the transport and docking of RNAs at their proper cellular addresses. In general these localization elements lie in the 3′ untranslated regions (UTRs) of RNAs, although signals for localization have also been identified in 5′ UTRs. Some RNA localization signals are modular, with separable elements mediating distinct steps in RNA localization. For instance, distinct signals in the 3′ UTR of MBP mRNA mediate RNA transport and RNA anchoring in the cell processes. Localization signals usually lie in regions with intricate secondary RNA structures. In several cases, such as *Xenopus Vg1* mRNA, and *Drosophila bcd* mRNA, redundant elements are dispersed over a larger RNA segment required for localization.

The Role of the Cytoskeleton and Other Cellular Organelles

In many cases, localized RNAs are transported as large granules, visible by light microscopy. The movement of these RNA granules, identified with fluorescently labeled RNAs or green fluorescent protein (GFP)-tagged proteins, has been studied in living cells by time-lapse video microscopy. Both the actin and microtubule cytoskeletons play important roles in RNA localization. These cytoskeletal elements provide scaffolds for directional transport of RNAs through the cytoplasm, and structural components of anchors at the site of localization. For instance, in *Xenopus* oocytes, microtubules and actin filaments perform distinct temporal and spatial roles in localizing *Vg1* mRNA: microtubules mediate transport of the RNA through the cell, and actin filaments are required for anchoring the RNA at the vegetal cortex. The endoplasmic reticulum (ER) also contributes to RNA localization in *Xenopus* oocytes.

Motor proteins are expected to contribute to these events. In budding yeast localization of *ASH1* mRNA to the daughter cell is actin-dependent, and requires the *SHE1* type V myosin. In oligodendrocytes, MBP RNA forms RNP granules that can be visualized in cultured oligodendrocytes. The transport of these RNA granules along microtubules to the cell processes requires kinesin. In *Drosophila*, two microtubule motor proteins mediate sorting of maternal mRNAs along a polarized microtubule cytoskeleton to their destinations in the oocyte. Kinesin I is required to localize oskar mRNA to the posterior pole of the oocyte, whereas cytoplasmic dynein is implicated in the localization of bicoid mRNA to the anterior oocyte pole. The actin cytoskeleton also plays a role in anchoring oskar mRNA at the posterior pole of the *Drosophila* oocyte.

Proteins that Bind RNA Localization Elements

RNA-binding proteins constitute one important component of the large RNA/protein complexes that transport localized RNAs. In some cases the same RNA-binding protein is used to localize RNAs in different species. ZBP-1 (zipcode-binding protein-1) is an RNA-binding protein identified in chicken fibroblasts that binds to the RNA localization signal of β-actin mRNA. The homologous protein in *Xenopus*, Vera (also known as Vg1 RBP (RNA-Binding protein)), binds to *Vg1* mRNA localization signals in oocytes. Some RNA-binding proteins mediate RNA localization in different types of cells within a species. Thus Staufen, a protein that binds double-stranded RNA (dsRNA), was identified for its role in localizing maternal RNAs in the *Drosophila* oocyte, and is also required for localizing *prospero* mRNA in dividing embryonic neuroblasts. Proteins that cycle in and out of the nucleus may function in very early steps in RNA localization pathways. Several heterogeneous nuclear RNP (hnRNP) proteins have been implicated in cytoplasmic RNA localization in *Xenopus* oocytes, *Drosophila* embryos, and mammalian oligodendrocytes.

Localization of RNA by Degradation

A final mechanism that yields spatial localization of RNAs is selective degradation of RNA coupled to protection in specialized regions of the cytoplasm. This type of localization mechanism is exemplified by Hsp83 mRNA in *Drosophila* embryos. In young embryos, maternally loaded Hsp83 mRNA is distributed uniformly. However, following egg activation there is degradation of the mRNA throughout the cytoplasm, except at the posterior pole, where it is protected. This selective degradation yields an embryo with posteriorly localized Hsp83 mRNA.

Further Reading

Barbarese E, Brumwell C, Kwon S, Cui H and Carson JH (1999) RNA on the road to myelin. *Journal of Neurocytology* 28(4–5): 263–270.

Bashirullah A, Cooperstock RL and Lipshitz HD (1998) RNA localization in development. *Annual Review of Biochemistry* 67: 335–394.

Bassell GJ, Oleynikov Y and Singer RH (1999) The travels of mRNAs through all cells large and small. *FASEB Journal* 13(3): 447–454.

Etkin LD and Lipshitz HD (1999) RNA localization. *FASEB Journal* 13(3): 419–420.

Gavis E (1997) Expeditions to the pole: RNA localization in *Xenopus* and *Drosophila*. *Trends in Cell Biology* 7: 485–492.

Hazelrigg T (1999) The destinies and destinations of RNAs. *Cell* 95: 451–460.

Lasko P (1999) RNA sorting in *Drosophila* oocytes and embryos. *FASEB Journal* 13(3): 421–433.

Lehmann R (1995) Cell–cell signaling, microtubules, and the loss of symmetry in the *Drosophila* oocyte. *Cell* 83: 353–356.

Macdonald PM and Smibert CA (1996) Translational regulation of maternal mRNAs. *Current Opinion in Genetics and Development* 6: 403–407.

Mowry KL and Cote CA (1999) RNA sorting in *Xenopus* oocytes and embryos. *FASEB Journal* 13(3): 435–445.

Oleynikov Y and Singer RH (1998) RNA localization: different zipcodes, same postman? *Trends in Cell Biology* 8: 381–383.

Schnapp BJ, Arn EA, Deshler JO and Highett MI (1997) RNA localization in *Xenopus* oocytes. *Seminars in Cell and Developmental Biology* 8: 529–540.

St Johnston D (1995) The intracellular localization of messenger RNAs. *Cell* 81: 161–170.

Wilhelm JE and Vale RD (1993) RNA on the move: the mRNA localization pathway. *Journal of Cell Biology* 123: 269–274.

See also:* Cell Lineage; Messenger RNA (mRNA); *Xenopus laevis

Subcloning

I Schildkraut

doi: 10.1006/rwgn.2001.1252

Subcloning is the process of dividing a large DNA fragment carried in a vector into smaller more manageable DNA fragments each carried independently in its own vector. DNA cloning often results in large DNA fragments that encode more than one gene. These large DNA fragments are subcloned in order to determine DNA sequence or study the effect of single genes or overexpress specific gene products.

***See also:* DNA Cloning; Vectors**

Substitution Mutations

***See:* Base Substitution Mutations; Gene Substitution; Mutagens**

Superinfection Immunity

E Kutter

doi: 10.1006/rwgn.2001.1256

Superinfection is the productive entry of a phage into a cell that is already infected with another phage. Infection of bacteria by certain phages interferes with the ability of certain other phages to reproduce or to contribute to the genetic composition of the progeny.

Several different types of superinfection immunity are observed.

First of all, cells carrying prophages are immune to superinfection by other phages that use the same repressor, since that repressor is already present in the cell in high enough quantity to prevent the lytic mode of infection for the incoming phage. Thus lambda prophages, for example, give their hosts the benefit of protection against lytic infection by other lambdoid prophages. Even though HKO22 does not belong to the same immunity group as lambda in this regard, it protects against superinfection by lambda in another way. As discussed in the article on antitermination (see Antitermination Factors), HKO22 makes a very special version of the N protein that, rather than being involved in its own antitermination process, directly interferes with the ability of the N protein of lambda and certain related phages to carry out their own antitermination. This alternative mechanism makes cells infected with HKO22 immune to infection by that group of lambdoid phages.

The pair of lambda *rex* genes render cells carrying lambda prophages immune to infection by a wide variety of other phages. In this case, the immunity is suicidal; the cell's membrane potential breaks down some time into infection by the superinfecting phage. The T4 rII A and B genes are able to overcome this effect of lambda *rex* genes. It is this phenomenon that rendered rII mutants unable to grow on lambda lysogens – a factor that was crucial to the elaborate fine-structure-genetics work of Seymour Benzer which established that the unit of recombination is the individual nucleotide (see T Phages).

Cells infected with bacteriophage T4 are immune not only to infection by most other phages but to other T4 phage arriving more than a few minutes after the T4. This immunity against late-arriving T4 is the consequence of a membrane protein encoded by the *imm* gene that helps block entry of the newly arriving phage DNA into the cell; it is related also to a process called 'superinfection exclusion.' The precise mechanism is not understood. The DNA of the phage attempting superinfection remains in the periplasmic space and is largely degraded. In the case of non-T-even phages attempting to superinfect a cell infected by a T-even phage, transcription is inhibited by the Alc protein, which blocks transcription of all cytosine-containing DNA, as discussed in the entry on T phages, and the incoming phage DNA is subject to degradation by endonucleases II and IV, produced by T-even phage to degrade the host DNA.

Phage P22 prophages have yet other mechanisms of producing immunity to superinfecting phages. In addition to the repressors needed to maintain the prophage state, P22 prophages express several genes which are very effective in keeping out homologous and heterologous superinfecting phages. The product of gene *sieA* interferes with DNA injection by P22 and related phages. Gp*sieB* causes the lytic cycle of certain other *Salmonella* phages (not including P22 itself) to abort early in infection. The product of gene *a1* interferes with adsorption by P22 and related phages by changing the structure of the lysogen's O antigen.

The variety of mechanisms found in the few phages studied to date for engendering superinfection exclusion and thus protecting resident phage from external competition makes it very likely that many more interesting mechanisms will be found as more phages are studied from the vast pool present in our biosphere.

See also: **Bacteriophages; Prophage, Prophage Induction**

Superrepressor

C A Royer

doi: 10.1006/rwgn.2001.1257

Superrepressor refers to a mutant form of a repressor protein that represses gene expression more efficiently than the wild-type. Repressor proteins are found in phage, bacteria, and eukaryotes and intervene in the control of gene expression at the level of transcription or translation, such that the structure–function mechanisms underlying the superrepressor phenotype can be quite varied.

Superrepressors were first identified in the context of the study of the control of prokaryotic operons coding for metabolic enzymes, such as the *lac* (lactose), *hut* (histidine), *gal* (galactose), *put* (proline), *src* (sucrose) and *nag* (n-acetlyglucosamine-6P) operons. The repressor proteins of these operons act by binding to specific DNA operator sequences found in or near the operon. RNA polymerase cannot bind or initiate transcription when the repressor is bound to the operator site. When the small molecule metabolites (inducers) of these operons (i.e., histidine in the case of the *hut* operon) are present in the medium, they interact with high affinity with the repressor proteins, inducing a conformational change such that the repressor's affinity for the operator site decreases significantly, thereby decreasing operator occupancy and allowing transcription to take place. In this context, superrepressor mutants typically involve amino acid mutations that abolish or at least significantly diminish the affinity of the repressor proteins for their inducers. They are termed un-inducible.

Another class of bacterial repressors for which superrepressors have been identified involves proteins that negatively regulate gene transcription in response to the binding of a corepressor molecule, typically the product or an intermediate in the biosynthetic pathway implicating the enzymes encoded by the operon in question. Thus, the *trp* repressor (trpR) in reponse to L-tryptophan represses transcription of genes whose products are enzymes involved in the synthetic pathways for aromatic amino acids. Another example is the *arg* repressor. Strains bearing strong superrepressor mutations in these proteins are typically tryptophan (or arginine) auxotrophs, since the enzymes responsible for the biosynthesis of these amino acids are never produced. The mechanisms by which these mutations lead to the superrepressor phenotype can be quite diverse and include increased operator affinity in the absence of corepressor or only when bound by the corepressor. Increases in DNA affinity in general can also lead to superrepression since operator affinity increases accordingly. Superrepressor mutants with altered repressor oligomerization and protein-folding properties have also been identified.

Certain repressors such as birA (biotin repressor) and putR also act as enzymes in the control of the biosynthetic pathway, catalyzing reactions on the inducer or corepressor molecule. Superrepressor mutants of these multifunctional proteins may exhibit altered catalytic properties, as well.

Translational superrepressors have been found in the case of the R17/MS2 RNA phage coat protein. Their superrepressor activity appears to arise from an increase in the size of the RNA site recognized leading to an increase in overall affinity. Superrepressors bearing deficiencies in coat assembly result in an increase in the amount of protein available for repression.

Eukaryotic superrepressors typically involve mutations that increase the ability of a repressor protein to undergo heterodimerization with a transcriptional activator. Transcriptional activation by the yeast transcriptional activator Gal4 is repressed through its interaction with a repressor protein, Gal80. In response to binding of galactose by Gal80 and phosphorylation of Gal4, the Gal4/Gal80 complex is destabilized in favor of a complex between Gal3 and Gal80. Superrepressor forms of Ga180 are galactose uninducible, and are deficient for interaction with Gal3.

Another very interesting group of eukaryotic superrepressors are the superrepressor mutants of I-κB which heterodimerizes with NF-κB, a transcriptional activator of immunoglobulin, and certain antiapoptotic genes. NF-κB is retained in the cytoplasm when complexed with I-κB. In response to a variety of extracellular signals, I-κB is phosphorylated and eventually degraded, resulting in the transport of NF-κB to the nucleus. Transfection of malignant cell lines with I-κB superrepressors can lead to dramatic decreases in their abnormal proliferation rates.

See also: **Repressor**

Suppression

R W Alexander and P Schimmel

doi: 10.1006/rwgn.2001.1258

Mutations in genes can be detrimental when they result in premature stop signals (nonsense errors), amino acid substitutions (missense errors), or shifts in the translational register (frameshift errors). Compensatory substitutions in translational components rescue these initial mutations by a mechanism known as genetic suppression. Many naturally occurring suppressors have been identified, and the most common are variants of transfer RNAs (tRNAs). These so-called 'suppressor tRNAs' typically contain substitutions in their anticodons that recognize error-inducing codons and allow insertion of an amino acid into the growing polypeptide chain. Other suppressors are variants of ribosomal RNA, ribosomal proteins, and termination factors. Interest in therapeutically induced suppression is high, given the numerous diseases caused by single nucleotide changes.

Types of Translational Errors

Mutations in genes can have a wide range of effects. The amino acid sequence of an encoded protein may not be changed at all, for example, if the resulting codon is read by an isoaccepting tRNA. Similarly, no phenotypic change is observed at the functional level if an amino acid change does not affect the structure or function of the encoded protein. In contrast, a single nucleotide substitution can have a drastic effect not only on the protein being produced, but also on cellular pathways that depend on the encoded protein.

Three types of translational errors can result from genetic point mutations. A substitution that changes a codon's specificity from one amino acid to another is a missense error. A substitution that changes an amino-acid-inserting (sense) codon to a stop codon is a nonsense error. Finally, insertion or deletion of one or more nucleotides can result in a translational frameshift. Frameshifts result in a new amino acid sequence and often lead to nonsense errors when the ribosome encounters a premature stop codon in the new reading frame.

Suppressor tRNAs

A transfer RNA that is able to suppress genetic mutations is the result of advantageous substitutions in the tRNA gene. Such substitutions are typically located in the anticodon of tRNA, such that a suppressor tRNA

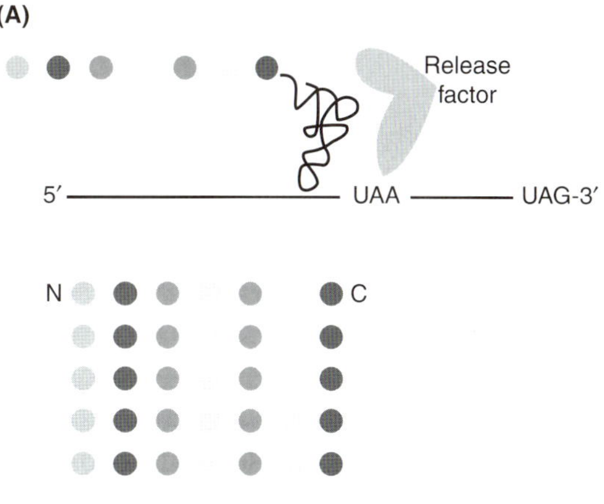

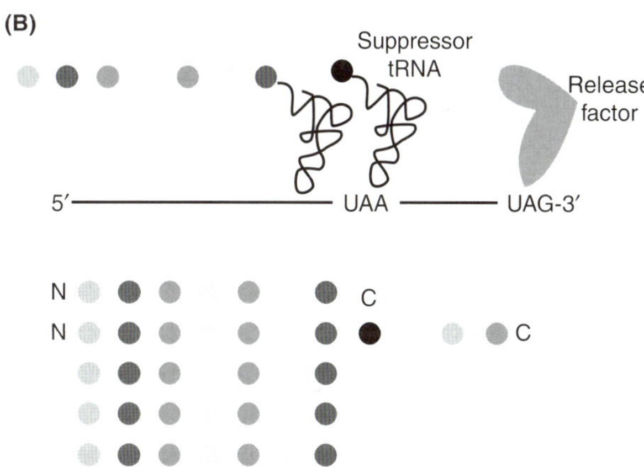

Figure 1 A suppressor tRNA rescues a nonsense mutation by inserting an amino acid at the position of the premature termination codon. Suppression generally occurs at low levels compared with wild-type production of the protein. (A) Premature termination; (B) nonsense suppression.

recognizes the mutated codon or unintended frameshift rather than its cognate codon. Suppressor tRNAs may be aminoacylated according to the amino acid specificity of their parent sequence, or they may be misacylated because of the anticodon change (for example, the *Escherichia coli su7* tRNA is derived from $tRNA^{Trp}$ but inserts glutamine). In either case the suppressor tRNA inserts its attached amino acid into the growing polypeptide chain at the location of mutation in the mRNA (**Figure 1**). As long as the inserted amino acid is not detrimental to the protein, the gene mutation is rescued. One example of this most common type of suppression is the *su2* suppressor tRNA in *E. coli*. This variant of $tRNA_2^{Gln}$ has a G to A substitution in its anticodon so that it recognizes a UAG stop codon instead of its cognate CAG codon. The *su2* tRNA inserts glutamine in place of the premature stop.

While missense and nonsense suppression occurs primarily through single nucleotide substitutions in the anticodon, tRNAs that suppress +1 frameshifts sometimes contain an extra base in the anticodon. These suppressors read a four-nucleotide codon, thereby restoring the correct translational frame. An example is the *E. coli* $tRNA^{Gly}$ suppressor that has a four-base anticodon.

Not all suppressor tRNAs contain anticodon substitutions, however. The *E. coli su9* nonsense suppressor is a variant of $tRNA^{Trp}$ that retains its wild-type CCA anticodon but has a G to A change in its D-arm. This substitution leads to a tRNA with increased thermal stability that recognizes both its cognate UGG codon and, through an unusual A:C pair in the third anticodon position, the UGA stop codon.

Several mechanisms combine to limit suppression-mediated amino acid insertion, which is typically between 5 and 50% of wild-type levels. Suppressor tRNAs are typically derived from minor isoacceptors, ensuring that translation of most sense codons is not reduced. Wild-type mRNAs often contain tandem stop codons, so even if one is suppressed the other will lead to termination. Finally, suppressor tRNAs must compete with termination factors for binding to stop codons. Together these features maintain the overall accuracy of protein synthesis.

Other Compensatory Changes Result in Suppression

When the ribosome reaches a stop codon, translational release factors facilitate hydrolysis of the fully synthesized protein from the peptidyl-tRNA. In addition to genetic suppression through altered tRNAs, variants of release factors have also been identified that produce suppression phenotypes. Mutations are primarily within the C-terminal regions of the release factors, and likely prevent sequence-dependent recognition of stop codons.

Certain ribosomal proteins and regions of rRNA have long been implicated in translational accuracy control. For example, mutations in small subunit proteins S4, S5, and S12 lead to ribosomes that are either hyperaccurate or error-prone. Errors in protein synthesis include increased levels of stop codon readthrough, frameshifting, and missense errors. Codon-specific suppression variants have also been identified within ribosomal components. A single C to A substitution within the small subunit rRNA (*E. coli* C1054A) results in UGA readthrough without affecting other termination events or causing missense or frameshift suppression. This mutation decreases the binding affinity of release factor 2 for the ribosome, resulting in the observed genetic suppression.

Finally, suppression of genetic mutations can be the result of compensatory changes in partner molecules. Mitochondria translate a limited number of proteins from a small genome that also contains a complete set of tRNA genes. A nucleotide substitution in the acceptor stem of yeast mitochondrial $tRNA^{Asp}$ was shown recently to be suppressed by a variant of aspartyl-tRNA synthetase (AspRS). The single amino acid subsitution in the nuclear-encoded AspRS enzyme is in a region known to contact the acceptor stem of its cognate $tRNA^{Asp}$. New contacts between the variant $tRNA^{Asp}$ and AspRS result in enhanced aminoacylation efficiency and genetic suppression of the tRNA defect.

Therapeutic Potential of Suppression

Recent advances in gene sequencing have revealed that numerous human genetic diseases are the result of nonsense or missense mutations. For example, substitutions in the tumor suppressor p53 are reported to be responsible for as many as half of human cancers. These p53 substitutions typically result in missense mutations within a critical DNA-binding region. Likewise, about 5% of individuals with cystic fibrosis carry a premature stop codon in the gene for cystic fibrosis transmembrane conductance regulator (CFTR). If even small amounts of functional protein could be produced in these cases, significant reduction in symptoms might be achieved. Many researchers are therefore actively working to develop suppressor tRNAs that could be used therapeutically. Several challenges to such gene therapy exist. As with all approaches to gene therapy, any suppressor tRNA must be transported into the affected cells. The suppressor tRNA must be transcribed and aminoacylated at high levels. Finally, suppression must be selective for the target gene's mRNA so authentic termination signals for other proteins are not read erroneously.

The use of aminoglycoside antibiotics has also been proposed as a treatment for some genetic diseases. Aminoglycosides interact with the decoding center of the ribosome and decrease translational accuracy by allowing readthrough of stop codons. At high levels of antibiotic this decrease in accuracy completely eliminates protein synthesis, while at low levels limited readthrough may provide enough of the deficient protein for a near-normal phenotype. Such an approach has shown potential in models of both cystic fibrosis and Duchenne muscular dystrophy, a disease in which 5–15% of patients carry a premature stop codon in the gene for dystrophin.

Further Reading

Atkinson J and Martin R (1994) Mutations to nonsense codons in human genetic disease: implications for gene therapy by nonsense suppressor tRNAs. *Nucleic Acids Research* 22: 1327–1334.

Bedwell DM, Kaenjak A and Benos DJ *et al.* (1997) Suppression of a CFTR premature stop mutation in a bronchial epithelial cell line. *Nature Medicine* 3: 1280–1284.

Moore B, Persson BC, Nelson CC, Gesteland RF and Atkins JF (2000) Quadruplet codons: implications for code expansion and the specification of translation step size. *Journal of Molecular Biology* 298: 195–209.

Murgola EJ (1995) Translational suppression: when two wrongs do make a right. In: Söll D and RajBhandary U (eds) *tRNA Structure, Biosynthesis, and Function*, pp. 491–509. Washington, DC: American Society for Microbiology Press.

See also: **Frameshift Mutation; Gene Therapy, Human; Mutation, Missense**

Suppressor Mutations

J Parker

doi: 10.1006/rwgn.2001.1259

Like all mutations, suppressor mutations are inheritable alterations in the sequence of the genetic material of an organism. What is distinctive about a suppressor mutation is that it reverses the phenotypic change caused by a previously existing mutation, without actually reversing the original mutation itself. The continued presence of the original mutation distinguishes a suppressor mutation from a true 'reverse mutation.' For this reason organisms containing the original mutation and the suppressor are sometimes referred to as 'pseudorevertants' to distinguish them from 'true' revertants. Although, in many instances, the phenotype found in the new double mutant is not identical to the wild-type phenotype, it is sufficiently normal to allow the organism to function under selective conditions. Because the suppressor mutations occur at a sites other than that of the original mutations they suppress, they are also called 'second-site mutations.' (It must be emphasized that the suppressor mutation in the absence of the original mutation is unlikely to yield an organism with a wild-type phenotype.)

Typically suppressor mutations have been classified as being of two types, depending on where in the genome they occur compared with the original mutation they are suppressing. If they occur in the same gene they are said to be intragenic suppressors. If they are in another gene, they are said to be intergenic suppressors. (The term 'extragenic suppressor' is also often used to refer to a suppressor mutation occurring in a gene other than the gene containing the original mutation.) Intragenic suppressors tend to restore the function of the gene containing the original mutation, a situation sometimes termed 'direct suppression.' While intergenic suppressors can also be direct, they may allow the organism to somehow bypass the original defect. This latter situation is termed 'indirect suppression.' While some suppressors can suppress only a specific mutation in a specific gene, others can suppress a number of different mutations (in one gene or related genes), and still others can suppress entire classes of mutations in many different genes.

Intragenic Suppression

If the suppressor mutation is within the same gene as the original suppressor it is said to be an intragenic suppressor. Intragenic suppression is direct in that the function of the originally mutated gene is restored.

One type of such mutations are intragenic frameshift suppressors, and their discovery and characterization helped elucidate the mechanism by which the genetic code was read. Mutations of the bacteriophage T4 *rII* genes induced by the acridine proflavin were found to be of two types, microinsertions (+) and microdeletions (−). It was discovered that some revertants of these mutants actually contained two mutations, the original *rII* mutation and an intragenic suppressor. When the latter was isolated by recombination, it was found to behave as a typical *rII* mutant but had the opposite sign (+/−) of the original mutation. This indicated that, during expression, the genetic code was read in a particular frame and, while the removal (or addition) of a single base pair caused a frameshift to a nonfunctional state, the nearby addition (or removal) could restore the correct reading frame and thereby the function of the gene. Also, whereas a single mutation or two mutations of the same sign (+ or −) within the gene led to loss of function, it was observed that three closely linked mutations of the same sign led to a normal, or nearly normal phenotype. This gave evidence that the code is read in groups of three bases. Please note that this type of suppressor is 'direct' in that the function of the gene is (at least partially) restored.

The other main type of intragenic suppressors are second site mutations within a gene that lead to an amino acid residue change in the protein product which compensates for a change brought about by the original mutation. Once again the restoration of the original phenotype may not be complete, but the doubly mutant protein has activity. Such mutants were discovered by Yanofsky in his studies on the colineraity of the *trpA* gene and its product, tryptophan synthetase. The replacement of the glycine at residue 211 by a glutamic acid residue leads to an inactive tryptophan synthetase, but the activity can be restored if the normal tyrosine at residue 175 is replaced by a cysteine. (The suppressor mutation in the absence of the original mutation leads to an inactive protein.) Such intragenic suppressors can give considerable insight into the functional/structural requirements of the protein for activity.

Intergenic Suppressors

If the suppressor mutation is in a gene other than that containing the original mutation it is said to be an intergenic suppressor (or an extragenic suppressor). These have proved both useful and interesting. Once again there are different types of intergenic suppressors. Many are indirect and allow the organism to bypass the original mutation. These bypass suppressors may activate a new pathway; if so they should be able to suppress essentially any mutation in the original gene (and also mutations in other genes in the original pathway). Such suppressor mutations may be found in regulatory genes of other pathways.

However, other intergenic suppressors may restore the function of the original pathway. If the original mutation led to a partially active gene product, intergenic suppressors may be found in that gene's regulatory genes. These then will be somewhat allele-specific, since many mutant alleles will have no activity. Other intergenic suppressor mutations may be in genes whose products interact with the product of the gene containing the original mutation. Mutations in the interacting protein may compensate for the change in the original protein. Such suppressor mutations would be expected to be very allele-specific, only restoring activity to a very limited number of mutations in the original gene.

Another type of intergenic suppressor suppresses mutations of a particular class rather than mutation in specific genes. These are informational suppressors that alter, to a limited extent, how the cell containing them reads the genetic code. These informational suppressors were also discovered during analysis of tryptophan synthetase mutants and T4 mutants. These suppressor mutations act directly and allow the cell containing them to make a functional product from the original mutated gene. These suppressors, then, are intergenic but direct.

The first suppressors of this type to be understood completely were nonsense suppressors, and these were investigated using nonsense mutations in the T4 *rII* genes. Cells with nonsense suppressors insert an amino acid at the site of a nonsense mutation, that is, they read nonsense (stop) codons as sense. Most of these suppressors are themselves mutant tRNAs which have been altered to respond to one (or more) of the stop codons (and compete with the mechanism of chain termination). The majority of these have mutations in that part of the tRNA which encodes the anticodon. The mutant tRNA is normally aminoacylated but now reads a nonsense codon and not the normal sense codon. Such mutations are possible in cells that have duplicate genes for the normal tRNA or at least alternative tRNAs which can still read the normal sense codon. Nonsense suppressors differ in the efficiency of suppression. Part of this is due to the fact that suppression will occur only if the amino acid they carry will lead to a return of activity of the full-length peptide. Part is also due to the fact that the normal translational termination machinery continues to function. Nonsense suppressors tend not to lead to a loss of termination at normal stop codons at the end of genes, and efficient nonsense suppressor tRNAs tend to suppress nonsense codons other than the one(s) the organism prefers. For instance, most *Escherichia coli* genes terminate with a UAA codon and most efficient suppressor tRNAs read UAG or UGA. Many different suppressor tRNA mutations have been isolated and others have also been constructed using *in vitro* genetic manipulations. Such tRNAs can be used to assess the activity of proteins with different amino acids at a particular residue.

Not all nonsense suppressor mutations need to be in genes encoding tRNAs. Such mutations have also been isolated in the genes encoding both the 16S ribosomal RNA and the 23S rRNA of bacteria. Presumably the bases that are changed in the mutants normally participate in translation termination.

In addition, mutant tRNAs have also been isolated that can suppress missense mutations and frameshift mutations. Indeed among the first intergenic, direct suppressors to be isolated was a mutant tRNA that suppressed a missense mutation in *trpA*.

The selection and characterization of suppressor mutations remains one of the most interesting and rewarding experimental genetic approaches to uncovering gene function. Such mutations can yield important information on the function or regulation of the gene containing the original mutation and also uncover new genes and even new pathways. While we have here emphasized the 'suppressor mutation' occurring in a organism containing an original mutation, important information can also be gained by studying the suppressor mutation in isolation. Some, particularly the intragenic mutations, are likely to give rise to phenotypes very like those of the mutation they suppress. However, others may lead to new and interesting phenotypes, disclosing the functions of the genes that contain them.

***See also:* Acridines; Mutation, Missense; Nonsense Mutation; Phenotype; Reverse Mutation; Suppressor tRNA; Yanofsky, Charles**

Suppressor tRNA

E J Murgola

doi: 10.1006/rwgn.2001.1261

The term 'suppressor tRNA' usually refers to a genetically altered or mutant tRNA that, because it translates a codon other than its normal (cognate) codon or is aminoacylated with an amino acid other than its normal amino acid, reverses, at least to some extent, the effect of a mutation in the gene for one or more proteins. Such translational suppression has been described as a "mistake upon a mistake." The latter mistake refers to a mutation in a protein-encoding gene, which results in some altered phenotype of the encoded protein. That mutation could be: (1) a missense change, converting one of that protein's codons to a codon for an amino acid that renders the protein inactive; (2) a nonsense mutation, i.e., a change of one of the sense codons to a termination codon (UGA, UAA, or UAG), causing premature termination of polypeptide synthesis and resulting in synthesis of a truncated protein; or (3) insertion or deletion of one or more nucleotides can result in a frameshift mutation that leads either to an early termination codon in the new frame or continued synthesis past the original normal termination codon until a new termination codon is reached in the new translational reading frame. The first mistake in the "mistake upon a mistake" expression refers to a change in specificity at some step in translation that reverses the effects of the missense, nonsense, or frameshift mutation in the protein-encoding reference gene, resulting in a change in the primary structure of the mutant protein, either back to the wild-type amino acid sequence or to one that confers on the protein some degree of normal activity.

Several types of mutant suppressor tRNAs have been characterized. The most common type arises from anticodon base changes that allow the mutant tRNAs to read codons other than their normal ones. Less frequent are mutations outside of the anticodon that result in decoding changes that lead to suppression, particularly of frameshift mutations. Such mutations have been found to occur in any one of the three tRNA arms, the anticodon arm (outside of the anticodon), the dihydroU arm, and the T-pseudoU-C arm, as well as the amino acid acceptor stem. Another kind of one-step suppressor tRNA mutation, one that is found infrequently but was predicted in the original hypothesis for missense suppression, is a mutation that changes the aminoacylation specificity of the tRNA but allows the retention of the normal decoding specificity. An example of this type is a lysine tRNA base change in the amino acid acceptor stem that allows the tRNA to be misacylated some of time with alanine while still decoding the lysine codons AAA and AAG. Some mutant tRNAs with anticodon changes that alter their decoding properties are also misacylated to some extent with noncognate amino acids.

Not limited to mutant suppressor tRNAs, translational suppression is a most effective way in which to examine the structure, function, and interactions of any translational macromolecule, as long as that molecule is involved in the specificity or accuracy of translation. Translational suppressor mutations have been found and characterized in the genes for translational molecules other than tRNAs, namely in the genes for elongation factors (bacterial EF-Tu and EF-G), termination factors (bacterial release factors RF1, RF2, and RF3, and yeast release factor eRF1), aminoacyl synthetases, and the ribosomal RNAs of both subunits of bacteria and yeast. Suppression of nonsense mutations has also been achieved by high expression of rRNA fragments from cloned segments of the bacterial 23S rRNA (large ribosomal subunit) gene, either in the sense orientation only, in the antisense orientation only, or in both orientations, depending on the segment examined.

Two special situations should be mentioned in which apparently mutant tRNAs (suppressor tRNAs) are normally present in the cells. First, some organisms carry, in addition to the tRNAs for a sense codon, another tRNA, acylatable with the same amino acid, but whose anticodon allows it to decode the termination codons UAA or UAG in certain contexts within a coding sequence. Second, from bacterial to human cells, a specialized translational mechanism employs a tRNA that inserts selenocysteine at high frequency at strategically located UGA codons within a protein-coding sequence. In bacteria, that special tRNA is acylated with serine, which is then converted in two steps to selenocysteine. The selenocysteyl-tRNA then interacts, not with elongation factor Tu, but rather with a special Tu-like elongation factor to position the special aminoacyl-tRNA at the UGA codon for peptide bond formation.

***See also:* Elongation Factors; Translation; Translational Control**

Symbionts, Genetics of

B D Dyer

doi: 10.1006/rwgn.2001.1491

Any change in the environment in which a genome is expressed may be expected to have some impact on the phenotypes of at least some of the genes. Alterations in phenotype, even if subtle, may have an effect on the overall fitness (ability to leave reproductive offspring) of an organism. Environmental parameters are not limited to the obvious ecological ones of temperature, moisture, sunlight, and pH but should also be considered to include the environment within organisms, within particular organs and tissues, and within cells. Indeed these may be among the most exacting of environments, especially when the additional factor of symbiosis is included.

Symbiosis is a close association of two or more organisms of different species such that the two leave more offspring (are more fit) in a particular environment as partners than as individuals. Lichens are a striking case in point. As a symbiont, a lichen can colonize the dry, bare surface of a grave stone, a habitat where the individual fungus and algae would not be able to survive let alone reproduce. In any intimate symbiosis the 'environment' includes the partners themselves and the phenotype is collective, a combination of all of their expressed genes.

Genetic consequences of symbioses (especially long-term intimate symbioses) may include:

1. A loss of genes (or gene function) that are redundant in one or the other partner.
2. A loss of genes (or gene function) that are no longer necessary in the new circumstances.
3. A transfer of genes from one partner to another such as via an exchange of viruses. Such exchanges may be lethal in some cases but in other cases may result in a shared coding for a multigenic structure or pathway and thus a sort of cementing of the relationship.
4. Loss and transfer of genes may ultimately result in an obligate relationship in which the partners can never again be free-living. The phenotype of the symbiosis is effectively a shared phenotype of two altered genomes. All kinds of herbivorous animals are striking examples in that none can digest cellulose and yet a major aspect of a herbivorous niche is the consumption of plants. Symbionts of herbivorous digestive systems (such as those of bovine rumens) provide a major part of the overall phenotype by digesting cellulose.

Parasitism may be seen as a variation of symbiosis, although one in which the reproduction of the associates is not coordinated. In a typical mutual symbiosis, reproduction of one partner is accompanied by reproduction in the other such that the ratio remains the same. In contrast, parasites often overrun their hosts and may be defined in part by this tendency. Parasites and symbionts may be seen as opposite extremes of a changeable continuum. Some symbioses may revert to more independent (and sometimes pathogenic) living if environmental conditions change. For example, green hydra will consume their algal symbionts if kept in the dark. Some parasites are lethal only under particular circumstances. For example, many intestinal microbes are endemic in some populations and not especially harmful but can become harmful in an unfamiliar host. Therefore, parasite genetics may also be viewed as a variation on the symbiotic theme and the points listed above are valid for parasites as well. For example, malaria parasites are truly obligate internal inhabitants of their host's blood cells, in part because of a loss of some of their metabolic capabilities and a reliance on those of the host. Mitochondria and chloroplasts may be considered as extreme examples of obligate symbionts with several additional consequences to their genetics.

See also: **Mitochondria; Predator–Prey and Parasite–Host Interactions**

Symbiosis Islands

C Ronson and J Sullivan

doi: 10.1006/rwgn.2001.1633

The symbiosis island of *Mesorhizobium loti* is a 501.8-kb chromosomally integrated element that transfers to nonsymbiotic mesorhizobia in the environment and converts them to symbionts able to nodulate and fix nitrogen with *Lotus* species. The island integrates into a phe-tRNA gene, reconstructing the gene at one (left) end of the island and producing a 17-bp direct repeat of the 3′ end of the tRNA gene at the other end. Integration and excision of the island are mediated by a phage P4-type integrase encoded just within the left end of the island (Sullivan and Ronson, 1998). The island has a mosaic structure suggesting that it evolved in a stepwise fashion via multiple recombination events. It contains nodulation and nitrogen fixation genes, including some which are spread across several replicons in other rhizobia, and a wide range of other genes. Such genes include those likely to be involved in transfer of the island, genes of unknown function found on symbiotic replicons in other rhizobia, genes with no homologs in current databases, several putative regulatory genes, genes encoding cell-membrane-associated components including porins, and an array of metabolic genes which may contribute to 'fine tuning' of nodule metabolism. The island is a member of an emerging class of acquired genetic elements that may be termed 'fitness islands' (Preston *et al.*, 1998). These conjugative genetic elements integrate chromosomally in a site-specific manner, and contribute to the diversification and adaptation of bacteria to environmental niches. Examples of such elements include many pathogenicity islands (Kaper and Hacker, 1999), the *clc* element conferring chlorocatechol degradation in *Pseudomonas* sp. (Ravatn *et al.*, 1998), and the SXT element conferring antibiotic resistance in *Vibrio cholerae* (Hochhut *et al.*, 2000).

References

Hochhut B, Marrero J and Waldor MK (2000) Mobilization of plasmids and chromosomal DNA mediated by the SXT element, a constin found in *Vibrio cholerae* O139. *Journal of Bacteriology* 182: 2043–2047.

Kaper JB and Hacker J (1999) *Pathogenicity Islands and Other Mobile Virulence Elements*. Washington, DC: American Society for Microbiology Press.

Preston GM, Haubold B and Rainey PB (1998) Bacterial genomics and adaptation to life on plants: implications for the evolution of pathogenicity and symbiosis. *Current Opinion in Microbiology* 1: 589–597.

Ravatn R, Studer S, Zehnder JB and van der Meer RJ (1998) Int-B13, an unusual site specific recombinase of the bacteriophage P4 integrase family, is responsible for chromosomal insertion of the 105-kb *clc* Element of *Pseudomonas* sp. Strain B13. *Journal of Bacteriology* 180: 5505–5514.

Sullivan JT and Ronson CW (1998) Evolution of rhizobia by acquisition of a 500-kb symbiosis island that integrates into a phe-tRNA gene. *Proceedings of the National Academy of Sciences, USA* 95: 5145–5149.

See also: *Lotus japonicus*; Nodulation Genes; Symbionts, Genetics of

Symbiosome

G Stacey

doi: 10.1006/rwgn.2001.1738

The symbiosome is defined as a membrane-bounded compartment containing one or more symbionts that is located in the cytoplasm of eukaryotic cells (**Figure 1**). Rhizobia have the ability to infect and establish a nitrogen-fixing symbiosis with a variety of legume plants. This symbiosis involves the

(A)

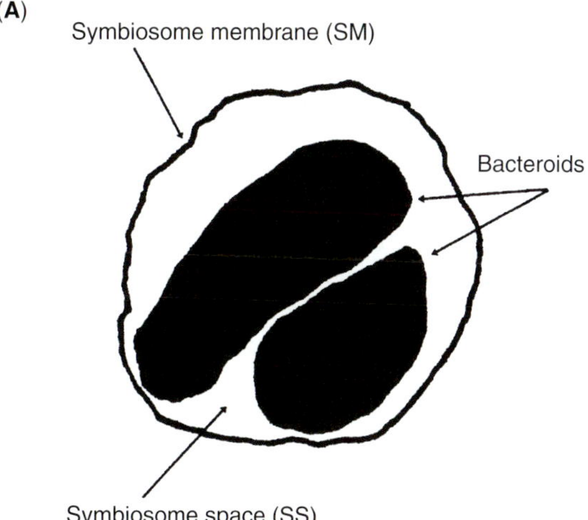

(B)

Figure 1 (A) Schematic drawing of a symbiosome showing the location of the symbiosome membrane (SM), symbiosome space (SS), and bacteroids. (B) Transmission electron micrograph showing a symbiosome within an infected cell of a *Lotus japonicus* root nodule. The SM, SS, and *Mesorhizobium loti* bacteroids should be easily identifiable by referring to the schematic diagram. (Drawing and photo kindly provided by Dr John Dunlap, University of Tennessee.)

intracellular colonization of root cortical cells contained in the nodule structure. The symbiosome is composed of the symbiosome membrane (SM) which surrounds a symbiosome space (SS) that contains the symbiont (in the case of rhizobia, the nitrogen-fixing symbiont, i.e., bacteroid). The term symbiosome is specific to stable, intracellular symbioses. For example, it should not be confused with an endosome resulting from endocytosis of bacteria.

The term symbiont has general meaning regardless of what system is being discussed. However, homology among many intracellular symbionts extends to the compartment (vacuole) that encloses such bacteria. For example, similar membrane-bounded, bacteria-containing structures are found in legume nodules, amoeba endosymbionts, intracellular *Legionella*, malaria, etc. Moreover, all intracellular symbionts in eukaryotic cells have similar problems to overcome. For example, how do such symbionts enter the cell, avoid lysosomes, obtain nutrients, etc. Terminology that draws attention to the similarity among intracellular symbionts could stimulate cooperative research and speed discovery. This is the reason why the term symbiosome was proposed and has been widely adopted.

Before 'symbiosome,' the intracellular compartment housing rhizobial bacteroids was given a variety of names; peribacteroid vacuoles, bacteroid-enclosing compartments, peribacteroid units, symbiotic vesicles, nifixasome, etc. Although all such terms were descriptive, they failed to draw attention to the homology of symbiosomes to similar structures in other intracellular symbionts.

Owing to the endosymbiotic theory for the evolution of mitochondria and chloroplasts, the suggestion has been made that symbiosomes be considered as quasi-organelles.

***See also:* *Rhizobium*; Symbionts, Genetics of**

Sympatric

E Mayr

doi: 10.1006/rwgn.2001.1469

'Sympatric' is the term used to describe species (or higher taxa) coexisting at the same locality. Such taxa may coexist in the same habitat or prefer different habitats at the same geographical location.

***See also:* Phylogeography; Speciation; Species**

Sympatric Speciation

***See:* Speciation**

Symplesiomorphy

E O Wiley

doi: 10.1006/rwgn.2001.1262

At any restricted part of the phylogeny of life, Hennig (1966) asserted that there were two kinds of homologous characters: apomorphies and plesiomorphies. In most cases a symplesiomorphy is a homologous character shared by two or more taxa and is hypothesized to have evolved before the common ancestor of these taxa. Symplesiomorphies are also known as shared primitive or plesiotypic characters. Symplesiomorphy is a relative term that is contrasted with an apomorphic homolog at this restricted level. Consider the characters 'having feathers on the body' shared by robins and hawks and 'having epidermal scales on the body' shared by crocodiles and lizards. At this restricted level, Hennig argumentation would lead to the hypothesis that 'having scales on the body' was the plesiomorphic homolog while 'having feathers on the body' was the apomorphic homolog. This is because the sister group of lepidosaurs (including lizards) and archosaurs (crocodilians and birds) has epidermal scales, indicating that the scales were already present in the common ancestor of all four taxa. All symplesiomorphies are synapomorphies at some higher level in the phylogeny that is more inclusive than the restricted level considered by the investigator. So, epidermal scales are synapomorphic at the level of a larger monophyletic group that includes all four or our taxa plus mammals, turtles, fish, etc. In the phylogenetic system symplesiomorphies cannot be used to corroborate monophyletic groups because they have already been used (actually or logically) to corroborate a larger, more inclusive, group. Although symplesiomorphies have been used to diagnose paraphyletic groups by some investigators, this use is suspect because classifications containing paraphyletic groups are logically inconsistent with the phylogeny as reconstructed by synapomorphies (Hull, 1964; Wiley, 1981). In other words, grouping by symplesiomorphy represents the use of a homolog to group at an inappropriate level of the phylogeny.

References

Hennig W (1966) *Phylogenetic Systematics*. Urbana, IL: University of Illinois Press.

Hull DL (1964) Consistency and monophyly. *Systematic Zoology* 13: 1–11.

Wiley EO (1981) Convex groups and consistent classifications. *Systematic Botany* 6: 346–358.

See also: **Apomorphy; Phylogeny; Plesiomorphy; Synapomorphy**

Synapomorphy

E O Wiley

doi: 10.1006/rwgn.2001.1263

Henning (1950, 1966) distinguished between three kinds of taxic homologs. A synapomorphy is a homologous character (evolutionary novelty) shared between two or more species or higher taxa whose presence in these taxa diagnoses a monophyletic group (clade). If we consider all of phylogeny (the complete descent pattern of all species, the tree of life), all taxic homologies are either synapomorphies or autapomorphies (evolutionary novelties of a single terminal species). At some restricted level that considers only part of the tree of life, synapomorphies are those homologous characters shared by two or more taxa that have evolved during the time span covered by the phylogenetic tree the investigator is attempting to discover. Other characters are either plesiomorphies (apomorphies that have evolved earlier) or autapomorphies (character states that diagnose single species). Thus, at a restricted part of the phylogeny the term synapomorphy is a relative term used to differentiate character states that are relevant to reconstructing phylogenetic history from characters that are not relevant. Because all shared taxic homologies are ultimately synapomorphies at some level in the phylogeny, the origin of synapomorphies is the same as the origins of the characters themselves (see Homology). Ultimately, synapomorphies arise as mutations that come to be fixed in species lineages as autapomorphies and become synapomorphies when the ancestral species speciates (see Ax, 1987; Haszprunar, 1991).

Synapomorphies diagnose monophyletic groups (clades) precisely because they are the evidence required to hypothesize that the descendant species shared a common ancestral species. In contrast, shared plesiomorphies are not evidence that can be used to diagnose a monophyletic group precisely because they evolved earlier than the origin of the common ancestor hypothesized to be shared by the members of the monophyletic group. Since all species and groups of species from monophyletic groups at some level in the phylogeny, the use of plesiomorphic characters to diagnose any group would represent the use of the same character at least twice, even though it had a singular origin (singular from the taxic viewpoint, arising in a single lineage).

Although the concept of synapomorphy had been used by many investigators before Hennig, the systematic use of synapomorphies as tools for reconstructing phylogenetic trees springs from his systematic application of the principle of grouping by synapomorphy, commonly known as Hennig argumentation or phylogenetic systematics. In its simplest form, Hennig argumentation requires a search for similarities within the group of study (the ingroup), formulation of hypotheses of character (using various criteria), and comparison of the distribution of these characters within groups that are closely related to the group of study (outgroups).

Given an initial hypothesis of homology, if the character state in question is uniformly found within the ingroup but not found in the closest relative (the sister group) or any other closely related groups, then the investigator may deduce that the character in question is a synapomorphy for the group as a whole. For example, tetrapod limbs are found in all major groups of tetrapod vertebrates but is not found in the sister group of tetrapods (certain fishes generally termed rhipidistians) or any other taxon closely related to the tetrapods or their sister group (coelacanths, lungfishes, other groups of rhipidistians, etc.). Thus, we may deduce that the tetrapod limb is a synapomorphy of a monophyletic group, Tetrapoda. This does not mean that all tetrapods must have tetrapod limbs. Certain lizards and snakes may have only embryonic vestiges or limb buds but lack limbs as adults. But, they are members of groups whose ancestors are hypothesized to have limbs. These seeming anomalies are the reason why synapomorphies are diagnostic of groups rather than defining groups.

Hypotheses of relationships within a group are argued in a similar manner. If some members of a group have one character and other members have a different but homologous character, then that character found in the sister group and other outgroups is the synapomorphy that diagnosed a monophyletic group within the group of study. For example, within the plant group composed of mosses and tracheophytes, tracheophytes have an independent sporophyte generation while in mosses the sporophyte is dependent on the gametophyte. The sister group of mosses and tracheophytes are the hornworts and in hornworts the sporophyte is dependent on the gametophyte. Thus, we can conclude that within the monophyletic group of land plants that has xylem and phloem (mosses + tracheophytes), the tracheophytes share the synapomorphy of an independent sporophyte generation. This hypothesis gains additional corroboration when we observe that liverworts (another outgroup, but not part of the sister group) also have saprophytes that are dependent on the gametophyte.

The final arbitrator of synapomorphy and thus homology is the test of congruence of many independent characters corroborating a particular phylogenetic tree (the congruence test: Patterson, 1988) (see Homology). This is because some identical characters are not homologies, but homoplasies. The assumption inherent in the congruence test is that the most parsimonious explanation of character distribution yields the maximum number of hypotheses of homology and the minimum number of ad hoc hypotheses of homoplasy (Farris, 1980).

References

Ax P (1987) *The Phylogenetic System*. New York: John Wiley.

Farris JS (1980) The information content of the phylogenetic system. *Systematic Zoology* 28: 483–519.

Haszprunar G (1991) The types of homology and their significance for evolutionary biology and phylogenetics. *Journal of Evolutionary Biology* 5: 13–24.

Hennig W (1950) *Grundzüge einer Theorie der phylogenetischen Systematik*. Berlin: Deutscher Zentralverlag.

Hennig W (1966) *Phylogenetic Systematics*. Urbana, IL: University of Illinois Press.

Patterson C (1988) Homology in classical and molecular biology. *Molecular Biology and Evolution* 5: 603–625.

***See also:* Apomorphy; Homology; Phylogeny**

Synapsis, Chromosomes

***See:* Chromosome Pairing, Synapsis**

Synapsis in DNA Transactions

M M Cox

doi: 10.1006/rwgn.2001.1264

The word synapsis is derived from the Greek word *sunapsis*, meaning point of contact. The term synapse is used to describe a cell–cell junction that allows a nerve impulse to pass from one nerve cell to another. In the study of DNA metabolism, the word synapse is also used to describe the point where two DNA molecules come together during recombination. Genetic recombination is any process that brings about an exchange of genetic information between two DNA molecules, and it can take several forms. Homologous genetic recombination, transposition, and site-specific recombination are the most common, and each of these is described in more detail elsewhere in this encyclopedia. In each case, at least two DNA molecules or two different segments of the same DNA molecule must be brought together at the point where the genetic exchange is to take place. This key step, which may precede any covalent chemistry, is referred to as synapsis. The synapsis may be either DNA-mediated or protein-mediated.

DNA-Mediated Synapsis

Homologous genetic recombination is a genetic exchange between any two DNA molecules (or segments of the same molecule) with a similar sequence. In principle, it can occur at any site on any DNA molecule. Thus, proteins that bind to a particular sequence on the DNA play no role in the synapsis step in this process. Instead, synapsis involves the alignment of similar sequences in the two DNA molecules, a process that requires direct DNA–DNA interaction. Proteins participate in this process as catalysts. Proteins that facilitate DNA–DNA alignment include the RecA protein in bacteria and its homologs: the Rad51 or Dmc1 proteins in eukaryotes, the Rad1 protein of Archaea, and related proteins produced by some viruses. These proteins (with the possible exception of Dmc1) form helical filaments on single-stranded DNA, with the bases of the DNA displayed in the major groove of the filament. The DNA is then aligned with homologous sequences in a second, duplex DNA, in a process sometimes called the search for homology. Recent studies indicate that the homology search involves base flipping, in which Watson–Crick interactions in the duplex are weakened and individual bases in the duplex are flipped out so that they can pair with the bases in the originally bound single strand. The homology search leading to synapsis is thus mediated by standard Watson–Crick base pairing (**Figure 1**). The sampling is very rapid. Once the correct alignment is found, there is an extensive transfer of one strand of the duplex to its new pairing partner.

Protein-Mediated Synapsis

Site-specific recombination and transposition both generally require the activity of at least one protein that binds to specific DNA sequences. This protein also plays a key role in bringing DNA molecules together in the right orientation for the genetic exchanges catalyzed in these reactions. Synapsis is mediated largely by protein–protein interactions. In

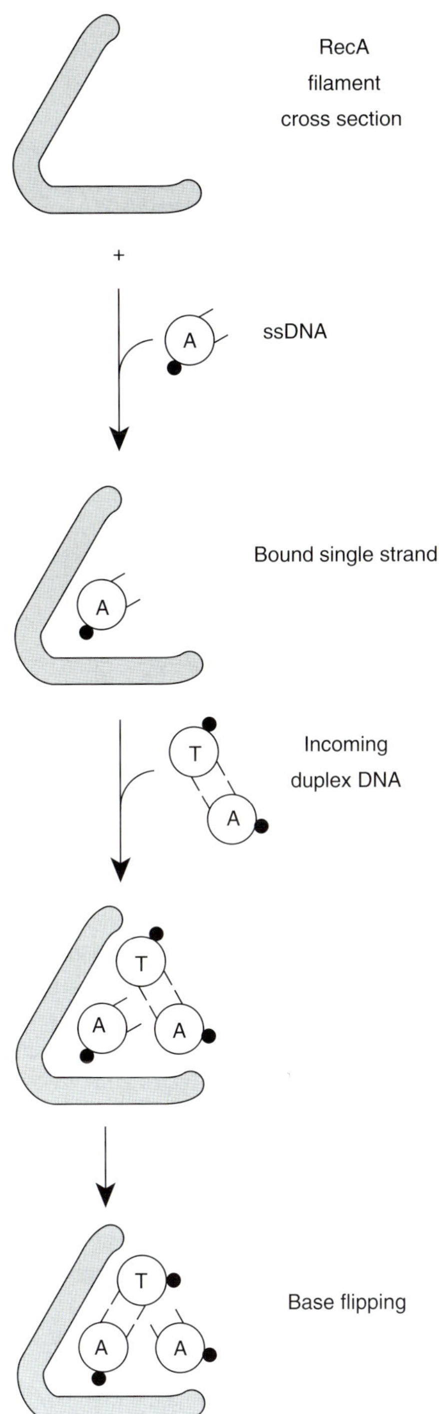

Figure 1 (left) DNA-mediated synapsis in homologous genetic recombination. The reaction is shown in cross section, with a RecA-bound single strand interacting with an incoming duplex DNA. Only one base (A) or base pair (A:T) from each DNA is shown. Base-flipping occurs within the duplex to allow synapsis mediated by a Watson–Crick interaction between the bound single strand and the base flipped out of the duplex. The small filled circles attached to each base are meant to represent the DNA backbone, and the duplex is thus shown approaching the single strand via its major groove. Several recent studies have provided strong evidence for a minor groove approach prior to base flipping, and this aspect of the DNA pairing mechanism is still considered controversial.

many of these systems, very elaborate protein–DNA complexes are formed with the DNA wrapped into the complex in a precise geometry. The architecture of the synaptic complex not only brings two DNA sites together for reaction, but can also determine the outcome of the reaction. In addition, formation of the synaptic complex is often a prerequisite for any covalent chemistry, preventing the occurrence of incomplete DNA cleavage or strand transfer reactions, which could be deleterious to chromosomal DNA.

Site-Specific Recombination

In conservative site-specific recombination, the genetic exchange occurs at specified sequences in the DNA which are recognized and bound by the recombinase enzyme and/or auxiliary proteins. There are two large classes of site-specific recombinases, the integrase class and the resolvase/invertase class.

In the integrase class, the simplest forms of the recombination sites consist of two protein-binding sites flanking a short sequence where the actual DNA recombination occurs. The recombinase proteins bind to a recombination site, then bring two bound recombination sites together in a synapse via protein–protein interactions (**Figure 2**). Recombination is then catalyzed by the recombinase and occurs within the complex.

For the resolvase/invertase class, synapsis not only brings two recombination sites together, but also determines the outcome of the reaction. Synapsis in these systems involves an elaborate complex with multiple proteins, with the DNA wrapped within and around the complex in a precise topology. In addition, the synaptic complex forms efficiently only when the DNA is negatively supercoiled. The synaptic complex acts as a topological filter. In the case of invertases, the architecture of the complex can form only with two recombination sites that are both on the same DNA molecule and inverted in orientation (**Figure 3A**). The result is that the reaction always leads to an inversion of DNA sequences between the recombination sites. Similarly, the architecture of the synaptic complex formed by resolvases allows them to catalyze recombination only between two recombination sites on the same DNA molecule that are in the same orientation, leading to a deletion of the intervening sequences (**Figure 3B**). Each of these systems is thus able to 'sense' the relative orientation of two recombination sites in a DNA molecule even though the sites may be separated by thousands of base pairs.

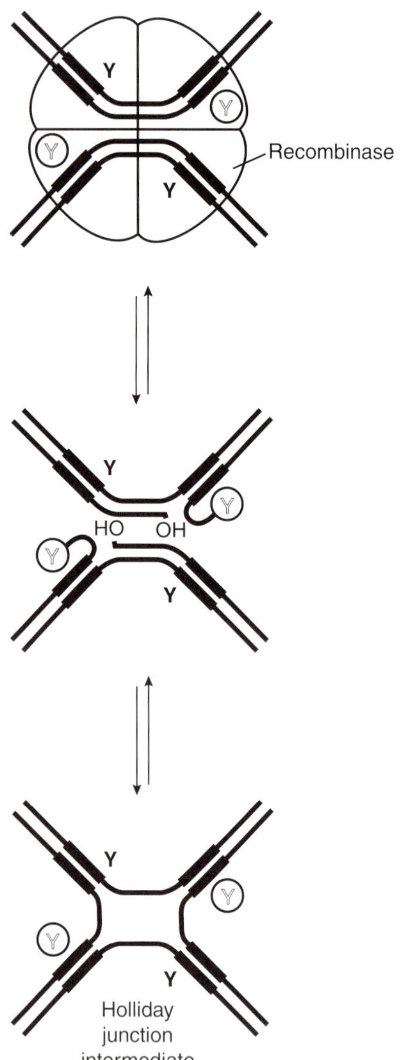

Figure 2 Protein-mediated synapsis in site-specific recombination by integrase class recombinases. Synaptic complexes generally include four recombinase proteins (as shown in the first panel only), but can be considerably more complex (e.g., the complexes formed by the bacteriophage λ integrase). The DNA-binding sites containing the base pairs specifically recognized by the recombinase proteins are indicated with thickened lines. Only the first few steps of the recombination reaction are shown, with the Holliday junction illustrated being a common intermediate in the reactions catalyzed by these enzymes. Integrase class recombinases have an active site tyrosine that forms a covalent intermediate with the DNA. These tyrosine residues are indicated by Y symbols. Only two of the four subunits promote formation and resolution of the covalent intermediates at any given time, and the active ones are circled. Resolution of the Holliday junction into recombinant DNA molecules involves the noncircled Y residues.

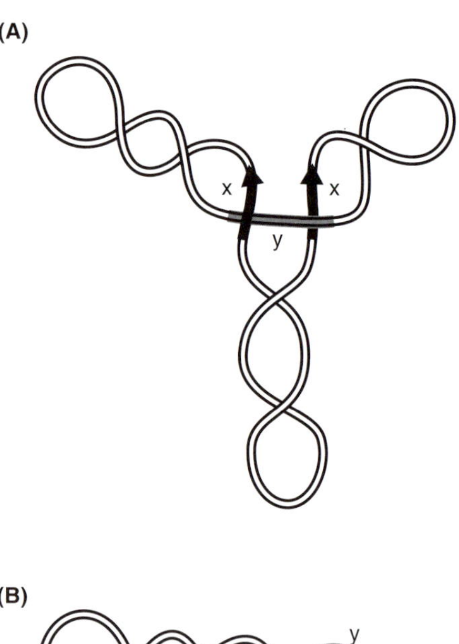

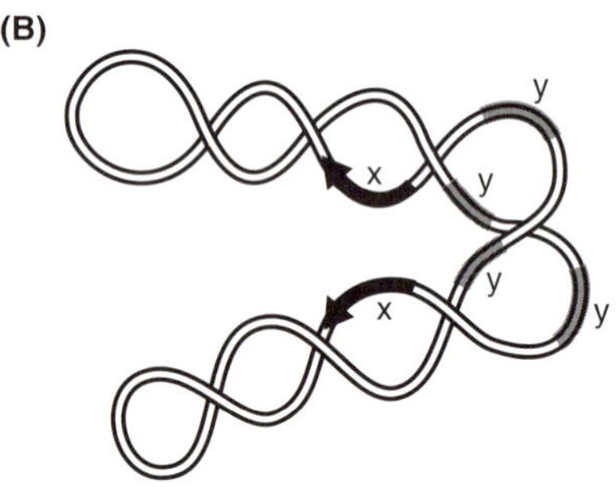

Figure 3 Synaptic complexes as topological filters: The invertase/resolvase class of site-specific recombination systems. Proteins are not shown to keep the figure uncluttered, although multiple recombinase and auxiliary proteins generally bind to the indicated sites and are essential to form and maintain the DNA architecture shown. (A) The likely architecture of the synaptic complex formed by an invertase system. This complex can readily form between two DNA sites only if the sites are in the opposite orientation and the DNA molecule is negatively supercoiled. (B) The likely architecture of a synaptic complex formed by a resolvase system. This complex restricts reaction to recombination sites that have the same orientation within a negatively supercoiled DNA molecule. The structures effectively filter out sites in the incorrect orientation even if they are thousands of base pairs apart. In each complex, the sites labeled x are those where the DNA rearrangement takes place, and these sites are positioned and held together by protein–protein interactions. The orientations of the x sites are indicated with arrows in both panels. The sites labeled y are places where additional recombinase proteins (which do not take place in the chemical steps) or other auxiliary proteins bind to maintain the overall DNA architecture. In the DNA beyond the protein-binding sites, the right-handed twisting represents the natural supertwisting of negatively supercoiled DNA.

Transposition

Transposons are discrete DNA segments that have the capacity to move between different chromosomal locations that may share no homology. Elaborate protein complexes are also used to bring about synapsis between a migrating transposon and a new target site in a host chromosome. The synaptic complex generally includes the target DNA as well as both ends of the transposon, or three sites altogether. The transposase enzyme that catalyzes the DNA splicing steps of the reaction is always a critical, and sometimes the only, protein component in the complex. There are many types of transposons. Transposition can involve a simple cut-and-paste movement from one site to another, or replication of the transposon so that a copy is left behind in the original location. Some transposons can migrate in either a replicative or nonreplicative mode, and the architecture of the synaptic complex can play a role in defining the mode employed. In some complex transposons, the architecture of the synaptic complex also helps ensure that the two transposon ends in the reaction are inverted relative to each other and thus are likely to have come from the same transposon. An example is the synaptic complex formed by the transposing bacteriophage Mu (**Figure 4**). The principle is the same as that employed in the invertase/resolvase systems described above, in which the complex architecture serves as a topological filter preventing the juxtaposition of sites that are not properly oriented on the chromosome.

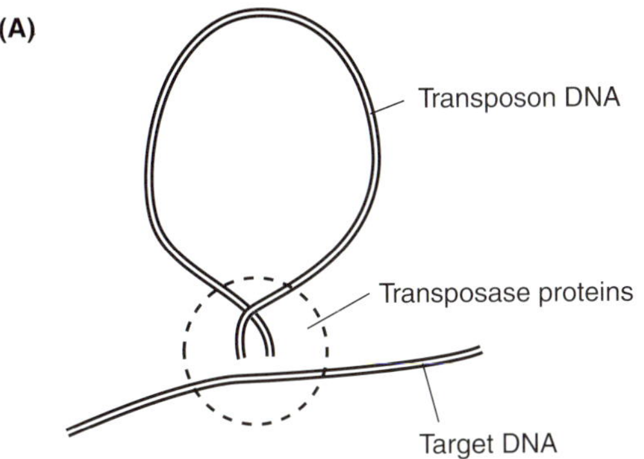

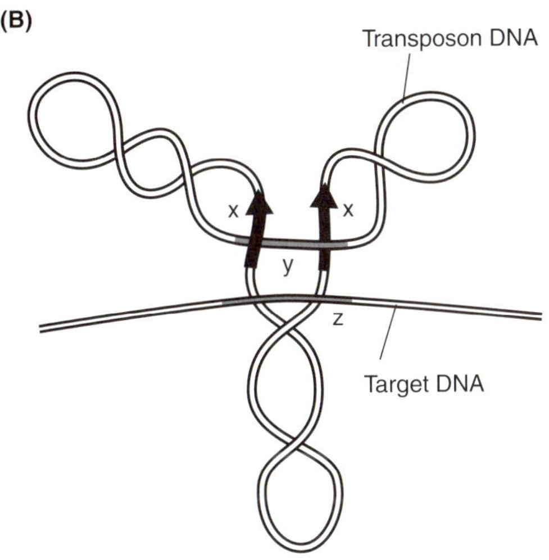

Figure 4 Transposon synaptic complexes. (A) A generic synaptic complex with the two transposon ends juxtaposed over a chromosomal DNA target site. (B) The synaptic complex formed by the transposing bacteriophage Mu. This structure is closely related to the complex formed by the invertase class of site-specific recombinases. There are inverted DNA-binding sites at the ends of the Mu transposon, and the complex helps ensure that the two ends brought together are from the same transposon. The sites labeled x are the sites where the Mu DNA is cleaved. The sites labeled y are places where additional proteins bind to help define the complex architecture. The site labeled z is the target site on a different DNA segment where the Mu transposon will be inserted. As in **Figure 3**, the proteins needed to form and maintain this architecture are not shown.

Further Reading

Aldaz H, Schuster E and Baker TA (1996) The interwoven architecture of the Mu transposase couples DNA synapsis to catalysis. *Cell* 85: 257–269.

Davies DR, Groyshin IY, Reznikoff WS and Rayment I (2000) Three-dimensional structure of the Tn5 synaptic complex transposition intermediate. *Science* 289: 77–85.

Grindley ND (1997) Site-specific recombination: synapsis and strand exchange revealed. *Current Biology* 7: R608–R612.

Guo F, Gopaul DN and van Duyne GD (1997) Structure of Cre recombinase complexed with DNA in a site-specific recombination synape. *Nature* 389: 4–6.

Gupta RC, Folta-Stogniew E, O'Malley S, Takahasi M and Radding CM (1999) Rapid exchange of A:T base pairs is essential for recognition of DNA homology by human rad51 recombination protein. *Molecular Cell* 4: 705–714.

Haber JE (1998) Meiosis: avoiding inappropriate relationships. *Current Biology* 8: R832–R835.

Pena CE, Kahlenberg JM and Hatfull GF (2000) Assembly and activation of site-specific recombination complexes. *Proceedings of the National Academy of Sciences, USA* 97: 7760–7765.

See also: **Genetic Recombination; Holliday Junction; Transposable Elements**

Synaptonemal Complex

P B Moens

doi: 10.1006/rwgn.2001.1265

The synaptonemal complex (SC) is two parallel aligned proteinaceous chromosome cores of a pair of homologous chromosomes at prophase of meiosis. The cores are in the order of 60 to 100 nm wide, lie about 100 nm apart, and are interconnected by transverse filaments. The chromatin is attached to the cores in a series of loops. Superficially, the structure of the SC is similar over a wide range of organisms from fungi, to protists, to plants, and to invertebrate and vertebrate animals. Although present at meiosis of most sexually reproducing organisms, it is notably absent in some species such as the fungus *Ustilago maydis*, the protist *Tetrahymena*,

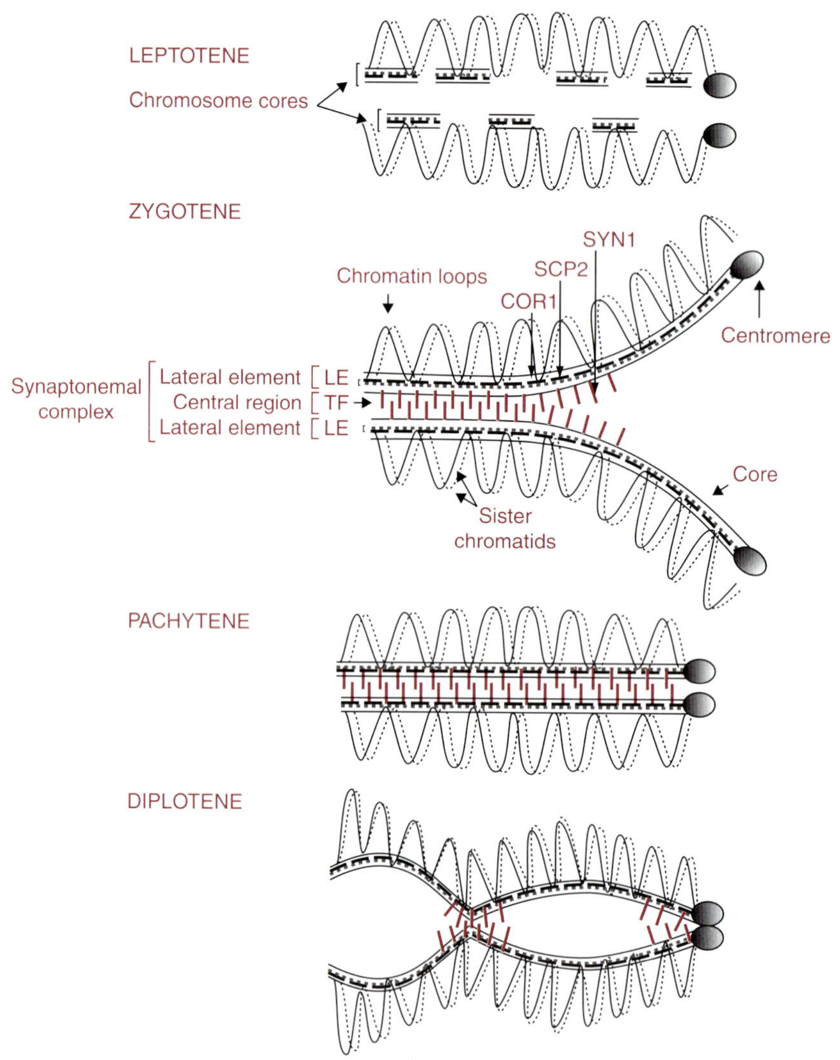

Figure 1 A diagrammatic representation of the development and structure of the synaptonemal complex at successive stages of meiotic prophase. The two chromatids of each homolog (black, wavy lines) are associated with a chromosome core which begins to form during leptotene. By zygotene, these cores become parallel aligned and commence 'zipping up' by means of the interaction of transverse filaments with the core proteins. The entire structure is termed the synaptonemal complex, which is fully formed by pachytene. In diplotene, the transverse filaments detach and the cores separate for most of their length but remain together at the chiasmata, the points of genetic recombination. The names of the stages and structures indicated in the diagram. The positions of three of the proteins that form the complex in mice, COR1 (long, dashed lines), SCP2 (dotted gray lines), and SYN1 (gray lines), are shown at the various stages of meiotic prophase.

and males of some species of fruit flies. It is only partially developed in the fission yeast *Schizosaccharomyces pombe*.

Historical Background

The SC was first reported in 1957 in a number of mammals and in an invertebrate species. Originally, the structure could be visualized only with the electron microscope and the date of the reports coincides with the perfection and commercialization of the electron microscope. It was intuitively assumed that this structure was of fundamental importance to the structure and behavior of chromosomes at meiosis, and this was soon after supported by the observation that male dipteran insects having genetic recombination possessed SCs while those that had no recombination lacked this structure. However, after a number of cases were discovered to have genetic recombination without SCs, their importance was downplayed in the literature. This trend was reversed in the 1970s with the discovery that recombination-associated nodules (RNs) are uniquely located at the SCs, implying that the SCs play a role in recombination. The SC regained full recognition as a main player in the meiotic process with the discoveries in the 1990s that recombinationally active proteins are located at the SCs and that they interact with SC components.

Development of the Synaptonemal Complex

The diagrams of **Figure 1** illustrate the development of the SC. At the leptotene stage of meiotic prophase, the chromosomes are unpaired and small segments of chromosome core appear in the nucleus as shown in **Figure 2**. The early unpaired core segments in the figure are visualized by immunofluorescent microscopy using antibodies against one of the core proteins and a secondary antibody that is conjugated with a green fluorochrome.

The short segments become joined into longer stretches of cores and simultaneously the cores of homologous chromosomes start to associate with each other, thereby forming the first SC segments at the zygotene stage of meiosis. The synapsis of the cores is related to the formation of transverse filaments between the cores. Proteins of the transverse filament are reacted with their respective antibodies linked to a red fluorescent secondary antibody. Where the cores have started to synapse in the zygotene nucleus, yellow segments appear due to the overlapping of the red and green fluorochromes. Once synapsis is complete, the antibody against the transverse filaments is present along the entire SCs as

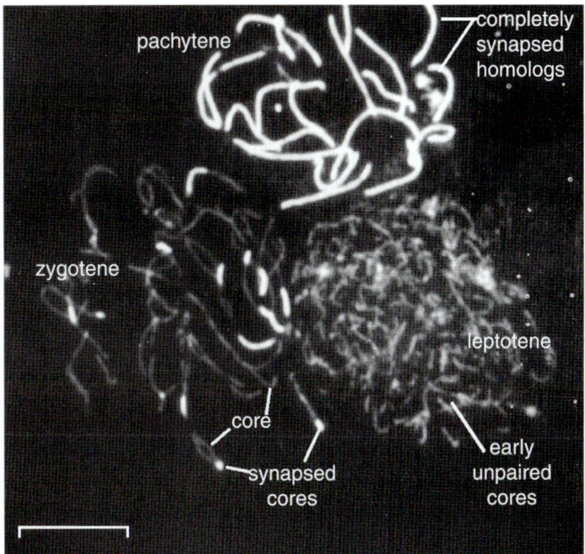

Figure 2 (See Plate 39) The meiotic chromosome cores/SCs of a mouse are visualized here by indirect immunofluorescence with antibodies against a core protein and against a synaptic protein which produces a yellow color when the two proteins are present at the same site. Three meiotic prophase nuclei in successive stages of development (leptotene, zygotene and pachytene) are indicated in the figure. Scale bar = 10 μm.

shown in **Figure 1** and in the pachytene nucleus of **Figure 2**.

During subsequent development, the chromosome cores start to move apart when the transverse filaments are removed. In the diplotene stage depicted in **Figures 1** and **3**, the separated cores are green fluorescent and the last remaining points of contact that have transverse filaments are yellow-colored. Some filament material still adheres to the separated cores at a few points. Where two cores show a sharp convergence, it is suspected that this is the site of a reciprocal recombination event which gives rise to a chiasma (**Figure 3**, *ch*).

Structural Components of the Synaptonemal Complex

Two of the structural components of the meiotic chromosome cores in mammals are a 30 kDa and a 190 kDa protein (**Figure 1**, COR1 and SCP2) which can form multimers with themselves and with each other resulting in long and relatively strong cores. The transverse filaments that lie between two cores during SC formation contain a 125 kDa protein (**Figure 1**, SYN1). This protein can be thought of as the interdigitating elements of a zipper that fastens the cores together in close (about 100 nm) parallel proximity.

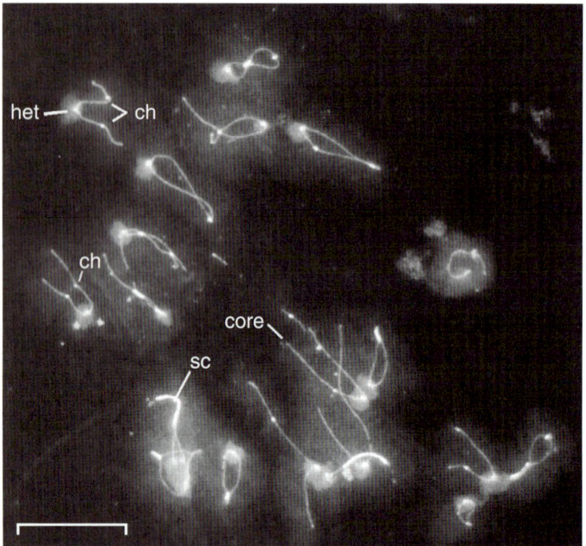

Figure 3 (See Plate 40) Near the end of prophase in this mouse nucleus, the homologous chromosomes and their cores separate wherever the synaptic protein is no longer present. The centromeric chromatin (*het*) is stained with DAPI. Some of the locations that are suspected of having a chiasma are designated *ch*. Scale bar = 10 μm.

These structural elements were identified by the use of antibodies raised against SCs. In insects and plants, the SCs are a dominant feature of the meiotic prophase nucleus, but their protein components have not yet been determined.

In the yeast *Saccharomyces cerevisiae*, an entirely different methodology has resulted in the identification of SC components. The products of a number of genes that affected the meiotic process were found to affect the formation of SCs. The gene products of *HOP1* and *RED1* are required for normal core formation, and the *ZIP1* product is present between the cores, where it functions in chromosome synapsis. Surprisingly, none of the SC components of mammals resembles those of yeast SC components, even though they are structurally and functionally similar. No decision can yet be made whether this can be attributed to evolutionary divergence or to polyphyletic origins.

Proteins Associated with the Synaptonemal Complex

To induce a double-strand break in the DNA for the formation of a joint molecule as a preliminary to recombination in *S. cerevisiae*, it has been estimated that approximately ten different proteins are clustered at the site of the break. Some of these proteins reside in distinct 100 nm foci on the SCs of yeast, plants, and animals. Antibodies against recombinationally active proteins, RAD51 and DMC1, detect approximately 300 foci at the cores/SCs of early mouse meiotic prophase nuclei. Other proteins may be a part of the complex but could be present in too low abundance to be detected with this technique. Proteins involved in later recombination functions, such as BLM and MLH1, are present in SC-associated foci at later stages of meiotic prophase. In addition to these recombinationally active proteins, there are foci at the SC that contain the checkpoint proteins ATR and hRAD1. The functions of these proteins in the regulation of the cell cycle following DNA damage have been reported for somatic cells but their functions in meiosis are not well understood.

Chromatin Attachment to the Synaptonemal Complex

The diagrammatic chromatin loops in **Figure 1** and the chromatin haloes in **Figure 3** illustrate the organization of the chromatin relative to the SC. Particularly evident is the intensely blue-stained centromeric heterochromatic chromatin in **Figure 3**. The average length of the loops varies among species: short (0.5 μm) for yeast, about 5 μm for various mammals, and long (20 μm) for some insects. The implication is that there are specific mechanisms that regulate loop size and attachment to the SC. Foreign DNA from bacteriophage lambda that is inserted into mouse DNA fails to attach to the core/SC, indicating that there is some recognition mechanism.

Further Reading

Moens PB, Pearlman RE, Heng HHQ and Traut W (1998) Chromosome cores and chromatin at meiotic prophase. *Current Topics in Developmental Biology* 37: 241–262.

Roeder SG (1997) Meiotic chromosomes: it takes two to tango. *Genes and Development* 11: 2600–2621.

See also: **Chromatin; Meiosis**

Syndactyly

M A Ferguson-Smith

doi: 10.1006/rwgn.2001.1266

Syndactyly refers to the complete or partial fusion of two or more fingers or toes. Severe forms of syndactyly involve bony fusion of digits, lesser forms include webbed fingers and toes. The condition is frequently familial.

Syngenic

L Silver

doi: 10.1006/rwgn.2001.1267

'Syngenic' means literally 'of the same genotype.' The term is used most frequently by immunologists to describe interactions between cells from the same inbred strain.

***See also:* Inbred Strain**

Synovial Sarcoma

C S Cooper

doi: 10.1006/rwgn.2001.1623

Synovial sarcoma is an aggressive soft tissue sarcoma that arises most commonly in young adults and adolescence, with around 200 new cases in the United Kingdom and 800 in the United States each year. Histologically biphasic and monophasic subtypes can be distinguished. Both these subtypes contain the diagnostic translocation t(X;18)(p11.2;q11.2) that results in the fusion of the *SYT* gene on chromosome 18 with either the *SSX1* or *SSX2* gene on chromosome X.

***See also:* Sarcomas; Translocation**

Synteny (Syntenic Genes)

L Silver

doi: 10.1006/rwgn.2001.1268

Synteny describes two or more genes or loci that have been mapped to the same linkage group. Conserved synteny refers to the situation where two linked loci in one species (such as the mouse) have homologs that are also linked in another species (such as humans).

***See also:* Linkage Map**

Systematics

See: Taxonomy, Numerical

Systemic Acquired Resistance (SAR)

J P Métraux

doi: 10.1006/rwgn.2001.1683

Plants defend themselves against pathogens by constitutive barriers and a number of inducible defense mechanisms deployed after contact with a pathogen. Often, a first infection with a fungal, bacterial, or viral pathogen induces resistance toward subsequent infections. To a certain extent, this plant immunization is analogous to immunization in animals. Induced resistance may be expressed locally at the site of infection as well as systemically, in uninfected parts of the plant. This phenomenon was termed 'systemic acquired resistance' (SAR) to emphasize the power of the plants to acquire resistance after an initial infection even in tissues remotely located from the first infection. This type of resistance is expressed by many plants against a wide variety of pathogens, including organisms unrelated to the inducing pathogen (**Table 1**). SAR is explained by the production of a signal released from the infected leaf and translocated to other parts of the plant, where it induces defense reactions. Nonpathogenic root-colonizing bacteria were also found to induce SAR in leaves. The biochemical nature of the changes induced in infected plants was intensely studied and led to the discovery of a number of proteins termed 'pathogenesis-related' (PR) proteins. It was also observed that the simple phenolic compound salicylic acid (SA) can induce PRs in tobacco and protect the plant against tobacco mosaic virus (TMV). Later, SA was shown to be produced by plants locally, at the site of infection, but also in the phloem sap as well as in uninfected systemic leaves, and SA was proposed as a possible endogenous signal for SAR. These observations opened the way for molecular investigations on induced resistance. This field evolved considerably when SAR was found to operate in the genetically tractable system *Arabidopsis thaliana*.

Reactions after a First Infection

Generally, the success of the induced defense mechanisms depends on the outcome of the race between the invading pathogen and the reactions of the plant. In compatible interactions, the virulent pathogen is often recognized too late and the plant will be infected. In the case of incompatible interactions, plants rapidly recognize the avirulent pathogen and the resistance mechanisms are efficiently blocking the invader. A

Table 1 Inducing agents and disease agents of plants

Plant	Inducer organism	Systemic protection against
Alfalfa	*Colletotrichum lindemuthianum*	*Colletotrichum lindemuthianum*
Arabidopsis thaliana	Turnip crinkle virus *Pseudomonas syringae* *Fusarium oxysporum* *Pseudomonas fluorescens WCS417*	*Pseudomonas syringae* Turnip crinkle virus *Pseudomonas syringae* *Erysiphe cichoracearum* *Botrytis cinerea* *Alternaria brassicicola*
Asparagus bean	Tobacco necrosis virus Tobacco rattle virus	*Fusarium oxysporum* f.sp. *raphani* *Pseudomonas syringae* pv. *tomato*
Barley	*Erysiphe graminis* f. sp. *hordei*	Tobacco necrosis virus
Bean	*Collectotrichum lindemuthianum* *Collectotrichum lagenarium* *Uromyces phaseoli* *Pseudomonas fluorescens*	*Erysiphe graminis* f. sp. *hordei* *Colletotrichum lindemuthianum* Tobacco necrosis virus *Pseudomonas syringae* pv. *phaseolicola*
Carnation	*Pseudomonas* sp.	*Fusarium*
Cucumber	*Colletotrichum lagenarium* *Pseudoperonospora cubensis* *Pseudomonas lachrymans* Tobacco necrosis virus *Pseudomonas putida* *Serratia marcescens*	*Colletotrichum lagenarium* *Cladosporium cucumerinum* *Fusarium oxysporum* *Pseudomonas lachrymans* *Sphaerotheca fuliginea* Tobacco necrosis virus *Colletotrichum orbiculare*
Muskmelon	*Colletotrichum lagenarium*	*Colletotrichum lagenarium*
Oilseed rape	*Leptosphaeria maculans*	*Leptosphaeria maculans*
Pearl millet	*Sclerospora graminicola*	*Sclerospora graminicola*
Potato	*Phytophthora infestans* *Phytophthora cryptogea*	*Phytophthora infestans*
Radish	*Pseudomonas fluorescens*	*Fusarium oxysporum* f. sp. *raphani* *Pseudomonas syringae* pv. *tomato* *Alternaria brassicola*
Red clover	Bean yellow mosaic virus	*Erysiphe polygoni*
Rice	*Pseudomonas syringae*	*Magnaporthe grisea*
Sicklepod	*Alternaria crassiae*	*Alternaria crassiae*
Soybean	*Colletotrichum lagenarium* *Colletotrichum truncatum*	*Colletotrichum truncatum*
Stylosanthes guianensis	*Colletotrichum gloeosporioides*	*Colletotrichum gloeosporioides*
Tobacco	Tobacco mosaic virus Tobacco necrosis virus *Thielaviopsis basicola* *Peronospora tabacina* *Pseudomonas syringae* *Pseudomonas fluorescens* CHAO	*Thielaviopsis basicola* *Phytophthora parasitica* *Peronospora tabacina* *Pseudomonas syringae* *Phytophthora parasitica* *Pseudomonas tabaci* Tobacco mosaic virus Tobacco necrosis virus
Tomato	*Phytophthora infestans*	*Phytophthora infestans*
Watermelon	*Fusarium oxysporum*	*Colletotrichum lagenarium*

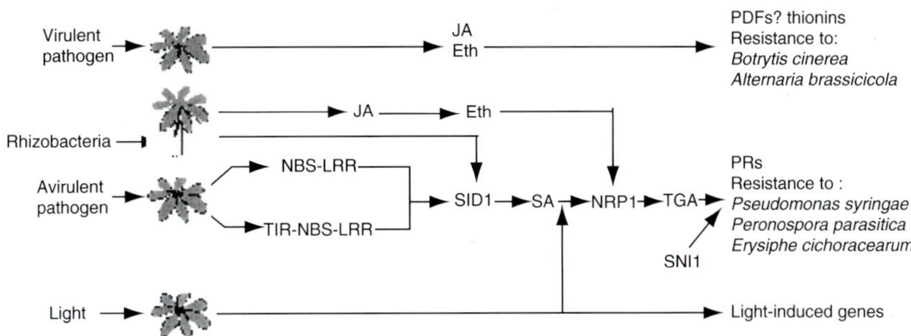

Figure 1 Schematic, simplified diagram of the signal transduction network operating in SAR. In this diagram arrows represent a flow of the information and proteins are ordered with respect to the sequence from incoming signals (left side) to the responses (right side).

Abbreviations: Eth: ethylene; JA: jasmomic acid; NBS-LRR: nucleotide-binding site leucine-rich-repeat protein; NPR: non-expresser of PR genes; NDR: non-race-specific disease resistance; PDFs: plant defensins; PRs: pathogenesis-related proteins; SID: salicylic acid induction deficient; SNI: suppressor of NPR1 inducible; TGA: basic leucine zipper (btP) transcription factor; TIR-NBS-LRR; toll-interleukin-receptor nucleotide-binding site leucine-rich-repeat protein.

first infection leads to many changes, some of which may eventually lead to the deployment of various barriers that can block the invading pathogen. These barriers include modifications of the cell wall such as deposition of lignin. The deposition of this phenolic polymer hinders pathogens in any of several ways: mechanical reinforcement of the cell wall, formation of a hydrophobic layer preventing diffusion of water and solutes, protection of the other cell wall components from the actions of hydrolytic enzymes of the pathogens. Antimicrobial secondary metabolites are produced *de novo* at the site of infection. These phytoalexins often have a broad unspecific activity and represent a toxic barrier to invaders. Activation of a programmed cell death similar to apoptosis in animals also prevents the spread of invaders. This hypersensitive reaction (HR) is genetically determined. It is mostly triggered when a product of a pathogen gene (avirulence gene) is recognized by the product of a host gene (resistance gene) during a so-called gene-for-gene interaction. The synthesis of novel proteins after pathogen attack is perhaps the most intensely documented reaction. These host-encoded, PR proteins are induced locally and systemically after pathogen infection. They occur in most plants where they have been looked for and have various biochemical activities (**Table 2**). Some of the PRs were found to be enzymes such as β-1,3-glucanases, chitinases, or proteinases capable of hydrolyzing the cell wall of invading fungal pathogens, while the function of others, for example PR-1, still remains unknown to date. Combinations of PRs (for example glucanase and chitinase) are likely to be most efficient, and different types of PRs are directed at different types of pathogens.

The sequence of reactions taking place in a leaf undergoing a first attack by a pathogen have been extensively studied using various mutants of *Arabidopsis* spp. After initial recognition of the pathogen by the plant, a cascade of early events is induced that includes ion fluxes, phosphorylation events, and generation of nitric oxide and active oxygen species. SA acts as a secondary signal molecule and is required for increased expression of resistance and various defense-related proteins such as the PRs. Depending on the inducing microorganism, the signal transduction pathway takes a different course according to the nature of the initial interaction (virulent versus avirulent pathogen, rhizobacteria). A further level of complexity exists among the incompatible interactions where the pathway shows a dependency on either one of two classes of leucine-rich-repeats proteins (LRRs; **Figure 1**). Resistance against a given pathogen might be activated via different signal transduction pathways. For example, infection with leaf pathogens which induce resistance to *Pseudomonas syringae* depends on a pathway involving SA, while rhizobacteria-induced SAR act via the plant hormone ethylene and jasmonic acid (**Figure 1**). The complexity of these signaling pathways is further illustrated by the occurrence of cross-talk or interference between pathways. For instance, both the induction of PR-1 and the resistance to *P. syringae* show a strong dependency on the light signal transduction pathway (**Figure 1**). Another example of cross-talk is given by the signaling pathway used by plants after rhizobacteria infection. In this case, the NPR1 protein is recruited, which otherwise is parts of the SA pathway.

Given the central role of SA in pathogen-induced signaling for induced resistance, studies have been

Table 2 The families of pathogenesis-related (PR) proteins

Family	Property
PR-1	Unknown
PR-2	β-1,3-Glucanase
PR-3	Chitinase (type I, II, IV, V, VI, VII)
PR-4	Chitinase (type I, II)
PR-5	Thaumatin-like
PR-6	Proteinase inhibitor
PR-7	Endoproteinase
PR-8	Chitinase (type III)
PR-9	Peroxidase
PR-10	RNase-like
PR-11	Chitinase (type I)
PR-12	Defensin
PR-13	Thionin
PR-14	Lipid-transfer protein

directed at understanding the regulation of its production and its molecular mode of action. SA is produced from phenylalanine via coumaric and benzoic acid, but the exact precursor of SA is still unknown, and the enzymes involved in SA biosynthesis have not yet been identified or isolated. More work is needed to understand the regulation of SA and its localization after pathogen attack both locally and systemically. Mutants impaired in SA biosynthesis might be a valuable alternative with which to discover enzymes involved in SA biosynthesis. The *sid1* and *sid2* mutants impaired in SA accumulation after pathogen attack represent interesting candidates, and the function of these genes is actively been pursued.

The mode of action of SA was investigated by searching for SA-binding proteins (SAPs). These SAPs include catalase and ascorbate peroxidase. The binding of SA to such H_2O_2-scavenging enzymes was hypothesized to lead to the formation of a phenolic radical involved in lipid peroxidation. Lipid peroxidation products can activate defense gene expression, providing a link between SA and defense. It remains to be shown that sufficient lipid peroxides are formed by such phenolic radicals in the right time-frame for the defense response to take place. Another SAP of unknown biochemical function shows a higher affinity for SA or related functional analogs such as 2, 6-dichloroisonicotinic acid (INA) or benzothiadiazole (BTH; BION®) than catalase, but its biological relevance remains to be determined.

Responses induced by SA include transcriptional activation of genes. For instance, a SA-inducible protein kinase (SIPK) belonging to the MAP kinase family has been identified in tobacco. This SIPK is also induced upon infection and is likely to be part of the chain of phosphorylation events taking place downstream of SA. A number of studies have focused on the upstream regulatory sequences of the *PR-1* gene, one of the culminating responses in SAR. One indispensable regulatory element for SA-induced *PR-1* gene expression is a consensus sequence (TGACG) for recognition with transcription factors of the bZIP protein family. TGA proteins belonging to the plant bZIP transcription factors were shown to bind to the TGACG box in the *PR-1* promoter of *A. thaliana*. TGAs were also shown to interact physically with NPR1, a 65-kDa ankyrin repeat-containing protein with homology to iκBα. A further level of regulation is provided by SNI1, which represses *PR* gene expression, presumably by direct binding to a specific DNA sequence or via a transcription factor. Regulation of *PR* gene expression also involves phosphorylated WRKY DNA-binding factors (**Figure 1**).

Systemic Signal for SAR

The importance of SA in SAR was provided by various correlative studies, but most compellingly by transgenic plants overexpressing a bacterial salicylate hydroxylase gene (the *NahG* gene). When expressed in the plant, this enzyme degrades SA effectively. Transgenic plants carrying the *NahG* gene are unable to display SAR. The role of SA as a systemic signal was critically assessed using leaf detachment assays or grafting experiments between NahG-expressors or other plants with altered SA levels and wild-type plants. These experiments all show that SA is necessary for the induction of pathogen-induced SAR but that a signal other than SA can be translocated to the upper leaves and induce resistance. However, SA produced in a lower leaf during infection can be transported to the upper leaf in sufficient amounts before appearance of systemic resistance in that leaf. In conclusion, SA as well as another putative systemic signal might be involved in systemic signaling during SAR.

Reaction in Systemic Tissue

The systemic responses can be clearly separated from the reactions taking place in the infected parts of the plant. For instance, the upper leaves of plants inoculated on the lower leaf have elevated levels of PRs. In this case, the systemic signal has triggered defense-related reactions before contact with the challenging pathogen. In contrast, other reactions such as changes in cell wall lignification were only detected after challenge infection of the upper leaf but with faster induction kinetics. Thus, the systemic signal has conditioned the tissue to respond faster. Evidence for conditioning has been provided using cultured plant cells. Defense reactions can be induced in cultured cells by treatment with elicitors, e.g., molecules derived directly from pathogens or released after the plant–pathogen interaction. Pretreatment with SA or functionally related inducers prior to exposure to an elicitor leads to potentiation of the elicitor-induced expression of defense-related phenylpropanoid genes such as phenylalanine ammonia-lyase (PAL) or 4-coumarate: CoA ligase. In the same tissue, the expression of other genes not directly related to defense such as mannitol dehydrogenase or anionic peroxidase is induced directly by SA. SA as well as functional analogs such as INA or BTH appear to have a dual function by inducing directly the expression or potentiating the expression of elicitor-induced genes. Similar observations were also made in cucumber hypocotyls and in whole *A. thaliana* plants and they are similar to the situation in a noninfected upper leaf of a plant infected on the lower leaf. Future experiments should now be designed to describe closely how the systemic signal induces conditioning in the induced leaves.

Prospects

Much progress has been achieved in the study of SAR in the last few years. An increasing number of new elements in the signal transduction pathway are being discovered and their number will undoubtedly increase with the advent of large-scale investigations of gene expression. Clearly, to understand how proteins encoded by the novel gene products operate and interact, interest is expected to move swiftly to the biochemical level for a functional understanding of SAR. Among the fascinating questions on SAR are those concerning the systemic signal of SAR, its regulation, and mode of action. SA has been implied in this process initially, and its role as a key signal in pathogen-induced SAR is well documented. Its function as a translocated systemic signal remains a matter of debate. In the near future, more will be learned on the regulation and localization of SA synthesis and its mode of action. The signal transduction involved in the regulation of the SAR response turns out to be far from a linear chain of events, and several pathways interact, partially leading to sets of responses targeted to specific pathogens.

Compounds such as SA represent interesting model structures for the development of non-antibiotic crop protectants, which trigger the natural potential for resistance in various plants. These chemicals could be considered as immunostimulants by analogy to certain drugs used in humans. A good example is BION®, a compound recently released on the market.

Further Reading

Ellis JPD and Pryor T (2000) Structure, function and evolution of plant disease resistance genes. *Current Opinions in Plant Science* 3: 278–284.

Genoud T and Metraux JP (1999) Crosstalk in plant cell signaling: structure and function of the genetic network. *Trends in Plant Science* 4: 503–507.

Glazebrook J (1999) Genes controlling expression of defense responses in *Arabidopsis*. *Current Opinion in Plant Science* 4: 280–286.

Grant M and Mansfield J (1999) Early events in host–pathogen interactions. *Current Opinion in Plant Biology* 2: 312–319.

Hammerschmidt R (1999) Phytoalexins: what have we learned after 60 years? *Annual Review of Phytopathology* 37: 285–306.

Malock K, Levine A, Eulgam T *et al.* (2000) The transcriptome of *Arabidopsis* during systematic acquired resistance. *Nature Genetics* 403–410.

Pieterse CMJ and van Loon LC (1999) Salicylic acid-independent plant defence pathways. *Trends in Plant Science* 4: 52–58.

Van Loon LC and Van Strien EA (1999) The families of pathogenesis-related proteins, their activities, and comparative analysis of PR-1 type proteins. *Physiological and Molecular Plant Pathology* 55: 85–97.

See also: *Arabidopsis thaliana*: The Premier Model Plant; *Rhizobium*; Signal Transduction

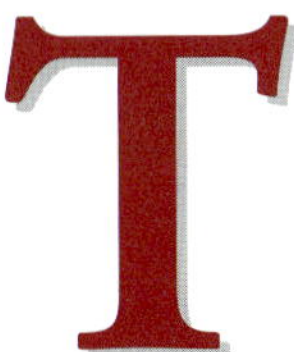

T Cell Receptor Gene Family

L Silver

doi: 10.1006/rwgn.2001.1275

The T cell receptor (Tcr) gene family is a member of the immunoglobulin gene superfamily. The Tcr gene family encodes polypeptides that are placed on the surface of the immune cells (called T cells) that provide the body's cellular immune response to foreign viruses and bacteria.

***See also:* Immunoglobulin Gene Superfamily**

t Haplotype

K Ardlie

doi: 10.1006/rwgn.2001.1287

History and Genetics

The *t* haplotypes of the house mouse are one of the best-known mammalian examples of meiotic drive. *t* haplotypes are a selfish variant of chromosome 17 that have evolved the ability to enhance their own transmission at the expense of the wild-type homolog. They were first discovered in 1927 and were originally referred to as '*t* mutations' or '*t* alleles' because they appeared to be mutant alleles of the *T* (*Brachyury*) locus on chromosome 17: *T*/+ animals have short tails while *T*/*t* animals are tailless (both +/+ and +/*t* mice have tails of normal length). These *t* mutations were thought to act in several aberrant pleiotropic ways, as they appeared to have effects on embryonic development, genetic recombination, male fertility, and Mendelian segregation. Subsequently, the combined use of formal genetic analysis and molecular markers revealed that they were instead an extended genetic entity with altered chromosomal structure and multiple independent loci responsible for the various phenotypes they express. This extended genetic entity became known as a *t* haplotype (also called the *t* complex).

t haplotypes can be found worldwide in natural populations of all subspecies of the house mouse, *Mus musculus*. Current knowledge of *t* haplotypes indicates that they are genetically complex, and comprise a 20 cM (30–40 Mbp) region of the proximal third of chromosome 17 (see **Figure 1**). They are defined relative to the wild-type homolog by a series of four major, nonoverlapping inversions. These act to suppress recombination across the entire region in +/*t* heterozygotes such that the integrity of the *t* haplotype is maintained, and they are typically transmitted as a single genetic entity. Within this region there are numerous independent loci which produce the characteristic effect on tail length, cause embryonic lethality and male sterility, and mediate the meiotic drive phenotype.

Transmission Ratio Distortion

The most characteristic feature of *t* haplotypes is their capacity to distort Mendelian segregation in their favor. This is known as transmission ratio distortion (TRD). Segregation is normal in +/*t* females who produce offspring in the expected 50:50 Mendelian ratios. In contrast, heterozygous +/*t* males will typically transmit the *t* haplotype to more than 90% of their offspring. Early studies demonstrated that meiosis is normal in these males, and that segregation distortion occurs postmeiotically. Equal numbers of + sperm and *t* sperm are produced, and thus the TRD phenotype is a consequence of the production of functionally inactivated wild-type sperm.

Although the mechanism of TRD has not been fully resolved, the molecular basis of segregation distortion is finally beginning to emerge. Multiple independent loci, the *t* complex distorters (*Tcd*) and a *t* complex responder (*Tcr*), interact to cause transmission bias in favor of the *t* haplotype (**Figure 1**). At least three, and perhaps as many as five, distorter loci have been identified: *Tcd-1, Tcd-2, Tcd-3*, and *Tcd-4* and *Tcd-5*. These vary in their individual strength and effectiveness, and show a cumulative effect on TRD such that the extent of distortion is dependent on the

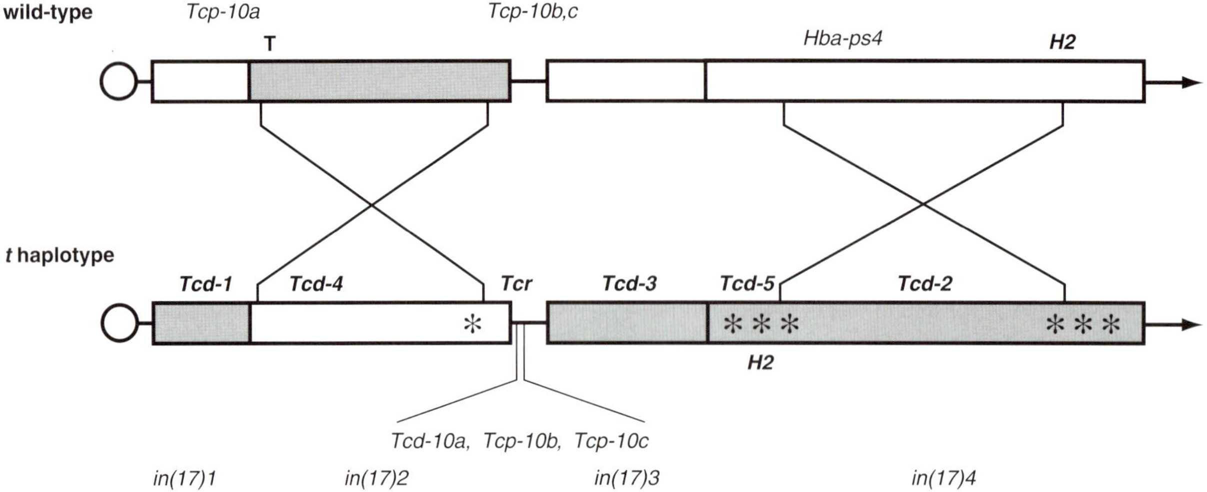

Figure 1 *t* haplotype and wild-type forms of mouse chromosome 17. Shaded boxes represent the four *t*-associated inversions, *in(17)1* through *in(17)4*. The second inversion, *in(17)2*, is believed to have arisen on the wild-type lineage, where it breaks up the *Tcp-10* gene family, and not within the *t* haplotype lineage. Several distorter loci, *Tcd-1* through *Tcd-3* (and possibly *Tcd-4* and *Tcd-5*), interact with the responder locus, *Tcr*, to cause the high transmission of the *t* haplotype. The locations of the *t*-associated lethal mutations are indicated by stars. The *H2* (or mouse major histocompatibility complex) is contained within the fourth inversion, and is displaced proximally relative to its position on the wild-type homolog.

absolute levels of expression of the individual *Tcd* genes. Although multiple loci play a role in the expression of the phenotype, only one (*Tcr*) is responsible for determining which of the two homologs is transmitted at high ratios.

Physiological and genetic studies of sperm have demonstrated that motility defects are observed in the wild-type sperm from $+/t$ males, and a model has been suggested in which the TRD phenotype might result from the expression of defective axonemal dyneins (the microtubule associated ATPases that control flagellar motility, providing locomotive force), producing impaired flagellar function of the + sperm. A gene encoding an axonemal dynein heavy chain has recently been identified as a molecular candidate for the most powerful of the distorter elements, *Tcd-2*. Recent studies of the *Tcd-1* region indicated that the distorter activity of *Tcd-1* is independent of any sterility effects and might be due to more than one independent distorter in the region. A candidate gene for *Tcr*, *Tcp-10b*, was one of the first candidates to be isolated and cloned; however targeted mutagenesis of *Tcp-10b* in *t* haplotypes failed to eliminate TRD.

Evolution

The accumulated evidence from studies addressing the origin of *t* haplotypes suggests that they evolved through the stepwise assembly of the four inversions. Phylogenetic analyses of the DNA sequences of two *t*-associated loci (an intron of the *Tcp-1* gene in the second inversion and the *Hba-ps4* pseudogene in the fourth inversion) suggest that inversion 2 is much older (~3 million years) than inversion 4 (~1.5 million years), and is likely to have been the first inversion to have arisen. Because the second inversion is so old, and can suppress recombination over a region adjacent to both the responder (*Tcr*) and distorter (*Tcd*) loci, it is assumed that this inversion was the primary event leading to the spread of *t* haplotypes. The remaining inversions probably accumulated subsequently, with each one 'locking in' additional *Tcd* loci, increasing the overall strength of TRD.

Despite this ancient age, sequence comparisons among independent *t* haplotypes show extremely reduced levels of nucleotide polymorphism in contrast to the high levels obtained in comparisons among independent wild-type chromosomes. Additionally, there are very few independent *t* haplotype lineages in contrast to the large number of wild-type ones that have arisen since their divergence. The finding that *t* haplotypes have diverged considerably from their wild-type homologs, yet have very low levels of variation among different *t* haplotypes, has been interpreted as evidence that all contemporary *t* haplotype may share a more recent common ancestor (dated from 100 000 to as recently as 10 000 years ago) which must have spread rapidly, perhaps due to drive, across all subspecies of *Mus musculus* in which *t* haplotypes are now found. In striking contrast with the low levels of polymorphism found at most *t* loci are the large number of recessive lethal alleles associated with *t* haplotypes. To date, 16 independent complementing lethal loci have been identified on *t* haplotypes present in different populations. These have presumably accumulated since the recent divergence of *t* haplotypes from a common ancestor. This empirical finding of a high diversity of recessive lethals has led several authors to speculate that recessive lethality may impart a selective advantage to the *t* chromosome.

Population Biology

An obvious consequence of the high transmission of *t* haplotypes is that they should increase in frequency and become fixed in natural populations. Yet, and like other meiotic drive systems in *Drosophila*, they remain as a polymorphism, suggesting that strong natural selection may act against drive systems in general. Two counterbalancing forces account for why *t* haplotypes have not become fixed. First, all males homozygous for *t* haplotypes (*t/t*) are completely sterile due to motility defects of all of their sperm. Second, most *t* haplotypes carry recessive lethal mutations. Because *t* haplotypes may carry different recessive lethals this results in two possible outcomes. All mice homozygous for the same lethal *t* haplotype (e.g., t^x/t^x) die early in gestation, while mice carrying two *t* haplotypes with different, complementing lethals (e.g., t^x/t^y) are viable, but male-sterile. Nevertheless, these counterbalancing forces do not account for why the frequencies of *t* haplotypes in natural populations are so low. Mathematical models indicate that for a lethal *t* haplotype with a TRD of 95%, a high equilibrium frequency of *t* haplotypes should result, such that about 77% of wild mice should be heterozygous for a *t* haplotype. All field studies, however, have found that far fewer wild mice, as few as 10–15%, actually carry *t* haplotypes.

Several forces have been proposed that might maintain a low frequency of *t* haplotypes. Theoretical studies show that strong selection to reduce the transmission bias of drive chromosomes will favor the spread of genes that suppress meiotic drive, and this kind of genetic suppression has been described for several drive systems in *Drosophila*. In contrast, the evidence for modifiers of TRD is mixed. There is some evidence for a general effect of genetic background in long-term laboratory studies. However, studies on transmission ratio in matings from wild mice have so far found no evidence for reduced TRDs, suggesting modifiers are not common in natural populations.

Why are modifiers of drive not prevalent in this system when they have evolved to counteract drive in many other systems? Evidence so far suggests that selection is acting on other components of fitness instead. Studies of field-inseminated litters have found that multiple matings of +/+ females with both +/*t* and +/+ males can effectively lower TRD in a given litter from 90% down to 20%. Selection has also been demonstrated to be acting against +/*t* heterozygotes in several studies. Mean litter size has been shown to be ~20% less for litters produced by either +/*t* males or +/*t* females relative to wild-type (+/+) litters, and this is likely to be due to a reduced viability of +/*t* embryos *in utero*. Additionally, a broad pattern emerging from empirical studies suggests a relationship between *t* haplotypes and population size and structure, with genetic drift or inbreeding also contributing to lowering *t* frequencies in larger populations. The combined effects of even a small reduction in TRD, with a 20% reduction in heterozygote fitness, and moderate levels of population subdivision can be shown to considerably lower *t* frequencies in simulation studies, and thus a combination of interacting population-level effects seems most likely to account for the low frequencies of this strong meiotic drive chromosome in natural populations.

See also:* Meiotic Drive, Mouse; *Mus musculus

T Phages

E Kutter and B Guttman

doi: 10.1006/rwgn.2001.1297

Bacteriophages have been major research tools for molecular biology; the history of phage research in the West is virtually a history of molecular biology itself. Many early advances in understanding the detailed biology of phage infection were made by Max Delbrück (1939), Salvador Luria, Alfred Hershey (Hershey and Chase, 1952) and their students such as A.H. Doermann, using a select group of phages, the 'T phages,' that are still extremely important. Until 1944, various laboratories had used many different kinds of phages and bacteria, often isolating them themselves, making it virtually impossible to compare results between laboratories. Delbruck set up a phage training course, regular (still-ongoing) meetings at Cold Spring Harbor, Long Island, and a "phage truce." He convinced the community of American phage investigators to focus on seven phages selected by Demerec and Fano (1945), grown on *Escherichia coli* strain B in nutrient broth at 37 °C.

These phages, numbered T1–T7 (T for 'type'), are all well-behaved in that they give clear, easily countable plaques with virtually 100% plating efficiency and show no confusing phenomena such as lysogeny. T2, T4, and T6 (the 'T-even' phages) happen to be closely related to each other, as are T3 and T7. Much of the early work focused on the T-even phages, the largest of the group and structurally the most complex, with a genome size of 169 kb and contractile tails that assign them to the family Myoviridae. T3 and T7 are Podoviridae, with short stubby tails. They are about a quarter the size of the T-even phages and distinguish themselves particularly by producing their own phage-directed RNA polymerase to transcribe their late genes. This polymerase and its associated, distinct promoters have been useful for cloning work, particularly when potentially very toxic gene products are involved. T5 belongs to the Siphoviridae, with a long, flexible noncontractile tail and an icosahedral head (90 nm in diameter); its genome is about two-thirds the size of T4's. The last of the group, T1, is also a member of the Siphoviridae; it has a 60-nm icosahedral head and a genome size of about 48.5 kb, and looks much like the temperate bacteriophage lambda. It has been studied much less than the others, mainly because it is so difficult to contain in the laboratory; unlike the other T phages, it survives drying, and thus often turns up in unexpected and undesired places. The T phages infect various strains of *E. coli* and *Shigella dysenteriae* to varying degrees. While all infect most common laboratory strains, T1, for example, only replicated in 2 of 290 clinical isolates of *E. coli*, and T7 also infects very few wild strains. Relatives of the various T phages can be found infecting most or all of the gram-negative bacteria, but none have yet been seen invading gram-positive bacteria or Archaea.

T2, T4, and T6 are so closely related that they can recombine with one another in a mixed infection, as do T3 and T7. The T-even phages show dominance over virtually all other phages in mixed infections, inhibiting their synthesis just as they do that of the host even when the other phages are well into their infection cycle. T-even phages further distinguish themselves by using an odd base, 5-hydroxymethylcytosine (HMC) rather than cytosine in their DNA. This substitution plays a key role in many aspects of the mechanisms the phage uses to efficiently subvert the host to its purposes and to avoid attack by most host restriction endonucleases.

Here we focus primarily on the T-even phages because of their central role in elucidating many fundamental processes and control mechanisms. For example, T2 was used to first demonstrate that viruses encode enzymes and that DNA is the genetic material. Other important advances using T2 and/or T4 include demonstrations of the colinearity of gene and protein; the nonoverlapping triplet nature of the code, with three specific triplets set aside to signal "terminate here"; the existence and properties of mRNA; the processes leading to the assembly of complex functional structures; the mechanism of DNA replication; and the occurrence of DNA restriction and modification. We will, however, begin by summarizing the properties of the three families of T-odd phages.

T-Odd Phages

Bacteriophages T7 and T3

Bacteriophage T7 has played several important roles in the development of molecular biology. It was the first of the larger phages to be fully sequenced – by I.J. Dunn and F.W. Studier in 1983 – and the functions of most of its genes were soon identified. It encodes its own RNA polymerase, which transcribes 10 times as fast as the host polymerase. This T7 enzyme has been used extensively in developing expression systems in which a cloned protein can be so overproduced that it forms as much as half of the total cell protein – a tremendous bonus for gene engineers.

T7 is the prototype of the Podoviridae phages, with a stubby, noncontractile external tail about 10 × 20 nm. The tail also has an intraviral portion that expands after attachement to the target cell, forming a complex organ for DNA transfer into the cell. The DNA transfer process always starts from the left end on the standard genomic map, aided by host-polymerase transcription of the first 19% of the genome from three strong promoters that are located within the first 750 bp. The genes in this 'first-step transfer' portion encode inactivators of the host restriction enzyme and of its deoxyguanosine triphosphate (dGTP) triphosphohydrolase, a protein kinase that also shuts off host-catalyzed transcription by a phosphorylation-independent mechanism, a DNA ligase and the new single-subunit RNA polymerase. This polymerase has significant homology with the *Saccharomyces cerevisiae* mitochondrial RNA polymerase. It is required to transcribe (and draw in) the remainder of the genome, transcribing first a cluster of genes involved primarily in DNA metabolism and then, from stronger promoters, the genes responsible for the phage capsid. The promoters for this polymerase consist of a highly conserved sequence between bases −17 and +6 relative to the transcription start site. There is little recognition for noncognate promoters between the polymerases from T7, T3, and the related *Salmonella* phage SP6 and *Klebsiella* phage K11, but changes in a single amino acid can interconvert the T3 and T7 specificities. T7 has 10 promoters for the middle genes, five for the late genes, and one to initiate replication.

T7 DNA replicates as a linear molecule and then forms concatamers using unreplicated 160-bp terminal repeats that are later duplicated during the packaging process. Growth of T7 and many of its relatives (but not T3) is inhibited by F plasmids. This inhibition involves specific interactions with the F-factor *pif* gene and causes inhibition of membrane functions and of all macromolecular synthesis. Several other prophages and resident plasmids can inhibit infection by T7 or by particular T7 mutants. The RNA polymerases and cognate promoters from T7 and several of its relatives have been used to generate tightly controlled, high-level expression vectors capable of producing as much as 50% of the cellular protein as a desired cloned product. The promoter sequences are rare enough that these can be engineered to work even in eukaryotic cells.

Bacteriophage T5

The DNA of T5 is about 121 kb long, with 10-kb terminal repeats, unique ends, and four nicks at precise sites in one strand. The DNA enters the cell in a two-step process. The left terminal repeat enters first, and pre-early genes it encodes completely shut off host replication, transcription, and translation, block host restriction systems, and degrade the host DNA to free bases and deoxyribonucleosides that are ejected from the cell. This first-step transfer DNA segment also encodes genes needed for the rapid entry of the rest of the genome once this process is complete, for shutoff of the pre-early genes, and for the orderly expression of early and then late genes from the rest of the genome. T5 encodes a variety of genes for enzymes of nucleotide metabolism and DNA synthesis and for modifiers of RNA polymerase. Circular DNA molecules with a single copy of the terminal repeat are found inside the cell; it appears to replicate in a rolling-circle mode. Precut genomes containing both terminal repeats are inserted into the preformed heads. About 25 kb of the genome, in three large blocks, is in principle deletable; this includes genes for tRNAs for all 20 amino acids as well as a number of open reading frames (ORFs). However, unless there is a compensating insertion, not more than 13.3 kb can be deleted and still have the DNA packaged.

Bacteriophage T1

T1 uses proteins of the iron-transport pathway to enter the cell, first binding reversibly to membrane protein T on A, then irreversibly to TonB. It is the only one of the T phages requiring an energized cell membrane for irreversible binding; its DNA entry is effected by a proton symport involving *tonB*. A transient fall in proton-motive force (PMF), ATP, and GTP inhibit the initiation of translation of host proteins while allowing phage transcription and translation to proceed. However, host transcription continues until the host DNA is degraded – a process that is tightly coupled to phage DNA synthesis (though not required to produce phage DNA). Initiation of phage DNA synthesis requires phage protiens, but elongation is carried out by the host pol III α-subunit, a mode of replication that is common for temperate but not for lytic phages. The latent period of T1 is only about 13 min, with a burst size of approximately 100. The DNA is packaged from concatamers by a headful mechanism. Early in the cycle (or in the absence of host DNA degradation), it can package host DNA to produce generalized transducing phages; these carry various parts of the host genome in predictable ratios. However, the transducing phage can be observed only with double *amber* mutant strains plated on nonpermissive recipients, since otherwise all of the transductants are killed by the large excess of viable, virulent phage.

T-Even Phages

The T-even phages include not only T2, T4, and T6 but also hundreds of other phages that have been isolated around the world: for example, from the sewers of Long Island, animals in Denver Zoo, and patients recovering from dysentery in the USSR and eastern Europe. They have been found in substantial quantity everywhere they have been sought, such as in habitats supporting the growth of *E. coli* and/or *Shigella* sp., and some have also been isolated for acinetobacters, aeromonads, pseudomonads, and vibrios. Many have been tabulated by Ackermann and Krisch (1997) and by Kutter *et al.* (1996). Although the early work was carried out with T2, the development of a detailed genetic system for T4 led to it becoming the primary object of study. T4 is the only member of the family to have been completely sequenced – a task that involved the coordinated efforts of the extensive T4 community – and has been studied so extensively that it is the subject of two entire books. All of the T-even phages share the same general gene organization and most of the same genes. The main known differences among them are in the tail tips and receptors recognized, in some modifications of the DNA, in the complements of genes for tRNAs and for internal proteins packaged in the capside, and in some additional nonessential genes. Each of the known phages in this family has been isolated only once from nature, emphasizing the extent and variability of the family.

The T4's spaceship-like capsid (**Figure 1**) carries 168 903 bp worth of genetic information, coding for about 300 genes. Only 161 of these have been functionally characterized (**Figure 2**). Genes of related function are largely clustered. About 40% of the genome codes for proteins involved in the phage's complex structure and its assembly: 24 for head morphogenesis, 10 of which encode structural components, and 26 used in the tail and tail fibers, with five additional proteins needed for tail assembly. The six tail fibers are attached to a hexagonal baseplate; the double tail shaft consists of a hollow tail tube built up on the baseplate and surrounded by the contractile sheath, which has a molecule of GTP bound to each subunit to provide the energy for contraction. With this complexity and its ability to largely carry out its assembly *in vitro*, T4 has served extensively as a model system for studying protein self-assembly and mediated-assembly processes. The T4 DNA is linear, but the molecules are circularly permuted, ending at many different sites in the genome, with a terminal redundancy of about 6%, so the genetic map is circular; the physical basis for this phenomenon is discussed in detail in Bacteriophage recombination (see Bacteriophage Recombination).

T4 rapidly directs the bacterial cell to stop making all of its own macromolecules – DNA, RNA and protein – and turns it into a factory for making more T4. This transition involves a finely orchestrated series of developmental steps; the times given in the following list refer to infection at 37 °C in rich medium under strong aeration, the conditions under which most studies have been carried out. (Recent work indicates that T-even phages also grow well anaerobically, as in the mammalian gut, but few details are available for anaerobic fermentation and virtually none for anaerobic respiration, where no glucose is present and molecules such as nitrate or fumarate act as electron receptors in the place of oxygen.)

To summarize the infection process (**Figure 3**):

1. As the T4 DNA is injected, the host RNA polymerase binds to nearly 50 strong promoter regions, leading to transcription of the immediate-early genes. The products of these genes are mainly

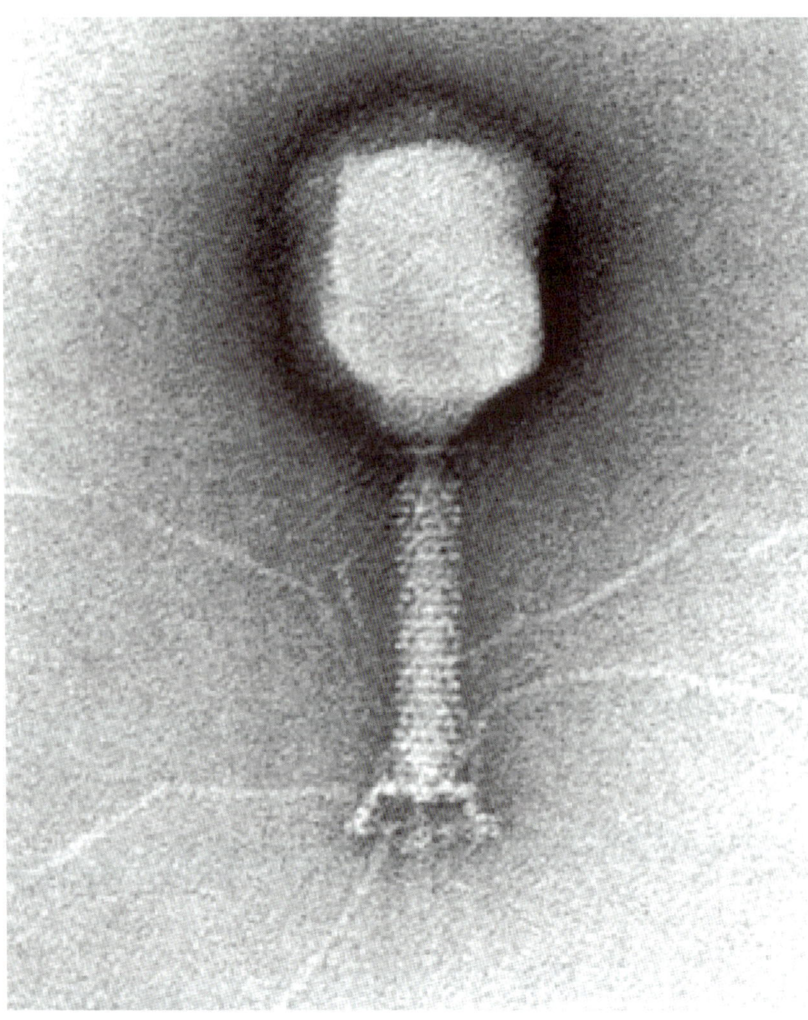

Figure 1 This electron micrograph of bacteriophage T4 was kindly donated to the phage archives at Evergreen State College by Michael Wurtz, Basel, Switzerland.

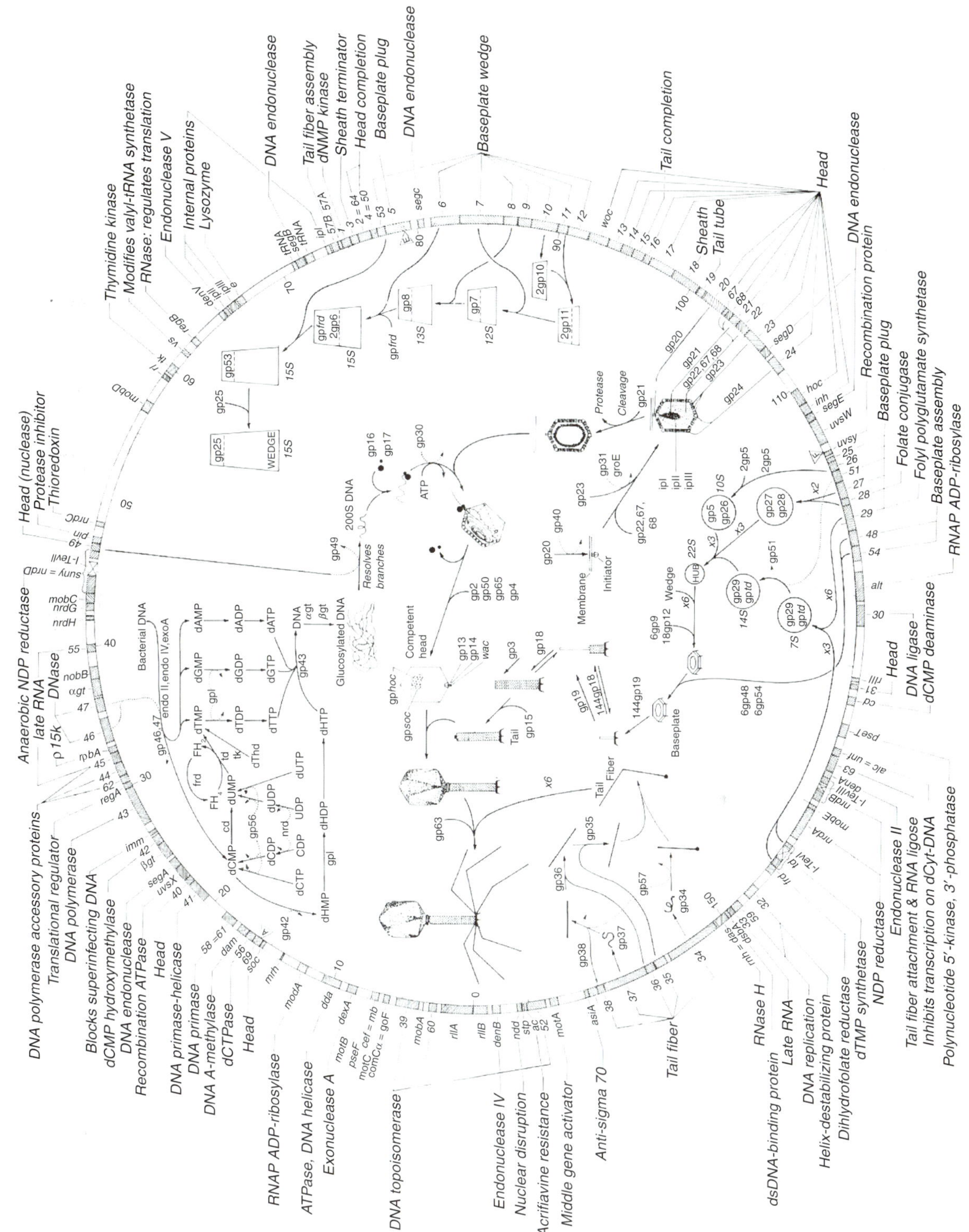

Figure 2 Genomic map of bacteriophage T4. Sizes and positions of all the characterized genes have been determined on the basis of the DNA sequence. In addition, there are a similar number of open reading frames, whose functions have not yet been determined; there are only a few kilobase pairs of apparently noncoding DNA in the genome. (Diagram prepared by Burton Guttman and Elizabeth Kutter, Evergreen State College.)

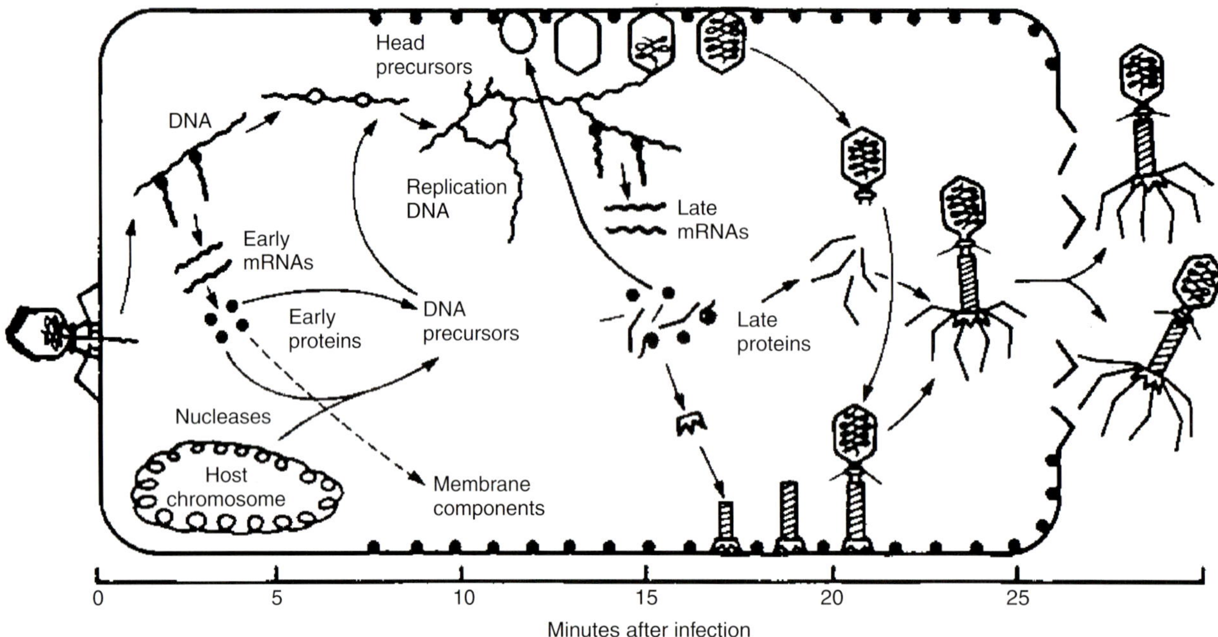

Figure 3 The normal course of T4 infection at low multiplicity of infection. The only difference at high multiplicity is that the time of lysis is delayed by several hours, and synthesis of DNA, structural proteins, and phages continues, albeit more slowly.

small proteins and are primarily involved with shutoff of host functions. Most are only made for the first 3–5 min after infection.

2. A second group of early proteins is made starting about 3 min after infection. Some of these delayed-early proteins help replicate T4 DNA. Others are nucleases that degrade the host DNA, and some are proteins that further modify the host RNA polymerase to allow recognition of the late genes and reduce early-gene transcription.
3. Phage DNA synthesis starts about 5 min after infection, mediated by an intricate complex of eight proteins. Nucleotides are efficiently fed into this 'replisome' by a second complex made up of nucleotide-synthesizing enzymes, most of which are phage-encoded. The daughter DNA molecules recombine extensively, resulting in a complicated, multibranched ball of replicating DNA that can include over a hundred genome equivalents.
4. Synthesis of late phage proteins, mostly those that constitute the phage capsid, starts about 7 min after infection. Meanwhile, synthesis of the T4 enzymes of DNA synthesis gradually stops. If anything blocks phage DNA synthesis, such as an antibiotic or a mutation in a gene essential for phage DNA replication, synthesis of many T4 early enzymes continues, as if trying to bypass the block by sheer numbers. Furthermore, no structural proteins are made; the system is regulated so that it does not make capsids until there are phage DNA molecules available to be packaged in them.
5. Phage heads, tails, and tail fibers assemble via independent pathways. The heads assemble bound to the cell membrane, assisted by chaperonins. Then each of them is packed with a head full of DNA from the replicating complex; single-strandard breaks are repaired and branches are resolved in the process. Tails and tail fibers are then added.
6. Cell lysis generally occurs about 25–30 min after low-multiplicity infection. Oxidative metabolism suddenly stops, and lysis is mediated by the action of T4 lysozyme, whose mechanism is similar to that of known eukaryotic lysozymes. The lysozyme begins to be synthesized relatively early in infection, but it has no access to the peptidoglycan layer until after the synthesis and proper assembly of a holin that is encoded by the *t* gene. The released phages (about 100–200 per cell) are then ready to start another cycle of infection.
7. If T4 is forewarned during its intracellular phase that there is something of a bacterial shortage, it has yet another growth strategy, 'lysis inhibition.' This process is initiated if another T-even phage tries to get into a cell already infected by T4, indicating an overabundance of phage relative to cells and signaling that the best strategy for reproduction is to delay lysis. Instead of lysing the cell after only half an hour, the virus somehow maintains the cell intact

for 4 – 6 h through a mechanism that depends on an interaction between a T4-encoded periplasmic protein, rI, and the *t*-encoded holin in response to the secondary infection. The phages are thus protected from a relatively inhospitable environment for a longer period of time, as well as allowing more phages to be made. Expansion of the phage population is clearly slower under these lysis-inhibition conditions than when a new round of infection is initiated every half-hour, but this is a more effective strategy when the bacterial population is limited. Most of the T-even phages show lysis inhibition, but the details of the timing and extent vary among the different phages and are affected to some degree by the specific host being infected. No such phenomenon has yet been reported for phages from non-T-even families.

One T4 particle is enough to cause a normal infection. When several T4 infect *E. coli* at the same time, they peacefully coexist, mutually complement any genetic defects they may have, recombine extensively with each other, and produce progeny with all possible combinations of the available genetic information. However, if more than 20 – 40 phages try to infect the same cell simultaneously, they damage the bacterial membrane so much that the cell just disintegrates; this phenomenon is called 'lysis from without.' T4 virions, like those of many other viruses, can remain viable for many years, unless they dry out, their DNA is damaged by radiation, their tail fibers get knocked off, or their DNA is released by osmotic shock before they encounter a susceptible host.

Foundations of Viral Genetics

Mutant Phages

No technical innovation has been more productive than the development of gentic analysis, which depends upon finding mutants and using them to elucidate the normal structure and operation of specific systems. Phage genetics began with the recognition of the existence of T2 and T4 plaque-morphology mutants. Turbid (*tu*) mutants make somewhat cloudy plaques; minute (*mi*) mutants make small plaques; and rapid-lysis (*r*) mutants make larger-than-normal plaques with sharp edges. Wild-type T4 cannot grow on a phage-resistant strain of *E. coli* B (B/4). Host-range (*h*) mutants have altered adsorption properties, so they can grow on B/4 and thus can form clear plaques on mixed B and B/4 indicator bacteria, where h^{+} phages make turbid plaques, since growth of the B/4 bacteria continues unimpeded. The ability to find such phage and bacterial mutants shows how specific the attachment of the phage to the cell surface is; the resistant bacteria do not adsorb the phage in question, because the requisite surface structures have been altered and the phage mutants have altered adsorption structures suited to the new bacterial morphology.

Hershey and Rotman (1948) used these various plaque morphology mutants to demonstrate genetic recombination in phages. R.S. Edgar and R.H. Epstein recognized that T4 could only be properly explored genetically if one could collect a large number of mutants that might, in principle, affect any gene. They therefore searched for, and found, temperature-sensitive (*ts*) mutants, able to grow at 30 °C but not at 42 °C. Since any protein of the phage might become inactivated at high temperature through a change in one of its amino acids, *ts* mutations might be obtained in any gene. However, they can only be observed if the absence of that gene product is very deleterious to the phage under the growth conditions being used. They also identified a second general type of conditionally lethal mutant, a host-dependent type that is able to grow in certain bacterial hosts but not others. Seymour Benzer had already shown that the *rII* mutants are host dependent, growing readily in *E. coli* strain B but not in strain K, although it soon became apparent that the critical factor is really that the K strain they used carried the unrelated temperate phage lambda. Epstein and Charles Steinberg (Epstein *et al.*, 1963) searched for *anti-rII* mutants that could grow in strain K but not in B, and found many of them. But in contrast to *rII* mutants, these mutants mapped in many genes around the T4 genome. It thus became evident that these mutations were of a general type that might occur in any gene. They were subsequently named *amber* (*am*) in honor of Harris Bernstein, the graduate student who helped them map large number of mutants (the German word 'Bernstein' means 'amber') Such *am* mutants have since been identified in many organisms and viruses; they involve mutation to a stop codon in the middle of the gene and thus to premature termination.

Genetic Fine-Structure Analysis

Benzer Seymour (1955) carried out a now-classic series of experiments with the *rII* mutants that revealed basic facts about genetic fine structure, which can only be investigated if one can detect very small numbers of recombinants among large numbers of progency. He took advantage of the fact that when two *rII* mutants are crossed, any wild-type recombinants are easily detectable because they alone will plate on K (λ), while all progeny plate on B. Most of the *rII* mutants carry point mutations, which revert (back-mutate) to wild-type at measurable rates. However, Benzer also found *rII* mutants that involved deletions: they do not revert to wild-type and they fail to recombine with two or more point mutants at different sites. If any two such deletions overlap – that is, delete some common stretch of the genome – they also will be unable to recombine. Eventually he was able to arrange 145 deletions in an unambiguous sequence. His experiments thus showed that the phage genome is topologically linear, as it should be if its information is simply encoded in a DNA molecule. Given the deletion map, it is then easy to map any point mutation by crossing it against the deletions. It is first localized roughly by crossing against sets of deletions; then its position relative to nearby point mutations is determined by standard crosses. Using this procedure, Benzer determined that the *rII* point mutations map at a large number of sites, some so close to each other that there must be changes at neighboring nucleotide pairs in the DNA. These experiments confirmed that the phage genome can be understood as a simple DNA molecule, with mutations being changes in any nucleotide.

Complementation and Operational Definition of a Gene

Mapping a series of mutations, even those very close to one another, still leaves open the question of boundaries between genes. Theoretically, a gene is best understood as the region that encodes a single polypeptide; operationally, there must be some way to delimit such a region. Benzer determined that the *rII* mutations actually fall into two neighboring genes, both required for growth in K(λ). Benzer infected K(λ) cells simultaneously with any two *rII* mutants, reasoning that if their defects are in different genes, each one can supply a function that the other lacks; they are able to complement each other and produce an infection yielding viable phage. If, on the other hand, both mutations fall in the same gene, the two phages together are no better off than each one by itself, and they cannot grow. This complementation test is thus an operational definition of a gene. When applied to the *rII* mutants, it allowed unambiguous assignment of each mutation to one of the two neighboring genes. (The well-defined boundary between them was then chosen as the origin of the T4 genetic map.) The same test can be used with any host-lethal mutants. Note that complementation is quite different from recombination. In recombination tests, the question is whether, and with what frequency, two genomes can recombine their information to produce a new genome; one must wait until the next generation to determine the answer. In complementation tests, the question is whether two genomes, each missing some functional unit, can mutually supply gene products (generally proteins) to produce a normal function and, thus, viable phage under otherwise nonpermissive conditions.

By using a large collection of *am* and *ts* mutants, Epstein et al. (1963) outlined the general structure of the T4 genome (**Figure 2**). The genome has since been found to contain 168 903 nt pairs; it is standardly drawn with its (arbitrary) zero point, the junction between the rIIA and rIIB genes, at 9 o'clock. It is largely organized by function, with the late or capsid-related genes falling predominantly into two large blocks. The genes for a few late proteins are interspersed in early regions; transcription and translation of such genes is subjected to unusual transcriptional and translational controls. The other regions contain primarily early genes, many of which are not essential under ordinary laboratory conditions and thus cannot be defined by *am* or *ts* mutations in the usual way. Many of these genes have been identified by mutations obtained by functional methods; they have names of one to four letters which usually relate to their function (such as '*e*' for lysozyme ('endolysin'), *denA* and *denB* for DNA endonucleases, or *rpbA* and *rpbB* for RNA polymerase-binding proteins). Note that genes that appear 'nonessential' under standard laboratory conditions may be necessary or at least highly beneficial under conditions in the natural environment, most commonly the mammalian gut.

Special Properties of T-Even Phages

The hydroxymethyl groups, like the methyl groups of thymine, are located in the major groove of the DNA helix, where they do not affect base-pairing but can be used as recognition signals. This is analagous to the 5-methylcytosine that is formed at specific sites after DNA synthesis in both prokaryotic and eukaryotic systems and acts as a control signal for many cellular processes. The use of HMC in place of C in T4 DNA facilitates the viral domination of the host in several ways:

1. T4 is immune to most bacterial restriction systems. Bacteria protect their own DNA from restriction endonucleases by marking it with methyl groups at the cleavage site; most of these enzymes modify a C residue and are also blocked by the hydroxymethyl group, so that they do not attack T4 DNA. *E. coli* does have a nuclease that specifically recognizes HMC as foreign, but T4 blocks this nuclease by glucosylating the HMC residues in its DNA. No *E. coli* enzyme has yet been observed that can attack the sugar-coated DNA,
2. T4 encodes cytosine-specific nucleases that degrade the host DNA but do not attack its own DNA.
3. T4 inhibits transcription of bacterial DNA by producing a small protein, gpAlc, which interacts with

both the RNA polymerase and DNA to terminate the elongation of transcription of all cytosine-containing DNA.

T4 could also ensure that it does not accidentally package host DNA by only packaging hmdC DNA. However, it has no such mechanism, and can therefore package host DNA and carry it to a new cell, acting as a transducing phage, as long as the degradation of host DNA is blocked by elimination of the genes that encode the specific T4 nucleases. This seems to work much more efficiently in complex phage mutants that use cytosine rather than HMC in their DNA and are missing gpAlc.

T4 also encodes several enzymes that are particularly useful in genetic engineering work, even though their value to T4 itself is still unclear:

1. A DNA ligase that can join two blunt-ended pieces of DNA. (The other known DNA ligases will only join DNA pieces that have complementary single-stranded ends, and thus can be held in register.)
2. An RNA ligase (whose main known function in the phage is to join the tail fibers to the tail, a process that seems to involve only proteins, not RNA). Though at least three T4 genes contain introns (the first to be demonstrated in eubacterial systems), the RNA ligase apparently plays no role in their splicing, which occurs autocatalytically.
3. A 3′-phosphatase, 5′-kinase which acts on DNA, RNA, a number of vitamins and cofactors, and a variety of other molecules. Mutations in this gene have no observable deleterious effects on the phage.

The tail fiber genes are generally the most diverse region between the various T-even phages, as might be expected considering their role in host range determination. Each tail fiber is made primarily of two very long proteins, gp34 and gp37, with two much smaller proteins, gp35 and gp36, forming the flexible joint between them. In T4, gp38 is added to the distal tip of gp37 and provides the specificity of binding to the receptor. Many of the other T-even phages have a totally different version of gp38 that is involved only in the assembly process; in those cases, the receptor binding site is near the distal end of gp37 itself. T4 phage particles initially absorb to the surfaces of sensitive cells through specific contact between the distal ends of the tail fibers and specific diglucosyl residues on the outer membrane lipopolysaccharides of the *E. coli* B cell surface or the outer membrane protein ompC on K strains. Each of the other T-even phages has its own specific receptors: membrane proteins OmpF and FadL for T2, Tsx = NupA for T6, OmpA for K3 and Ox2, OmpF for TuIa, OmpC for TuIb. The reversible initial binding is quickly followed by irreversible attachment by means of gp12, a short tail fiber that extends down from each vertex of the baseplate. The gp 12 binding leads to an allosteric hexagon-to-star transition in the arrangement of the baseplate proteins and thence in the relationship between the tail sheath monomers. Using energy from bound GTP, the sheath contracts while the baseplate stays bound to the cell surface, forcing the central, noncontracting core of the tail fiber through the membrane of the cell.

Introns in Genes of T-Even Phages

A large fraction of eukaryotic genes are now known to be fragmented by the insertion of one or more non-translated intervening sequences, or 'introns,' within their coding sequences, which must then be excised from the primary transcripts. However, such complexities were considered a purely eukaryotic phenomenon until the report by Chu *et al.* (1984) of a 1-kb intron within the thymidylate synthase (*td*) gene of bacteriophage T4. Two additional T4 genes have since been shown to also contain introns: those for the aerobic and anaerobic nucleotide reductases, genes *nrdB* and *nrdD*.

The observation of introns in T4 raises obvious questions about splicing mechanisms, especially once it was observed that these introns could also be spliced out when cloned into *E. coli* in the absence of other T4 genes. Does this indicate that *E. coli* carries splicing machinery and thus presumably also contains introns? Additional research in the laboratories of Belfort, Shub, and Chu has shown that the T4 introns are, in fact, self-splicing and are capable of assuming a secondary structure virtually identical to that of the eukaryotic type I self-splicing introns. The splicing, in all cases, occurs via the same mechanism, involving a series of transesterifications, or phosphoester bond transfers, with the RNA functioning as an enzyme. The ease of doing genetic analysis in T4 facilitates studying the role of various intron and exon sequences in the self-splicing reaction. It is not yet clear whether the structural homology between the T4 and eukaryotic introns reflects some ancient evolutionary origin of such splicing or a later transfer of introns from eukaryotes to T4; in either case, many interesting questions are opened up.

Phage T4 seems to be a never-ending source of novelties for molecular biologists, and the latest of them has been described by Huang *et al.*, 1988, who found a mechanism in which a 50-base segment of the information in one gene is skipped over during translation. Gene 60 encodes an 18-kDa subunit of the DNA topoisomerase, which is involved in phage DNA replication. There is strong evidence that the extra sequence is not removed as an intron. The extra bases, which are bracketed by a direct repeat of 5 nt pairs, appear to be pushed out in a kind of hairpin loop so that the codons on either side of it are brought together. Though the structure at this point is unusual, a ribosome can presumably move right through it, translating the messenger properly while ignoring all the nucleotides in the loop. This segment of gene 60 was inserted into the N-terminal coding sequence of the β-galactosidase gene of *E. coli*, where it also was neither excised nor translated. The fused genes showed comparable levels of enzyme activity to a fusion without the extra 50 bases, indicating that the looped-out sequence has little effect on translation of the messenger. This ribosomal bypass region has been found only in gene 60 of T4 and a few of the other T-even phages; there is nothing like it in the comparable gene of phages T2, T6, and most other family members. In those cases, T4's gene 60 is actually fused with the gene for another topoisomerase subunit, coded in T4 by gp39. In T4, genes 39 and 60 are separated by several hundred base pairs that are noncoding except for what appears to be the residue of one of the homing endonuclease genes described here.

Conclusions

Bacteriophages, especially the large, complex phages discussed here, have long been a major focus of molecular biology. An amazing amount of basic biological information has been uncovered with the T-even coliphages alone. Although much of the excitement of molecular biology has now shifted to eukaryotic systems, many investigators continue to work with phages and continue to astonish their colleagues with discoveries of previously undreamt-of mechanisms and processes. Phage systems, which are relatively easy to handle and involve inexpensive materials and short time scales, remain excellent material for working out the details of many kinds of complex mechanisms and for training young investigators. One major advantage is the degree of genetic understanding of the phage-host system and the ability to combine genetic, physical, and biochemical tools in attacking a problem. But this line of work reemphasizes an important point about biological research: that simply knowing the structure of a DNA molecule is not enough, because the sequence of nucleotides tells little about the function of that sequence, even though it may yield important clues. Biology is something more than chemistry. A gene is not merely a segment of a DNA molecule; it is a meaningful segment, which must be expressed and regulated, often through complex mechanisms, and there is no way to know those mechanisms a priori just by doing chemical experiments. This has again been reemphasized with the discovery of the folded-out intron in gene 60. Molecular biology has been fruitful primarily because it combines chemical work with biological – especially genetic – studies. And much of its fascination lies in its promise of another surprise after every experiment.

Further Reading

Karam J *et al.* (eds) (1994) *Molecular Biology of Bacteriophage T4.*

Webster R and Granoff A (eds) (1994) *Encyclopedia of Virology.* London: Academic Press

References

Ackermann and Krisch (1997) *Archives of Virology* 142: 2329–2345.

Benzer S (1955) Fine structure of a genetic region in bacteriophage. *Proceedings of the National Academy of Sciences, USA* 41: 344–354.

Chu FK, Maley GF, Maley F *et al.* (1984) Intervening sequence is the thymidylats synthesis gene of bacteriophage T4. *Proceedings of the National Academy of Sciences, USA.* 1(10): 3049–3053.

Demerec M and Fano U (1945) Bacteriophage-resistant mutants in *Escherichia coli*. *Genetics* 30: 119–136.

Ellis EL and Delbrück M (1939) The growth of bacteriophage. *Journal of General Physiology* 22: 365–384.

Epstein RH, Bolle A and Steinberg C *et al.* (1963) Physiological studies of conditional lethal mutants of bacteriophage T4D. *Cold Spring Harbor Symposia on Quantitative Biology* 28: 375–392.

Hershey AD and Chase M (1952) Independent functions of viral protein and nucleic acid in growth of bacteriophage. *Journal of General Physiology* 36: 39 – 56.

Hershey AD and Rotman R (1948) Linkage among genes controlling inhibition of lysis in a bacterial virus. *Proceedings of the National Academy of Sciences, USA* 34: 89–96.

Huang WM, Ao S-Z, Casjens S *et al.* (1988) A persistent untranslated sequence within T4 DNA topoisomerase gene 60. *Science* 239: 1005–1012.

Kutter E, Gachechiladze K, Poglazov A *et al.* (1996) Evolution of T4-Related Phages. *Virus Genes* 11: 285–297.

See also: **Bacteriophages; Lysogeny**

TAL Gene Family (*SCL*)

R Baer

doi: 10.1006/rwgn.2001.1625

The *TAL* family consists of three proto-oncogenes (*TAL1*, *TAL2*, and *LYL1*) that were identified through the analysis of tumor-specific chromosomal translocations associated with human T-cell acute lymphoblastic leukemia (T-ALL). Each of these genes encodes a polypeptide that harbors the basic helix–loop–helix domain (bHLH), a DNA-binding motif common to many eukaryotic transcription factors. Since the bHLH domains of TAL1, TAL2, and LYL1 are more closely related to one another than to those of other proteins, they constitute a discrete subgroup within the larger family of bHLH proteins (**Figure 1**). Nevertheless, each of the *TAL* genes has a different pattern of tissue-specific expression during normal development. Also, while gene targeting has shown that the mouse Tal1 protein (also called Scl) is essential for the formation of all blood cell lineages, similar studies have not revealed overt defects of hemapoiesis in mice devoid of Tal2 or Lyl1. Thus, despite their common role in leukemogenesis and the striking amino acid sequence homology of their respective bHLH domains, it appears that each of the *TAL* genes has assumed distinct functions during mammalian development.

TAL Proteins Serve as DNA-Binding Transcription Factors

The bHLH motif is a structural domain of 50–60 amino acids that forms two amphipathic α-helices separated by an intervening loop (**Figure 1**). The bHLH domains of two separate polypeptides can interact to form a dimer that binds DNA as a parallel, left-handed four-helix bundle. Sequence-specific DNA recognition is mediated primarily by a stretch of basic amino acids that reside near the N-terminal flank of each dimerized bHLH motif (**Figure 1**). Although some bHLH proteins can form homodimers, the TAL proteins only bind DNA upon heterodimerization with the 'E proteins,' a distinct subgroup of bHLH proteins encoded by the *E2A*, *E2-2*, and *HEB* genes. These heterodimers (e.g., TAL1/E2A) have been shown to bind DNA in a sequence-specific fashion and to modulate the transcription of reporter genes that contain a cognate recognition sequence. Thus, at the biochemical level, the TAL proteins appear to function as transcription factors.

The bHLH domains of TAL1, TAL2, and LYL1 also interact with the LIM domains of the LIM-only oncoproteins LMO1 and LMO2. This interaction allows for the assembly of a larger DNA-binding complex, which includes not only a bHLH heterodimer such as TAL1/E2A but also a member of the GATA transcription factor family. One such oligomeric complex (E2A/TAL1/LMO2/LBD/GATA-1) has been observed in erythroid cells as well as in leukemic T cells derived from patients with T-ALL. This complex binds DNA in a bidentate fashion in which the E2A/TAL1 heterodimer contacts its recognition sequence on DNA (the E-box), while

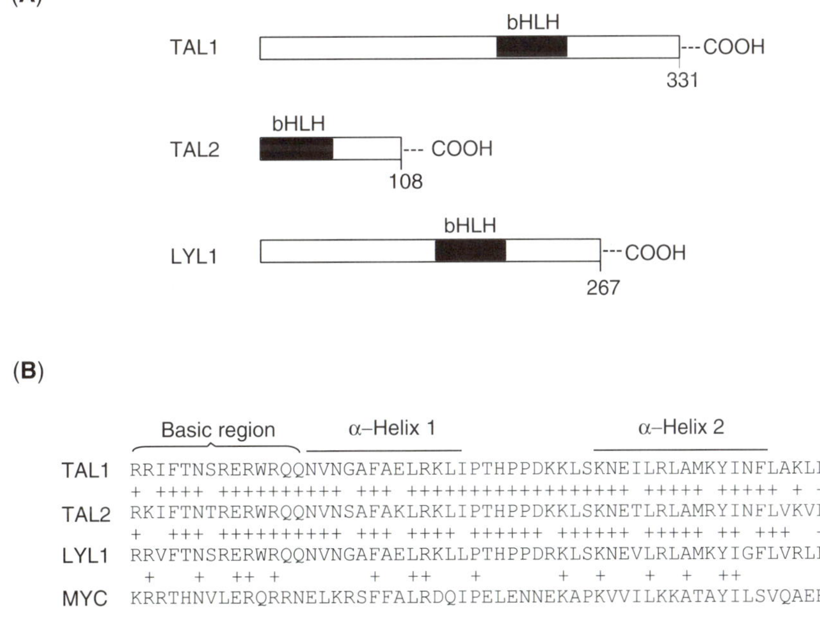

Figure 1 TAL family of basic helix–loop–helix (bHLH) proteins. (A) Schematic of the *TAL1*, *TAL2*, and *LYL1* gene products. Shaded bars represent the DNA-binding bHLH domains. (B) An alignment of amino acid sequences from the bHLH domains. Whereas the bHLH motifs of TAL1, TAL2, and LYL1 share more than 85% amino acid identity, they are less similar to the corresponding motifs of other bHLH proteins (e.g., MYC).

GATA-1 binds a neighboring GATA sequence (see **Figure 1** in *LMO* Family of LIM-only Genes).

Role of *TAL* Genes in T-Cell Leukemia

Certain lymphoid malignancies are characterized by common chromosome abnormalities that can be found in almost all affected patients. For example, more than 95% of patients with Burkitt's lymphoma have chromosome translocations that activate the *MYC* proto-oncogene, while most cases of follicular B-cell lymphoma feature a translocation that activates the *BCL2* proto-oncogene. In contrast, cytogenetic studies have not uncovered a common chromosomal defect associated with T-ALL. Instead, a series of rare, but recurrent, chromosome translocations are found in T-ALL patients. Each results in the transposition of a proto-oncogene into the T-cell receptor (TCR) locus on either chromosome 7 (TCR β-chain) or 14 (TCR α/δ-chain). Of the nine proto-oncogenes known to be activated in this manner in T-ALL, three encode the members of the TAL family and two encode the LIM-only proteins with which they are known to interact (i.e., LMO1 and LMO2). Thus, although the chromosomal abnormalities associated with T-ALL are diverse, many of them target proteins within the same biochemical pathway. Moreover, malignant activation of the *TAL1* gene is especially frequent in T-ALL. While chromosomal translocations involving *TAL1* are observed in only 3% of cases, an additional 25% of patients harbor local rearrangements of the *TAL1* gene and nearly half of all pediatric cases show evidence of tumor-specific *TAL1* activation. As such, *TAL1* represents the most commonly activated proto-oncogene known to be involved in T-ALL.

TAL1, *TAL2*, and *LYL1* are not expressed during T-cell development. In contrast, the translocated alleles of these genes are actively transcribed in T-ALL cells, suggesting that ectopic expression of any one of these genes in the T-cell lineage is potentially leukemogenic. This has been confirmed using mouse models in which targeted expression of a *Tal1* transgene in thymocytes results in the formation of clonal T-cell leukemias after a long latency. The exact mechanisms by which ectopic expression of the *TAL* genes elicits T-cell tumorigenesis are not clear. DNA-binding protein complexes containing, for example, TAL1 might promote leukemogenesis by altering the normal pattern of gene expression in T-lineage cells. Alternatively, the *TAL* gene products might serve as dominant-negative inhibitors of the E proteins, which normally function as homodimeric bHLH transcription factors during lymphoid development. By disrupting these E-protein homodimers, ectopic TAL1 might impair T-cell maturation and thereby increase the likelihood of leukemic transformation.

Further Reading

Baer R, Hwang LY and Bash RO (1997) Transcription factors of the bHLH and LIM families: synergistic mediators of T cell acute leukemia? *Current Topics in Microbiology and Immunology* 220: 55–65.

Rabbitts TH (1998) *LMO* T-cell translocation oncogenes typify genes activated by chromosomal translocations that alter transcription and developmental processes. *Genes and Development* 12: 2651–2657.

See also: **Leukemia, Acute; *LMO* Family of LIM-only Genes; Oncogenes; Transcription; Translocation**

Tandem Repeats

S T Lovett

doi: 10.1006/rwgn.2001.1269

Repeated DNA segments are sometimes found adjacent to each other in a direct orientation. Schematically, if 'ABCD' represents an ordered genetic sequence, a direct tandem repeat could consist of 'ABCBCD,' with segment 'BC' being the repeat unit. Such tandem repeats can be of various lengths ranging from several nucleotides to entire groups of genes. Tandem repeats can occur both in coding and noncoding DNA sequences. In certain instances, large numbers of these DNA repeats can found in direct orientation in repeat arrays. For example, the nucleolar organizer region of many organisms including *Xenopus*, *Drosophila*, and humans carries hundreds of rRNA genes in a tandem repeat array.

One prominent characteristic of tandem repeats is their instability. Tandem repeats are prone to both increases (such as 'ABCBCD' to 'ABCBCBCD') and decreases ('ABCBCD' to 'ABCD') in the number of repeat elements. Such rearrangements are believed to occur by homologous recombination between the repeated DNA segments (including unequal crossing-over between chromosomes) or by slipped misalignment of the repeat sequences during DNA replication. Repeats at a variety of genetic loci in humans are variable enough among individuals (so-called variable number tandem repeats or VNTR) that they have been used as molecular 'fingerprints' for forensic purposes. For certain loci in humans, rearrangements between tandem repeats can lead to genetic disease. For example, in Huntington disease, fragile X syndrome, or myotonic muscular dystrophy, expansion of a trinucleotide repeat array to a larger size, within their respective loci, is associated with the disease phenotypes. Certain thalassemias are caused by deletion between tandem globin genes.

How are tandem repeats formed? During replication, slipped misalignment at short repeated DNA sequences flanking a single-copy gene segment can cause the repeat and the gene segment to be duplicated. Alternatively, mispairing and unequal crossing-over between chromosomes can initially produce the duplication. Once a duplication has been established, it is free to undergo further rearrangements, as discussed above, including deletion or expansion in repeat number. Genetic selection may sometimes play a role in driving expansion of genes into tandem arrays. For instance, in bacteria, selection for increased resistance to certain antibiotics can result in the amplification of a drug-resistance gene as direct repeats in a tandem array. Selection for increased gene copy may also explain the existence of tandem arrays of rRNA or histone genes in certain organisms. Once a gene is duplicated, it is free to diverge in sequence from its partner and, therefore, tandem repeats may play an important role in evolution. This is probably the case for the globin loci of humans where diverged globin genes, expressed differentially in development, are found in a tandem array. Nevertheless, despite the usefulness of certain tandem duplications, many tandem-repeated DNA sequences are in noncoding DNA with no known function. Tandem duplication may be, like genetic mutation, a source of genomic flux that can occasionally be beneficial to the organism.

***See also:* Evolution of Gene Families; Globin Genes, Human**

Targeted Mutagenesis, Mouse

L Silver

doi: 10.1006/rwgn.2001.1271

The first method developed for creating transgenic mice – direct injection of DNA into embryonic nuclei – is very powerful, but it has two serious limitations. First, it can only be used to add, not subtract, genetic material. Second, the insertion of genetic material cannot be targeted to particular genomic locations. In genetic terms, this means that transgenic mice produced by embryonic nuclear injection are only useful for the analysis of dominant phenotypes.

By 1989, a second independent transgenic technology – targeted mutagenesis or gene targeting – was developed to circumvent these limitations. Targeted mutagenesis provides researchers with the ability to eliminate, or knockout, any cloned gene. The same technology can even be used to replace single amino acids or larger regions of a gene to obtain an allele with an altered function. This ultimate tool of genetic engineering can be used in experiments with two different kinds of objectives. First, as means for determining gene function by examining the phenotypic consequences incurred in a developing embryo or animal that does not express a particular gene, or expresses an alternative form of that gene. Second, as a means for creating mouse models for human diseases, like cystic fibrosis, that are caused by the loss of gene function.

The targeted mutagenesis technology is technically more demanding and more complex than nuclear injection technology, and its development was dependent on two critical advances in cell culture that occurred during the 1980s. The first was the establishment of *in vitro* conditions that allow researchers to place mouse embryos, at the blastocyst stage, into culture where they continue to divide without differentiating. These cultured cells are called embryonic stem (ES) cells. ES cells appear to be similar to cells from the inner cell mass (ICM) in that they retain totipotency. It is possible to grow cultures containing many millions of ES cells from a single embryo, and then recover a handful of cells from this culture for injection back into the blastocoele cavity of a normal embryo, where the ES cells can attach to the ICM, divide, and contribute to all of the tissues in the adult mouse that develops out of that embryo. Most importantly for geneticists, the ES cells even contribute to the germline of these chimeric mice so that genes present in the ES cell genome can be passed on to future generations.

The second critical advance necessary for the development of the targeted mutagenesis technology came with the establishment of a protocol for homologous recombination in ES cells. When mouse cells are transfected (the word used to describe DNA transformation of mammalian cells) with mouse-derived DNA, chromosomal integration almost always occurs at random locations other than the site from which the DNA was derived. However, very occasionally, the added DNA will 'find' its endogenous homolog and replace it by a process of 'homologous recombination.' The frequency of homologous recombination events as a fraction of total integrants is on the order of 10^{-3} to 10^{-5}.

If ES cells were simply transfected with cloned fragments of mouse DNA, homologous recombination events would not cause any genomic changes. But, the methods of recombinant DNA technology provide researchers with tools for modifying cloned genes so that they are no longer functional. Now, when homologous recombination occurs with one of these specially designed knockout constructs, the endogenous wild-type gene is replaced by a nonfunctional allele. Finally, in order to make use of the homologous recombination technology, researchers needed to develop special protocols for identifying and recovering the very rare cells in which these events took place.

One, although not the only, appeal of the gene targeting technology is the ability to create mouse models for particular human diseases. But, in essence, gene targeting can provide investigators with powerful tools to study any cloned gene. While patterns of RNA and protein expression provide clues to the stages and tissues in which genes are active, it is only with mutations that a true understanding of function can be obtained.

Creating Gene Knockouts

Once a particular gene has been cloned and characterized, the steps involved in obtaining a mouse with a *null* mutation in the corresponding locus can be outlined briefly as follows.

1. Design and construct an appropriate targeting vector in which the gene of interest has been disrupted with a positive selectable marker; in the most commonly used protocol, a negative selectable marker is also added at a position that flanks the gene sequence. The most commonly used positive selectable marker is the neomycin resistance (*neo*) gene, and the most commonly used negative selectable marker is the thymidine kinase (*tk*) gene.
2. Introduce the targeting vector into a culture of embryonic stem (ES) cells (usually derived from the 129 strain), then select for those cells in which the internal positive selectable marker has become integrated into the genome without the flanking negative selectable marker.
3. Screen for clones that have integrated the vector by homologous recombination rather than by the more common nonhomologous recombination in random genomic sites.
4. Once 'targeted clones' have been identified, produce chimeric embryos through the injection of the mutated ES cells into the inner cavity of a blastocyst (usually of the B6 strain), and place these chimeric embryos back into foster mothers who bring them to term.

The experiment is deemed a success if the ES cells successfully enter the germline of the chimeric animals as demonstrated by breeding. If the disrupted gene is indeed transmitted through the germline, the first generation of offspring from the chimeric founder will include heterozygous animals that can be intercrossed to produce a second generation with individuals homozygous for the mutated gene.

See *also:* Chimera; Embryonic Stem Cells; Knockout

TATA Box

doi: 10.1006/rwgn.2001.2045

The TATA box is a conserved A:T-rich sequence of nucleotides found about 25 bp upstream of the startpoint in a eukaryotic RNA polymerase II transcription unit; it is involved in positioning the enzyme for correct initiation.

See *also:* RNA Polymerase; Transcription

Taxonomy, Evolutionary

E Mayr

doi: 10.1006/rwgn.2001.1466

The biodiversity on earth, running to many millions of species, is so enormous that it would not be possible to study it if it were not classified. Endeavors at ordering this diversity were made many years ago by the Greeks (Theophrastus). From the sixteenth century to the time of Linnaeus the main interest was in medicinal plants, and here correct identification was of the utmost importance. The 'downward' classification, employed by Linnaeus, in which the mass of unidentified species was divided at every step into smaller groups by employing divisional logic (dichotomy), rather quickly led to identification: animals are either cold-blooded or warm-blooded; if warm-blooded they have either hair or feathers; if they have feathers they can either fly or not (e.g., the ostrich), etc. This was a method of identification but not of classification. It placed far too much weight on single characters and often led to very unnatural groupings, like placing the whales among the fishes.

In the period from *c.* 1770 to 1830 ‘downward’ classification was replaced by ‘upward’ classification. This consists of the construction of classes of similar species in which the classes are arranged in a hierarchy by the degree of their similarity to each other. Such an arrangement is called a classification. The definition of classification given by a dictionary is approximately as follows: a classification is the systematic arrangement of entities into groups or classes, according to the degree of their similarity or relationship.

This concept of classification, based on similarity, is widely used in human affairs. Books in a library or goods in a store or any other heterogeneous mass of entities are classified according to the principle of similarity. However, applying this universal method to biodiversity ran into difficulties. Different authors often disagreed on what was most similar. Worse still, some authors fell back on relying on a few conspicuous characters. As a result, for the next 150 years there continued to be much argument as to what would be the best classification.

It had long been recognized by perceptive philosophers that similarity alone is not always sufficient for a good classification. If similarity or difference among groups of entities is caused by a particular factor, this causal factor must also be taken into consideration in a classification. Darwin applied this principle to the classification of biodiversity. He realized that, according to the theory of common descent, the descendants of a particular ancestor would tend to be more similar to each other than they would be to unrelated species. More importantly, if owing to superficial similarity, an unrelated species is included in a taxon, a detailed character analysis would reveal that it had not descended from the common ancestor of the other species of the taxon. Such a superficially similar species is then removed from the taxon. Such an analysis is referred to as cladistic analysis (see below).

Darwin presented his new ideas on classification in *On the Origin of Species* (Darwin, 1859). Since then evolutionary taxonomists have more or less adopted his principles. They are best designated as the method of Darwinian classification.

Darwinian Classification

Darwinian classification employs two sets of criteria in the classification of biodiversity, degree of difference, and genealogy. Darwin emphasized repeatedly, verbally and in his correspondence, that genealogy alone cannot produce a good classification. The crucially important aspect of a Darwinian classification is that in the first step the classes of similar species are determined (‘classification’) and in the second step (‘cladistic analysis’) all species are removed from these taxa that are not clearly descendants of the nearest common ancestor. A taxon that consists exclusively of all descendants of the nearest common ancestor is called a monophyletic taxon. Haeckel, 1866 defines the term monophyletic taxon as consisting of the descendants of the nearest common ancestor.

Cladistic Analysis

Hennig (1966) clearly recognized that only derived (apomorph) characters can be used to determine branching pattern, not ancestral (patristic, plesiomorph) characters. Followers of Darwinian classification use, likewise, cladistic analysis in order to determine whether or not a taxon delimited by them is monophyletic. There have been earlier authors who appreciated the importance of this principle, but Hennig was the first to articulate it clearly. It is a legitimate method for the weighting of characters. The use of cladistic analysis does not make a classification cladistic. The most successful recent applications of cladistic analysis were made by testing additional classifications for strict monophyly.

Objective of Classification

The purpose of any classification is to serve as an information storage and retrieval system. Every taxon is relatively homogeneous and all included species share some well-defined attributes: all mammals have a mammalian jaw articulation, are warm-blooded, are hairy, and suckle their young with milk. Furthermore, almost any Darwinian taxon is adapted for a particular niche or adaptive zone. Hence, almost invariably, the taxa of a Darwinian classification have an ecological significance. Particularly helpful is the arrangement of the taxa in a hierarchical pyramid: species, genus, family, order, class, phylum. The more different two species are, the higher the (higher) taxon to which they belong. Cat and dog belong to different families, but to the same order (Carnivora). Cat and yeast belong to different kingdoms (Animalia versus Fungi).

The Darwinian classification is ideally suited to fulfill the functions of a classification and is, therefore, also used when, for special purposes, other ordering systems are used simultaneously.

Cladification

In 1950 Hennig proposed a new ordering system for organisms. It was based entirely on the branching pattern of the tree of descent and was called by him ‘phylogenetic systematics.’ This term was rather misleading because the differences arising during the divergence of the various lineages are as much part of phylogeny as their branching. Hennig's system, consisting of an ordering of the branches, was therefore renamed cladification, and taxonomists practicing it are called cladists.

Method of cladification

Cladification is not a method of classification, it does not establish 'classes of similar and/or related species.' Rather, it recognizes phyletic lineages, branches of the phyletic tree (clades or cladons). A cladon consists of the stem species that gave rise to the branch and all of its descendants. As a result, for instance, the mammals are combined into a cladon with their reptilian ancestors, the Therapsida and Pelycosauria. Taking these two groups out of the Reptilia makes the Reptilia for a cladist a 'paraphyletic' group (see Paraphyly). The remaining Reptilia are then no longer a valid taxon. By contrast, for a Darwinian taxonomist, birds and mammals are the only species that answer to the diagnostic characters of birds and mammals. Apart from the fact that a taxon must be monophyletic, its descent does not determine its classification. Cladification is most helpful whenever questions of phylogeny are involved. It sheds light on the time of origin of particular characters. For instance, it shows that nest building did not originate with the birds, because it occurs already in other taxa of the branch of the reptiles, thearchosaurians (thecodonts, dinosaurs, crocodiles), from which birds are descended. Cladification permits inferences on phylogeny from an analysis of the characters of living forms without the use of fossil material.

In recent years numerous cladists have suggested that the establishment of a cladification (a cladogram) would make the Darwinian classification superfluous; however, this is not the case. As already described above, the virtues of the Darwinian system for information retrieval and its importance in ecology make its preservation indispensable. Classification and cladification have different objectives and can exist side by side.

Incompatibilities between Cladifications and Traditional Classifications

In view of the evident merits of the method of cladification, the question is often raised why so many taxonomists still use the Darwinian classification as their preferred ordering system. The main reason is that the two systems have entirely different objectives and cladification is unable to produce a classification as traditionally understood. It is impossible to convert a cladogram into a Darwinian classification. Therefore, it is not correct to say that establishing a cladification makes a classification superfluous. An analysis of the incompatibility of the two methods is the best way to demonstrate how different the two systems of ordering are. Among the numerous incompatibilities between the two systems the following may be listed:

1. The method of combining all descendants of a stem species into a single cladon results in great heterogeneity. For instance, the synapsid stem species gave rise to the pelycosaurs, therapsids, and mammals, a highly heterogeneous cladon. By contrast, the taxa in a classification are relatively homogeneous, a property which is very important for information retrieval.
2. Cladification only uses derived characters. This is indeed a necessity in cladistic analysis, but in a classification one must use the totality of all characters, derived as well as ancestral ones. Indeed, its ancestral characters are often the most diagnostic characters of a taxon.
3. Cladists tend to assume that characters originate uniquely, hence a study of the distribution of newly derived characters permits the construction of unequivocal cladograms. This assumption overlooks the frequency of parallelophyly. Parallelophyly is the independent origin of the same character in two related phyletic lineages owing to their possession of a similar ancestral genotype. Parallelophyly is one of the major causes for the frequency of homoplasy. The irregular distribution of stalked eyes in the acalypteran flies is an example of homoplasy owing to parallelophyly.
4. Autapomorphic characters, that is characters that evolved in only one of two sister groups, are largely neglected in cladification. This prevents giving proper weight to the differences between sister groups. One sister group often deserves a much higher categorical rank than the other.
5. The stem species that gives rise to a new cladon usually has only one (or very few) of the derived features (synapomorphies) that later characterize the cladon. Others gradually accumulate during its further evolution. Even though the stem species belongs to the cladon, it may lack most of the apomorphic characters which later become diagnostic for this cladon.
6. Since degree of difference is, in principle, ignored and likewise the use of ancestral characters, cladification has no adequate method for the ranking of cladons. Both of Hennig's criteria, geological age and equal ranking of sister groups, have proven unworkable and cladists, to achieve some sort of ranking, developed a new method that makes use of degree of difference, an approach expressly rejected by Hennig. This method, called sequencing, is vulnerable to various difficulties.

Owing to their difficulties with ranking, cladists cannot construct a hierarchy of cladons that would reflect the degree of difference among higher taxa. This is a grave weakness, since a hierarchical arrangement is the most important property of any efficient classification.

7. The earlier portions of a highly derived cladon are usually members of a well-established taxon in a Darwinian classification. The dinosaurs, for instance, are part of the Reptilia in such a classification. In a cladification the dinosaurs forming with birds and crocodilians the Archosauria cladon are removed from the Reptilia, which thereby become 'paraphyletic' and are no longer a valid taxon. All those fossil taxa that have given rise to a derived (descendant) taxon become paraphyletic in a cladification and must be broken up and renamed. The adoption of the principle of paraphyly thus results in the destruction of the majority of traditional taxa, particularly of fossil taxa. Thus, cladification is in conflict with the highest objective of classification, namely, stability. Other provisions of cladification likewise are in conflict with stability.
8. Equally inimical to stability is the custom of cladists to give an entirely new meaning to traditional terms. For instance, phylogeny when introduced by Haeckel, referred to both of its components, cladogenesis and anagenesis. But Hennig restricted it to the former. Likewise, the term 'monophyletic' referred for 100 years to a taxon that was derived from the nearest common ancestor. For Hennig, the term describes the mode of descent of a branch (see Monophyly, Holophyly).

Other weaknesses of the methods of cladification have been pointed out in the recent literature (Cronquist, 1987; Hedberg, 1995; Knox, 1998). It is for this reason that the Darwinian classification continues to be so widely adopted. It represents and classifies organic diversity better than cladification, which restricts itself to the study of branching patterns.

Other Systems of Ordering Species

A number of taxonomic methods have been proposed that are not specifically evolutionary.

Special Purpose Classifications

These are usually based on a single characteristic, like diploid versus polyploid plants, or the traditional arrangement of plants under the headings trees, shrubs, herbs, and grasses. Such special classifications are not evolutionary; also they have such a low information content that they cannot be used for broader generalizations.

Phenetics

A system of classification based entirely on degree of similarity (or difference). Some taxonomists thought that by taking enough characters, preferably more than one hundred, one was sure to come up with taxa that were clearly the descendants of the nearest common ancestor. But this method never became popular, not only because it was very laborious, but also because it was usually impossible to find so many reliable differences. Also it encounters great difficulties owing to homoplasy, mosaic evolution, and the absence of criteria for character weighting (Mayr and Ashlock, 1991). However, Sibley and Ahlquist's (1983) classification of birds based on the DNA difference of the taxa is essentially a phenetic method.

References

Cronquist A (1987) A botanical critique of cladism. *Botanical Review* 53: 1–52.

Darwin C (1859) *On the Origin of Species by Means of Natural Selection or the Preservation of Favoured Races in the Struggle for Life*, 1st edn. London: John Murray.

Haeckel E (1866) *Generelle Morphologie der Organismen*. Berlin: Georg Reiner.

Hedberg O (1995) Cladistics in taxonomic botany: master or servant? *Taxon* 43: 3–11.

Hennig W (1966) *Phylogenetic Systematics*. Urbana, IL: University of Illinois Press.

Knox EB (1998) The use of hierarchies as organizational models in systematics. *Biological Journal of the Linnean Society* 63: 1–49.

Mayr E and Ashlock PD (1991) *Principles of Systematic Zoology*, 2nd edn, pp. 195–205. New York: McGraw Hill.

Sibley CG and Ahlquist JE (1983) The phylogeny and classification of birds based on the data of DNA–DNA hybridization. *Current Ornithology* 1: 245–292.

See also: **Cladistics; Holophyly; Monophyly; Paraphyly; Phenetics; Phylogeny; Taxonomy, Numerical**

Taxonomy, Numerical

P H A Sneath

doi: 10.1006/rwgn.2001.1272

Numerical taxonomy is the grouping of taxonomic units by numerical methods into groups on the basis of their properties. It requires the information about taxonomic entities to be converted into numerical quantities which can then be analyzed by appropriate algorithms. It includes the drawing of phylogenetic inferences to the extent that this is possible. Therefore, the term broadly covers much of systematics, and includes the distinct activities of classification (ordering of entities into groups) and identification (assignment of additional entities to their correct groups).

Attempts to quantitate biological relationships go back many years, and isolated advances were made in fields such as statistics, psychometrics, and anthropology. However, no consistent scheme for choosing properties, for coding and comparing the data numerically, and for constructing a classification or identification scheme had been developed. These problems were addressed in the late 1950s and early 1960s, and resulted in the welding together of many single themes into a coherent plan of action. This program was made feasible by the advent of digital computers. Numerical taxonomy has also been referred to as taxometrics, or Adansonian taxonomy after the early French botanist Adanson who first addressed the problem of differential weighting of characters.

The philosophical bases of numerical taxonomy rest on the empirical tradition of statistics and on the theory of predictivity propounded by philosophers of science such as Whewell and Mill. The groups formed by numerical taxonomy are 'natural groups' in the philosophical sense, though not necessarily natural in the commonly used sense of phylogenetic groups. Thus, the groups of chemical elements in the periodic table, such as halogens or alkaline metals, are natural in the philosophical sense although their members are not related by common ancestry. The relationships between the entities are highly multivariate, because numerous properties are considered. The groups thus produced are not defined by the invariable possession of certain of the properties, that is they are not monothetic. They are instead polythetic, which means that they share many properties, but no property must of necessity be present in members of a group. The groups thus formed can accommodate a limited number of exceptional properties.

One of the key concepts in numerical taxonomy is predictivity. In essence this means that groups of entities can predict correctly the most likely situation in new, as yet unanalyzed, members. Thus, a new bird will be expected to have feathers and a new mammal will be expected to have hairs, though exceptions may occur (for example whales do not have hair). Despite some limitations the concept of predictivity allows one to test classifications. Those that make more correct predictions are superior to those that make less.

The first problem is the choice of entities to be studied, and especially the choice of properties. In most biological work this seldom causes difficulties. Thus, the length of an insect will usually be considered relevant but not the day of the week the specimen was collected. There are still some conceptual problems, such as how to separate size from shape of organisms, and how best to combine information that is partly contradictory. Numerical taxonomy does require deliberate choices of properties, and one can see that this will depend on the aims of the classification, whether to express for example morphological or molecular detail.

The second most important question is to decide what properties are to be compared across entities. It seems obvious that one should compare length of wing in one entity with length of wing in another, and not with length of foot. In biology this is usually called the concept of homology: homologous comparisons should be made. Yet this leads to some serious problems. Homology is usually taken to mean that the properties are the same by ancestry, and wing and foot on the gross level have different ancestries. But to be sure of homologies one must first know the phylogeny, and this is either not certain, or should not be prejudged. Instead, character complexes that share the most similarity (in some sense) are taken as homologous. Thus, wings are more similar to other wings than they are to feet.

These issues are best shown by molecular sequence data, where there may be several families of proteins, globins for example. One wishes to compare orthologous sequences, i.e., those that have the strongest matches, and by implication the closest evolutionary relationships (aided perhaps by physiological evidence). Thus, one would compare one myoglobin with another myoglobin and one hemoglobin with another hemoglobin. The existence of several subfamilies of sequences (such as α and β globins) and of pseudogenes that are not expressed can lead to difficulties. Once the comparable sequences have been chosen there is still the problem of which sites should be compared, because of insertions and deletions in the sequences. This is again a question of homology, homology of sites, and in general it is solved by searching for that alignment between two sequences that gives the greatest number of matches along the entire sequence. The great success of molecular classification shows that this approach is sound, although there may be some areas that present difficulty.

A third problem is what weight should be given to different properties or characters. Thus, should character 1 be considered ten times as important as character 2, or perhaps only one-tenth? The answer is in general based on the amount of information given by a character. Complex characters should be broken down into several unit characters, each of which is given the same weight. The justification is that each unit of information should contribute unit weight. This broad principle implies that in numerical taxonomy characters are equally weighted as far as possible, although minor deviations from the rule have little effect on the final outcome. The principle prevents wildly different weights from producing serious distortion of the findings.

Numerical taxonomy became embroiled in a dispute about evolution. The view has grown that biological groupings should above all reflect phylogeny. It is of course perfectly permissible to aim for the best phylogenies, and numerous well-founded algorithms have been devised for this. But insistence on phylogeny does not obviate the need to consider character choice, homology, and character weighting. The dispute centered about the concepts of cladistics, i.e., of how to recognize phyla or clades. These ideas were introduced by followers of the German zoologist Hennig, and in their strict form this required phylogenetic groups to be formed exclusively from identical properties that were derived by descent from a common ancestor. Such properties clearly cannot be known before one knows the phylogeny, which is what is to be determined. Strict cladistics is therefore not an operational method. Cladistic ideas have since become more complicated, and cladistics is now becoming largely a synonym for some form of phylogenetic analysis.

Steps in Numerical Taxonomy

The steps in numerical taxonomy are as follows:

1. The entities to be studied are chosen, together with the properties that are to be employed. The properties are then coded in numerical form.
2. Similarities between the entities are calculated.
3. The salient taxonomic structure is determined from the similarities and is summarized in the form of groups of entities.
4. The groups are treated as successively inclusive groupings using criteria such as taxonomic rank or phylogenetic age.
5. The data are reorganized to give identification systems for new, unknown, entities.

These steps must be carried out in the order given above. One cannot, for example, choose the best characters for identification before the groups have been determined. Many studies only continue to step 3, and at present few continue to step 5. More details are given below.

Step 1

The entities to be studied, t in number, may be of many kinds – species, genera, populations, individuals, and molecular sequences. Therefore, these are termed operational taxonomic units (OTUs). The properties are termed characters, each with their character states. Thus length of leaf is a character and 11 cm is a character state of leaf length. Molecular properties are usually sites in protein or nucleotide sequences, and their character states are amino acids or nucleotides. Character complexes are complex properties that can be broken down to single characters. Thus, leaf shape is a character complex. At this stage decisions on homology must be taken to ensure that comparisons will be among the correct characters. It is usual to employ as many characters as are feasible, covering a wide range of properties that are considered relevant, because the reliability of the groupings generally increases with the number of characters.

The character states are then coded in a suitable numerical form. Characters are either qualitative (presence–absence, coded 1 or 0) multistate (e.g., amino acids, coded as one of the 20 alternatives), or quantitative (e.g., length of leaf, coded in cm). The latter require scaling, because the units in which they are measured must be controlled, otherwise their effect on the analysis becomes indeterminate. Thus, 5 cm could be scored as 50 mm or 0.05 m or even 0.00005 km. Some rational solution is needed, and this is usually by ranging them between 0 and 1 (for the smallest and largest measurement in the OTUs, respectively) or else by standardizing them to zero mean and standard deviation of 1. The OTUs and characters form a rectangular matrix of n rows of unit characters with t columns of OTUs, whose entries are the coded and scaled character states.

Step 2

The similarities between the OTUs are calculated using one of the coefficients of similarity or dissimilarity. A simple coefficient is the proportion of matches, m, in a set of n qualitative characters. For example, 25 matches in 32 characters gives 78.1% similarity. This can also be expressed as dissimilarity of $1-m$, and it can be represented in the alternative form of a distance of 0.219. The similarity of identical OTUs can be given as 1.0 or 100 %, or as distance of zero. These values yield a square similarity matrix of size $t \times t$ (though usually only the lower triangular half is recorded). The relationships can also be represented in space, where the positions of the OTUs are points in an imaginary space of n dimensions. There are numerous coefficients of similarity, which are chosen to reflect various desired types of relationship.

Certain experimental techniques yield the equivalent of similarity matrices directly. Thus, a table of serological cross-reactions, or of nucleic acid hybridization, records similarity between organisms from physicochemical reactions. The entries are not character states of OTUs.

Step 3

A table of similarities does not make evident the relationships between groups of OTUs. Two main classes of method are available for elucidating the taxonomic structure. The first leads to tree-like diagrams, or dendrograms, in which the OTUs are situated at the tips of the branches. The second yields plots of the positions of the OTUs in a simplified space, usually two- or three-dimensional diagrams.

In the first group are algorithms for cluster analysis and for phylogenetic reconstruction. These differ in the assumptions about the cause of the observed relationships. Most cluster methods search the similarity matrix for the most similar OTUs and group them together, and then successively add the next most similar OTUs until all have joined. Various criteria are used for the joining process. The smallest distance between any OTU of a cluster to any OTU in another cluster gives straggly clusters (single linkage analysis). The average distance between all members of one cluster and all of another cluster is the criterion in average linkage analyses (the best known is the unweighted pair group method with averages, UPGMA). The similarity level at which branches join forms one axis of the tree and the OTUs are given in order of joining along the other axis. The tips of the tree are all at the same level, i.e., at similarity of 100% or distance of zero. Such a tree is termed a phenogram, and the relationships express similarity without phylogenetic assumptions. Methods for reconstructing phylogeny rely on assumptions (often very complex) on the way evolution has proceeded. The basic principle is that evolutionary change has been as small as possible to yield the observed relationships between OTUs. This may be described loosely as the principle of evolutionary parsimony. The term parsimony is also used in more restricted senses, so that a most parsimonious tree is different from a minimum distance tree or a maximum likelihood tree, though all of these rest on the broad principle of minimum evolutionary change. These techniques also add OTUs successively to give a tree-like diagram, where the branches are phylogenetic groups (phyla or clades). Some methods bypass the similarity matrix itself, though similarities are implied in some form. Because different rates of evolution are taken into account the tips are not all at the same level. The resultant dendrogram is a phylogenetic tree or cladogram. Furthermore, the position representing the earliest point in time, which corresponds to the most recent common ancestor of the OTUs, is often uncertain, and such a tree is termed an unrooted tree. Further analysis may be needed to determine the root, for example by including a distant OTU that is believed to belong to a different clade than all the other OTUs. The diagram then becomes a rooted tree.

A method that is less often used relies on evolutionary compatibility between pairs of characters, known as clique analysis. The character states of a pair of characters may allow representation on a phylogenetic tree such that no repeated mutations, or back mutations, are required to account for the observed data. Such a pair is termed compatible, and the groups of mutually compatible characters are taken to indicate the clades. The criterion of parsimony here is parsimony of unnecessary mutations.

One pervasive problem is that unlike most methods of cluster analysis, many techniques for phylogenetic reconstruction cannot be guaranteed to find the optimal tree. The reason is that the number of alternative tree topologies grows very rapidly with increasing numbers of OTUs. It is then infeasible to test every topology, even with powerful computers, and computational short-cuts still leave many alternatives to test. For example, there are over 8×10^{38} topologies for 30 OTUs. Furthermore, there are often many trees with the same greatest optimality, so that although they differ little, there may be no criteria to choose among them. It is evident that phylogenetic reconstructions must always be regarded to some extent as approximate.

The second group of methods is known as ordination analysis. They reduce the many dimensions represented by the similarity matrix to a few dimensions that express as much as possible of the observed variation. This yields plots in two or three dimensions. The OTUs are represented by points on such diagrams, and clusters of closely related OTUs can be seen by eye. Well known techniques are principal component and principal coordinate analysis. If ordination is to be useful, a high proportion of the variation must be expressed in the first two or three dimensions. There is always a danger that clusters of OTUs that are quite separate in multidimensional space, and are easily revealed by cluster analysis to be distinct, will be overlapped in ordination plots.

If a similarity matrix is rearranged to bring highly similar OTUs together, the cells of the matrix can be shaded in different intensities according to the similarity values. Such shaded similarity matrices are occasionally useful in interpreting taxonomic structure.

These various methods for structure emphasize different aspects of relationship, and inevitably lead to some loss of information. Their choice depends largely on the aims of the investigator.

Step 4

The taxonomic structure can now be represented formally as taxonomic or phylogenetic groups of OTUs, with appropriate indications of their status. Thus taxonomic ranks can be defined and named. There are some problems, because objective criteria for rank or for cladal status are not well developed, and nomenclature can be controversial. Good scientific judgement in the light of other knowledge is therefore indispensable. The final groupings can then be described in various forms, such as tables of common character states or age of origin.

Step 5

The relevant information is now available for producing an identification system whereby further, unknown, members of the groups can be identified with their correct group. Various strategies are employed. One is to construct a diagnostic key, preferably by one of the algorithms for this. Such a key is similar to 'rule-based systems' in information retrieval, and the groups are usually treated as monothetic (i.e., it is assumed all new members of a group will possess the character states given in the key). Another strategy is to treat the groups and the unknown as having position in a phenetic space defined by the characters that distinguish groups. The unknown is then identified with the group to which it is nearest (i.e., most similar). A simple form of such a system is a diagnostic table, which can be compared with the unknown to find the closest match. Such methods are very similar to 'expert systems' in information retrieval. The groups are still treated as polythetic; therefore, the correct identity is not excluded by an occasional atypical or missing character state of the unknown. Also, the probability of a correct identification can usually be calculated. This strategy is close to discriminant functions in statistics. Numerical identification has been most used in microbiology but it is being applied to many fields where identification, recognition, or diagnosis is required.

Additional Techniques

Criteria of quality are needed for numerical techniques. Thus, statistical sampling theory leads to estimates of how accurate a similarity value is likely to be. The extent to which a dendrogram represents a similarity matrix can be measured by the cophenetic correlation coefficient, and there are techniques to assess the agreement or congruence between different data sets. Related to this are methods to combine data in the form of consensus trees. Repeated random sampling can estimate the reliability of clades in a cladogram (bootstrap analysis).

Current work in biology is largely concentrated on reconstructing phylogeny from molecular sequences, using algorithms that are highly specialized to proteins or nucleotide sequences. However, it should be remembered that molecular or genomic data are not necessarily phylogenetic. The distinction between phenetic and cladistic relationships is based on the methods employed, and phenetic analyses can be made from genomic data.

Other Applications

The commonest types of analysis, as described above, are those that group organisms together; this is termed Q analysis. However, one can group the characters; this is termed R analysis. This is useful in several ways. The grouping of characters can reveal complexes that covary, and provide insight into developmental genetics. Similarly, grouping geographical areas according to their biota can be of assistance in biogeography.

Numerical taxonomy has been successfully applied in a wide range of disciplines. It has been adapted to problems in ecology, morphometrics, epidemiology, and geographical variation. Its concepts are extensively used in genomic analysis and information retrieval. Most of these applications have had to face basic questions, such as homology, choice of characters, and character weighting. It has thus led many disciplines to re-examine and redefine their aims and assumptions.

Further Reading

Hillis DM, Moritz C and Mable BK (eds) (1996) *Molecular Systematics*, 2nd edn. Sunderland, MA: Sinauer Associates.

Jardine N and Sibson R (1971) *Mathematical Taxonomy*. Chichester, UK: John Wiley.

Sneath PHA and Sokal RR(1973) *Numerical Taxonomy*. San Francisco, CA: WH Freeman.

Wiley EO (1981) *Phylogenetics*. New York: John Wiley.

***See also:* Phenetics; Phenogram; Phylogeny; Taxonomy, Evolutionary**

Tay–Sachs Disease

M M Kaback

doi: 10.1006/rwgn.2001.1273

Tay–Sachs disease (TSD) is a progressive, uniformly fatal, neurodegenerative disorder of infancy – the acute infantile form of the G_{M2} gangliosidoses, one subgroup of the lysosomal storage disorders. TSD is named for the two physicians who first described the condition in the 1880s, Warren Tay, a British ophthalmologist, and Bernard Sachs, a neurologist in New York City.

Inheritance, Distribution, and Frequency

TSD is inherited with an autosomal recessive pattern of transmission. Heterozygote carriers are entirely normal. TSD has been described in infants of all racial and ethnic groups, but historically has been identified predominantly among children of Central/Eastern European Jewish ancestry (Ashkenazim). The heterozygote frequency for TSD among Ashkenazi Jews is between 1/25 and 1/30 individuals, with a disease incidence of about 1 in 3000 births (1/27 × 1/27 × 1/4). Among general non-Jewish populations, the TSD carrier rate is approximately 1 in 300, making the disease incidence approximately 1 in 360 000 births (1/300 × 1/300 × 1/4). Certain non-Jewish isolates with increased TSD have been found among the Pennsylvania-Dutch, the Cajuns of Louisiana, and some French Canadians from Quebec, Canada. Other non-Jewish isolates with TSD in China, Japan, and Morocco, also have been identified, all probable examples of genetic founder effect and drift.

Pathogenesis

TSD results from the progressive intralysosomal accumulation of G_{M2} ganglioside, a normal component of neuronal membranes. The defect in TSD is deficient activity of G_{M2} gangliosidase, the enzyme required to catalyze the intralysosomal hydrolytic cleavage of the terminal N-acetylgalactosamine from G_{M2} ganglioside. This enzyme is also named hexosaminidase A (HEX A) when its activity is assayed with colorometric or fluorogenic artificial substrates. In the absence of this hydrolysis, G_{M2} remains intact and, with continued normal biosynthesis, progressively accumulates within neuronal lysosomes. Increasing storage of G_{M2} leads to progressive engorgement of cytoplasmic lysosomes, forming the characteristic membranous cytoplasmic bodies ('onion skin lesions') on electron microscopy. Cytoplasmic engorgement disrupts normal neuronal cell function as reflected by the increasing neurologic symptomatology (weakness, blindness, seizures, etc.) and ultimately leads to neuronal cell death.

Clinical Description

Although the disease process begins in the fetal nervous system early in gestation, the affected infant appears entirely normal at birth and remains so throughout the first 4–6 months of life. Motor weakness (e.g., floppiness or poor head control) or an 'increased startle response' to sharp sounds may be the first difficulties observed by parents. Further, wandering eye movements at about this age may lead to specialist referral, where the characteristic 'cherry red spot' in the fovea of the maculae is seen, thus leading to the diagnosis. From 6–12 months, there is progressive deterioration, with increasing weakness, loss of motor and developmental milestones if previously gained (e.g., rollover, sit alone) and failure to gain new ones (e.g., stand, coast, walk, talk). Physical findings reveal only fundoscopic changes, absence of liver, spleen, or other organ enlargement, and evidence of upper motor neuron dysfunction (hyperreflexia, sustained ankle clonus, pathologic startle response, etc.). Diminished vision is evident and progressive seizures usually begin around 12–14 months. Deterioration continues and by 16–18 months decerebrate posturing, blindness, and complete loss of meaningful interaction with the immediate environment is apparent. Management issues subsequently include feeding, hydration, airway care, skin care, seizure control, and maintenance of bowel and bladder function. The child will remain in this 'chronic vegetative state' until death between 2 and 5 years of age, usually the result of acute aspiration or overwhelming infection, secondary to pneumonia.

Diagnosis

With artificial substrates, HEX A and HEX B are easily quantified in serum, leukocytes, cultured skin fibroblasts, or amniotic fluid cell samples from suspected patients or fetuses at risk for TSD. A complete or near complete deficiency of HEX A activity (in the presence of normal or increased HEX B) is diagnostic of TSD in a symptomatic infant or an affected fetus. Heterozygotes have HEX A levels of approximately 50% of the control level. Where specific mutations are known to segregate in a family, PCR-based DNA mutation analysis can be used both for diagnosis and carrier identification.

Molecular Genetics

The 40-kb, 14-exon gene for TSD is located on chromosome 15q23 and directs the synthesis of the α-subunit of G_{M2} gangliosidase (HEX A). This enzyme is a heterodimer comprised of one α-subunit and one β-subunit (derived from the HEX B gene on chromosome 5q13). Mutations in the α-subunit gene are associated with TSD and its later onset variants, while β-subunit gene mutations account for Sandhoff disease and its variants. Mutations in a third gene, the G_{M2} activator gene on chromosome 5q, can also lead to insufficient G_{M2} degradation and a rare form of G_{M2} gangliosidosis, 'activator-deficient TSD.' More than 100 α-subunit (HEX A) gene mutations have been identified to date.

Treatment

No specific treatment for TSD is presently available. Attempts to introduce G_{M2} gangliosidase into the central nervous system by purified enzyme infusions, by cellular transfusions or by bone marrow transplantation have uniformly failed to date. Traversing the blood–brain barrier – with enzymatic protein, cellular elements, or gene-carrying vectors – remains the major obstacle to therapeutic breakthrough. The creation of knockout mouse models for both TSD and Sandhoff disease provides important new avenues for such studies. Most recently, research efforts have been directed to minimizing the accumulation of G_{M2} by inhibiting its biosynthesis with nojirimycin derivatives, relatively nonspecific inhibitors of glycosphingolipid production. Some have hope that this may prove beneficial. Lastly, while widely discussed, vector-mediated gene therapy remains a hope for the future, and will require major further research breakthroughs if ever to become a reality.

Prevention

Major strides have been made to prevent the birth of infants affected with TSD. Community-based education, voluntary carrier testing, and genetic counseling programs in Jewish communities throughout the world have lead to a greater than 90% reduction in the incidence of TSD in Ashkenazi Jewish populations. Carrier testing (HEX A and /or DNA based testing) enables the identification of at-risk couples (both partners: heterozygotes) before the birth of affected offspring. Such couples, with comprehensive genetic counseling, may choose to monitor each pregnancy by amniocentesis or chorionic villus sampling and interrupt (abort) those pregnancies where the fetus is affected (25% risk with each pregnancy). Other options include adoption, use of noncarrier sperm or ovum donors, preimplantation testing of embryos after *in vitro* fertilization, taking their chances, or use of carrier status information in marriage or mating decisions (as carried out by certain ultra-orthodox Jewish groups). Until effective therapeutic breakthroughs occur, preventative approaches will remain the mainstay in the management of this dreaded disorder.

Further Reading

Desnick R and Kaback M (eds) (2001) *Tay–Sachs Disease: From Clinical Description to Molecular Defect.* New York: Academic Press.

Gravel RA, Clark JTT, Kaback MM *et al.* (1997) The G_{M2} gangliosidoses. In: Scriver CR, Beaudet AL, Sly WS and Valle D (eds), *Metabolic and Molecular Bases of Inherited Disease*, 7th edn. pp. 2839–2879. New York: McGraw-Hill.

Kaplan F (1998) Tay–Sachs disease carrier screen: A model for prevention of genetic disease. *Genetic Testing* 2: 271–293.

***See also:* Gene Therapy, Human; Genetic Counseling**

T-Box Genes

V E Papaioannou

doi: 10.1006/rwgn.2001.1274

History

A mutation in the *Brachyury* or *T* locus, first described in 1927, was one of the earliest recognized developmental genes in mice. Embryological defects caused by mutations at this locus cause embryonic death in homozygotes and defects in tail development in heterozygotes (hence the locus name *Brachyury* which means short tail, symbolized as *T* for tail). In 1990, the gene itself was cloned and found to be a novel transcription factor, and it was not long thereafter that the existence of a family of *T*-related genes was demonstrated by the discovery of genes in both *Drosophila* and mice, with sequence homology to *T*. Discovery and exploration of the family, which was called the T-box gene family, has proceeded by leaps and bounds and T-box genes have been found in species as divergent as the roundworm, *Caenorhabditis elegans*, and *Homo sapiens*, as well as many species in between (**Figure 1**).

Defining Features of the Family

The proteins encoded by the T-box family genes are all putative transcription factors. The defining feature of the gene family is a region of DNA sequence homology that encodes a polypeptide region named the T-box, extending across 180 to 190 amino acid residues. The gene products of several members of the T-box gene family have been shown to have a domain of specific DNA binding activity which includes the T-box region, leading to the hypothesis that DNA binding is conserved among all proteins containing the T-box polypeptide domain, even though polypeptides diverge widely outside the region encoded by the T-box. DNA-binding activity, along with the nuclear localization of the gene products, suggested that these proteins act as transcriptional regulators of other genes. Indeed, transcriptional regulation has been demonstrated for several T-box gene products, and it

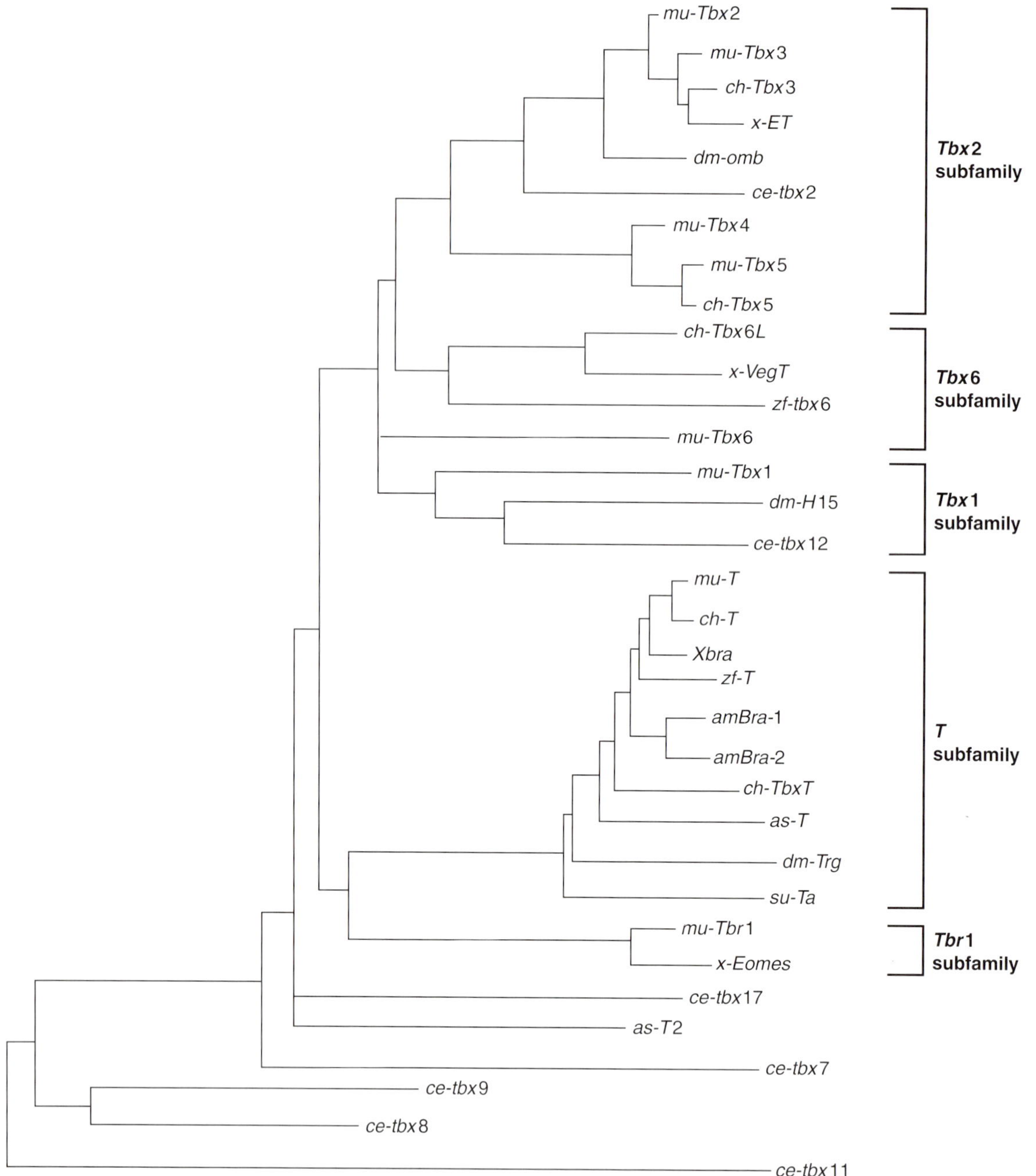

Figure 1 A phylogenetic tree of the T-box gene family constructed using the neighbor-joining algorithm based on Poisson-corrected distances between amino acid sequences. The length of the horizontal lines is proportional to evolutionary distance. T-box subfamilies are grouped and indicated by brackets. Five *Caenorhabditis elegans* T-box genes and one ascidian gene at the bottom of the tree have yet to be classified into particular subfamilies. Eight known human genes are not included but are closely related to their mouse orthologs. *as*, Ascidian; *am*, amphioxus; *ce*, *C. elegans*; *dm*, *Drosophila*; *ch*, chick; *mu*, mouse; *x*, *Xenopus*; *zf*, zebrafish. (Reproduced with permission from Papaioannou and Silver, 1998.)

seems very likely to be a common feature of all members of the family. A productive area of future research will be the elucidation of the nature of this transcriptional control and the discovery of the specific genes that are regulated by T-box genes.

Phylogenetic Analysis

Phylogenetic analysis provides a powerful tool for dissecting the evolution of gene families, and for predicting and understanding functional relationships among different family members. Phylogenetic analysis of the T-box gene family (**Figure 1**) reveals that this is an ancient gene family. Its initial expansion from a single progenitor sequence appears to have occurred at the outset of metazoan evolution and further expansions have occurred by gene duplication along individual evolutionary lineages. Together with gene expression studies, phylogenetic comparisons also provide evidence for the existence of T-box gene subfamilies whose more recently duplicated members retain similar or overlapping patterns of gene expression, most likely correlated with conserved function as well. For example, the *Tbx2* and *Tbx3* genes are members of an ancient vertebrate subfamily that expanded prior to the divergence of bony fish and tetrapods. These two genes have expression patterns that are broadly similar, both within and between species, although minor temporal and spatial differences in expression may reflect divergence of function that could have occurred since the separation of the genes.

Phylogenetics also allows the identification of what are likely to be orthologs of the same gene in different species. Orthologs are defined as direct descendants from a single ancestral gene that was present in the genome of the common ancestor of the species under analysis, for example, the *Brachyury* orthologs found in many species (**Figure 1**, *T* subfamily). Analysis of the T-box family tree can direct the search for new T-box genes by predicting the existence of orthologs in species where they have yet to be discovered, thus hastening gene discovery.

Role in Development

T-box genes have been discovered primarily through screens designed to detect genes with embryonic expression or function. Although several T-box genes are known to be expressed in adult tissues, their widespread expression in embryonic tissues, particularly in areas of inductive tissue interactions, emphasizes what is almost certainly a family feature: a major role for T-box genes in embryonic development.

One highly effective means of ascertaining the function of a gene is to find or create mutations in that gene in order to study the effects of its disruption. There is currently only a handful of known mutations in T-box genes. In addition to the well-studied *Brachyury* mutations in the mouse which affect the development of posterior structures including the tail, mutant alleles of the *Drosophila* ortholog, *Trg*, constitute a series of alleles with effects of varying severity on the development of posterior structures. Similarly, mutant alleles of the zebrafish *Brachyury* ortholog, *no tail*, also show effects on the development of posterior structures, illustrating comparable functions of orthologs in widely divergent species. Among the other T-box genes, there are spontaneous mutations at the *Drosophila omb* locus, and in two human genes, *TBX3* and *TBX5*. The human mutations are of considerable interest in that they are responsible for autosomal, dominant, developmental syndromes known as ulnar–mammary syndrome and Holt–Oram syndrome, respectively. The ulnar–mammary syndrome is characterized by limb defects and abnormalities of apocrine glands including the mammary glands, while the Holt–Oram syndrome is characterized by cardiac septal defects and abnormalities of the forelimbs.

A mutation in the mouse gene *Tbx6* has been produced by targeted mutagenesis, a technique by which specific mutations can be created at will. Homozygous mutant embryos have severe defects in the specification and differentiation of the somites. Although the head region forms normally, neck somites are misshapen and more posterior somites fail to form at all. Instead, two ectopic neural tubes are present in place of the posterior somites. This mutant phenotype of three parallel neural tubes and no posterior somites, as well as the phenotypes of all other known mutations in T-box family genes, indicates a critical role for these genes in the specification and differentiation of tissues and structures during embryonic development.

Even this small number of mutants has been extremely valuable in elucidating the nature of T-box gene function. Future mutational analysis, particularly by targeted mutagenesis, holds the key to understanding individual T-box gene function and the functional significance of the family as a whole.

Further Reading

Herrmann BG (1995) The mouse *Brachyury* (*T*) gene. *Seminars in Developmental Biology* 6: 385–394.

Papaioannou VE and Silver LM (1998) The T-box gene family. *BioEssays* 20: 9–19.

***See also:* Gene Family; Transcription Factor**

Telomerase

T M Picknett and S Brenner

doi: 10.1006/rwgn.2001.1806

Telomerase is a ribonucleoprotein complex that maintains chromosome ends. It is a cellular reverse transcriptase composed of both RNA and proteins that employs its internal RNA component as a template for the synthesis of telomeric DNA. It stabilizes telomere length by adding hexameric (TTAGGG)n repeats onto telomeric ends of chromosomes. After adding six bases, the enzyme is believed to pause while it repositions (translocates) the template RNA in order to synthesize the subsequent 6 bp repeat. This extension of the 3′ DNA template end in turn allows replication of the 5′ end of the lagging strand. It thus compensates for the continued erosion of telomeres and has been referred to as a 'cellular immortalizing enzyme.'

See also: Reverse Transcriptase

Telomeres

J R S Fincham

doi: 10.1006/rwgn.2001.1276

The Need for Telomeres

Telomeres are found at the ends of chromosomes; they provide the answer to two problems of chromosome management. First, there has to be something to distinguish true chromosome ends from the accidental ends resulting from chromosome breakage. There is much evidence from studies of the effect of radiation on chromosomes that broken ends are prone to indiscriminate rejoining, with the possibility of segmental rearrangement. Presumably they provide substrates for double-stranded DNA ligase, and they may also be subject to erosion by exonuclease. True chromosome termini must be sealed in some way to protect them against these hazards.

Second, there has to be a way of completing DNA replication. DNA polymerase extends DNA strands from their 3′ ends, and so the two strands of double-stranded DNA are replicated in opposite directions. The synthesis of one follows the replication fork and can be continuous, but the synthesis of the other runs 'backwards' in piecemeal fashion, each piece having to be initiated afresh by an RNA primer that is subsequently removed. The consequence is that the DNA strand with a 5′ terminus cannot be fully replicated by the regular mechanism, since there is no 3′ end to prime the filling of the gap left by removal of the last RNA primer (**Figure 1**). Indeed, there is evidence that the 5′-terminated strand, already shorter, may be further shortened, leaving an even longer single-stranded 3′ 'tail.' So, without some additional end-replication mechanism, one would expect the chromosome to get a little shorter with each cycle of replication. Cell viability depends on the constant replenishment of terminal sequences.

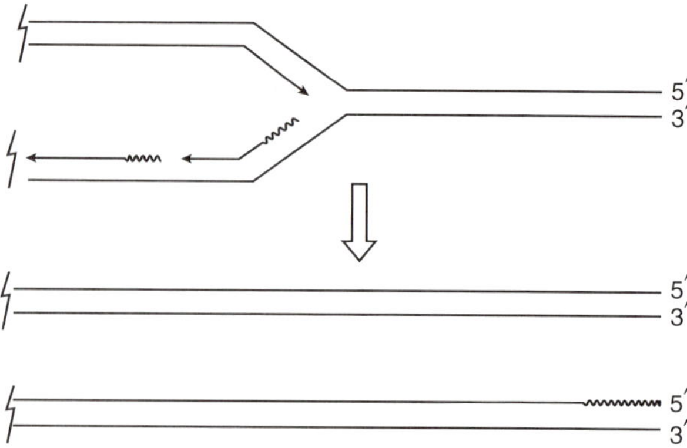

Figure 1 The end-replication problem. As the DNA replication fork progresses, one of the two parental strands can be replicated continuously, primed by its own 3′ end; the other (the lower in this figure) has to be replicated 'backwards' in patches, the replication of each patch being primed separately by primase RNA (shown here as a wavy line). When the replication fork reaches the chromosome end, the 5′-terminated strand of one daughter chromatid is incomplete, since the final RNA primer cannot be replaced by DNA.

Terminal Repeats

With the notable exception of *Drosophila* (see below), chromosomes, so far as they have been sequenced, have short tandem repeats at their extremities. These repeats are characteristically rich in G–C base pairs, with the Gs predominantly in the 3′-terminated strand, sometimes called the G-strand. The G-rich repeats are more or less constant within organisms but variable between organisms: TTGGGG in the ciliate *Tetrahymena*, TG, TGG, or TGGG in *Saccharomyces*, and TTAGGG in human and mouse, to give just a few examples. The total length of this region of simple repeats ranges from just 20 base pairs in some ciliates to more than 100 kb in mouse, it can also vary within species, especially with ageing (see below). Adjoining the simple repetitive telomeric sequence, there is generally a region, called the subtelomere, of less regular and generally longer repeats.

Subtelomeric Structure

In *Saccharomyces* yeast, where all chromosomes are now completely sequenced, the terminal TG1–3 repeats are usually flanked on the inside by up to four tandemly arranged copies of sequences called Y′, which come in two main sizes, 5.2 and 6.7 kb, related by an internal deletion. They contain open reading frames of uncertain function. Inside the Y′ sequences is a segment called X, variable in length among yeast telomeres but with a core sequence of about 0.5 kb which is common to all. It is thought to have a role in the positioning of the telomeres at the periphery of the nucleus. Inside the X segment are further repeats, variable in length and sequence from one chromosome to another, before the gene-containing interior of the chromosome is reached. Further copies of the TG1-3 repeats may be interspersed among the Y′ sequences (**Figure 2**).

The X/Y′ subtelomeric sequence tends to attract certain DNA-binding proteins which serve as foci for the formation of transcriptionally silent chromatin structure (see below).

The telomeres of other organisms are similar to those of *Saccharomyces* in the general sense that their terminal G/C-rich repetitive sequences are flanked by other kinds of repeats, but these show no consensus between organisms.

Telomerase Function

Terminal repeats are replenished through the activity of telomerase, an enzyme first isolated from *Tetrahymena*, which is a particularly rich source. Like other ciliates, *Tetrahymena* has two kinds of cell nuclei: micronuclei, containing the total genome in single copy, and macronuclei, in which those genes currently active are excised in fragments and amplified to high copy number. Consequently, this organism has an exceptionally large number of chromosome ends to look after. In spite of the peculiarities of the organism, the *Tetrahymena* telomerase system appears to be typical of eukaryotic organisms generally.

Telomerase is a reverse transcriptase which carries its own single-stranded RNA molecule to serve as a template for DNA synthesis. The total length of the telomerase RNA varies widely between organisms (159 bases in *Tetrahymena*, about 500 bases in mammals, and 1.5 kilobases in *Saccharomyces*), but the crucial section in each case is a sequence including the complement of the repetitive G-rich telomere sequence. Thus in *Tetrahymena*, the primase se quence 3′-AACCCC-5′ corresponds to the telomere 5′-TTGGGG-3′, and in *Saccharomyces* 3′-CACACCC-5′ corresponds to 5′-GTGTGGG-3′ in the telomere. The telomerase binds to the chromosome end and, using its own RNA sequence as a template and the 3′-terminus at the chromosome end as primer, synthesizes an additional terminal repeat. The enzyme then moves to the new end and adds another repeat copy, and so on (**Figure 3**).

After a number of such sequential additions to the 3′-terminated telomere strand, the extension can be made double-stranded by ordinary RNA-primed 'backwards' replication (**Figure 3**). The removal of the final RNA primer will result in a short single-strand gap and a consequent shortening of the G-strand, but the length already gained through telomerase action will generally be more than enough to compensate for this loss as well as for shortening resulting from the previous round of replication. An alternative method of second strand synthesis also seems possible; if the first strand folded back on itself, as it might do through G–T base pairing, it could prime the synthesis of its own complement.

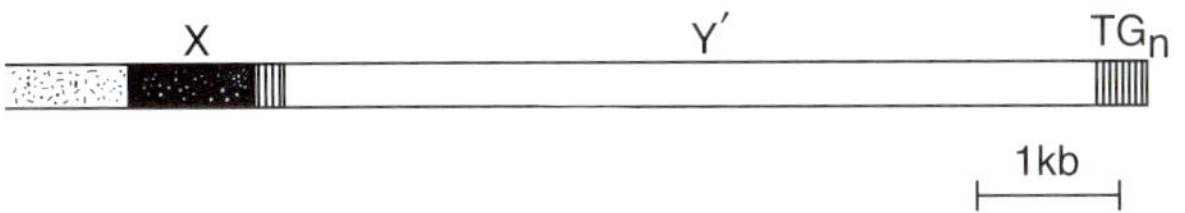

Figure 2 The general structure of the telomeric DNA of the budding yeast *Saccharomyces cerevisiae*. Closely-spaced vertical lines represent the short terminal repeats (TG_n on the 3′-terminated strand), sometimes present also in the subtelomeric DNA. Open and filled boxes represent Y′ and X sequences and internal chromosome sequence is stippled. The number of Y′ elements varies between 0 and 4, but only one is shown here.

Telomere Binding Proteins

Proteins binding at telomeres probably serve several functions. The most obvious is protection of the DNA

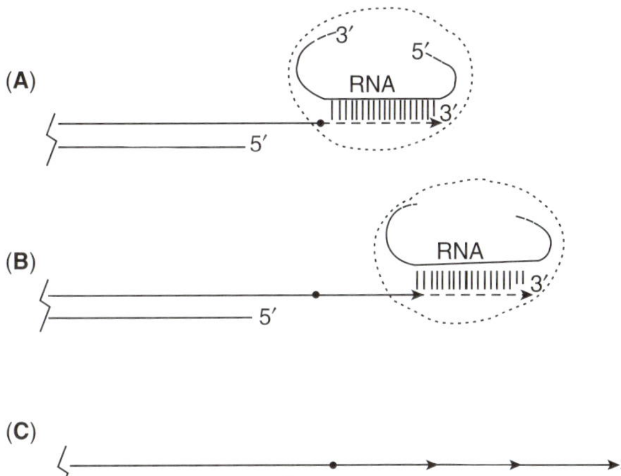

Figure 3 The proposed mechanism for the maintenance of telomeres. (A) Telomerase protein (represented by the dotted ellipse) binds to the 3′-terminated G/C-rich single strand which extends beyond the shortened 5′-terminated strand, and a new repeat sequence is added, with the single-stranded telomerase RNA acting as a template. (B) The telomerase shifts to the new 3′ end, and another repeat sequence is added. (C) After a certain number of repeats have been added (three depicted here), the extended sequence is made double-stranded by RNA-primed repliction, shown as a dashed line; the RNA primer is shown as a short wavy line.

termini against attack by exonuclease. Another, about which little can be said as yet, is to do with the positioning of the chromosomes in the nucleus. It has been known for many years that, especially in meiotic cells, chromosome ends often appear to be attached to the nuclear envelope in clusters, giving the so-called 'bouquet' appearance to the prophase chromosomes. It has been suggested that this could be a part of the process bringing homologous chromosomes together for meiotic pairing.

One idea about the protection of chromosome termini is that the G-rich nature of the telomere DNA permits the formation of tetrameric associations between guanine residues, perhaps resulting in hairpin loops in the DNA chain. Such structures might be protective against exonuclease attack in themselves and might also provide sites for protein binding. Tetra-G-binding proteins do exist, but whether they function especially at telomeres is not yet clear.

Most is known about the telomere binding proteins of the yeast *Saccharomyces*. Not all of them are known to be necessary for telomere function. The *Saccharomyces SIR* genes were first identified as necessary for the maintenance of the 'silent' (i.e., nontranscribed) state of auxiliary genes in the yeast mating-type switching system. At least some of the proteins encoded by them bind not only to the silent mating type 'cassette' loci but also to the subtelomeric repeats of yeast telomeres and help maintain a state of the chromatin that silences, in a clonal fashion, genes artificially inserted within it (an example of position-effect variegation). When *SIR3* is overexpressed, this 'silent' form of chromatin, which is akin to the heterochromatin of higher eukaryotes, can spread along the chromosome, extending the silenced region. It is not clear, however, that this property of telomeres has anything to do with the essential telomere function of preserving chromosome ends.

The DNA binding protein encoded by the essential gene *RAP1* is more clearly concerned with telomere function. It binds at many sites in the genome and acts as a transcription regulator, but it is particularly concentrated at telomeres and functions in the regulation of the number of simple repeats. The increasing amount of bound RAP1p as the telomere elongates appear to inhibit further elongation. Disruption of the binding, either by truncation of the RAP1 polypeptide chain, or by alteration of the telomere repeats by engineering mutations in the *TER1* gene which encodes the RNA template, results in a lengthening of the telomeric sequence and a faster rate of turnover of the terminal repeats (Krauskopf and Blackburn, 1998). At least one other gene *RIF1*, appears to cooperate with *RAP1* in limiting telomerase function. RIF1 protein binds to RAP1 protein, and temperature-sensitive alleles of *RIF1* cause telomere elongation at the restrictive temperature.

Proteins that may be analogous to the yeast telomere binding proteins have been identified in mouse and human cells. The genes *TRF1* and *TRF2* both encode proteins that bind to the TTAGGG terminal repeats, and there is strong evidence that TRF1, like yeast RAP1, regulates telomere length. A mutational deficiency in the protein results in telomere lengthening, and its overproduction to telomere shortening. It thus seems to be functionally similar to yeast RAP1, but there is no sequence similarity between the two proteins.

The yeast gene *EST2*, so called because some *est* mutants have Extremely Short Telomeres, encodes the protein component of the telomerase itself. Telomere-shortening *est* mutants in general suffer early senescence but give rise to occasional clones of cells with restored growth. Analysis of the telomeres of these revived clones has shown that they had been regenerated not by regrowth of the standard telomeric repeats, but by recruitment of additional subtelomeric Y′ elements. It seems that the revived cells have built up a surplus of Y′ elements by some amplification process and transferred them to chromosome termini, probably by recombination between the few TG_n repeats still present at the termini and TG_n repeats intercalated among the donor Y′ elements (Lundblad and Blackburn, 1993). This example shows that organisms are not necessarily dependent on one particular kind of repeat for maintaining their chromosome ends: other kinds of renewable sequence may be able to substitute. This principle is still more strikingly illustrated by the telomeres of *Drosophila*.

Drosophila Telomeres

Surprisingly, exploration of the ends of *Drosophila* chromosomes has revealed no short G/C-rich terminal repeats of the kind found in other organisms. Instead, the termini are composed largely of a long repetitive element (HeT-A), interspersed at some termini with another such element (TART). HeT-A and TART are respectively 6 kb and 5.1 kb in length and are allied to the LINE (long interspersed element) class of retroelements. TART has open reading frames encoding a reverse transcriptase and a protein with similarities to gag, typical of retroelements. HeT-A encodes no reverse transcriptase, but reveals its relationship to a retroelement by possessing a gag-like open reading frame. Both have poly-A sequences at their 3′ ends, as one would expect of sequences propagated by reverse transcription from mRNA. Next to the HeT-A/TART elements on the centromere side is a region of about 10 kb consisting of shorter repeats of between 0.5 and 1.8 kb (**Figure 4A**).

The current hypothesis for *Drosophila* telomere maintenance is that HeT-A and TART RNA transcripts are reverse-transcribed into DNA single strands, the 3′-poly(A) tails of which are brought into alignment with, and then ligated to 5′-termini at the chromosome ends. The added sequence is then made double-stranded by DNA synthesis primed from the previous chromosome end (**Figure 4B**). The gag-like DNA binding protein is essential for the addition process and may be involved in the alignment step (**Figure 4B**).

The addition of a HeT-A or TART element is a very substantial elongation by comparison with the small additions typically catalyzed by telomerase. One added HeT-A or TART copy should sustain chromosome replication for many fly generations.

Whereas previously-known LINE elements, such as the I element of Drosophila which is responsible for one kind of hybrid dysgenesis, are inserted all over the genome, HeT-A and TART seem to be confined to chromosome ends. LINEs, like other transposons and retrotransposons, are generally considered to be 'selfish' elements, but HeT-A and TART appear to have been recruited to perform an essential function for their host organism.

The subtelomeric regions of *Drosophila* chromosomes are like those of other organisms in consisting of the order of tens of kilobases of tandemly repetitive sequence of no ascribed function. As in *Saccharomyces*, this repetitive sequence may provide a substrate for formation of heterochromatin, since genes artificially placed within it tend to be 'silenced,' i.e., not transcribed.

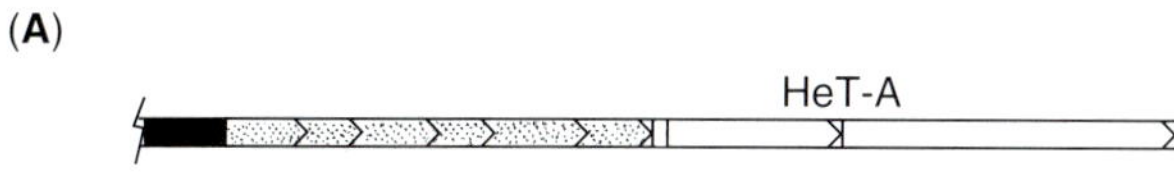

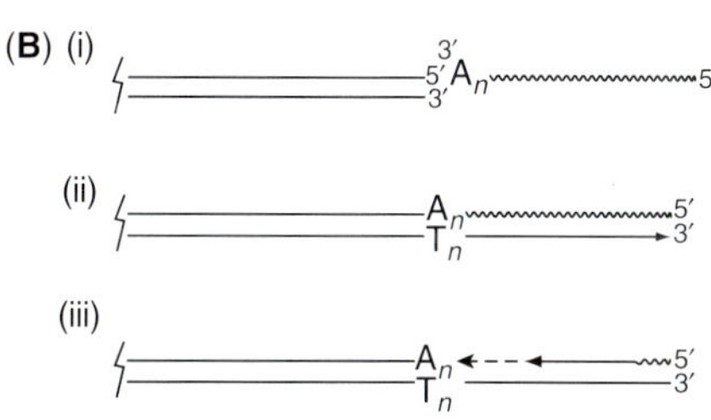

Figure 4 (A) The DNA structure of *Drosophila* telomeres. The terminal tandemly arranged Het-A/TART retroelements are shown as open boxes; the subterminal element has been shortened by many cycles of replication. Stippled boxes represent shorter subtelomeric tandem repeats of variable length, and the black section non-telomeric chromosomal DNA. (B) Proposed mechanism of Het-A/TART addition; (i) RNA (wavy line) is reverse-transcribed from Het-A or TART, with the 3′-terminal poly-A tract brought into alignment with the chromosome end, probably with the aid of the gag protein encoded in both elements; (ii) the 3′-terminated DNA strand is extended by reverse transcriptase (encoded in TART) using the RNA as template, so adding a single-strand DNA copy of one of the retroelements; (iii) the RNA is removed, and second DNA strand synthesis is primed from the distal end by primase-synthesized RNA.

Telomeres, Aging, and Cancer

The chromosomes of embryonic cells are adequately equipped with telomeric repeated sequences and generally (a notable exception is *Drosophila*) possess telomerase activity. But as cells begin to differentiate, their telomerase usually decreases to undetectable levels. This is one reason why somatic cells have a limited lifespan. The telomeres of cells in culture decrease in length with successive transfers (**Figure 5**), and when they fall below a critical minimum length, cell division stops.

On the other hand, cancer cells, or cells with cancerous potential, are 'immortalized' and can grow indefinitely. A common, though not quite universal,

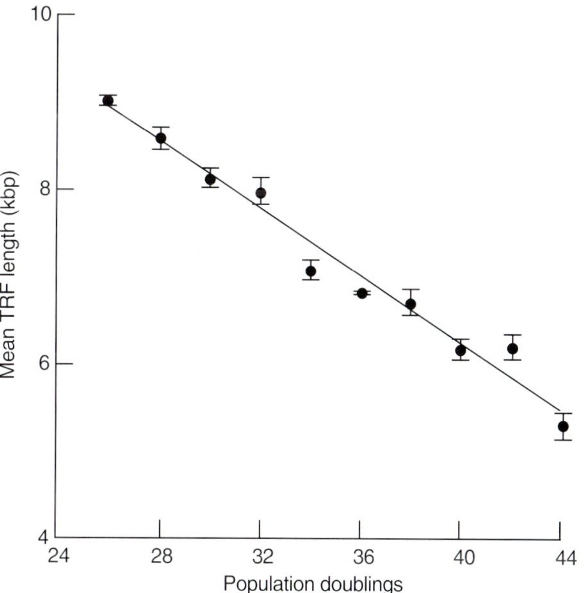

Figure 5 Decrease in telomere length with time of culture of cells from human vascular tissue. The size of a terminal restriction fragment (TRF) – i.e., the narrowing distance between a fixed subterminal restriction site and the chromosome end – is plotted (with standard errors) against number of cell generations (population doublings). (From Chang and Harley, 1995.)

characteristic of immortalized cells is that they have at least detectable levels of telomerase. Very significantly, the lifetime of at least some nonimmortalized human cell lines in culture can be extended, perhaps indefinitely, by transfection with DNA vectors encoding telomerase protein (Bodnar *et al.*, 1998), though other studies have shown that extra telomerase alone is not sufficient to immortalize all cell types. An increased amount of telomerase is not usually considered to be a main cause of cancer, and neither is the failure to maintain telomerase levels considered to be a main cause of programmed cell death (apoptosis). But it does seem that upregulation of telomerase production usually accompanies oncogenesis, and is perhaps necessary for it.

Further Reading

Blackburn EH and Greider CW (eds) (1995) *Telomeres*. Plainview, NY: Cold Spring Harbor Laboratory Press.

Chadwick DJ and Cardew G (eds) (1997) *Telomeres and Telomerase (*Ciba Symposium). Chichester, UK: John Wiley.

Greider CW (1996) Telomere length regulation. *Annual Review of Biochemistry* 65: 337–365.

Kipling D (1995) *The Telomere*. Oxford: Oxford University Press.

Louis EJ The chromosome ends of *Saccharomyces cerevisiae*. *Yeast* 11: 1553–1573.

Mason JM and Biessman H (1995) The unusual telomeres of *Drosophila*. *Trends in Genetics* 11: 58–62.

van Steensel B and de Lange T (1997) Control of telomere length by the human telomeric protein TRF1. *Nature* 385: 740–743.

Zakian VA (1996) Structure, function and replication of *Saccharomyces cerevisiae* telomeres. *Annual Review of Genetics* 30: 141–172.

Zakian VA (1995) Telomere function: lessons from yeast. *Trends in Cell Biology* 6: 29–33.

References

Bodnar AG, Ovelette M, Frolks M, Holt SE *et al.* (1998) Extension of life-span by introduction of telomerase into normal human cells. *Science* 279: 349–352.

Chang E and Harley CB (1995) Telomere length and replicative aging in human vascular tissues. *Proceedings of the National Academy of Sciences, USA* 92: 11190–11194.

Krauskopf A and Blackburn EH (1998) *Rap1* protein regulates telomere turnover in yeast. *Proceedings of the National Academy of Sciences, USA* 95: 12486–12491.

Lundblad V and Blackburn EH (1993) An alternative pathway for yeast telomere maintenance rescues est1 senescence. *Cell* 73: 347–360.

See also: **Aging, Genetics of; Chromosome Structure; DNA Replication; Heterochromatin; Mating-Type Genes and their Switching in Yeasts; Position Effects; Retroposon; Reverse Transcription; SINE**

Temperate Phage

E Thomas

doi: 10.1006/rwgn.2001.1277

Temperate phages are bacteriophages that can sometimes coexist with their host for extended periods of time, during which time the host and its internal phage multiply in synchrony (unlike lytic or virulent phages). Instead, temperate phages have two lifecycles to choose from. They can undergo a lytic life cycle (see Virulent Phage for more detail) or a lysogenic lifecycle. The lysogenic lifecycle is unique to temperate phages.

The first step in any phage life cycle is entry of the phage into the host cell. After the virus has successfully attached itself to the outside of the bacterial cell it inserts its genome (either DNA or RNA) into the cell. In the lytic life cycle, the genome is immediately used to begin replication of the virus. In the lysogenic lifecycle, the viral genome will remain dormant within the bacterial cell, either as a plasmid or incorporated into the host's genome, replicating only when the host genome is replicated; the regulatory mechanisms involved are discussed elsewhere. The lysogenized phage can later enter the lytic life cycle; this process is called induction. Upon induction, viral genes will be transcribed and translated, new progeny phage will be made, and the cell will be lysed to release the new phage into the environment. Lysogeny presumably provides temperate phage with a selective advantage, because they can delay the lytic life cycle until conditions are favorable.

See also: **Bacteriophages; Lysogeny; Virulent Phage**

Temperature-Sensitive Mutant

B S Guttman

doi: 10.1006/rwgn.2001.1278

A temperature-sensitive mutation creates a phenotype that varies with the temperature. The classic temperature-sensitive (*ts*) mutants were obtained by R.S. Edgar as part of a program to map the phage T4 genome extensively; at the same time, host-dependent (*amber*) mutations were found by R.H. Epstein and his associates for the same purpose (see Epstein *et al.*, 1963). (Both types of mutants are called 'conditional lethals'; Edgar, 1966.) The *ts* mutants of T4 were defined as those that are able to form plaques at 25 °C but not at 42 °C (Edgar and Lielausis, 1964). A set of *ts* mutants can be mapped relative to one another by crosses in which bacteria are infected simultaneously by two mutants at the low (permissive) temperature, so they can produce recombinants. Progeny phage are plated and incubated at low temperature to count total phage and at high temperature to count recombinants. The *ts* mutants are also mapped relative to classical mutants and *amber* mutants.

Edgar (1966) pointed out that temperature-sensitive mutants had been obtained much earlier in other organisms. Some classical alleles are also temperature sensitive. For instance, some alleles for fur color in Siamese cats and mice cause the expression of color only in the cooler tissues at the extremities of the feet, ears, nose, and tail.

References

Edgar RS (1966) Conditional lethals. In: Cairns J, Stent GS and Watson JD (eds) *Phage and the Origins of Molecular Biology*, pp. 166–172. Plainview, NY: Cold Spring Harbor Laboratory Press.

Edgar RS and Lielausis I (1964) Temperature sensitive mutants of bacteriophage T4D: their isolation and genetic characterization. *Genetics* 49: 649–662.

Epstein RH, Bolle A, Steinberg C *et al.* (1963) Physiological Studies of conditioned lethal mutants of bacteriophage T4D. *Cold Spring Harbor Symposia on Quantitative Biology* 28: 375–392.

See *also*: Conditional Lethality; Plaques; T Phages

Template

M M Hingorani

doi: 10.1006/rwgn.2001.1279

A template is a single-stranded DNA or RNA polymer that is used to direct synthesis of another polymer such as DNA, RNA, or protein. DNA is used as a template molecule for DNA replication, DNA repair, as well as for transcription. DNA polymerases use template DNA by covalently linking deoxyribonucleoside 5′-triphosphates that base pair with template DNA to form a new, complementary DNA strand. The polymerase 'reads' the template in the 3′→5′ direction, while synthesizing DNA in the 5′→3′ direction to form the antiparallel double-stranded DNA product. This process is essential for duplication of genomic DNA, which precedes cell division. RNA polymerases use template DNA similarly to create complementary strands of RNA, which are then used as messenger RNA and translated into proteins or used as RNA components of cellular machinery.

Template DNA is prepared for replication and transcription by the unwinding action of enzymes known as helicases. These enzymes utilize energy from molecules such as ATP to destabilize the hydrogen bonds between base pairs, which leads to separation of the two strands of the duplex. Single-stranded DNA can quickly reanneal to duplex form, therefore most cellular organisms have special single-stranded DNA binding proteins that stabilize the template until it is copied or transcribed. Excision of damaged DNA or DNA nuclease activity can also generate single-stranded DNA. This template is converted to double-stranded DNA during the processes of DNA repair or recombination.

RNA is also used as a template molecule for synthesis of both RNA and DNA. Similar to the DNA-dependent polymerases described above, RNA-dependent polymerases recognize and bind single-stranded RNA template and covalently link complementary ribonucleotides to form RNA polymers. RNA-dependent DNA polymerases known as reverse transcriptases use single-stranded RNA template to form complementary strands of DNA. Retroviruses such as human immunodeficiency virus (HIV) use these enzymes to convert their genomic RNA into single-stranded DNA, which is used as a template for transcription, and into double-stranded DNA which is integrated into the host organism DNA during infection. RNA-dependent RNA polymerases also use RNA as a template. RNA viruses such as the influenza virus and poliovirus use RNA polymerases to transcribe as well as to replicate their genomic RNA.

As well as its role in nucleic acid metabolism, single-stranded RNA is used as template during protein synthesis. Messenger RNA (mRNA), a scrupulous copy transcribed from the coding DNA template or RNA template, is in turn translated into amino acid polymers by the translation machinery in the cell. During translation, groups of three consecutive bases on mRNA (codons) are recognized and paired with corresponding amino acids, and the amino acids are successively linked to form polypeptides. Thus, the mRNA template is 'read' in the 5′→3′ direction, codon by codon leading to a faithful copy of the information in the form of a protein.

By using a relay system of templates from replication to transcription to translation, living organisms can faithfully maintain their genetic information over successive generations as well as utilize this genetic information accurately for the process of living.

See also: **DNA Repair; DNA Replication; Polymerase Chain Reaction (PCR); Transcription; Translation**

Terminal Redundancy

doi: 10.1006/rwgn.2001.2046

Terminal redundancy refers to the situation where certain phages have a duplication at one end of the other end. This allows them to circularize when they enter the host cell.

See also: **Bacteriophages**

Termination Codon

See: **Amber Codon; Nonsense Codon; Ochre Codon; Opal Codon**

Termination Factors

See: **Release (Termination) Factors**

Terminator

doi: 10.1006/rwgn.2001.2048

A terminator is a DNA sequence at the end of a transcript that causes RNA polymerase to stop transcription.

See also: **RNA Polymerase; Transcription**

Test Cross

J Merriam

doi: 10.1006/rwgn.2001.1281

Mendel originated the test cross as a cross of a hybrid individual by a purebred, or homozygous for the recessive trait(s) segregating in the hybrid. Mendel proposed that a hybrid individual would make two kinds of gametes, each equally frequent, for each heterozygous character, resulting in two types for a monohybrid, four types for a dihybrid, or eight types for a trihybrid. These types could only be inferred from the results of hybrid × hybrid crosses, however. To test each hybrid for the gamete types it produces more directly, Mendel introduced the idea of crossing to a purebred for recessive traits as the phenotype of each progeny would then reflect the gamete type contributed by the hybrid parent. Test crossing a dihybrid, for example, resulted in 1:1:1:1 ratios, confirming the gamete types that underlie the 9:3:3:1 ratios observed from the self-cross of the dihybrid. Currently test crosses are widely used to simplify the analysis of progeny from parents that are heterozygous for many genes, for example, in linkage and gene mapping studies.

See also: **Backcross; Linkage Map; Punnett Square**

Testes Determining Locus

L Silver

doi: 10.1006/rwgn.2001.1282

The mammalian testes determining locus (symbolized as *Tdy*, because of its location on the Y chromosome) is also called the testes determining factor (or TDF) in humans. This single gene present on the Y chromosome of all male mammals is necessary and sufficient for the development of the fetus along a male pathway of differentiation in both germ cell and somatic cell tissues. The locus was identified through genetic studies of people who carry a Y chromosome but are developmentally female (such people have a deletion over the *TDF* gene), and other people who do not carry the Y chromosome but are developmentally male (such people have a copy of the *TDF* gene translocated to another chromosome).

See also: **Sex Determination, Human; Y Chromosome (Human)**

Tetrad Analysis

J R S Fincham

doi: 10.1006/rwgn.2001.1283

Meiotic Tetrads: Advantages and Availability

In most eukaryotic organisms that have been studied genetically, the segregation of alleles at meiosis can be studied only in randomized meiotic products. On the female side, only one product of each meiotic cell survives to form the egg nucleus, and in males, although all products are potentially viable, they can be recovered in their original tetrads only rarely. In many fungi, however, and some algae such as *Chlamydomonas*, analysis of whole tetrads is possible. Tetrad analysis permits the confirmation of rules of meiosis which are believed to apply universally but are less directly demonstrable in randomized meiotic products and can be confirmed by microscopy only with especially favorable material.

Some of the same advantages can be obtained by half-tetrad analysis, that is the recovery of two out of the four postmeiotic chromosome copies in the same meiotic product; this is possible in *Drosophila melanogaster* through the use of attached-X chromosomes.

Tetrads and Octads in Fungi

The meiotic products of the ascomycete fungi are ascospores, initially held together within the ascus, which is a sac formed from the cell wall of the meiotic mother cell. The budding yeast *Saccharomyces cerevisiae* and

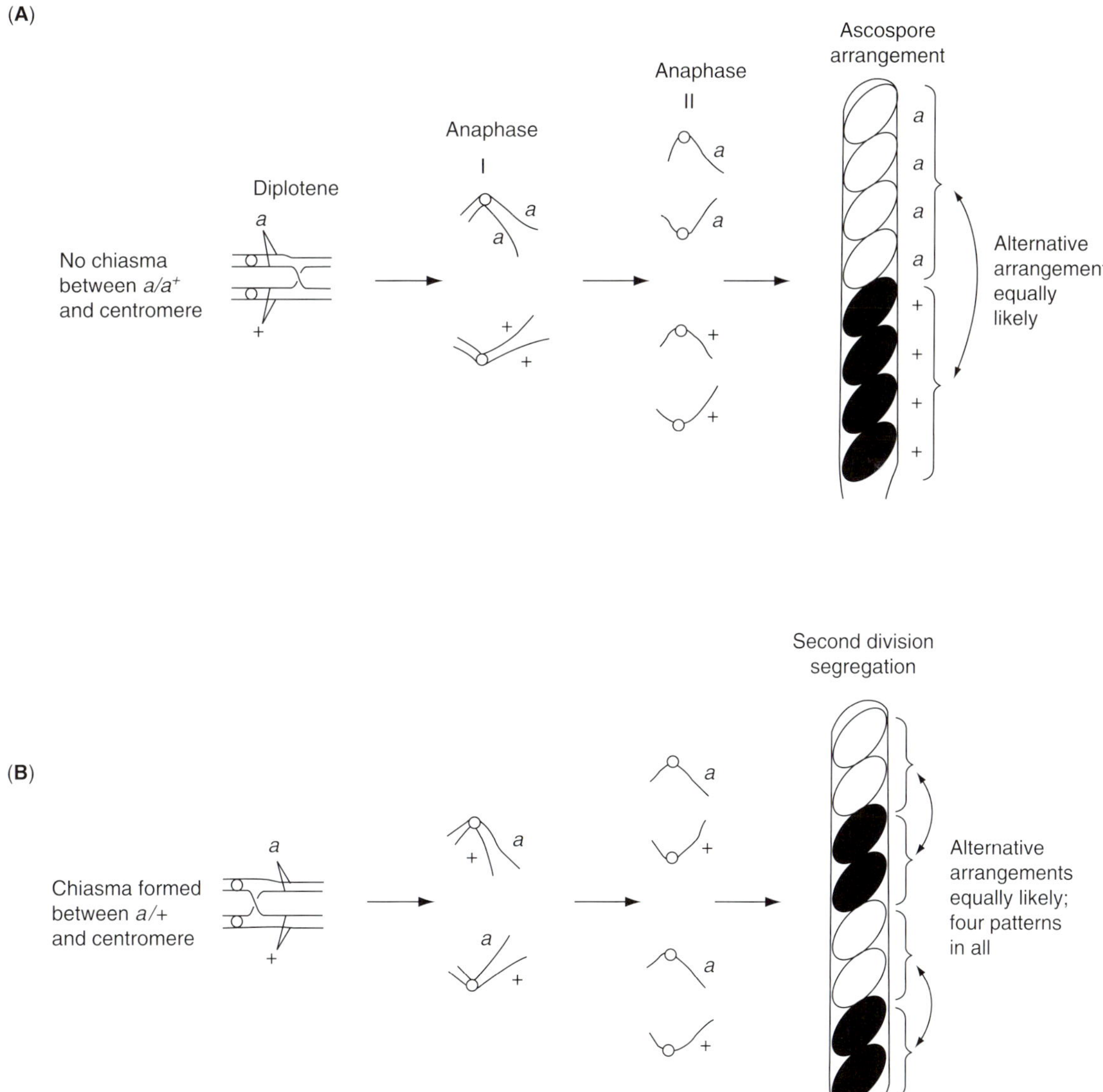

Figure 1 The 4:4 segregation of a pale-ascospore mutation (shown as *a*) in an ascomycete fungus with ordered asci, such as *Sordaria fimicola*. (A) No crossing over between the gene and the centromere results in segregation at the first division of meiosis. (B) A crossover between the gene and the centromere results in segregation at the second division. (From Fincham JRS (1983) *Genetics*. Bristol, UK: John Wright & Sons.)

the fission yeast *Schizosaccharomyces pombe* have four ascospores in each ascus. Most of the filamentous species, however, (e.g., *Neurospora*, *Sordaria*, and *Ascobolus*) have eight – a tetrad of spore pairs resulting from a mitotic division following meiosis.

In all these species it is relatively easy to make crosses between genetically distinct strains, dissect out ascospores before their discharge from the asci, and grow them into separate haploid cultures. Tetrad analysis has also been performed in the basidiomycete fungi, particularly in the mushroom group (agarics, e.g., *Coprinus* and *Schizophyllum* spp.), but not so commonly as in the Ascomycetes.

Confirmation of 1:1 Segregation (and Exceptions)

Simple Mendelism and the chromosome theory predict that meiosis in a heterozygote, with alleles *A* and *a*, should result in two *A* and two *a* products in every tetrad (or 4:4 in eight-spored asci). This prediction is most easily checked when, as in *Sordaria* and *Ascobolus*, mutant alleles are available that affect ascospore color and are therefore scorable in undissected asci. In *Sa. cerevisiae* (the most intensively studied yeast) such directly visible markers are not available. The asci have to be dissected and the ascospores grown, but this can now be done sufficiently rapidly for analysis of hundreds or thousands of tetrads.

The data show that the simple Mendelian rule holds in the great majority of asci, but not universally. A minority of tetrads, usually of the order of 1% or 0.1% in the filamentous species but typically ranging between 1 and 10% in *Saccharomyces*, show 3:1 or 1:3 ratios of spores or spore pairs, an anomaly now attributed to gene conversion (Gene Conversion).

Another exception is the occasional segregation of the allelic difference at the mitotic division following meiosis, giving mismatched sister spores and 5:3, 3:5, or (with two mismatches in the ascus) aberrant 4:4 ratios in the eight-spored species, or 50:50 mosaic single-spore colonies in yeast. This postmeiotic segregation can be explained by gene conversion affecting half-chromatids (single DNA strands) at the first prophase of meiosis.

Although tetrad analysis provided the means of detecting exceptions to the 2:2 rule, gene conversion will not be considered in this article, which will proceed on the assumption that the rule of 2:2 segregation always holds.

Ordered Tetrads

The long narrow asci of such Ascomycetes as *Neurospora* and *Sordaria* species are ordered in the sense that the positions of the spores reflect the two divisions of meiosis. The first division spindle is oriented lengthwise along the ascus. The second division spindles are (with some exceptions, notably in *So. brevicollis*) arranged end-to-end without overlap, so that alleles separated from one another at the first division end up in spores in different halves of the ascus. The postmeiotic mitotic spindles are also nonoverlapping, so each product of meiosis is represented by a pair of adjacent spores. The occurrence of spore pairs carrying different alleles in the same half of the ascus indicates second division segregation, the result of crossing over between the gene locus and the chromosome centromere (First and Second Division Segregation). Second division segregation frequency generally approaches a maximum of two-thirds for genes far from their centromeres (**Figure 1**).

Representing the allelic difference as $+$, *a* (for wild-type and mutant alleles), and writing the constitutions of spore pairs in order from the tip to the base of the ascus, there are two equally frequent first division segregation patterns, $+ + a\ a$ and $a\ a + +$, and four equally frequent second division segregation patterns, $+ a + a$, $+ a\ a +$, $a + a +$ and $a + + a$ (**Figure 1**). These statistical equalities show that the orientation of the first division bivalents and second division dyads on the division spindles is essentially random.

Asci in *Ascobolus* and *Saccharomyces* are more or less spherical and not ordered, so that information on first versus second division segregation has to be obtained in other ways (see below).

Two-Locus Segregations: Independent Assortment

When two crossed strains differ at two gene loci, there are (neglecting order within the ascus) three possible ascus types: parental ditype (PD), nonparental ditype (NPD), and tetratype (T). A good example is provided by ascospore color mutants, yellow (y) and buff (b) in *Sordaria* spp. Wild-type ascospores are nearly black, and double mutant ascospores are white. We can symbolize a buff × yellow cross as $b\ y^+ \times b^+\ y$, the $+$ superscripts indicating the wild-type alleles. Then PD asci will have two buff ($b\ y^+$) and two yellow ($b^+\ y$) spore pairs, NPD asci will have two black ($b^+\ y^+$) and two white ($b\ y$) spore pairs, and T asci will have one spore pair of each color (**Figure 2**). If the two segregating loci are unlinked, PD and NPD asci should be statistically equal in frequency. If PD asci are significantly more frequent than NPD, that is *prima facie* evidence for linkage.

If the loci of the two genes are not linked, the frequency of T asci depends on the two second division segregation frequencies, which we will call p and q.

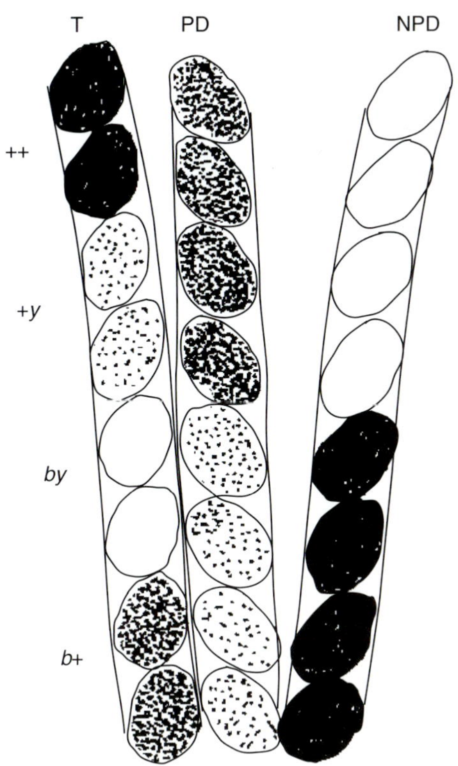

Figure 2 The three ascus types resulting from a cross between two pale-ascospore mutants in *Sordaria*. *b*, buff; *y*, yellow; *b y*, the double mutant, is white. PD, parental ditype; NPD, nonparental ditype; T, tetratype. Buff and yellow are distinguished from wild-type (black) by dark and light stippling, respectively. Wild-type alleles are shown as +.

Tetratypes must always result when one locus segregates at the first division and the other at the second: the overall frequency is $p(1-q)+q(1-p)$. They will also result from one half of the cases where both segregate at the second division, i.e., $pq/2$. One can easily show this by writing down all the equally probable possibilities. PD and NPD asci will each result from one half of the cases where both loci segregate at the first division, and one quarter of the cases where both segregate at the second division. Thus:

$$\mathrm{f(PD)} = \mathrm{f(NPD)} = (1-p)(1-q)/2 + pq/4 \quad \text{and}$$
$$\mathrm{f(T)} = p(1-q) + q(1-p) + pq/2$$

With three independently segregating allelic differences, a^+/a, b^+/b and c^+/c, it is in principle possible, by determining the tetratype frequencies from the three crosses $a^+\ b \times a\ b^+$, $a^+\ c \times a\ c^+$ and $b^+\ c \times b\ c^+$, to evaluate all three second division frequencies from three simultaneous equations. But if second division segregation frequencies are required, it is simpler to find a mutation that virtually always segregates at the first division. Then the tetratype frequency in a cross of that mutant to any other will be the second division segregation frequency of that other mutant.

Linked Segregations

When two segregating loci are linked, the three tetrad types give information about crossing over between the two. PD asci are most simply interpreted as absence of crossing-over, although they can also result from two-strand double crossovers, with the second crossover involving the same two chromatids as the first, so cancelling its effect. (As a first approximation, one can neglect the possibility of three or more crossovers in the same interval.)

Tetratypes result mainly from single crossovers. The observation that, except for the most distant linkages, the majority of recombinant products occur in tetratype asci is the most direct genetic evidence that crossing-over occurs after the chromosomes have divided, and involves only one chromatid from each chromosome. Tetratype tetrads also confirm that crossing-over is reciprocal, with *Ab* and *aB* always produced together. Except in a case of very close linkage, tetratype asci also result from three-strand double crossovers – that is with one chromatid crossing over twice, two chromatids involved once each and the fourth one not at all.

Four-strand doubles, with the second crossover occurring between the two chromatids not involved in the first (**Figure 3**), result in nonparental ditype asci.

Strand Relationships in Double Crossovers

As **Figure 3** shows, there are four kinds of double crossover, but their relative frequences cannot be obtained from two-point crosses because two-strand doubles are indistinguishable from no crossing-over and the two kinds of three-strand doubles both look like singles. With three linked allelic differences, however, the four types can all be distinguished when one crossover falls within each interval (**Figure 4**).

Extensive data, mostly from *Saccharomyces* and *Neurospora*, support the generalization that the four types of double crossover are equally probable, and that each chromatid has a 50% chance of being involved in each crossover, whether or not it is involved in another crossover. This is called absence of chromatid interference.

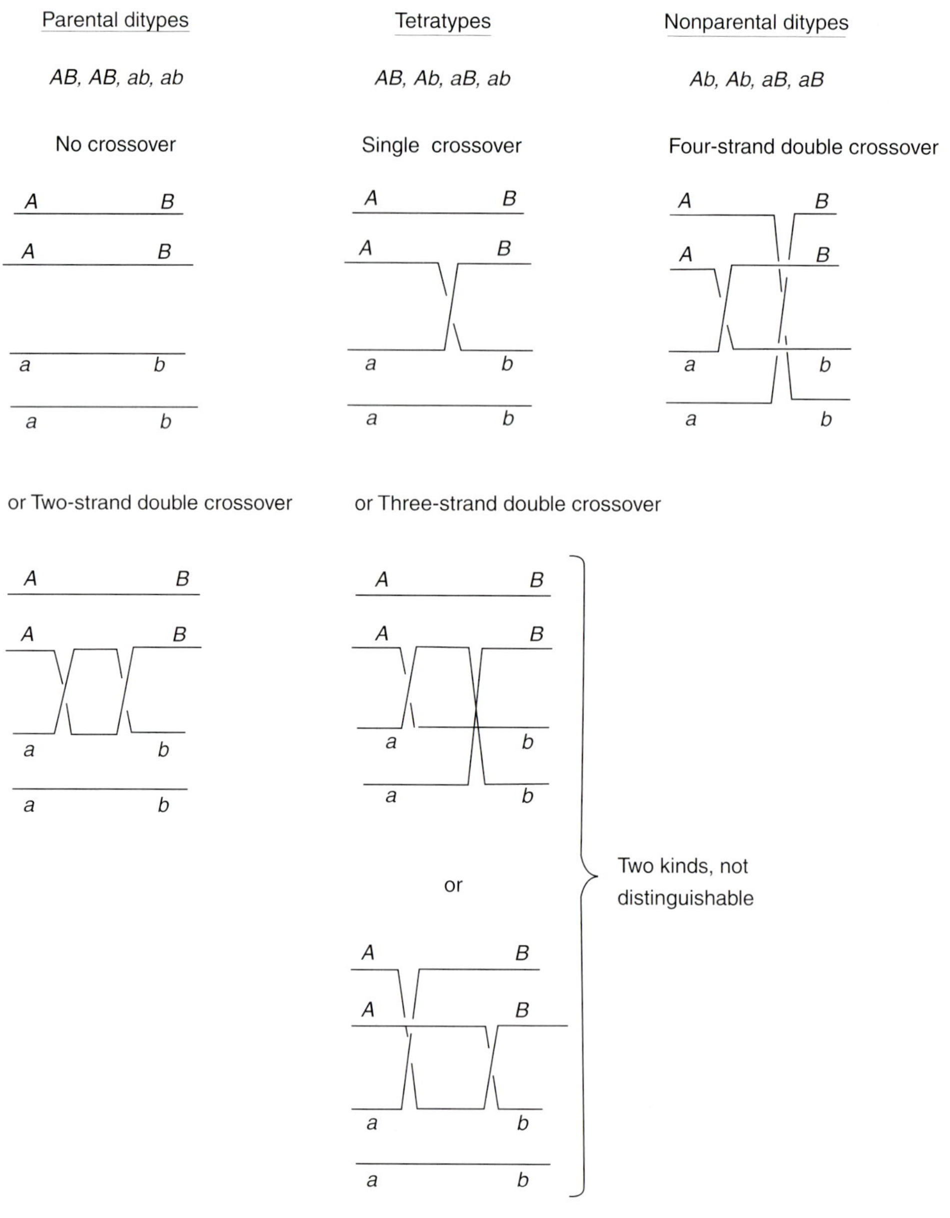

Figure 3 The explanation of the three tetrad classes, PD, NPD, and T, when the two genes concerned are linked. The cross is *A b* × *a B*, with *a* and *b* two linked mutations and *A* and *B* the corresponding wild-type alleles.

Distinguishing between Independence and Distant Linkage

When loci marked by allelic differences (markers) are very far apart on the same chromosome, the frequency of nonparental ditype tetrads may approach that of parental ditypes, and, by this criterion, the loci may be judged to be unlinked. However, this situation will arise only when the crossover frequency is so high that the *A*/*a* and *B*/*b* pairs of alleles are distributed among the four products of meiosis virtually independently of each other and at random. The result of the four meiotic products each receiving one random *A* allele and one random *B* allele will be a PD:NPD:T ratio of 1:1:4.[1]

In other words, a crossover frequency high enough to bring the NPD and PD frequencies close to equality will generate twice as many tetratypes as ditypes. With unlinked loci, the frequency of tetratypes can

[1]Given that four meiotic products carry two *A* and two *a* alleles, in arbitrary order *A A a a*, there are six ways of distributing two B and two *b* alleles among them at random: *AB AB ab ab*, *Ab Ab aB aB*, *AB Ab aB ab*, *Ab AB ab aB*, *AB Ab ab aB*, and *Ab AB aB ab*.

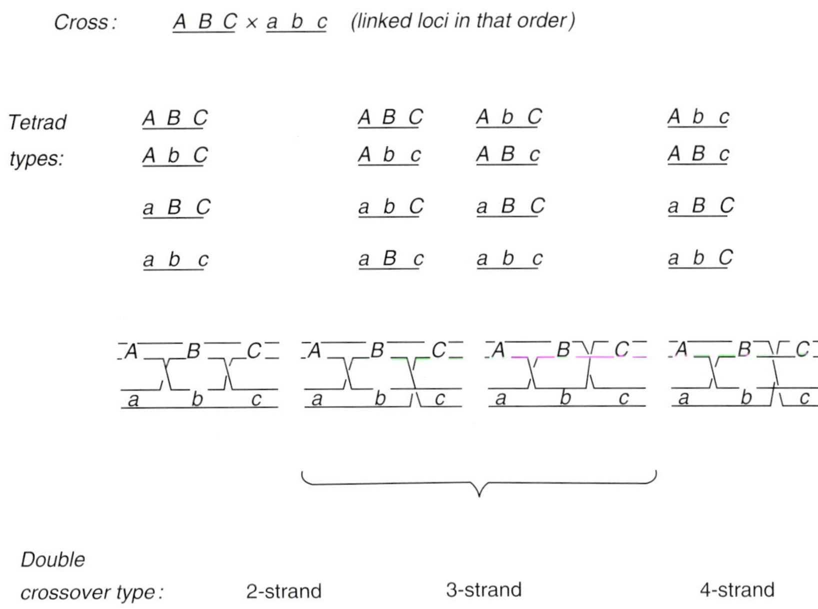

Figure 4 The detection of four kinds of double crossover (two-strand, four-strand, and two kinds of three-strand) by tetrad analysis. They occur with equal probabilities. The three linked marker mutations are here shown as *a*, *b*, *c* and the corresponding wild-type alleles as *A*, *B*, *C*.

take any value between two-thirds and zero, depending on the second division segregation frequencies of the loci concerned (see equation above). So, while a tetratype frequency of two-thirds is in itself ambiguous, a tetratype frequency of significantly less than two-thirds of the total can be taken as evidence that the two loci are unlinked.

Further Reading

Fincham JRS and Day PR (1961) *Fungal Genetics*. Oxford: Blackwell Scientific Publications.

See also: **Attached-X and other Compound Chromosomes; Chromatid; Chromatid Interference; Crossing-Over; First and Second Division Segregation; Gene Conversion; Meiosis; Mendel's Laws; *Neurospora crassa*; Postmeiotic Segregation**

Tetraparental Mouse

L Silver

doi: 10.1006/rwgn.2001.1284

The cells contained within the early mammalian embryo (from zygote to eight-cell morula) are all undifferentiated and all capable of directing the development of an entire individual animal by themselves, even though they typically participate in the creation of an animal with their sister cells. A bizarre consequence of this 'totipotency' of cleavage stage cells is the formation of chimeras. The term chimera comes from Greek mythology and is used to designate an embryo or animal that is composed of cells from two or more different origins. (The mythological Chimera is a composite of a lion, goat, and serpent.)

The production of chimeric mice was first reported in 1961 by the Polish embryologist Tarkowski. He accomplished this feat by first taking the zona pellucida off two cleavage stage embryos to obtain denuded cell masses that are naturally sticky. When two denuded embryos are pushed together, they form a single chimeric cell mass that is capable of undergoing normal development within the female reproductive tract. When the two embryos are obtained from different females mated to different males, the resulting animal has four parents and is considered to be tetraparental. It is also possible to produce hexaparental animals that are derived from a combination of three embryos. The production of chimeric mice is an essential component of the targeted mutagenesis technology that has revolutionized the use of the mouse as a model organism for studying human diseases.

See also: **Chimera; Targeted Mutagenesis, Mouse**

Tetratype

J R S Fincham

doi: 10.1006/rwgn.2001.1285

This term refers to the type of meiotic tetrad of haploid products formed in a diploid, doubly heterozygous meiotic cell (*A/a B/b*) that contains all four possible combinations of alleles: *A B*, *A b*, *a B*, and *a b*. For formation of a tetratype tetrad at least one of the two allelic differences must segregate at the second division of meiosis. When the *A* and *B* loci are linked on the same chromosome, a tetratype is most simply explained as due to a single crossover between the two, but can also result from certain kinds of double or multiple crossing-over.

***See also:* Tetrad Analysis**

Thalassemias

D J Weatherall

doi: 10.1006/rwgn.2001.1286

The thalassemias are a group of inherited disorders of hemoglobin. They were first reported independently from the United States and Italy in 1925. The word 'thalassemia,' derived from Greek roots for 'the sea' and 'blood,' was invented under the mistaken belief that these diseases were confined to the Mediterranean region. It was later discovered that they are the commonest single human gene disorders and have a widespread distribution in many countries of the world (**Figure 1**).

Different Types of Thalassemia

The thalassemias result from inherited defects in the synthesis of the globin chains of hemoglobin. Humans have different hemoglobins at various stages of development. Normal adults have a major hemoglobin (Hb) called HbA, comprising about 97% of the total, and a minor component, HbA_2 which accounts for 2–3%. The main hemoglobin in fetal life is HbF, traces of which are found in normal adults. There are three embryonic hemoglobins. All these different hemoglobins are tetramers of two pairs of unlike globin chains. Adult and fetal hemoglobins have α chains associated with β (HbA, $\alpha_2\beta_2$), δ (HbA_2, $\alpha_2\delta_2$), or γ chains (HbF, $\alpha_2\gamma_2$), whereas in the embryo there are different α-like chains called ζ chains and distinct β-like chains called ε chains. Each individual globin chain has a heme moiety attached to it, to which oxygen is bound.

There are two common types of thalassemia, α- and β-thalassemia, which result from defective synthesis of α or β chains. There are rarer forms in which both δ and β chain, or ε, γ, δ, and γ chain production, is defective, called $\delta\beta$- or $\varepsilon\gamma\delta\beta$-thalassemia, respectively.

The thalassemias are inherited in a Mendelian recessive fashion. The severe, homozygous form of the disease is called thalassemia major while the carrier state, in which only one defective globin gene is inherited, is called the trait. The disease is very heterogeneous from the clinical viewpoint and many patients are encountered who fall in between these extremes; these disorders are called 'thalassemia intermedia.'

Molecular Pathology

Most of the thalassemias result from mutations which involve either the α- or β-globin genes.

α-Thalassemia

The genetics of α-thalassemia is complicated because normal humans receive two α genes from each parent, a genotype which is written $\alpha\alpha/\alpha\alpha$. There are two main classes of α-thalassemia. First, there are the α^o-thalassemias, in which both α genes are deleted, that is all or part of the gene is missing; the homozygous state is written --/--, and the heterozygous state --/$\alpha\alpha$. On the other hand, in the α^+-thalassemias only one of the α genes is lost; the homozygous and heterozygous states are designated -α/-α and, -α/$\alpha\alpha$, respectively. Sometimes α^+-thalassemia results from a mutation which inactivates the α-globin gene rather than deleting it. In this case the heterozygous state is written $\alpha^T\alpha/\alpha\alpha$.

β-Thalassemia

Over 200 different mutations of the β-globin genes have been found in patients with β-thalassemia. They may affect gene function at any level between transcription, processing of the primary messenger RNA transcript, translation, or posttranslational stability of the gene product. Rarely, β-thalassemia, like α-thalassemia, may result from a partial or complete deletion of the β-globin gene. Some of these mutations cause an absence of β-chain production and the resulting disease is called β^o-thalassemia, while others cause a reduced output of β chains, β^+-thalassemia. Some of the latter forms are extremely mild and may not be identifiable in carriers; most heterozygotes for β-thalassemia have very mild anemia and an elevated level of HbA_2.

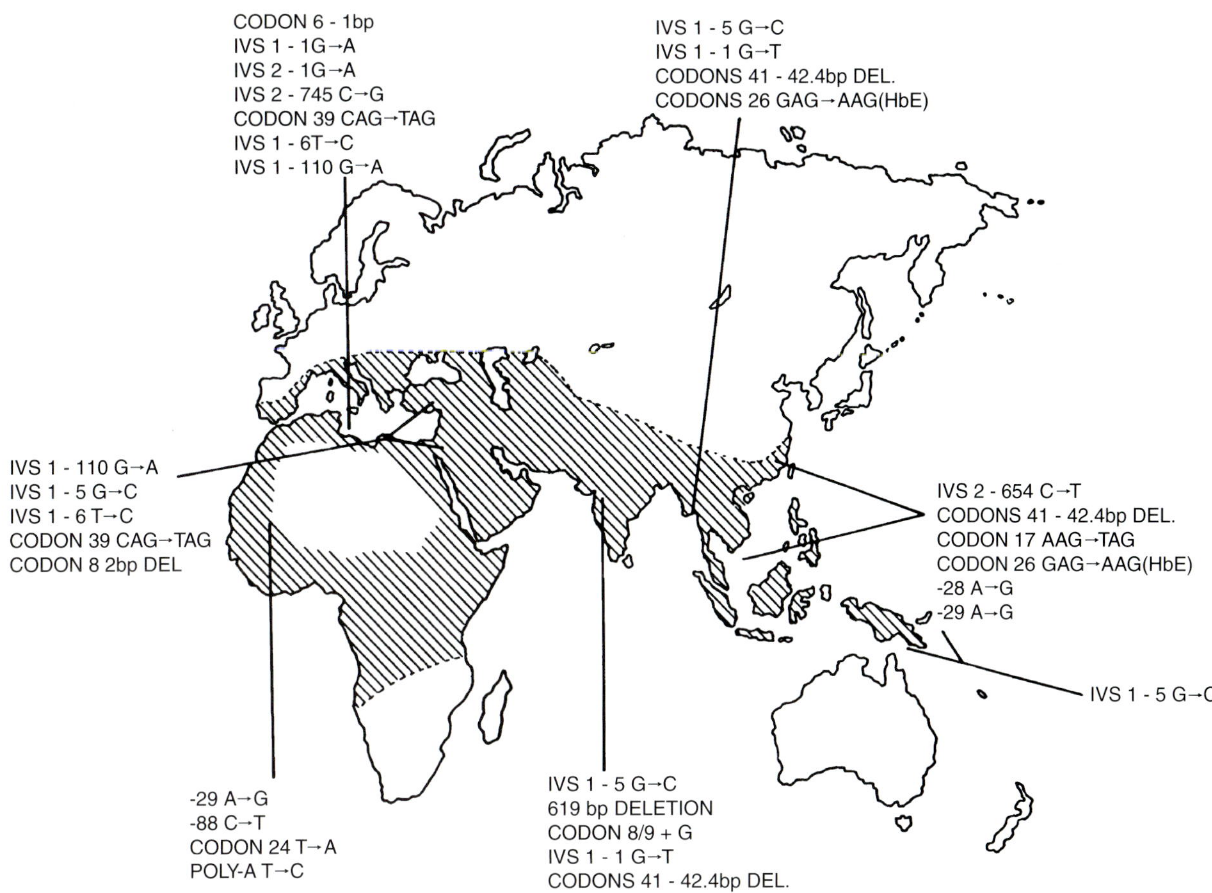

Figure 1 The world distribution of the β-thalassemias. Each population has a different set of mutations. These are described either by the nucleotide base position in introns (IVS 1 or 2) or in the particular codons in exons. Mutations that are given the prefix – are those in the 5′ noncoding regions of the β-globin genes. Those marked poly(A) are mutations in the 3′ noncoding regions.

The hallmark of all the thalassemias is imbalanced globin chain production. In the β-thalassemias this results in an excess of α chains, which precipitate in the red cell precursors, leading to their damage in the bone marrow and shortening the survival of their progeny in the peripheral blood. The pathology of the α-thalassemias is different. In the face of defective α chain production excess γ chains produced in fetal life form γ_4 molecules, while in adults excess β chains form β_4 molecules; these homotetramers are called Hb Bart's (γ_4) and H (β_4), respectively. They do not give up oxygen at normal physiological tensions and are also unstable. This leads to a shortened red cell survival and hence anemia, and patients are further disadvantaged because the high oxygen affinity of the homotetramers leads to reduced oxygen delivery to the tissues.

Clinical Features

The homozygous state for α^0-thalassaemia, that is the loss of all four α globin genes, results in the stillbirth of a hydropic fetus, usually late in pregnancy. These infants are anemic and edematous and show all the features of severe intrauterine hypoxia. Pregnancies carrying these babies are complicated by a high frequency of toxemia and difficulties in delivery, particularly because of enormously enlarged placentas. Individuals who have lost three of their four α genes (-α/--) have a condition called hemoglobin H disease which is associated with moderate anemia and enlargement of the spleen. Persons who have lost two or one of their α globin genes are not incapacitated, but of course may pass on the defective chromosomes to their children.

The homozyous or compound heterozygous (the inheritance of two different alleles) states for severe forms of β-thalassemia are characterized by severe anemia which is manifest during the first year of life when the switch from γ- to β-globin chain production occurs. If these children are not given regular blood transfusions they usually die within a few months. If they are inadequately transfused they become growth retarded, develop a curious mongoloid facial appearance, have gross skeletal deformities due to overgrowth of the bone marrow, and a variety of other complications (**Figure 2**). Children who are well transfused grow and develop normally but if they do not receive drugs to remove the excess iron gained by transfusion they die of the effects of iron overload, which involves particularly the liver, endocrine glands, and heart. Some of the milder forms of β-thalassemia are compatible with relatively normal development without regular blood transfusions, despite a variable degree of anemia.

Reasons for Clinical Variability

Particularly in the case of the β-thalassemias, there is remarkable variability in the clinical severity of the disorder. Several factors have been identified. First, children with severe forms of β-thalassemia produce variable amounts of fetal hemoglobin after the first year of life. All normal adults produce small amounts of fetal hemoglobin in some of their red cell precursors; in β-thalassemia these cells come under intense selection because part of the excess of α chains, which destroy red cell precursors, are bound to γ chains to produce hemoglobin F. It is now clear that one of the major factors in the clinical variability of β-thalassemia is a genetically determined ability to produce unusually high levels of fetal hemoglobin. A second factor which has been clearly identified is that the coinheritance of α-thalassemia will ameliorate the β-thalassemia. This remarkable experiment of nature provides clearcut evidence that it is the imbalance of globin chain production, and the excess of α chains, that is the major reason why β-thalassemia is so severe. Patients who are fortunate enough to inherit both types of thalassemia are less severely affected because the reduction of α chains caused by the α-thalassemia gene decreases the overall degree of globin chain imbalance and hence red cell production in the bone marrow is more effective.

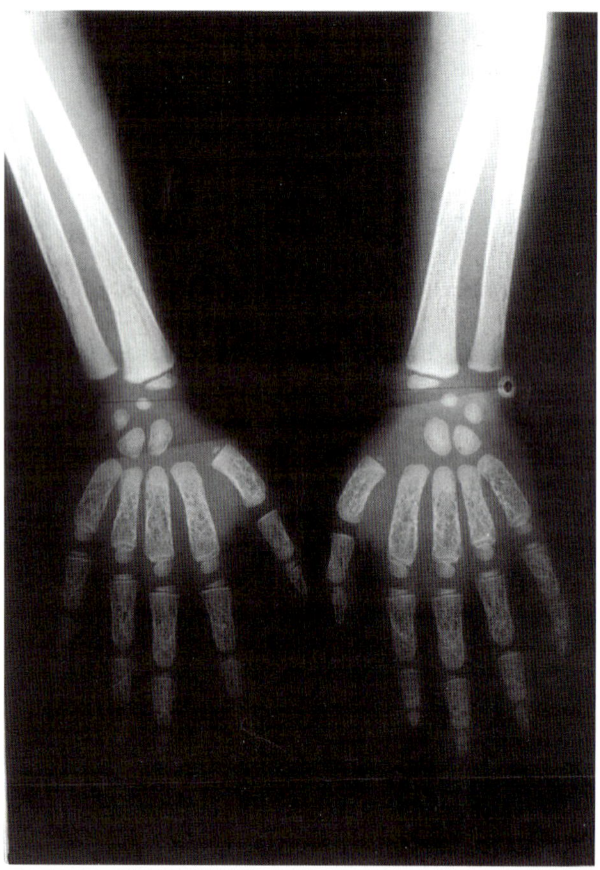

Figure 2 An X-ray photograph of the hands of a child with severe β-thalassemia showing the marked thinning of the bones of the hands due to expansion of bone marrow.

Coinheritance of Thalassemia with Hemoglobin Variants

Although there are many structural hemoglobin variants, most of them are rare and only three, hemoglobins S, C, and E, reach high frequencies. Hence it is not uncommon for a person with β-thalassemia to coinherit a gene for one of these variants. The compound heterozygous state for β-thalassemia and the sickle cell gene, sickle cell β-thalassemia, results in a clinical picture very like sickle cell anemia (Sickle Cell Anemia). On the other hand, the inheritance of β-thalassemia together with hemoglobin E, a hemoglobin variant which is produced at a reduced rate and hence is associated with a mild β-thalassemia phenotype, produces a severe form of thalassemia which is usually, but not always, transfusion-dependent. Hemoglobin E β-thalassemia is one of the commonest forms of severe thalassemia, and is becoming a major public health problem in parts of India, and further east, particularly in Thailand and Indonesia.

Distribution and Population Genetics

The thalassemias occur at a particularly high frequency in a band stretching from the Mediterranean region, through the Middle East and Indian subcontinent into south-east Asia where they are distributed in a vertical line from China, through the Malaysian peninsula and into the island population of Indonesia.

Each population has its own particular varieties of α- or β-thalassemia, which suggests that they have arisen by mutation and that the gene frequency has been increased by a local selective process. There is good evidence that the milder forms of α^+-thalassemia are protective against *Plasmodium falciparum* malaria. Although it has not yet been formally proved, it seems very likely that this will also be the case for carriers of β-thalassemia.

Control and Treatment

All the thalassemias can be identified in the carrier state, and most forms can be diagnosed in the fetus, and thus it is possible to offer counseling and prenatal diagnosis for parents who wish to terminate pregnancies carrying babies with severe forms of the disease. This approach has resulted in a major reduction in the birth of new cases in some of the Mediterranean islands and in other countries.

The only definitive form of treatment for thalassemia is bone marrow transplantation, which is only possible when there is a matching donor relative. Symptomatic treatment involves regular blood transfusion and the use of iron-chelating drugs to remove the excess iron which results from transfused blood. Children with β-thalassemia who are adequately transfused and chelated grow and develop normally and in some cases are now able to have children of their own. They need expert care because they are prone to a variety of complications, including blood-borne infections, notably hepatitis C and human immunodeficiency virus (HIV), endocrine damage leading to growth retardation and bone disease, and the side effects of chelating agents.

Future therapeutic efforts are being directed at trying to stimulate the production of fetal hemoglobin production, or at somatic gene therapy, directed at replacing defective α- or β-globin genes.

Further Reading

Cao A and Rosatelli MC (1993) Screening and prenatal diagnosis of the haemoglobinopathies. *Clinical Haematology* 6: 263–286.

Olivieri NF, Nathan DG, MacMillan JH *et al.* (1994) Survival of medically treated patients with homozygous β thalassemia. *New England Journal of Medicine* 331: 574–578.

Weatherall DJ (2000) The thalassemias. In: Stamatoyannopoulos G, Perlmutter RM, Marjerus PW and Varmus H (eds) *Molecular Basis of Blood Diseases*, 3rd edn. Philadelphia, PA: WB Saunders.

Weatherall DJ and Clegg JB (1996) Thalassaemia: a global public health problem. *Nature Medicine* 2: 847–849.

Weatherall DJ and Clegg JB (1999) *The Thalassaemia Syndromes*, 4th edn. Oxford: Blackwell Scientific Publications.

Weatherall DJ, Clegg JB, Higgs DR and Wood WG (1999) The hemoglobinopathies. In: Scriver CR, Beaudet AL, Sly WS, Valle D, Childs B and Vogelstein B (eds) *The Metabolic and Molecular Bases of Inherited Disease*, 8th edn. New York: McGraw-Hill.

See also: **Genetic Counseling; Sickle Cell Anemia**

Thermophilic Bacteria

K M Noll

doi: 10.1006/rwgn.2001.1288

Thermophiles or heat-loving organisms, are generally defined as those organisms that grow optimally (T_{opt}) above 55 °C. Although some animals can tolerate brief exposure to these temperatures, all organisms that require these temperatures for growth are prokaryotic, i.e., lack a membrane-defined nucleus. Thermophiles can be further described as extreme thermophiles ($T_{opt} > 65\,^{\circ}C$) or hyperthermophiles ($T_{opt} > 80\,^{\circ}C$) and most of the latter are archaebacteria. Since most of the currently known thermophiles have only been in culture for 20 years or less, the pace of development of genetic methods to study them has lagged behind that of mesophilic (moderate temperature) microbes. Genetics, as considered in this discussion, concerns the ability of thermophilic microbes to transfer genetic information among cells, to express those acquired traits, and to pass this information on to progeny. This discussion will be limited to those organisms belonging to the domain Bacteria. A comprehensive presentation of the genetics of Archaea is provided in a separate article (see Archaea, Genetics of).

Why the Interest in Thermophiles?

Since their discovery, thermophilic microbes have engendered interest in their ability to withstand high temperatures. From both applied and theoretical perspectives, these organisms provide unique opportunities to understand and exploit their biochemical capacities. The intrinsic stability of their cellular components, particularly their proteins, has attracted interest in using them as sources of high-temperature biological catalysts. The most spectacular examples of this are the thermostable DNA polymerases isolated from thermophilic microbes for use in the polymerase chain reaction (PCR). The first such enzyme was obtained from the extremely thermophilic bacterium *Thermus aquaticus*. Since complex biochemical transformations may require many enzymes to work in concert to produce desirable endproducts, it may be desirable in these situations to employ whole organisms engineered to carry out the needed biotransformations. One goal of developing genetic methods with thermophiles is to allow the engineering of whole cells to facilitate the development of green technologies such as bioremediation and generation of useful products from waste biomass.

More recently, it has become apparent that organisms very much like modern thermophiles may have played a pivotal role in the early evolution of life. Current evolutionary studies of living organisms suggest that the earliest ancestors of all life were hyperthermophiles. In addition, as complete genome sequences of several thermophiles have become available, evidence has accumulated suggesting that genetic information has been transferred among thermophiles. The impact this has had upon the evolution of life may be profound and will continue to be an important topic of study.

Challenges of Thermophile Genetics

In Vivo versus *In Vitro* Genetics

Using recombinant DNA methods, it is often possible for one to remove a gene of interest from an organism and study it in another organism (such as *Escherichia coli*) that is easier to handle in the laboratory. This *in vitro* genetics has made possible the study of many enzymes from thermophiles by expressing the genes encoding those enzymes in hosts such as *E. coli*. In particular, this has allowed investigators to study in detail the molecular basis of the thermostability exhibited by these enzymes and to alter their properties to suit given needs in biotechnological applications. Although these *in vitro* genetic studies account for the great majority of studies of thermophile enzymes, they do not involve true genetic manipulations of the thermophilic organisms themselves. The actual manipulation of DNA within the thermophiles (*in vivo* genetics) is much less common due to technological constraints and our general lack of knowledge of these microbes. It is these manipulations that will be the focus of the following discussion.

Transfer of DNA Among Thermophiles

Genetic analysis of thermophilic bacteria is limited to only a few species: *Bacillus stearothermophilus* ($T_{opt} \sim 60°C$), *Thermus thermophilus* ($T_{opt} \sim 70°C$), *Thermoanaerobacter* species ($T_{opt} \sim 60°C$), and *Thermotoga* species ($T_{opt} \sim 80°C$). In all cases, the genetic tools available are much less sophisticated than those used to study mesophiles like *E. coli*. Three methods are used to introduce DNA into bacteria: transduction, conjugation, and transformation. Transduction is DNA transfer mediated by a bacterial virus (bacteriophage) that contains a segment of genomic DNA removed from its previous host. Although bacteriophage that infect some thermophiles are known, none are known to be capable of transferring chromosomal genes to an infected host. Conjugal transfer of DNA between cells involves cell-to-cell contact during the transfer process and has been observed among some *Thermus* strains in the laboratory. No other thermophiles have been shown to do so. Finally, transformation, the process by which naked DNA is taken up by cells, occurs in all of the thermophiles listed above. However, only *Thermus* species undergo 'natural' transformation whereby naked DNA is taken up by cells during growth without pretreatment of those cells in the laboratory. The other organisms must be forced to take up DNA by prior chemical treatment of the cells or by driving the DNA into cells using electrical force (electroporation).

State of Genetics for Specific Thermophiles

Strictly Anaerobic Thermophiles

In the cases of *Thermotoga* and *Thermoanaerobacter*, investigators have only recently demonstrated DNA uptake and expression of genes encoding selectable phenotypes. A major impediment to the use of selectable markers is the instability of the selective agents, typically antibiotics, in the growth medium. For strict anaerobes, the combination of high temperature and the reactive compounds needed to exclude oxygen from growth media can often inactivate the drugs. Additionally, many of the proteins that confer resistance to antibiotics in mesophiles are themselves unstable at high temperatures. The antibiotic kanamycin has been found to be of great use with thermophiles since it is generally stable to heat in culture medium and there is a gene that encodes a heat-stable protein that confers resistance to it. This selectable phenotype is used in many thermophiles including species of these two strictly anaerobic groups. Both *Thermoanaerobacter* and *Thermotoga* have been transformed to kanamycin resistance using artificial methods. However, these methods have not as yet been used to examine the physiology or molecular biology of these organisms.

Bacillus stearothermophilus

Bacillus stearothermophilus has been artificially transformed with plasmid DNA containing the kanamycin-resistance gene to allow stable maintenance of the plasmid in the cell population. Genes from other organisms have also been introduced into this organism where their encoded proteins were expressed. Mutants of *B. stearothermophilus* have been constructed by introducing into them a transposon that integrated into a resident plasmid encoding the ability to degrade phenol. The mutation introduced into this plasmid caused those cells to produce catechol, a chemical useful for a variety of industrial processes.

Thermus thermophilus

Arguably the most genetically tractable thermophilic bacterium is *Thermus thermophilus*. As noted above, this organism, like many *Thermus* species, can take up DNA during growth without prior chemical treatment of cells. The mechanism by which transformation occurs is not yet known, though it is currently the subject of investigation. Unlike most naturally transformable organisms, *T. thermophilus* is capable of taking up DNA at all phases of growth in the laboratory. Further, it is very efficient in doing so, with transformation efficiencies for uptake of chromosomal genes of up to 12% of the cells in a culture taking up added DNA. *Thermus* cells can be transformed by *T. thermophilus* DNA that has been cloned into *E. coli*. This allows investigators to simply mix *E. coli* cells harboring cloned *T. thermophilus* DNA with *T. thermophilus* cells, heat the mixture (killing the *E. coli*), and then incubate the mixture under conditions that only allow genetically transformed *T. thermophilus* cells to grow. Thus *Thermus* genes can be readily removed, altered in *E. coli*, and then returned to *T. thermophilus* cells to observe the effects of the alterations.

A variety of plasmid vectors have been constructed to allow manipulation of *T. thermophilus* genes in *Thermus* itself. Plasmids capable of replication in *T. thermophilus* contain the gene conferring resistance to kanamycin. Other plasmids cannot replicate in *T. thermophilus*, but contain portions of chromosomal DNA so that they can recombine into defined sites within the *T. thermophilus* genome. These allow stable maintenance of cloned genes in single copies in the chromosome. In addition, expression of these genes can be placed under the control of the promoter where the plasmid integrates.

Recent work has shown that DNA can be transferred by conjugation from *T. thermophilus* strain HB8 to a closely related strain, HB27. Strain HB8 harbors a region of its chromosomal DNA similar in sequence to the F plasmid of *E. coli* which allows conjugal transfer of chromosomal genes after it integrates into the chromosome of the donor cell. It is thought that this DNA element allows strain HB8 to transfer copies of its chromosomal genes to recipient cells. There is as yet no evidence for the existence of this mobilizable DNA element as a free plasmid in cells. This element is adjacent to genes allowing strain HB8 to grow under anaerobic conditions by nitrate respiration and the transfer of this trait to strain HB27, which cannot respire nitrate, is being used to examine the conjugation mechanism. Mutations are being made in suspected conjugation genes by site-directed mutagenesis to systematically assign functions to these genes.

Further Reading

Mai V and Wiegel J (1999) Recombinant DNA applications in thermophiles. In: Demain AL and Davies JE (eds) *ASM Manual of Industrial Microbiology and Biotechnology*, 2nd edn, pp. 511–519. Washington, DC: ASM Press.

Noll KM and Vargas M (1997) Recent advances in genetic analyses of hyperthermophilic Archaea and Bacteria. *Archives of Microbiology* 168: 73–80.

See *also*: Archaea, Genetics of

Theta(θ) Replication

E Thomas

doi: 10.1006/rwgn.2001.1289

Theta replication is the primary type of replication used to replicate circular genomes (such as those of bacteria and many viruses and plasmids). These circular genomes each have one origin of replication. Theta replication proceeds bidirectionally from this origin, creating a replication bubble. Replication occurs in both directions simultaneously because there are two strands; on each strand, the replication fork moves from 5′ to 3′. The replication bubble grows during DNA replication (see **Figure 1**), eventually splitting off as a new double-stranded copy of the genome when the forks traveling in opposite directions merge. The structure formed during this process resembles the Greek letter theta (θ), hence the name of the model.

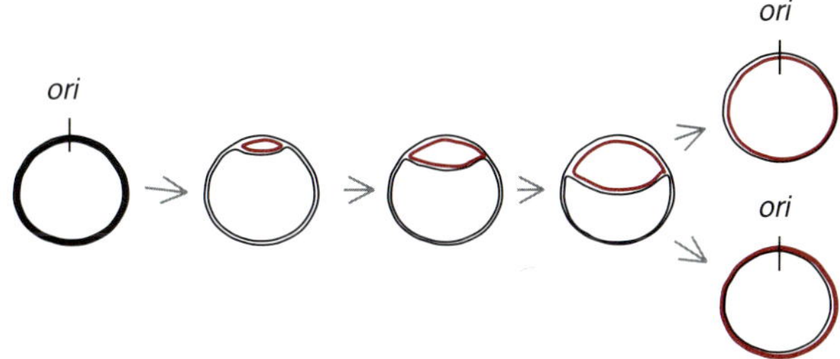

Figure 1 Theta replication.

In **Figure 1**, the replicating genome is shown as having only a single replication bubble. In reality, additional replication bubbles can be initiated soon after the parental origin of replication has been duplicated; this is how a bacterium that takes 40 min to replicate its DNA can still reproduce with a doubling time of only 20 min, and it means that the copy number of genes near the origin can be several-fold higher than that of genes on the opposite side of the circle.

See *also*: Replication

Three-Point Cross (Test-Cross)

J R S Fincham

doi: 10.1006/rwgn.2001.1290

This is the standard method in classical genetics of ordering linked genes. Basically, it involves determining the relative frequencies of the eight possible products of meiosis in a triply heterozygous diploid, say *A/a B/b C/c*, where *A*, *B*, and *C* are linked, or suspected of being linked. If the organism is a haploid fungus or alga, the products of meiosis, which will be spores of one kind or another, can be individually germinated and scored directly. In the case of diploid organisms, the triple heterozygote is crossed to a homozygous triple recessive *a/a b/b c/c*, so that the eight kinds of meiotic product are represented by eight distinguishable phenotypes in the test-cross progeny.

If *A*, *B*, *C* are unlinked, the eight types of meiotic product will, apart from sampling error and any differences in viability, be equally frequent (see Independent Segregation). If all are linked, defining two adjacent intervals in the linear order of the chromosome, two types of reciprocally constituted products will represent noncrossovers (parental types), two types will result from crossing-over in one interval, two from crossing-over in the other, and two from crossing-over in both. **Table 1** sets out the possibilities, with some hypothetical frequencies for illustration. We would conclude from these data that the triple heterozygote inherited *A B C* from one parent and *a b c* from the other, and that the loci were in the order as written. The order of the loci is usually obvious from the observation that two of the eight types are conspicuously less frequent than the others and can therefore be identified as the double crossover types. It is important that both putative double crossover classes should agree in their low frequency, since a low frequency of just one of them could be due to its low viability. Approximately equal frequencies of reciprocally constituted classes give reassurance that the conclusions regarding linkage are not being seriously distorted by viability differences.

The recombination percentages (uncorrected linkage map distances) between the loci, taken two at a time, are calculated by summing the recombinant classes 3, 4, 7 and 8 for *A–B* and classes 5, 6, 7 and 8 for *B–C*: 36 and 25 respectively, based on the numbers in **Table 1**. Unless the loci are sufficiently close together for there to be no double crossovers (classes 7 and 8 both zero), the recombination frequency between *A* and *C* ($3 + 4 + 5 + 6$, totalling 51 here) will always be less than the sum of *A–B* and *B–C*. Because of double and multiple crossovers, recombination frequency does not increase linearly with map distance. True map distance, measured in centimorgans, is 100 times the mean crossover frequency per meiotic product or (because each crossover involves only two out of four chromatids in the meiotic bivalent) 50 times the total mean number of crossovers per meiotic cell. This is equal to the percentage recombination between the ends of the chromosome interval only when double crossovers do not occur.

Three-point data give information about crossover interference. If crossovers are formed independently, with no effect of one on the chance of formation of another, the probability of simultaneous crossing over in both of the marked intervals should be the product of the probabilities in each interval separately. A lower double crossover frequency than predicted on this basis is an indication of crossover interference, the intensity of which is measured by the extent to which the coefficient of coincidence (observed/predicted double crossover frequency) falls below unity.

Table 1 A typical result of a three-point test-cross: *ABC/abc* × *abc/abc*

Diploid phenotypes or meiotic products[a]		Frequency	Interpretation
1	*A B C*	44 %	No crossover
2	*a b c*		
3	*A b c*	31 %	Single crossover between *A* and *B*
4	*a B C*		
5	*A B c*	20 %	Single crossover between *B* and *C*
6	*a b C*		
7	*A b C*	5 %	Double crossover, *A–B* and *B–C*
8	*a B c*		

[a] It is the phenotypes of the diploid progeny of the cross that are scored, but they represent the haploid meiotic products (germ cells) of the triply homozygous parent *A B C/a b c*, since the other parent contributes only the recessive alleles *a*, *b*, *c*.

Given the illustrative numbers in **Table 1**, we would expect, if there were no interference, the frequency of doubles to be $0.36 \times 0.25 = 0.09$ of the total, as compared with the 0.05 observed. The coefficient of coincidence is therefore 5/9 or 0.56, which is a fairly typical value. Generally, interference is stronger over shorter distances.

See also: **Independent Segregation; Interference, Genetic; Map Distance, Unit; Mapping Function; Meiotic Product**

Threonine

J Read and S Brenner

doi: 10.1006/rwgn.2001.1659

Threonine (Thr or T) is one of the 20 amino acids commonly found in proteins. Its side-chain contains an OH group allowing it to form H-bonds with water. It is therefore classed as a polar amino acid. Its chemical structure is given in **Figure 1**.

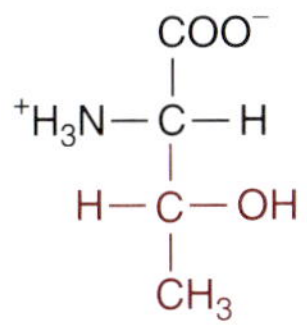

$$
\begin{array}{c}
\mathrm{COO^-} \\
| \\
{}^{+}\mathrm{H_3N}-\mathrm{C}-\mathrm{H} \\
| \\
\mathrm{H}-\mathrm{C}-\mathrm{OH} \\
| \\
\mathrm{CH_3}
\end{array}
$$

Figure 1 Threonine.

See also: **Amino Acids; Proteins and Protein Structure**

Threshold Characters

P Sham

doi: 10.1006/rwgn.2001.1291

Threshold characters refer to discrete (i.e., discontinuous) traits that are not inherited in a Mendelian fashion but in a manner more similar to continuous traits. In other words, despite being discontinuous in nature, their inheritance is polygenic rather than monogenic. An example of a threshold character is litter size in mice, pigs, and cattle. Although litter size itself must be a whole number, the factors that control litter size are quantitative, namely the levels of circulating gonadotrophic hormones.

The simplest form of discrete trait is one that is either present or absent. The liability-threshold model describes how polygenic effects (which are cumulative and therefore continuous) can lead to a dichotomous character. This model assumes the presence of an unmeasured continuous variable that, if it could be measured, would be appropriate for quantitative genetic analysis. Such an underlying continuous variable is known as the liability. If the liability of an individual exceeds a certain threshold value, then the trait will be present in the individual. Conversely, if the liability falls below the threshold, then the trait will be absent.

Since liability is defined as an unmeasured variable, its frequency distribution in the population is unknown. In order to apply standard quantitative genetic theory to liability, it is usual to assume that it has normal distribution, and that its joint distribution in members of a family has a multivariate normal distribution. However, it is also possible to incorporate the effect of one or more major loci into the liability. A model that incorporates both polygenic and major locus effects on liability is called a 'mixed model.' The correlation in liability between any pair of relatives is determined by the degree of the relationship and the underlying genetic model. Empirical data on the frequencies of the trait in the general population and in relatives of probands with and without the trait, or on pedigrees ascertained via probands, can be fitted to the liability-threshold models to obtain estimates of the threshold and of the variance components of the liability. In this way it is possible to test for the presence of a major locus effect (complex segregation analysis), and obtain an estimate of the heritability of the liability.

The validity of a liability-threshold model can be assessed by goodness-of-fit statistics of empirical data involving pedigrees containing relative pairs of different classes (e.g., identical and fraternal twins, full and half siblings). However, rejection of the model may mean that complicating factors (such as assortative mating) have not been accounted for, rather than failure of the basic liability-threshold model.

See also: **Continuous Variation; Heritability; Multifactorial Inheritance**

Thymine

R L Somerville

doi: 10.1006/rwgn.2001.1292

Thymine is a pyrimidine (molecular formula, $C_5H_6N_2O_2$) found primarily within DNA in the form of a deoxynucleotidyl residue, paired with adenine. Thymine is also found in trace quantities within transfer RNA. Chemically, thymine can be considered to be a derivative of uracil, and is sometimes referred to as 5-methyluracil. Because thymine is a critically important constituent of DNA, considerable effort has been put into the design of drugs that might selectively inhibit thymine biosynthesis, thereby blocking DNA replication, especially in rapidly dividing malignant cells. Examples of success in this area include the drugs fluorouracil, methotrexate, and aminopterin, each of which directly or indirectly block the attachment of the methyl substituent to the pyrimidine ring of the thymine precursor deoxyuridylic acid.

See *also:* Pyrimidine

Thymine Dimer

doi: 10.1006/rwgn.2001.2050

A thymine dimer is a cross-linked pair of adjacent thymine residues in DNA, which results from damage induced by ultraviolet radiation.

See *also:* DNA Repair; Thymine

Ti Plasmids

M Van Montagu

doi: 10.1006/rwgn.2001.1293

Ti plasmids are large, often more than 200 kb long, catabolic plasmids harbored by *Agrobacterium tumefaciens* strains. A Ti plasmid can be transferred by conjugation to most *Agrobacterium* and some *Rhizobium* species. A major characteristic of a Ti plasmid is that it contains, the *vir* or virulence genes, which enable a copy of one or more segments (T-DNA) of the Ti plasmid be transferred into plant cells, where it can become integrated into the plant genome. The genes encoded by the T-DNA are under eukaryotic control and can be expressed in a plant background. This can result in a plant cell proliferation (crown gall formation) and the synthesis and secretion of a specific metabolite, of no use for the plant. These metabolites, called opines, are condensation products of amino acids, such as arginine and lysine, and abundant plant metabolites such as pyruvic acid, ketoglutaric acid, succinate, and mannose.

Ti plasmids encode also functions that allow the *Agrobacterium* to catabolize these opines. So Ti plasmids bring to their host the capacity to engineer plant tissue to proliferate (hence the name 'tumor inducing') and force these tissues to synthesize large amounts of compounds that can only be catabolized by Ti-harbouring bacteria.

The discovery of the Ti plamids and the understanding of the mechanism of T-DNA transfer was a start for altering these plasmids and turning them into vectors for plant gene engineering. These transgenic plants have now been commercialized and are grown worldwide (on more than 45 million hectares in the year 2000), all having been constructed with the help of the Ti plasmid as gene vector.

See *also:* *Agrobacterium*; Plasmids; Transfer of Genetic Information from *Agrobacterium tumefaciens* to Plants

Tissue Culture

J W Pollard

doi: 10.1006/rwgn.2001.1527

In the modern context, tissue culture usually refers to the long-term culture of dispersed animal or plant cells rather than the short-term incubation of organs or tissues. However, its evolution has been to mimic complex cellular interactions *in vitro* in a manipulatable fashion and had its origins in the maintenance of organs or tissues *ex vivo*. As such, its history goes back to the beginning of biology. However, it was only at the end of the nineteenth century that Roux demonstrated the viability of cells outside of the body in physiologic saline, and not until 1907 when Harrison showed the outgrowth of neurons from explanted tissue, that tissue culture can be truly considered to begin. These explant cultures were grown in a lymph clot, a technique replaced by the plasma clot and perfected by Carrel. Carrel and Burrows showed the growth-promoting activities of a chick embryo extract and using this technique, with rigid control of sterility, Carrel maintained a strain of cells for over 34 years. This technique became the mainstay of cell culture. It was used by investigators such as Rous who transferred cells back into chickens and showed that they formed tumors, a phenomenon that was later shown to be due to the Rous sarcoma virus (RSV) named after him. Thus, cancer cell biology was born. However, it was only towards the end of the 1940s when manipulatable cell lines were established that tissue culture, as we know it, can be said to have started.

Milestones in Development of Tissue Culture

Many of the cell lines originally isolated in the late 1940s and early 1950s are still in use today, illustrating their long-term culturability. These include the L-cell line that Earle showed could be dispersed and plated as single cells to grow up as clones, HeLa cells derived from a human cervical tumor, and Chinese hamster ovary (CHO) cells from a disaggregated ovary. A key parallel event was the development of defined culture media. These were originally developed from physiological salt solutions defined by Earle and Ham, although it was Eagle who systemically tested many reagents and developed the complex media with over 25 ingredients that bears his name. In fact, many media available today bear these pioneers' names, and even those media developed for specific cell types such as Iscove's medium for hemapoietic cells, are based on these original media formulations. These defined media, however, are not sufficient to support the growth of cells but must be supplemented with serum, usually fetal bovine. In addition, some cells need the products of replication-inactivated (usually by treatment with mitomycin C or γ-irradiation) feeder fibroblastic cell layers. It is thought that the serum or feeder cells provide the source of adhesion molecules, growth factors and carriers such as transferrin. It was the observation in the middle of the 1950s that nerve cells required nerve growth factor (NGF) for outgrowth and maintenance of their viability that allowed the development of more defined media for the support of differentiated cells. Innumerable growth factors have since been isolated such as epidermal growth factor (EGF), platelet-derived growth factor (PDGF) for epithelial and fibroblastic cells, respectively, as well as those for hemapoietic cells including macrophage and granulocyte colony stimulating factors. Thus several cell types can now be cultured in completely defined media. However, despite the identification of numerous cell-type-specific growth factors, the ability of cells to grow in the complete absence of serum is still the exception rather than the rule.

Cells are often cultured from tissue explants or from disaggregated embryos. These cells are anchorage dependent and retain a normal diploid chromosome complement. The primary cultures can be propagated to form secondary cultures. But Hayflick observed that cultured human fibroblasts have a finite replicative life span, the 'Hayflick limit' after which they cannot be passaged. This cellular senescence has proven to be a fertile ground for the study of aging. In mouse fibroblasts, a similar replicative life span exists. However, the cultures usually go into a crisis when most cells cannot be propagated but some outgrow as morphologically transformed cells. These are usually aneuploid and are immortal. They also display less stringent growth requirements including the loss of contact inhibition of growth, a lower requirement for serum growth factors, anchorage independence, and often an ability to form tumors in nude (immunology compromised) mice. These are also the characteristics of cultured cells derived from tumors. Similar changes can be induced in normal primary cells by transformation of receptive cells with RNA and DNA tumor viruses such as RSV, polyoma, and simian virus 40 (SV40). These transformed cells and the events leading from the mortal to the immortal tumorigenic state have been major areas of study for cancer biology. Indeed, a major assay for the tumorigenic phenotype is the ability of cells to grow as foci of morphologically transformed cells, or as colonies in semisoft media containing agar or methylcellulose. These characteristics are not shared by normal mortal cells.

Almost all cells that can be continuously cultured are neoplastically transformed and genetically altered. An exception to this are embryonic stem (ES) cells. These are derived from blastocysts and represent uncommitted cells form the inner cell mass (the cells that will form the embryo proper). ES cells can be cultured indefinitely either on feeder cells or in the presence of the misnamed growth factor, leukemia inhibitory factor (LIF). They show many characteristics of transformed cells including anchorage independence, loss of contact inhibition, and the ability to form tumors *in vivo*. However, when reintroduced into a blastocyst, they take on normal cell fates and can contribute to all tissues including the germline. These cells are also unusual in that they permit homologous recombination into their nuclear DNA of introduced genes thus allowing the ablation or mutation of specific genes. By reintroduction of these mutated cells into the blastocyst, mutations can be introduced into the mouse germline. This technology of switching between cultured cell and living organism has dramatically enhanced the study of mammalian development. Furthermore, the pluripotent nature of ES cells has been exploited in culture by inducing differentiation thereby allowing the study of this process. Differentiation *in vitro* is now becoming common and many lineages have cell culture systems suitable to study the biochemistry of differentiation. These include myogenic, adipogenic, neurogenic, and hemapoietic lineages. To date, however, maintenance and differentiation of epithelially derived cell types has been difficult.

Cell culture has also been invaluable for genetics. The first cloning from single cells allowed the establishment of genetically homogenous cell lines and the ability to identify mutation and purify these free from other cells. This search for mutations was particularly fertile in Puck's CHO cell line that is, in fact, a proline autotroph. It was in this cell line that the first temperature-sensitive and, therefore recessive, mutation was identified. This was a paradox since mammalian cells are diploid making the identification of recessive mutations theoretically almost impossible. However, it appears that despite its almost diploid chromosome number, the CHO genome has large areas of functional hemizygosity. This cell line is easily mutagenized and consequently large numbers of mutants were isolated in this and some other cell lines. The availability of mutants led to attempts to rescue the mutations by complementation with DNA in a manner similar to that pioneered in bacterial genetics. DNA transformation was first achieved by Graham and Van der Erb using adenovirus DNA and this soon became routine with both plasmid and chromosomal DNA. A similar transfer of genetic information could also be achieved into cells with a variety of viral vectors. Consequently, DNA transformation of cultured cells could be used to isolate genes by complementation, to study structure–function relationship of gene products and to study gene regulation in promoter assays. It also became the basis of the commercial use of cell culture for the production of biologicals.

Somatic cell mutants allowed the study of cells hybrids. In this case, fusion is made between cells carrying particular selectable markers. The first metabolic selection was the HAT (hypoxanthine, aminopterin, and thymidine) selection technique developed by Szybalaski and exploited by Littlefield, that kills cells that lack hypoxanthine guanine phosphoribosyl transferase (HGPRT) and thymidine kinase (TK). This is because the aminopterin blocks *de novo* synthesis of purines and conversion of dUMP to dTMP forcing the use of salvage pathways. Consequently, parent cells that are $TK^+/HGPRT^-$ and $HGPRT^+/TK^-$ can be hybridized and complementing hybrids selected by incubation in HAT media. Somatic cell hybrids can be created both within and between species. Fusion can be spontaneous or promoted by inactivated Sendai virus or, more commonly now, by incubation in polyethylene glycol. This allows complementation assays to be performed and questions to be addressed such as the dominance of the neoplastic phenotype. Controversies over some of the results led to the realization that chromosomes were being lost from heterokaryons. Thus, in human–hamster hybrids, human chromosomes are preferentially lost. This enabled the establishment of somatic cell hybrids that contained single human chromosomes to be used for rapid chromosomal mapping of genes; a technique still in use today. The immortalization of normal B cells by fusion with myeloma cells led to hybrids that produce a single antibody specificity. These cells can be propagated indefinitely producing a monoclonal antibody of defined type and directed to a single epitope. These cultures can be scaled up for industrial production of monoclonal antibodies.

Concomitant with the development of animal cell culture, plant cell culture was also developed. It lagged early on because of the lack of appropriate media but, by the 1930s, suitable media were developed by White and others. In contrast to animal cell culture, plant cells can grow in defined media. A major advance occurred when whole plants were generated from a single cell, originally performed in carrot, by Kato and Takeuchi, but now routine in many species. This allowed for clonal propagation of plants as did the regeneration of buds, shoots, and roots in culture. Thus, in many ways, plant cell culture is in advance of animal cell culture since development and differentiation can be studied, as can the sophisticated interactions between parasites and hosts in culture.

Culture Methods

Isolated cells were originally cultured on glass, hence *in vitro* literally meaning 'in glass.' However, these methods have been superseded by use of specially treated plasticware available from many different suppliers. Flasks or tissue culture plates are usually designated by their surface area, e.g., 25 cm^2, 75 cm^2 flasks or their diameter, 30 mm, 60 mm plates, etc. Generally, cells are grown in a modification of Eagle's medium, e.g., α-minimal essential medium (αMEM) or Dulbecco's modified Eagle's medium (DMEM) supplemented with 5% to 20% serum. Usually the medium is bicarbonate buffered since this physiological buffering seems to enhance the growth of cells, and thus the cells require a CO_2 containing atmosphere and usually contain phenol red as a PH indicator. Cells are subcultured by trypsinization to dispense the cells or by scraping for trypsin-resistant cells. Thus, they can be propagated from flask to flask. Cells can also be frozen at temperatures below $-70\,^{\circ}C$ in media containing serum and dimethyl sulfoxide in a state from which they can be effectively resurrected. Thus, clones or cell lines can be laid down in storage for use many years later.

Surface culture of cells is limited for the large-scale production of cells. Some transformed cells can be adapted to growth in suspension. Suspension culture is usually in spinner flasks that contain a magnetic stirrer bar and these can be scaled up to 10 liters or so. After that, for industrial production, various fermenter designs have been developed. However, most cells require substrate attachment to grow, limiting the large-scale production of cells or their products. Consequently, the design of culture flasks has been modified to contain spirals or racks of plastic to enhance surface area, or cells have been grown in suspension culture attached to small beads known as microcarriers. Given the ability to transfer DNA into mammalian cells and to obtain stable expression of gene products, such large-scale techniques have become an essential component of the biotechnology industry to produce recombinant products on a commercial scale. At the other end of the scale, hemapoietic stem cells can be grown in small colonies, usually in semisolid medium, and differentiation induced by application of particular growth factors. These methods have been used to define lineage relationships. These culture methods have also provided invaluable assays for the isolation and eventual cloning of hemapoietic growth factors, many of which are now in commercial production as therapeutics using the large-scale culture techniques described above.

Further Reading

Celis JC (1998) *Cell Biology: A Laboratory Handbook*. San Diego, CA: Academic Press.

Paul J (1975) *Cell and Tissue Culture*. Edinburgh, UK: Churchill Livingstone.

Pollard JW and Walker JM (1994) *Plant Cell and Tissue Culture*. Totowa, NJ: Humana Press.

Pollard JW and Walker JM (1997) *Basic Cell Culture Protocols*. Totowa, NJ: Humana Press.

Spector DL, Goldman RD and Leinwand LA (1998) *Cells: A Laboratory Manual*. Plainview, NY: Cold Spring Harbor Laboratory Press.

***See also:* Biotechnology; Cell Lineage**

Tjio, Joe-Hin

M A Hultén

doi: 10.1006/rwgn.2001.1294

Joe-Hin Tjio (1919–) is best known for his part in the discovery in 1956 that the diploid chromosome number in humans is 46 and not 48 as had been dogma since 1912. This discovery was made at the Institute of Genetics, University of Lund (Sweden) in collaboration with Albert Levan.

The successful enumeration of the correct diploid chromosome number was the product of many factors, among which were Tjio's acknowledged technical skill, the use of an amenable tissue, and application of techniques that had been developed for plant and animal cytogenetics. The crucial study was performed on fetal lung fibroblasts, cultured *in vitro* by Rune Grubb at the University's Department of Microbiology, which proved to be an excellent source material for chromosome preparations. These cells were induced to arrest at metaphase (the best stage for counting chromosome number) by use of the spindle poison, colchicine; Albert Levan had pioneered this approach in studies of plant chromosomes. The seminal paper by Tjio and Levan, entitled 'The chromosome number of man,' was published in *Hereditas* in 1956 (Tjio and Levan, 1956). This publication marks the start point of the discipline of clinical cytogenetics.

Joe-Hin Tjio was born on 11 February 1919 in Indonesia where he grew up. He trained as an agronomist from 1936 to 1940, then took up a position as a cytogeneticist at the Botanical Institute of Bogor in Indonesia. His initial foray into cytogenetics came to an abrupt halt following the invasion of the country by the Japanese and his internment for the remainder of the war. After the war, Tjio moved to Europe. He worked as a research assistant in laboratories in Denmark and Sweden (including that of Albert Levan), before taking up a position as head of cytogenetics in Zaragosa (Spain) where he remained from 1948 to 1957. This move to Spain did not, however, end his association with Albert Levan. Tjio made regular study visits to Lund and it was on one of these, in 1955, that he carried out the work resulting in the identification of the human diploid chromosome number.

Following the publication of the correct chromosome number and presentation of the observations at the First International Human Genetics Congress in Copenhagen, Tjio received a number of invitations to work in the United States. Although initially reluctant to move, in 1957 Tjio joined T.T. Puck at the University of Colorado. While there, he completed his PhD entitled 'The somatic chromosomes of man.' In 1959, Tjio moved to the National Institutes of Health, initially at the National Institute of Arthritis and Metabolic Diseases, and subsequently, at the National Institute of Diabetes and Digestive and Kidney Diseases. At the latter, he was head of the cytogenetics section, a position he held until retirement in 1992. He continued to work after retirement until 1997.

The continuing thread through Joe-Hin Tjio's career was the study of chromosomes. His initial publications in 1948 were on plant chromosomes. A series of publications followed, many with Albert Levan, in which the effects of chemicals on plant chromosomes were described. In 1954, with encouragement from Levan, he turned his attention to mammalian chromosomes, initially using mouse ascite tumors and later human cells. This interest developed further in the United States, and with T.T. Puck and A. Robinson he published a number of papers highlighting chromosome abnormalities in constitutional genetic defects. He later became especially interested in the Philadelphia chromosome (a marker chromosome resulting from a reciprocal translocation involving chromosomes 9 and 22) in chronic myeloid leukemia. For many years he applied his cytogenetic expertise in studies on autoimmunity in mouse model systems.

Joe-Hin Tjio received many awards and honors for his scientific endeavors. The Kennedy International Award from the Joseph P. Kennedy foundation that he received in 1962 is probably the most prestigious. This award recognized his important contribution to our understanding of genetically determined mental retardation.

Further Reading

Tjio J-H and Levan A (1956) The chromosome number of man. *Hereditas* 42: 1–6.

***See also:* Levan, Albert; Painter, Theophilus Schickel**

Tm

***See:* Melting Temperature (Tm)**

TNF

***See:* Tumor Necrosis Factor (TNF)**

Topoisomerases

E Kutter

doi: 10.1006/rwgn.2001.1295

The processes of DNA transcription and replication both require separating the two strands of the double-helical DNA molecule, and they happen at tremendous speed. For example *Escherichia coli* DNA replication unwinds about 100 000 base pairs per minute. This potentially leads to overcoiling of the DNA into tight knots, with the two strands wrapped too tightly around each other. Topoisomerases are proteins that solve this problem by cutting the DNA backbone, letting the cut ends twist past each other to a more relaxed configuration, and then resealing the phosphodiester bonds in the backbone to again form an intact double helix. DNA topoisomerases bond covalently to the DNA phosphate as they break the phosphodiester linkage between neighboring nucleotides, storing the energy in that bond to use in the process of resealing the bond. They are thus in effect reversible nucleases that break and reseal the DNA molecule rapidly and efficiently, with no added energy – they cannot let go until they repair the break they have caused.

There are two families of DNA topoisomerases. Members of the topoisomerase I family break only one strand, causing a nick in the DNA and letting the two free ends rotate relative to each other around the phosphodiester linkage in the other strand, driven by the stress of any supercoiling of the DNA. Members of the topoisomerase II family are primarily responsible for resolving the potential tangle where two DNA double helices cross each other. They cut both strands of one such molecule at the same time, binding to both free ends and thus forming a sort of double protein-lined gate. The other double helix can be passed through this gate, while the topoisomerase still keeps both ends of the cut strands close to each other and ready to reseal. Type II topoisomerases are useful, for example, for separating the two replicated strands of circular DNA molecules. Some of them need extra ATP energy to make the molecules relax. In eukaryotic cells, the type II enzymes are mainly found during periods of DNA replication, so they are useful targets for anticancer drugs. Bacteria also have another form of type II topoisomerases, called gyrases, that are able to put in extra superhelical turns. Many enzymes that work on DNA can only work properly if the DNA has a bit of extra twist – a so-called ‘supercoil’ – that forces the two strands of the DNA to separate over a short region. Gyrases can generate such supercoils.

***See also:* Replication; Transcription**

Tortoiseshell Coloring

M F Lyon

doi: 10.1006/rwgn.2001.1296

A well-known feature of the tortoiseshell cat is that almost all such animals are female. This is a result of X chromosome inactivation. In all female mammals one of the two X chromosomes in every cell becomes genetically inactive early in development. Either X chromosome may become inactive, and in each cell once the choice has been made the same X chromosome remains inactive throughout the further multiplication of that cell line in development. Thus, in the tissues of the adult there may be large clumps of cells with the same X chromosome active. If the two X chromosomes carry different alleles of an X-linked gene, the animal will be a mosaic of two types of cells, and if the gene concerned affects some visible feature, such as coat color, then this mosaicism will be visible as variegation.

In cats there is an X-linked gene that determines a ginger coat versus nonginger (i.e., black or tabby). A tortoiseshell cat carries an allele for ginger on one X chromosome and an allele for black or tabby on the other. As a result the coat is variegated with patches of ginger and of black or tabby. Male cats have only a single X chromosome (being chromosomally XY) and can thus only be either ginger or nonginger. Similar X-linked coat color genes occur in other species, including the Syrian hamster and the mouse. In addition, a patchy effect can result from genes affecting other characteristics, such as coat texture. The tabby pattern in the cat is not due to X chromosome inactivation as the gene is autosomal, but a gene called tabby in the mouse is X-linked and gives a striped effect in heterozygotes due to X-inactivation. With appropriate techniques of histology or cell culture the presence of two types of cells can be shown in heterozygotes for many X-linked genes. The shape and size of the patches depends on patterns of cell growth and cell mingling during development. In tortoiseshell cats the patches are larger if the cat carries an autosomal gene for white spotting because this gene reduces the number and distance of migration of pigment cells, so that descendants of a single cell occupy a larger area.

It is possible to use the distribution of patches to determine whether particular structures arise from a single cell (monoclonal) or many cells (polyclonal) in development; e.g., intestinal crypts arise from a single cell, so that all cells of a crypt have the same X chromosome active, and intestinal villi arise from more than one cell, and hence show variegation. Similarly, X chromosome inactivation can be used to show whether tumors arise from one or many cells.

Occasionally male tortoiseshell cats are found and these usually result from a chromosomal anomaly. If an animal is chromosomally XXY it is male, due to the male-determining effect of the Y chromosome, but having two X chromosomes it undergoes X chromosome inactivation, as in an XX female. Thus, if it is heterozygous for ginger and nonginger it will show tortoiseshell coloring. A similar pattern can also be produced by other developmental events that cause the animal to be a mixture of two types of cells. One such event is a somatic mutation, occurring early in development, so that some cells carry the mutation and others not. Another possibility is the accidental or experimental fusion of two early embryos to form a single individual. Since the pattern depends on the multiplication and movement of cells after the two-cell populations form, then different types of event, such as mutation or embryo fusion, produce similar patterns. In some species patterns similar to tortoiseshell can be produced by autosomal genes, if these lead to somatic mutation or to epigenetic developmental changes in gene expression, as in the tabby cat.

See *also*: Coat Color Mutations, Animals; X-Chromosome Inactivation

Trans, Cis Configurations

See: *Cis–Trans* Configurations

Trans-Acting Factors

D S Latchman

doi: 10.1006/rwgn.2001.1298

Trans-acting factors are regulatory proteins which act to control gene transcription. They are therefore also known as transcription factors. In eukaryotes, these factors play a key role in the regulation of gene expression by determining which genes are transcribed in a particular tissue or in response to a given stimulus.

To produce their effects *trans*-acting transcription factors will, in general, require the ability to bind directly or indirectly to DNA and then to influence gene transcription either positively or negatively. Each of these aspects will be considered in turn.

DNA Binding

Detailed analysis of a number of different transcription factors has indicated that they have a modular structure in which specific regions of the molecule are responsible for binding to the DNA while other regions produce a stimulatory or inhibitory effect on transcription. Studies on the DNA binding regions of different transcription factors have revealed several distinct structural elements which can produce DNA binding.

Indeed transcription factors are frequently classified on the basis of their DNA binding domains.

Well-characterized DNA binding domains include the helix–turn–helix motif found in the homeobox transcription factors, the two cysteine–two histidine zinc finger which is found, for example, in the Sp transcription factor family, the multicysteine zinc finger which is found in the steroid-thyroid hormone receptor family, the Ets domain, and the basic DNA binding domain.

This last example is of particular interest since factors containing the basic DNA binding domain can only bind to DNA once they have formed transcription factor dimers. Hence, factors containing the basic binding domain are further subgrouped according to the nature of the dimerization motif which they contain. Thus some of these factors contain a helix–loop–helix motif which mediates dimerization while others undergo dimerization via the so-called leucine zipper motif which contains a regular array of leucine residues.

Thus a wide variety of DNA binding domains (which in some cases have associated dimerization domains) allow *trans*-acting transcription factors to bind to their appropriate DNA sequences within target genes.

Activation of Transcription

Many transcription factors contain, in addition to the DNA binding domain, specific regions that are necessary for the activation of transcription. Such regions were identified on the basis of their ability to stimulate transcription when linked to the DNA binding domain of a completely unrelated factor. These regions are known as activation domains. As with DNA binding domains, a number of distinct types of activation domain have been identified. They are classified on the basis that they are rich respectively in acidic amino acids, glutamine residues, or proline residues.

These activation domains appear to function by interacting with components of the basal transcriptional complex. This is a complex of RNA polymerase II and various transcription factors such as TFIIB and TFIID which assembles at the gene promoter and is essential for transcription to occur. Activation domains have been shown to interact either directly with specific components of this complex or indirectly by interacting with so-called coactivator molecules which then interact with the basal complex itself. Whatever the case, such interactions appear to result in enhanced transcription either by stimulating the rate of transcription factor complex assembly or by stimulating the level of its activity.

Hence, following binding to their appropriate DNA binding site via the DNA binding domain, the activation domains of specific activating transcription factors can interact with the basal transcriptional complex so as to stimulate transcription. In this manner the binding of specific transcription factors can stimulate gene transcription.

Repression of Transcription

Although it was originally thought that most eukaryotic transcription factors acted by stimulating transcription, it has now become clear that a wide variety of factors act by inhibiting the transcription of specific genes and that such inhibitory transcription factors may be at least as important as stimulatory factors.

The earliest examples of such inhibitory transcription factors were shown to act by interfering with the activity of a positively acting factor thereby blocking its stimulatory effect on transcription. This could be achieved, for example, by preventing the positively acting factor from binding to DNA either via the negatively acting factor binding to its DNA binding site or by the formation of a non-DNA binding protein–protein complex between the positively acting factor and the negatively acting factor. Alternatively, the negatively acting factor could act by interacting with the positively acting factor to block the activity of its activation domain in a phenomenon known as 'quenching.'

It has now become clear, however, that a class of inhibitory transcription factors exists which can directly inhibit transcription even in the absence of a positively acting factor. These factors can thus reduce the basal level of transcription below that observed even in the absence of any activating molecule and appear to function by interacting either directly or indirectly with the basal transcriptional complex so as to reduce its activity. They thus constitute the antithesis of the activating molecules discussed in the previous section and possess defined inhibitory domains which are responsible for their effects and which, like activation domains, can function when transferred to the DNA binding domain of another molecule.

Hence the balance between binding of transcriptional activators and transcriptional repressors to the regulatory region of a particular gene will determine its rate of transcription in any particular situation. Clearly however, in order for a particular gene to respond to specific signals or to be regulated in a cell type specific manner, the balance between these activating and repressing molecules must change in different situations. The mechanisms which are used to regulate transcription factor activity are discussed in the next section.

Regulation of Transcription Factors

Transcription factors can be regulated at two levels, namely, the regulation of transcription factor synthesis and the regulation of transcription factor activity.

Regulation of Synthesis

In a number of different situations a transcription factor is regulated by being synthesized in one particular tissue or cell type and not in other tissues. The most dramatic example of this concerns the MyoD transcription factor which is synthesized only in skeletal muscle cells. Thus in this case the overexpression of the MyoD factor in undifferentiated fibroblast cells is sufficient to convert them to skeletal muscle cells, indicating that this factor is critical in the induction of muscle-specific gene expression.

Regulation of Transcription Factor Activity

Although the regulation of transcription factor synthesis is an important control point, it cannot be the only regulatory mechanism that controls transcription factor activity. Thus if this were the case, the enhanced synthesis of a transcription factor in response to a particular stimulus would be controlled by enhanced transcription of its corresponding gene, which in turn would require the *de novo* synthesis of further transcription factors, so resulting in the need for new transcription of these genes and so on. Therefore it is necessary to have an additional mechanism which allows *de novo* gene transcription by the activation of pre-existing transcription factors.

Such activation of pre-existing transcription factors can occur via a number of different mechanisms which can involve ligand binding, alterations in protein–protein interaction and transcription factor phosphorylation. Thus, for example, in the case of the steroid receptors the inactive receptor is associated with an inhibitory heat-shock protein hsp90. Following binding of the steroid hormone ligand, hsp90 dissociates and moves to the nucleus where it can bind to its appropriate response element and switch on transcription.

Regulation of transcription factor activity by phosphorylation is seen in the case of the CREB factor which binds to the cyclic AMP response element (CRE) and plays a critical role in the regulation of transcription in response to cyclic AMP. Thus, following treatment with cyclic AMP the CREB factor becomes phosphorylated on a particular serine residue. This phosphorylation prevents the binding of CREB to another protein, CBP, which does not bind to unphosphorylated CREB.

This CBP factor appears to play a critical role in the activation of transcription. Thus this factor is able to bind to specific components of the basal transcriptional complex thereby linking CREB to this complex and allowing stimulation of its activity following cyclic AMP treatment. In addition however CBP has been shown to possess histone acetyltransferase activity. Such enhanced acetylation of histones has been shown to occur in regions of DNA that are active or potentially active in transcription and to be involved in the open chromatin structure characteristic of such regions. It is therefore possible that the binding of CBP to CREB recruits it to the DNA and allows it to produce changes in the chromatin structure which lead to enhanced transcription.

Hence, in a specific cell type or in response to a specific stimulus, specific transcription factors are either synthesized or become activated following posttranslational modification. The binding of these transcription factors to their appropriate recognition sequences thus produces specific patterns of gene transcription in specific cell types or in response to specific stimuli.

Further Reading

Latchman DS (1998) *Gene Regulation: A Eukaryotic Perspective*, 3rd edn. Cheltenham, UK: Stanley Thorne.

Latchman DS (1998) *Eukaryotic Transcription Factors*, 3rd edn. London: Academic Press.

***See also*: Transcription**

Transcribed Spacer

doi: 10.1006/rwgn.2001.2053

A transcribed spacer is part of an rRNA transcriptional unit that is transcribed but subsequently discarded during maturation. It does not ultimately give rise to part of rRNA.

***See also*: Nontranscribed Spacer; Transcription**

Transcription

C Kane

doi: 10.1006/rwgn.2001.1299

Transcription Decodes the Genetic Information and Is a Complex Process

Transcription is the process of physically decoding the genetic information into an RNA that can be used by the cell, either directly or as a template for protein synthesis. RNA polymerase is the enzyme that carries out the synthesis of RNA from a nucleic acid template, and the complexity of the enzyme's polypeptide composition depends on the organism. Likewise the complexity of the cellular machinery that regulates RNA synthesis depends on the organism, and on the necessity for timing the synthesis of particular RNAs in particular locations to particular levels. Transcription is but one of many processes that influence the overall regulation of gene expression in the cell, and cells and their viruses have developed multiple strategies to modulate the transcription machinery. In addition to protein components that have an impact on the behavior of RNA polymerase, the template and nascent transcript can also be involved in regulation, through nucleic acid sequences and structures and/or by virtue of the association of the DNA and the RNA with other proteins.

In developing a mechanistic model for transcription regulation, experimentalists have used both genetics and biochemistry. Most recently these models have been highlighted by the availability of three-dimensional structures for RNA polymerases from prokaryotes and eukaryotes (**Figure 1**). However, transcription is a cyclic process, and in addition to regulating the actual catalytic properties of RNA polymerase, there is also regulation in finding the proper gene for transcription (promoter recognition), in initiating the synthesis of the RNA chain (both promoter escape and promoter clearance), in transitioning into a stable transcript elongation process and modulating elongation across a transcription unit, and in terminating the synthesis of the transcript at the proper location (**Figure 2**). The polymerase is then available to start the cycle again. For transcripts synthesized by prokaryotic RNA polymerases, there is coordination between synthesis and translation of the messenger RNA. In eukaryotes, there is coordination between synthesis and a variety of processing reactions for the transcript: capping (for RNA polymerase II), splicing, and 3′ end maturation. Transcription also must be coordinated with DNA metabolic processes such as replication, recombination, and repair.

Transcription Templates

Transcription is most commonly discussed in the context of a double-stranded DNA template being copied into RNA by a DNA-dependent RNA polymerase. The majority of this entry will focus on that category of transcription. However, there are a number of viruses that utilize RNA as a template for transcription carried out by RNA-dependent RNA polymerases. There are many mechanistic similarities between the DNA- and RNA-templated reactions, but the RNA-dependent RNA polymerases often contain a subunit with sequences closely related to reverse transcriptases (i.e., that copy RNA into DNA). In addition, much less is known about how these latter enzymes recognize specific genes and the proper sites at which to begin transcription. Next to nothing is known about their regulation during elongation, or about the process of termination.

In neither prokaryotic nor eukaryotic cells is the template highly purified nucleic acid. In eukaryotes, positively charged histone and nonhistone chromosomal proteins are associated with the template to create the chromatin that is found in the nucleus, and these proteins participate in structuring the chromosomes as well as in the regulation of both DNA and

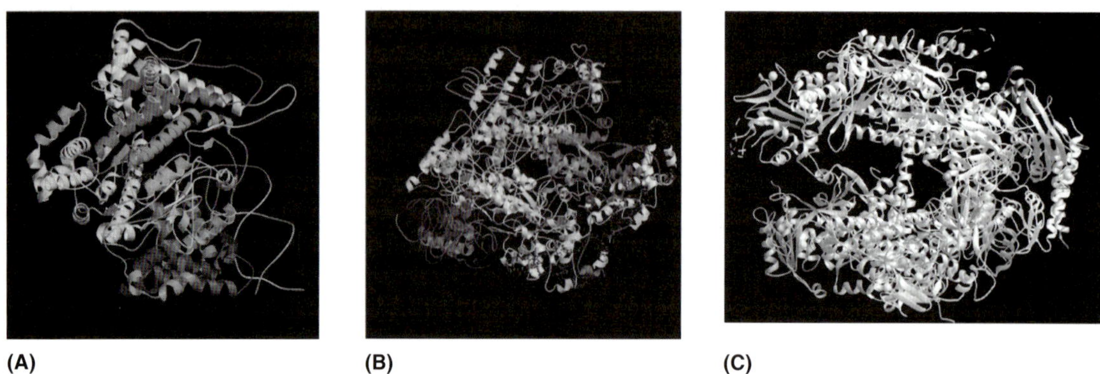

Figure 1 (See Plate 41) Crystal structures have been determined for (A) T7 RNA polymerase (2.4 Å, 1QLN; Cheetham and Steitz, 1999; image from Brookhaven Protein Database); (B) *Thermus aquaticus* core RNA polymerase (3.3 Å, 1DDQ, Zhang *et al.*, 1999; image from Brookhaven Protein Database); and (C) *Saccharomyces cerevisiae* RNA polymerase II (3.0 Å, 1EN0; Cramer *et al.*, 2000; image courtesy of Patrick Cramer and David Bushnell).

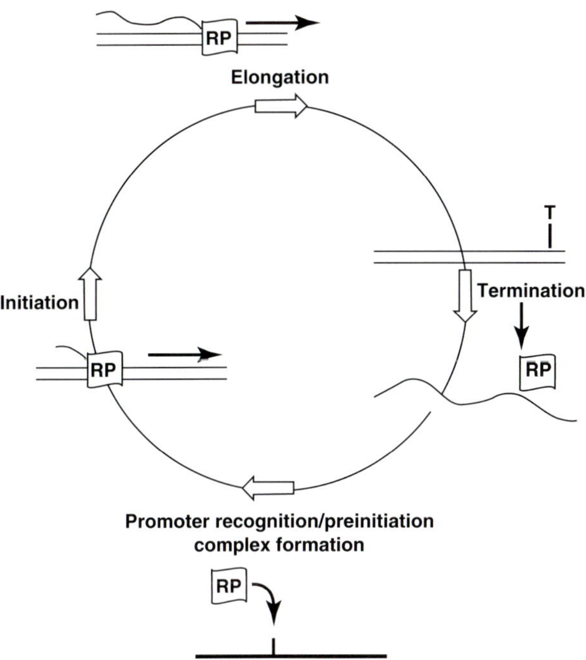

Figure 2 Transcription cycle. Representation of RNA polymerase (RP) during the process of transcription. The dark parallel lines represent the DNA template. The thin line represents the nascent transcript. T refers to a termination site on the DNA template. Details of the transcription cycle are given in the text.

RNA metabolism (see Chromosome). In prokaryotes, there are also basic proteins associated with the DNA to create a chromosome, but the structure is distinct from that in eukaryotes. There are no histones, although there are histone-like proteins. The prokaryotic chromosomes are condensed into a nucleoid structure, and are associated with the inner cell membrane. Mutations in the histone-like proteins have a significant impact on transcription, and thus, as in eukaryotes, the naturally occurring template should not be considered as simply free DNA.

RNA Polymerases

The DNA-dependent RNA polymerases range in subunit complexity from those of the single subunit polymerases of bacteriophages (bacterial viruses) to the 12–15 subunit complexes associated into an active enzyme necessary for catalysis in the eukaryotic nucleus (**Table 1**). Many phages utilize the host cell's polymerase as well as their own encoded RNA polymerase, as is the case with phages (e.g., T7) that encode the smaller single subunit RNA polymerases. Some of these polymerases have become major reagents in biotechnology because they are now relatively easy to overproduce in bacterial cells, they need no additional protein cofactors to recognize a promoter region, and they synthesize transcripts at least twice as fast as the bacterial host RNA polymerases. Curiously, the single subunit bacteriophage RNA polymerase itself can recognize the correct gene and start at the correct site on purified DNA, whereas the multisubunit eukaryotic nuclear polymerases require a very large number of accessory protein factors in order to locate the proper sites on the template for transcription. Bacterial RNA polymerases contain four subunits of three nonidentical proteins in the catalytic core of the enzyme, and this core complex requires one additional accessory factor (called sigma) that both enables the recognition of specific sequences that promote transcription and reduces the binding to nonpromoter sites on the DNA. There are several different sigma factors in each species of bacteria; one in each species is considered to be the primary factor for most transcription, whereas the others are utilized for special regulated events such as the stress response, motility, differentiation, and spore formation.

The genomes of some bacteriophages, such as *Bacillus subtilis* phage SP01 and coliphage T4, encode their own sigma factors that redirect the bacterial RNA polymerase catalytic core for transcription of the viral DNA templates. The catalytic cores of the bacterial enzymes remain the same for all this transcription. Also, many bacteriophages utilize other virally encoded and host-encoded regulatory proteins to modify subsequent steps in the transcription cycle after promoter recognition.

Bacteriophage N4 uses both the host polymerase and two other polymerases encoded by its own genome (**Table 1**). One of the viral polymerases is a very large single subunit enzyme that, in contrast to the smaller single subunit bacteriophage-encoded polymerases, needs a host cofactor, ssb, a single-stranded DNA-binding protein. The ssb protein allows this 320 kDa polymerase to recognize an unusual hairpin structure in the viral DNA. This allows synthesis of further viral RNAs, resulting in replication and assembly of new viral particles.

Eukaryotic viruses, like their prokaryotic counterparts, use the entire spectrum of transcription strategies, from encoding their own polymerases (e.g., vaccinia, baculovirus) to using the host-cell machinery. Several eukaryotic viruses encode RNA-dependent RNA polymerases as well (e.g., hepatitis C virus, poliovirus). Virally encoded protein factors also modulate transcription in the host cell at many positions in the transcription cycle.

Eukaryotic RNA polymerases can be found in the nucleus and in cellular organelles. The polymerase found in mitochondria and one polymerase found in chloroplasts have similarity to the small, single

Table 1 DNA-dependent RNA polymerase subunit variation

Source	Number of subunits in core	Sizes (kDa)
Bacteriophage T3, T7, SP6[a]	1	~100
Bacteriophage N4[a,b]	1	320
	3	40, 30, 15
Eubacteria, e.g., *Escherichia coli*	4	150, 145, 45
Archaebacteria, e.g., *Sulfolobolus acidocaldarius*	13	122, 101, 44, 30, 27, 13.8, 12, 11.8, 10, 9.7, 9.7, 7.5, 5.5
Saccharomyces cerevisiae RNAP I	14	190, 135, 49, 43, 40, 34.5, 27, 23, 19, 14.5, 14, 12.2, 10, 10
Saccharomyces cerevisiae RNAP II	12	191, 140, 35, 25.4, 25, 19, 17.9, 16.5, 14.2, 13, 8.3, 7.7
Saccharomyces cerevisiae RNAP III	13–17	160, 128, 82, 53, 40, 37, 34, 31, 27, 25, 23, 19, 17, 14.5, 11, 10, 10
Human and yeast mitochondria	1	140
Spinach and mustard chloroplast[c]	1	110
	4	154, 120, 78, 38
	~13	141, 110, 36, 26 and others not yet fully characterized
Vaccinia virus	8	147, 132, 35, 30, 22, 19, 18, 7
Baculovirus[a]	4	98, 55, 53, 46

Examples are given of several classes of DNA-dependent RNA polymerases.
[a]Indicates that the virus uses the host RNA polymerase as well as its own encoded RNA polymerase for temporal regulation of transcription.
[b]Indicates that the virus encodes more than one RNA polymerase.
[c]Indicates that more than one RNA polymerase is used.

subunit bacteriophage RNA polymerases; chloroplasts have two other RNA polymerases, one of which is very similar to the bacterial enzyme (**Table 1**). The nuclear RNA polymerases come in three distinct complexes isolated biochemically, and they were originally categorized by the class of genes each transcribes. More recently they have been defined by the subunits and accessory protein factors that act on them to regulate RNA synthesis. Termed RNA polymerases I, II, and III (or A, B, and C, respectively), each has over a dozen subunits (**Table 1**). RNA polymerase I transcribes the genes that encode the structural RNAs for the subunits of the ribosome. RNA polymerase II transcribes the genes that encode proteins as well as a subset of small RNAs. RNA polymerase III transcribes the genes encoding ribosomal 5S RNA, tRNAs, and a subset of other small RNAs.

In *Saccharomyces cerevisiae* where the subunits have all been cloned and characterized by sequence, five subunits are shared by all three polymerase complexes. In addition, there is sequence similarity among four other subunits that are unique to each of the three polymerases. Three of the four subunits that are related by sequence similarities are also related in sequence to the three different subunits in the bacterial catalytic core RNA polymerase. Biochemical and genetic studies suggest that these three subunits in the eukaryotic polymerases are probably functionally homologous to the bacterial subunits as well, and thus they are liable to be the catalytic core of the enzymes. However, none of the eukaryotic nuclear polymerases has been reconstituted in active form from individually isolated subunits, and thus the function of the individual subunits can at best only be inferred from the genetics and from biochemical analysis of mutant polymerases.

These studies, coupled with the recently published crystal structures of T7 RNA polymerase, yeast RNA polymerase II, and a thermophilic bacterial RNA polymerase, have provided great insight into function and predictions of function for individual subunits. There are some overall similarities but also significant differences among the enzymes, especially in the potential route for the template DNA and the nascent transcript within ternary complexes. The structures of the purified polymerases also differ significantly from the structure of cocrystals with nucleic acids, and there are major conformational changes in going from free enzyme to transcriptionally engaged enzyme. A number of mechanistic models have been built based on the structures of the purified enzymes, assimilating both mutational information and biochemical cross-linking of the DNA and RNA with polymerase subunits. In addition, there are lower resolution structures of elongation complexes for both yeast RNA polymerase II and *E. coli* RNA polymerase that inform the models. However, not all experimental results are consistent with the models that predict the position of the nucleic acids; the large conformational differences between the free enzyme and enzyme associated with DNA or with DNA and RNA probably explain these discrepancies. Nonetheless, the models are extremely useful in making predictions for further analyzing the molecular mechanisms during transcription.

In both prokaryotes and eukaryotes, the two largest subunits form the catalytic center of the polymerase. The individual nucleotide precursors cross-link to each of these subunits, as do the template DNA and the nascent RNA transcript. These subunits are also contacted by proteins that are regulatory for initiation and for elongation, and they are important in recognition of proper termination sequences. The third subunit of the bacterial core RNA polymerase, α, is important for proper structural integrity of the polymerase, and it also contacts DNA and mediates signals from activator and repressor proteins when the polymerase is bound at the promoter. In eukaryotes, the α-subunit equivalent has also been shown to be important for the structural integrity of RNA polymerase II in yeast, although there is no information as yet about its contacts or function with other regulatory proteins.

Two of the smaller subunits of yeast RNA polymerase II, 4 and 7, are important in mediating the stress response in cells, and *in vitro* they are essential for efficient promoter recognition and initiation. Subunit 7 is encoded by an essential gene, whereas a deletion of the gene encoding subunit 4 renders the cells temperature-sensitive and results in a weakened association of subunit 7 with the polymerase. The association is so inefficient that both subunits 4 and 7 are missing from polymerase biochemically isolated from cells in which the gene encoding only subunit 4 has been deleted.

No function has been ascribed to most of the subunits that are common to all three yeast nuclear RNA polymerases (subunits 5, 6, 8, 10, and 12), although subunit 6 exhibits genetic interactions with a factor involved in regulating transcript elongation by RNA polymerase II and also is important for the assembly and structural integrity of RNA polymerases I and II. Each polymerase also has a small subunit that is related but nonidentical in sequence, subunits A12.2, B12.6, and C11, in RNA polymerases I, II, and III, respectively. This subunit has an inferred role in transcript elongation because it mediates an RNA hydrolysis reaction carried out by the polymerase when elongation stalls during transcription.

Archaebacterial DNA-dependent RNA polymerases are also multisubunit complexes, and they resemble eukaryotic polymerases more than prokaryotic polymerases. However, they require fewer accessory protein factors for promoter recognition, and they also contain subunits that are related in sequence to a transcript elongation factor (TFIIS) for yeast RNA polymerase II.

Transcription Cycle

The cell utilizes the many reactions and interactions during transcription to provide exquisite temporal, spatial, and quantitative control over the synthesis of RNA. At any point in the transcription cycle (**Figure 2**), a step may be rate-limiting and thus a target for control. The step that is rate-limiting may also change depending on the needs of the cell. Experiments looking at transcription tend to focus on one specific aspect or another, or simply on whether the levels of transcript go up or down in the presence of various treatments or in different types of mutant cells. Detailing the mechanistic steps in transcription assists in appreciating and understanding the complexity and power of the multiple layers of regulation.

Finding the Proper Gene to Transcribe: Promoter Recognition and Preinitiation Complex Formation

In a complex genome, how does RNA polymerase find the location at which RNA synthesis needs to begin for a particular gene? This search-and-locate mission is referred to as 'promoter recognition' because the DNA sequences 'promote' specific transcription. As mentioned above, the single subunit bacteriophage RNA polymerases can find the location and the transcription start site for specific genes without the assistance of any other proteins. Nearly all other polymerases require one or more accessory factors to effect promoter recognition. Bacterial RNA polymerases need the accessory factor referred to as sigma to locate the desired promoter and to position the polymerase at the correct start site. Eukaryotic nuclear RNA polymerases need a large constellation of accessory proteins that work in sequence and in concert to recruit the polymerase to the proper location and position it for accurate initiation; this macromolecular complex bound to the DNA at the promoter has been referred to as the 'preinitiation complex.'

When the polymerase binds to the double-stranded DNA promoter region, it forms what has been termed a 'closed complex,' because the DNA remains in a regular, closed duplex. For bacterial RNA polymerase, conformational isomerization through several physically distinct closed complexes has been observed, and these isomerizations precede the opening of the DNA in the promoter region to expose the single-stranded DNA in an 'open complex' that can now be 'read' by RNA polymerase. The open complex can be distinguished from the closed complex by physical and biochemical properties. Open complex formation by bacteriophage and bacterial RNA polymerases occurs *in vitro* without any infusion of energy. However, the eukaryotic preinitiation complex requires the energy of ATP hydrolysis to open the DNA. In the open complex the polymerase is poised to receive nucleoside triphosphate substrates and begin RNA synthesis. A very large amount of published work has gone into examining the regulation of preinitiation complex formation, and closed-to-open complex formation. There are a number of rate-limiting steps in both the binding and isomerization reactions as well as in the association and dissociation of regulatory factors. The detailed kinetics of the rate-limiting steps have been analyzed *in vitro*, and genetic analysis has been invaluable in identification of regulatory factors and specific nucleic acid sequences essential in the regulation of promoter binding and preinitiation complex formation. There is still much to learn.

Regulation of Promoter Recognition and Preinitiation Complex Formation

Multiple sequence-specific DNA-binding proteins regulate promoter recognition and preinitiation complex formation. Other protein factors that do not bind DNA regulate these processes through protein–protein interactions. Hundreds of proteins in prokaryotes and eukaryotes (such as activators, repressors, and insulators, as well as enhancer-binding proteins and general transcription factors (GTFs)) have been catalogued based on their presence or absence in this phase of the transcription cycle. There are many ways in which activators, repressors, enhancers, and insulators work and, in large part, the regulation focuses upon a specific mechanistic step that is rate-limiting. In addition, the activity of these factors may be modulated by posttranslational modifications such as phosphorylation, glycosylation, acetylation, or methylation, as well as the chromosomal environment and chromatin structure surrounding the promoter. In addition, many of these regulatory proteins also have effects on initiation and elongation reactions. Thus, studies of each step in transcription in isolation can reveal the individual important details of the mechanism, but understanding the overall regulation requires a broader view of the transcription cycle.

Initiating RNA Synthesis

In the past, initiation of transcription was defined as the formation of the first phosphodiester bond, or dinucleotide synthesis. In this view, elongation started with the third nucleotide incorporated into a transcript. However, increased understanding of the different stages in the transcription cycle has redefined initiation to include events that continue into the initial transcribed region. The 'initial transcribed complexes' are those engaged in the initiation process that have not yet entered elongation. The formation of a fully processive elongation complex follows the completion of the initiation reaction and release of the accessory initiation factors, and promoter escape has become the operational term for the process that occurs prior to the establishment of the stable elongation complex.

Promoter escape

The 'switch' that dictates the release of the initiation factors from the polymerase and permits its transition into productive elongation is not known. 'Abortive initiation' is characterized by the repetitive synthesis of small oligomeric transcripts near the start site of transcription, and these abortive products are released from the transcribing complex without release of the polymerase or initiation factors. Rather, the polymerase starts again at the correct initiation site and begins synthesis anew. Sequences in the promoter region and in the not yet transcribed downstream region can influence the length of the abortive products and the efficiency of their production. In contrast to the predictions of some models for transcription, the length and efficiency of production of the abortive products is not correlated with the predicted thermodynamic stabilities of potential RNA–DNA hybrids in the initial transcription complexes. Nonetheless, promoter escape is clearly rate-limiting for some promoters and not for others.

What events characterize promoter escape? For bacterial RNA polymerase, the release of the sigma factor marks the end of the initiation process. The sigma factor is released when transcripts are between 9 and 16 nucleotides long, and the length depends on the promoter and the early transcribed sequence. For eukaryotic RNA polymerases, there is not yet a discrete event that marks the end of initiation, but the transition occurs within a broader range, when the transcript is between 8 and 40 nucleotides long. Several accessory protein factors influence the transition, and regulatory proteins also have an impact on the promoter escape process. For RNA polymerase II, the phosphorylation of the largest subunit is correlated with promoter escape, and several different protein kinases may be involved in this transition. However, the specific phosphorylation sites are not known.

Promoter clearance

The term 'promoter clearance' refers to the point in the reaction when the polymerase and associated accessory factors have moved away from the sequences necessary for promoter recognition and

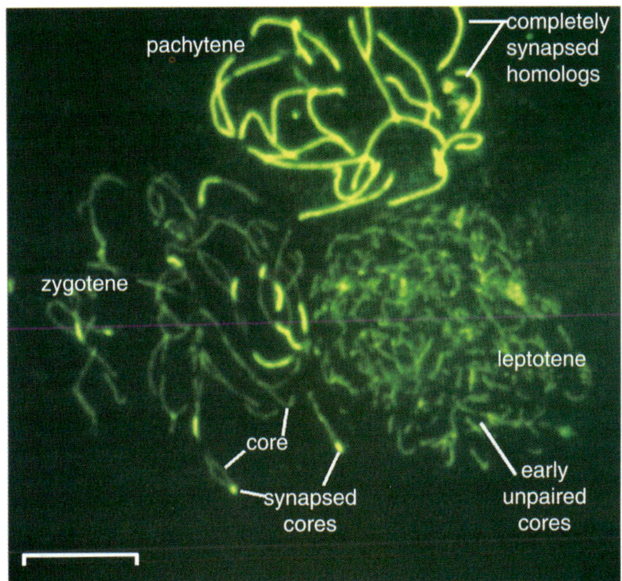

Plate 39 Synaptonemal Complex. The meiotic chromosome cores/SCs of a mouse are visualized here by indirect immunofluorescence with antibodies against a core protein and against a synaptic protein which produces a yellow color when the two proteins are present at the same site. Three meiotic prophase nuclei in successive stages of development (leptotene, zygotene and pachytene) are indicated in the figure. Scale bar = 10 μm.

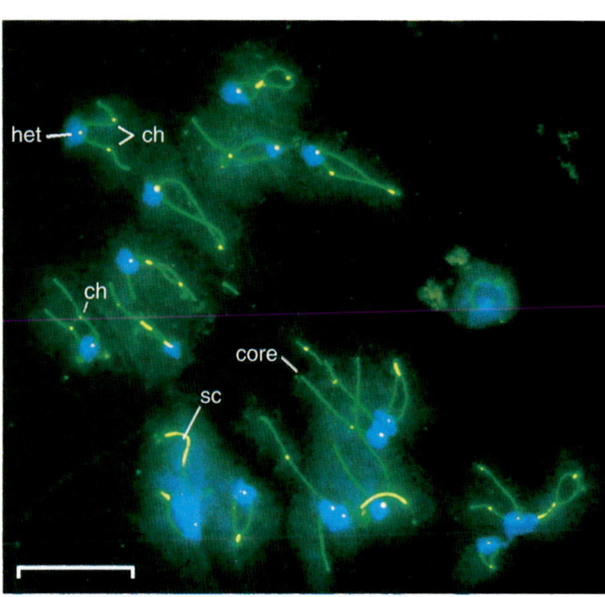

Plate 40 Synaptonemal Complex. Near the end of prophase in the mouse nucleus, the homologous chromosomes and their cores separate wherever the synaptic protein is no linger present. The centromeric chromatin (*het*) is stained with DAPI. Some of the locations that are suspected of having a chaisma are designated *ch.* Scale bar = 10 μm.

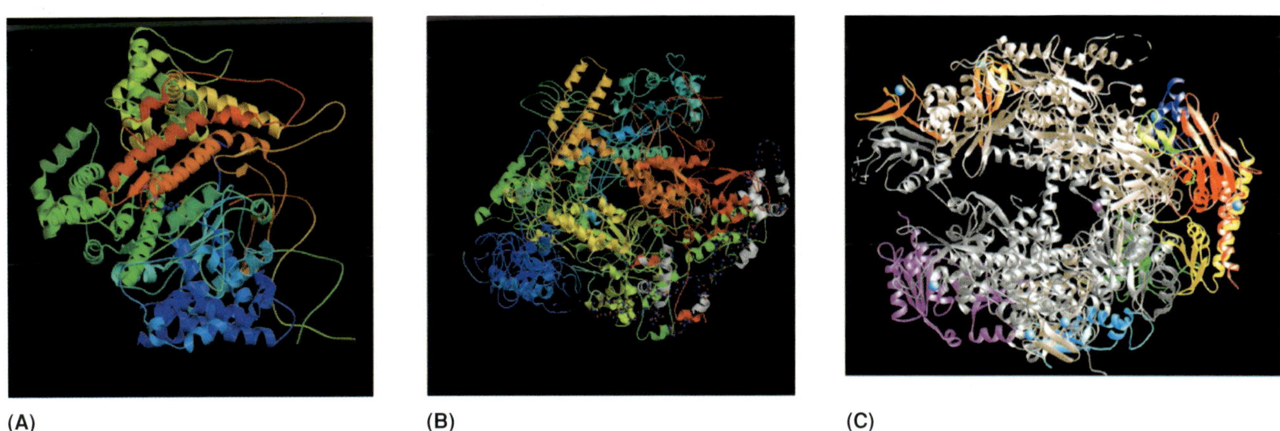

Plate 41 Transcription. Crystal structures have been determined for (A) T7 RNA polymerase (2.4 Å, IQLN; Cheetham and Steitz, 1999; image from Brookhaven Protein Database); (B) *Thermus aquaticus* core RNA polymerase (3.3 Å, IDDQ, Zhang *et al*., 1999; image from Brookhaven Protein Database); and (C) *Saccharomyces cerevisiae* RNA polymerase II (3.0 Å, IENO; Cramer *et al.,* 2000; image Courtesy of Patrick Cramer and David Bushnell).

Plate 42 Transposable Elements in Plants. The excision of the *Ac* transposable element at the *P* locus of *Zea mays* resulting in a variegated color pattern in the pericarp of the kernels.

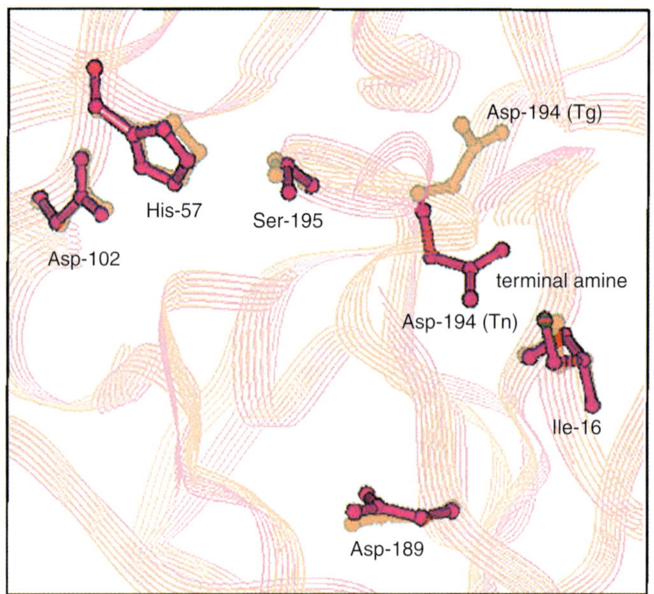

Plate 43 Trypsin. Arrangement of key residues in trypsinogen (Tg) and trypsin (Tn). The arrangement of the catalytic triad is essentially identical in the zymogen (Peach) and mature forms (Pink) of the enzyme. The major difference is the position of Asp 194. This residue rotates about the backbone and forms a salt bridge with he N-terminus of the mature enzyme. The salt bridge formed between Asp 194 and the N-terminus of the mature enzyme causes a subtle change in backbone structure and forms a site on the enzyme that stabilizes the transition state.

preinitiation complex formation. Thus, the 'cleared' promoter becomes accessible to another RNA polymerase molecule with its appropriate attendant proteins. Clearance can occur without escape, as occurs for heat shock genes in *Drosophila*. Thus, stalling near the start site of transcription can have a variety of effects, including promoter occlusion, although it can also poise the polymerase to respond rapidly to cellular needs that are essential for its survival.

Regulation of the initiation reaction

The transition from initiation to elongation requires the establishment of a 'ternary elongation complex,' which physically includes the RNA polymerase, the template, and the nascent transcript in a stable complex, that is able to continue the incorporation of nucleotides as progressive movement is made across the template. In addition to *cis*-acting promoter sequences and early transcribed sequences, there are *trans*-acting protein factors that influence the ability of the polymerase to move from initiation to elongation. The sigma factors and the GreA and GreB factors in bacteria can have an impact on the position and extent of abortive initiation, and the location at which the polymerase enters productive elongation. In more specific examples the 'catabolite regulatory protein' affects promoter escape in the *E. coli* maltose operon, and UTP concentration regulates promoter escape in the pyrimidine biosynthetic operon.

For eukaryotic RNA polymerase II, the general transcription factors TFIIE and TFIIH are very influential in the transition from preinitiation complex formation to elongation. The kinase activity of TFIIH acts upon RNA polymerase II changing its phosphorylation state, but the helicase activities of this factor are also critical in this transition. There are also situations, as with the heat shock genes of *Drosophila* and yeast, where the polymerase does not enter productive elongation until gene-specific regulatory factors have an impact on the initial transcribing complex. Clearly, initiation is a series of sequential steps that are regulated in the cell.

Elongation

Once the action proximal to the promoter has been resolved, the polymerase (and occasionally some associated proteins) can continue with its catalytic function, the synthesis of the RNA. However, elongating the nascent transcript is not a monotonic process of fixed rate. Rather, a large amount of cellular regulation influences the progress of the polymerase across a transcription unit. There are a variety of blocks to elongation that impede the polymerase as it makes its way across a transcription unit (see Table 1 in Lin and Lynch (1996), for eukaryotic blocks to elongation), and these regulated stops have a significant impact on the rate of transcription of any gene. At any nucleotide position, the polymerase can do one of four things (**Table 2**). Of course, it can **continue elongation**, incorporating the next nucleotides without detectable hesitation. Alternatively, the polymerase can 'pause,' and eventually begin transcription again. The signals that cause pausing are not clear, although pausing occurs at very specific locations all along a transcription unit. Pausing does not seem to occur when the templates are synthetic homopolymers (such as poly (dA) or poly (dC)) or alternating copolymeric templates (such as poly (dAT) or poly (dCG)). The amount of the polymerase that can be 'trapped' in the pause and the time spent pausing varies for each site and with different polymerases. There are few rules that predict a pause site or the duration of the pause. It is known that sequences both upstream and downstream of the pause site can influence the reaction, as can accessory protein factors. Some pauses have consecutive Ts in the nontranscribed DNA strand, or potential hairpin secondary structures upstream of the pause site. However, these are not characteristic of all the sequences that pause polymerases. Clearly the frequency and duration of pauses are attractive targets for regulation. Protein factors that influence polymerase pausing are discussed below.

During elongation, a polymerase also can 'arrest.' Arrest is an operational definition for an *in vitro* observation wherein the polymerase stops transcription upon encountering a block to elongation.

Table 2 Options for RNA polymerase during elongation

Elongate	Continue incorporation of nucleotide substrates
Pause	Halt during elongation, but resume nucleotide incorporation in a finite period of time without the need for an accessory factor. Accessory factors can influence the efficiency or duration of the pause
Arrest	Halt during elongation, but cannot resume synthesis without the assistance of accessory protein factor. Catalytic center of the polymerase and the 3′ end of the transcript physically separate, creating a requirement for transcript hydrolysis to generate a new 3′ end in the catalytic center
Terminate	Halt during elongation, release the transcript and the template; a terminator can be factor-independent or factor-dependent

However, in contrast to a pause site, an arrested polymerase will not spontaneously resume elongation in the presence of nucleotide substrates; rather the stalled and arrested elongation complex requires an accessory protein factor. That factor stimulates a transcript hydrolysis activity that resides in the polymerase. This unusual activity results in cleavage of the 3′ end of the transcript, release of a small RNA oligonucleotide (from 2 to 20 nucleotides long), retention of the 5′ portion of the transcript, and the resumption of elongation across the site that originally blocked progress of the polymerase (**Figure 3**). This cleavage activity is necessary because, when arrested, the polymerase's catalytic center is displaced from the 3′ end of the transcript. Lacking a 3′ hydroxyl group in the active site, the polymerase can no longer incorporate nucleotide substrates. To create a new 3′ end, the catalytic site hydrolyzes the transcript somewhat 5′ to the displaced end of the transcript. This creates a new 3′ end in the catalytic site that is in proper sequence register with the template; the polymerase can resume elongation and pass the original site of arrest (**Figure 3**). This cleavage reaction by the polymerase alone is very inefficient. Thus, factors that stimulate the cleavage reaction were isolated because they could stimulate the overall elongation reaction of the polymerase. Both prokaryotes and eukaryotes have such factors. Even the single-subunit bacteriophage RNA polymerase carries out the cleavage reaction, apparently without a cofactor. The mechanistic details that characterize the transcript cleavage reaction are under intense study, as the regulation of this reaction can strongly influence the efficiency of the elongation reaction.

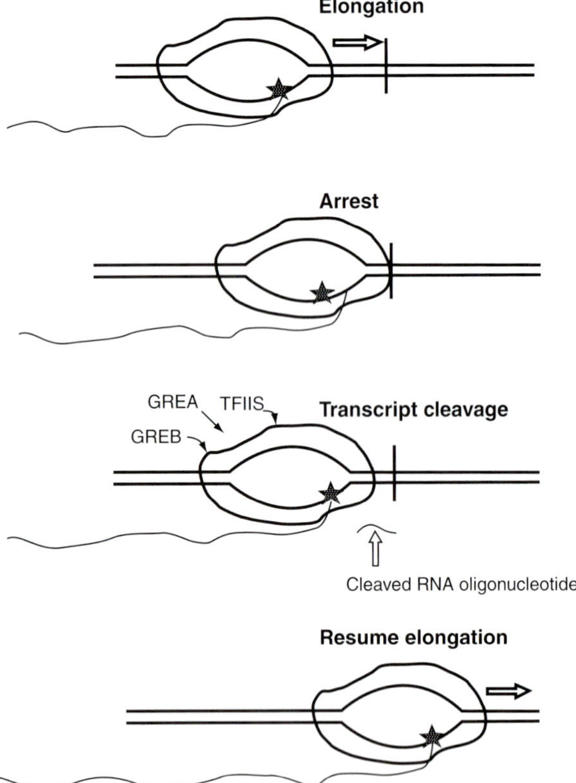

Figure 3 Transcript cleavage reaction in arrested elongation ternary complexes. RNA polymerase moving along the DNA template encounters a block to elongation and arrests transcription. The catalytic center of the polymerase moves away from the 3′ end of the transcript. The catalytic center is stimulated by an accessory protein (GreA or GreB in bacteria, TFIIS in eukaryotes) to hydrolyze an oligonucleotide from the 3′ end of the nascent transcript. The hydrolysis results in a new 3′-OH in register with the catalytic center, and nucleotide incorporation can resume for continuing the elongation process. The star represents the catalytic center of the polymerase. The perpendicular line represents the block to elongation.

Finally, the elongating polymerase may 'terminate' (described below).

Regulation during elongation

The multiple reactions that contribute to elongation (pausing, arresting, elongating) highlight the variety of targets for regulation. In eukaryotes, there are several proteins that reduce the pausing of RNA polymerase II *in vitro*. These include the general transcription factor TFIIF, which is also needed for the formation of the preinitiation complex. In addition, elongin and HMG14 and HMG17 have an effect on pausing. The TFIIS/SII protein of eukaryotes and the GreA and GreB proteins of bacteria stimulate the cleavage reaction necessary to reverse the arrested state of RNA polymerase II and bacterial RNA polymerase, respectively. The TFIIS protein is quite specific for RNA polymerase II in promoting readthrough of arrested polymerases, but it can also function with RNA polymerase I under some conditions to stimulate the cleavage reaction *in vitro*. The N-TEFb and DSIF protein complexes negatively regulate elongation by RNA polymerase II, and their action is overcome by a protein complex, P-TEFb, that phosphorylates the polymerase.

There are also many proteins that regulate elongation across specific transcription units, as illustrated by the following examples. The Tat protein of the human immunodeficiency virus stimulates more efficient elongation across the viral genome. The G2R and A18R proteins of vaccinia virus appear to modulate elongation for temporal control of viral RNA synthesis. Several cellular eukaryotic transcription units are controlled in a tissue-specific manner by regulating elongation, e.g., adenosine deaminase and, retinoic acid receptor isoforms. In prokaryotes, factors specific to the purB, Bgl, and S10 operons attenuate transcription elongation. Other proteins affect elongation by affecting termination, and examples of such proteins are discussed below.

Proteins that modulate the phosphorylation state of RNA polymerase II also regulate the elongation reaction, although the mechanism of this effect is unclear. The exchange of a protein complex (mediator) that promotes initiation with one(s) that facilitates elongation (elongator) is also involved. A highly phosphorylated polymerase is engaged in elongation; however, the pattern of phosphorylation and the particular kinase(s) involved are not well defined. There also may be the need for the sequential action of distinct kinases to generate an elongation complex capable of transcribing over long distances. Dephosphorylation is carried out by an enzyme that is essential in yeast, and is thought to be involved in the cycling of the polymerase after termination. It is likely that the extent and pattern of phosphorylation of RNA polymerase II changes during the elongation process, and that the combined effects of kinases and phosphatases have a significant regulatory impact *in vivo* during pausing, arrest, and in the coordination of RNA and DNA metabolism coincident with transcription.

In eukaryotes, chromosomal structure also affects the elongation reaction, and there are several interactions between chromatin remodeling machines and elongation factors inferred from genetics. In addition, proteins genetically involved in chromatin structure changes *in vivo* (such as the Spt4p and Spt5p proteins from yeast and their mammalian homologs) have been shown subsequently to have an impact on elongation *in vitro*. The Swi/Snf chromatin remodeling complex has also been linked *in vivo* and *in vitro* to elongation control.

Clearly, when considering the transcript elongation reaction, one must think about not only the mechanism of the catalytic synthesis but also the variety of behaviors of RNA polymerase during that synthesis (elongation, pausing, arrest, transcript cleavage). Overlaying the regulation are protein factors that alter each of these processes or change the template conformation.

Termination

Termination involves release of both the RNA and the template by RNA polymerase (**Figure 2**). It is crucial that this termination process is tightly regulated, since RNA polymerase is a completely processive enzyme, incorporating nucleotides until termination occurs. Once the polymerase has terminated transcription, it cannot pick up where it left off. This contrasts with DNA polymerases, distributive enzymes that dissociate from and reassociate with a template–primer to continue extension of a replicating strand of DNA.

Bacterial and bacteriophage RNA polymerases recognize specific sites as 'factor-independent' terminators, and bacterial polymerases also recognize a distinct set of sites only in the presence of accessory termination protein factors such as the rho protein, giving 'factor-dependent' sites of termination. The factor-independent termination sites include a T-rich sequence in the nontranscribed strand downstream of a potential hairpin secondary structure in the RNA. The T-rich sequence in the DNA is thought to signal the polymerase to halt transcription, and the hairpin structure is thought to contribute to the release reaction that dissociates the RNA from the polymerase. Curiously, T-rich sequences are part of the signal that causes pausing, arrest, and sometimes termination for eukaryotic RNA polymerases as well, reinforcing the notion of conserved signals during transcription. There are also a number of 'antitermination' proteins that regulate the recognition of terminators, and these are known to be especially important for the temporal regulation of bacteriophage production. In prokaryotes, the translation apparatus also regulates terminator recognition within a transcription unit in a process called 'attenuation' (see Attenuation), which is significant in several amino acid biosynthetic operons.

In eukaryotes, the termination reaction is best understood for RNA polymerase III, for which a T-run of four or five nucleotides in the nontemplate DNA strand seems sufficient to signal termination. For RNA polymerase II, all primary transcripts extend beyond the sequence that will become the 3′ end of the mature transcript ultimately formed in a processing reaction. The termination reaction is poorly understood, but it is directly coupled to or coordinated with the 3′ end processing events. The termination event is not random, but falls in a specific delimited region of the transcription unit. Termination by RNA polymerase I must be particularly precise. The ribosomal RNA genes (rDNA) are arranged in clusters of repeats, yet each transcription unit is independent of the others. Thus, RNA polymerase I terminates at specific locations that fall between the transcribed repeats, and this termination event in *S. cerevisiae* is dependent upon the Reb-1 protein. Like RNA polymerase II, RNA polymerase I transcribes well beyond the location of what will become the mature 3′ end of the rRNA, and this 'extra' transcription may be involved in regulating preinitiation complex formation for the next copy of rDNA downstream in the repeats.

Regulation during termination

Accurate and efficient termination prevents interference with transcription regulation of downstream genes. That is, if transcription continued unhindered,

Figure 4 Colliding polymerases. RNA polymerase ternary elongation complexes (A and B) are copying opposite strands of the DNA template and moving toward each other. Each might interfere with the progress of the other, and the chicken α-globin gene cluster uses this mechanism for regulation of transcription termination.

transcription regulation of a downstream gene would be disrupted. Occasionally this 'promoter occlusion' is used to the cell's advantage. In addition, when two transcription units are oriented facing each other in opposite directions on the chromosome, the termination reaction for gene A may also be regulatory for gene B (**Figure 4**). However, when the DNA replication machinery encounters the transcribing RNA polymerase, termination does not always occur. Surprisingly, these two molecular machines can pass each other nearly unhindered, although how this occurs is not understood.

In prokaryotes, the rho termination factor is essential for the proper utilization of factor-dependent termination sites in both cellular and bacterial virus transcription units. The rho factor is a hexamer of identical subunits that binds RNA, has RNA helicase activity, and hydrolyzes ATP during the termination reaction. There are rho utilization sites (rut sites) in rho-dependent termination regions, but the exact molecular nature of the sites and their recognition is incompletely defined. In a mechanism distinct from that utilized by the rho protein, the *B. subtilis* TRAP complex directs termination during tryptophan biosynthesis, in a manner quite distinct from the ribosome-mediated attenuation for tryptophan synthesis seen in *E. coli*.

Rho-dependent termination sites and factor-independent termination sites are ignored by prokaryotic RNA polymerases in the presence of antitermination factors such as the bacteriophage lambda N and Q proteins and similar proteins from related bacterial viruses. The antitermination function of N requires several host proteins and N-utilization-substances (Nus), first identified genetically. These include NusA, NusB, the rho factor, NusE (ribosomal protein S10), the α-subunit of RNA polymerase, as well as NusG. In a distinct mechanism, the psu protein encoded by phage P4 promotes antitermination which allows its partner phage, P2, to complete its replication. Cellular antitermination activity has been detected genetically in the bacterial ribosomal operons, where readthrough of transcription is regulatory and dependent on sites with sequence similarity to those needed for effective N function, but the host protein that carries out this reaction has not been identified as yet.

There are also mutant bacterial RNA polymerases that are considered 'hyperterminators' and 'hypoterminators,' and these mutations lie in the two largest subunits of the catalytic core of the enzyme. The change in termination efficiency holds for both factor-independent and rho-dependent terminators. However, the termination efficiency is unrelated to the elongation rate when factor-independent terminators are studied, whereas the termination efficiency for rho-dependent terminators seems to be related to the rate of elongation when the polymerase encounters the termination sequences. The mechanistic explanation for these differences is not known.

In eukaryotes, the La protein enters into termination regulation for RNA polymerase III. The La protein binds to U-rich 3′ ends of RNA, characteristic of the sequence for terminated RNAs for RNA polymerase III. Whether La is a termination factor that promotes release or is a stability factor for terminated RNAs is still under study. Genetic analyses of the largest subunits of RNA polymerase III very clearly delineate regions of the polymerase important for the recognition of T-rich regions of the non-transcribed DNA strand in the termination reaction. An analysis of similar regions of the largest subunits of RNA polymerases I and II has not been done, but the terminators for these two polymerases are also less well defined. In these cases, it may be more difficult to select individual mutations in the polymerase that have an impact on termination since there is a coupling of 3′ end processing with the termination reaction.

Termination is the final step in the synthesis of a productive RNA transcript from a transcription unit, but it is the essential part of the cycle that permits the polymerase to move from one template sequence to another so that the process of transcription can begin again.

Summary

Transcription makes an accurate copy of the information found within the DNA (or RNA) template so that the cell can decode the genetic information and allow it to be used in synthesizing protein, as well as structural, regulatory, and catalytic RNAs. Transcription is a multistep process with a large number of rate-limiting steps, both catalytic and stoichiometric. It should be emphasized that transcription is a cycle wherein the proteins involved are cycled again and again to promote synthesis of the proper RNAs at the proper place and time and in appropriate amounts in the cell. Transcription is one of many processes that the cell uses to regulate gene expression, and the coordination of transcription with other RNA and DNA metabolic processes emphasizes the importance of accurate regulation of these events to ensure survival of the organism.

Further Reading

Adhya S (ed.) (1996) *Methods in Enzymology*, vols 273 and 274, *RNA Polymerase and Associated Factors*. San Diego, CA: Academic Press.

Archambault J and Friesen JD (1993) Genetics of eukaryotic RNA polymerases I, II and III. *Microbiological Reviews* 57: 703–724.

Cheetham GM and Steitz TA (1999) Structure of a transcribing T7 RNA polymerase initiation complex. *Science* 286: 2305–2309.

Cheetham GM and Steitz TA (2000) Insights into transcription: structure and function of single subunit DNA-dependent RNA polymerases. *Current Opinion in Structural Biology* 10: 117–123.

Cramer P, Bushnell DA, Fu J *et al.* (2000) Architecture of RNA polymerase II and implications for the transcription mechanism. *Science* 288: 640–649.

Lai MC (1998) Cellular factors in the transcription and replication of viral RNA genomes: a parallel to DNA-dependent RNA transcription. *Virology* 244: 1–12.

Lewin B (1997) *Genes VI*, Chs 3 and 6. Oxford: Oxford University Press.

Lin ECC and Lynch AS (eds) (1996) *Regulation of gene expression in* Escherichia coli. Austin, TX: RG Landes.

Stillman B (ed.) (1998) Mechanisms of transcription. *Cold Spring Harbor Symbosia on Quantitative Biology*, 63.

Uptain SM, Kane CM and Chamberlin MJ (1997) Basic mechanisms of transcript elongation and its regulation. *Annual Review of Biochemistry* 66: 117–172.

Zhang G, Campbell EA, Minakhin L, Richter C, Severinov K and Darst SA (1999) Crystal structure of *Thermus aquaticus* core RNA polymerase at 3.3 A solution. *Cell* 98: 811–824.

Reference

Uptain SM, Kane CM and Chamberlin MJ (1997) Basic mechanisms of transcript elongation and its regulation. *Annual Review of Biochemistry* 66: 117–172.

See also: Attenuation; Chromosome; Gene Expression; Ribosomal RNA (rRNA); RNA Polymerase

Transcription Factor

See: Bacterial Transcription Factors, Transcription

Transduction

B S Guttman

doi: 10.1006/rwgn.2001.1300

Transduction is the term used to describe the transfer of genetic material from one cell to another by means of a virus. As microbial genetics developed, it became clear that bacteria can exchange genetic information in two ways: through transformation, in which a cell picks up naked DNA from the medium, and through conjugation, in which two cells come into contact temporarily and a copy of some genes are transferred from one to the other. The requirement for cell–cell contact was demonstrated by Davis, who put cells of two strains of *Escherichia coli*, normally capable of conjugation, on opposite sides of a filter, thus preventing their contact; no recombination occurred, thus showing that recombination was not due to transformation. When Norton Zinder discovered recombination between strains LT2 and LT22 of *Salmonella typhimurium*, he repeated the Davis experiment, expecting to find no recombination. But recombination did occur, showing that conjugation was not involved; adding deoxyribonuclease to the medium did not prevent recombination, so the mechanism was not transformation. Zinder then showed that strain LT22 was lysogenic for a phage, designated P22, and that P22 virions were carrying genes into LT2 cells. Zinder called the phenomenon 'transduction.'

Transduction can take two distinct forms, generalized and specialized. In generalized transduction, as in the P22 case, the phage (either temperate or virulent) multiplies in a host cell while breaking the host DNA into fragments. While most virions include copies of the phage genome, some 'pseudovirions' encapsidate fragments of host DNA. When a second strain of bacteria, carrying suitable genetic markers, is then infected at a low multiplicity of infection, some of the bacteria receive the transducing particles; instead of becoming infected, they receive a fragment of bacterial DNA, which can then recombine with their own DNA and render them recombinant. Generalized transducing phage can carry any host genes.

Specialized transduction, in contrast, occurs only with a temperate phage whose prophage occupies a specific site in the host genome. The best-known case is with phage lambda, whose prophage inserts between the *gal* (galactose metabolism) and *bio* (biotin biosynthesis) genes. When the prophage is excised from the genome and begins lytic multiplication, it usually recombines with a precise reversal of the insertion crossover, so a burst of normal lambda phage is produced.

Sometimes, however, the excision crossover occurs at the wrong point, so the excised lambda genome leaves some of its own genes behind and carries some of the host genes from one side or the other. This phenomenon was first studied by using the *gal* genes; the resulting transducing phage carried *gal* DNA instead of some critical genes needed for phage multiplication and so were known as λ*dg* (defective, galactose-transducing). Later it was discovered that lambda phage could also carry *bio* genes, and some of these phage are not defective in their ability to multiply.

Transduction of either type is particularly useful for fine-structure genetic analysis. The general sequence of bacterial genes can be determined with relative ease through conjugation experiments, but such experiments cannot determine the sequence of a series of very close markers. Various experimental designs can be used for fine-structure mapping. If two close markers are transduced together, they are said to be cotransduced, and the frequencies of cotransduction can indicate the relative sequence of three or more contiguous markers, since those closest to each other should be cotransduced most frequently. One can also determine the sequence of a series of markers affecting a single function ($X_1, X_2, \ldots, X_n$) relative to a nearby outside marker; for illustration, we will use the marker *leu* (leucine biosynthesis). One then sets up a pair of reciprocal crosses such as the following, where the recipient cells are made leu^- and the transducing phage are grown on leu^+ cells:

$$\begin{array}{ll} \text{donor:} & \underline{leu^+ \; x_1 \; +} \\ \text{recipient:} & \underline{leu \quad + \; x_2} \end{array}$$

and

$$\begin{array}{ll} \text{donor:} & \underline{leu^+ \quad + x_2} \\ \text{recipient:} & \underline{leu \quad x_1 \; +} \end{array}$$

One then measures the frequency of leu^+ transductants and the frequency of $leu^+ \; X^+$ transductants. The experiment depends on the principle that the more crossover events are required to produce a given recombinant, the rarer that recombinant will be. Notice that in the first cross, one pair of crossovers is needed to bring the leu^+ marker into the recipient and a second pair of crossovers is necessary to obtain the wild-type X_2^+. In the second cross, however, a single pair of crossovers can bring in both the leu^+ and X_1^+ markers. Thus, if the sequence is indeed leu–X_1–X_2 more of the leu^+ recombinants should also be X^+ in the second cross than in the first.

See *also*: Bacterial Transformation; Conjugation, Bacterial; Specialized Transduction

Transfection

R J Redfield

doi: 10.1006/rwgn.2001.1301

The term 'transfection' is used differently by bacterial geneticists and animal cell biologists. Both usages describe processes where a DNA fragment is introduced into a cell and the gene or genes it carries are expressed. In bacteria this general process is called transformation, and transfection refers to the special case where bacterial cells are transformed with DNA from an infectious bacterial virus (a bacteriophage), to produce an infected cell. In tissue culture of animal cells, transfection refers to any process that artificially introduces DNA into cultured cells.

See *also*: Bacterial Transformation; Tissue Culture

Transfer of Genetic Information from *Agrobacterium tumefaciens* to Plants

B Hohn

doi: 10.1006/rwgn.2001.1638

Agrobacterium tumefaciens, a soil-borne phytopathogenic bacterium, transfers a segment of its Ti (tumor-inducing) plasmid, called T-DNA (transferred DNA), to plants (**Figure 1**). Virulence proteins, coded for by the virulence region also localized on the Ti plasmid, mediate this transfer. They are involved in generation, translocation, protection, and nuclear localization of the T-DNA. Finally the bacterial DNA integrates into plant chromosomal DNA in a random fashion. Genes located on the integrated T-DNA carry eukaryotic expression signals. Their expression yields enzymes providing unique nutrients for the bacterium and turning on mitotic activity of the transformed cell. The result is a tumor, also called crown gall. Molecular understanding of this process allows the use of this bacterium as a general tool for generating transgenic plants.

Events in the Bacterium

In free-living agrobacteria most of the virulence genes, which are organized in operons, are inactive.

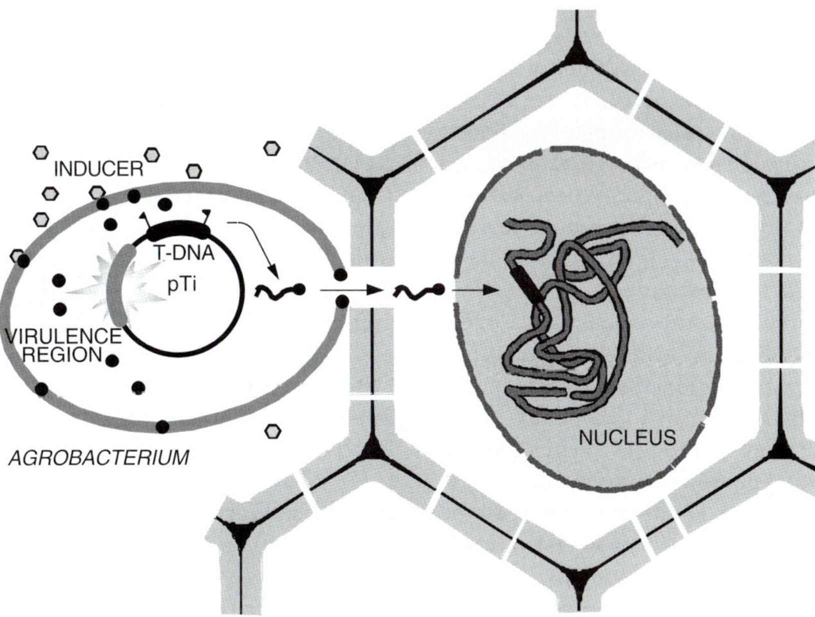

Figure 1 Scheme of T-DNA transfer from *Agrobacterium tumefaciens* to a plant cell. (Adapted from Rossi *et al.*, 1998, with modifications.)

Perception by the bacteria of signals emanating from wounded plant cells leads to a unique molecular dialogue. Wounded plant cells secrete phenolic compounds and sugars, which are sensed by the bacterial virulence protein VirA (**Table 1**) and interpreted as proximity of a plant cell ready to be invaded. In a cascade of events following attachment of the bacteria to plant cells, the receipt of the plant signal is turned into a general transcriptional activation of all virulence genes. The VirA protein thereby becomes autophosphorylated and subsequently transfers the phosphate group to the virulence transcription factor VirG in order to activate it. This leads to general transcription of all virulence genes, involving specific regions upstream of the virulence promoters. As a consequence virulence proteins are produced. These are involved, apart from the signal sensing/transcriptional regulation activities, in generating a transferable T-DNA unit and accomplishing its transfer to the plant cell with the help of a transfer machine.

The T-DNA on the Ti plasmid is flanked by two almost identical sequence elements in direct orientation. These so-called border sequences are substrates for the site-specific nicking enzyme composed of VirD1 and VirD2. Upon cleavage, the catalytic subunit VirD2 remains covalently attached to the 5′ terminus of the emerging single-stranded T-DNA. This protein–DNA complex travels, in an unknown fashion, to the plant cell. It thereby uses the transfer apparatus composed of the exocellular pilus, the mating channel, and cytoplasmic membrane ATPases. Despite intensive research in this area, the exact composition of this machine is not known. Also the local connection of the functional units and, most importantly, the use by the T-DNA-protein complex of the transfer machinery awaits elucidation.

Events in the Plant

After the T-DNA–VirD2 complex has reached a plant cell's interior, by whatever mechanism, it receives a

Table 1 Functions of Ti-plasmid encoded virulence proteins. Chromosomally located virulence genes are not included. In addition to the general Ti-plasmid encoded virulence proteins, there exist additional strain specific proteins

Protein	Function
VirA	Sensor/transmitter of plant signal
VirG	Transcriptional activator of virulence genes
VirD1, VirD2	Enzyme complex generating a site-specific nick into the border sequence
VirC	Enhancement of virulence
VirE2	Single-stranded DNA binding protein; activity in the bacterium is dependent on the presence of VirE1
VirB1-11, VirD4	Components of transfer machine

protective coat of VirE2 protein molecules that most likely reaches the plant cell independently of the T-DNA complex. This virus-like particle has a compact, yet flexible structure. It enters the plant nucleus at nuclear pores by virtue of nuclear localization signals contained in virulence proteins. Specific transporters ferry the T-DNA complex across these huge macromolecular machines.

The 'ultimate goal' of the pathogenic organism *A. tumefaciens* is to subvert the metabolism of the infected plant cell to serve the invader's need and to produce compounds only the inciting pathogen can use. For this to be guaranteed in a genetically stable way, the T-DNA complex integrates. To elucidate the mechanism of this integration step *in vitro* as well as *in vivo* analyses have been employed. Whereas the former approach yielded information on known bacterial- and plant-specific proteins involved in this process, analysis of mutants of the model plant *Arabidopsis thaliana* impaired in transformation leads to the identification of plant proteins specifically involved in this process.

The tumorous phenotype of plants infected with *Agrobacterium tumefaciens* originally suggested oncogenic principles at work that might yield information on human cancer. Once the bacterially initiated transformation principle was established, this suggested relationship of course could no longer be maintained. Indeed, the tumor-inducing principle (TIP) could be explained by the presence and expression of genes involved in the hormone balance of plant cells. The T-DNA was shown to contain genes coding for enzymes involved in the biosynthesis of the plant hormones auxin and cytokinin. Overexpression of these genes leads to uncontrolled proliferation of transformed cells, ultimately yielding the phenotype of tumors. The underlying pathogenic mechanism was found to be the overreplication of transformed nuclear DNA and specifically of T-DNA genes involved in the production and secretion of specific secondary metabolites called opines. These biochemicals, specific for agrobacterial strains inducing their synthesis, are used exclusively by the bacterial species responsible for the respective tumors as their sole nitrogen, carbon, and energy source. This special bacterium–plant relationship thus represents a microcosm of specific exploitation of plant resources by a sophisticated plant pathogen.

Use of *Agrobacterium tumefaciens* for Generation of Transgenic Plants

During the study of the mechanisms underlying T-DNA-mediated plant transformation, it became obvious that there was an opportunity to exploit this system for generating transgenic plants. In particular, the fact that for the transfer no T-DNA sequences other than the borders are required suggested the process of inserting genes of interest into the T-DNA, which the bacterium faithfully inserted into the plant. In many plant species *Agrobacterium*-mediated plant transformation is the method of choice for the generation of transgenic plants, as in many cases only one or a few copies of (almost) complete DNA units are integrated. Successful transformation has been achieved not only for dicotyledonous plants, the natural hosts producing tumors upon infection, but also for some monocotyledonous plants, to which the agriculturally important cereals such as wheat, maize, and rice belong. Thereby the capacity of the transformed cells to regenerate frequently into fertile, transgenic plants has been the critical step. In addition to crop improvement, however, plant transgenesis is a key factor in the establishment of modern genetics and genomics.

Further Reading

Chilton MD, Drummond MH, Merio DJ *et al.* (1977) Stable incorporation of plasmid DNA into higher plant cells: the molecular basis of crown gall tumorigenesis. *Cell* 11: 263–271.

Crouzet P and Hohn B (2001) Transgenic plants. In: *Encyclopedia of Life Science*, http://www.else.net.

Gelvin SB (2000) *Agrobacterium* and plant genes involved in T-DNA transfer and integration. *Annual Review of Plant Physiology and Plant Molecular Biology* 51: 223–256.

Horsch RB, Fraley RT, Rogers SG *et al.* (1984) Inheritance of functional foreign genes in plants. *Science* 223: 496–498.

Rossi L, Tinland B and Hohn B (1998) Role of the virulence proteins of *Agrobacterium tumefaciens* in the plant cell. In: Spaink H, Hooykaas P and Kondorosi A (eds) *The Rhizobiaceae*, pp. 303–320. Dordrecht: Kluwer.

Tzfira T and Citovsky V (2000) From host recognition to T-DNA integration: the function of bacterial and plant genes in the *Agrobacterium*–plant cell interaction. *Molecular Plant Pathology* 1: 201–212.

Zupan J, Muth TR, Draper O and Zambryski P (2000) The transfer of DNA from *Agrobacterium tumefaciens* into plants: a feast of fundamental insights. *Plant Journal* 23(1): 11–28.

***See also:* Ti Plasmids**

Transfer RNA (tRNA)

A Liljas

doi: 10.1006/rwgn.2001.1302

The transfer RNAs are the central molecules in translation or protein biosynthesis. They are the adaptors that mediate the translation of the genetic message into proteins, which are the principal gene products of cells.

History

When the structure of DNA and the basics of protein synthesis were first clarified tRNA molecules were unknown. Crick pointed out that there was a significant problems to understand how a polypeptide could be assembled from an RNA template since there was no stereochemical complementarity between the codons and the amino acids. He suggested that adaptors, small RNA molecules that could be charged with specific amino acids by enzymes, would decode the messenger RNA (mRNA) by complementarity. These adaptors could thus participate in incorporating the amino acids into a growing polypeptide. Subsequently these adaptors were identified and are now known as the tRNA molecules.

Structure

When the first nucleotide sequence of a tRNA was determined the possibilities for base-paired secondary structures was examined. Among the three possibilities suggested only one, the classical cloverleaf (**Figure 1**), was consistent with subsequently determined sequences. Here the stem regions contain four to seven base pairs. The cloverleaf is arranged in such a way that the 5′ and 3′ termini are base-paired to each other. The three leafs are formed by three base-paired regions each making a loop. The middle loop contains the anticodon of the tRNA. In 1973 the first three-dimensional structures of tRNA were determined. Here a number of surprises became apparent. First, the secondary structure of the clover leaf was confirmed, but it was found to be folded into the shape of an L (**Figure 2**). Second, the anticodon was found at one end and the 3′ acceptor, the CCA sequence, at the opposite end, approximately 80 Å apart. This meant that the anticodon has no possibility of interacting with the amino acid. It also means that when the tRNA incorporates the amino acid into the growing polypeptide on the ribosome the mRNA and the decoding site is far from the site for peptidyl transfer.

Synthesis

The genes for the tRNA molecules are dispersed in the genomes. There are sometimes several genes for a certain tRNA and the genes frequently are found as clusters of several tRNA genes. Some of them are located among genes for ribosomal RNA molecules. The genes for the tRNA molecules do not code for the final functional molecule. Thus the transcript contains sequences that are removed by specialized tRNA processing nucleases. Likewise the 3′ terminal residues, CCA, are added by an enzyme to eucaryal tRNA precursors.

In addition the mature tRNA molecule is extensively modified. Thus some modifications are so

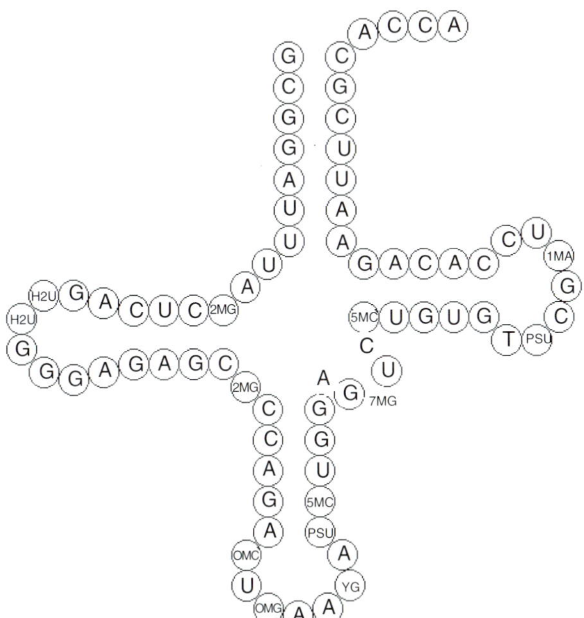

Figure 1 The classical cloverleaf two-dimensional structure of phenylalanine tRNA from yeast. (Figure made by Maria Selmer.)

Figure 2 The three-dimensional structure of a tRNA (Phe tRNA from yeast). (Figure made by Maria Selmer.)

typical that they have given the names to the parts of the structure they belong to. Thus a pseudouridine (Ψ) has given the name to the Ψ loop and a 5,6-dehydrouridine (D) has given the name to the D-loop. The anticodon loop is also frequently modified.

Codon–Anticodon Relationships

The universal genetic code has 64 words or codons, three of which designate stop and are not normally read by tRNAs. Since there are 20 different amino acids in the regular protein the code is degenerate. Thus there are six codons that correspond to serine and arginine while there is only one codon for several other amino acids. This situation is handled differently in different organisms. The codons used must be correlated to the set of tRNAs expressed by the organism. In some organisms the codon usage is limited to a small set of tRNAs (minimally 20) while in other species there are tRNAs corresponding to most codons. Thus the codon usage is different for different organisms.

The anticodon, normally positions 34–36 of the tRNA, interacts with the codons of the mRNA primarily by Watson–Crick base-pairing. However, the first position of an anticodon of a tRNA can base-pair with different nucleotides at the third position of a codon. Thus noncanonical base pairs are formed in the third or the so-called wobble position of the codon. This may also involve modified bases of the tRNA. This allows a tRNA with a certain anticodon to read several codons; thus a limited set of tRNAs can read a larger set of codons.

The tRNA Synthetàses

The enzymes that charge the tRNAs with amino acids, the tRNA synthetases, are specific for one amino acid each. Thus there are normally 20 tRNA synthetases in an organism even though there are some deviations. These enzymes are specific for one amino acid and the corresponding tRNA or a family of tRNAs. There are two classes of tRNA synthetases, ten of each class. These two classes have entirely different structures, recognize the tRNA in different ways and charge the tRNA on the 2′ (class I) and 3′ (class II) hydroxyls of the terminal ribose of the tRNA.

The fidelity of translation primarily depends on the synthetases. They are enzymes that recognize two different substrates with utmost accuracy. First the correct amino acid has to be bound by the enzyme and activated by an ATP molecule. In the second step the correct tRNA is bound to the enzyme and the amino acid has to be transferred to it. Since the distance between the amino acid and the anticodon is large not all tRNA synthetases contact the anticodon, but select the correct tRNA using specific sequences along the lengths of the tRNA molecules.

The Adaptor Molecule in Protein Synthesis

The charged aminoacyl tRNA is readily protected by a protein that is very abundant in the cell and that carries the tRNA to the ribosome. In procaryotes it is called elongation factor Tu (EF-Tu) and in Eukarya and Archaea elongation factor 1 which is composed of several subunits. This elongation factor is activated by binding a GTP molecule to bind the aminoacyl-tRNA The complex between the elongation factor and the tRNA primarily concerns the acceptor stem and the amino acid that becomes well protected from hydrolysis by the protein.

The ribosome has three main sites for tRNAs: the A-, P-, and E-sites. They span the space between the small and large subunits. Here the long distance between the two functional ends of the tRNA becomes necessary. The mRNA is bound to the small ribosomal subunit, thus the decoding is performed there. The large subunit contains the peptidyl transferase site where the acceptor ends of two tRNAs come together to transfer the nascent peptide to the incoming amino acid. To be able to start the synthesis of a protein a special tRNA is required to read the first AUG codon. This is called the initiator tRNA. It binds to initiation factor 2 (IF2) which brings it into the ribosomal P-site before the factor dissociates.

Complexes between tRNA and the elongation factor Tu (or EF-1) bind to the ribosome in an initial selection between cognate and noncognate tRNAs. Cognate, and less frequently near-cognate, tRNAs cause the elongation factor to hydrolyze its GTP molecule. This results in the dissociation of the factor from the tRNA and the ribosome. The tRNA molecule is now free to place its amino acid in the peptidyl transfer site. In a proofreading step the fidelity of the codon–anticodon interaction is improved by the dissociation of near-cognate tRNAs before the nascent peptide is attached. In a subsequent step the peptidyl tRNA is translocated from the A-site to the P-site, a step that is mediated by elongation factor G (EF-G) or in Eukarya and Archaea EF-2. The deacylated tRNA is moved from the P- to the E-site before it dissociates.

tRNA Mimicry

Several proteins participating in protein biosynthesis interact with the ribosomal sites for tRNA. The structures of several of those are being unraveled. So far the eukaryotic termination factor 1 (eRF1), that causes the hydrolysis of the peptide when a stop codon is encountered in the A-site, has a tRNA like structure and may bind to the A-site like a tRNA. The ribosome recycling factor is an excellent mimic of a tRNA. It may bind to the A-site after termination, to dissociate the mRNA from the ribosome and cause the subunits to dissociate from each other. The ribosomal translocase, EF-G or EF-2, specifically mimics the complex containing EF-Tu and a tRNA. Thus parts of that protein mimic the tRNA. During translocation it has been observed to place itself in part of the A-site.

Further Reading

Söll D and RajBhandary UL (1995) *tRNA: Structure, Biosynthesis, and Function*. Washington, DC: American Society for Microbiology Press.

Spirin AS (1999) *Ribosomes*. New York: Plenum Press.

***See also:* Adaptor Hypothesis; Amino Acids; Aminoacyl-tRNA Synthetases; Anticodons; AUG Codons; Elongation Factors; Genetic Code; Messenger RNA (mRNA); Protein Synthesis; Ribosomes; Translation; Wobble Hypothesis**

Transformation

doi: 10.1006/rwgn.2001.2054

1. Transformation in bacteria or eukaryotic cells is the acquisition of new genetic markers by incorporation of added DNA.
2. Transformation of eukaryotic cells is the conversion to an unrestrained growth pattern.

***See also:* Bacterial Transformation; Marker**

Transgenes

J Austin

doi: 10.1006/rwgn.2001.1304

Transgenes are exogenous DNA sequences introduced into the genome of an organism. These transgenes may include genes from the same organism or novel genes from a completely different organism. The resulting plant, animal, or microorganism is said to be transformed. Transformation occurs naturally in organisms such as bacteria, which can take up DNA from their surrounding environment. In addition, techniques have been developed to introduce and maintain transgenes in plants, animals, and bacteria. Transgenes can be used to analyze or alter the function of a known gene. In other cases, introduction of transgenic DNA has been used to add new functions to an organism, such as the expression of a protein normally not present in that organism. In addition to the application of transgenes in research, transgenic DNA has many potential medical applications, including the creation of DNA-based vaccines and gene therapy.

The process of transformation was first identified in 1928 by Frederick Griffith, in experiments using two different strains of streptococcal bacteria. It had been demonstrated previously that injection of virulent S (smooth) strain bacterial cells could kill mice, but that injection of cells from the nonvirulent R (rough) strain did not. When heat-killed S strain cells were mixed with live R strain cells and injected into mice, however, the mice also died, indicating that in some way the dead S strain cells had been able to transform the nonvirulent R strain cells into virulent S strain cells. In subsequent experiments, when DNA isolated from S strain cells was injected into mice along with live R strain cells, the injected animals were found to contain a mixture of both R and S strain bacterial cells, demonstrating that transformation of nonvirulent R strain bacteria into virulent S strain bacteria was mediated by the S strain DNA.

Initial experiments in the transformation of multicellular animals took place in *Drosophila melanogaster* and in mice. In *Drosophila*, transformation is carried out using a transposable element to transport the transgenic DNA into the *Drosophila* genome. Initial transformation experiments in *Drosophila* made use of the *Drosophila rosy* gene, which is required for normal red eye color. In order to create transformed flies, the wild-type *rosy* gene was cloned into the middle of a *Drosophila* transposable element and the transgene/transposon construct was then injected into *Drosophila* embryos containing a mutant *rosy* gene. Flies in which the transposon successfully excised from the plasmid and then inserted into the genome could be identified based on their change in eye color, from rosy to the normal red. The first transgenic mice were created by microinjection of transgene DNA into fertilized mouse eggs. One transgene used in these early experiments contained the promoter for the metallothionein gene, which is activated by increased levels of heavy metals, fused to the gene encoding human growth hormone. Using this transgene, the animals that developed from the transformed embryos could be identified based on their increased size relative to normal mice when fed either cadmium or zinc.

A critical first step in creating a transgenic organism is to get the transgene DNA into the organism. In bacteria, transformation is carried out by mixing transgenic DNA with bacterial cells treated to increase their ability to take up DNA. A variety of other methods have been used to introduce DNA into plant and animals cells including DNA injection, electroporation, and microparticle bombardment. A different type of approach involves the use of viruses: the transgene is cloned into the viral genome and is then introduced into cells by viral infection. If the transgene-containing virus is competent to carry out multiple rounds of infection, the transgene may be spread from cell to cell by the virus. Alternatively, if the transgenic virus is unable to infect cells on its own and requires the presence of a helper virus for infection, the transgene will be transmitted to a more limited group of cells.

Typically, transformation is carried out in such a way that while many individuals are exposed to the transgenic DNA only a small percentage are actually transformed. Therefore, a means of distinguishing the successful transformants from the background of untransformed individuals is essential. One approach often used is to include a selectable marker, such as a gene that provides antibiotic resistance, in the transgenic DNA; in other cases, transformed individuals are identified based on rescue of a mutant phenotype, such as described above for the *Drosophila rosy* gene. An increasingly important approach is the use of reporter genes such as *LacZ* or green fluorescent protein (GFP) that can be used to both identify transformants and to monitor when and where the transgene is expressed *in vivo*.

In some cases transformation is a transient event; for example, when transgenic DNA is introduced into tissue culture cells, most of it is lost after several rounds of cell division. The creation of stable transformants requires a means of maintaining the transgenic DNA through multiple rounds of cell division. For multicellular animals, there is the additional requirement that germ cells must be transformed in order to ensure that the transforming DNA is transmitted to the next generation (in the case of multicellular plants, transformation can be maintained either by transformation of the germline or by clonal propagation of transformed cells).

One way to produce stable transformation is by insertion of the transgene into the genomic DNA. In many cases where this occurs, the site of transgene insertion is random. In species such as the yeast *Saccharomyces cerevisiae* and *Tetrahymena*, however, transgene insertion occurs by homologous recombination in which the site of transgene insertion is based on homology between transgene and genomic sequences. In *S. cerevisiae*, stable transformation can also occur when transgenic DNA is cloned into Cen plasmids which are faithfully segregated to both daughter cells during each cell division due to the presence of yeast centromere sequences. A different approach to transformation is observed in the nematode *Caenorhabditis elegans*, where injection of transgenic DNA into the *C. elegans* germline can result in formation of an extrachromosomal DNA array containing hundreds of tandemly repeated copies of the transgene.

Transgenes have played a critical role in research. They have been used to clone genes by phenotypic rescue, in which transgenes are tested for their ability to rescue the mutant phenotype of individual genes. Using transgenes that fuse gene regulatory sequences to reporter genes such as β-galactosidase or GFP, it is possible to observe directly gene expression patterns. Another important use of transgenes has been the creation of constructs that are targeted to a specific gene based on sequence homology and can be used to alter or knockout gene function. Increasingly, these approaches are now being applied to the creation of genetically altered plants and animals. One example of this growing transgenic technology is the use of mammalian embryonic stem cells, which are totipotent and therefore have the capacity to form any tissue type.

See also: **Embryonic Stem Cells; Genetic Engineering; Transfer of Genetic Information from *Agrobacterium tumefaciens* to Plants; Transgenic Animals**

Transgenic Animals

F Costantini

doi: 10.1006/rwgn.2001.1305

A transgenic animal is one carrying an experimentally introduced gene or DNA segment (transgene) in many or all of its cells. The transgene is usually acquired by gene transfer at an early embryonic stage or by transmission from a transgenic parent. Transgenes may be inserted at random sites in the host genome, or they may be targeted to specific loci via homologous recombination. Many (but not all) transgenes express a gene product, which in some cases alters the phenotype of the animal. Transgenic mice are widely used in biomedical research, and the application of similar technologies to other species is of potential importance in agriculture.

History of Transgenic Animals

Transgenic mice were the first transgenic animals to be produced, due to the widespread use of this species as a model mammalian organism for genetic and embryological studies. This important development resulted from a convergence of technical advances in several fields, including the culture of preimplantation embryos *in vitro*, embryo transfer, recombinant DNA methods, and the ability to transfer genes into cultured cells. The first transgenic mice were produced in the mid-1970s, by the viral infection of preimplantation embryos, which resulted in stable integration of the viral DNA into the host genome and transmission through the germline. The first transgenic mice carrying an exogenous eukaryotic gene were produced in 1980 by a different method, the microinjection of purified recombinant DNA into a pronucleus of the fertilized egg. This method has since been used extensively in mice, and also adapted for transgenesis in many other mammalian species. Later in the 1980s, the development of embryonic stem (ES) cell lines that could colonize the germline in chimeric mice provided another route for the introduction of transgenes, as well as for targeted mutagenesis of endogenous genes. However, ES cells able to colonize the germline have not been produced from other mammals. In the late 1990s, techniques for cloning sheep, cows, and mice via nuclear transfer from fetal or adult somatic cells into the oocyte provided a new potential route for transgenesis.

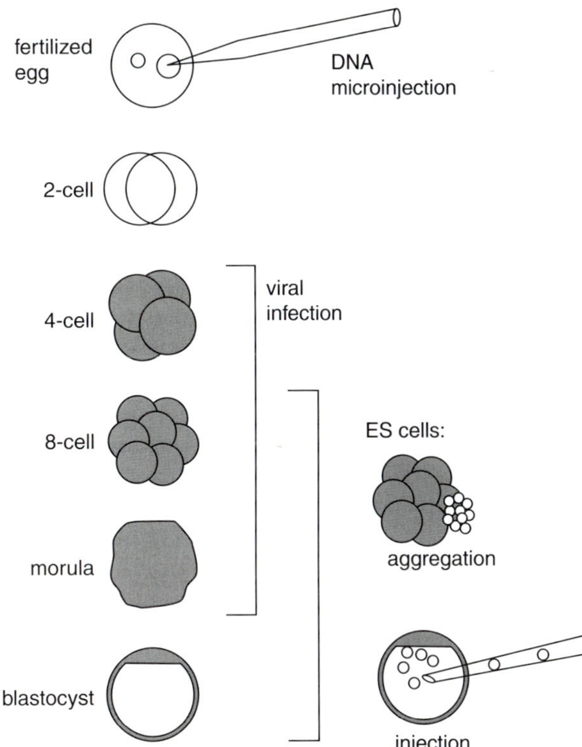

Figure 1 Stages of preimplantation mouse embryogenesis at which transgenes are introduced. Microinjection of transgene DNA is normally performed at the one-cell (fertilized egg) stage. A fine, hollow glass needle containing a DNA solution is inserted through the zona pellucida (outer glycoprotein envelope) and plasma membrane of the egg into one of the two pronuclei. A small volume of the DNA solution is injected. Infection with retroviral vectors is usually performed between the four-cell and morula stages: the zona pellucida is removed and the embryo is exposed to a preparation of infectious viral particles (not shown). Transgenesis via embryoic stem (ES) cells is performed at the eight-cell or blastocyst stage: several ES cells containing the transgene are aggregated with an eight-cell embryo, or injected into the cavity of a blastocyst. The ES cells then intermingle with the host embryonic cells, giving rise to a chimeric mouse.

Methods for Production of Transgenic Animals, and Genetic and Physical Properties of Transgenes

In order to introduce a transgene into most or all the cells of the animal, including the germ cells, transgenesis is normally performed at very early stages of embryogenesis, ranging from the oocyte to the balstocyst (**Figure 1**). Oocytes, fertilized eggs, or preimplantation embryos are recovered from female animals and maintained in culture for several hours to several days. After introduction of the transgene, the embryos are reimplanted into the reproductive tract of a 'foster mother,' where they can develop *in utero*.

Transgenesis via Pronuclear Microinjection of DNA

Pronuclear microinjection of DNA into fertilized eggs is the most widely used method for introducing transgenes when site-specific insertion into the host genome is not required. The eggs are viewed with a microscope and, using a glass micropipette controlled by a micromanipulator, a small volume of a DNA solution containing several hundred transgene molecules is introduced into one or both pronuclei of each egg. One or more copies of the transgene may integrate stably into a host chromosome, resulting in mitotic transmission to all cells in the developing embryo. Integration of the transgene occurs in approximately 10–40% of injected mouse eggs. Therefore, each animal born from a microinjected embryo must be screened to determine whether it is transgenic. For this purpose, genomic DNA is isolated from a tissue sample (typically the tail tip) and polymerase chain reaction (PCR) and/or Southern blot analyses are performed to detect the transgene and determine its physical integrity. An animal that develops from the injected egg and carries the transgene is called a 'founder,' and it can be mated with normal animals to transmit the transgene and derive a 'transgenic line.' Some transgenic founders are mosaics that carry the transgene in only a fraction of their cells, presumably because it integrated into the genome after the first round of DNA replication. Such mosaic animals are recognized because they transmit the transgene to fewer than the normal 50% of offspring, or because they express the transgene in only a fraction of cells. Except when inherited from a mosaic founder, a transgene displays Mendelian transmission. It is stably inherited over many generations, implying that it is permanently integrated at a single genomic locus. The loci of transgene insertions can be determined by genetic mapping or by *in situ* hybridization to metaphase chromosomes. Transgenic lines can be maintained in either a hemizygous or a homozygous state, unless the transgene insertion has induced a recessive mutation in an essential host gene (see section 'Targeted mutations,' below).

Transgenes usually insert at a random site in the genome, through an unknown mechanism involving illegitimate recombination (**Figure 2**). This allows DNA from any source (eukaryotic, prokaryotic, viral, etc.) to be introduced into the mammalian genome. Frequently, there is a deletion or rearrangement of host DNA at the site of insertion. In some cases, copies of the transgene insert at two or more unlinked loci in a single founder. Each insertion locus can contain a single copy of the transgene or, more typically, a head-to-tail tandem array of tens or even hundreds of transgene copies. Even when DNA from the same species is injected, the frequency of insertion into the host genome through homologous recombination is extremely low, about 10^{-4}. However, microinjected DNA molecules readily recombine with each other through a homologous mechanism. This allows a large gene to be injected in several overlapping fragments, which can recombine to reconstruct the gene before or during the integration process. The size of a transgene appears to be limited only by breakage during handling and microinjection of the DNA, and cloned DNA segments several hundred kb in length (e.g., yeast artificial chromosomes) can be successfully introduced. Similar methods have been used to introduce transgenes into a wide variety of mammalian species, including rats, rabbits, sheep, goats, pigs, and cows. However, for unknown reasons the frequency of transgenesis is much lower in large mammals than in mice and rats.

Transgenesis via Infection with Retroviral Vectors

Retroviral vectors take advantage of the natural ability of retroviruses to enter the cell and integrate into the genome in a single copy. A retroviral vector is produced by inserting the transgene in place of part of the viral genome, and a preparation of infectious viral particles is produced by introducing the recombinant virus into tissue culture cells. Mouse embryos at the cleavage or morula stage are then infected with the virus (**Figure 1**), resulting in retroviral DNA integration in one or more cells (**Figure 2**). The site of insertion is essentially random, and the founder mice are usually mosaics. The size of the transgene is limited to 8–9 kb, due to packaging limitations of the retroviral particle.

Transgenesis via Embryonic Stem Cells

Embryonic stem cells are pluripotent cell lines derived from early mouse embryos. ES cells can be cultured *in vitro*, and retain the ability to contribute to all somatic and germ cell lineages when introduced into an early embryo. Therefore, they represent an important route for the introduction of transgenes. The main advantage of ES cells for transgenesis is that they can be transfected with DNA and then subject to selection and screening procedures to identify clones of cells in which the transgene has inserted at a specific site in the host genome. This permits the generation of targeted mutations in specific endogenous genes (e.g., gene knockouts), as well as the insertion of transgenes at specific sites in the genome.

Transgenes for use in ES cells are designed to include a selectable marker gene allowing positive selection during cell culture, e.g., a neomycin resistance gene. The transgene DNA is introduced into a large population of cultured ES cells ($\sim 10^7$ cells) by any of several methods for DNA transfection, such as electroporation or lipofection. Clones of ES cells that carry the transgene, and therefore can grow in the presence of the selective agent, are isolated. These clones can then be screened to identify rare clones in which the transgene has inserted at a specific target site, as in gene knockout experiments (**Figure 2**). Retroviral vectors can also be used to introduce exogenous genes or to induce random insertional mutations in ES cells.

Once a transgenic ES cell clone with the desired properties has been identified, it is used to produce chimeric mice. Several ES cells are microinjected into the cavity of a blastocyst-stage embryo or aggregated with a cleavage-stage embryo (**Figure 1**), and several such embryos are implanted into the uterus of a foster mother mouse and allowed to develop to term. This results in the development of chimeric animals (i.e.,

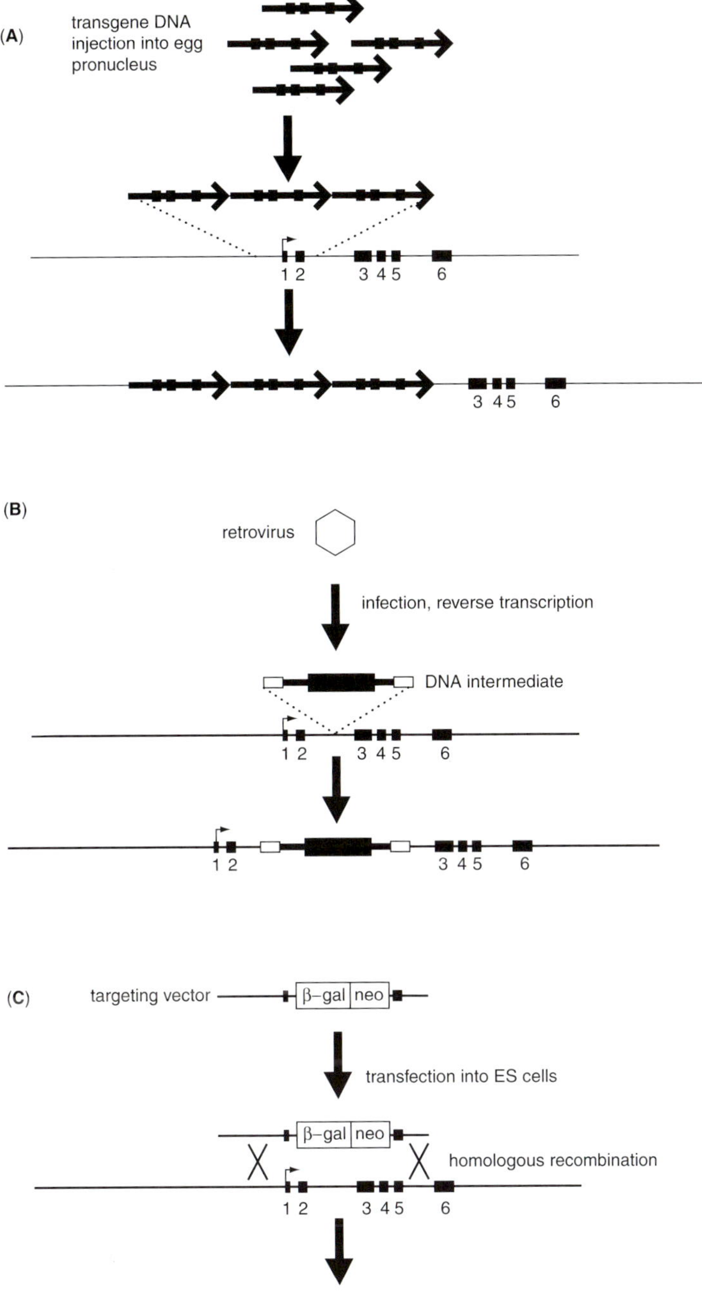

Figure 2 Modes of transgene insertion into the genome. (A) DNA molecules microinjected into the egg pronucleus (shown by arrows, with black boxes representing exons of the transgene) usually integrate in a multicopy, head-to-tail tandem array at a random site in the host genome. In the example illustrated, the transgene has, by chance, disrupted a host gene (whose six exons are indicated by boxes), resulting in deletion of the first two exons. (B) Retroviral vectors also insert into the host genome at random sites, but they do not cause deletion of host sequences. In the example shown, the retroviral vector, which carries a transgene (long black box) in between the long terminal repeats (open boxes), has by chance inserted into an intron of the same host gene. (C) Targeting vectors are designed to integrate into a specific host gene via homologous recombination. The vector shown contains two segments of DNA derived from the host gene, allowing it to undergo site-specific integration via homologous recombination with its target gene. In the example shown, the targeting vector is specifically designed to delete exons 2–4 of the host gene, replacing them with a *neo* gene (a selectable marker) and a β-galactosidase reporter gene, which will thus be expressed under the control of the promoter of the target gene (small arrow).

those composed of a mixture of cells from two sources, in this case the host embryo and the donor ES cells) in which the cells derived from donor ES cells carry the transgene. The host embryos are obtained from a mouse strain genetically distinct from the ES cells, typically carrying different hair color genes so that the chimeric mice can be identified by their mixed hair pigmentation. Chimeras in which the ES cells have contributed to the germline will transmit the transgene to some or all of their offspring, which are then bred to perpetuate the transgenic line.

Transgenesis via Nuclear Transfer

Somatic cells derived from embryonic, fetal, or adult tissues of the same species are cultured and transfected with the transgene, and a clone of transgenic cells is identified. A transgenic cell is then fused with an oocyte whose nucleus has been previously removed, resulting in transfer of the transgenic somatic cell nucleus in to the oocyte. The oocyte is then implanted into the reproductive tract of a female animal to allow it to develop.

As in the case of ES cells, donor cells are first selected for insertion of the transgene, and can be screened for the desired transgene insertion site, copy number, etc. Because the host oocyte is enucleated, the transgenic animal that develops carries the entire genome of the donor cell, including the transgene, in every cell.

Expression of Transgenes

Transgenes that are introduced by pronuclear microinjection are designed to function after insertion into diverse sites in the host genome. The expression of such a transgene depends on several factors, including the regulatory elements (i.e., the sequences that regulate transcription, RNA processing, and translation) included in the transgene, the site of insertion into the host genome, and the number of copies of the transgene. Some transgenes consist of a genomic DNA segment including a gene (exons and introns) together with a certain extent of the natural 5′ and 3′ flanking DNA. The expression of such a 'genomic' transgene depends on whether the DNA segment includes all the regulatory elements that normally regulate the gene's expression. In mammals, regulatory elements sometimes reside >100 kb away from the gene, so that it may be necessary to transfer a very large segment of genomic DNA to ensure proper expression. More frequently, a transgene consists of a cDNA clone (providing the coding sequences) joined to a heterologous promoter, enhancer(s), intron, and polyadenylation signals to create an artificial 'cDNA transgene.' This approach can be used to express essentially any gene in any cell type or tissue for which appropriate regulatory sequences have been defined, although the level of expression may be lower than that obtained with a genomic transgene. The pattern of expression dictated by the regulatory sequences in a genomic or cDNA transgene may be overridden by the influence of neighboring host DNA, a consequence of random insertion into the genome. Such 'position effects' can silence a transgene in all cells or in a fraction of cells, or alter its level or pattern of expression. There is in general a positive correlation between transgene copy number and expression level, although this relationship can be masked by position effects.

In addition to regulatory elements that can direct expression in specific cell types, tissues, or temporal patterns, sequences can be included in a transgene to make its expression conditional on administration of a drug, a change in temperature, or other experimental manipulations. For example, promoters that can be regulated by administration of antibiotics, hormones, or metal ions have been used extensively to turn transgenes on and off at will during development of the embryo or during the life of the adult animal.

Through the use of ES cells (and potentially through nuclear transfer from somatic cells), the problem of position effects can be circumvented by targeting the desired coding sequence to a specific locus, where it falls under the control of the natural regulatory mechanisms at that locus. For example, to express a foreign protein in red blood cells, the appropriate coding sequences might be fused to the regulatory elements of a β-globin gene (encoding the β chain of hemoglobin) and introduced into random sites in the mouse genome by pronuclear microinjection. Alternatively, the coding sequences could be inserted, via homologous recombination, to replace the coding sequences of the mouse β-globin gene in ES cells, which would then be used to produce germline chimeric mice.

Applications of Transgenic Animals

Transgenesis is an extremely powerful tool for the genetic analysis and manipulation of mice and other animals. As defined above, a transgene is an experimentally introduced DNA segment carried in the genome of a host animal. A transgene can be designed to encode a new gene product in the transgenic animal, or it can be introduced with the intent of altering or disrupting a host gene at its site of insertion. In many cases, a transgene will do both, e.g., disrupt an endogenous gene while expressing a new gene product. Thus, the applications of transgenesis take advantage of its ability to induce both loss-of-function and gain-of-function genetic alterations.

Targeted Mutations, Random Insertional Mutations, and Gene Traps

Targeted mutations are caused by the insertion of a transgene into a specific, predetermined host gene via homologous recombination. These include 'knock-outs,' which are designed to eliminate expression of a host gene, and 'knock-ins,' which are designed to modify a host gene without blocking its expression. Targeted mutagenesis, whose aim is generally to study the function of a known gene, is described in detail elsewhere in this volume.

A different and complementary application of transgenesis is random insertional mutagenesis, which is generally conducted with the aim of identifying new genes. Because of the random site of insertion of most transgenes, a host gene is sometimes disrupted, resulting in a lethal or phenotypically visible mutation in approximately 5–10% of transgenic mouse lines. Virtually all insertional mutations identified in transgenic animals are recessive, and can be propagated in the heterozygous state even if they are lethal in the homozygous state. In contrast to mutations induced by chemicals or radiation, the molecular analysis of insertional mutations is relatively easy, because the transgene provides a tag that can be used to clone DNA from the mutant locus. In mammals, it is not generally feasible to screen sufficient numbers of transgenic animals to identify mutations in a specific gene, or those causing a preordained phenotypic defect. Nevertheless, random insertional mutations in mice, which are often an unanticipated byproduct of other transgenic experiments, have led to the discovery of many important genes.

A different approach to the use of random insertional mutagenesis for gene discovery is 'gene trapping.' Here, a transgene encoding an easily detected reporter protein (e.g., the enzyme β-galactosidase), but lacking its own promoter and/or or enhancer, is introduced at random sites in the genome. In mice, gene trapping is usually performed in ES cells, which are then used to produce chimeric mice. When the gene trap vector lands in a host gene, it is expressed under the control of the host gene's regulatory apparatus. Many animals with different gene trap insertions are screened to identify those that express the reporter gene at particular anatomic sites or development stages. The inserted gene trap vector provides a tag to clone the gene that has been 'trapped,' and also often generates a loss-of-function mutation.

Cell-Type-Specific or Conditional Gene Disruption

In addition to heritable mutations, which are present in every cell in the animal throughout its development, it is sometimes useful to disrupt a gene in only a specific cell type or at a specific time. This can be accomplished in mice by introducing a transgene encoding a site-specific recombinase (e.g., Cre recombinase of bacteriophage P1) under the control of cell-type-specific or inducible regulatory elements. Animals carrying this transgene are crossed with a second transgenic strain in which the target gene has been modified by the insertion of recognition sites for the recombinase (e.g., loxP sites). In cells that express the recombinase transgene, recombination at the target locus is specifically induced. Depending on the placement of the recognition sites, this can result in deletion, silencing, or activation of a target gene.

Genetic Rescue of Mutations

When attempting to identify the gene responsible for a mutant phenotype in the mouse, transgenesis is often used to test individual candidate genes. A transgenic strain is generated by microinjecting DNA encoding the wild-type allele of the candidate gene, which inserts at a different locus than the mutant gene. The transgenic strain is then crossed with the mutant strain, to test whether the transgene can correct (or 'rescue') the phenotypic defect in mutant animals. This approach is also useful when attempting to positionally clone a mutant gene that is believed to be located within a large region of DNA, but whose identity is unknown. Large segments of wild-type DNA (often cloned as yeast or bacterial artificial chromosomes) within the genomic region of interest are tested for their ability to rescue the mutation. If a successful rescue is observed, smaller clones of DNA are tested to narrow in on the gene.

Mapping Transcriptional Regulatory Elements

One of the earliest applications of transgenesis was to delimit the regulatory DNA sequences (e.g., promoter elements and enhancers) required to direct correct tissue-specific or developmental-stage-specific expression of eukaryotic genes. The rationale of such experiments is that if a transgene displays a consistent pattern of expression when inserted at several different sites in the host genome, then that pattern must be dictated by *cis*-acting regulatory elements included in the transgene DNA. A transgene whose RNA or protein product can be distinguished from endogenous gene products is used to produce several transgenic animals or lines, and the expression pattern is characterized. Often a reporter gene encoding an easily detected protein is inserted in place of the coding sequences of the transgene. The expression of several transgenes containing varying amounts of 5′ or 3′ flanking DNA is compared to deduce the location of *cis*-acting regulatory elements. Such regulatory elements can then be used to direct the expression of other transgenes.

Gene targeting can also be used to define the regulatory elements of a gene, for example, by deleting a segment of flanking DNA from a host gene, and examining the consequences for the gene's expression.

Ectopic Expression or Overexpression of Normal Gene Products

A frequent application of transgenesis involves intentionally altering the developmental or tissue-specific pattern of expression of a normal gene product. This is usually accomplished by placing the coding sequences of one gene under the control of promoter and enhancer elements of a second gene, and introducing the hybrid transgene into an animal. Alternatively, the coding sequences can be targeted, via homologous recombination, to a new locus in the genome, where they come under the control of regulatory elements of a different gene. These approaches have been used extensively to study the roles of proteins such as growth factors, receptors, and transcription factors during animal development. Two examples are shown in **Figures 3** and **4**. **Figure 3** illustrates one of the earliest examples of genetic engineering in mice, in which growth hormone coding DNA sequences were expressed under the control of the promoter of the metallothionein gene, resulting in expression in and secretion from ectopic tissues, and abnormal growth. **Figure 4** illustrates an experiment in which a mouse homeobox cDNA, *Hoxd4*, was placed under the control of the promoter of a different homoebox gene, *Hoxa1*, resulting in expression of the Hoxd4 protein in a more anterior region of the embryo than normal, and causing a 'homeotic' transformation of certain bones in the axial skeleton.

Figure 3 Abnormal growth of a transgenic mouse expressing growth hormone under the control of the mouse metallothionein promoter. Growth hormone coding sequences were fused to the metallothionein promoter, resulting in abnormal expression of growth hormone in ectopic tissues such as liver and intestine, and its consequent overproduction in the transgenic animals. Transgenic mice (left) grew two to three times as fast as controls, and reached a size of up to twice normal. (Photograph courtesy of Dr. Ralph L. Brinster.)

Expressing Dominant Gain-of-Function or Dominant-Negative Mutant Gene Products

Certain diseases and developmental defects in humans and animals, including many forms of cancer, are

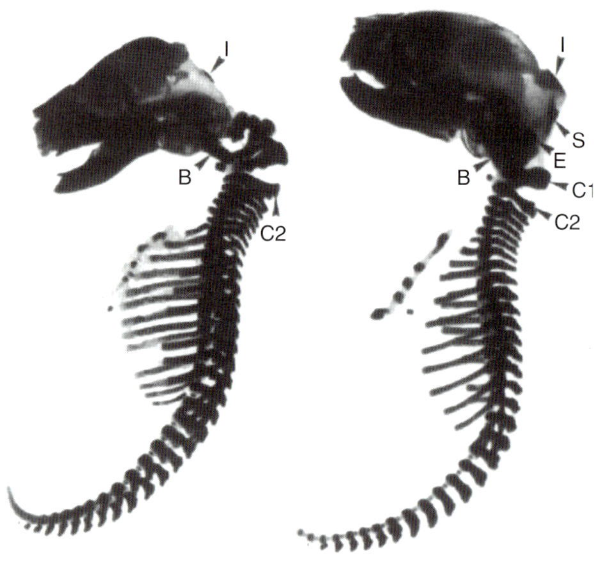

Figure 4 The ectopic expression of a homeobox transcription factor in transgenic mice causes a homeotic transformation of the occipital bones of the skull. To study the role of *Hox* genes in anterior–posterior patterning of the body plan, the *Hoxd4* coding sequences were fused to the promoter of the *Hoxa1* gene, which is expressed more anteriorly than *Hoxd4* in the mesoderm of the developing embryo. The photograph shows the axial skeletons of a transgenic (left) and a normal newborn mouse (right). The ectopic expression of *Hoxd4* caused an absence of the supraoccipital (S) and exoccipital (E) bones and a reduction in size of the interparietal bone (I) of the skull. Furthermore, the transgenic mouse contained ectopic bony structures located anteriorly to the first cervical vertebra (C1) and fused ventrally to the basioccipital bone (B). (Photograph courtesy of Dr. Thomas Lufkin.)

caused by dominantly acting mutations, which may encode a modified gene product or cause inappropriate expression of a normal gene product. Animal models of such diseases can be generated by introducing the mutant gene as a transgene. For example, many mouse models of cancer have been generated by introducing dominantly acting oncogenes with appropriate regulatory elements to target their expression to a specific tissue or cell type. Similarly, the individual roles of viral gene products in pathogenesis can be examined by expressing viral genes in transgenic animals.

Dominant-negative mutations are those that cause a mutant protein to interfere with the function of its wild-type counterpart in a heterozygote. An example is an enzyme that can bind its substrate but is catalytically inactive, and therefore competes with the wild-type enzyme for substrate. Transgenes encoding dominant-negative mutant proteins can be used to block the function of a host gene product. This approach is particularly useful in cases where it is not feasible to generate a loss-of-function mutation, or where it is desirable to block gene function in only a limited population of cells in the animal.

Genetic Markers of Specific Cell Types or Lineages

Transgenic animals that express a reporter gene product (e.g., β-galactosidase or GFP) in a specific cell type, cell lineage, or anatomical region are useful in a wide variety of biological experiments. Transgenic strains of this type are commonly produced in three ways: (1) by introducing a transgene in which the reporter gene is controlled by regulatory elements from a gene with the desired expression pattern, (2) by identifying a strain carrying a randomly inserted reporter gene with the desired expression pattern, i.e., a 'gene trap,' or (3) by targeting the reporter gene into the locus of a specific gene, and thereby placing it under the control of regulatory elements at that locus. An example of the expression of a β-galactosidase reporter gene in a specific region of the developing nervous system is shown in **Figure 5**.

Genetic Ablation of Specific Cell Lineages

Instead of marking the cells expressing a specific gene, it is also possible to ablate those cells, and all of their descendants, in the developing animal. This can be used to analyze cell lineage relationships or to generate disease models. Cell lineage ablation is accomplished by introducing a transgene encoding a toxin (e.g., diphtheria toxin A fragment) that will kill the cells in which it is expressed, but will not harm surrounding cells.

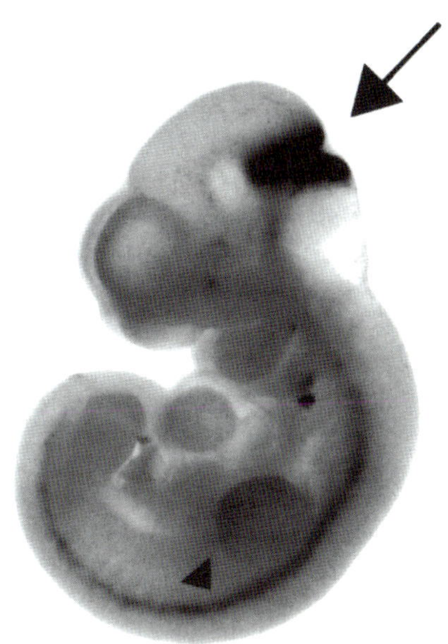

Figure 5 β-galactosidase staining reveals the expression of a transgene at the mid-hindbrain junction of an E10.5 day transgenic mouse embryo. The embryo carried a transgene encoding β-galactosidase under the control of the *Engrailed-2* enhancer and the promoter of the *hsp68* heat-shock gene. *Engrailed-2* is a homeobox gene required for normal brain development. The arrow points to the mid-hindbrain junction, a site of *Engrailed-2* expression, while the arrowhead indicates expression of the transgene in the spinal cord, which is due to the activity of the *hsp68* promoter. (Photograph courtesy of Dr. Alexandra Joyner.)

Applications in Agriculture and the Pharmaceutical Industry

Transgenic animal models of human disease can be useful for preclinical drug testing. Animals engineered to be susceptible to human viruses, by introduction of viral receptors or other host range determinants, can also be used for testing human vaccines.

Transgenic animals can serve as 'factories' that, in some cases, may produce large amounts of proteins more efficiently than alternative expression systems such as bacteria, yeast, or mammalian cell cultures. Transgenic mice have been engineered to express human antibodies (which are superior to murine antibodies for use as drugs) by introducing large segments of human DNA encoding human immunoglobulin genes, and breeding these transgenic animals with strains in which the endogenous immunoglobulin loci are mutated. In transgenic large animals such as cows or sheep, proteins of pharmaceutical value can be produced in large quantity in milk (and later purified) by introducing the appropriate gene under the control of regulatory elements that direct expression in the mammary glands.

Transgenesis can in principle be used to alter many phenotypic properties that may increase the value of agriculturally important animals. These include growth rate, fat composition, milk production, and hair texture. It may also be possible to modify domestic animals such as pigs to make them more suitable as organ donors for human transplant patients.

Further Reading

Grosveld F and Kollias G (eds) (1992) *Transgenic Animals*. San Diego, CA: Academic Press.

Hanahan D (1989) Transgenic mice as probes into complex systems. *Science* 246: 1265–1275.

Hogan B, Beddington R, Costantini F and Lacy E (1994) *Manipulating the Mouse Embryo: A Laboratory Manual*, 2nd edn. Plainview, NY: Cold Spring Harbor Laboratory Press.

Jaenisch R (1988) Transgenic animals. *Science* 240: 1468–1474.

Palmiter RD and Brinster RL (1986) Germline transformation of mice. *Annual Review of Genetics* 20: 465–499.

See also: Chimera; Embryonic Stem Cells; Knockout

Transient Polymorphism

M A Asmussen

doi: 10.1006/rwgn.2001.1306

A transient polymorphism is one where the genetic variation (polymorphism) currently found in a population at a certain genetic locus is expected to be of limited duration, and will ultimately be eliminated over time by the evolutionary force(s) acting on the population. The end result in such cases will necessarily be a monomorphic population with only one genetic type (allele and genotype) at the locus in question.

The biological significance of transient polymorphisms naturally depends strongly on the number of generations for which the existing genetic variation will be maintained. If this is a very long time, say hundreds or thousands of generations, then, for all practical purposes, the observed variation may be effectively permanent. Moreover, in such cases, environmental or other relevant conditions are apt to change during such a long time span, and along with it, the evolutionary forces acting in the population on the genotypes at this locus. Once conditions change, genetic variation may either be maintained or eliminated even faster under the new evolutionary regime. Genetic variation would thus be apt to be maintained in such a population as long as the current conditions prevail, which is all one can ordinarily hope to predict.

Because of the critical impact of the time frame of transient polymorphisms, it is important for evolutionary biologists to take this into account when deriving and interpreting the biological conditions under which genetic variation will be maintained in a population. In particular, these conditions may be much broader than suggested by the standard ones based simply on when there will be a stable polymorphism (i.e., a locally stable equilibrium at which genetic variation will be maintained at the locus in question). The time-dependent dynamics of the system must also be examined under the conditions in which genetic variation will ultimately be lost to determine if and when long-lasting, transient polymorphisms (and effectively permanent genetic variation) will be maintained.

An example of a long-term, but transient polymorphism is provided by the case of a recessive lethal allele at a single, diallelic autosomal locus within a diploid population of organisms. Suppose, in particular, that a locus has the two alleles, A_1 and A_2, where the three genotypes (A_1A_1, A_1A_2, and A_2A_2) have the relative viabilities of 1, 1, and 0, respectively. This means that genotypes with one or two copies of the dominant A_1 allele survive equally well from birth to reproduction, while homozygotes for the recessive A_2 allele are lethal, with none surviving to reproduce. Suppose further that this selection is the only evolutionary force acting on this locus in the population. In particular, this is a random mating population with no mutation to new alleles at this locus, migration, or stochastic effects from genetic drift present.

Under these evolutionary conditions, natural selection will steadily reduce the frequency of the deleterious A_2 allele to 0. Genetic variation will thus always ultimately be eliminated at this locus, as long as these conditions prevail; however, this selection regime can result in a long-lasting, transient polymorphism, with the frequency of the recessive lethal allele remaining above 0.001 for thousands of generations, as shown in **Figure 1**. The reason for this slow decline is that once the deleterious allele is at low frequencies only rarely will individuals carry two such rare alleles; once this happens, the recessive lethal allele is effectively shielded from selection, since almost all the individuals in the population will be normal, with full fitness, as they carry at least one normal, dominant allele.

The allele frequency trajectories in **Figure 1** are based on the recursion equation giving the new frequency of a recessive lethal allele A_2 (q') after one generation of selection, in terms of its previous frequency (q). This is given by:

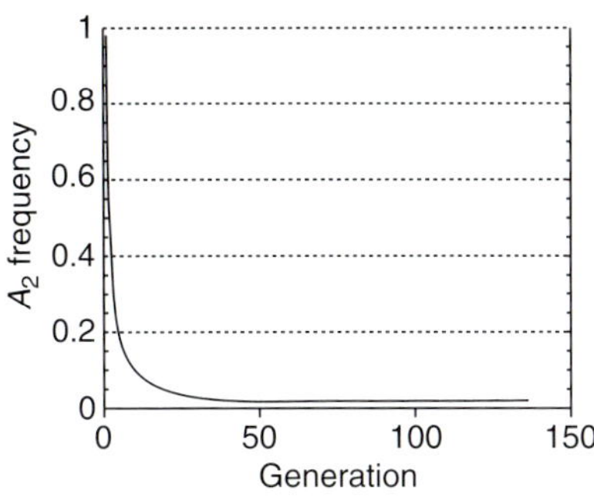

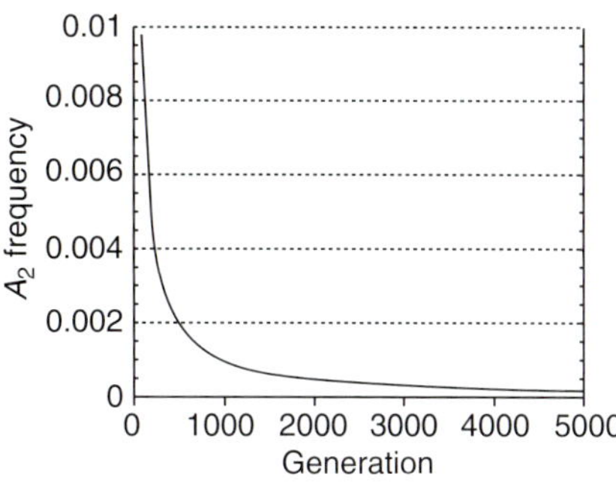

Figure 1 Trajectory through time in generations of the frequency of a recessive lethal allele A_2, showing how its frequency will only slowly be reduced below 0.001 by natural selection, resulting in a long-lasting, transient polymorphism.

$$q' = \frac{q}{1+q}$$

This recursion equation can be solved to obtain an explicit dynamical solution for the frequency q_t of the deleterious allele A_2 after any number of generations of selection, t:

$$q_t = \frac{q_0}{1+tq_0}$$

where q_0 is its initial frequency. From this solution we see clearly that, over the course of many generations, natural selection will by itself slowly but steadily reduce the frequency of a recessive lethal allele to 0.

Further Reading

Hartl DL and Clark AG (1997) *Principles of Population Genetics*, 3rd edn. Sunderland, MA: Sinauer Associates.

Hedrick PW (2000) *Genetics of Populations*, 2nd edn. Sudbury, MA: Jones & Bartlett.

See also: **Balanced Polymorphism; Polymorphism; Population Genetics**

Transition

J H Miller

doi: 10.1006/rwgn.2001.1307

Transition is the term that describes the change of a purine to a purine on the same strand of nucleic acid, and of a pyrimidine to a pyrimidine on the complementary strand. Changes from A:T to G:C, T:A to C:G, G:C to A:T, and C:G to T:A are transitions.

See also: **Transversion Mutation**

Translation

A Liljas

doi: 10.1006/rwgn.2001.1308

A genome may contain 500–50 000 genes that are translated to the corresponding proteins. In most genomes the hereditary material is in the form of DNA; however, certain viruses contain genomic RNA. The process of translation is usually preceded by transcription, and the product of translation, a protein, is frequently transported into a different compartment of the cell from that where it has been synthesized.

Genetic Code

The genetic code is the universal dictionary by which genetic information is translated into the functional machinery of living organisms, the proteins. The words or 'codons' of the genetic message are three nucleotides long. Since there are four different nucleotides used in messenger RNA (mRNA), this results in a dictionary of 64 words. There are 20 amino acids that are normally used in proteins and which are translated. In addition the translation needs a definition of 'start' and 'stop.' The start codon defines the start of translation as well as the reading frame (the sequence of nucleotide triplets) that is to be translated. The start or initiator codon is identical to the methionine codon. Special mechanisms are used to identify the correct initiation site; in addition there are three stop codons. Thus 61 codons are available for 20 amino acids, and hence the genetic code is degenerate. In the case of leucine, serine, and arginine, there are as many as six codons, whereas methionine and tryptophan have only one codon.

The universal genetic code deviates slightly in mitochondria, where a few codons are translated in alternative ways. The most prevalent are methionine and tryptophan, which have two codons instead of the usual one. Different organisms use the degenerate genetic code differently. The usage of the codons is coupled to the availability to tRNAs that can translate them. Thus the codon usage can differ to the extent that a gene that is transferred from one organism to another cannot be translated unless the new organism is supplemented with extra tRNAs.

Transcription

Genomic DNA cannot be translated but has to be copied or transcribed into RNA by different RNA polymerases. Here the classic mechanism discovered by Watson and Crick applies. One strand of the double-stranded DNA (the negative strand) is copied through Watson–Crick base-pairing to a positive strand of RNA. The process of transcription is in all cases strongly regulated. Some proteins are synthesized in large numbers, whereas others are only present in a few copies per cell. Again some proteins are synthesized during a brief period in the life of the cell, whereas others are produced more or less continuously.

In eukarya, transcription is performed in the nucleus and the transcript is transported into the cytoplasm to be translated. Transcription and translation in mitochondria and chloroplasts is performed in these cellular organelles. In the case of eubacteria and archaea, the whole process is performed in the cytoplasm. The eubacterial transcripts frequently contain several genes controlled by one operator; the mRNA is polycistronic.

Processing of Transcribed RNA

A number of transcribed RNAs are never translated but have the same cellular functions as RNA. These are primarily ribosomal RNA (rRNA) and transfer RNA (tRNA). The transcribed RNA, called the 'primary transcript,' frequently has to be processed to become mRNA. Several different processes are involved; the processes in eukarya differ from those in eubacteria. The primary transcripts normally contain long or short regions, which are not to be translated. They form so-called introns, while the translated regions form exons. The splicing machinery removes the introns by cutting and ligation. Eukaryotic mRNAs are also modified by the addition of a poly(A) tail at the 3′ end of the message.

In eukarya the primary transcripts are also frequently edited to become mRNAs. This is sometimes done by changes of U to C or vice versa. More extensive editing occurs of mitochondria from trypanosomes, where the mRNAs are extensively modified by large enzymatic particles that use templates called 'guide RNAs.'

Reading Frame and Usage of Genetic Code

The initiator AUG codon defines the reading frame of a mRNA. Translation proceeds from this start in steps of three nucleotides (one codon). The frequent occurrence of termination codons out of frame prevents translation in the wrong frame for other than short stretches. However, there are mRNAs which for correct translation need a change of reading frame. This is the case for *Escherichia coli* termination or release factor-2 (RF2). The readthrough of a stop codon requires a tRNA that would decode a stop (nonsense) codon as a sense codon and incorporate a specific amino acid. Such tRNAs are called 'suppressor tRNAs.'

A few proteins in eubacteria and eukarya contain seleno-cystein (Se-Cys). This is not incorporated by a posttranslational modification as in other cases of nonstandard amino acids. Se-Cys is incorporated instead during translation in response to one of the stop codons. The mechanism for this involves a special tRNA ($tRNA^{Sec}$) which reads the stop codon. A set of enzymes has specific functions in this system. One of them is a special form of elongation factor Tu called SelB that uniquely binds $tRNA^{Sec}$. SelB has the property of identifying a specific secondary structure of the mRNA that precedes the stop codon that corresponds to Se-Cys. This leads to the suppression of the stop codon and the incorporation of Se-Cys.

Translation on Ribosomes

The process of translation occurs on the ribosome in the cytoplasm or in the cellular organelles, mitochondria and chloroplasts. The ribosome is a complex of a few large rRNA molecules and between 50 and 90 different proteins. It is made of two subunits (large and small) with different functions that dissociate from each other during part of the process. Translation is traditionally divided into three steps: initiation, elongation, and termination. A fourth step, ribosome recycling, also belongs to the process. Soluble protein factors catalyze the process by binding to the ribosome transiently. More than 10 factors participate in eubacterial translation, whereas a considerably larger number participate in eukaryal translation.

Initiation

Translation is initiated by the binding of a messenger RNA (mRNA) to the ribosomal small subunit. In this process, the initiation (methionine) codon is selected and bound at the ribosomal decoding site. Subsequently the initiator methionyl-tRNA and the large ribosomal subunit are bound to this initiation complex. Eubacterial initiation is stimulated by three initiation factors, IF-1, IF-2, and IF-3. In eukaryotes a much larger number of initiation factors participate.

Elongation

In each cycle of elongation, one amino acid is incorporated into the nascent peptide. There are three elongation factors in eubacteria, which catalyze two of the basic steps in translation: the binding of an aminoacyl-tRNA to the A-site and the translocation of the peptidyl-tRNA from the A-site to the P-site. During translocation the tRNAs and the mRNA are moved to expose the next codon in the ribosomal A-site. However, the central event in elongation, peptidyl transfer, is a spontaneous process where no protein factor is needed.

The recognition of the codon by the anticodon of the tRNA is a process that is done in several steps. In the initial selection, the anticodon of the aminoacyl-tRNA in complex with elongation factor Tu (EF-Tu) and GTP is matched against the codon in the A-site of the ribosome. In the case of a good match, the ribosome induces EF-Tu to hydrolyze its bound GTP to GDP and phosphate. The EF-Tu·GDP complex has a conformation that has low affinity for the aminoacyl-tRNA and the ribosome; accordingly it dissociates. The aminoacyl moiety of the tRNA, when bound to EF-Tu, is located far from the peptidyl transfer center and, after the disassociation of EF-Tu, has to reorient itself into the A-site of the ribosome, while retaining the interaction with its codon. This process coincides with the proofreading of the anticodon of the tRNA by the codon of the mRNA. An incorrect (noncognate) match of the anticodon to the codon increases the likelihood that the aminoacyl-tRNA will dissociate before its amino acid has reached the peptidyl transfer site of the ribosome.

Peptidyl transfer is catalyzed by the rRNA of the large subunit without direct assistance of ribosomal proteins or translation factors. A completely conserved nucleotide, A2451 of the *E. coli* 23S rRNA, serves as a general base during peptide bond formation. Once the aminoacyl moiety reaches the A-site of the peptidyl transfer site, the peptide on the peptidyl-tRNA in the P-site can be transferred to it. This leads to a peptidyl-tRNA in the A-site and a deacylated tRNA in the P-site.

The final step of elongation is the translocation of the peptidyl-tRNA from the A-site to the P-site and the movement of the mRNA by three nucleotides so that the next codon is exposed in the A-site. EF-G, which catalyzes this process, binds to the ribosome in complex with GTP. After translocation is performed, it dissociates in complex with GDP. A surprising finding is that the ternary complex of EF-Tu with GTP and aminoacyl tRNA has the same shape as EF-G. It remains possible that EF-G, when it dissociates from the ribosome, leaves an imprint into which this ternary complex fits nicely.

Termination

The termination of protein synthesis depends on the exposure of one of the three stop codons (UAA, UAG, and UGA) in the decoding part of the A-site. In eubacteria two release factors, RF1 and RF2, participate to decode the stop codons and hydrolyze the completed peptide from the P-site tRNA. In eukarya they correspond to a single decoding factor, eRF1. The crystal structure of eRF1 indicates that these factors may perform their function by mimicking tRNA. The termination factor RF3 in all cases catalyzes the dissociation of the decoding factors from the ribosome.

Ribosome Recycling

The ribosome-recycling factor (RRF) removes mRNA from the ribosome so that the ribosome is available to synthesize new protein from new mRNAs. It performs this role together with EF-G. An amazing observation is that RRF also closely mimics tRNA. This suggests that RRF binds to a tRNA binding site, possibly the A-site, and is translocated from it by EF-G. This then leads to the dissociation of the mRNA from the ribosome and the subunits from each other.

Inhibition of Translation

Translation can be inhibited generally by a large group of antibiotic inhibitors. Translational repression also occurs and has been observed for some eubacterial ribosomal proteins. When these specific proteins have been synthesized in excess over rRNA they bind to a specific region of their own polycistronic mRNA and prevent further synthesis of any of the proteins encoded by this mRNA.

Transport of Product

The translated protein is frequently targeted to another compartment of the cell distinct from the site of synthesis. This is the case for membrane proteins in general, but certain proteins, functioning in different cellular compartments, are synthesized in the cytoplasm and subsequently transported to their final destination by various different transport systems. One of the best studied involves the signal recognition particle (SRP), which sorts proteins according to their different destinations. The protein to be transported has an N-terminal sequence of amino acids (tag) that is recognized by SRP, which assists the protein in passing through the cytoplasmic membrane. The SRP is composed of an RNA molecule and a number of proteins.

Molecular System of Translation

The ribosome is essential to the process of translation. In translation, different molecules bind to the ribosome and proteins are produced. mRNA is vital, because it contains the message to be translated. The tRNA molecules are also absolutely essential components of the system, since they are the ultimate tools of the translation. They are the adaptors that in one end read the codons of the message and at the opposite end incorporate the amino acid that corresponds to the codon into the growing polypeptide chain. The factors participating in translation, different in number depending on the type of cell, catalyze the different steps of translation. In their absence, the process becomes so slow that the life of the cell is impossible. It is interesting that several of the protein factors, EF-G, RRF, and eRF1, imitate tRNA. This suggests that they bind to a tRNA binding site on the ribosome to perform their catalytic function.

The ribosome itself has a number of functional sites, primarily the decoding site and the peptidyl transfer sites. It is remarkable that these sites are made up of RNA without any direct participation of ribosomal or factor proteins. Thus the ribosome is a ribozyme. This observation, together with the fact that tRNA and mRNA are the only additional essential components in translation, suggests that the early translation system could have been constructed entirely of RNA. This is consistent with the idea that the prebiotic world was an RNA world.

Further Reading

Spirin AS (1999) *Ribosomes*. New York: Kluwer Academic Publishers.

Garrett RA, Douthwaite SR, Liljas A, Matheson AT, Moore PB and Noller HF (eds) (2000) *The Ribosome: Structure, Function, Antibiotics and Cellular Interactions*. Washington, DC: ASM Press.

Carter AP, Clemons WM, Brodersen DE *et al.* (2000) Functional insights from the structure of the 30S ribosomal subunit and its interactions with antibiotics. *Nature* 407: 920–930.

Nissen P, Hansen J, Ban N, Moore PB and Steitz TA (2000) The structural basis of ribosome activity in peptide bond synthesis. *Science* 289: 920–930.

Yupupov M *et al.* (2001) Crystal structure of the ribosome at 5.5 Å resolution. *Science* (in press).

***See also:* Genetic Code; Messenger RNA (mRNA); Ribosomes; Transcription; Transfer RNA (tRNA)**

Translational Control

E J Murgola

doi: 10.1006/rwgn.2001.1309

Regulation of expression of a protein-encoding gene can occur at several molecular levels between the DNA and the accumulation of a functional protein: transcription, mRNA processing and stability, translation (ribosome-dependent, mRNA-programmed protein synthesis), and protein modification and turnover. Translational control usually refers to regulation of expression at the level of translation. It is properly used precisely to mean modulation of the efficiency of mRNA translation at the initiation, elongation, or termination stages of polypeptide synthesis, although it is sometimes used more broadly to include translation-coupled regulation of mRNA stability. Translational control allows cells to respond rapidly to changes in physiological conditions. This is especially important in organisms with a nuclear barrier between the sites of mRNA synthesis and translation, as well as a considerable time lag due to the interval between activation of nuclear pathways for the synthesis of mRNA and its transport to the cytoplasm. Because it involves virtually instantaneous recruitment and action of regulatory macromolecules, translational control is particularly well suited to regulation of the structural and temporal aspects of cell proliferation, developmental processes, and cell differentiation, and for integrating the various metabolic pathways in the cell. A distinction can be made between 'global' and 'selective' translational control. The former affects the entire population of mRNAs in a cell, switching their translation on or off or modulating it by degrees in unison. This type is usually achieved by adjustments in the activity of general components of the protein synthetic machinery acting in a nonspecific manner. By contrast, selective controls affect a subset of the mRNAs in a cell, sometimes even just a single species.

Cis-acting elements of mRNA often interact with *trans*-acting molecules (mostly, but not exclusively, proteins) to achieve activation or deactivation of mRNA translation. For example, the iron-responsive element (IRE), a sequence- and structure-specific negative regulatory element, is found in the 5′-untranslated region (5′UTR) of the mRNA for ferritin, the iron storage protein. The IRE regulates translation of ferritin mRNA in accordance with changes in the level of cellular iron. This regulation is mediated by a *trans*-acting iron repressor protein (IRP) that binds to the IRE. At low iron concentration, IRP binds to the 5′UTR of the ferritin mRNA and prevents ferritin translation; at higher iron levels, the IRP is saturated with iron and falls off the ferritin mRNA. Release of the IRP from the ferritin mRNA leads to efficient translation of the mRNA. Translation of mRNAs for ribosomal proteins and *Drosophila* sperm tail proteins is also regulated by 5′UTRs.

In one mechanism for repressing translation, termed 'translational masking,' the RNA is sequestered into translationally silent mRNP particles. Translational masking was first described for the phenomenon, during gametogenesis, in which mRNA is stored for use at a later stage in development. Several 'masking' phosphoproteins bind to the mRNA with relatively little sequence specificity and, by preventing mobilization of the mRNA to ribosomes, inhibit translation. Specific sequences in the 3′-untranslated region (3′ UTR) are necessary for unmasking at specific stages of development.

The length of an mRNA's poly(A) tail seems to be important for modulating the level of its translation. During oogenesis and embryogenesis, the translational activity of specific mRNAs is regulated by poly(A) length, and sequence elements in the 3′UTR have been shown to control polyadenylation. For some mRNAs, a long poly(A) tail activates translation but a short one does not. The 3′UTR of some developmentally regulated genes is also known to be involved in repression of mRNA translation by interaction with *trans*-acting factors, as, for example, in the case of the 15-lipoxygenase (*Lox*) gene. The Lox protein is required for the breakdown of internal membranes in mature reticulocytes. *Lox* RNA is synthesized in bone marrow, but translation of the RNA is repressed until reticulocytes reach the blood. The repression is achieved by binding of a 48-kDa protein specifically to a repeat motif in the 3′UTR of *Lox* mRNA.

A novel mechanism of translational regulation by 3′UTR sequences is displayed by the interaction of the *lin-14* and *lin-4* genes, which regulate the timing of developmental events in *Caenorhabditis elegans*. Surprisingly, *lin-4* encodes not a protein but two small RNA transcripts with partial complementarity to sequences within the *lin-14* mRNA 3′UTR. Translational repression of *lin-14* mRNA may be caused by the binding of the *lin-4* antisense RNAs to a repeated sequence motif in the 3′UTR of the *lin-14* mRNA.

The majority of known cases of translational control operate at the initiation stage of polypeptide synthesis, but examples are known for both the elongation and termination stages. Particularly notable are mechanisms involving frameshifting to produce longer or shorter proteins for specific functions and termination codon readthrough to produce a longer protein with a specific function. Finally, it should be mentioned that translational control of transcription is observed in many cases of transcription attenuation, where the termination or continuation of RNA polymerase is determined by the procession and positioning of the ribosome while translating a leader region of the nascent RNA transcript.

***See also:* Gene Expression; Translation**

Translocation

C V Beechey and A G Searle

doi: 10.1006/rwgn.2001.1310

Translocation is the transfer of a length of genomic material to a new site, either in a chromosome or chromatid. It is, therefore, a type of 'structural aberration.' When within a chromosome (intrachange) it is generally known as a 'shift.' When between chromosomes (interchange) it may be a 'reciprocal translocation' or a nonreciprocal 'insertion' of material. Reciprocal translocations are symmetrical when the centric part of one broken chromosome combines with the acentric part of the other, or asymmetrical when the centric parts combine with each other, as do the acentric parts. A symmetrical exchange is usually viable, but an asymmetrical one, with its dicentric and acentric products, is usually cell-lethal. In a 'whole-arm translocation,' all or nearly all of a chromosome arm from a metacentric is interchanged with that from a telocentric or another metacentric. Closely related to this is the 'Robertsonian translocation,' or 'centric fusion,' in which the long arms of two acrocentrics unite to form a metacentric, with loss of a centric fragment. The reverse process of 'centric fission,' or dissociation, leads to the formation of two acrocentrics from one metacentric. Each is important in evolution, as is reciprocal translocation.

Reciprocal Translocation (Symbol *T* in Mouse, t in Humans)

Induction in Male Germ Cells

Reciprocal translocations were first studied in *Drosophila* by H.J. Muller and coworkers, who induced them by X-irradiation of spermatozoa. This has been done in the mouse too, where work has concentrated on products of treated spermatogonial stem cells, namely, primary spermatocytes and subsequent progeny. Translocations have also been studied in many other animals and plants.

Meiotic Effects

Translocations are revealed in spermatocytes at diakinesis/metaphase I by their characteristic multivalent configurations, typically rings or chains of four elements (because of the association between nonhomologous chromosomes) among the normal bivalents (see **Figure 1**). F_1 progeny which carry a translocation generally show 'semi-sterility' on mating to normal mice: litter size is approximately halved because about half the gametes produced by a translocation heterozygote will be unbalanced (see **Figure 2**), with death *in utero* of resultant zygotes. This chromosomal imbalance may arise from normal (alternate or adjacent-1) disjunction, which leads to duplications and deficiencies of segments distal to the translocation breakpoint, or from the rarer adjacent-2 disjunction, in which proximal (centric) regions are duplicated and deficient in the gamete.

X–Autosome Translocations

In male mice, translocations between an autosome and a sex chromosome cause sterility of carriers through breakdown of spermatogenesis. Female carriers of X–autosome translocations have small litters due to 'semi-sterility,' although there could also be effects on oogenesis. Some X–autosome translocations in which autosomal coat color genes are moved on to the X have thrown light on X-inactivation. If this process is random, cells in which the normal X is inactivated will express the autosomal color gene but if the translocated X is affected, inactivation may spread into the autosomal segment to suppress expression of the color gene and produce variegation. In most X–autosome translocations inactivation is random but in *T(X;16) 16H* the normal X is inactive in all cells. This has led to its extensive use in studies on sex determination.

Marker Chromosomes

Meiotic effects

Some autosomal translocations are male-sterile in the mouse through defective spermatogenesis, especially when one breakpoint is near the centromere and the other is distal, so that long and short 'marker chromosomes' tend to be formed. This may lead to failures of pairing at meiosis, to spermatogenic breakdown, and to the production by fertile females of tertiary (or partial) trisomic and monosomic zygotes, which sometimes survive.

Experimental use

Translocations which give long and/or short chromosome markers in somatic cells and which are fertile have been used extensively in transplantation and chimerism research since no special pretreatment is needed to reveal the transplanted cells, e.g., the very short T6 marker in the translocation *T(14;15) 6Ca* (see **Figure 1**). Unequal translocations have also proved useful in the fine mapping of loci by *in situ* hybridization since probe signals can easily be detected as being distal or proximal to the breakpoint.

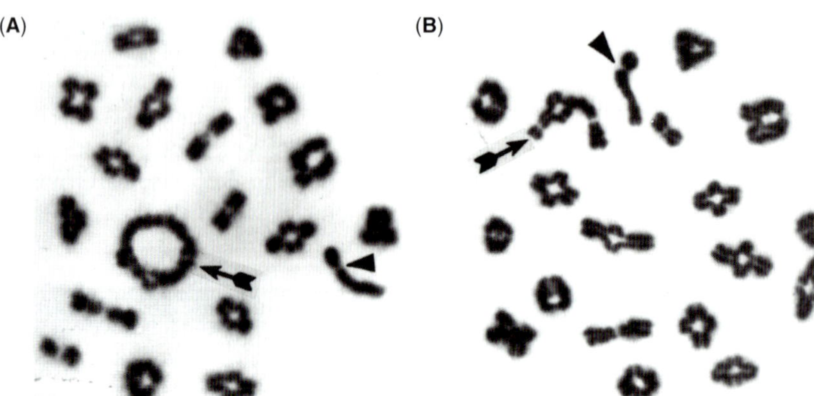

Figure 1 Quadrivalents observed amongst bivalents in metaphase preparations from male mice heterozygous for reciprocal translocations. (A) Ring of four chromosomes (arrowed). (B) Chain of four chromosomes from a mouse carrying the small T6Ca chromosome (arrowed) referred to in the section 'Experimental use.' In both preparations the unequal X/Y bivalent is shown with an arrowhead marking the junction of the large X and small Y chromosomes. (Photograph courtesy of E.P. Evans.)

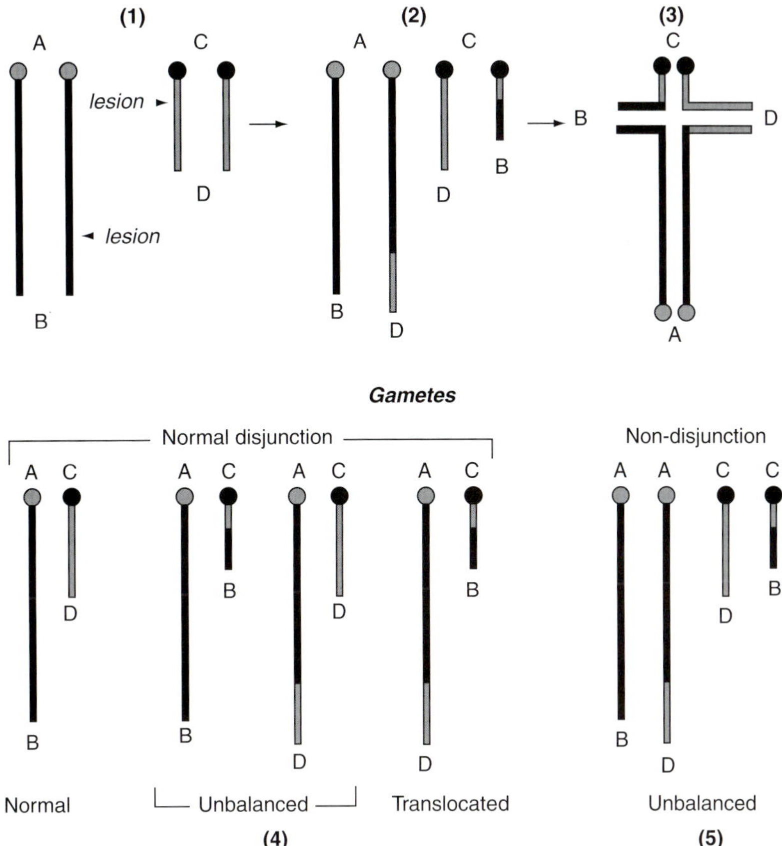

Figure 2 Gametic products of a reciprocal translocation (RT). (1) Initial lesions in nonhomologous chromosomes (may occur at diploid or haploid stage); (2) exchange of distal regions to give RT; (3) resultant cross-configuration at meiotic pachytene which gives characteristic rings or chains at diakinesis/metaphase I; (4) haploid products of normal disjunction, half unbalanced; (5) products of rarer nondisjunction, all unbalanced. Note that if an unbalanced gamete AB.CB combines with another unbalanced gamete AD.CD then a balanced zygote heterozygous for the translocation results, but it has uniparental partial disomy (see section 'Genomic Imprinting') for the chromosomes concerned.

Locating Breakpoints on Linkage Maps

Translocation breakpoints can be mapped because the heterozygote behaves as if it had a dominant gene for semi-sterility at that point. Some translocations have a phenotypic effect associated with a breakpoint. Thus *T(2; 8)26H* in the mouse is completely linked to the non-agouti locus with a distinctive dark agouti phenotype in the homozygote. With some translocations the homozygote is lethal.

Genomic Imprinting

Offspring with both copies of a specific chromosome region derived from one parent only (uniparental partial disomy) can be generated by intercrossing heterozygotes for reciprocal translocations (see **Figure 2**). Such offspring can be recognized by the use of marker genes on the chromosome regions concerned. However, for certain chromosome regions and translocation breakpoints this expected complementation of unbalanced gametes to form normal offspring results in abnormalities instead. This is one manifestation of 'genomic imprinting,' which means that, for certain genetic factors, passage through both parental germlines is necessary for normal development. Work with many mouse translocations (both reciprocal and Robertsonian) have helped us to understand this phenomenon and the chromosome regions involved.

Insertion (Symbol *Is* in Mouse)

These seem to be rarer than reciprocal translocations, though some apparent duplications could really be insertions, as demonstrated by chromosome painting techniques. Only three have been described in the laboratory mouse, of which the most interesting is Cattanach's translocation, an inverted insertion of about one-third of chromosome 7 into the X chromosome with symbol *Is(In7;X)1 Ct*. This has facilitated studies on X-inactivation, imprinting, and gene dosage.

Shift (Transposition)

Only one chromosomal shift has been described in the mouse. Otherwise known as sex-reversed (*Sxr*) this causes a part of the proximal end of the Y chromosome to move to its distal end. Its effects include the partial sex reversal of XX and XO females to males with small testes. It has proved useful in studies on sex determination.

Robertsonian Translocation (Symbol *Rb* in Mouse)

Nondisjunctional Effects

There is no evidence for the induction of Robertsonian translocations by ionizing radiation but they are a common type of aberration in animals, especially mammals. Because their heterozygotes tend to undergo nondisjunction, they lead to the formation of monosomic and trisomic zygotes (see **Figure 3**). All the former and most of the latter die *in utero*, but in humans and mice some trisomics survive beyond birth, usually with severe abnormalities. Thus heterozygotes for a Robertsonian involving human chromosome 21 may have offspring with trisomy 21, which causes Down syndrome. Since mouse chromosome 16 shows extensive genetic homology with human 21, Robertsonians for the former have been used to generate mice trisomic for 16 as models for this syndrome. Higher rates of nondisjunction are found in mice having two Robertsonians with an arm in common,

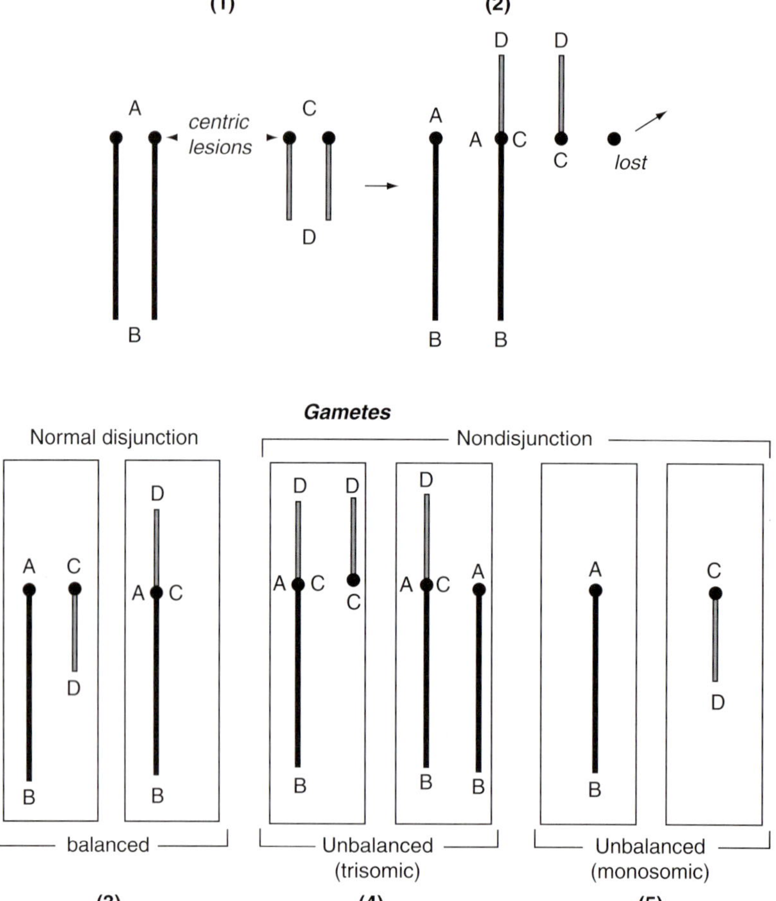

Figure 3 Robertsonian translocation and its gametic consequences. (1) Initial centric lesions in nonhomologous acrocentric chromosomes AB and CD; (2) formation of metacentric with loss of small product (genetically inert); (3) haploid balanced gametes from normal disjunction in Robertsonian heterozygote: 1:1 ratio of gametes with both acrocentrics or the metacentric alone; (4) trisomic unbalanced gametes from nondisjunction, with two copies of chromosome arm AB or of chromosome arm CD; (5) monosomic unbalanced gametes, with absence of chromosomes AB or CD. Note that in intercrosses of Robertsonian heterozygotes a trisomic unbalanced gamete (shown in 4) may combine with a monosomic one (shown in 5) to give a balanced zygote heterozygous for the translocation but with uniparental disomy for AB or CD (see section 'Use for Imprinting Studies').

i.e., monobrachial homology; a mouse stock with tribrachial homology generates 100% nondisjunction (Beechey and Searle, 1991).

Use for Imprinting Studies

When Robertsonian heterozygotes are intercrossed, complementation of unbalanced gametes can result in offspring with both copies of a specific chromosome inherited from the same parent (uniparental disomy). As with reciprocal translocations abnormal phenotypes result from uniparental disomy for certain chromosomes.

Robertsonian Translocations in Wild Mammals

Mice

Robertsonian variation is studied in many wild and domestic mammals, particularly mice and shrews. The rate of chromosomal evolution in both species is high. The known distribution of Robertsonian races in the mouse stretches from North Africa to Scotland but is concentrated in mountainous regions around Switzerland, home of the tobacco mouse (Gropp and Winking, 1981). This is homozygous for seven centric fusions, so has a diploid complement of 14 metacentric and 12 acrocentric chromosomes, in contrast to the 40 acrocentrics found in the house mouse. Hybrids between tobacco and laboratory mice have reduced fertility, with high but variable frequencies of nondisjunction, but these Robertsonians have been transferred successfully into laboratory strains.

Shrews

In the common shrew (*Sorex araneus*) 52 karyotypic races, differing in the arrangements of acrocentrics and metacentrics, have now been described. Apparently they are all derived through Robertsonian fusions from an all-acrocentric ancestral stock, although whole-arm translocation may be responsible for some metacentric arrangements (Searle *et al.*, 1990). They have been used mainly to study evolutionary problems, such as the links between chromosomal variation and speciation and the significance of hybrid zones.

Further Reading

Beechey CV and Evans EP (1996) Numerical variants and structural rearrangements. In: Lyon MF, Rastan S and Brown SDM (eds) *Genetic Variants and Strains of the Laboratory Mouse*, 3rd edn, vol. 2, pp. 1452–1506. Oxford: Oxford University Press.

Daniel A (ed.) (1988) *The Cytogenetics of Mammalian Autosomal Rearrangements*. New York: Alan R. Liss.

Dyban AP and Baranov VS (1987) *Cytogenetics of Mammalian Embryonic Development*. Oxford: Clarendon Press.

Epstein CJ (1986) *The Consequences of Chromosome Imbalance*. Cambridge: Cambridge University Press.

Sankaranarayanan K (1982) *Genetic Effects of Ionising Radiation in Multicellular Eukaryotes and the Assessment of Genetic Radiation Hazards in Man*. Amsterdam: Elsevier Biomedical Press.

References

Beechey CV and Searle AG (1991) Aneuploidy induction in mice: construction and use of a tester stock with 100% nondisjunction. *Cytogenetics and Cell Genetics* 56: 2–8.

Gropp A and Winking H (1981) Robertsonian translocations: cytology, meioses, segregation patterns and histological consequences of heterozygosity. *Zoological Society of London Symposia* 47: 141–181.

Searle JB, Hubner R, Wallace BMN and Garagna S (1990) Robertsonian variation in wild mice and shrews. *Chromosomes Today* 10: 253–263.

***See also:* Cattanach's Translocation; Down Syndrome; Homology; Mouse; Nondisjunction; Radiation Genetics, Mouse; Robertsonian Translocation; Sex Reversal; Spermatogenesis, Mouse; X-Chromosome Inactivation**

Transmissible Spongiform Encephalopathy

M A Ferguson-Smith

doi: 10.1006/rwgn.2001.1746

Transmissible spongiform encephalopathies (TSEs) comprise an unusual group of invariably fatal neurogenerative diseases which affect both humans and animals. They are unusual in that they are not transmitted by conventional microorganisms such as viruses. The infective agent is believed to be a normal host protein, the prion protein, which has undergone a posttranslational modification that changes the conformation of the protein and renders it resistant to enzyme degradation. Natural transmission of the protease-resistant form of the prion protein in food, or its inoculation parenterally, for example in contaminated hormones, are known routes of infection. Mutation of the prion protein gene can lead to familial forms of TSE in humans and it has been shown that in these cases transmission to experimental animals can occur by inoculation of infected brain material from deceased family members. The disease is thus unusual in that it is both genetic and infective. Another important characteristic is that the incubation period after infection is measured in years rather than days.

TSEs affect the central nervous system and produce characteristic neuropathological features, including vacuolation of the gray matter, loss of neurons, astrocytosis, and the occurrence of amyloid plaques composed largely of abnormal prion protein. There is no evidence of any immune response which would be expected if the disease was due to a viral pathogen or due to an autoimmune reaction. Treatments that inactivate viruses, such as heat, irradiation, formaldehyde, and DNA and RNA nucleases, fail to inactivate TSE agents.

The following lists the various TSEs that have been recognized in animals and humans. In animals:

Scrapie in sheep and goats.
Chronic wasting diseases in mule deer and elk.
Transmissible mink encephalopathy in farmed mink.
Bovine spongiform encephalopathy (BSE) in cattle, exotic ungulates, and carnivores, including cats.

In humans:

Sporadic Creutzfeldt–Jakob disease (CJD).
Kuru (see Kuru).
Iatrogenic CJD (from transplants and hormones).
Gerstmann–Sträussler–Scheinker (GSS) syndrome.
Fatal familial insomnia.
Variant CJD (from BSE).

See *also:* Kuru; Spongiform Encephalopathies (Transmissible), Genetic Aspects of

Transmission Genetics

J R Fabian

doi: 10.1006/rwgn.2001.1311

Transmission genetics is often referred to as Mendelian genetics or classical genetics. Genetic crosses and pedigrees are studied in transmission genetics in order to reveal the mode of inheritance for a trait. The mode of inheritance describes whether a gene influencing a trait is dominant or recessive, and if the trait is transmitted in an autosomal, sex-linked, or maternal fashion. The mode of inheritance is most easily identified for traits that are determined by the action of a single gene, which are also known as Mendelian traits. In contrast, polygenic traits reflect the combined activities of more than one gene, and the inheritance of these traits is more difficult to trace. An extremely important consideration in transmission genetics is to sort out the degree to which the variation in an observed trait (phenotype) is due to the genetic constitution of an individual (genotype) and how much is attributable to and arises from influences in the environment.

In the past century, several major achievements in genetics blossomed from work in transmission genetics. These successes include the discovery that genes are the unit of inheritance, that genes reside on chromosomes, and that mutation in the DNA sequences of genes can alter their function. The tools and concepts of transmission are used today in a wide range of applications including agriculture, biotechnology, medicine, genetic mapping, genetic counseling, and evolutionary studies.

Selective Breeding

Domestication of Animals

In 1906, 22 years after the death of Gregor Mendel, William Bateson offered the name 'genetics' for the new branch of biology concerned with the study of heredity. Despite its recent scientific origin, work in transmission genetics began in prehistoric times with the development of agriculture. As early as 8000 BC humans observed the transmission of biological traits from parent to offspring as they bred and domesticated animals such as horses, oxen, and camels. Heredity was important to early breeders, and had a major impact on the value of their livestock. For example, a prize-winning ram was priceless if its advantages were passed on to its offspring, but would have little value if they were not. Without an understanding of the mechanism of heredity it was difficult for breeders to accurately predict the outcomes of matings. As a result, many of the early breeding methods often involved a curious mixture of folklore, magic, and science.

Acquired Traits

The story of Jacob in the book of Genesis (30: 27–42) is an example of the use of folklore and magic in the breeding of animals. Jacob desired to increase the number of rare speckled and spotted animals in his father-in-law's flock, in order to keep these animals for his own profit. He tried to improve his chances for success by using his own form of magic to influence the coat color patterns in offspring. To do this he displayed wood rods decorated with black and white speckled patterns at the watering hole where the animals mated. According to the story, Jacob's breeding technique resulted in large numbers of speckled and spotted animals. The modern genetic explanation of Jacob's success is that many of the solid colored animals in Jacob's flock were carriers of recessive alleles for spotting and speckling. Since the speckled and spotted coat patterns are recessive traits, these traits will appear in offspring of parents that are both carriers.

Up until the twentieth century, the spotted and speckled patterns in Jacob's flocks would likely be explained as an acquired trait or character. Support for the inheritance acquired characteristics waned in the late nineteenth century as a result of the work of embryologists. They demonstrated the existence of both germ cells and somatic cells. Germ cells were shown to carry the genetic material transmitted to future generations, and this material was not readily influenced by environmental or life experiences.

Cultivation of Plants

Despite the long history of animal husbandry, the first key insights into the mechanism of heredity did not come from animal breeding, but rather from botany. Mendel's model for heredity stemmed from his observations of dominance, segregation and independent assortment in experiments using the garden pea. Mendel's work was in turn the culmination of an ancient history of applied genetics in plants. As early as 8000 BC, humans in various regions of the world began cultivating and selectively breeding plants such as maize, wheat, barley, rice, and the date palm. While some crops of relatively minor importance have been domesticated since the Middle Ages, the majority of modern crop plants (and domesticated animals) were brought under human management during prehistoric times. The selective breeding strategy used for crop development was not complex, but it was very effective. Farmers would collect seeds from those plants they liked the most and grow them more frequently than plants that were less desirable. Evidence of the powerful effects of long-term selective breeding on crop development is evident in the common cabbage-like vegetables that are very popular today. These plants include kale, cabbage, kohlrabi, cauliflower, broccoli, and Brussels sprouts. While all of these plants look very different, they are in fact members of the same species, *Barsac olearia*.

Plant Hybrids

The powerful effects of selective breeding that resulted in plants such as the cabbage-like vegetables, gradually reaches its limit as plants become inbred and genetic variation decreases. One of the greatest challenges to early plant breeders was to introduce new traits into crops in order to improve their productivity. Some crop variants appear spontaneously as a result of mutations, where others result from cross-pollination. Artificial pollination of plants has an ancient history as evidenced by plants such as the date palm. The date palm is a type of plant with separate male and female individuals in a population. Cultivation of the date palm occurred as early as 4000 BC and ancient Assyrian art dating from 800 BC demonstrates pollination ceremonies involved a curious combination of science and magic. Despite this early use of artificial pollination in plant breeding, the first systematic study of plant hybrids was not carried out until late in the eighteenth century. Cross-pollination studies became popular with academics and plant breeders in the early nineteenth century who were interested in vegetative vigour and potential economic significance of hybrid plants. In 1866, Mendel's principles were presented in a paper titled 'Experiments in plant hybridization.' Unfortunately, it took 34 years for the scientific community to appreciate Mendel's contribution to biology. Modern genetics began with the study of trait transmission and the production of plant hybrids.

Mendelian Inheritance

Mendelian Traits

Mendelian genetics displaced several erroneous ideas about heredity. Mendel used the common garden pea (*Pisum sativum*) as his experimental organism. *Pisum* proved to be a remarkably tractable organism for Mendel's genetic studies. Pea plants have a relatively short generation time when compared to other plants such as fruit trees, which were among those plants used by Mendel's predecessors. An additional advantage of *Pisum* was the wide range of variable traits available for use in cross-pollination studies. Furthermore, pea plants self-fertilize when left alone. This allowed Mendel to produce purebred (homozygous) parental strains by self-crossing plants for several generations. Mendel studied traits influenced by the action of single genes and these traits are now referred to as Mendelian traits. The single-gene traits studied by Mendel produced easily visible and distinguishable phenotypes such as round versus wrinkled seed, yellow versus green seedpod, and tall versus dwarf stem. These phenotypes were readily recognizable in subsequent generations and were not complicated by in-between forms in progeny. Mendel studied the transmission of seven different single-gene traits in hybridization experiments using purebred parental strains of pea plants.

Reciprocal Crosses

Since ancient times a major question regarding the mechanism of heredity was whether males and females have an equal contribution to the traits of offspring. Since the golden age of Greece, philosophers and scientists debated whether the genetic contribution of one sex exceeded the contribution of the other sex. Mendel proposed that the genetic contribution of both sexes was equal. Mendel proposed that organisms inherit pairs of unit factors (genes) for each trait, with one member of each pair derived from each parent. Mendel's most conclusive evidence for an equal genetic contribution for each sex came from his use of reciprocal crosses. For example, Mendel observed that crosses could be designed that used the pollen from a tall plant pollinating a dwarf plant, or vice versa. He obtained the same results with all seven traits in his study and concluded that the transmission of these traits was not sex-dependent.

Traits Do Not Blend

The blending theory of inheritance was a deeply entrenched misconception that was eventually resolved by Mendelian genetics. The blending model proposed that maternal and paternal genetic material mixed together after fertilization, just like two different colored liquids in a cup. Blending theories gained support from observations based on continuously varying traits, such as skin color and height, where the physical appearance (phenotype) was intermediate of each parent. Blending theories failed to explain the behavior of discontinuous traits, or discrete traits, that consisted of only two contrasting phenotypes, with no intermediate phenotypes between. These discrete traits were not altered in offspring and could skip generations. By studying discrete traits, now known as Mendelian traits, Mendel's work demonstrated that genes are stable entities that are inherited in pairs. Mendel's results also showed that in hybrid organisms dominant versions of genes, or alleles, could mask the presence of recessive alleles. Recessive alleles, such as those for speckled coat patterns in Jacob's flock, are therefore hidden in hybrids but are stable and can be transmitted to future generations.

Statistical Approach

Mendel's experiments involved 287 crosses of 70 different purebred plants and used approximately 28 000 pea plants. One of Mendel's major contributions to genetics was his methodical approach. His approach involved setting up clearly defined crosses of plants with readily distinguishable variables, and then applying mathematical analysis (statistics) in order interpret his results. Mendel's method of counting large numbers of progeny followed by statistical analysis became widely applied in biological research and helped to earn early research in transmission genetics the nickname 'bean bag' genetics. Despite earning this undesirable nickname, the statistical analysis of trait transmission has been instrumental to the mapping of genes, evolutionary studies, and models used for population genetics.

Monohybrid Cross and Test Cross

Mendel's cross-hybridization studies involved purebred plants that differed with regard to a single contrasting trait. Purebred, homozygous, parental stocks were crossed and the offspring of this cross are called F_1 hybrids, or monohybrids. In the F_1 generation, all of the hybrids resembled the parent with the dominant trait. The genotype of these monohybrid, or heterozygous, plants can be represented as genotype *Aa*, with the uppercase letter representing the dominant allele and the lowercase letter representing the recessive allele. The F_1 hybrid plants were next self-fertilized ($Aa \times Aa$) and this cross is known as a monohybrid cross. In the offspring of monohybrid crosses, or F_2 generation, Mendel repeatedly observed a phenotype ratio of three plants with the dominant phenotype to one plant with the recessive phenotype (3:1 phenotype ratio) in the F_2 generation. Mendel predicted that the plants with a dominant phenotype in the F_2 generation were of mixed genotypes with some being homozygous dominant genotype *AA* and others being heterozygous genotype *Aa*. In order to determine the genotypes of plants with dominant phenotypes in the F_2 generation Mendel devised the test cross.

The test cross takes the organism with a dominant phenotype but unknown genotype and crosses it to a homozygous recessive individual with a known genotype *aa*. In a test cross with a plant of genotype *AA* all offspring will have the dominant phenotype and will have the heterozygous genotype *Aa*. However, if a plant with genotype *Aa* is used in a test cross, then the genotypes of 50% of the offspring will have the genotype *Aa* and display the dominant trait. The other 50% will be display the recessive phenotype since they will have the homozygous recessive genotype *aa*. Mendel's test cross method is still used today in breeding procedures with plants and animals in order to determine the genotype of plants with dominant phenotypes.

Independent Assortment

Mendel also studied crosses where he followed the segregation of two separate pairs of contrasting traits. The initial cross involved two homozygous parents that differed in two different traits represented by the cross $AABB \times aabb$. The F_1 offspring of this cross were dihybrid plants of the genotype *AaBb*. Mendel performed a dihybrid cross and examined the phenotypes and genotypes of F_2 plants. Mendel observed that each pair of traits was inherited independently. He observed a 3:1 ratio of dominant to recessive trait when the *A* and *B* pairs of traits were considered separately, as independent crosses ($Aa \times Aa$ and $Bb \times Bb$). When considered together in one cross ($AaBb \times AaBb$), the combinations of traits appeared in the phenotype ratio of 9/16 with both dominant traits, 3/16 with one dominant trait, 3/16 with the other, and

1/16 with both recessive traits. This ratio is designated as Mendel's 9:3:3:1 dihybrid ratio and is based on probability events involving segregation, independent assortment, and random fertilization.

Meiosis

Within a few years of the rediscovery of Mendel's principles of heredity, the fields of cytology and genetics were brought together through the work of Walter Sutton and Theodor Boveri. The work of these investigators and others introduced the concept that chromosomes in somatic cells exist as definite pairs. In contrast, gametes were found to be haploid and thus contained only one member of each chromosome pair. These observations were incorporated into a chromosome theory of heredity, which states that chromosomes are the carriers of genes and serve as the basis for Mendelian mechanisms of segregation and independent assortment. Both segregation and independent assortment are explained by events involving distribution and sorting of homologous chromosomes during meiosis I. Independent assortment occurs when homologous pairs of chromosomes randomly align during metaphase I of meiosis and segregation results from the separation of homologous pairs during anaphase I.

Linkage

As early as 1903, Sutton and Boveri suggested that there are likely many more genes than there are chromosomes. This indicated that if chromosomes were involved in heredity, then each chromosomes would have to contain many genes linked together, like beads on a string. These predictions based on cytology were supported by studies in transmission genetics. Specifically, it was observed that in certain dihybrid crosses some traits were inherited together and were therefore not transmitted according to the law of independent assortment. Mendel's principle of independent assortment states that the distribution of one pair of genes into gametes is independent of the distribution of another pair. Genes located on the same chromosome are said to be linked genes and tend to be inherited as a group. Linkage of genes on chromosomes is usually not complete. New combinations of linked genes occur as a result of crossing-over during prophase I of meiosis. Crossing-over results in a reshuffling of the allele repertoire for a particular chromosome and adds to the genetic variation of sexually reproducing organisms. The degree of crossing-over between two loci on a single chromosome is proportional to the distance between the two loci. This correlation provided the basis for the construction of the first chromosome maps. Genetic mapping therefore owes its origins to experiments in transmission genetics, or 'bean bag' genetics, conducted by two *Drosophila* geneticists, Thomas H. Morgan and Alfred H. Sturtevant. These experiments involved scoring the phenotypes of offspring and calculating recombination frequencies of linked genes. Along with providing the basis of genetic mapping, the work of Sturtevant and Morgan also confirmed the chromosomal theory of heredity because they established that chromosomes contain genes in a linear order, and these genes were the units of inheritance observed by Mendel.

Non-Mendelian Inheritance

Interactions among Alleles

In the early twentieth century, researchers sought to confirm and extend Mendel's observations of heredity in *Pisum* using different organisms. In addition to the observation of genetic linkage, researchers also began to encounter other examples of non-Mendelian inheritance. Non-Mendelian patterns of inheritance were identified when crosses yielded a modified version of Mendel's 3:1 phenotype ratio in the F_2 generation of a monohybrid cross. In some cases the altered phenotype ratios in the F_2 generation reflected different types of dominance relationships among the alleles of a gene. This is observed for phenotypes resulting from incomplete dominance and codominance. Codominance and incomplete dominance yield unique phenotypes for heterozygous offspring (*Aa*). Incomplete dominance results in heterozygotes with intermediate phenotypes, as in the case of snapdragons when parents with red flowers and white flowers are crossed resulting in heterozygous offspring with pink flowers. Codominance occurs when both alleles show dominance, as in the case of the AB blood type ($I^A\ I^B$) in humans. Furthermore, the human ABO blood groups represent another deviation from Mendelian simplicity since there are more than two alleles (A, B, and O) for this particular trait. Deviations from Mendelian inheritance are also observed in traits with phenotypes having variable penetrance and expressivity. In these cases, individuals with the same allele combination can produce different degrees of a phenotype in different individuals. An example of a trait that is incompletely penetrant is polydactyly, which is a dominant trait causing extra fingers and toes. In some cases, the dominant trait of polydactyly can skip generations due to incomplete penetrance and then reappear in future generations.

Interactions among Genes

Not long after the rediscovery of Mendel's work, research in transmission genetics revealed that discrete traits, are often regulated by more than gene. This indicated that the role of genes in determining phenotype was more sophisticated than Mendel's observations had suggested. We know now that in reality few phenotypes result from the action of a single gene acting alone. Phenotypes result from the combined actions of the many genes and their protein products. The protein products of many genes functionally interact in metabolic pathways and regulatory processes, or physically interact as structural proteins within and between cells.

Epistasis is a type of functional interaction between nonallelic gene pairs. In epistasis, a homologous pair of recessive alleles can override the phenotypic input of separate locus. Epistasis was first observed for genes that participate as part of a pathway involving many gene products, such as those controlling coat color and patterns in animals. An example of epistasis is albinism. In albinism, the homozygous pair of recessive albino alleles for albinism overrides the function of other nonallelic gene pairs involved in pigmentation.

Effect of Sex on Phenotype

Another type of non-Mendelian inheritance involves traits that are affected by the sex of an organism. In cases of sex-limited traits, expression is exclusively limited to one sex. In sex-limited traits, the expression genes are modified by an individual's sex hormones. Examples of sex-limited inheritance include genes influencing the heaviness of beard growth in humans and genes influencing sex-limited differences in tail and neck plumage in domestic fowl. In comparison, sex-influenced traits are influenced by the sex of the organism, but are not limited to one sex or the other. Examples of sex-influenced traits include pattern baldness in humans, horn formation in sheep, and certain coat patterns in cattle.

Mendelian Traits in Man

Mendel's Laws Are Universal

Although the mechanisms of heredity were initially demonstrated using the garden pea plant, these principles were rapidly confirmed in a variety of different organisms. Mendelian inheritance was demonstrated in 1902 in poultry and mice. In 1905, W.E. Castle, a chief pioneer in genetics, introduced *Drosophila* as an experimental organism for genetic studies. In 1902, Sir Archibald Garrod made the observation that the human disease alkaptonuria was caused by a block in a metabolic reaction sequence. Garrod hypothesized that the metabolic block was due to a congenital deficiency of a specific enzyme, and he suggested that the trait appeared to be inherited as a recessive trait. Mendelian inheritance was officially extended to humans in 1903, when albinism became the first human trait classified as a Mendelian recessive trait, with normally pigmented skin inherited as a dominant trait. By 1910, other human traits such as brachydactyly (digit malformation) and the ABO blood groups were shown to be genetically determined. Additional examples of dominant and recessive Mendelian traits are listed in **Table 1**.

Table 1 A representative list of Mendelian traits in humans

Recessive traits	Dominant traits
Albinism	Achondroplasia
Alkaptonuria	Brachydactyly
Ataxia telangiectasia	Ehler–Danlos syndrome
Cystic fibrosis	Huntington disease
Duchenne muscular dystrophy	
Galactosemia	Hypercholesterolemia
Hemophilia	Marfan syndrome
Lesch–Nyhan syndrome	Neurofibromatosis
Phenylketonuria	Osteogenesis imperfecta
Sickle-cell anemia	Phenylthiocarbamide tasting (PTC)
Straight hair line	Porphyria
	Widow's peak hairline

Early Genetic Mapping

Genetic mapping of human traits has its origins in 1911, with the assignment of the gene resulting in colorblindness (a recessive trait) to the X chromosome when it was observed that trait was inherited by sons from mothers who saw colors normally. Other disorders that affected males only were also mapped to the X chromosome. For X-linked disorders, females are protected by a normal copy of the gene on their second X chromosome, while males only have one copy of the X chromosome.

Somatic Cell Hybrids

The other 22 pairs of human chromosomes were virtually uncharted until late in the 1960s. The first breakthrough in mapping genes to autosomes resulted from studies using somatic cell hybrids, which are cell lines made by fusing mouse and human cells, and contain only a few copies different human chromosomes. Advances in the 1970s ultimately led to our modern methods in gene mapping. The first of these advances was the development of specific stains that produced banding patterns, making it easier for researchers to identify human chromosomes in hybrid cells.

Genetic Markers and DNA Sequencing

A major advance in genetic mapping occurred in the 1970s with the development of recombinant

DNA techniques. Recombinant DNA technology led to new mapping strategies based on using DNA variations as markers on chromosomes and to the technique of *in situ* hybridization. DNA sequencing was first introduced in the 1970s and major advances in the technique, including automated DNA sequencing in the 1980s, allowed the determination of the order of bases in a strand of DNA and ultimately revealed the molecular structure genes on chromosomes.

Mendelian Trait Database

Many disorders in humans are inherited as simple dominant or recessive Mendelian traits, including some 3500 disease genes. Most of the Mendelian disorders are rare and recessive traits occur more frequently when offspring result from matings between related individuals. Dominant disorders sometimes appear in families with no history of the trait and these are often cases resulting from spontaneous mutations in the germline of a parent. A database known as Online Mendelian Inheritance in Man (OMIM) is available and contains information on Mendelian traits in humans. The OMIM database that contains over 11 000 phenotypes in humans that are presumed to represent a trait caused by a single gene. Over 6000 of the entries in the OMIM database represent mapped genetic loci, and this number will likely increase rapidly as work in the human genome project draws to completion.

Modes of Inheritance

Pedigrees

Researchers today still use transmission genetics and the methods used by Mendel to study the inheritance of a trait. The first step with many experimental organisms begins with designing genetic crosses to study the inheritance of a phenotype. The results of these crosses are most significant when many of these designed crosses can be set up to yield large numbers of offspring for statistical analysis. However, how do we study trait transmission in organisms such as humans, where designed crosses and large numbers of offspring are not practical or available? The answer is that the inheritance of a phenotype in humans can be determined by pedigree analysis. Pedigrees are charts that depict family relationships and phenotypes. In pedigrees, males are represented by squares and circles represent females presenting family tree information in a chart known as a pedigree. In pedigrees, shaded symbols are used to represent individuals with a particular phenotype. Examples of pedigrees demonstrating two different modes of inheritance are shown in **Figure 1**

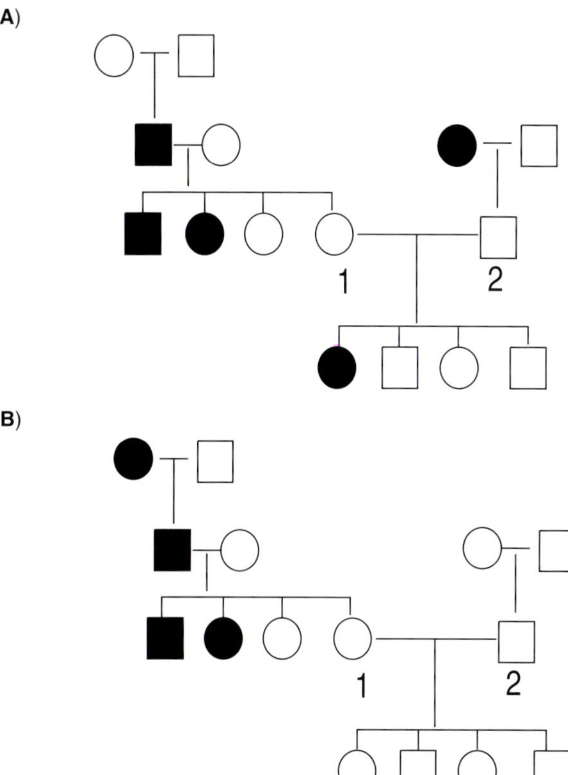

Figure 1 Examples of pedigrees showing different modes of inheritance. Males are represented by squares and females by circles. Horizontal lines indicate parents, vertical lines show generations, and elevated horizontal lines depict siblings. Symbols for individuals displaying a phenotype are shaded. (A) This pedigree is consistent with an autosomal recessive mode of inheritance. Individuals labeled 1 and 2 are both heterozygotes for the recessive trait (*Aa* × *Aa*). (B) This pedigree is consistent with an autosomal dominant mode of inheritance. Individuals labeled 1 and 2 do not have the phenotype and are therefore homozygous recessive (*aa* × *aa*).

Standard Modes of Inheritance

There are six standard modes of inheritance that can be reliably determined by careful pedigree analysis. The modes of inheritance reflect whether a trait is dominant or recessive, and the chromosomal linkage of the trait in question. When first attempting to eliminate modes of inheritance it is easiest to initially assume that the trait displayed in the pedigree is due to the action of a single gene, and that the trait shows complete penetrance and uniform expressivity. The six standard modes of inheritance and examples of diseases are (1) autosomal recessive (cystic fibrosis, Tay–Sachs disease), (2) autosomal dominant (achondroplasia, neurofibromatosis), (3) X-linked recessive (hemophilia, colorblindness), (4) X-linked dominant (congential generalized hypertrichosis, Rett syndrome), (5) Y-linked (genes involved in male fertility and development), and (6) mitochondrial inheritance (leber optic atrophy, Leigh syndrome). In each case, characteristic patterns in pedigrees are used to eliminate other modes of inheritance. Important characteristics used to determine inheritance patterns are presented in **Table 2**.

Polygenic and Multifactorial Traits

Traits that result from the activities of more than one gene are polygenic traits. Unlike single-gene traits, which produce discrete phenotypes, polygenic traits produce a range of phenotypes. In a population, the distribution of phenotypic classes produced by polygenic trait follows a bell-shaped curve. Polygenic traits are often multifactorial, since the resulting phenotypes are influenced to a certain degree by the environment. Human behavior and many diseases are multifactorial and it is often difficult to determine how much of the phenotype is genetically determined. A list of representative multifactorial disorders in humans is presented in **Table 3**.

The recurrence risk of Mendelian traits in a family can be predicted by establishing the mode of inheritance using pedigree analysis. However, it is much more difficult to predict the recurrence risks for polygenic traits and to do so geneticists must use a variety of information from family and population studies. The human genome project will help with the diagnosis of many polygenic disorders as many of the genes that predispose people to these illnesses are identified. Using data from the human genome project, insight into key genes controlling multifactorial traits such as intelligence and hypertension by analysis of genome data using quantitative trait loci (QTL) algorithms. QTL analysis can reveal these genes by detecting loci that account for as little as 1% of the observed variance in a trait. The Human Genome Project has the potential to radically change medicine, since someday the diagnosis of many multifactorial disorders may begin at birth; decades before an individual experience the first symptoms.

Hereditarianism

Complex Traits

In the early 1900s the public was introduced to Mendelian inheritance. At that time there was little distinction made between the inheritance of Mendelian traits and complex traits. Hereditarianism is the concept that all human traits are controlled solely by genetic inheritance and ignores the contribution of the environment. Many single-gene traits, such as those studied by Mendel, conform to hereditarian analysis since they are relatively resistant to environmental influences. Hereditarianism viewed human personality traits as Mendelian traits, and presented them in contrasting pairs such as politeness versus bluntness and obedience versus disobedience. Hereditarianism did not acknowledge the possibility that not all familial

Table 2 Mode of inheritance and pedigree analysis

Mode of inheritance	Some characteristic patterns in pedigree
Autosomal recessive	• Affected offspring usually born to unaffected parents • Chance of affected offspring is 25% for children of carriers • If both parents are affected, all children will exhibit trait • Affects either sex • Increased incidence with parental consanguinity
Autosomal dominant	• Affected individual has at least one affected parent[a] • Children with one affected parent have 50% risk of being affected • Affects either sex
X-linked recessive	• Affects almost exclusively males • Not transmitted from father to son • If female inherits, father must have trait
X-linked dominant	• All daughters of affected fathers exhibit the trait • All sons of an unaffected mother will not have trait
Y-linked	• Females never exhibit trait • Son always has same phenotype as father
Mitochondrial inheritance	• All children of an affected mother inherit the disorder • None of the children of an affected father inherit the disorder

[a]May not apply in the cases of non-penetrance or spontaneous mutation.

Table 3 Representative examples of multifactorial human genetic disorders

Multifactorial Disorders
Breast cancer
Bipolar affective disorder
Cleft palate
Dyslexia
Diabetes mellitus
Hypertension
Neural tube defects
Schizophrenia
Seizure disorders

traits are biologically inherited, and that even inherited traits can have complex causes.

Eugenics

In 1911, geneticist Reginald Crundall Punnett was not alone when he warned that the knowledge of heredity in humans was "at present too slight and too uncertain to base legislation upon." Nevertheless, by the early twentieth century hereditarianism became a part of American popular and political culture in the form of eugenic ideology. Eugenicists argued that society pays a high price for the birth of 'socially inadequate' people. Eugenicists warned that undesirable traits such as pauperism, feeblemindness, alcholism, rebelliousness, nomadism, criminality, and prostitution were spreading in the general population. The goal of eugenic programs was to institute social policies that promoted certain human matings (positive eugenics) while discouraging others (negative eugenics). Social policies supported by eugenicists included restriction in marriage laws, immigration restrictions, and sterilization laws.

Sterilization law

An example of a dangerous eugenic social policy that is still valid is the 1927 Supreme Court ruling in the case of Buck *v.* Bell. The case involved 17-year-old Carrie Buck, who was chosen as the first person to be sterilized under Eugenical Sterilization Act passed in 1924 by the State of Virginia. Carrie had a child but was not married, and her mother was a resident at an asylum. A lower court decreed that Carrie was "the probable potential parent of socially inadequate offspring" and that her sterilization would be a benefit to society, since "experience has shown that heredity plays an important part in the transmission of insanity, imbecility, etc." The US Supreme Court upheld the lower court opinion and in the decision rendered Justice Holmes issued the infamous phrase that "three generations of imbeciles are enough." Carrie Buck and more than 60 000 Americans in institutions for the mentally ill were involuntarily sterilized following the 1927 decision. The practice of sterilization for the mentally ill continued into the mid-1970s and the Buck *v.* Bell precedent allowing sterilization of the 'feebleminded' has never been overuled.

Conclusions and Prospects

It is essential to examine both heredity and variation when tracing the passage of traits from generation to generation. One aspect of heredity that has been apparent since prehistoric times is that the offspring of sexually reproducing organisms are not exact duplicates of their parents – instead they usually vary in many traits. Plant and animal breeders have been able to harness this genetic variation and using controlled breeding they have accentuated certain desired traits in offspring over many generations.

The principles of transmission genetics have enduring practical applications in the design of selective breeding strategies for agriculture. Hybrid corn provides yields incomparable to that of inbred varieties. Poultry farmers can separate male and female chicks upon hatching through careful use stocks with sex-linked traits influencing plumage. Transgenic farm animals can produce rare human proteins for pharmaceutical use in their milk (or semen in the case of boars) as a result of an innovative technique that crosses species barriers and streamlines gene transmission by inserting foreign DNA directly into the germline.

Mankind has applied the principles of transmission genetics since primitive times by the introduction of agriculture and selective breeding. Genetics is still a young science and our knowledge of heredity has been radically altered within just a few generations. Our fascination of heredity has had a dark past as evidenced by the eugenic policies in the early twentieth century. The genome project is an exciting endeavor that is rapidly transforming genetics into an information science. It is important to consider the rights of individuals and to remember the lessons of the past when using genetic information in the future. Overall, both science and society have been radically transformed by research related to transmission genetics.

Further Reading

Mayr E (1982) *The Growth of Biological Thought: Diversity, Evolution, and Inheritance.* Cambridge, MA: Belknap Press of Harvard University Press.

Olby RC (1997) Mendel, Mendelism and genetics. In Blumberg RB (ed.) *Mendel Web* (http://www.netspace.org/MendelWeb/, Edition 97.1, 1997)

Online Mendelian Inheritance in Man, OMIM(TM). McKusick–Nathans Institute for Genetic Medicine, Johns Hopkins University (Baltimore, MD) and National Center for Biotechnology Information, National Library of Medicine (Bethesda, MD), 2000. http://www.ncbi.nlm.nih. gov/omim/

Orel V (1996) Heredity before Mendel. In *Gregor Mendel: The First Geneticist*. Oxford: Oxford University Press. This chapter can also be viewed at *Mendel Web* (edited by RB Blumberg) (http://www.netspace.org/MendelWeb/, Edition 97.1, 1997)

Ridley M (1999) *Genome: The Autobiography of a Species in 23 Chapters*. New York: HarperCollins.

Sturtevant AH (1965) *A History of Genetics: Modern Perspectives in Biology*. New York: Harper and Row.

***See also:* Breeding of Animals; Ethics and Genetics; Mendelian Genetics; Mendelian Inheritance; Multifactorial Inheritance; Transgenic Animals**

Transposable Elements

R H A Plasterk

doi: 10.1006/rwgn.2001.1316

Transposable elements, or transposons, are discrete segments of DNA that can move within genomes. They were discovered in maize by Barbara McClintock, as the cause of unstable mutations (the instability resulting from excision of the transposon), and have since been found in every organism that was analyzed in any detail.

In the first approximation transposons are probably best viewed as molecular parasites, segments of genetic material that can ensure their own replication (albeit with the aid of multiple host factors). Note that this does not exclude that some transposons may have acquired a function that is beneficial to the host organism; in ecological terms one could say that in those cases strict parasitism has turned into symbiosis. A well-known example in the present time is the rapid spread of antibiotic resistance genes via transposable elements. A more hypothetical example is the integration of the RAG transposon into a predecessor of the immunoglobulin genes; this triggered the development of the current repertoire of segments of vertebrate immunoglobulin genes, which are combined through V(D)J recombination, a descendant of RAG transposon excision.

There are bacterial transposons that encode factors for conjugative pili, and thus ensure their spread between cells. One step further along these lines are transposons where the state after leaving one cell and before entering the other cell has become stabilized. Thus bacteriophage Mu can be viewed as a transposon, but it is also a perfectly fine member of the family of lambdoid temperate bacteriophages. Similarly retroviruses such as MoMLV and MMTV can be viewed as retrotransposons that can be packaged.

Not all genetic elements that move within or between genomes are considered transposons. In general, the ability to integrate into several different positions in the genome distinguishes transposons from other elements. Alternatively mobile elements may move by site-specific recombination, as is the case for the well-studied bacteriophage lambda (see Site-Specific Recombination).

Structure of Transposons

Transposons are mostly flanked by short direct repeats of host DNA, which result from duplication of the target DNA during the integration reaction (see below). The termini of the transposons themselves are often inverted repeats; some level of symmetry should be no surprise if one realizes that the mechanistic events that integrate each of the two transposon ends are usually identical. Transposons minimally encode their own transposase, the protein(s) required for the transposition reaction (see below). In addition they can encode e.g., antibiotic resistance genes, in the case of many bacterial transposons, and structural virion genes, e.g., in the case of retroviruses and bacteriophage Mu. Bacterial transposons such as Tn10 and Tn5 are composite elements: they consist of two insertion sequences (IS10 for Tn10, and IS50 for Tn5) flanking a unique middle segment, which encodes the transposase; IS sequences are also found as separate mobile elements in the genome.

Bacterial transposase proteins often show a *cis*-preference, meaning that the protein acts preferentially on the nearest transposon termini it encounters after it has been synthesized, those of the actual encoding element itself. In eukaryotes where there is spatial separation between cytoplasmic transposase protein synthesis and nuclear transposase activity, there can be no preference for any transposase to act on the encoding copy of the transposon. Hence, in the common situation where a genome contains multiple copies of a given transposon, there is no selective disadvantage for a given copy to lose the transposase gene, as long as it can move using the transposase encoded by other copies. Copies encoding an active transposase are called autonomous; mutant derivatives (usually deletions) that have lost their own transposase genes are called nonautonomous. The first class of transposons, discovered by McClintock, owes its double name to this: Ac/Ds for the autonomous activator *Ac*, and the nonautonomous dissociator *Ds*.

Classification

For an extensive overview of classes of transposons readers are referred to the monographs mentioned below. A classification by host organism makes little sense, since it has been discovered that families of transposons that share the most important features of their mechanism of jumping, and also often show extensive sequence similarity can be encountered in plants, as well as animals, fungi, and prokaryotes. A more meaningful distinction is by the mechanism of jumping, and (often related) the structure of the element.

One distinction is between elements that transpose through a a DNA intermediate and those that have an RNA intermediate. The former are called DNA transposons, the latter retrotransposons.

DNA Transposons

DNA transposons come again in two types. There are those that jump via a simple cut-and-paste mechanism. Those include, to name some of the best-studied transposons, bacterial transposons Tn7, Tn10, the P element of *Drosophila*, the Tcl transposon (and related mariner transposon) from *Caenorhabditis elegans* and other organisms, and Tam and Ac/Ds elements in plants. In all cases the transposition reaction is initiated by double-strand DNA breaks at the transposon termini, after which the excised element can move to a new genomic target and reintegrate. Note that for these transposons the transposition process itself does not result in a rise in transposon copy number. For the transposon to multiply within a genome, it must thus depend on additional features. One possibility is that the transposon preferentially transposes from replicated DNA into a part of the genome that has not yet been replicated, which results in duplication of the transposon in one of the two daughter cells. Another mechanism that seems to be responsible for transposon duplication in some cases is templated repair of the donor site after transposon excision: a transposon excises, and the break left in the donor DNA is repaired using as repair template either the sister chromatid or the homologous chromosome. This repair replication will then often insert a new copy of the transposon into its old position.

The other class of DNA transposons is characterized by a transposition process that is intrinsically replicative. Examples are bacterial transposons Mu, gamma-delta, and Tn3. These transposons are never excised from their original position in the genome. Instead breaks occur only at the 3′ ends of the two transposon ends. These are then fused to a new target in the genome; this reaction creates a forked structure at each transposon end, which is very similar to a replication fork, and can indeed be zipped open by replication of the transposon. Note that, as shown below, the difference between the two classes of transposons, replicative and nonreplicative, is smaller than one might have thought.

Retrotransposons

These transposons do not excise, and do not even undergo single-strand DNA breaks at their termini to initiate transposition. They are transcribed. The resulting RNA can be reverse transcribed into DNA, by the enzyme reverse transcriptase (RT), which is usually encoded by the pol (for polymerase) gene of the transposon. There are two classes of retrotransposons: LTR and non-LTR. LTR stands for long terminal repeat. LTR transposons are first transcribed into an RNA that contains part of each LTR at each terminus. Then via a reverse transcription process that includes a complex series of template jumps between the two termini, a genomic cDNA is generated that has complete LTR copies at each end. This DNA copy is then integrated into a new chromosomal DNA target. This integration reaction is catalyzed by the transposon encoded integrase protein. Well-known LTR retroelements are the retroviruses MoMLV, MMTV, and HIV, the yeast transposons Ty1 and Ty3, Copia of *Drosophila melanogaster*, Tal of *Arabidopsis thaliana*, and IAP of mice.

Non-LTR retroelements are, e.g., the LINES and SINES that make up a considerable part of the human genome, and the I element of *Drosophila*. SINES do not encode RT and integrase, they probably use the transposition machinery of LINES (also see Retroposon).

Mechanism of Transposition

Chemical Steps in Transposition

While the above classification might suggest otherwise, the chemical steps in transposition are universal. In all cases the transposon is inserted into its new target DNA by a pair of nucleophylic attacks of the two 3′ hydroxyls at each terminus of the transposon on two phosphodiester bonds in the opposite strands of the target DNA. Since these two phosphodiester bonds are always a few positions apart, there is a stagger in the target DNA, which causes the target duplication characteristic for all transposons (see **Figure 1**). It is important to note that the target DNA is thus never actually cut: the in-line attacks of the transposon 3′ hydroxyls remove and replace the original hydroxyl groups at the target. This implies that no exogenous energy is required to ligate the

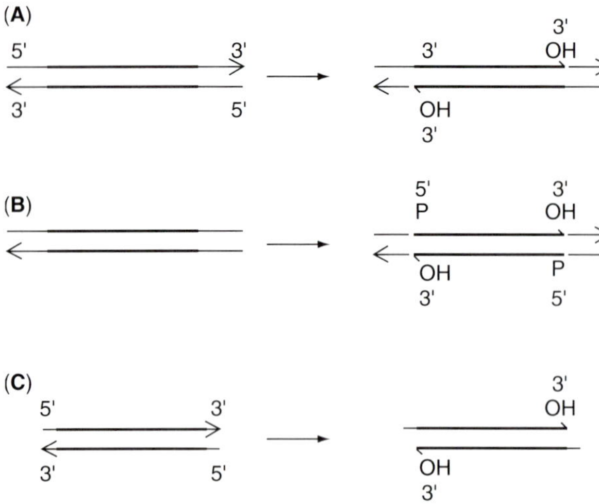

Figure 1 The donor cleavage. (A) Cutting of 3′ ends only (e.g., Mu, Tn3). (B) Cutting of both ends (e.g., Tn10, Tn7, P element, Tc1). (C) Retrovirus donor cleavage.

transposons to its new target, since the energy of the cleaved bonds is retained in the new bonds. How then can replicative and nonreplicative transposons all integrate via this one universal reaction? All reactions are initiated by a simple hydrolysis that releases the 3′ transposon terminus from its flank. (Note that this is even true for most LTR retrotransposons: the products of the reverse transcription of e.g. HIV are two base pairs longer at each end than the integrated copies; two nucleotides are removed from the 3′ end by a hydrolysis.) The difference between replicative and nonreplicative elements is what happens to the other strands. These are cut for simple cut-and-paste transposons, while they are not for, e.g., phage Mu. Then in both cases the newly generated 3′ ends are fused to the target DNA; the difference is that the replicative elements still have the complete flanking DNA dangling on, while the cut-and-paste transposons have at most a few nucleotides (as result of a short stagger with which the transposon was excised). The point is illustrated in **Figure 1B**, and discussed in more detail in Sherratt (1995).

The standard technical term for the hydrolysis that releases the 3′ hydroxyl groups from their flanking DNA is donor cleavage. The reaction in which these 3′ hydroxyls are fused to the new target DNA is called strand transfer. Donor cleavage and strand transfer are not exactly the same as excision and integration: e.g., retrotransposons never excise, but as mentioned above most of them need a donor cleavage to remove a few nucleotides from the double-stranded linear transposon DNA before integration can occur. Integration is almost identical to strand transfer, except that the strand transfer reaction *per se* only fuses the 3′ ends of the transposon to its new target, so that subsequent DNA repair is required before the transposon is fully integrated.

Target Choice

While transposons are in the first approximation distinguished from site-specific recombination systems by their ability to integrate randomly, there are actually varying degrees of freedom in target choice, but there hardly ever is complete randomness. Some transposons have absolute target requirements (e.g., the Tc1/mariner transposons need a TA dinucleotide at the target). Most transposons have some preference for a few nucleotides at the target DNA; for Mu, Tn10, and Tc1 a preferred integration consensus could be found. These consensuses are loose enough that the transposon will still integrate into most genes at many positions. Some yeast retroelements prefer to integrate into promoter regions. The bacterial Tn7 transposon has two modes of integration: site-specific as well as (at a lower frequency) random. Apart from general target preferences, there can be other restraints, such as the preference for P elements and for plant transposons to integrate into the region of the chromosome it excised from; apparently there is only limited diffusion of the excised transposon, either in three dimensions, or by scanning along the chromosome after excision.

The Transposase Proteins

Both the donor cleavage and the strand transfer reactions are catalyzed by the transposase. Starting from the assumption that transposons are selfish DNA, one could expect that in principle transposons encode their own transposase, and this is indeed what is found. In all cases there is a separate domain that specifically binds to the termini of the transposon (usually it binds to a region of approximately 20 base pairs, which is a few base pairs removed from the actual transposon end). Then there is a catalytic domain. Interestingly, these catalytic domains show a clear conservation at two levels: the overall folds of the transposase of Mu (A), resembles that of retroviral integrases (and also that of the ruvC protein involved in resolution of intermediates in homologous recombination). Also, a characteristic triad of amino acid residues is observed, with a conserved spacing: the so-called DD(35)E motif; these residues are at the heart of the catalytic site of these phosphoryl transferase proteins, and they are thought to coordinate the divalent cations (commonly Mg^{2+}) that are required for the reaction.

Some analysis has been done concerning the stoichiometry of the transposase complex; it is clear that

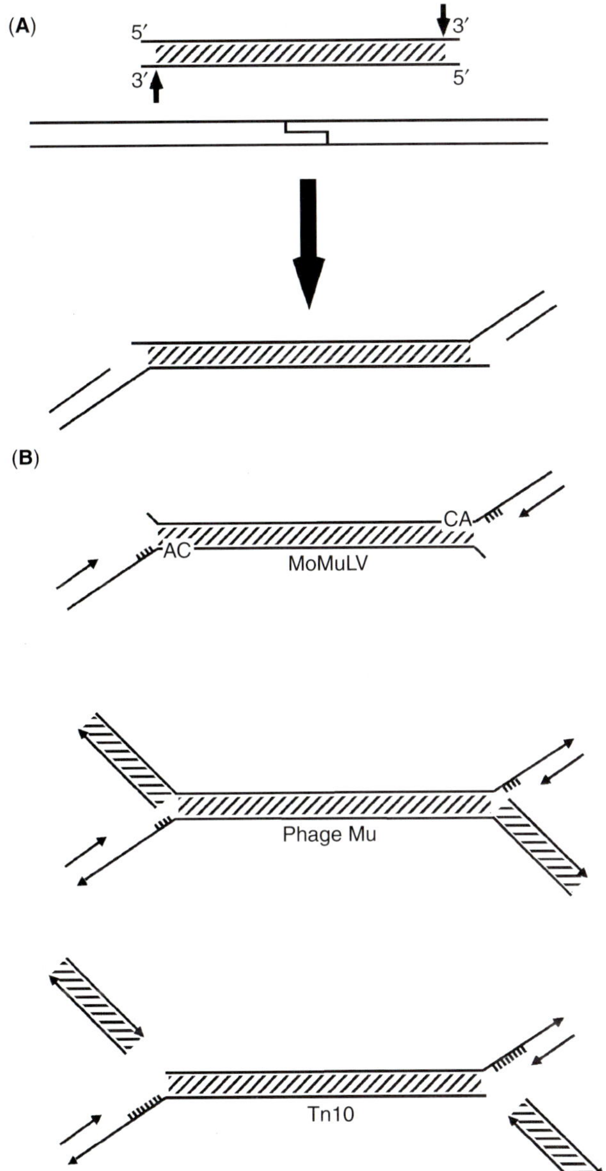

Figure 2 Donor cleavage and strand transfer. (A) Here shown for an integrating retrovirus are the donor cleavage (as in **Figure 1**), indicated by small arrows, followed by strand transfer. Note that the stagger shown in the target DNA results in short single-strand sequences in the target DNA that flank the integrated element; after repair replication these will become short target duplications. (B) The product of strand transfer shown for a retrovirus, a replicative DNA transposon (note that this structure needs to be resolved by DNA replication through the Mu transposon), and a cut-and-paste transposon, Tn10. In the latter case the two ends of the broken donor DNA are still shown in the picture; in reality they may be far removed from the integration site, but they are shown to emphasize the similarity in the reaction mechanisms. (See Sherratt, 1995.)

at least two, possibly four, transposase subunits are required for the complete reaction. The two ends of the transposon are actually complexed during the donor cleavage as well as the strand transfer reaction. It has been shown that cutting at one transposon ends depends on proper recognition also at the other end; in the strand transfer reaction, it is the precise spatial organization of the two transposon termini which determines the precise distance between the phosphodiester bonds at the target DNA that are being cleaved. Small differences in the multimeric integrase complex can thus explain why the MLV genome is always flanked by four and the HIV genome by five base pair target duplications.

There is limited knowledge of the host proteins required in transposition. What is known is almost all based on bacterial transposons, where small basic proteins are sometimes required or at least strongly stimulate transposition.

Regulation of Transposition

All organisms contain transposons in their genomes, and it is clear that the frequency of transposition must be controlled. The population biology is as for any parasite: the transposon can not replicate too aggressively, since it should not affect the fitness of the host too much. The regulation of transposition is still an open area for research. The P element of *Drosophila* encodes its own repressor; an alternative splice in the transposase transcript determines whether an active transposase or a repressor protein is synthesized. This alternative splice ensures that the transposon can jump in the germ line, but remains silent in somatic cells. Interestingly, the opposite is found for the Tc1 element of the nematode *C. elegans*, which is active in somatic cells of all strains, but is kept quiet in the germline of some strains; the nature of the silencing genes (mut genes) is to be determined. Plant transposable elements are also regulated at the substrate level by DNA methylation. Especially with the aid of ongoing genome projects, it should now become feasible to identify the genes that control transposon activity.

Transposons as Tools

Transposons generate DNA fusions. This can be exploited in experimental genetics in many ways: e.g., interruption by a transposon can inactivate a gene, or insertion next to a gene can activate a gene. After transposon insertion mutagenesis, the transposon can be used as a tag to recover the mutant gene relatively easily. An enhancerless reporter gene can be included in a transposon, so that after massive transposon insertion, one can screen for organisms in which the reporter is expressed in an interesting fashion (e.g., tissue specific, or only conditionally); these experiments are referred to as enhancer traps or gene traps. Other applications focus less on the integration site as such; transposons can be used for transgenesis: to ensure that a precisely defined DNA segment is integrated in single copy (but at an unknown position in the genome). Transposons have also found numerous applications in gene mapping and sequencing projects.

Further Reading

Berg D and Howe MM (eds) (1989) *Mobile DNA*. Washington, DC: American Society for Microbiology Press.

Craig N (ed.) (2001) *Mobile DNA II*. Washington, DC: American Society for Microbiology Press.

Haren L, Ton-Hoang B and Chandlers M (1999) Integrating DNA: transposases and retroviral integrases. *Annual Review of Microbiology* 53: 245–281.

Saedler H and Gierl A (eds) (1996). *Current Topics in Microbiology and Immunology*, no. 204, *Transposable Elements*. Berlin: Springer-Verlag.

Turlan C and Chandler M (2000) Playing second fiddle: second strand processing and liberation of transposable elements from donor DNA. *Trends in Microbiology* 8: 268–274.

Reference

Sherratt DJ (ed.) (1995) *Mobile Genetic Elements*. Oxford: IRL Press.

See also: **Insertion Sequence; P Elements; Phage Mu; Retrotransposons; Retroviruses; Transposons as Tools**

Transposable Elements in Plants

H-A Becker, H Saedler, and W-E Lönnig

doi: 10.1006/rwgn.2001.1639

The concept of transposable elements (TEs) was first published in 1948 by Barbara McClintock as a result of combined genetic and cytological studies in maize. In contrast to the linear and relatively stable arrangement of genes in linkage groups, transposable elements seemed to possess the ability to change their position in the genome. They were even able to jump from one chromosome to another. After the nature of transposable elements had been revealed by molecular analysis and it had become clear that transposable elements are ubiquitous in prokaryotic and eukaryotic organisms, McClintock's discovery and work were finally rewarded by the Nobel Prize in 1983. TEs have, in fact, been found in every organism analyzed so far. In plants they can contribute more than 80% of the total DNA of the genome. In general, plants serve as hosts for elements that are structurally and functionally similar to those found in yeast or mammals. However, there are differences in the genomic organization of plants and in the heterogeneity of their families (see below).

In prokaryotes (where the number of transposable elements is much lower) there are clear advantages for the host that carrys, for example, antibiotic-resistant elements. Despite the often enormous contribution of these elements to DNA in plants, it is still unclear whether there are long-term benefits for the host species carrying them or whether they must be regarded as mainly 'selfish DNA.'

Discovery of Transposable Elements in Maize

McClintock's successful studies in the 1940s were based on classical genetic presuppositions. For example, she concentrated on suitable genetic traits such as genetic loci associated with plant pigments or distinguished morphological characters belonging to an ideal monitoring organ for genetic analysis such as the maize karyopsis. Although transposable elements are normally known to cause unstable mutations by insertion into marker genes, the first evidence of transposition came from a genetic constellation where a defective nonautonomous element able to cause aberrant transposition events was activated in *trans* by an autonomous element of the same family.

Although the molecular basis of transposition was not known for another 30 years, McClintock made three observations that enabled her to devise the concept of transposable elements. McClintock termed her first observations such as 'unstable loci' *c* (colorless aleurone) and *wx* (*waxy*). Normal loci known to be stable during propagation suddenly became unstable. All the mutable loci were located on the short arm of chromosome 9. They became unstable because a locus termed *Ds* (*Dissociation*) caused chromosome breakage, which resulted in the loss of the distal chromosome fragment carrying the markers in subsequent cell divisions. The second observation was that chromosome breakage only occurred in the presence of a controlling element, which McClintock termed *Ac* (*Activator*). The third key to the idea of transposition was the proof that *Ac* and *Ds* could both move from their loci to other sites in the genome. They both displayed Mendelian inheritance; however, they changed their location on the chromosomes by a transposition event (McClintock, 1987). Later, molecular analysis revealed the *Ds* at the proximal position of the short arm of chromosome 9 to be a defective nonautonomous element of the *Ac/Ds* family called double *Ds*. One *Ds* copy of this family had jumped into another *Ds* and the resulting element often causes chromosome breaks and rarely transposes. The *Ac* element is the autonomous master element of the family and mobilizes nonautonomous *Ds* elements in *trans*. Transposons of the *Ac/Ds* family are DNA elements (Class II transposons) (see '*Ac/Ds* superfamily in maize and *Antirrhinum*' below), which transpose by a cut-and-paste mechanism.

Overview and Classification

The transposition mechanisms of TEs in plants are more diverse than those of prokaryotic elements. In plants, a transposition event leads to a mobile element insertion at a new acceptor site in the genome. However, there are differences between the various types of elements concerning the generation of insertion sequences. Two main classes, comprising elements with similar mechanisms of transposition, can be distinguished (**Figure 1**). Class I elements are retrotransposons which transpose via an RNA intermediate; therefore, transcription is the first step to generate the inserted copy. The RNA is retrotranscribed in the next step by reverse transcriptase action into extrachromosomal DNA and this DNA is finally inserted at the target site. Class II elements transpose via a 'cut-and-paste' mechanism; the DNA copy is excised at the donor locus and inserted at the acceptor site. In Class I elements the life cycle is replicative and increases the number of elements per cell. In Class II elements the mechanism is nonreplicative and the number of elements is not increased by the transposition mechanism *per se*. The elements of the *Ac/Ds* family described above belong to Class II (see below). Class I and Class II elements can be subdivided into different groups displaying similar structures or similar modes of transposition. Both classes comprise autonomous and nonautonomous elements. Nonautonomous transposable elements have to be mobilized by autonomous ones in *trans*. In plants Class I retrotransposable elements occur in greater numbers. In contrast to nonreplicative Class II elements, Class I elements give rise to stable mutations because they are not excised from their donor site. Autonomous Class II elements create unstable mutations and a gene function may be restored by their excision.

Class I: Retrotransposons

A clear correlation between structure and mode of transposition can be observed for all transposable elements. Whilst retrotransposition always starts by transcription of the element, the full retroviral life cycle generates their characteristic long terminal repeats (LTRs). The presence of these repeats is also the basis for a further classification of retrotransposable elements: Although all retrotransposons transpose via an RNA intermediate, some of these elements follow the mechanism of the retroviral replication, yet others do not. In plants the first retroelements were described some 15 years ago. The LTR-elements *cin1* and *Bs1* as well as the non-LTR element *cin4* have all been isolated from maize. LTR-elements have been detected in algae, bryophytes, pteridophytes, gymnosperms, and angiosperms. Non-LTR elements have also been found throughout the plant kingdom. Looking at the elements sequenced so far, the vast majority of plant retroelements are nonautonomous. Only a few actively transposing retroelements have been isolated until now.

LTR-Retrotransposons in Plants

In comparison with other retroelements, LTR-retrotransposable elements are the most retrovirus-like TEs. They encode the *gag*, *pol*, *RT/RNaseH*, and the *int* (integrase) sequences constituting genes

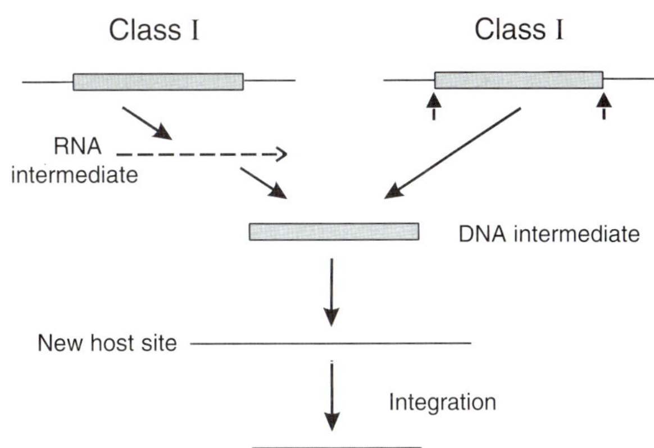

Figure 1 Transposition of Class I and Class II elements results in the integration of a mobile DNA sequence at a new host acceptor site. Class II elements transpose via a cut-and-paste mechanism. The inserted sequence has to be cut out prior to another insertion. Class I elements generate the inserted sequence by transcription in the first step and retrotransposition in the second.

characteristic for retroviruses; however, the *env* genes are missing. (If a coding sequence similar to the *env* (envelope) can be detected, the element is generally classified as a retrovirus.) In yeast, retrovirus-like particles are formed by the *Ty1 gag*, which encodes for the structural proteins of the virion core. However, as the *env* homolog is missing, they do not give rise to infectious viruses. Thus, the elements might have originated from retroviruses either by loss of the *env* gene, or the incorporation of this gene into an LTR-element might have led to a retrovirus. Because of the missing *env* gene, LTR-retrotransposable elements can be regarded as retroviruses that are unable to leave their host cellular units.

In yeast and *Drosophila* these elements have been classified into two groups according to the differences in their structural organization as well as homology of their sequences: *Ty1/copia* and *Ty3/gypsy*. The structural organization and mechanism of transposition are identical to that of retroviruses. Even the arrangement of retroviral proteins is similar in both groups. The *gag* and the *protease* genes precede the reverse transcriptase and the *RNaseH* gene. However, they differ in the location of the *int* (integrase) gene (**Figure 2**) (Kunze *et al.*, 1997).

The transposition mechanism of LTR transposons and retroviruses is thought to be the same. In all cases retrotransposition starts with the transcription of the element from a promoter located in the 5′ LTR. The primary transcript is then reversely transcribed into double-stranded copy-DNA and finally integrated into the host genome by integrase action. This enzymatic activity creates a staggered nick at the new host site. The gaps resulting from ligation of the double-stranded element to the single-stranded staggered host DNA are filled in by cellular repair enzymes. Therefore, the inserted copy shows LTRs (caused by the retroviral-like replication) which are bordered by short direct repeats (caused by integration into the host DNA) that are usually 5 bp long. The length of these repeats is due to the integrase, which sets the staggered cuts.

In plants elements of the *Ty1/copia* group seem to be ubiquitous. They have been identified in algae, ferns, gymnosperms, monocots, and dicots. Sequence analysis has revealed that most of the elements identified seem to be inactive: They either carry defects in the protein-coding regions, but can still be mobilized by a different reverse transcriptase activity in *trans*, or they have accumulated mutations in the LTRs after integration, which totally inhibit mobilization.

The first functional and autonomously transposing *Ty1/copia* elements to be isolated in plants were the *Tnt1* and *Tto1* retrotransposons from tobacco. *Tnt1* has a size of 5.3 kb, 610 bp LTRs and generates 5 bp target-site duplications upon integration. There are more than 100 *Tnt1* copies in the tobacco genome. The polyprotein encoding sequence is similar to that for the *Drosophila copia* element (sequence homology and length). At the protein level homologies range from 29 to 42% (*gag*, *prot*, *endo*, and *RT/RNaseH* domains). The number of elements may vary in the extreme. *Bs1* in maize, for example, is present in only one to five copies, but *BARE-1* in barley exists in more than 50 000 copies per genome, so constituting a considerable part of the genome.

Elements of the *Ty3/gypsy* group have been isolated from plants as well, but so far they have been investigated in less detail than the *Ty1/copia* elements. The *Ty3/gypsy* group has been detected in gymnosperms and angiosperms and is also expected to occur throughout the plant kingdom. The first *Ty3/gypsy* member to be identified was the *del1-46* TE from *Lilium henryi* (see **Figure 2**). It occurs in about 13 000 copies in the nuclear DNA. In contrast to their similar structural organization, protein homology with the *Ty1/gypsy* group is rather limited. The most obvious structural difference is the position of the *int* coding region. It is located upstream of the *RNaseH* in the *Ty3/gypsy* elements. The number of certain *Ty3/gypsy* transposons may be as high as the *Ty1/copia* elements. In maize, for instance, there are about 20 000 copies of *Cinful-1* which is 8.6 kb long and terminated by 586 bp LTRs (Kumar and Bennetzen, 1999).

Owing to their transposition mechanism, the retrotransposon copy number is enlarged with every transcript eventually converted into a cDNA. There are many retrotransposon elements that have been isolated from plants. In general, plant genomes seem to be surprisingly tolerant to increasing copy numbers of TE sequences until they finally contribute significant amounts of DNA to the host genome. Whereas the number of *Ty1* elements is, for example, rather limited in yeast (some 40 copies), an estimated 1 000 000 LTR-retrotransposons occur in *Vicia faba*.

LINEs and SINEs: Non-LTR-Retrotransposons in Plants

The non-LTR-retrotransposons constitute a second group of retrotransposable elements. The lack of LTRs distinguishes them structurally and mechanistically from LTR elements (the latter following the retroviral replication cycle). LINEs (long interspersed nuclear elements) and SINEs (short interspersed nuclear elements) occur in the genome as part of the repetitive DNA that is not arranged in tandem repeats but is interspersed.

Interestingly, LINEs partly encode the same proteins as LTR-transposable elements: *gag* and *RT/*

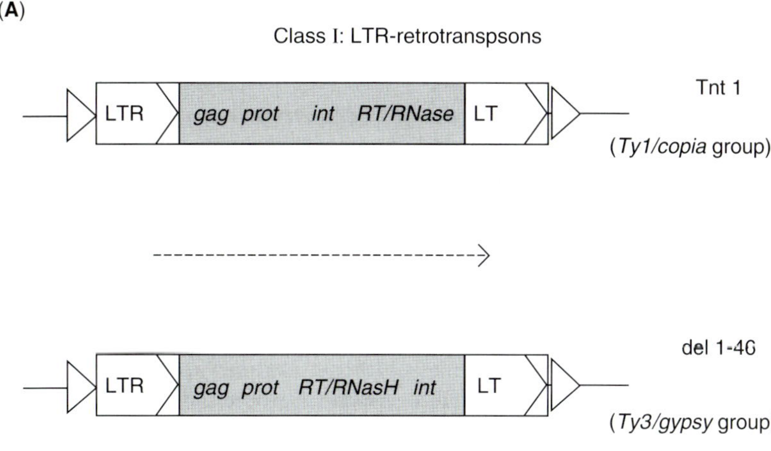

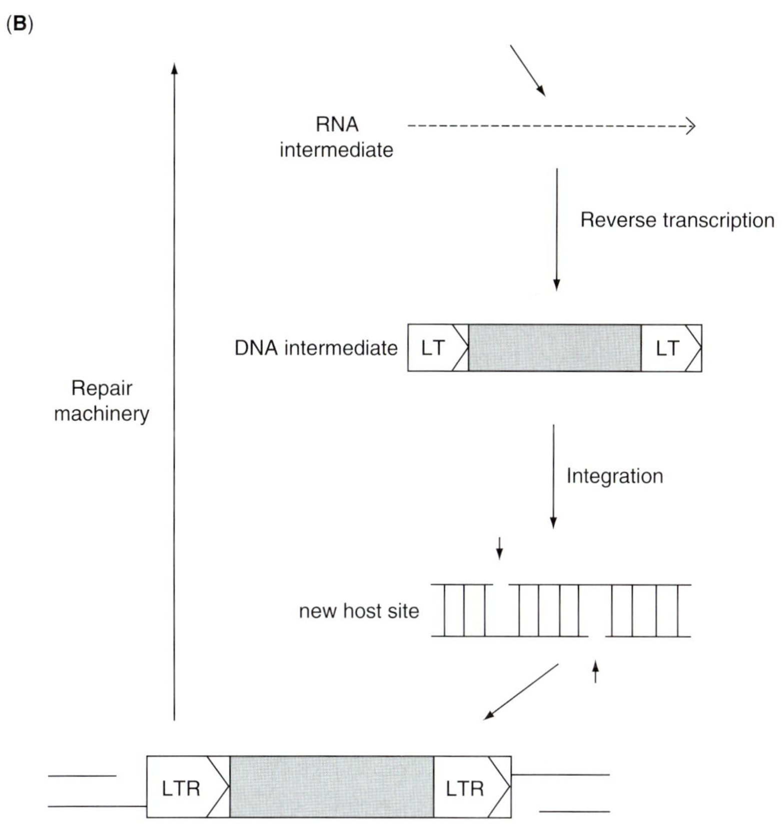

Figure 2 (A) *TntI* from tobacco (belonging to the *TyI/copia* group) and *delI-46* from *Lilium henryi* (a member of the *Ty3/gypsy* group) display a similar structural organization. LTRs (long terminal repeats) mark the termini of the integrated elements. Target site duplications are indicated by triangles outside the LTRs; however, location of the *int* (integrase) differs. (B) Retrotransposition of LTR-retrotransposons starts by transcription which is followed by reverse transcription. The new copy is inserted by integrase action. The host repair machinery fills in the gaps resulting from the insertion process.

RNaseH. Instead of *int* (integrase), a coding sequence for an endonuclease (EN) seems to be responsible for the insertion into the host genome. The terminal poly(A) tract, the missing LTRs, and the flanking direct inverted repeats indicate that, despite the partly identical protein functions, a different mechanism for transposition compared to LTR-bordered transposons is used by these elements. Hints about the mode of transposition were obtained by the isolation of the first plant non-LTR retrotransposon.

The first plant LINE *Cin4–1* was detected as an insertion in the 3′ untranslated region of the *A1* gene in maize. Full-length *Cin4* elements are 4 kb in size and present in 50 to 100 copies per genome. Further isolated *Cin4* elements showed identical 3′ ends, but heterogeneously truncated 5′ ends. Based on the observation that short regions of homology could be identified between the sequenced *Cin4* 5′ ends and the adjoining target site duplication, a model for LINE transposition as depicted in **Figure 3** was developed (Schwarz-Sommer *et al.*, 1987). Transcription starts at a promoter at the 5′ end of the element. LINEs carry RNA-PolII and SINEs carry RNA-PolIII promoters. These promoters are transposed as well, thus ensuring further mobility of an element that has already moved in the genome. In the next step reverse transcription

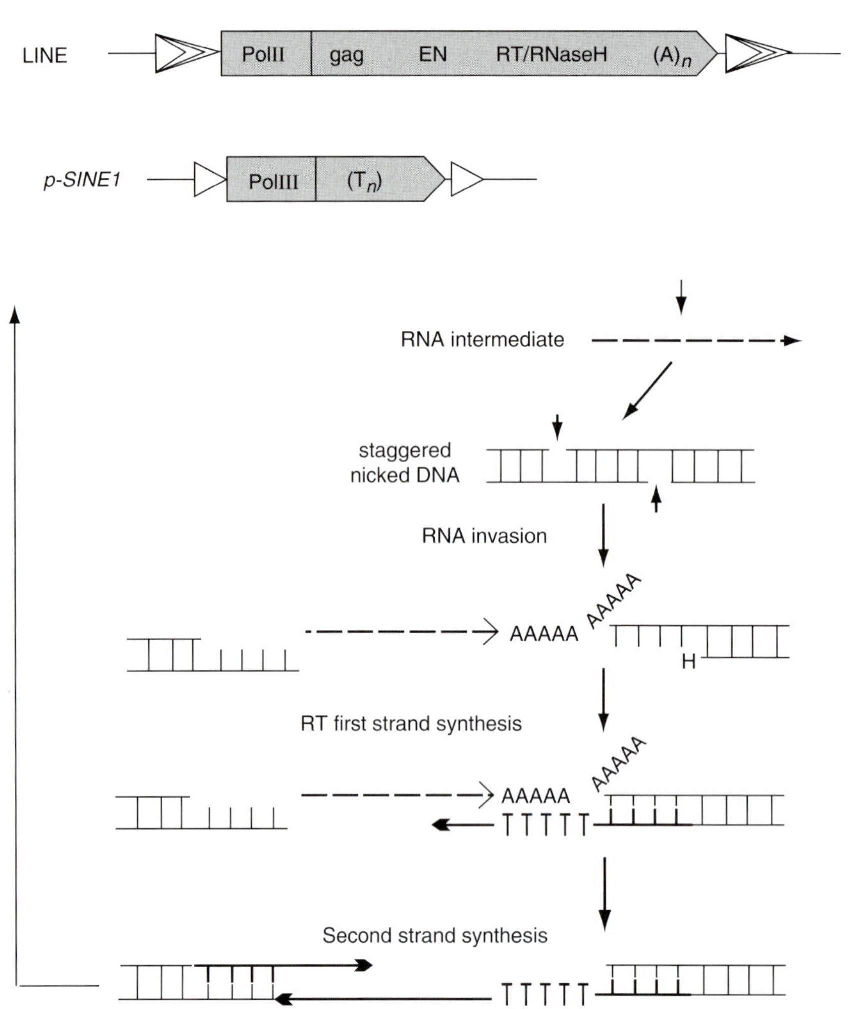

Figure 3 (A) Typical LINE structure and *p-SINE1* from rice. The LINE encodes for the Gag (coat protein), the RT/RNaseH (reverse transcriptase/RNaseH), and the EN (endonuclease). The varying target site duplications are marked by nested and non-nested triangles. A hypothetical transposition model for non-LTR retrotransposons starts by transcription from Po I I or Po I II promoters. (B) In the second step the endonuclease (En) cuts the host acceptor site. Retrotranscription is primed by RNA invasion using the free 3′-hydroxyl end in the host DNA as a primer. The host DNA of the opposite strand serves as primer for the second strand synthesis.

starts at the 3′ end of the element. The transcription is primed using a free 3′ hydroxyl end which results from endonuclease (EN) activity in the genomic DNA. The second strand is finally synthesized using the free 3′ OH end in the opposite genome strand as a primer for the cellular repair machinery. Owing to the mechanism of integration by short homologies between the 5′ end of the element and the host DNA at the staggered nick, the length of the target side duplication varies from about 3 to 16 bp. In contrast, LTR elements (see 'LTR-Retrotransposons in plants,' above) and Class II elements (see below) produce discrete target side duplications (generally 5 bp in the case of LTR elements).

Of special interest concerning quantitative effects is the *del2* LINE in the lily. A 4.5 kb unit is present in 250 000 copies in *Lilium henryi*. It constitutes 4% of the lily genome. The element is found in many monocotyledons but interestingly in only four of eight lily species investigated. The vast number of copies and its presence in only some lily genomes illustrates the astonishing plasticity of the plant genome. In particular LINEs contribute a major part to the repetitive DNA in plants.

SINEs are relatively small elements. In contrast to LINEs they only give rise to transcripts that do not encode for proteins necessary for transposition; therefore, they are also termed retrogenes. For mobilization these functions have to be supplied in *trans*. However, the *cis*-acting determinants and structures are very similar to those in LINEs. A promoter that is part of the transcript, and a 3′ end that terminates in poly(A) or A- or T-rich sequences as well as target site duplications in the host DNA all indicate that the transposition mechanism is similar to that of LINE elements. Most probably the necessary *trans*-acting functions are provided by other retroelements present in the host genome. The promoter regions show motifs similar to those known from transfer RNA (tRNA) genes. The promoter, usually an RNA-polIII promoter that is part of the transcript, distinguishes them from intronless pseudogenes.

In rice, a family of related sequences has been identified that occurs in more than 100 copies in the genome: the *p*-SINE*1* elements are on average 125 bp long, flanked by 14–15 bp target site duplications and they contain an RNA-PolIII promoter (**Figure 3**). However, they terminate in a T-rich pyrimidine tract at the 3′-ends. Similar elements have been found in *Craterostigma plantagineum* where the insertions are flanked by 12–17 bp target site duplications and terminate in an oligo(T) tract as well. The size varies between 0.65 and 0.9 kb, which is unusually large for SINE-like elements.

Given that the model for the transposition mechanism of LINEs and SINEs is correct, the question arises as to what extent normal genomic transcripts could be retrotransposed as well. If a poly(A) tail is sufficient to prime integration and reverse transcription, any poly(A) tail could serve as a substrate for retroelement enzymatic activities. As a matter of fact, DNA sequences resulting from such events can be found in the plant genome. They are termed processed pseudogenes. In contrast to the genes they originated from they display neither an internal promoter, nor any intron sequences, yet they possess the poly(A) tail. They generally represent processed RNA transcripts converted into DNA.

Therefore, after integration, these byproducts of retroelement activity in the genome are immobilized and cannot be transcribed anymore. Extending the line of reduction from complex autonomous elements like retroviruses to LTR-transposable elements to non-LTR-transposable elements (LINEs), and no protein-encoding SINEs, these processed pseudogenes are one-way products of accidentally chosen templates by *trans*-acting reverse transcriptase activities in the genome.

Genome Distribution and Chromosomal Organization

Two types of repetitive DNA that are structurally different and dissimilarly arranged constitute a major part of the genetic material in higher plants. Tandemly arranged sequences build up satellite and microsatellite DNAs. Interspersed sequences widely distributed in the genome mainly consist of retrotransposable elements. In maize they constitute up to more than 80% of the genome. Besides their individual structure, which is correlated to their mode of transposition, their arrangement at the chromosome level is of special interest.

Although the elements are often widely interspersed, investigations in *Beta* species by fluorescent *in situ* hybridization (FISH) disclosed a nonrandom distribution for LINE elements. They are located in discrete clusters. Homogenization of certain areas (by loss of TEs) might be a genomic mechanism to limit the presence of retroelements to specific chromosomal locations. Alternatively, preferential insertions at specific positions might concentrate the elements. Such a tendency has been shown for *Zepp*, a non-LTR retroelement in *Chlorella vulgaris*, which preferentially integrates into other *Zepp* copies (Schmidt, 1999).

Ty1-copia elements are dispersed in euchromatin. In some species they are unevenly distributed. The preferred regions differ for various elements and separate species. Some families are clustered in or absent from chromosomal areas such as centromeres, telomeres, or paracentric regions. In euchromatin (the location of most genes), the elements occur mainly in spacer DNA (intergenic regions). *Drosophila* specific retroelements are detected at telomere or centromere sites, yet in the Gramineae the *cereba* and *Ty3/copia* sequences cluster at centromere regions. To date, no known function can be correlated with these different occurrences.

Class II: DNA Transposons

Since the discovery of DNA transposons in maize more than 50 years ago, the elements of this class have been investigated in detail. They transpose via a nonreplicative cut-and-paste mechanism. Owing to the fact that the transposon is cut out at the donor site, the number of Class II transposons does not increase during transposition. Yet, they can multiply when they transpose from an already replicated site into one that is replicated later. In general, the number of DNA transposons is limited from a few copies up to several hundred in the genome. The elements encode for only one or a few proteins constituting *trans*-acting factors sufficient for transposition to occur. However, the structure of the termini (the *cis*-determinants for transpositional activity) are more complicated than the corresponding retroelement structures. Current models for transposition of Class II elements emphasize the necessity of concerted cuts at both ends to avoid chromosome breakage. The DNA-based *cis*-requirements for exision seem to be more intricate than those for retroelements. The termini contain two types of *cis*-acting signals: terminal inverted repeats (TIRs), which determine the position to be cut, and subterminal motifs, which seem to guarantee the alignment of the ends in an appropriate arrangement for the cut-and-paste mechanism.

Ac/Ds Superfamily in Maize and *Antirrhinum*

The *Ac/Ds* family comprises some hundred elements per genome. Most of them are inactive or nonautonomous. Nonautonomous elements are derived from the autonomous *Ac* by, for example, base substitutions or deletions in the protein-coding region. In general, only a few *Ac* master elements exist per genome. *Ac* is 4565 bp long and the central part encodes for the 807 amino acid transposase (TPase) (see **Figure 4**). The TPase recognizes two different types of *cis*-acting signals in the *Ac* termini via a bipartite DNA-binding domain. The outermost ends of *Ac* are marked by 11 bp TIRs. There are multiple 3–4 bp-long DNA-binding motifs which are internally located to the TIRs and that are bound by the TPase as well. In spite of the absence of sequence homology, both the 11 bp TIRs and the short motifs are recognized (Becker and Kunze, 1997). According to the transposom hypothesis, the subterminal motifs are required for the correct alignment of both ends by TPase–TPase interactions. Concerted cuts at both TIRs might then be set by TPase molecules binding to the TIRs as well. During integration a staggered incision is set into the host DNA and the element is ligated to the single strands of the locus. The gaps are finally filled in by the actions of host repair enzymes. 8 bp target site duplications are created by integration. After excision a more or less perfect track of the TE activity in the form of 'footprints' can be detected, generated by imprecise excision and repair of TE Class II visits (see 'Transposable elements and DNA diversity' and **Figure 4B**).

The TPase is acting in *trans* on the *Ac* (*activator*) element as well as on the nonautonomous *Ds* elements. In some cases *Ds* elements can be clearly identified as decendants of autonomous elements. Although in other instances, for example, *Ds1*, 11 bp TIRs and subterminal motifs can also be identified, sequence comparison indicates *Ds1* is not derived from an *Ac* element.

The presence of many different TEs in the same plant species raises the question whether (and if so, to what extent) TEs are related within and between various organisms. The *Ac/Ds* transposons, comprising several hundred elements in maize, belong to the *Ac* superfamily. All these elements generate 8 bp target site duplications and display similarities in their TIR sequences. In maize, the TEs *Ds1*, *Bg*, and *rDT* belong to this superfamily. Further members are found in *Chlamydomonas*, *Petunia*, *Pisum*, and *Petroselinum*. Also, one of the genetically and molecularly best-characterized transposons, *Tam3* of *Antirrhinum majus*, is a member of this superfamily. The *Tam3* (transposon Antirrhinum majus) DNA consisting of 3629 bp is bordered by 12 bp TIRs, and contains an open reading frame coding for 749 amino acids. An overall 30% similarity is found between the *Tam3* TPase and the *Ac* TPase (807 amino acids long) (in some conserved regions the similarity rises to about 65%). These findings indicate that homologies between elements from different species are not only limited to the *cis*-acting determinants for transposition at the DNA-level, but also extend to the proteins necessary for transposition. The homologies at the DNA and protein level indicate common patterns and blueprints and thus clear systematic relationships.

En/Spm in Maize: CACTA Superfamily

The CACTA superfamily of TEs is defined by similarities in their outermost TIR sequences. Elements belonging to this family have been identified in maize, rice, *Antirrhinum* (*Tam1*), *Glycine max*, and *Pisum*. The *En/Spm* element is the most thoroughly

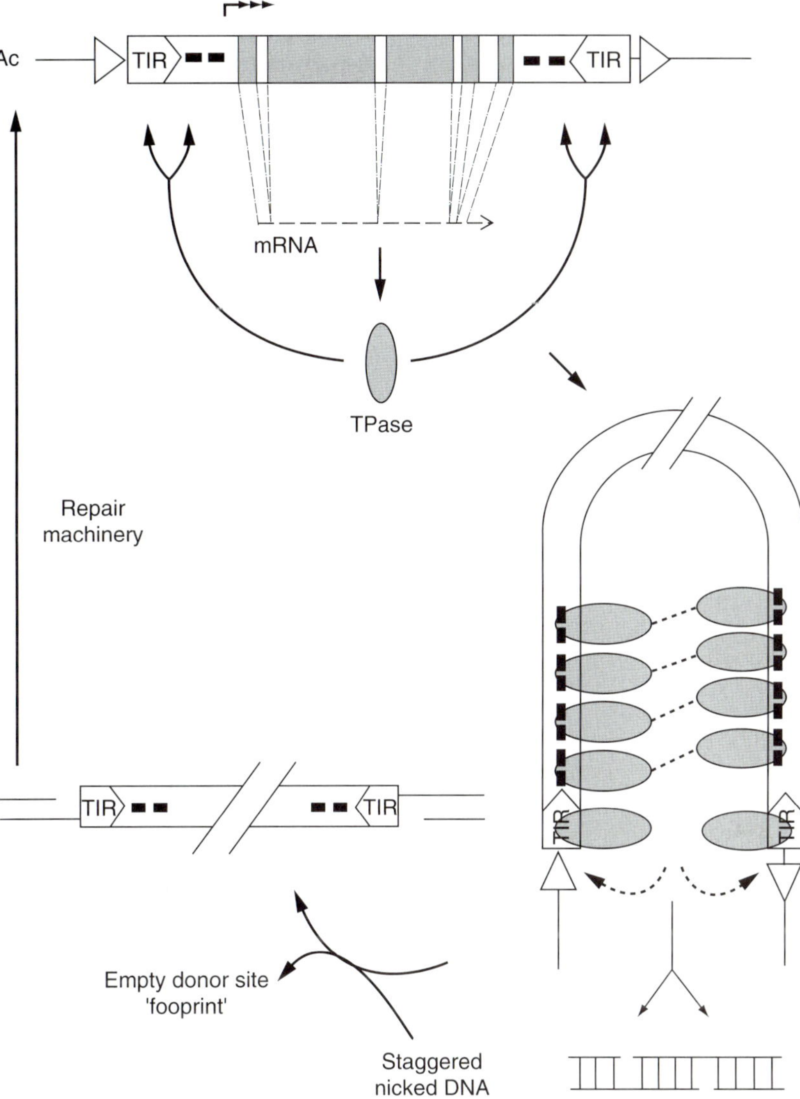

Figure 4 The *Ac*-encoded TPase (transposase) recognizes the TIRs (terminal inverted repeats) and short subterminal motifs (indicated by black bars). The target site duplications at the outermost ends of the TIRs are depicted as triangles. The three black arrows mark transcription start sites. In the hypothetical transposom (a complex arrangement of the *Ac* termini and TPase molecules) the TPase binds to the short subterminal motifs. Thus the ends are brought together in correct alignment. TPase molecules that have been shown to recognize the TIRs then might cut the termini. The host repair machinery fills in the gaps that result from the first steps of integration. At the donor site a more or less perfect target site duplication is left as a footprint.

analyzed member of the CACTA superfamily. It was independently discovered by Peterson in 1953 who termed the element *Enhancer* (*En*) and McClintock in 1954 who called this transposon *Suppressor of mutation* (*Spm*). *En*/*Spm* as depicted in **Figure 5** is the autonomous master element that carries all the necessary *cis*- and *trans*-determinants for transposition. As in *Ac*; the central protein-encoding part of the 8287 bp element is dispensable for *cis*-acting transposition signals. The *cis*-determinants for transposition consist of perfect 13 bp-long TIRs and 12 bp subterminal sequence motifs. Hence, the basic organization of *Ac* and *En*/*Spm* is similar. However, on *En*/*Spm* integration, only 3 bp-long target site duplications are generated which are characteristic for all the members of this family. In further contrast to *Ac*, *En*/*Spm* encodes for two proteins. A primary transcript is subjected to alternative splicing. The two mRNAs of 2.5 kb and 6 kb are transcribed into two proteins, the 67 kDa TNPA and the 132 kDa TNPD

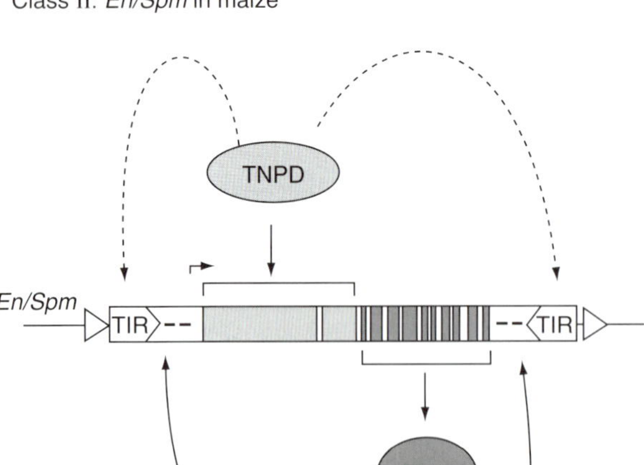

Figure 5 The ends of the *En/Spm* element are marked by 13 bp TIRs. Target site duplications in the host DNA are indicated by triangles. The element encodes for the TNPA and the TNPD (transposon proteins A and D, respectively). Transcription start site is marked by the small arrow. The corresponding mRNAs are generated by alternative splicing. TNPA recognizes the subterminal motifs depicted as black bars. Although being essential for transposition, specific DNA-binding has not yet been demonstrated for TNPD. A hypothetical transposon TNPD recognizes the TIRs and cuts just outside the termini.

(transposon proteins A and D). Both functions are necessary for transposition to occur. The quantitatively dominating TNPA binds to the subterminal 12 bp motifs and presumably works as a glue between the termini and ensures the correct alignment of the TIRs. Whereas TNPA recognizes subterminal motifs, the 132 kDa TNPD presumably binds the 13 bp TIRs that are not recognized by TNPA. Both proteins are required for transposition as shown in tobacco as a heterologous host.

In *Antirrhinum*, a close relative of maize, *En/Spm* is the 15164 bp-long *Tam1* element that is bordered by 13 bp TIRs and that also generates 3 bp target site duplications. *Tam1* encodes for two putative proteins TNP1 and TNP2 which, in all probability, constitute the functional equivalents of TNPA and TNPD.

Mutator Elements: An Unusual Class II Family

Mutator (*Mu*) elements were originally identified by their property of an abnormally high mutation rate, hence the name. Although the independently isolated autonomous elements seem to be virtually unvarying and hence form just one subfamily, the nonautonomous elements have been classified into six subfamilies because of often totally unrelated internal sequences between the families and strong homologies within them (**Figure 6**). Thus, in comparison with other Class II elements like *Ac/Ds* or *En/Spm*, members of the Mutator family display an unusual amount of sequence diversity. However, similar TIRs border autonomous and nonautonomous elements alike. To date, large numbers of *Mu* elements have been isolated and sequenced.

The autonomous element is termed *MuDR*. It is 4942 bp long and is bordered by *c.* 220 bp-long TIRs containing the *Mu* promoters. The first 180 bp of the TIRs of *MuDr* elements show 99% identity. Two major transcripts have been identified, a 2.8 kb *mudrA* and a 1.0 kb *mudrB* product. Deletions in the *mudrA*-coding region abolish *Mu* activity. The 823 amino acid MUR-A product contains a sequence motif which is also found in transposase proteins from nine prokaryotic IS (insection sequence) elements. Hence MUR-A is supposed to be the transposase of this family. A clear function for the *mudrB* transcript has not been defined up to now. For *mudrB* alternative splicing has been reported: either two or three small introns are spliced out. The longer mRNA encodes a 207-amino acid protein and the shorter mRNA a polypetide consisting of 167 amino acids (**Figure 6**).

Besides their sequence diversity, another interesting feature of *Mu* elements is their unusual transposition behavior. In contrast to other Class II elements, which transpose only via a cut-and-paste mechanism, two different types of mechanisms have to be postulated for *Mu*. Although the number of elements can increase threefold per generation in low copy stocks, the germinal reversion rate is extremely low. Yet, new *Mu* insertions can rise to a frequency of 10–15 copies per gamete per generation. A simple cut-and-paste mechanism does not seem to be sufficient to explain this transposition behavior. Furthermore, somatic revertant sectors show characteristic footprints resembling those of other Class II TEs. *Mutator* seems to use two different types of transposition depending on tissue and developmental stage of the host cell. The somatic effects can be explained by a cut-and-paste mechanism (as shown for *Ac* in **Figure 4**). However, during development of the germ cells, *Mu* transposition was found to be duplicative: The double-strand breaks at the exision site trigger a gene conversion-like gap repair, which may start either from the sister chromatid or from the homologous chromosome as a template. Such a transposition event finally leads to a duplication of the element. Despite the fact that the exact function for MUR-B is still unknown, *in situ* immunolocalization of germinal tissues has revealed an accumulation pattern of MUR-B that would be consistent with a possible MUR-B function in switching between the two alternative pathways: a gene-conversion-like gap repair in germinal cells and a fill-in/religation repair belonging to the cut-and-paste mechanism in somatic tissues (for a review see Kunze *et al.*, 1997).

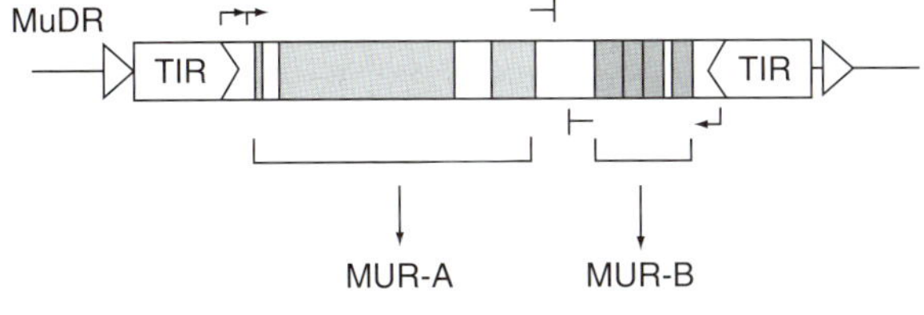

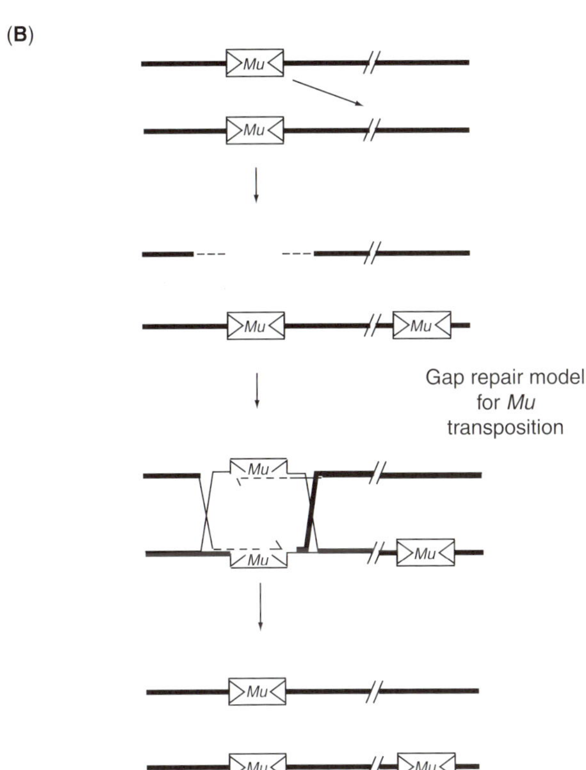

Figure 6 (A) The ends of the autonomous MuDR element are marked by 220 bp TIRs. The host DNA exhibits 9 bp target side duplications as indicated by triangles. Transcription start sides are marked by the arrows above the element. Two convergent transcripts encode for MUR-A (the presumed transposase) MUR-B is possibly involved in switching between the two types of transposition mechanisms as described in the text. (B) The gap repair model for *Mu* transposition ultimately leads to sequence duplication of the *Mu* element. The double-stranded gap left behind after transposition is widened at the donor site by exonuclease activity. A homologous genomic sequence serves as a template for the gap repair. After completion of the repair process, the element has been duplicated.

Regulation of Transposition

Any transposition event has the potential to be more or less harmful for the genome affected. Hence, a limitation of transposition frequency might be advantageous for both, the hosts and the TEs. Yet, if there is any benefit by downregulation of TE activities, how would such an advantage concur with the enormous numbers of retroelements mentioned above indicating an impressive tolerance and plasticity of the plant genome for repetitive DNA?

One possibility is that benefits from downregulation may not necessarily limit the overall number of elements but only the number of transposition events in time, compensating for potential damage by distributing rare events over long periods. As mentioned above, plant genomes contain vast amounts of retroelements which occur as diverse but related sequences. However, active elements are either absent or their number is extremely limited. The transposition rate of retrotransposons is generally controlled at the level of transcription. Normally transcription rate is low and transposition has been detected for just a few plant retroelements.

However, some methods of regulation have already been demonstrated experimentally: generally, in *Tnt1* (belonging to the *Ty1/copia* group) transcription is low, but can be stimulated by microbial induction and abiotic factors like wounding and freezing. The promoter of the *Tnt1* element in tobacco is located in the 5′ LTR and the corresponding regulatory sequences could be identified in its U3 region. Activation by genomic stress has likewise been established for Class II elements such as *Ac* and *En/Spm*, which also display a tight regulation of the transposition rate.

Moreover, the *Ac/Ds* and *En/Spm* TE families can epigenetically be *in*activated. This effect is found to be associated with cytosine methylation of the elements upstream of their promoters or regions nearby. Inactive *Ac* elements are hypermethylated; however, they can be activated *in trans* by an active *Ac* element which supplies the TPase. Reactivation seems to be correlated with partial demethylation and stimulation of transcription. Similar effects have been found for *En/Spm*. Here different states of inactivation can be observed. Cryptic inactive elements are not transiently reactivated by an active *En/Spm* in the genome, yet silent inactive elements are activated by the TNPA protein supplied *in trans*. The protein binds in a methylation-sensitive manner to the promoter of *En/Spm*. Like *Ac*, *En/Spm* encodes for a protein, TNP A, which maintains the unmethylated state (Becker and Kunze, 1996).

On the other hand, reactivation can be induced by gamma irradiation or chromosome breakage, leading to mass activations of transposable elements comparable to the phenomenon of hybrid dysgenesis by the *Drosophila P* elements.

Apart from such stress activations there are other factors influencing time and role of transposition. Transposition of *Ac*, for instance, seems to be dependent on the number of elements in the genome. A negative dosage effect can be observed for certain genetic backgrounds: The more autonomous elements occur in the genome, the less frequent and more delayed transposition events are in development. Thus, an increasing number of elements can be correlated with a reduction of transpositon.

Transposons as Tools: Transposon Tagging and Reverse Genetics

The classical transposon tagging strategy has been widely used in bacteria and plants (for a review in plants, see Kunze *et al.*, 1997). A quick isolation of an inactivated gene is guaranteed by the use of an inserted transposon sequence as a tag to isolate the neighboring host sequences. This classical strategy has been extended into the era of functional genomics by more quantitative approaches. The aim changed from the isolation of a specific gene to quantitative mutagenesis of the genome. Saturation mutagenesis by either TEs or T-DNA transferred from *Agrobacterium tumefaciens* to the plant has successfully been applied for this purpose.

Figure 7 (See Plate 42) The excision of the *Ac* transposable element at the *P* locus of *Zea mays* resulting in a variegated colour pattern in the pericarp of the kernels.

Additionally, quantitative mutagenesis of the genome allows reverse genetics. A gene of interest whose function is yet unknown can be investigated when an individual carrying a suitable insertion has been identified. The use of PCR primers against the TE and the gene of interest allows the identification of such insertion mutants. The PCR screens are usually performed by analyzing the DNA pools of many plants. Hence, quick isolation of a specific insertion mutant is possible. In the model plant *Arabidopsis thaliana* the well-analyzed *En/Spm* system has been used successfully for this purpose. Reverse genetic strategies based on endogenous TEs have also been applied efficaciously in *Antirrhinum*, *Petunia*, and maize.

Quantitative Contributions to the Genetic Material

The vast number of retrotransposons and their impressive contribution to the total amount of nuclear DNA in plants reflects the astonishing tolerance and plasticity of the plant genome. Whereas the copy number of Class II elements is usually limited to a maximum of several hundred TEs in maize, thousands of different Class I element families are estimated to account for 70–85% of its nuclear DNA. In comparison *Saccharomyces cerevisiae* contains only five LTR retrotransposon families, which contribute 3% of the genome. In *Arabidopsis* there are several hundred families of LTR and non-LTR elements accounting for 14% of its genome. In contrast, in mammals LINEs and SINEs are the dominating groups comprising up to 100 000 copies and 35% of the nuclear genome, whereas elements of the *Ty1/copia* group are either not present at all or occur only in small numbers. Plants again possess elements belonging to the *Ty1-copia* and *Ty3-gypsy* groups as well as LINE-like and SINE-like elements. Ordinarily, plant TEs surpass those of the other kingdoms in diversity and abundance (Kumar and Bennetzen, 1999).

Transposable Elements and DNA Diversity

Owing to their peculiarity to insert at new genomic sites, Class I and Class II TEs have the capacity to generate mutations and to constitute homologous sequences at nonallelic loci. Both are known to be essential for the generation of genetic diversity. Homologous recombination between such nonallelic copies leads to chromosomal rearrangements which comprise deletions, duplications, inversions, and translocations. The genetic diversity is further increased by alterations in gene expression, for example, when elements insert into regulatory regions (usually leading either to reductions in or to losses of function of a pathway). Unstable mutations observed for Class II elements reflect insertions and excisions at the DNA level. Additional sequence diversity is generated upon excision (also found for the P elements of *Drosophila*): At the donor site a more or less precise target site duplication is left behind. Whereas target site duplications lead to additional bases and/or amino acids, the excision mechanism often deletes or adds bases and thus creates imprecise footprints (Saedler and Nevers, 1985; Kunze *et al.*, 1997). Multiple different TE footprints result in altered host sequences: over 90% of more than 800 analyzed *Ds* excision products of maize *Waxy* alleles revealed mutant sequences not restoring the wild-type. However, the sequence deviations proved to be 'surprisingly nonrandom' (Scott *et al.*, 1996): Depending on the allele and the insertion site, some 37–88% carry a predominant footprint and even the less prevalent footprints are often similar to the prevalent ones. Moreover, in 1–6% of the excisions, no footprint was formed. Only the rest in between seems to consist of random sequences.

Interestingly, sequence deviations leading to the addition of one to three amino acids to a Waxy protein did not abolish protein function: although the new alleles encode proteins with reduced enzymatic activities, wild-type function is still largely maintained. The question whether and to what extent such results detected in cultivated plants (whose existence depend on human care and interest) can be extrapolated to the origin of plant species in the wild leads us to our next TE topic.

Transposons and Evolution

The present discussion about transposons and evolution is one of the most interesting and undoubtedly also the most controversial topic(s) of TE research. Seeing that often about one third (and sometimes up to 70–85%) of the total nuclear DNA amount can consist of transposon sequences in plants (not to mention the similar situation in animals and humans), the question is, of course, what (if any) is the biological function of TEs?

In contrast to bacteria, where a range of clear-cut functions has been detected and described in detail (see entries on related topics), it is not possible to make a similar sweeping statement for the TEs of eukaryotes. In the latter the search for a general biological function has, in fact, until now resulted in no more than several alluring, yet contradictary hypotheses without much hard scientific evidence behind them (for reviews see Kunze *et al.*, 1997, Lönnig and Saedler, 1997).

The two main hypotheses on the existence of TEs may be presented as follows:

1. Transposons as evolutionary tools. TEs are thought to have a general biological function by essentially contributing to the origin of eukaryotic species and (perhaps also) higher systematic categories (Nevers *et al.*, 1986; McClintock, 1987). At present, the reason for this hypothesis consists mainly of the fact that TEs have an enormous mutagenic potential: As mentioned above, nearly the whole range of mutations is covered by TE activities. Even the generation of intron-like sequences and ectopic expression of genes belongs to the possible mutagenic effects of plant TEs. Moreover, TEs can generate mutations at such accelerated rates that no other known natural mutagenic agency can compete with transposons. So the basic inference is: DNA variation is necessary in evolution. TEs produce DNA variation. Thus TEs are important in evolution. In sum, transposon existence and spread is due to natural selection of superior phenotypes caused by TE-induced advantageous mutations.
2. Transposons as parasites. The second hypothesis was first formulated by Doolittle and Sapienza (1980) and Orgel and Crick (1980): TEs have no general biological function. They exist only for their own sake and not for the organism's. As transposons represent at least a slight energetic burden for the organisms harboring them and since many TE activities are clearly destructive, they essentially consist of 'sefish DNA' which may also be labeled 'parasitic DNA' or even the 'ultimate parasite' (Orgel and Crick, 1980). Their only 'function' is survival in their environments where they can spread as long as they do not become too harmful for their hosts. TEs can replicate and spread because they happen to be in surroundings in which DNA replication is part and parcel of the regular key events of cell division (for several further points see below, as well as the review by Kunze *et al.*, 1997). However, the hypothesis does not exclude occasional contributions to the origin of useful variation in nature.

Concerning hypothesis (1), it may be conceded that TEs have a place in the microevolution of wild populations (which has yet to be tested) (Lönnig and Saedler, 1997). We must, nevertheless, distinguish between 'necessary' and 'sufficient' causes for explaining a phenomenon as the origin of the plant (and animal) world: is the origin of species and higher systematic categories fully explained by natural selection of TE-induced mutations as well as by the DNA sequence variation produced by the rest of the mutation processes?

It was no less a person than Charles Darwin himself who provided the following sufficiency test for his theory (Darwin, 1859): "If it could be demonstrated that any complex organ existed, which could not possibly have been formed by numerous, successive, slight modifications, my theory would absolutely break down." However, Darwin stated that he could "not find out such a case." Yet, the question whether the situation has changed in the interim of some 150 years of biological research is answered in the affirmative by several biologists. Michael J. Behe (Behe, 1996, 2000) has refined Darwin's statement by introducing and defining his concept of "irreducibly complex systems," specifying:

> By *irreducibly complex* I mean a single system composed of several well-matched, interacting parts that contribute to the basic function, wherein the removal of any one of the parts causes the system to effectively cease functioning.

Behe and like-minded researchers are convinced that they have detected several such systems at the biochemical level (origin of the cilium, the flagellum, blood clotting, vesicular transport, and further examples, see Behe, 1996, 2000). Lönnig (1998), suggests that irreducible complexity may also be found at the anatomical level (in combination with biochemical systems) in angiosperms as, for instance, in the trap mechanism(s) of *Utricularia* and several other carnivorous plants.

Supposing that such systems exist, could the non-Darwinian (more or less saltational) view of TEs as evolutionary tools as favored by many TE researchers (for a review, see Kunze *et al.,* 1997) help solve the problem in a naturalistic way?

For a balanced answer, we have to consider briefly what protagonists of hypothesis (2), i.e., transposons as 'parasites,' can further object to as a general evolutionary role of TEs:

1. Since "low mutation rates are necessary for life as we know it" (Alberts *et al.,* 1994, p. 243, with a review of the evidence), the unusually frequent movements of activated TEs can be life-threatening for the natural populations affected by them. High mutation rates result in 'error catastrophe' leading to the extinction of a population (for the details of population genetics, see ReMine, 1993).
2. Apart from very few exceptions, almost all TE insertions into coding sequences cause losses of gene functions.
3. Even the footprints leading to an addition of one to three amino acids of a protein have to be mainly classified as regressive evolution, i.e., protein function is reduced (for examples, see Lönnig and Saedler, 1997).
4. The hierarchy of gene functions has to be considered. Transposon activities in genes coding for histones, ubiquitin, actin, many tRNAs, and other ultraconservative and conservative parts of the genome have to be classified as nearly always 'parasitic.'
5. Yet, even in the plant genome's more redundant parts (flower color, plant height, form of leaf margins, etc.), TE activities can be disadvantageous. For instance, loss of the *Nivea* gene function (one of the basic functions in the anthocyanin pathway) affects not only flower color, but also lowers resistance to stresses, such as UV light, cold, pathogens, and mechanical damage. (As for the problems of gene duplications and exon shuffling by TEs, as well as a possible synthesis between different views, see Kunze *et al.,* 1997.)
6. The probability that activated TEs will simultaneously generate several independent, advantageous mutations in different parts of the genome, saltationally resulting in irreducibly complex structures or organs appears to be very low.
7. Besides the problem of irreducible complexity and pertaining more generally to the origin of species, the following points should be considered. The *Lilium* case mentioned above already hints at the fact that there is neither a correlation between the number and kinds of TEs and the number of species and genera formations within plant families, nor is there a strict connection between overall species (and higher systematic category) complexity and the DNA amount (C-value paradox).

Concerning the question whether TE activities could solve the origin of irreducibly complex systems and organs in particular and the generation of species in general, you are invited to judge for yourself whether the facts and arguments presented so far suggest a direction to the answer(s) of the problems raised (for further reading see Starlinger, 1993; Kunze *et al.,* 1997; Lönnig, 2001).

References

Alberts B, Bray D, Lewis J *et al.* (1994) *Molecular Biology of the Cell*, 3rd edn. New York: Garland Publishing.

Becker H-A and Kunze R (1996) Maize nuclear protein binding sites in the subterminal regions of transposable element *Activator. Molecular and General Genetics* 251: 428–435.

Becker H-A and Kunze R (1997) Maize *Activator* transposase has a bipartite DNA binding domain that recognizes subterminal motifs and the terminal inverted repeats. *Molecular and General Genetics* 254: 219–230.

Behe MJ (1996) *Darwin's Black Box: The Biochemical Challenge to Evolution*. New York: Free Press.

Behe MJ (2000) Self-organization and irreducibly complex systems: A reply to Shanks and Joplin. *Philosophy of Science* 67: 155–162. (See also "Behe responds to critics": http: // www.discovery.org/crsc/fellows/MichaelBehe/index.html/. Here the reader finds several well thought-out contributions to the recent controversy about 'irreducibly complex structures.')

Darwin C (1859) *On the Origin of Species by Means of Natural Selection or the Preservation of Favoured Races in the Struggle for Life*. London: John Murray. (1967: Everyman's Library No. 811, reprint of the 6th edn of 1872.)

Doolittle WF and Sapienza C (1980) Selfish genes, the phenotype paradigm and genome evolution. *Nature* 284: 601–603.

Kumar A and Bennetzen JL (1999) Plant retrotransposons. *Annual Review of Genetics* 33: 479–532.

Kunze R, Saedler H and Lönnig W-E (1997) Plant transposable elements. *Advances in Botanical Research* 27: 331–470.

Lönnig, W-E (1998) *Zehn Paradebeispiele gegen Zufalls-Evolution*, 2nd edn. Köln: Naturwissenschaftlicher Verlag Köln.

Lönnig WE (2001) Natural selection. In: Craighead WE and Nemeroff CB (eds) *The Corsini Encyclopedia of Psychology and Behavioral Sciences*, (3rd edn) vol. 3, pp. 1008–1016. New York: John Wiley.

Lönnig W-E and Saedler H (1997) Plant transposons: contributors to evolution? *Gene* 205: 245–253.

McClintock B (1987) (Anthology of all her important papers) In: Moore JA (ed.) *The Discovery and Characterization of Transposable Elements: Genes, Cells and Organisms*. New York: Garland Publishing.

Nevers P, Shepherd NS and Saedler H (1986) Plant transposable elements. *Advances in Botanical Research* 12: 103–203.

Orgel LE and Crick FHC (1980) Selfish DNA: the ultimate parasite. *Nature* 284: 604–607.

ReMine WJ (1993) *The Biotic Message*. St Paul, MN: St Paul Science Publishers.

Saedler H and Nevers P (1985) Transposition in plants: a molecular model. *EMBO Journal* 4: 585–590.

Schmidt T (1999) LINEs, SINEs and repetitive DNA: non-LTR retrotransposons in plant genomes. *Plant Molecular Biology* 40: 903–910.

Schwarz-Sommer Z, Leclerq L, Göbel E and Saedler H (1987) *Cin*4, an insert altering the structure of the A1 gene in *Zea mays*, exhibits properties of nonviral retrotransposons. *EMBO Journal* 6: 3873–3880.

Scott L, Lafoe D and Weil CF (1996) Ajacent sequences influence DNA repair accompanying transposon excision in maize. *Genetics* 142: 237–246.

Starlinger P (1993) What do we still need to know about transposable elements? *Gene* 135: 251–255.

***See also:* McClintock, Barbara; Retrotransposons; Transposable Elements; Transposons as Tools**

Transposase

N D F Grindley

doi: 10.1006/rwgn.2001.1313

Transposase is the transposon-encoded protein that is responsible the element's transposition. The transposase recognizes and binds to both ends of its cognate transposon, brings the two ends together in a synaptic complex, and cuts the DNA at the two 3′ ends (and in some cases at the 5′ ends as well). In a single combined cleavage-ligation step, the transposase then inserts the 3′ transposon ends into target DNA.

***See also:* Insertion Sequence; P Elements; Transposable Elements; Transposable Elements in Plants; Transposons as Tools**

Transposon Excision

D B Haniford

doi: 10.1006/rwgn.2001.0436

Transposons are mobile genetic elements. Their movement within and between DNA moleculas can result in a wide range of genome rearrangements including insertions, deletions, inversions, duplications, replicon fusions, and probably chromosomal translocations. In addition they are important elements in the spread of antibiotic resistance genes in bacteria. They are extremely abundant, being found in almost all organisms. They can constitute a significant percentage of the total genomic DNA of a species.

A large number of different types of transposons have been identified. Of these a fairly large subset transpose by a mechanism in which the transposon is simply excised from the flanking 'donor' DNA by a pair of double-strand breaks, one at each transposon end. The excised transposon intermediate is then inserted into a new site. The new location usually has no sequence relationship to the transposon or the donor site. Excision and insertion steps are catalyzed by one or more transposon-encoded proteins called transposases.

Three different strategies for generating the excised transposon intermediate have been documented. Two separate transposase proteins can be involved in making the double strand break at each end. For example, in Tn7 (a bacterial transposon) the proteins TnsB and TnsA each are responsible for cleaving a different DNA strand at and just outside of the transposon end, respectively.

However, other transposons such as the bacterial transposons Tn10 and Tn5 encode only a single transposase protein with a single active site. The excision reaction takes place within a nucleoprotein complex in which there are only two molecules of the transposase present. In this situation excision takes place by a mechanism in which a transposon end hairpin intermediate is formed. Here, the single active site of one transposase monomer first introduces a nick to expose a 3′OH group at the transposon terminus. Then the same active site is used to catalyze the joining of this 3′OH terminus to a phosphate group on the opposite strand of the same end. This generates the transposon end hairpin and severs the final connection between the transposon and the flanking donor DNA in one chemical step. The hairpin end must then be cleaved to reexpose the 3′OH group for joining to the target DNA in the final step. It is likely that a number of plant transposons also use this mechanism of excision.

In the third excision mechanism the transposase first makes a nick at only one transposon end. Then the exposed 3′OH terminus is joined to a phosphate group on the same DNA strand but just outside of the second end. This generates an excision intermediate in which the cleaved strand forms a covalently closed circle and the two transposon ends are held together by a single-strand bridge. Double-strand circles are generated from this by an as yet undefined mechanism. Then the transposase introduces a pair of single-strand cleavages at the abutting transposon ends. This opens up the circle generating a linear form of the excised transposon that can be directly inserted into a new target site. This mechanism is used by the bacterial transposon IS911 and other members of the IS3 family. It is not understood what constraints are responsible for the evolution of this seemingly complex excision pathway.

Interestingly, the hairpin mechanism for transposon excision is also used in the formation of double-strand DNA breaks in V(D)J recombination. This is the process whereby antigen receptor genes are pieced together from separate coding segments in developing T and B cells in the immune systems of jawed vertebrates. Furthermore, the proteins that catalyze this double-strand break reaction, RAG1 and RAG2, have been shown to catalyze DNA transposition reactions *in vitro* via the hairpin mechanism. The use of a common mechanism for double-strand break formation in V(D)J recombination and bacterial DNA transposition suggests that the V(D)J recombination system evolved from an ancient bacterial transposon.

It will be interesting to see if other mechanisms for carrying out transposon excision are used and what constraints favor these mechanisms.

Further Reading

Kennedy AK, Guhathakurta A, Kleckner N and Haniford DB (1998) Tn10 transposition via a DNA hairpin intermediate. *Cell* 95: 125–134.

McBlane JF, van Gent DC, Ramsden DA *et al.* (1995) Cleavage at a V(D) J recombination signal requires only RAG1 and RAG2 proteins and occurs in two steps. *Cell* 83: 387–395.

Sarnovsky RJ, May EW and Craig NL (1996) The Tn7 transposase is a heteromeric complex in which DNA breakage and joining activities are distributed between different gene products. *EMBO Journal* 15: 6348–6361.

Ton-Hoang B, Betermier M, Polard P and Chandler M (1997) Assembly of a strong promoter following IS911 circularization and the role of circles in transposition. *EMBO Journal* 16: 3357–3371.

Turlan C and Chandler M (2000) Playing second fiddle: second-strand processing and liberation of transposable elements from donor DNA. *Trends in Microbiology* 8: 268–274.

***See also:* Insertion Sequence; Transposable Elements**

Transposons as Tools

S A des Etages, A Kumar, and M Snyder

doi: 10.1006/rwgn.2001.1523

Transposons are ubiquitous mobile DNA elements that can relocate within the genome of their hosts. These elements generally have either inverted or direct repeats at their termini and encode enzymes important for transposition. Movement occurs either by duplication and insertion of the new copy into the genome (replicative) or by excision of an existing copy and insertion at a new site (nonreplicative). Insertion of the transposon causes duplication of the target site and therefore these elements are flanked by direct repeats. Retrotransposons are a class of transposons that contain long terminal repeats and use an RNA intermediate for transposition. These elements encode reverse transcriptase and integrase, enzymes required for their duplication and transposition. Early observations by McClintock and others indicated that transposons could mediate transfer of genetic material within and between species. The application of transposons as molecular biology tools did not begin until the 1970s, coincident with our increased understanding of

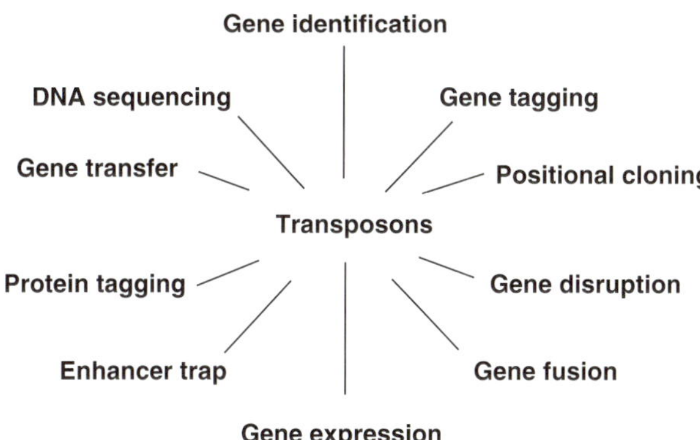

Figure 1 Common uses for transposons.

transposon structure and movement and the ability to manipulate these elements.

Transposons as Molecular Biology Tools

Transposable elements have emerged as versatile, powerful, and informative biological tools since their discovery 40 years ago. Current applications of transposon technology include gene identification and tagging, positional cloning, gene disruption, generating reporter fusions and enhancer traps, DNA sequencing, altering protein localization or gene expression, introducing protein tags, and introducing new genes (**Figure 1**). Prior to their use as tools, transposons are generally modified in order to control their replication and to allow detection of their presence. Typically, the endogenous transposase is deleted from the transposon and expressed in *trans* when movement is required. If not already present, a suitable selectable marker gene (usually encoding antibiotic resistance or a metabolic enzyme) is inserted to allow identification of transposon-bearing organisms. Further modifications are made depending on the specific transposon and the goals of the application.

The methods of introduction of transposons into their target sites have evolved considerably over the years. Traditionally, *in vivo* transposition was achieved by crossing strains that lack the transposon to those that contain the transposon and a functional transposase. A more commonly used method is to transform a vector containing the modified transposon into the cells of interest and induce expression of the transposase gene in *trans* to achieve transposition. While random insertions occur, the method is not ideal if the organism of choice is not easily transformed, lacks a manipulable transposon system, or is not compatible with the expressed transposase. These problems can be overcome using *in vitro* transposition. In the presence of the appropriate transposase, the transposon (e.g., Tn5, Mu) randomly inserts into the target DNA which is subsequently transformed into the cells. *In vitro* transposition allows the creation of large libraries of randomly tagged clones that can simplify DNA sequencing, gene mapping, and mutagenesis. Tools such as genetic footprinting were developed as a result of the ability to quickly and economically generate large numbers of transposon mutagenized clones. Most recently, Tn5 transposition complexes are formed *in vitro* and introduced into cells via electroporation. *In vitro* transposition will allow researchers to apply transposon technology to answer questions in a larger number of organisms.

Gene Identification and Tagging

Transposons are used to identify genes of interest in a wide variety of organisms that include viruses, bacteria, fungi, flies, plants, worms, and mice. Historically, the *Drosophila melanogaster* transposon known as the P element has been used to isolate genes of interest to researchers. When female flies from laboratory strains are crossed to male flies from natural populations, the resulting progeny exhibit a variety of mutant phenotypes. This phenomenon, known as hybrid dysgenesis, is the result of transposition of the P element introduced into the cross by the male flies. Progeny flies exhibiting a phenotype of interest are selected, and DNA from these flies is isolated. Since the linkage of the P element to the mutant gene serves as a tag, radioactively labeled P element sequence is used as a probe to identify the DNA fragments that contain a P element. The DNA encoding the gene of interest is present in a subset of this population and once identified, is cloned and sequenced.

Today, P elements are used to identify *Drosophila* genes by inducing movement within the germline and screening for the phenotype of interest. These elements have been modified to encode markers that affect eye color to allow identification of flies that contain a P element insertion. Upon mutagenesis with these modified elements, flies that exhibit the phenotype of interest as well as the eye color encoded by the P element are selected. The sequence of the DNA flanking the insertion element can be identified by inverse PCR. If the P element contains a bacterial selectable marker and origin of replication, the flanking region can be cloned by plasmid rescue in *Escherichia coli*.

Transposons are used to circumvent the challenges associated with identifying genes in large eukaryotic genomes. Random transposon mutagenesis in intron-rich genomes does not yield many gene mutations since the transposons often insert into introns and are spliced out of the mRNA. Springer and colleagues developed a *Dissociation* (*Ds*) transposon-based tool for gene trapping in *Arabidopsis thalania* that bypasses this problem. The transposon contains a small intron, three splice site acceptors, a reporter gene, and a splice donor site in the order listed. Insertion of this transposon into an intron leads to alternative splicing and expression of the reporter gene. The presence of the reporter gene also disrupts the coding sequence of the tagged gene and might therefore generate a phenotype that allows identification of strains bearing mutations in the gene of interest.

Transposons are used to identify genes that have tissue-specific or developmentally induced expression patterns. Transposons designed for this purpose usually contain a reporter gene that is fused to a weak promoter (see 'reporter fusions and gene expression'). The reporter gene is not expressed unless there is an enhancer nearby. This tool allows identification of genes on the basis of their expression patterns. This approach has been used in worms, flies, and plants to identify genes that are tissue or developmentally regulated.

Markers for Gene Mapping and Cloning

Transposons are used as markers for mapping and positional cloning of genes in maize and *Arabidopsis thalania*. A collection of *Arabidopsis* plants that contain *Ds* transposon insertions encoding the kanamycin resistance gene is being constructed. After 1000–2000 genomic transposon insertions have been mapped, they will become valuable tools for fine-scale mapping and cloning of gene mutations. There are currently enough molecular markers in *Arabidopsis* to allow preliminary mapping of a gene mutation within 10 to 20 cM. Transposon tagging will generate approximately 20 mapped transposons within a 10 to 20 cM region. This will facilitate further mapping since tagged plant lines will have several transposons that bear dominant genetic markers throughout the genome. The strain containing the mutation to be mapped can be crossed to the appropriate tagged lines and progeny in which the phenotype of interest is linked to the dominant marker selected. The region containing the mutation can be mapped to the sequence between two mapped transposons on the basis of linkage. Since a recombination event will have occurred between the mutation of interest and the transposon, the recombination breakpoint will facilitate cloning of the gene of interest.

Gene Mutagenesis

Once a gene has been identified, researchers can use transposons to produce disruptions or small mutations within its coding sequence. Transposon insertions can be generated in the gene by inducing transposon movement through genetic crosses or other methods and selecting or screening for the mutation of interest. If the organism of choice lacks a manipulable transposon system, *in vitro* transcription or electroporation can be used to introduce the transposon. Disruptions, insertions, and small deletions within the coding sequence are the most common mutagenic effects. Disruptive alleles are the result of transposon insertion while deletions and small insertions are caused by imprecise excision of nonreplicative transposons.

Disruption of a gene of interest is made easier if there is a transposon in a nearby chromosomal location. Many eukaryotic transposons such as P elements in *Drosophila*, and *Ac/Ds* elements in maize and *Arabidopsis* tend to reinsert very close (within 200 kb) to the initial site when mobilized. Thus one can begin with a strain in which a transposon is located near the gene of interest, induce transposase expression and generate a collection of mutations a subset of which will contain disruptions in the gene of interest. Alternatively, if the gene has been cloned into a stable extrachromosomal DNA element, it can be mutagenized with transposons in bacterial cells and reintroduced into its host by transformation or gene transfer. The advantage of many transposon-based disruption alleles is that induction of a second round of transposon movement causes reversion to the wild-type allele. This feature is often used to verify that the phenotype observed is the result of the transposon insertion.

Transposons are used to generate multiple insertional alleles in a gene of interest. Depending on the location of a transposon insertion within a gene, it can cause complete or partial inactivation of the gene, or alter its activity or the conditions under which it is active. Generation of a large number of insertional alleles in a gene of interest is useful in dissecting gene function since the effect of truncating a protein at different positions can then be assessed. In addition, many nonreplicative transposons such as fly P elements, plant *Ac/Ds*, yeast Ty, and worm Tc1 excise imprecisely and cause deletions of various sizes and insertions. Deletions occur when flanking DNA is removed during excision and insertions occur when transposon sequences are not completely excised. If there is a transposon near a gene of interest, one can generate many different mutations within the gene and analyze the phenotypes of the resulting strains.

If the phenotype of a mutation in the gene of interest is not known, mutagenized strains can be screened to identify those that contain mutations in the gene of interest. For example, in *Caenorhabditis elegans* large mutant libraries that contain Tc1 mutagenized genes have been generated. Tc1 is a ubiquitous transposon whose excision usually leads to deletions. After inducing Tc1 excision, the mutagenized worms are subcultured and allowed to produce progeny. Every subculture will contain siblings that share the same mutation. Genomic DNAs from the progeny of Tc1 mutagenized strains are pooled and screened for mutations in the gene of interest by PCR. Since Tc1 excision usually leads to deletions, the fragment produced by the mutant allele would be smaller than that made by the wild-type allele and favored in PCR reactions. Once a positive result is obtained, genomic DNA derived from each subculture is recovered and screened to identify which subculture contains the worms of interest.

Transposons are used to mutagenize genes of interest in a wide variety of organisms that include viruses, bacteria, fungi, plants, worms, and mice. Cloned genomic DNA or cDNA from many organisms can be mutagenized in bacterial cells and reintroduced to assess the phenotypic effects of the resulting mutations. Endogenous transposons such as P elements in flies, Ty in yeast, Ac/*Ds* in plants, or Tc1 in worms can also be used to mutagenize their hosts.

Reporter Fusions and Gene Expression

Transposons facilitate gene expression studies by generating reporter gene fusions. Reporter genes encode proteins (for example, the β-galactosidase gene (*lacZ*)) whose presence can be easily detected using colorimetric or fluorescence assays. Transposon insertions that lead to production of reporter proteins are tremendously useful in determining tissue-specific expression patterns from multicellular organisms and the conditions under which genes are expressed in unicellular organisms.

Reporter fusions can be generated in two ways: the first method involves expression of a fusion protein that contains sequences from the endogenous protein as well as the reporter, while the second involves expression of an intact reporter protein from endogenous regulatory sequences. Production of a fusion protein occurs when the reporter gene is missing a start codon (ATG) and therefore cannot be expressed unless it is downstream of and in frame with the start codon for the endogenous gene. A transposon is modified so that a reporter gene that lacks a start codon is adjacent to one end. Insertion of this transposon into the coding region of a gene in the appropriate reading frame results in expression of a fusion protein. Expression of the reporter gene would then correlate with the expression of flanking genomic DNA, enabling the establishment of an expression profile for a given gene. This method has been used to study gene expression patterns in yeast, plants, and flies.

Transposons can also be used for enhancer trapping. In this application, the transposon is modified so that it contains a weak promoter and an intact reporter gene whose expression is dependent on insertion near a chromosomal enhancer. The expression of the reporter gene reflects the expression pattern conferred by the enhancer. The transposon does not have to insert within the coding sequence of a gene to be expressed and does not generate a fusion protein (**Figure 2**). The approach is identical to that used to identify genes on the basis of their expression patterns (see above). It helps researchers to identify regulatory elements and determine the conditions under which they are active, and is valuable in identifying enhancers involved in tissue and developmentally regulated gene expression.

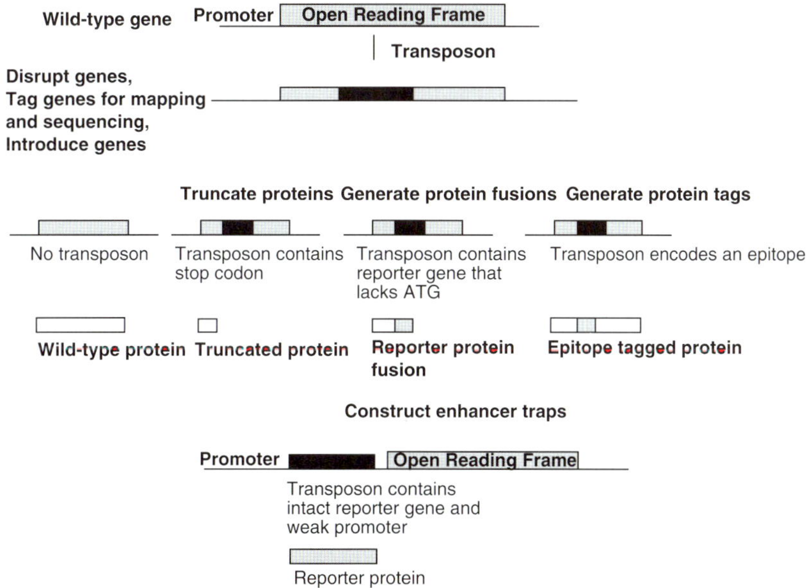

Figure 2 Multiple applications of transposons: transposon generation of gene disruptions, gene tags, truncated proteins, reporter fusions, protein tags and enhancer traps. Gray box, wild-type gene; black box, transposon; open box, wild-type protein; hatched box, reporter protein; lined box, protein tag.

A popular modification of the traditional enhancer trap experiment involves cloning a promoter into a transposon and then using it to cause inappropriate expression of nearby genes. Expression pattern data is extremely useful in dissecting the functions of phenotypically silent genes. This was demonstrated by Rorth and colleagues who used a P element that contained the yeast *GAL* promoter to carry out gain-of-function screens in *Drosophila*. Upon insertion of this P element, expression of the tagged gene was under the control of the *GAL* promoter. Mating these tagged lines to files that express the Gal4p activator resulted in controlled overexpression. P elements tend to insert at the 5′ ends of genes and lead to production of full-length transcripts. In some cases, the P element inserts downstream of a start site and in the opposite orientation. This leads to the expression of antisense transcripts resulting in loss-of-function mutations.

Cellular localization of fusion proteins provides important clues about gene function. Transposons can be used to alter protein localization by placing specific localization signals within the transposon. The fusion proteins encoded by the transposon-tagged genes are then targeted on the basis of the transposon-mediated signal. Researchers constructed a transposon that contains a yeast promoter and the DNA binding domain of the Gal4p protein. Insertion of this transposon often leads to the generation of Gal4p fusion proteins that localize to the nucleus and bind the *GAL* promoter. The transcriptional activity of the tagged protein can then be assessed in one-hybrid assays. Transposon tagging can therefore allow direct detection of a target protein or localize it to facilitate further analysis of protein function.

Transposons can therefore be used to generate fusion proteins or reporter proteins that reflect endogenous gene expression, identify enhancers and alter gene expression levels or cellular localization.

Protein Tags

Transposons are also used to generate protein tags that can be used for easy *in vivo* and *in vitro* protein detection (**Figure 2**). For this approach to be successful, the sequence encoding the tag must be inserted such that the reading frame of the transposon-bearing gene is maintained through the tag; the resulting protein can be visualized by detecting the presence of the tag. Common tags include green fluorescent protein (GFP), and the hemagglutinin and myc epitopes. Epitope-tagging a protein also allows researchers to learn if a protein is a part of a larger complex through coimmunoprecipitation studies. The tagged protein can be identified using an antibody to the tag and immunoblot analysis, and purification may be aided by the use of affinity columns that contain antibody to the epitope.

Gene Transfer

Transposons can also be used to introduce genes into organisms – an approach currently being exploited to create transgenic strains of zebrafish and flies. Homologous recombination does not occur in these organisms making it difficult to incorporate genes into the chromosome. Transposon insertion is a highly efficient method of introducing genes into the genome. In the case of zebrafish, researchers clone a gene of interest into a transposon, which is subsequently used to mutagenize zebrafish embryos. A subset of the embryos develop into fish that can transmit the transposon insertion through the germline. The resulting transgenic organisms are used to dissect gene function during development.

DNA Sequencing

Transposons are currently being used to facilitate large-scale DNA sequencing in a cost-efficient and accurate manner. When using a transposon-based approach to genome sequencing, large clones are broken into smaller redundant and overlapping clones that are subsequently subjected to transposon mutagenesis. The transposition conditions are controlled so that there is approximately one insertion every three kilobases. After mapping the locations of the transposons, clones are aligned so that a group of clones that together represent a transposon insertion at every 300 bases within a region are identified (see **Figure 3**). The DNA on both sides of the transposon is sequenced using primers that are specific to the ends of the transposon. Since an identical priming site is used for all the sequencing reactions, the cost of synthesizing primers is very low and there is no lag time spent waiting for new sequence data in order to design new primers. Multiple transposon insertions decrease the length of DNA to be sequenced in one run thereby increasing the probability of obtaining accurate data in a single sequencing run. Additionally, random transposon insertion into cosmids and other large clones facilitates rapid sequencing and does not require mapping or subcloning. As testimony to the practicality of this approach, transposons have already been used to help sequence large amounts of bacterial, insect, and human DNA.

Combining Transposon Applications

All the transposon applications discussed have been used in isolation from each other. Using individual

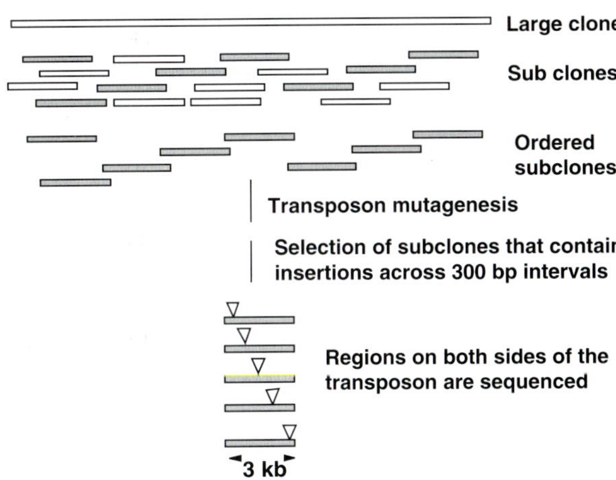

Figure 3 Transposons in DNA sequencing. Transposons are depicted as open triangles.

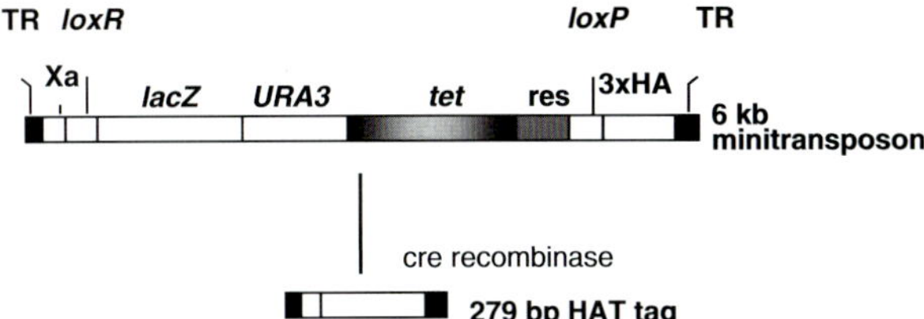

Figure 4 A multipurpose transposon. TR, terminal repeats; Xa, protease cleavage site; *loxR* and *loxP*, sites that recombine in the presence of *cre* recombinase; *lacZ*, β-galactosidase gene; *URA3*, yeast selectable marker that allows growth in the absence of uracil; *tet*, bacterial selectable marker that encodes the tetracycline resistance gene; res, transposon resolvase site required for resolution of co-integrates; 3×HA, three copies of the hemagglutinin epitope.

procedures, a gene is often tagged, sequenced, and distrupted, and reporter fusions generated. It is possible, however, to use a single transposon to attain all of these goals in a minimal number of steps. A multipurpose mini-Tn3 transposon designed for this purpose is shown in **Figure 4**. This transposon contains a *lacZ* reporter gene that is missing its promoter sequences and initiator methionine codon near one end of the transposon. The *lacZ* gene is usually only expressed if it is downstream of and in frame with a promoter and an initiator methionine codon. This engineered mobile element contains yeast and bacterial selectable markers that allow selection for its presence in both organisms. The reporter gene and selectable markers are separated from the terminal repeats by *loxR* and *loxP* sites, and there are three copies of the hemagglutinin (HAT) epitope (a convenient protein tag) between the *loxP* site and the terminal repeat. In the presence of cre recombinase, recombination occurs between the *lox* sites, resulting in excision of the intervening region. The remaining transposon sequence encodes three copies of the hemagglutinin epitope that are in the same reading frame as the recently removed reporter gene; this reduced transposon is transcribed and translated as part of the surrounding gene.

This highly modified transposon is currently being used for the large-scale analysis of gene function in yeast. Specifically, a yeast genomic DNA library is mutagenized with the multipurpose transposon described above. Strains carrying reporter fusions are used for quantitative and qualitative measurements of gene expression. Strains bearing HAT-tagged genes allow application of protein detection by immunoblot analysis, immunoprecipitation, and immunolocalization. Finally, strains bearing insertion and disruption alleles are being used for phenotypic analysis. Further modification of this transposon by replacement of the HAT sequence with that of GFP will allow *in vivo* protein localization studies. Additionally, the insertion of a protease cleavage site near the HAT tag will enable controlled degradation of the tagged protein, thereby increasing the versatility and efficacy of this transposon for further studies.

Conclusion

Transposable elements have clearly emerged as a versatile and informative tool in modern molecular biology. These elements usually lead to the development of new biological tools. As evidence of this fact, consider the ever-increasing collection of *Drosophila* and *Saccharomyces* strains that bear transposon-mediated single gene disruptions. These libraries have proven invaluable in determining gene function in these model organisms. The recent development of a transposon that can mutagenize *Mycobacteria tuberculosis* will drastically speed the identification of proteins required for the virility of this organism. Transposons are also being widely used in medical diagnostics since the presence of their conserved sequences provides a primer site for amplification of the nearby DNA and consequent identification of the infecting organism. As our understanding of transposons in different model systems continues to grow, additional uses are likely in the future analysis of their respective organism.

Further Reading

Burns N, Grimwade B, Ross-Macdonald PB *et al.* (1994) Large-scale analysis of gene expression, protein localization, and gene disruption in *Saccharomyces cerevisiae*. *Genes and Development* 8: 1087–1105.

Engels WR (1996) P elements in *Drosophila*. In: Saedler H and Gierl A (eds) *Transposable Elements*, pp. 103–123. Berlin: Springer-Verlag.

Parinov S and Sundaresan V (2000) Functional genomics in *Arabidopsis*: large-scale insertional mutagenesis complements the genome sequencing project. *Current Opinion in Biotechnology* 11: 157–161.

***See also:* Insertion Sequence; P Elements; Phage Mu; Transposable Elements; Transposable Elements in Plants; Transposase**

Trans-Splicing

T Blumenthal

doi: 10.1006/rwgn.2001.1334

Most instances of RNA splicing involve removal of internal sequences of a single molecule and splicing together of the two surrounding sequences. In contrast, *trans*-splicing results in the splicing of two originally separate RNA molecules. *Trans*-splicing can take several different forms: (1) the splicing of a so-called spliced leader (SL) onto the 5′ ends of mRNAs, which occurs in trypanosomes, euglena, roundworms, flatworms, and primitive chordates (2) group II splicing of separate RNAs in certain organellar systems, and (3) situations in which separate RNAs undergo *trans*-splicing by group-I-dependent, group-II-dependent, or spliceosome-dependent mechanisms, either because they have been engineered to do so or because of a rare, poorly understood, low-frequency event. Because these three kinds of *trans*-splicing are unrelated processes and are grouped here only because they are each classified as '*trans*-splicing,' they will be considered separately.

Spliced Leader Addition

In this kind of *trans*-splicing, a short donor RNA contributes its 5′ end to form the 5′ end of an mRNA. The spliced leader (SL) replaces the 5′ sequences of the pre-mRNA, and the reaction is catalyzed by most of the same machinery that catalyzes nuclear intron removal. That is, these are spliceosome-catalyzed reactions. The donor in *trans*-splicing is itself a small nuclear ribonucleoprotein particle. It is comprised of a short RNA, the SL RNA, which is 100–135 nucleotides in length, and several bound proteins. The first 21 to 51 nucleotides of the SL RNA are transferred to a recipient RNA by *trans*-splicing. The SL RNA is folded into a three-stem/loop structure with a conventional Sm protein-binding site located between the second and third stem. The SL snRNP contains the Sm proteins also found on U1, U2, U4, and U5 snRNPs, as well as some unique proteins that have not been found on other snRNPs. The spliced leader itself is immediately followed by a conventional 5′ splice site that acts as the donor in *trans*-splicing. The recipient in *trans*-splicing is a standard pre-mRNA in most respects. However it differs from most pre-mRNAs by beginning with an intron-like sequence, sometimes called an outron, instead of the usual exon at the 5′ end. The outron ends with a conventional 3′ splice site that acts as the *trans*-splice acceptor. The 5′ splice site on the SL RNA interacts with a branch point in the outron to form a Y-branched intermediate that is subsequently resolved by splicing of the short SL to the 3′ splice site at the end of the outron. The resulting products are (1) the SL spliced to the first exon of the acceptor RNA, and (2) the outron branched to the downstream portion of the SL RNA. The latter is presumably debranched and the nucleotides recycled as with the lariat byproducts of *cis*-splicing or intron removal, the more familiar nuclear splicing event. *Trans*-splicing is catalyzed by most of the same snRNPs as catalyze *cis*-splicing. One exception though is the U1 snRNP responsible for recognition and choice of the 5′ splice site. In *trans*-splicing U1 plays no role since the 5′ splice site is present on a snRNP already. In fact it is base paired in all known SL snRNPs to the SL itself in a short helix reminiscent of the U1 RNA/5′ splice site helix. However, this base pairing is not required for *trans*-splicing *in vitro* or *in vivo*.

In trypanosomes and at least some nematodes (and possibly some flatworms) many or all of the acceptor molecules are synthesized as polycistronic precursors. Each pre-mRNA contains RNA copies of several genes, and in these cases *trans*-splicing is used to resolve the polycistronic precursor into mature monocistronic mRNAs. In addition, 3′ end formation occurs just upstream (generally about 100–400 nt upstream) which results in a polyadenylated upstream mRNA and an SL-containing downstream mRNA. In these cases, the *trans*-splicing reaction follows the same course as described above except the branching occurs at a branch point between genes rather than near the 5′ end of the mRNA. In trypanosomes, there is only a single SL RNA, which is used for *trans*-splicing both at the 5′ ends and at internal sites in polycistronic mRNAs. In the nematode *Caenorhabditis elegans*, about 25% of genes are transcribed as parts of polycistronic precursors containing two to more than five genes. There is a special SL RNA, called SL2, which is used for *trans*-splicing at *trans*-splice sites between genes in these polycistronic pre-mRNAs. SL2 RNA has a secondary structure similar to the SL RNAs described above, but its sequence is different.

In *C. elegans*, many polycistronic precursors must undergo SL1-*trans*-splicing at their 5′ ends, intron removal throughout, and SL2-*trans*-splicing at internal *trans*-splice sites between genes. How are these different processes accomplished with specificity? Not all the players in the reaction are known yet, but it is clear that an intron or synthetic intron-like RNA can serve as an outron if placed at the 5′ end of a pre-mRNA. Furthermore an outron can be excised as an intron if a 5′ splice site is placed within it. Thus the context of a 3′ splice site, rather than any particular sequence, determines whether it is subjected to *trans*- or *cis*-splicing. In general, a 3′ splice site near the 5′ end of a pre-mRNA, with no upstream 5′ splice site, will be *trans*-spliced. This can be most easily understood by envisioning a spliceosome beginning to form around a 3′ splice site; if an upstream U1 snRNP bound to a 5′ splice site pairs with it, then *cis*-splicing occurs, whereas if no upstream site is found, then the SL snRNP provides a 5′ splice site in *trans*.

The rules for SL2 *trans*-splicing at internal sites in polycistronic pre-mRNAs are less clear. In trypanosomes the downstream *trans*-splicing event determines the location of upstream 3′ end formation. However, in worms the events are largely independent, although interference with 3′ end formation does affect the SL2 specificity of *trans*-splicing. In the only operon studied so far, a 22-nucleotide U-rich sequence about 30 nt downstream of the 3′ end formation site has been shown to be required for utilization of SL2. It is not yet known what *trans*-acting factors interact with this sequence. The sequence of the remainder of the intercistronic region is not required for *trans*-splicing.

Group II *Trans*-Splicing

Group II introns occur in plant mitochondria and chloroplasts. Most exist between adjacent exons and their removal by *cis*-splicing is dependent on a complex secondary structure containing six stem/loop domains. However, in some instances, especially in the *nad1*, *nad2*, and *nad5* genes of higher plant mitochondria, the exons have become rearranged. In these cases the individual pieces of the genes are transcribed separately. The transcripts can then form the analogous stems by intermolecular base pairing and the splicing occurs in *trans* just as if it were occurring within a single transcript. In some cases one of the transcripts contains no exon sequences; apparently its only purpose is to bring the correct exons together in a trimolecular stem–loop structure to allow the correct *trans*-splicing to occur.

Other Instances of *Trans*-Splicing

It is well established that autocatalytic Group I splicing can be engineered to occur in *trans*, as can Group II intron splicing that normally occurs in *cis*. Furthermore the eukaryotic nuclear mRNA splicing machinery can splice together two separate RNAs by conventional mechanisms *in vitro*. This reaction is relatively efficient when spliceosomes are offered two substrates, one of which contains only a 3′ splice site while the other has only a 5′ splice site. In these cases, splice sites normally used for *cis*-splicing are used in *trans*. A low level of *trans*-splicing occurs *in vivo* as well. *Trans*-splicing has been detected in a variety of cells, always cases of splicing between two different mRNAs at *cis*-splice sites. These have been detected by RT-PCR, which greatly amplifies rare products, and by isolation of single rare cDNA clones. Nevertheless, there have now been numerous reports that can be explained only by *trans*-splicing having occurred. Most of these examples have occurred in mammalian cells. So far, it is not clear what has brought the two exons from separate mRNAs together. One possibility would be formation of an RNA double helix or other tertiary structure involving the two molecules which could artificially bring the two splice sites into proximity. It has been possible to force *trans*-splicing to occur in mammalian systems by engineering two molecules in which portions of the 'intron' sequences can anneal. In these cases it appears as if the splicing machinery is 'fooled' into believing the 5′ and 3′ splice sites are on the same molecule and so it splices them together, creating a hybrid molecule. This is splicing in *trans*, but it is presumably mechanistically identical to normal splicing, since conventional 5′ and 3′ splice sites are used. The significance of these rare events is unclear, since there are no cases in which the products of these *trans*-spliced chimeric RNAs have been shown to function. Presumably this sort of *trans*-splicing is just an unintended consequence of normal nuclear pre-mRNA processing events.

See also: **Pre-mRNA Splicing**

Transvection

doi: 10.1006/rwgn.2001.2057

Transvection is the ability of a locus to influence activity of an allele on the other homolog only when the two chromosomes are synapsed.

See also: **Synapsis in DNA Transactions**

Transversion Mutation

C Beamish

doi: 10.1006/rwgn.2001.1335

Transversion mutation is a specific kind of point mutation, one in which a single purine is substituted for a pyrimidine or vice versa. As the result of a transversion mutation, the mutated position in the gene may for example have an adenine where it had a thymine or cytosine. Transversions are much less common than transition mutations – the other form of point substitution mutations, in which one of the two purines or pyrimidines is substituted for the other – because the generation of transversions during replication requires much greater distortion of the double helix than does the production of transition mutations. The general shape of the base mistakenly inserted into the base pair to produce a transition mutation is conserved and there mainly is a change in electron distribution; this may happen simply because the base in the template strand is, for example, in a rare enol state rather than its more common keto form at the moment of replication. On the other hand, the base pair formed during the production of transversion mutations is either much larger (involving two purines) or much smaller (with two pyrimidines) than the standard base pair.

Interestingly, the genetic code has evolved in such a way that transversion mutations are much more likely than are transition mutations to lead to substituting an amino acid with very different properties and thus to significant changes in the properties of the protein, due to the relationships between the sequence patterns of the codons for the various amino acids. Because of the degree of degeneracy in the third nucleotide of most codons (except those for tryptophan and methionine), a transversion in the third nucleotide is less likely to affect the organism than a transversion in the second or first nucleotides.

A classic example of a transversion leading to a major change in protein properties is found in the sickle cell mutation in human hemoglobin. In those with sickle cell anemia, a thymine is substituted for an adenine in the second position of the sixth codon of the gene for the β subunit, leading to the incorporation of valine (a hydrophobic amino acid) rather than glutamic acid (which is very hydrophilic) at that position. As a consequence of this change in each of the two β subunits, the individual hemoglobin tetramers stick to each other to form very long chains that lead to the characteristic sickle shape of the cells. In the case of sickle cell trait, where only one of the two hemoglobin alleles carries the sickle-cell mutation, only half of the subunits are mutated so no long chains are formed.

See also: **Invariants, Phylogenetic; Sickle Cell Anemia; Transition**

Trees

N Saitou

doi: 10.1006/rwgn.2001.1479

In evolutionary biology, phylogenetic trees of organisms or genes are often called 'trees.' Mathematically, a tree is defined in terms of graph theory as follows: all the nodes are connected via edges, and there is only one path to connect any two nodes. Therefore, a network is not a tree because there is more than one path (route) to connect one node with another. Nodes are divided into external and internal ones. The former are also called operational taxonomic units (OTUs) in evolutionary biology, or leaves in computer science. There are five OTUs (1–5) and four internal nodes

(A)

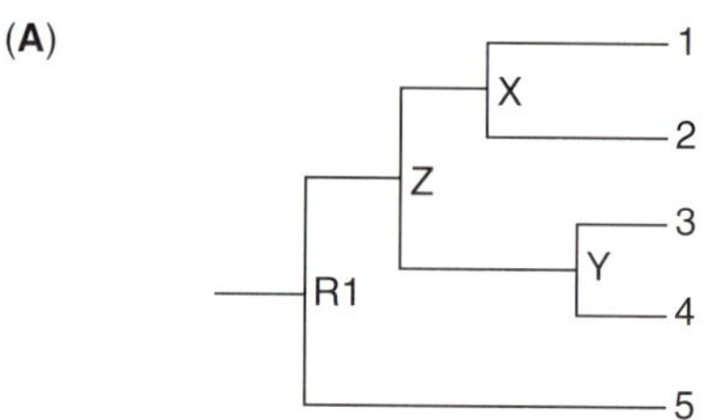

(B)

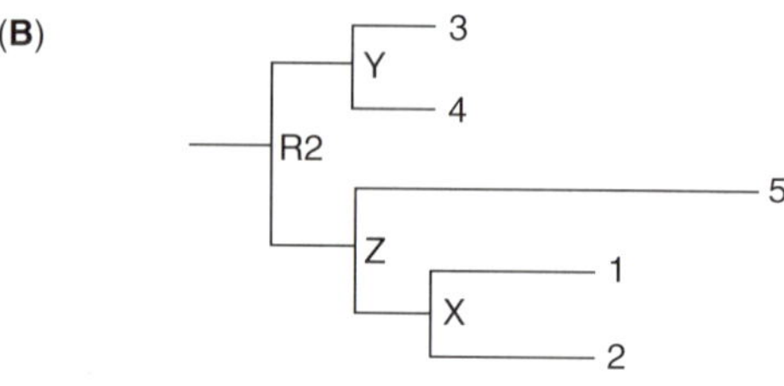

(C)

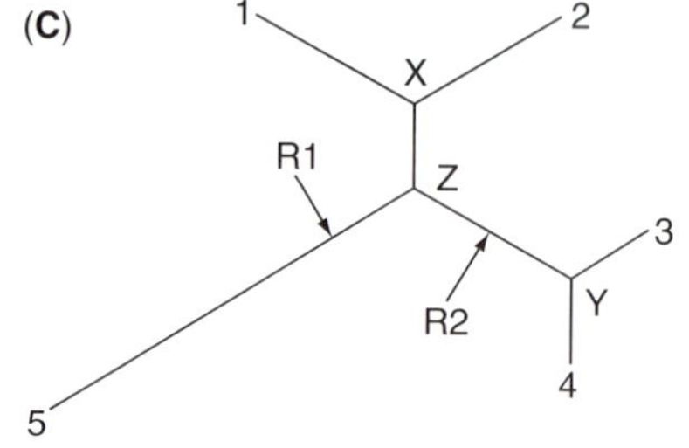

Figure 1 Tree types. (A, B) Rooted; (C) unrooted.

(X, Y, Z, and R1) in the tree of **Figure 1A**. Edges are usually called branches in evolutionary biology, and not only topological relationship (how to connect nodes) but also distance value is often added to branches. Branches can also be divided into external and internal ones. An external branch connects an external node to an internal node (e.g., branch 1–X of **Figure 1**), while an internal branch connects two internal nodes (e.g., branch X–Z of **Figure 1**).

A tree can be either rooted or unrooted. A rooted tree has a special node called the root that is defined as the position, R, of the common ancestor (see **Figures 1A** and **1B**). There will be a unique path from the root to any other node, and the direction of this is of course that of time in evolution. A phylogenetic tree in an ordinary sense is a rooted tree. Unfortunately, however, many methods for building phylogenetic trees produce unrooted trees, such as the tree shown in **Figure 1C**. An unrooted tree can be converted to a rooted tree if the position of the root is specified. Trees of **Figures 1A** and **1B** were thus produced from the unrooted tree of **Figure 1C**. Rooted/unrooted trees are also called directed/undirected trees in mathematics.

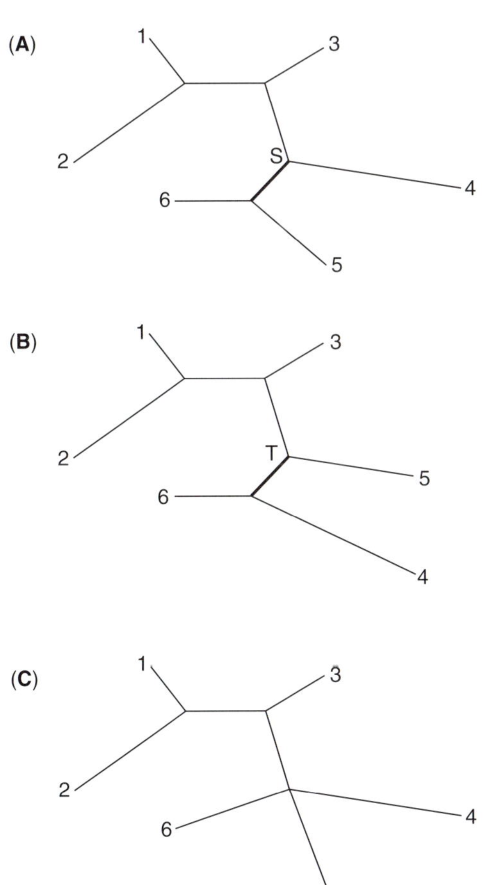

Figure 2 Tree types. (A, B) Bifurcating; (C) multifurcating.

This relation between rooted and unrooted trees is used for the 'outgroup' method of rooting as follows. When we are interested in determining the phylogenetic relationship among the n sequences, we will add one (or more) sequence that is known to be an outgroup relative to the n sequences. The obtained unrooted tree for the $n + 1$ sequences can easily be converted into a rooted tree of n sequences. Sequence 5 corresponds to the outgroup in the tree of **Figure 1C** when the root is R1, and the tree of **Figure 1A** is then obtained. When the root is R2, sequences 3 and 4 are considered to be the outgroup to sequences 1, 2, and 5, and we obtain the tree of **Figure 1B**.

The number of possible tree topologies rapidly increases with an increasing number of OTUs. The general equation for the number of possible topologies for bifurcating unrooted trees (Tn) for n OTUs is given by:

$$Tn = (2n - 5)!/[2^{n-3}(n - 3)!]$$

If we apply this equation, there are 221 643 095 476 699 771 875 possible tree topologies for 20 OTUs. It is clear that the search for the true phylogenetic tree of many sequences is a very difficult problem. This is why so many methods have been proposed for building phylogenetic trees.

Other important concepts in trees are bifurcating trees and multifurcating trees. The trees shown in **Figure 1** and those in **Figures 2A** and **2B** are all bifurcating ones, while the tree in **Figure 2C** is multifurcating. Theoretically, multifurcating trees can be considered as bifurcating trees in which some branches have zero length. For example, the tree of **Figure 2C** can be equated to those of **Figures 2A** and **2B** when branches S and T of **Figures 2A** and **2B**, respectively are zeros. This relationship is used to produce 'consensus' trees. Let us compare the tree structures of **Figures 2A** and **2B**. There are slight differences between them, and if we ignore branches S and T, we obtain **Figure 2C**. This tree can be considered as the consensus tree of those in **Figures 2A** and **2B**.

If the evolutionary rate is constant, we obtain a particular type of rooted tree, which can be called a 'clock' tree. This example is shown in **Figure 3A**. When there is heterogeneity of evolutionary rates in different lineages of a tree, a non-clock tree is obtained, as in **Figure 3B**. It should be noted, however, that trees that look like a clock tree can be constructed if we assume constancy of evolutionary rate, even if constancy does not hold in reality. Unweighted pair group method with arithmetic mean (UPGMA) is such a method for producing a clock-like tree. Many

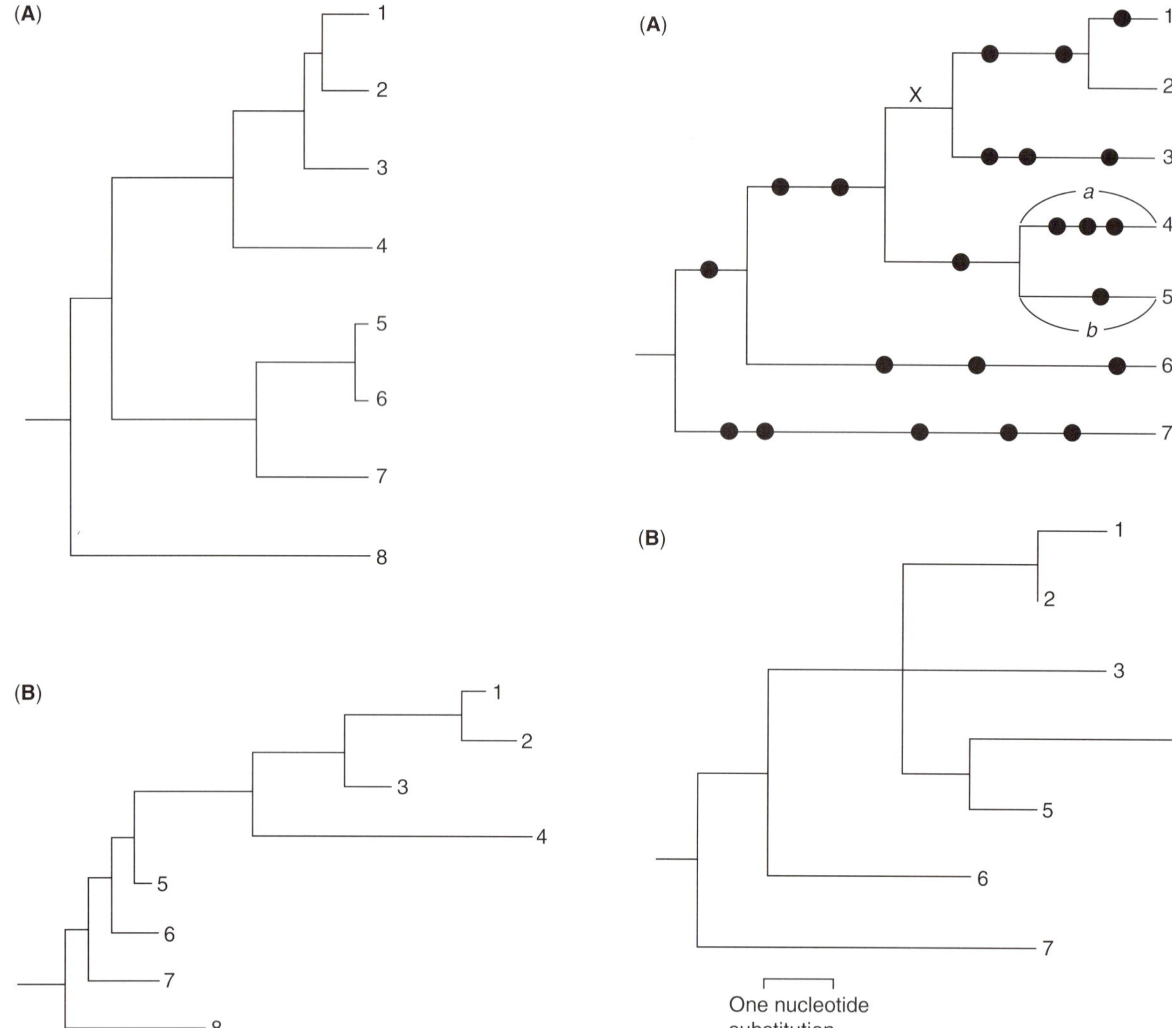

Figure 3 Rooted trees. (A) Clock tree; (B) non-clock tree.

Figure 4 Phylogenetic trees. (A) 'Expected'; (B) 'realized.'

other tree-making methods usually produce trees without assuming constancy of evolutionary rate. However, they only produce unrooted trees, unlike UPGMA which always produces rooted trees.

Ideally, branch lengths of a phylogenetic tree are proportional to the physical time since divergence. Thus branches *a* and *b* of **Figure 4A** should be the same length. We call this type of rooted tree the 'expected tree.' Both species and gene trees have their expected trees, but their properties are somewhat different from each other. An expected gene tree directly reflects the history of DNA replications, while an expected species tree is a gross simplification of the course of differentiation of populations. Therefore, the speciation time is not always clear.

The genealogical relationship of genes, or expected gene tree, is independent from the mutation process. However, mutation events are essential for the reconstruction of phylogenetic trees. Thus we can at best estimate a gene tree according to the mutation events realized on its expected gene tree. We call this ideal reconstruction of the gene tree as the 'realized' gene tree (**Figure 4B**), while the reconstructed one from observed data is called the 'estimated' gene tree. Branch lengths of realized and estimated genes tree are proportional to mutational events. These mutational events are not necessarily proportional to physical time. By definition, expected gene trees are strictly bifurcating, while realized and estimated gene trees may be multifurcating. This is because of the possibility of no mutation in a certain branch, such as branch X of **Figure 4A**.

A species tree reconstructed from observed data is called an 'estimated' species tree, while there is no realized species tree. It should also be noted that both expected and realized trees are rooted, while estimated trees are often unrooted due to the limitations of available information.

See also: Gene Trees; Genetic Distance; Phylogeny; Species Trees; Taxonomy, Numerical

Trichome Development, Genetics of

M Hülskamp and J Mathur

doi: 10.1006/rwgn.2001.1678

Plant hairs, also called trichomes, are specialized epidermal cells. On aerial organs trichomes have a protective role against insects or sun. In *Arabidopsis thaliana* trichomes are easily accessible and have become a genetic model system for the analysis of pattern formation and cell differentiation.

Genetic Dissection of Trichome Development in *Arabidopsis*

In *Arabidopsis*, trichomes are unicellular, branched cells that are regularly distributed on most aerial surfaces. Systematic screens for trichome mutants in *Arabidopsis* revealed 37 complementation groups. The analysis of these mutants enabled the dissection of trichome development into distinct, genetically controlled steps (**Figure 1**): (1) initiation, (2) endoreduplication, (3) differentiation, (4) branching, (5) expansion, and (6) maturation.

Trichome Initiation

On leaves, trichomes are initiated at the base in a field of dividing epidermal cells. The incipient trichome cells are separated by three to four epidermal cells and show a characteristic spacing pattern. Trichome patterning does not seem to involve cell lineage. Rather it is thought to be based on a mechanism where initially equivalent epidermal cells compete with each other via cell–cell interactions (**Figure 2A**). According to the current models *GLABRA1 (GL1)*, a MYB-related transcription factor that is expressed in developing trichomes, and *TRANSPARENT TESTA GLABRA1 (TTG1)*, a WD40 protein, function as positive regulators of trichome development. Epidermal cells surrounding a young trichome are inhibited from becoming trichomes by the negative regulator TRIPTYCHON (TRY) probably by downregulating the two positive regulators GL1 and TTG. *TRY* encodes a MYB-related protein lacking the activation domain and is assumed to execute its inhibitory function by directly moving to neighboring cells.

Endoreduplication

Incipient trichome cells stop cell divisions but proceed, on an average through four cycles of DNA replication (called endoreduplication). The number of endoreduplication cycles in trichomes is controlled by two genetic pathways. One pathway depends on the plant hormone gibberellin (GA). Mutants deficient in GA biosynthesis lack trichomes and a mutant, *spindly (spy)*, that results in a constitutive activation of the GA signal transduction pathway displays trichomes with an increased DNA content (**Figure 2B**). In addition three genes control trichome endoreduplication in a GA independent pathway. Strikingly, two of them, *GL1* and *TRY*, also play a role during trichome patterning, with *GL1* promoting and *TRY* inhibiting additional endoreduplication cycles. In addition, the *GL3* gene is required as a positive regulator. *gl3* mutants undergo only three cycles of endoreduplication and since this phenotype can not be rescued by any other overreplicating mutant *GL3* is assumed to act upstream of all other known genes.

Differentiation Mutants

Five genes, *GLABRA2*, *ROOT HAIRLESS1 (RHL1)*, *RHL2*, *RHL3*, and *ECTOPIC ROOT HAIR3 (ERH3)*, appear to function early during trichome differentiation in the regulation of genes acting later. The corresponding mutants show a wide range of trichome phenotypes: trichome size and branching is generally reduced, and mutant trichomes often lack papillae on their surface. These phenotypic aspects resemble the single mutant phenotypes of other trichome morphogenesis mutants and it is therefore believed that the differentiation genes are required to integrate the function of later-acting trichome morphogenesis genes. Consistent with this idea is the finding that the cloning of the *GLABRA2* gene revealed that it encodes a protein with sequence similarity to homeodomain transcription factors.

Branching Mutants

Fifteen genes have been identified that function as positive or negative regulators of branch number. They fall into two groups. One group establishes a connection between the DNA content and branch number (**Figure 2B,C**). Accordingly, mutants with a reduced DNA content, e.g. *glabra3*, have fewer branches while mutants with an increased DNA content, e.g. *triptychon*, show more branches. Since changes in

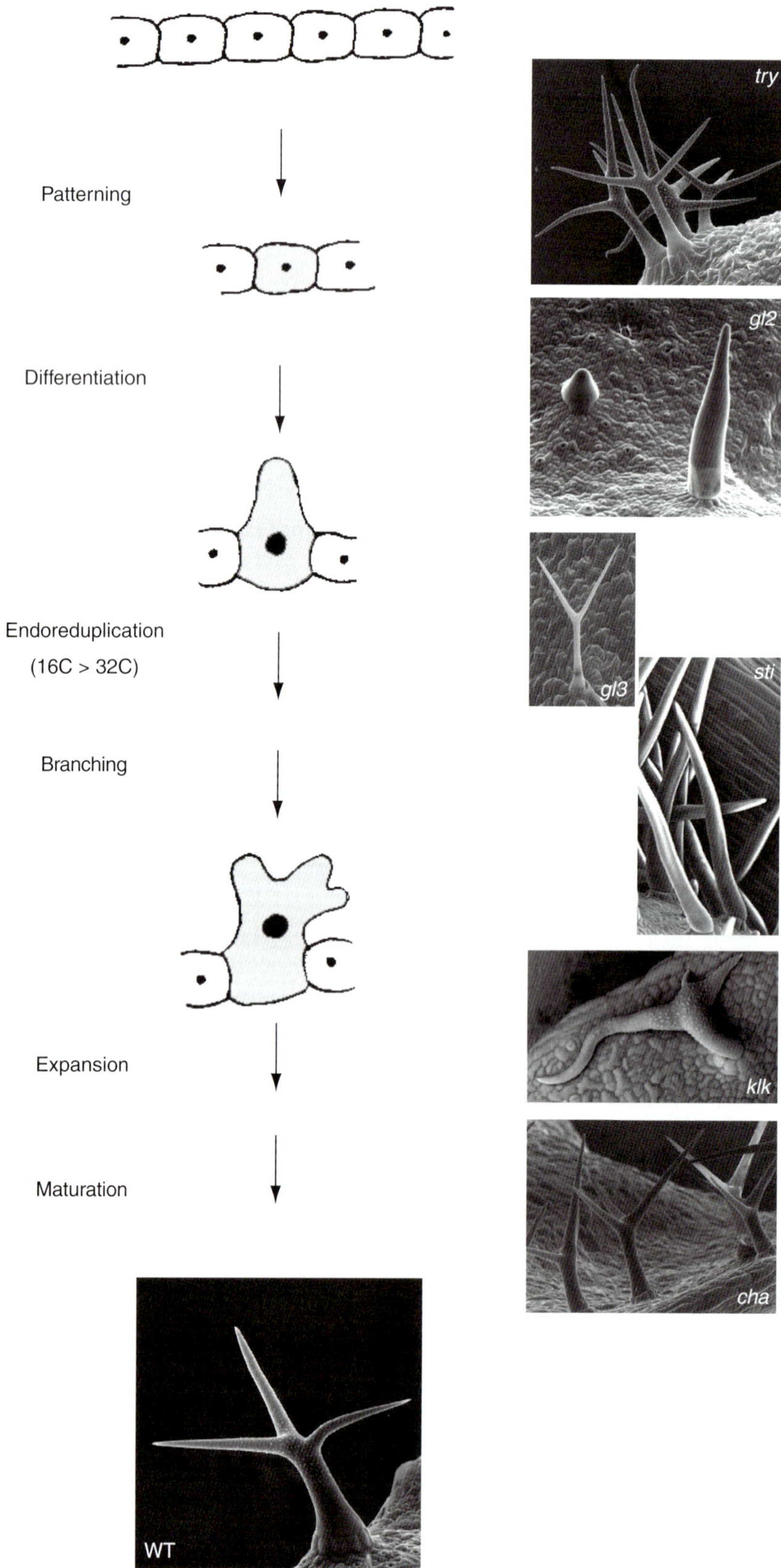

Figure 1 Trichome development mutants. Left: Schematic illustration of developmental steps during trichome formation. Right: Examples of mutants affecting various developmental steps. Abbreviations: *try (triptychon)*, *gl2 (glabra2)*, *gl3 (glabra3)*, *sti (stichel)*, *klk (klunker)* (a member of the distorted class of mutants), *cha (chabli)*.

(A)

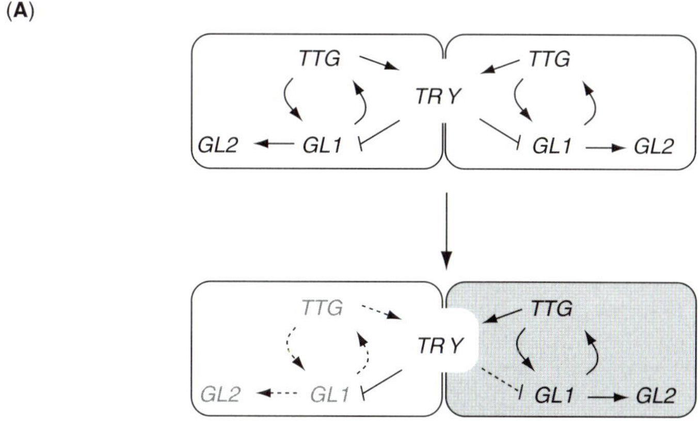

(B)

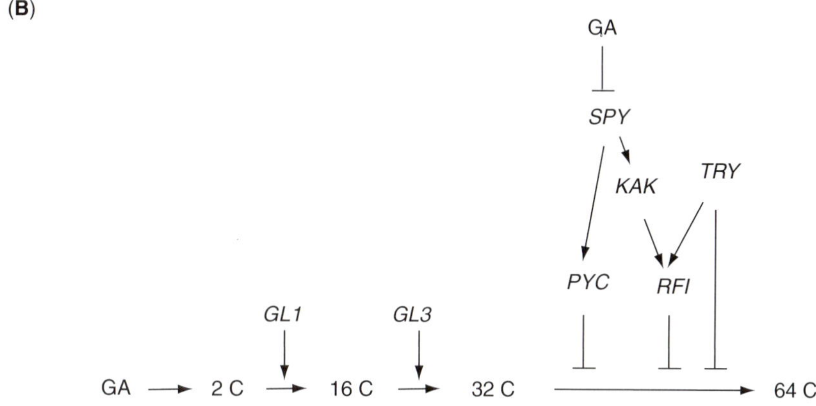

(C)

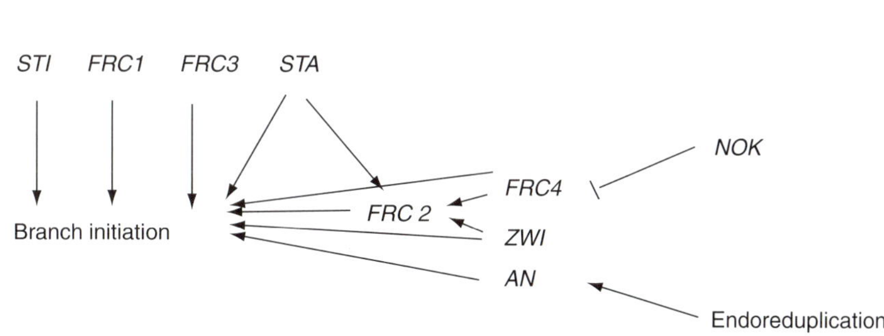

Figure 2 Genetic models of trichome development. (A) Trichome cell selection. The genetic model postulates that *GLABRA1 (GL1)* and *TRANSPARENT TESTA GLABRA1 (TTG)* form a positive regulatory loop and trichome development is initiated by trichome differentiation genes such as *GLABRA2 (GL2)*. Cell–cell communication is mediated by *TRIPTYCHON (TRY)* which is activated by *TTG* and downregulates *GL1*. As a result cells compete with each other to become a trichome cell. In the upper situation both cells are in an equilibrium. Below, the right, shaded cell has gained higher concentrations of *GL1* and *TTG* and suppresses trichome development in the left cell. (B) Endoreduplication. The number of endoreduplication cycles is controlled by positive and negative regulators. Arrows indicate positive regulation events, blunted bars indicate negative regulatory events. Abbreviations: GA (gibberellin), *GL1 (GLABRA1)*, *GLABRA3 (GL3)*, *SPY (SPINDLY)*, *PYC (POLYCHOM)*, *KAK (KAKTUS)*, *RFI (RASTAFARI)*, *TRIPTYCHON (TRY)*. (C) Branching. The number of branches is controlled by several independent pathways. Abbreviations: *STI (STICHEL)*, *FRC1 (FURCA1)*, *FRC2 (FURCA2)*, *FRC3 (FURCA3)*, *FRC4 (FURCA4)*, *STA (STACHEL)*, *ZWI (ZWICHEL)*, *AN (ANGUSTIFOLIA)*, *NOK (NOEK)*.

the DNA content in the wild-type background by using inhibitors of DNA replication or in tetraploid plants also result in a correlation between the DNA content and branch number it is unlikely that the mutants have two separate roles in the two processes. This suggests that either cell growth or cell size controls branch number.

In the second group of mutants the DNA content is like in the wild-type. Genetically, they seem to act largely in independent pathways (**Figure 2C**). Only *FURCA2* and *STACHEL* seem to function redundantly and downstream of *ZWICHEL*, *FURCA4*, and *NOEK*. The DNA content-related pathway seems to be mediated by *ANGUSTIFOLIA*. Presently the underlying molecular mechanisms are unknown. Only the *ZWICHEL* gene has been cloned. The *ZWICHEL* gene encodes a member of the kinesin superfamily of motor proteins that contains a calmodulin-binding site. It is therefore assumed to be either involved in the transport of important intracellular components or in the reorganization of microtubules prior to branch initiation. A role of the microtubules in branch formation is also suggested by drug inhibitor experiments: destabilization of microtubules results in unbranched trichomes and the stabilization of microtubules can trigger branch formation in the unbranched *stichel* mutant.

Trichome Expansion

Eight genes, grouped in the *DISTORTED* class, are required to maintain the directionality of trichome cell expansion. Development of trichomes in *distorted* mutants is nearly normal until branch initiation while later growth is irregular resulting in mature trichomes displaying a twisted and distorted phenotype. Although none of these genes is cloned yet, an analysis of the cytoskeleton in these mutants suggests that the *DISTORTED* genes play a role in the organization of the actin cytoskeleton. All *distorted* mutants show strong abnormalities in the organization of the actin cytoskeleton. The biological relevance of the actin cytoskeleton in the expansion growth has been independently demonstrated with drug inhibitors. Drugs interfering with the actin organization result in a phenotype indistinguishable from the *distorted* mutants.

Trichome Maturation

During trichome maturation the cell wall thickens and small papilla are formed. This step is affected in five mutants, *under developed trichome (udt)*, *trichome birefringence (tbr)*, *chablis (cha)*, *chardonnay (cdo)*, and *retsina (rts)*. These mutants appear transparent and may even collapse at some point. Only the *tbr* mutant has been studied in some detail and was shown to be affected in cellulose deposition.

Further Reading

Hülskamp M (2000) How plants split hairs. *Current Biology* 10: R308–R310.

Hülskamp M, Folkers U and Schnittger A (1999) Trichome development in *Arabidopsis thaliana*. *International Review of Cytology* 186: 147–178.

Marks MD (1997) Molecular genetic analysis of trichome development in *Arabidopsis*. *Annual Review of Plant Physiology and Plant Molecular Biology* 48: 137–163.

Oppenheimer D (1998) Genetics of plant cell shape. *Current Opinion in Plant Biology* 1: 520–524.

Szymanski DB, Lloyd AM and Marks MD (2000) Progress in the molecular genetic analysis of trichome initiation and morphogenesis in *Arabidopsis*. *Trends in Plant Sciences* 5: 53.

See also: *Arabidopsis thaliana*: The Premier Model Plant; Cell Lineage

Trinucleotide Repeats: Dynamic DNA and Human Disease

V Brown and S T Warren

doi: 10.1006/rwgn.2001.1337

For many years a handful of heritable disorders puzzled geneticists by showing a tendency of the disease phenotype to become more severe or have earlier age-of-onset as the disease is passed on through subsequent generations in a family (Wells and Warren, 1998). This has been termed genetic anticipation, which is similar to the Sherman paradox described in fragile X syndrome, where the likelihood of having an affected child increases through subsequent generations of a pedigree (Warren and Sherman, 2000). In 1991, through research on spinal bulbar muscular atrophy and the fragile X syndrome, scientists discovered of a new class of genetic mutation termed trinucleotide repeat expansions or dynamic mutations (La Spada *et al.*, 1991; Warren and Sherman, 2000). Understanding this novel type of mutation revealed in molecular terms the underlying mechanism of both genetic anticipation and the Sherman paradox. To date, at least 24 neurological diseases and 17 non-neurologic genetic diseases show evidence of genetic anticipation (Wells and Warren, 1998) and at least 20 genetic disorders have been linked to mutations in trinucleotide repeat tracts (**Table 1**).

Table 1 Human disorders caused by trinucleotide repeats

Repeat	Locale	Normal	Affected	Disorder	Inheritance	Gene	Locus	Protein	Result
CGG	5′UTR	6–52 repeats	52–230 (pre) >230–2000 (full)	Fragile X syndrome	X-linked dominant	*FMR1 FRAXA*	Xq27.3	RNA-BP	CpG DNA methylated, transcriptionally silenced, mental retardation
GCC	5′UTR	7–35	130–150 (pre) 230–750 (full)	Fragile X, mental retardation	X-linked dominant	*FMR2 FRAXE*	Xq28	Transcription factor	CpG DNA methylated, transcriptionally silenced, mild mental impairment
CGG	5′UTR	11	80 (pre) 100–1000 (full)	Jacobsen syndrome (FRA11B)	Expansion carrier, recessive	*CBL2*	11q23.3	Proto-oncogene	11q23->qter loss, mild mental impairment
CTG	3′UTR	5–37	50–80 (pre) >80–3000 (full)	Myotonic dystrophy (DM)	Autosomal dominant	*DMPK*	19q13	Ser-Thr protein kinase	Impaired nuclear export of mRNA
GAA	Intron	<39	80 (pre) >112–1700 (full)	Friedreich's ataxia (FA)	Autosomal recessive	*X25 FRDA1*	9q13–21.1	Mitochondrial iron-binding protein	Low expression of mature mRNA, impaired iron transport
CAG	Coding Gln	11–33	38–66	Spinobulbar muscular atrophy or Kennedy's disease (SBMA)	X-linked recessive	*AR*	Xq12–21	Androgen receptor	Mildly androgen insensitive (loss of function), neuronal loss: spinal cord, cranial nerves
CAG	Coding Gln	6–39	36–121	Huntington disease (HD)	Autosomal dominant	*IT15*	4p16.3	Huntingtin	Intracellular protein aggregates, neuronal loss: striatum, cerebral cortex
CAG	Coding Gln	7–34	51–88	Dentatorubral pallidoluysian atrophy or Haw River syndrome	Autosomal dominant	*DRPLA B27*	12p13	Atrophin-1 or drplap	Nuclear protein aggregates, neuronal loss: dentate nucleus, neocortex
CAG	Coding Gln	6–39	41–81	Spinocerebellar ataxia type 1 (SCA 1)	Autosomal dominant	*SCA1*	6p23	Ataxin-1	Nuclear protein aggregates, neuronal loss: Purkinje cells, brainstem
CAG	Coding Gln	14–31	35–64	Spinocerebellar ataxia type 2	Autosomal dominant	*SCA2*	12q24.1	Ataxin-2	No protein aggregate detected

Repeat	Locale	Normal	Affected	Disorder	Inheritance	Gene	Locus	Protein	Result
CAG	Coding Gln	12–41	40–84	Spinocerebellar ataxia type 3 or Machado–Joseph disease (SCA3/MJD)	Autosomal dominant	*SCA3 MJD1*	14q32.1	Ataxin-3	Nuclear/perinuclear protein aggregates, neuronal loss: basal ganglia, brainstem
CAG	Coding Gln	7–18	20–23 (EA2) 21–27 (SCA6)	Spinocerebellar ataxia type 6 (SCA 6) or episodic ataxia type 2	Autosomal dominant	*CACNA1A*	19p13	α1A-voltage dependent Ca^{2+} channel subunit	Cytoplasmic protein aggregates, neuronal loss: Purkinje cells
CAG	Coding Gln	7–17	38–130	Spinocerebellar ataxia type 7 (SCA 7)	Autosomal dominant	*SCA7*	3p12–13	Ataxin-7	Nuclear protein aggregates, neuronal loss: cerebellum, inferior olive
$(CTA)_{3-17}$ CTG_n	3′UTR	<100	107–250 >>250 (?)	Spinocerebellar ataxia type 8, schizophrenia, or BPAD (allelic variant?)	Autosomal dominant	*SCA 8* (antisense) *KLHL1* (sense)	13q21	Antisense RNA Sense RNA-BP	Spastic, ataxic dysarthria, cerebellar atrophy
CAG	Coding Gln	7–28	>65	Spinocerebellar ataxia type 12 (SCA 12)	Autosomal dominant	*PP2A* (putative)	11q22–24	Regulatory subunit of protein phosphatase 2A	Tremor, ataxia, dementia, damaged basal ganglia and cerebellum
GCG	Coding Ala	6–7	8–13 (AD)	Oculopharyngeal muscular dystrophy	Autosomal dominant	*PABP2*	14q11.2	Poly-A RNA-BP, adenylation factor	Nuclear filament protein accumulation
			7 (AR) homozygote		Autosomal recessive				

Two common techniques are used to identify repeat expansions in known genes: Southern blotting and polymerase chain reaction (PCR). Southern blots of genomic DNA are hybridized with radioactive probes that will bind to the unique DNA sequence in or near the repeat region in question. The radioactive probe identifies the DNA fragment harboring the repeat in question and the size of that fragment reflects the number of triplet repeats at that locus. PCR uses primers that bind to the unique sequences on each flank of a repeat and amplifies the DNA fragment containing the repeat in question. The size of the amplified DNA product reflects the size of the repeated region (Wells and Warren, 1998; Warren and Sherman, 2000). To locate new genes affected by repeat expansion, genome-wide screens are performed in search of large trinucleotide tracts of DNA. Tagged synthetic oligonucleotides of any desired repeat sequence (for example: $(GAC)_{17}$) bind any DNA fragment containing GAC/CAG repeats. By virtue of the oligonucleotide tag, repeat-containing DNA fragments are tracked down, cloned, and analyzed for unique flanking sequences. The unique flanking sequence provides information about the genomic location of the large repeat tracts. The method of finding trinucleotide repeat expansions without previous knowledge of genomic location is named repeat expansion detection (RED) (Vincent *et al.,* 2000).

Repeated triplet sequences occur naturally within expressed genes. The 'dynamic' repeats show intergenerational instability; the number of triplet units found in a child's gene is different than the parental alleles. However, this instability is typical only in families with the disorder. In most normal families, repeats are stable. Although most known dynamic triplet repeats are GC-rich, the GAA expansion in Friedreich's ataxia implies that any triplet combination may be subject to dynamic instability. Each trinucleotide tract has a characteristic range of polymorphic alleles in the normal population (**Table 1**) (Wells and Warren, 1998; Cummings and Zoghbi, 2000). Massive expansions occur in 5′ untranslated regions, introns, or 3′ untranslated regions, which are noncoding regions of genes that are transcribed into RNA but not translated into protein. The CGG repeat in fragile X syndrome, the GAA repeat in Friedreich's ataxia, and the CTG repeat in myotonic dystrophy can each expand by hundreds to thousands of repeats during transmission from parent to child. In contrast, even moderate expansions are sufficient to cause disease when they occur in the coding regions of genes as seen with the CAG repeats in the spinocerebellar ataxias. The trinucleotide repeat diseases can therefore be subdivided into two categories: coding (moderate) expansions and noncoding (massive) expansions.

Trinucleotide Repeat Mutations in Coding Regions of Genes

In the coding region of a gene, even moderate expansions of roughly 15–100 repeats can be pathogenic (**Table 1**). Each added DNA triplet encodes an extra amino acid, which potentially disrupts the protein structure and function. As the repeat expands beyond a certain threshold, protein function is pathologically altered. With the exception of oculopharyngeal muscular dystrophy, all of the known coding repeat expansion mutations code for polyglutamine (polyQ) stretches, and almost universally cause neurologic disorders (SBMA, HD, SCA1, SCA3/MJD, SCA6, SCA7, SCA12). Polyglutamine diseases are dominantly inherited. CAG expansions are defined as altered-function alleles that cause a hallmark cellular phenotype: intracellular protein aggregates containing the mutant polyQ protein, ubiquitin, transglutaminase, and heat-shock proteins (Kaytor and Warren, 1999). Although each of these proteins harboring expanded polyQ tracts are found throughout the brain, a particular mutant protein causes only selective degeneration of a unique subset of neurons (**Table 1**). Interacting proteins that are expressed in a specific subset of cells and interact with a specific mutant polyQ protein may impart neuronal specificity. In support of this hypothesis, huntingtin, ataxin-1, atrophin-1, and the SBMA androgen receptor protein are each associated with a different cell-specific partner (Wells and Warren, 1998). Another view is that cell-specific proteases release pathogenic peptides from specific polyQ protein substrates. These peptides seed intracellular protein aggregates which may either catalyze or be a consequence of neuronal death. Cell culture experiments have established that polyQ peptides are more toxic than polyQ stretches within the context of a full-length huntingtin protein, and indeed caspases are sufficient to facilitate polyQ peptide toxicity (Ferrigno and Silver, 2000). One view is that polyQ stretches impede protein degradation by the proteasome such that mutant protein overwhelms the normal protein turnover machinery of the cell. The consequent disruption of general protein metabolism would then lead to cell death. Cellular ubiquitin is a 'tag' that targets proteins for degradation by the proteasome (Alves-Rodrigues *et al.,* 1998). Cummings and coworkers studied pathogenesis in cells that could not ubiquitinate mutant expanded ataxin-1 cells. Surprisingly, in transgenic mice lacking ubiquitin ligase in brain Purkinje cells expressing mutant ataxin-1, there was no correlation between nuclear inclusions (protein aggregates) and pathogenicity. In fact, the brains of mice with reduced protein aggregates had more neurodegeneration (Cummings and Zoghbi, 2000).

This has fueled debate over whether protein aggregation in neurons may somehow protect neurons, and not catalyze cell death.

Trinucleotide Repeat Mutations in Noncoding Regions of Genes

Trinucleotide repeat mutations are not always found in coding regions of genes; they have also been detected within introns as well as within noncoding exons of the 5′- and 3′-untranslated regions in mature mRNA. Moderate expansions of trinucleotide tracts that do not code for amino acids are often asymptomatic, as seen in fragile X syndrome, myotonic dystrophy, and Friedreich's ataxia. The moderate expansions generate 'pre-mutation' alleles – highly unstable trinucleotide tracts that may undergo massive expansions, adding hundreds to thousands of triplet repeats during transmission from parent to offspring. These massive expansions can become pathogenic by silencing transcription, inhibiting translation, altering mRNA splicing, or impeding RNA export from the nucleus (Wells and Warren, 1998). In fragile X syndrome, as the triplet repeat expands into the premutation range, more disease-causing 'full-mutation' alleles arise through transmission to offspring, which explains the increased penetrance of fragile X syndrome in successive generations and thus the Sherman paradox. The full-mutation CGG tract, which is near the promoter, becomes a target for DNA methylation thereby silencing transcription (Cummings and Zoghbi, 2000). In myotonic dystrophy, the transcribed expanded CUG repeat in the DMPK mRNA 3′ UTR is believed to sequester CUG-binding proteins, thereby squelching proteins away from other cellular mRNAs and impairing general cellular RNA splicing and export (Cummings and Zoghbi, 2000). Massive GAA expansions in intron 1 of the X25 gene are believed to impair transcription or possibly splicing of the expanded intron, reducing the cellular level of mature frataxin (Cummings and Zoghbi, 2000).

DNA Expansion Mechanism

DNA expansions that occur within trinucleotide tracts are attributable to the unique biochemical difficulties of performing DNA replication within long stretches of repeat DNA. Mounting *in vitro* evidence suggests that triplets are deleted or added to long repetitive tracts as the cellular machinery tries to replicate through DNA hairpins formed by triplet repeat sequences (Warren and Sherman, 2000). During replication of the genome, DNA synthesis on the lagging strand of template requires small bits of DNA, Okazaki fragments, to be synthesized along the template and then joined into one continuous strand (**Figure 1**) (Zubay, 1993). In order to join Okazaki fragments, the

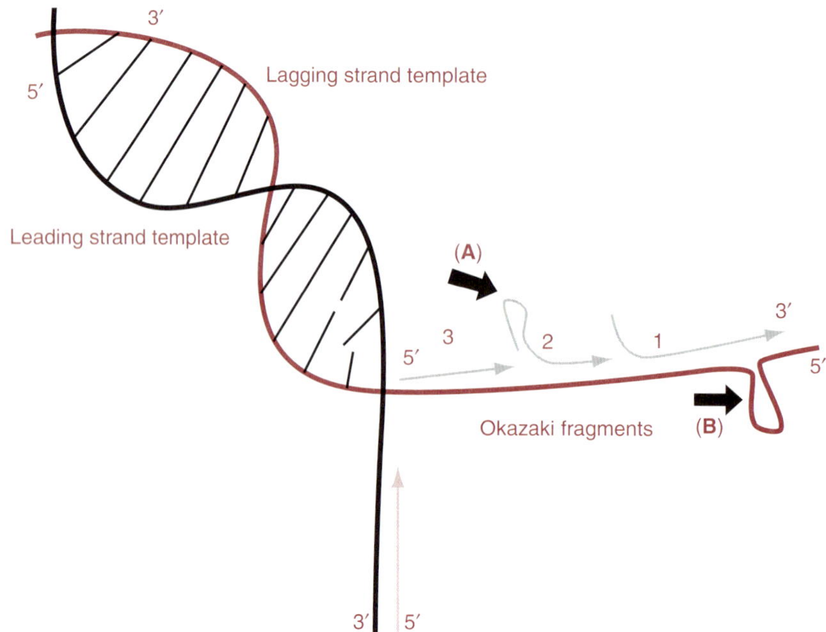

Figure 1 DNA replication. Polymerization proceeds from 5′ to 3′ for the newly synthesized DNA strands. On the lagging strand, synthesis proceeds 5′ to 3′ for each Okazaki fragment, but overall lagging strand synthesis proceeds 3′ to 5′ as the fragments extend, meet, and are ligated together. Okazaki fragments are shown in chronological order of synthesis indicated by 1, 2, 3.

upstream fragment extends downstream (3′) to the next fragment and displaces a small portion of it as a single strand 'flap' of DNA (**Figure 1, 1**). Flap endonuclease 1, FEN1, is an enzyme responsible for removing the 'flap' of single-standed DNA (Warren and Sherman, 2000). This allows end-to-end ligation of the Okazaki fragments and conserves the wild-type DNA sequence. In trinucleotide repeat tracts, the single-strand flap DNA may form a stable hairpin structure and remain in the DNA unremoved by FEN1 (**Figure 1, 2**). Upon ligation to the neighboring Okazaki fragment, the displaced hairpin region will be incorporated in the new DNA strand, effectively expanding the repeat region. The model predicts that DNA replication through hairpins in Okazaki fragments will cause an insertion of triplet units (**Figure 1A**), whereas hairpins formed in the template strand of DNA will cause triplet deletions (**Figure 1B**). In support of this model, yeast mutants defective for the FEN1 homolog RAD27 show increased trinucleotide expansion rates (reviewed in Warren and Sherman, 2000). Other supporting data show that CGG trinucleotide repeats which harbor AGG interruptions are more stable than pure CGG tracts. The AGG interruptions may act as landmarks on the DNA that facilitate proper primer/template alignment during replication, or may destabilize transient hairpins in the DNA. Interestingly, many of these dynamic mutations show gender-bias instability, meaning expansion or contraction occurs only during inheritance through the female (fragile X, Friedreich's ataxia, myotonic dystrophy) or through the male (spinocerebellar ataxias). This 'meiotic drive' however is often accompanied by somatic mosaicism, where nongametic cells harbor a spectrum of different allele sizes (Wells and Warren, 1998).

For diseases with genetic anticipation (polyQ diseases, fragile X syndrome, myotonic dystrophy, Friedreich's ataxia) a negative correlation between triplet repeat length and disease age-of-onset coupled with a positive correlation between repeat length and disease severity reveals a story much like the Sherman paradox (McInnis, 1996). Again, by elucidating the mechanism of trinucleotide expansions through transmission to offspring, this anticipation can be defined at a molecular level. For coding mutations, as the CAG repeats expand the polyQ stretches became longer, gaining the dominant function of expanded polyQ.

Both coding and non-coding dynamic triplet repeat genes have a characteristic range of 'normal' repeat, allowing some degree of heterogeneity in the unaffected population (**Table 1**). PABP2, the polyA-binding protein 2 gene for oculopharyngeal muscular dystrophy, does not tolerate even small expansions or contractions. Expansion of the GCG repeat in PABP2 becomes pathogenic after addition of just one repeat (recessive form) or three repeats (dominant form) coding for polyalanine (Brais *et al.*, 1998).

Discovering the mutational mechanism that underlies trinucleotide repeat expansion has allowed us to define genetic anticipation phenomena in molecular terms. We have insight into an entire class of heritable disorders, and can associate phenotype with underlying genotype of trinucleotide repeat disease. As we learn more about each of these diseases we are also gaining insight into genomic polymorphisms, chromatin structure, DNA replication, methylation, transcription, translation, and protein degradation. Trinucleotide repeats have thus opened a new window into dynamic research on dynamic DNA.

References

Alves-Rodrigues A, Gregori L and Figueiredo-Pereira ME (1998) Ubiquitin, cellular inclusions and their role in neurodegeneration. *Trends in Neurological Science* 21(12): 516–520.

Brais B, Bouchard JP and Xie Y-G *et al.* (1998) Short GCG expansions in the PABP2 gene cause oculopharyngeal muscular dystrophy. *Nature Genetics* 18: 164–167.

Cummings CJ and Zoghbi HY (2000) Fourteen and counting: unraveling trinucleotide repeat diseases. *Human Molecular Genetics* 9(6): 909–916.

Ferrigno P and Silver PA (2000) Polyglutamine expansions: proteolysis, chaperones, and the dangers of promiscuity. *Neuron* 26(1): 9–12.

Kaytor MD and Warren ST (1999) Aberrant protein deposition and neurological disease. *Journal of Biological Chemistry* 274: 37507–37510.

La Spada AR, Wilson EM, Lubahn DB, Harding AE and Fishbeck KH (1991) Androgen receptor gene mutations in X-linked spinal and bulbar muscular atrophy. *Nature* 352: 77–79.

McInnis MG (1996) Anticipation: an old idea in new genes. *American Journal of Human Genetics* 59: 973–979.

Vincent JB, Paterson AD, Strong E, Petronis A and Kennedy JL (2000) The unstable trinucleotide repeat story of major psychosis. *American Journal of Medical Genetics* 97(1): 77–97.

Warren ST and Sherman SL (2000) The fragile X syndrome. In: Scriver CR, Beadet AL, Sly WS and Valle D (eds) *The Metabolic and Molecular Basis of Inherited Disease*, vol. 2, pp. 1257–1289. New York: McGraw-Hill.

Wells RD and Warren ST (1998) *Genetic Instabilities and Hereditary Neurologic Diseases*. San Diego, CA: Academic Press.

Zubay G (1993) *Biochemistry*. Dubuque, IA: Wm C. Brown.

***See also:* Fragile X Syndrome; Muscular Dystrophies; Okazaki Fragment**

Triplet Code, Genetic Evidence

E J Murgola

doi: 10.1006/rwgn.2001.1336

The direct demonstration that the genetic code is a triplet code came only when the relationship between the nucleotide sequence of a segment of DNA and the amino acid sequence of the polypeptide encoded by it was shown to be three nucleotides to one amino acid. Although procedures for determining the amino acid sequence of small polypeptides were extant in the late 1950s, DNA-sequencing protocols were not generally available until the late 1970s. By that time the triplet nature of the code had become established through the interpretation of the results of both genetic and biochemical experiments.

Published in 1961, a brilliant series of genetic experiments clearly indicated that the genetic code is a triplet code. Proflavin, a derivative of acridine, is highly mutagenic for replicating genomes. In the experiments alluded to, addition of proflavin to a bacterial culture infected with phage T4 yielded many mutant phage progeny. In general, the resulting mutant phage characteristically did not revert to the wild-type phenotype after subsequent treatment with base-substitution mutagens, but did do so after treatment with mutagenic acridines. It was proposed that acridine mutations and the majority of spontaneous mutations are not caused by base substitutions but by addition or deletion of nucleotides, resulting in disruption of synthesis of the protein encoded by the mutated gene. That disruption is best explained by supposing that the sequence of letters comprising the genetic code is read in one direction, from a specific starting point, in groups of three or whatever the coding ratio might be. Consequently, the addition or removal of a letter would displace the reading mechanism by one letter from the site of alteration onward so that every triplet thereafter would be misread relative to the wild-type gene. Reversion of acridine-induced mutations is nearly always by suppressor mutations at a different but closely linked site, mutations that are assumed to work by adding a base where one has been deleted, and vice versa, so that the proper wild-type number of nucleotides (letters in the sequence) would be restored.

The 1961 study began with the isolation of independent, spontaneous reversions of a proflavin-induced mutation in the *rII* region of phage T4 and the subsequent demonstration in recombination experiments that the reversions were second-site, suppressor mutations as hypothesized, and that, after separation from the original mutation, each displayed the mutant phenotype. Once the suppressor mutations were isolated, they could be treated like any new *rII* mutations. First, they were mapped and all were found to be located close to and on either side of the original *rII* mutation that they suppressed. Second, reversions of the separated suppressor mutations were obtained and the revertant strains were analyzed in the same way as before. Once again the reversions were found to be due to suppressor mutations, that is, to suppressors of suppressors! When these 'second-generation suppressors' were isolated, they proved, like the first-generation suppressors, to have the typical suppressor phenotype and to map in the same segment of the *rII* region. They, in turn, were reverted and their suppressors, that is, suppressors of suppressors of suppressors, or third-generation suppressors, were isolated. In this way some 80 new *rII* mutations of the spontaneously reverting, acridine type were obtained.

The investigators, assuming that the original acridine-induced mutation resulted from, for example, the insertion of an extra nucleotide, and that the reversions were due to a compensating deletion of a nucleotide at other sites close to it, assigned arbitrarily a plus sign (+) to the original mutation and a minus sign (−) to its suppressors. Similarly all the suppressors of those suppressors could be labeled as 'plus' mutations, while the third generation suppressors would be 'minus' mutations. It did not matter whether the original mutation was an insertion or a deletion; the important point was that a mutation and its suppressor had opposite signs. According to this convention, some of the 80 new *rII* mutations were plus and some were minus. The question addressed next was: What happens if new combinations of these mutations are constructed by recombination? First, any combination of a plus and a minus mutation gave a wild-type or pseudo-wild-type phenotype. Second, as expected, combinations of two pluses (++) or two minuses (−−) always yielded the mutant phenotype. On the other hand, if the coding ratio is really 3, or a multiple of 3, the addition or substraction of three nucleotides should not throw the reading mechanism out of alignment with the code. This was verified experimentally. (Of course the reading between the outermost insertions or deletions will be different from wild-type, but for a particular gene segment this may not affect the protein product.) The conclusion of this study was that the coding ratio is probably 3, or, if more than one nucleotide is added or removed at a time, some multiple of 3.

The triplet nature of the code was supported by other mutational studies, utilizing, independently, tobacco mosaic virus (TMV) coat protein and the tryptophan synthetase alpha chain (encoded by the *trpA* gene) of *Escherichia coli*. Strong support for a triplet code came from correlation of amino acid substitutions with the type of action of specific mutagens (TMV), from analyses of amino acid substitutions in forward mutations and subsequent nonwild-type reversions (*trpA*), and particularly intracodon recombination in *trpA*. Biochemical confirmation of the triplet nature of the code and specific assignment of codons to their amino acids came from translation of defined RNA polymers, transcription and translation of DNA polymers, and finally from ribosome/triplet aminoacyl-tRNA-binding assays.

***See also:* Genetic Code; Universal Genetic Code**

Triploidy

P L Pearson

doi: 10.1006/rwgn.2001.1338

Triploidy is the term referring to the presence of a complete extra haploid set of chromosomes in an organism or cell line resulting in a $3n$ number of chromosomes where n is the haploid chromosome number for the species concerned. Triploids are both euploid and polyploid in that they contain a completely balanced extra set of chromosomes. They are rarely found in a viable state in wild animal populations but can occur in plant communities and are widely used in both commercial fruit and fish production. As discussed below, triploid organisms are invariably infertile.

Triploidy in Humans

In humans triploid embryos occur with chromosome constitutions of either 69, XXX, 69, XXY or 69, XYY. Triploidy is one of the commonest chromosome aberrations in humans and ends in spontaneous abortion in nearly all cases constitutes an estimated 1% of all recognized human conceptions and 17% of all chromosomally abnormal abortuses. Approximately 50 cases that came to birth but died almost directly afterwards are described in the literature.

Two distinct triploid fetal phenotypes have been recognized: one in which a relatively well developed fetus occurs together with an extremely large cystic placenta known as a partial hydatidiform mole and a second characterized by a grossly retarded embryo and placenta. Other severe malformations frequently present in both forms include cardiac defects, cleft palate, and skeletal defects. Chromosome and DNA marker studies have demonstrated that placental and fetal enlargement in triploid conceptions occurs when two paternal and one maternal set of chromosomes are present (diandry) and the phenotypic form with a reduced fetal and placental development arises in the presence of two maternal and one paternal set of chromosomes (digyny). Various lines of evidence, principally from mouse embryo nuclear transfer experiments, suggest that variations in fetal development associated with an excess of either paternal or maternal genome material are related to differences in parent specific imprinting patterns induced during gamete formation.

Diandrous triploidy most frequently arises through the fertilization of a single oocyte by two spermatozoa and digynous triploidy by complete failure of chromosome segregation in either the maternal first or second meiotic division giving rise to a diploid egg with two maternal sets of chromosomes.

The 69, XXX and 69, XXY triploids occur equally frequently with the 69, XYY form being rarely observed suggesting a much reduced viability of the 69, XYY triploid relative to the other two.

Other Organisms

Triploidy is encountered occasionally in natural populations of flowering plants containing diploid ($2n$) and tetraploid ($4n$) plants. It is presumed that such triploids arise by natural crosses between diploid and tetraploid plants in the same population. However, there are many examples of triploid strains of cultivated plants that have been induced artifically by crossing diploid and tetraploid parental strains. Unlike human triploids, such triploid plants appear to be morphologically normal, but are characterized by being completely infertile and can only be propagated vegetatively. Their infertility arises during gamete formation. Typically during meiosis the three homologs of each chromosome join and cross-over to produce a trivalent at the first meiotic division. The resulting chromosome segregation from each trivalent is completely random and it is extremely unlikely that a sufficiently large number of genetically balanced gametes can be produced to provide fertility. This phenomenon of triploid sterility was widely studied in the 1930s and 1940s in various plant species with notable contributions from Darlington and Mather in triploid *Hyacinthus* (hyacinth) and Dermen in *Petunia* by studying chromosome segregation in pollen.

One of the most famous and ancient examples of a triploid plant species is the cultivated banana characterized by its widely used and fleshy seedless fruit. The cultivated banana is believed to have been derived from a cross between a diploid species *Musa acuminata* and the tetraploid species *M. balbisiana*, both of which produce seeded fruit, some 1000 years ago in southeast Asia. This gave rise to a sterile triploid plant with large seedless fruit and enormous food-producing properties. Propagation of the cultivated banana occurs by dividing its root system. There are now more than 600 varieties of cultivated banana, including the plantain, which have been introduced into the majority of tropical countries. Although the original seeded wild species are still available, they are considered to be so inferior that they are only eaten in times of famine when the cultivated banana crop fails.

Modern plant breeders have adopted the same strategy of using the triploid status to produce seedless fruit for various fruit varieties of which currently the most important and fashionable ones are grapes, watermelons, and citrus fruits. Breeders have developed sophisticated methods of vegetative propagation including *in vitro* tissue culture and vegetative regeneration from endosperm tissue via somatic embryogenesis.

Although viable triploidy is rarely found in the animal kingdom, examples of natural newt populations were described the early 1940s in which triploids occurred together with diploid and tetraploid animals. In 1978 it was demonstrated that triploid newts could be induced experimentally by fertilizing heat-shocked eggs from diploid mothers; it was discovered that the heat shock treatment caused retention of the second polar body to produce a diploid egg. Similar strategies have been used to induce triploidy in fish for commercial reasons. Triploid salmon are sterile and demonstrate a steady and long-term growth combined with an improved flesh quality by comparison to fertile salmon and have a further advantage that they can be harvested at any time of the year. Triploid grass carp have been introduced into rivers and ponds in the United States for the purpose of weed control without the fear of reproduction and spreading in an uncontrolled fashion.

***See also:* Aneuploid; Chromosome Number; Euploid**

Trisomy

P L Pearson

doi: 10.1006/rwgn.2001.1339

Trisomy is the term that refers to the occurrence of an extra chromosome in a diploid ($2n$) organism. Trisomy is an example of aneuploidy in that the presence of the extra chromosome induces a genetic and numerical imbalance of the genome of the organism concerned.

Trisomy of a chromosome frequently leads to induction of a specific phenotype for the chromosome concerned. In a series of classical studies carried out between 1923 and 1931, Blakeslee demonstrated specific phenotypic effects associated with trisomy of each type of chromosome in *Datura stramonium*, the Jimson weed. The haploid chromosome number of diploid Jimson weed is 12 and Blakeslee observed 12 different characteristic phenotypes expressed as changes in the general size and shape of the seed capsules and the structure of hooks carried on their surface. Further, Blakeslee noted that the phenotypic changes induced by trisomy were significantly larger than those caused by induction of either triploidy or tetraploidy and concluded that the genetic imbalance of trisomy was the major cause of the phenotypic changes.

These studies came to the notice of the scientific community and in 1933 the Dutch ophthalmologist Waardenberg speculated that Down syndrome in humans might also be caused by the occurrence of an extra specific chromosome, an opinion that was later endorsed by Penrose.

Various improvements in human chromosome preparations lead to the accurate counting of human chromosomes by Tjio and Levan in 1956. In 1959, Lejeune and colleagues identified an extra chromosome caused by trisomy of a specific chromosome in patients with Down syndrome. The chromosome involved was subsequently referred to as chromosome 21 and Down syndrome as trisomy 21 syndrome. This observation was confirmed by Pat Jacobs in the same year, who also observed trisomy of the X-chromosome in Klinefelter syndrome patients and monosomy X in Turner syndrome. In the following year, trisomy of chromosome 18 giving rise to Edwards syndrome and of trisomy 13 associated with Patau syndrome were described. The respective birth frequencies of trisomies 21, 18, and 13 have been estimated to be 1/700, 1/7000, and 1/12 000 respectively. The short form of the ISDN human chromosome nomenclature refers to trisomy 21 as 47, XX, +21 in the case of female Down syndrome patients and 47, XY, +21 for male Down syndrome patients. Complete trisomies of autosomes other than 21, 18, and 13 do not appear to be compatible with life after birth in humans.

However, there are occasional descriptions in the literature of mosaic trisomies being found for several other chromosomes in liveborn individuals, including trisomies for chromosomes 8 and 9. Approximately 5% of Down syndrome patients do not arise from a complete extra chromosome 21, but are caused by unbalanced translocations resulting in three copies of all or part of the long arm of chromosome 21. The most common form of translocation is a Robertsonian translocation formed by centric fusion involving the long arm of 21 and the long arm of another acrocentric chromosome, usually 14 or 13. The short arms of the translocation chromosomes are eliminated resulting in a phenotypically normal translocation carrier with 45 chromosomes in place of 46. Missegregation of the translocation in gametes of the translocation carriers results in Down syndrome children with 46 chromosomes with three copies of 21q. Other forms of translocation in unbalanced state can give rise to partial trisomy in which just part of a given chromosome is trisomic. Specific phenotypes of partial trisomies for chromosomes besides 21, 18, and 13 have been described. In particular partial trisomies involving the short arm of chromosome 9 occur regularly and have a very characteristic phenotype. Partial trisomies can also arise from meiotic recombination within the inversion loop in carriers of pericentric inversions, resulting in a recombinant chromosome with a partial monosomy at the end of one arm of the chromosome and partial trisomy at the end of the other arm. Interstitial duplications have also been described which result in partial trisomy and an associated clinical phenotype. Investigation of the size and position of partial trisomies has led to the recognition that only part of the chromosome is necessary and permits defining a minimal essential region to produce a trisomic phenotype. Gene mapping studies of such cases has helped identify which genes are involved in causing genetic imbalance. In the case of chromosome 21 the recent sequencing of the entire chromosome shows that probably far fewer genes are involved in causing Down syndrome than originally believed.

Many independent studies indicate that the liveborn incidence of Down syndrome must constitute only a small fraction of the presumed disease incidence at conception. Chromosome studies on spontaneous abortions, which represent between 15–25% of all recognized pregnancies, show that at least 80% of Down syndrome pregnancies must spontaneously abort. Trisomy 21 is involved in approximately 10% of all trisomic spontaneous abortions. Importantly, trisomy has been observed for all chromosomes in spontaneous abortions in varying frequencies and demonstrates a general increase with maternal age in the abortus population. Astonishingly, monosomy X (Turner syndrome) is the only total chromosome monosomy found in either human liveborns or spontaneous abortions. However, studies in the mouse on fetal wastage associated with chromosomal abnormalities carried out by Alfred Gropp in the early 1970s clearly show that monosomies are involved in conception at an equal frequency to trisomies (the theoretical expectation); autosomal monosomic conceptions never come to term and are lost at a much earlier stage of pregnancy. We may assume that the same is true for humans but monosomic embryos are lost too early in the pregnancy to be even recognized. The inevitable conclusion is that the frequency of chromosomal aneuploidies at conception, including trisomies, is much higher than can be deduced from frequencies in either liveborns or spontaneous abortions. Further, the frequency of chromosome abnormalities at conception can be expected to rise rapidly with increasing maternal age such that the majority of embryos are chromosomally abnormal in women approaching their 40th year. This expectation is confirmed by recent investigations on the chromosome status of IVF embryos which demonstrate that nearly all embryos derived from women ≥ 37 years are chromosomally abnormal.

Geneticists have been fascinated for decades with the factors determining the enormous increases in chromosome nondisjunction in the human female with increasing age. One possible factor, first described by Edwards and Henderson by direct examination of mouse meiotic oocytes some 33 years ago, is that of a reduced meiotic crossing-over (chiasma forming) resulting in homologous chromosomes failing to segregate normally into daughter cells at the first meiotic division. This phenomenon has also been shown to take place in humans by using genetic marker segregation analysis; the level of genetic recombination between genes located on a particular chromosome is a direct measure of the number of chiasmata occurring on that chromosome during the preceding meiosis. It appears that the specific chromosome 21s involved in giving rise to Down syndrome children exhibit a much reduced level of genetic recombination than chromosome 21s from normal children. Interestingly, the reduced recombination frequency is not just confined to chromosome 21 and genetic marker analysis of all chromosomes in Down syndrome children shows a genome-wide reduction in genetic recombination. This leads to the conclusion that there is a subpopulation of oocytes with a strongly reduced level of recombination over all chromosomes which predisposes the oocyte to nondisjunction of chromosome 21. The central problem is how the occurrence of a subpopulation of oocytes with reduced recombination could be released during ovulation with increasing frequency with advancing maternal age. Henderson and Edwards advanced the 'production line theory' to explain this: the theory assumes that oocytes which were the first to differentiate during development of the embryonic ovary had a higher chromosome recombination frequency than those which differentiated later; oocytes were subsequently released during monthly ovulation in the order of their embryonic differentiation according to the maxim, first in – first out, last in – last out. Many questions arising from the production line theory remain unanswered, including, how can the order of oocyte differentiation induce the postulated differences in recombination frequency. In 1996 Lamb *et al.* introduced their two-hit model of nondisjunction in which the first hit was a reduced recombination in particular oocytes and the second hit a generalized ovarian aging related to maternal age. This model explains the rare occurrence of Down syndrome babies to young mothers to be caused primarily by a reduced recombination and the much higher incidence in older mothers by a combination of both ovarian aging and reduced chromosome recombination.

See also: **Aneuploid; Down Syndrome; Gene Mapping; Klinefelter Syndrome; Monosomy; Nondisjunction; Trisomy 18; Turner Syndrome**

Trisomy 18

J C Carey

doi: 10.1006/rwgn.2001.0396

Trisomy 18 syndrome, also known as Edwards syndrome, was originally described by Professor John Edwards of Oxford University and his colleagues in a single case report published in 1960 (Edwards *et al.*, 1960). Other case descriptions in North America soon followed and this syndromic pattern became established by the mid-1960s. Since then, hundreds of case reports and several series have been published throughout the world. The trisomy 18 syndrome is due to an extra copy of chromosome 18 and represents the third most common autosomal syndrome behind trisomy 21/Down syndrome and the deletion 22q11 syndrome. Based on a number of population studies performed in different areas of North America, Europe, and Australia, the prevalence at birth in liveborn infants ranges from 1 in 3600 to 1 in 8500 (summarized by Embleton *et al.*, 1996). The most accurate estimate of the frequency of trisomy 18 in live births with only a minimal influence by prenatal screening is from the Utah study which documents a prevalence of just under 1 in 6000.

The pattern of malformation, i.e., the syndrome, consists of a recognizable constellation of major and minor anomalies, a predisposition to increased neonatal and infant mortality, and a significant neurodevelopmental and motor disability in the older surviving children. The most consistent findings include the presence of prenatal growth deficiency (low weight for gestational age), recognizable craniofacial features (high forehead, short palpebral fissures, small face for cranium, external ear anomalies, and micrognathia), a distinctive hand posturing (overriding fingers, camptodactyly, and nail hypoplasia), a short sternal length, and foot deformities. While the facial features are not as distinctive as in the newborn with Down syndrome, the clinical diagnosis in the neonate with trisomy 18 is relatively straightforward for the clinician with training in clinical genetics, dysmorphology, and neonatology. The hand findings are particularly unique as the combination of overriding fingers with the second on the third, accompanied by the camptodactyly is observed in only a few other congenital malformation syndromes. In addition to the prenatal growth deficiency and pattern of minor anomalies, infants with trisomy 18 are at risk for a number of medically significant structural defects: over 90% of children will have a cardiovascular malformation; other congenital anomalies that occur in varying proportions of 10–50% of patients include omphalocoele, tracheal esophageal fistula, radial aplasia, and talipes equinovarus. Defects of the external ear are common and usually comprise a small ear that is broad with unraveling to the helix and is frequently accompanied by fusion to the scalp skin (crypotia). About 5% of infants will have an open neural tube defect, usually meningomyelocele. A summary of these and other features of trisomy 18 are available in pediatric and genetic texts.

The cardiovascular malformations, like in most chromosome syndromes, are nonrandom and relatively specific. In large series, about 90% of trisomy 18 patients with a heart defect have a ventricular septal defect with polyvalvular disease. Usually the valvular dysplasia involves two or three of the heart valves and occasionally produces the tetralogy of Fallot. About 10% of patients with trisomy 18 syndrome will have a more complicated cardiac malformation, such as a double outlet right ventricle or endocardial cushion defect.

The most significant manifestation of infants with trisomy 18 syndrome is the increased neonatal and infant mortality. This particular feature of the syndrome is well known to perinatologist, neonatologists, and pediatricians. Since the mid-1980s, five population-based survival studies have been performed in various parts of the United Kingdom, Australia, and the United States. The actual figures are remarkably similar in all of the investigations and indicate that about 50% of newborns with trisomy 18 die by 7 days of age. Approximately

80–90% have succumbed by 6 months of age with over 90% having died by 12 months of age. It is this increased infant mortality that has often led physicians to regard trisomy 18 as a 'lethal condition.' However, it is important to emphasize that about 5% of children will survive the first year of life. The exact reason for death in children with this syndrome is not completely clearcut. Recent investigations into the natural history of trisomy 18 have indicated that central apnea or its presence in combination with other medical problems is the primary cause of the high infant mortality. Aspiration events, upper airway obstruction, hypoventilation, and the heart defects accompany central apnea as part of the multifactorial complex of findings that are related to this increased infant mortality. Other medical problems that complicate morbidity in infancy include feeding difficulties, gastroesophageal reflux, and the potential for congestive heart failure related to the cardiovascular malformations. The majority of infants with trisomy 18 who survive the newborn period are not able to feed by mouth and require tube feedings. Placement of a gastrostomy tube becomes a consideration for older infants.

The other major manifestation of trisomy 18 is the neurodevelopmental disability. As mentioned, less than 10% of children survive the first year of life. However, once a child with trisomy 18 is older than 12 months, the individual appears to have passed an important threshold regarding the occurrence of central apnea and the aspiration events mentioned above. (In later childhood if there is a serious illness or demise, it is usually due to a more specific medical reason or complication such as pneumonia and pulmonary hypertension.) All older infants and children with trisomy 18 exhibit a significant but nonregressive neurodevelopmental and motor disability. While all children progress in their milestones, they do so quite slowly. Toddlers and older children have enough developmental involvement that they are not able to walk unassisted or use verbal expression. A study by Baty *et al.* (1994a, b) reviewed the developmental and medical records of 62 children with trisomy 18. Analysis of these developmental evaluations showed that on average the age equivalent performance skills were between 6 and 10 months of age regardless of chronological age. Although all individuals with trisomy 18 in the study were functioning at an age of severe to profound developmental lag, the children did achieve many skills of early childhood and continued to learn throughout life: a number of older children with trisomy could use a walker and several were able to feed themselves and understand cause and effect. Of note, there is one child in the medical literature with trisomy 18 who was able to walk unsupported. Investigations on this child showed that he had full trisomy 18 without mosaicism. Because of the developmental progress in older infants and young children, referral for early intervention programs is recommended. Due to the increased infant mortality and degree of developmental disability, the ethical issues surrounding the management of infants and children with trisomy 18 are quite complicated, yet seldom discussed. As mentioned above, a small but definite proportion of infants with trisomy 18 will be alive at 12 months of age and it is usually not possible in the newborn period to predict survival accurately. These issues are discussed in more detail in a recent review by the author (Carey, 2001). Other complications of older children include hearing loss, scoliosis, and increased risk for Wilms' tumor. The author has outlined suggestions for routine health supervision (Carey, 2001) in persons with trisomy 18.

Trisomy 18 is frequently recognized in the prenatal setting because of the high occurrence of ultrasound abnormalities that are seen in second trimester fetuses with this condition. In addition, prenatal triple screen will show the presence of a low value of all three parameters and such population programs will detect about 60% of second trimester fetuses with trisomy 18. Thus, the dilemma of discussing the condition and its natural history in the prenatal scenario is a common occurrence in these times.

Trisomy 18 is usually due to a complete trisomy, i.e., three copies of chromosome 18. Over 90% of infants in recent population series who had the Edwards syndrome phenotype have complete trisomy, while about 8% have either mosaicism or a partial 18q trisomy. In full trisomy 18, the extra chromosome is presumably present due to a nondysjunctional event in meiosis. The error in nondysjunction predominantly occurs in oogenesis and is evenly divided between meiosis I and II. This is in contrast to all other human trisomies where the error is usually in maternal meiosis I. Thus, the biology of nondysjunction in trisomy 18 is unique. However, as in the case of the other common autosomal trisomes, i.e., 21 and 13, there is a maternal age affect. As in other chromosome syndromes, prenatal diagnosis with amniocentesis or chorionic villous sampling in future pregnancies is routinely offered to families who have had a child with trisomy 18.

Families of infants and children with trisomy 18 in Utah formed an international lay advocacy group in 1980. This group, called the Support Organizational for Trisomy 18, 13, and Related Disorders (SOFT), now includes thousands of families from all over the world. SOFT publishes a newsletter six times a year, holds an annual conference, and connects families to each other for support. The contact address is SOFT, 2982 South Union Street, Rochester, NY 14624, USA,

1-800-716-SOFT, and the web page is www.trisomy.org. The contact address for SOFT UK is through Tudor Lodge Redwood, Ross on Wye, Herefordshire HR9 5UD, UK (tel: 01-989-67480). Another organization entitled Chromosome 18 Registry and Research Society, San Antonio, Texas, USA (http://www.chr18.uthscsa.edu) focuses on research aspects of chromosome disorders involving chromosome 18.

References

Baty BJ Blackburn BL and Carey JC (1994a) Natural history of trisomy 18 and trisomy 13: I. Growth, physical assessment, medical histories, survival, and recurrence risk. *American Journal of Medical Genetics* 49: 175–188.

Baty BJ, Jorde LB, Blackburn BL and Carey JC (1994b) Natural history of trisomy 18 and trisomy 13: II. Psychomotor development. *American Journal of Medical Genetics* 49: 189–194.

Carey JC (2001) The Common Medically Serious Classical Trisomy Syndromes. In: Cassidy S and Allanson J (eds) *Management of Common Genetic Syndromes*. New York: John Wiley.

Chromosome 18 Registry and Research Society, San Antonio, Texas. http://www.chr18.uthscsa.edu

Embleton ND, Wyllie JP, Wright MJ, Burn J and Hunter S (1996) Natural history of trisomy 18. *Archives of Disease in Childhood* 75: 48–41.

Edwards JH, Harnden DG, Cameron AH, Crosse VM and Wolff OH (1960) A new trisomic syndrome. *Lancet* 1: 787–788.

SOFT. http://www.trisomy.org

See *also:* Trisomy

Triticum Species (Wheat)

J Dvořák

doi: 10.1006/rwgn.2001.1672

Evolution and Domestication

Evolution

Triticum (wheat) comprises six biological species at the diploid, tetraploid, and hexaploid levels (**Table 1**). The polyploid *Triticum* species originated by hybridization between *Triticum* and the neighboring genus *Aegilops* (goatgrass), as shown schematically in **Figure 1**. The tetraploid species, *T. turgidum* (genomes AABB) and *T. timopheevii* (genomes AAGG), are polyphyletic. The A genomes of both species were contributed by *T. urartu*. The B and G genomes are closely related to the genome of *Ae. speltoides* (S genome). The designation of these genomes as B and G rather than as modified S genomes has been retained for historical reasons. Since *T. turgidum* is older than *T. timopheevii*, the B genome has diverged more from the S genome of *Ae. speltoides* than the G genome. The G genome is virtually identical to the S genome at the molecular level but it differs from it, as well as from the B genome, by major structural chromosome rearrangements. Hexaploid *T. aestivum* originated some 6000–7000 years ago by the hybridization of tetraploid wheat, most likely cultivated emmer (*T. turgidum* subsp. *dicoccon*), with *Ae. tauschii*. *Ae. tauschii* subsp. *strangulata* in Transcaucasia was the principal source of the wheat D genome gene pool but, since several hybridization events were responsible for the formation of this gene pool, it cannot be excluded that *Ae. tauschii* from other geographical regions, participated in shaping it. Hexaploid *T. zhukovskyi* originated recently by interspecific hybridization of cultivated *T. timopheevii* with cultivated *T. monococcum*.

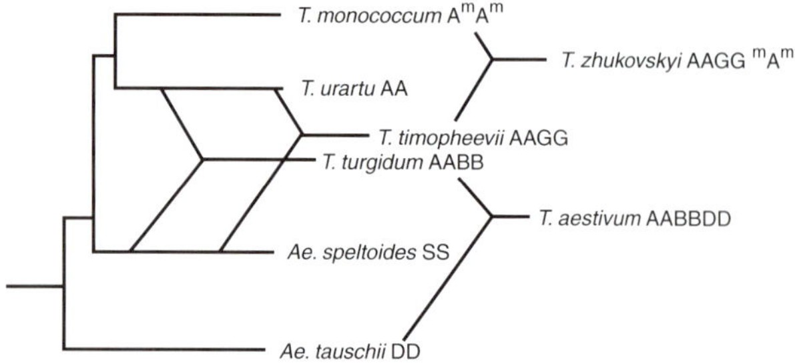

Figure 1 Evolution of the polyploid complex of *Triticum*.

Table 2 summarizes genome relationships in the *Triticum–Aegilops* alliance. Genomes designated by the same capital letter share homologous chromosomes. Different superscripts attached to a common capital letter mark slightly differentiated (modified) versions of a basic genome. Diploid sources of genomes designated as X and Y are currently uncertain, although some evidence suggests that the X genome evolved from an ancient S genome. **Table 2** also summarizes relationships among cytoplasms (plasmons) of the species in the *Triticum–Aegilops* alliance.

Domestication

Cultivated *T. monococcum*, *T. turgidum*, and *T. timopheevii* originated by domestication of wild progenitors (**Table 1**). Einkorn wheat, *T. monococcum*, was domesticated in the Karacadag mountains of southeastern Turkey about 10 000 years ago. Emmer wheat was domesticated at a currently unknown site either at the same time as *T. monococcum* or slightly later. *T. timopheevii* was probably domesticated in Transcaucasia. There is no evidence that hexaploid wheat ever existed in the wild. The semi-wild *T. aestivum* subsp. *tibetanum* (**Table 1**) is a weedy race of unknown origin. Hulled, brittle-rachis forms of *T. aestivum* have often been considered ancestral to the free-threshing

Table 1 Ploidy, domestication status, and spike characteristics of *Triticum* species and subspecies

Ploidy	Status	Spike	Species	Subspecies
2×	wild	hulled, brittle	*T. monococcum*	*aegilopoides* (wild einkorn wheat)
	cult.	hulled, nonbrittle	*T. monococcum*	*monococcum* (cultivated einkorn wheat)
	wild	hulled, brittle	*T. urartu*	
4×	wild	hulled, brittle	*T. turgidum*	*dicoccoides* (wild emmer wheat)
	cult.	hulled, nonbrittle	*T. turgidum*	*dicoccon* (cultivated emmer wheat) *Paleocolchieum*
		naked, nonbrittle	*T. turgidum*	*durum* (durum), *turgidum* (pollard wheat), *turanicum* (Khorassan wheat), *polonicum* (Polish wheat), *carthlicum* (Persian wheat), *isphahanicum*
	wild	hulled, brittle	*T. timopheevii*	*armeniaeum* (syn. *araraticum*)
	cult.	hulled, nonbrittle	*T. timopheevii*	*timopheevii*
6×	cult.	hulled, brittle	*T. aestivum*	*macha*, *tibetanum* (Tibetan wheat)
		hulled, partially brittle	*T. aestivum*	*spelta* (spelt), *vavilovii*, *yunanense* (Yunan wheat)
		naked, nonbrittle	*T. aestivum*	*aestivum* (bread wheat), *compactum* (club wheat), *sphaerococcum* (Indian dwarf wheat), *petropavlovskyi* (Chinese rice wheat)
	cult.	hulled, nonbrittle	*T. zhukovskyi*	

forms (**Table 1**). However, both molecular and archeological evidence suggest that modern hulled forms of *T. aestivum* were probably derived by hybridization between free-threshing hexaploid wheats and hulled domesticated or wild emmer wheat.

Cytogenetic Structure

Genomes within the *Triticum–Aegilops* alliance are large ranging from 4.1 pg (4024 Mb) of DNA per 1 C nucleus in *Ae. tauschii* (Arumuganathan and Earle, 1991) to 6–7 pg of DNA per 1 C nucleus in a number of other diploid species (Bennett, 1972; Arumuganathan and Earle, 1991). The 1C DNA content of *T. aestivum* was estimated to be 16.5 pg (15 966 Mb) (Arumuganathan and Earle, 1991) and 18 pg (Bennett, 1972). The relative sizes of the three *T. aestivum* genomes are B > A > D (**Figure 2**). Repeated nucleotide sequences comprise a minimum of 83% of the *T. aestivum* genomes (Flavell *et al.*, 1974).

The basic chromosome number x is 7 in all genomes of the *Triticum–Aegilops* alliance. Accessory chromosomes have been reported only in *Aegilops mutica* (syn. *Amblyopyrum muticum*) and *Ae. speltoides*.

Triticum aestivum cultivar Chinese Spring has been shown to possess a 'primitive' chromosome structure and has been extensively used in genetic studies and the development of cytogenetic stocks. Its idiogram (**Figure 2**) was adopted as the wheat standard. The ability to replace a chromosome in one wheat genome by a chromosome from another wheat genome (as in the nullisomic–tetrasomic lines) led to the identification of related (homoeologous) chromosomes in *T. aestivum* A, B, and D genomes (**Figure 2**). To reflect homoeologous relationships among wheat chromosomes, the 21 chromosomes of *T. aestivum* have been assigned to seven groups of three homoeologous chromosomes, one from each genome. For example, chromosomes 1A, 1B, and 1D are the A-, B- and D-genome chromosomes, respectively, of wheat homoeologous group 1 (**Figure 2**). Wheat homoeologous groups are the standard for the assignment of homoeologous relationships among chromosomes across the tribe Triticeae.

The regions of wheat chromosomes containing constitutive heterochromatin can be revealed by the C-banding procedure followed by staining with Giemsa (heterochromatin stains dark). The pattern of C-bands facilitates unequivocal identification of each of the wheat chromosomes (**Figure 2**). Chromosomes of the B genome are most heavily heterochromatic (**Figure 2**) with the principal heterochromatic sequence being the Ag-satellite, based on a trinucleotide motif $(GAA)_n$. Wheat chromosome arms are designated S for the short arm and L for the long arm (wheat chromosome arm designations p for short and q for long, that have been used in other genomes, did not find a broad acceptance and were abandoned after a few years). The subdivision of wheat chromosome arms into smaller segments based on C-banding is shown in **Figure 2**.

The majority of Chinese Spring chromosomes belong to a single homoeologous group. An exception are the structurally rearranged chromosomes 4A, 5A,

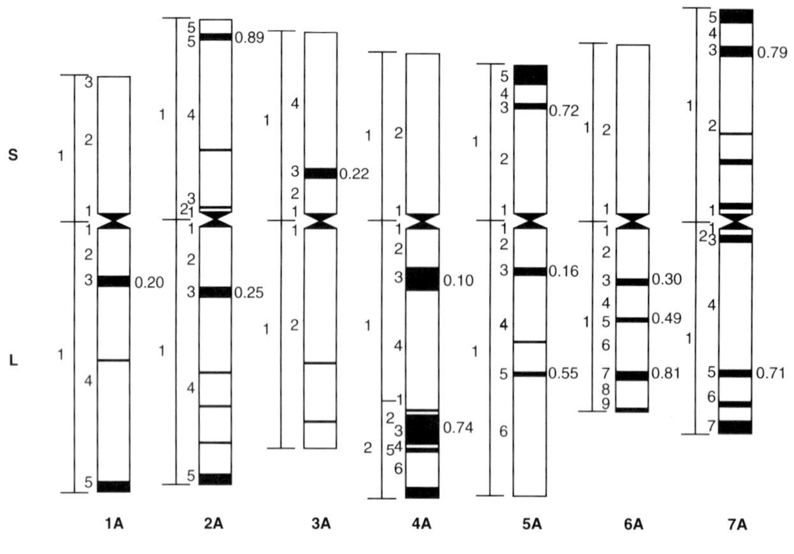
S
L
1A
2A
3A
4A
5A
6A
7A

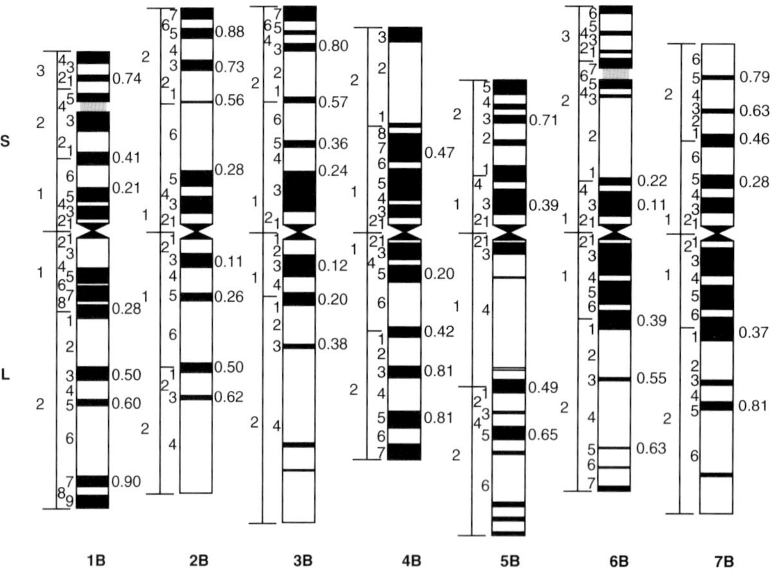
S
L
1B
2B
3B
4B
5B
6B
7B

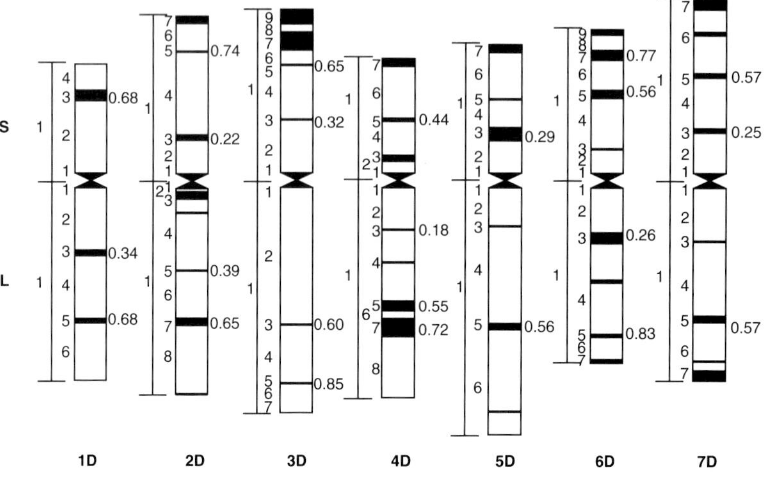
S
L
1D
2D
3D
4D
5D
6D
7D

Figure 2 (Opposite) Idiogram of the *Triticum aestivum* cultivar Chinese Spring complement. Chromosomes 1A to 7A, 1B to 7B, and 1D to 7D belong to the A, B, and D genomes, respectively. Chromosomes designated by the same Arabic numeral are homoeologous. Chromosome arms are designated S for short and L for long. Except for chromosome 4A, S arms are homoeologous with each other and L arms are homoeologous with each other in the seven homoeologous groups. Each arm is divided into one to three regions, numbered 1 through 3, delineated by major C-bands (numbers to the left of chromosomes). The regions are further subdivided into C-positive (solid rectangles) and C-negative (opened rectangles) bands. The first band in each region is a C-positive band. Bands within a region are numbered in the proximal to distal direction. The position of major C-bands in a chromosome arm is characterized by the fraction length (numbers to the right of chromosomes) of the arm (Endo and Gill, 1996). (Courtesy of B. S. Gill.)

Table 2 Genome and plasmon constitution of species in the *Triticum–Aegilops* alliance

Species	Plasmon[a]	Genome[b]
T. monococcum	A^2	A^m
T. urartu	?	A
Ae. speltoides	S,G,G^2	S
Ae. searsii	S^v	S^{se}
Ae. bicornis	S^b	S^b
Ae. sharonensis	S^l	S^l
Ae. longissima	S^{l2}	S^l
Ae. uniaristata	N	N
Ae. comosa (incl. *Ae. heldreichii*)	M, M^h	M
Ae. caudata	C	C
Ae. umbellulata	U	U
Ae. tauschii	D	D
Ae. mutica (syn. *Amblyopyrum muticum*)	T,T^2	T
T. turgidum	B	AB (B is related to S)
T. aestivum	B	ABD
T. timopheevii	G	AG (G is related to S)
T. zhukovskyi	G	AGA^m
Ae. cylindrica	D	DC
Ae. ventricosa	D	DN
Ae. crassa 4×	D^2	D^cX
Ae. crassa 6×	D^2	D^cXD
Ae. vavilovii	D^2	D^cXS^{se}
Ae. juvenalis	D^2	D^cXU
Ae. triuncialis	U, C^2	UC
Ae. columnaris	U^2	UY
Ae. neglecta 4× (syn. *Ae. triaristata*)	U	UY
Ae. neglecta 6× (syn. *Ae. recta*)	U	UYN
Ae. geniculata (syn. *Ae. ovata*)	M° (T)	UM°
Ae. biuncialis	U	UM°
Ae. kotschyi	S^v	US^l
Ae. peregrina (syn. *Ae. variabilis*)	S^v	US^l

[a]Superscripts indicate minor differentiation of a plasmon from the basic type (for more details see Wang et al., 1997).
[b]Modified from Dvořák (1998).

and 7B. The first step in these rearrangements was fixation of a 4A–5A translocation in einkorn wheat prior to the divergence of *T. monococcum* and *T. urartu*. The remaining rearrangements were fixed during the evolution of *T. turgidum* subsp. *dicoccoides*. These involved a pericentric inversion in 4A which converted the long arm into the short arm, a paracentric inversion in the 4AL arm, and a reciprocal translocation between 7BS and the rearranged 4AL arm. Translocations fixed during the evolution of *T. turgidum* subsp. *dicoccoides* differ from those fixed during the evolution of *T. timopheevii* subsp. *armeniaeum*.

Although additional small terminal translocations and inversions in the A and B genomes may have been fixed during the evolution of the A, B, and D genomes of *T. aestivum*, and have escaped molecular detection, the order of loci in wheat homoeologous chromosomes is largely colinear.

The genomes in the *Triticum–Aegilops* alliance possess either one or two pairs of nucleolar organizing regions (NORs). In the chromosome complement of *T. aestivum*, NORs are on chromosome arms 1AS (*Nor9*), 1BS (*Nor1*), 6BS (*Nor2*), and 5DS (*Nor3*). These multigene loci contain several hundred to several thousand repeated 18S-5.8S-26S rRNA gene units arranged in tandem and separated by nontranscribed spacers. Although additional minor loci were detected by *in situ* hybridization in wheat, they may not contain complete gene units and no evidence exists that they function as NORs. Also, comparative mapping in Triticeae showed that the multigene *Nor* loci and multigene loci encoding 5S rRNA have occasionally transposed into new locations during the evolution of wheat and other Triticeae genomes without perturbation of the colinearity of surrounding chromosome regions. The genetic mechanism of these transposition events is not known.

Cytogenetic Stocks and their Use

Aneuploids and Alien Addition and Substitution Lines

Because of hexaploidy, *T. aestivum* is exceptionally tolerant of aneuploidy and sets of monosomics, tetrasomics, ditelosomics, and double ditelosomics for each of the 21 chromosomes of Chinese Spring have been developed (Sears, 1954; Kimber and Sears, 1968). Since limited polymorphism exists in Chinese Spring wheat, Chinese Spring aneuploid stocks are not absolutely isogenic. That fact must be considered in experimental designs and interpretation of results. The definitions and designations of the various types of wheat aneuploids and alien addition and substitution lines and structural chromosome variants are compiled in **Table 3**.

Monosomics hold a central position among wheat aneuploid stocks since they facilitate development of other cytogenetic stocks and have played a critical role in gene mapping in wheat. Selfed monosomics segregate for nullisomics in their progeny. Theoretically, half of the microspores or megaspores produced on a monosomic plant should receive the monosome and half should not. However, since the univalent is lost in about 50% of meioses, only about 25% of the microspores and megaspores acquire the monosome; about 75% are nullisomic. Since nullisomic eggs are functional in *T. aestivum*, a monosomic produces about 25% euploid eggs and 75% nullisomic eggs. However, the transmission of nullisomy via pollen is adversely affected by competition between monosomic and nullisomic pollen grains. Sears, (1954) reported the frequency of nullisomic progeny from selfed monosomics to range from 0.9% for monosomic 5D to 7.6% for monosomic 3B. From these frequencies, he inferred that only 4% of the pollen grains involved in fertilization are nullisomic while 96% are euploid. Hence, a selfed monosomic plant is, on average, expected to produce 24% euploid, 73% monosomic, and 3% nullisomic zygotes.

Irregular disjunction of the univalent in monosomics occasionally results in centromere misdivision, leading to the occurrence of telocentric chromosomes (telosomes) and isochromosomes (isosomes) in the progeny. Sears (1954) and Sears and Sears (1979) developed ditelosomic and double ditelosomic stocks for all 42 chromosome arms of *T. aestivum* cv. Chinese Spring and isosomic stocks for most of the arms. Monotelosomics can be produced by crossing ditelosomics with corresponding monosomics. Monotelosomics are viable and fertile and have the advantage over monosomics in that the aneuploid chromosome is identified easily with a microscope, which can be critical in crossing schemes in wheat. Monosomy has been transferred from Chinese Spring to other *T. aestivum* varieties by recurrent backcrossing, thereby facilitating chromosome manipulation in other genetic backgrounds.

Some cultivars of *T. aestivum* (Chinese Spring being one of them) possess genes for high interspecific crossability, *Kr1* (5B), *kr2* (5A), *kr3* (5D), and *kr4* (1A). Such wheat genotypes can be used to produce hybrids with virtually any species in the tribe Triticeae. By backcrossing amphiploids produced from interspecific hybrids, individual chromosomes of *Aegilops*, rye, barley, *Lophopyrum*, *Thinopyrum*, and other species have been added to the wheat chromosome complement. These genetic stocks are called monosomic or disomic alien addition lines (**Table 3**). Alien addition lines can be used as the initial material in substituting an alien chromosome for a specific wheat homoeolog (alien substitution lines) (**Table 3**).

Some wheat and alien chromosomes carry gametocidal genes. An example of a wheat gametocidal gene is the *Pollen killer* (*Ki*) locus on chromosome arm 6BL. In the *Kiki* heterozygote, male gametophytes not carrying the *Ki* allele are aborted; only pollen grains having the *Ki* allele are able to function, resulting in an extreme form of segregation distortion. The breakage of wheat chromosomes, being a frequent result of the activity of gametocidal genes, has been exploited in the development of wheat deletion stocks. An example is the breakage of wheat

Table 3 Types, designation, sporophytic chromosome numbers, and meiotic pairing configurations of wheat aneuploids, alien addition and substitution lines, and structural variants

Name	Designation (group 1 and 2 chromosomes are used as examples)	Sporophytic chromosome no.	Meiotic pairing[a]
Disomic	D1A	42	$21''$
Monosomic	M1A	41	$21'' + 1'$
Nullisomic	N1A	40	$20''$
Trisomic	Tri1A	43	$20'' + 1'''$
Tetrasomic	T1A	44	$20'' + 1''''$
Nullisomic–tetrasomic	N1A-T1B	42	$19'' + 1''''$
Monotelosomic	Mt1AS	40 + t	$20'' + t'$
Ditelosomic	Dt1AS	40 + 2t	$20'' + t''$
Double monotelosomic	dMt1A	40 + 2t	$20'' + t' + t'$
Double ditelosomic	dDt1A	40 + 4t	$20'' + t'' + t''$
Ditelo-monotelosomic	Dt1AS-Mt1AL	40 + 3t	$20'' + t'' + t'$
Monoisosomic	Mi1AS	40 + i	$20'' + i'$
Diisosomic	Di1AS	40 + 2i	$20'' + i''$
Monotelodisomic	MtD1AS	41 + t	$20'' + t1''$
Double monotelo-trisomic	dMitri1A	41 + 2t	$20'' + (t + t)\ 1'''$
Monoisodisomic	MiD1AS	41 + i	$20'' + i1''$
Double monoiso-trisomic	dMitri1A	41 + 2i	$20'' + (i + i)\ 1'''$
Double monosomic	dM1A-M2A	40	$19'' + 1' + 1'$
Double monotelo-disomic	dMtD1AS-MtD2AS	40 + 2t	$19'' + t1'' + t1''$
Monosomic addition	MA1R[b]	43	$21'' + 1'$
Disomic addition	DA1R	44	$22''$
Monotelosomic addition	MtA1RS	42 + t	$21'' + t'$
Ditelosomic addition	DtA1RS	42 + 2t	$21'' + t''$
Monosomic substitution	MS1R(1A)	41	$20'' + 1'$
Disomic substitution	DS1R(1A)	42	$21''$
Substitution double monosomic	SdM1R(1A)	42	$20'' + 1' + 1'$
Intervarietal disomic substitution	DS1ACnn[c] (1ACS)	42	$21''$
Terminal translocation	T1AS1AL-1BL[d, e]	42	$21''$
Terminal translocation (explicit description)	T1AS1AL1.4::1BL1.2	42	$21''$
Intercalarly translocation	T1AS1AL-1DL-1AL	42	$21''$
Intercalarly translocation (explicit description)	T1AS1AL1.4::1DL1.2::1AL	42	$21''$
Deletion	del1AS-1	42	$21''$

[a]$'$, $''$, $'''$, and $''''$ indicate univalent, bivalent, trivalent, and quadrivalent, respectively. Arabic numerals indicate the number of complete chromosomes present; t, a telosome; i, an isosome. Telosomes or isosomes for the opposite arms of a chromosome in a single muttiralent are placed in parentheses.

[b]Rye chromosome 1R.

[c]Cnn: *T. aestivum* cultivar Cheyenne; CS: cultivar Chinese Spring.

[d]Centromere is indicated by '.'

[e]Translocation breakpoint is indicated by '::'

Modified from Kimber and Sears (1968).

chromosomes in plants with the monosomic addition of *Ae. cylindrica* chromosome 2C. Gametophytes lacking the chromosome suffer chromosome breakage, and the aberrant chromosomes are often transmitted to progeny. Over 400 stocks with terminal deletions have been isolated in the genetic background of Chinese Spring and the location of the breakpoints on the chromosome arms determined (Endo and Gill, 1996). These deletion stocks are a powerful tool for gene mapping in wheat.

Compared to *T. aestivum*, tetraploid *T. turgidum* is far less tolerant of aneuploidy. *T. turgidum* monosomics are difficult to produce, have poor vigor and fertility, and the monosomic state is poorly transmitted to progeny. Therefore, a set of disomic substitution lines of the D-genome chromosomes for their

A- and B-genome homoeologs, developed in the genetic background of durum variety Langdon, is used instead of monosomics in mapping studies at the tetraploid level.

Gene Mapping with Monosomics

Synteny mapping with *T. aestivum* monosomics (or monotelosomics) exploits the altered segregation ratio that characterizes progeny of a monosomic compared to a disomic. In the monosomic portion of progeny from a cross between a monosomic female and euploid male, the monosome is contributed by the male parent. If the male has a recessive allele on the chromosome the monosomic F_1 will express the recessive allele. In practice, a homozygous recessive (*aa*) male is crossed with each of the 21 possible monosomics. Only one of the F_1 progeny will express the recessive phenotype, indicating the chromosome on which the locus is located. The entire F_2 progeny derived from F_1 will show the recessive phenotype, provided that the phenotype of rare nullisomics is the same as that of the hemizygous (*a-*) or homozygous (*aa*) plants. If it is not, the F_2 progeny will not be uniform. Nevertheless, even in this instance, the F_2 progeny will show a vast excess of recessive phenotypes.

If the genotypes of the parents are reversed, the male parent being homozygous for the dominant allele (*AA*) and the monosomic female hemizygous for the recessive allele (*a–*), the hemizygous (*A–*) monosomic F_1 progeny from the cross usually expresses the dominant phenotype and is phenotypically similar to the disomic (*AA*). In the F_2 generation, most plants are either homozygous *AA* or hemizygous *A–*. Since nullisomics are usually rare, segregation in the F_2 generation is greatly distorted in favor of the dominant class. In synteny mapping schemes, a homozygous dominant (*AA*) male is crossed with the entire set of 21 possible monosomics. All F_1 monosomic progenies from these crosses are usually phenotypically identical. In the F_2 generation, 20 progenies will segregate in the classical monohybrid phenotypic 3:1 ratio but one will show an excess of dominant phenotypes, indicating the syntenic group in which the locus resides. In some crosses, the superiority of the euploid pollen over the nullisomic pollen (see 'Aneuploids and alien addition and substitution lines') is weak. If the frequency of nullisomics approaches 0.25, it will be concluded that a normal Mendelian 3 *AA*:1 *aa* segregation has occurred, if the phenotype of the nullisomic (– –) is the same as that of the recessive homozygote (*aa*). Meiotic examination of the recessive class in each cross is needed in such cases. All recessive plants in the critical progeny will be found to be nullisomic. This outcome identifies the syntenic group in which the locus resides.

Intervarietal Substitution Lines

In an intervarietal disomic substitution (DS) line (**Table 2**), a single chromosome pair from wheat variety A (chromosome donor) is substituted for the homologous pair in wheat variety B (chromosome recipient). To develop an intervarietal disomic substitution line, variety A is crossed as a male with variety B monosomic for the targeted chromosome. The monosomic F_1 is selected cytologically. The monosome, which is contributed by the donor variety, cannot recombine because of the absence of a homolog. The F_1 monosomic is backcrossed to the monosomic of line B, and again a monosomic is selected in the progeny. This backcross is repeated six or more times, thereby producing a monosomic intervarietal substitution of a specific chromosome of line A in the genetic background of line B. An intervarietal disomic substitution line is then produced by selfing. Any wheat genotype can be used as a source of a chromosome but the choice of a chromosome recipient is limited by the availability of a set of monosomics in the specific genetic background.

Alien disomic substitution lines, such as the D-genome disomic substitution lines in the Langdon genetic background, can be used instead of monosomics in the production of intervarietal DS lines. In this scenario, an alien disomic substitution line is used as a recurrent parent instead of a monosomic. In each backcross generation, the male parent is monosomic for two homoeologous chromosomes (double monosomic). The absence of recombination between homoeologous monosomes ensures that an intact wheat chromosome is ultimately substituted.

Intervarietal disomic substitution lines partition the genome of a donor variety into individual chromosomes in the nearly isogenic background of a recipient variety and thus provide a powerful tool for gene mapping. If the genetic background of the recipient variety has been fully restored by backcrossing, all phenotypic differences between a disomic substitution line and the recipient variety are owing the activity of genes located on the substituted chromosome.

A locus placed into a syntenic group by analysis of disomic substitution lines can be further mapped by employing disomic recombinant substitution lines (disomic RSLs) for that specific chromosome. A disomic RSL is a line in which a single pair of recombined chromosomes is substituted into the genetic background of a recipient variety. In a population of disomic RSLs, each line is homozygous for a pair of recombined chromosomes. To develop a disomic RSL mapping population, a line with disomic substitution of a specific chromosome of line A in the genetic background of line B is crossed with line B. F_1 progeny is crossed as a male with the corresponding monosomic of line B. Monosomic progeny harbor recombined (A/B) monosomes. Homozygous disomic RSLs are produced by selfing. Because of isogenicity of the genetic background, populations of disomic RSLs can be used for mapping of genes with minor effects and genes affecting quantitative traits.

Genetic Transmission

All polyploid species in the *Triticum–Aegilops* alliance are allopolyploid. Their chromosome complements are composed of either two or three pairs of related genomes. Artificial allopolyploids invariably show some heterogenetic chromosome pairing (pairing between homoeologous chromosomes at meiosis I). In marked contrast, chromosomes pair only homogenetically (only homologs pair) in natural polyploids. Because of this, natural allopolyploids show strictly disomic inheritance. Studies of aneuploids and aneuploid interspecific hybrids showed that heterogenetic chromosome pairing in wheat is prevented by the activity of a completely dominant gene, *Ph1* (pairing homoeologous), on the long arm of chromosome 5B. If the locus is absent, heterogenetic chromosome pairing occurs. Additional suppressors of heterogenetic pairing have been detected in the A and D genomes. Of these weaker loci, the best characterized is the *Ph2* locus on the short arm of chromosome 3D. While *Ph2* has been found to exist in *Ae. tauschii*, a diploid, where it plays an unknown role, the evolutionary source of *Ph1* is currently unknown, since no diploid species has so far been found to compensate fully for the absence of *Ph1* in *Ph1*-deficient interspecific hybrids. Interestingly, accessory chromosomes of *Ae. speltoides* and *Ae. mutica* exert a pairing effect on homoeologous chromosomes similar to that of *Ph1*.

Genotypes of virtually all polyploid *Aegilops* species were found to suppress pairing of homoeologous chromosomes to some degree, indicating that disomic inheritance in these species is a result of genetic suppression of heterogenetic chromosome pairing. Since none of these species possess the *Ph1* gene, genetic suppression of heterogenetic chromosome pairing in *Aegilops* must employ different genes.

The suppression of heterogenetic chromosome pairing by *Ph1*, *Ph2*, and other loci is opposed by a number of genes that either promote heterogenetic chromosome pairing or inhibit the expression of suppressors. *Ae. speltoides* and *Ae. mutica* are polymorphic for major genes inhibiting *Ph1* activity.

The mechanism by which *Ph1* and other genes with a similar function regulate heterogenetic chromosome pairing is currently unknown. It has been suggested that *Ph1* regulates premeiotic association of chromosomes and that premeiotic associations, mediated by the centromere–spindle interactions, predetermine meiotic pairing pattern. Studies of recombination between chromosomes composed of homologous and homoeologous segments showed that *Ph1* precludes recombination in homoeologous segments even if the centromere and telomere are simultaneously homologous, which is inconsistent with these hypotheses.

The activity of *Ph1* also affects meiosis I pairing of homologous chromosomes by detecting heterozygosity in homologous chromosome pairs in wheat intervarietal F_1 hybrids. Partial suppression of crossovers in F_1 plants from intervarietal crosses reduces the regularity of chromosome disjunction and results in the presence of aneuploids in F_2 and early selfing generations.

Mating Systems

All species in the *Triticum–Aegilops* alliance, except for *Ae. mutica* and *Ae. speltoides*, are naturally self-pollinating. The outcrossing rates vary among the self-pollinating species. In wheat, the outcrossing rate is typically about 1% in field conditions.

Recombination between Homoeologous Chromosomes

The *Ph1* locus prevents recombination and meiosis I pairing between homoeologous chromosomes not only in tetraploid and hexaploid wheat but also in wheat haploids and interspecific hybrids. As a result, *Ph1* represents a potent barrier for the introgression of genes from related species into wheat. The initial strategy to incorporate alien genes into wheat chromosomes relied on the production of translocations between alien and wheat chromosomes by irradiation. The elucidation of the genetic control of heterogenetic chromosome pairing in wheat facilitated the development of techniques for introgression of alien genes by recombination between homoeologous chromosomes allowed by nullisomy for chromosome 5B or by homozygosity or hemizygosity for a recessive mutation of *Ph1*. Several *ph1* mutations exist in *T. aestivum*; the *ph1b* deletion mutation has been most extensively used. Only one mutant, *ph1c*, exists in *T. turgidum*.

Genetic Mapping

Mapping of traits in wheat has to a large extent relied on natural variation since induced mutations are rare in polyploid wheats. Mutations induced by ionizing radiation are most often large deletions. Many genes controlling isozymes, disease resistance, environmental stress tolerance, morphological traits, and other types of genetic markers have been placed into syntenic groups by monosomic, nullisomic–tetrasomic, and ditelosomic analyses, and analyses of alien disomic addition and alien and intervarietal disomic substitution lines. A compilation of mapped wheat genes can be accessed in the Wheat Gene Catalog in GrainGenes (http://wheat.pw. usda.gov).

Linkage maps employing RFLP and simple sequence repeat (SSR) markers have been developed for *T. aestivum* and *T. turgidum* and many of these maps have been compiled in GrainGenes (http://wheat.pw.usda.gov). SSR markers, most based on dinucleotide motifs, are highly polymorphic in *T. aestivum*, and most primer sets amplify DNA from only a single genome.

Extensive deletion maps have been constructed for all 21 chromosomes of common wheat. Deletion mapping is an efficient means of placing molecular markers into bins delineated by the breakpoints of terminal deletions and is the backbone of the expressed sequence tag (EST) mapping in wheat.

In wheat genomes, over 30% of loci are duplicated or multiplicated. That fact must be considered in comparative mapping and other uses of wheat genetic maps.

A notable characteristic of the wheat linkage maps is their great distortion relative to physical-type maps (such as the deletion maps) employing the same markers. These distortions reflect the fact that crossovers are preferentially localized at the ends of the chromosomes while large proximal regions of chromosomes are largely devoid of crossovers. These proximal euchromatic regions of the wheat chromosomes also tend to be poor in gene content. Genes tend to be clustered in gene-rich islands in wheat chromosomes. The locations of these islands are currently being investigated.

References

Arumuganathan K and Earle ED (1991) Nuclear DNA content of some important plant species. *Plant Molecular Biology Reporter* 9: 208–218.

Bennett MD (1972) Nuclear DNA content and minimum generation time in herbaceous plants. *Proceedings of the Royal Society of London, Series B* 181: 109–135.

Dvorak J (1998) Genome analysis in the *Triticum–Aegilops* alliance. In: Slinkard AE (ed.) *9th International Wheat Genetics Symposium*, pp. 8–11. Saskatoon Canada: University Extension Press, University of Saskatchewan.

Endo TR and Gill BS (1996) The deletion stocks of common wheat. *Journal of Heredity* 87: 295–307.

Flavell RB, Bennett MD, Smith JB and Smith DB (1974) Genome size and the proportion of repeated nucleotide sequence DNA in plants. *Biochemical Genetics* 12: 257–269.

Kimber G and Sears ER (1968) Nomenclature for the description of aneuploids in the Triyieinae. In: Finky KW (ed.) *3rd International Wheat Genetics Symposium*, pp. 468–473. Canberra, Australia: Australian Academy of Science.

Sears ER (1954) The aneuploids of common wheat. *Research Bulletin of the University of Missouri Agricultural Experiments Station* 572: 1–59.

Sears ER and Sears LMS (1979) The telocentric chromosomes of common wheat, In: Ramanujam S (ed.) *5th International Wheat Genetics Symposium*, pp. 389–407. New Delhi: Indian Society of Genetics and Plant Breeding, Indian Agricultural Research Institute.

Wang GZ, Miyashita NT and Tsunewaki K (1997) Plasmon analyses of *Triticum* (wheat) and *Aegilops*: PCR-single-strand conformational polymorphism (PCR-SSCP) analyses of organellar DNAs. *Proceedings of the National Academy of Sciences, USA* 94: 14570–14577.

See also: **Aneuploid; Grasses, Synteny, Evolution, and Molecular Systematics; *Hordeum* Species; Polyploidy**

Trivial Equilibrium

M A Asmussen

doi: 10.1006/rwgn.2001.1340

A 'trivial equilibrium' in evolutionary biology generally refers to an equilibrium state in which a population is monomorphic, that is has only one version (allele) of a certain gene. This state is also called a 'fixation equilibrium' or 'boundary equilibrium.' When there are only two alleles present at a particular genetic locus, the stability of the allele frequency equilibrium where the frequency of one of these two alleles is 0 determines whether or not that allele can be maintained in the population. If the fixation equilibrium is unstable where the frequency of allele *A*, for instance, is 0, then the frequency of allele *A* will always move away from 0, and thus increase when it is rare, thereby insuring that it can never be eliminated as long as the current conditions prevail. If such an equilibrium is locally stable, however, the frequency of allele *A* will approach 0, and it will be lost, whenever its frequency starts or becomes sufficiently low.

The stability of the two trivial equilibria, or fixation states, where only one of two alleles is present, can also give useful insight into the conditions under which both alleles (and thus genetic variation) are maintained in the population. In particular, when fixation is unstable for both alleles at a diallelic genetic locus (when either allele has a frequency of 0), then neither allele can be lost, and we say we have a

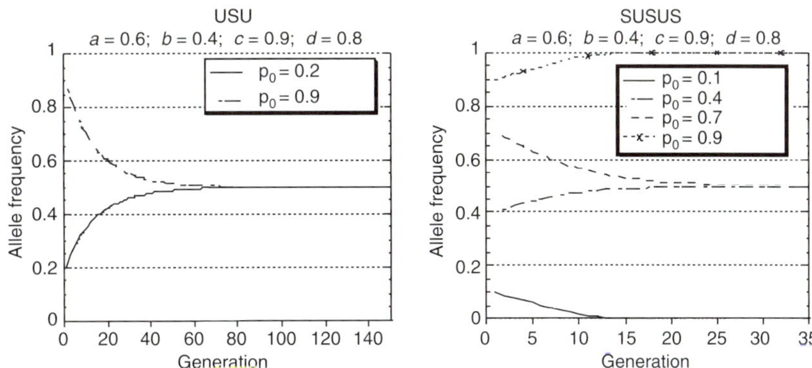

Figure 1 Sample allele frequency trajectories as a function of time in generations under a form of frequency-dependent selection for various initial allele frequencies (p_0). Left: fitness conditions giving a unstable–stable–unstable (USU) equilibrium pattern with a protected polymorphism; right: a stable–unstable–stable–unstable–stable (SUSUS) equilibrium pattern without a protected polymorphism. The first and last entries of 'unstable' (U) or 'stable' (S) refer to the stability of the two fixation equilibria with allele frequencies of 0 and 1. The intermediate entries refer to the stability of the internal, polymorphic equilibria with allele frequencies greater than 0 and less than 1. The fitness parameters (a,b,c,d) give the values of the pairwise fitnesses for the symmetric pairwise interaction model defined in Table 5 of Asmussen and Basnayake (1990).

'protected polymorphism.' A protected polymorphism thus ensures that permanent genetic variation will be maintained in the population. Because of this useful practical application, in analyzing more complicated population genetic models, where it is difficult if not impossible to derive the full analytic conditions for the maintenance of genetic variation, researchers often rely instead on determining the conditions for a protected polymorphism.

In interpreting protected polymorphism results, however, it should be realized that a protected polymorphism is only sufficient, and not always necessary, to retain a population's existing allelic variation. This is because it is sometimes possible to have a situation where an internal, polymorphic equilibrium is stable along with one or both of the two trivial equilibria. Such a combination of simultaneously stable internal and fixation equilibria occasionally arises in more complicated models of natural selection, such as frequency-dependent selection, where the fitnesses of the various genotypes vary with the changing genetic composition of the population in which they are found.

In such cases the evolutionary fate of existing genetic variation depends on the initial genetic composition of the population; from some initial allelic frequencies (e.g., those near the stable fixation equilibrium) the population will converge to the stable fixation equilibrium and lose all genetic variation, while from other initial frequencies it will converge to a stable internal equilibrium and maintain that variation. Genetic variation can thus be maintained in such populations under certain initial conditions, even in the absence of a protected polymorphism. In these populations, the conditions for a protected polymorphism underestimate the full conditions under which allelic variation is maintained.

Figure 1 illustrates this type of situation in the symmetric pairwise interaction model of frequency-dependent selection introduced by Asmussen and Basnayake (1990). In this model both alleles are maintained via a protected polymorphism whenever the fitnesses in like × like interactions (homozygote × like homozygote) are lower than those in like × unlike interactions (homozygote × heterozygote). Given the right initial conditions, however, both alleles can also be maintained in the population when fitnesses in homozygote × like homozygote interactions are less than a 1:2 weighted average of the fitnesses in homozygote × unlike homozygote and heterozygote × heterozygote interactions.

In the left panel of Figure 1, the genotypic fitnesses in the pairwise interactions among the three genotypes are such that there is a protected polymorphism; fixation for both alleles is unstable, and there is a single, stable polymorphic equilibrium, with each allele having a frequency of 0.5, to which the population converges from all initially polymorphic states. Genetic variation is thus always maintained under these conditions. Genetic variation can also be maintained in this model under the selection conditions shown in the right panel, however, where both fixation equilibria are actually stable. Here there is most definitely not a protected polymorphism, since when either allele is rare it is lost from the population; however, a permanent polymorphism may still be reached, since for intermediate initial frequencies the population converges over time to a stable internal equilibrium at which both alleles are maintained at a frequency of 0.5. The fitness conditions giving a protected polymorphism thus underestimate the potential for permanent genetic variation under this type of (frequency-dependent) selection.

References

Asmussen MA and Basnayake E (1990) Frequency-dependent selection: the high potential for permanent genetic variation in the diallelic, pairwise interaction model. *Genetics* 125: 215–230.

See *also*: Equilibrium; Frequency-Dependent Selection; Polymorphism

tRNA

See: Transfer RNA (tRNA)

Trophoblast

J L Rinkenberger and Z Werb

doi: 10.1006/rwgn.2001.1341

The individualization of the trophectoderm from the inner cell mass during the late morula stage is the first differentiation event of the developing mammalian embryo. Trophoblast cells are derived from the trophectoderm and mediate embryonic implantation, the process by which an embryo initiates and maintains contact with the maternal uterine epithelia or stroma. Trophoblast precursors are diploid, mononuclear stem cells that differentiate to form nonproliferating trophoblast and giant cells. Giant cells are characterized by their large size, increased phagocytosis, and the endoreduplication of their DNA. In humans, giant cells fuse with each other to form multinucleate syncytiotrophoblasts.

Implantation begins with the apposition of the trophectoderm to the uterine epithelia and continues with the formation of adhesive contacts between them. In some mammals, adhesion may be followed by the invasion of the uterine epithelium by the trophoblast cells during the formation of the placenta. There are two main types of placentation, epitheliochorial and hemochorial. In the epitheliochorial, or yolk sac placenta, the uterine epithelia and trophoblast maintain intimate contacts through the interdigitation of microvilli without invasion of the uterine epithelial layer by the trophoblast. Nutrient, waste, and gas exchange occurs across these cell layers. In the hemochorial or chorioallantoic placenta, the trophoblast invades the uterine epithelial cell layer by the intrusion of the trophoblast into the uterine stroma, or the displacement of the uterine epithelium by the trophoblast.

Most ungulates, like the porcine, bovine, and ovine, have an epitheliochorial placenta. The attachment of a blastocyst is followed by adhesion and the formation of microvilli which interdigitate the trophoblast cells with the uterine epithelium. The trophoblast, or chorion, expands to cover the entire surface of the uterine epithelia surrounding the implantation site, and the maternal and fetal blood vessels form beside the uterine epithelia and trophoblast cell layers, respectively. Although the distance between the vessels may be as little as 2 μm, the integrity of the uterine epithelial cell layer is not breached by the trophoblast. There may be specialized areas of the epitheliochorial placenta for the exchange of gases or other nutrients. The majority of marsupials also exhibit epitheliochorial placentation, even though the gestation period of marsupials is short compared with other mammals, and most development of the offspring is supported by lactation after birth, rather than *in utero* by placentation.

In rodents and humans a hemochorial placenta forms when the trophoblast invades the uterine stroma. In the rat and mouse, the mural trophectoderm cells that are not in contact with the inner cell mass attach to the uterine epithelia. They become highly phagocytic and engulf the apoptotic uterine epithelial cell layer as they invade into the uterine stroma. Invasion of the stroma is facilitated by the production of serine, matrix metallo- and cysteine proteinases by the trophoblast cells. After the invasion of the stroma, the mural trophectoderm cells differentiate to form the primary trophoblast giant cell layer. The polar trophectoderm cells remain as diploid stem cells that continue to proliferate and form the ectoplacental cone. Some of these cells will become giant cells, while others become the extraembryonic ectoderm, which ultimately forms the chorion of the placenta.

Upon terminal differentiation to giant cells, the trophoblast cells expand in size and endoreduplicate their DNA. During endoreduplication multiple rounds of DNA replication occur without intervening mitosis, resulting in the accumulation of extra chromatids. In murine trophoblast giant cells, the downregulation of a transcription factor called Snail may promote endoreduplication.

The trophoblast cells produce hormones, like placental lactogen, that help to maintain pregnancy, and factors that promote maternal angiogenesis. Before the placenta forms, the embryonic trophoblast cells directly contact maternal blood sinuses, and in combination with Reichert's membrane and the yolk sac, serves as a primitive placenta to enable nutrient and gas exchange. The definitive placenta forms when the allantois, which is derived from the embryonic mesoderm, attaches to the chorion. Fetal blood vessels, including the umbilicial cord vessels, form from this mesoderm and invade the chorion and labyrinthine layers of the placenta, where nutrient and gas exchange occur.

After the mouse placenta forms, the trophoblast giant cells and diploid trophoblast cells are found in the labyrinthine and spongiotrophoblast layers. In the labyrinth, the trophoblast cells will contact the fetal blood vessels and maternal blood sinuses, whereas in the spongiotrophoblast layer the maternal blood sinuses and stromal cells are in close apposition. In humans the cytotrophoblast cells invade and fuse with the maternal spiral arterioles coming in direct contact with maternal blood, whereas the syncytiotrophoblast layer is bathed in maternal blood. The embryo proper does not come in direct contact with maternal blood. The trophoblast giant cell layer isolates the embryo early in pregnancy, and maintains a barrier between fetal and maternal blood vessels after the placenta forms.

Female mammals inherit one copy of the X chromosome from each parent, and one copy is transcriptionally inactivated in the trophectoderm at the blastocyst stage. In mice the trophoblast cells selectively inactivate the paternally inherited X chromosome, whereas either the maternal or paternal X chromosome is inactivated in the embryonic tissue. Genomic imprinting and X chromosome inactivation prevent the complete development of mouse embryos created from parthenogenetic, gynogenetic, and androgenetic embryos. In parthenogenotes and gynogenotes, which contain only maternal chromosomes, the extraembryonic tissues fail to develop properly. In androgenotes with only paternal chromosomes the embryonic tissues do not form.

The proper development of the trophoblast is essential to embryo implantation and normal development. In humans, implantation failure is a frequent cause of infertility. In the disease of pregnancy, preeclampsia, the cytotrophoblast exhibits shallow invasion of the maternal spiral arterioles resulting in poor placenta development, fetal growth retardation and, often, fetal death. Choriocarcinoma, a rare form of cancer, is a malignant growth of the trophoblast in the absence of an embryo, that is highly invasive and may cause maternal death.

***See also:* Embryonic Development, Mouse; Embryonic Stem Cells**

Trypsin

T T Baird Jr and C S Craik

doi: 10.1006/rwgn.2001.1342

Trypsin, the canonical serine protease, has often been used to investigate the relationship between the structure of proteins and their respective functions. This relationship is one that exists in a delicate balance. It has been refined through evolution and is a significant challenge to understand. Historically, various and sometimes harsh, chemical and biochemical methods have been used to dissect this relationship. However, with modern molecular biology and biophysical technology, one may examine and explore this relationship in an intentional, systematic, and more selective manner. The coupling of molecular cloning with techniques such as X-ray crystallography has provided a means to directly examine the various relationships that exist between the amino acids of a given protein. Once the high resolution three-dimensional structure of an enzyme has been determined, the potential role(s) of individual amino acids in a protein often becomes more obvious, and many times new or unexpected features are discovered. Furthermore, the understanding of these relationships may be used to alter pre-existing functions or to introduce new functions into a protein in a predictable manner, i.e., protein engineering. The proteolytic enzyme trypsin has played a fundamental role in our current understanding of the relationship between the structures of the serine proteases and their activities. A common three-dimensional structure and catalytic mechanism has been conserved among the eukaryotic and prokaryotic members of this large family of enzymes. However, the diverse activities of the serine proteases that range from digestion to fertilization are the results of different sets of amino acids that are used by each enzyme for its specific function. This article profiles various approaches to study trypsin and further our understanding of enzyme structure and function, particularly in the serine proteases.

Background

Trypsin (enzyme classification number 3.4.21.4) was first described in the late 1800s as a proteolytic activity present in pancreatic secretions. Subsequent studies revealed that this enzyme specifically hydrolyzed peptide bonds C-terminal to the amino acid residues of Lys (lysine) and Arg (arginine) 10^9 times faster than hydrolysis by hydroxide ion. Since its initial discovery, trypsin has been identified in all animals, including insects, fish, and mammals. Trypsin from each source can differ slightly in activity, but the natural substrate for the enzyme is generally any peptide that contains Lys or Arg. The specificity of trypsin allows it to serve both digestive and regulatory functions. As a digestive agent, it degrades large polypeptides into smaller fragments. As a regulatory protease, it activates other proteins through limited proteolysis at specific Lys or Arg bonds.

A Model for Protein Engineering

Trypsin has been extensively studied both mechanistically and structurally producing a vast amount of information regarding how its structure relates to certain aspects of its function. This large informational database makes this enzyme an ideal candidate for examining protein structure and function in detail. Additional features of this enzyme have also made it a desirable candidate for protein engineering. It is relatively small (molecular weight = 25 kDa) and it is monomeric. The protein can be overexpressed in bacteria or yeast and its primary structure can be readily altered through the use of recombinant DNA technology. The enzyme is very robust and stable for long periods of time at low temperature, either lyophilized or in solution at low pH.

Structural Aspects

Detailed analysis of various high-resolution crystal structures of trypsin provided the initial understanding of how the three-dimensional structure relates to its function. The crystal structures of trypsin from several sources have been solved under a wide range of conditions in the presence and absence of both small molecule and macromolecular inhibitors. Analyses of these structures have identified key amino acids that are important for activation, catalysis, and substrate recognition.

Zymogen activation

In all natural systems, trypsin is expressed in an inactive form, or zymogen, known as trypsinogen. In mammals, trypsinogen is expressed in the pancreas and then secreted into the duodenum where it is activated by the highly selective protease, enteropeptidase. Enteropeptidase recognizes a specific sequence (Asp-Asp-Asp-Asp-Lys) at the amino-terminus of try psinogen and cleaves the peptide bond C-terminal to the Lys residue to produce the active enzyme. A comparison of the high-resolution crystal structures of trypsinogen and trypsin revealed that after removal of this specific N-terminal segment, a buried aspartate (Asp194) rotates about the polypeptide backbone and forms a salt bridge with the new N-terminus at Ile16 (**Figure 1**). This interaction orders a site on the polypeptide that is essential for stabilizing the negatively charged transition state of the substrate. Once this region is stabilized, trypsin is irreversibly activated.

Catalysis

Catalytic activity in trypsin proceeds by the utilization of an elegant arrangement of three amino acids that constitute the 'catalytic triad.' Using a numbering system based on chymotrypsinogen, this essential triumvirate consists of serine 195, histidine 57, and aspartate 102. These three amino acids are strictly conserved among all members of the trypsin family. Several features identified via crystallographic analysis suggested that the functional moieties of the triad do not act independently, but instead work in concert to facilitate peptide and ester bond hydrolysis. The crystal structures show that the δ-oxygen of Asp102 accepts a hydrogen bond from the δ-1-nitrogen of His57 (**Figure 2A**). This interaction highly polarizes the His such that there is a proton on the δ-1-nitrogen and not on the 2-nitrogen. Consequently, the ε-2-nitrogen faces Ser195 and acts as a general base to increase the nucleophilic character of the hydroxyl group of Ser195. In the currently accepted mechanism

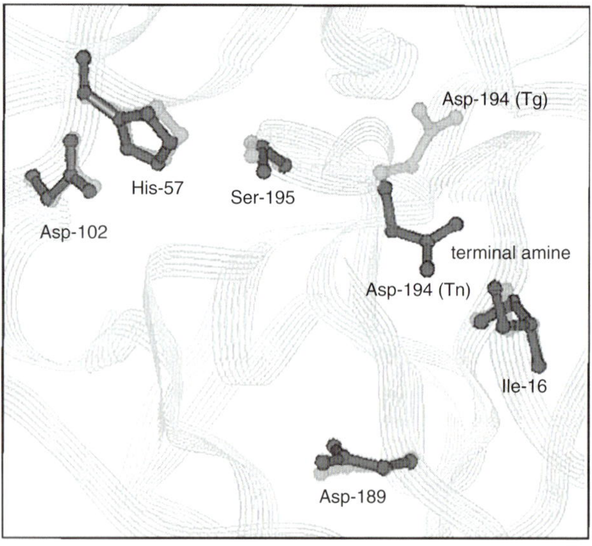

Figure 1 (See Plate 43) Arrangement of key residues in trypsinogen (Tg) and trypsin (Tn). The arrangement of the catalytic triad is essentially identical in the zymogen (Peach) and mature forms (Pink) of the enzyme. The major difference is the position of Asp194. This residue rotates about the backbone and forms a salt bridge with the N-terminus of the mature enzyme. The salt bridge formed between Asp194 and the N-terminus of the mature enzyme causes a subtle change in backbone structure and forms a site on the enzyme that stabilizes the transition state.

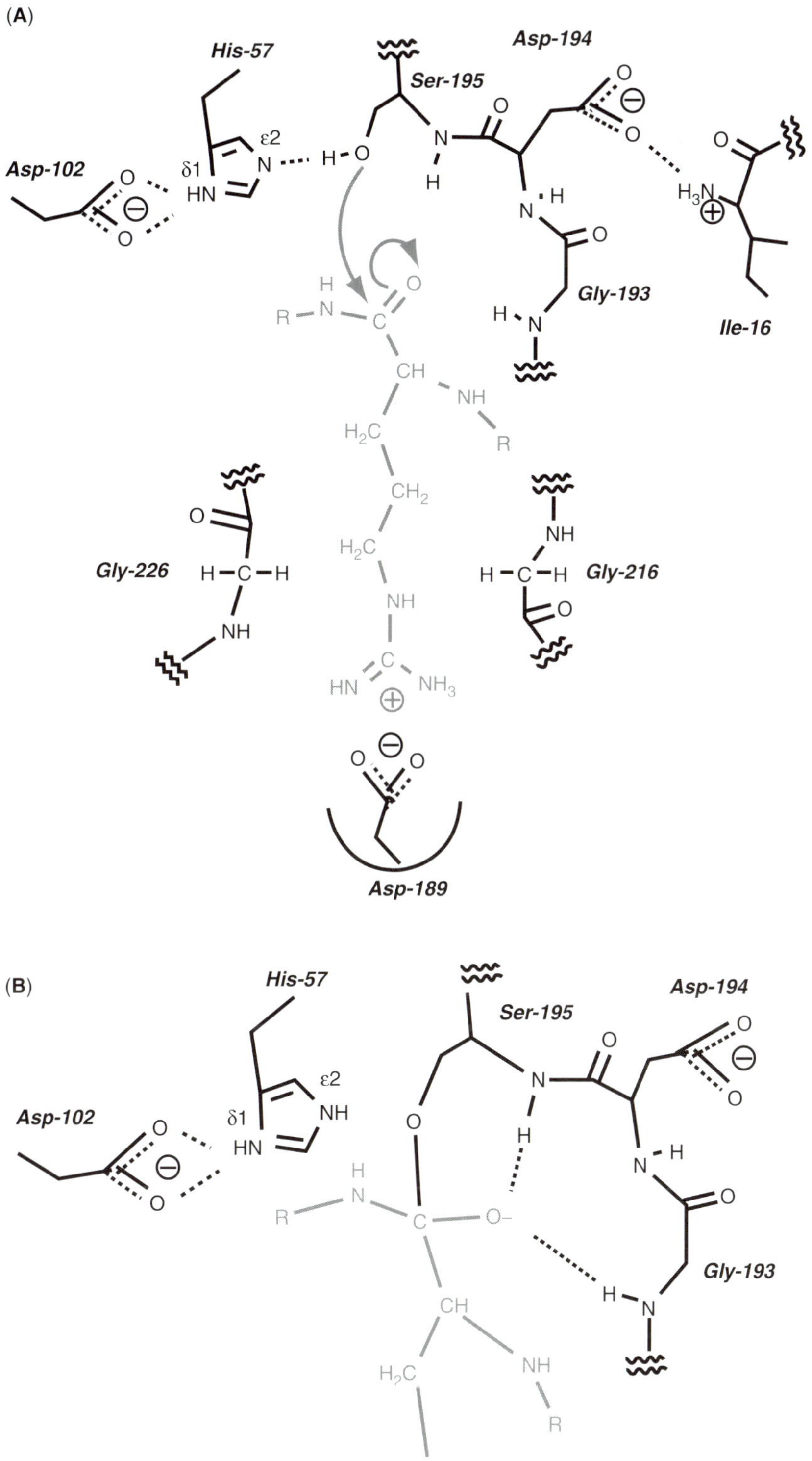

(A)

(B)

Figure 2 Schematic representation of the structure and mechanism of the trypsin active site. (A) Key features of trypsin: the arrangement of Asp102, His57, and Ser195 is essential for catalysis. The δ-oxygen of Asp102 accepts a hydrogen bond from the δ1-nitrogen of His57. The ε2-nitrogen faces Ser195 and acts as a general base to increase the nucleophilic character of the hydroxyl group of Ser195 which attacks the carbonyl carbon of the scissile peptide bond. The base of the S1 site is occupied by Asp189 and facilitates Lys and Arg recognition primarily through electrostatic interaction. (B) Stabilization of the transition state: after nucleophilic attack of Ser195 on the carbonyl carbon, the backbone amide hydrogens of Gly193 and Ser195 stabilize the negatively charged tetrahedral intermediate formed in peptide hydrolysis.

of catalysis, after the initial complex between trypsin and a substrate is formed, the hydroxyl oxygen of Ser195 attacks the carbonyl carbon of the scissile bond. The histidine then acts as a general acid and donates the proton abstracted from Ser195 to the newly formed amine or alcohol group. The first product then dissociates and a covalent acyl–enzyme complex is simultaneously formed (**Figure 2B**). Deacylation occurs through the same mechanism except that solvent provides the attacking nucleophile. The serine and histidine residues had been shown to be essential for catalysis by chemical modification experiments. Defining the role of the third member of the catalytic triad required more sophisticated analytical methods.

Substrate recognition

The understanding of substrate recognition in trypsin has been facilitated by co-crystal structures of trypsin complexed with macromolecular inhibitors such as pancreatic trypsin inhibitor. The structural basis for recognition of Arg and Lys residues is a clearly defined region in the protein referred to as the S1 site. An Asp residue (Asp189) occupies the base of this site and forms favorable interactions with the positively charged Arg and Lys side chains of bound substrates (see **Figure 2**). The cocrystal structures also revealed multiple interactions at the protease–inhibitor interface that define the extended substrate-binding pocket. These additional interactions increase the catalytic efficiency of trypsin toward peptide substrates. Residues lining the sides of the S1 site also affect substrate recognition. Glycines at positions 216 and 226 (Gly216 and Gly226) lie on opposite walls of the pocket and interact with the aliphatic portion of the long side chains of Lys and Arg. Large hydrophobic residues do not bind productively in the S1 site and negatively charged residues are chemically incompatible. Similar, but less obvious, principles are involved in the recognition of extended peptide substrates.

Mutational Analysis

Structural analysis of native trypsin provided a framework to further clarify the existing biochemical data and to assign specific roles for individual amino acids from a physical frame of reference. However, the proposed roles for various amino acids could not be tested directly until the mid-1980s when advances in recombinant DNA technology allowed their direct replacement.

Catalysis

Further support for the importance of each of the residues in the catalytic triad was obtained by replacing Ser195 and His57 of rat trypsin with alanines. These substitutions led to a 10^4-fold decrease in activity. The role of the active site aspartic acid was evaluated by replacement with an asparagine (Asn). Kinetic and structural analysis of the Asp102 → Asn variant of trypsin revealed that the role of Asp102 is to maintain the imidazole ring of His in the correct three-dimensional arrangement. Substituting any residue in the catalytic triad dramatically reduces activity. Such deleterious effects verified that each component of the triad is an indispensable part of the catalytic machinery.

Substrate recognition

Both mutational analysis and genetic selection have been used to generate trypsin variants that address the role of Asp189 in trypsin-catalyzed reactions. Kinetic analysis of these variants such as Asp-189→Ser demonstrated that the presence of a negative charge at the base of the specificity pocket is essential for catalysis. Variants that did not contain the negative charge were 10^5-fold lower in activity toward Arg or Lys containing substrates. This conclusion is further substantiated by the observation that activity was partially restored if acetate were added to the assay mixture. The crystal structure of this variant reveals that the acetate occupies the base of the specificity pocket, facilitating interaction between the variant trypsin and a substrate molecule, similar to Asp189 in native trypsin. However, substitution of Asp189 with the positively charged Lys did not result in a reversal of charge recognition. Clearly, Asp189 is not the only residue involved in substrate recognition.

Substrate specificity

Trypsin has been used to explore the structural features that govern substrate specificity among the serine proteases. For example, the arrangement of the catalytic residues and the positioning of the S1 site are remarkably similar between trypsin and chymotrypsin, but the residue at the base of the specificity pocket differs. Chymotrypsin has a Ser instead of an Asp at position 189 that facilitates association of the large hydrophobic residues typically recognized by chymotrypsin. When this residue in trypsin was substituted with Ser, the catalytic efficiency of the variant enzyme was markedly reduced against substrates containing Lys and Arg, but no increase in activity was observed against typical chymotrypsin substrates. Therefore, there are other factors governing the specificity of the protease toward its substrate. In fact, to achieve chymotrypsin-like substrate specificity, two surface loops in trypsin must be replaced by the corresponding loops in chymotrypsin. The new 'hybrid' trypsin variant exhibited an acylation rate constant equal to that of chymotrypsin toward phenylalanine amide substrates, but still was 10^3-fold lower in overall efficiency relative to wild-type chymotrypsin. This result demonstrated that portions of the enzyme distal to the active site could play essential roles in affecting catalysis. Structural comparisons of the hybrid enzyme with native trypsin and chymotrypsin illustrated that other features, such as the backbone conformation of Gly216, are important to substrate specificity. These features were altered to increase the activity of the hybrid trypsin toward typical chymotrypsin substrates.

Altering Protein Function

A goal in studying protein structure–function relationships is to understand the basic principles of structure that lead to specific function(s). The intent is to develop the capability to design a protein *de novo* that possesses a predicted function. Information from the structures of other proteins coupled with that from the numerous crystal and cocrystal structures of trypsin have been used as a guide for designing new regulatory functions into trypsin. In certain cases at least, the existing structural and biochemical information may be used to alter predictably the function of a protein by making subtle changes in its structure. For example, an enzyme whose activity could be regulated by transition metals was made. A variant of trypsin was designed that utilized the tendency of histidines to coordinate transition metals in proteins. Through molecular modeling, an Arg residue near His57 was selectively substituted with a His (Arg96→His). The expectation was that in the presence of certain transition metals, the two His residues would coordinate the metal, precluding use of His57 in catalysis. Indeed, in the presence of copper, nickel, or zinc, the proteolytic activity of this variant was arrested. Furthermore, the activity could be restored by the addition of a strong metal chelator supporting the assumption that the metal was indeed acting in the predicted manner. Structural analysis of the variant trypsin verified that a metal-regulated protease had been constructed.

A similar design scheme was developed in which metal binding was used to alter the substrate specificity of trypsin. Two residues in the extended binding pocket of trypsin were substituted with histidines that could act as metal ligands. Additionally, a histidine in the substrate was positioned such that if it completed the metal coordination polyhedron, it would register the scissile bond at the cleavage position in the protein. A substrate atypical for trypsin, containing tyrosine and a correctly positioned histidine, was used to test the design. The two protein ligands and one substrate ligand coordinated the metal thereby forcing the tyrosine residue into the specificity pocket. This strategic placement of a histidine in the substrate that facilitated productive binding of the substrate on the enzyme, resulted in a form of 'substrate assisted' catalysis. Not only was activity observed, but the efficiency was also increased through further engineered substitutions. By replacing the Asp in the specificity pocket with a Ser residue, which is more characteristic of chymotrypsin, the serine protease that typically recognizes tyrosine as a substrate, a metal dependent proteolytic activity was achieved.

Conclusion

Investigations with trypsin have demonstrated that information obtained through various studies may be used to define principles and relationships of protein structure that lead to function. Each new technology provides not only a deeper understanding of how this particular enzyme works, but also new means to explore novel mechanisms that enzymes in general may employ. The activities that have been engineered into trypsin may already be used in closely related family members. Trypsin will continue to serve as a powerful and effective 'divining rod' to discover examples of these activities.

***See also:* Proteins and Protein Structure**

Tryptophan

J Read and S Brenner

doi: 10.1006/rwgn.2001.2078

Tryptophan (Trp or W) is one of the 20 amino acids commonly found in proteins. Its nonpolar, aromatic side chain is only slightly soluble in water. Tryptophan mapping is a technique that exploits the fluorescent nature of tryptophan residues to determine the chemical environment of a specific tryptophan in a protein. Its chemical structure is given in **Figure 1**.

Figure 1 Tryptophan.

***See also:* Amino Acids; Proteins and Protein Structure**

Tryptophan Operon

C Yanofsky

doi: 10.1006/rwgn.2001.1343

The *trp* Operon and Tryptophan Biosynthesis

All organisms that synthesize tryptophan use the same sequence of biochemical reactions. In *Escherichia coli*, these reactions are catalyzed by enzymes formed from polypeptide chains encoded in the structural genes of a single transcriptional unit, the *trp* operon. This article will describe the organization of the *trp* operon of *E. coli*, the pathway and enzymes of tryptophan biosynthesis, and the regulatory mechanisms this organism employs to regulate *trp* operon expression.

Gene Arrangement in the *trp* Operon of *Escherichia coli*

The *trp* operon of *E. coli* and some other enteric microorganisms contains five major structural genes, designated *trpE* through *trpA* (**Figure 1**). These five genes encode five polypeptides bearing the seven functional domains that are necessary for tryptophan formation. The genetic segments corresponding to two pairs of functional domains are fused, yielding

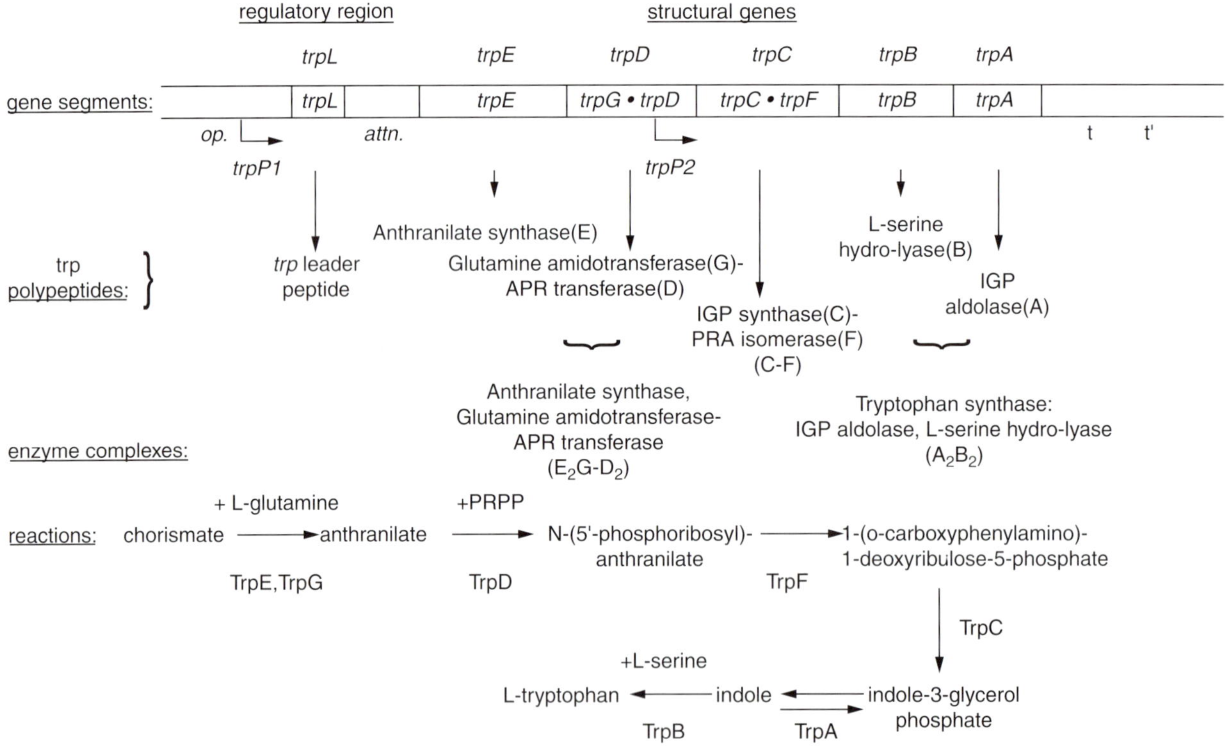

Figure 1 The *trp* operon of *Escherichia coli*, its specified polypeptides, the enzyme complexes they form, the reactions in tryptophan biosynthesis, and the polypeptide or polypeptide domain responsible for catalysis of each reaction. The operon consists of a transcription regulatory region followed by five structural genes and tandem sites of transcription termination (*t* and *t'*). The principal promoter (*trpP1*) overlaps multiple operators (*op.*) at which the tryptophan-activated *trp* repressor can bind and inhibit transcription initiation. Following the promoter, there is a transcribed regulatory leader region containing the coding region (*trpL*) for a 14-residue peptide. Transcription may either terminate at a regulated site of transcription termination, the attenuator (*attn.*), located in this leader region, or proceed into the structural genes of the operon. Two of the structural genes, *trpD* and *trpC*, consist of fused genetic segments. Each genetic segment specifies a polypeptide domain that can catalyze one of the tryptophan biosynthetic reactions. There is an internal promoter (*trpP2*) near the distal end of *trpD*. TrpA through TrpG (and A through G) refer to the polypeptide domains responsible for catalysis of the indicated reactions. Four of the five *trp* polypeptides form enzyme complexes. APR transferase, anthranilate phosphoribosyl transferase; PRA isomerase, phosphoribosyl anthranilate isomerase; IGP synthase, indoleglycerol phosphate synthase; IGP aldolase, indoleglycerol phosphate aldolase; PRPP, 5-phosphoribosyl-1-pyrophosphate.

the bifunctional polypeptides TrpG-TrpD and TrpC-TrpF. Dissection studies with these fused polypeptides have established that each domain is a more or less independent functional unit.

The Pathway and Enzymes of Tryptophan Biosynthesis

The pathway of tryptophan biosynthesis proceeds from chorismate, the common precursor of the three aromatic amino acids. Chorismate also serves as precursor of several minor aromatic metabolites, including p-aminobenzoic acid, a component of folic acid. The biochemical reactions proceeding from chorismate to tryptophan, and the seven polypeptide domains catalyzing these reactions, are illustrated in **Figure 1**. The synthesis of anthranilate from chorismate, and phosphoribosyl anthranilate from anthranilate, are catalyzed by a tetrameric enzyme complex consisting of two TrpE and two TrpG-TrpD polypeptides. Although L-glutamine is the preferred amino group donor during the synthesis of anthranilate (o-aminobenzoate) from chorismate, ammonia may be used as alternative source of this amino group by the complex or by the TrpE polypeptide alone. Glutamine utilization requires the TrpG glutamine amidotransferase domain. In the conversion of anthranilate to phosphoribosyl anthranilate, by the TrpD domain, 5-phosphoribosyl-1-pyrophosphate (PRPP) contributes the side chain of phosphoribosyl anthranilate. Phosphoribosyl anthranilate is then rearranged by anthranilate phosphoribosyl transferase, the TrpF domain, to form 1-(o-carboxyphenylamino)-1-deoxyribulose-5-phosphate (CdRP). The carboxyl group of CdRP is then removed and the pyrrole ring of the indole moiety is formed, yielding the next intermediate in the pathway, indole-3-glycerol phosphate. The latter reaction is catalyzed by indoleglycerol phosphate synthase, the TrpC domain. Indole glycerol phosphate is then converted to indole by the TrpA polypeptide of the tryptophan synthase enzyme complex; this tetrameric complex consists of two molecules each of TrpA and TrpB. Finally, indole is condensed with a pyridoxal phosphate derivative of L-serine, to form L-tryptophan; the final reaction is catalyzed by the TrpB polypeptide of the tryptophan synthase complex.

Synthesis of tryptophan from chorismate requires the products of four additional biosynthetic pathways, the compounds L-glutamine, phosphoribosyl-1-pyrophosphate, L-serine, and pyridoxal phosphate. Glutamine provides the amino group of anthranilate, phosporibosyl pyrophosphate is the source of two carbon atoms of the pyrrole ring of indole, L-serine provides the alanyl side chain of tryptophan, and pyridoxal phosphate is the coenzyme essential for activation of L-serine during catalysis of the final reaction in tryptophan formation.

Structure/Function Studies with the Tryptophan Biosynthetic Enzymes

The mechanism of enzymatic catalysis of each of the tryptophan biosynthetic reactions has been investigated and appreciable information has been gathered on the key active site residues in each biosynthetic protein or protein domain. The three-dimensional structure of the tryptophan synthase enzyme complex of *Salmonella typhimurium* has been determined, as well as the structures of complexes containing mutant protein variants. The three-dimensional structure of the bifunctional phosphoribosyl anthranilate isomerase-indoleglycerol phosphate synthase of *E. coli* has also been determined. These structures have revealed that the TrpA, TrpC, and TrpF polypeptide domains have similar structures of the α/β TIM barrel type, raising the possibility that they evolved from one another or from a common ancestor. Structural studies with the tryptophan synthase enzyme complex have shown that a tunnel connects the active site of the TrpA polypeptide to the active site of the TrpB polypeptide. Indole, generated in the TrpA active site, travels through this tunnel to the TrpB active site, where it is condensed with serine. Studies with this enzyme complex have also revealed features of the complex that explain the mutual activation of each polypeptide upon complex formation with the heterologous polypeptide.

Regulation of Expression of the *trp* Operon of *Escherichia coli*

The five structural genes of the *trp* operon are preceded by a transcription regulatory region consisting of a promoter/operator, at which transcription initation is regulated, and a transcribed leader segment, within which transcription termination is regulated. Initation at the *trp* promoter is regulated by the tryptophan-activated *trp* repressor protein; the extent of repression varies in response to changes in the intracellular concentration of free tryptophan. Repression regulates operon expression over about an 80-fold range. Polymerase molecules that have initiated transcription at the *trp* promoter and escaped repression are subject to a second regulatory mechanism, transcription attenuation. The latter mechanism determines whether or not transcription will terminate at a site located in the distal portion of the leader region. This decision is influenced by the intracellular concentration of tryptophan-charged $tRNA^{Trp}$. When the $Trp\text{-}tRNA^{Trp}$ concentration is high, transcription terminates in the leader region. When $tRNA^{Trp}$ is mostly uncharged, which occurs when cells experience a severe tryptophan deficiency, termination is avoided and transcription proceeds to the end of the operon. Transcription attenuation in the *trp* operon of *E. coli* regulates transcription of the structural genes of the operon over about an eightfold range. The combined action of repression and attenuation regulates transcription of the structural genes of the operon over about a 600-fold range. There is an internal promoter located in the distal portion of *trpD* (**Figure 1**). Transcription initiation at this promoter is unregulated and proceeds at a frequency less than 10% that attributable to the principal promoter. Tandem sites of transcription termination are located following the *trpA* structural gene; the first is protein-factor-independent, a so-called intrinsic terminator, while the second required the protein Rho. Completion of transcription of the operon yields a polycistronic messenger RNA. Ribosomes can initiate translation at any of the five major ribosome binding sites on this polycistronic messenger.

The *trp* promoter region of *E. coli* contains three operators that can bind *trp* repressor. Operator-bound repressor inhibits transcription initiation. The *trp* repressor also regulates transcription initiation in several other operons concerned with tryptophan metabolism. The three-dimensional structures of the *trp* aporepressor (aporepressor lacks bound tryptophan), the *trp* repressor, and the *trp* repressor–operator complex, have been determined. These structures have revealed the features of this protein that are responsible for its activation by tryptophan and its recognition of specific operators.

The transcribed leader region of the *trp* operon of *E. coli* is about 160 bp in length. As mentioned, this genetic segment encodes an mRNA segment that can cause transcription termination in the leader region. The transcript of the leader region can fold to form three RNA structures, termed terminator, antiterminator, and transcription pause structure. The terminator and antiterminator are alternative RNA structures, i.e., they have a sequence of nucleotides in common, thus either, but not both, can exist at one time. When cells are deficient in charged $tRNA^{Trp}$ the antiterminator forms; this precludes formation of the terminator. When cells have adequate levels of charged $tRNA^{Trp}$, the terminator forms and transcription terminates in the leader region. A deficiency of charged $tRNA^{Trp}$ is sensed during attempted translation of tandem Trp codons in a 14-residue leader peptide coding region, *trpL*, located near the 5′ end of the *trp* operon transcript. Coupling of transcription and translation, essential to this mechanism of attenuation, is achieved by the formation of the transcription pause structure, located near the 5′ end of the transcript. Polymerase pausing allows a ribosome to bind to the transcript and initiate synthesis of the leader peptide. The movement of this ribosome then releases the paused transcription complex, and transcription and translation proceed in unison.

Two of the *trp* polypeptides, the products of genes *trpE* and *trpA*, lack tryptophan, therefore they are synthesized preferentially during severe tryptophan starvation. An additional regulatory feature, translational coupling, insures equimolar synthesis of the polypeptide products of two pairs of adjacent genes, *trpE* and *trpD*, and *trpB* and *trpA*. As mentioned, the products of these genes form enzyme complexes. The enzyme complex catalyzing the first two reactions in the pathway is feedback-inhibited by tryptophan. The tryptophan binding site is located in the TrpE polypeptide.

The use of two transcription regulatory mechanisms and feedback inhibition of anthranilate synthase activity allows *E. coli* to regulate tryptophan biosynthesis efficiently in response to changes in the availability of tryptophan and the rate of protein synthesis.

Further Reading

Crawford IP (1989) Evolution of a biosynthetic pathway: the tryptophan paradigm. *Annual Review of Microbiology* 43: 567–600.

Miles EW (1995) Tryptophan synthase: structure, function, and protein engineering. *Subcellular Biochemistry* 24: 207–254.

Yanofsky C (1984) Comparison of regulatory and structural regions of genes of tryptophan metabolism. *Molecular Biology and Evolution* 1: 143–161.

Yanofsky C and Crawford IP (1989) The tryptophan operon of *Escherichia coli*. In: Neidhardt FC, Ingraham JL, Low KB *et al.* (eds) Escherichia coli *and* Salmonella: *Cellular and Molecular Biology*, vol. 2, pp. 1453–1472. Washington, DC: American Society for Microbiology Press.

Yanofsky C, Platt T, Crawford IP *et al.* (1981) The complete nucleotide sequence of the tryptophan operon of *Escherichia coli*. *Nucleic Acids Research* 9: 6647–6668.

***See also:* *Escherichia coli*; Operon**

Tumor Antigens Encoded by Simian Virus 40

J M Pipas

doi: 10.1006/rwgn.2001.1624

Simian virus 40 (SV40) is a small (45 nm) DNA-containing virus that establishes a lifelong, harmless persistent infection in the kidney of its natural host, the Rhesus maqaque. SV40 causes tumors in some rodents and confers tumorigenic properties to cell types from many species. Because it grows well in established monkey kidney cell lines and efficiently transforms rodent cell lines in culture, SV40 has been studied extensively as a model for viral productive infection and as a probe to understand molecular mechanisms of tumorigenesis.

SV40 is a member of the polyomavirus subfamily of the Papovavirus family. The other subfamily includes the papillomaviruses. Polyomaviruses are characterized by small icosahedral nonenveloped virions and circular, double-stranded DNA genomes. Some members of the polyomavirus subfamily such as murine polyomavirus or simian lymphotrophic polyomavirus, are tumorigenic in their natural hosts. Other members of the subfamily, such as murine K virus and budgerigar fledging disease virus, are important pathogens. Two human polyomaviruses, BKV and JCV, have been identified and both are closely related to SV40. Both BKV and JCV establish lifelong, harmless persistent infections of the kidney in most humans. JCV can undergo a productive infection of the brain in immunocompromized individuals and is the causative agent of progressive multifocal leukoencephalopathy, an AIDS-associated dementia.

Infectious Cycle of SV40

The SV40 virion consists of a single molecule of circular, double-stranded DNA consisting of 5243 base pairs complexed with cellular chromatin and three viral-encoded proteins termed VP1, VP2, and VP3. The viral genome also encodes three proteins that are not present in the mature virion. One of these, the agnoprotein, is poorly understood but is thought to be involved in virion assembly and/or release from cells once the infectious cycle is complete. The other two proteins termed large tumor antigen (T antigen) and small tumor antigen (t antigen), play central roles in regulating the infectious cycle and are responsible for the tumorigenic properties of SV40.

The circular viral genome is divided into two transcriptional units. The early promoter produces a single primary transcript that is differentially spliced to yield two mRNAs, one encoding large T antigen (T antigen) and the other small t antigen (t antigen). Expression of these messenger RNAs (mRNAs) requires only the cellular transcription apparatus and consequently the SV40 early promoter is frequently used in mammalian expression vectors to drive transcription of heterologous genes. The agnoproteins, VP1, VP2, and VP3, are encoded by differentially spliced mRNAs derived from the viral late promoter. The SV40 late promoter is inactive in most cell types. Large T antigen is a potent activator of the SV40 late promoter. Hence, transcription of the virion proteins requires the prior expression of T antigen in infected cells.

The infectious cycle starts when an SV40 virion attaches to a susceptible cell. Most studies have been done on cultures of established African green monkey kidney cells. Internalized virions are thought to be transported to the nucleus with uncoating of the viral chromatin occurring either during this transport or subsequent to arrival in the nucleus. The cellular transcription apparatus then drives expression from the SV40 early promoter resulting in expression of large T antigen and small t antigen. T/t expression is followed by an increase in the transcription of a number of cellular genes, many of which are involved in nucleotide metabolism (thymidine kinase), DNA replication (histones, DNA polymerase), and cell growth (rRNAs). Approximately 24 h postinfection, the infected cells enter S phase and cellular DNA is replicated. Shortly after this, viral DNA replication is initiated and continues throughout infection. Transcription from the viral late promoter begins at about the same time as viral DNA replication, followed by expression of the virion proteins and the coordinated assembly of progeny virions. Cell death occurs about 96 h postinfection with approximately 300 infectious progeny being released from each infected cell.

Not all cell types are permissive for SV40 infection. For example, human fibroblasts are semipermissive with viral replication being restricted to a few cells in the population. Rodent fibroblasts are nonpermissive and thus no progeny viruses are produced following infection of these cells.

Role of Large T Antigen in Viral DNA Replication

Large T antigen is the only viral protein directly required for viral DNA replication. The remaining proteins are recruited from the cellular replication apparatus. Viral DNA replication initiates at a unique site termed the origin of replication (*ori*) and proceeds bidirectionally around the genome. The minimal *ori* is a 64 bp fragment that contains three important elements: (1) a 21 bp imperfect palindrome; (2) a central region consisting of four repeats of the pentanucleotide GAGGC; and (3) an AT-rich region. Large T antigen possesses a DNA-binding domain that recognizes the GAGGC pentanucleotide. In addition, T antigen possesses an ATP-binding and hydrolysis domain. To initiate viral replication, T antigen monomers bind to the pentanucleotide and this is followed by the ATP-dependent assembly of a double hexamer structure that cooperatively forms around the central portion of *ori*. The formation of the T antigen double hexamer leads to a distortion of the AT-rich region of *ori* that is essential for initiation of replication. T antigen then recruits the cellular replication apparatus to *ori* by forming direct associations with DNA polymerase, primase, RPA, and topoisomerase. Following initiation, T antigen serves as a DNA helicase with each of the two hexamers hydrolyzing ATP.

The replication functions of T antigen are both positively and negatively regulated by phosphorylation. Phosphorylation at T124 is necessary for the cooperative assembly of double hexamers at *ori*, thus T antigen molecules not modified at this site are defective for replication. On the other hand, the phosphorylation of several serine residues antagonizes double hexamer formation.

Regulation of Gene Expression by the Large and Small T Antigens

Both large and small T antigens are transcriptional regulators. Large T antigen also acts as a transcriptional repressor. T antigen recognition sequences (GAGGC elements) are present in the early promoter and when T antigen binds to these sequences early region transcription is abated. Thus, T antigen regulates its own levels via an autoregulatory feedback loop. This is clearly shown during SV40 productive infections where T antigen levels rise during the first 24 h postinfection, but then reach a steady-state level. Mutation of the T antigen binding sequences in the early promoter eliminate this autoregulation and lead to constituitively high levels of T antigen.

Large T antigen is also a transcriptional activator. For example, large T antigen is necessary for activation of the viral late promoter, allowing expression of the virion structural proteins. Large T antigen also activates expression of a number of cellular genes, many of which are required to drive the cells into and through the cell cycle (see below). Transcriptional activation does not require the DNA-binding activity of T antigen. Rather, T antigen associates with the basal transcriptional apparatus and, by mechanisms that are still unclear, this association leads to transcriptional activation. For example, T antigen binds directly to TBP and to the transcriptional adapters CBP/p300 as well as to a number of transcription factors.

Small t antigen also activates the transcription of cellular genes, including cyclinD and cyclinA. Small t antigen-mediated activation of cyclinD transcription is indirect and requires small t antigen interaction with the cellular phosphatase pp2A. This association leads to inhibition of pp2A activity, which in turn leads to increased cyclinD transcription signaled by activation of the MAP kinase pathway. The mechanism by which small t antigen transactivates the cyclinA gene is not clear, but this action does not require interaction with pp2A. Rather, the small t antigen J domain, a conserved domain in all members of the DnaJ class of molecular chaperones, is necessary to activate cyclin A expression.

Viral Tumorigenesis

SV40 normally grows in growth-arrested, terminally differentiated cells; however, a number of cellular proteins required for viral replication are only expressed during S phase. Thus, a successful infection requires that SV40 drives the cells into the cell cycle. In permissive cells, this is followed by viral DNA replication, late gene expression, virion assembly, and cell death. Thus, tumor cells rarely, if ever, result from a SV40 productive infection. On the other hand, rodent cells are nonpermissive for SV40 replication. For example, SV40 infection of mouse cells results in T antigen expression and the subsequent driving of the infected cells into the cell cycle, but viral DNA replication and late gene expression are blocked. This results in a population of cells that remain in the cell cycle as long as T antigen is present. However, since no viral replication occurs T antigen is diluted as the cells divide, and eventually the culture returns to its normal growth-arrested state. This phenomenon is termed abortive transformation.

In rare instances, SV40 DNA becomes integrated in the cellular chromosome in such a manner that allows continuous T antigen expression. Such cells are permanently transformed and display a variety of altered properties including the ability to: (1) grow in the absence of serum or specific growth factors; (2) overgrow a monolayer of normal growth-arrested cells; (3) grow in the absence of anchorage to a substrate; and (4) form tumors in animals. The expression of both large and small T antigen is required for full transformation by SV40.

Small t antigen does not induce transformation by itself, but rather cooperates with large T antigen to induce the fully transformed phenotype. This function is clearly linked to small t antigen inhibition of pp2A activity and the consequent increase in cyclinD levels. Small t antigen-mediated activation of the cyclinA promoter, which is independent of its action on pp2A, also contributes to transformation.

Large T antigen is necessary for transformation and is sufficient to confer some aspects of the transformed phenotype. Large T antigen effects transformation by blocking the Rb and p53 tumor suppressor pathways. The Rb family of tumor suppressors (pRb, p107, and p130) block cell proliferation by inhibiting the action of the E2F family of transcription factors. T antigen binds to Rb proteins resulting in the release of E2F from Rb-mediated repression. This results in the activation of E2F-responsive genes and the consequent progression through the cell cycle. One cellular response to this unscheduled entry into S phase is the activation of the p53 tumor suppressor pathway. The subsequent activation of p53-responsive genes results in cell cycle arrest and apoptosis. SV40 large T antigen circumvents this defense by binding to p53 and preventing it from activating its target genes.

Clearly SV40 tumorigenesis is complicated requiring the activation and inhibition of multiple cellular pathways. Studies with animal models suggest that tumorigenesis may be even more complicated than originally thought. T antigen expression in some tissues, for example the β islet cells of the pancreas, is sufficient to drive cell proliferation resulting in hyperplasia, but is not sufficient for progression to carcinoma. This progression requires the subsequent mutation of, as yet unidentified, cellular genes. Nevertheless, the T antigens are a powerful model for understanding how the perturbation of specific cellular pathways contributes to cancer.

Future Directions

The DNA tumor viruses, including SV40, continue to provide surprises and insight into the mechanisms of cellular growth control. Future studies will be aimed at obtaining a detailed understanding of the biochemical mechanisms used by the T antigens to alter their cellular targets, and at discerning which specific aspects of the transformed phenotype are altered by T antigen action on each target. The next few years should prove very exciting in this regard.

Further Reading

Cole CN (1996) Polyomaviridae: the viruses and their replication. In: Fields BN, Knipe DM and Howley PM (eds) *Fields' Virology*, 3rd edn, pp. 1997–2043. Philadelphia, PA: Lippincott–Raven.

See also: **Origin (*ori*); Transformation; Virus**

Tumor Necrosis Factor (TNF)

T H Rabbitts

doi: 10.1006/rwgn.2001.1627

Tumor necrosis factor (TNF; also called cachectin, macrophage cytotoxin, necrosin, lymphotoxin) is a mediator of inflammatory responses. It is made by macrophages and activated monocytes, eosinophils, and NK cells. Its function in tumor biology is complex and it can have a cytotoxic and cytostatic effect. It also has an effect on angiogenesis and has a negative effect leading to cachexia in cancer patients (thereby its other name cachectin).

The gene is on human chromosome 6p2 and mouse chromosome 17. It has four exons and produces an mRNA encoding a proprotein of 233 amino acids, cleaved to 157 amino acids in its mature form.

See also: **Angiogenesis**

Tumor Suppressor Genes

C Caldas and A R Venkitaraman

doi: 10.1006/rwgn.2001.1345

Normal cellular genes whose inactivation by mutation predisposes to cancer are termed 'tumor suppressors.' Generally speaking, the protein products of tumor suppressor genes work in processes that prevent the transformation of normal cells into cancer cells. Inactivation of both copies of a tumor suppressor gene vitiates the protective function of its product, predisposing to transformation. Germline mutations in certain tumor suppressor genes give rise to inherited cancer predisposition syndromes. In predisposed individuals, the second copy of the gene is inactivated in tumors by somatic mutations.

The first tumor suppressor gene to be characterized was *Rb*, associated with the rare cancer retinoblastoma. Retinoblastoma can exhibit a familial or nonfamilial (sporadic) pattern of incidence. In the 1970s, Knudson proposed that both types could be the result of a 'two hit' mutational process inactivating a protective gene, a notion later proven with the identification of *Rb*. Familial cases inherit one defective copy of *Rb* in the germline, and inactivation of the remaining copy suffices to trigger retinoblastoma. In sporadic cases, both copies of *Rb* must be inactivated by somatic mutation in the tumor cells.

Two Classes

Broadly speaking, the protein products of tumor suppressor genes can be classified on the basis of their cellular function as 'gatekeepers' (which regulate cell division, death, or lifespan) or 'caretakers' (which preserve genetic stability) (see **Figure 1**).

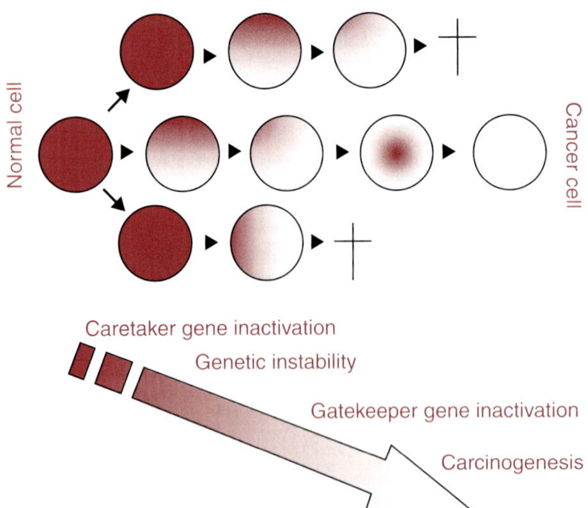

Figure 1 An overview of the events leading to tumorigenesis following the inactivation of tumor suppressor genes. Top: Multiple genetic alterations are selected during the evolution of a cancer cell from a normal cell. The loss of tumor suppressor genes belonging to the caretaker and gatekeeper classes facilitates the evolution of a cancer cell.

Inactivation of gatekeeper genes facilitates, through a variety of mechanisms, the unrestrained growth typical of cancer cells. Mutations in several different gatekeeper genes may be required for the neoplastic transformation of a cell. Gatekeeper gene products often participate in the control of cell cycle (for example, the *Rb* gene product) or in the signals that regulate proliferation (for example, the *APC* gene product).

In contrast, caretaker gene mutations lead indirectly to carcinogenesis. Inactivation of a caretaker gene typically induces an increase in the rate of genetic mutation, thereafter favoring somatic mutations in gatekeeper genes. Often, caretaker gene products participate in DNA repair or the pathways that maintain chromosome stability. Examples include the breast cancer susceptibility genes *BRCA1* and *BRCA2*, or the DNA repair genes mutated in different complementation groups of the disorder xeroderma pigmentosum.

The Biological Basis of Tumor Suppression

Space does not permit a comprehensive discussion of the functions of known tumor suppressor genes. Nonetheless, it is instructive to examine some of the biological processes in which they participate.

The division of mammalian cells is normally triggered or limited by signaling pathways transduced by ligand-receptor interactions at the cell surface. Several tumor suppressor genes of the gatekeeper class serve to regulate these pathways. For example, inactivation of the *APC* tumor suppressor dysregulates a signaling pathway initiated by the *Wnt* receptor.

Many tumor suppressors of the gatekeeper class are key regulators of progression through the cell cycle. Of particular importance in human tumors is the disregulation, by tumor suppressor inactivation, of the events that induce progression from the G_1 to the S-phase. Normally, a complex series of inhibitory interactions governed by the Rb and p53 tumor suppressor proteins prevents inappropriate G_1 to S progression. It has been estimated that this pathway is inactivated by mutation in over 80% of human tumors.

The elimination of abnormal, damaged, or unwanted cells is accomplished by a mechanism of programmed cell death (apoptosis), whose initiation and execution are controlled by several tumor suppressor gene products belonging to the gatekeeper class. Inactivation of these genes prevents or impedes apoptosis, prolonging the survival of aberrant cells that may evolve into tumors.

Normal human cells have a finite lifespan, and in culture, attain a state of replicative quiescence (termed 'senescence') after a limited number of cell divisions. Many tumor suppressor genes of the gatekeeper class (for example, *p53* and *p16*) participate in senescence induction. Their inactivation eliminates one important control preventing the unlimited cell proliferation typical of cancer cells.

Multiple mechanisms governed by tumor suppressor genes belonging to the caretaker class are responsible for the maintenance of genetic stability. They include the mechanisms that ensure that a cell can sense, signal, and repair damage to DNA that either arises during processes such as replication or is induced by exogenous agents such as UV radiation. For example, inactivation of the tumor suppressor gene *Msh2*, involved in the correction of mismatched DNA bases, increases the frequency of mutations throughout the genome and is associated with hereditary predisposition to colorectal cancer. Moreover, genes whose products participate in the mechanisms that ensure the correct segregation of duplicated chromosomes to daughter cells during mitosis may also behave as caretakers, although this remains to be firmly established.

Finally, it is important to appreciate that tumor suppressor genes may perform multiple cellular functions whose inactivation is relevant to carcinogenesis. An important example is the *p53* tumor suppressor, which has been implicated in pathways for G_1–S checkpoint control, in DNA damage sensing, in DNA repair, and in programed cell death by apoptosis. The breast cancer genes *BRCA1* and *BRCA2* have been implicated in the regulation of transcription as well as DNA repair. The *APC* tumor suppressor participates in intracellular signaling through the *Wnt* pathway, but also in the regulation of mitosis. In these and other examples, it is overly simplistic to ascribe tumor suppression to a single biological function of the mutant gene.

Approaches to Identification

The majority of 'classical' tumor suppressor genes have been characterized through the identification of germline mutations associated with predisposition to human cancer. Most of these genes have been first isolated using linkage analysis in rare, large families with highly penetrant autosomal dominant genetic predisposition to cancer. *Rb* is the first example of a tumor suppressor gene identified by this approach. These inherited cancer syndromes account for less than 5% of the global cancer burden. A different strategy will be required to identify high-prevalence, low-penetrance predisposition genes, mostly using association studies and linkage disequilibrium. The proportion of cancers attributable to these genes is probably much higher than 5%.

Tumor suppressor genes are recessive at the cellular level and therefore inactivation of both alleles is required. This is more often accomplished by mutation of one allele and deletion of the second allele. The second allele in some cases is targeted by deletion (homozygous deletions), methylation with consequent loss of expression, or mutation. Finally some mutations act as dominant negative, and effectively a single event inactivates both alleles.

p53 was the first human tumor suppressor gene identified by mutational analysis of sporadic tumors, and since then several others have been described. Classic tumor suppressor genes are defined by mutation in both familial and sporadic forms of cancer. An increasing number of candidate tumor suppressor genes are identified by somatic mutations and have not been associated with genetic predisposition. Examples include *BUB1*, *BUBR1*, *TGF-βRII*, *Axin*, *DPC4*, *p300*, and *PPARγ*.

The most frequent mechanism of inactivation of the second allele of a tumor suppressor gene is allelic deletion, and therefore loss of specific chromosomal regions occurs frequently in human neoplasia. The classic method used to detect these allelic losses was Southern blotting, with restriction fragment length polymorphism (RFLP) and, later, variable-number tandem repeat (VNTR) probes. The advent of PCR and the mapping of very informative microsatellite markers has facilitated significantly the screening for loss of heterozygosity (LOH). More recently, single-nucleotide polymorphisms (SNPs) have also been used in LOH studies. PCR-based methods can also be used for whole-genome screens for homozygous deletions, and several tumor suppressor genes have been cloned using homozygous deletion mapping (for example, *p16*, *DPC4*, *PTEN*, and *hSNF5*). Detection of chromosomal copy number loss can also be done using hybridization-based approaches: comparative genomic hybridization (CGH) and array-based CGH. Tumor suppressor genes have also been identified with functional approaches, chromosomal transfer-based complementation, and using mouse genetics.

Examples

An exhaustive review of all tumor suppressor genes identified to date and their function is beyond the space available here, and therefore a list of tumor suppressor genes is presented (**Table I**) and a selected group of tumor suppressor pathways are discussed in more detail.

Rb–p53 Pathway: Cell Cycle Control at G_1–S Transition

The *Rb* gene, first identified in a search for the gene mutated in familial retinoblastoma, encodes a member of a small family of proteins that has a critical role in the control of progression from the G_1- to the S-phase of the cell cycle. It has been estimated that the pathway controlled by *Rb* is defective in over 85% of all cancers. *Rb* itself is mutated not only in retinoblastoma but also in a variety of sporadic tumors arising in different tissues.

Rb and related proteins regulate the E2-F family of transcription factors at the G_1–S transition. Activation of the heterodimeric E2-F complex, which contains one of the E2-F subunits (E2-F1 to −5) bound to the DP protein, suffices to initiate the transcription of a number of genes essential for entry into the S-phase, including DNA polymerase, enzymes involved in nucleotide synthesis, and cyclin E. Prior to S-phase entry, E2-F is held in an inactive state bound to Rb. Passage through the G_1–S transition is triggered by cyclin-dependent protein kinases (CDKs) that form active complexes with cyclins D, E, or A. When the cyclin D-CDK complex is activated in response to mitogenic growth regulatory signals, it hyperphosphorylates Rb. Hyperphosphorylated Rb is released from its association with E2-F subunits, leaving them free to transactivate genes required for S-phase entry. Cyclin E, itself a target for transactivation, initiates a positive feedback circuit by promoting the assembly of cyclin E-CDK complexes that can also hyperphosphorylate Rb.

Table 1 Tumor suppressor genes in human cancer.

Tumor suppressor gene	Chromosomal location	Gene function	Germline mutations in hereditary cancer syndromes	Somatic mutations in sporadic cancers
RB1	13q14	Transcriptional regulator of cell cycle (G_1–S)	Familial retinoblastoma	Retinoblastoma, osteosarcoma, lung cancer
WT1	11p13	Transcriptional regulator	Wilms' tumor	Nephroblastoma, AML
p53	17q11	Guardian of genome	Li–Fraumeni syndrome	50% of human cancers
hCHK2	22q12.1	G_2–M checkpoint enforcement	Li–Fraumeni syndrome	Unknown
NF1	17q11	Ras-GAP activity	Von Recklinghausen neurofibromatosis	Neurofibroma, sarcoma, glioma
NF2	22q12	ERM protein/cytoskeletal regulator	Neurofibromatosis type 2	Schwannomas, meningiomas
VHL	3p25	Regulator of proteolysis; interacts with elongins	Von Hippel–Lindau syndrome	Renal cell carcinoma, pheocromocytoma, hemangioma
APC	5q21	Negative regulator of *Wnt* signaling; association with microtubule cytoskeleton	Familial adenomatous polyposis	Colorectal cancer
INK4a (*CDKI2A*, *p16*, *MTS1*)	9p21	Cyclin-dependent kinase inhibitor	FAMM	Pancreatic cancer; melanoma; esophageal carcinoma
PTC	9q22.3	Receptor for sonic hedgehog	Gorlin syndrome	Basal cell carcinoma, medulloblastoma
BRCA1	17q21	DNA repair; transcriptional regulator	Familial breast cancer	Breast and ovarian carcinoma (rare)
BRCA2	13q12.3	DNA repair; transcriptional regulator	Familial breast cancer	Breast and ovarian carcinoma; pancreatic carcinoma (rare)
DPC4	18q21.1	Regulator of TGF pathway	Juvenile polyposis	Pancreatic and colorectal carcinoma; hamartoma
PTEN	10q23	Dual-specificity phosphatase; regulation of PI3K /AKT pathway	Cowden syndrome	Glioblastoma; prostate and breast carcinoma
TSC1/*TSC2*	9q34/16p13.3	Associates with TSC1/TSC2; a putative GTPase-activating protein	Tuberous sclerosis	
LKB1	19p13	Serine/threonine kinase	Peutz–Jeghers syndrome	
E-cadherin (*CDH1*)	16q22.1	Cell adhesion and *Wnt* signaling	Hereditary diffuse gastric cancer	Gastric (diffuse), breast (lobular), and gynecologic carcinoma
hMSH2	2p22	DNA MMR	HNPCC	Colorectal carcinoma; MMR-deficient tumors
hMLH1	3p21	DNA MMR	HNPCC	Colorectal carcinoma; MMR-deficient tumors
hPMS1	2q31	DNA MMR	HNPCC	Colorectal carcinoma; MMR-deficient tumors
hPMS2	7p22	DNA MMR	HNPCC	Colorectal carcinoma; MMR-deficient tumors
hMSH6	2p16	DNA MMR	HNPCC	Colorectal carcinoma; MMR-deficient tumors
hMSH3	5q11-q12	DNA MMR	HNPCC	Colorectal carcinoma; MMR-deficient tumors
CBP/*p300*	16p13.3/22q13	Transcriptional regulator; acetylase	Rubinstein–Taybi syndrome (CBP)	Colorectal, pancreatic, gastric and breast carcinoma (p300)

Table I *(Continued)*

Tumor suppressor gene	Chromosomal location	Gene function	Germline mutations in hereditary cancer syndromes	Somatic mutations in sporadic cancers
hSNF5	22q11.2	SWI/SNF multiprotein complex	Hereditary MRTs	Sporadic MRTs; chroroid plexus carcinomas
PPARγ	3p25	Nuclear hormone receptor family of transcription factors	–	Colorectal carcinoma
Axin 1	16p13.3	*Wnt* signaling	–	Liver carcinoma
Axin 2 (Conductin)	17q23-q24	*Wnt* signaling	–	Colorectal carcinoma
EXT1/EXT2	8q24.11-q24.13/ 11p12-p11	Glycosyltransferase activity (heparan sulfate metabolism)	Familial exostoses	
ATM	11q23	DNA damage/repair	Ataxia telangiectasia (autosomal recessive)	B-CLL
XP(A-H)	Multiple loci	DNA excision repair	Xeroderma pigmentosum (autosomal recessive)	
NBS1	8q21	DNA double-stranded break repair	Nijmegen breakage syndrome (autosomal recessive)	
BLM	15q26.1	DNA helicase	Bloom syndrome (autosomal recessive)	
FANCA-H	Multiple loci	UV damage/repair	Fanconi anemia (autosomal recessive)	

FAMM: familial atypical mole melanoma syndrome; AML: acute myeloid leukemia; HNPCC: hereditary nonpolyposis colorectal cancer; MMR: mismatch repair; MRT: malignant rhabdoid tumor.

Cyclin-CDK activity is inhibited by two families of inhibitory proteins (CDK-Is), whose prototypic members are the p21 or $p16^{INK4A}$ proteins, respectively. The p21 family includes the CDK-Is p27KIP-1 and p57; whereas the p16 family includes $p19^{ARF}$/$p14^{ARF}$, an alternatively spliced form, and $p15^{INK4B}$. By preventing Rb hyperphosphorylation, CDK-Is antagonize the cascade of events that culminates in S-phase entry. CDK-Is of the p21 family are generally induced through the activation of the *p53* tumor suppressor, which serves to integrate the cellular response to metabolic or genotoxic stress. p53 protein is normally present only at very low concentrations in cells, due to its rapid proteolysis induced by association with a negative regulator, Mdm-2. Following exposure to stress, p53 activity becomes elevated through a variety of mechanisms. Levels of p53 protein are increased, following release from Mdm-2 association. Moreover, posttranslational modifications of p53 such as phosphorylation are induced that activate the protein. Active p53 functions as a sequence-specific activator of transcription. A number of different targets have been identified that mediate cell-cycle arrest, DNA repair, and programed cells death by apoptosis. CDK-Is such as p21 are important targets for p53 transactivation. When levels of p21 protein are increased following p53 activation, they bind to and inactivate cyclin-CDK complexes, preventing S-phase entry and cell-cycle progression. Thus, the G1–S transition is rendered sensitive to DNA damage and other cellular stresses through the activation of the p53 pathway.

The multifaceted control mechanisms that govern transition from the G1- to the S-phase of cell cycle can be perturbed by cancer-associated mutations that affect the genes encoding the cyclin-CDK complexes or the CDK-Is and other components of the p53 pathway, besides Rb. The net effect of these mutation-induced changes is to effectively remove the controls that prevent unfettered entry into the S-phase, enabling unrestrained cell proliferation.

cyclin D1 is often overexpressed in human cancers following gene amplification or translocation to an active chromosomal locus. Indeed, the chromosomal region (11q13) that contains the *cyclin D1* gene is amplified in a diverse group of human cancers, including head and neck tumors (>40%), breast cancer (~15%), and small cell lung tumors (10%). Similarly, the *CDK4* gene encoding the CDK that binds to cyclin D1 can also undergo cancer-associated amplification or mutation. That overexpression of *cyclin D1* in transgenic mice promotes carcinogenesis suggests a direct role for *cyclin D1* amplification in tumor cell proliferation.

Genes encoding the CDK-Is are frequent targets of tumor-associated mutations that result in the inactivation or deletion of their encoded proteins. For instance, the *p16* gene is mutated in familial cases of melanoma, in over 40% of sporadic pancreatic carcinomas, and in about one-third of (sporadic) tumors of the esophagus. The locus encoding *p16* undergoes homozygous deletion in about half of all gliomas, in 40% of pancreatic carcinomas, and, less frequently, in acute lymphocytic leukemias, ovarian cancer, and lung carcinomas. Consistent with these observations, mice deficient in *p16* spontaneously develop a number of different tumors.

Over half of all human tumors contain mutations in *p53*. Typically these mutations cause a missense alteration in one allele, resulting in the synthesis of a faulty p53 protein, with the second *p53* allele often inactivated by gross chromosomal rearrangements such as deletions. Over 90% of the missense mutations found in human cancers affect the DNA-binding domain of p53, encoded in the middle third of the protein. Many result in the loss of sequence-specific DNA binding, or transactivation, by the p53 molecule. Since p53 works as a tetrameric complex, mutant p53 can poison the function of the wild-type molecule in a *trans*-dominant manner. This property may explain why humans who inherit a single defective *p53* allele (Li–Fraumeni syndrome) are predisposed to a variety of different tumors.

It is important to emphasize that the transactivation capacity of p53 is important not only for correct regulation of the G_1–S transition, but also for many other functions whose inactivation is relevant to tumorigenesis. Thus, although disregulation of the *Rb* pathway by *p53* mutations is likely to make an important contribution to transformation, it is unlikely – in isolation – to explain the multifaceted role of p53 in tumor suppression.

APC and *E-cadherin*: *Wnt* Signaling, Cell Adhesion, and Human Tumorigenesis

APC mutations were initially identified in the rare hereditary form of colorectal cancer familial adenomatous polyposis (FAP). Patients with FAP develop hundreds to thousands of adenomatous polyps during the first three decades of life and almost inevitably some of these polyps will progress to invasive carcinoma. *APC* mutations have been described subsequently in the large majority of sporadic colorectal cancers.

The APC protein binds to and acts to prevent β-catenin accumulation. This finding suggests a function of APC in both cadherin-based cellular adhesion and *Wnt* signaling. The canonical *Wnt* signaling pathway has β-catenin at its heart. In unstimulated cells, free cytoplasmic β-catenin is destabilized by a multiprotein complex containing the APC tumor suppressor, Axin and glycogen synthase kinase-3β. The phosphorlation of β-catenin earmarks it for ubiquination and subsequent degradation by the proteasome. In contrast, when cells are stimulated by Wnt ligands, the cytoplasmic protein Dishevelled is recruited to the membrane, where directly or indirectly it binds to Frizzled, the seven transmembrane Wnt receptor. The mechanism by which Dishevelled inhibits the Axin complex is not completely understood, but the net result is stabilization of β-catenin. β-catenin is released from the Axin complex and translocated into the nucleus, where it binds to TCF proteins and stimulates transcription of Wnt target genes (including *c-Myc* and *cyclin D*). In addition to its role in *Wnt* signaling, β-catenin binds to the homotypic adhesion molecule E-cadherin and links the cadherin junctions to the actin cytoskeleton to mediate cell adhesion. The current model of APC's function in tumor suppression proposes that its main role is to shuttle β-catenin from the nucleus and the cytoplasm to the junctional compartment of epithelial cells. Here β-catenin is delivered either to the Axin complex for degradation or to E-cadherin to integrate adherens junctions. A role for APC in spindle function has also been suggested and could explain the chromosomal instability observed in APC mutant colorectal cancer cells.

E-cadherin is an homophilic cell adhesion molecule and integrates the *Wnt* signaling pathway. Somatic mutations of the *E-cadherin* gene, *CDH1*, were originally identified in diffuse gastric, lobular breast and gynecological carcinomas. Subsequently using a combination of linkage and candidate-gene analysis, truncating germline mutations of *CDH1* were identified in kindred with autosomal dominant predisposition to diffuse gastric cancer, affecting predominantly young individuals. The mechanism trough which E-cadherin inactivation initiates tumorigenesis is poorly understood, but it is tempting to speculate that it might be the result of increasing the cytoplasmic pool of β-catenin.

Breast Cancer Genes *BRCA1* and *BRCA2*: New Paradigms in Tumor Suppression

About one-tenth of all breast cancer cases exhibit a familial pattern of inheritance. Of these familial cases, germline mutations in either *BRCA1* or *BRCA2* occur in between 20 and 60%. Mutations in *BRCA1* or *BRCA2* are not a feature of nonfamilial (sporadic) breast cancer. *BRCA1* and *BRCA2* were first identified in 1994–95 through the analysis of families exhibiting a predisposition to early-onset breast cancer.

Founder mutations affecting these genes occur in Iceland and amongst the Ashkenazim, where they confer a highly penetrant risk of breast, ovarian, and other cancers (including cancers of the male breast, pancreas, and prostate).

The cellular functions of the proteins encoded by these large genes remain uncertain. In meiotic cells, colocalization of BRCA1 and BRCA2 proteins to the synaptonemal complexes of developing axial elements has been reported, consistent with a role in meiotic recombination, a process that is initiated by DNA double-stranded breakage. Similarly, there is increasing evidence that BRCA1 and BRCA2 are essential in mitotic cells for the repair of DNA double-stranded breaks by homologous recombination. Targeted disruption of the murine homologs of *BRCA1* or *BRCA2* gives rise to genotoxin hypersensitivity and chromosomal instability suggestive of defective DNA double-stranded break repair. Furthermore, homology-directed repair of double-stranded DNA breaks introduced into chromosomal substrates is impaired by the disruption of BRCA1 or BRCA2 pathways, although pathways for repair by nonhomologous end-joining remain unaffected.

BRCA2 interacts directly, and at a high stoichiometry, with Rad51, the mammalian homolog of the RecA protein essential for DNA repair by recombination. It has therefore been proposed that BRCA2 works to control the activity or availability of Rad51, although the precise molecular mechanism remains to be defined. The interaction of BRCA1 with Rad51 is less well characterized, although both proteins colocalize – along with BRCA2 – to discrete nuclear foci following DNA damage.

There is good evidence that *BRCA* genes work as caretakers of genetic stability. Cells that harbor disruptions in *BRCA1* or *BRCA2* accumulate aberrations in chromosome structure reminiscent of diseases such as Bloom syndrome or Fanconi's anemia, where chromosomal instability is associated with cancer predisposition. This is likely to increase the frequency of gross chromosomal rearrangements such as translocations and deletions throughout the genome.

Homozygous inactivation of BRCA1 or BRCA2 results in the failure of cell proliferation through the activation of cell-cycle checkpoints responsive to DNA damage. There is evidence that checkpoints operative during mitosis may be of particular relevance in preventing the proliferation of BRCA2-deficient cells. It therefore seems likely that inactivation of these checkpoints will be an important step in the transformation of cells lacking the *BRCA* genes.

It is unclear why carcinogenesis in individuals who inherit one mutant allele of *BRCA1* or *BRCA2* should exhibit a predilection for specific tissues such as the breast or ovary. Both BRCA1 and BRCA2 are expressed in many different cell types and have been implicated in biological functions common to all tissues. Several possible explanations have been proposed, but none is as yet supported by conclusive evidence.

Mismatch Repair Genes

The identification of the mismatch repair (MMR) genes is probably the single example where a molecular phenotype combined with cross-species comparison helped in cloning a novel class of cancer genes. Cancers from individuals belonging to kindred with hereditary nonpolyposis colorectal cancer (HNPCC) displayed evidence of replication errors upon microsatellite analysis of colorectal cancer. These replication errors were similar to those identified in yeast with defective MMR genes, homologous to bacterial MutS and MutL. Human homologs of bacterial/yeast MMR genes were cloned and shown to be responsible for the cancer predisposition in HNPCC. Loss of MMR genes, as with other genes with caretaker function, predisposes to cancer by increasing the DNA mutation rate, thereby increasing the chance of inactivation of gatekeeper genes such as *APC*.

Von Hippel–Lindau Gene

The von Hippel–Lindau (VHL) gene product forms part of a complex that targets proteins for degradation by proteolysis. One of the VHL targets for proteolysis is hypoxia-inducible factor-1 (HIF1), which in VHL-negative tumor cells becomes stabilized irrespective of oxygen concentration, leading to induction of vascular endothelial growth factor (VEGF). Other growth-regulatory molecules are also regulated by proteolysis and potential VHL targets.

Mutations in Cancer Pathogenesis

Cancer arises as a result of successive rounds of mutation and clonal selection, leading to the progressive conversion of normal human cells into cancer cells (see **Figure 1**). Several independent lines of evidence support the contention that cancer is a multistep process. Measurements of age-dependent human cancer incidence have shown that the rate of tumor development is proportional to the 4th–6th power of elapsed time, implicating four to six rate-limiting, independent stochastic events. Estimates of the number of genetic alterations derived from the study of these alterations in colorectal tumorigenesis reveal that carcinomas arise from a minimum of five or more events. Remarkably, in experiments involving the introduction of genes in human cells, a minimum of four distinct pathways must be disrupted to convert a normal cell into a tumor cell: the mitogen-response pathway, the telomere maintenance pathway, the retinoblastoma pathway, and the p53 pathway. In normal cells genomic integrity at the sequence and karyotypic levels is maintained by complex systems of DNA-monitoring and -repair enzymes and of checkpoint enforcement, and therefore mutations are rare. Disruption of one or more of these systems results in genomic instability, an enabling characteristic which allows mutations at the rate required for malignant transformation.

Clonal evolution through mutation results in the acquisition of capabilities that underlie the malignant transformation process: self-sufficiency in growth signals, insensitivity to antigrowth signals, evasion of apoptosis, evasion of senescence, ability to induce and sustain angiogenesis, and ability to invade and metastasize. Inactivation of tumor suppressor genes may underlie the acquisition of one or more of these six essential malignant characteristics (gatekeeper genes) and/or result in the enabling genetic instability (caretaker genes). Tumor suppressor genes integrate these molecular pathways that have evolved to maintain cellular homeostasis. It is not clear how many genes have to be mutated in order to disrupt a pathway, and inactivation of some genes can have consequences in several pathways (*p53* regulates genomic stability, the cell cycle, and apoptosis). In pancreatic carcinoma the *Rb* pathway is abrogated by inactivation of the *p16* gene (through point mutation, homozygous deletion, and methylation) in over 95% of cases, and disruption of other members of the pathway (*Rb, cyclin D*, and *CDK4*) is almost never seen. This contrasts with the situation in breast cancer, where *p16* mutations are rare (as are mutations in *Rb*) and *cyclin D* amplification is relatively common. This might indicate that different tissues have different gatekeepers and therefore require different mutations for tumor initiation and progression.

Conclusions

Much of the recent progress that has illuminated the biological functions of tumor suppressor genes and their role in carcinogenesis has moved in steps from identification of the genes by linkage and mutation studies in humans to the analysis of mouse strains in which genes have been disrupted by targeting. The recent availability of the human genomic DNA sequence and the anticipated availability of the murine genomic sequence in the near future should considerably facilitate both identification and biological analysis. Moreover, the widespread availability of sequence information coupled to the use of new and better techniques to manipulate the genome of organisms ranging from mice to fruit flies promises to make possible global searches for genes whose disruption or alteration confers increased cancer predisposition. It will be particularly interesting and important if these studies result in fresh insights into the question of why germline tumor suppressor gene mutations are generally associated with a tissue-specific – rather than ubiquitous – increase in cancer risk. A better understanding of this problem, perhaps through the identification of tissue-specific gatekeeper genes for carcinogenesis, will be an important step in the evolution of new strategies for cancer treatment and prevention.

Further Reading

Fearon ER and Vogelstein BA (1990) Genetic model for colorectal tumorigenesis. *Cell* 61: 759–767.

Kinzler KW and Vogelstein B (1997) Cancer-susceptibility genes: gatekeepers and caretakers. *Nature* 386: 761–763.

Knudson AG (1985) Hereditary cancer, oncogenes, and antioncogenes. *Cancer Research* 45: 1437–1443.

Nowell PC (1976) The clonal evolution of tumor cell populations. *Science* 194: 23–28.

Weinberg RA (1989) Oncogenes, antioncogenes, and the molecular bases of multistep carcinogenesis. *Cancer Research* 49: 3713–3721.

***See also:* BRCA1/BRCA2; Breast Cancer; Cancer Susceptibility; Oncogenes**

Turner Syndrome

M A Ferguson-Smith

doi: 10.1006/rwgn.2001.1346

Although the syndrome described by Turner in 1938 comprised infantilism, congenital webbed neck, and cubitus valgus in women, it has since been recognized that the important feature of Turner syndrome is abnormal ovarian development. In affected infants the ovaries appear normal and contain oogonia, but by adolescence primordial germ cells are virtually absent and the ovaries are replaced by thin streaks of ovarian stroma. These 'streak' gonads are incapable of producing sufficient estrogens for feminization and so the vulva remains infantile, breasts and pubic and axillary hair fail to develop, and there is primary amenorrhea. Short stature is invariable, and the final height is usually no more than 140 cm. Webbed neck and cubitus valgus are only two of a number of associated congenital abnormalities that may or may not be present. These associated malformations are so characteristic that they are often termed 'Turner stigmata.' The important ones are shield chest, webbed neck, cubitus valgus, peripheral lymphedema at birth, short IVth metacarpals, hypoplastic nails, multiple pigmented naevi, atrial septal defect, bicuspid aortic valve, and coarctation of the aorta. The condition presents in infancy with malformations, in childhood with short stature, and in adolescence with primary amenorrhea. Plasma and urinary gonadotrophins are elevated at puberty to postmenopausal levels. Substitution therapy with estrogens allows the adult to develop secondary sex characteristics and to live a comparatively satisfactory, although sterile, married life. Pregnancy can be achieved by ovum donation, cyclical hormone therapy, and IVF. Moderate improvement in stature (up to an additional 5 cm) is possible by prolonged treatment with growth hormone started in childhood and continued through adolescence.

The more frequent occurrence of coarctation of the aorta in males prompted Polani and colleagues in 1954 to examine the sex chromatin status of three patients with Turner syndrome and coarctation of the aorta. Sex chromatin was found to be absent and this was interpreted as evidence that the patients were sex-reversed males. A study of the frequency of color blindness in a large series of patients tended to confirm this as the frequency found corresponded with the frequency in normal males. However, in 1959 Ford and colleagues showed that the sex chromosome complement consisted of a single X chromosome and no Y. While most patients with the complete Turner phenotype and associated malformations have been shown to have a 45,X karyotype, a wide variety of variants of Turner syndrome are described with structural abnormalities of the X or Y chromosomes. Those patients with deletions of the short arm of the X chromosome tend to have short stature and Turner stigmata, whereas those with long-arm deletions of the X tend to have few Turner stigmata and stature within the normal range. The fact that normal XY males do not have Turner stigmata and short stature is explained by the existence of X/Y homologous loci present in a double dose. In other words, the disabilities associated with Turner syndrome are the result of haploinsufficiency of X/Y homologous loci.

The more common structural sex chromosome abnormalities found in Turner patients include long-arm X isochromosome, long-arm Y isochromosome, short-arm deletions of the X chromosome and ring-X chromosomes. These structural abnormalities are invariably associated with mosaicism for 45,X cells. 46,XX/45,X, and 46,XY/45,X are more common forms of mosaicism found in patients with features of Turner syndrome. The presence of a normal cell line clearly modifies the phenotype. 46,XX/45,X cases may menstruate, and 46,XY/45,X cases may show variable degrees of masculinization. Some XY/X mosaics may have asymmetrial sex differentiation with a 'streak' gonad on one side and a rudimentary testis on the other; in these cases Mullerian derivatives are repressed on the side of the testis, and the external genitalia are ambiguous.

In Turner patients with structural X-chromosome aberrations the abnormal X is preferentially inactivated. This is reflected in the size of the sex chromatin body. Isochromosomes for the long arm of the X have measurably larger sex chromatin bodies than normal, while X-chromosome deletions have smaller than normal sex chromatin. Among cases of 45,X Turner syndrome two classes are recognized depending on the parental origin of the X chromosome. Those with a paternally derived X chromosome have better social communication and cognitive skills than those in whom the single X is maternally derived. The findings are explained on the basis of differential imprinting of maternal and paternal X-linked genes influencing behavior.

The finding that a substantial proportion of early spontaneous abortions have a 45,X karyotype indicates that most conceptions with Turner syndrome are inviable. In fact, it has been estimated that 97% of 45,X conceptions are lost early in pregnancy. Many of those that continue in pregnancy can be recognized by ultrasound examination. The most important feature is cystic hygroma resulting from hypoplasia of the lymphatic system, a precursor not only of the webbed neck evident after birth but also of the associated cardiac malformations and peripheral lymphedema. The frequency of Turner syndrome among female livebirths is believed to be in the order of 1 in 3000.

***See also:* Imprinting, Genomic; Klinefelter Syndrome; Sex Chromatin; Sex Determination, Human; X-Chromosome Inactivation**

Twisting Number

doi: 10.1006/rwgn.2001.2059

The twisting number is the total number of base pairs divided by the number of base pairs per turn of a DNA double helix.

***See also:* DNA Structure**

TY Elements

See: Transposons as Tools

Tyrosine

J Read and S Brenner

doi: 10.1006/rwgn.2001.2079

Tyrosine (Tyr or Y) is one of the 20 amino acids commonly found in proteins. Its side-chain contains a hydrophilic hydroxyl group attached to an extremely hydrophobic benzene ring, making its chemical properties somewhat ambiguous. Its chemical structure is given in **Figure 1**.

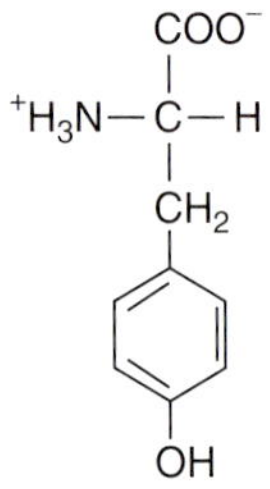

COO^-
$^+H_3N-C-H$
CH_2
OH

Figure 1 Tyrosine.

***See also:* Amino Acids; Proteins and Protein Structure**

Ubiquitin

A Varshavsky

doi: 10.1006/rwgn.2001.1348

Ubiquitin (Ub) is a 76-residue protein that exists in cells either free or covalently conjugated to other proteins. Ub is present in all eukaryotes but apparently absent from prokaryotes. Among eukaryotes, Ub is one of the most conserved proteins. For example, the sequences of yeast and mammalian Ub differ by two residues out of 76. Ub is a component of a multipathway intracellular proteolytic system called the Ub system, or the Ub–proteasome system. Most of the Ub-dependent pathways involve processive degradation of Ub-conjugated (ubiquitylated) proteins by the 26S proteasome, an ATP-dependent multisubunit protease. (Note: Ub whose C-terminal (glycine 76) carboxyl group is covalently linked to another compound is called the ubiquityl moiety, with the derivative terms being ubiquitylation, ubiquitylated. The abbreviation Ub refers to both free ubiquitin and the ubiquityl moiety.)

Ub is conjugated to other proteins (including other Ub molecules) through an amide bond, called the isopeptide bond, between the C-terminal residue (glycine 76) of Ub and the ε-amino group of a lysine residue in a substrate protein. Ub is activated for the conjugation reaction by a Ub-activating enzyme (E1), which couples ATP hydrolysis to the formation of a high-energy thioester bond between glycine 76 of Ub and a specific cysteine residue of the E1 enzyme (**Figure 1**). The E1-linked Ub moiety is moved, in a transesterification reaction, from E1 to a specific cysteine residue of a Ub-conjugating enzyme (E2), and from there to a lysine residue of an ultimate substrate protein, yielding a Ub–protein conjugate (**Figure 1**). This latter step requires participation of another component, called E3 or recognin. One function of E3 is to select a protein for ubiquitylation through an interaction with the protein's degradation signal.

Some ubiquitylated proteins (for example, histone H2A in mammalian chromosomes) bear a single Ub moiety and, in that state, appear to be metabolically stable. However, most of the other Ub conjugates contain a substrate-linked multi-Ub chain, in which the C-terminal glycine of Ub is linked to an internal lysine of an adjacent Ub, resulting in a chain of Ub–Ub conjugates containing two or more Ub moieties. One function of a substrate-linked multi-Ub chain is to facilitate the substrate's degradation by the 26S proteasome, in part through the binding of the multi-Ub chain to a specific component of the proteasome.

The covalent bond between Ub and other proteins can be cleaved: every eukaryotic cell contains multiple, ATP-independent proteases that recognize a Ub moiety and cleave at the Ub–adduct junction. The multiplicity of these Ub-specific proteases stems in part from the diversity of their targets, which include Ub precursors (linear, DNA-encoded fusions of Ub to other proteins, including Ub itself); Ub adducts with small nucleophiles such as glutathione; and either free or substrate-linked multi-Ub chains.

Degradation signals, or degrons, are features of proteins that confer metabolic instability. In most cases, Ub is a secondary degradation signal, in that it is linked to a protein that bears one of several primary degradation signals recognized by the Ub system. Degradation signals can be active constitutively or conditionally. Signals of the latter class, found in many regulatory proteins, including cyclins and transcription factors, are controlled through phosphorylation or interactions with other proteins whose binding may either shield a degron or activate it by providing a missing determinant.

Ub-dependent degradation signals consist of two major determinants: an amino acid sequence or a conformational feature that is specific for a given degron, and a lysine residue, the latter being the site of ubiquitylation. An example of the first determinant is a destabilizing N-terminal residue of a short-lived protein, which is recognized by a specific E2–E3 targeting complex. The set of amino acid residues that are destabilizing in a given cell yields a rule, called the N-end rule, which relates the *in vivo* half-life of a protein to the identity of its N-terminal residue. The corresponding Ub-dependent system, called the N-end rule pathway, is one of several proteolytic pathways of the Ub system. Among the other Ub-dependent

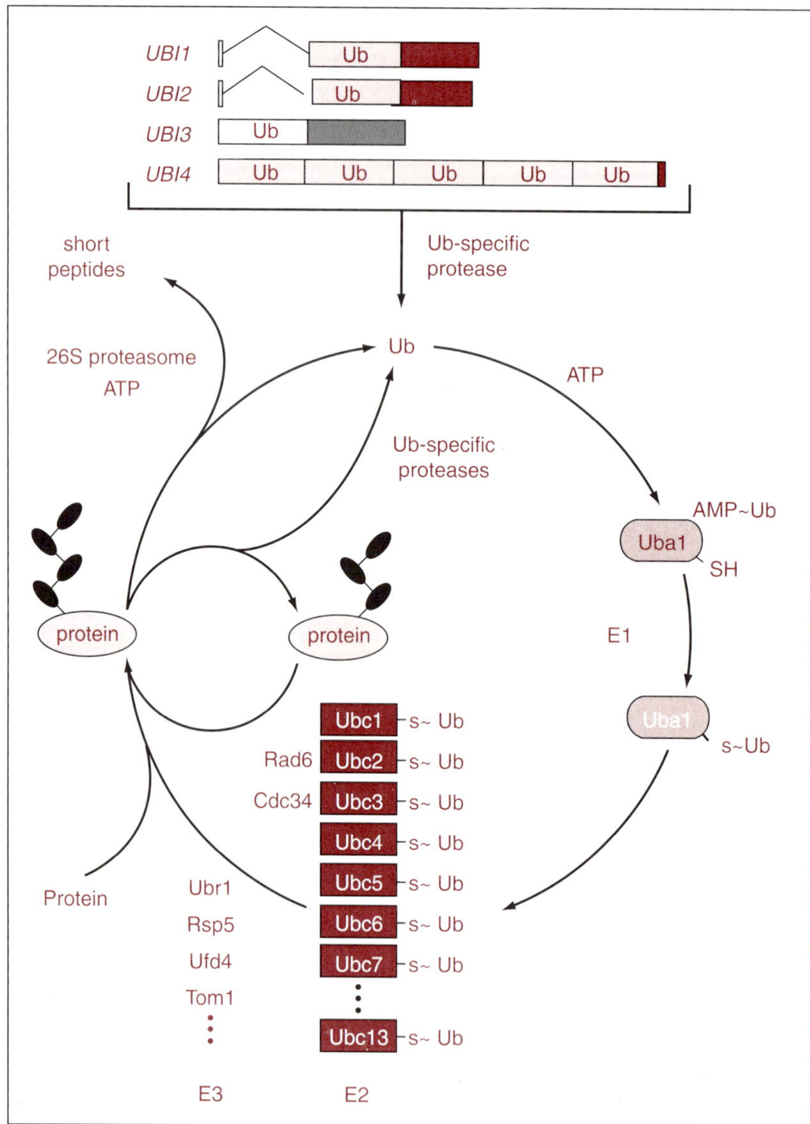

Figure 1 Ubiquitin pathways.

pathways are the systems whose large E3–E2 complexes target for degradation a class of conditionally unstable regulatory proteins called cyclins. The resulting periodic changes in the concentrations of cyclins (which function as subunits of cyclin-dependent kinases) drive and regulate the cell cycle. For example, mitotic cyclins are destroyed specifically at the end of mitosis, in part through the recognition of a sequence motif in these cyclins called the 'destruction box.'

Most, though not all, of the damaged and otherwise abnormal cytosolic and nuclear proteins are recognized and selectively destroyed by the Ub system, apparently through the exposure of their normally buried degrons. Moreover, proteins which are translocated from the cytosol across the endoplasmic reticulum (ER) membrane into the ER but fail to fold properly can be selectively transported back to the cytosol for their degradation by the Ub system. In addition, the conjugation of Ub to some membrane proteins acts as a signal for their endocytosis and delivery to lysosomes, in contrast to the more common function of Ub as a signal for the proteasome-mediated proteolysis in the cytosol. Under certain conditions, Ub can also function as a molecular chaperone: the transient, cotranslational linkage of Ub to specific ribosomal proteins was found to be essential for the efficient biogenesis of ribosomes. Eukaryotic cells also contain several homologs of Ub, called Ub-like proteins, which, similarly to Ub, exist either free or conjugated to other proteins. The conjugation of Ub-like proteins to their substrates involves distinct sets of conjugating enzymes and, in contrast to Ub conjugation, appears to have functions different from proteolysis.

The vast functional range of Ub stems from the fact that a large fraction of intracellular proteins (cyclins, transcription factors, components of signal transduction pathways, damaged proteins) are physiological substrates of the Ub system. This system has been shown to play major roles in a legion of biological processes: the cell cycle, cell growth and differentiation, embryogenesis, apoptosis, signal transduction, DNA repair, regulation of DNA transcription, replication and segregation, transmembrane transport, endocytosis, stress responses, antigen presentation, other aspects of the immune response, and functions of the nervous system, including circadian rhythms, axon guidance, and acquisition of memories. A number of tumor suppressors and proto-oncoproteins are specific components of the Ub system. Viruses often target Ub-dependent proteolysis to bypass or suppress the host's immune response. The Ub system and its perturbations have been implicated in the pathogenesis of cancer, senescence, bacterial and viral infections, specific genetic syndromes, and major neurodegenerative diseases.

Further Reading

Hershko A, Ciechanever A and Varshavsky A (2000) The ubiquitin system. *Nature Medicine* 6: 1073–1081.

Hicke L (1997) Ubiquitin-dependent internalization and downregulation of plasma membrane proteins. *FASEB Journal* 84: 277–287.

Peters J-M, Harris JR and Finley D (eds) (1998) *Ubiquitin and the Biology of the Cell.* New York: Plenum Press.

See also: Cell Cycle; Cyclin-Dependent Kinases; Mitosis; Proteolysis; Proteome

Umber Mutation

See: Nonsense Mutation; Start, Stop Codons

Underdominance

M A Asmussen

doi: 10.1006/rwgn.2001.1349

'Underdominance' or 'heterozygote inferiority' refers to cases of natural selection in diploid organisms where the fitness of genetic heterozygotes, carrying two different forms (alleles) of a gene (e.g., A_1A_2), is strictly less than that of both of the corresponding homozygotes, which carry two copies of one of those alleles (e.g., A_1A_1 and A_2A_2). This situation is perhaps most commonly found in interspecific hybrids formed by the mating of two distinct biological species. In such cases, each parental species may be fixed for one allele at a certain genetic locus that has proved advantageous in its typical habitat, and all individuals of that species are accordingly homozygous for that allele; hybrids carrying a mixed combination of the two parental alleles sometimes have reduced viability or fertility in the two parental environments and thus show underdominance there. In other, intermediate or alternative environments, however, the hybrids may instead have an intermediate fitness, or even enjoy a fitness advantage over both parental species, since the relative fitness of genotypes is often habitat or environment specific.

Evolutionary Outcome under Constant Underdominant Selection

The evolutionary outcome from underdominance is readily predicted if there are only two alleles at a certain locus, e.g., A_1 and A_2, and heterozygotes have a constant fitness disadvantage through a lower viability (rate of survival from birth to reproduction) than either of the associated homozygous genotypes. If no other evolutionary forces such as mutation, nonrandom mating, migration, or genetic drift act on this genetic locus, constant underdominant selection will always eliminate one of the two alleles from the population, with the frequency of one allele monotonically declining to zero, and that of the other monotonically increasing to 1.

Which allele is lost (frequency becomes zero) and which fixed (frequency becomes 1) is readily predicted and depends on the relative fitnesses of the three genotypes, as well as upon the initial allele frequencies in the population. This works as follows: the relative fitness values determine a threshold allele frequency for each allele, such that populations with frequencies initially below that value will lose that allele, while, in populations with higher initial allele frequencies, the frequency of that allele will increase to 1 and the other allele will be lost. In general, an allele increases in frequency under constant viability selection if and only if it has the higher average, or marginal fitness, which equals the weighted average of the fitnesses of the two genotypes that carry the allele, with the weights given by the frequency of the other allele in that genotype. The marginal fitness of allele A_1, for instance, equals the fitness of A_1A_1 weighted by the frequency of A_1, plus the fitness of A_1A_2 weighted by the frequency of A_2. In underdominance, the marginal fitnesses of the two alleles vary in their relative magnitudes, such that each allele enjoys the advantage when it is at sufficiently high frequencies, but not at lower frequencies.

The threshold allele frequency determining when each allele is fixed or eliminated by selection can be specified precisely if we introduce some notation and let the viabilities of the two homozygotes, A_1A_1 and A_2A_2, be $1 + s$ and $1 + t$, respectively, relative to a value of 1 for A_1A_2 heterozygotes. The critical, threshold frequency for allele A_1 is then $t/(s + t)$. The frequency of A_1 will steadily decrease to zero (and that of the alternative allele A_2 will increase to 1) if it is initially below this threshold value, and A_1 will steadily increase to 1 (and the alternative allele A_2 will decrease to zero) if it starts above $t/(s + t)$. The frequency of allele A_1 will remain at $t/(s + t)$ if it starts exactly there. This critical allele frequency is an example of an unstable equilibrium and unstable polymorphism, since the population always moves away from this polymorphic equilibrium value if it is perturbed from it. (Since there are only two alleles at this locus, their frequencies add to 1, implying that the corresponding threshold frequency for the alternate allele A_2 is $s/(s + t)$.)

Figure 1 shows underdominant allele frequency dynamics. This represents a numerical example in which the relative fitnesses of the three genotypes (A_1A_1, A_1A_2, and A_2A_2) are 1.1, 1.0, and 1.2, respectively, so that $s = 0.1$ and $t = 0.2$. The threshold frequency for allele A_1 is thus $t/(s + t) = 0.2/(0.1 + 0.2) = 2/3$. As shown by the upper and lower allele frequency trajectories in **Figure 1** the frequency of allele A_1 steadily increases to 1 if its frequency starts above 2/3 (e.g., from 0.7), and steadily declines to zero if its frequency is initially below two-thirds (e.g., from 0.6). The frequency of allele A_1 remains at 2/3 should it start at precisely that value.

Note that in this case the threshold frequency for allele A_1 is more than 0.5, which means that its frequency declines to 0 (and that of the alternate allele increases to 1) for more than half the possible initial values; in this example, for all allele frequencies less than 2/3. This asymmetrical disadvantage to allele A_1 will occur whenever A_1A_1 homozygotes have the lower fitness (of the two homozygotes), as it does here (1.1 versus 1.2). If A_1A_1 homozygotes instead have the higher fitness, then the threshold frequency for allele A_1 is below 0.5 and its frequency will increase to 1 (and that of the alternate allele decrease to 0) for more than half its possible initial values.

Figure 1 was computed using the recursion equation predicting the new frequency of allele A_1 (p') after one generation of underdominant selection, in terms of its previous frequency (p) and the fitness advantage (s and t) of the two homozygotes. This is given by:

$$p' = p\left[\frac{1 + sp}{1 + sp^2 + tq^2}\right]$$

where $q = 1 - p$ is the frequency of the alternate allele A_2.

Lastly, it should be emphasized that **Figure 1** gives just one numerical example showing how the threshold frequency determining which allele is lost is determined by the relative fitnesses of the three genotypes. The strength of selection, which depends on the magnitude of the differences among the relative fitnesses, also strongly affects the time until equilibrium is reached and the relevant allele is lost from the population.

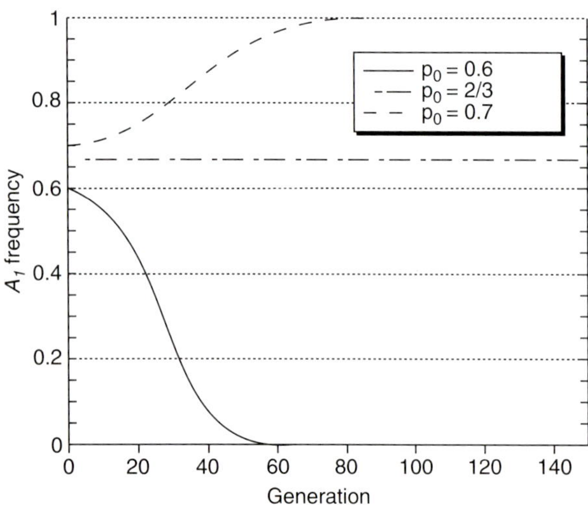

Figure 1 Trajectory through time in generations of the frequency of allele A_1, subject to constant underdominant selection (see text), for three initial allele frequencies (p_0). The alternate allele does the reverse, decreasing in frequency to zero when allele A_1 increases to 1, and vice versa.

Further Reading

Hartl DL and Clark AG (1997) *Principles of Population Genetics*, 3rd edn, ch. 6. Sunderland, MA: Sinauer Associates.

Hedrick PW (2000) *Genetics of Populations*, 2nd edn, ch. 3. Sudbury, MA: Jones & Bartlett.

See also: **Balanced Polymorphism; Fitness**

Underwinding

doi: 10.1006/rwgn.2001.2060

Underwinding of DNA is produced by negative supercoiling, when the double helix itself is coiled in the opposite sense from the intertwined strands.

See also: **DNA Supercoiling; Negative Supercoiling; Overwinding**

Undirectional Replication

doi: 10.1006/rwgn.2001.2062

Undirectional replication occurs with the movement of a single replication fork from a given origin.

See also: Replication; Replication Fork

Unequal Crossing Over

J R S Fincham

doi: 10.1006/rwgn.2001.1350

Definition

Unequal crossing over occurs as a result of pairing between similar chromosome segments residing at different chromosome loci. This occurs most readily when tandemly repetitious segments are present close together on the same chromosome, since in this case only a slight mutual slippage in normal meiotic pairing is required. Unequal crossing-over in this situation yields products differing only in that one has an extra tandem repeat at the expense of the other.

Bar in *Drosophila*

The classical example is the *Bar* mutant of *Drosophila melanogaster*, investigated in the 1920s and 1930s by Sturtevant and Bridges. *Bar* is a tandem duplication of a segment of the X chromosome comprising about six polytene chromosome bands. In the male or the homozygous female it causes an extreme narrowing of the eye; when heterozygous with wild-type in the female it has a less extreme effect. Stocks of *Bar* are not completely stable, mutating to wild-type or a more extreme mutant allele called *Double-bar*, with a frequency of the order of 1 in 1000 flies. Examination of the polytene chromosomes shows that the apparently wild-type revertants had lost the duplicated segment and that *Double-bar* was a tandem triplication. The explanation in terms of out-of-phase meiotic pairing of the repeats and unequal crossing over is shown in **Figure 1**.

From his original experiment, Sturtevant concluded that the wild-type and double-bar derivatives resulted from unequal crossing-over. This was made virtually certain by the finding that they nearly all showed recombination of allelic differences at two loci, *fused* (*fu*) and *forked* (*f*), closely placed on each side of the *Bar* locus (**Figure 1**). Though resulting from out-of-phase pairing, the unequal crossovers still had the normal crossover property of interference – reducing the probability of crossing-over in an adjacent chromosome interval.

Unequal Exchanges between Sister Chromatids in *Drosophila*?

In *Drosophila*, as in other eukaryotes, certain genes are naturally tandemly repeated to high copy number; the ribosomal RNA-encoding (rDNA) sequences are the prime example. *Drosophila* rDNA occurs in about 400 copies, divided about equally between the X and Y chromosomes. The *bobbed* (*bb*) series of mutations, detectable through their effect on bristles, are drastic reductions in rDNA copies. A *bobbed* fly stock with a special Y chromosome (Y*bb*–) with mutant *bobbed*, and an X chromosome with another reduced copy number *bb* allele (X*bb*) shows instability of rDNA copy number on the X chromosome of male flies. New alleles, either with magnification of copy number up to wild-type levels or further diminution, arise with high frequency (although the diminished alleles are recovered less often because of their low viability). Unequal meiotic crossing-over between homologous sequences appears to be ruled out here, since homolog exchange does not occur in *Drosophila* males. Perhaps the most plausible hypothesis is unequal sister-strand exchange within the single X, both at meiosis and mitosis. It is not known what feature of the Y*bb*– chromosome induces this instability in the X.

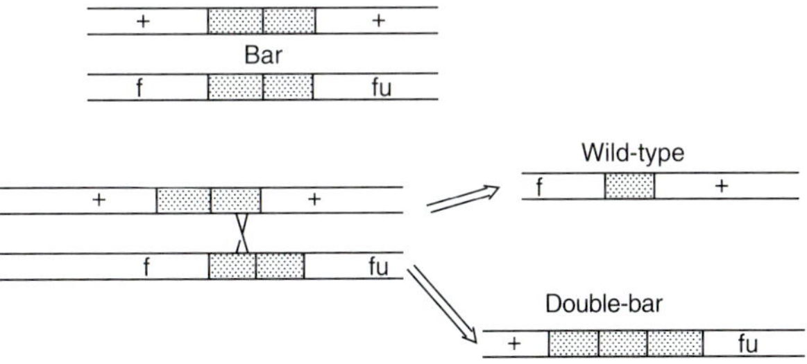

Figure 1 The reversion of the X-linked *Bar* mutant of *Drosophila melanogaster* to wild-type, and the origin of the more extreme mutant *Double-bar*, by slippage in the pairing of the *Bar* duplication and unequal crossing-over. The females whose progeny were screened were homozygous *Bar* and heterozygous with respect to two closely-placed flanking mutations, *fused* (*fu*) and *forked* (*f*). The non*Bar* ('wild-type') and *double-Bar* derivatives were, with very few exceptions, reciprocally-constituted crossovers with respect to the flanking markers.

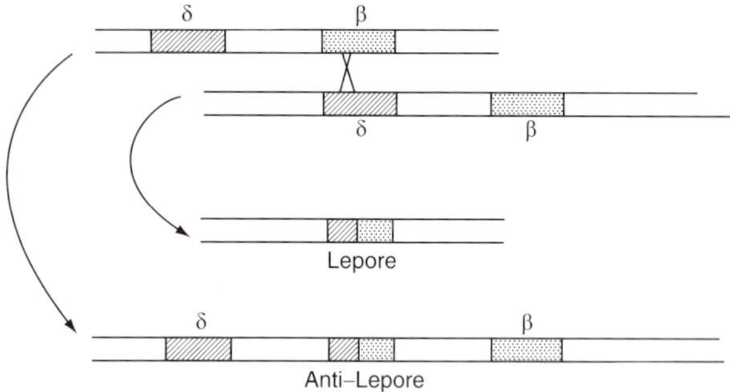

Figure 2 The presumed modes of origin of the human mutant globins Lepore and anti-Lepore by out-of register pairing and crossing-over between the β- and δ-globin genes. The two kinds of variant globin were identified in different families.

Unequal Crossing-Over in Yeast

In more recent years, the new techniques of DNA manipulation and transformation have been used to 'engineer' tandem duplications in the budding yeast, *Saccharomyces cerevisiae*. Here the frequency of unequal crossing-over is much higher than in *Drosophila*; in one study, about one-fourth of all the crossovers in the region of a tandem repeat were unequal, with one participating chromatid gaining the gene copy that the other had lost. In yeast the situation is complicated by the fact that some of the gains and losses of repeats are due to conversion events without crossing-over.

The Human β-Globin Gene Cluster

In humans, as in other mammals, the genes encoding the globin polypeptides of hemoglobins occur in tandemly arranged clusters, one for α-related and one for β-related globins. The human β-globin cluster consists, in order of linkage, of the genes ε (embryonic globin), Gγ, Aγ (two closely similar forms of fetal globin), φβ (a degenerate and nonfunctional β-related pseudogene), and δ and β (δ and β adult globins). It is speculated that this tandem array of homologous genes resulted from unequal crossing-over, either meiotic or mitotic, in the course of vertebrate evolution, although the initial generation of two copies from one must have occurred without much homology. Following the generation of repeats, the initially redundant extra copies would seem to have either diverged in the times of their expression and, to different extents, in their coding sequences, or decayed by mutation to loss-of-function, as in the pseudogenes.

This gene arrangement still entails some small risk of expansion and deletion by unequal exchange. One abnormal, though functional type of hemoglobin, called hemoglobin Lepore, is the result of the replacement of the normally separate δ and β genes by a single δ–β chimera, with the N-terminal sequence of δ joined to the C-terminal part of β. The reciprocal condition, called anti-Lepore, has also been found, with the N-terminus of β joined to the C-terminus of δ, and normal copies of the β and δ genes retained (**Figure 2**). The β and δ genes, which are relatively close together and more than 90% identical, are presumably more prone to this kind of mistake than the other members of the cluster.

Further Reading

Bridges CB (1936) The *Bar* 'gene' is a duplication. *Science* 83: 210.

Hawley ES and Tartof K (1985) A two-stage model for the control of rDNA magnification. *Genetics* 109: 691–700.

Jackson JA and Fink GR (1985) Meiotic recombination between duplicated genetic elements in *Saccharomyces cerevisiae*. *Genetics* 109: 303–332.

Sturtevant AH (1961) Further study of the so-called mutation at the *Bar* locus of *Drosophila*. Reprinted in *Genetics and Evolution – Selected Papers of A. H. Sturtevant*, pp. 107–114. San Francisco, CA: WH Freeman.

***See also:* Globin Genes, Human; rDNA Amplification; Tandem Repeats**

Uniparental Inheritance

A C Ferguson-Smith

doi: 10.1006/rwgn.2001.1351

When offspring inherit their genotype from only one parent, this is known as uniparental inheritance. This term can be applied to a wide range of genetic events, some examples of which are given below. Much of what follows addresses uniparental inheritance in diploid organisms. Uniparental inheritance in plants will not be considered here. Nonetheless, a substantial proportion of the earth's biomass is composed of organisms with a predominantly haploid life cycle such as fungi and algae and these deserve a mention. For example, the fission yeast *Schizosaccharomyces pombe* normally reproduces asexually during vegetative growth, thus propagating the haploid state by uniparental inheritance. This yeast is also, however, capable of sexual reproduction to produce diploid cells. In response to starvation, these diploids will sporulate and give rise to haploid cells again by meiosis. When nutrients are plentiful, budding yeast strains such as the baker's yeast *Saccharomyces cerevisiae* prefer to proliferate as diploid cells. On starvation they too will undergo meiosis and can subsequently either proliferate asexually as haploids or sexually by mating with cells of the opposite mating type to form diploids. Yeast haploids are a valuable genetic tool for the identification and analysis of mutant genes in key cellular pathways such as cell-cycle control. In the life cycle of the unicellular freshwater alga, *Chlamydomonas*, uniparental inheritance is evident through the asexual reproductive activity of the haploid cells. Once again in adverse environmental conditions, some of these cells are transformed into gametes and a pair fuse to form a diploid zygote. In favorable conditions after this sexual phase, meiosis occurs which gives rise to a new haploid generation.

Mitochondria are organelles which occupy a substantial cytoplasmic portion of the eukaryotic cell. They are responsible for completing the energy conversion used to drive cellular reactions. These organelles contain DNA which in mammals is about 10^{-5} times the size of the nuclear genome and are capable of carrying out their own DNA replication, transcription and protein synthesis. The mitochondrial genome in humans encodes tRNAs, rRNAs and 13 other polypeptides. Mitochondrial genes undergo non-Mendelian cytoplasmic inheritance which, in higher mammals is uniparental because the egg contributes much more cytoplasm to the zygote than does the sperm. Hence, this uniparental inheritance is maternal in origin.

Parthenogenesis in Invertebrates

Parthenogenesis can be defined as the production of an embryo from a female gamete without any genetic contribution from a male gamete, with or without the eventual development into an adult. It is distinct from asexual reproduction since it involves the production of egg cells. Parthenogenesis is a normal method of reproduction in many lower organisms, but does not lead to viable mammalian offspring. Parthenogenetic development can proceed by various routes depending on whether meiosis has occurred or has been supressed, in which case the egg develops as as result of mitotic divisions. Whenever sex is determined by chromosome constitution, parthenogenetic offspring, in the absence of effective meiosis, all will be, mostly female. In birds, however (see below), the offspring are male as in this case females are the heterogametic sex. In bees, males originate by haploid parthenogenesis while diploid females are produced by fertilization in the normal way. Other aphids, such as greenfly (Hemiptera) have generations which alternate between parthenogenesis and fertilization, so called cyclical parthenogenesis. The formation of female parthenogenetic offspring is widespread among many order of insects. For example in *Drosophila parthenogenetica*, a small proportion of eggs laid by virgin females develop to produce viable adults. Another example is the parthenogenetic grasshopper *Warramaba virgo*, a species which consists of females only. Parthenogenesis is also successful in some Crustaceae such as the brine shrimp, *Artemia salina*.

Uniparental Inheritance in Vertebrates

As organism complexity increases, experimentally induced parthenogenesis has been used to test the developmental potential of parthenogenetically produced vertebrates. For example, in several species of amphibia including *Rana japonica*, *R. nigromaculata*, and *R. pipiens*, viable parthenogenetic adults have been described. Parthenogenetic fish have also been reported and in these abnormal situations it is thought that the egg activation may result from infection with the phycomycete *Ichthyonophonus hoferi*. Surprisingly, for some reptiles of the order *Squamata* the natural mode of reproduction is through parthenogenesis in some unisexual populations, and through fertilization in populations with two sexes.

In warm-blooded vertebrates, parthenogenetic offspring are less viable and there is no known instance of parthenogenesis as the normal mode of reproduction. Nonetheless, in some breeds of chicken and turkey, parthenogenetic development has been reported. In the latter case, the parthenogenetically produced male turkeys (in birds, males are the homogametic sex) were small with reduced fertility. In mammals, spontaneously occuring cleavage divisions in oocytes have been described, notably in the LT/Sv strain of mice in which parthenogenesis occurred regularly in a small percentage of virgin females resulting in development to the blastocyst stage. A high incidence of ovarian teratomas are also seen in this mouse strain. These benign teratomas are derived from germ cells and consist of both differentiated and undifferentiated cells. In humans, benign teratomas (dermoid cysts) account for 15–20% of ovarian neoplasia and there have been several reports of differentiated human tissue within these teratomas.

The inability of activated mammalian eggs to develop into viable term fetuses is most likely due to the phenomenon of genomic imprinting. Genomic imprinting is a process that causes some genes to be expressed solely from one parental allele. Some imprinted genes are expressed from the maternal chromosome and others from the paternally inherited homolog. This results in a requirement for both parental genomes in order for development to proceed normally. The embryological consequences of genomic imprinting have been studied experimentally in mice. Uniparental conceptuses were generated by egg activation to make parthenogenones, or via pronuclear transplantation to make bimaternal gynogenones or bipaternal androgenones. Gynogenetic and parthenogenetic conceptuses die around midgestation and exhibit growth retardation and poor development of extraembryonic tissues. If these embryos are given normal extraembryonic components, development will proceed only slightly further, indicating that the lethality is not solely due to the failure of the placenta but that a diploid maternal contribution to the embryo itself is incompatable with life. Androgenetic embryos fare less well than their bimaternal counterparts. These embryos rarely survive gastrulation and usually the conceptus resembles a mass of extraembryonic tissue in the absence of embryonic components. These findings suggest that a maternal genome is required for the development of embryonic components and a paternal genome for the development of extraembryonic tissues, at least at these early stages. Androgenetic conceptuses are reminiscent of the complete hydatidiform molar pregnancy in humans which has a diploid paternal genotype in the absence of a maternal genetic component. A small number of complete moles have been reported which contain differentiated embryonic cells. The vast majority of hydatidiform moles are diploid and are likely to have arisen through duplication of the paternal genome (homozygous) or by dispermy (heterozygous). The majority of partial moles are triploid with the extra genome set being paternal in origin.

Mouse chimeras made with normal cells and either parthenogetic or androgenetic cells allow the study of the later developmental potential of these experimental cells during development. The presence of normal cells rescues the lethality observed in the parthenogenetic embryos. The resulting chimeras are small, but morphologically normal and fertile. Parthenote cells are predominantly found in tissues of ectodermal and neurectodermal origin and are not well-represented in mesodermal derivatives. In contrast, androgenetic chimaeras develop to term only if their contribution is low. Resulting chimeras show dysmorphology of the axial skeleton and embryos show growth enhancement. The androgenetic cells do not favour neurectodermal lineages but prefer to contribute to mesoderm. Both androgenetic and parthenogenetic cells can contribute to the germline. These studies suggest that imprinted genes have roles to play in regulating growth and in the development of mesodermal and neurectodermal lineages. This is consistent with the functions of the small number of imprinted genes identified to date.

Successful androgenesis has been reported to occur naturally in interspecfic hybrids of the Sicilian stick insect (*Bacillus ressius*). In contrast with the failure of mammalian androgenesis, androgenetic development can be induced to occur in lower organisms including some vertebrates, notably fish. In fish research, androgenesis has been used successfully to generate homozygous lines of fish. In addition, the fish species *Oncorhyncus mykiss* and *Oreochromis niloticus* have been recovered from cryopreserved sperm, via sperm–sperm fusion followed by fertilization of irradiated eggs. The resulting fish survive and appear normal suggesting that if genomic imprinting occurs in these androgenetic organisms, it is not essential for development.

Mammalian Cloning

A clone is any collection of cells that are descendants of a single ancestor cell. In the case of the cloning of a whole complex organism, that single ancestor is usually an enucleated egg which, instead of being fertilized, has received a diploid nucleus from a somatic cell. This can be considered as another form of uniparental inheritance as offspring are produced asexually without a contribution from two parents. Such nuclear transplantations in vertebrates were carried out in frogs in the 1980s in order to test the ability of nuclei from differentiated cells to support normal development. These studies, which addressed the question of whether irreversible changes in genes accompany differentiation, resulted in development to adulthood in a small number of cases. In general, in frogs, the later the developmental stage of the nucleus used, the more limited the developmental potential of the embryo. These studies suggested that there was limited reversibility of the differentiated state.

Until recently, mammalian cloning from fully differentiated cell types was even less successful. In cattle and sheep, a series of technical advances allowed transplanted nuclei from undifferentiated embryonic cells to give rise to viable offspring. More recently, donor nuclei from differentiated adult cells were also found to be capable of making a sheep clone after nuclear transplantation into an egg. This significant result indicates that the differentiated mammalian genome can be ‘reprogrammed’ to support development. This approach has now been used to produce mouse clones at a frequency of about 2% from adult somatic cell nuclei injected into enucleated eggs. The frequency of successful mammalian cloning is very low and not all somatic cell nuclei are capable of producing adult clones. This historical achievement provides a valuable model system in which to address key questions regarding the regulatory mechanisms of genome programming and reprogramming. In addition, it opens wide the debate over ethical issues surrounding the application of this technology, notably in ‘reproductive cloning.’ Nonetheless, the benefits of cloning technology are great, for example as applied to cell and tissue therapy – 'therapeutic cloning,' as well as the preservation and propagation of endangered species. Mammalian cloning is an example of how uniparental inheritance has moved from nature into the laboratory with profound implications for genetics and biomedicine.

Further Reading

Mittwoch U (1978) Parthenogenesis. *Journal of Medical Genetics* 15: 165–181.

***See also:* Androgenone; Imprinting, Genomic; Mitochondria; Parthenogenesis, Mammalian**

Unique DNA

E Kutter

doi: 10.1006/rwgn.2001.1352

Unique DNA is a stretch of DNA that is present in only a single copy in a cell. This includes most DNA in bacterial cells and most of the DNA that is expressed as mRNA in eukaryotic cells. However, as reported by Roy Britten and David Kohne in 1968, most eukaryotic cells also have classes of DNA that are present in many copies in the genome. In reassociation kinetics studies, where the DNA was fragmented into pieces a few hundred nucleotides long, denatured, and allowed to reanneal, at least three general populations of DNA were seen. There are sections that reanneal very rapidly, indicating that they must be present in over a million copies per cell to be able to find a mate so quickly. This is now termed highly repetitive DNA. Segments that anneal as slowly as one would expect from the size of the genome if each DNA fragment had to find a unique mate are termed unique DNA. Fragments that reanneal at intermediate rates, as if a few hundred nearly identical copies were present in the genome are called (moderately or middle repetitive DNA). The amount of unique DNA tends to be fairly similar between different related species, even though the total DNA content may vary widely; in amphibians and plants, the total DNA in the haploid genome can vary by over two orders of magnitude, to nearly 10^{12} base pairs! In most higher eukaryotes, including humans, it appears that only a few percent of the DNA encodes proteins.

Much of the highly repetitive DNA has an average base composition significantly different from that of the bulk DNA of the cell, and thus shows up as a separate, or 'satellite,' peak during density-gradient centrifugation. This satellite DNA hybridizes mainly to the heterochromatic centromere regions of the chromosome and consists of short sequences repeated a very large number of times. Much of the other rapidly annealing DNA may be related to large numbers of copies of various (generally defective) viruses present in the DNA. Moderately repetitive DNA may also be clustered as tandem repeats which can include functional genes, such as the ribosomal RNA and tRNA genes.

***See also:* Repetitive (DNA) Sequence**

Universal Genetic Code

E Kutter

doi: 10.1006/rwgn.2001.1353

Throughout the whole of the prokaryotic, plant, and animal kingdoms, the same codons are used for the same amino acids, with very few exceptions – the code is thus *almost* universal. However, the genetic code in mitochondria has several significant differences from the common universal code (**Table 1**).

In the mitochondria, the third codon thus is less important in selection, reducing the number of tRNAs needed in this compact organelle.

The second known variation from the universal genetic code is that ciliated protozoans read AGA and AGG as additional Stop signals rather than as Arg.

Table 1 Genetic code in mitochondria

Codon	Common code	Mitochondrial code
AUA	Ile	Met
AGA	Arg	Stop
AGG	Arg	Stop
UGA	Stop	Trp

***See also:* Genetic Code; Start, Stop Codons**

Unscheduled DNA Synthesis

doi: 10.1006/rwgn.2001.2064

Unscheduled DNA synthesis is any DNA synthesis occurring outside the S phase in the eukaryotic cell.

***See also:* Cell Cycle; S Phase**

Unstable Equilibrium

See: **Equilibrium**

Up, Down Mutations

See: **Alleles; Promoters**

Upstream

doi: 10.1006/rwgn.2001.2066

The term 'upstream' identifies sequences preceding in the opposite direction from expression.

***See also:* Downstream; Gene Expression**

Upstream, Downstream Site

J Parker

doi: 10.1006/rwgn.2001.1357

The terms 'upstream' and 'downstream' are used to refer to the location of sites on the DNA relative to other sites in the same gene. The first step in the expression of a gene is transcription, the production of an RNA copy of the gene using one strand of the DNA as a template. For a given gene it is always the same DNA strand that is used as a template and transcription starts at a specific site on the double-stranded DNA molecule and proceeds in only one direction. The RNA molecule is synthesized from its $5'$ to its $3'$ end. For genes whose products are proteins, translation also begins at a specific sequence on this RNA transcript and proceeds unidirectionally in the $5'$ to $3'$ direction. This common directionality for both transcription and translation means that from the point of view of expression, one end of the gene is its beginning (the end where transcription begins) and the RNA polymerases and ribosomes flow unidirectionally from this end toward the other end of the gene, something like a stream.

This has led to the practice of referring to specific DNA sequences as 'upstream' or 'downstream' as compared to other sequences in a gene (or in relationship to the gene itself). For instance, the promoter, a site involved in transcription initiation, is upstream of the coding sequences of a gene, but may be downstream of some regulatory sites on the DNA. Please note that a site could be upstream with regard to some sequences and downstream with regard to others. In addition, although some sites invariably have a fixed orientation relative to other sites, others do not. For example, in a gene encoding a protein, the sequence that encodes the start codon is of necessity upstream from the sequence encoding the stop codon. However, the position of certain regulatory sequences, such as an operator, may be either upstream or downstream of other regulatory sequences, such as the promoter. Such differences in position may have important consequences for gene regulation. For sequences called enhancers, which are involved in transcription, their position relative to the gene(s) whose expression they influence seems less important. However, such sites tend to be exceptional.

Many sequences upstream of the encoding region of a gene play important roles in the regulation of gene expression. These include the promoter, various sites that make up the promoter (e.g., TATA box and GC box), and sites where regulatory proteins bind (e.g., operators and activator-binding sites). Another site found in yeast is the upstream activator site (UAS) which functions somewhat like an enhancer but must be upstream of the beginning of transcription. Often these regulatory sites may be simply referred to as 'upstream sites.' Of course, there are also sites that are invariably downstream of the coding region of a gene, such as transcription terminators and, in eukaryotes, the sites encoding sequences involved in the polyadenylation of messenger RNA.

***See also:* Enhancers; Operators; Promoters; Transcription; Translation**

Uracil

R L Somerville

doi: 10.1006/rwgn.2001.1358

Uracil (U; molecular formula, $C_4H_4N_2O_2$) is a naturally occurring representative of the heterocyclic nitrogen-containing aromatic bases termed pyrimidines. Although found predominantly in RNA, uracil is also a major constituent of the DNA of certain bacterial viruses. In every other species, uracil is a transient constituent of DNA, arising through the random deamination of cytosine residues. Efficient surveillance and repair systems exist within most cells to prevent uracil residues from accumulating within DNA. Because the base-pairing properties of uracil are identical to those of thymine, replication of DNA containing a G:U base pair would give rise to a daughter molecule containing an A:U base pair and a granddaughter molecule containing an A:T base pair. If the original G:C base pair happened to be indispensable, the failure to replace U would be lethal.

***See also:* Pyrimidine**

URF

doi: 10.1006/rwgn.2001.2067

A URF (unidentified reading frame) is an open reading frame that is presumed to code for protein, but for which no product has been found.

***See also:* Reading Frame**

UTP (Uridine Triphosphate)

E J Murgola

doi: 10.1006/rwgn.2001.1359

Uridine-5′-triphosphate (UTP) is an energy-rich, activated precursor for RNA synthesis. It is synthesized in the cell by phosphorylation of uridine diphosphate (UDP), catalyzed by a nucleoside diphosphate kinase, with adenosine triphosphate (ATP) as the phosphate donor:

$$\mathrm{UDP} + \mathrm{ATP} \leftrightharpoons \mathrm{UTP} + \mathrm{ADP}$$

For the synthesis of deoxyuridine triphosphate (dUTP), a precursor of DNA, the 2′ hydroxyl group of the ribose moiety of guanosine triphosphate (GTP) is replaced by a hydrogen atom. The final step in this conversion is catalyzed by ribonucleotide reductase.

***See also:* RNA**

UVR Genes

***See:* Excision Repair**

V Gene

doi: 10.1006/rwgn.2001.2068

A V gene (variable gene) is a sequence coding for the major part of the variable (N-terminal) region of an immunoglobulin chain.

***See also:* Immunoglobulin Gene Superfamily; Variable Region**

Valine

E J Murgola

doi: 10.1006/rwgn.2001.1362

Valine is one of the 20 amino acids commonly found in proteins. Its abbreviation is Val and its single letter designation is V. As one of the essential amino acids in humans, it is not synthesized by the body and so must be provided in an individual's diet.

The chemical structure of valine is given in **Figure 1**.

```
         COOH
          |
  H2N—— C —— H
          |
         CH
        /    \
     H3C      CH3
```

Figure 1 Valine.

***See also:* Amino Acids; Proteins and Protein Structure**

Variable Codons

***See:* Codons, Invariable; Genetic Code**

Variable Region

doi: 10.1006/rwgn.2001.2069

Variable regions are regions in the amino acid sequence of both heavy and light chains of immunoglobulins with great diversity of sequence. They are associated with the antigen-binding areas.

***See also:* Constant Regions; Immunoglobulin Gene Superfamily; V Gene**

Variegation

doi: 10.1006/rwgn.2001.2070

Variegation of phenotype is a phenomenon caused by a change in genotype during somatic development.

***See also:* Somatic Mutation**

Vascular Endothelial Growth Factor (VEGF)

J Laurén

doi: 10.1006/rwgn.2001.1715

Vascular endothelial growth factor (VEGF) is essential for the embryonic differentiation and growth of endothelial cells and for tumor angiogenesis. VEGF expression is induced by hypoxia, activated oncogenes, and a variety of cytokines. Five different isoforms of VEGF, generated by alternative splicing, bind to two endothelial tyrosine-kinase receptors, VEGFR-1 (flt-1) and VEGFR-2 (flk-1/KDR). VEGF-induced intracellular signaling results in increased cell proliferation, migration, and inhibition of apoptosis.

VEGFs have great potential in the induction of therapeutic angiogenesis in ischemic diseases and blocking their signal transduction is a promising approach for the inhibition of tumor angiogenesis.

***See also:* Angiogenesis; Signal Transduction**

Vectors

W E Jack

doi: 10.1006/rwgn.2001.1364

Vectors were the first DNA tools used in genetic engineering, and continue to be cornerstones of this technology. A vector is a DNA segment that can replicate independent of the chromosome and be transferred between hosts. DNA fragments integrated into the vector are coreplicated, making multiple identical copies, or clones, of the target sequence. Accordingly, these elements are often referred to as cloning vectors or vehicles. Vectors have been developed and adapted for a wide range of uses. Two primary uses are (1) to isolate, identify and archive fragments of a larger genome and (2) to selectively express proteins encoded by specific genes.

Vector Characteristics

Several vector characteristics are central to their function as tools in molecular biology. The first feature is autonomous replication. This autonomy facilitates purification of the vector and its integrated DNA insert, paving the way for characterization and manipulation of the DNA sequence. Such manipulations frequently include schemes for expression and/or mutagenesis of associated gene-coding regions. This autonomy also confers the ability to have multiple copies of the DNA insert in the cell, improving DNA yields and increasing the gene dosage.

A second central feature is the availability of mechanisms to insert desired DNA fragments into the vector. Ideally, precisely defined DNA fragments can be readily integrated into the vector without disrupting essential vector functions. If this step can be accomplished with high efficiency, screening for the desired construct is minimized. Traditionally, restriction endonucleases have been used to cleave both vector and insert, which are then joined by DNA ligase to form the desired clone. In this regard, vectors with a large number of unique restriction endonuclease sites provide the greatest flexibility in accommodating inserts, which can be flanked by any of a variety of restriction sites. When these sites are clustered in the vector, this cluster is referred to as a multiple cloning site (MCS). The advent of PCR amplification has dramatically altered the landscape of cloning in that precise insert regions can be amplified, adding flanking restriction sites if necessary to facilitate attachment to the vector. Recently, enzymes that function in recombination, such as lambda int/xis and Cre recombinase, have also been used to transfer DNA segments between vectors.

A final feature is the ability to select for hosts containing the vector. Introducing purified DNA into cells, a process called transformation or transfection, is inefficient in many cell types. Accordingly, transformation is generally followed by a selection in which untransformed cells are killed. Selection often employs cytotoxic agents, such as antibiotics, to kill untransformed cells, relying on drug resistance genes on the vector to confer survival to transformed cells. Selection can also rely on metabolic enzymes on the plasmid that complement either naturally-occurring or introduced defects in the host cell.

In addition to these basic features, DNA sequences on the vector can act to extend the host range or control expression of cloned gene-coding regions, providing an opportunity to selectively express a desired protein in a variety of novel contexts, as when tissue- or cell cycle-specific expression is desired. Still other sequences can provide means for producing and selecting protein variants with desired properties.

Two broad categories of vectors can be defined. The first is composed of DNA derived from naturally-occurring elements (plasmids) in the cell. These elements can differ widely in size and complexity and are most often circular. The second utilizes the genomes of viruses infecting bacteria or higher organisms such as fungi, plants, insects or mammals. These are referred to as viral vectors, or as bacteriophage (phage) vectors when the host is a bacterium.

Plasmid Vectors

While plasmids have been reported in eukaryotes such as yeast, their presence is largely limited to bacteria where their existence is widespread. Accordingly, the discussion here focuses on the bacterial systems, especially on *Escherichia coli*, the most widely used bacterial system for manipulating DNA. The wealth of knowledge and experience concerning DNA DNA manipulations in *E. coli* makes it a starting point for almost all *in vitro* DNA manipulations, no matter what the intended use or fate of the construct DNA.

One obvious utility of plasmids is the ability to separate that DNA from cellular chromosomal DNA, permitting a vast array of *in vitro* manipulations.

Plasmid purification protocols rely both on the relatively small size and closed circular structure of plasmids to effect separation. The ability to perform such purifications rapidly on a small scale is of great advantage in analysis.

The origin of replication is the key determinant of plasmid properties. This DNA element controls the host range, copy number and compatibility of the plasmid. Replication initiates within that element, and typically involves host proteins, frequently in association with plasmid-encoded proteins. Despite using host proteins, the mode of replication can differ from that of the host chromosome, allowing independent accumulation of the plasmid. In one manifestation of this effect, the replication of certain plasmids is independent of new protein synthesis (relaxed control), unlike host chromosomal replication (stringent control). As a result, in the presence of protein synthesis inhibitors, plasmid replication continues in the absence of chromosomal replication, resulting in a selective amplification of plasmid DNA.

The type of origin of replication also dictates the copy number, or number of plasmids per cell. Copy numbers range from one to thousands per cell, and can be affected by growth conditions. A high copy number is an advantage in increasing the yield of DNA, and can also elevate gene expression by the concomitant increase in gene dosage. However, if the expressed genes are toxic to the cell, lower copy number vectors are advised.

The origin of replication determines whether the plasmid will replicate in a single host, or in multiple hosts. Bacterial plasmids do not, however, replicate in eukaryotic cells. When there is a need to replicate in disparate hosts, multiple replication origins can be included in a single shuttle vector. Thus, constructs that can be more readily assembled in a bacterial cloning system can still be used with more complex genetic systems.

An additional feature arising from replication origin interactions is plasmid incompatibility: the inability of plasmids with similar replication origins to be stably maintained in the same cell. When incompatible plasmids are introduced into the same cell, the plasmids compete for survival in subsequent generations of cells, eventually segregating to leave two populations of cells, each containing only one type of plasmid. When multiple types of plasmids are desired in the same cell, they ideally will have compatible replication origins, and each carry a different selectable marker.

Viral Vectors

Viruses are distinguished from plasmids in that the viral DNA is encapsulated by a protein coat. This viral capsid not only protects the DNA, but also provides an efficient mechanism to deliver viral DNA to the cell. This packaging facilitates isolation of the viral DNA via purification of the viral capsids. In general, viral vectors have larger genomes, encoding proteins involved in packaging viral DNA, and in some cases proteins involved in viral replication, transcription and/or translation.

The diverse biology of viral vectors has been exploited for specialized applications. For example, one class of bacteriophages packages a circular single strand of the viral genome. DNA isolated from the resulting viral particles yields an excellent template for *in vitro* DNA replication, useful in DNA sequencing and oligonucleotide-directed mutagenesis. Other viruses can integrate their genome into the host chromosome, providing stable transgenes for phenotypic analysis.

A number of eukaryotic viral vectors have been constructed, although development has lagged behind phage vectors because of the more complex systems they represent. Although different elements govern replication, transcription and translation, the basic features of viral vectors mimic those of phage vectors. Viral vectors have played an essential role in elucidating gene function, as well as the function of regulatory sequences, and promise to play a prominent role in the development and implementation of gene therapy.

Library Construction in Vectors

A fundamental use of vectors is to isolate and store smaller segments of a larger genome. Such fragmentation simplifies DNA sequence determination and analysis of associated gene-coding regions. Such collections, or libraries, are created by fragmenting a genome, and combining each of the ensuing fragments with a vector backbone. This genomic library will ideally have at least one representation of all DNA sequences. Even better, there will be multiple representations, each bounded by different borders to increase the probability that a single clone can be isolated containing all the relevant contiguous genetic elements, such as those associated with a single gene-coding region. Correct clones can be identified by a number of methods, with one common method being hybridization with a probe sharing DNA sequence similarity with the desired region.

In higher organisms, study of gene-coding regions is facilitated by constructing cDNA libraries, formed by making double-stranded DNA copies of the mRNA found in the cell. These cDNA libraries simplify analysis in that sequences outside the coding regions are not represented.

One final library, an expression library, bears mentioning. This type of library joins coding sequences from the genome with a specific coding sequence expressed on the vector, producing a hybrid protein containing at least a portion of the protein encoded in the genome. Identification of this expressed protein, for example by interaction with a specific antibody, identifies the underlying gene-coding region, and through this correlation opens avenues to identify and isolate the complete protein-coding region.

As can be imagined, a critical feature of these libraries is the size of the insert that can be cloned. High copy number vectors (> 20 copies per cell) can easily accommodate inserts of 3–8 kb, a reasonable range for a gene-coding region (a 60 kDa protein-coding region is about 1.6 kb in length). Larger inserts tend to be less stable, particularly in high copy number vectors, presumably due to the selective advantage of replicating smaller plasmids, including those with spontaneous deletions. This bias against large inserts is also seen in phage vectors that do not limit the amount of DNA packaged in the capsid.

However, screening, mapping and DNA sequence compilation often benefits from having large inserts. Many viral cloning vectors allow larger inserts, with the length limited by the cavity within the viral capsid. Insert sizes in these viral vectors can be increased by deleting nonessential viral genes. In an extreme example, cosmid vectors derived from bacteriophage lambda can package inserts of 30–45 kb, only slightly smaller than the 48 kb wild-type genome. Single copy plasmids such as bacterial artificial chromosomes (BAC) and yeast artificial chromosomes (YAC) can maintain even larger inserts (350–1000 kb).

Vectors for Protein Expression

A specialized class of vectors are designed for protein expression. Vectors in this class promote high yields of a desired protein in either bacterial or eukaryotic systems, although usually not both due to differences in transcriptional and translational signals. It should additionally be noted that a number of eukaryotic proteins require posttranslational modifications which are lacking in bacterial systems, and must be expressed in eukaryotic systems to ensure full functional operation.

High yields in either prokaryotic and eukaryotic expression systems can be fostered by multicopy vectors that increase the gene dosage, and by juxtaposition of transcriptional and translational control sequences next to the gene-coding sequence. Inducible transcriptional promoters are often used, and in the case of eukaryotes promoters can also be controlled in a cell-cycle or tissue-specific manner, providing means to control the timing and localization of protein expression. This is useful not only in producing large amounts of protein for biochemical analysis, but also in establishing regulatory pathways and phenotypic associations for particular gene products.

One variety of expression vectors modifies the expressed protein to include extra amino acid sequences at one or both termini. These added sequences can target the resulting fusion protein to specific cellular compartments, or can act as detection tags in defining spatial and temporal patterns of expression. Still other fusions act as affinity tags to assist in purification of the attached protein.

Conclusions

In the short history of recombinant DNA, vectors have played a major role in advances in technology. Without a doubt, refinement and development of new vectors will continue to pace the rapidly expanding field of molecular biology.

Further Reading

Balbas P and Bolivar F (1990) Design and construction of expression plasmid vectors in *Escherichia coli*. In: Goeddel DV (ed.) *Methods in Enzymology*, vol. 185, pp. 14–37. San Diego, CA: Academic Press.

Colosimo A, Goncz KK, Holmes AR *et al.* (2000) Transfer and expression of foreign genes in mammalian cells. *Biotechniques* 29: 314–333.

Feinbaum R (1998) Introduction to plasmid biology. In: Ausubel FM, Brent R, Kingston RE *et al.* (eds) *Current Protocols in Molecular Biology*. pp. 1.5.1–1.5.17. New York: John Wiley.

See also: **Cloning Vectors; Plasmids; Recombinant DNA; Transduction**

Velocardiofacial Syndrome

See: **DiGeorge Syndrome**

Vertical Transmission

D E Wilcox

doi: 10.1006/rwgn.2001.1365

Vertical transmission occurs when a trait or a disease is passed down through several generations, directly from an affected individual to affected descendants in successive generations. It is typically seen in autosomal dominant inheritance. In this pattern of inheritance, both sexes can be affected and, in turn, transmit the trait or disease to both males and females. It is also seen in mitochondrial inheritance but here, although both males and females can be affected, only females transmit the trait. This is because only eggs and not sperm transmit mitochondria to the zygote.

Example pedigrees may be seen at http: //www.gla.ac.uk/medicalgenetics/encyclopedia.htm

See also: **Mitochondrial Inheritance**

Viroids

C Beamish and E Kutter

doi: 10.1006/rwgn.2001.1367

Viroids might be considered 'naked viruses' – viruses without their space ships. They are infectious circular RNA molecules only about 300 nucleotides long, with no protein coats. Cells somehow take them up, replicate them extensively and then release them. Sometimes they harm their host cell, sometimes not. They cause a variety of problematic diseases in higher plants. For example, in potato spindle-tuber disease, a viroid causes the potatoes to become cracked, gnarled, and elongated. Viroids also damage coconut palms, hops, peaches, cucumbers, and avocados. They are transmitted between plants mechanically or through pollen or ovules. In addition to their large economic impact on plant crops, these RNAs are of great interest to molecular biologists because they are the smallest and simplest replicating molecules known, and because it is possible that they are a sort of living fossils, reflecting precellular evolution in a hypothetical RNA world.

Viroids were discovered in 1971 by T.O. Diener, a plant pathologist. It is still not clear how they cause disease, especially since they may cause severe problems in one plant and no particular symptoms in a related species. They encode no proteins. Perhaps they bind to something in the cell and disrupt some regulatory mechanism. The RNA of many viroids contains sections of nucleotide sequence complementary to key regions at the boundaries of RNA introns; maybe that is how they damage cells. They also have nucleotide sequences similar to some seen in transposons and retroviruses.

Two quite distinct groups of viroids have been identified. Group A viroids (such as peach latent mosaic viroid) replicate in chloroplasts, while group B viroids multiply in the nucleus of the host cell. Viroid replication involves either symmetric or asymmetric rolling circle replication mediated by an RNA-directed RNA polymerase, long thought to be unique to plant cells. (Now it seems that there may be a similar animal-cell enzyme that is involved in replicating hepatitis delta virus RNA.) The infecting circular monomer is copied to make a long linear multimeric minus strand. In the symmetric mode, the multimeric minus strand is cut and sealed to make minus circular monomers. These then act as templates to form the plus-strand monomers by the same set of three reactions. All group B viroids studied to date use the symmetric mode. In the asymmetric mode, seen for group A viroids, the minus strand serves directly as a template to make linear multimeric plus strands. These are cleaved through autocatalysis involving a so-called 'hammerhead' ribozyme structure and sealed into closed circles. Such hammerhead structures have also been seen in other RNA molecules that have enzyme-like properties. No hammerheads are seen in the group B viroids and no self cleavage has yet been observed, but neither has a cellular enzyme yet been identified that is responsible for their cleavage. Thus, a still-unidentified form of autocatalysis may be involved.

Further Reading

Diener TO (1991) Subviral pathogens of plants: viroids and viroid-like satellite RNAs. *FASEB Journal* 5 (13): 2808–2813.

Scott A (1985) *Pirates of the Cell*. Oxford: Blackwell.

Sanger HL (1989) Viroid function: viroid replication. In Diener TO (ed.) *The Viroids*, pp. 117–166. New York: Plenum Press.

See also: **Rolling Circle Replication**

Virology

See: **Virus**

Virulent Phage

E Thomas

doi: 10.1006/rwgn.2001.1370

A virulent phage will always substantially alter the physiology of the cell it infects and lyse its bacterial host at the completion of its infection cycle. Because of this distinguishing characteristic, virulent phages are also known as lytic phages. Coliphages T4 and T7 are classic example of virulent phages. The life cycle of virulent phages (**Figure 1**) is initiated with adsorption of the phage to some specific set of macromolecules it uses as receptors on a bacterial host. Once adsorption has occurred, the viral DNA enters the

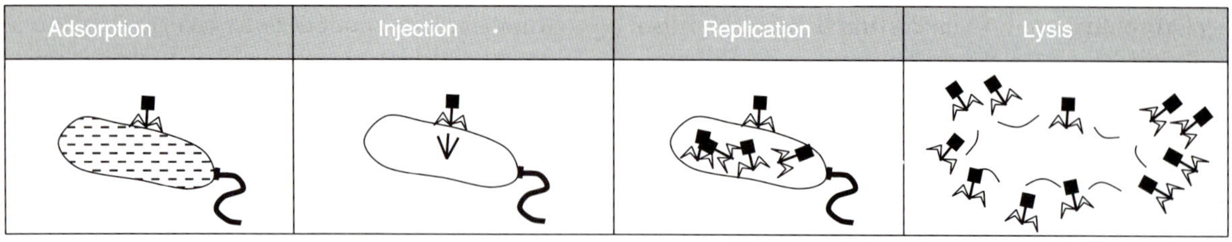

Adsorption	Injection	Replication	Lysis

Figure 1 The life cycle of virulent phages.

host and the viral genes are transcribed and translated in a specific preprogrammed sequence. The resulting viral proteins shut down the host's cellular machinery and begin building new viruses. A lysozyme or other endolysin is produced and, once a phage-encoded porin gives it access across the inner membrane to the bacterial cell wall, the wall is degraded and the new phages are released into the surrounding environment. Under optimal conditions, it takes about 25 min for T4 to complete this life cycle.

See also: **Lysis; T Phages**

Virus

B S Guttman

doi: 10.1006/rwgn.2001.1371

A virus can be any of a variety of minute, particulate, infectious agents consisting of protein and nucleic acid, each capable of multiplying in the cells of a specific type of organism. Each type of virus causes a characteristic infection, which commonly culminates in destruction of the host cells and liberation of more viruses. Viruses of all types of organisms (bacteria, fungi, algae, protists, plants, animals) have been identified.

As microbiologists of the late nineteenth and early twentieth centuries gradually sorted out the entities of their domain, they depended on what we would now consider quite primitive tools. Microscopes of the time were good enough to reveal bacteria of various forms, but the theoretical limit of resolution of a microscope using visible light, about 0.2 μm, kept them from observing any smaller particles. However, the porcelain filters that were commonly used to sterilize solutions provided a useful criterion of size. Thus, when Iwanowski demonstrated, in 1892, that infected tobacco plants carried the infectious agent in their sap, he could not observe the agent with a microscope, but, by demonstrating that it passed through a filter that retained bacteria, he showed it to be a previously unknown life form. In 1899, Beijerinck independently discovered the same phenomenon and called it a *contagium vivum fluidum* – a contagious living fluid. Similar infectious agents of other diseases were identified during the same period (foot-and-mouth disease by Loeffler and Frosch in 1898, yellow fever by Reed and his colleagues in 1900) as filterable viruses, from the Latin *virus*, meaning a poison or stench.

The study of viruses was advanced considerably by the discovery of bacterial viruses, or bacteriophages (phages; see article on Bacteriophages), by Twort in 1915 and by d'Herelle (see D'Herelle, Félix) in 1917. As described by Evans (1952):

> When one adds a small amount of bacterial virus to a vigorously growing bacterial culture, nothing occurs immediately. Then, suddenly, the suspension begins to foam, as materials inside the bacterial cell are liberated into the medium, and within a short time the heavy mass of growing bacteria has been replaced by floating shreds of debris that settle slowly to the bottom of the containing vessel. The clear bluish supernatant fluid now contains a hundred fold multiplication of the original virus inoculum.

Phage multiplication is easy to study because of plaque formation: an appropriately diluted suspension of phage is mixed with susceptible bacteria in a few milliliters of melted agar, and the mixture is poured over the layer of nutrient medium in a petri plate. After overnight incubation, the plate is covered with a continuous layer, or 'lawn,' of bacteria which is interrupted by small, circular clearings called 'plaques,' where bacteria have been killed. This phenomenon reveals at least two important points. First, bacteria are killed at discrete points on the plate, not in widespread killing spread nebulously throughout the culture, thus showing that the infectious agents are particulate. Second, the number of plaques formed is directly proportional to the amount of the original suspension added to the plate, making a convenient way to determine the concentration, or titer, of infectious particles: the number of plaques formed multiplied by the dilution factor. One can also count viruses by electron microscopy and thus determine the plating efficiency: the fraction of particles in a preparation that are capable of forming plaques. For many phages, the plating efficiency is essentially 1, but for other viruses it may be quite low.

In a series of classic experiments, Ellis and Delbrück (1939) and Delbrück (1940) showed the value of the plaque method and used it to demonstrate that bacteriophages multiply in steps, now demonstrated by a single-step growth curve. Consider each plaque to be formed by an infective center; the number of such centers stays constant for about 25–30 min after phages and bacteria have been mixed in a flask, and then it rises dramatically, typically by a factor of about 100. This shows that the phages multiply within bacteria that remain intact for a time and then suddenly burst (lyse). Until 25–30 min, the infective centers are intact, infected bacteria which burst on the plate, each yielding a single plaque; after that time, more and more of the infective centers are free phage particles that have been liberated by cells bursting in the infection flask. Experiments with viruses of other cells, including plants and animals, give comparable results, except that infected cells do not commonly burst so suddenly and dramatically but rather disintegrate gradually, liberating the virus particles that have been formed within. The understanding generated by these experiments has defined much of our thinking about viruses in general.

Nature of Viruses

The most often-asked question about viruses is "Are they alive?" This is a semantic problem that could only be answered by addressing the more complicated question of what the criteria for life are. Instead of addressing that issue, we will adopt the now-well-accepted viewpoint of Evans (1952) that viruses are entities with their own distinctive characteristics and are quite distinct from other entities called organisms. In Lwoff's words, "viruses are to be considered viruses because viruses are viruses." As Lwoff and Tournier emphasized, there are no entities intermediate between viruses and organisms. Certain kinds of infectious bacteria (rickettsias, chlamydias) show virus-like features such as intracellular multiplication and extreme metabolic dependency on their hosts, but they are still recognizable as bacteria. The differences between viruses and organisms are easily shown by the following comparison.

1. An organism is always a cell or collection of cells. No virus has such a structure. A virus is a particle, called a 'virion,' which consists of a nucleic acid genome enclosed in a protein covering: (a) The virion contains only one kind of nucleic acid – either DNA or RNA – whereas every cell needs both kinds to function. Viruses reproduce solely using the information from this one nucleic acid, whereas organisms, including infectious organisms, reproduce through an integrated action of their nucleic acid constituents; (b) Cells grow by enlargement and binary fission. No virus grows in this way. The virion is merely a vehicle for transporting the nucleic acid genome to another host cell. The genome enters the host cell and begins an infection, which results in production of a large number of new virions.
2. Viral genomes do not contain the information for any kind of apparatus to generate high-potential energy – what Lwoff called a 'Lipmann system.' The virus is thus totally dependent upon its host cell for a chemiosmotic potential, for ATP, and for any other source of energy.
3. A virus makes use of its host's protein-synthesizing apparatus: its ribosomes, transfer RNAs, and other factors. Some viral genomes encode special tRNAs, but no virus supplies the entire protein-synthetic system. Again, it is absolutely dependent upon its host.

Characteristics of Virions

Individual virus particles (virions) have distinctive features that help to identify them in electron micrographs, but all virions share certain basic features. Each virion consists of a nucleic acid genome (see Genome; Nucleic Acid) and a protective covering of protein called the 'capsid'; the combination of nucleic acid and protein makes the nucleocapsid, which in many cases is the entire virion. In other cases, however, the nucleocapsid is enclosed by an envelope, or 'peplos,' made of a somewhat modified cell membrane from the cell in which the virion was made.

Crick and Watson (1956) pointed out that even a small virus has too much capsid protein to be encoded by its genome if the protein were one unique sequence. Instead, they argued, a capsid must be made of small subunits (protomers) that combine to form the large (multimeric) capsid. Caspar and Klug (1962), in considering design principles for virus structure, combined this principle of subassembly with a principle of self-assembly: the protomeric units should assemble themselves spontaneously into the correct (that is, lowest energy) form without the need for additional structural information from outside. If identical protomers are to assemble themselves into a structure, the bonds between all subunits must be identical and all subunits must bear the same geometrical relationship to one another; the resulting structure must therefore be symmetrical. The laws of crystallography limit the possible modes of symmetry to two classes: helical or cubic. In helical virions, the nucleic acid associates with the capsid protein units to form a helix coated with protein. In cubic virions, the nucleic acid is wound up rather like a ball of thread inside a closed shell of protein subunits. The capsid of a cubic virion is therefore a surface crystal of protein.

Helical Capsids

The helical capsid, exemplified by tobacco mosaic virus, is easiest to describe (**Figure 1**). The capsid is a large multimer of a single type of polypeptide enclosing the RNA genome in a groove. Protomeric units associate with one another to initially form a disk, which soon is transformed into the beginnings of a helix. Protomers assemble along the RNA until it is entirely enclosed, thus determining the length of the capsid. This type of capsid is therefore very similar to other large, helical protein structures such as the flagella and pili of bacteria.

Figure 1 The helical structure of tobacco mosaic virus. The nucleocapsid consists of many identical protein molecules enclosing the RNA genome.

Cubic Capsids

Electron microscopy shows that all cubic capsids actually have the form of an icosahedron, with 20 equilateral triangular faces; this becomes a dodecahedron in the limit for the smallest viruses. In fact, the principle of construction was discovered by R. Buckminster Fuller, in facing the challenge of designing easily assembled structures ("assembly by child," as he put it). A plane can be paved with identical equilateral-triangle tiles; in this plane, six tiles meet at many points of sixfold symmetry. The plane can be bent into a third dimension, so it can start to enclose a space, by removing a wedge of tiles touching one point and connecting the remaining tiles, thus creating a vertex with fivefold symmetry. Completely enclosing a volume requires 12 fivefold vertices (making a dodecahedron), but more generally the space between vertices is filled in with varying numbers of units of sixfold symmetry, thus creating an icosahedron. This is the basis of Fuller's geodesic dome. Just as geodesic domes can vary in size, actual cubic capsids vary in number of protomers. All possible icosadeltahedrons can be defined by a number T, the triangulation number, where the number of subunits is $20T$. T is given by the rule $T = Pf^2$, where f is any integer and $P = h^2 + hk + k^2$, where h and k are any two integers with no common factor. Virions of different sizes are known to have T-values of 1, 3, 4, 7, 9, 16, 25, and 81. When negatively stained and examined by electron microscopy, the surface of an icosahedral nucleocapsid shows distinct units called capsomers; those at the vertices are pentons, made of five protomers, and the rest are hexons, made of six protomers.

Types of Virus–Host Interactions

Virus–host interactions are exemplifying by lytic virus multiplication: One or more virions infect a host cell, which is then converted into a factory for the synthesis of new viruses. Virions accumulate in the cell, which eventually disintegrates and scatters its contents.

However, this is not the only possible type of virus-host interaction. Many viruses establish a state of lysogeny (see Lysogeny), in which the viral genome remains in a stable condition (a provirus) inside the cell. Lysogenized cells may multiply indefinitely, like other cells, each member of the clone retaining its own copy of the provirus.

A number of bacterial viruses are known to multiply in their hosts in a nondestructive way. A copy of the viral genome remains inside the cell, replicating at a low rate and directing the synthesis of viral proteins; but, instead of lysing, the cell remains intact and new virions are extruded from the cell surface. Such cells can apparently continue to produce virions indefinitely.

Common Pattern of Lytic Multiplication

All viruses whose lytic cycles have been studied show very similar patterns of virus multiplication. The method of infection varies, but, once a viral genome is established in the host cytoplasm, a series of early genes (see Early Genes (in Phage Genomes)) are expressed. These genes may encode proteins that disrupt host activities, as by hydrolyzing the host genome and stopping translation of host mRNAs, and their

Table 1 Known types of viral genomes

	Description	Example
DNA genomes		
Single-stranded	Circular	Phages: ϕX174, M13
	Linear	Parvoviruses
Double-stranded	Linear	Many viruses
	Linear with nicks	Phage T5
	Circular	Papovaviruses
RNA genomes		
Single-stranded (linear)	Positive strand	Picornaviruses
	Positive strand segmented	Brome mosaic virus
	Negative strand	Rhabdoviruses; some paramyxoviruses
	Negative strand segmented	Some paramyxoviruses
Double-stranded (linear)	Segmented	Reoviruses

protein products may be new enzymes essential for the replication of the viral genome. Synthesis of these proteins ceases by the middle of the infection period and a set of late genes (see Late Genes) is then turned on. The late proteins encoded by these genes are primarily structural proteins of the virion; their synthesis continues throughout the rest of the infection period, and the lysed cell contains capsid proteins and nucleic acids that have not formed virions.

Types of Viral Genomes

Each type of virus has a genome of either DNA or RNA, but no virus carries both. The genome may be either single- or double-stranded, and may take a variety of forms. Viral nucleic acid strands are also designated either positive (plus) or negative (minus), where the viral messenger RNA is defined as being positive. Finally, some viruses carry segmented genomes made of separate nucleic acid molecules, and a complete genome requires all of the segments; in this case, however, a successful infection can result from simultaneous infection by several virions that contribute all of the segments in combination, even if no one virion has a complete genome. **Table 1** shows the types of genomes that have been identified.

When viruses with double-stranded genomes infect, transcription can occur as it does in a normal cell, although special transcriptases may be required. However, infection by a single-stranded genome requires the formation of a double-stranded intermediate called a 'replicative form' (RF). For instance, the single-stranded DNA genome of the phage ϕX174 is a plus strand. Upon infection, it is converted by the bacterial DNA polymerase into a double-stranded RF, whose negative strand is then used as the template strand for transcription of mRNA and also as the template for the production of new positive-strand genomes. Similarly, when a virus with a positive-strand RNA genome infects, that RNA is itself capable of being a messenger, and one of the early products made by translation of this mRNA is an RNA replicase, an RNA-dependent RNA polymerase which converts some of the infecting strands into RFs. As with a single-stranded DNA, these RFs are the sources of new mRNA molecules and new positive-strand genomes.

Infection by a virus with a negative-strand RNA genome requires a transcriptase associated with the virion; this enzyme transcribes the genome to produce positive mRNA, which is then translated into proteins. Among these proteins is a replicase, which converts the infecting genomes into RFs, from which new negative-strand genomes are formed.

The most unusual and complex form of replication occurs among the retroviruses, such as human immunodeficiency virus (HIV) and Rous sarcoma virus (RSV). They carry positive-strand RNA genomes along with RNA-dependent DNA polymerase, a so-called reverse transcriptase. Upon infection, this polymerase converts the RNA genome into a unique double-stranded form (+RNA combined with −DNA). After removal of the RNA strand by a ribonuclease activity of the same enzyme, the remaining negative DNA strand is replicated to form a double-stranded proviral DNA. Furthermore, the proviral DNA then forms a circular molecule, which integrates itself into the host chromosomes. This DNA can then

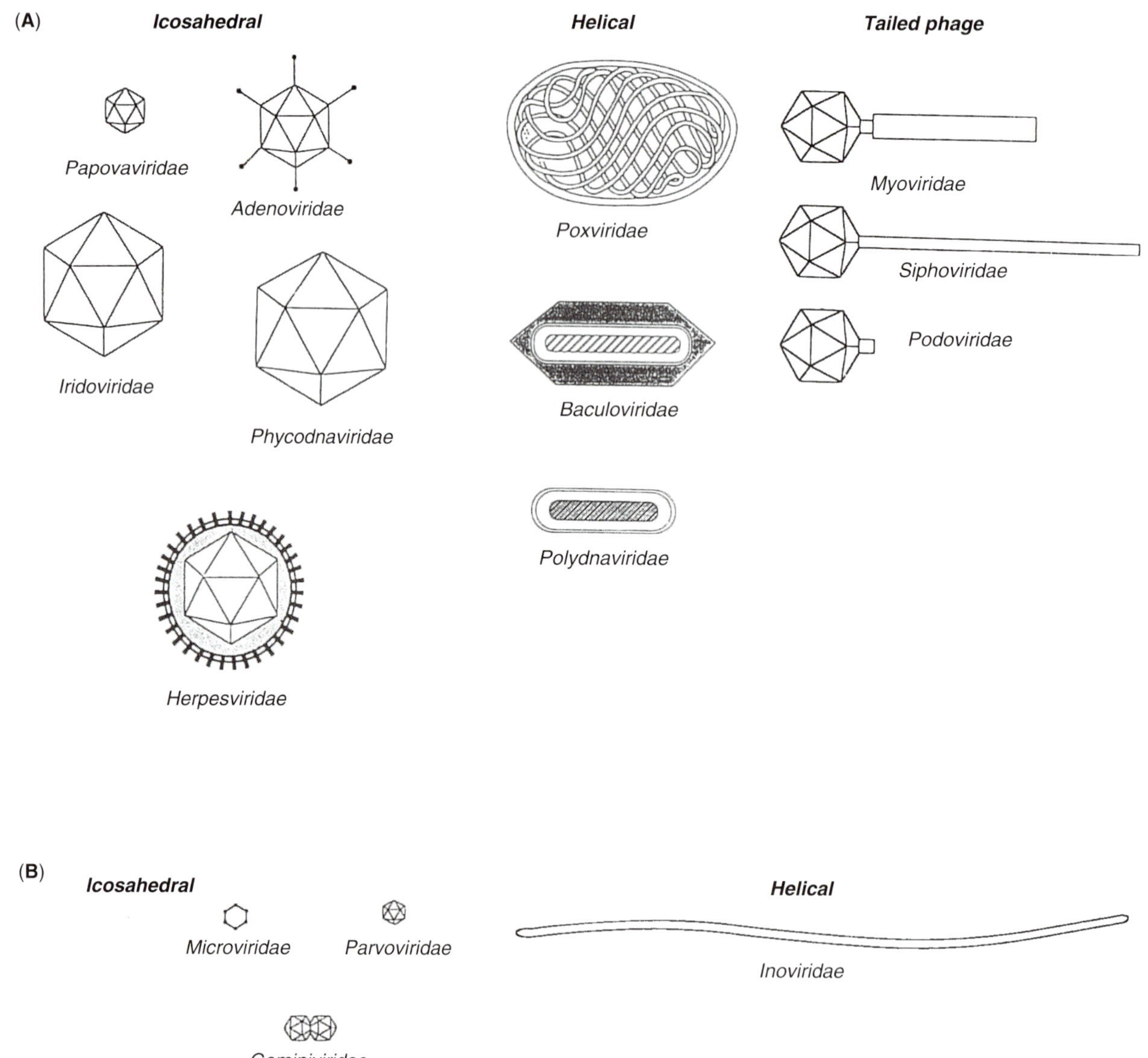

Figure 2 Forms of representative DNA viruses. (A) Viruses with double-stranded DNA; (B) viruses with single-stranded DNA.

be transcribed into new RNA strands, which serve either as mRNAs or as genomes for new viruses. Cells infected in this manner then take on new characteristics of their own and may be transformed into tumor cells.

Classification of Viruses

The general scheme of virus classification proposed by Lwoff and Tournier, based on the characteristics of the virion, has been considerably modified to take account of other features. The Lwoff–Tournier system divides viruses into riboviruses or deoxyviruses as their nucleic acid is RNA or DNA; then into helical or cubic classes, depending on symmetry of the nucleocapsid, thus defining four classes: Ribocubica, Ribohelica, Deoxycubica, and Deoxyhelica. Finally, the nucleocapsid may be naked or enveloped, creating a tertiary division and defining eight orders. However, this scheme leaves no room for the common bacteriophages, which have icosahedral heads and helical tails, nor for some viruses with complex – and sometimes still unknown – structures. The classification currently being developed by an international committee is explained by Lwoff and Tournier (1966). **Figures 2** and **3** show some major families of viruses, defined by a combination of nucleocapsid form, presence or absence of an envelope, and type of nucleic acid.

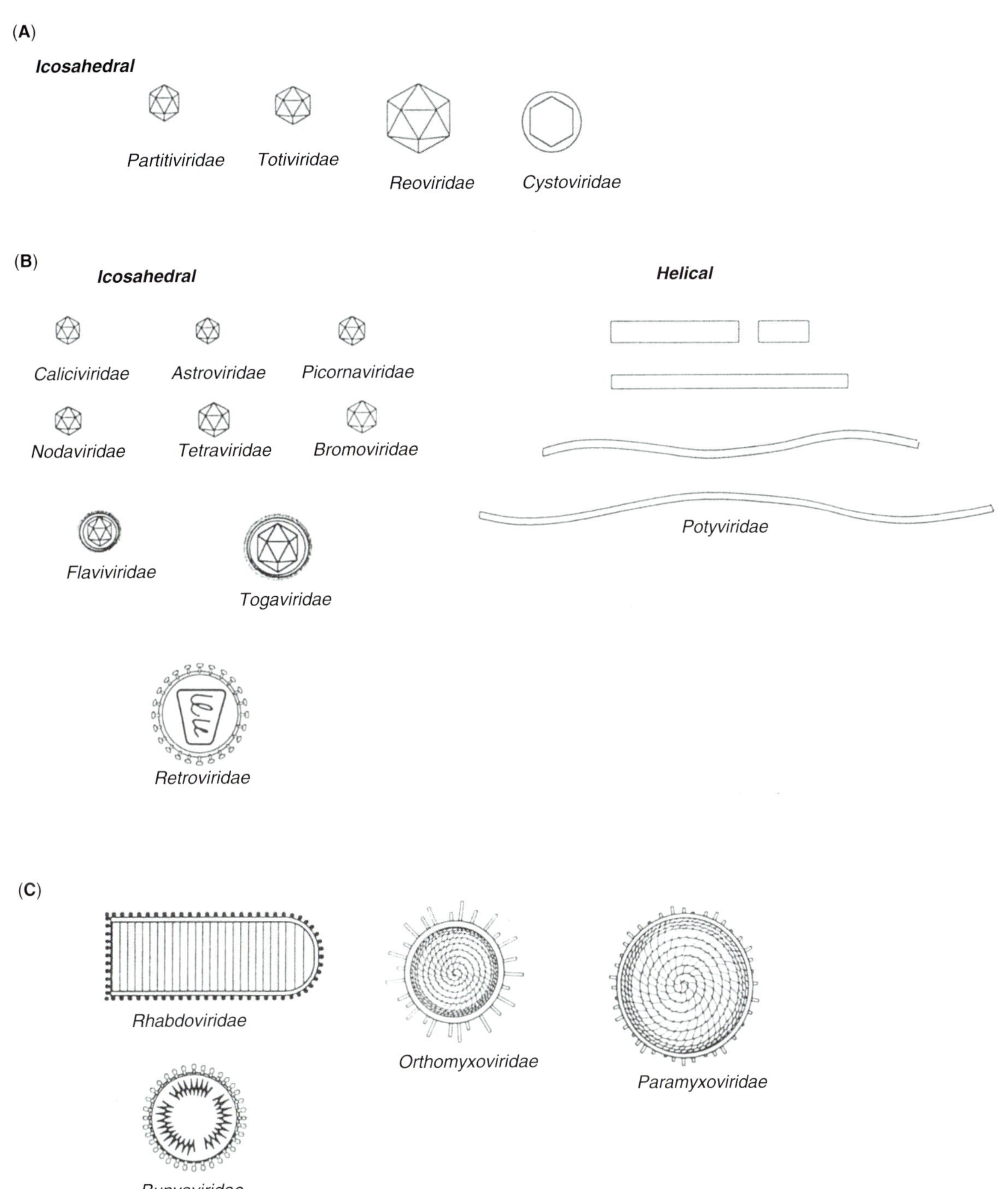

Figure 3 Forms of representative RNA viruses. (A) Viruses with double-stranded RNA; (B) viruses with single-stranded RNA (+ strand); (C) viruses with single-stranded DNA.

References

Caspar DLD and Klug A (1962) Physical principles in the construction of regular viruses. *Cold Spring Harbor Symposia on Quantitative Biology* 27: 1.

Crick FHC and Watson JD (1956) The structure of small viruses. *Nature* 177: 473–475.

Delbrück M (1940) The growth of bacteriophage and lysis of the host. *Journal of General Physiology* 23: 643–660.

Ellis EL and Delbrück M (1939) The growth of bacteriophage. *Journal of General Physiology* 22: 365–384.

Evans EA (1952) *Biochemical Studies of Bacterial Viruses*. Chicago, IL: University Press of Chicago.

Lwoff A and Tournier P (1966) The classification of viruses. *Annual Review of Microbiology* 20: 46–74.

Murphy FA, Fauquet CM, Bishop DHL, Ghabrial SA, Jarvis AW and Martelli GP (1995) *The Classification and Nomenclature of Viruses*. New York: Springer-Verlag.

See also: **Bacteriophages; D'Herelle, Félix; Early Genes (in Phage Genomes); Genome; Late Genes; Lysogeny; Nucleic Acid**

Viruses of the Archaea

D Prangishvili and W Zillig

doi: 10.1006/rwgn.2001.1516

In contrast to the large number of bacteriophages and viruses of eukaryotes known to date, only about two dozen viruses of the Archaea have so far been identified and studied in some detail. Several of these viruses have unique morphologies, although their genome structures and virus–host relationships show certain similarities to either bacterial or eukaryotic viruses, furnishing evidence for the primeval existence of common ancestral modules.

Viruses of Euryarchaeota

The morphotypes of archaeal viruses reflect the division of the domain Archaea into two kingdoms, the Euryarchaeota and the Crenarchaeota. All but two viruses of euryarchaeotes are typical head-and-tail phages, including virions with contractile and noncontractile tails, thus belonging to the families Myoviridae and Siphovoridae. All have double-stranded DNA genomes. Circular permutation and terminal redundancy of the genomes of some phages indicate a headful mechanism of packaging from concatemeric precursors.

Both temperate and lytic viruses were found. The prophage of *Halobacterium* phage ϕH persists as a circular episome, similar to the prophage form of the coliphage P1, rather than being integrated in the host's chromosome. The regulation of lysogeny has features resembling the regulation of lysogeny in the lambda phage of *Escherichia coli*. The promoters of the genes encoding an early lytic product that is necessary for the expression of late genes and the repressor of that transcript are situated back to back, in a manner similar to that of *cI* and *cro* in lambda, and transcription of the two genes is mutually exclusive.

About 25% of the genome of the haloarchaeophage ϕH and the complete genome of the *Methanobacterium* phage ΨM2 have been sequenced. Similarities with bacteriophages were again found in the genome organization. Several open reading frames (ORFs) of the *Methanobacterium* phage show significant similarities to genes encoding structural proteins, proteins involved in packaging DNA into capsids, and a site-specific recombinase of bacteriophages that infect *Bacillus* and other gram-positive hosts.

There are two examples of euryarchaeal viruses that have morphologies different from those of tailed phages: His 1, a virus infecting *Haloarcula hispanica*, and a virus-like particle produced by *Methanococcus voltae* strain A3 both have a spindle-like shape.

Viruses of Crenarchaeota

Viruses have been described for only two genera of the kingdom Crenarchaeota, the hyperthermophile *Thermoproteus* and the extreme thermophile *Sulfolobus*. All of these viruses have unique morphotypes and have been assigned to four novel virus families: Fuselloviridae (the spindle-shaped enveloped viruses SSV1, SSV2, and SSV3 of *Sulfolobus*), Rudiviridae (the stiff rod-shaped, nonenveloped viruses SIRV1 and SIRV2 of *Sulfolobus*), Lipothrixviridae (the filamentous enveloped viruses TTV1, TTV2, and TTV3 of *Thermoproteus*, DAFV of *Acidianus*, and SIFV of *Sulfolobus*), and Guttaviridae (the droplet-shaped virus SNDV of *Sulfolobus*). Typical virus particles are shown in **Figure 1**.

Only viruses TTV1 and TTV4 are lytic. The fuselloviruses are temperate and the rest are present in their hosts in more or less stable carrier state. Possibly this strategy helps them to escape prolonged direct confrontation with the harsh natural environment, with temperatures up to 100 °C, and, for viruses of the acidophilic *Sulfolobus*, pH values down to 1.5. However, due to significant inhibition of the growth of host cells, plaque tests could be established for all viruses infecting *Sulfolobus*, except SNDV.

In fusellovirus lysogens the viral genome is integrated specifically into the host genome by means of a virally encoded integrase and is also present as a plasmid copy. As in the case of some temperate bacteriophages, integration occurs in a tRNA gene of the host. Similar to bacterial lysogens, virus production can be induced by UV irradiation or by mitomycin C treatment, apparently resulting from an SOS-like response of the host cells. This response includes activation of a short transcription unit which is situated between two large 'back-to-back' transcription units, similar to the C1 gene of the lambdoid bacterial phage 186.

The rudivirus SIRV1 shows unusual behavior in new hosts. The virus genome varies by extensive accumulation of point mutations with a rate of about 10^{-3}

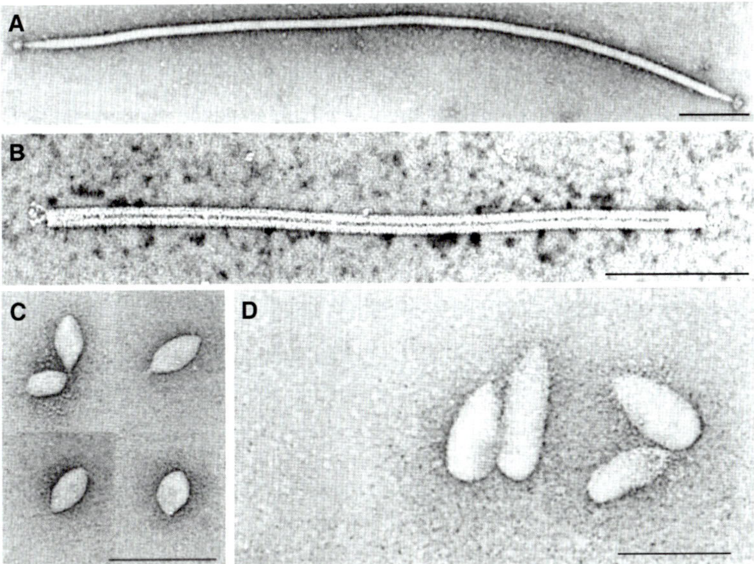

Figure 1 Electron micrographs of representatives of the four families of viruses of the Crenarchaeota. (A) Lipothrixvirus SIFV; (B) rudivirus SIRV2; (C) fusellovirus SSV1; (D) guttavirus SIFV. Scale bars = 200 nm. (Reprinted from Prangishvili *et al.*, 2001 with permission from Elsevier Science.)

substitutions per nucleotide per replication cycle – unprecedented for DNA viruses and approaching values seen for the most rapidly mutating RNA viruses. Accumulation of point mutations eventually leads to the selection of conditionally stable virus variants, coinciding with the recovery of high fidelity replication. Such stable variants of SIRV1 produce further variants when infecting a new host, demonstrating that stability of the viral genome in certain hosts does not exclude the potential to vary. SIRV2, which has a similar but 3.2 kb longer genome, remains stable in the same hosts.

The virus TTV1 shows genetic variability indicative of an undefined recombination mechanism. The variance arises from the regrouping of homologous specific sequences between two nonadjacent reading frames.

The genomes of all crenarchaeal viruses isolated to date consist of double-stranded DNA. In members of the Rudiviridae and the Lipothrixviridae it is linear, and in members of the two other families covalently closed circular. The circular DNA of SSV1 was shown to be positively supercoiled. The termini of the linear genome of the lipothrixvirus SIFV are modified in an as yet uncharacterized manner, and those of the rudiviruses are covalently closed – the two DNA strands form a continuous polynucleotide chain. Such structure is characteristic for linear genomes of eukaryotic poxviruses, *Chlorella* viruses, and African swine fever virus. The genomes of rudiviruses share with these genomes one more characteristic feature, long inverted terminal repeats. Presumably the mode of DNA replication is similar in rudiviruses and these eukaryotic viruses.

The complete genomes of both rudiviruses and of the fuselloviruses SSV1 and SSV2, and more than 90% of the genomes of the lipothrixviruses TTV1 and SIFV have been sequenced. Except for the latter two, the sequences of members of the same families are highly homologous to each other. The two rudiviruses share 16 ORFs with the lipothrixvirus SIFV, indicating that the two virus families may be related. Paralleling the unique morphotypes of the viruses, only a very limited number of their ORFs show any similarity to proteins from other viruses or from organisms. On the basis of sequence similarity, only the SSV1 gene encoding an integrase and the SIRV1 genes encoding a dUTPase and a Holliday junction resolvase could be unambiguously identified. All three genes have been functionally expressed in *E. coli*.

The virus SSV1 has proven to be a useful model for studying transcription in the Archaea. During the analysis of viral transcripts and promoters it was found that the promoter sequences contained TATA-boxes resembling the promoters of eukaryotic RNA-polymerase II rather than those of bacteria.

Evolutionary Considerations

Studies on archaeal viruses allow insight into virus evolution. Conservation of a characteristic virion morphology and sequence similarity of some of the genes indicates that euryarchaeal viruses could share ancestry with tailed bacteriophages. Certain characteristics of the genomes as well as virus–host relationships of crenarchaeal fuselloviruses indicate a common ancestry with lambdoid bacteriophages. Crenarchaeal rudiviruses, sharing with poxviruses, *Chlorella* viruses, and African swine fever virus peculiarities of genome organization and replication, seem to be related to them. Considering biochemical barriers between the three domains of life, e.g., the incompatibility of the archaeal and the bacterial transcription systems, and between different life styles, direct spreading of viruses from one domain to the others appears unlikely. It seems more plausible to assume common ancestors prior to the divergence of the lineages of their hosts.

Further Reading

Pfister P, Wasserfallen A, Stettler R and Leisinger T (1998) Molecular analysis of *Methanobacterium* phage ΨM2. *Molecular Microbiology* 30: 233–244.

Prangishvili D, Stedman K and Zillig W (2001) Viruses of the extremely thermophilic archaeaon *Sulfolobus*. *Trends in Microbiology* 9: 39–43.

Stedman K, Schleper C, Rumpf E and Zillig W (1999) Genetic requirements for the function of the archaeal virus SSV1 in *Sulfolobus solfataricus*: construction and testing viral shuttle vectors. *Genetics* 152: 1397–1405.

Stolt P and Zillig W (1994) Gene regulation in halophage ϕH, more than promoters. *Systematic and Applied Microbiology* 16: 591–596.

Zillig W, Prangishvili D, Schleper C *et al.* (1996) Viruses, plasmids and other genetic elements of thermophilic and hyperthermophilic archaea. *FEMS Microbiological Reviews* 18: 225–236

***See also:* Archaea, Genetics of; Virus**

Visconti–Delbrück Hypothesis

F W Stahl

doi: 10.1006/rwgn.2001.1373

The Visconti–Delbrück hypothesis was offered by N. Visconti and M. Delbrück in 1953 to rationalize the linkage relations observed in genetic crosses conducted with the T-even phages, T2 and T4.

Phage crosses usually involve the mixing of about 1×10^8 bacteria (in 1 ml) with about 7×10^8 phage particles of each of (usually) two different genotypes. Phage particles are given time to adsorb to the bacteria and inject their DNA, thereby entering the 'vegetative' state. Phage particles that fail to adsorb to cells are eliminated, and the infected cells are diluted so that progeny phage particles released when the cells lyse do not readsorb to bacteria or bacterial debris. The progeny phage are assayed by plaque formation, and the genotypes of the particles determined by either the morphology of the plaques generated or by their ability to grow under one or another condition. The fraction of the progeny that is recombinant for a given pair of markers defines the recombinant frequency for those markers. For phages T2 and T4, early studies suggested three 'linkage groups.' Crosses involving markers in the same linkage group gave convincingly less than 50% recombinants, and the frequencies observed allowed the construction of linear maps based on the rule that markers giving the largest frequencies of recombinants be placed farthest apart on the map. However, crosses with markers on different linkage groups gave 40–45% recombinants, significantly less than the Mendelian expectation of 50%.

The linkage data departed from that usually observed with eukaryotic genetic crosses in an additional way: Frequencies of double recombinants for adjacent intervals were higher than (negative interference), rather than equal to (no interference) or lower than (positive interference), that expected if exchanges were randomly distributed among the linkage groups of different progeny particles. Furthermore, simultaneous infection by three appropriately marked genotypes of phage produced a progeny in which some of the particles had markers derived from each of the three infecting genotypes ('triparental recombinants').

In an effort to put these novel data into a meiotic framework, Visconti and Delbrück proposed that a phage cross should be interpreted within a 'population genetics' framework. They proposed that the vegetative phage within an infected cell paired ('mated') with each other. During a mating, the linkage groups are randomly assorted, and markers within the linkage groups are recombined by reciprocal exchanges that do not interfere with each other. Since there is no evidence of mating types or sexes in a phage population, the vegetative phage were assumed to mate with each other at random with respect to their genotype, with the consequence that half the matings are unproductive of recombinants. Under this assumption, one round of mating would result in 25% recombination for unlinked markers. To account for the observed 40–45% recombination between markers on different linkage groups, the progeny particles must have had several such matings, with partners chosen at random for each mating. Since mating was assumed to be contemporaneous with replication, this is equivalent to saying that progeny particles derive from lineages that have experienced several matings on the average. These assumptions explained the formation of triparental recombinants as having arisen in successive biparental matings. They explained the negative interference as a consequence of the unequal numbers of productive matings experienced by the different lineages. The model, formulated algebraically (see below), accounted for most of the data it was intended to explain. An apparent weakness in the model, however, related to the assumption of reciprocal exchange. Crosses with T2 had shown that complementary recombinants emerge from individual infected bacteria in numbers that are uncorrelated. At face value, this suggests that the recombination process is not reciprocal, i.e., the exchange process does not result in the formation of complementary recombinants in the same event. Visconti and Delbrück argued that the vegetative phage that have emerged from a mating could, by chance, enjoy differing opportunities for replication, obscuring the reciprocality of recombination which they had assumed. However, subsequent considerations revealed that the assumption of reciprocality played no role in the form of the final equations obtained.

The elementary Visconti–Delbrück equation relating recombinant frequency R to the probability of recombination in a single mating p is

$$R = 2f(1-f)(1-e^{-mp})$$

where m is the average number of matings per lineage and f is the fraction of one of the two parental genotypes in the infecting phage population. When the two infecting types are equal, as was the usual intention, the expression simplified to $R = \frac{1}{2}(1-e^{-mp})$. These equations suppose that matings are Poisson-distributed among lineages.

The mean number of matings, m, per lineage was estimated to be about 5 by letting p be 0.5 for markers that gave $R = 0.45$. With m set at 5, the observed R values were converted to p values that showed no crossover interference. The Visconti–Delbrück theory met similar success with linkage data from phage lambda, when m was set at 1.

For T-even phage, the mating theory required modification when it was demonstrated that the 'three linkage groups' were, in fact, well separated segments of a single, circular linkage map. When the theory was adjusted for this circularity, the data were best explained by the assumption that there was exactly one exchange per mating, rendering the concept of mating superfluous. The negative interference in the data, which had been the primary justification for the notion of mating, proved to be the result of several contributing factors: (1) The classical definition of interference is inappropriate for a circular map; (2) as envisioned by Visconti and Delbrück, some vegetative phage are withdrawn from the mating pool (becoming infectious particles) before others, further contributing to heterogeneity in opportunities for recombination; (3) when a mixture containing equal numbers of two phage genotypes is added to cells to initiate a cross, some cells, by chance, are infected by unequal numbers of the two types. This 'finite input' effect introduces further heterogeneity in recombination opportunities.

For T4 phage, the mating theory was dealt its final blow with the demonstration that separate regions of the linkage map behave independently. The demonstration made use of intragenic crosses employing markers (call them *1* and *2*) in both the *r* and the *e* genes. One parent was the double mutant *r1 e1*; the other was *r2 e2*. The frequencies of r^+ and of e^+ recombinants were measured in the absence and in the presence of a third infecting phage, which was *e1e2* and was deleted for the *r* gene. In the presence of the third phage, the reduction in frequency of e^+ progeny phage was greater than was the reduction in frequency of the r^+ progeny phage. Apparently, the *e1* and *e2* mutant genes indulged in fruitless interactions with the *e1e2* double mutant gene, while the *r1* and *r2* genes interacted to produce r^+ as if the third phage was not present in the cross.

See also: Interference, Genetic; T Phages

Vitamins

S Reichheld and T M Finan

doi: 10.1006/rwgn.2001.1374

Vitamins are organic molecules that are required in very small amounts by some, but not all living organisms. After ingestion, vitamins are converted to active derivatives, in which form they act primarily as cofactors for enzymatic reactions. Vitamins such as A and D are also involved in regulating gene expression, while others have antioxidant capabilities. Vitamins differ from other metabolic intermediates, because they either cannot be synthesized by the organism or are present in such low quantities that they must be acquired externally. Historically, vitamins have been classified by their solubility: The B vitamins and vitamin C are soluble in water, whereas vitamins A, D, E, and K are soluble in organic solvents.

Vitamins Act Catalytically as Cofactors in General Cell Metabolism

Most vitamins or their active metabolites are needed as cofactors in the synthesis, catabolism, and modification of organic compounds. They interact, covalently and noncovalently, with enzymes and are necessary for their activity.

Biotin is a key cofactor required by the enzymes acetyl coenzyme A (-CoA) carboxylase (EC 6.4.1.2), pyruvate carboxylase (EC 6.4.1.1), transcarboxylase (EC 2.1.3.1), methylmalonyl-CoA decarboxylase (EC 4.1.1.41), and oxaloacetate decarboxylase (EC 4.1.1.3), each of which catalyze carboxyl transfer reactions. The widespread distribution of biotin in foodstuffs and the ability to absorb some of the biotin synthesized by intestinal bacteria makes biotin deficiency in humans rare. However, deficiency has been observed in humans and experimental animals on diets containing large amounts of uncooked egg white. The glycoprotein avidin, found in abundant quantities in egg white, binds biotin with high affinity, thereby preventing its absorption through the gastrointestinal wall.

Another vitamin directly involved in the synthesis and degradation of organic compounds is vitamin B_5 (pantothenic acid). This compound becomes the functional moiety of coenzyme A and of the acyl carrier protein (see **Figure 1**). Vitamin B_1 (thiamine) is necessary for energy-yielding metabolism, as its active form thiamine pyrophosphate is the prosthetic group for some enzymes such as pyruvate dehydrogenase (EC 1.2.4.1) and 2-oxoglutarate dehydrogenase (EC 1.2.4.2) that are involved in oxidative decarboxylation.

Many dehydrogenases use nicotinamide adenine dinucleotide (NAD^+) or nicotinamide adenine dinucleotide phosphate ($NADP^+$). Niacin is a precursor in the synthesis of both of these cofactors. Other enzymes, designated flavoproteins, use the riboflavin (vitamin B_2)-derived oxidation–reduction cofactors, flavin mononucleotide (FMN) and flavin adenine dinucleotide (FAD).

Once translated, many proteins must be modified to be enzymatically or structurally active. Of significance in this regard is the action of vitamin C (ascorbic acid) as a cofactor for the hydroxylation of proline residues in collagen. This reaction is important for the maintenance of connective tissue, with deficiency causing related symptoms, including muscle fatigue, easily bruised skin, swollen gums, osteoporosis, and poor wound healing. Vitamin B_6, in phosphorylated form, is a necessary cofactor for transamination reactions required for amino acid synthesis and in the breakdown of the glucose storage polymer glycogen. Vitamin K is involved in converting blood coagulation precursor proteins to an active conformation.

Folic Acid and Vitamin B_{12} are Critical for DNA Synthesis

Sufficient amounts of folic acid and cobalamin (Vitamin B_{12}) are required for nucleotide synthesis. Folate derivatives directly associate as cofactors to enzymes that synthesize purines and the pyrimidine thymidine, while vitamin B_{12} is necessary to ensure that enough folate-derived cofactor is present to support nucleotide biosynthesis.

Folic acid deficiency leads to megaloblastic and macrocytic anemia, a hallmark of improper DNA replication and cell division in rapidly dividing hematopoetic cells. This condition is attributable to insufficient thymidine synthesis, as purines can normally be obtained in the diet. 5,10-Methylenetetrahydrofolate, a cofactor derived from folic acid, acts as a methyl donor for thymidylate synthase (EC 2.1.1.45), which converts deoxyuridine monophosphate (dUMP) to thymidine monophosphate (TMP). A lack of folic acid causes an accumulation of dUMP as well as a TMP deficiency. In folate-deficient cells, RNA transcription and subsequent translation is normal, but normal DNA replication is impeded. Consequently, the cytoplasm is able to expand, but the growth and division of the nucleus lags behind, preventing cell division. An overabundance of dUMP coupled with a deficiency of TMP also increases the likelihood of misincorporation of dUMP in place of TMP, which in turn increases DNA instability and the probability of chromosome breakage.

One of the functions of vitamin B_{12} is to act as a cofactor for the enzyme methionine synthase (EC 2.1.1.14), which catalyzes the conversion of homocysteine to methionine and in the process demethylates 5-methyltetrahydrofolate. Methyltetrahydrofolate tetrahydrofolate (THF) can be recycled and converted to such cofactors as 5,10-methylenetetrahydrofolate to be used for thymidine biosynthesis. Vitamin B_{12} deficiency leads to an accumulation of homocysteine and, more importantly, THF is not recycled, mimicking the effects of folate deficiency.

Vitamins E and C Act as Antioxidants

Reactive oxygen species (ROS) cause extensive damage to DNA and membrane lipids. Enzymatic reactions catalyzed by superoxide dismutase, peroxidase, and catalase counteract the effects of ROS. Protection against ROS, such as from the highly reactive peroxyl radicals, is also accomplished by antioxidant molecules such as vitamins E and C. These react with the initial oxygen species to form less reactive species, which are readily quenched by molecules other than DNA and lipids.

Vitamin E consists of a mixture of related compounds named tocopherols, all of which are soluble in lipids. Owing to their lipophilic nature, they accumulate in cell membranes and adipose tissue. By behaving as a lipid peroxyl radical scavenger, vitamin

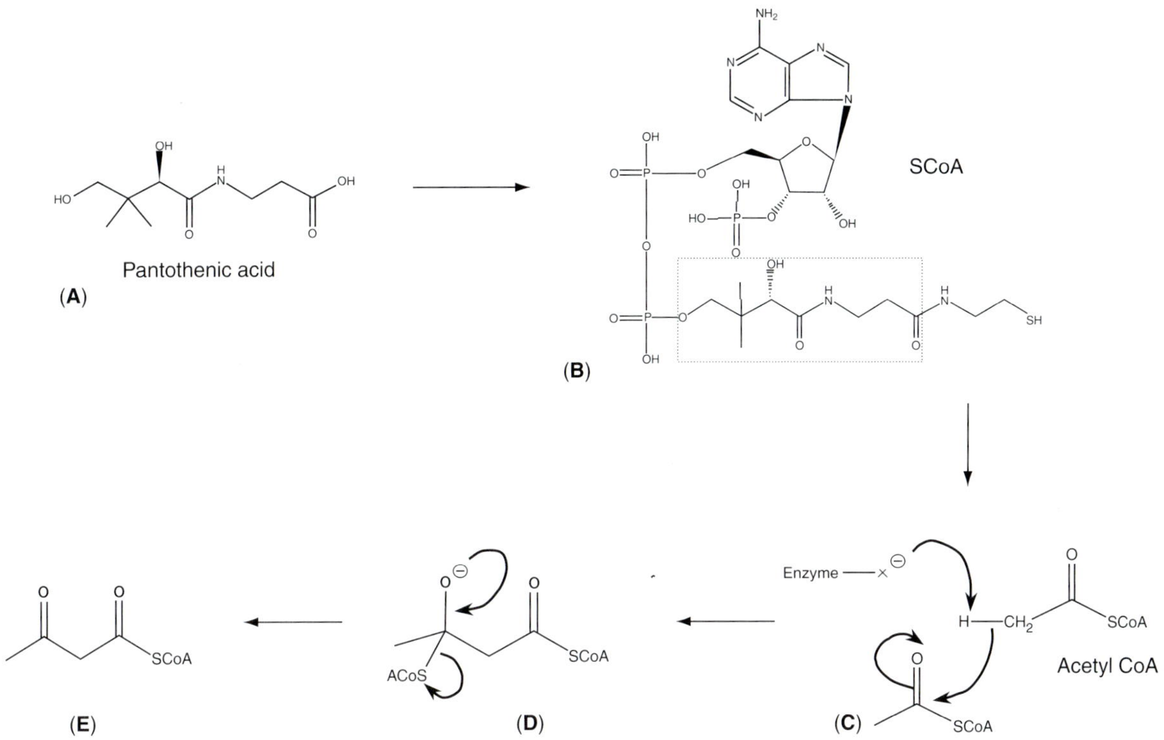

Figure 1 Pantothenic acid is an essential component of coenzyme A (SCoA). (A, B) The vitamin pantothenic acid (A) is incorporated as a functional component of coenzyme A (B); the dashed box (B) shows the substituents of coenzyme A that were derived from pantothenic acid. (C, D) Coenzyme A can be attached to acetyl-CoA and other carbonyl species to act as a good leaving group in carbon–carbon bond formation. A basic residue (X) of an enzyme can deprotonate the α-carbon of acetyl-CoA. The carbon ion created can attack the electrophilic center of the carbonyl group of another acetyl-CoA, creating a carbon–carbon bond. The curved arrows show the direction of electron flow in the reaction mechanism. (E) The product of the substitution reaction.

E prevents the peroxidation of polyunsaturated membrane fatty acids. Once reacted, the radical form of vitamin E can be converted back to α-tocopherol in a redox reaction, enhancing its antioxidant capabilities. Deficiency of vitamin E can cause red blood cell instability, but there are no major vitamin E diseases, because this substance is present in most food sources.

Vitamin C (ascorbic acid) is water soluble and acts as a free radical scavenger in the cytoplasm and organelles of cells. Ascorbic acid is able to react with the α-tocopheryl radical as well as donate electrons to ROS. This accomplishes two important tasks: the recycling of vitamin E and the prevention of the tocopheryl radical from starting a phospholipid peroxidation chain reaction.

Vitamin A Regulates Cell Growth and Development

A member of the vitamin A family of molecules, retinoic acid, affects DNA replication and cell division. Like the other vitamin A molecules, retinol and retinal, retinoic acid is also derived from carotene. Retinal is an integral part of the membrane-bound light receptors rhodopsin and iodopsin. However, retinoic acid acts as a signaling molecule by binding ligand-dependent nuclear receptor proteins in a manner comparable with the behavior of steroid hormones.

Retinoic acid aggregates with retinoic acid receptors (RARs) and retinoic acid X receptors (RXRs), allowing dimeric complexes of receptors to bind recognition sequences in the promoter region of target genes. RAR is able to heterodimerize with RXR, while RXRs can homodimerize or heterodimerize with other members of the nuclear receptor superfamily. DNA binding can activate transcription or block the binding by other transcription factors, both of which contribute to repressing cell replication and thymidine uptake.

11-*cis*-retinal is bound to the photopigments rhodopsin, in rod cells, and iodopsin, in cone cells. Light absorption by the opsins stimulates a series of conformational changes of the bound retinoid. Each conformational change makes the association between retinal and the protein progressively less stable. This process culminates with the dissociation of 11-retinal in a *trans* conformation. The dissociation causes closing of sodium channels, hyperpolarizing the cell membrane. The change in membrane potential is transmitted as a nervous impulse along the optic neurons.

Vitamin D is Important for Calcium and Phosphorus Metabolism

The most important derivative of vitamin D, the biologically active form calcitrol (1,25-dihydroxyvitamin D_3) acts similarly to retinoic acid (see Vitamin A regulates cell growth and development), by binding the nuclear vitamin D receptor (VDR). The binding to response elements in promoter regions by vitamin D-bound VDR dimers induces transcription of genes involved in increasing the intestinal absorption and kidney resorption of calcium and phosphorus. A deficiency reduces overall calcium and phosphorus levels, leading to the childhood disease rickets, characterized by the incomplete mineralization of bones. An analogous condition in adults is osteomalacia, which is the result of the demineralization of mature bones. Deficiencies are rare because of the capacity of skin cells to synthesize vitamin D in a light-dependent reaction.

The Evolution of Vitamin C Synthesis

Most mammals, birds, amphibians, and reptiles are able to synthesize ascorbic acid. However, guinea pigs and primates (including humans), have lost the ability to synthesize this vitamin. This phenomenon is caused by mutations in the gene encoding the enzyme L-gulonolactone oxidase (EC 1.1.3.8), which is responsible for converting gulonolactone to ascorbic acid. It is generally believed that this deficiency was maintained because adequate dietary intake removed a selective advantage.

Future Research

There is still much to be understood about the biosynthesis and mechanism of action of vitamins. The role of antioxidant vitamins in preventing cancer and possibly increasing longevity is not fully understood. Further research on vitamin gene regulation and the metabolic pathways effected will help to elucidate vitamins' role in development. Ultimately, what is learned can be applied to enhance dietary intake and subsequently better human life.

Further Reading

Bender DA (1992) *Nutritional Biochemistry of the Vitamins.* Cambridge: Cambridge University Press.

Friedrich W (1988) *Vitamins.* Hawthorne, NY: Walter de Gruyter.

Gardner DG and Chen S (1999) Retinoids and cell growth in the cardiovascular system. *Life Sciences* 65: 1607–1613.

Rachez C and Freedman LP (2000) Mechanisms of gene regulation by vitamin D_3 receptor: a network of coactivator interactions. *Gene* 246: 9–21.

See also: **Inherited Rickets; Nutritional Mutations**

VNTR (Variable Number of Tandem Repeats)

See: **Minisatellite**

Von Gierke disease

A Burchell

doi: 10.1006/rwgn.2001.1378

Glycogen storage diseases are metabolic disorders resulting in storage of abnormal amounts and/or forms of glycogen. Von Gierke disease is a glycogen storage disease caused by defective liver glucose-6-phosphatase activity. The disease causing mutation(s) can either be in the gene coding for the liver glucose-6-phosphatase enzyme (G6PC) or in the gene coding for the endoplasmic reticulum substrate and/or product transport proteins of the glucose-6-phosphatase system (see **Figure 1**).

History

In 1929, the pathologist von Gierke carried out an autopsy on a child who had died of influenza. He noticed a very large liver that stained positively for glycogen. He sent tissue samples to Schoenheimer, a biochemist, who showed that the glycogen levels were 36% of the dry weight of the liver. In subsequent years, more patients were described with abnormal liver glycogen storage. It took over 20 years for the first enzyme defect in glycogen storage disease to be delineated. In 1952, Carl and Gerty Cori found abnormally low glucose-6-phosphatase enzyme activity in livers of some (but not all) patients with glycogen storage disease. Glucose-6-phosphatase deficiency was therefore termed von Gierke disease. Since 1952, virtually all proteins involved in the synthesis or

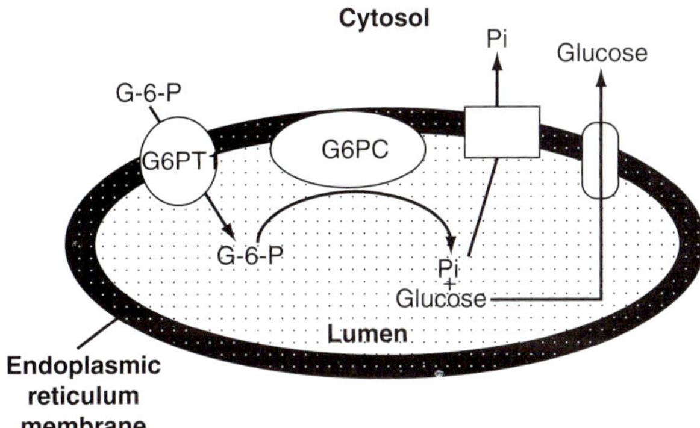

Figure 1 A schematic representation of the human liver glucose-6-phosphatase system. G6PC, glucose-6-phosphatase enzyme; G6PT1, endoplasmic reticulum glucose-6-phosphate transport protein; Pi, inorganic phosphate; G-6-P, glucose-6-phosphate.

degradation of glycogen (see **Figure 2**) and its regulation have been found to cause forms of glycogen storage disease. The glycogen storage diseases are now usually either named after the enzyme that is defective or numbered in the order in which the enzymatic defects were identified. Von Gierke disease was the first glycogen storage disease to be delineated and it is now more commonly called type 1 glycogen storage disease or glucose-6-phosphatase deficiency.

Function of Liver Glycogen

Glucose is the primary source of energy for most mammalian cells. Most tissues cannot make sufficient glucose to meet their metabolic needs. Blood glucose levels must stay within a narrow range to maintain normal metabolic function in brain and other tissues. It is, therefore, advantageous to an individual to have the ability to store glucose at times of plenty (for example, after a meal) in a compact macromolecular form, which can be rapidly broken down and released into the bloodstream at times of need. In human liver, glycogen is the storage form of glucose.

Role of Liver Glucose-6-Phosphatase

Whenever blood glucose levels fall, or at times of stress, the liver releases glucose into the bloodstream. In addition to breaking down glycogen to form glucose, the liver can also synthesize new glucose via the pathway called gluconeogenesis (see **Figure 2**). Glucose-6-phosphatase is the final step of both gluconeogenesis and glycogen breakdown. Glucose-6-phosphatase breaks down glucose-6-phosphate to glucose and inorganic phosphate and is the only enzyme that is capable of forming significant amounts of glucose in the body. The major role of liver glucose-6-phosphatase is therefore to produce glucose for use by other tissues.

In patients with type 1 glycogen storage disease significant amounts of glucose cannot be made by either pathway. In contrast, in most other types of glycogen storage disease only glycogen breakdown is affected, and glucose can still be made via the gluconeogenic pathway.

The Liver Glucose-6-Phosphatase System

Liver glucose-6-phosphatase is an integral membrane protein and its active site is inside the lumen of the endoplasmic reticulum, whereas all the other enzymes that produce or use glucose-6-phosphate are in the cytoplasm (see **Figure 2**). This means that the substrates and products of glucose-6-phosphatase must cross the endoplasmic reticulum membrane (see **Figure 1**). The substrate and product transport proteins are facilitative transporters that transport their substrates down a concentration gradient. The two common forms of type 1 glycogen storage disease are termed type 1a and type 1b glycogen storage disease

Molecular Bases of Type 1a Glycogen Storage Disease: Glucose-6-Phosphatase Enzyme (G6PC) Deficiency

A human liver glucose-6-phosphatase enzyme has been cloned, and the human glucose-6-phosphatase gene (*G6PC1*), which has five exons, is located on chromosome 17 at q21. To date, over 40 different mutations in the glucose-6-phosphatase enzyme gene have been found in patients with type 1a glycogen storage disease. Mutations have been found throughout all five exons and at the intron–exon boundaries. Two other highly related genes, *G6PC2* and *G6PC3*, have been found recently. The three *G6PC* genes have different tissue-specific patterns of expression. Only *G6PC1*, the gene mutated in type 1a glycogen storage disease, is expressed at significant levels in human liver.

Molecular Bases of Type 1b Glycogen Storage Disease: G6PT1 Deficiency

The gene mutated in type 1b glycogen storage disease has been mapped to chromosome 11q23. The gene has nine exons spanning a region of approximately 4 kb. To date, mutations have been found in all the exons except exon 7. The gene is differentially spliced in a

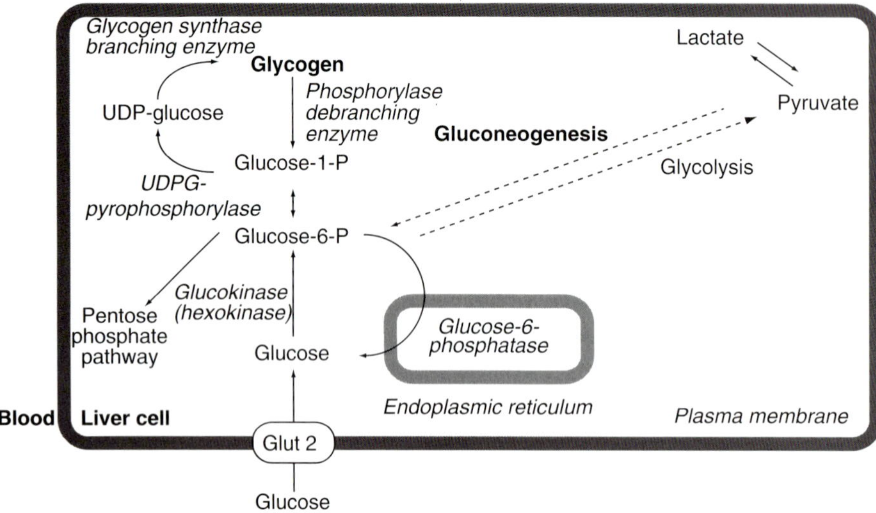

Figure 2 Schematic representation of the pathways of liver glucose production.

tissue-specific manner and exon 7 is not expressed in liver. The 1b gene is expressed more widely than the *G6PC1* gene, which may explain why patients with type 1b glycogen storage disease often have additional symptoms to those with type 1a glycogen storage disease. For example, patients with type 1b glycogen storage disease often have neutropenia, and *G6PT1* (but not *G6PC1*) is expressed in neutrophils.

Clinical Presentation of Glycogen Storage Disease

Type 1a

The clinical manifestations of type 1a glycogen storage disease are many and varied, including the logical effects of defective glucose production, e.g., growth retardation, hepatomegaly, fasting hypoglycemia, lactic acidemia, hyperuricemia, and hyperlipidemia. The severity of individual symptoms vary greatly among patients, who may be virtually asymptomatic in rare cases. These diseases normally present in childhood but, surprisingly, some present in adulthood. Long-term complications include gout, hepatic adenomas, hepatomas, and renal disease.

Type 1b

Type 1b glycogen storage disease is often more severe than type 1a glycogen storage disease. Type 1b glycogen storage disease has a similar clinical course to type 1a glycogen storage disease, with the additional findings of neutropenia and impaired neutrophil function resulting in recurrent bacterial infections. Oral and intestinal mucosa ulceration commonly occur, and cases of chronic inflammatory bowel disease have been reported.

Management

In the past, many patients with type 1 glycogen storage disease died, and prognosis was guarded in those who survived. In the past 20 years, major progress has been made in managing this disorder. Current treatment of type 1 glycogen storage disease involves the nocturnal nasogastric infusion of glucose and/or oral uncooked cornstarch. Early diagnosis and initiation of treatment has improved the prognosis, with normal growth and pubertal development and reduced risk of gout in adult patients.

Further Reading

Cori GT and Cori CF (1952) Glucose-6-phosphate of the liver in glycogen storage disease. *Journal of Biological Chemistry* 199: 661–667.

See *also:* Glucose 6-Phosphate Dehydrogenase (G6PD) Deficiency

Von Hippel–Lindau Disease

E Maher

doi: 10.1006/rwgn.2001.1629

Germline mutations in the VHL tumor suppressor gene cause von Hippel–Lindau disease – a dominantly inherited familial cancer syndrome characterized by the development of vascular tumors in the retina and central nervous system (hemangioblastomas), clear cell renal cell carcinomas, pheochromocytomas, pancreatic islet cell, and other tumors. Tumors from von Hippel–Lindau patients have somatic inactivation of the wild-type allele and somatic inactivation of both alleles occurs in most sporadic clear cell renal cell carcinomas and hemangioblastomas. The VHL gene product plays a critical role in regulating proteosomal degradation of the α subunits of the hypoxia-inducible transcription factors HIF-1 and HIF2.

See also: **Tumor Suppressor Genes**

Von Willebrand Disease

M A Ferguson-Smith

doi: 10.1006/rwgn.2001.1379

Von Willebrand disease is the commonest inherited bleeding disorder in humans. Deficiency of Von Willebrand factor (vWF) causes defective platelet adhesion and abnormal Factor VIII function. In its effects it thus mimics hemophilia A. Unlike hemophilia A, the condition is inherited as an autosomal trait, usually autosomal dominant, but a severe autosomal recessive form is also recognized due to homozygosity of the gene. The vWF gene has been sequenced and many mutations associated with the disease have been identified.

See also: **Hemophilia**

W Chromosome

D W Burt

doi: 10.1006/rwgn.2001.1384

Avian and mammalian sex chromosomes evolved independently (Fridolfsson *et al.*, 1998; Nanda *et al.*, 1999, 2000) and should therefore have fundamentally different sex-determining genes. The sex chromosomes in birds are designated Z and W. The female is the heteromorphic (ZW) sex and the male homomorphic (ZZ). The average avian Z chromosome is a medium-sized macrochromosome. The W chromosome of most modern birds is a microchromosome, considerably smaller than the Z and largely heterochromatic. The Z and W chromosomes are derived from a common ancestral chromosome (Fridolfsson *et al.*, 1998; Shetty *et al.*, 1999). Heterochromatinization, deletion, and rearrangement of the W chromosome have contributed to the evolution of highly differentiated sex chromosomes in modern birds. In primitive birds, such as ratites, the W chromosome resembles the Z in size and morphology, and most likely in gene content (Fridolfsson *et al.*, 1998; Ogawa *et al.*, 1998). It is not clear in birds, whether the Z or W chromosome determines sex.

References

Fridolfsson A-K, Cheng H, Copeland NG *et al.* (1998) Evolution of the avian sex chromosomes from an ancestral pair of autosomes. *Proceedings of the National Academy of Sciences, USA* 95: 8147–8152.

Nanda I, Shan Z, Schartl M *et al.* (1999) 300 million years of conserved synteny between chicken Z and human chromosome 9. *Nature Genetics* 21: 258–259.

Nanda I, Zend-Ajusch E, Shan Z *et al.* (2000) Conserved synteny between the chicken Z sex chromosome and human chromosome 9 includes the male regulatory gene DMRT1: a comparative (re)view on avian sex determination. *Cytogenetics and Cell Genetics* 89: 67–78.

Ogawa A, Murata K and Mizuno S (1998). The location of Z- and W-linked marker genes and sequences on the homomorphic sex chromosomes of the ostrich and the emu. *Proceedings of the National Academy of Sciences, USA* 95: 4415–4418.

Shetty S, Griffin DK and Marshall-Graves JA (1999) Comparative painting reveals strong chromosome homology over 80 million years of bird evolution. *Chromosome Research* 7: 289–295.

***See also:* Microchromosomes; Sex Determination, Human**

W (White Spotting) Locus

A Bernstein, J Shulman, and G Caruana

doi: 10.1006/rwgn.2001.1390

Mice with mutations at either the dominant *white spotting* (*W*) locus on chromosome 5 or the *Steel* (*Sl*) locus on chromosome 10 are severely anemic, mast-cell deficient, sterile, and lack neural crest-derived melanocytes and interstitial cells of Cajal in the gut. Reciprocal bone marrow transplantation and chimeric embryo studies with these mutant mice and normal littermates established that the cellular deficits in *W* mutant mice are due to a cell autonomous, intrinsic defect expressed by the cells in the affected lineages. In contrast, the *Sl* defect resides in the microenvironment in which these cells develop during embryogenesis and function in the adult. This complementary nature of the phenotypic abnormalities in *W* and *Sl* mutant mice was explained by the discovery that *W* encodes the Kit receptor tyrosine kinase, while *Sl* encodes its membrane-bound ligand, Steel factor (SLF), also referred to as Kit ligand or stem cell factor. Furthermore, Kit is expressed in early cells of all the lineages affected in *W* and *Sl* mice while SLF is expressed in cells that are located immediately contiguous to Kit-expressing cells.

Both the murine and human Kit receptors are 970 amino acids in length with nine potential sites for N-linked glycosylation in their extracellular domains. Like other receptor tyrosine kinases (RTKs), Kit has a large extracellular domain with five immunoglobulin (Ig)-like domains involved in ligand binding and receptor dimerization and perhaps heterodimerization with other receptors. The receptor has a single transmembrane domain and a cytoplasmic tyrosine kinase domain split in two by a kinase-insert domain into an ATP-binding region and a phosphotransferase domain. As with other RTKs, ligand binding induces rapid receptor dimerization and autophosphorylation at 8 tyrosine residues in the cytoplasmic domain of the receptor. These phosphotyrosines, in the context of the 3 C-terminal amino acids, then serve as high-affinity docking sites for a number of SH2-containing signalling molecules that lie downstream of Kit.

In mice and humans alternative splice variants of c-*kit* exist. Two of these differ by an in-frame deletion or insertion of 12 base pairs in the extracellular domain just prior to the transmembrane domain, encoding four amino acids Gly-Asn-Asn-Lys (GNNK). In addition, in humans two other Kit variants have been described which involve an in-frame insertion or deletion of 3 bp encoding a serine (S) at residue 715 located in the kinase insert of the cytoplasmic domain. All isoforms are co-expressed in a number of cell types with the GNNK- and S^{+-} isoforms being more prominently expressed. At present the roles of the various splice variants is not known; however, it has been shown that the absence of the GNNK motif results in low levels of constitutive phosphorylation of the receptor in the absence of ligand.

Naturally occurring *W* mutations result in partial or complete loss of function of Kit tyrosine kinase activity, either as the result of mutations (usually point substitutions; **Figure 1**) in the cytoplasmic domain, or regulatory mutations (large deletions or chromosomal inversions in the W^{57} and W^{bd} alleles, respectively) which reduce or eliminate the levels of c-*kit* transcripts.

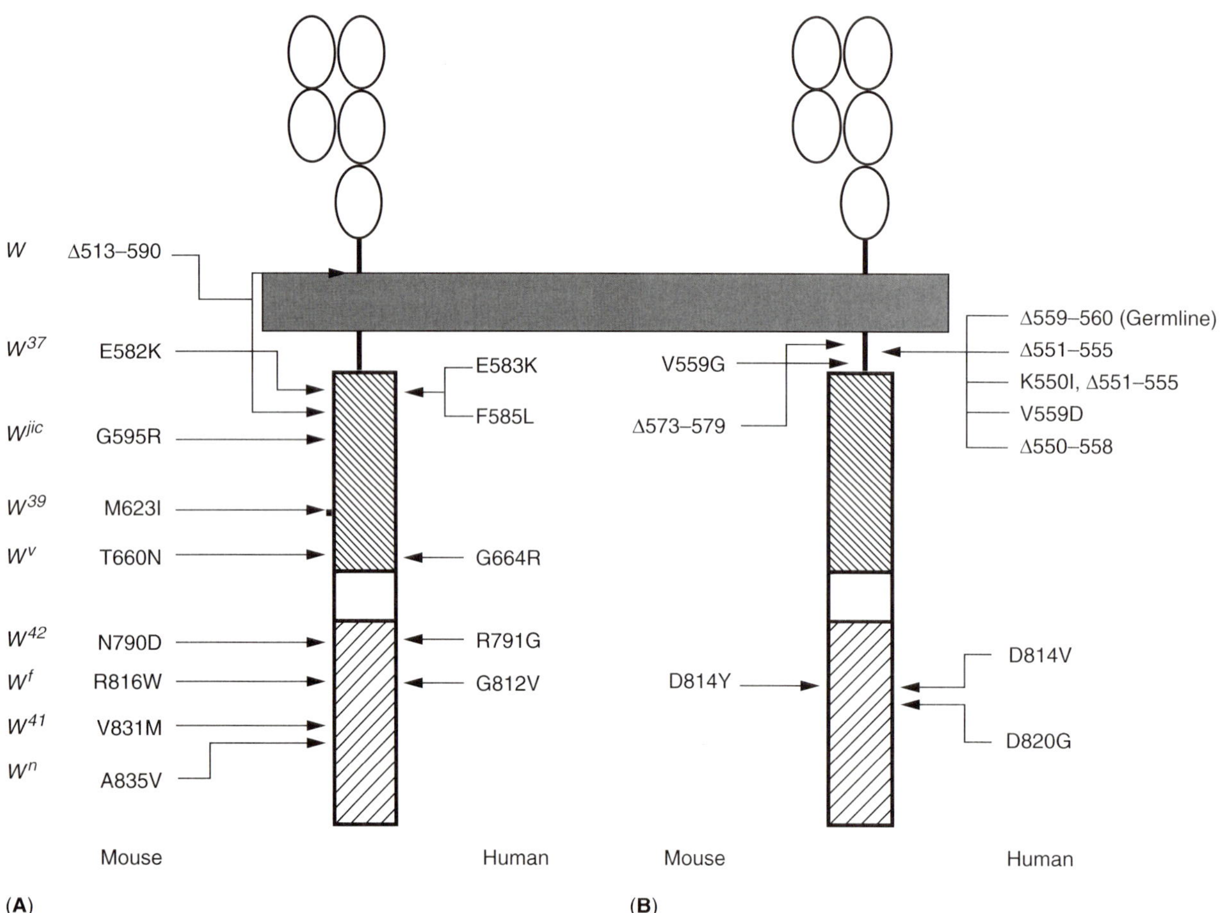

Figure 1 Schematic illustration of loss (A) and gain (B) of function mutations of the Kit receptor. The extracellular domain of Kit comprises five Ig-like repeats (circles) and the cytoplasmic domain consists of two kinase domains (hatched areas). Loss of function murine *W* and human piebaldism mutations are depicted on the Kit receptor on the left. Murine and human gain of function mutations are denoted on the Kit receptor on the right. All murine gain of function mutations have been found in mastocytoma cell lines. In humans, mutations located in the juxtamembrane domain have been reported in patients with gastrointestinal stromal tumors (GISTs), while mutations located in the kinase domain have been identified in patients with mastocytosis, an associated myelodysplastic disorder and urticaria pigmentosa.

The human c-*kit* gene maps to human chromosome 4q11–21. In both mice and humans, the genes for three RTKs, PDGFRα, Kit, and flk-1, are physically linked on chromosomes 5 and 4, respectively. Dominant loss of function mutations in human Kit (either large deletions or point mutations in the cytoplasmic domain of the receptor) are associated with a rare human genetic pigmentation disorder known as piebald trait (**Figure 1**). Interestingly, there are several reports in the literature of mental retardation associated with piebaldism, consistent with the high levels of c-*kit* and *Sl* expression in the brain, and the spatial learning and memory deficits in *Sl* mutant mice.

Gain of function or activating mutations in the Kit receptor have also been described in mastocytoma cell lines and in some primary human leukemias, particularly mastocytosis (**Figure 1**). These mutations result in constitutive (ligand-independent) receptor signaling and/or qualitative changes in the downstream signaling pathway controlled by this receptor.

Activating mutations in c-*kit* have also been described in gastrointestinal stromal tumors (GIST). These mutations, like the mutations observed in leukemias, are usually acquired (i.e., somatic), but there is one report of a family with a predisposition to GIST associated with a germline, activating mutation in c-*kit* (**Figure 1**).

Further reading

Reith AD and Bernstein A (1991) Molecular biology of the *W* and *Steel* loci. In: Davies K and Tilghman S (eds), *Genome Analysis*, vol. 3, *Genes and Phenotypes*, pp. 105–133. Plainview, NY: Cold Spring Harbor Laboratory Press.

Galli SJ, Zsebo KM and Geissler EN (1994) The kit ligand, stem cell factor. *Advances in Immunology* 55: 1.

***See also:* Piebald Trait; *Steel* Locus**

WAF1

W S El-Deiry

doi: 10.1006/rwgn.2001.1630

Wild-type p53-activated fragment 1 (*WAF1*) is a gene isolated from subtractive hybridization screening for transcriptional targets of the p53 tumor suppressor protein as candidate mediators of cell growth suppression. The *WAF1* gene (also known as *CIP1*, *SDI1*, *MDA6*, *CDKN1A*) is located on human chromosome 6p21 and encodes a 21-kDa protein (p21WAF1) that functions as a cyclin-dependent kinase inhibitor, which causes cell-cycle arrest, primarily in G_1. p21WAF1 also interacts with the proliferating cell nuclear antigen, resulting in inhibition of DNA synthesis. Transcriptional activation of p21WAF1 by p53 permits maintenance of DNA damage-induced or other stress-induced checkpoint responses until repair has taken place, so that cells can survive and maintain genetic fidelity.

***See also:* Tumor Suppressor Genes**

Wahlund Effect

M A Asmussen

doi: 10.1006/rwgn.2001.1448

The 'Wahlund effect' or 'Wahlund principle' refers to the unexpected, but often major, statistical effect of population subdivision or geographical structure on the observed genetic structure of a population. Such subdivision and geographical structure are quite common in nature, particularly for organisms whose habitat is discontinuous, such as a starfish population distributed across a series of tidepools, a population of trees distributed around a string of lakes, or the insect or bird populations that live in such disjoint clusters of trees.

The Wahlund effect in these and other subdivided populations is most commonly manifested as deviations in genotypic frequencies from the Hardy–Weinberg (HW) equilibrium values expected in the absence of all evolutionary forces such as natural selection, nonrandom mating, migration, or random genetic drift, etc. Since a test for statistically significant deviations from HW frequencies is often used as a critical, preliminary screen for the presence of such nonrandomizing evolutionary forces, it is essential to realize that such deviations may be due to the Wahlund effect from population subdivision alone, in the absence of all other evolutionary forces.

What is the so-called Wahlund effect and how does it come about? This is a statistical phenomenon that produces deviations from HW frequencies in a population as a whole when it is made up of multiple subpopulations with distinct allele frequencies, even though each of the individual subpopulations is at HW frequencies. The same is true of a total population sample based on surveys of a set of multiple sample sites (e.g., populations of trees around a string of different lakes), not all of which have the same allele frequency. A graphical illustration of this type of situation is shown in **Figure 1**, where the inner circles could represent, for instance, a series of tidepools (subpopulations), across which a population of a marine species is distributed.

More specifically, if a population or sample is made up of two or more parts, each of which is in HW proportions, the overall population will have a deficit of heterozygotes (and an excess of homozygotes) with respect to HW proportions if there is any variation in allele frequency across the parts. These deviations from HW equilibrium are solely a statistical consequence of the subdivision itself. When this happens, one may erroneously think a population is subject to some nonrandomizing evolutionary force such as natural selection or assortative mating, even when no such force is acting on that population. Because of this rather unexpected, purely statistical effect from subdivision itself, it is critical for properly interpreting tests for deviations from HW frequencies, to know whether the data are from a single, unified population, or from one composed of multiple parts, with distinct allele frequencies.

A Numerical Example

Let us first illustrate the Wahlund effect through a simple numerical example of a diploid insect population comprising two subpopulations associated with the host plants distributed around two nearby lakes. Suppose no nonrandomizing evolutionary forces such as natural selection or migration are acting on the insects at either lake for a given autosomal locus with two alleles, A_1 and A_2; the genotypic frequencies for the insects at each lake (subpopulation) should thus be at HW equilibrium at this locus. The 400 individuals in this insect population as a whole are made up of 100 insects from lake (subpopulation) 1 and 300 from lake (subpopulation) 2. The genotypic data for each subpopulation of insects are as shown in **Table 1**, where # denotes number, and p_1 and p_2 are the respective frequencies of allele A_1 in lake (subpopulation) 1 and 2.

The first point to verify is that each subpopulation of insects is indeed in HW proportions. The HW equilibrium frequencies for a diallelic, autosomal genetic locus with an allele frequency of p for allele A_1 are $p^2 =$ freq.(A_1A_1), $2p\ (1-p) =$ freq.(A_1A_2), and $(1-p)^2 =$ freq.(A_2A_2), where 'freq.' denotes 'the frequency of.' Applying these formulas with the allele frequencies at each lake (0.4 and 0.8) shows that they are indeed each in HW equilibrium for their respective allele frequency (**Table 2**).

Since the insect subpopulations at the two lakes have different allele frequencies (0.4 versus 0.8), Wahlund's principle now tells us that the overall population of 400 insects will not be in HW proportions, but will have a deficit of heterozygote genotypes with two different alleles (A_1A_2), and an excess of each homozygote genotype with two like alleles (A_1A_1 and A_2A_2), relative to the expected HW frequencies, p^2, $2p(1-p)$, and $(1-p)^2$, based on the allele frequency, $p = p^T$, in the insect population as a whole. Let us verify that this is indeed the case. First, we find that the allele frequency in the overall population of 400 insects is

$$p^T = \text{freq.}(A_1) = \text{freq.}(A_1A_1) + \frac{1}{2}\text{freq.}(A_1A_2)$$
$$= 0.52 + \frac{1}{2}(0.36) = 0.7$$

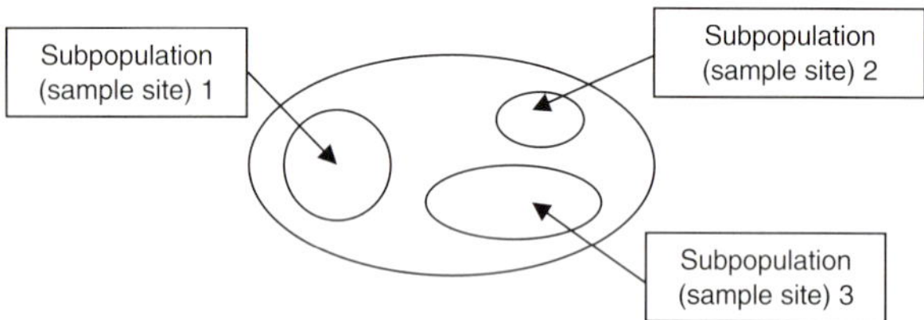

Figure 1 A schematic showing a population comprising multiple (three) subpopulations, or a population sample comprising multiple (three) study sites.

Table 1 Numerical example of a population of insects distributed across two lakes (subpopulations)

	Frequency (A_1)	$\#A_1A_1$	$\#A_1A_2$	$\#A_2A_2$	Total #
Subpopulation 1	$p_1 = 0.4$	16	48	36	100
Subpopulation 2	$p_2 = 0.8$	192	96	12	300
Total population		208	144	48	400
Genotypic frequencies in total population	$p^T = 0.7$	0.52	0.36	0.12	1.0

Table 2 Comparing observed genotype counts in subpopulations 1 and 2 to their Hardy–Weinberg values

Genotype	$\#A_1A_1$	$\#A_1A_2$	$\#A_2A_2$
Subpopulation 1	$16 = (0.4)^2 100$	$48 = 2(0.4)(0.6)100$	$36 = (0.6)^2 100$
Subpopulation 2	$192 = (0.8)^2 300$	$96 = 2(0.8)(0.2)300$	$12 = (0.2)^2 300$

This value can also be obtained directly, by counting and dividing the number of A_1 alleles in the overall population [2(208) + 144 = 560] by the total number of alleles in this diploid insect population [2(400) = 800], which gives 560/800 = 0.7. The expected HW frequencies in the overall, combined insect population are thus

$$(0.7)^2 = 0.49 \quad 2(0.7)(0.3) = 0.42 \quad (0.3)^2 = 0.09$$

for genotypes A_1A_1, A_1A_2, and A_2A_2, respectively.

Subtracting these expected frequencies from the observed values in Table 1 shows that the observed genotypic frequencies in the overall population of insects indeed deviate from HW expectations, as shown in **Table 3**. In particular, there is a deficit of heterozygotes and an excess of each type of homozygote in the overall population of 400 insects, which is caused not by any nonrandomizing evolutionary forces in this population, but solely as a result of its population subdivision, as predicted from the Wahlund principle.

Formal Specification of the Wahlund Effect for Diallelic Autosomal Loci

If a diploid population (or sample) is made up of multiple subpopulations (or sample sites), each of which is in HW frequencies for its allele frequency, the population (or sample) as a whole may *not* be in HW frequencies: the frequency of each homozygous genotype in the overall population as a whole will exceed its expected HW proportion by an amount equal to the variance in allele frequency at that genetic locus across the subpopulations (sample sites). There will be a corresponding deficit of heterozygotes in the population as a whole equal to twice the variance in allele frequency across each part.

To formalize this statement, suppose a diploid population (or sample) is comprised of k subpopulations (or sample sites), each in HW frequencies at a given diallelic, autosomal locus (**Table 4**).

The allele and genotypic frequencies in the overall (total) population, denoted here by a superscript ^{T}, are the weighted arithmetic averages of those in the k subpopulations:

$$p^T = \text{freq.}(A_1) = \sum_{i=1}^{k} c_i p_i = \bar{p}$$

$$P^T = \text{freq.}(A_1A_1) = \sum_{i=1}^{k} c_i (p_i)^2$$

$$H^T = \text{freq.}(A_1A_2) = \sum_{i=1}^{k} c_i[2p_i(1 - p_i)]$$

$$Q^T = \text{freq.}(A_2A_2) = \sum_{i=1}^{k} c_i(1 - p_i)^2$$

where each summation is over the relevant frequency in each of the k subpopulations, with the value for subpopulation i weighted by $c_i = N_i/N^T$, the fraction of the total population located in that component. These formulas can be easily verified by simply 'counting' alleles and genotypes to calculate the four

Table 3 Deviation from Hardy–Weinberg genotypic frequencies in the numerical example of a subdivided insect population given in **Table 1**

Total population	A_1A_1	A_1A_2	A_2A_2	**Total**
Observed frequencies	0.52	0.36	0.12	1.0
Expected (HW) frequencies	0.49	0.42	0.09	1.0
Observed − HW frequencies	0.03	−0.06	0.03	

Table 4 A subdivided diploid population made up of k subpopulations

Subpopulation	**Freq.(A_1)**	**Freq.(A_1A_1)**	**Freq.(A_1A_2)**	**Freq.(A_2A_2)**	**#Individuals**
1	p_1	$(p_1)^2$	$2p_1(1 - p_1)$	$(1 - p_1)^2$	N_1
2	p_2	$(p_2)^2$	$2p_2(1 - p_2)$	$(1 - p_2)^2$	N_2
.	.	.	.	.	.
.	.	.	.	.	.
k	p_k	$(p_k)^2$	$2p_k(1 - p_k)$	$(1 - p_k)^2$	N_k
Overall population	p^T	P^T	H^T	Q^T	$N^T = \sum_{i=1}^{k} N_i$

frequencies as the number of each allele (or genotype) divided by the total number of that allele (or genotype) in the population as a whole. For example, in subpopulation i the number of A_1 alleles is $p_i(2N_i)$ and the number of A_1A_1 individuals is $(p_i)^2 N_i$.

Wahlund's Formula

The genotypic frequencies in the overall population specified in **Table 4** deviate from HW proportions as follows:

$$\begin{aligned} P^T &= \text{freq.}(A_1A_1) = (\bar{p})^2 + Var(p) \\ H^T &= \text{freq.}(A_1A_2) = 2\bar{p}(1-\bar{p}) - 2Var(p) \\ Q^T &= \text{freq.}(A_2A_2) = (1-\bar{p})^2 + Var(p) \end{aligned} \tag{1}$$

where

$$\bar{p} = \sum_{i=1}^{k} c_i p_i \tag{2}$$

is the average or mean allele frequency across the k subpopulations, which equals p^T the frequency of allele A_1 in the population as a whole.

$$Var(p) = \sum_{i=1}^{k} c_i (p_i)^2 - p^2 \tag{3}$$

is the variance in allele frequency across the subpopulations, formally defined as $\sum_{i=1}^{k} c_i(p_i - \bar{p})^2$, where p_i is the frequency of allele A_1 in subpopulation i, and $c_i = N_i/N^T$ is the fraction of the total population located in subpopulation i.

These formulas show that a subdivided population or sample as a whole will deviate from HW frequencies if and only if there is variation in allele frequencies across its parts (i.e., not all parts of the population or sample have the same allele frequency). When this occurs, there will be an excess of each type of homozygote (A_1A_1 and A_2A_2) and a deficit of heterozygotes (A_1A_2) in proportion to the variance in allele frequency at the locus across the subpopulations that comprise the overall population.

Check of our Numerical Example

To apply Wahlund's formula, we need the mean, $\bar{p}$, and variance, *Var(p)*, in allele frequency across the two subpopulations of insects specified in **Table 1**. We already calculated the allele frequency in the overall insect population to be 0.7, which also equals the mean across the subpopulations at the two lakes. This equivalence can be verified by calculating the mean allele frequency directly from (2), as the weighted average of the allele frequencies in the two subpopulations, with each value weighted by the fraction of the total population which is in that subpopulation, which yields:

$$\bar{p} = \sum_{i=1}^{2} c_i p_i = (1/4)(0.4) + (3/4)(0.8) = 0.7$$

The variance in allele frequency across the two lakes can now be calculated from (3) as

$$\begin{aligned} Var(p) &= \sum_{i=1}^{2} c_i(p_i)^2 - \bar{p}^2 = (1/4)(0.4)^2 \\ &\quad + (3/4)(0.8)^2 - (0.7)^2 \\ &= 0.04 + 0.48 - 0.49 = 0.03 \end{aligned}$$

Based on these findings, the Wahlund formulas in (1) tell us that the overall population will have an excess of each of the two homozygote genotypes (A_1A_1 and A_2A_2) equal to the variance in allele frequency across the subpopulations (0.03), and a deficit of heterozygotes (A_1A_2) equal to twice this variance (0.06), as indeed is the case, as shown from the direct calculations reported in **Table 3**.

As a further test of the analytic formulas for the Wahlund effect in (1), we can substitute the mean allele frequency (0.7) and variance (0.03) into these three equations and verify that the predicted genotypic frequencies,

$$\begin{aligned} P^T &= \text{freq.}(A_1A_1) = (\bar{p})^2 + Var(p) = (0.7)^2 + 0.03 \\ &= 0.49 + 0.03 = 0.52 \end{aligned}$$

$$\begin{aligned} H^T &= \text{freq.}(A_1A_2) = 2\bar{p}(1-p) - 2Var(p) \\ &= 2(0.7)(0.3) - 2(0.03) = 0.42 - 0.06 = 0.36 \end{aligned}$$

$$\begin{aligned} Q^T &= \text{freq.}(A_2A_2) = (1-\bar{p})^2 + Var(p) \\ &= (0.3)^2 + 0.03 = 0.09 + 0.03 = 0.12 \end{aligned}$$

In fact, all agree with the observed frequencies in the overall insect population as a whole (see **Table 3**).

Practical Implications

Returning to the biology of our numerical example, one would clearly be drawing a false biological inference were one to take the observed departure from HW in this insect population as indicative of the presence of nonrandom mating, natural selection or some other nonrandomizing evolutionary force. In fact, the observed deviations are solely a consequence of its being composed of multiple parts (lake subpopulations) with different allele frequencies, on which no such forces are acting. One must therefore be very careful to check for evidence of population subdivision when interpreting the results of tests for HW equilibrium, since departures from HW can be generated solely as a statistical consequence of the data being from multiple components rather than from a unified whole.

It should also be emphasized that the Wahlund principle simply ensures that there will be a deviation from HW frequencies in the population as a whole if it is composed of multiple parts not all of which have the same allele frequency. It says nothing about whether the resulting excess of homozygotes and deficit of heterozygotes is statistically significant. This can be determined by performing a standard χ^2 test comparing the expected and observed numbers of each genotype.

Derivation of the Wahlund Formulas

Having a departure from HW genotypic frequencies in the total population even when its subpopulations are at HW genotypic proportions for their allele frequencies is not an accident, but a general feature of population subdivision known as the Wahlund effect. We can fairly easily derive the general formulas in (1) showing exactly what these deviations will be. Suppose, as in the general statement of the Wahlund principle based on **Table 4**, that a diploid population is made up of k subpopulations, each in HW frequencies for its allele frequency at a given, diallelic autosomal locus. The first question is, what are the allele and genotype frequencies in the overall population in terms of the corresponding frequencies in the individual subpopulations?

As in our simple numerical example in **Table 1**, the allele and genotypic frequencies in the population as a whole are all simply the weighted averages of the corresponding frequencies in the subpopulations, with the frequency in subpopulation i weighted by the fraction of the total population that is located in that part. For example, recalling that a diploid population of N individuals has $2N$ alleles at each genetic locus, we calculate, using the notation of **Table 4**, that the frequencies of allele A_1 and genotype A_1A_1 in the overall population are

$$p^T = \text{freq.}(A_1) = \frac{\#A_1 \text{ alleles}}{\text{total } \# \text{ alleles}} = \frac{p_1(2N_1) + p_2(2N_2) + \cdots p_k(2N_k)}{2N^T} = \sum_{i=1}^{k} c_i p_i = \bar{p}$$

and

$$P^T = \text{freq.}(A_1A_1) = \frac{\#A_1A_1 \text{ individuals}}{\text{total } \# \text{ individuals}} = \frac{P_1N_1 + P_2N_2 + \cdots + P_kN_k}{N^T} = \sum_{i=1}^{k} c_i P_i$$

respectively, where in each case p_i and P_i are the frequencies of the A_1 allele and A_1A_1 homozygotes, respectively, in subpopulation i, and $c_i = N_i/N^T$ is the fraction of the total population located in that subpopulation. Note that this direct derivation shows why each part must be weighted by its relative census size, its proportion of the whole in terms of number of individuals, which is why all our Wahlund formulas involve weighted averages, rather than simple arithmetic averages of the values in each component.

In the special case where each subpopulation is in HW frequencies, the frequency of A_1A_1 homozygotes in subpopulation i is $P_i = (p_i)^2$, so that

$$P^T = \sum_{i=1}^{k} c_i(p_i)^2 = (\bar{p})^2 + Var(p)$$

where the final equality follows immediately from a simple rearrangement of the computational formula for the variance across subpopulations in (3). The frequency of heterozygotes in the total population can also be easily expressed in terms of the mean and variance in allele frequency using the same principles:

$$\begin{aligned} H^T &= \text{freq.}(A_1A_2) = \sum_{i=1}^{k} c_i[2p_i(1-p_i)] \\ &= 2\sum_{i=1}^{k} c_i p_i - 2\sum_{i=1}^{k} c_i(p_i)^2 \\ &= 2\bar{p} - 2[(\bar{p})^2 + Var(p)] = 2\bar{p}(1-\bar{p}) - 2Var(p) \end{aligned}$$

The frequency Q^T of A_2A_2 homozygotes in the overall population as a whole given in (1) can be derived by similar reasoning, or alternatively, by applying the basic relation, freq. $(A_2A_2) = 1 -$ freq.$(A_1A_1) -$ freq.(A_1A_2), which yields

$$\begin{aligned} Q^T &= 1 - P^T - H^T \\ &= 1 - [\bar{p}^2 + Var(p) + 2\bar{p}(1-\bar{p}) - 2Var(p)] \\ &= (1-\bar{p})^2 + Var(p) \end{aligned}$$

These formulas show that under the Wahlund effect, an overall population or sample will deviate from HW frequencies whenever there is a nonzero variance in allele frequency across the subpopulations (sample sites) of which it is composed, which will be the case whenever the components do not all have the same allele frequency.

Wahlund Effect with Multiple Alleles

A Wahlund effect is also found for autosomal loci with multiple alleles (i.e., more than two), but with some modification from what we found above for diallelic loci. For a multiallelic locus with three or more alleles (A_1, A_2, A_3, ...) there will be an excess of each homozygous genotype A_jA_j and an overall deficit of heterozygotes, but not necessarily for each specific heterozygous genotype A_mA_n. Assuming each genotype is in HW frequencies within each of k subpopulations (or sample sites), the genotypic frequencies in the overall population will be

$$\text{homozygotes: freq.}(A_jA_j) = (\bar{p}_j)^2 + Var(p_j) \quad \text{for all alleles } A_j$$

$$\text{heterozygotes: freq.}(A_mA_n) = 2\bar{p}_m\bar{p}_n + 2\text{cov}(p_m, p_n) \quad \text{for all } m \neq n$$

where p_j here denotes the frequency of allele A_j, with $\bar{p}_j$ its frequency in the population as a whole, and cov (p_m, p_n) is the covariance in the two allele frequencies across the k subpopulations. The covariance of two quantities x and y across multiple subpopulations is calculated as:

$$\text{cov}(x, y) = \sum_{i=1}^{k}(x_i - \bar{x})(y_i - \bar{y}) = \sum_{i=1}^{k} x_iy_i - \overline{xy}$$

where x_i and y_i are the respective values of x and y in subpopulation i, and $\bar{x}$ and $\bar{y}$ are their mean values across the k subpopulations.

Conclusion

A population or population sample will deviate from Hardy–Weinberg frequencies even in the absence of all nonrandomizing evolutionary forces if it is made up of multiple subpopulations or sample sites, not all of which have the same allele frequencies. It is therefore crucial to know the geographical structure of a population or sample to properly interpret the results from standard population genetic tests for a deviation from HW frequencies. In particular, a statistically significant deviation from HW frequencies may simply be due to the statistical effects of population subdivision alone, and not be indicative of the presence of major evolutionary forces such as natural selection, nonrandom mating, or migration.

***See also:* Allele Frequency; Hardy–Weinberg Law**

Wallace, Alfred Russel

K N Gracy and J H Miller

doi: 10.1006/rwgn.2001.1381

Alfred Russel Wallace (1823–1913) is best known for his independent formulation of the theory of evolution. His work on the subject was published with that of Charles Darwin, in the *Proceedings of the Linnean Society* in 1858 (Darwin and Wallace, 1858).

Wallace was born in 1823 in Monmouthshire, Wales, where he grew up in strained financial circumstances and with limited formal education. He spent his early career as a surveyor in England and Wales, but developed an enthusiastic interest in natural history, cartography, and biology. Unable to find work due to political uprisings, Wallace traveled to Brazil in 1848. There, with his naturalist friend Henry Walter Bates, he began collecting entomological specimens. He went on to map several unexplored regions of the Amazon River basin and documented the language and culture of the indigenous people there. His articles, books, and maps on the Amazon and Rio Negro won him acclaim from the Royal Geographical Society and established him as a serious naturalist.

Wallace's next trip took him to the Malay Archipelago, where he spent 8 years collecting specimens and documenting his research. During this time he formulated a theory of the origin of new species, which he published in the *Annals and Magazine of Natural History* in 1855 (Wallace, 1855). His idea that new species arose progressively from parent species due to environmental pressures was honed in an 1858 manuscript, which he sent to Charles Darwin for his opinion. Darwin published extracts of his own writing along with Wallace's manuscript in an article entitled 'On the tendency of species to form varieties; and on the perpetuation of varieties and species by natural means of selection' in the *Journal of the Proceedings of the Linnean Society* in 1858 (Darwin and Wallace, 1858). The resultant interest spurred Darwin to publish *On the Origin of Species* in November of 1859 (Darwin, 1859).

Wallace went on to publish many books on the Malay Archipelago that provided additional support for the theory of evolution. Because the species he found in Australia were more evolutionarily primitive than those found in nearby Asia, Wallace proposed that the two continents had become geographically separated quite early in the species' development. The boundary dividing the fauna of the two regions is now known as Wallace's line.

Wallace's most noted works are *Contributions to the Theory of Natural Selection* (Wallace, 1870), *Geographical Distribution of Animals* (Wallace, 1876), and *Island Life* (Wallace, 1881). He received many awards, including the Royal Society of London's Royal Medal (1868), the Darwin Medal (1890), the Linnean Society of London's Gold Medal (1892), the Royal Geographical Society's Founder's Medal (1892), and the Linnean Society of London's Darwin–Wallace Medal (1908). Wallace died in 1913 at the age of 91.

References

Darwin C (1859) *On the Origin of Species by Means of Natural Selection or the Preservation of Favoured Races in the Struggle for Life*. London: John Murray.

Darwin C and Wallace AR (1858) On the tendency of species to form varieties; and on the perpetuation of varieties and species by natural means of selection. *Journal of the Proceedings of the Linnean Society: Zoology* 3(9): 53–62.

Wallace AR (1855) On the law which has regulated the introduction of new species. *Annals and Magazine of Natural History* 16(2): 184–196.

Wallace AR (1870) *Contributions to the Theory of Natural Selection: A Series of Essays*. London: Macmillan.

Wallace AR (1876) *The Geographical Distribution of Animals, with a Study of the Relations of Living and Extinct Faunas as Elucidating the Past Changes of the Earth's Surface*. London: Macmillan.

Wallace AR (1881) *Island Life; or, the Phenomena and Causes of Insular Faunas and Floras, Including a Revision and Attempted Solution of the Problem of Geological Climates*. New York: Harper & Brothers.

***See also:* Darwin, Charles; Evolution; Lamarckism**

Watson, James Dewey

N C Comfort

doi: 10.1006/rwgn.2001.1382

Science has been the dominant intellectual enterprise of the twentieth century. Molecular genetics is the most important science of that century's second half. James Dewey Watson (1928–) – first as scientist, then writer, then administrator – has shaped the development of molecular genetics more than anyone else.

Born on 6 April 1928 in Chicago, the second child of James D. Watson, a businessman, and Jean Mitchell, Watson grew up poor. In an autobiographical note, Watson recalled three pillars of his early intellectual and moral life: the power of knowledge to drive out superstition, i.e., religion; ornithology; and the Democratic party. He attended Horace Mann Grammar School and for two years South Shore High School. At 14, he appeared on the popular *Quiz Kids* radio program, produced in Chicago. The next year, 1943, he went to college at the University of Chicago, at the time of president Robert Hutchins's extraordinary experiment in liberal education. Hutchins's program, Watson later reflected, gave him intellectual tools that proved essential to his success: a commitment to reading original sources and a lack of respect for tradition and authority. Though Watson did not find a particularly strong biology program at Chicago, the great geneticist Sewall Wright was teaching there. In 1946 Watson took Wright's course in physiological genetics. That, combined with having just read *What Is Life?*, the physicist Erwin Schrödinger's slim volume of speculations which became a cult favorite among the founders of molecular genetics, set Watson's intellectual course. He would dedicate himself to the search for the gene.

In 1947, Watson entered graduate school at Indiana University in Bloomington. Besides a scholarship – offered on condition that Watson forget about studying birds – Indiana held three attractions for him: Hermann Joseph Muller, a *Drosophila* geneticist and recent Nobel laureate for his discovery that X-rays cause mutations; Tracy Sonneborn, a microbial geneticist with exciting, controversial evidence for cytoplasmic inheritance; and Salvador Luria. An Italian emigré, Luria had worked in the United States throughout the recent war, collaborating with the German expatriate physicist Max Delbrück, then at Vanderbilt University, on the genetics of bacteriophage, or bacterial viruses. Phage had the most elementary genetic system known. Experiments were quick, inexpensive, and extremely sensitive – one could detect a single mutant virus among many millions. Watson worked in Luria's laboratory for his PhD, which he received in 1950. The only other person in the laboratory was Renato Dulbecco, a postdoctoral fellow who later developed animal-cell culture techniques that won him a share of the 1975 Nobel Prize in Physiology or Medicine.

In the summer of 1948, Watson and Dulbecco accompanied Luria to what had become the phage geneticists' traditional summer destination. Most summers since 1941, Luria had gone to the laboratories at Cold Spring Harbor, midway out the north shore of Long Island, New York, to do and talk science with Delbrück and others. In 1945, Delbrück had initiated there a small, intensive postgraduate course on phage genetics – a hands-on tutorial in his and Luria's research. The phage course became the training ground for many of that first generation of American molecular biologists. It established the intellectual style (ruthless, irreverent) and social style (informal, loyal, whimsical) that has dominated molecular genetics ever since. Watson did not enroll in the course but he did do some experiments, and he drank beer and played softball with the phage course students.

In 1950, Watson, as a Merck Fellow of the National Research Council, went to the laboratory of Hermann Kalckar, a biochemist at the University of Copenhagen and a graduate of the first phage course. Watson could hardly have been less interested in biochemistry. Within the year, he managed to transfer his fellowship money from Copenhagen to the Cavendish Laboratory at Cambridge University, home to the most distinguished tradition in experimental physics.

At that time the Cavendish Professor was Sir Lawrence Bragg, who, with his father, Sir William, had developed X-ray crystallography, beginning in 1912. In X-ray crystallography, a chemical is crystallized, its atoms thus packed into a rigid structure and its molecular form repeated many times in three dimensions. The crystal is then placed in an X-ray beam. The crystal scatters the beam and a collector (at the time, a photographic plate) placed behind the crystal records the scattering pattern. From this pattern, by laborious calculations, the structure of atoms that produced the pattern can be computed.

On arriving in Cambridge in September 1951, Watson met Francis Crick, 12 years his senior and an all-but-dissertation physicist working on the structure of hemoglobin, in a unit at the Cavendish funded by the Medical Research Council. Crick and Watson, each bored with his assigned project, found in each other someone as excited as himself by the gene, which by now to them meant DNA. Both men preferred talking and thinking to experimental work. Late in the fall of 1952, Watson attended a talk by Rosalind Franklin, a skilled crystallographer at King's College London, in a laboratory where Maurice Wilkins also worked. Watson, with Franklin's data in his head, devised with Crick a first model for the structure of DNA. They presented the model to the Cavendish and King's groups, where Franklin showed with scorn that it was wrong. Watson had misremembered the amount of water in the interstices of the DNA molecule. But in February of 1953 Watson and Crick built a model of elegance and precision: they knew instantly it was correct.

Deoxyribonucleic acid is a twisted zipper. The binding of the zipper (the fabric at the edge that gives it structure) is a pair of backbones, composed of alternating molecules of phosphate and ribose, a simple sugar. To each ribose is bound one of the zipper's teeth: the bases. The bases are of four kinds, adenine, cytosine, guanine, or thymine, abbreviated A, C, G, and T, respectively. Up one side of the zipper the sequence of bases can be anything, but when the two strands zip together, A always binds with T, and C always with G. Thus, given one strand, the sequence of bases on the other strand is determined. The whole molecule coils about a common axis, forming a double helix.

The unique pairings mean that each chain serves as a template for assembling a new double helix. This provides a copying mechanism and explains how genetic information is passed on through generations. Watson and Crick's first paper on the double helix, an 800-word note to *Nature*, was published on 25 April 1953. A second paper, also published in *Nature*, on 30 May, explored the genetic implications of the structure. It was soon clear that genetic information was encoded in the sequence of bases along the chain. A gene was a sequence of DNA.

After two further fellowships, at Caltech and then back to the Cavendish for a year, Watson landed an assistant professorship at Harvard University, in 1956. Promotions followed rapidly, aided no doubt by Watson receiving the Lasker and Eli Lilly awards in 1960, and by 1961 he was a full professor. In 1962, he was elected to the National Academy of Sciences and he, Crick, and Wilkins shared the Nobel Prize in Physiology or Medicine. Franklin had died in 1958, aged 37, her strong career in science cut short by cancer.

As a professor, Watson showed the same integrity and awkwardness that characterized his science. Horace Freeland Judson described his classroom style as mumbling to the shoes of the students in the front row, his own shoes untied, delivering a thrilling, unrehearsed analysis of current research. Watson's Harvard laboratory produced some important results – in 1960, for example, François Gros discovered messenger RNA, nearly simultaneously with François Jacob and Sydney Brenner – and students who went on to distinguished careers in science. But as a laboratory head, Watson refused to add his own name to papers published by his students. Consequently, the publication record underestimates his influence on the field. He helped set the current style among principal investigators, doing science by thinking and talking, and letting his students be his hands at the laboratory bench.

By age 35, Watson had risen to the top of his profession. In the mid-1960s, he began a second career as a writer. In 1965, he published *The Molecular Biology of the Gene*, the first textbook of molecular biology and one with a style appropriate to the tradition-bashing bravura of his cohort. Watson replaced the monochromatic descriptive subheads conventional in biology texts ('The chromosomes,' 'Mitosis,' etc.) with declarative subheads that spell out an argument for the molecular vision of life: 'Mitosis maintains the parental chromosome number'; 'The cell theory is universally applicable.' The book became a classic. Soon thereafter, Watson began sending friends and colleagues a draft manuscript entitled *Honest Jim* (a play on Kingsley Amis's *Lucky Jim*). A breathless, novelistic account of his and Crick's discovery of the double helix, complete with a search for treasure, girl-chasing, back-stabbing, and the ultimate victory, all set in romanticized Cambridge. The manuscript infuriated Watson's colleagues. Crick was outraged, and led the attack. Harvard University Press had signed up the book, but withdrew. *Honest Jim* was published as *The Double Helix* with Athenaeum in 1968 and went on to be perhaps the most widely read scientific memoir ever, selling over a million copies, and is still in print today.

That same year, Watson began a third career, when he assumed the directorship of Cold Spring Harbor Laboratory of Quantitative Biology. It was a labor of love. Watson's old summer haunt was in serious financial trouble, although its previous director, John Cairns, had brought it back from the brink of bankruptcy. Retaining his Harvard appointment until 1976, Watson commuted to Cold Spring Harbor and set about reviving it. He dropped 'of Quantitative Biology' from the name, bolstered the revenue-producing summer courses and meetings, and invigorated the science program. The shambling, blurting Watson proved an extraordinary fund-raiser. In 1971, he capitalized on President Nixon's 'War on Cancer,' winning for the Lab a 5-year, 5-million-dollar grant from the National Cancer Institute. Cancer research dominated Cold Spring Harbor science for 25 years and remains a central focus. Watson also began for the first time in the Lab's 90-year history to build an endowment. All this enabled the hiring of more staff, construction of buildings, paving of roads, even installation of air-conditioning. Between 1968 and 1994, the scientific staff grew from 20 to more than 200, the total staff from 47 to nearly 600. In 2000, the grounds are manicured and decorated with sculpture, the façades by turns charming and stylish. The common wisdom is that the mature Jim Watson made Cold Spring Harbor into a place where the young Jim Watson could never have flourished.

Watson also became a leading spokesman for molecular genetic research. He was vocal and visible in the 1970s, during extensive and passionate debates over the safety of recombinant DNA research. He was a signer of the 1974 letter to *Science* magazine that called for a moratorium on recombinant DNA research; 10 months later, he participated actively in the international meeting on recombinant DNA technology at the Asilomar conference centre in California, where it was voted that the research should proceed, albeit carefully. By the late 1970s, Watson had become persuaded that the dangers associated with recombinant DNA were negligible; he became one of the most outspoken critics of restrictions on the research.

From 1989 to 1992, Watson served as the first director of the National Center for Human Genome Research, the National Institutes of Health arm of the Human Genome Project. First suggested in the 1970s, the genome project sought to discover, map, and sequence the complete set of human genes. Watson established the genome project as a major national and international effort, secured Congressional support, and established a crucial precedent by setting aside at first 3% and later 5% of the Center's budget for scholarly study of the many complex ethical, legal, and social issues raised by the project. On 26 June 2000, at a ceremony in the East Room of the White House, the completion of the Human Genome Project was announced. Watson was present.

See *also*: Delbrück, Max; DNA, History of; Luria, Salvador

Watson-Crick Model

See: DNA Structure

WHO Classification of Leukemia

B Bain

doi: 10.1006/rwgn.2001.1710

In the late 1990s, multiple-expert groups working under the aegis of the World Health Organization (WHO) drew up consensus proposals for the classification of neoplastic disorders of hemapoietic and lymphoid tissues. These were first published, in outline, in 1999. Classifications were proposed for acute myeloid leukemia (AML), the myelodysplastic syndromes (MDS), myeloproliferative disorders (MPD), lymphoid neoplasms (including acute lymphoblastic leukemia/lymphoma, non-Hodgkin's lymphoma, posttransplant lymphoproliferative disorders, and Hodgkin's disease), mast cell diseases and histiocytic and dendritic neoplasms. Some, but not all, of these classifications incorporate cytogenetic/molecular genetic findings. Thus, within the AML group there are specific categories for AML associated with t(15;17)(q22;q11–12), t(8;21)(q22;q22), inv(16)(p13q22), t(16;q22), and translocations with an 11q23 breakpoint. Similarly, within the MPD classification, chronic myeloid leukemia associated with t(9;22)(q34;q11) is specifically identified. Within the MDS group, the 5q− syndrome is specifically recognized. Similarly, the WHO classification of the acute and chronic leukemias makes use of the results of molecular genetic analysis. Specifically, the fusion genes and *MLL* abnormalities associated with the abovementioned cytogenetic abnormalities are incorporated in the classification. Four cytogenetic/molecular genetic subtypes of B-lineage ALL are also recognized, those associated with t(9;22), rearrangements with an 11q23 breakpoint involving the *MLL* gene, t(1;19)(q23;p13), and t(12;21)(p12;q22). It is recommended that pathologists diagnosing these neoplasms should be familiar with the types and significance of associated genetic abnormalities and that genetic analysis should form part of (or an addendum to) the pathology report whenever feasible. Cases of acute and chronic leukemia not falling into these cytogenetic/molecular genetic groups are classified on the basis of hematological features.

The WHO classification of lymphoid neoplasms with a mature phenotype also incorporates cytogenetic/molecular genetic entities, although the association with rearrangement of specific oncogenes is sometimes implied rather than stated. Genetic subtypes specifically recognized include: follicular lymphoma, mantle cell lymphoma, Burkitt's lymphoma (associated with t(8;14)(q24;q32) and *MYC* rearrangement), and anaplastic large cell lymphoma (associated with t(2;5)(p23;q35) and *ALK* rearrangement). Other neoplasms, including some specifically associated with recurring cytogenetic and molecular genetic abnormalities, are diagnosed on the basis on cytology, histology, and immunophenotype.

See also:* FAB Classification of Leukemia; Leukemia; Leukemia, Acute; Leukemia, Chronic; MIC and MIC-M Classifications of Leukemia; *MLL

Whole Organism Cloning

J B Gurdon

doi: 10.1006/rwgn.2001.1704

Whole animal cloning, or nuclear transplantation, as it used to be called, involves the formation of a complete individual by replacing the nucleus of an egg with that of a body (somatic) cell. Thus it is possible, in principle, to make huge numbers of genetically identical individuals by transplanting nuclei from one embryo or adult into numerous eggs. The genetically identical individuals so formed are called a clone. The production of multiple genetically identical organisms is well known in plants, where vegetative propagation is routine, and in humans in the case of identical twins.

The first multiple genetically identical adult animals were produced by nuclear transplantation in *Xenopus* in the late 1950s. The original purpose of these experiments was not to produce clones of individuals, but was to determine whether the process of development and cell differentiation involved any loss or stable change to the genome of cells as they differentiate from the egg. If the nuclear material of an egg could be replaced by that of a somatic or differentiated cell and still produce, in combination with egg cytoplasm, a normal individual, then clearly the genetic material of the donor cell must be equivalent to that of the egg and sperm nuclei and therefore the process of development and differentiation could not involve any change in the genome. This was the conclusion reached by transplanting nuclei from embryonic, and later differentiated cells. The production of sexually mature male and female frogs from the nuclei of the intestinal epithelium proved the principle of the constancy of the genome during development and cell differentiation. This was the first direct evidence that cell differentiation involves the selective activation and repression of genes that are retained in a potentially active state in all kinds of cells.

The methods used for cloning vary according to the species. In all species it is best to use unfertilized eggs as recipients. In frogs, the egg nucleus is situated at a visible point on the surface and can be eliminated by ultraviolet radiation which does not penetrate enough to damage the egg cytoplasm. Also in frogs, the transfer of a nucleus is achieved by sucking a donor cell into a small pipette so as to break the cytoplasm but not the nucleus of the cell and then to inject the broken cell into the irradiated egg. In all such experiments genetic markers are used to prove that the resulting cloned individuals arose from the activity of the transplanted nucleus and not from a failure to kill the nucleus of the unfertilized egg. In mammals, where great cloning success has been achieved in recent years, different methods have to be used. Irradiation is too damaging for such small eggs, and the nucleus has to be removed by sucking it out with a fine pipette. According to the species of mammal, various types of egg activation, for example electrical, are undertaken to prepare the egg. The nucleus is transplanted either by direct injection with a micropipette, by virus fusion to the egg, or by other means.

The first successful cloning of a mammal to produce a normal adult was achieved in 1997 by Wilmut *et al.*, who produced 'Dolly' the sheep by transplanting a nucleus from a primary culture of cells originating from an adult mammary gland. Dolly carried the genetic markers of the donor and not recipient strains. The cloned embryos were transferred back into a foster mother of the recipient strain, and were in due course born. One lamb, Dolly, survived to become a sexually mature adult sheep. Since that time, cloning success has been achieved in several mammal species including mice, goats, and cows, in addition to sheep. Furthermore, normal adult cloned animals have been produced from various different tissues and also from established lines of cells including embryonic stem cells. Thus, it is now clear that the principle of the constancy of the genome applies not only during cell differentiation but also to cells of old animals and to cells that have been cultured for many passages in the laboratory.

The success of cloning in mammals offers a number of practical benefits. For example, it is possible to generate multiple copies of any one animal that happens to carry a particularly desirable genetic constitution. Furthermore, embryonic stem cells can be genetically transformed and cells of the desired genetic constitution can be used to generate fertile adult animals and so create a novel genetic strain.

The principle of the constancy of the genome during life and during the propagation of cell lines is well established. There are, of course, particular exceptions, as in the case of antibody-forming cells where genetic recombination occurs to create the great variety of antibodies that can by synthesized. There is also the point that the mitochondria and their genes may differ according to the recipient eggs used, although it is not generally thought that mitochondrial genes contribute to the diversity of cell types. It is only because of the constancy of the genome during development that whole animal cloning is possible.

The success rate of cloning is surprisingly low. Using nuclei from embryonic cells of frogs, as many as 50% of all nuclear transfers can become normal adults. However, as nuclei are taken from increasingly differentiated or aged cells, the success rate becomes greatly reduced. Currently the frequency with which a normal adult can be produced by nuclear transfer from differentiated an adult cells is between 0.1 and 1.0% of all nuclei transferred. For many purposes this does not matter very much. However, there is considerable scientific interest in the mechanism of gene reprogramming that must accompany a successful cloning experiment. The switching off of genes no longer needed, for example, of muscle genes in the intestinal epithelium or skin, appears to be accompanied by modifications of chromosome structure (e.g., DNA-associated proteins). It is remarkable that these modifications can be reversed at all by nuclear transfer, but the process is not efficient, and it is at present obscure why this should be so. Some but not all workers in the field believe that the stage of the donor cell cycle is important. There is much current interest in trying to improve the success rate of cloning and to determine the molecular mechanism by which gene reprograming is brought about through the action of egg cytoplasm. The failure of the great majority of nuclear transplant experiments occurs progressively, in both frogs and mammals, during their development from egg to adult. The death rate of newborns is particularly high in mammals. One possible reason for this is that the unnatural process of combining a somatic cell nucleus with an unfertilized egg leads to defective chromosome replication during early development such that it is a rare event for the totipotent state of the original donor nucleus to be successfully maintained through to adulthood.

There is no scientific reason why the cloning technology that has been developed successfully for several species of mammals could not be applied to humans. In nearly all countries where legislation concerns itself with such things, it is illegal to carry out experiments aimed at human reproductive cloning. An entirely different objective is to use cloning from somatic cells to prepare tissues for transplantation. For example, it is theoretically possible to transplant nuclei from adult skin cells into unfertilized eggs and so generate embryos. Cells from these embryos or embryonic stem cells can be prepared and directed by the use of appropriate growth factors to differentiate in other directions, such as into a cornea for the eye. Experiments with human cells aimed at understanding and facilitating such technology is legally permitted in many countries and should be encouraged. If consideration is ever given to the reproductive cloning of humans, it should be remembered that the very great majority of nuclear transplant experiments yield defective embryos that die at or before birth. Furthermore, it is not yet known whether cloning from aged somatic cells would cause a predisposition to cancer.

Further Reading

Colman A (1999) *Cloning* 1: 185–200.

Gurdon JB (1962) *Journal of Heredity* 53: 4–9.

Gurdon JB (1977) *Proceedings of the Royal Society Series B* 198: 211–247.

Gurdon JB and Coleman A (2000) *Nature* 402: 743–746.

***See also:* Embryonic Stem Cells; Ethics and Genetics; Nuclear Transfer**

Wild-Type (WT)

L Silver

doi: 10.1006/rwgn.2001.1385

'Wild-type' is the term used traditionally to define the 'normal' allele at a particular locus. It is based on the assumption that organisms from 'the wild' (meaning a natural environment) will all carry a normal non-mutant allele at a locus for which mutant alleles have been uncovered in laboratory strains. Although this definition is useful in the context of experimental crosses, it suffers from the fact that at many loci, there are multiple normal alleles in a population (i.e., the locus is polymorphic). From the perspective of population genetics, a wild-type allele is one that is present at a frequency of at least 1% in a population, based on the assumption that truly deleterious alleles could never reach this frequency. However, some disease-causing alleles (like the sickle cell mutation) are maintained at wild-type levels in certain populations because of an effect of heterozygous advantage (in the case of sickle cell, heteozygotes are resistant to contracting malaria).

***See also:* Balanced Polymorphism; Sickle Cell Anemia**

Wilms' Tumor

N Hastie

doi: 10.1006/rwgn.2001.1386

Wilms' tumor (nephroblastoma) of the kidney is a striking example of the way that cancer may arise through the disruption of a normal developmental process. This tumour affects around 1:10 000 children worldwide, usually during the first 5 years of life. Wilms' tumors are thought to arise from metanephric mesenchymal stem cells that would normally differentiate to form components of the nephron. Hence, these stem cells would normally differentiate and, in doing so, exit the cell cycle. In the malignant situation these cells fail to exit the cycle and there is uncontrolled proliferation, leading to the formation of tumors which often contain disorganized counterparts of normal kidney structures.

The majority of Wilms' tumors arise sporadically and do not involve inherited or constitutional genetic change. However, major insights into the genetic basis of Wilms' tumor have come from the study of the rare cases in which Wilms' tumor is one feature of a contiguous gene syndrome, the WAGR syndrome. Children with WAGR suffer from Wilms' tumor, aniridia (no iris), genitourinary abnormalities, and mental retardation. In all cases the children have constitutional hemizygous deletions of chromosome 11p13. Mapping within these deletions led to the isolation of the first Wilms' tumor predisposition gene, *WT1*. *WT1* is classified as a tumor suppressor gene because in order for tumors to develop in these children the remaining allele is always mutated during kidney development. Thus, complete loss of WT1 function is required for tumorigenesis. It is now known that the *WT1* gene plays a vital role in the normal development of the kidney and gonads; homozygous *wt1* null mice completely lack these two organs.

Perhaps the most fascinating aspect of *WT1* from the genetic perspective is the way that different mutations in the gene can lead to different phenotypes. Whereas in WAGR, hemizygous deletion of *WT1* can result in mild genital anomalies and Wilms' tumor (once the wild-type allele is mutated), dominant mutations in the gene can lead to a different spectrum of phenotypes.

The *WT1* gene encodes a protein(s) with four zinc fingers that has all the hallmarks of a transcription factor. Through a combination of alternative splicing, RNA editing and alternative translational start sites, 16 slightly different WT1 isoforms can be produced in mammals. One of these alternative splices leads to the inclusion (+KTS) or exclusion (−KTS) of three amino acids, lysine, threonine and serine, between zinc fingers three and four. This alternative splice is conserved throughout vertebrate evolution.

One syndrome arising through dominantly acting *WT1* mutations is the Denys–Drash syndrome (DDS). Children with DDS not only suffer from Wilms' tumor but also have severe hypertension associated with kidney disease and a spectrum of gonadal abnormalities, ranging from no gonads to XY sex reversal. All these children have heterozygous mutations that disrupt the DNA binding, zinc finger domain. The most likely explanation is that these mutant proteins act in a dominant-negative way, interfering with the function of the normal protein encoded by the wildtype allele.

Another syndrome, Frasier syndrome, is characterized by glomerular disease and XY sex reversal in the absence of Wilms' tumor. Remarkably, all the children with Frasier syndrome have a heterozygous mutation that inhibits the alternative splice which would normally allow the production of the WT1 + KTS isoform. Thus these severe phenotypes arise because there is an alteration in the ratio of the +KTS to −KTS isoforms.

Hence these human genetic data suggest that these two isoforms of WT1, differing by just three amino acids, have very different functions and that the correct balance between these functions is vital. It is extremely gratifying that this conclusion is borne out by biochemical studies. There is a body of evidence that WT1 is a transcription factor that can influence the expression of other genes as a transcriptional activator, coactivator or repressor. However, recent evidence suggests that WT1 may also play a role in RNA splicing. Furthermore, the data suggest that the −KTS isoforms are more likely to play a role in transcription, whereas the +KTS isoform may function to regulate RNA processing. These data raise the intriguing possibility that WT1, through its different isoforms, coordinates tissue-specific and developmental-specific regulation of transcription and alternative splicing. Now the major goal is to identify the key target genes regulated by WT1 at the transcriptional and post-transcriptional level. It is likely that some of these may also play crucial roles in Wilms' tumors and other urogenital conditions.

The *WT1* gene plays a role in only 10–15% of all Wilms' tumors. Studies of rare families in which Wilms' tumor predisposition is inherited have resulted in the mapping of two familial Wilms' tumor loci to chromosomes 17q and 19q. When these genes are identified it is hoped that they will be shown to work in the same pathway as WT1.

There is another fascinating syndrome in which Wilms' tumor may be a feature, but not involving the *WT1* gene. This is the so-called Beckwith–Weidemann syndrome (BWS) which illustrates the remarkable phenomenon of genomic imprinting. Children with BWS always suffer from gigantism, have disproportionate overgrowth of certain organs, and have renal dysplasia along with several less penetrant phenotypes. Of children with BWS, 5–10% develop embryonal malignancies, particularly Wilms' tumor, but more rarely rhabdomyosarcomas and hepatoblastomas. In a few percent of cases, BWS is a familial trait, the responsible locus mapping to chromosome 11p15. Within these families, however, children only develop the condition if they inherit the mutation from the mother.

A few percent of BWS cases carry balanced translocations involving 11p15, always inherited from the mother. Furthermore, a significant proportion of cases have paternal isodisomy of chromosome 11p15 in the presence or absence of the maternal chromosome 11. Finally, in 30% of Wilms' tumors there is loss of heterozygosity of chromosome 11p15 markers, but always involving loss of the maternal allele.

Thus it has been posited that BWS and the associated tumors may arise either through a double dose of a growth promoting gene expressed normally from just the paternal 11p15 allele or loss of expression of a maternally expressed growth suppressing gene mapping to 11p15. A strong candidate for the former is the insulin-like growth factor (IGF_2) gene and for the latter the p57 KIP_2 gene, encoding a cell cycle inhibitor. Recent studies with animal models suggest that BWS may involve both these genes.

In conclusion, the study of Wilms' tumor and the genes predisposing to this condition has illuminated several biological issues: the relationship of cancer to development; the link between kidney and gonad development; the way that human mutation can provide insight into gene function; how different mutations in a gene may lead to very different phenotypic consequences; and into the process of genomic imprinting.

Further Reading

Hastie ND (1994) The genetics of Wilms' tumor: a case of disrupted development. *Annual Review of Genetics* 523–558.

Hastie ND (1997) Disomy and disease resolved? *Nature* 389: 785–787.

Van Heyningen V (1997) Sugar and spice and all things nice. *Nature Genetics* 17: 367–368.

See also: **Balanced Translocation; Genetic Diseases; *Igf2* Locus**

Wilson Disease

D W Cox

doi: 10.1006/rwgn.2001.1388

Wilson disease (hepatolenticular degeneration) is a disorder of copper transport. Copper from the diet is not eliminated adequately from the liver or incorporated into the copper plasma protein, ceruloplasmin. As a result, the copper accumulates in liver and brain and causes damage to these organs. The age of onset varies from 3 to 50 years, and the copper deposition can cause neurological disease or destruction of the liver. The gene maps to human chromosome 13q14 and was cloned in 1993. The gene encodes a membrane copper transporter which is critical for elimination of copper from the liver. The disease is treated with chelating agents such as penicillamine or trien, or is prevented from absorption by high doses of oral zinc. Treatment with these agents is successful, providing this is started before organs have been permanently damaged.

***See also:* Genetic Diseases**

Wilson, Edmund Beecher

K N Gracy and J H Miller

doi: 10.1006/rwgn.2001.1387

Edmund Beecher Wilson (1856–1937), recognized for his early work in establishing the significance of the sex chromosome, was born in Geneva, Illinois, USA. He graduated from Yale University in 1878 with a PhB, and received a PhD from Johns Hopkins University in 1881. After further training abroad, he returned to the USA to teach at the Massachusetts Institute of Technology and Bryn Mawr College. He then moved to Columbia University, where he spent the rest of his career (1894–1928) as a professor of zoology.

Wilson's early work focused on embryology and experimental morphology. This then led to an interest in cell lineage. His work tracing the development of specialized tissue from precursor cells included the study of internal cellular organization. In 1896, Wilson published *The Cell in Development and Inheritance* (Wilson, 1896), a work that emphasized the role of the cell in biological processes.

Wilson's research then turned to sex determination and heredity. The rediscovery of Mendel's work in 1900 strongly influenced Wilson's studies, and he began to examine chromosomes and their role in sex determination. A series of papers, the first published in 1905, explored these findings and served to shape the growing field of genetics.

In later years, Wilson expounded on the idea that the function of chromosomes went beyond sex determination, and his writing helped to shape the theory of chromosomes as carriers of all hereditary information. The geneticists T.H. Morgan and Hermann Muller were part of the Zoology Department at Columbia University at that time and were influenced by Wilson's work. Wilson went on to write other books, including *The Physical Basis of Life* (Wilson, 1923). He retired from Columbia University in 1928 and died in New York in 1939.

References

Wilson EB (1923) *The Physical Basis of Life*. New Haven, CT: Yale University Press.

Wilson EB (1896) *The Cell in Development and Inheritance*. New York: Macmillan.

***See also:* Morgan, Thomas Hunt; Muller, Hermann J**

Winding

***See:* DNA Supercoiling**

Wiskott–Aldrich Syndrome

M A Ferguson-Smith

doi: 10.1006/rwgn.2001.1389

Wiskott–Aldrich Syndrome is a rare X-linked recessive disorder characterized by eczema and a severe reduction in blood platelets (thrombocytopenia) which are abnormally small. Those affected are subject to life-threatening infections due to a severe immune defect involving both B and T lymphocytes. Bone marrow transplantation has been used to treat the condition.

***See also:* Sex Linkage**

Wobble Hypothesis

R W Alexander and P Schimmel

doi: 10.1006/rwgn.2001.1391

Even before the genetic code was fully elucidated, Francis Crick proposed his 'wobble hypothesis' to explain how a single transfer RNA (tRNA) molecule could decipher more than one codon. He rationalized that the first two codon positions require standard pairings with bases in the tRNA anticodon to assign amino acid specificity unambiguously. That is, guanine (G) pairs with cytidine (C) and uracil (U) pairs with adenine (A). Equivalent pairs such as inosine (I):C and thymine (T):A are also permitted in the first two positions. (Inosine is deaminated adenine, and is often found as the 5′-nucleotide of a tRNA anticodon.) Crick suggested that the third codon position has enough freedom or ‘wobble’ to allow nonstandard base pairs. Because base pairing is antiparallel, this ‘wobble position’ corresponds to the 5′-nucleotide of the anticodon.

Of all the possible pairs at the wobble position, only those close in geometry to the standard pairs are likely to occur, due to the structural constraints provided by the first two base pairs. Crick therefore predicted that G:U, I:U, and I:A pairs would be permitted in the third position (**Table 1**), while other base pairs (such as U:U) would bring the glycosyl bonds of the nucleotides too close together.

The wobble hypothesis correctly predicts the minimal number of tRNAs required to fully determine the genetic code and accounts for the observed degeneracy in the code. There are 64 (4^3) possible combinations of trinucleotide codons, of which 61 code for amino acids and three are termination signals. If wobble is permitted in the third position, the codons UUA and UUG can both be recognized by a single tRNA containing a UAA anticodon. That is, either the standard A:U or the wobble G:U would occur. These particular codons therefore cannot specify two different amino acids, and indeed they both specify leucine. Similarly, if an amino acid is specified by codons XYN (where N represents any nucleotide), a minimum of two tRNAs is necessary to translate the four codons. Both XYU and XYC can be recognized by an anticodon with G in the first position, while XYA and XYG can pair with an anticodon containing U in the wobble position. If inosine is the 5′-nucleotide of the anticodon, wobble pairing allows recognition of codons ending in U, C, and A. The XYG codon of this family can then be recognized by a tRNA with a C in the first position.

Table 1 Base pairs allowed at the third codon position according to the wobble hypothesis

Base on the tRNA anticodon	Bases recognized on the mRNA codon
U	A
	G
C	G
A	U
G	C
	U
I	C
	U
	A

The 20 amino acids of the genetic code are therefore specified by a minimum of 32 tRNAs, with a unique initiator tRNA also necessary. Most organisms actually have more than 32 cytoplasmic tRNAs, including some with the same anticodon sequence. All tRNAs that specify a particular amino acid are called isoacceptors, and all members of an isoaccepting family are aminoacylated by the same aminoacyl-tRNA synthetase.

Mitochondria have ‘nonstandard’ genetic codes, and this is evident in their codon–anticodon interactions. In addition to the wobble pairs permitted according to Crick’s hypothesis, some mitochondrial tRNAs with U in the first position of the anticodon decode all four members of the XYN family. These tRNAs read only the first two bases of the codon and ignore the third. This ‘two-base’ reading reduces the number of tRNAs necessary to translate the genetic code. Indeed, mitochondria require only 24 tRNAs.

Further Reading

Breitenberger CA and RajBhandary UL (1985) Some highlights of mitochondrial research based on analyses of *Neurospora crassa* mitochondrial DNA. *Trends in Biochemical Sciences* 10: 478–483.

Crick FHC (1966) Codon–anticodon pairing: the wobble hypothesis. *Journal of Molecular Biology* 19: 548–555.

***See also:* Genetic Code; Mitochondrial Genome**

Wright, Sewall

J F Crow

doi: 10.1006/rwgn.2001.1449

Sewall Wright (1889–1988) was the leading population and quantitative geneticist in the United States. Along with the British geneticists R.A. Fisher and J.B.S. Haldane, he founded the subject of population genetics and the mathematical theory of evolution, and for many years the three of them dominated the field.

Wright was precocious as a child; he could extract cube roots before he started school. He attended tiny Lombard College, where his father was on the faculty, and got interested in genetics by reading the *Encyclopaedia Britannica*. He spent a postgraduate year at the University of Illinois and then joined William Castle at Harvard where he received his doctorate in 1915. He then spent 10 years as senior animal husbandryman at the United States Department of Agriculture, before moving to the University of Chicago. Upon retirement from Chicago he joined the faculty at the University of Wisconsin and remained there until his death.

Wright is best known among genetics students for his inbreeding coefficient. This gives the proportion by which inbreeding has reduced the average heterozygosity, relative to a noninbred individual. Wright provided a simple algorithm for calculating the inbreeding coefficient for an individual with any pedigree, no matter how complicated. This has become the starting-point for the study of inbreeding in animal, plant, and human populations. Alternatively, the inbreeding coefficient is a measure of the probability that two alleles in an individual are identical by descent, that is, both derived from the same ancestral gene or one from the other. In recent years Wright's correlation definition has been almost entirely replaced by the probabilistic definition based on identity by descent. The algorithm is the same.

Wright invented a statistical procedure called 'path analysis.' This assigns a path coefficient, a standardized partial regression coefficient, to each step in a causal diagram, and their relative sizes reflect the influence of that step. As with his inbreeding coefficient, Wright provided simple algorithmic rules for calculating path coefficients. In particular, Wright used this method to separate genetic components from each other and from environmental influences. The method has been widely used in the study of livestock breeding, and forms the basis for standard heritability calculations. More recently path analysis has become popular in the social sciences.

Wright started at Harvard with research on guinea pigs and these studies continued through most of his life. He studied many developmental problems and his analyses of coat color interactions were especially deep. On moving from Harvard to the United States Department of Agriculture in 1915, he inherited a guinea pig colony with a long history of inbreeding. In 1922 he published a thorough analysis of the effects of inbreeding and crossbreeding. He also used these methods to study the history of livestock breeds, especially British Shorthorn cattle. He was able to apportion the inbreeding coefficient between effects of finite population size and nonrandom mating. The former turned out to be much more important. Later he extended this idea to natural populations and developed his F statistics, extensively used to study genetic differentiation in geographically structured populations.

On moving to the University of Chicago in 1926, Wright turned his research attention to problems of evolution. In 1931, he published his most famous paper, 'Evolution in Mendelian populations' (Wright, 1931). He included the effects of mutation, selection, migration, and random processes in one general equation. In later years the work was made more general and rigorous, but the essential ideas remained the same.

Out of this grew Wright's shifting-balance theory, and he spent much of the rest of his long life elaborating and defending it. The theory is based on (1) the finding in his guinea pigs that gene interactions were common and often unpredictable, (2) the observation that inbreeding and random gene frequency drift are important in small populations and affect important, fitness-related characters, and (3) the history of British Shorthorn cattle. Some herds, for no obvious reason, were particularly good, and bulls exported from these herds were used to upgrade the whole breed. These observations formed the model for Wright's theory of evolution, in which he envisaged a large population divided into small, partially isolated groups, analogous to Shorthorn herds.

The shifting-balance process was viewed as happening in three phases. Phase I consists of differentiation among numerous small subpopulations by random processes. Among them one or more will drift into a happy combination of genes producing increased fitness. In Phase II, when a threshold frequency is passed by random processes, natural selection will carry the subpopulation to a higher fitness, an adaptive peak in Wright's terminology. In Phase III, because of its excess growth and migration, individuals from the successful subpopulation diffuse through the population. Continuing in this way the entire species is upgraded and the process can start over again.

In Wright's view, a major barrier to evolutionary progress is that natural selection in a large population has no way to move a population from one well-adapted state to a better one if this involves passing through states that are less well adapted. This will often be the case if fitness is strongly dependent on epistatic interactions. Wright regarded his shifting-balance process, a combination of random drift and intergroup selection by differential migration, as a way of getting over this difficulty.

The shifting-balance theory has enjoyed great popularity among biologists from the time of its first presentation in the early 1930s and throughout Wright's life. Its popularity was aided by its being espoused by two influential disciples, J.L. Lush, America's leading student of livestock breeding, and Th. Dobzhansky, one of the best-known evolutionists. The theory was not universally accepted, however. In particular, R.A. Fisher believed that the kind of epistatic hang-up that Wright envisaged was not likely to be important. It would almost always be possible, he said, for some gene frequency changes to improve fitness. For him, a large unstructured population in which random noise is less important insures systematic evolutionary progress, and there is no need for any special Wright process. Wright's view has also been criticized for requiring a rather special combination of subpopulation size and migration rates, and for the fact that the population has low average fitness during the shifting-balance process. The issue was never settled during Wright's lifetime and current opinion seems to be drifting in the direction of the Fisher view. Unfortunately, the problem is almost insoluble and is not likely to be settled by any single or small number of experiments. Nor is it likely that the answer will be a clean either/or. Possibly both are correct.

After retiring from the University of Chicago at age 65, Wright moved to the University of Wisconsin where he continued to work actively well into his nineties. His four-volume set of books summarizing a lifetime of work by himself and reviewing the work of many others, were all written after his formal retirement. He died from a fracture, the result of slipping on an icy sidewalk during his regular walk, only a little more than a year away from his 100th birthday.

Further Reading

Provine WB (1986) *Sewall Wright and Evolutionary Biology.* Chicago, IL: University of Chicago Press.

Wright S (1968, 1969, 1977, 1978) *Evolution and the Genetics of Populations,* 4 vols. Chicago, IL: University of Chicago Press.

Reference

Wright S (1931) Evolution in Mendelian populations. *Genetics* 16: 97–159.

***See also:* Fitness Landscape; Shifting Balance Theory of Evolution**

WT

***See:* Wild-Type (WT)**

X Chromosome

Y Boyd

doi: 10.1006/rwgn.2001.1392

The human X chromosome contains around 150 000 000 base pairs (150 Mb) of DNA, approximately 5% of the genetic content of each cell. There are estimated to be around 3000 to 5000 genes carried on the X chromosome and several hundred of these have been associated with clinical disease. Lists of genes and genetic diseases that have been mapped to the human X chromosome can be found on the Genome Database (GDB) website. Two features of the X chromosome have made it particularly amenable to study. The first feature is that caused by the difference in X chromosome content of females and males, with XX females inheriting one X chromosome from each parent and XY males inheriting their single X chromosome from their mothers and the sex-determining Y chromosome from their fathers. Pedigrees for X-linked traits, such as color blindness and hemophilia, therefore exhibit a distinctive inheritance pattern, with the trait being manifested only in males and being passed on through their unaffected daughters to their grandsons. The second catalyst for research on X-chromosome genes was the discovery that the chromosome contained the locus encoding hypoxanthine phosphoribosyl transferase (HPRT), a non-essential enzyme in the purine salvage pathway. Since HPRT can be subject to both forward and reverse chemical selection, the presence or absence of the entire X chromosome, or portions of the X chromosome containing HPRT, can be selected for in cell culture systems. This has led to the production of somatic cell hybrid panels that contained different portions of the X chromosome and which have been extensively used to map X-chromosomal genes.

The X chromosome has also been the focus of research efforts because of its association with X-inactivation (see X-Chromosome Inactivation) the phenomenon whereby most genes on one of the two X chromosomes selected at random are silenced in early female development. Recent advances in molecular biology have brought considerable insight into the mechanisms behind this phenomenon, which is thought to have arisen as a means of ensuring that both XX females and XY males have a single dose of the products of X-linked genes in all their somatic cells.

Evolutionary Origin of the X Chromosome

Examination of the sex chromosomes of nonmammalian vertebrates such as snakes has led to the conclusion that the mammalian X and Y chromosomes were derived from a pair of autosomal homologs which initially differed at only one, or a few loci, that were required for sex determination. The difference in size and shape of the X and Y chromosome, known as heteromorphism, is thought to have arisen through recombination suppression and loss of genes from the sex-determining Y chromosome. There are still

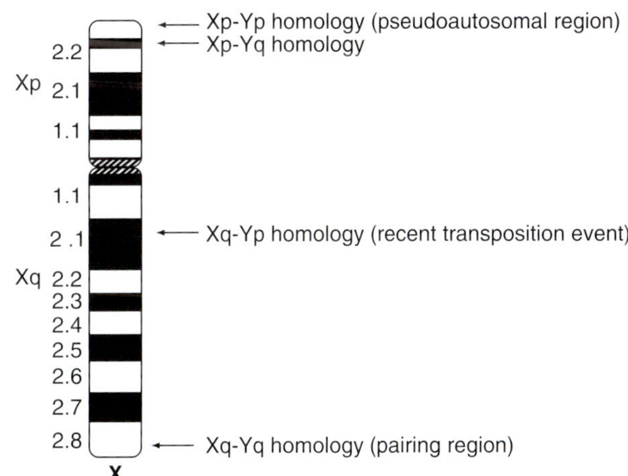

Figure 1 The human X chromosome represented as a Giemsa-banded ideogram. Xp (X chromosome short arm) and Xq (X chromosome long arm) are divided into stained and unstained regions that are referred to by a standard nomenclature (numbers given to the right of the chromosome). The four main regions of homology between the human X and Y chromosomes are indicated by the arrow. There are also much smaller regions of homology represented by single genes.

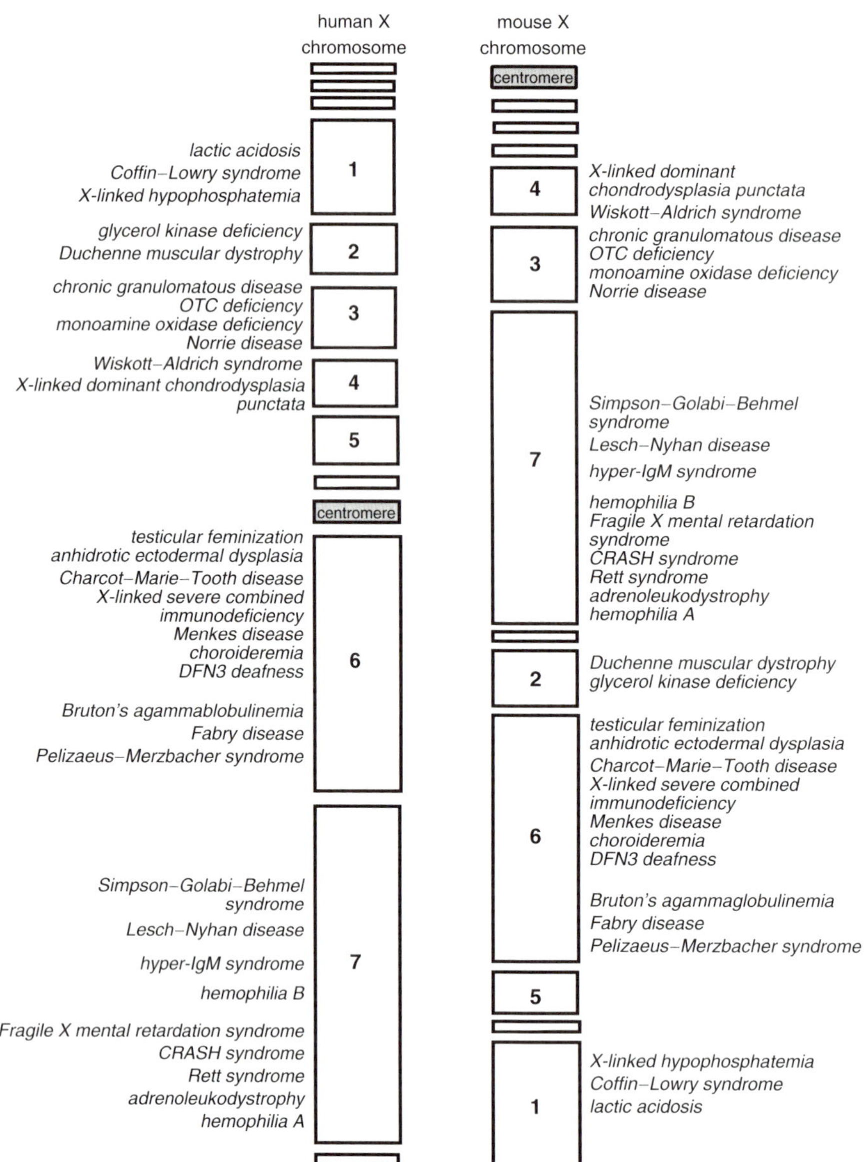

Figure 2 Comparative map of the human and mouse X chromosomes. Each rectangular block represents a chromosomal region, known as a conserved segment or homologous block, which contains the same genes in the same order on both chromosomes. The seven largest blocks have been given equivalent numbers on the human and mouse X chromosomes. It is not known whether there is any sequence homology between the human and mouse centromeres. It can be seen that, whereas there are only two large conserved segments (blocks 6 and 7) that are shared between the human X chromosome long arm and the mouse X chromosome, the human X chromosome short arm is composed of several smaller conserved segments. The position of genes responsible for human X-linked disorders, which have either natural or engineered mouse models, are positioned to the side of each chromosome.

some regions of homology between the human X and Y chromosomes, including the 5 Mb region pseudo-autosomal region where the two chromosomes pair during meiosis (**Figure 1**). The ancestral sex chromosomes of mammals are probably represented by the region of approximately 120 Mb covering the entire human X long arm and the centromeric region of the short arm which is homologous to the X chromosome found in noneutherian mammals such as marsupials. The additional material on the X chromosome short arm of eutherian (placental mammals), is autosomal in marsupials and is thought to have been transposed at intervals during evolution onto the X chromosome. X and Y chromosomes of differing sizes are also present in insects and worms. Birds have a reciprocal system where females are the heterogametic sex (i.e., making two different types of gametes) and carry two different sex chromosomes this time called Z and W and males have a ZZ complement. The sex chromosomes of birds are thought to have evolved independently from those of mammals as genes that are Z-linked in birds have been mapped to a range of human and mouse autosomes.

Conservation of X Linkage in Mammals

In the late 1960s Susumu Ohno predicted that genes that were X-linked in one mammalian species would also be X-linked in another to avoid the imbalance in gene dosage that would occur if X chromosome genes were moved onto an autosome. This prediction has been shown to be true and has helped to identify animal models for human X-linked diseases, which can be recognized by their unusual inheritance pattern. However, although the same genes lie on the X chromosome in all mammals, mapping experiments have revealed that groups of loci in the same order, known as conserved segments or homologous blocks, have been rearranged with respect to each other during evolution. A good example of this is illustrated by the comparative map of the human and mouse X chromosomes which comprises a series of conserved segments that range in size from 100 000 to 50 000 000 base pairs (**Figure 2**). The pseudoautosomal region, which contains genes that are expressed on the active X, the inactive X and the Y chromosome and are therefore not subject to dosage compensation, is the only region on humans and mouse that does not have a conserved gene content.

Human X-Linked Disease

The most common human syndromes associated with the X chromosome are anomalies in sex chromosome number that arise through nondisjunction at meiosis. Turner syndrome occurs in approximately 1 in 2000 female births and is caused by the loss of an entire chromosome leading to an XO karyotype. To explain why the presence of a single X chromosome is deleterious in XO females but not in XY males, it has been proposed that Turner syndrome is caused by a single, not double, dose of one or more of the few genes that normally escapes from X-inactivation. This is in tune with the observation that mice with an XO karyotype do not have an overt phenotype and that there are fewer mouse genes reported that escape X-inactivation. An additional X chromosome is present in the 1 in 600 males that are Klinefelter syndrome patients, who have an XXY karyotype. More rarely, females have also been identified with XXX and XXXX complements. Mutations, or rearrangements, in genes that are important in primary or secondary sex determination can give rise to females with an XY chromosome complement and males with an XX complement.

Mutations in single X-linked genes are fully expressed in males and give rise to 'sex-linked' disorders, for example, Duchenne muscular dystrophy which has an incidence of around 1 in 3000 males and the fragile X-linked mental retardation syndrome which has an incidence of around 1 in 10 000 males. As a result of the random inactivation of one of their two X chromosomes in early development, all females are mosaics of two populations of cells and the relative numbers of cells in these two populations will differ between individuals. Often females heterozygous for a mutated gene are completely unaffected as the population of cells expressing the nonmutated allele either provides a sufficient quantity of normal gene product, or, during development or lineage differentiation, predominates over the population of cells carrying the mutant allele. However, some female carriers for X-linked 'recessive' diseases manifest some disease symptoms because of a natural skew in favor of cells with the mutated X as the active chromosome. Very occasionally, carrier females may manifest the same severity of disorder as that seen in males. Mutations in X-linked genes may also give rise to X-linked 'dominant' disorders found only in females and in these instances it is assumed that affected males die before birth. The most common example of an X-linked dominant is Rett syndrome, a severe progressive neurological disorder affecting approximately 1 in 20 000 females, which has recently been associated with mutations in the gene encoding methyl-CpG-binding protein.

Further Reading

Boyd Y, Blair HJ, Cunliffe P, Masson WK and Reed V (2000) A phenotype map of the mouse X chromosome: models for human X-linked disease. *Genome Research* 10: 277–292.

Genomic Database (GDB) http://www.gdb.org/gdb.

Lahn BT and Page DC (1999) Four evolutionary strata on the human X chromosome. *Science* 286: 964–967.

Miller JR (1990) *X-Linked Traits: A Catalog of Loci in Non-Human Mammals.* Cambridge: Cambridge University Press.

Ohno S (1969) Evolution of sex chromosomes in mammals. *Annual Review of Genetics* 3: 495–521.

***See also:* Ohno's Law; Sex Determination, Human; Sex Linkage; W Chromosome; X-Chromosome Inactivation; Z Chromosome**

X-Chromosome Inactivation

M F Lyon

doi: 10.1006/rwgn.2001.1511

In mammals sex is determined by the X and Y chromosomes, females being chromosomally XX and males XY. In females one of the two X chromosomes in each cell becomes genetically inactive and untranscribed early in development and remains so throughout life. This is termed X-chromosome inactivation (XCI). The result is that the effective dosages of products of X-linked genes are equal in males and females. The X chromosome is typically large and with many genes unconnected with sex, whereas the Y chromosome is typically much smaller and carries orthologs of only a few of the X-linked genes. Without XCI males and females would thus differ in levels of X-linked gene products. X-chromosome inactivation is thus a form of dosage compensation. The existence of XCI was first suggested by Mary Lyon in 1961. For a time this suggestion was known as the 'Lyon hypothesis,' and the inactive X chromosome was said to be 'lyonized.' These terms are now outdated, however.

Either X chromosome can be inactivated in different cells in the embryo proper of eutherian mammals. Once the choice has been made the same X chromosome remains inactive in the descendants of each cell throughout life. By contrast in the extraembryonic membranes of eutherian mammals, and in all cells of marsupials, the paternally derived X chromosome is inactivated in all cells. In the female germ cells the inactive X chromosome is reactivated as the cells approach meiosis, whereas in the male germ cell the single X chromosome becomes inactive. Rare individuals are found with supernumerary or missing X chromosomes such as XXY males, or XXX or XO females. In these individuals a single X chromosome remains active no matter how many are present. Thus, there is a counting mechanism ensuring that a single X chromosome remains active per two autosome sets.

When the X chromosome becomes inactive it takes on a set of characteristic properties. It replicates its DNA late in S-phase, and it remains condensed during interphase and forms the sex chromatin body against the nuclear membrane. It shows hypoacetylation of lysine residues in histones, and in somatic cells of eutherian mammals, the cytosines in CpG islands of promoter regions of housekeeping genes are differentially methylated. In these respects the inactive X chromosome (Xi) behaves like heterochromatin, whereas the active X chromosome (Xa) in the same cell behaves as euchromatin. In addition a specific protein, histone macro H2A1, is concentrated in the Xi.

X-chromosome inactivation in marsupials differs from that in eutherians not only in preferential paternal X inactivation but also in the lack of differential methylation. Late replication and hypoacetylation of histones are seen as in eutherians. The inactivation is both less complete and less stable than in eutherians.

The mechanism of XCI remains unknown but there have been recent major advances in knowledge. The initiation of inactivation in early development requires the presence of the X-inactivation center (XIC) on the X chromosome. Segments of X chromosome lacking an XIC through translocation or deletion do not undergo XCI. A single gene *Xist* (X inactive specific transcript) located at the XIC is essential for XCI. Gene knockouts show that it is needed for both random and paternal XCI, but is not needed for spermatogenesis (when the single X chromosome becomes inactive). Knockout of the promoter and first exon prevented initiation of inactivation of the affected X chromosome but counting of X chromosomes still occurred. Knockout of the region 3′ to exon 6 prevented counting. Insertion of transgenes for *Xist* showed that a 40-kb cosmid including the *Xist* gene was sufficient for counting and inactivation. *Xist* RNA can coat the autosome and inactivate autosomal genes and hence X-specific sequences are not essential for *Xist* function. Before the onset of XCI *Xist* is transcribed from both X chromosomes but the transcripts are unstable. At XCI the transcript from the incipient Xi becomes stabilized and its RNA accumulates over the entire length of the Xi and appears to coat it. The allele on the Xa is then silenced. Methylation of cytosines at certain sites in *Xist* is required to maintain its silence on the Xa. Methylation is also thought to form the imprint that prevents the maternally derived X chromosome becoming inactive in the extraembryonic membranes. *Xist* has not yet been found in marsupials. It may be that they have lost the gene or it may be present but very poorly conserved.

It is not clear whether *Xist* has a role in maintaining the inactive state once initiated. Loss of *Xist* activity after the X chromosome has become inactive does not necessarily lead to reactivation. There is good evidence that differential methylation of cytosines on the Xi is important in stabilizing inactivation. Lack of differential methylation of the Xi in marsupials is thought to explain the lower stability of XCI in this group. In addition, late replication of the Xi is thought to provide a stabilizing mechanism in both eutherians and marsupials.

How *Xist* RNA brings about the conversion of the Xi from the active euchromatic to the inactive heterochromatic state remains unknown. Since late replication of DNA and hypoacetylation of histones are both found in marsupials as well as eutherians they are thought to be part of the fundamental mechanism. During the process of initiation of XCI late replication appears earlier than hypoacetylation, suggesting that induction of a delay in replication may be the first step leading on to hypoacetylation, condensation, and inactivation. However, other possibilities remain open.

***See also:* Sex Chromosomes; X Chromosome**

Xenology

W Fitch

doi: 10.1006/rwgn.2001.1393

Xenology is the condition of having been, in its history, transferred, not from parent to offspring, but from one species to another (horizontal transfer). It does not include the transfer between organelles and the nucleus.

***See also:* Horizontal Transfer**

Xenopus laevis

R P Elinson

doi: 10.1006/rwgn.2001.1699

The South African clawed frog, *Xenopus laevis* (**Figure 1**), is a model organism for the analysis of vertebrate development. Its long generation time of 1.5 years precludes generation of mutants for genetic analysis. Nonetheless, the availability of large numbers of large eggs and the ease of manipulating experimentally the early embryos allow the identification and characterization of genes that are important for development.

While amphibians have long been used for embryological research, *X. laevis* has been the frog of choice since 1960. An important reason for this choice is that *X. laevis* is the easiest amphibian to maintain as a breeding colony. Unlike most other frogs, it remains aquatic as an adult, it tolerates dirty water, and it does not require live food. Females can be injected with

Figure 1 *Xenopus laevis.* (From: *Amphibians: Guidelines for Breeding, Care, and Management of Laboratory Animals* (1974) National Academy of Sciences, USA.)

hormones to induce egg-laying, every few months over many years, and tadpoles are easily raised to become reproductive adults.

Although it takes too long to do crosses to characterize genes by mutation, the combination of molecular and embryological techniques in *X. laevis* allows genes to be analyzed. The embryological advantages include the numbers of eggs, their large size, the speed of development, and the hardiness of the embryos. An adult female will spawn hundreds of eggs at a time, each 1.3 mm in diameter. By 3 days after fertilization, many organ systems have formed and the embryo begins to swim. It is simple to obtain large numbers of synchronously developing embryos and to dissect out particular parts of them. Similarly it is easy to inject molecules into the embryos and to perform microsurgery. Pieces of embryo can be transplanted to other embryos or joined together as recombinants in culture, requiring only simple salt solutions and antibiotics.

In place of mutagenesis to find genes of interest, those working with *X. laevis* use various criteria to select genes from cDNA libraries. Criteria may include: expression of the gene at one stage or in one tissue of the embryo and not in another, homology to a gene of interest in another model organism such as the fruit fly *Drosophila* or the nematode *Caenorhabditis elegans*, or an activity in the embryo itself. The activity of any cDNA is easily assayed by injecting its corresponding RNA into the dividing embryo and examining the effect on development. The simplicity of the RNA injection assay allows molecular screens for genes in place of mutant screens. For example, the dorsal axis is the defining feature of the vertebrates and consists of the central nervous system, the backbone, and the body musculature. Use of molecular approaches on *X. laevis* embryos identified genes important for the development of the dorsal axis. These include *Siamois*, *goosecoid*, *chordin*, *noggin*, and *Cerberus*, most of which play important roles in mammalian development.

The strengths of *X. laevis* as a model system can be seen by the variety of investigations that can be done. Signaling between two tissues in development can be detected by the simple microsurgical procedure of making a recombinant in culture between those two tissues. Basic life processes can be analyzed using preparations from *X. laevis* eggs. For example, extracts made from eggs or early embryos continue to exhibit features of cell division in the test tube, and molecules can be added to the extracts to test their role in the cell cycle. In addition, the *X. laevis* egg itself can serve as a tiny, 1-μl test tube to examine activities of molecules. For instance, RNAs for proteins that serve as ion channels can be injected into eggs, and the eggs will now use those channels. In this way, the chloride channel involved in cystic fibrosis was analyzed.

Finally, new genes can be introduced into *X. laevis* by mixing cDNAs with sperm nuclei. The sperm nuclei are injected into unfertilized eggs to start development. This transgenic procedure permits more extensive testing of genes in *X. laevis* development as well as the establishment of genetic lines that carry a gene of interest. *X. laevis* and its relative *X. tropicalis*, which has a simpler genome, provide valuable opportunities for the study of vertebrate development and cell biology.

***See also:* Developmental Genetics**

Xeroderma Pigmentosum

N G J Jaspers

doi: 10.1006/rwgn.2001.1394

Xeroderma pigmentosum (XP) is the classical human recessive disorder caused by defective nucleotide excision repair of DNA damage, including pyrimidine dimers induced by UV radiation. Sun-exposed skin of XP patients appears parchment-like and hyperpigmented and has an over 1000× increased risk of skin cancer. In the most severely affected cases there is progressive neuronal degeneration as well. Clinical management is largely restricted to stringent sunlight protection measures. Seven different genes are involved. Of these, *XPA*, *XPC*, and *XPE* are required for DNA lesion recognition; the XPB and XPD gene products are helicases mediating local strand unwinding and *XPF* and *XPG* specify structure-specific endonucleases performing strand incision on either side of the lesion. There exists an additional relatively mild 'variant' form of XP caused by defective DNA polymerase η (eta), a translesion polymerase that can replicate DNA templates containing UV damage.

***See also:* DNA Polymerase η (Eta); Excision Repair; Pyrimidine Dimers**

XIST

S M Gartler

doi: 10.1006/rwgn.2001.1395

All organisms with XX:XY systems of sex determination generally exhibit equal levels of X-linked gene expression in XX females and XY males. This phenomenon, known as dosage compensation, is achieved in mammals by inactivation of one X chromosome in female cells (Lyon, 1961). As a result, the two identical X chromosomes in the same cell are distinguishable from one another. Almost from the inception of the X-chromosome inactivation hypothesis, it was realized that this unequal expression of the Xs in the same cell could be best explained by a single initiation site on the X chromosome from which inactivation would spread. Studies of X:autosome translocations supported this idea in that only one X chromosome segment of a rearrangement could be inactivated, and studies of the X chromosome break points from different translocations permitted an approximate location of the initiation site. In addition, heritable variation in X chromosome susceptibility to inactivation permitted an approximate mapping of the initiation site, the results of which agreed with the results from the translocation studies. The initiation site was named the X inactivation center (XIC), but little progress in identifying the specific gene involved was made until the late 1980s.

Discovery

Realizing that the controlling gene might have the unique property of being expressed only from the inactive X chromosome, in 1989, Willard and his group began a systematic search for X-linked genes that escape inactivation concentrating on the region believed to contain the XIC. In 1991 they discovered a gene that was expressed from the inactive X alone and named it X inactive specific transcript (*XIST/Xist*;

Brown *et al.*, 1991). (Human genes are depicted in italicized capital letters, such as *XIST*, while mouse genes are depicted in italicized capital and lower-case letters, such as *Xist*. For simplicity, the all capitals version will be used here.) Both the human and mouse *XIST* transcripts have no significant open reading frame and they remain in the nucleus, with the processed transcript coating the inactive X. Prior to inactivation, *XIST* is expressed from both X chromosomes, but in an unstable form which does not coat either X chromosome. A critical point in the inactivation process is the stabilization of the *XIST* transcript from one X, leading to coating of that chromosome and its inactivation. At this stage, *XIST* transcription from the other X ceases.

Developmental Studies

If *XIST* is the gene that initiates X inactivation, its expression should precede the other features characteristic of X inactivation, such as widespread promoter methylation, hypoacetylation of histones, late replication, and histone macroH2A1.2 association. This timing was verified, as *XIST* expression is first detected at the four-cell stage in the mouse (Kay *et al.*, 1993) before any obvious signs of X inactivation, and it continues to be expressed through later developmental stages and into adult life. In the mouse, the early expression of *XIST* is imprinted, with the paternal allele being exclusively expressed in extraembryonic tissues and, accordingly, the paternal X chromosome is exclusively inactivated. Later, in the embryo proper, *XIST* expression and inactivation are random.

Requirement of *XIST* for X-Inactivation

Deletion of *XIST*, including the promoter, prevents that chromosome from becoming inactivated, which indicates that *XIST* is required in *cis* for inactivation to occur (Penny *et al.*, 1996). Transfection of *XIST* in multiple copies onto a murine autosome can bring about some of the characteristics of X-inactivation on that autosome (Lee *et al.*, 1996). These results along with the findings from developmental studies demonstrate that *XIST* is necessary and probably sufficient for initiating X-inactivation.

Determining the Active versus the Inactive X

X-inactivation occurs normally only in cells with more than one structurally normal X chromosome. In cells with multiple X chromosomes, but a normal number of autosomes, only one X remains active. In polyploid cells, however, more than one X can remain active. This means that the initiating factor must be involved in 'counting' the number of X chromosomes relative to the autosome content. Furthermore, the paternal X in the mouse is selectively inactivated in extraembryonic cells, and in the embryo proper, where inactivation is normally random, a choice must be made in each cell as to which chromosome will remain active or be inactivated. This is the 'choice' aspect of X-inactivation.

Some light has been shed on these questions with *XIST* knockout experiments in mice. As mentioned earlier, when the 5′ region of *XIST*, including the promoter, is deleted, no XIST RNA is produced and the chromosome cannot be inactivated. The normal X is inactivated in about half the cells and in the remaining cells both the deleted and normal Xs are active. The cells with both Xs active are not viable in the embryo. This outcome implies that in about half the cells, the deleted X was chosen for inactivation, but could not be inactivated because of the 5′ deletion. Most importantly, this result suggests that the 5′ end of *XIST* may not be of value in the 'counting and choice' functions of X-inactivation. In contrast to the 5′ deletion, a 3′ deletion of *XIST* insures that the deleted X will be inactivated; this is true even if there is only a single X in the cell (Clerc and Avner, 1998). This latter observation indicates that the 'counting' function can be destroyed without affecting inactivation.

In humans a mutation in the *XIST* promoter appears to alter the probability of inactivation of the chromosome carrying the mutation. If the interpretation is true, it would mean that in humans, in contrast to mice, the 5′ end of *XIST* has a role in counting and choice. Alternative interpretations of these data are possible and these will have to be tested before the human results can be considered to be contrary to the murine findings.

The recent discovery of a gene, *TSIX*, which is antisense to *XIST* leads to an explanation of the different functions of the 5′ and 3′ ends of *XIST* (Lee *et al.*, 1999). Like *XIST*, *TSIX* produces an untranslated nuclear RNA. It has its own promoter that is deleted in the 3′ *XIST* deletion referred to above, and it is transcribed at low levels from both Xs prior to inactivation. As the inactivation process begins, in an *in vitro* embryonic culture system, *TSIX* transcription is shut down on the chromosome to be inactivated, while on the other X, low-level *TSIX* and *XIST* expression are maintained. The next step appears to be an enhancement and stabilization of *XIST* expression on the X that does not express *TSIX*. *XIST* is then shut down on the other X (destined to be the active X). When the differentiation process leading to an active X and an inactive X in the same cell is complete, *TSIX* transcription is silenced on both Xs and *XIST* transcription and accumulation occurs only from the inactive X. This series of events implies that *TSIX* expression directly blocks *XIST* accumulation in *cis* and that silencing *TSIX* on one X is all that is necessary to produce an inactive X chromosome. The establishment of the active X chromosome requires that *XIST* be shut down prior to *TSIX* silencing so that *XIST* will not accumulate.

Control of *TSIX* and *XIST* Expression

Since the promoters of both genes are rich in CpG dinucleotides, promoter methylation appears to be a possible means for switching transcription on and off. The *XIST* promoter has been well studied, and in fully differentiated female cells the *XIST* promoter on the active X chromosome is hypermethylated, while the promoter on the inactive X is hypomethylated. In mature sperm, the *XIST* promoter is hypermethylated, while in the oocyte it is hypomethylated. Soon after fertilization, however, there is a general wave of demethylation that includes *XIST*, and current methylation studies of early mouse embryonic stages show no *XIST* methylation from the eight-cell stage through blastula formation. We expect no *XIST* promoter methylation on the inactive X as *XIST* is expressed on that chromosome. On the active X, however, *XIST* is silenced, and in fully differentiated *in vitro* cultured embryonic cells and *in vivo* somatic cells the *XIST* promoter on the active X chromosome is fully methylated. Apparently, the initial shutdown of *XIST* on the active X is not determined by methylation, although methylation of the promoter must occur soon after silencing. What the initial silencing factor for *XIST* is remains to be determined. As mentioned earlier, the key to inactivating an X is silencing *TSIX* and limited evidence suggests that promoter methylation of *TSIX* may be involved.

Aneuploidy and X-Inactivation

Another important question is: what are the details of the inactivation process in cells with a normal complement of autosomes but multiple X chromosomes? Most workers assume that one X is first randomly selected to be active and that the remaining Xs are inactivated by a default mechanism. The fact that a single X with the 3′ *XIST* deletion can be inactivated could be considered as support for this idea. That is, the 3′ region contains the site for 'marking' and/or 'counting' the active X and its deletion allows its inactivation. It may be that observations of polyploid cells will also have bearing on this 'marking' event. In a triploid with two Xs and a Y, both Xs remain active, but in a triploid with three Xs, either one or two Xs may be inactivated. That inactivation does not occur in a triploid cell with two Xs and a Y suggests that the 'counting' mechanism must involve not only the number of X chromosomes, but an autosomal:X chromosome ratio, as in *Drosophila* and *Caenorhabditis*. An excess of autosomes relative to X chromosomes would act to prevent inactivation as in normal males and in triploids with two Xs and a Y, while an equal or lower autosomal:X chromosome ratio, as in normal females or triploids with three Xs, results in inactivation. The absence of inactivation in the triploid XXY cell also argues against inactivation being initiated by pairing or interaction between X chromosomes.

The primary experimental system used to obtain much of the developmental information on *XIST* is the murine embryonic cell (ES) culture system. XX ES cells in an undifferentiated state represent a pre-X-inactivation embryonic state; upon induction of differentiation one X becomes inactivated, mimicking the process *in vivo*. With respect to the comments just made on ploidy and X-inactivation, it would be very valuable if ES cultures with more than two X chromosomes and various levels of polyploidy were available so that the questions just raised might be better explored.

How does *XIST* Work?

At the outset it must be stated that we have no definitive answer to this most important question. There are numerous observations and interpretations, however, that can help develop a working model.

As noted, XIST RNA binds to the inactive X. If the chromatin is digested away, XIST RNA remains bound in the nucleus, suggesting that it is not directly in contact with chromosomal DNA, but rather bound to protein components of the nuclear matrix, some of which must be the chromosomal scaffold. The amount of XIST RNA in the nucleus has been calculated to be insufficient to cover all the coding regions of the inactive X, which implies that XIST RNA does not interact with individual genes. Careful cytogenetic studies have shown that XIST RNA exhibits a banded structure on the murine inactive X chromosome, with the RNA being excluded from regions of constitutive heterochromatin.

Silencing of the X is a complex phenomenon including XIST RNA, promoter methylation, hypoacetylation, late replication, histone macroH2A1.2 association, and undoubtedly other as yet undiscovered factors. The interrelationships and possible interactions of these various silencing factors have yet to be fully worked out, but it is already clear that, at least, some of these factors have a considerable degree of independence. For example, the *XIST* gene can be deleted from the inactive X of a somatic cell, and inactivation at individual loci is still maintained. Inactivation is not as stable when XIST RNA is not properly localized, however, as is the case in the transformed cells from which *XIST* was deleted. In such cases it is likely that promoter hypermethylation would be sufficient for silencing. It is also possible that in the presence of *XIST*, but with promoter hypomethylation and advanced replication, that a particular gene on the inactive X can escape inactivation.

XIST RNA has been shown to colocalize with the histone variant, macroH2A1.2, very early in development (Costanzi *et al.,* 2000). It is likely that direct or indirect interactions with other proteins will be found in the near future. One possible mode of action of XIST RNA is that interactions with various proteins could bring about facultative heterochromatinization of the inactive X, thus acting as an initial silencing agent. This could also mark the chromosome for further modifications by additional silencing factors, such as methylation and histone hypoacetylation. Since neither XIST RNA nor macroH2A1.2 are critical for maintaining inactivation, it is also possible that XIST is not directly involved in silencing, but acts as an early developmental mark for compartmentalizing the inactive X and permitting modification by silencing agents such as methylation, hypoacetylation, and late replication.

Spreading of the X-Inactivation Signal

Another important aspect of X-inactivation is how the inactivation signal is spread from the XIC throughout the X. With the discovery of *XIST* and its role in initiation of X-inactivation, the question of spreading is one of how XIST RNA is spread in *cis* along the inactive X. As discussed above, XIST RNA is likely to interact with the X chromosome via protein intermediates. In fact, macroH2A1.2, mentioned above, may be such a protein. Since XIST RNA covers a large part of the inactive X and may, under certain conditions, bind to autosomal segments, it is unlikely that the XIST RNA protein complex binds to a DNA sequence unique to the X chromosome. Lyon (2000) suggested that such an XIST-protein complex might bind to a long interspersed repeat element, LINE-1(L1), which are more frequent on the X chromosome, especially the younger L1 elements, than on autosomes. L1 elements are retrotransposons unique to mammals (Burton *et al.,* 1986). This hypothesis remains to be confirmed.

XIST Expression in Male Meiosis

XIST RNA has also been detected in male meiosis, at a time when the X is going through precocious condensation. Eventually all the sperm chromosomes become highly condensed. It was natural to think that *XIST* expression plays a role in this condensation, but the level of *XIST* expression in spermatogenesis is extremely low and could cover only a small fraction of the X chromosome. It would be interesting to know if *TSIX* is also expressed at this stage, thereby preventing any accumulation of XIST RNA.

Future Questions

Although the discovery of *XIST* has led to rapid progress in our understanding of the X-inactivation process, significant questions remain. What does XIST RNA actually do? What is the nature of the autosomal signal in the initiation process? What is the nature of the mechanism for the switching on and off of *TSIX* and *XIST* on the active and inactive X chromosomes.

Finally, we can consider the question of whether X-inactivation resulting from stable *XIST* expression can be initiated at more than one time in development. X-inactivation is initiated at an early and developmentally specific stage. Its occurrence appears to depend on differentiation, because the absence of dosage compensation is tolerated in early undifferentiated states. Can inactivation be induced at a later developmental stage in cells that have more than one active X chromosome, such as in a male tumor or in a somatic cell culture system? *XIST* gene expression has been reactivated on the active X in transformed somatic cells, but without any detectable silencing effect on the rest of the chromosome. This may mean that the complete inactivation process is restricted to a specific early developmental stage. In contrast, there is evidence of sex chromatin (Atkin and Baker, 1992) and *XIST* expression (Looijenga *et al.,* 1997) in male germ cell tumors with more than one X, although demonstration of inactivation of genes on the 'inactivated X' has not been reported. It is unlikely that these germ cell tumors would have been present at the time of normal embryonic inactivation; if this work is correct, therefore, it would mean that the inactivation process is not restricted in developmental time and place.

References

Atkin NB and Baker MC (1992) X-chromatin, sex chromosomes, and ploidy in 37 germ cell tumors of the testes. *Cancer Genetics and Cytogenetics* 59: 54–56.

Brown CJ, Ballabio A, Rupert JL *et al.* (1991) A gene from the region of the human X inactivation centre is expressed exclusively from the inactive X chromosome. *Nature* 349: 38–44.

Burton FH, Loeb DD, Voliva CF *et al.* (1986) Conservation throughout mammals and extensive protein-encoding capacity of the highly repeated DNA long interspersed repeat sequence one. *Journal of Molecular Biology* 187: 291–304.

Clerc P and Avner P (1998) Role of the region 3′ to Xist exon 6 in the counting process of X-chromosome inactivation. *Nature Genetics* 19: 249–253.

Costanzi C, Stein P, Worrad DMS *et al.* (2000) Histone macroH2A1 is concentrated on the inactive X chromosome of female preimplantation mouse embryos. *Development* 127: 2283–2289.

Kay GF, Penny GD, Patel D *et al.* (1993) Expression of *Xist* during mouse development suggests a role in the initiation of X chromosome inactivation. *Cell* 72: 171–182.

Lee JT, Strauss WM, Dausman JA and Jaenisch R (1996). A 450 kb transgene displays properties of the mammalian X-inactivation center. *Cell* 86: 83–94.

Lee JT, Davidow LS and Warshawsky D (1999) *Tsix*, a gene antisence to *Xist* at the X-inactivation centre. *Nature Genetics* 21: 400–404.

Looijenga LH, Gillis AJ, van Gurp RJ, Verkerk AJ and Oosterhuis JW (1997) X inactivation in human testicular tumors: XIST expression and androgen receptor methylation status. *American Journal of Pathology* 151: 581–590.

Lyon MF (1961) Gene action in the X-chromosome of the mouse (*Mus musculus* L). *Nature* 190: 372–373.

Lyon MF (2000) Line-1 elements and X chromosome inactivation: a function for 'junk' DNA? *Proceedings of the National Academy of Sciences, USA* 97: 6248–6249.

Penny GD, Kay GF, Sheardown SA, Rastan S and Brockdorff N (1996) Requirement for Xist in X chromosome inactivation. *Nature* 379: 131–137.

***See also:* LINE; X-Chromosome Inactivation**

X-Ray Crystallography

T M Picknett and S Brenner

doi: 10.1006/rwgn.2001.1396

X-ray crystallography is a technique for determining the three-dimensional structure of molecules, including complex biological macromolecules such as proteins and nucleic acids. It is a powerful tool in the elucidation of the three-dimensional structure of a molecule at atomic resolution. Data is collected by diffracting X-rays from a single crystal, which has an ordered, regularly repeating arrangement of atoms. Based on the diffraction pattern obtained from X-ray scattering off the periodic assembly of molecules or atoms in the crystal, the electron density can be reconstructed.

The use of X-ray diffraction patterns in the study of molecular structure dates to the early part of the twentieth century. In 1901, Röntgen received the first Nobel Prize for Physics for the discovery of X-rays, and by 1912 the discovery of X-ray diffraction in crystals by Von Laue, Friedrich, and Knipping, and the research of Bragg and Bragg on the structure of crystals (for which the 1915 Nobel Prize for Physics was awarded) laid the foundation for the field of X-ray crystallography. In 1953, X-ray diffraction patterns from ordered fibers of DNA were instrumental in Watson and Crick's discovery of the double-helical structure of DNA.

***See also:* DNA, History of**

Y Chromosome (Human)

N A Affara

doi: 10.1006/rwgn.2001.1400

The Y chromosome has been studied extensively in humans and mice where in both species it is required for testis determination and spermatogenesis. In humans the chromosome also contributes to normal somatic development. In this brief article, the organization and gene content of the human Y chromosome will be considered with particular reference to the unique features that have shaped its evolution. The mammalian sex chromosomes evolved from an ancestral pair of autosomes. The dominant male-determining gene arose on the proto-Y and became genetically sequestered from the rest of the genome through the suppression of recombination over most of the length of the proto-Y with the proto-X. This suppression of recombination has led to the rapid degeneration of genes on the Y chromosome and, consequently, a genetic imbalance in gene dosage between males and females for homologous genes carried on the progenitor sex chromosomes and retained as functional loci on the modern X chromosome. The evolution of dosage compensation has restored equivalence of gene expression between males and females for genes on the X chromosome and, in most mammals, this is achieved by random X-inactivation in the female.

Thus, the Y chromosome differs from other mammalian chromosomes in two fundamental ways. First, as mentioned above, it is the only chromosome that does not recombine along the majority of its length and, second, it is present only in one sex, the male. The evolution of the genetic functions, and therefore the gene content, of the Y chromosome are believed to be a reflection of these basic properties. Two principal theories for the evolution of genes on the Y chromosome have been formulated each based on one of its two distinctive properties. First, it is theorized that the absence of recombination prevents the segregation of deleterious mutations from advantageous mutations, thus leading to an inevitable deterioration in the genetic content of the nonrecombining regions of the Y chromosome (NRY). The lack of Y-linked genetic functions, which originally inspired this theory over 80 years ago, is still the best known feature of the Y chromosome, after sex-determination. Second, the male-specific nature of the Y chromosome may promote the accumulation of male-enhancing/female-damaging alleles, sexually antagonistic (SA) alleles, leading to the Y chromosome becoming a specialized male chromosome; for example, genes involved in spermatogenesis or the promotion of fertilization success of the male. These two theories are not mutually exclusive but are contradictory, in the sense that the first proposes a loss of genetic information while the second proposes a gain of genetic information.

Sequence Content of the Human Y Chromosome

Despite the morphological and sequence divergence of the Y chromosome from its ancestral homolog the X, all mammals retain a small region of strict X–Y homology to permit pairing and correct segregation of the sex chromosomes during male gametogenesis. This segment is known as the pseudoautosomal region (PAR) and pairing region and in humans there are two; a major one (PAR1; 2.6 Mb) found at the extremities of the X and Y short arms and a lesser region (PAR2; 300–400 Kb) found at the extremities of the X and Y long arms (see **Figure 1**). In humans, there is an obligatory crossover in PAR1 at each male meiosis indicating that recombination between the sex chromosomes is unrestricted in this region. Recombination

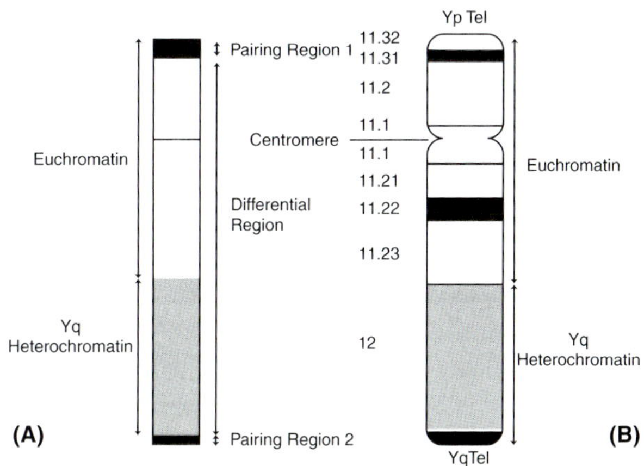

Figure 1 (A) Bipartite model of the Y chromosome. (B) Standard idiogram of the Y chromosome.

also occurs in the PAR2, but at a much lower frequency. The PAR regions show considerable variation in gene content between different groups of species. The evolution of the sex chromosomes and the establishment of distinct PARs is believed to have occurred both by the acquisition of regions from autosomes onto the X with subsequent recombination onto the Y and the loss of material from the male-determining Y chromosome through rearrangements and degeneration: the addition–attrition model of sex chromosome evolution.

Within the nonrecombining portions of the sex chromosomes several other blocks of homology and X–Y homologous genes have been described. By studying the patterns of homology in different species, the events leading to discrete blocks of sequence conservation can be reconstructed. Consequently, regions of X–Y homology defined on the modern human sex chromosomes represent either the ancient remnants of the ancestral pair of autosomes, or reflect exchanges of material with the X chromosome that have occurred more recently during evolution. These NRY X–Y homologous regions have arisen through transfers mediated by PAR recombination and subsequent rearrangements on the Y (e.g., inversions) and by direct duplication and transposition from the X into the NRY. There is evidence to suggest that sequences have also been recruited directly to the Y chromosome from autosomes through duplicative transpositions; for example the *DAZ* genes. The result of these different mechanisms of sequence recruitment to the Y has been to create a chromosome where X–Y homologies, Y-autosome homologies, and Y-specific sequences are interspersed within the euchromatic portion of the chromosome.

Whilst most genes on the X chromosome are subject to inactivation on one of the X chromosomes in each female cell, this does not apply to genes found to be homologous between the X and Y chromosomes where both X and Y copies are functional. These genes escape X-inactivation and are, therefore, expressed in diploid dose in both males and females. This applies to almost all genes within the PARs and to X–Y homologous genes located in the NRY. Deficits of genes in this latter category lead to the features of Turner syndrome (XO females, and females and males with partial deletions of the X and Y), suggesting that they are required in diploid dose in both males and females. Other subtle aspects of somatic phenotype may also be influenced by such genes.

The third prominent sequence feature of the Y chromosome is the block of Y long arm heterochromatin that accounts for almost half of the 50 Mb of the human Y. This is composed of two tandem repeat sequences and contains virtually no single-copy DNA sequences. Amongst individual males, there is extreme polymorphism in the size of the Yq heterochromatin, ranging from its complete absence to the occupation of almost half of the Y chromosome DNA content. As no clinical consequences are associated with the absence of Yq heterochromatin, it is believed that no critical genes reside in this part of the chromosome.

Amplification of Sequences on the Y Chromosome

Sequences related to the highly amplified Yq heterochromatin repeats are found in other regions of the genome but at much lower copy number. This indicates that these sequences have become amplified once placed onto the Y chromosome. This amplification of sequences on the Y chromosome is a distinguishing feature of DNA sequences on this chromosome. Much of the sequence content in the euchromatic NRY consists of amplified sequence and gene families. It would appear that the absence of recombination may remove restraints upon copy number and, where this does not compromise chromosome function, amplification is tolerated. It may also be that there is selection for the amplification of genes coding for particular functions; for example, the *DAZ* and *RBM* genes believed to be important for successful spermatogenesis. There may be several reasons for this selection for amplification. First, there may be selection for multiple copies of a gene because this results in increased amounts of suboptimal gene product arising from accumulated deleterious mutations. Second, it may be that subtle variants of the same amplified gene have been selected that act synergistically. Third, it may be that the inability to reconstruct defective genes through recombination drives gene amplification resulting in reduced selection against any member of that gene family that suffers an alteration. One or a combination of these possibilities may lie behind the emergence of amplified sequence and gene families on the Y chromosome.

Model of the Y Chromosome

A bipartite model of the human Y chromosome emerges from the above observations. The key features are as follows:

1. The Y has two pairing regions that are strictly homologous with the X where there is recombination, and hence genetic exchange between the two sex chromosomes.
2. The Y has a nonrecombining region (NRY) composed of the Yq heterochromatin and a euchromatic

segment containing an interspersed arrangement of Y-specific, X–Y homologous and Y-autosome homologous sequences.

3. The Y is populated by amplified sequence and gene families that are likely to have arisen from the nonrecombining status of the majority of the chromosome.

This model is summarized in **Figure 1A**.

The analysis of genetic functions encoded by the Y chromosome and its gene content has been facilitated by, first, extensive deletion mapping and, second, cloning of the entire euchromatin (some 30 Mb of DNA) in a series of overlapping yeast, P1, and bacterial artificial chromosomes. Deletion mapping has exploited structural abnormalities of the Y chromosome and a wide range of single-copy cloned DNA and STS (sequence-tagged site) markers to score the presence or absence of Y chromosome regions in different individuals. This has allowed the correlation of a series of deletion intervals with the phenotypes possessed by individuals carrying any particular Y chromosome abnormality. By mapping cloned markers and STSs back onto the physical clone maps, it has been possible to determine which set of clones cover each of the defined deletion intervals associated with a genetic function defined by a phenotype. These clones provide the basis for determining the gene content of deletion intervals through a variety of molecular analyses. **Figure 2** summarizes the deletion map, the location of genetic functions assigned to the Y, the blocks of homology with the X, and the genes and pseudogenes that have been mapped to the chromosome.

Genes and Phenotypes Mapped to the Y Chromosome

The following is a brief summary of the genes and phenotypes that have been assigned to the human Y chromosome. These will be listed starting at the Yp telomere.

PAR1

The only definitive phenotype that has been assigned to the PAR1 is short stature associated with Turner syndrome. Several genes have been assigned to this region. These are:

1. *SHOX* (a homeodomain-containing gene) where mutations have been shown to be present in individuals with idiopathic short stature, suggesting its involvement in Turner syndrome.
2. *CSF2RA* encodes the α chain of the GM-CFS (granulocyte–macrophage colony-stimulating factor) receptor heterodimer and its frequent deletion in the M2 subtype of acute myeloid leukemia suggests its involvement in this malignancy.
3. *IL3RA* encodes the α subunit of the receptor for the cytokine IL-3 that promotes growth of hemapoietic cells. This receptor shares the β subunit with the GM-CFS and IL-5 receptors.
4. *ANT3* encodes an ADP/ATP translocase and is believed to be involved in energy metabolism.
5. *ASMT* encodes the enzyme acetylserotonin methyltransferase and catalyses the final step in the synthesis of serotonin in the retina and pineal gland. There have been suggestions that this gene may be involved in affective disorders.
6. *XE7* encodes a ubiquitously expressed protein of unknown function.
7. *MIC2* encodes a surface antigen expressed on all cells except spermatozoa. It is adjacent to a related pseudogene, *MIC2R*, that has probably arisen by gene duplication.
8. On the Y, *XG* represents a nonfunctional copy of its X homolog. Only the first three exons are present in the Y-linked sequence. The gene on the X encodes the red blood cell antigen XG.

Yp NRY Euchromatin

A number of genes have been mapped to the nonrecombining region of the Y chromosome short arm. The majority of these are X–Y homologous genes but do not participate in genetic exchange with their X counterparts. Three major genetic functions have been assigned to Yp by deletion mapping; sex determination in distal Yp close to the boundary with the PAR1, the locus (or loci) involved in the lymphoedemic anomalies of Turner syndrome to the human-specific block of X–Y homology between Yp11.2 and Xq21.3 and the locus causing gonadoblastoma (*GBY*) to proximal Yp. The genes on Yp are:

1. *SRY* encodes a transcription factor belonging to the HMG gene family. It has been shown that its function is determination of male gonadal development (*TDF*).
2. *RPS4Y* encodes a ribosomal protein and it has been suggested that haploinsufficiency of this protein may contribute to aspects of the Turner phenotype. This remains controversial.
3. *ZFY* encodes a zinc finger protein with potential to function as a transcription factor. The function of this gene remains unknown.
4. *PCDHY* encodes a protocadherin gene that resembles cadherin neuronal receptors. The gene is expressed in the brain and may be involved in forming neuronal networks.

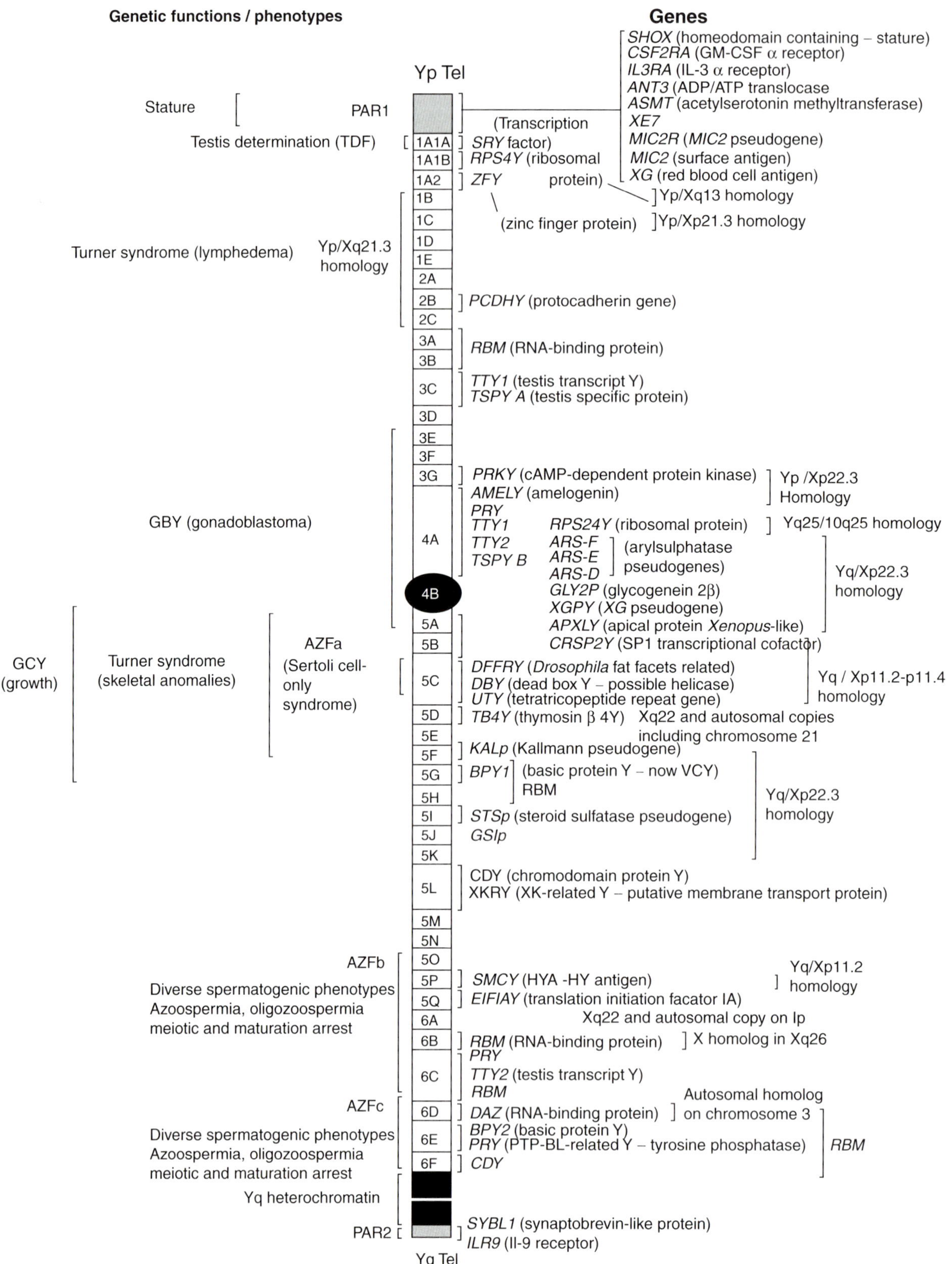

Figure 2 The human Y chromosome.

5. *RBM* encodes an RNA-binding protein and is part of an amplified gene family of 30–50 genes on the Y. Copies on Yq are involved in germ cell differentiation.
6. *TTY1* and *TTY2* encode testis-specific transcripts with no evident open reading frames. More than one locus exists on the Y for both transcripts.
7. *TSPY* encodes a testis-specific protein with some homology to the *SET* oncogene, a nuclear phosphoprotein. It is believed to be involved in germ cell differentiation and it has been suggested that it may play a role in the development of germ cell tumors; a possible candidate for *GBY*. The gene is part of a tandemly amplified gene family with two clusters (*TSPYA* and *TSPYB*) on Yp (see **Figure 2**).
8. *PRKY* encodes a protein kinase related to the cAMP-dependent kinases. Its function is unknown.
9. *AMELY* encodes amelogenin, a constituent of tooth enamel and may contribute to tooth size in males. Both *PRKY* and *AMELY* map into a second block of homology on Yp with the Xp 22.3 region of the X chromosome.
10. *PRY* (*PTP-BL* related) encodes a tyrosine phosphatase and is also present at more than one locus on the Y.

Yq NRY Euchromatin

Several genes have been mapped to the long arm of the human Y chromosome. Two features are evident when the gene content of Yq is considered. First, it is noticeable that at least three loci associated with germ cell development and male infertility have been mapped to Yq. This supports the idea that genes controlling spermatogenesis will accumulate on the Y. Second, there is an accumulation of pseudogenes that are homologous to functional genes mapping to the X chromosome. Two further pseudogenes, *ASSP6* of the argininosuccinate synthetase gene family and *ACTP2* of the actin gene family, have also been assigned to Yq11. The long arm also contains several copies of sequences homologous to retroviruses. Deletion analysis has assigned phenotypes for male infertility (*AZFa*, *AZFb*, and *AZFc*), Turner syndrome skeletal anomalies, and growth (*GCY*; including tooth size) to Yq. The following have been mapped to the Yq euchromatin:

1. A series of nonfunctional pseudogenes with homology to functional homologs mapping to Xp. These are: ribosomal protein *RPS24Y*, arylsulfatases *ARSFY*, *ARSEY*, *ARSDY*, glycogenein 2 *GLY2P*, *XG* pseudogene *XGPY*, apical protein *Xenopus*-like *APXLY*, and an SP1 transcription cofactor *CRSP2Y* (formerly known as *EXLM1Y*).
2. *DFFRY* (also known as *USP9Y*) encodes a ubiquitin-specific protease and is homologous to the *Drosophila* developmental gene, *faf*.
3. *DBY* encodes a potential RNA helicase that may be involved in mRNA translation. Both *DFFRY* and *DBY* are removed by deletions resulting in the *AZFa* male infertility phenotype primarily characterized by Sertoli cell-only syndrome. One or both of these are, therefore, likely to underpin the *AZFa* phenotype.
4. *UTY* encodes a tetratricopeptide repeat gene that may have a role in transcriptional repression.
5. *TB4Y* is homologous to thymosin β and has homologs on Xq22 and various autosomes.
6. *KALp* encodes a nonfunctional copy of the X-linked *KAL* gene that is responsible for Kallmann syndrome (anosmia and hypogonadism). The sequence of this gene resembles that of cell adhesion molecules and is believed to have a role in neuronal cell migration.
7. *BPY1* (basic protein on the Y) encodes a basic protein of unknown function. The gene is now known as *VCY* (variable charge protein on the Y) and has been shown to have homologous genes in Xp22.3. There are two Y copies. *BPY2* is unrelated but potentially encodes a different basic protein.
8. *STSp* is a nonfunctional copy of the steroid sulfatase gene located in Xp22.3. Deletion of the X gene leads to the skin condition X-linked ichthyosis. Closely linked to this is the pseudogene *GS1p*, homologous to an X-linked gene in Xp22.3 of unknown function.
9. *CDY* encodes a protein containing a chromodomain and may be involved in remodeling chromatin during the maturation stages of spermatogenesis. There are at least two loci for this gene on the Y.
10. *XKRY* encodes a protein related to XK, a putative membrane transport protein.
11. *SMCY* encodes the male-specific HY antigen (*HYA*) expressed on the surface of male cells. This gene has a homolog mapping to Xp11.2.
12. *EIF1AY* encodes a translation initiation and elongation factor with homologs in Xq22 and chromosome 1p.
13. *DAZ* encodes another RNA-binding protein and has been suggested as a candidate for the AZFc infertility phenotypes. Although *DAZ* and *RBM* are strong candidates for the AZFc and AZFb phenotypes, respectively, it can be seen that both intervals contain loci for a number of other genes.

It should, therefore, be kept in mind that combinations of these genes may underpin the spermatogenic phenotypes associated with these regions.

Par2

The PAR2 is of recent evolutionary origin having appeared after the divergence of chimpanzees and hominids. Two genes have been mapped to this minor pairing region at the Yq telomere. The first, SYBL1, encodes a synaptobrevin-like protein that may have a function is synaptic vesicle docking. Unlike other potentially functional X–Y homologous genes, the Y copy is specifically inactivated. The second, ILR9, encodes the interleukin 9 receptor.

Conclusion

The sex chromosomes evolved from a common ancestral pair of homologs as a chromosomal basis for sex determination emerged. The suppression of recombination outside the regions pairing with the X chromosome has created unique conditions on the differential portion of the Y, leading to the rapid degeneration of its sequence and gene content. This genetic isolation has driven Y chromosome evolution resulting in an accumulation of repeated sequence and gene families, defective pseudogenes, and functional genes shared with the X and autosomes. There is a continuous process of addition and attrition of sequences and genes on the Y chromosome, creating a rapidly changing genetic content. The absence of recombination on and presence only in the male of the chromosome have produced evolutionary pressures predicted to lead to the accumulation of male-specific functions and dimorphisms on the Y chromosome. Many of the genes and genotypes assigned to the Y support this prediction as they have been shown to be associated with male-specific traits such as male sexual development, greater stature, and spermatogenesis. It is expected that the concept of sexual selection (the selection for genes on the Y that confer an advantage in male competition for fertilization success) may account for the presence of other functional genes on the Y chromosome once their biological function(s) have been clarified.

See also: **Sex Determination, Human; X Chromosome**

Y Linkage

See: **Linkage; Y Chromosome (Human)**

YAC (Yeast Artificial Chromosome)

E J Louis

doi: 10.1006/rwgn.2001.1398

Yeast artificial chromosomes (YACs) were originally constructed in order to study chromosome behavior in mitosis and meiosis without the complications of manipulating and destabilizing native chromosomes. This allows for the alteration of structures on a nonessential chromosome to study their effects. There are three essential components for chromosome maintenance and stability (**Figure 1**): a functioning centromere (CEN); origin of replication (autonomous replication sequence, ARS); and telomeres (TEL) at the ends. A technical barrier to building YACs was creating telomeres. CENs and ARSs could be cloned on small plasmids in *Escherichia coli* and their function tested by shuttling into yeast. As linear DNA molecules cannot be maintained in *E. coli*, the TEL component was constructed as inverted repeats of telomere sequence that could resolve into functioning telomeres when moved into yeast. Tetrahymena telomere repeats function as telomeres in yeast and are used in most YAC constructs.

YACs that contain the three essential components and are of sufficient size behave as normal chromosomes. They replicate and segregate properly during mitosis and meiosis and are affected similarly by mutations that alter native chromosome behavior. Using YACs we now know that centromere function is severely impaired on very short YACs (less than 20 kb), which may be due to antagonism between

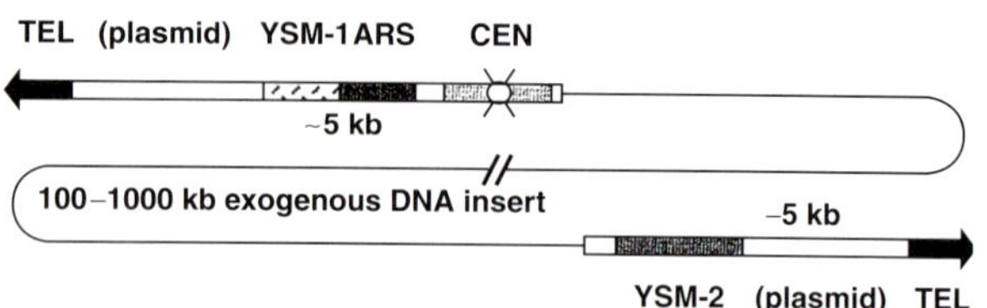

Figure 1 Basic structure of a yeast artificial chromosome. TEL, *Tetrahymena* telomere-derived sequences; (plasmid) sequences derived from bacterial cloning vector such as pBR322; YSM-1 and YSM-2, yeast genes for selecting yeast host transformants, generally prototrophic markers; ARS, yeast autonomously replicating sequence; CEN, yeast centromere DNA.

cetromere and telomere functions. Using markers along a pair of homologous YACs it has been shown that crossovers near the telomeres are not sufficient to guarantee proper segregation in meiosis, while those that are more internal are sufficient. YACs were also used to demonstrate distributive segregation, an alternative segregation mechanism originally described in *Drosophila*. This occurs in the absence of a homolog or absence of crossovers. YACs continue to be used to study the function of chromosome components, and as an assay for segregation problems in different mutant backgrounds and treatments.

It was quickly recognized that YACs could provide a new cloning vehicle for very large contiguous fragments. These could be much bigger than those cloned using the conventional cloning vectors of the time. Technically this is more difficult than other cloning methods as it requires ligation of three molecules and uptake of very large fragments by yeast. Despite this, several large YAC libraries of various genomes, including humans, have been constructed. These have proven very useful for physical characterization, as well as positional cloning of genes of interest. The ability of telomeres from other organisms to function as telomeres in yeast has led to the cloning of large terminal fragments of chromosomes as 'half-YACs.'

The advantages of using YACs go beyond the ability to clone large fragments. The high levels of homologous recombination in yeast and the developments in the ability to alter any sequence in any fashion allow specific mutations to be made in the sequences in the YAC without having to resort to subcloning. In many cases these altered YACs can be directly moved back into the organism or cell type of origin for assaying phenotypes. Recombination between YACs that partially overlap can be used to generate new YACs of larger size, allowing for the building up of contiguous fragments larger than the original library inserts. This is particularly useful when the genomic organization of a gene or region of interest covers a larger than insert size. YACs in excess of 2 megabases have been constructed in this fashion.

The disadvantages of YACs are threefold. The first disadvantage is the presence of a significant number of chimeric YACs that are due to coligation of fragments from different parts of the genome being cloned. A second disadvantage is the instability of certain sequences in yeast such as the alphoid repeats from human centromeres. Finally, YACs, as originally constructed, are not amenable to easy separation from yeast genomic DNA for further analysis and manipulation. For these reasons, the use of YACs for construction of a library in genome projects is being replaced by alternatives such as BACs (bacterial artificial chromosomes). There is a recent resurgence in the use of YACs due to a technique that combines the advantages of yeast recombination with the ease of manipulation of BACs. Transformation-associated recombination can be used to target specific regions of genomic DNA for cloning as a YAC. The YAC vector can incorporate bacterial sequences that allow a large circular YAC to be shuttled into *E. coli* as a BAC. This removes two of the problems associated with YACs and may solve the stability problem by utilizing an alternative host.

***See also:* BAC (Bacterial Artificial Chromosome); *Saccharomyces* Chromosomes**

Yanofsky, Charles

D R Helinski

doi: 10.1006/rwgn.2001.1399

Charles Yanofsky was born in New York City on 17 April 1925. He received his BS degree from City College of New York and PhD degree from Yale University under the guidance of David Bonner. His academic career began with his appointment as an Assistant Professor at Western Reserve University Medical School. He moved to Stanford University in 1958 where he is currently Herzstein Professor of Biological Sciences. His early pioneering studies with *Neurospora crassa* involved suppressor mutations that restored the ability of this organism to form an active enzyme in a mutant that previously produced an inactive protein. Considerably later and working with the A protein subunit of tryptophan synthetase in *Escherichia coli*, he established that suppression causes mistakes in amino acid incorporation with the net result that a specific amino acid in the mutant protein that is responsible for loss of protein activity is replaced by an amino acid that restores enzymatic activity. It was later shown that these suppressor mutant strains contained altered tRNAs. The switch from *Neurospora* to *E. coli* allowed Charles Yanofsky to carry out fine-structure genetic analysis of a large number of isolated tryptophan-requiring A protein mutants of *E. coli* and to isolate sufficient quantities of the corresponding mutant proteins for sequence analysis. By aligning the order of mutations on the genetic map with the order of positions of the amino acid changes in the corresponding A protein mutants, direct evidence was provided that gene structure and protein structure are colinear in bacteria. The further analysis of different amino acid changes at the same codon position in the A protein provided *in vivo* verification of the genetic code deduced from *in vitro* studies. The genetic and biochemical analysis of the tryptophan synthetase system in *E. coli* subsequently led to the discovery of attenuation as a major regulatory mechanism that controls the level of transcription of the tryptophan operon in response to the cellular level of tryptophanyl-tRNA. These fundamental contributions of Charles Yanofsky have been key to the rapid development of molecular genetics and our basic understanding of mechanisms controlling the flow of information from gene to protein.

See *also*: Tryptophan Operon

Yeast Plasmids

S L Forsburg

doi: 10.1006/rwgn.2001.1402

Yeasts are simple, single-celled eukaryotes that provide outstanding model systems for understanding basic cell biology. The ability to manipulate yeast cells in the laboratory depends upon the ability to transform them with engineered plasmids and to maintain these plasmids within the cell. Naturally occurring yeast plasmids provided the original template used to design laboratory vectors. These have been further developed so that modern laboratory plasmids provide a variety of sophisticated features. These plasmids can be used as research tools to study yeast biology, or as practical tools to manipulate yeast cells.

Naturally Occurring Plasmids

Yeast cells have naturally occurring double-stranded circular DNA plasmids. In order to persist in a population of growing cells, these plasmids must replicate, and they must segregate to both daughter cells during cell division. In the well-studied budding yeast *Saccharomyces cerevisiae*, a naturally occurring plasmid called 2-micron is present at up to 100 copies per cell. This plasmid provides a useful example of the strategies required to replicate and transmit an extrachromosomal element. The 2-micron plasmid provides no apparent selective advantage or disadvantage to the cell that harbors it. It is maintained in the nucleus and its DNA is packaged in the same way as normal chromatin. It contains four genes, a unique origin of DNA replication required for its replication once per cell cycle, and a partitioning system, still not completely understood, that ensures its transmission to both cells during cell division.

How can a plasmid that is replicated only once per cell cycle achieve such high copy number? The structure of the 2-micron plasmid provides a clue. It contains two tracts of repeated sequence that separate the molecule into two halves. If these homologous regions are aligned with one another and recombination occurs between them, the net effect is to flip the orientation of one half of the plasmid relative to the other. If a bidirectional replication fork is proceeding around the plasmid at the time of recombination, with the two halves separated by the homologous recombination region, then this event effectively reverses the orientation of one of the forks relative to the other. That is, the forks follow one another, rather than converge. This rearrangement allows amplification by additional replication of the plasmid. Another recombination event restores the original orientation, and when the forks finally converge, replication ceases. The enzyme responsible for these rearrangements is encoded in the plasmid genome. Similar plasmids have been isolated from a number of yeast species, and they all appear to employ the recombination method for amplification. The existence of these natural plasmids provides yeast geneticists with useful molecular tools.

Engineered Plasmids

In order for yeast plasmids to be useful in the laboratory, they require several features. First, there must be a means of preparing large quantities of pure plasmid DNA. For this purpose, recombinant yeast plasmids are built as yeast/*Escherichia coli* shuttle vectors that contain a bacterial origin of replication and a bacterial selective marker, such as the β-lactamase gene that confers ampicillin resistance. With these components, large amounts of the recombinant plasmid can be manipulated in and purified from *E. coli* cultures. The second requirement is that the plasmid be maintained in the yeast cell. Thus, it requires a yeast origin of replication. For *S. cerevisiae* plasmids, the 2-micron origin isolated from the naturally occurring plasmid is commonly used. However, chromosomal replication origins can also be employed. In fact, the ability of a fragment of chromosomal DNA to support plasmid maintenance as an autonomously replicating sequence (ARS) is one of the definitions of a chromosomal replication origin in yeast. Because the ARS elements from the *S. cerevisiae* chromosome are compact, on the order of 100 base pairs of DNA, they are easily added to a plasmid. The centromeres from *S. cerevisiae* are also sufficiently compact to be encompassed on a plasmid, again on the order of a few hundred base pairs, so that an ARS ± CEN-containing plasmid can be maintained and transmitted through mitosis and meiosis as a circular minichromosome.

Not all yeasts provide such handy cellular components. The fission yeast *Schizosaccharomyces pombe* is another popular experimental system. However, naturally occurring plasmids have not been studied in this organism, so there is no native equivalent to the 2-micron origin. Sequence fragments with ARS function from the fission yeast genome have been cloned based on their ability to support plasmid maintenance and these are commonly used in plasmid constructions. These fission yeast ARS elements are somewhat larger than the replication origins in *Sa. cerevisiae*, typically over a kilobase of DNA. However, the centromeres of the fission yeast approach 100 kilobases of DNA, and are far too large to be included on any plasmid. Therefore, most fission yeast plasmids rely upon a *Sc. pombe* ARS element and random segregation; as a result, they suffer a relatively high frequency of loss compared to plasmids in *Sa. cerevisiae*.

A third necessary feature is a selectable marker, so that yeast cells containing the plasmid ('transformants') can be distinguished from those that do not. Because yeasts are eukaryotes, they are not sensitive to antibacterial drugs such as ampicillin. Instead, the plasmids contain wild-type yeast genes to complement nutrient-requiring yeast mutants. For example, *Sa. cerevisiae* cells lacking a functional *URA3* gene are unable to grow in the absence of exogenous uracil. However, if the active *URA3* gene is included on a plasmid, cells that take up the plasmid and maintain it will grow in the absence of uracil ('complementation'). *URA3* thus provides a positive selection for plasmid-containing strains. Similarly, in *Sc. pombe*, the $ura4^{+}$ gene on the plasmid will complement a strain with a *ura4* deficiency. Since many standard laboratory yeast strains carry multiple auxotrophic mutations, and yeast origins do not suffer from replication interference, it is possible to transform a single strain with several plasmids that differ only in their selectable markers.

A final requirement for yeast plasmids is one of size. Because simple methods of plasmid purification are more difficult with large molecules, most workable plasmids are at most 20 to 30 kb. Larger plasmids exist, but are refractory to simple manipulation in *E. coli* and difficult to transform into yeast. Instead, they are manipulated in the yeast and moved from strain to strain using classical genetics.

Different Plasmids Have Different Uses

Additional plasmid features depend upon their intended use. First, there are simple cloning vectors designed to maintain genomic DNA fragments. These typically add a set of useful restriction enzyme sites to the basic plasmid backbone described in the preceding section. Such plasmids are often used to construct genomic DNA libraries, in which each plasmid contains a random fragment of the genome and the pool of plasmids represents the entire genome. These libraries are useful for cloning genes by complementation of a mutant strain. A second class of plasmids allows expression of a cloned gene under controlled conditions. These require a regulated promoter. A number of different yeast promoters that can be turned on and off in response to particular growth conditions have been isolated from both *Sa. cerevisiae* and *Sc. pombe*. Such expression plasmids can be used not only to express yeast genes, but also to express genes from other species in yeast cells. By fusing the cloned fragment to targeting signals, such as secretion signals, a heterologous protein can be produced and secreted or otherwise directed to specific cellular compartments. Yeast cells can therefore be used as factories to produce large amounts of heterologous recombinant proteins.

A specialized subset of plasmids are those designed to integrate into the yeast genome and be maintained as part of a chromosome rather than as free episomes in the cell; in a formal sense, these are not yeast plasmids at all, because they are only maintained as true plasmids in *E. coli*. Simply removing the yeast replication origin will prevent a plasmid from being maintained efficiently as an episome in the yeast cell. Because yeast cells are proficient at homologous recombination, an integrating plasmid is likely to insert at a position in the chromosome that matches some sequences on the plasmid, such as the marker, or the cloned gene of interest. Integration ensures that the plasmid will be present in single copy in the cell, and eliminates concerns about copy number variation and inefficient transmission through the cell cycle. An integrated plasmid is relatively stable, so that it is likely to be maintained even in the absence of selection, and it can be moved from strain to strain genetically as any other chromosomal marker.

Manipulating Cells with Plasmids

Once a plasmid is constructed, it can be used in a variety of experiments. First, the yeast cells must be induced to take up the plasmid in a process referred to as transformation, usually involving chemical treatment or electroporation. The transformed cells are plated under selective conditions so that only those cells that have successfully taken up and established the plasmid and its marker will grow. Subsequently, the gene(s) contained on the plasmid can be analyzed for ability to complement mutations in the host strain, or for phenotypes associated with overproduction. Using plasmid libraries of genomic DNA or cDNA transformed into mutant strains, the investigator can clone genes and suppressors by complementation. For example, by transforming the plasmid library into a host strain that contains a temperature-sensitive mutation in an interesting gene, and selecting for growth at the restrictive temperature, plasmids that contain the wild-type copy of the mutant gene or a suppressor of the mutant can be isolated and subsequently characterized. If a yeast plasmid contains an equivalent gene from human cells (for example), the ability of that gene to replace the mutated yeast gene can be assessed. Such cross-complementation has been used to isolate homologous genes from different species.

However, presence of a plasmid-borne gene in the cell may have unanticipated effects. Expression of a toxic gene on a plasmid may provide a negative selection that counters the positive selection of the plasmid marker. This reduces the efficiency of plasmid maintenance and attenuates viability of the host strain. Such toxic phenotypes also confer a genetic selection for random mutations that reduce expression or otherwise modify the responsible gene or the host strain to ameliorate the effects. This adaptability can be turned to the investigator's advantage. An example is a method called the 'plasmid shuffle,' which exploits the ability to select against a plasmid. In this technique, expression of a gene on a plasmid is essential for viability of the cell. The investigator transforms the strain with a second plasmid containing a mutant derivative of the same gene, and a different selectable marker. If the mutant derivative is still functional, and the missing nutrient is provided, then the cell no longer relies upon the first plasmid for viability. This provides away to assess the function of mutations *in vivo*.

Integrating plasmids provide an opportunity to manipulate the yeast chromosome directly. As described in the previous section, integration relies upon the yeast cell's proficient homologous recombination. This can be exploited to target insertion of DNA to a particular locus. Depending on the exact construction, an integrating plasmid may be used to replace a genomic copy of a gene of interest, to insert a mutation or an epitope tag, or to physically link a chromosomal locus to a selectable plasmid marker. This can be important for subsequent genetic analysis.

Plasmids in yeast thus provide the ability to identify unknown genes and examine their function, to manipulate the yeast chromosome, and to program the yeast cell to produce any protein of interest. These episomes provide essential laboratory tools, as well as important models for studying extrachromosomal elements. Without them, yeast genetics would never have developed as a powerful model system for understanding eukaryotic cell biology.

Further Reading

Broach JR and Volkert FC (1991) Circular DNA plasmids of yeasts. In: Broach JR, Pringle JR and Jones EW (eds) *The Molecular and Cellular Biology of the Yeast* Saccharomyces: *Genome Dynamics, Protein Synthesis, and Energetics*, pp. 297–332. Plainview, NY: Cold Spring Harbor Laboratory Press.

See *also*: *Saccharomyces cerevisiae* (Brewer's Yeast); *Saccharomyces* Chromosomes; *Schizosaccharomyces pombe*, the Principal Subject of Fission Yeast Genetics; Transposable Elements

Yeast Two-Hybrid System

J Read and S Brenner

doi: 10.1006/rwgn.2001.2093

The yeast two-hybrid system is a valuable tool used to identify interacting proteins. The protein of interest is expressed in yeast as a fusion to the DNA-binding domain of a transcription factor lacking a transcription activation domain. The DNA-binding fusion protein is generally called the bait. The yeast strain also contains one or more reporter genes with binding sites for the DNA-binding domain. To identify proteins that interact with the bait, a plasmid library that expresses cDNA-encoded proteins fused to a transcription activation domain is introduced into the strain. Interaction of a cDNA-encoded protein with the bait results in activation of the reporter genes, allowing cells containing the interactors to be identified.

See *also*: cDNA; DNA-Binding Proteins; Reporter Gene

Z

Z Chromosome

D W Burt

doi: 10.1006/rwgn.2001.1404

Avian and mammalian sex chromosomes evolved independently (Fridolfsson *et al.,* 1998; Nanda *et al.,* 1999, 2000) and should therefore have fundamentally different sex-determining genes. The sex chromosomes in birds are designated Z and W: the female is the heteromorphic (ZW) and the male the homomorphic (ZZ) sex. The average avian Z chromosome is a medium-sized macrochromosome. It is not clear in birds whether the Z or W chromosome determines sex.

References

Fridolfsson A-K, Cheng H, Copeland NG *et al.* (1998) Evolution of the avian sex chromosomes from an ancestral pair of autosomes. *Proceedings of the National Academy of Sciences, USA* 95: 8147–8152.

Nanda I, Shan Z, Schartl M *et al.* (1999) 300 million years of conserved synteny between chicken Z and human chromosome 9. *Nature Genetics* 21: 258–259.

Nanda I, Zend-Ajusch E, Shan Z *et al.* (2000) Conserved synteny between the chicken Z sex chromosome and human chromosome 9 includes the male regulatory gene DMRT1: a comparative (re)view on avian sex determination. *Cytogenetics and Cell Genetics* 89: 67–78.

***See also:* Sex Determination, Human; W Chromosome**

Z DNA

J H Miller

doi: 10.1006/rwgn.2001.1405

A zig-zag-like structure of the DNA chain that is observed in GC-rich segments of DNA which form left-handed helices.

Zinc Finger Proteins

***See:* DNA-Binding Proteins**

Zoo Blot

doi: 10.1006/rwgn.2001.2073

A zoo blot is a technique using Southern blotting to evaluate the ability of a DNA probe from one species to hybridize with genomic DNA from a variety of other species.

***See also:* Southern Blotting**

Zygote

P M Wassarman

doi: 10.1006/rwgn.2001.1407

The zygote, or fertilized egg, is a single cell produced by fusion of female and male germ cells, that is, the unfertilized egg and sperm, respectively. Since germ cells undergo meiotic divisions to a haploid state (n) during oogenesis and spermatogenesis, fusion of the unfertilized egg and sperm (fertilization) restores a diploid ($2n$) number of chromosomes to the zygote. In mammals, the second meiotic division of the egg, with separation of chromatids, occurs shortly after fusion with sperm. At an appropriate time after fertilization, the zygote begins to divide mitotically, eventually giving rise to a multicellular organism that exhibits all of the characteristics of the species.

Nuclei contributed to the zygote by the unfertilized egg and sperm are called female and male pronuclei, respectively. In mice, the female pronucleus forms at $\sim$7.5 h and the male pronucleus at $\sim$5.5 h after fusion of the unfertilized egg and sperm. The two pronuclei must come together near the center of the zygote and form a single diploid nucleus. The timing of nuclear formation varies greatly from one species to another; for example, it takes $\leq$1 h in sea urchins and

$\geq$ 12 h in mice. In fact, in mice, pronuclei approach each other, but do not actually fuse to become a diploid nucleus. Rather, pronuclear membranes disappear and chromosomes assemble on a spindle. DNA replication occurs ~14.5 h after fertilization, as the pronuclei migrate toward the center of the zygote. In mice, the first cleavage division occurs at ~20 h after fertilization, when chromosomes are assembled on a spindle.

In many animals, sperm contribute a centriole to the zygote and this organelle helps to organize the first mitotic spindle on which the chromosomes are arranged. In this respect, the sperm centriole acts as a microtuble-organizing center in the zygote. On the other hand, sperm contribute very few of the large number of mitochondria found in the zygote ($\leq$ 0.01%), ensuring that mitochondrial DNA is maternally inherited.

The zygote is inactive with respect to nascent transcription of genomic DNA, although translation of maternal transcripts takes place. The onset of transcription is delayed until after the first cleavage division in mammals and after the first 12 cleavage divisions in some nonmammals. Presumably, this period of transcriptional inactivity exhibited by the zygote provides time to remodel parental chromosomes.

In mammals, genomes derived from the unfertilized egg and sperm appear not to be equivalent. In some cases, only the maternally derived allele of a particular gene is active, whereas, in others, only the paternally derived allele is active ('genetic imprinting'). Some of these genes are absolutely essential for normal development. Apparently, as a result of this nonequivalence of pronuclei, parthenogenetic (bimaternal), gynogenetic (bimaternal), and androgenetic (bipaternal) mammalian zygotes cannot give rise to normal fetuses and live births.

Further Reading

Gilbert SF (1997) *Developmental Biology*, 5th edn, p. 918. Sunderland, MA: Sinauer Associates.

Hogan B, Beddington R, Costantini F and Lacy E. (1994) *Manipulating the Mouse Embryo*, 2nd edn, p. 497. Plainview, NY: Cold Spring Harbor Laboratory Press.

See *also*: Fertilization

Zygotic Lethal Gene

R K Herman

doi: 10.1006/rwgn.2001.1408

A zygotic lethal gene is a gene that leads invariably or almost invariably to the death of an organism prior to the reproductive stage. 'Zygotic' in this case refers to the organism that develops from a single-celled diploid zygote. For some zygotic lethal genes, a rare individual carrying the gene in a dose that is normally lethal survives to the reproductive stage. Such rare survivors are called 'escapers.' Zygotic lethal genes are to be distinguished from the following: gametic or haplophasic lethal genes, which exert their effects in haploid gametes; from sterile genes, which allow their bearers to reach reproductive maturity but render them sterile; and from maternal-effect or paternal-effect lethal genes, which kill the progeny of affected individuals.

Many zygotic lethal genes result in developmental arrest and death at a particular stage during development, referred to as the lethal phase. Zygotic lethal genes may thus be classified as embryonic or postembryonic, depending on their lethal phase. Examples of postembryonic zygotic lethal genes in insects include larval, pupal and early adult lethals.

A dominant zygotic lethal gene is lethal when present as a single copy per cell, even when a wild-type allele of the same gene is also present in the diploid cells. Such lethal genes are rarely studied because they cannot be propagated by breeding. Recessive zygotic lethal genes are lethal only when they are present in the homozygous or hemizygous condition. Individuals that are heterozygous for the lethal gene are viable because the wild-type allele is dominant to the lethal allele. A recessive lethal gene can be maintained and propagated in a heterozygous stock. One-fourth of the progeny produced from the mating of heterozygous lethal parents are expected to be homozygous for the lethal gene and exhibit the lethal phenotype. Two-thirds of the surviving progeny are expected to be heterozygous for the lethal gene and can be used to propagate the lethal gene. In rare cases, a recessive zygotic lethal gene may confer a visible phenotype dominantly, in which case individuals heterozygous for the lethal can be readily identified. For most lethal genes, however, the heterozygous lethal organisms are indistinguishable from the homozygous wild-type organisms, unless a closely linked marker gene conferring a visible phenotype is used to track either the lethal-bearing chromosome or its non-lethal homolog. Chromosomal rearrangements are often used to suppress recombination between the lethal gene and the visible tag.

X-linked zygotic lethal genes have long (since 1912) been recognized in the fruit fly *Drosophila melanogaster* because the progeny of a heterozygous mother mated to a wild-type male exhibit an altered sex ratio: half of the sons are hemizygous for the zygotic lethal gene and inviable, whereas all of the daughters receive a dominant wild-type allele from their father. Half of the daughters will be heterozygous for the recessive lethal gene; they can be identified by the altered sex ratio of their progeny.

The phenotype conferred by a conditional zygotic lethal gene can be influenced by changes in the growth conditions of the organisms carrying it or in the genotypic background in which the lethal gene is embedded. An example of a zygotic lethal gene influenced by growth conditions is a gene that causes lethality only when the organism carrying it is raised at an elevated (non-permissive or restrictive) temperature. In this case, the conditional lethal gene can be propagated in homozygous stocks maintained at permissive conditions. A shift to restrictive conditions permits analysis of the lethal phenotype.

Most recessive zygotic lethal genes differ from their wild-type alleles by having reduced or no wild-type gene activity. Such genes are referred to as vital or essential because wild-type gene activity, even if provided by a single wild-type gene per cell, is required for development of the organism to the reproductive stage. Very approximate estimates of numbers of essential genes have been made for organisms that have been intensively studied genetically. For example, it has been estimated that the fruit fly *D. melanogaster* and the nematode *Caenorhabditis elegans* each have 5000 essential genes and that the mouse *Mus musculus* has 5000–10 000 essential genes. For all three of these organisms, it is estimated that there are many more genes that are active and unessential. Any essential function provided by two or more genes redundantly would not have been counted as essential in these estimates because a loss-of-function allele of one gene within an overlapping set would not have been recessive lethal.

In *Drosophila* and *C. elegans*, the earliest stages of embryogenesis are controlled largely by genes expressed in the mother rather than in the embryo. The products of such maternal-effect genes are stored in the oocyte prior to fertilization and are needed for early embryogenesis, before the activities of many zygotic genes are required.

In general, the essential physiological role played by a zygotic lethal gene cannot be deduced simply from an analysis of its lethal phenotype. A molecular analysis of the gene and its products is usually required, as well as other methodologies. Essential genes may be required at more than one stage of development, including stages that in normal development occur after the lethal phase. The role of an essential gene in stages subsequent to the lethal phase may be studied in genetic mosaics, in which only some cells of an organism are homozygous for the lethal gene. Genetic mosaics of *D. melanogaster* have been used to show that only a small proportion of genes represented by recessive zygotic lethal alleles are essential for the viability of all cells of the animal.

***See also:* Balanced Translocation; Chimera; Maternal Effect**

Index

NOTE

Bold page number locators refer to complete articles on the various topics covered by this encyclopedia. Illustrations and tables are indicated by *italic* page numbers.

Text is located by page numbers in normal print.

Cross references, prefixed by *see* and *see also*, are also listed at the end of each article.

A

a1 gene 1895
5A2 gene 603
aadA gene 336
Aaronson S A **126–128**
aat site 1435–1436
ABA *see* abscisic acid (ABA)
abaA gene 110, *110f*
abc-1 gene 301
ABCB1 *see* P-glycoprotein
abd-A (abdominalA) gene *960f*
Abd-B (AbdominalB) gene 958, *960f*, 961
Abelson murine leukemia virus 250
Abies alba 840
ABL1 gene **250–251**
ABL2 gene 251
abl oncogene 1484
ABL-related gene 251
abnormal placental morphogenesis 989–990
ABO blood groups 187, 356, 1507
abortive transduction **1–2**
abscisic acid (ABA) 1481
Acacia 1083
Ac (Activator) gene 2021, 2029
Ac (Activator) locus 1161
Acanthamoeba castellanii 1740
Accipiter 408
Ac/Ds superfamily 2026
ACeDB database 254
acentric fragment **2**
Acetospora 1882
N-2-acetyl-2-aminofluorene 557
N-acetyl glucosamine 1131
N-acetylmuramic acid 1131
ACF protein *342t*, 1737
achaete-scute genes 1317, *1318f*, 1321
Achard, Emile 90
achondroplasia **2–3**, 591
  cause 2
  clinical features 2
  growth factors 900
  hypochondroplasia 2
  medical problems 2–3
  missense mutations 1277
  mutation 1271
  single-gene inheritance 1836
  thanatophoric dysplasia 2
achromatopsia 421–422
Achromobacter 1478
acquired traits 2009–2010
Acrasiomycota 1883
acridines **3**
acrocentric chromosome **3**
acrosome **3–4**, 689, 1872
  composition 3
  exocytosis 4
  function 3
  spermiogenesis 3, 1875
Actinidia deliciosa 1421–1422
Actinomycetes 1882
active site **4**
acute lymphoblastic leukemia (ALL) *see* leukemia, acute
acute myelogenous leukemia (AML) *see* leukemia, acute
acy-1/sgs-1 gene 317
ada gene 36
Adalia 730
Adams J M **205–207**
Adams T H **106–111**
adaptive landscapes **4–7**, *6f*
  shifting balance theory of evolution 1825–1826, *1826f*
  types 5–6, *6f*
  Wright, Sewall 4
adaptor hypothesis **7**, 59, 78
additive genetic variance **7–9**
adenine (A) **9**, 540, *540f*, 572, 1358
adenocarcinomas (ADCs) **9–12**
  colorectal adenocarcinomas (CRCs) 10, *10t*
  genetic abnormalities *12t*
  lung adenocarcinomas 11
adenoma **12**
adenomatous polyposis coli (APC) **12–14**, 422
  beta (β)-catenin 13
  function 13–14
  homologs 13
  molecular organization 13
  pathology 14
adenosine 14, 1360
adenosine diphosphate (ADP) 14
adenosine monophosphate (AMP) 14
adenosine phosphates **14**
adenosine triphosphate (ATP) 14, **115–116**, 324–325
  cAMP nucleotide 258
  characteristics 115
  function 116
  heat shock proteins (Hsps) 915
  histidine biosynthesis 943
  mitochondrion 1379
  production 116
  see also cyclic AMP (cAMP)
adenoviruses **15–18**
  adenovirus recombinants 17–18
  E family proteins 15–17
  oncogenic transformation 17
  replication cycle 15–17
  structure 15, *15f*
  transcription factors 16
  transcription map *17f*
  Virus Associated I (VAI) 16
adenylate cyclase 258, 260
adenylic acid 1360
Adh gene 187, 863
Adhya S **1652–1657**
adipsin 429
adjacent/alternate disjunction **18–19**, *19f*
adrenal steroidogenic enzymes *446t*
adrenodoxin *446t*
adrenodoxin reductase *446t*
adrenogenital syndrome *see* congenital adrenal hyperplasia (CAH)
adult T-cell lymphoma/leukemia (ATLL) 1094
advanced intercross lines (AILs) **19–20**, *20f*
Aedes egypti 246
Aegilops 2060
Aegilops speltoides 1511
Aegilops squarrosa 1511
Aequorea victoria (jellyfish) 311, 485
Aeschynomeneae 1083
af (afila) gene 1471
Affara N A **2155–2160**
aflatoxins **20–21**
  aflatoxin B_1 20–21, 557
  Aspergillus flavus 20
  Aspergillus parasiticus 20
  carcinogenicity 21
  structure *21f*
AFRC Institute of Plant Science Research (UK) 1179
African trypanosomiasis 1736
agamospecies 1861, 1866–1867
AG gene 715
aging **21–23**
  chromosome nondisjunction 2057
  evolutionary theory 21–22
  Hayflick limit 1967
  humans 22
  Hutchinson-Guilford syndrome 22
  mechanisms 23
  model systems
    Caenorhabditis elegans 22
    Drosophila melanogaster 22
    Methuselah gene 22
    Saccharomyces cerevisiae 22
  tissue culture 1967
  Werner syndrome (WS) 22
  see also telomeres
agouti **23**
Agouti (a) gene 23, 398, 400, 527, 1470
agouti (*a*) locus 1599, 1871
Agouti protein 23
agr gene family 1886
agriculture
  cultivation of plants 2009
  domestication of animals 2008–2009
  Phaseolus vulgaris (Beans) 1445
  polyploidy 1510–1511
  transgenic animals 1997–1998
Agrobacterium **23–25**, 973
  characteristics 24

Agrobacterium (*continued*)
  classification 24
  genetic engineering 24
  plasmids 24
Agrobacterium radiobacter 24, 492
Agrobacterium rhizogenes 24, 1122
Agrobacterium rubi 24
Agrobacterium tumefaciens 24, 454, 491, 885, 1162, 1475, 1966, **1984–1986**
  Arabidopsis thaliana 891, 1781
  DNA cloning 546
  genetic information transfer 1984–1986, *1985f*
  genome sequencing 859
  Lotus japonicus 1122
  Lycopersicon esculentum (tomato) 1126
  seed development 1781
  T-DNA **1984–1986**, *1985f*
  transgenic plants 1984, 1986
  transposons 2030
  virulence proteins *1985t*
Agrobacterium vitis 24
Agr proteins 1886
AIB1 gene 425
aidB gene 36
Aitchison J D **1352–1356**
Aizawa S-I **711–712**
alanine **25**, *25f*
alarmones 281–282
Alavidze A **175–179**
Albertson D G **956–957**
albinism **25–27**, 401, 1468
  autosomal inheritance 131–132
  Chediak–Higashi syndrome 26
  clinical features 25–26
  founder effect 27
  Hermansky–Pudlak syndrome 26
  history 25
  melanin 26
  ocular albinism (OA) 25
  oculocutaneous albinism (OCA) 25, *26f*
  Prader–Willi syndrome 26
albino (c) gene 398
albino (*c*) locus 1599, 1871
Albrecht K H **1253–1254, 1816–1819**
albumins 1783–1784
alcoholism **27–29**
  animal studies 27–28
  clinical findings 28
  QTL mapping 28
aldosterone synthase *446t*
Alexander R W **1897–1899, 2140–2141**
alfalfa *1914t*
Alifano P **943–947**
alignment problem **29–35**
  alignment parameters 31–32
  BLAST program 33
  BLOSUM weight matrix 32, *32f*
  Dayhoff PAM weight matrix 32
  dot matrix plots 32, *33f*
  dynamic programming 33–34
  Hidden Markov models (HMMs) 33
  sequence alignment *29f*, 29–30, *30f*
  sequence changes 29
alkA gene 36, 1664
alkaptonuria **35–36**
  clinical features 35
  metabolism defect 35, *35f*
  pathology 36
  single-gene inheritance 1836
  treatment 36
alkB gene 36
alkyltransferases **36–37**
  crystallographic structure 37
  DNA binding 37
  DNA repair 36–37
  helix–turn–helix motif 37
  occurrence 37
ALL *see* leukemia, acute
allele frequency **37**, 908
  effective population number 602
  equilibrium *645t*, 645–646, *646f*, 646–648
  evolution 665
  gene flow 785
  genetic distance 828
  Hardy–Weinberg law 37, *912t*, 912–914
  mutant allele **1268**
  population substructure 1520–1522
  shifting balance theory of evolution 1825–1826, *1826f*
  speciation 665
  stationarity 647–648
  trivial equilibrium *2069f*
  underdominance 2093–2094, *2094f*
alleles **38–39**
  codominant 38
  codominance 1017
  definition 38
  diploid organisms 38
  dominant 38
  hemizygote 38
  heterozygote 38, 934–935
  homozygote 38
  incomplete dominance 1016–1017
  meiosis 38
  molecular mechanisms 38
  mutant allele **1268**
  recessive 38
  types 38
allelic association *see* linkage disequilibrium
allelic exclusion **39**
Allium test 1094
allohexaploid 537
allopatric **39**, 665
Allorhizobium 24
allostery **39–40**
  allosteric proteins 39
  allosteric transition 39
  hemoglobin 40
  Monod, Jacques 1238
allotetraploid 537, 1164, 1510
allotypes **40**
allozyme electrophoresis 968
Alnus glutinosa 840
Alnus rubra 840
alopecia 1027–1028
alpha (α_1)-antitrypsin deficiency (α1AT) **40–41**
  chromosome 14 41
  disease associations 41
  genetic variation 41
  occurrence 41
  plasma proteins 40
  serine protease inhibitor family (serpins) 40
  therapy 41
alpha (α)-catenin 620
alpha (α)-ENaC ion channel 315
alpha (α)-fetoprotein (AFP) **42**
alpha (α) globin gene 878
alpha (α)-helix 1569, *1569f*
alpha (α)-sarcin 611
Alternaria brassicicola 1914t
Alternaria crassiae 1914t
alternation of gene expression **42–46**
  DNA rearrangement 43, *44t*
  IS element insertion 45, *45f*
  site-specific inversion
    FimB/FimE-mediated inversion 43
    Hin-mediated inversion 43
    Piv-mediated inversion 43, *45f*
    Type 4 pilin 43
alternative splicing **47–48**
  binding process 47
  Drosophila melanogaster 47
  mammals 48
  pre-mRNA splicing 47–48
  splicing variations 47–48
  Sxl gene 48
Altmann, Richard 553
Altman, Sidney 1730, 1751
altruism *see* Hamilton's theory
Alu family **48**
Alzheimer's disease **48–49**, 313, 317, 318
  apo gene family 49
  chromosome 10 49
  chromosome 12 49
  chromosome 14 49
  chromosome 19 49
  human behavioral genetics 211
  pathology 48–49
  presenilin gene family 49
Amadori rearrangement 943
Aman P **1287**
amber (*am*) mutants 1927
amber codon **49**, 1887
amber mutation **49**, 245
amber suppressors **50**
amBra-1 gene *1944f*
amBra-2 gene *1944f*
Ambrose M **36–37, 1650–1651**
Ambystoma mexicanum 1304
Ambystoma tigrinum 1304
AMELY gene 2159
American Phage Group 554
American trypanosomiasis 1736
American Type Culture Collection (ATCC) 850
Ames, Bruce **50–51**
  contributions 50
  Gold, Lois Swirsky 51
  mutagenicity test 1712
  reversion assays 1710
Ames dwarf (*df*) mouse mutation 592
Ames G F **50–51**
Ames test **51–54**, 1710, 1712
  histidine operon 52
  rat liver S9 53
  reversion assays 1710
  Salmonella typhimurium tester strains 51, *52f*, *52t*
  techniques 52
  test result guidelines 53–54
  uvrB gene 52
AMH gene 1814
Amhr2 gene 1818
amino acids **54–56**, *55f*, *56f*
  alanine **25**, *25f*
  arginine **95**, *95f*
  asparagine **105–106**, *106f*
  cysteine **507**, *507f*
  glutamic acid **884**
  glutamine **884**
  glycine **884**
  histidine **943**
  isoleucine 1055
  joining *1568f*

amino acids (*continued*)
leucine **1088**, *1088f*
lysine 972, **1130**, *1130f*
methionine **1189**
phenylalanine **1447**
proline **1549**
protein structure 1568
sense codons 522
serine **1809**, *1809f*
sources 55–56
structure *55f*, *56f*, 1568, *1568f*
threonine **1965**
tryptophan **2075**
tyrosine **2090**
uses 56
valine **2103**, *2103f*
amino acid substitution **57–58**
impact 57
Kimura correction 1062
location 57
Yanofsky, Charles 57
aminoacylation 601
aminoacyl-tRNA 57, **58**, 601–602
Bacillus subtilis 141
elongation 609, 610
guanosine triphosphate (GTP) 901
nutritional mutations 1363
aminoacyl-tRNA synthetases **59–60**
adaptor hypothesis 59
amino acid transformations 59
function 59
tryptophan (Trp) 59
5-aminoimidazole 4-carboxamide ribonucleotide (AICAR or ZMP) 943–945
5-aminoimidazole-4-carboxamide riboside-5′-triphosphate (ZTP) 945
aminomethyl coamarin acetic acid (AMCA) 1002
aminopterin **60**
aminopurine (2-aminopurine) 191, 196
amino terminus **58**
ami operon 1079
amitosis 1135
amniocentesis **60–61**, 1540
amphibians 331, 390
amphidiploid 537, 1164
amphiregulin (AR) 626, 649
amplicons **61–62**, 580, 963
amplified fragment length polymorphism (AFLP) 831, 972
amyotrophic lateral sclerosis 313
Anacardiaceae 238
ana gene 930
anagenesis **62**
analogous similarities 969
analogy **62–63**, 969
ancestral inheritance theory **63–64**
Galton, Francis **63–64**
mathematical-statistical procedures 63
pangenesis 63
Pearson, Karl 64
ancestral polymorphisms 974
anchorage-independent growth **65**
anchor locus **64–65**
Anderson, M. 852
Anderson P **38–39**
Anderson, Thomas 180
androgenesis 1000, 1419, 2098
androgen insensitivity 1822
androgenone **65**
androgen receptor gene 425
aneuploid **65–66**, 346, 980
compound chromosomes 117
deleterious effects 65–66
effects on humans 66
effects on plants 66
meiosis 66
mitosis 66
nondisjunction 66
Angelman syndrome 346, 634, 1000, 1150, 1179
angiogenesis **66–73**
capillary blood vessel configuration *67f*
capillary blood vessel growth 66–67
nonneoplastic diseases 72–73
pathologic angiogenesis 67
physiologic angiogenesis 67
research history 67–68
tumor angiogenesis 68
clinical trials 72
endogenous inhibitors 70–72, *71t*
endogenous stimulators *68t*
exogenous inhibitors 72
human breast cancer *70f*
mechanisms 68–69
metastasis 69–70
neovascularization 69
tumor growth 69
vascular endothelial growth factor (VEGF) 68
angiosarcoma 1768, *1769t*
angustifolia gene 2048
annealing **73–74**
anonymous locus **74**
Anopheles 1511
Anraku, Yasuhiro 1565
Anstee, D. 1713
ANT3 gene 2157
ANT-C complex 962
ANT gene 716
antibiotic resistance **74–76**
bypass mechanism 75
decreased uptake 74–75
enzymatic modification 74
evolutionary origins 76
genes 1042–1044, 1687
genetics **1687**
molecular basis 75
plasmids 75
resistance plasmids **1686**
target sites 75
antibiotic-resistance mutants **76–78**
coumarins 76–77
erythromycin 77
Escherichia coli 76
fusidic acid 77
kasugamycin 77
kirromycin 77
occurrence 76
quinolones 77
rifampicin 77
streptomycin 77–78
antibody **78**
see also C genes; immunoglobulin
anticipation 593, 982
anticodons **78–79**
adaptor hypothesis 78
composition 78–79
decoding process 78–79
elongation factors 609
wobble hypothesis 79
antigen **79**
antigenic variation **79–81**
purpose 79
antigenic variation (*continued*)
random causes 80
specific causes 80
anti-Mullerian hormone (AMH) 1812, 1816
anti-oncogenes **81–82**
Knudson, Alfred 81
tumor formation 81
tumor suppressor genes *81t*, 81–82
two-hit model 81
antiparallel **82**
Antirrhinum 531
Antirrhinum majus 713, 2026
antisense DNA **82–83**
applications 83
aptamer binding 83
gene suppression 82–83
oligonucleotides 82–83
antisense RNA **83–84**
function 83
gene regulation 83
structure 84
antitermination factors **84–85**
attenuation 85
mechanisms 84–85
phage HK022 85
phage λ 84–85
RNA polymerases 84
transcript regulation 84
antitermination proteins **85**
antithrombin III deficiency 357
Antoun H **1477–1480**
Antp (Antennapedia) gene 958, *960f*, 962
ant^{R} mutation 1223
AP180 protein 1313
AP1 gene 715
AP2 gene 715
AP3 gene 716
APC *see* adenomatous polyposis coli (APC)
Apc (Adenomatous polyposis coli) gene 1245
APC (Adenomatous Polyposis Coli) gene 9, *81t*, 263, 1769, *1770t*, 2082, *2084–2085t*
AP endonucleases **85**
Apert syndrome *see* craniosynostosis
Apicomplexa 1379, 1882
apicoplasts 1379
aplasia 448
Aplysia 210
APOBEC-1 1737, *1737f*
Apodemus (European field mice) 688
apoenzyme 625
apo gene family 49
apolipoprotein (ApoB) mRNA 1737, *1737f*
apomorphy **85**, 965–967, 968, 1904, 1905
apoptosis **86**, 313, 566, 595
Brenner, Sydney 245
Caenorhabditis elegans 314, *315f*
chromatin 313
DNA 313
pattern formation 1426
proteolysis 1574
tumor suppressor genes 82
see also pattern formation
aporepressors **86**
aptamers 1006
apterous gene 1097
apurinic or apyrimidinic (AP) site *562f*
apx-1 gene 617–618, *618f*
APXLY gene 2159
Aquadro C F **790–792**
Aquifex aeolicus 150t, 192

Aquificales 149
Arabidopsis 1084
cell lineage 309
gene number 796
mitochondria 1217
paternal inheritance 1421
root *1754f*
Arabidopsis Biological Resource Center (ABRC) 852
Arabidopsis Centre (England) 852
Arabidopsis Genome Initiative 89
Arabidopsis griffithiana 86
Arabidopsis Pin-formed 1 (PIN1) gene 1472
Arabidopsis thaliana **86–87, 87–90,** *88f*, 2035
Arabidopsis Genome Initiative 89, 238
Arabidopsis griffithiana 86
The Arabidopsis Information Resource (TAIR) 90
Arabidopsis wallichii 86
background 87
Brassicaceae 86, 238
Cardaminopsis 86
contribution to genetic research 88
Crucihimalaica 86
developmental genetics 531
dispersed transposable sequences 861
DNA cloning 545
EST number *366t*
evolution 86–87
flower development, genetics *714f*
gene function 89–90
genetic map 88
genome organization 860
genome sequencing 86, 89, 859
history 87
homeobox genes 962
model organism 86, 87, 1454–1455, 1474–1475
molecular map 88
molecular phylogeny 86
morphology 86
Olimarabidopsis 86
photosynthesis 1461, 1462
plant genetics 87
plant hormones 1480
polyploidy 1510
seed development 1780–1782
seedling development *1455f*
systemic acquired resistance (SAR) 1913, *1914t*
trichomes **2045–2048**
Arabidopsis wallichii 86
arabinose **90**
araC gene 1652
Arachis hypogaea (peanut) 1083
arachnodactyly **90**
AraC protein 543
ara operon 1652
Arber, Werner **90–91**, 182
Nobel Prize 90, 1290, 1846
restriction enzymes 90
Archaea **91–95**
conjugation 94
Crenarchaeota 92
Crenarchaeum symbiosum 92
diversity 92
DNA transfer 93–94
Euryarcheota 92
genetic mechanisms 92
genome sequences
Archaeoglobus fulgidus 92
Methanobacterium thermoautotrophicum 92
Archaea (*continued*)
Methanococcus jannaschii 92
Pyrococcus horikoshii 92
growth studies 93
Haloarculae hispanica 94
Haloferax mediterranei 94
halophiles 93
heat shock proteins (Hsps) 915
histidine genes *946f*
history 91
hyperthermophile 92
Methanococcus voltae 93
occurrence 91
phage growth 94
physiology 92
plasmid vectors 94
RNA polymerase 1746
Sulfolobus acidocaldarium 92
Sulfolobus islandicum 94
Sulfolobus solfataricus 94
viruses **2114–2116**, *2115f*
see also bacteria; eukaryotes; thermophilic bacteria
Archaeoglobus fulgidus 92
Ardlie K **1165–1167, 1919–1921**
ARG1 gene 892
argA gene 1624, *1624f*
AR gene *2049t*
ARG gene 251
arginine **95**, *95f*
arg repressor 1896
ArgR protein 1040, *1040t*
Aristotle 964
Arkin M **1010–1014, 1279–1284**
ARL2 gene 892
Armoracia rusticana (horseradish) 237
Arnold J **832–834**
Arnott, Struther 555
ARO1 gene 1765
arrowhead gene 1097
ARSDY gene 2159
ARSEY gene 2159
ARSFY gene 2159
Artemia salina 2097
Arthrobacter 1478
Arthrobacter luteus 1694t
artificial chromosomes, bacterial *see* BACs (bacterial artificial chromosomes)
artificial chromosomes, yeast *see* YACs (yeast artificial chromosomes)
artificial selection 96–101
correlated responses 97
gene frequencies 97
selection experiments *98f*, 98–101, *99f*, *100f*
selection response 96–97
uses 101
Artzt K **590–591**
Ascaris megalocephala 236
Ascobolus **101–104**, 636, 738, 1954
Ascobolaceae 101
Ascobolus immersus 101, *103f*, **104–105**
Ascobolus nidulans 813
characteristics 101, *102f*
gametes 749
germination 103
recombination models 1642
sexual reproduction 101–102
usefulness in genetic research 103–104
vegetative state 102–103
Ascobolus immersus **104–105**
Ascomycetes 749, 1882
ascorbic acid 2118
ascus **105**
Ashworth A **239–241**
Asilomar Conference **105**, 1638, 2135
Askin's tumor 672
ASMT gene 2157
Asmussen M A **644–646, 1998–1999, 2068–2070, 2093–2094, 2127–2132**
asnA gene *1383f*
asnC gene *1383f*
ASO (allele-specific oligonucleotide) **105**
asparagine **105–106**, *106f*
asparagus bean *1914t*
aspartic acid **106**
Aspergillus 738, 749
Aspergillus flavus 20
Aspergillus nidulans **106–111**, 405, 535, 1019
aneuploidy 107–108
cell cycle 111
characteristics 106
development 109–110, *110f*
diploidy 107–108
Emericella nidulans 106
heterokaryosis 107
history 106
life cycle 106, *107f*
mapping 109
meiosis 108
metabolic regulation 110
mitosis 108, *108f*, *108t*
mutagenesis 108
Neurospora crassa 106
nitrate assimilation 110, *110f*
resources 109
Saccharomyces cerevisiae 106
transformation 108
Trichocomaceae 106
usefulness in genetic research 111
Aspergillus parasiticus 20
assortative mating **111–113**, 1410
negative assortative mating 112–113
positive assortative mating 111–112
assortment **113**
as-T2 gene *1944f*
as-T gene *1944f*
Astragalus 1083
ataxia telangiectasia (AT) **113–115**, 568, *569–570t*
clinical features 113, 568
immune system deficiency 568
leukemia, chronic 1093
malignancy development 114
mutational studies 114
Nijmegen breakage syndrome 114
sensitivity to ionizing radiation 113, 568
tumorigenesis 114
ataxia-telangiectasia mutated (*ATM*) gene *see ATM* gene
ATF1 gene *1772t*
ATF2 gene 1772
Atkinson, D.E. 943
AT-like disorder 571
ATLL (adult T-cell lymphoma/leukemia) *see* leukemia, acute
ATM gene 11, 113, 263, 568, 1093, *2084–2085t*
atonal gene 1321
ATP *see* adenosine triphosphate (ATP)
atp1 gene 238
atpB gene 238
ATP-binding cassette (ABC) transporters 1430
atp gene 1221, 1461
Atractomorpha similis 860

ATR protein 1912
attached-X chromosomes *see* compound chromosomes
attachment sites **120–122**
  att sites 120
    function 120–121
    location 121, *122f*
    structure *120f*, 120–121
  bacteriophage λ *120f*, 120–121
  Campbell, Allan 120
  Campbell Model 120
  Escherichia coli 120–121
  excision 120
  site-specific recombination 120–122
  specialized transduction 1858–1859
*att*B site 1435–1436
attenuation **122–123**
  gene regulation 122
  RNA polymerases 122, 123
  RNA synthesis 122
  termination 122
  transcription attenuation 122, **123–125**
    mechanisms 122, 123–125
    trp operon 122
  transcription termination 123, *124f*
    antiterminator proteins 124–125
    Bacillus subtilis 124
    bacteriophage λ 125
    his operon 123
    RNA binding proteins 123
    terminator proteins 124
    translation 123
    tRNA-directed antitermination 125
    trp operon 123
*att*P site 1435–1436
att sites **116**, 120, 174, *174f*
Audreau, Alice 1237
Aufderheide K J **1411–1414**
AUG codons **125–126**
Austin J **1989–1990**
autapomorphy 965, 1905
autocrine mode **126–128**
  growth factor activities 126–127
  paracrine mode 127
  tumorigenesis 126
  vascular endothelial growth factor (VEGF) 127
autogenous control **128**
autoimmune diseases **128**
autonomous controlling element **128**
autonomously replicating sequences (ARS) 1382
autoradiography **128**
autoregulation **128–129**
  DNA-binding proteins 129
  Drosophila melanogaster 129
  negative autoregulation 129
  positive autoregulation 129
autosomal dominant *2014t*
autosomal inheritance **129–133**
  albinism 131–132
  autosomal codominant expression 132
  autosomal dominant inheritance *130f*, 130–131
  autosomal recessive inheritance 131–132, *132f*
  chromosome 14 130
autosomal recessive *2014t*
autosomes **133**
autotetraploids 1510
AUX1 gene 892
auxin 491, 1480–1481
auxin-resistant 6 (axr6) mutation 1475
auxotroph **133**, 218, 1362
Avery, Oswald 540, 553, 654, 815, 1491–1492
Avian Genetic Stock Collection (domestic chicken) 854
avian myelocytomatosis virus (MC29) *see* MC29 (avian myelocytomatosis virus)
avir (avirulence) allele 1535
Avise J C **1466–1467**
Axin gene 2083
Axin 1 gene *2084–2085t*
Axin 2 gene *2084–2085t*
Axinella mexicana 92
Axolotl Colony 854
azidothymidine (3′-deoxy-3′-azidothymidine, AZT) 192, 586
Azobacter 1328
Azobacter vinelandii 1067, 1365
azoospermia 1023
Azorhizobium 1328
Azorhizobium caulinodans 707, 1329
Azospirillum barsilence 1479
Azospirillum brasiliense 946f

B

B gene 1045
B27 gene *2049t*
B6 *see* C57BL/6
B6 mouse strain 250
Bacillus Genetic Stock Center 850
Bacillus megaterium 1131
Bacillus natto 144
Bacillus ressius 2098
Bacillus stearothermophilus 1962
Bacillus subtilis **135–144**, *150t*, 1382, 1478, 1882
  Bacillus natto 144
  background 136
  bacteria 148
  base composition 192
  catabolite repression 140, *283f*, 283–284
  cell division 303
  chemotaxis 138, *139f*
  Chi sequences 328
  codon usage 141–142
  codon usage bias *403t*, 404–405
  cotransformation 472
  cytoplasm organization 141
  elongation factors 141
  envelope 136, *136f*
  gene transfer 142–143
  genome sequence 140–142, 859
  histidine operon *946f*
  industrial uses 143–144
  metabolism 138–140
  microbial genetics 1191
  model organisms 136
  phylogeny 143
  protein secretion 138
  quorum-sensing 137–138
    Staphylococcus aureus 138
    Streptococcus pneumoniae 138
  replication 142
  restriction and modification 1692
  ribosome lattices 142
  sigma factors 1832
  sporulation *136f*, 136–137, *137f*
  SubtiList online database 141
  transcription 142
  transcription termination 124
  translation 142
Bacillus thuringiensis 25, 511
backcross **144–145**, *444f*, 911, 971–973, 1289, 1400
  see also congenic strain
background selection 145–146
  Charlesworth, Brian 145
  Charlesworth, Deborah 145
  Drosophila melanogaster 145
  mutation rates 145
  polymorphism 145
  population genetics 145
  process 145
  selective sweep 1804
  Wright–Fisher population model 145
back mutation **1275**
BACs (bacterial artificial chromosomes) **135**, 547, 980–981, 1394–1395
  chromosome mapping 362–363
  chromosome walking 375–376
  Human Genome Project 135
  Lotus japonicus 1122
  microbial genomics 1201
  plasmid vectors 135
  vectors 869–870
bacteria **146–151**
  aminoacyl-tRNA binding 610
  Bacillus subtilis 148
  bioluminescence 311
  Borrelia burgdorferi 148f
  cAMP nucleotide 260
  carbohydrate phosphotransferase system (PTS) 281–282
  catabolite repression 260, *282f*, 282–283, *283f*
  cell division genetics 297–298
  characteristics 146–147, *147f*, *148f*
  chromosome dimer resolution 351–353
  chromosomes 149–150, *150t*
    circular 149
    Haemophilus influenzae 149
    linear 149
    open reading frames *150t*
    plasmids 149
  cis-acting proteins 380
  developmental genetics 531
  diversity 147–149
  DNA cloning 547, 548, 549
  elongation factors 610
  Epulopsicum fishelsoni 147
  Escherichia coli 148
  gene transfer
    conjugation 151
    transduction 151
    transformation 151
  heat shock proteins (Hsps) 915
  histidine genes *946f*
  linkage 1105
  Luria, Salvador E. 1123
  Micrococcus luteus 147f
  Myxococcus xanthus 147
  Proteus vulgaris 147f
  RNA phages 1744–1746
  RNA polymerase 1746–1747
  taxonomy
    Aquificales 149
    Chlamydia 149
    Cyanobacteria (blue-green algae) 149
    Cytophagales 149
    Deinococcus/Thermus 149
    Fusobacteria 149
    gram-positive Bacteria 149
    green non-sulfur Bacteria 149

bacteria (*continued*)
green sulfur Bacteria 149
Nitrospira 149
Planctomyces 149
Proteobacteria (Purple Bacteria) 149
Spirochetes 149
Thermotagales 149
Thiomargarita namibiensis 147
translocation 610
bacterial artificial chromosome (BAC) *see* BACs (bacterial artificial chromosomes)
bacterial genes **151–156**
gene names 152
gene organization 154–155
domain structure 154–155
frameshifting 154
inteins 154
intervening sequence 154
introns 154
lac operon *153f*, 154
operons 154
overlapping 154
gene sequences 155–156
BLAST program 156
codon usage 155
DDBJ (DNA Data Bank of Japan) 155
DNA polymerases 155
EMBL (European Molecular Biology Laboratories) 155
GenBank 155
modules 156
NCBI web site 156
open reading frames 155
sequence similarity 155
gene types 152
mutations 152
nomenclature 151–152
phenotypes 152
structure 152–154
cis/trans tests 153
cistron 153
complementation 153
genetic code 153
protein-coding genes 152–153, *153f*
RNA genes 153
bacterial genetics **156–163**
circular DNA molecules 162
genetic analysis 160–162
genetic variation 156
gene transfer *159t*, 159–160
homologous sequences 162
importance of *Escherichia coli* 654–656
mutations *157t*, 157–159
effects of mutations 157
enrichment 159
isolation of mutants *157t*, 157–158
screening 158–159
selection 158
wild-type strains 157
transposable elements 162
insertion sequence (IS) 162
transposons 162
see also plasmids
bacterial transcription factors **163–165**
RNA polymerases 163–165
sigma factors
anti-sigma factors 163–164
function 163
regulation 163–164
transcription activators 164–165
catabolite repressor protein (CRP) 164
bacterial transcription factors (*continued*)
cyclic AMP (cAMP) 164
merP promoter 164
sigma^{54}-holoenzyme 164–165
transcription repressors 165
Cyt repressor 165
LacI protein 165
lac repressor 165
bacterial transformation **165–168**
artificial competence 168
chemical competence 168
competence 165–168
DNA uptake 165–168
electroporation 168
natural competence
distribution 166
importance 166
mechanisms 167
recombination 167
regulation 167
specificity 167–168
bacteriocines 1124
bacteriophage λ 173–176, 260–262, 638, 656, 926, 998, 1029, 1040, 1080
attachment sites *120f*, 120–121
meiotic hot spots 976–977
transcription termination 125
bacteriophage M13 656, 1012
bacteriophage Mu 174, *174f*
bacteriophage recombination **168–177**
bacteriophage recombination
hot spots 174–175
intron mobility 175
history 168–169
homologous recombination 169–173
concatameric DNA 169–171
DNA replication 169–172, *170f*, *171f*
Okazaki fragments 172
phage λ 171–174
phage T4 169–173, *170f*, *172f*
phage T7 170–172, *171f*
recombination-dependent replication (RDR) 171–172, *172f*
hot sports 172–173
intron mobility 173
Luria, Salvador E. 169
Phage and the Origins of Molecular Biology 169
site-specific recombination 173–175, *174f*
att sites 174, *174f*
bacteriophage Mu 174, *174f*
transposons 174
T-phages 169
bacteriophages **179–186**
bacteriophage λ 260–262
behavior 181
characteristics 179
classification *181f*, 181–182
Caudoviridae 181
Myoviridae 181
Podoviridae 181
Siphoviridae 181
coliphage PRD1 184
composition 180
cyanophages 183
definition 179
D'Herelle, Félix 511–512
distribution 179
DNA (deoxyribonucleic acid) 926
early genes 599
evolution 185
genome sequence on-line database 182
bacteriophages (*continued*)
gram-positive Bacteria 183
phage ϕ29 184
SPO1 183–184
helper phage 917
Hershey, Alfred 925–926
heteroduplex DNA 931–932
history
Anderson, Thomas 180
Chase, Martha 180
Delbrück, Max 180
D'Herelle, Félix 179
Ellis, Emory 180
Hershey, Alfred 180
Hershey–Chase Experiment 181
Luria, Salvador E. 1123
plaque method 179–180
single-step growth curve method 180
study techniques 179–180
Twort, Frederick 179
immunity 998
linkage 1105
morphology 181, *181f*
phage P1 182
phages P2 and P4 182–183
phage P22 183
phage ϕ6 184
phage T4 184–186
Pittsburgh Bacteriophage Institute 182
replication 917
temperate bacteriophages 184
Clostridium botulinum 185
Cornebacterium diphtheriae 185
diphtheria toxin 185
pyrogenic staphylococcal enterotoxins 185
pyrogenic streptococcal enterotoxins 185
Shiga-like toxins 184
see also Archaea; filamentous bacteriophages; phage Mu; T phages
bacteriophage therapy **175–179**
advantages 178
current research
Eliava Institute of Bacteriophages, Microbiology and Virology 176–177
Hirszfeld Institute of Immunology and Experimental Therapy, Polish Academy of Sciences 177
Intestiphage 176
Phagobioderm 177
Piophage 176
Tbilisi, Georgia 176–177
therapeutic process 177–178
D'Herelle, Félix 175–178
history 175–178
precautions 178
Bacterium subtilis 1078
Bacterium tumefaciens 491
baculovirus system **186**
Baer R **1930–1932**
Baert J-L **661–663**
Bafico A **126–128**
Bain B **680–681, 1190–1191, 2135–2136**
Baird T T Jr **2071–2075**
balanced polymorphism **186–188**
frequency-dependent selection 187–188
multiple-niche polymorphish 188
mutation–drift balance 187
overdominance
ABO blood groups 187
Drosophila melanogaster 187

balanced polymorphism (*continued*)
Gaucher's disease 187
glucose 6-phosphate dehydrogenase (G6PD) deficiency 187
Niemann–Pick disease 187
sickle-cell anemia 187
Tay–Sachs disease 187
selection–migration balance 187
transient polymorphism 187
balanced translocation **188–190**
definition 188
distribution 188
humans 188–189
medical problems
chromosome 22 189
chromosome 9 189
chronic myelogenous leukemia (CML) 189
Philadelphia chromosome 189
reproduction 189
meiosis 189, *189f*
alternate segregation 189
quadrivalent 189
BALB/c mouse **190**, 1482, 1636
Balbiani rings (BR) **190–191**, 1513
Ballou, Jon 459
Balmain A **191**
Baltimore, David 590, 1702
Bambusoideae 890
BARE-1 gene 2022
Bar gene 973
barley *see Hordeum* sp.
Barley Genetic Stock Center 853
barley mild mosaic virus (BaMMV) 973
barley yellow dwarf virus (BYDV) 973
barley yellow mosaic virus (BaYMV) 973
Bar mutant 2095, *2095f*
Barnes W M **1499–1503**
Barrand M A **1430–1433**
Barratt, R. 849
Barr body **191**, 629, 1054, 1809–1810
Barr, Murray 1809
Barsac oleria 2009
Barsh G S **23**, **397–401**
Bartel D P **1751–1752**
basal cell carcinoma **191**
base analog mutagens **191–192**
aminopurine (2-aminopurine) 191
azidothymidine (3′-deoxy-3′-azidothymidine, AZT) 192
bromouracil (5-bromouracil) 191
diaminopurine (2,6-diaminopurine) 191
hydroxyaminopurine (6-hydroxyaminopurine) 191
base composition **192**
base pairing 192
codon bias 192
eukaryotes 192
GC content 192
prokaryotes 192
base excision repair (BER) 560, *562f*, 888, 1663–1664, *1664f*
see also DNA repair; repair mechanisms
base pair **193**
base pairing/mispairing **193–197**
consequences 197
Crick, Francis 193–197
DNA polymerases 193–197
ionization 193–197
research techniques 195–196
chemical synthesis 195
NMR spectroscopy 195
base pairing/mispairing (*continued*)
X-ray crystallography 195
structure 193–197, *196–195f*
tautomerization 193–197
Watson–Crick model 193–197
Watson, James 193–197
wobble hypothesis *194–195f*, 196
bases **198–199**, *199f*
base substitution mutations **197–198**
cause 197–198
missense mutation 198
nonsense mutation 198
silent mutation 198
transition mutations 197, *198f*
transversion mutations 197, *198f*
basic helix–loop–helix (bHLH) proteins *1931f*
Basidiomycetes 749, 1882
Bates, Henry Walter 2132
Bateson, William 353, 512, 521, 962, 2008
early genetic research 1173
genes 759
Mendel's Principles of Hereditry 701
Batten disease 316
Battus philenor 731
Baumeister R **1320–1322**
Baur, Erwin **199–203**
Antirrhinum genetics 200
contributions
Antirrhinum genetics 199
neo-Darwinism 199
plant chimeras 199
plant virology 199
plastid genetics 199
eugenics 202
infectious chlorosis 199
Kaiser-Wilhelm-Institut für Züchtungsforschung Müncheberg 202
Lupinus angustifolius 201
Lupinus luteus 201
neo-Darwinism 200–201
non-Mendelian inheritance 199–200
plant breeding projects 201
plant chimeras 200
publications
Bibliotheca Genetica 201
Der Züchter (Theoretical and Applied Genetics) 201
Die Wissenschaftlichen Grundlagen der Pflanzenzüchtung (The Scientific Basis of Plant Breeding) 201
Einführung in die experimentelle Vererbungslehre (Introduction to Experimental Genetics) 201
Handbuch der Vererbungswissenschaft (Handbook of Genetics) 201
Zeitschrift für induktive Abstammungs- und Vererbungslehre (Molecular and General Genetics) 201
seed collections 201
Bax gene 274
BAX gene 423
Bayesian analysis **203–205**
Bayes theorem 203
linkage map 204
logarithm of the odds (LOD) score 204
null hypothesis 204
Bayesian hierarchical model 1159
bcd (bicoid) gene *960f*, 961
B-cell chronic lymphocytic leukemia (CLL) *see* leukemia, chronic
B-cell precursor acute lymphoblastic leukemia (BCP-ALL) *see* leukemia, acute
B-cell prolymphocytic leukemia (B-PLL) *see* leukemia, chronic
B-cells 953, 998, 1251
BcgI 1011
B chromosomes 1807
BCL-2 family 314
BCL-2 gene 1208
Bcl-2 gene family **205–207**
apoptosis regulation 205–207
Bcl-2 homology domains
Bax subfamily 208
Bcl-2 subfamily 205
structure 208
biological roles 208
carcinoid tumors 274
caspase 205
caspase activators 207
discovery 205
function 205
organelle interaction 207
pathways
CED-3 protein 205
CED-4 protein 205
CED-9 protein 205
EGL-1protein 205
regulation mechanisms 208
Bcr/Abl fusion protein 208
BCR/ABL oncogene **207–208**, 1208, *1208f*, 1450, *1450f*
acute lymphoblastic leukemia (ALL) 208
acute myelogenous leukemia (AML) 208
chromosome 22 207–208
chromosome 9 207–208
chronic myelogenous leukemia (CML) 207–208
distribution 208
history 207–208
mechanisms 208
non-Hodgkin's lymphoma (NHL) 208
Philadelphia chromosome 207–208
BCRF1 gene 643
Beadle, George 419, 1080, 1323
genes 760
Nobel Prize 1070
bead theory **208–209**
Beagle, HMS 513–514, 1291
Beamish C **2042**, **2107**
bean *1914t*
bean yellow mosaic virus *1914t*
Beaudet, Arthur 1179
Becker, Fred 67
Becker H-A **2020–2033**
Becker muscular dystrophy 587, 1262, *1263f*
Beckingham K M **591**, **827–828**, **981**
Beckwith–Wiedemann syndrome **209**, 996, 1000, 1770, 2139
Beechey C V **2003–2007**
Beggs J D **1536–1539**
behavioral genetics **209–212**
Benzer, Seymour
clock mutants 210
Drosophila melanogaster 210
neurogenetics 210
Brenner, Sydney 211
Caenorhabditis elegans 211

behavioral genetics (*continued*)
eugenics 209
future research 212
Galton, Francis 209
human behavioral genetics 211
intelligence quotient 210
molecular behavioral genetics 210–211
Aplysia 210
Drosophila melanogaster 210
Scheller, Richard 210
Plato 209
twin studies 210
Beijerinck, M.W. 521
BEL1 gene 716
Belfort M **1047–1052**
Belgian Coordinated Collections of Microorganisms (BCCM) 851
Bell, Julia 354
Benfey P N **1474–1477**
Benne, Rob 1741
Bennetzen J L **863–865**
Bentham, George 1083
Benzer, Seymour **212**, 383, 655, 1314, 1927
bacteriophage genetics 212
cistron 212, 383
clock mutants 210
countercurrent apparatus 1314
Delbrück, Max 212
deletion mapping 795
Drosophila melanogaster 210, 212, 386, 1314–1315
fate mapping 1315
genes 760
hot spots 977
Lwoff, André 212
neurogenetics 210
techniques 1314
X chromosome mutagenesis 1314
benzo(a)pyrene diolepoxide 557
Berenbaum, M. R. 1535
Berg, Paul 877
Bergstrom D E **911–912**, **1648–1649**
Berk A J **15–18**, **954**, **1029**
Berlyn M K **846–854**
Bernardi A **544–550**
Berns A **1470**
Bernstein A **1887–1888**, **2125–2127**
Bernstein, Harris 1927
best linear unbiased prediction (BLUP) method 1799, 1792
Bestor, Timothy 630
beta (β)-amyloid peptide 1–42, 317
beta (β)-catenin 620, 13
beta (β) cellulin (BTC) 627
beta (β)-ENaC ion channel 315
beta (β)-fold motif 1655
beta (β)-galactosidase **212–214**, 303, 312, *331f*, 334, 819, 1020, 1069, 1070
applications 213
binding sites 213
lac operon 212
lactose 1070
lactose interaction 213
lacZ gene 212
metal requirements 213
*o*NPG 213
*p*NPG 213
structure 213
see also chimera
beta (β) globin gene 878, 2096
beta (β)-glucosidase 753
beta (β)-strand 1569–1570, *1570f*
BF gene *430t*, *431t*
Bg gene 2026
bgl operon 1079
BH3 family 314
Bhattacharya S S **1697**
Bhattacharyya's distance 828, *829f*
bHLH proteins *see* basic helix–loop–helix (bHLH) proteins
BHRF1 gene 643
Bibliotheca Genetica 201
bidirectional replication **214**
Biek, Donald 1624
BIME (bacterial interspersed mosaic elements) 214
bim gene 208
binomial distribution **215–216**
biochemical genetics **216–218**
classical approach 216–217
definition 216
genome sequences 217
modern approach 217
tryptophan (Trp) pathway *216f*, 217
bio gene 1436, 1983–1984
bioinformatics 1201
biolistics 973
biological clocks 386–390, 390
biological diversity 458
biological markers *see* β-galactosidase; green fluorescent protein (GFP)
bioluminescence *see* green fluorescent protein (GFP)
BioMolecular Engineering Research Center (BMERC) 519
bio operon 1858–1859, *1859f*
biosynthetic pathways **218–224**, 655
classical genetic approaches
chemical analogs 220
complementation 220
crossfeeding 221
epistasis 221–222
functional domains 221
genetic mapping 221
hypersensitivity 220
mutagenesis 219
mutant enrichment 219
plasmids 220
potential problems 222
simple screening 218–219
transposon insertions 220
transposons 219
enzyme purification 222
genomic libraries 222
genomics
applications 223
genome sequencing 223
regulation 224
reverse genetics 222
small molecules 218–224, *219f*
biotechnology **224–227**
background 224–227
contributions
drugs 225, *226f*
gene therapy 226
plants 225–226, *227f*
Penicillium chrysogenum 225
Pseudomonas syringae 226
techniques 225
biotin 2118
bipolar affective disorder *2015t*
Bipolaris maydis 1223
birA repressor 1896
Birchler, J. 849
Bird, Adrian 629
birth defects *see* congenital disorders
Biston betularia 1534
Black D L **47–48**
Blank R D **29–35**, **91–95**, **203–205**, **246–248**, **258–260**, **814–819**, **1407–1410**, **1549–1551**, **2165**
Blashko's lines 600
Blastocladiales 748
blastomeres *615f*, 616–618, *620f*
BLAST program 33, 156, 1197
Blender Experiment 554, 925
blending theory 2010
blepharochalasis 499
Bleuler, E. 1774
blindness 269
Blk protein 1883
BLM gene 229–230, *2084–2085t*
BLM protein 1646, 1912
Blobel, Günter 1553
blood group chimeras **227–228**
Blood-Group Serum Unit 701
blood groups polymorphism 1507
blood group systems *228t*, **228–229**
Bloomington *Drosophila* Stock Collection 847
Bloom's syndrome (BS) **229–230**, *569–570t*, 571
chromosomal instability 1285
clinical features 229–230
complications 230
diagnosis 230
recombination nodules (RNs) 1646
BLOSUM weight matrix 32, *32f*
Blumenthal T **2040–2041**
blunt-end ligation **230**
bmall gene 389
BMERC (BioMolecular Engineering Research Center) 519
Bmp5 (se) gene 524–525
bobbed (bb) mutant 2095
bodenlos (bdl) mutation 1475
Bohr, Niels 522
bom (basis of mobility) gene 450–451
Bombay blood group phenotype **230**
Bombyx mori **231–233**
Bombyx mandarina 231
classical genetic approaches 231–232
cloning 232
cytogenetics 231
expressed sequence tags (ESTs) 232
gene regulation 232
genetic applications 232
genetic resources 231
linkage maps 232
mulberry silkworm 231
SilkBase database 232
Bonhoeffer, Friedrich 257
Bonhomme F **1259–1261**, **1261–1262**
Bonhomme, François 1248
bootstrapping *see* trees
Bordetella 1556
Bordetella pertussis 452, 454–455
Bordet, Jules 512
Borges-Walmsley M I 249–260
Born, Max 522
Borrelia burgdorferi 80, *148f*, *150t*, 192, 404, 859, 1198
Borrelia hermsii 80
Bos taurus 366t
Botrytis cinerea 1914t
Botstein, David 185, 357
bottleneck effect **233–235**, 640, 840, 1802
effective population number 602
evolution 666

bottleneck effect (*continued*)
  evolutionary biology 516
  evolutionary factors 234
  founder effect 233
  genetic effects 233–234
  linkage disequilibrium 1105
  population genetics 233
boundary equilibrium 2068
Boursot P **1259–1261**
Boveri, Theodor **235–236**, 618, 1177, 1408, 2011
  chromosome studies 236
  Zur Frage der Entstehungen maligner Tumoren 236
bovine spongiform encephalopathy (BSE) 1068, 1879, 1880, 2008
Boyd Y **2145–2147**
Boyer, Herbert 1638
Boynton, John 1462
B-prolymphocytic leukemia (B-PLL) **236**
  see also leukemia
BPY1 gene 2159
BPY2 gene 2159
Braaten B **1442–1444**
brachydactyly **236–237**
Brachydanio rerio 1320
brachyury locus **237**, 1943
  DiGeorge syndrome 237
  Holt–Oram syndrome 237
  T-box genes 237
  T (Brachyury) locus 590
Bradyrhizobium japonicum 709, 885, 1329, 1334
bradytroph 133
Bragg, Lawrence 2134
Bragg, William 2134
Brahma chromatin remodeling complex *342t*
Brammer W J **1858–1860**
branch migration **237**, 1756–1758, *1758f*
Brassica 1422
Brassicaceae 86, 112, **237–238**
  Cruciferae 237
  molecular genetics 238
  mustard family 237–238
    Arabidopsis thaliana 238
    Armoracia rusticana (horseradish) 237
    Brassica juncea (Chinese mustard) 237
    Brassica napa (canola oil) 238
    Brassica nigra (black mustard) 237
    Brassica oleracea 237
    Brassica rapa (turnip) 238
    Raphanus sativus (radish) 237
    Sinapis alba (white mustard) 237
  phylogenetic analysis 238
    Anacardiaceae 238
    Capparaceae 238
    Malvaceae 238
    Onagraceae 238
    Rutaceae 238
    Sapindaceae 238
  see also Arabidopsis thaliana
Brassica juncea (Chinese mustard) 237
Brassica napa (canola oil) 238
Brassica nigra (black mustard) 237
Brassica oleracea 237
Brassica rapa (turnip) 238
brassinosteroids (BRs) **238–239**, 1481
BRCA1/BRCA2 genes **239–241**
  BRCA1 gene 10, *81t*, 114, 239–240, 262, 571, 1300
BRCA1/BRCA2 genes (*continued*)
  BRCA1 protein 239–240
  BRCA2 gene 10, *81t*, 240, 262, 564, 571, 1300
  BRCA2 protein 240
  breast cancer 242
  Breast Cancer Information Core (BIC) database 240
  breast/ovarian cancer syndrome 239
  clinical management 239
  DNA repair 241
  double-strand break repair 241
  founder effect 240, 242–243
  mouse studies 240
  mutational studies 240
  pathology 239
  transcription regulation 241
BRCA1 gene 496, 2082, *2084–2085t*, 2086–2087
BRCA2 gene 2082, *2084–2085t*, 2086–2087
BrdU labeling 339
break-copy/break-join **241–242**
breast cancer **242–243**, 263, *2015t*
  androgen receptor mutation 243
  BRCA1 gene 242
  BRCA2 gene 242
  chromosome 13q12 242
  chromosome 17q21 242
  Cowden's disease 243
  family history 242
  founder effect 242–243
  heredity 262
  Li–Fraumeni syndrome (LFS) 243
  screening 243
Breast Cancer Information Core (BIC) database 240
breast/ovarian cancer syndrome *569–570t*
  BRCA1 gene 242, 571
  BRCA2 gene 571
breeding of animals **243–244**
  backcross 244
  best linear unbiased prediction (BLUP) method 1792
  inbred strain 243–244
  incross 244
  intercross 244
  Mendel's laws 244
  outcross 243–244
  selection index 1792
  selection intensity **1793–1794**
Brembs B **906–910**
Brennan P J **1306, 1394**
Brenner C **661–663**
Brenner S **7, 86, 191, 193, 557, 663**, 700, **821–822, 884, 917, 983–984, 984, 989, 1055, 1095, 1189, 1377, 1379–1381, 1532–1533, 1587, 1710, 1710–1711, 1730, 1732–1733, 1768, 1834, 1838, 1945–1946, 1965, 2045–2048, 2154, 2075, 2090, 2164**
Brenner, Sydney **244–246**, 366, 815
  awards 246
  Caenorhabditis elegans 211, 245, 251, 299, 1309
  colinearity 419
  contributions
    amber mutation 245
    frameshift mutation 244
    molecular biology 244
    mRNA (messenger ribonucleic acid) 245
Brenner, Sydney (*continued*)
    ochre mutation 245
  gene sequencing 246
  Human Genome Project 245
  Jacob, François 244, 1059
  Molecular Sciences Institute (MSI) 246
brewer's yeast *see Saccharomyces cerevisiae*
Brewin N J **1334–1335**
BRG1 chromatin remodeling complex *342t*
Bridges B A **1676–1677, 1853–1855**
Bridges, Calvin Blackman 344, 353–354, 847, 962, 1240
Brilliant M H **1469–1470**
Brinker A **324–325**
Brink, R. Alexander 629
brlA gene 110, *110f*
5-bromo-4-chloro-3-indolyl-β-D-galactosidase 1069
bromouracil (5-bromouracil) 191, 196
Bronfenbrenner, Jacques 925
Brookfield J **1233–1234**
Broughton W **903**
Browman K E **27–29**
brown (b) gene 398
brown (*b*) locus 1599, 1871
Brown, Louise 689
Brown M S **682–683**
Brown T A **1855–1857**
Brown V **2048–2053**
BRs *see* brassinosteroids (BRs)
46BR syndrome 564
Bruce M E **1879–1882**
Bruck syndrome 1397
Bruessow, Harald 185
Bs1 gene 2022
BspMI 1011
BTAK gene 425
BUB1 gene 2083
BUBR1 gene 2083
Buck *v.* Bell 2015
Budarf M L **188–190, 190–192**
Buffon, Georges 1071
Bull, J. 1453
Bülow L **333–334**
buoyant density **246**
Burchell A **2120–2122**
Burgess R R **1831–1834**
Bürglin T R **958–962, 962–963**
Burkholderia 1478
Burkholderia cepacia 1480
Burkitt, Denis 246
Burkitt's lymphoma (BL) **246–248**, 643
  Aedes egypti 246
  B-cell non-Hodgkin's lymphoma 247
  Burkitt, Denis 246
  carcinogens 271
  cytogenetic analysis 247
  epidemiology 246
  Epstein–Barr virus (EBV) 247
  etiology 247
  gene deregulation 247
  histology 247
  human immunodeficiency virus (HIV) 247
  immunoglobulin gene superfamily 247
Bürkle A **1533**
Burn J **533–535**
Burt D W **1203, 2125, 2165**
BWS gene 1770
BX-C complex 962
BZLF1 gene 643

C

C1qA gene 429
C1qB gene 429
C1qC gene 429
C1q protein 429
C1r protein 429
C1s protein 429
C2 gene *430t*, *431t*
C2 protein 429
C3AR1 gene *430t*, *431t*
C3b protein 429
C3 gene *430t*, *431t*
C3 protein 429, 432
C4A gene *430t*, *431t*, 447
C4B gene *430t*, *431t*, 447
C4-binding protein 433
C4BPA gene *430t*, *431t*
C4BPB gene *430t*, *431t*
C4 protein 432
C57BL/6 mouse strain **250**
C5 gene *430t*, *431t*
C5 protein 432
C5R1 gene *430t*, *431t*
C6 gene *430t*, *431t*
C6 protein 429, 432
C7 gene *430t*, *431t*
C7 protein 429, 432
C8A gene *430t*, *431t*
C8B gene *430t*, *431t*
C8G gene *430t*, *431t*
C8 protein 432
C9 gene *430t*, *431t*
C9 protein 432
CAAT box **250**
c-ABL oncogene 207–208, **250–251**
c-abl oncogene 1484
c-Abl protein 208, 251
cac gene 1317
CACNA1A gene *2050t*
CACTA superfamily 2026–2027
cadherin 620
cadherin superfamily 1320
Caedobacter taeniospiralis 1413
Caenorhabditis briggsae 255
Caenorhabditis elegans **251–256**
  aging 22
  anatomy *252f*, 253
  apoptosis 314, *315f*
  base composition 192
  behavioral genetics 211
  cell biology 256
  cell death 314–318, *315f*
  cell division **298–302**, 304
    asymmetric 301, 306–308
    coordination 301
    cytokinesis 301
    genetic studies 299–300
    mitosis 300–301
    mitotic spindle 300–301
    reverse genetic studies 299–300
  cell lineage 254, 302, *304f*, 304–308, 314
    embryonic lineage *305f*
    gene 306–308
    homeotic cells 306
    larval stages (L1-L4) 306, *307f*
    mutations 306, *307f*
  centromere 860
  chromosome mapping 366–367, *367t*
  codon usage bias 405
  degenerins 314–316
  development 256, *300f*, 314
  developmental genetics **531–532**

Caenorhabditis elegans (*continued*)
  dispersed transposable sequences 861
  distal spermatheca 1373
  DNA cloning 545
  DNA methylation 633
  DNA transposons 2017
  embryonic development **612–621**, 622
    analysis methodology 614–615
    anatomy 613, *613f*
    anterior/posterior (A/P) axis 615–617, *616f*
    blastomeres *615f*, 616–618
    cell diversity 617
    cell lineage 613, *615f*
    dorso/ventral (D/V) axis 617–620, *618f*
    embryogenesis 613–614, *614f*
    epithelial cells 619–620
    first cell cycle 619–620
    gastrulation 617–618
    germline lineage 618–619
    left/right (L/R) axis 617
    mechanisms 619
    mosaic development 621
    muscle fibers 620, *620f*
  epithelial cells 620
  EST number *366t*
  gene number 796
  genes 255–256
  genome 251, 253, 254–255, 299, 1314
  genome sequencing 859
  genome size *367t*
  green fluorescent protein (GFP) 312–313
  growth 253–254
  heterochronic genes 928, 928–930, *929f*
  heterochronic mutation 928–930, *929f*
  heterotrimeric G proteins 934
  holocentric chromosomes 956–957
  homeobox genes 959
  homeotic genes 963
  ion channels 314–315
  kinases 1063
  life cycle 252–253, *253f*, 256
  model organism 253–254, 298–299, 314, 612, 1309
  mutagenesis 255–256
  necrosis *315f*
  Nematoda 251–252
  neurobiology 256
  neurodegeneration 314–318
  neurogenetics **1309–1314**
    advantages as model organism 1309
    AWA chemosensory neuron 1313
    axon outgrowth 1311
    brain disorders 1313
    cell-ablation studies 1309
    cell lineage 1310, *1311f*
    founder cells 1310, *1311f*
    guanosine triphosphatase (GTPase) 1311
    helix–loop–helix proteins 1310
    human homologs 1313
    LIM domain 1310
    nervous system 1310, *1310f*
    neuronal cell fate 1310
    neuronal development 1310–1312
    neurotransmission 1312–1313
    POU domain 1310
    sensory adaptation 1313
    sensory behavior 1313
    SNARE complex 1312–1313
  neuronal specification 1320

Caenorhabditis elegans (*continued*)
  neurons 314–318
  olfaction 1369
  oogenesis **1373–1374**
    cytoplasm accumulation 1373
    diakinesis 1373
    distal spermatheca 1373
    fertilization 1374
    germline development 1373
    GLD-1 protein 1373
    gonad 1373
    meiosis 1373
    meiotic cell division 1374
    mitogen-activated kinases (MAPK) 1374
    morphology 1373
    mRNA (messenger ribonucleic acid) 1373
    mutational studies 1374
    ovulation 1374
    pachytene stage 1373
    process 1373
    protein synthesis 1373
    regulation 1373
    sheath cells 1373
    syncytium 1373
  paternal inheritance 1422
  postgenomic tools 255
  regulatory RNA 1658
  replication errors *1676t*
  reproductive biology 1873
    hermaphrodite 1873
    mutants 1873–1874
    sexual system *253f*
  research areas 255–256
    cell biology 256
    developmental biology 256
    life cycle 256
    neurobiology 256
  resources 254
  reverse genetics 1706
  RNA degradation 1750
  RNA interference 1743
  self-fertilization 254, 1805
  sex linkage 1819
  sheath cells 1373
  size 254, 299
  spermatogenesis **1873–1874**, *1874f*
  Tc1 transposon 2036
  transgenes 1990
  translation control 2003
  transparency 254, 298, 314
  WWW server 254
  zygotic lethal gene 2167
Caenorhabditis Genetics Center (CGC) 254, 853
café au lait spots 1307
cagA gene 916–917
CAH *see* congenital adrenal hyperplasia (CAH)
Cairns, John **256–258**, 1651
  Cold Spring Harbor Laboratory 257
  DNA polymerases 257
  Escherichia coli 257
  Hershey, Alfred 257
  Matters of Life and Death 258
Cajanus (pigeon pea) 1083
Caldas C **916–917**, **2081–2088**
CAL gene 715
Calvin cycle 1463
cAMP *see* cyclic AMP (cAMP)
Campbell A **260–262**, **856–857**
Campbell, Allan 120

Campbell Model 120, **260–262**, *261f*
Camper S **591–593**
cAMP nucleotide
catabolite gene activator protein (CAP) 279–280
catabolite repression 260
cell signaling **258–260**
chemoattractant 260
Dictyostelium discoideum 260
eukaryotes 258–260
microorganisms 281–283
protein kinase A (PKA) 258–259
Saccharomyces cerevisiae 258–259
signal pathway *259f*
see also CAP (CRP)
campomelic dysplasia 1812, 1817, 1822
Canaani E **1227–1228**
Canavalia (sword or jack bean) 1083
cancer
carcinogens 271–272
hereditary syndromes 262–263
nucleolus organizing region (NOR) 349
telomeres 1949–1950
see also carcinoid tumors
Cancer Chromosome Laboratory 1095
Cancer Genome Project 1773
cancer susceptibility **262–263**
see also carcinogens; plasmacytomas (mouse)
Candida 738
Candida albicans 1154
candidate gene **263–264**
Candido E P M **914–915**
Canis familiaris see dogs
Canis latrans 1866
Canis lupus 1866
Cap **270**
CAP *see* catabolite gene activator protein (CAP)
capacitation 689, 691
CAP (CRP) **270**, 1653
see also catabolite gene activator protein (CAP)
capillary electrophoresis 758–759
Capparaceae 238
capsid **271**, 554, 696, 2105, 2109
carbohydrate phosphotransferase system (PTS) 281–282
carcinogens **271–272**
Burkitt's lymphoma 271
cell lines 310
chemicals 272
epigenetic 272
genotoxic 272
ionizing radiation 271
retroviruses 271–272
smoking 272
UV radiation 271
viruses 271–272
see also cancer susceptibility
carcinoid tumors **272–274**
Bcl-2 gene family 274
carcinoid syndrome 273
cyclin 274
histology 273
karyotypes 274
lungs 272, 273
p53 protein 273–274
peptides 273
ploidy 273
telomeres 274
carcinoid tumors (*continued*)
tumorlets 272
types 273
see also cancer
carcinoma *see* basal cell carcinoma
Cardaminopsis 86
CA repeats **250**
Carex 322
Carey J C **2058–2060**
carnation *1914t*
Carneiro F **916–917**
Carpenter A T C **116–120**
Carr D **733**
carrier **274–275**
cystic fibrosis 275
detection 275
frequency 274–275
genetic counseling 275
sickle cell anemia 275
Tay–Sachs disease 275
Carroll D **558**, **778–780**, **841–845**, **996–998**, **1610–1612**, **1637–1639**
Carruthers, Marvin 555
Carter, T.C. 1871
Cartron, J-P. 1713
Caruana G **1887–1888**, **2125–2127**
Case S M **513–517**
Cashel M **1891–1892**
Caspersson, T. 348
Casselton L A **1151–1153**
cassette model **275–278**, 278
cassette mutagenesis *see in vitro* mutagenesis
Castle, William E. **278**, 397, 590, 912, 2012
catabolite gene activator protein (CAP) **278–280**, 1070
cAMP nucleotide 279–280
CAP–DNA complex 280, *280f*
discovery 279
Escherichia coli 279
helix–turn–helix motif 280
lac operon 278–279
structure *279f*, 280
see also CAP (CRP); catabolite repression
catabolite repression **281–284**
Bacillus subtilis *283f*, 283–284
bacteria 260
cAMP nucleotide 260
enteric bacteria *282f*, 282–283
lac operon 279
Saccharomyces cerevisiae 284
see also catabolite gene activator protein (CAP)
catabolite repressor protein (CRP) 164, 544
CATCH 22 534
Catovsky D **236**, **903–904**
cats *see* feline genetics; tortoiseshell coloring
Cattanach B M **284–286**
Cattanach's translocation **284–286**, 2005
chromosomal breakpoints 285
inheritance 285
phenotypes 285
position effect variegation 285–286
sex chromosome aneuploids and homozygotes 285
uses 286
see also translocation
caudal (cad) gene 959
Caudoviridae 181
Caulobacter crescentus 303
C-banding 348
CBL2 gene *2049t*
CBP/p300 gene *2084–2085t*
CBP transcription factor 1973
Cbx2 gene 1818
CC *see* commitment complex (CC)
C/c antigen 651
c (colorless aleurone) gene 2020
C/c polypeptide 651
Cd36 gene 1608
CD4+ T cells 533, 979, 1251
CD59 gene *430t*, *431t*
CD5+ B cells 1093
CD8+ T cells 641, 1251
cda (chablis) gene 2048
cdc13$^+$ gene 500
CDC2a gene 1755
cdc2$^+$ gene 500
Cdc2 protein 291–292
Cdc45 protein 1384
Cdc6 protein 1384
cdc gene 298, 500
CDH1 gene *2084–2085t*, 2086
CDK *see* cyclin-dependent kinases
CDK12A gene *2084–2085t*
CDK1 protein 502
CDK2 protein 502
CDK4 gene 1772, 2085
CDK4 protein 502–502
CDK5 protein 504
CDK6 genes 1093
CDK6 protein 502–503
CDK7 protein 503–504
CDKN6/P16 gene 1772
cDNA *see* complementary DNA (cDNA)
cdo (chardonnay) gene 2048
cd-tbx9 gene *1944f*
CDY gene 2160
Cech, Thomas 1730, 1751
ced-1 gene 314
ced-2 gene 314
ced-3 gene 205, 314
ced-3(lf) gene 316
CED-3 protein 205, 314
ced-4 gene 205, 314
ced-4(lf) gene 316
CED-4 protein 205, 314
ced-5 gene 314
ced-6 gene 314
ced-7 gene 314
ced-9 gene 205
CED-9 protein 205, 314
ced-10 gene 314
ced-12 gene 314
ced-14 gene 1097
ced genes 255
C. elegans see Caenorhabditis elegans
Celera 356, 364–365, 367
cell adhesion molecules 1320
cell culture *see* tissue culture
cell cycle **286–296**
Cdc2 protein 291–292
chromosome segregation 288, *290f*
cyclin/CDK oscillator 292–294, *293f*
cyclins 291–294
cytokinesis 288–290, *291f*
DNA checkpoint controls 295–296
DNA replication 287, 294
external conditions 294–295
interphase 287–288
mitosis 288
regulation *503f*
Schizosaccharomyces pombe *292f*
stopping and starting 294–296
timing 290–296

cell cycle (*continued*)
  ubiquitin 2092
  yeast *292f*
  see also cell division genetics; chromatid; chromosome; cytokinesis; kinetochore; mitosis
cell death 313–314
  Caenorhabditis elegans 313, 314–318, *315f*
  neurons 314–318
  see also apoptosis; necrosis
cell determination **296–297**
  cell signaling *277*
  environmental stimuli 276, 304
  homeotic cells 306
  intrinsic mechanisms 304
  mating type locus 276
  yeast 275–277
  see also cell lineage; embryonic stem (ES) cells
cell differentiation
  cell lines 311
  chromosome break 350
  yeast 1153–1154
cell division
  asymmetric 303, 306–308
  Bacillus subtilis 303
  branching tree 303, *303f*
  Caenorhabditis elegans 304, 305, 306–308
  Caulobacter crescentus 303
  chromosome dimer resolution 353
  diversifying 303
  proliferative 303
  Saccharomyces cerevisiae 303
  types 303
  Xer recombination 353
cell division genetics **297–298**
  bacteria 297–298
  Caenorhabditis elegans 299–300
  cdc gene 298
  Drosophila melanogaster 298
  eukaryotes 297
  Saccharomyces cerevisiae 298, 1153
  Schizosaccharomyces pombe 298, 1153
  see also cell cycle; pattern formation
cell lineage **302–310**
  Caenorhabditis elegans 302, *304f*, 304–308, *305f*, 314
  chimera 303, 331
  clonal analysis 302–303, *303f*
  compartmentalization **426–427**
  direct observation 302
  Drosophila melanogaster 302, 303, *303f*, *308f*, 308–309, **1415**
  evolution 309
  history 302
  plants 309
  vertebrates 309
  see also cell determination; parasegment; pattern formation
cell lines **310–311**
  carcinogens 310–311
  cell differentiation 311
  cell signaling 311
  epithelial cells 310
  fibroblasts 310
  stem cells 310
  see also tissue culture
cell markers *see* green fluorescent protein (GFP)
cell/neuron degeneration **313–318**
cells
  creation 286–287
  cytokinesis 288–290, *291f*
  degeneration **313–318**
  interphase 287–288
  microtubule 287, 288, *290f*
  mitosis 288, *289f*
  senescence 310, 311
cell signaling
  cAMP nucleotide **258–260**
  cell determination *277*
  cell lines 311
  yeast *276–277*
cell specialization *see* cell determination
cenancestor **318**
CEN gene 715
CENP-A protein 1065
CENP-B protein 1065
CENP-C protein 1065
CENP-E protein 1065, 1225
CENP-F protein 1065
Center for Genetic Research Information (Japan) 850
centimorgan (cM) **319**, 699, 904, 1140, 1141
centiray (cR) 360
Centre d'Etude du Polymorphisme Humaine (CEPH) 358, *358t*
centric fusion **319–320**
  see also Robertsonian translocation
centrioles **320**, 2166
centromere **320–323**, *841f*, 996
  acentric fragment **2**
  acrocentric chromosome **3**
  chromatid 321–322
  chromosome 10 (human) 322
  chromosome bridge 350–351
  compound chromosomes 116–117
  Darlington, Cyril Dean 512
  DNA 322, *322f*
  DNA probes 1003
  double-minute chromosomes (*dmin*) 580
  Drosophila melanogaster 322
  first and second division segregation 698, *699f*
  holocentric chromosomes 956–957
  isochromosome 1054
  kinetochores 321, 1064
  linkage 1104
  metacentric chromosome 1189
  mitosis 321–322, 1225
  mitotic spindle 321–322
  neocentromere 322
  Saccharomyces cerevisiae 322, *322f*
  Schizosaccharomyces pombre 322f
  structure 321, *322f*
  tetrad analysis 682
  Y chromosome (human) *322f*
  yeast *322f*
  see also chromosome banding
CEPH *see* Centre d'Etude du Polymorphisme Humaine (CEPH)
cephalosporins 1890
Ceratis capitata 985
Ceratonia (carob) 1083
Cerberus gene 2150
Cercis (judas tree, redbud) 1083
cereba gene 2026
Cervus canadensis 1295
ce-tbx2 gene *1944f*
ce-tbx7 gene *1944f*
ce-tbx8 gene *1944f*
ce-tbx11 gene *1944f*
ce-tbx12 gene *1944f*
ce-tbx17 gene *1944f*
CF *see* cystic fibrosis
CF gene 359
CFTR *see* transmembrane conductance regulator (CFTR)
Cftr gene 1245
CFTR protein 507
C genes **249**
CGH *see* comparative genomic hybridization (CGH)
CGN1 gene *430t*, *431t*
Chaganti R **235–236**
Chagas disease 1736
chain initiation, elongation, and termination **323–324**
Chakraborty R **233–235**
Chalfie M **311–313**
Chandler M **1029–1037**
Chandley A C **1255**
Changeux, J.-P. 39
chaperonins **324–325**
  see also heat shock proteins; proteins
character **325**, 1416
character state **325**
C-Ha-ras oncogene 1484
Chargaff, Erwin 325
Chargaff's rules **325**, 477
Charlesworth, Brian 145
Charlesworth, Deborah 145
charon phages *see* vectors
Chase, Martha 554, 815, 925
  bacteriophages 180, 654
CHD4 chromatin remodeling complex *342t*
Chediak–Higashi syndrome 26, 400
Chedin F **1731–1732**
che genes 255
chemoaffinity theory 1319
chemoheterotrophs 147
chemolithoautotrophs 147
chemotaxis 711
Chetelat R **1125–1127**
chiasma **328–330**, *329f*, 512, 927
  adjacent/alternate disjunction 18, *18–19f*
  chromatid 328
  chromosome 328–329
  chromosome break 330
  holocentric chromosomes 957
  interference 329
  localized 329
  meiosis 536, 875
  multiple 329
  recombination nodules (RNs) 330
  resolution *329f*, 329–330
chimera **330–333**
  129 inbred strain 1016
  amphibians 331
  cell lineage 331
  embryogenesis 331
  Ford, Charles 723
  hermaphrodite 925
  mosaic 332–333
  mouse *331f*, 331–332, *332f*
  parthenogenesis 2098
  recombinant DNA 1637
  teraparental mouse **1957**
  uniparental inheritance 2098
  see also β-galactosidase; interference
chimeric genes and proteins **333–334**
  Escherichia coli 333
  gene fusion 333

chimeric genes and proteins (*continued*)
  linker regions 333
chirality 55 *see* handedness
Chironomidae 763
Chironomus 985, 1511
Chironomus tentans 190
Chi sequences **325–328**
  Bacillus subtilis 328
  double-strand break repair 1668
  Escherichia coli 977
  eukaryotes 328
  genetic properties 326
  Klebsiella pneumoniae 328
  Lactococcus lactis 328
  phage λ *327f*
  RecBCD enzyme 325 328, *327f*, 1623
  Salmonella typhimurium 328
  see also RecBCD enzyme; recombination
Chisholm A **296–297**, **302–310**
chi-square test 729–730, 751, 1104, 1589, 2131
CHK2 gene 1770
Chlamydia 149, 1199, 1556
Chlamydia trachomatis 150t, 1892
Chlamydomonas 535, 1953, 2026
  RNA degradation 1750
  uniparental inheritance 2097
Chlamydomonas Genetics Center (CGC) 851
Chlamydomonas reinhardtii **334–337**, 748
  cells 334, *335f*
  chloroplast DNA 336
  chloroplasts biogenesis 336–337
  flagella 337
  life cycle 334–335
  mitochondrial DNA 336, 1219
  mitochondrial inheritance 1222
  model organism 334, 336
  nuclear and chloroplast transformation 336
  nuclear genes 335–336
  photosynthesis 335, 1461
  see also chloroplasts; photosynthesis
CHLC *see* Co-operative Human Linkage Centre (CHLC)
Chloealtis conspersa 1646, *1646f*
chloramphenicol 1686, 1729, 1890
Chlorella vulgaris 2025
Chloridoideae 890
chlorophyll 335, 337
chloroplast DNA *see* ct DNA
chloroplasts **337–338**, 1379, 1466
  biogenesis 336–337
  DNA recombination 338
  genome 335–336, 337–338
  maternal inheritance 338
  mRNA 338
  shared coding 338
  see also Chlamydomonas reinhardtii; mitochondria; symbionts
CHN/TEC gene 1772
CHO cell line 1968
cholesterol 445
chondrosarcoma 1768, *1769t*
Chong S **1565–1567**
chordin gene 2150
chorion villus sampling (CVS) 1540
Chorthippus parallelus 840
Chory J **238–239**
CHRAC chromatin remodeling complex *342t*
Christmas disease **338**, 918
  see also hemophilia
chromatid 321–322, **338–339**
  chiasma 328
  expression 340–341
  first and second division segregation 698, *699f*
  genetic recombination *841f*
  harlequin chromosomes 339
  meiosis 339–340
  mitosis 339, 1225
  nondisjunction 1346
  recombination 339
  sister chromatid exchange (SCE) **1774**
  unequal crossing over 339
  uninemes 373
  X-chromosome inactivation 343
  see also cell cycle; chromosome; chromosome break; crossing-over; meiosis; mitosis; unequal crossing-over
chromatid interference **339–340**
  see also interference
chromatin **342–343**, 1524, 1524–1525
  apoptosis 313
  chromatin loops 374
  DNA 340
  DNA methylation 632–633
  expression 340, 340–341
  histone genes 950
  histones 340, *342f*
  holocentric chromosomes 957
  imprinting, genomic 1001
  kinetochore 1064
  minichromosome 1205
  necrosis 313
  nucleosome 340, 340
  remodeling complexes *342t*, 342–343
  replication 343, *632f*
  structure 340, *341f*, 811–812, *812f*, 976
  synaptonemal complex 1912
  transcription 340
chromomere **343–344**, *1074f*
chromosome **344–345**
  centromere **320**, *321f*
  chiasma 328–329
  Drosophila melanogaster 344
  end-replication problem 1946, *1946f*
  eukaryotes 344
  gene 344
  integration 260–262
  laboratory preparation 345
  meiosis 344
  mitosis 288, *290f*, 300–301, 344, 1225–1226
  muntjac deer 320, *321f*
  recombination 261–262
  Robertsonian translocation 1752
  satellited chromosome **1773–1774**
  structure 512
  synteny 359
  telomeres **1946–1950**
  visualization 369
  see also cell cycle; centric fusion; centromere; chromatid; double-minute chromosomes; telomeres
chromosome 1 753
chromosome 4 1582
chromosome 5 9, 486
chromosome 6 1582
chromosome 7 (mouse) 2005
chromosome 8 2056
chromosome 9 189, 207–208, 250, 1450, 2056
chromosome 10 49, 322, 2125
chromosome 11 878, 2121
chromosome 12 9, 49
chromosome 13 10, 66, 240, 242, 583, 1420, 2057
chromosome 14 41, 49, 130, 583
chromosome 15 1421, 1450
chromosome 16 878
chromosome 17 10, 239, 242, 359–360, 986, 988, 1166, 1185, 1308, 1582, 2121
chromosome 17 (mouse) 1919, *1920f*
chromosome 18 10, 66, 2056, 2058
chromosome 19 49
chromosome 21 583, 1086, 2056
chromosome 22 189, 207–208
chromosome aberrations **345–348**
  aneuploidy 346
  Angelman syndrome 346
  digyny 345
  dispermy 345
  DNA sequence map 364–365
  Down syndrome 346, 348
  human chromosomes 980
  insertional translocations 348
  inversions 348
  irradiation 347
  isochromosome 348
  Klinefelter syndrome 346
  monosomy 346
  nullisomy 346
  numerical aberrations 345, 345–346
  Patau syndrome 346, 348
  point mutation 345
  Prader–Willi syndrome 346
  reciprocal transloactions 347
  Robertsonian translocation 347–348
  structural aberrations 347–348
  tetraploidy 346
  triploidy 345
  trisomy 346
  trisomy 13 346
  trisomy 18 346
  trisomy 21 346
  Turner syndrome 346
  uniparental disomy (UPD) 346
  see also Down syndrome; Klinefelter syndrome; Patau syndrome; polyploidy; trisomy 18
chromosome banding **348–350**, 354
  applications 349
  C-banding 348
  CREST labeling 349
  Drosophila melanogaster 348
  euchromatic bands 349
  functions 349
  G-banding 349
  heterochromatic bands 348–349
  kinetochores 349
  nucleolus organizing region (NOR) 348, 349
  Q-banding 348, 349
  R-banding 349
  T-banding 349
  see also centromere; heterochromatin; kinetochore
chromosome break **350**
  cell differentiation 350
  chiasma 330
  mitosis 350

chromosome break (*continued*)
Nijmegen syndrome 351
nonhomologous end-jointing (NHEJ) 351
repair 350
Saccharomyces cerevisiae 330
Schizosaccharomyces pombe 330
see also chromatid
chromosome breakpoint maps *363t*
chromosome bridge **350–351**
chromosome dimer resolution **351–353**
dif recombination site 351
Escherichia coli 351–353
homologous recombinations 351–353, *352f*
recombination reaction 351–352
Xer recombination 351–353
see also Holliday junction
chromosome imprinting 629
Angelman syndrome 634
conflict theory 635
gametogenesis 635
Peromyscus polionotus 635
Prader–Willi syndrome 634
spermatogenesis 634
chromosome mapping **353–368**
BACs (bacterial artificial chromosomes) 362–363
Caenorhabditis elegans 366–367, *367t*
clone contig map 363–364
comparative mapping studies 366–367
Danio rerio 367, *367t*
dbEST database 364
DNA sequencing 355–356
Drosophila melanogaster 366–367, *367t*
Escherichia coli 366, *367t*
expressed sequence tags (ESTs) 364, *365f*
Fugu rubripes rubripes 367, *367t*
Gene Mapping Conferences 355
genetic disease genes 364–365
Haldane mapping function 357
Human Genome Project 355–356
humans 354–356
Kosambi mapping function 357
linkage 353–354
long-range restriction maps 363
man *364f*
mouse 367, *367t*
Online Mendelian Inheritance in Man (OMIM) 365
phage T4 *1925f*
physical mapping *363t*, 363–365
polymerase chain reaction (PCR) 354
radiation hybrid mapping 354, 360
restriction fragment length polymorphism (RFLP) 357
Saccharomyces cerevisiae *367t*
Sanger Centre 360
single-nucleotide polymorphisms (SNPs) 359
in situ hybridization (ISH) 361–362
somatic cell hybrid analysis 359–360
Stanford Human Genome Center 360
transcript maps 364
Whitehead Institute 360
see also physical mapping
chromosome movement **368**
see also meiosis; mitosis
chromosome number **368**, 1969
see also polyploidy
chromosome painting *368f*, **368–369**, 964, 1002–1004
chromosome pairing, synapsis **369–372**, *370f*
bivalents 369, *370f*
bouquet stage 369–370, *371f*
genetic aspects 371–372
history 369
initiation 371
inversion loops 371, *371f*
see also meiosis; somatic pairing; synaptonemal complex
chromosome rearrangement 956
chromosome scaffold **372–373**, 373–374
see also mitosis
chromosome structure **373–375**
chromatin loops 374
chromosomal fibers 373
chromosome condensation 374
chromosome scaffold 373–374
nucleolar organizer region (NOR) 374
packing ratio 373, 374
perichromosomal material 374–375
position effects 1524
scaffold-attached region (SAR) 374
solenoids 373
uninemes 373
see also telomeres
chromosome walking **375–376**, *375f*
computer analysis 376
contig 376
chronic myelogenous leukemia (CML) *see* leukemia, chronic
chronic nonspherocytic hemolytic anemia (CNSHA) 882
CHS1 gene 27
CHS protein 27
ch-T gene *1944f*
ch-Tbx3 gene *1944f*
ch-Tbx5 gene *1944f*
ch-Tbx6L gene *1944f*
ch-TbxT gene *1944f*
Churchill G A **638–641**
Churchward G **455–456**
Cicer arientum 1470
Cicer (chickpea) 1083
Cicereae 1083
Cichlidae 1862
cI gene 1131
ciliates
amitosis 1135
Euplotes 1135
Euplotes crassus 1133
macronuclear development **1133–1135**
macronucleus 1133, **1135–1136**
micronucleus 1133, 1135
Oxytricha 1135
Oxytricha fallax 1134
Oxytricha trifallax 1134
Paramecium 1135
Stylonychia 1135
Tetrahymena 1135
Tetrahymena thermophila 1133
Cin4–1 gene 2024
Cinful-1 gene 2022
CINH gene *430t*, *431t*
CIQA gene *430t*, *431t*
CIQB gene *430t*, *431t*
CIQC gene *430t*, *431t*
CIQR gene *430t*, *431t*
circular chromosome 377
circular linkage map **376–377**
CIR gene *430t*, *431t*
cis-acting locus **377–380**
enhancers and silencers 378
cis-acting locus (*continued*)
detection 379
function 378–379
structure 379
locus control region (LCR) 378
promoters 379–380
detection 380
function 379
structure 379
scaffold/matrix attachment region (S/MAR) 377, 378
detection 378
function 378
structure 378
see also transcription
cis-acting proteins **380–382**, 948
bacteria 380
classification 380
enhancers 624
mechanism 381–382
multiple binding sites 381
protein instability 381–382
protein sequestration 381
role 381
see also promoters; transcription
CisA protein 380
cis-dominance **382–383**
CIS gene *430t*, *431t*
cis-platin 557
cis–trans configurations **383**
cis–trans test 153, 383, 658, 760, 1194
cistron 155, 212, **383–384**, 655, 1194–1195
see also gene
Citrobacter 1439
Citrobacter freundii *1694t*
CJD (Creutzfeldt–Jakob disease) *see* Creutzfeldt–Jakob disease (CJD)
c-*Kit* proto-oncogene 398
c-kit (W) gene 525, 2126
CL *see* cutis laxa
clade 265, **384**, 1905
see also parsimony
cladistics **384**
see also phylogeny
cladists 1936
cladogenesis **384**
cladograms **384–385**, 1936
see also character
cladon 1929
Cladosporium cucumerinum *1914t*
Clark A G **873**, **1513–1519**, **1601–1602**
Clark, A. John 1614, 1623, 1635, 1647
Clark S G **1318–1320**
classical genetics *see* transmission genetics
class switching **385**
Clavibacter michiganensis 1850
clavulanic acid 1890
clay hypothesis 1389
CLCN5 gene 1028
Cleary M A **578**, **911**, **1088**, **1106**, **1631**
cleavage *see* nuclease; restriction endonuclease
Cleaver, James 565–566
cleft lip and cleft palate **385**
epistasis 640, *640t*
Fogh–Andersen, Poul 385
Fraser–Juriloff model *386f*
susceptibility *386f*
cleft palate and cleft palate *2015t*
clf1 gene 640
clf2 gene 640

clinical genetics **386**
see also genetic counseling; genetic diseases
CLK protein 389
Clock (Clk) gene 387
clock mutants **386–390**
Drosophila melanogaster 388f
feedback loops 387, *388f*, 389
model systems
bread mold 389
Cyanobacteria 389
mouse *388f*, 389
Neurospora crassa 389
Synechococcus 389
clonal analysis 302–303, *303f*
clone **390**
clone contig maps *363t*, 363–364
cloned organisms **390–391**, 656
amphibians 390
Dolly (sheep) 391
Drosophila melanogaster 585
embryo transfer 612
humans 391
mouse 391
mouse genetics 1248
nuclear transfer 390–391
Cloning Vector Collection, National Institute of Genetics 852
cloning vectors **391–392**
lac mutants 1069
lac operon 1070
shuttle vector **1828**
vectors 2104
closed reading frame **392**
Clostridium botulinum 185
Clostridium perfringens 177
CLU gene *430t*, *431t*
CLV1 gene 715
CLV2 gene 715
CLV3 gene 715
cM *see* centimorgan (cM)
c-*met* proto-oncogene 1185
CML *see* leukemia, chronic
C.M. Ricks Tomato Genetics Resource Center (TGRC) 852, 1126–1127
cms-T strain 1223
coalescent **392–397**
ancestral process
neutral case 392–393
selection case 393–394
ancestral recombination graph 395–396
ancestral selection graph 393, *394f*
coalescent process 392
coalescent tree *393f*
embedded genealogy *394f*
genealogy robustness 394–395
interference 396
migration and subdivision 396
recombination *395f*, 395–396
strong selection 396
ultimate ancestor 394
varying population size 395
see also trees
coat color mutations, animals **397–401**
albinism 401
human pigmentation 401
loci 398
melanin and other pigments 399–400
mouse 397–398
number of mutations and genes 398
pigment granule trafficking 400
pigment type-switching 400–401
types 398
coat color mutations, animals (*continued*)
white spotting 398–399
see also piebald trait; tortoiseshell coloring
cobalamin 2118
cob gene 1221
COBRA gene 1755
coccids 1420–1421
Coccobacillus acridiorum 511
Cochliobolus sativa 973
Cockayne syndrome (CS) 564, 568, *569–570t*
clinical features 568
defective DNA repair 568
nucleotide excision repair (NER) genes 1285
sunlight sensitivity 568
cod genes 255
coding sequences **401–402**
see also introns
coding strand **402**
codominance **402**
codons **406**
anticodons 78
AUG codons **125–126**
bacterial genes 153
DNA cloning 545
elongation 609
insertion sequence (IS) 1036
mRNA (messenger ribonucleic acid) 1184
neutral mutation 1325
non-Darwinian evolution 1344
nonsense codon **1350**, 1366
Ochoa, Severo 1366
ochre codon **1366**
ochre mutation **1366**
opal codon **1375**
open reading frame (ORF) 1375
start, stop codons **1886–1887**
termination codon 1366
universal genetic code 2099
upstream, downstream sites 2100
see also genetic code
codons, invariable **406**, *406f*
codon usage bias **402–406**, *403t*
Bacillus subtilis 403t, 404–405
Caenorhabditis elegans 405
Drosophila melanogaster 403t, 405
Escherichia coli 403t, 404
fine-scale variations 404
fitness 405
gene identification 405
genome G+C content 402–404
Helicobacter pylori 405
heterologous gene expression 405–406
humans *403t*
introgenomic regions 404
leading vs. lagging strands 404
mutation biases 402, *403t*
Mycoplasma capricolum 403t, 405
natural selection 404
Saccharomyces cerevisiae 403t, 405
Salmonella typhimurium 405
Streptomyces coelicolor 403t, 405
Coe, E.H. 849
Coe E H Jr **1780–1787**
coefficient of coincidence 1302, 1303
coelenterates (jellyfish) 311
coenzymes 625
coevolution **407–412**
adaptation costs 409
classification 407
coevolution (*continued*)
competing species 408
consequences 411–412
consumers and victims 408–409, 409–410
Ficus 407
levels of selection 409
methods of studying 407–408
mutualism 411
pairwise and diffuse 409
Red Queen coevolution 410
virulence and avirulence 410–411
CO gene 715
cog (congenital goiter) mutation 592–593
cognate tRNAs **412**
Cohen, Stanley 1624, 1638
cohesive ends **412–413**
dsDNA phages 412
phage λ 412, *413f*
Cohn, Melvin 1019
coincidence, coefficient of **413–414**, *414f*
coincidental evolution *see* concerted evolution
coisogenic strain **414**
see also inbred strain
COL1A gene 604, 1772, *1772t*
Col2a1 gene 1817
COL5A1 gene 603
Col B factor 415
colchicine 1094
cold-sensitive mutant **417**
Cold Spring Harbor 1400, 2135
Cold Spring Harbor Laboratory of Quantitative Biology 257, 2135
Col E1 factor 415
ColE1 plasmid 418, 656, 1381, 1486, 1488
Cole–Carpenter syndrome 1397
Coleman N **672, 1715**
col factors **415–417**
distribution 415
eco-evolutionary dynamics 416
invasion dynamics 416
resistance to colicins 415–416
Col Ia factor 415
colicins **417–418**
ColE1 418
col factors **415–416**
coliconogency and lysogeny 417–418
discovery 417
diversity and action 417
epidemiology 418
gene-cloning vectors 418
plasmids 418
receptors 418
colinearity **418–420**
demonstration 419–420
description 418–419
early history 419
colinearity principle 795, 1095
coliphage PRD1 184
coliphage T4 2107–2108
coliphage T7 2107
Collection of the Institut Pasteur (CIP) 851
Colletotrichum coccodes 1850
Colletotrichum gloeosporioides 1914t
Colletotrichum lagenarium 1914t
Colletotrichum lindemuthianum 1914t
Colletotrichum orbiculare 1914t
Colletotrichum truncatum 1914t
Collins V P **878**
collodion baby 994
Col M factor 415
colony hybridization **420**

color blindness **420–422**, *421t*, 1835
achromatopsia 421–422
red–green 421
tritanopia 421
colorectal cancer **422–423**
epidemiology 422
genetic basis 422–423
M-FISH *362f*
risk factors 422
color vision 421
Comfort N C **2133–2135**
commaless code **423**
commensal **424**
commitment complex (CC) 1538
Committee for the Standardized Karyotype of the Dog 267
comparative genomic hybridization (CGH) 362–363, **424–426**, 1003–1004
cancer research 424–425, *1609f*, 1609–1610
Hodgkin's disease 953
methodology 424
miroarrays 425
rats *1609f*, 1609–1610
compartmentalization **426–427**
borders 427
Drosophila melanogaster 426–427
genetic control of development 426–427
units of growth 427
see also pattern formation
compatibility group **427**
complementary DNA (cDNA) **286**, **433**, 651
see also DNA cloning; reverse transcription
complementation **434**
complementation test **434**
complement loci **427–433**
C3, C4, and C5 432
Cl1 and lectins 429
complement control proteins 432–433
factor D 429
factor I 429
polymorphisms 432
serine proteases 429
terminal complement components 432
complement system 427–433, *428f*
gene locations *430t*
gene structures *431t*
complete penetrance *see* penetrance
complex locus **434**
complex traits **434–435**
compound chromosomes **116–120**
attached-X chromosomes 117–119
centromere 116–117
construction 117, *119f*
crosses 117, *118f*
Drosophila melanogaster 116–119
genetic consequences 119
isochromosomes 119, *119f*
types 117–119, *119f*
viability 117
concatemer **435**
concatemer (genomes) **435–436**
concatenated circles **436**
concerted evolution **436–441**, *437f*
mechanisms 438–440, *439f*
nonallelic sites 437–438
phases 436
regulation 440
repetitive DNA sequences 1673
tandemly repeated genes 436–437, *438f*
concerted evolution (*continued*)
see also gene conversion; gene family; tandem repeats; unequal crossing-over
c-onc gene *1250f*, 1252
concordance **441**
see also linkage; pleiotropy
conditional lethality **441–443**
dominance 441
Escherichia coli 441–442
nonsense mutations 442
phage T4 *443f*
temperature-sensitive (*ts*) mutant 441–442
uses 442–443
see also temperature-sensitive (*ts*) mutant
confined placental mosaicism (CPM) 1241
conflict theory 635
congenic strain **443–445**, *444f*
see also backcross; inbred strain
congenital absence of vas deferens (CAVD) 1024–1025, *1025f*
congenital adrenal hyperplasia (CAH) **445–447**, 1809
3β-hydroxysteroid dehydrogenase 447
aldosterone deficiency 447
congenital lipoid adrenal hyperplasia 447
hypertensive form with virilism 447
hypertensive form without virilism 447
virilizing adrenal hyperplasia 447
congenital disorders **448–449**
environmental factors 449
genetic causes 448–449
malformations 448–449
multiple malformations 449
single malformations 448
terminology 448
without malformations 449
see also genetic diseases
congenital muscular dystrophy 1264
congruence test 968, 970
conjugation **449–453**
bacterial 450, **453–455**
donors 451–452, 454
Escherichia coli 450, 453
evolution 454–455
genes *453f*
history 452–453
recipients 452, 454
resistance plasmids **1686**
stages *452f*
type IV secretion systems 452
conjugative plasmids 450–451
Escherichia coli 657
Hfr strains 657–658
Paramecium tetraurelia 1412
plasmid 657
protozoan conjugation 450
conjugative transposition *455f*, **455–456**
conjugative transposons 455
host range 455
mechanism of conjugation 456
mechanism of transposition 455–456
conjunction test 967, 970
Connor J M **385–386**, **1428–1429**
conplastic **456**
consanguinity **456–457**
consensus sequence **457–458**
conservation genetics **458–462**
captivity 460
extinctions 459
conservation genetics (*continued*)
genetic diversity 459–460
genetic markers 461
islands 460
methodology 461–462
mutations 460
small population sizes 459
taxonomy 460–461
see also inbreeding depression; population genetics
conservative recombination **462**
conserved synteny *see* synteny
consomic **462**
Constantini F **611–612**, **1990–1998**
constant regions 249, **462**
constitutive expression **462–463**
constitutive heterochromatin *see* heterochromatin
constitutive mutations **463**
see also gene expression
Consultative Group for International Agricultural Research (CGIAR) 852
contig **463**
chromosome walking 376
continuous variation **463–464**
contractile ring **464**
contraselection *see* counterselection
controlling elements **464–465**
see also transposons
Convention on Trade in Endangered Species (CITES) 461
convergent similarities 969–970
convergert evolution **465**
conversion gradient **465–468**
gradient 465
heteroduplex rejection *466f*
Holliday junction *466f*
meiotic gene conversion 465, *466f*
mismatch repair 466, *467f*
Saccharomyces cerevisiae 466–467
Cooke A **128**
Co-operative Human Linkage Centre (CHLC) 358, *358t*, 1107
Cooper C S **712–713**, **1768–1773**, **1913**
coordinate regulation **468**
cop gene 1455
COPII protein 1562
Coprinus 1398, 1954
Coprinus cinereus 1153
copy-choice hypothesis **468**
COR1 protein *1910f*, 1911
cordycepin **468**
core particle **468–469**
see also ribosomal RNA (rRNA)
corepressor **469**, 1678
Cori, Carl 1067, 1365, 2120
Cori, Gerty 1067, 1365, 2120
Cormier-Daire V **2–3**, **236–237**
Cornebacterium diphtheriae 185
correlated response **469–471**
applications 470
evolution *470f*, 471
limitations of quantitative theory 470
quantitative genetic theory 469–470
Correns, Carl Erich 199, 353, 521, 1137, 1178
corticotropin-releasing hormone (CRH) 445
cortisol 445–446
cosmids **471**, 547
cotransformation **471–472**
Bacillus subtilis 472
competent population 472

cotransformation (*continued*)
congression 472
dispersive mechanisms 471
DNA disruption 471
eukaryotes 472
genetic mapping 472
integrative recombination 471
mismatch repair 471–472
Cotterman, Charles 995
Coughlin B C **798–800**
Coulson A **1706–1709**
coumarins 76–77
counterselection **472–473**
counting model *see* mapping function
covarion model of molecular evolution **473–477**
biochemical basis 474
Hidden Markov model 474, 475–476, *476f*
Kimura model 474–475, *475f*
tests 476
transition matrices 475
covarions 406, 473
Cowden's disease 243
COX-2 *see* cyclooxygenase-2 (COX-2)
Cox, David 360
Cox D W **40–41, 753–757, 2139–2140**
cox gene 1221
Cox M M **1906–1909**
Coyne J **1679–1686, 1825–1828**
cozymase 1365
CpG islands **477**, 631–632, 1386
adenocarcinomas (ADCs) 10
hemophilia 918
X-chromosome inactivation 2148
cR *see* centiray (cR)
CR1 gene *430t*, *431t*
CR1 protein 433
CR2 gene *430t*, *431t*
CR2 protein 433
Crabbe J C **27–29**
c-raf oncogene 1484
Craik C S **2071–2075**
Crandall K A **1465–1466**
craniofrontonasal syndrome 479
craniosynostosis **478–481**, *480t*
Apert syndrome 478, *479f*
classification 478
cranial sutures *478f*
craniofrontonasal syndrome 479
Crouzon syndrome 478
Muenke syndrome 479
mutations 479–481, *480f*, *480t*
Pfeiffer syndrome 478–479
Saethre–Chotzen syndrome 479
Crassostrea virginica 790
Craterostigma plantagineum 2025
CREB trancription factor 1973
Cre/*lox* - transgenics **481–486**
chromosome rearrangements 484–485
conditional mutations 483
Cre/*loxP* technology 332–333, 1707
Cre/*lox* recombination 482–484, *483f*, *484f*
Cre protein regulation 485–486
deletions, duplications, and inversions 484–485
gain of function 483–484
genomic targeting of DNA 485
green fluorescent protein (GFP) 485
loss of function 484
marker recycling 482
modified Cre proteins 485–486
Cre/*lox* - transgenics (*continued*)
phage P1 481
point mutations 482–483
reverse genetics 1707
selectable marker genes 482
translocations 485
Crenarchaeota 92
Crenarchaeum symbiosum 92
Cre recombinase 1040, *1040t*
CREST labeling 349
Creutzfeldt–Jakob disease (CJD) 595, 1068, 1879, 2008
see also Gerstmann–Sträussler disease (GSD)
Crick, Francis Harry Compton **486**, 540, 554, 815, 1178
adaptor hypothesis **7**
frameshift mutation 728
Nobel Prize 2134
RNA world 1751
Watson, James Dewey 2134
cri-du-chat syndrome **486–487**, 1086
Crithidia 1741
Cro protein 1131
cross **487**
see also parental
crossed-strand DNA junctions 1756, *1757f*
crossing-over **488**, 957, 1046
first and second division segregation 699
genetic recombination 841
hot spots 976
isomerization 1055–1056, *1056f*
see also chromatid; unequal crossing over
cross-linking agents 551
crossover suppressor **488–489**, *489f*
see also inversion
Crouse, Helen 629
Crouzon syndrome 478, 900
Crow, James F. **489–491**
Basic Concepts in Population, Evolutionary, and Quantitative Genetics 490
Genetics Notes 490
An Introduction to Population Genetics Theory 490
Crow J F **602–603, 706–707, 838–839, 906, 933, 1045–1046, 1063–1064, 1255–1256, 1276, 1400–1401, 2141–2143**
crown gall tumors *491f*, **491–493**
dissemination and control 492–493
history 491–492
oncogene transmission 492
opine genes 492
T-DNA transfer mechanism 493
Crow T J **1774–1776**
CRP (cAMP receptor protein) 1070
CRP (cysteine-rich protein) family 1098
crp gene 1069, 1193
CRSP2Y gene 2159
CRT *see* corticotropin-releasing hormone (CRH)
Cruciferae 237
cruciform DNA **493–495**, *494f*
branch migration 494–495
four-way DNA junction 493–494, *494f*
protein interactions 495
supercoiled DNA 495
Crucihimalaica 86
Cruz-Reyes J **1741–1743**
Cruzan M B **4–7**
cry1 gene 1456–1457
CRY2 gene 715
CRY protein 387–388
cryptic satellite **495**
cryptic splice sites and cryptic splicing **495–497**
clinical significance 496
detection 496
RNA products 496
cryptochromes 1456–1457
cryptorchidism 1023
CSA gene 563, 568
CSB gene 563, 568
CSF2RA gene 2157
Csink A K **1851–1853**
c-ski gene 1846
CTCF protein 634–635
ctDNA **497**
CTP *see* cytidine triphosphate (CTP)
Cuculus canorus 409
cucumber *1914t*
Culex quinquefasciatus 766
cutis hyperelastica 497
cutis laxa (CL) **497–499**, *498f*, 599
acquired 499
blepharochalasis 499
pathogenesis 499
primary 497–498
secondary 498–499
Cuvier, Georges 1072
C value **249**, 249–250, 1807
C-value paradox **249–250**, 1807
CVS *see* chorion villus sampling (CVS)
cya gene 1193
Cyamopsis (guar gum) 1083
Cyanobacteria (blue-green algae) 149, 1378
cyanophages 183
cybrid 1357
CYC1AT gene 1755
CYC gene 716
cycle (cyc) gene 387
cyclic 3′,5′-AMP *see* cAMP nucleotide
cyclical parthenogenesis 1419
cyclic AMP (cAMP) **499–500**, 1069, 1131
bacterial transcription factors 164
olfaction 1369
regulatory genes 1653
see also adenosine triphosphate (ATP)
cyclin/CDK oscillator 292–294, *293f*
cyclin D1 gene 1772, 2085
cyclin-dependent kinases **500–506**, *503f*
assembly factors 504–505
CDK1 protein 502
CDK2 protein 502
CDK4 protein 502
CDK5 protein 504
CDK6 protein 502
CDK7 protein 503–504
CDK–activating kinase (CAK) 502
CDK dysfunction and disease 506
CDK protein family 502–504
CDK–protein interaction sites *504f*
CDK regulation 504–506
CDKs in complex with cyclin A *501f*
CDKs in metazoans 502–503
CDKs in yeasts 502
CKI binding 505
cyclin binding 504
discovery 500
function 501
localization 505
Pho85 502
phosphorylation 505, 505–506

cyclin-dependent kinases (*continued*)
  protein degradation 506
  structure and activity 501–502
cyclin D gene 2086
cyclins
  carcinoid tumors 274
  cell cycle 291–294
  ubiquitin 2092
  yeasts 291–292
cyclooxygenase-2 (COX-2) 423
Cyclops strenuus 860
cyk-1 gene 301
cyk-3 gene 301
Cynidium 763
CYP11A (P-450$_{\text{scc}}$) *446t*
CYP11B1 gene 447
CYP11B1 (P-450$_{\text{c11}}$) *446t*
CYP11B2 *446t*
CYP11B2 gene 447
CYP17 gene 447
CYP17 (P-450$_{\text{c17}}$) *446t*
CYP1A1 gene 263
CYP21 *446t*
CYP21 gene 447
CYP21P (P-450$_{\text{c21}}$) *446t*
CYP21P gene 447
CYP27P1 gene 1027
CYP2E1 gene 263
cysteine **507**, *507f*
cystic fibrosis **507–509**, 1024
  carrier 275
  Caucasians 509
  gene 359, 507
  genetic diagnosis 508
  genotype–phenotype correlation 508
  mutations 507–508
  prenatal diagnosis 1540
  transmembrane conductance regulator (CFTR) 1431, 1565
  treatment 509
cystic fibrosis transmembrane conductance regulator *see* CFTR
Cystoviridae 1744–1746
cytidine 1360
cytidine triphosphate (CTP) **497**
cytidylic acid 1360
cytochrome *c* 473
cytogenetics **509**
cytokinesis 288–290, *291f*, **509**
  Caenorhabditis elegans 301
  see also cell cycle
cytokinin 491, 1481
Cytophagales 149
cytoplasm **509–510**
cytoplasmic genes **510**
cytoplasmic inheritance **510**
cytoplasmic male sterile (CMS) mutant 1223
cytoplast 1357
cytosine (C) **510**, 540, 572, 631, *631f*, 1358
cytoskeleton **510**
cytosol **510**

D

d1 gene 1395
Dab1 gene 1650
Dab1 protein 1650
D$_{\text{A}}$ distance 830
daf genes 22, 255
daf-12 gene 928
DAF gene *430t*, *431t*
DAF protein 433
Dalbergieae 1083
Dale, Henry 1365
Dalgaard J Z **1153–1157**
Danaus chrysippus 731
Danchin A **135–144**
Daniels G **227–228**, **228–229**, **230**, **588–589**, **1125**
Danio rerio 309, 531
  chromosome mapping 367, *367t*
  genome size *367t*
D antigen 650
Daphnia 410
Darlington, Cyril Dean **512–513**
  centromeres 512
  chromosome structure 512
  evolutionary biology 513
  The Evolution of Genetic Systems 512
  meiosis research 512
  mitosis research 512
  population genetics 513
  Recent Advances in Cytology 512
Darvasi A **19–20**
Darwin, Charles **513–517**
  criticism 516
  The Effects of Cross and Self-Fertilization in the Vegetable Kingdom 933
  evidence for evolution 515
  evolution 664
  evolutionary biology 516
  evolutionary synthesis 516
  genes 759
  Huxley, Thomas Henry 984
  influences 514–515
  major concepts 515
  natural selection 514–515, 1291
  On the Origin of Species 514–515, 743, 907, 984, 1004, 1071, 1344, 1597
  pangenesis 516
  philosophical implications 516
  reaction to *On the Origin of Species* 516
  speciation 1860–1861
  taxonomy, evolutionary 1935
  theory of evolution 513
  The Variations of Animals and Plants under Domestication 516
  Wallace, Alfred Russel 2132
Darwin, Leonard 700
Darwin's finches (*Geospiza* spp.) 672
Datura stramonium 492, 2056
Davisson M T **876–877**, **1336–1340**
Davisson, Muriel T. 1337
Dawkins, Richard 1165
Dawson, W.D. 849
Dax1 gene 1813, 1817–1818
DAX1 gene 1813, 1817–1818
Dayhoff PAM weight matrix 32
DAZ gene 2156–2160, *2158f*
dbEST database 364, *366t*
Dby gene *1253f*, 1254
DBY gene 2159
DCC (Deleted in Colon Carcinoma) gene 10
Dce gene 1713
dce gene 1713
dCE gene 1713
DCP1 gene 1748
DDBJ (DNA Data Bank of Japan) 155, 517
Dean, D.H. 850
De Barsey syndrome 498
decapentaplegic *(dpp)* gene 427
decay-accelerating factor (DAF) *see DAF* gene; DAF protein
Deep Vent DNA polymerase 1502
*defective endosperm** B30 mutant 1786
DEF gene 716
deficiency *see* deletion mutation
deg-1(d) gene 316
deg-1 gene 316
deg-3 gene 316–317
deg-3(u662) gene 316–317
degenerate code 522
degrees of freedom **729–730**
De Gregorio L **1087–1088**
de Haan, H. 1337
5,6-dehydrouridine 1988
Deinococcus radiodurans 859, 1200
Deinococcus/Thermus 149
del1–46 gene 2022
de Launoit Y **661–663**
Delbank 1600
Delbrück, Max 257, **522–524**, 554, 1921
  American Phage Group 522
  bacteriophages 180, 522, 2109
  Benzer, Seymour 212
  Drosophila melanogaster 522
  Dulbecco, Renato 589
  Escherichia coli 654
  Hershey, Alfred 522
  Luria–Delbrück experiment 522, 1124, **1124–1125**
  Luria, Salvador E. 522
  lysogeny 1130
  Nobel Prize 523, 925, 1123
  The Phage Course 522
  phototropism 522
  radiobiological studies 522
  selection process 1709
  standardization in phage biology 522
  T phages 522
  Viscounti–Delbrück hypothesis 2116
  Watson, James Dewey 2133
deletion **524**
  indel 1017
  occurrence
    excision 524
    intramolecular transposition 524
    replication errors 524
    specialized recombination 524
  resolvase-mediated deletion **1688–1692**
deletion mapping **524**, 795, *795f*
deletion mapping (mouse) **524–528**
  background 524
  deletion complexes 527
  embryonic stem (ES) cells 527
  fine-structure mapping 527
  functional maps 527
  manifestations 525
  sex-reversed rearrangement 1253–1254
  uses 524
    codominant genetic markers 526, *526f*
    complementation 525, 527
    genetic banks 526
    genetic mapping 525, 527
    microsatellites 526
    pseudodominance test *525f*, 525–526
    restriction fragment length polymorphism (RFLP) 526
    simple sequence length polymorphism (SSLP) 526
deletion mutation **528**
delta (δ) globin gene 878, 2096
Delta gene 1317, *1318f*
DeLuca D C **625–626**
De Lucia, Paula 257
Demant P **1606–1637**
Demerec, M. 850

demes **528**, *528t*
  definition 528
  isolation by distance 528
  panmictic demes 528
  shifting balance theory of evolution 1825
demethylases 542
Dénarié J **1330–1332**
denaturation (proteins) **529**
  amphipathic 529
  biological consequences 529
  function 529
  heat shock proteins (Hsps) 529
  mechanisms
    chemical 529
    cooling 529
    heating 529
  molten globule 529
  reversiblity 529
dentatorubal pallidoluysian atrophy (DRPLA) 983, 1421, *2049t*
Dent's disease 1028
Denys–Drash syndrome (DDS) 1814, 1818, 1822, 2138
Derbyshire K M **380–382**
derepression **530**
dermal neurofibromas 1307
Der Züchter (Theoretical and Applied Genetics) 201
des Etages S A **2034–2040**
Desmoideae 1083
desmosomes 1289
Desnick R J **681**
destruction box 2092
Desulfovibrio desulfuricans 1694t
Desulfurococcus mobilis 92
det gene 1455
detoxification **530**
deuteranopia *see* color blindness
Deutsche Sammlung von Mikroorganismen und Zellkulturen (DSMZ) 851
developmental genetics **530–531**
  Caenorhabditis elegans **531–532**
    axis development 532
    Drosophila melanogaster 532
    embryogenesis 532
    fate decisions 532
    methodology 531–532
  experimental intervention 530
  fate mapping 530
  findings 531
  genetic markers 530
  genetic systems 531
  impact on developmental biology 530
  methods 531
  molecular cloning 530
  Morgan, Thomas Hunt 530
  strategy 530
developmental genetics (mouse) *see* embryonic development
deviation from Mendelian inheritance (DMI) 1167
Devos K M **863–865**
de Vries, Hugo Marie 353, 516, 521–522, 1137, 1178
  ancestral inheritance theory 63
  pangenes 521
Dfd (Deformed) gene *960f*
DFFRY gene 2159
DF gene *430t*, *431t*
D gene 1713
D'Herelle, Félix **511–512**, 522
  bacteriophage studies 175–178, 179, 511–512, 2108
  fermentation studies 511
diabetes mellitus 1150, *2015t*
Diablo/Smac 207
diaminopurine (2,6-diaminopurine) 191
diauxy 281
dicentric chromosome **533**
Diceros bicornis 859
DICH gene 716
dichopatric 39, 1861
Dickerson, Richard 555
Dickinson, Alan 1879
Dictyostelium **533**, *606f*, 763, 1707
  heterochronic mutation 930
  homeobox genes 961
  kinases 1063
Dictyostelium discoideum 260, 405, 934, *1732f*
Didelphys virginiana 1408
dideoxy sequencing *see* DNA sequencing
Diener, T.O. 2107
diethyl pyrocarbonate (DEP) 551
Die Wissenschaftlichen Grundlagen der Pflanzenzüchtung (The Scientific Basis of Plant Breeding) 201
differentially methylated domain (DMD) 903, 996
differential segment **533**
dif recombination site 351–353, *352f*
DiGeorge syndrome 361, **533–535**
  brachyury locus 237
  CATCH 22 534
  CD4+ T cells 533
  chromosomal deletions 533
  clinical features 534, *534f*
  fluorescence microscopy (FISH) 535
  velocardiofacial syndrome 534
digyny 345
1,25-dihydroxycholcalciferol 1027
1,25-dihydroxyvitamin D 1027–1028
dilute (d) gene 398
dilute (*d*) locus 1599, 1871
dinA gene 1854
dinB gene 1651, 1854
Diplococcus pneumoniae 1491
diploid androgenesis 989
diploidy **535–537**, 908–910
  artificial diploidy
    Aspergillus nidulans 535, *536f*
    Drosophila melanogaster 536f
    haploid fungi 535
    instability 535–536
    mitotic crossing-over 536, *536f*
    Schizosaccharomyces pombe 535
    Ustilago maydis 535
  haploid phase
    algae 535
    animals 535
    bacteria 535
    fungi 535
    plants 535
    Saccharomyces cerevisiae 535
  homologous chromosomes 963–964
  homologs 964
  homozygosity 970
  imprinting, genomic 1000
  infertility 1021
  meiosis 536
Dipodomys heermani tularensis 860
Dipodomys ordii monoensis 860
Diptera 763
diphtheria toxin 185
directed deletion *see* gene rearrangement
directed mutagenesis *see* complement loci
directed mutation **537**
direct repeats **537**
disassortative mating *see* assortative mating
discontinuous replication **537**
discordance **537**
disequilibrium *see* gametic disequilibrium; linkage disequilibrium
disjunction **537–538**
dislet gene 1097
dispermy 345, 989
disruptive selection **538**
distal **538**
distal muscular dystrophy *1263f*, 1264
distorted genes 2048
divergent evolution **539**
divergent transcription **539**
dlim1 gene 1096
dlim3 gene 1097
dLMO gene 1097
D-loop **539–540**, 1988
  genetic diversity 539–540
  haplotype analysis 539–540
  Meselson–Radding model 1181
  mitochondrial DNA (mtDNA) 539
    control region 539
    H-strand replication 539
    L-strand replication 539
    *ori*H 539
    *ori*L 539
    replication *539f*
  nucleotide variations 539
  RuvAB enzyme 1756, *1757f*
$(dm)^2$ distance 830
dm-H15 gene *1944f*
dm-omb gene *1944f*
DMC1 protein 1645, *1645f*, *1646f*, 1912
Dmd^{mdx} gene 1244
DMPK gene 1838, *2049t*
DMR1 (mesodermal tissue silencer element) 996
DMRT gene 1811
DMRT1 gene 1813
dm-Trg gene *1944f*
DNA A 1
dnaA gene 1382
DnaA protein *1383f*, 1486
DNA binding *see trans*-acting factors
DNA-binding proteins **542–544**
  alkyltransferases 37
  binding interactions 542–543
  cisplatin adducts 544
  DNA-binding domains (DBDs) 542–543, *808f*, 808–809, *809f*
  DNA organization
    Dps proteins 542, 544
    factor for inversion stimulation (FIS) 542
    H1 histone 542
    H5 histone 542
    histones 542
    H-NS proteins 542
    HU proteins 542
  DNA regulation
    activators 542
    demethylases 542
    DNA polymerases 542
    invertases 542
    phage λ 542
    processive single-strand exonucleases 542

DNA-binding proteins (*continued*)
repressors 542
resolving enzymes 542
restriction endonucleases 542
RNA polymerases 542
T7 exonucleases 542
function 542
protein motifs 542, *542t*
AraC protein 543
bZip motif 543
H1 histone 543, *543f*
H5 histone 543, *543f*
helix–loop–helix proteins 543
helix–turn–helix motif 542
HMG domain proteins 542
lac repressor 542
POU domain 543
sequence specificity 542–543
winged helix domain 543
pyrimidine dimers 544
sequence selectivity 543–544
binding energy 543–544
single-stranded DNA-binding proteins (SSBs) **1839–1842**
single-stranded nicks 544
structure distortions 544
bending 544
catabolite repressor protein (CRP) 544
histones 544
HMG domain proteins 544, *544f*
IHF (integration host factor) protein 544
lac repressor 544
mechanisms 544
TATA binding protein 544
see also DNA structure
DNA cloning **544–550**
animals 546
cellular cloning 544
definition 544
gene therapy 546
history 544–545
insert 545
knock-in 546
knockout 546
ligation step 547–548, *548f*
molecular cloning 546–548, *547f*, *548f*
mRNA (messenger ribonucleic acid) 545
plants 546
process 546, *548f*
purpose 545–546
recombinant DNA 1637
reproductive cloning 544
screening methods 549
selection of the transformants 549
shuttle vectors 545
somatic cloning 544, 550
therapeutic cloning 550
transformation 548–549
uses 550
vectors 545, *548f*
see also cDNA
DNA code *see* genetic code
DNA damage *see* DNA repair
DNA Data Bank of Japan (DDBJ) 517
DNA denaturation **550–553**
biological reasons 551
lac operon *552f*
methods 550–551, *551f*
renaturation 550
single-stranded state 551
stacking energies 550
DNA (deoxyribonucleic acid) 516, **540–541**, 654
adenine (A) 540, *540f*
apoptosis 313
aptamers 1006
bacteriophages 926
binding 951, 1039–1040
centromere 322, *322f*
checkpoint controls 295–296, 298
chemical structure 540, *540f*
uracil (U) 541
chloroplasts 335–336, 1421–1422
chromatin 340, *341f*
chromosomal fibers 373
cis-acting locus **377–380**
cis-acting proteins **380–382**
cis-dominance **382–383**
cis–trans configurations *383*
cleavage 1033, 1038
coding sequences 401–402
cohesive ends **412–413**
C values 249–250
cytosine (C) 540
damage 271, *272*
DNAzymes 1006
double-helical structure 540, *541f*
electron microscopy (EM) *607f*
evolution 666–668
evolutionary rate 671
fingerprinting 516
function 541
guanine (G) 540
helicases 915–916
histone 633
history see DNA, history of
Holliday junction 954–955
Holliday's model 668–669, 955–956
Human Genome Project 981
hybridization **556**, 968, 1380
insertion sequence 1029
ligation 1038
locus 378
mitochondria 335–336, 1421–1422
molecular genetics 540
necrosis 313
nucleosome 633
nucleotide 540
packing ratio 373
recombination **558**
repair 257, **558–564**, **564–571**, 656
replication 287, 294, 295–296, 298, 541, **571–572**, 630–632, *632f*
restriction fragment length polymorphism (RFLP) 357
scaffold-attached region (SAR) *372–373*
sequences 934–935, 973–974, 998
deletion mapping (mouse) 526
epigenetics 628
lampbrush chromosomes 1075
upstream, downstream sites 2100
see also DNA probes
sequencing 516, **572**
silencing 636
solenoids 373
structure 257, 325
synthesis **577**, 1067
thymine (T) 540
virus 2109, *2112f*
Watson, James Dewey 2134
see also DNA structure
DNA duplex 377
DNA Genome Atlas 519, *520f*
dnaG gene 1543
DNA glycosylases 560, *562f*, 566–567
DNA, history of **553–556**
chemical characterization 553–554
DNA structure 554–555
early days 553
DNA hybridization **556**
DNA inversion system **938–942**
enhancers 939–941
FIS (factor for inversion stimulation) 939–941, *942f*
gamma delta ($\gamma\delta$) resolvase 941
gene expression 938, *939f*
Gin invertase 938–942, *939f*, 1843, *1844t*
helix–turn–helix motif 941, *942f*
Hin invertase 938–942, *939f*, *942f*, 1843, *1844t*
invertase recombination sites *939f*
invertasome 939–941, *940f*
mechanisms 941–942
regulation of inversion reaction 941
resolvase/invertase family 938
serine family 938
site-specific inversion reaction 939–941, *940f*
site-specific recombination 1843
structure 941, *942f*
topology of DNA strands 941
tyrosine integrase family 938
DNA lesions **556–557**
altered bases 556
carcinogens 557
causes 556
chemicals 557
depurination 557
depyrimidination 557
DNA replication 556
ionizing radiation 557
mutagenesis 556
UV radiation 557
DNA libraries 1013, 953
DNA ligases **557**, *562f*
DNA ligation 1012
DNA marker **557**
DNA methylation 628–637, 1443, *1444f*
5-methylcytosine (m^5C) *631f*, 631–632
chromatin remodeling 632–633
cytosine 631, *631f*
DNA replication 631–633, *632f*
endogenous parasites 636
epigenetic regulation 628–637
gametogenesis 635
gene regulation 631–633
heterochromatin 927
imprinting, genomic 1001
m^5CpG 636
MeCP2 633
methylated DNA binding domain (MBD) 633
silencing 636
spermatogenesis 635
TpG 636
DNA modification 557
dnaN gene 1384
DNA photolyases 559
DNA polymerase η (eta) **558**
DNA polymerases **558**, *562f*, 1382
bacterial genes 155
base pairing/mispairing 193–197
Cairns, John 257
DNA cloning 546
DNA regulation 542
epigenetics 630

DNA polymerases (*continued*)
excision repair 674–675
function 558
Kornberg, Arthur 558
Kornberg enzyme 257
nuclease 1358
proofreading 558
replication errors 1676
retroposon 1698
single-stranded DNA-binding proteins (SSBs) 1840–1841
SOS function 558
types
DNA polymerase I 257, 1677
DNA polymerase II 257, 1854
DNA polymerase III 257, 1676
DNA polymerase IV 1622, 1651, 1677, 1854
DNA polymerase V 1622, 1651, 1677, 1854
in vitro mutagenesis 1010–1012
see also polymerase chain reaction (PCR)
DNA probes 1002–1003
centromeric 1003
chromosome painting 1002–1004
telomeric 1003
DNA recombination **558**
illegitimate recombination 558
legitimate recombination 558
DNA repair **558–564, 564–571**
base excision repair (BER) 560, 1651, 1663–1664, *1664f*
apurinic or apyrimidinic (AP) site 560, *562f*
DNA glycosylases 560, *562f*, 566–567
DNA ligases 560, *562f*
DNA polymerases 560, *562f*, 1664
glycosylase repair 888, 1664, 1670
hereditary diseases 564, 566
oxidative damage repair **1669–1671**
8-oxoguanine *1670f*, 1671
repair synthesis 560
uracil (U) 566, 1664
BRCA1/BRCA2 genes 241
cyclobutane pyrimidine dimer 559, *560f*
defective DNA repair 563–564
cancer predisposition 563–564
Cockayne syndrome (CS) 568
transcription-coupled repair (TCR) 568
xeroderma pigmentosum (XP) 563–564, 567
enzymatic photoreactivation 565
chromophores 559, *560f*
DNA photolyases 559, *560f*
excision repair 565–566, 673
exogenous damage 559
genetic recombination 843–844, *844f*
glycosylase repair 888
hereditary diseases 565, 567–571, *569–570t*
ataxia telangiectasia (AT) 568, *569–570t*
AT-like disorder 571
Bloom's syndrome (BS) *569–570t*, 571
breast/ovarian cancer syndrome *569–570t*, 571
Cockayne syndrome (CS) 568, *569–570t*
DNA repair (*continued*)
Fanconi anemia (FA) *569–570t*
Li–Fraumeni syndrome (LFS) *569–570t*, 571
Lynch syndrome *569–570t*
Nijmegen breakage syndrome *569–570t*, 571
recQ homologs 568, 571
Rothmund–Thompson syndrome (RTS) *569–570t*, 571
trichothiodystrophy (TTD) 568, *569–570t*
Werner syndrome (WS) *569–570t*, 571
xeroderma pigmentosum (XP) 567–568, *569–570t*
homologous recombinations 351–353
mechanisms 559
base excision repair (BER) 560, *562f*, 566
enzymatic photoreactivation 559, 565
excision repair 560, 565–566
mismatch excision repair (MMR) 560–562, 566
nucleotide excision repair (NER) 562–563, 566
polynucleotide ligase 565
repair of alkylated sites 559–560, 565
strand-specific excision repair 563
mismatch excision repair (MMR) 560–562, 566, 1651
cancer predisposition 563
DNA ligation 562
hereditary diseases 566
hereditary nonpolyposis colon cancer (HNPCC) 563, 566
methylation 561–562, *562f*
Mut system *562f*
repair synthesis 562
spontaneous mutations 566
strand discrimination 561–562
mouse embryonic stem cells
gene replacement 564
genetic framework 564
mutagenic specificity 1267
nuclease 1358
nucleotide excision repair (NER) *563f*, 562–563, 566, 1651
apoptosis 566
bimodal incision 563
cancer predisposition 563
cyclobutane pyrimidine dimer 562
eukaryotes 563
excision repair pathways 566
helicases 566
oligonucleotide fragments *562f*, 562–563
prokaryotes 563
RNA polymerases 566
strand discrimination 562
structure-specific endonucleases 563
transcription 563, 566
transcription-coupled repair (TCR) 566
oxidative damage repair *1669f*, **1669–1671**
RecBCD enzyme 1621–1622, *1622f*
recombinational repair **1649**, 1651
regulation **1650–1651**
ADA-dependent adaptive response 1651
OxyR-mediated response 1651
DNA repair (*continued*)
SOS repair 1651
SoxRS-mediated response 1651
repair mechanisms **1661–1669**
repair of alkylated sites 559–560, 565
alkyltransferases **36–37**
hereditary diseases 564
O^4-alkylthymine 560
O^6-alkylguanine 560
O^6-alkylguanine-DNA alkyltransferase 560
O^6-methylguanine-DNA methyltransferase (O^6-MGT) *561f*
repair of strand breaks 564, 567
double-strand break repair model *580f*, **580–582**
double-strand breaks 564, 567, 1668, *1668f*
genetic recombination 567
recombinational repair 564
SOS repair 1651, **1853–1855**
spontaneous damage 558–559, 565
strand-specific excision repair
Cockayne syndrome (CS) 564
RNA polymerases 563
transcription 563
see also template
DNA replication **571–572**, *2052f*
bidirectional synthesis 572
semiconservative replication 572
see also template
DNase 1357
DNase-1 624, 1076
DNA sequencing **572**
chromosome aberrations 364–365
Human Genome Project 355–356
mapping *363t*
nucleotides and nucleosides 1360
Sanger, Fred 572
technique 572
DNA shuffling 1014
DNA Structural Atlas of Genomes 519, *520f*
DNA structure **572–574**, 1524
A-form (A-DNA) 573, 1730–1731, *1731f*
bases 572, *573f*
B-form (B-DNA) 573, *574f*, 1376, 1730–1731, *1731f*
cruciforms 574
deoxyribose sugar ring 572
double-helical structure 572–573
gene expression 574
handedness **1730–1731**, *1731f*
junctions 574
phosphate group 572
structural variations 574
supercoiling 574, **575–577**
Z-form (Z-DNA) 573, 1730–1731, *1731f*
DNA supercoiling **575–577**
mathematical calculations 575
free energy 575
linking number 575
superhelix density 575
twist 575
writhe 575
physical properties 575, *576f*
topoisomerases 576
topoisomers 575, *576f*
topological origin 575
transcription 576

DNA synthesis **577**
DNA polymerases
DNA polymerase α 577
DNA polymerase β 577
DNA polymerase γ 577
DNA polymerase I 577
DNA polymerase II 577
DNA polymerase III 577
occurrence 577
Okazaki fragments 577
process 577
DNAase 541
DNAzymes 1006
Dobberstein, Berhard 1553
Dobzhansky, Theodosius 513, **577–578**, 1466
balanced polymorphism 578
contributions 577–578
Drosophila melanogaster 578
eugenics 578
evolution 663
Genetics and the Origin of the Species 201, 578
population genetics 578, 1516
population structure 578
postzygotic isolation 1685
reproductive isolation 1679
shifting balance theory of evolution 2142
Dobzhansky–Muller model 1685, *1685f*
Doermann, A. H. 1921
dogs **264–270**
breeds 264, 265, 267
clades 265
evolution 264–265
genes 269–270
genetic diseases 268–269
genetic mapping 267–269
genome 267
mitochondrial DNA 265, *266f*
phenotypic diversity 265–266
radiation hybrids 270
segregation of alleles *268f*
wolf–dog hybridization 265
Dolly, the sheep 391, 2137
domestication of animals 2008–2009
dominance **578**
dominant lethals 525
Donahue, Jerry 554
Donelson J E **79–81**
Donnai D **597–598**
Dorit R L **1004–1008**
dosage compensation **579**, 927
epigenetics 634
possible mechanisms 579
sex linkage 1819
transcription 579
X-chromosome inactivation 579, 2148
X-linked gene products 579
Y chromosome (human) 2155
Double-bar mutant 2095, *2095f*
double-minute chromosomes (*dmin*) 344, 580, 963, 1123
amplicons 580
gene amplification 580, 766
double-strand break repair model *580f*, **580–582**, *775–777*, *776f*
BRCA1/BRCA2 genes 241
genetic recombination 843–844, *844f*
process *580f*, *581f*, 581–582
reciprocal recombination 1632, *1633f*
recombination models 1643, 1668, *1668f*
resolution of Holliday junction 582
single-strand annealing 1838–1839, *1839f*
trans-heteroduplex 582
doubletime (dbt) gene 387
Dougherty, Ellsworth 251
Douglas K **591–593**
Dove W F **489–491**
Down, John Langdon 583
Downs D M **216–218**
downstream **584**
downstream promoter element (DPE) 1550
Down syndrome 346, **583–584**, 2056–2057
amniocentesis 61
aneuploidy 66
clinical features 583
cognitive impairment 583
cytogenetics 583
dementia 583
Down, John Langdon 583
frequency 583
Lejeune, Jerome 1085
life expectancy 583
meiosis 1164
mutation 1271
nondisjunction 1346
prenatal diagnosis 1540, 1541
quality of life 583
research 584
Robertsonian translocation 348
trisomy 21 583–584
see also chromosome aberrations
Doyle J J **1081–1085**
DPC4 gene 2083, *2084–2085t*
DPC4/SMAD4 gene *81t*
DPE *see* downstream promoter element (DPE)
DpnI gene 1494
DpnII gene 1494
Dps proteins 542, 544
dpy genes 255
Drake, John W. 257, 1676
Drake J W **1277–1279**
Drake's rule 1676, *1676t*
Drd1a gene 1244
Dresser M E **605–608**
Dressler, David 877
Driscoll M **313–318**
Drlica K **1485–1490**
Dronamraju K R **904–906**
Drosophila 763
heterochronic mutation 930
reverse genetics 1708
Drosophila arizonae 1683t
Drosophila auraria 1683t
Drosophila biauraria 1683t
Drosophila buzatti 1683t
Drosophila heteroneura 1683t
Drosophila koepferae 1683t
Drosophila littoralis 1683t
Drosophila lummei 1683t
Drosophila madeirensis 1683t
Drosophila mauritiana 1405, 1592, *1682f*, *1683t*
Drosophila melanogaster 522, **584–585**, 765, *1683t*
ACF protein *342t*
aging 22
alternative splicing 47
anterior/posterior (A/P) axis 615
autoregulation 129
background selection 145
balanced polymorphism 187
Bar mutant 2095, *2095f*
Drosophila melanogaster (continued)
Benzer, Seymour 210
bobbed (bb) mutant 2095
Brahma chromatin remodeling complex *342t*
cell division genetics 298
cell lineage 302, 303, *303f*, *308f*, 308–309
centric fusion 319
centromere 322, 860
CHRAC chromatin remodeling complex *342t*
chromosome 344
chromosome banding 348
chromosome mapping 366–367, *367t*
circadian feedback loop *388f*
clock mutants 386–389
codon usage bias *403t*, 405
compound chromosomes 116–119
developmental genetics 530, 531
developmental genetics (*Caenorhabditis elegans*) 532
dispersed transposable sequences 861
disruptive selection 538
DNA cloning 545
DNA methylation 633
DNA transposons 2017
Dobzhansky, Theodosius 578
Double-bar mutant 2095, *2095f*
dynamic mutations 595
embryonic development 613, 622
epigenetic regulation 636
EST number *366t*
evolution 666
fecundity selection 1296, *1296t*
frequency-dependent selection 730
gene mapping 793
gene number 796
genetic research 584
genetic stock collections and centers 847–848
genome 249
genome sequencing 585, 859
genome size *367t*
half-tetrad analysis 1953
heat shock proteins (Hsps) 915
heterochromatin 349, 927
heterotrimeric G proteins 934
histone repeats *438f*
homeobox genes 959–961
homeotic genes 636
homeotic mutation 962–963
homologous chromosomes 963
horizontal transfer 975
hybrid dysgenesis 984–985, *1404f*
importance 585
independent assortment 1018
introns 1052
ISWI chromatin remodeling complex *342t*
kinases 1063
kinetochore 1064
Lewis, Edward 1095
model organisms 584, 653, 1080
molecular drive 1234
Morgan, Thomas Hunt 353, 584, 1239
Muller, Hermann Joseph 1256
mutant varieties
coitus interruptus 1315
drop-dead 1314
ether-a-go-go 1314
fruitless 1316
inactive 1316
sevenless 1316

Drosophila melanogaster (*continued*)
Shaker 1316
stuck 1315
neurogenetics **1314–1318**
behavioral screeing 1314
Benzer, Seymour 1314–1315
circadian rhythms 1315
complex behavioral phenotypes 1316
countercurrent apparatus 1314
fate mapping 1315
genetic screening 1314
learning/memory 1315
mutant varieties 1314–1315
mutational studies 1314–1315
neurogenic genes 1317
neuronal guidance 1318
Notch signal transduction pathway 1317, *1318f*
sexual behavior 1315
simple behavioral phenotypes 1315–1316
X chromosome mutagenesis 1314, *1315f*
neuronal specification 1320
NURF chromatin remodeling complex *342t*
parasegment **1415**
paternal inheritance 1422
P elements **1403–1407**, 2035
physical mapping 354
polytene chromosomes 1511, 1512–1513
population biology 1405
position effects *1525f*, 1525–1526
proteome 1577
radiation genetics 1512–1513
rare male mating advantage *732*, *732f*
reciprocal translocations 2003–2004
repetitive DNA sequences 860
replication errors *1676t*
retrotransposons *1699f*
reverse genetics 1706
RNA-binding protein domains 1734
selective sweep 1804
selfish DNA 1808
sex linkage 1819
sexual behavior
bisexual behavior 1316–1317
courtship 1316
lovesong patter 1317
somatic pairing *1851f*, 1851–1852
Staufen protein 1734
subcellular RNA localization 1893
telomeres 861, 1948–1949, *1949f*
transgenes 1989
transposable elements 585
viability selection 1295, *1295t*
zygotic lethal gene 2166
Drosophila Mid-America Center (*Drosophila* Species Collection) 847
Drosophila mojavensis 1683t
Drosophila montana 1683t
Drosophila nubialis 1683t
Drosophila parthenogenetica 2097
Drosophila persimilis 1683t
Drosophila pseudoobscura 112, 836, 1516, *1522t*, *1683t*
Drosophila sechellia 1682f, *1683t*
Drosophila silvestris 1683t
Drosophila simulans 861, 952, 1405, 1592, *1683t*, 1804
Drosophila Stock Center (Mexico) 847
Drosophila subobscura 1683t
Drosophila texana 1683t
Drosophila virilis 249, 319, 985, *1683t*
Drosophila willistoni 1405, *1522t*
DRPLA *see* dentatorubal pallidoluysian atrophy (DRPLA)
DRPLA gene *2049t*
drug resistance **585–586**
altered resistance patterns 586
antibiotics 586
Enterococcus faecalis 586
Mycobacterium tuberculosis 586
Pseudomonas aeruginosa 586
resistance plasmids 586
retroviruses 586
Staphylococcus aureus 586
transposable elements 586
vancomycin 586
Ds1 gene 2026
Ds (Dissociation) gene 2020
Ds (Dissociation) locus 1161
dsDNA phages 412
D-serine deaminase activator protein 380
dsx (doublesex) gene 1813
Duchenne, Guillaume-Benjamin-Amand 586
Duchenne muscular dystrophy (DMD) **586–587**, 1262, *1263f*
clinical features 586
gene *362f*
genetic counseling 587
Meryon, Edward 586
mosaicism 587
muscle membrane proteins 586, *587f*
prenatal diagnosis 587
single-gene inheritance 1836
treatment 587
X chromosome 2147
Dudley, H.W. 1365
Duffy antigen receptor for chemokines (DARC) 588
Duffy blood group system **588–589**
alloantigens 588
Duffy antigen receptor for chemokines (DARC) 588, *589f*
Duffy glycoprotein
blood cells 588
chemokines 588
gene expression 588
malaria 589
Plasmodium falciparum 589
Plasmodium vivax 588–589, 589
frequency 588
phenotypes 588, *588t*
Dulbecco, Renato **589–590**
bacteriophage multiplicity reactivation 589
Baltimore, David 590
Delbrück, Max 589
gene therapy (human) 816
human genome sequence 590
Nobel Prize 590
photoreactivation 589
poliomyelitis virus 589
Salk Institute 590
Temin, Howard 590
tumor viruses 589
early genes 590
late genes 590
polyomavirus 589
Vogt, Marguerite 589
Watson, James Dewey 2133
dunce gene 210
Dunckley T **1748–1751**
Dunn, I. J. 1922
Dunn, L.C. 278, **590–591**
developmental biology 591
developmental genetics 590
Glueckshon-Waelsch, Salome 591
Mendel's laws 590
mouse genetic nomenclature 1337
pseudoalleles 590
T (Brachyury) locus 590
Dussoix, Daisy 91
Dutrilleaux, Bernard 1086
Dvořák J **2060–2068**
dwarfism **591**
achondroplasia **2–3**, 591
gibberellin 591
mouse **591–593**
Ames dwarf (*df*) mutation 592
cog (congenital goiter) mutation 592–593
growth hormone (GH) 592
growth regulation 591
hyt (hypothyroid) mutation 592
lit (little) mutation 592
Pit1 (pituitary specific transcription factor1) gene 592
Snell dwarf (*dw*) mutation 592
Tgn (Thyroglobulin) gene 592–593
thyroid hormone (TH) 592
Dyer M J S **337–338, 953–954,** 979, **1091–1092, 1093–1094, 1217–1219, 1824, 1902**
dyf genes 255
DYN-1 protein 1313
dynamic mutations **593–597**
anticipation 593
change in function 593
DNA repeat sequence expansion 593, *594t*
diseases 594, *594t*
fragile X syndrome (FRAXA) 593
myotonic dystrophy 593
polyglutamine diseases 594
spinobulbar muscular atrophy (SBMA) 593
trinucleotide repeat disorders 594
gain of function 595, *596f*
copy number threshold 595
Drosophila melanogaster 595
transglutamination 595
loss of function 595, *596f*
FRAXA 595, *596f*
FRAXE 595, *596f*
Friedreich ataxia (*FRDA*) 595, *596f*
myoclonic epilepsy (*EPM1*) 595, *596f*
mechanisms 594–595
molecular pathways 595
myotonic dystrophy 596
non-Mendelian inheritance 593
dyslexia *2015t*
dysmorphology **597–598**
diagnostic steps 597
function of diagnosis 598
new syndromes 597–598
study of birth defects and syndromes 597
dysplasia 448
dystrophin **598**
Becker muscular dystrophy 598
Duchenne muscular dystrophy (DMD) 586, *587f*, 598

E

E6E7 gene 310
Eacker S M **1420–1422**

Eα gene 1640
Eanes W F **145–146**
early genes **599**, 1029, 2110
eat genes 255
EBERs protein 643, 1289
Eberwine J **1350–1352**
Eβ gene 1640
EBNA **599**
  see also Epstein–Barr virus (EBV)
EBNA1 protein 642–644, 1289
EBNA2 protein 642–643
EBNA3A protein 642–644
EBNA3B protein 642–644
EBNA3C protein 642–644
EBNA5 protein 642
EBNA6 protein 642
EBNA-LP protein 642
E-cadherin gene 2086
Eckhart W **589–590**
Ecocyc database 1201
E. coli see Escherichia coli
E. coli Genetic Stock Center 847, 849–850
Ectocarpus 748
ectoderm *see* developmental genetics
ectodermal dysplasias **599–600**
  anhidrotic 600
  autosomal anhidrotic 600
  Blashko's lines 600
  characteristics 599–600
  hidrotic 600
  hypohidrotic 600
ectopic expression 1996, *1996f*
EDA1 gene 600
Edgar, R.S. 1927, 1950–1951
editing and proofreading **601–602**
  aminoacylation 601
  aminoacyl-tRNA 601–602
  proofreading 601
  ribosome 601–602
  translation 601
Ednrb (s) gene 398, 527
EDS *see* Ehlers-Danlos syndrome
Edwards A W F **700**
Edwards, John 2058
Edwards R **1008–1009**
Edwards syndrome 66, 2056, **2058–2060**
E/e antigen 651
Eeken J C J **190–191, 1359–2952, 1511–1513**
E/e polypeptide 651
EF-1α elongation factor 611
EF-1β elongation factor 611
EF-1γ elongation factor 611
EF-1δ elongation factor 611
EF-2 elongation factor 611
EF-3 elongation factor 611
effective population number **602–603**
  allele frequency 602
  genetic drift 602
  mathematical calculation 602
  Wright, Sewall 602
EF-G elongation factor 609, 610, 901, 1722, 1901
EF-G-GTP elongation factor 610
EF-G protein factor 1988, 2002
efp gene 610
EF-Ts elongation factor 610
EF-Tu elongation factor 79, 323, 610, 901, 1988
  editing and proofreading 601
  elongation factors 609
  ribosomal RNA (rRNA) 1722
  ribosomes 1728
  translational suppression 1901
EF-Tu-GDP elongation factor 610
Eg5 protein 1225
Egel R **1776–1779**
EGF *see* epidermal growth factor (EGF)
EGF growth factor 626
Eggler A L **1839–1842**
egl-1 gene 205
EGL-1 protein 205, 314
egl genes 255
Ehlers–Danlos syndrome (EDS) 497, 498, **603–605**
  characteristics 603–605
  collagen III 603–604
  ectodermal dysplasias 599
  EDS types I/II *603f*, 603–605
  EDS type III 605
  EDS type IV 603–604, *604f*
  EDS type V 605
  EDS type VI 604
  EDS type VII 604
  EDS type VIII 605
  EDS type IX 605
Ehrlich, P. R. 411
Ehrman L **731–732**
EIAF gene 1772
Eicher E M **1816–1819, 1253–1254**
Eickbush T H **1699–1701**
EIF1AY gene 2160
eIF1 initiation factor 1029
Eif2&y gene *1253f*, 1254
eIF2 initiation factor 1029
eIF3 initiation factor 1029
eIF4 initiation factor 1029
eIF5 initiation factor 1029
Eigen's paradox 1393
Einführung in die experimentelle Vererbungslehre (Introduction to Experimental Genetics) 201
Eisensmith R C **1447–1449**
El-Deiry W S **2127**
electrofusion 1356
electron microscopy (EM) **605–608**, 614
  scanning electron microscopy (SEM) *607f*, 608
  techniques *606f*, 606–608, *607f*, *608f*
  transmission electron microscopy (TEM) 605
electrophoresis
  see gel electrophoresis; pulsed field gel electrophoresis (PFGE)
electroporation 549, **609**
electrospray ionization (ESI) 1576
Eleutherodactylus 762
Elgerdy N A **272–274**
Eliava, George 176
Eliava Institute of Bacteriophages, Microbiology and Virology 176–177
Elinson R P **2149–2150**
ELISA *see* enzyme-linked immunosorbent assay (ELISA)
Ellis, Emory 180, 522
elongation **609**
  aminoacyl-tRNA 609
  antibiotics, effect of
    alpha (α)-sarcin 611
    fusidic acid 611
    kanamycin 611
    kirromycin 611
    ricin 611
    sordarin 611
    tetracycline 611
    thiostrepton 611
  bacteria 610
elongation (*continued*)
  codon 610
  elongation factors **610–611**
    EF-1α 611
    EF-1β 611
    EF-1γ 611
    EF-1δ 611
    EF-2 611
    EF-3 611
    EF-G 609, 610, 901, 1722, 1729
    EF-G-GTP 610
    EF-Ts 610
    EF-Tu 601, 609, 901, 1722, 1728
    EF-Tu-GDP 610
  guanosine diphosphate (GDP) 609, 610
  guanosine triphosphate (GTP) 609, 610
  mRNA (messenger ribonucleic acid) 609, 610
  peptidyl-tRNA 609, 610
  ribosome 601–602, 609, 610
  RNA polymerases 609
  transcription 609
  translation 609
  translocation 610
  tRNA 609
ELP1 locus 1817
ELP2 locus 1817
ELP3 locus 1817
elt-1 gene 619
Eμ-myc oncogene 1484
emb genes 255, 1475
EMBL (European Molecular Biology Laboratories) 155, 517
embryogenesis
  Caenorhabditis elegans 613–614
  chimera 331
  imprinting, genomic 1000
  mutations 1422
  Pax genes 1426
  in vitro fertilization (IVF) 1009
embryonic development
  Caenorhabditis elegans **612–621**
  mouse **621–623**
embryonic stem (ES) cells **623**
  129 inbred strain 1016
  cell lines 310
  deletion mapping (mouse) 527
  gene trapping 819
  LMO family of genes 1118
  mouse *331f*, 332
  tissue culture 1967
  transgenicanimals1991,1992–1994,*1993f*
  see also cell determination
embryo transfer **611–612**
Emericella nidulans 106
Emery A E H **586–587, 1262–1264**
Emery–Dreifuss muscular dystrophy 1262–1263, *1263f*
Emmerson, Peter 1623
emphysema 497–498
ems (empty spiracles) gene 959
en-1 (engrailed-1) protein 1100
ENaC family 315
Encyclopedia of the Mouse Genome 1108
end-1 gene 619
end labeling **623**
endocytosis 974, 1573
endoderm *see* developmental genetics
endo gene 2022
Endo I enzyme 1760
endometriosis 1024
endonucleases **623**, 1357
  epigenetics 636

endonucleases (*continued*)
histone genes 949
hot spots 976
intron homing 1050–1051
endoplasmic reticulum (ER) 1378, 1558–1561
endoreplication 1511–1512
endosymbiont 973–974
endothelin 3 gene 1469
Endo VII enzyme 1760
end-product inhibition **623**
En (Ehancer) gene 2027
Engelsberg, E. 1652
Engels W R **1403–1407**
Enhanced Microbial Genomes Library 519
enhancers **624–625**
cis-acting proteins 624
detection 379
DNA inversion system 939–941
enhanceosome 624
function 378–379, 624
gene expression 624
gene regulation 807
histone genes 949
mechanisms 624
structure 379
transcription factors 624
upstream, downstream sites 2100
enol gene 888
ENODs 1335
En/Spm element 2029
Entamoeba histolytica 974, 405
Enterobacter 74, 1439
Enterobacter aerogenes 1382
Enterobacteriaceae 75
Enterococcus spp. 75
Enterococcus faecalis 586
env gene 979, 1251, 1704, 2022
enzymatic misincorporation 1010
enzymatic photoreactivation 559
enzyme-linked immunosorbent assay (ELISA) 334
enzymes **625–626**
applications 626
classification 625
control mechanisms 625–626
inhibitors 625
isoenzymes 625
isozymes 625
restriction endonucleases 625
Ephrussi, Boris 359
epidermal growth factor (EGF) **626–628**, 1967
biological importance 627
erbB receptors 626–627, 648–649, 1306
ligand family 627
peptide growth factors 626–628
role in oncogenesis 648–649
Src family tyrosine kinases 1884
epidermolysis bullosa 599, 1540
epidermolytic hyperkeratosis (EH) 993, *994f*
Epifagus 337
epigenetics **628–637**, 1064
Angelman syndrome 634
chromosome imprinting 629
CpG islands 631–632
DNA methylation 628–637, *631f*
dosage compensation 634
gene regulation 628–629
historical background 629–630
non-Mendelian inheritance 636
Prader–Willi syndrome 634
epigenetics (*continued*)
silencing 635–636
Spm transposon 636
X chromosome inactivation 634
X-inactivation center (*XIC*) 634
epiregulin (EPR) 627
episodic ataxia type 2 *2050t*
episome **638**
see also bacteriophages; plasmids
epistasis **638–641**, 906
adaptive landscapes 5
biochemical 638–639
birth defect analysis 640
cleft lip and palate *640t*
disease trait analysis 640
fitness landscape 706
heritability 921
population genetics 640
shifting balance theory of evolution 1826
statistical 639–640
biometrical genetics 639, *639t*
quantitative trait analysis 639
Simpson's paradox 639
epithelial cells 310
epsilon (ε) globin gene 878, 2096
Epstein–Barr virus determined nuclear antigen
see EBNA; Epstein–Barr virus (EBV)
Epstein–Barr virus (EBV) 271, **641–644**
B-cells 641–644
Burkitt's lymphoma (BL) 247
CD8+ T cells 641
disease associations
Burkitt's lymphoma (BL) 643
Hodgkin's disease 644
mononucleosis 641
nasopharyngeal carcinoma (NPC) 643
growth transformation proteins 642–643
herpesvirus subfamily 641
immune responses 643–644
lymphocryptovirus 641
monoclonal antibodies 1236
nasopharyngeal carcinoma (NPC) 1289
non-Hodgkin's lymphoma (NHL) 1347
Reed–Sternberg cells **1649**
T-cells 641
viral strategy 643
Epstein, R. H. 1927
Epulopsicum fishelsoni 147
equilibrium **644–646**
allele frequency *645t*, 645–646, *646f*
mathematical criteria 644
population **646–648**
absence of selection 646
allele frequency 646–648
genetic drift 647–648
Hardy–Weinberg Law 646
Mutation–selection balance 647–648
overdominance 647
selection 647
stationarity 647–648
stable 648, 644–646
trivial equilibrium **2068–2070**
unstable 648, 645
ER *see* endoplasmic reticulum
erbB-1 (EGFR) 648–649, 1306
erbB-1 gene 648–649
erbB-2 (HER-2/neu) protein 627, 649, 1306
erbB-2 (HER-2/neu) gene 648–649
erbB-3 (HER-3) protein 627, 649, 1306
erbB-3 (HER-3) gene 648–649
erbB-4 (HER-4) protein 627, 649, 1306
erbB-4 (HER-4) gene 648–649
erbB family **648–649**
erbB receptors 626–627, 1306
eRF1 protein factor 2002
ERG gene 1772
erb3 gene 2045
ERIC (Enterobacterial Repetitive Intergenic Consensus) 214
Erinaceus 840
error catastrophe **649–650**
E.R. Sears Wheat Genetic Stock Collection 853
Erwinia 1439, 1556, 1850
Erwinia carotovora 1382
Erysiphe cichoracearum 1914t
Erysiphe graminis 973, *1914t*
Erysiphe polygoni 1914t
erythroblastosis fetalis **650–653**
historical background 650
preimplantation determination 652–653
prenatal determination 652
Rh antigen system
C/c antigen 651
D antigen 650
E/e antigen 651
molecular studies 651
Rh polypeptides 651–652
structure 651
Rh polypeptides 651–652
erythromycin 77, 1687, 1729, 1890
ES cells *see* embryonic stem (ES) cells
Escherichia coli 176, **653–659**, 711, 1382, 1423
abortive transduction 1–2
alternation of gene expression 43
attachment sites 120–121
bacteria 148, *150t*
bacteriophage λ 260–262
base composition 192
catabolite gene activator protein (CAP) 279
chaperonins 324
characteristics 653, *654f*
chimeric genes 333
Chi sequences **325–328**, 977
chromosome 257
chromosome dimer resolution 351–353
chromosome mapping 366, *367t*
circular linkage map 377
cloning 656
codon usage bias *403t*, 404
col factors **415–416**
colicins **417–418**
conjugation 450, 453, 657
contributions to bacterial genetics 654–656
core particle 468–469
dif recombination site 351–353, *352f*
dispersed transposable sequences 861
DNA cloning 545
editing and proofreading 601–602
elongation 610
excision repair 674, *674f*
gene expression 1029
gene regulation 655
genetic code 655
genetic exchange 656–658
genetic structure 659
gene transpositions 658–659
genome sequencing 859
genome size *367t*
helix–turn–helix motif 280

Escherichia coli (*continued*)
heteroduplex DNA 931
Hfr strain 936–938
histidine biosynthesis 943
histidine operon *946f*
illegitimate recombination 998
induction 1019–1020
inheritance 656
integrase family 1040
lac mutants 1069
lac operon 1070
leader sequence 1079
meiotic hot spots 977
microbial genetics 1192, 1199
missense mutations 1277
model organisms 653–655
mutagenesis 257
mutational studies 656
Mut system 932
nutritional mutations 1362
phage λ integration and excision **1434**
phage Mu 1439
"phage truce" 1921
photoreactivation *1458f*, **1458–1459**
plasmids 657–659, 1486, 1488
pleiotropy 1491
polymerase I 1497
primase 1542, *1543f*, *1547t*
protein secretion systems 1556
rearrangements 1613
RecA protein 931
RecBCD enzyme 325–328
rec genes **1614–1618**
reckless DNA degradation 1635
recombination pathways 1647
replication errors *1676t*
restriction endonucleases *1694t*
ribosomal RNA (rRNA) 1719
ribosomes 1724
RNA polymerase *1547t*
Salmonella 1767
sexduction **1823–1824**
short bacterial repeated elements 214
sigma factors 1831–1832, *1832t*
single-stranded DNA-binding proteins (SSBs) **1839–1842**
stringent response 1891
temperature-sensitive (*ts*) mutant 441–442
transcription attenuation 122
tryptophan operon *2076f*, 2076–2077, 2077–2078
vectors 2104
in vitro mutagenesis 1010
Yanofsky, Charles 2161
Escherichia faecalis 451, 452
Escherich, Theodor 653
ESI *see* electrospray ionization (ESI)
E(spl) gene 1317, *1318f*
Esposito, D. 351
ESS *see* evolutionarily stable strategies (ESS)
EST *see* expressed sequence tags (ESTs)
EST2 gene 1948
established cell lines *see* cell lines; tissue culture
Estelle M **1480–1481**
ethics **659–661**
eugenics 659–660
gene therapy 660
in medicine 660
nonhuman genetics 661
in reproduction 660
in research 660
ethics (*continued*)
social issues 661
ethylene 1481
ethylmethanesulfonate 255, 557
ets-1 gene 661
Ets family **661–663**
biological importance 662
cancer, role in 662
domains 662
ets-1 gene 661
ets genes 662
phosphorylation 662
regulation
DNA binding 662
gene expression 662
signal transduction pathways 662
transcription factors 662
ets genes 662
ETV-1 gene 1772
ETV6(TEL) gene 1772
euchromatic bands 349
euchromatin **663**, 927, 932
Eugenical Sterilization Act (Virginia) 2015
eugenics 578, 659–660, 2015
Baur, Erwin 202
behavioral genetics 209
Galton, Francis 742, 747
Euglena 535
eukaryotes **663**
cAMP signaling 258–260
cell cycle regulation *503f*
cell division genetics 297
Chi sequences 328
chromosome 344
cotransfomation 472
DNA cloning 547, 549
DNA repair 563
elongation factors 611
enhancers 624
excision repair 673
genetic code *822t*
genomes 249
heat shock proteins (Hsps) 915
heteroduplex DNA 931
histidine genes *946f*
histone genes 949–951
horizontal transfer 973–975
kinases 1063–1064
open reading frame (ORF) 1375
photosynthesis 337
protein secretion systems *1554f*, 1558–1564
Rad51protein 931
regulatory RNA 1658
RNA polymerase 1746
spliceosomal introns 1052
transcription 250
transcription regulation 949–951
ubiquitin 2091
eukaryotic genes **663**
see also introns
eukaryotic primases *1547t*
euploid **663**
see also aneuploid
Euplotes 1135
Euplotes crassus 1133
European Collaborative Interspecific Mouse Backcross Mapping Panel 912
European *Drosophila* Stock Center 847
European Molecular Biology Laboratories (EMBL) 517
Euryarcheota 92
eusociality 907
Evans E P **722–723**
Evans T C Jr **1565–1567**
evolution **663–666**
cell lineage 309
C-value paradox 250
Dobzhansky, Theodosius 663
dogs 264–265
Drosophila melanogaster 666
duplication mechanisms 667–668, *668f*
extinction 665
fitness 664
fossil record 665
gene families **666–669**
gene fusion 333–334
genetic drift 665
genetic exchange 668
Holliday's model 668–669
macroevolution 664
mechanisms 1004
microbial evolution 666
microevolution 664
paternal inheritance 1423–1424
photosynthesis 337
phyletic evolution 665
polyploidy 1510
punctuated equilibrium 665–666
selective breeding 1004
speciation 664
t haplotype 1920–1921
theory of evolution 664, 1004
thermophilic bacteria 1962
transposons in plants 2031–2032
in vitro evolution 1004–1008
evolutionarily stable strategies (ESS) **669–671**
cooperative strategies 670
game theory 669–670
Nash equilibrium 669
prisoner's dilemma 669–670
tit for tat 670
evolutionary biology
bottlenecks 516
Darlington, Cyril Dean 513
Darwin, Charles 516
DNA fingerprinting 516
DNA sequencing 516
founder effect 516
genetic drift 516
genetic technologies 516
evolutionary rate **671–672**
DNA (deoxyribonucleic acid) 671
genetic change 671–672
molecular clock 672
neutral mutation 671–672
phenotypic change 671–672
RNA (ribonucleic acid) 671
selective neutrality 1801–1802
evolutionary synthesis 516
Evx gene 959
Ewens, Warren 1517
Ewens W J **7–9**, **733–738**, **855–856**
Ewing's tumor 662, **672**, 713, 1768, *1769t*
see also Ets family
EWS gene 1772, *1772t*
exchange **673**
see also crossing-over; segmental interchange; unequal crossing-over
exchange pairing *see* exchange; segmental interchange
excinuclease 673–675
excision repair **673–675**
DNA damage 673

excision repair (*continued*)
DNA repair 673
Escherichia coli 674, *674f*
excinuclease 673–675
function 673
human *674f*, 674–675
mechanisms 673–675
xeroderma pigmentosum (XP) 674–675
see also DNA repair; introns
excision repair pathways 566
excitotoxicity 317
exd (extradenticle) gene 961
EXLM1Y gene 2159
Exner, Frank 1123
exogenous genes 545
exons 1052
exon skipping 496
exonucleases **675**, 1010, 1357
"Experiments in plant hybridization" (Mendel, 1866) 2009
expressed sequence tags (ESTs) 232, 364, *365f*
Arabidopsis thaliana 366t
Bos taurus 366t
Caenorhabditis elegans 366t
Drosophila melanogaster 366t
Glycine max 366t
human linkage maps 1108
humans *366t*
mouse *366t*
rat *366t*
expression vector **675**
see also vectors
expressivity **675–676**
see also penetrance
EXT1/EXT2 gene *2084–2085t*
extracellular polysaccharide (EPS) 45
eyes 1427

F

F_1 generation **680**
F_1 hybrid **680**
haplotype 911
inbred strain 680, 1243
intercross 680
mouse genetics 1248
mule 1255
reciprocal cross 1631
FAB classification of leukemia **680–681**
acute lymphoblastic leukemia (ALL) 680
acute myelogenous leukemia (AML) 680
chronic myelogenous leukemia (CML) 680
French-American-British (FAB) cooperative group 680
non-Hodgkin's lymphoma (NHL) 680
see also MIC and MIC-M classification of leukemia; WHO classification of leukemia
Fabian J R **2008–2016**
Fabry disease **681**, 756
α galactosidase A deficiency 681
clinical features 681
facioscapulohumeral muscular dystrophy *1263f*, 1264
fackel mutation 1475
factor B 429
factor crosses *see* genetic crosses
factor H 433
facultative heterochromatin *see* heterochromatin
FAD *see* flavin-adenine dinucleotide (FAD)
Fagus sylvatica 840
familial adenomatous polyposis (FAP) 9, 12, 422, 1769
familial fatal insomnia (FFI) 1881, 2008
see also Gerstmann–Sträussler disease (GSD)
familial hypercholesterolemia (FH) **682–683**
clinical features 682
low density lipoprotein (LDL) 682–683
familial retinoblastoma 1770
FANCA-H gene *2084–2085t*
Fanconi's anemia (FA) *569–570t*, **683**, 1285
Fane B A **1450–1454**
Fankhauser C **1454–1458**
FAP *see* familial adenomatous polyposis (FAP)
FASS gene 1476
fass mutation 1476
FASTA files 519, 1197
fate map 683
favism **683**
clinical features 683
fava bean 683
glucose 6-phosphate dehydrogenase (G6PD) deficiency 683, 882
Vicia faba 683
FBA1 gene 886
FBNI gene 1145
FBP11 gene 716
FBP7 gene 716
FCA gene 715
F-duction **683**
see also sexduction
Federoff, Nina 636
Fedoroff N **1161–1162**
Feiss M **412–413**
Feenstra, William 88
Feldblyum T V **865–872**
feline genetics **684–685**
characteristics 684
evolution 684
molecular phylogeny *685f*
species 684, *685f*
subspecies 685
survival threats 685
Felsenfeld, Gary 634
Felsenstein, Joseph 490, 1517
female carriers **685–688**
analysis of risk 687
Becker muscular dystrophy 686, *686f*
carrier tests 687
homozygosity 688
nonrandom X-inactivation 688
obligate carriers 686
Turner syndrome 688
X–autosome translocation 688
X-chromosome recessive disorders 685, *686f*
FEN-1 protein 1546–1547
feral **688**
fer genes 255, 1874, *1874f*
Ferguson-Smith A C **999–1001, 2096–2099**
Ferguson-Smith M A **60–61, 345–348, 368, 911, 924–925, 980, 996, 1002–1004, 1047, 1054, 1061, 1065–1066, 1068, 1189, 1239, 1361, 1362, 1539–1541, 1773–1774, 1809–1810, 1822–1823, 1912, 2007–2008, 2088–2089, 2123, 2140**
Fermi, Enrico 1123
Fersht A R **529**
fertilization **688–690**
acrosome 689, *692f*, 693
fertilization (*continued*)
capacitation 689, 691
fusion process 694
Golgi body 689, 693
Graafian follicle 691
intracytoplasmic sperm injection (ICSI) 689
mammalian **690–695**, *692f*
meiosis 689
oogenesis 689, 691
process 689, 691
Sertoli cells 691
spermatogenesis 689, 691
sperm binding process 692–693
syngamy 689
in vitro fertilization 689–690
zona pellucida 689, 692, *692f*, *693f*
zygote 689, 695
Festetics, Imre (Emerich) 1173
fetal aminopterin syndrome 60
FEV gene 1772
F (fertility) factor 453, 451, *452f*, **677–680**, 1080
bacterial genetics 679
conjugative plasmid 677, *677f*
conjugative transfer 677, *679f*
DNA metabolism 678
episome 638
Hfr strains 679
sexduction **1823–1824**
structure *677f*, 677–678, *678f*
F (fertility) plasmid 936–938
ffh gene 1555
ffs gene 1555
FGF *see* fibroblast growth factor (FGF)
FGF23 gene 1029
FGF-3 gene 126
FGF-4 gene 126
FGF-5 gene 126
FGFR1 gene 479–481, *480t*
FGFR1 protein *480f*
FGFR2 gene 479–481, *480t*
FGFR2 protein *480f*
FGFR3 gene 479–481, *480t*
FGFR3 protein *480f*
Fgr protein 1883
FHA gene 715
FHIT gene 1123
fibroblast growth factor (FGF) 480
fibroblasts 310
fibroids 1110
fibromatoses 1769, *1769t*, *1770t*
fibrous proteins 1572
Ficedula albicaulis 1296
Ficedula hypoleuca 1296
Ficus 407
filamentous bacteriophages **695–698**
capsids 696
characteristics 696, *696f*
cholera toxin (CT) 695
cloning vectors 697
Ff coliphages 695
mechanisms 696
uses 697
Vibrio cholerae phage 695
filial (F) generation **698**
see also F_1 hybrid; inbred strain
Filipowicz W **813–814**
filter hybridization **698**
fimA gene 800
fim gene 1443
Finan T M **2117–2120**

Fincham J R S **113**, **319–320**, **320–323**, **535–537**, **663**, **673**, **698–700**, **748–749**, **793–796**, **963–964**, **1018–1019**, **1787–1790**, **1819**, **1946–1950**, **1952–1957**, **1958**, **1964–1965**, **2095–2096**
fingerprinting **698**, 746–747
finO gene 678
first and second division segregation **698–700**
  centimorgan (cM) 699
  centromere 698, *699f*
  chromatids 698, *699f*
  crossovers 699
  meiosis 698, *699f*
  tetrad analysis 699, *700f*
first-cousins marriages 457
Firth H **1420**
FIS (factor for inversion stimulation) 121, 542, 544, *1040t*, 1040–1041, 1435
FISH *see* fluorescent *in situ* hybridization (FISH)
Fisher–Race nomenclature system 651
Fisher, Ronald A. 516, **700–701**, 1713
  Blood-Group Serum Unit 701
  contributions
    evolutionary genetics 700
    mathematical genetics 700
    statistics 700
  The Correlation between Relatives on the Supposition of Mendelian Inheritance 701
  evolution 664
  fitness 703
  Ford, E.B. 701
  fundamental theorem of natural selection 733, 1294
  The Genetical Theory of Natural Selection 701
  genetic drift 832
  Haldane, J.B.S. 701
  heritability 921
  Human Genome Project 701
  human linkage maps 701
  mathematical models 1516
  Penrose, L.S. 701
  population genetics 640, 905, 1514, 1518
  quantitative inheritance 1595
  quantitative trait 1597
  shifting balance theory of evolution 2142
  Theory of Inbreeding 701
Fisher's exact test 751
Fisher's theory of equal investment in the sexes 1820
FIS protein *1383f*
Fitch W **318**, **325**, **556**, **1017**, **1053–1054**, **1379–1381**, **1394**, **1411**, **1429**, **1696–1697**, **2149**
Fitch, Walter 473
fitness **702–707**
  adaptation 704
  adaptedness 705
  components 703–704
  Darwinian fitness 703
  definition 702
  Fisher, R.A. 703
  genotypic fitness 705
  history 702
  inclusive fitness 704
  Malthusian fitness 703
  mathematical calculations 703
  mutation load **1276**
fitness (*continued*)
  natural selection 702
  On the Origin of Species 702
  population fitness 705–706
  propensity interpretation 705
  Red Queen hypothesis 706
  sex ratios 1820
  shifting balance theory of evolution 1825–1826, *1826f*
  Spencer, Herbert 702
  Verhulstian fitness 703
  Wright, Sewall 703
fitness islands 1903
fitness landscape *706f*, **706–707**, 707
  adaptive surface 706
  shifting balance theory of evolution 706
  Wright, Sewall 706
fixation equilibrium 2068
fixation probability **709–710**, *710f*
  selection limit 1794
  selective neutrality 1800
fixed model 1801, *1801f*
fix genes **707–709**, *708t*
  fix gene cluster
    fixABCX gene *708t*
    fixD gene *708t*
    fixF gene *708t*
    fixGHIS gene 708, *708t*
    fixK gene 708, *708t*
    fixLJ gene 708, *708t*
    fixNOQP gene *708t*
    fixR gene *708t*
    fixT gene 708, *708t*
    fixU gene *708t*
    fixW gene *708t*
    fixY gene *708t*
    fixZ gene *708t*
  Klebsiella pneumoniae 707
  nifA gene 708
  nitrogen fixation 707
  phosphorylation 708
  promoters 707
  proteins
    FixJ protein 708, *708t*
    FixK protein 708, *708t*
    FixL protein 708, *708t*
  regulatory systems 708, *709f*
  Rhizobium 1715
  Sinorhizobium meliloti 707
  see also nif genes
FKHR gene 1772, *1772t*
fl2 mutant 1786
F-lac gene 451
flagella **711–712**
  chemotaxis 711
  Chlamydomonas reinhardtii 337
  flagellar system *712f*
  flagellin genes 711
  gene regulation 711
  motility 711
fla genes 711
flavin-adenine dinucleotide (FAD) 1456
Flavobacterium okeanokoites 1694t
flavonoid inducers 1333
FLC gene 715
Fleming, Alexander 1131
Flemming, Walther 1172
flg genes 711
flh genes 711
FLI1 (Friend leukemia virus integration 1) oncogene **712–713**, 1772, *1772t*
fli genes 43, 711
fljA gene *939f*
fljB gene 938, *939f*
fljC gene 938
flj genes 43, 711
FLO gene 715
Flor, H. H. 1535
floury2 mutant 1786
floury3 mutant 1786
flower development, genetics 713–717
  gene hierarchy 713, *714f*
  genes *714f*
  meristem formation 715
  model plants
    Antirrhinum majus 713
    Arabidopsis thaliana 714f
  organ development 715–716
  structure 713, *714f*
  timing regulation 713–715
flow-sorted chromosomes 361–362
Flp recombinase **717–721**, 1040, *1040t*
  Futcher model 720, *721f*
  Holliday junction 719
  integrase family 717
  mechanisms 718, *719f*
  plasmid physiology 720
  plasmid replication 720, *721f*
  plasmids 717, *718f*
  site-specific recombination 717
  structure 719, *720f*
flu gene 110, *110f*
fluorescein-conjugated dextran 302
fluorescein isothiocyanate (FITC) 1002
fluorescent *in situ* hybridization (FISH) 269–270, 361, *362f*, 535, **700**, 1002–1004
  applications 1003–1004
  chromosome painting 1002–1004
  DNA probes 1002–1003
  homologous chromosomes 963
  leukemia, acute 1091
  mitosis 1224
  somatic pairing 1853
  uterine leiomyoma 1110
  see also in situ hybridization (ISH)
fluorescently labeled oligomers 551
fluorochromes 348
fluorophores 312
FlyBase 847
FMR1 gene 726, *2049t*
FMR2 gene *2049t*, 595
FMRI gene 595
FMRI protein 1735
FMS oncogene **721–722**
Fogh-Andersen, Poul 385
foldback DNA **722**
Fol, Hermann 553
folic acid 2118
Folkman J **66–73**
follicle stimulating hormone (FSH) 1021–1022, 1875
follicular lymphoma **722**
Fontana, W. 1323
Food and Drug Administration (FDA) 818
footprinting **722**
Ford, Charles **722–723**
  human chromosome number 722–723
  mammalian cytogenetics 723
  nomenclature systems 723
Ford, E.B. 187, 701
Ford J **564–571**
Forejt J **1345–1347**
formic acid 1010
Forsburg S L **2162–2164**
Forterre, Patrick 186

- forward mutations **724**
- fosmid **724**
- Fossella J A **675–676**, **987**, **1016–1017**, **1429**, **1631–1632**
- fossil record 665
- Foster P L **20–21**, **197–198**, **256–258**, **1279**
- founder effect 594, **724–726**
  - albinism 27
  - bottleneck effect 233, 724
  - disease mutations 724, *725t*
  - evolution 665
  - evolutionary biology 516
  - genetic colonization 823
  - hemophilia 920
  - homozygosity mapping 725
  - identity by descent 725
  - speciation 665
  - uses 725
- founder principle *see* bottleneck effect
- four-point metric **1189**
- *FPA* gene 715
- F plasmid *603f*, 657, 1486, 1823
- fragile chromosome site **726**
- fragile X syndrome (FRAXA) 593, **726–728**, *2049t*
  - clinical features *726t*, 726–727
  - cytogenetics 727, *727f*
  - female carriers 687
  - FMRI protein 1735
  - mosaicism 727
  - Sherman paradox 728
  - single-gene inheritance 1837
  - treatment 727
  - X chromosome 2147
- fragmentation maps *363t*
- Frame M **1160**, **1824**, **1824–1825**, **1825**, **1846**, **1883–1885**
- frameshift mutation **728–729**
  - Brenner, Sydney 244
  - coding mechanisms 728
  - nucleic acid sequence alteration 728
- Frampton J **1287**
- Frankel, Sir Otto 459
- *Frankia* 1334
- Franklin, Rosalind 540, 554, 1730, 2134
- Fraser, Bruce 554
- Fraser C M **1196–1203**
- Fraser G R **1168–1171**
- Fraser–Juriloff model *386f*
- Frasier's syndrome 1814, 1818, 2139
- *FRAXA* gene *2049t*
- *FRAXE* gene *2049t*
- *FRDA1* gene *2049t*
- Fredericq, Pierre 417
- Fredga K **1094–1095**
- Fredriksson S **333–334**
- freedom, degrees of *see* degrees of freedom
- French-American-British (FAB) cooperative group 680, 1190–1191
- French Muscular Dystrophy Association 355
- frequency-dependent selection **730–731**
  - Batesian mimicry 730–731
  - competition 730
  - fitness 730, 731
  - Mullerian mimicry 731
  - overdominance 731
  - predator–prey interaction 730
  - rare male mating advantage 730, **731–732**, *732f*
  - self-incompatibility 730
- *frequency (frq)* gene 389
- Friedberg E C **558–564**
- Friedreich ataxia (*FRDA*) 595, *596f*, *2049t*
- *Fritillaria* 319–320
- fruit fly *see Drosophila*
- F^*_{ST} distance 829
- *FT* gene 715
- FtsK protein 353
- *ftsY* gene 1555
- *Ftzf1* locus 1817
- *ftz (fushi tarazu)* gene 958, *960f*
- *Fucus* 748
- *Fugu rubripes rubripes* 367, *367t*, 860
- *FUL* gene 715
- Fuller, Watson 555
- functional genomics **733**
- fundamental theorem of natural selection **733–738**
  - classical interpretation 734–736
  - modern interpretation 736–738
  - *see also* Fisher, Ronald A.
- fungal genetics **738**
- Fungal Genetics Stock Center 847, 849
- fungi **738**, 1882
- *furca2* gene 2048
- *furca4* gene 2048
- *Fusarium* 1850
- *Fusarium oxysporum 1914t*
- *fus* gene 610, 1455
- fusidic acid 77, 611
- fusion gene *see* gene fusion
- fusion proteins **739–740**
  - uses 740
  - *see also* green fluorescent protein (GFP)
- *Fusobacteria* 149
- *FUS/TLS-CHOP* fusion genes **1287**, 1772, *1772t*
- Futcher, Bruce 720
- Futcher model 720, *721f*
- Futuyma D J **407–412**
- *Fv1* gene 1250
- *Fv4* gene 1250
- Fyn protein 1883

G

- G_1 phase **741**
- G_2 phase **741**
- G6PC *see* liver glucose-6-phosphatase enzyme
- G6PC gene family 2121
- G6PD deficiency *see* glucose 6-phosphate dehydrogenase (>746PD) deficiency
- *G7c* gene 1640
- Gabor Miklos G L **859–863**
- *GA1* gene 715
- *Gag* gene 979
- *gag* gene 1704, 2021
- gain of function 982–983, *2126f*, 2127
- Gal80 repressor protein 1896
- galactosemia **741–742**
  - etiology 741
  - mechanisms 741
  - metabolic disorders 1187
  - mutational studies 742
  - treatment 742
- Galegeae 1083
- *gal* gene 1436, 1983–1984
- Gallant J **649–650**
- Gall, Joe G. 348, 361
- *gal* operon 1029, 1858–1859, *1859f*, 1896
- $G\alpha_s$ 317
- *GAL* promoter 2038
- Galton, Francis **742–748**, 1297–1298
  - ancestral inheritance theory **63–64**
- Galton, Francis (*continued*)
  - *The Art of Travel* 743
  - behavioral genetics 209
  - biometry 742
  - contributions
    - behavioral genetics 1298
    - fingerprinting 746–747
    - human variation 1298
    - measurement techniques 1298
    - statistical techniques 742
  - Darwin, Charles 743, 1298
  - *English Men of Science: Their Nature and Nurture* 1298
  - eugenics 742, 747, 1298
  - *Hereditary Genius* 743, 1298
  - heredity studies 743–744, *744t*
  - *Inquiries into Human Faculty and its Development* 747
  - mathematical–statistical procedures 63
  - Mendel, Gregor Johann 1170
  - *Natural Inheritance* 743
  - natural selection 746–747
  - nature–nurture controversy 742
  - pedigrees 744
  - Royal Geographical Society 743
  - statistical techniques 745, *745f*, *746f*
  - twin studies 744, 1298
- G.A. Marx Collection (pea) 853
- *Gambusia* 730
- gametes **748–749**
  - algae 748
  - fungi 749
  - mammalian **750**
  - protozoa 749
  - vegetative cells 748–749
- game theory 669–670
- gametic disequilibrium **750–752**
  - *see* linkage disequilibrium
  - chi-square test 751
  - definition 750
  - Fisher's exact test 751
  - gene mapping 751–752
  - relationship to recombination 750
  - *see also* linkage disequilibrium
- gametogenesis **752**
  - chromosome imprinting 635
  - DNA methylation 635
  - germ cell 875–876
  - haploid number 911
- *gam* gene 1625
- gamma (γ) -ENaC ion channel 315
- gamma (γ) distribution **752**
- gamma (γ) globin gene 878
- gamma delta ($\gamma\delta$) transposon 1688, *1689f*
- gamma delta ($\gamma\delta$) resolvase 941
- Gamow, George 821
- Ganetzky, Barry 490
- GAPase 1308
- GAP (*Ras* GTPase activating protein) **752–753**
- gargoylism 983
- Garland, T. H. 1535
- Garrels J I **1575–1578**
- Garrod, Sir Archibald 2012
- Gartler S M **2150–2154**
- GAs *see* gibberellins (GAs)
- *Gasterosteus aculeatus* 408
- *GATA4* gene 1814
- GATA4 protein 1814, 1818
- Gaucher's disease **753–757**
  - balanced polymorphism 187

Gaucher's disease (*continued*)
  beta (β)-glucosidase 753
  chromosome 1q21 753
  clinical features 754–755
  definition 753
  frequency 753
  genetics 753
  genotype-phenotype correlations 754
  mutational studies 754
  pathogenesis 756–757
  pathology 753–754
  treatment 755–756
Gavalas A **978–979**
Gavrias V **106–111**
Gazdar A F **9–12**
G-banding (Giemsa banding) 345, 349, **757**, **876–877**, 996
Gefter, Malcolm 257
Gehring, Walter 958
2D gel *see* two-dimensional polyacrylamide gel electrophoresis
gel electrophoresis **757–759**
  capillary electrophoresis 758–759
  Molecular Cloning: A Laboratory Manual 759
  Ohm's Law 758
  polyacrylamide gel electrophoresis (PAGE) 758
  Tiselius, Arne 757
gemmules 521
GenBank 89, 155, 364, 517
gender verification 1810
gene **759–761**
  alleles 760
  Bateson, William 759
  Beadle, George 760
  Benzer, Seymour 760
  chromosome 344
  Darwin, Charles 759
  definition 759–761
  function 760
  humans 364
  Johannsen, W. 759
  Mendel, Gregor Johann 759
  open reading frames 760
  operational definition 1928
  sequencing 760
  Sturtevant, A.H. 760
  see also cistron
gene action **761**
gene amplification 580, **761–771**
  bacteria 764
  Culex quinquefasciatus 766
  developmental 762–763
  Drosophila melanogaster 765
  frequency 767–768
  germline cells 762–763
  invertebrates 765
  mammalian 766–767
  phage λ 764
  plants 766
  protozoans 765
  Saccharomyces cerevisiae 764
  somatic cells 763
  sporadic 764–768
  vertebrates 766
  yeast 764
gene cassettes **771–774**
  59-be (59-base element) 771
  antibiotic resistance genes 1043–1044
  associated genes 772
  attI site 771
  integrons 773, 1041–1044, *1042f*, *1044f*
gene cassettes (*continued*)
  open reading frames 771, 1043–1044
  recombination sites *772*, *773f*
  structure *771–772*, *773f*
  Vibrio cholerae 772
  see also integrons
gene conversion **774–778**
  aberrant segregation 775
  conversion gradient 775
  crossing-over 775
  double-strand break repair model 775–777, *776f*
  frequency 775
  Holliday junction *776f*
  hot spots 777
  meiosis 774, *774f*
  mismatch repair 932
  mitosis 774
  molecular drive 1234
  nonreciprocal exchange 778
  nonreciprocal recombination 774
  reciprocal recombination 1632, *1633f*
  recombination models 1642
  segregation ratios 774
  single-strand gap repair model *777f*, 778
  see also concerted evolution
gene dosage **778**
gene duplication **778–780**
  dispersed duplication 779–780
  gene amplification 780
  gene families 779, *779f*
  pseudogenes 779
  tandem duplication 779, *779f*
gene expression **780–782**
  DNA inversion system 938, *939f*
  DNA replication 541, 630–632
  enhancers 624
  epigenetic regulation 628–630
  Ets family 662
  green fluorescent protein (GFP) 311–312
  homeobox genes 978
  insertion sequence (IS) 1030
  integrons 1043–1044
  introns 1052
  Jacob, François 1059
  lac operon 1070
  posttranslation steps 782
  transcription 780–781
  translation 782
gene family **783–785**
  concerted evolution 784–785
  superfamilies *783f*, 783–784
  tandem families 784
  unequal crossover 784–785
  see also concerted evolution; T-box genes
gene flow **785–790**, *1522t*
  allele frequency 785
  behavior 787
  Crassostrea virginica 790
  definition 785
  direct measurement 787
  dispersal 786–787
  genetic drift 785
  Geukensia demissa 786
  Hardy–Weinberg law 785
  homogeneous populations 786
  inference
    evolutionary equilibrium 789–790
    F_{ST} 787–788
    heterogeneity of estimates 790
    migration number 788
gene flow (*continued*)
    private alleles 789, *789t*
  mating system 787
  Mytilus edulis 786, *786f*
  natural selection 786
  Pinus flexilis (limber pine) 788–789, *788f*
  Semibalanus balanoides 786
gene frequency **790–792**
  allele frequency 790–792
  mutations 791–792
  sampling techniques 790–791
gene fusion **739**
  chimeric genes 333
  evolution 333–334
  history 739
  uses 739
gene insertion **792**
gene interaction **792–793**
gene library **793**
gene linkage *see* linkage
gene mapping **793–796**
  see chromosome mapping
  colinearity principle 795
  crossing-over 793, *794f*
  deletion mapping 795, *795f*
  Drosophila melanogaster 793, 584
  flanking markers 793–794, *794f*
  gene conversion 793
  haplotype 912
  history 353–355
  hybrid sterility (mouse) 985–986
  Neurospora 795
  recombination frequency 793
  recombination within genes 793
  Saccharomyces cerevisiae 793
  in situ hybridization (ISH) 1003
  see also centimorgan (cM)
Gene Mapping Conferences 355
gene number **796–797**
  accuracy 796
  Arabidopsis 796
  Caenorhabditis elegans 796
  determination 796
  Drosophila melanogaster 796
  Pseudomonas aeruginosa 796
  Saccharomyces cerevisiae 796
gene pool **797**
  balance theory 797
  classical theory 797
  Hardy–Weinberg law 797
  panmictic population 797
gene product **797**
general transcription factor (GTF) 379
gene rearrangement **798–800**, **800–803**
  eukaryotes **798–800**
    antibodies 799, *799f*
    antigenic variation 798
    Drosophila melanogaster 798
    mating regulation 798
    Saccharomyces cerevisiae 798
    T-cell antigens 799
    transposons 798
    trypanosomes 798
    vertebrate immune system 799
  prokaryotes **800–803**
    Anabaena 801, *802f*
    antigenic variation 800
    Bacillus subtilis 803
    cyanobacteria 801
    developmentally regulated rearrangements 801–803
    Escherichia coli 800

gene rearrangement (*continued*)
excision process 802, *802f*
Hfr strains 801
insertion sequence (IS) 801
negative regulation 801
phage λ 801
plasmids 801
recombination 800
rRNA 800
Salmonella 800
sporulation 803
stochastic rearrangement 800–801
gene regulation **803–813**
attenuation 943
chromatin structure 811–812, *812f*
C-terminal domain (CTD) 804, *804f*
DNA methylation 631–633
DNA replication 541, 630–632
enhancers 807
eukaryotes 803
histidine operon 943
lac operon 1070
prokaryotes 803
purpose 803
RNA polymerases 803, *804f*
DNA-binding domains (DBDs) *808f*, 808–809, *809f*
promoter structure *805f*, 807, *807f*
TATA box 805, *805f*, *806f*
transcription factors 805, 807, *808f*
transcription mechanism 805, *806f*
transcription 803, *809f*, 809–811, *810f*, *811f*
gene sequencing *see* DNA sequencing
gene silencing **813–814**, 1530
Ascobolus nidulans 813
Caenorhabditis elegans 813
Drosophila melanogaster 814
epigenetic modification 813
function 813
methylation 813
Neurospora crassa 813
posttranscriptional gene silencing (PTGS) 813
RNA interference **1743–1744**
RNA (ribonucleic acid) 813
transcriptional gene silencing (TGS) 813
Genesis 30:27–42 2009–2010
genes, specific 2096
5A2 gene 603
a1 gene 1895
aadA gene 336
abaA gene 110, *110f*
abc-1 gene 301
abd-A (abdominalA) gene *960f*
Abd-B (AbdominalB) gene 958, *960f*, 961
ABL1 gene **250–251**
ABL2 gene 251
ABL-related gene 251
Ac (Activator) gene 2021
achaete-scute genes 1317, *1318f*, 1321
acy-1/sgs-1 gene 317
ada gene 36
Adh gene 187
af (afila) gene 1471
AG gene 715
agouti (a) gene 398
Agouti gene 23
agr gene family 1886
AIB1 gene 425
aidB gene 36
albino (c) gene 398
genes, specific (*continued*)
alkA gene 36, 1664
alkB gene 36
alpha (α) globin gene 878
amBra-1 gene *1944f*
amBra-2 gene *1944f*
AMELY gene 2159
AMH gene 1814
Ambr2 gene 1818
ana gene 930
angustifolia gene 2048
ANT3 gene 2157
ANT gene 716
Antp (Antennapedia) gene 958, *960f*
AP1 gene 715
AP2 gene 715
AP3 gene 716
APC (adenomatous polyposis coli) gene *81t*, 9–10
Apc gene 1245
APC gene 2082, *2084–2085t*
apterous gene 1097
apx-1 gene 617–618, *618f*
APXLY gene 2159
AR gene *2049t*
araC gene 1652
ARG1 gene 892
argA gene 1624, *1624f*
ARG gene 251
ARL2 gene 892
ARO1 gene 1765
arrowhead gene 1097
ARSDY gene 2159
ARSEY gene 2159
ARSFY gene 2159
ASMT gene 2157
asnA gene *1383f*
asnC gene *1383f*
as-T2 gene *1944f*
as-T gene *1944f*
ATF1 gene *1772t*
ATF2 gene 1772
ATM gene 11, 113, 263, 568, 1093, *2084–2085t*
atonal gene 1321
atp1 gene 238
atpB gene 238
atp gene 1221, 1461
AUX1 gene 892
Axin 1 gene *2084–2085t*
Axin 2 gene *2084–2085t*
Axin gene 2083
B27 gene *2049t*
BARE-1 gene 2022
Bar gene 973
Bax gene 274
BAX gene 423
bcd (bicoid) gene *960f*, 961
BCL2 gene 1208
BCRF1 gene 643
BEL1 gene 716
beta (β) globin gene 878, 2096
B gene 1045
Bg gene 2026
BHRF1 gene 643
bim gene 208
bio gene 1436, 1983–1984
BLM gene 229–230, *2084–2085t*
bmall gene 389
Bmp5 (se) gene 524–525
BPY1 gene 2159
BPY2 gene 2159
genes, specific (*continued*)
BRCA1 gene 10, *81t*, 2082, *2084–2085t*, 2086–2087
BRCA2 gene 10, *81t*, 2082, *2084–2085t*, 2086–2087
brlA gene 110, *110f*
brown (b) gene 398
Bs1 gene 2022
BTAK gene 425
BUB1 gene 2083
BUBR1 gene 2083
BWS gene 1770
BZLF1 gene 643
C1qA gene 429
C1qB gene 429
C1qC gene 429
C2 gene *430t*, *431t*
C3AR1 gene *430t*, *431t*
C3 gene *430t*, *431t*
C4A gene *430t*, *431t*, 447
C4B gene *430t*, *431t*, 447
C4BPA gene *430t*, *431t*
C4BPB gene *430t*, *431t*
C5 gene *430t*, *431t*
C5R1 gene *430t*, *431t*
C6 gene *430t*, *431t*
C7 gene *430t*, *431t*
C8A gene *430t*, *431t*
C8B gene *430t*, *431t*
C8G gene *430t*, *431t*
C9 gene *430t*, *431t*
cac gene 1317
CACNA1A gene *2050t*
cagA gene 916–917
CAL gene 715
caudal (cad) gene 959
CBL2 gene *2049t*
CBP/p300 gene *2084–2085t*
Cbx2 gene 1818
c (colorless aleurone) gene 2020
Cd36 gene 1608
CD59 gene *430t*, *431t*
cda (chablis) gene 2048
cdc13$^+$ gene 500
CDC2a gene 1755
cdc2$^+$ gene 500
cdc gene 500, 298
CDH1 gene *2084–2085t*, 2086
CDK4 gene 1772, 2085
CDK6 genes 1093
CDK12A gene *2084–2085t*
CDKN6/P16 gene 1772
cdo (chardonnay) gene 2048
cd-tbx9 gene *1944f*
CDY gene 2160
ced-1 gene 314
ced-2 gene 314
ced-3 gene 205, 314
ced-3(lf) gene 316
ced-4 gene 205, 314
ced-4(lf) gene 316
ced-5 gene 314
ced-6 gene 314
ced-7 gene 314
ced-9 gene 205
ced-10 gene 314
ced-12 gene 314
ceh-14 gene 1097
ced genes 255
CEN gene 715
Cerberus gene 2150
cereba gene 2026
ce-tbx2 gene *1944f*

genes, specific (*continued*)
ce-tbx7 gene *1944f*
ce-tbx8 gene *1944f*
ce-tbx11 gene *1944f*
ce-tbx12 gene *1944f*
ce-tbx17 gene *1944f*
CF gene 359
Cftr gene 1245
che genes 255
CHK2 gene 1770
CHN/TEC gene 1772
chordin gene 2150
CHS1 gene 27
ch-Tbx3 gene *1944f*
ch-Tbx5 gene *1944f*
ch-Tbx6L gene *1944f*
ch-TbxT gene *1944f*
ch-T gene *1944f*
cI gene 1131
Cin4-1 gene 2024
Cinful-1 gene 2022
CINH gene *430t*, *431t*
CIQA gene *430t*, *431t*
CIQB gene *430t*, *431t*
CIQC gene *430t*, *431t*
CIQR gene *430t*, *431t*
CIR gene *430t*, *431t*
CIS gene *430t*, *431t*
c-kit (W) gene 525, 2126
CLCN5 gene 1028
clf1 gene 640
clf2 gene 640
Clock (Clk) gene 389
CLU gene *430t*, *431t*
CLV1 gene 715
CLV2 gene 715
CLV3 gene 715
cms-T strain 1223
cob gene 1221
COBRA gene 1755
cod genes 255
CO gene 715
COL1A gene 604
Col2a1 gene 1817
COL5A1 gene 603
COLA1 gene 1772, *1772t*
cox gene 1221
CR1 gene *430t*, *431t*
CR2 gene *430t*, *431t*
crp gene 1069, 1193
CRSP2Y gene 2159
cry1 gene 1456–1457
CRY2 gene 715
CSA gene 563, 568
CSB gene 563, 568
CSF2RA gene 2157
cya gene 1193
CYC1AT gene 1755
CYC gene 716
cyclin D1 gene 1772, 2085
cyclin D gene 2086
cyk-1 gene 301
cyk-3 gene 301
CYP11B1 gene 447
CYP11B2 gene 447
CYP17 gene 447
CYP1A1 gene 263
CYP21 gene 447
CYP21P gene 447
CYP27P1 gene 1027
CYP2E1 gene 263
d1 gene 1395
Dab1 gene 1650

genes, specific (*continued*)
daf-12 gene 928
DAF gene *430t*, *431t*
daf genes 255, 22
Dax1 gene 1813, 1817–1818
DAX1 gene 1813, 1817–1818
DAZ gene 2156, 2160
dbt (doubletime) gene 387
Dby gene *1253f*, 1254
DBY gene 2159
DCC (deleted in colon carcinoma) gene 10
Dce gene 1713
dce gene 1713
dCE gene 1713
DCP1 gene 1748
DEF gene 716
deg-1(d) gene 316
deg-1 gene 316
deg-3 gene 316–317
deg-3(u662) gene 316–317
del1–46 gene 2022
Delta gene 1317, *1318f*
delta (δ) globin gene 878, 2096
det gene 1455
Dfd (Deformed) gene *960f*
DFFRY gene 2159
D gene 1713
DICH gene 716
dilute (d) gene 398
dinA gene 1854
dinB gene 1651, 1854
dislet gene 1097
distorted genes 2048
dlim1 gene 1096
dlim3 gene 1097
dLMO gene 1097
Dmd^{mdx} gene 1244
dm-H15 gene *1944f*
dm-omb gene *1944f*
DMPK gene 1838, *2049t*
DMRT1 gene 1813
dm-Trg gene *1944f*
dnaG gene 1543
DPC4 gene 2083, *2084–2085t*
DPC4/SMAD4 gene *81t*
DpnI gene 1494
DpnII gene 1494
dpp (decapentaplegic) gene *427*
dpy genes 255
Drd1a gene 1244
DRPLA gene *2049t*
Ds1 gene 2026
Ds (Dissociation) gene 2020
dsx (doublesex) gene 1813
dunce gene 210
dyf genes 255
E6E7 gene 310
Eα gene 1640
eat genes 255
Eβ gene 1640
E-cadherin gene 2086
EDA1 gene 600
Ednrb (s) gene 527
efp gene 610
egl-1 gene 205
egl genes 255
EIAF gene 1772
EIF1AY gene 2160
Eif2&y gene *1253f*, 1254
elt-1 gene 619
emb genes 255
ems (empty spiracles) gene 959

genes, specific (*continued*)
end-1 gene 619
endo gene 2022
endothelin 3 gene 1469
En (Ehancer) gene 2027
eno1 gene 888
env gene 979, 1251, 1704, 2022
epsilon (ε) globin gene 878, 2096
erbB-1 gene 648–649
erbB-2 (HER-2/neu) gene 648–649
erbB-3 (HER-3) gene 648–649
erbB-4 (HER-4) gene 648–649
ERG gene 1772
erh3 gene 2045
E(spl) gene 1317, *1318f*
EST2 gene 1948
ets-1 gene 661
ets genes 662
ETV-1 gene 1772
ETV6(TEL) gene 1772
Evx gene 959
EWS gene 1772, *1772t*
exd (extradenticle) gene 961
EXLM1Y gene 2159
EXT1/EXT2 gene *2084–2085t*
FANCA-H gene *2084–2085t*
FASS gene 1476
FBA1 gene 886
FBN1 gene 1145
FBP11 gene 716
FBP7 gene 716
FCA gene 715
fer genes 255, 1874, *1874f*
FEV gene 1772
ffh gene 1555
ffs gene 1555
FGF23 gene 1029
FGF-3 gene 126
FGF-4 gene 126
FGF-5 gene 126
FGFR1 gene 479–481, *480t*
FGFR2 gene 479–481, *480t*
FGFR3 gene 479–481, *480t*
FHA gene 715
FHIT gene 1123
fimA gene 800
fim gene 1443
finO gene 678
fix genes 707–709, *708t*
FKHR gene 1772, *1772t*
fla genes 711
FLC gene 715
flg genes 711
flh genes 711
fli genes 711
fljA gene *939f*
fljB gene 938, *939f*
fljC gene 938
flj genes 711
FLO gene 715
flu gene 110, *110f*
FMR1 gene 726, *2049t*
FMR2 gene 595, *2049t*
FMRI gene 595
FPA gene 715
FRAXA gene *2049t*
FRAXE gene *2049t*
FRDA1 gene *2049t*
frq (frequency) gene 389
FT gene 715
ftsY gene 1555
ftz (fushi tarazu) gene 958, *960f*
FUL gene 715

genes, specific (*continued*)
furca2 gene 2048
furca4 gene 2048
fus gene 610
FUS/TLS-CHOP fusion genes 1772, *1772t*
Fv1 gene 1250
Fv4 gene 1250
G7c gene 1640
GA1 gene 715
Gag gene 979
gag gene 1704, 2021
gal gene 1436, 1983–1984
gam gene 1625
gamma (γ) globin gene 878
GATA4 gene 1814
gfp gene 739
Ghrhr (GHRH receptor) gene 592
gidA gene *1383f*
gidB gene *1383f*
gin gene *939f*
gl1 gene 2045
gl2 gene 2045
gl3 gene 2045
glb1 gene 1785
glb2 gene 1785
gli-3 gene 1101
GLK1 gene 885
Aγ globin gene 2096
Gγ globin gene 2096
GLO gene 716
glossy15 gene 930
glp-1 gene 617–618, *618f*
GLY2P gene 2159
GNOM gene 1475
gon genes 255
goosecoid gene 2150
GpsieB gene 1895
GS1p gene 2159
Gy1–Gy5 gene family 1785
Hba-ps4 pseudogene 1921
HBT (HOBBIT) gene 1755
hCHK2 gene *2084–2085t*
Hd1 gene 1395
Hd3a gene 1395
Hd6 gene 1395
hex genes 1666
Hfr gene 453
(hh) hedgehog gene 427
hifA gene 977
hifB gene 977
him-10 gene 301
him genes 255
hin gene *939f*
hisA gene *946f*
hisB gene 943–945, *946f*
hisC gene 944, *946f*
hisD gene 943–944, *946f*
hisE gene *946f*
hisF gene 943–944, *946f*
hisG gene 943, *946f*
hisH gene 943–944, *946f*
hisI gene 943–945, *946f*
hisP gene 945
his genes 943–945, *946f*
HMGIC gene 1110, *1112f*
HMGI(Y) gene 1110, *1112f*
hMLH1 gene 423, *2084–2085t*
hmp-1 gene 620
hmp-2 gene 620
hmr-1 gene 620
hMRE11 gene 114
hMSH2 gene 423, *2084–2085t*

genes, specific (*continued*)
hMSH3 gene *2084–2085t*
hMSH6 gene 423, *2084–2085t*
ho gene 1761
HO gene 277, 1156, 1761
Hoxa1 gene 1996
Hoxd4 gene 1996
Hox gene 1814
hPMS1 gene 423, *2084–2085t*
hPMS2 gene 423, *2084–2085t*
Hprt gene 978, 1087
HPS gene 27
hSNF5 gene *2084–2085t*
hth (homothorax) gene 961
HXK1 gene 885
HXK2 gene 885
iceA gene 916–917
Igf2 gene 996
Igf2r gene 996
IL3RA gene 2157
ILR9 gene 2160
imm gene 1895
INK4a gene *2084–2085t*
int22h gene 918
int gene 1859
intI gene 1041–1045, *1042f*, *1044f*
int (integrase) gene 1687, 1859, 2021
Is11 gene 1097
Is12 gene 1097
IT15 gene 982, *2049t*
ITGAM gene *430t*, *431t*
ITGAX gene *430t*, *431t*
ITGB2 gene *430t*, *431t*
kaiA gene 389–390
kaiB gene 389–390
KaiC gene 389–390
KALp gene 2159
ki (pollen killer) gene 2064
KK-λ gene 600
KLHL1 (sense) gene *2050t*
Knotted 1 gene 962
Kr1 gene 2064
kr2 gene 2064
kr3 gene 2064
kr4 gene 2064
lab (labial) gene *960f*, 961
lacA gene *153f*, 1070
lac gene 453, 655
lacI gene 977, 1069, 1491, 1652, *1653f*
lacL gene 1070
lacO gene 1069, 1070
lacP gene 1070
lacY gene *153f*, 154, 1069, 1070, 1193–1194, 1491
lacZ gene *153f*, 154, 1069, 1070, 1193–1194, 1491, 1990, 2039
lambda (λ) *rex* gene 1895
LD gene 715
LEC gene 1476
Le gene 1179
Lep^{ob} gene 1244
let-502 gene 620
let-805 gene *620f*
let-99 gene 301
lethal yellow gene 23
lexA gene 1651, 1853–1854
LFY gene 715
Lhx1 gene 1096, 1818
Lhx2 gene 1097
Lhx3 gene 1097
Lhx4 gene 1097
Lhx5 gene 1096
Lhx6 gene 1097

genes, specific (*continued*)
Lhx8 gene 1097
Lhx9 gene 1097
Lim1 gene 1814, 1818
lim-4 gene 1097
lim-6 gene 1097
lim-7 gene 1097
lin-4 gene 928–930, 2003
lin-5 gene 301
lin-6 gene 301
lin-11 gene 1096
lin-12 gene 306, *307f*, 963, 1313
lin-14 gene 306, *307f*, 928–930, *929f*, 2003
lin-26 gene 317
lin-28 gene 928–930
lin-29 gene 928–930
lin-32 gene 1310
lin-42 gene 928, 532
lin genes 255
lit-1 gene 617
LKB1 gene *2084–2085t*
LMO1–4 gene 1097
Lmp2 gene 1640
Lmx1a gene 1097
Lmx1b gene 1097, 1101
Lox gene 2003
loxP gene 482
LPP gene 1110, *1113f*, *1116f*
LUG gene 715
LYL1 gene 1930–1932
lys gene 1786
M33 gene 1814, 1818
mab-3 gene 532, 1813
mab genes 255
MADS-box genes 715
MAOA gene 211
MASP1 gene *430t*, *431t*
MASP2 gene *430t*, *431t*
mast cell growth factor (Mgf) gene 398
Mat1 gene 962
mat1 gene 1154
mat1 locus 1154
Mat2 gene 962
*MAT*a gene 276–278, 1152, 1154, 1398
MATα gene 276–278, 1152, 1154, *1155f*, 1398
matR gene 238
MBL2 gene *430t*, *431t*
Mc1r gene 400
mClk gene 389
MCP gene *430t*, *431t*
Mrc1 (melanocortin receptor 1) gene 1470
mCry gene 389
MDM2 proto-oncogene 1772
MDR1 gene 1431
MDR3 gene 1431
mec-3 gene 1096, 1311, 1322
mec-4(d) gene 314–316
mec-4 gene 314–316
mec-6 gene 317
mecA gene 75
mec gene 255, 1687
MEDEA gene 1476
MEN1 gene 1257
mes-2 gene 619
mes-6 gene 619
metallothionein gene 978
Methuselah gene 22, 214
mex-1 gene 619
mex-3 gene 619
Mgf gene 525, 1469

genes, specific (*continued*)
MIC2 gene 2157
mig genes 255
mioC gene *1383f*
Mis gene 1818
MITF gene 1469
MJD1 gene *2050t*
Mla gene 973
MLH1 gene 1285
Mlo gene 973
mob gene 454
mom gene 1440–1441
MONOPTEROS gene 1475, 1755
mot genes 711
MRE11 gene 571
MRP6 gene 1580–1581
Msh2 gene 2082
MSH2 gene 1285
MSH3 gene 1285
MSH6 gene 1285
MSX2 gene *480t*
MTCP1 gene 114, 1094
mTim gene 389
MTS1 gene *2084–2085t*
mudrA gene 2028
mudrB gene 2028
MuDR gene 2028
Mu (Mutator) gene 2028
mu-Tbr1 gene *1944f*
mu-Tbx1 gene *1944f*
mu-Tbx2 gene *1944f*
mu-Tbx3 gene *1944f*
mu-Tbx4 gene *1944f*
mu-Tbx5 gene *1944f*
mu-Tbx6 gene *1944f*
mutD gene 1286
mu-T gene *1944f*
mut genes 1666
mutH gene 1286
mutL gene 1286
mutM gene 1286
mutS gene 1286
mutT gene 1286
mutY gene 1286
Myc gene 1470
Myo5a (d) gene 524–525
Myo5a gene 398, 1469
nad1 gene 2041
nad2 gene 2041
nad5 gene 2041
nad gene 1221, 1224
NahG gene 1916
NBS1 gene 571, *2084–2085t*
neo gene 482
NF1 gene *81t*, 1770, *2084–2085t*
NF2 gene *81t*, *2084–2085t*
niaA gene 110, *110f*
nifB gene 1328
nifD gene 1328
nifE gene 1328
nifH gene 1328
nifK gene 1328
nifN gene 1328
nif genes 802
niiA gene 110, *110f*
N-myc gene 780
nodABC gene 1330
nodABCMFE genes 1332
nodD gene 1330
nod genes 1715, 1330
nodIJ genes 1332
noek gene 2048
noggin gene 2150

genes, specific (*continued*)
nonA gene 1317
no tail gene 1945
Notch gene 1315, *1318f*
nph1 gene 1457
nph4 gene 1457
nrdB gene 1929
nrdD gene 1929
NTRK3 gene 1772
numb gene 1322
OA1 gene 27
Oct-1 gene 959, 962
Oct-2 gene 959, 962
odr-10 gene 1313
ogt gene 36
oriT gene 450–451, 456
osm genes 255
$p14^{arf}$ gene 1772
p16 gene 10, *81t*, 2082, *2084–2085t*
$p16^{INKA}$ gene 1123
p300 gene 2083
p53 gene 10, 69, *81t*, 423, 571, 1093, 1770, 2082, 2083, *2084–2085t*
PABP2 gene *2050t*, 2053
PAN gene 716
pap gene 1443
parB gene 1488
par genes 616–618
pat-2 gene *620f*
pat-3 gene *620f*
Pax1 gene 1427
Pax2 gene 1427
Pax3 gene 1427
PAX3 gene 1715, 1772, *1772t*
Pax5 gene 1427
Pax6 gene 1427
PAX7–FKHR fusion gene 425
Pax7 gene 1427
PAX7 gene 1715, 1772, *1772t*
Pax9 gene 1427
pb (proboscopedia) gene *960f*
PCDHY gene 2159
Pcg gene 1814
Pctr1 gene 1483
Pctr2 gene 1483
per gene 210–211, 389, 1316
pet gene 1461
PEX gene 1028–1029
PFC gene *430t*, *431t*
PFK1 gene 886
PFK26 gene 886
PFK27 gene 886
PFK2 gene 886
P gene 26
p (pink-eyed dilution) gene 398, 527, 1469
PGI1 gene 885
PGK1 gene 886
Pgk gene 477, 978
PGM gene 892
PGM1 gene 886
ph1b gene 2067
ph1c gene 2067
Ph1 gene 2067
pha-4 gene 619
Phex gene 1028
phi beta (ϕβ) globin gene 2096
PHYA gene 1455
Pib gene 1395
pie-1 gene 619
piebald (s) gene 398
PI gene 716
pilE gene 1443

genes, specific (*continued*)
pilS gene 1443
PIN1 (Arabidopsis pin-formed) 1 gene 1472
Pit1 (pituitary specific transcription factor1) gene 959
PLE gene 716
PMS1 gene 1285
PMS2 gene 1285
Pol gene 979
pol gene 1704, 2021
pop-1 gene 617
pos-1 gene 619
PP2A (putative) gene *2050t*
PPARγ gene 2083, *2084–2085t*
prcd gene 269
PRKY gene 2159
prlD gene 1555
Prop1 (Prophet of Pit1) gene 592
prospero gene 1322
protease gene 2022
prot gene 2022
PRY gene 2159
psaA gene *1461t*
psaB gene *1461t*
psaC gene *1461t*
PsaD gene *1461t*
PsaE gene *1461t*
PsaF gene *1461t*
PsaG gene *1461t*
PsaK gene *1461t*
psal gene *1461t*
PsaL gene *1461t*
PsaM gene *1461t*
PsaN gene *1461t*
psb gene 1461
PTC gene *81t*, *2084–2085t*
PTEN gene *2084–2085t*
ptl gene *453f*
ptr gene 1624, *1624f*
Pvt-1 gene 1483
PYK1 gene 887
PYK2 gene 887
Q gene 1045
RAD10 gene 1838
RAD1 gene 1838
RAD26 gene 563
rad27 gene 595
RAD51L1 gene 1110, *1112f*, 1113
RADIALIS gene 716
RAP1 gene 1948
RAS1 gene 1603
RAS2 gene 22, 1603
RB1 gene 1770, *2084–2085t*
rbcL gene 238
Rb gene 2081, 2083
RB gene 274, 1123
RBI gene *81t*
RBM gene 2156, 2159
Rbm gene cluster *1253f*, 1254
rde gene 930
rDT gene 2026
recA gene 1651, 1853–1854
recA/RAD51 gene family 1110
recB gene 1623, *1624f*, 1647
recC gene 1623, *1624f*, 1647
recD gene 1624, *1624f*
recE gene 1647
rec genes 938
recQ gene 568
recT gene 1647
red gene 1625
relA gene 1891

genes, specific (*continued*)
Reln (reelin) gene 1649
repA gene 382
Rex gene 979
R gene 1045
RhCcEe gene 651, 1713–1714
RHD gene 1713–1714
RhD gene 1714
RHG gene 892
rhl1 gene 2045
rhl2 gene 2045
rhl3 gene 2045
RhoAy1 gene *1253f*, 1254
RhoAy2 gene *1253f*, 1254
Rmcf gene 1250
rosy gene 1989
rpl gene 1461
rpoD gene 1543
rpo gene 1461
RPS24Y gene 2159
RPS4Y gene 2159
rps gene 1461
rpsL gene 1891
rpsU gene 1543
rrs gene 1891
RT/RNaseH gene 2021
rts (retsina) gene 2048
RuvC gene 1032
ruv genes 1648
ry gene 795
SABRE gene 1755
Saccaromyces SIR genes 1948
SAS gene 1773
sbc genes 1648
SCA1 gene *2049t*
SCA2 gene *2049t*
SCA3 gene *2050t*
SCA7 gene *2050t*
SCA 8 (antisense) gene *2050t*
SCARECROW (SCR) gene 1755
SCR gene 1476
scr gene 892
Scr (sex combs reduced) gene *960f*
sdh gene 1221
secA gene 1555
secB gene 1554
sel-12 gene 1313
SEP1 gene 716
SEP2 gene 716
SEP3 gene 716
sey gene 1427
Sf1 gene 1814, 1817
SF1 gene 1813–1814
SFTPA1 gene *430t*, *431t*
SFTPA2 gene *430t*, *431t*
SFTPD gene *430t*, *431t*
SGS1 gene 22
sh gene 1395
shh (sonic hedgehog) gene 1100
short ear (se) gene 398
SHOX gene 2157
SHR (SHORT-ROOT) 1476, 1755
shr gene 892
Siamois gene 2150
sieA gene 1895
Sinc gene 1879
SIR2 gene 22
SIX5 gene 1838
skn-1 gene 618
SMAD4 gene 423
Smcy gene *1253f*, 1254
SMCY gene 2160
SOC1 gene 715

genes, specific (*continued*)
Sox9 gene 1813, 1817
SOX9 gene 1812–1813, 1817
SOX10 gene 1469
spd-2 gene 300–301
spe genes 255, 1874, *1874f*
sp gene 1427
Spl7 gene 1395
Spm (suppressor of mutation) gene 2027
Spo11 gene 350
spoT gene 1892
SQUA gene 715
sry gene 1812
Sry gene 1812, 1816–1817
SRY gene 1812, *1815f*, 1816, 1822–1823, 2159
SSX1 gene 1771, *1772t*
SSX2 gene 1771, *1772t*
stachel gene 2048
STM (shootmeristemless) gene 962, 1475
STSp gene 2159
SuH gene 1317, *1318f*
SUPERMAN gene 1473
sup genes 255
su-Ta gene *1944f*
SVP gene 715
swi1 gene 1156
swi3 gene 1156
swi7 gene 1156
Sxl gene 48, 129
Sxr gene cluster *1253f*, 1253–1254
SYBL1 gene 2160
sym (symbiosis) gene 1471
SYT gene 1771, *1772t*
T7 polymerase gene 975–976
tag gene 1664
TAL1 gene 1930–1932
TAL2 gene 1930–1932
Tam1 gene 2026
tau gene 389
Tax gene 979
TB4Y gene 2159
Tbr1 gene subfamily *1944f*
tbr (trichome birefringence) gene 2048
Tbx1 gene subfamily *1944f*
Tbx2 gene 1945
Tbx2 gene subfamily *1944f*
Tbx3 gene 1945
TBX3 gene 1945
TBX5 gene 1945
Tbx6 gene subfamily *1944f*
Tcd-1 gene 1919, *1920f*
Tcd-2 gene 1919, *1920f*
Tcd-3 gene 1919, *1920f*
Tcd-4 gene 1919, *1920f*
Tcd-5 gene 1919, *1920f*
Tcd gene 1919
TCL1 gene 114, 1094
Tcp-10b gene 1920
Tcp-10 gene *1920f*
Tcr gene 1919, *1920f*
TDF gene 1154, 1816
TDH1 gene 886
TDH2 gene 886
TDH3 gene 886
Tdy gene 1154, 1816–1817
Teopod1 gene 930
Teopod2 gene 930
Teopod3 gene 930
TER1 gene 1948
TFL1 gene 715
T gene subfamily *1944f*
TGFβIIR gene 423, 2083

genes, specific (*continued*)
Tbx6 gene 1945
thyA gene 1624, *1624f*
thymidylate synthase (*td*) gene 1929
tniA gene 1045
tnpA gene 1689
tnpR gene 1688–1689
Tnt1 gene 2022, 2029
tonB gene 1923
TP53 gene 1123
tpfQ/I gene 43, *45f*
TPI1 gene 886
tra gene 678
transformer gene 1316
TRF1 gene 1948
TRF2 gene 1948
triptychon gene 2045
Trp53 gene 1245
trpA gene 57, *218f*, 219, 419, *419f*, 655, 1900, 2078
trpB gene *218f*, 2078
trpC gene *218f*
trpD gene 218, *218f*, 2078
trpE gene *218f*, 2078
trp gene 453
TSC1 gene *81t*
TSC1/TSC2 gene *2084–2085t*
TSC2 gene *81t*
tsf gene 610
ts gene 442
TSH receptor (Tshr) gene 592
TSIX gene 2151
Tspy gene *1253f*, 1254
TSPY gene 2159
TTD-A gene 568
ttg1 gene 2045
Tto1 gene 2022
ttx-3 gene 1097
TTY1 gene 2159
TTY2 gene 2159
tufA gene 1221
tuf gene 610
T-*urf13* gene 1223
TWIST gene *480t*
Ty1/copia gene 2022, 2025, 2029
Ty1 gag gene 2022
Ty3/copia gene 2026
Ty3/gypsy gene 2022
Tyr (c) gene 527
TYR gene 26
Tyr (tyrosinase) gene 398
Tyrp 1 (tyrosinase-related) protein 1 gene 398
Tyrp 1 (b) gene 527
TYRP1 gene 27
Ube1y gene *1253f*, 1254
Ubx (ultrabithorax) gene 958, *960f*, 961, 962, 1852
udt (underdeveloped trichome) gene 2048
UFO gene 716
umuC gene 1622, 1651, 1854
umuD gene 1622, 1651, 1854
un gene 1427
unc-4 gene 961, 1312
unc-6 gene 532
unc-13 gene 1312–1313
unc-18 gene 1312–1313
unc-25 gene 1312
unc-26 gene 1313
unc-30 gene 1312
unc-34 gene 1311
unc-47 gene 1312

genes, specific (*continued*)
- *unc-52* gene *620f*
- *unc-54* gene 317, 1750
- *unc-73* gene 1311
- *unc-86* gene *307f*, 308, 959, 1310, 1322
- *unc-104* gene 1312
- *uncB* gene *1383f*
- *uncI* gene *1383f*
- *unc* genes 255
- *URA3* gene 2163
- *ura4* gene 2163
- *Usp9y* gene *1253f*, 1254
- *USP9Y* gene 2159
- *Uty* gene *1253f*, 1254
- *UTY* gene 2159
- *uvrA* gene 1854
- *uvrB* gene 1854
- *uvrD* gene 1286, 1854
- *vacA* gene 916–917
- *vanA* gene 75
- *VHL* gene *81t*, *2084–2085t*, 2087
- *vir* gene 493
- *VMA1* gene 1565
- *wac* gene *443f*
- *WAF1* gene **2127**
- *wc1* gene 389
- *wc2* gene 389
- *wetA* gene 110, *110f*
- w^+ gene *1525f*, 1525–1526
- *wg (wingless)* gene 427
- *white collar* genes 389
- *WIG* gene 715
- *Wnt4* gene 1815
- *Wt1* gene 1814
- *WT1* gene *2084–2085t*
- *wx (waxy)* gene 2020
- X25 gene *2049t*
- *Xa1* gene 1395
- *Xa21* gene 1395
- *Xbra* gene *1944f*
- *XE7* gene 2157
- *x-Eomes* gene *1944f*
- *x-ET* gene *1944f*
- *XG* gene 2157
- *XGPY* gene 2159
- *xisA* gene 802
- *xis* gene 1859
- *XIST* gene 2150
- *Xist* gene 634, 2148
- *XKRY* gene 2160
- *XPA* gene 2150
- *XP(A-H)* gene *2084–2085t*
- *XPB* gene 568, 2150
- *XPC* gene 568, 2150
- *XPD* gene 568, 2150
- *XPE* gene 568, 2150
- *XPF* gene 2150
- *XPG* gene 568, 2150
- *XRN1* gene 1748
- *x-VegT* gene *1944f*
- *xylB* gene 977
- *ycf* gene 1461
- *Yd2* gene 973
- *zen (zerknüllt)* gene *960f*
- *zen2 (zerknüllt-related)* gene *960f*
- zeta (ζ) globin gene 878
- *Z (Zebra)* gene 643
- *zf-tbx6* gene *1944f*
- *zf-T* gene *1944f*
- *Zfy1* gene 1253, *1253f*
- *Zfy2* gene *1253f*, 1254
- *ZFY* gene 2159
- *ZNF217* gene 425

genes, specific (*continued*)
- *zwichel* gene 2048
- *ZWILLE* gene 1475–1476
- *zyg-1* gene 300
- *zyg-9* gene 301

gene substitution **814**
- gene frequency 814
- random genetic drift 814
- substitution rate 814

gene therapy (human) **814–819**
- applications 817–818
- cancer, role in 817
- Dulbecco, Renato 816
- ethics 660
- Food and Drug Administration (FDA) 818
- Gene Therapy Subcommittee of the Office of Recombinant DNA Advisory Committee (RAC) 818
- gene therapy trials 818
- genetic transformation 815
- gene transfer 816
- historical medical advances 815
- history of medical treatment 815
- Lesch–Nyhan syndrome (LNS) 816
- methodology 816
- National Institutes of Health (NIH) 818
- retrovirus vectors 816–817
- usefulness 818
- vector systems 817, *817t*
- viral vectors 816
- viruses 816

Gene Therapy Subcommittee of the Office of Recombinant DNA Advisory Committee (RAC) 818

Genethon Human Genome Research Centre 1107

genetic anticipation 2048

genetic code 516, 655, *821f*, **821–822**, *822f*, 1999–2000
- adenine (A) 821
- coding problems 821
- cytosine (C) 821
- eukaryotes *822f*
- Gamow, George 821
- guanine (G) 821
- Khorana, Har Gobind 1061
- mRNA 821
- organellar organisms *822f*
- prokaryotes *822f*
- ribosome 821
- thymine (T) 821
- triplet code **2054–2055**
- tRNA 821
- *see also* codons

genetic colonization **822–823**
- climate change 823
- founder effect 823
- invading species 823
- metapopulations 823
- *Mytilus edulis* 823
- *Mytilus galloprovincialis* 823

genetic correlation **823–825**
- calculations 823–824
- cause 824
- estimations 824
- linkage disequilibrium 824
- pleiotropy 824

genetic counseling **825–826**, 920
- auditing 826
- carrier 275
- components 825, *825t*

genetic counseling (*continued*)
- cri-du-chat syndrome 487
- definition 825
- ethical issues 825
- Tay–Sachs disease 1943
- *see also* clinical genetics; prenatal diagnosis

genetic covariance **826–827**

genetic crosses **681–682**
- 1-factor crosses 681
- 2-factor crosses 681
- 3-factor crosses 682
- Mendel's laws 681
- tetrad analysis 682

genetic diseases 263–264, **827–828**
- achondroplasia 827
- dogs 268
- dominant mutations 827
- recessive mutations 827
- treatment 827
- *in vitro* fertilization (IVF) 1009
- *see also* clinical genetics; congenital disorders; phenylketonuria

genetic distance **828–832**
- allele frequency 828
- amplified fragment length polymorphism (AFLP) 831
- Bhattacharyya's distance 828, *829f*
- D_A distance 830, *831f*
- $(dm)^2$ distance 830, *831f*
- evolutionary relationships *830f*, 831
- F*ST distance 829
- function 828
- gene trees 830–831, *831f*
- microsatellites 830
- random amplified polymorphic DNA (RAPD) 831
- restriction fragment length polymorphism (RFLP) 831–832
- Rogers' distance 828
- standard genetic distance 829–830
- stepwise mutation model 830, *830f*

genetic drift **832–834**
- allele frequency 832–833, *833f*
- bottleneck effect 602
- *Drosophila melanogaster* 832–833, *833f*
- effective population number 602
- evolution 665, 667
- evolutionary biology 516
- Fisher, Ronald A. 832
- gene flow 785
- Hardy–Weinberg law 913
- hitchhiking effect 952
- neutral theory 833
- non-Darwinian evolution 1343
- population equilibrium 647–648
- shifting balance theory of evolution 832–833
- speciation 665
- Wright, Sewall 832

genetic engineering **834**
- genetic recombination 844–845
- *Hordeum* sp. 973
- plants 661
- selective breeding 1799

genetic equilibrium **834–835**
- Hardy–Weinberg law 834–835
- mutation–selection balance 835
- overdominance 835

genetic homeostasis **835–838**
- definition 835
- *Drosophila* 836
- fluctuating asymmetry 837

genetic homeostasis (*continued*)
  inbred strain 837–838
  individuals 836
  populations 835–836
genetic load **838–839**
  balance theory 839
  classical theory 839
  Haldane–Muller principle 838
  Muller, H.J. 838
  population genetics 839
  types 838–839
genetic mapping 263–264
  androgenone 65
  genetic recombination *841f*, 842
  microsatellites 268
  Morgan, Thomas Hunt 1239
genetic marker **839**
genetic material **839**
genetic migration **839–840**
genetic nomenclature *see* nomenclature, genetic
genetic ratios **840**
genetic recombination **841–845**
  centromere *841f*
  chromatids *841f*
  crossovers 841
  definitions 841, *841f*, *842f*
  DNA repair 843–844, *844f*
  DNA structure 842–843, *843f*
  double-strand break repair 843–844, *844f*
  exchange 673
  gene targeting 845
  genetic engineering 844–845
  genetic mapping *841f*, 842
  genetic markers 842
  heteroduplex DNA 842–843, *843f*
  meiosis 841
  postmeiotic segregation 843
  recombination frequency 842
genetic redundancy **845–846**
  cumulative benefit 845
  error buffering 845–846
  functional shift 845
  genetic duplication 845
  pleiotropy 846
  regulatory elements 846
genetics **856–857**
  Mendel, Gregor Johann 856
  physiological genetics 857
  population genetics 857
  transmission genetics 856–857
  word origin 2008
genetic stock collections and centers **846–854**
  Arabidopsis Biological Resource Center (ABRC) 852
  Arabidopsis Centre (England) 852
  Avian Genetic Stock Collection (domestic chicken) 854
  Axolotl Colony 854
  Bacillus Genetic Stock Center 850
  Barley Genetic Stock Center 853
  Belgian Coordinated Collections of Microorganisms (BCCM) 851
  Caenorhabditis Genetics Center (CGC) 853
  Chlamydomonas Genetics Center (CGC) 851
  Cloning Vector Collection, National Institute of Genetics 852
  Collection of the Institut Pasteur (CIP) 851
genetic stock collections and centers (*continued*)
  Deutsche Sammlung von Mikroorganismen und Zellkulturen (DSMZ) 851
  Drosophila
    Bloomington *Drosophila* Stock Collection 847
    Drosophila Mid-America Center (*Drosophila* Species Collection) 847
    Drosophila Stock Center (Mexico) 847
    European *Drosophila* Stock Center 847
    FlyBase 847
    Indian *Drosophila* Stock Centre 847
    Moscow Regional *Drosophila melanogaster* Stock Center 847
    National Institute of Genetics (Japan) 847
    online databases 847–848
    P Insertion Mutant Stock Centre 847
    University of Arizona 848
  early genetic stock centers
    E. coli Genetic Stock Center 847
    Fungal Genetics Stock Center 847
    Jackson Laboratory 847
    National Institutes of Health (NIH) National Center for Research Resources 847
    National Science Foundation (NSF) Living Stock Collections Program 847
    U.S. Department of Agriculture 847
  Escherichia coli
    American Type Culture Collection (ATCC) 850
    Center for Genetic Research Information (Japan) 850
    E. coli Genetic Stock Center 849–850
    Genetic Strains Research Center (Japan) 850
    Microbial Genetics Center (Japan) 850
    online databases 850
    Phabagen 850
  Fungal Genetics Stock Center 849
  G.A. Marx Collection (pea) 853
  Institute of Physical and Chemical Research (RIKEN) 852
  International Mycological Institute (IMI) 851
  Japan Collection of Microorganisms (JCM) 852
  maize
    Maize Genetics Cooperation Stock Center 849
    Maize Genome Database (Maize DB) 849
    online databases 849
  Microbial Germplasm Database (MGD) 852
  mouse
    Jackson Laboratory 849
    Mouse Genome Informatics (MGI) website 849
    MRC Mammalian Genetics Unit (UK) 849
    Oak Ridge National Laboratory Mutant Mouse Collection 849
    online databases 849
genetic stock collections and centers (*continued*)
    PeroBase 849
    Peromyscus (deer mouse) Genetic Stock Center 849
  National Collection of Type Cultures (NCTC) 851–852
  National Culture Collection of Bacteria of the Netherlands 850
  online databases 847, *848t*
  plant germplasm collections
    Consultative Group for International Agricultural Research (CGIAR) 852
    International Board for Plant Genetic Resources 852
    National Seed Storage Laboratory 852
    Rice Genetic Cooperative 852
    USDA National Genetic Resources Collection 852
  Pseudomonas Genetic Stock Center 850–851
  rice
    International Rice Genetic Cooperative (RGC) 853
    Oryzabase web site 853
    RiceGenes 853
  Salmonella
    Salmonella Genetic Stock Center (SGSC) 850
    Salmonella Reference Collection (SARC) 850
  SolGenes database 853
  tomato
    C.M. Ricks Tomato Genetics Resource Center (TGRC) 852
    USDA National Plant Germplasm System 852
  wheat
    E.R. Sears Wheat Genetic Stock Collection 853
    GrainGenes 853
    GRIN database 853
  yeast
    National Collection of Yeast Cultures (NYCY) (UK) 851
    online databases 851
    Peterhof Genetic Collection of Yeasts (PGC) (Russia) 851
    Saccharomyces Genome Database 851
    Yeast Genetic Stock Center 851
  zebrafish
    International Resource Center for Zebrafish 853
    University of Oregon 853
    ZFIN (Zebrafish Information Network) 853
Genetic Strains Research Center (Japan) 850
genetic technologies
  DNA fingerprinting 516
  DNA sequencing 516
  restriction fragment length polymorphism analysis 516
genetic translation **854–855**
  eukaryotes 855
  mechanisms 854
  prokaryotes 854–855
  Shine–Dalgarno sequence 855
genetic variation **855–856**
  measurement techniques 855–856
  Mendelian hereditary system 855

gene trapping **819**
  beta (β)-galactosidase 819
  embryonic stem (ES) cells 819
  insertional mutagenesis 819
gene trees **819–821**, *820f*, **1869–1870**, *1870f*
  gene genealogies 820, 1870
  genetic distance 830–831, *831f*
  orthology 819
  paralogy 819
  species tree 819, *1870f*
Genisteae 1083
Gennaro M L **1485–1490**
genome **857–859**
  chloroplasts 335–336, 337–338
  coding components 858
  CpG islands **477**
  definition 857
  DNA sequencing 858
  eukaryotes 249, 858
  G+C content 402–404, 477
  gene quantity 858
  mammals 269–270
  mitochondria 335–336
  organization 858
  size 249, 249–250, **865**
    Arabidopsis thaliana 857
    Caenorhabditis elegans 857
    Drosophila melanogaster 857
    Escherichia coli 857
    Homo sapiens 858
    Mus musculus musculus 858
    Saccharomyces cerevisiae 857
    simian virus 40 (SV40) 857
  vertebrates 249
Genome Data Base 1107, 2145
genome databases **517–521**
  archival databases
    DDBJ (DNA Data Bank of Japan) 517
    EMBL (European Molecular Biology Laboratories) 517
    GenBank 517
    NCBI web site 517
    plasmids 517
  bioinformatic databases
    BMERC (BioMolecular Engineering Research Center) 519
    DNA Genome Atlas 519, *520f*
    DNA Structural Atlas of Genomes 519, *520f*
    KEGG (Kyoto Encyclopedia of Genes and Genomes) 519
  GenBank 517
  links to sequenced genomes
    Enhanced Microbial Genomes Library 519
    FASTA files 519
    GOLD web site 519
    INFOBIOGEN web site 519
    NCBI web site 519
    NIAID (National Institute of Allergy and Infectious Diseases) 519
    U.S. Department of Energy (DOE) 519
  major sequencing centers
    NCBI web site 517
    Sanger Centre 517
    TIGR web site 517
    University of Oklahoma 519
    Washington University 519
  sequenced genomes 517
genome databases (*continued*)
  web site locations 517–519, *518t*
genome organization **859–863**
  ants 860
  bacteria 859
  centromere 860
  ciliates 860
  crustaceans 860
  dispersed transposable sequences 861
  fish 860
  gene families 861–862
  gene numbers 861
  gene size 862
  genome sequencing 859
  orphan genes 862
  placental mammals 859
  plants 860
  pseudogenes 862
  repetitive DNA sequences 860
  telomeres 860–861
  yeast 860
genome relationships **863–865**
  Adh gene 863
  applications 863
  Arabidopsis genome 863
  colinearity principle 863
  comparative genetic maps 863, *864f*
  DNA sequence analysis 863
  maize genome 863, *864f*
  Poaceae 863, *864f*
  process 863
  rice genome 863, *864f*
genome size **865**
genomic imprinting *see* imprinting, genomic
genomic library 545, **865–872**
  definition 865–866
  DNA fragmentation 871
  future libraries 872
  history 865–866
  Human Genome Project 866
  insert preparation 870
  library size 870
  requirements 866
  vectors 866–870, *867t*
    BACs (bacterial artificial chromosomes) 869–870
    cosmids 869
    P1 vectors 869
    PACs (P1 artificial chromosomes) 869
    phage λ *868f*, 868–869
    plasmids *867f*, 867–868
    YACs (yeast artificial chromosomes) 870
genomic position effects **1523–1530**
genomics **872**
genotype **873**
genotypic frequency **873**, *912t*, 912–914
Genzyme Company 756
Gepts P **1444–1445**
German J **229–230**
germ cell **873–876**
  description 873–874
  gametogenesis 875–876
  germline development 874, *874f*
  germline lineage 873
  meiosis 874–875
  totipotency 876
Gerstmann–Sträussler disease (GSD) **900**, 1881, 2008
  see also spongiform encephalopathies (transmissible)
Geukensia demissa 786
GFP *see* green fluorescent protein (GFP)
gfp gene 739
Gherardi E **1185–1187**
ghost gene phenomenon 1589
GHRH (growth hormone releasing hormone) 592
Ghrhr (GHRH receptor) gene 592
Giannelli F **338**, **917–920**
Giardia lamblia 405
Giardina's body 762
Gibberella 1481
gibberellins (GAs) 591, 1481
gidA gene *1383f*
gidB gene *1383f*
Giemsa banding *see* G-banding
Gigaspora margarita 1122, 1479
Gilbert, Walter 572, **877–878**
  Berg, Paul 877
  Dressler, David 877
  Maxam, Allan 878
  Maxam–Gilbert method 878
  Müller-Hill, Benno 877
  Nobel Prize 877
  protein synthesis 877
  RNA world 1751
  rolling-circle replication 877
  Sanger, Fred 877
Gillespie, John 1517
Gill G N **1096–1099**
Gilson M A **1047–1052**
gin gene *939f*
Gin invertase **938–942**
gl1 gene 2045
gl2 gene 2045
gl3 gene 2045
Glasgow A C **42–46**
glb1 gene 1785
glb2 gene 1785
Gleditsia (honey locust) 1083
gli-3 gene 1101
GLIMMER 1197
glioma **878**
GLK1 gene 885
Aγ globin gene 2096
Gγ globin gene 2096
globin genes (human) **878–881**
  chromosome 11 878
  chromosome 16 878
  cluster organization 878–879, *879f*
  evolution 880
  mutations 880–881
  regulation 879–880
  restriction fragment length polymorphism (RFLP) 880
  sickle cell anemia 1828
  structure 878, 879
globulins 1783–1784
GLO gene 716
Glomus intraradices 1122
Glomus mosseae 1479
glomus tumors 1000
glossy15 gene 930
glp-1 gene 617–618, *618f*
glp operon 1079
gluconeogenesis 2121, *2122f*
glucose 6-phosphate dehydrogenase (G6PD) deficiency 683, **881–884**
  acute hemolytic anemia (AHA) 882
  balanced polymorphism 187
  characteristics 882
  chronic nonspherocytic hemolytic anemia (CNSHA) 882

glucose 6-phosphate dehydrogenase (G6PD) deficiency (*continued*)
- clinical features 882
- evolution 881
- favism 882
- function 881
- housekeeping genes 881
- molecular genetics 881
- *Plasmodium falciparum* 882
- population genetics 882, *883f*

Gluecksohn-Waelsch, Salome 591
glutamic acid **884**
glutamine **884**
glutelins 1783–1784
GLY2P gene 2159
glycine **884**
Glycine max (soybean) *366t*, **884–885**, 1083, *1914t*, 2026
glycinin (*Glycine max*) 1784
glycogen 259
glycolysis **885–888**, *886f*
- enzyme deficiencies *887t*, 887–888
- glycolytic enzymes 885
- humans 887–888
- process 885–887
- *Saccharomyces cerevisiae* 885

glycosomes 1379
glycosylase repair **888–889**
- base excision repair (BER) 888
- BER pathway 889
- bifunctional 888
- crystallographic structure 889
- DNA repair 888
- function 888
- monofunctional 888

glycosylases 1494
G_{M2} ganglioside 1941–1943
GMP *see* guanosine monophosphate phosphodiesterase (GMP)
GNOM gene 1475
gnom mutation 1475
Gold, Lois Swirsky 51
Goldman M **382–383, 978**
Goldschmidt-Clermont M **1459–1465**
Goldschmidt, Richard 235
Goldsmith M **231–233**
Goldstein J L **682–683**
GOLD web site 519
Golgi body 3, 689, 693, 1378, 1558–1564
Golub E S **1235–1237**
gonadal mosaicism 131
gonadotropin-releasing hormone (GnRH) 1875
gon genes 255
gonorrhea 80
Goodman, Corey 1315
Goodman M F **193–197**
Goodrich J A **1549–1551**
Goodship J **533–535**
goosecoid gene 2150
Gordon D M **415–417**
Gosling, Raymond 554
Goss, Stephen 360
Gottesman II **1297–1301**
gp32 protein 1839
gp43 protein 1841
gp59 protein 1841
gp61 protein 1841
G-protein-coupled receptors (GPCRs) 1368, 1884
GpsieB gene 1895
Graafian follicle 691
GrainGenes 853, 2068
gram-positive bacteria 149, 183
grana 1379
Grant C S **1037–1038**
grasses **889–890**
- classification 890
- colinearity 890
- evolutionary history 890
- genomes 890
- Gramineae 889
- Poaceae 889

Gratia, André 417
Gratia J-P **417–418**
Graves J A M **1810–1811**
Graves, Jenny 1367
gravitropism **890–894**
- *Arabidopsis thaliana* 891–894
- auxin 892–893
- function 890
- future studies 894
- mechanisms 891–894
- mutational studies 892–893

Grawitz tumor 1659
Gray, Asa 1863
Gray M W **1215–1217, 1219–1220, 1220–1222, 1222–1224, 1377–1379**
green fluorescent protein (GFP) 303, **311–313**, 614, 1990
- bacteria 311
- *Caenorhabditis elegans* 312–313
- coelenterates (jellyfish) 311
- gene expression 311–312
- mitosis 1224
- modified Cre proteins 485
- mutations 312
- properties 312
- structure 312
- uses 312

Green, Howard 359–360
Green, Margaret C. 1337
Green, M.M. 383
green non-sulfur bacteria 149
green sulfur bacteria 149
Greene M I **1306, 1394**
Gregersen N **1187–1188**
Greig cephalopolysyndactyly (GCPS) 1101
Greig syndrome 911
Greighton, Harriet 1161
Gresshoff P **884–885**
Griep M A **1542–1545, 1546–1548**
Griffith, Frederick 553, 654, 1491, 1989
Grigliatti T **1523–1530**
GRIN database 853, 1127
Grindley N D F **524, 1038, 1041, 1054, 1687–1688, 1688–1692, 1842–1846, 1857–1858, 2033**
GroE (group I) chaperonins 324–325
Grollman A P **1669–1671**
Gros, François 877, 2134
Grosschedl R **624–625**
group selection **755–899**
- concepts 755–895
- epistasis *897f*, 897–898, *898f*
- examples *895f*, 895–896, *896f*, *897f*
- genetic structure 896–898
- individual selection 755–895

growth factors **899–900**
- achondroplasia 900
- Crouzon syndrome 900
- families 899
  - epidermal growth factor (EGF) 899
  - fibroblast growth factor (fgf) family 899
  - transforming growth factor α (TGF-α) 899
- function 899–900
- mechanisms 899
- mutational studies 900
- superfamilies 899
- usefulness in genetic research 900

Grubb, Rune 1094, 1969
Grunberg-Manago, Marianne 1365
Gruneberg, Hans 1337
Gruss P **1426–1428**
GS1p gene 2159
GT–AG rule **900**
GTF *see* general transcription factor (GTF)
guanine (G) 540, 572, **901**, 1358
guanosine 1360
guanosine 5′-diphosphate 3′-diphosphate (ppGpp) 945
guanosine diphosphate (GDP) 609, 610, 934, 1603, *1603f*
guanosine monophosphate phosphodiesterase (GMP) 269
guanosine triphosphatase (GTPase) 1308, 1311, 1603, *1603f*
guanosine triphosphate (GTP) **900–901**
- heterotrimeric G proteins 934
- hydrolysis 601, 609, 610
- RAS proteins 1603, *1603f*

guanylic acid 1360
guide RNA (gRNA) 1741–1742, *1742f*
Gurdon J B **2136–2138**
gurke mutation 1475
Gutmann, Antoinette 1130–1131
Guttman B S **244–246, 325, 383–384, 663–666, 728–729, 1130–1131, 1132, 1146, 1182–1184, 1358–1359, 1360–1361, 1437–1438, 1458–1459, 1549, 1850–1851, 1921–1930, 1950–1951, 1983–1984, 2108–2114**
Guttman, Burton *1925f*
Gy1–Gy5 gene family 1785
Gymnodinium acidotum 607f
gynogenesis 1419
gynogenone **901–902**
gyrase **902**

H

H19 gene cluster 209, **903**, 996
H1 histone 340, 343, 542
H2A/H2B dimers 343
H2A histone 340, *342f*
H2B histone 340, *342f*
H2 locus *see* major histocompatibility complex (MHC)
H2 protein *1920f*
H3 histone 340, *342f*
H4 histone 340, *342f*
H5 histone 542
hadulins **903**
Haeckel, Ernst 63, 199
Haemophilus influenzae 150t, 1423
- bacteria 149
- genome sequencing 517, 859
- histidine operon *946f*
- hot spots 977
- restriction endonucleases 1290
- Smith, Hamilton 1847

Haig, David 635
hairpin **903**

hairy cell leukemia (HCL) **903–904**
 see also leukemia, acute; leukemia, chronic
Haites N **242–243, 1370–1372**
Haldane, John Burdon Sanderson 354, 356–357, 357, 516, *904f*, **904–906**, 985
 The Biochemistry of Genetics 905
 The Causes of Evolution 905
 centimorgan (cM) 904
 Daedalus or Science and the Future 905
 evolution 664
 Fisher, R.A. 701
 fundamental theorem of natural selection 734
 Haldane–Muller principle 906
 Haldane's rule 904–905
 Heredity and Politics 905
 human genetics 905
 linkage map 904
 mouse genetics 1247
 My Friend, Mr. Leaky 905
 natural selection 905, 1326
 New Paths in Genetics 904, 905
 population genetics 905, 1514
 Possible Worlds and Other Essays 905
 reproductive isolation 1681
 Science and the Supernatural 905
Haldane–Muller principle 838, **906**, 1276
Haldane's mapping function *see* mapping function
Haldane's rule 904–905, 985, 1681
Halford S E **1693–1696**
Hall, Jeff 1315
Hall R M **771–774, 1041–1045**
haloacetaldehydes 551
Haloarcula hispanica 94, 2114
Halobacterium 2114
Haloferax mediterranei 94
Haloferax volcanii 92
halophiles 93
Hamerton J L **1407–1410**
Hamilton's theory **906–910**
 coefficient of relatedness 908–910, *909f*
 eusociality 907
 fitness 907–910
 inclusive fitness 908
 kin selection 908–910
 mathematical calculations 908
Hamilton, William Donald 907
Hanawalt P **564–571**
Hanawalt, Philip 565
Handbuch der Vererbungswissenschaft (Handbook of Genetics) 201
handedness (left/right) **910–911**
 amino acids 55
 behavioral 911
 helical 910–911
 intelligence 1046
 laterality 910
 molecular 910
Haniford D B **2033–2034**
Hansen, Emil Christian 1761
haploid 910
 infertility 1021
haploid number **911**
haploinsufficiency **911**
haplotype **911–912**
 D-loop 539
 gene mapping 911–912
 genetic diversity 539–540
 haplotype number 539
haplotype (*continued*)
 t haplotype 912
Harada H **422–423**
Hardies S C **1671–1675**
Hardy, Godfrey H. 912, 1514
Hardy–Weinberg law **912–914**
 allele frequency **37**, *912t*, 912–914, 935
 binomial distribution 215
 gene flow 785
 gene pool 797
 genetic drift 913
 genetic equilibrium 834–835
 genotype frequency *912t*, 912–914
 mathematical calculations 912
 mating table 912, *912t*
 meiosis 913
 population genetics 914
 population substructure 1519
 random mating 913
harlequin chromosomes 339, **914**, 1774
harlequin fetus 994, *995f*
Harper J C **688–690**
Harris, Henry 359, 360
Harrison C J **1449–1450**
Harrison D J **12**
Harris R **825–826**
Hartl, Dan 490
Hartl D L **111–113, 995–996**
Hartl F U **324–325**
Hartmann, Max 235
Hasegawa M **1157–1160**
Haselkorn R **800–803**
Hasleton P S **272–274**
Hastie N **2138–2139**
Hastings P J **237, 241–242, 580–582, 677–680, 683, 930–932, 936–938, 954–955, 955–956, 1055–1057, 1146–1147, 1181–1182, 1495, 1614–1618, 1623–1631, 1635–1636, 1642–1644, 1647–1648, 1756–1759, 1759–1760, 1823–1824, 1838–1839**
HAT *see* histone acetyl transferases (HAT); hypoxanthine-aminopterin-thymidine (HAT)
Hatfull, Graham 182
Hawkey P M **74–76**
Hawley R S **368**
Haw River syndrome *2049t*
Hayes, William 657, 679
Hayflick limit 1967
Hazelrigg T **1892–1895**
Hba-ps4 pseudogene 1921
hBRM chromatin remodeling complex *342t*
hCHK2 gene *2084–2085t*
Hck protein 1883
Hd1 gene 1395
Hd3a gene 1395
Hd6 gene 1395
HDAC *see* histone deacetylases (HDAC)
HDGS *see* homology-dependent gene silencing (HDGS)
Heath J K **899–900**
heat shock proteins (Hsps) **914–915**
 denaturation (proteins) 529
 Drosophila melanogaster 915
 function 915
 Hsp60 (GroEL) 915
 Hsp70 915
 mechanisms 915
 molecular chaperones 915
heavy chains *see* immunoglobulin gene superfamily
Hedgecock, George G. 491
hedgehog *(hh)* gene 427
HEI *see* hybrid element insertion (HEI)
Heidelberg J F **1196**
Heitz, Emil 1513
Helianthus 1807
Helianthus annuus 492, *1683t*
Helianthus petiolarus 1683t
helicases **915–916**
 DnaB 916
 DNA repair 566
 eIF4a 916
 PriA 916
 RecBCD enzyme 916
 Rho 916
 SV40 T-antigen 916
 UvrAB protein 916
Helicobacter pylori 150t, 192, 405, 452, **916–917**, 1423, 1693
 characteristics 916
 genetic studies 916–917
 pathogenicity island (PAI) 916
Heliconius 731
Heliconius cydno 1534
Heliconius erato 1534
Heliconius melpomene 1534
Helinski D R **2161–2162**
helix–loop–helix proteins *see* DNA-binding proteins
helix–turn–helix motif 280, **917**, *1036f*
 alkyltransferases 37
 DNA inversion system 941, *942f*
 homeodomain 958
 regulatory genes 1655
 resolvase-mediated deletion 1688
Helminthosporium solani 1850
helper phage **917**
hemagglutinatin virus of Japan (HVJ) 1356
hemagglutinin (H) 1026
Hemiptera 2097
hemizygote **917**
hemoglobin *see* globin genes (human)
hemoglobin anti-Lepore 2096, *2096f*
hemoglobin Lepore 2096, *2096f*
hemophilia 354, **917–920**
 blood coagulation defect 917–919
 factor V 919
 factor VII gene 918
 factor VIII gene 918
 factor IX gene 918
 factor X gene 918
 founder effect 920
 gene synthesis 918–919
 genetic counseling 920
 hemophilia A 918–920
 hemophilia B 918–920
 Leyden-type 919
 population genetics 918
 protein C 918
 treatment 920
 von Willebrand factor 919
 see also Christmas disease
Hemophilus influenzae 658
Henderson, C.R. 1797
Hendrix, Roger 182
Hengstler J G **51–54**
Hening, W. 384
Henkin T M **123–125**
Henning, W. 1935
Henslow, John Stevens 513
heparin-binding growth factor (HB-EGF) 627
hepatitis B virus 271

hepatocyte growth factor/scatter factor (HGF/SF) 1185
hereditarianism 2014–2015
hereditary diseases **920**
hereditary neoplasia **920–921**
hereditary nonpolyposis colorectal cancer (HNPCC) 10, 422, 563, 566, 1285, 1286
heritability **921–924**
  definition 921
  epistasis 921
  estimation 922–923
    computer modeling 923
    parent/offspring 922
    selection response 923
    siblings 922–923
    twins 923
  heritability values *922t*
  mathematical calculations 921–923
  uses 923
Herman R K **6–66, 2166–2167**
Hermansky–Pudlak syndrome 26, 1469
hermaphrodite **924–925**
  cause 925
  characteristics 924
  chimerism 925
  chromosomes 925
  Klinefelter syndrome 925, 1066
  lateral hermaphrodites 924–925
  mosaicism 925
  Mullerian ducts 925
  Wolffian ducts 924–925
Herpes virus primase *1547t*
herpesvirus subfamily 641
Herrick, James 1828
Hershey, Alfred 257, 412, 522, 554, 815, **925–926**, 1921
  American Phage Group 925
  bacteriophages 180, 654, 925–926
  Nobel Prize 523, 925, 1123
Hershey–Chase Experiment 181, 925
Herskowitz I **275–278**
Hertwig, Oskar 553
Hesketh R **1240, 1602–1607**
Heslop-Harrison J S **350–351, 512–513, 1509–1511**
heteroallele **926**
  complementation 926
  history 926
  recombination 926
heterobivalent recombinases 1040–1041
heterochromatic bands 348–349
  chromosome banding 349
heterochromatin 349, **926–927**
  constitutive 926–927
  DNA methylation 927
  euchromatin 927
  facultative 926–927
  inactive X chromosome 927
  mealybugs 927
  meiosis 927
  X-chromosome inactivation 927
  Y chromosome (human) 2156
  see also chromosome banding
heterochronic mutation **927–930**
  developmental timing
    Caenorhabditis elegans 928–930, *929f*
    Dictyostelium 930
    Drosophila 930
    humans 930
    maize 930
  heterochronic genes 928–930, *929f*
  heterochrony 930
heterochronic mutation (*continued*)
  see also neoteny
Heteroda avenae 973
heteroduplex DNA **930–931**
  bacteriophages 931–932
  Escherichia coli 931
  eukaryotes 931
  genetic recombination 842–843, *843f*
  Holliday's model 955–956
  hybrid DNA 930–931
  integrase family 1039
  integrons 1042
  meiosis 931
  mismatch repair *931f*, 931–932, 955–956
  mitosis 931
  Mut system 932
  postmeiotic segregation 931, 956
  Rad51protein 931
  RecA protein 931
  recombination models 1642
  replication 931, *931f*
  RuvAB enzyme 1756
  Saccharomyces cerevisiae 931
heterogenote **932**
heterokaryon **932**, 1323
heteropyknosis **932–933**
heterosis **933**
  dominance 933
  history 933
  hybrid vigor 933, 987
  overdominance 933
heterotrimeric G proteins **933–934**
  7-TM receptors 933–934
  guanosine diphosphate (GDP) 934
  guanosine triphosphate (GTP) 934
  helical domain 934
  signal transduction 934
  structural analysis 934
heterozygote **934–936**
  DNA sequences 934–935
  Hardy–Weinberg law 935
  heterozygote advantage 935–936
  phenotypic effects 935
  polymorphism 935
  recessive alleles 935
  reproductive fitness 935
  single nucleotide polymorphism (SNP) 935
hex genes 1666
Heyting C **372–373**
HF1 gene *430t, 431t*
Hfr gene 451, 453
Hfr strain 657–658, **936–938**
  conjugative transfer *937f*, 937–938
  Escherichia coli 936–938
  F (fertility) factor 678, *678f*
  F (fertility) plasmid 936–938
  formation *936f*, 936–937
  sexduction **1823–1824**
  transposable elements 936–937
Hidden Markov models (HMMs) 33, 1197
Hieter P **95–96**
hifA gene 977
hifB gene 977
Hill K L **79–81, 798–800**
Hill W G **96–101, 538, 823–825, 921–924, 1792–1793, 1793–1794, 1794–1795, 1796–1799**
him-10 gene 301
him genes 255
hin gene *939f*
Hingorani M M **1497–1499, 1951–1952**
Hin invertase **938–942**, *942f*
hinny 1255
Hiraizumi, Yuichiro 490
Hirschsprung disease **942**, 1468–1469, 1469, 1696
Hirszfeld Institute of Immunology and Experimental Therapy, Polish Academy of Sciences 177
hisA gene *946f*
hisB gene 943–945, *946f*
hisC gene 944, *946f*
hisD gene 943–944, *946f*
hisE gene *946f*
hisF gene 943–944, *946f*
hisG gene 943, *946f*
hisH gene 943–944, *946f*
hisI gene 943–945, *946f*
hisP gene 945
his genes 943–945, *946f*
his operon 123, 943
histidine **943**
histidine operon **943–947**
  Amadori rearrangement 943
  Ames test 52
  biosynthesis 943–945, *944f*
    attenuation 945, *946f*
    mechanisms 943–945
    polarity 947
    promoters 945
    regulation 945
    tRNA 945
  feedback inhibition 943
  gene organization 945, *946f*
  gene regulation 943
  histidine genes 945, *946f*
  leader region *946f*
  model system 943
  Salmonella 1766
  structure *944f*
histidyl-tRNA 1078
histocompatibility **947–948**, 999
histone acetyl transferases (HAT) 341
histone deacetylases (HDAC) 341, 633
histone genes **948–952**
  C-terminal histone-fold domain 949–951
  enhancers 949
  histones
    B4 948
    H2A 948–951
    H2B 948–951
    H3 948–951
    H4 948–951
    H5 949
  historical background 948
  mRNA (messenger ribonucleic acid) 949
  N-terminal domain 949–950
  nucleosome **1360**
  organization 948
  posttranscription regulation 949
  promoters 948–949
  transcriptional control 949–951
histone proteins *438f*
histones 340, *342f*, 633, **952**
hISWI chromatin remodeling complex *342t*
hitchhiking effect **952–953**, 1804
HIV-1 reverse transcriptase 1498
HMG domain proteins 542, 544, 1812, 1816
HMGIC gene 1110, *1112f*
HMGI(Y) gene 1110, *1112f*
HMG protein family 1111, *1112f*
hMLH1 gene 423, *2084–2085t*
HML locus 1154, *1155f*
hmp-1 gene 620

hmp-2 gene 620
hmr-1 gene 620
hMRE11 gene 351, 114
HMR locus 1154, *1155f*
hMSH2 gene 423, *2084–2085t*
hMSH3 gene *2084–2085t*
hMSH6 gene 423, *2084–2085t*
HNS protein 542, *1383f*
ho gene 1761
HO gene 277, 1156, 1761
Hobart M J **427, 433**
hobbit (hb) mutation 1476
HOBBIT (HBT) gene 1476, 1755
HobH protein *1383f*
Hodgkin J **128–129, 251–256, 530–531, 531–532, 724, 752, 792–793, 796–797, 865, 900, 910–911, 1065–1066, 1234–1235, 1275, 1279, 1335–1336, 1424–1426, 1743–1744, 1774, 1805**
Hodgkin's disease 644, **953–954**
 chromosome locations 953
 etiology 953
 pathophysiology 953–954
 see also Reed-Sternberg cells
Hodgkin, Thomas 953
Hogness Box **954**
Hohn B **1984–1986**
Holley, Robert W. 1061, 1328
Holliday junction 493, *466f*, *954f*, **954–955**, 1038–1039, *1039f*
 branch migration **237**, 1756–1757, *1758f*
 DNA, history of 555
 DNA repair *1622f*
 double-strand break repair model *580f*, *581f*, 582, *776f*
 Endo I enzyme 1760
 endonucleolytic cleavage 1759, *1760f*
 Endo VII enzyme 1760
 Flp recombinase 719
 homologous recombinations 352
 hot spots 976
 isomerization 955, 1055–1056, *1056f*
 Meselson–Radding model *1181f*, 1182
 Rap enzyme 1760
 RecBCD enzyme 1625, *1626f*
 reciprocal recombination 1632, *1633f*
 recombination models 1642
 resolvase/invertase family 1688
 Rus enzyme 1760
 RuvAB enzyme 1756, *1757f*
 RuvC enzyme 1759–1760
 site-specific recombination 1845
 synapsis (DNA transactions) *1908f*
 see also chromosome dimer resolution
Holliday, Robin 629, 955, 1756, 1759
Holliday's model **955–956**
 evolution 668–669
 heteroduplex DNA 955–956
 Holliday junction 955–956
 hybrid DNA formation *955f*, 955–956
Holloway, B. 851
holocentric chromosomes **956–957**
 Caenorhabditis elegans 956–957
 centromere 956–957
 chromosome rearrangement 956
 crossovers 957
 kinetic activity 956
 kinetochore 956–957
 meiosis 956–957, *957f*
 organization 956
 orientation 957, *957f*
 spindle microtubules 956–957
HO locus 1154
holoenzyme 625, 1497–1498
holophyly **958**
 see also cladogram; monophyly
Holsinger K E **528, 646–648, 797, 912–914, 1291–1297, 1410–1411**
Holweck, Fernand 1123
HOM-C complex 962–963
homeobox **958–962**
 homeobox genes 958–962, *960f*
  Antp (Antennapedia) gene 962
  Bar cluster 959
  BEL class 961
  caudal (cad) gene 959
  ceh19 cluster 959
  CUP class 961
  cut class 960
  Dll cluster 959
  empty spiracles (ems) gene 959
  en cluster 959
  evolution 961
  function 961–962
  gene expression 978
  HD-ZIP class 961
  Hex cluster 959
  Hlx cluster 959
  Hox cluster 958–959, *959f*, 962–963, 978
  Hox genes 978–979
  Hox-like cluster 959
  IRO class 961
  KNOX class 961, 962
  LIM class 960
  M-ATYP class 961
  MEIS class 961
  Mox cluster 959
  msh cluster 959
  NEC cluster 959
  NK cluster 959
  ParaHox cluster 959
  PBC class 961
  POU class 959, 962
  prd class 959
  prd-like class 960
  pros (prospero) class 961
  SO/SIX (sine oculis/Six) class 961
  spatial colinearity 978
  TALE class 961, 962
  temporal colinearity 978
  TGIF class 961
  Ubx (ultrabithorax) gene 962
  Xnot cluster 959
  ZF (zinc finger) class 960
  see also Hox genes
 homeodomain 958
  DNA binding 958
  helix–turn–helix motif 958–959
  Hox genes 979
  POU domain 959
  prd domain 959
  structure 958, *959f*
 homeotic genes 636, 666, 958, **962**
 transcription factors 958–962, 978
homeosis 962, 970
homeotic genes
 MLL gene 1228
 sex determination 1814
homeotic mutation **962–963**, 979
 ANT-C complex 962
 BX-C complex 962
 Drosophila melanogaster 962–963
 HOM-C complex 962–963
homing endonucleases 1358
homochirality 1392
Homo erectus 1519
homogeneously staining regions (*hsr*) 580, 766, **963**, 1123
homokaryon 1323
homologous chromosomes **963–964**
 meiosis 875
homologous recombinations *352f*
 DNA repair 351–353
homologs **964**
homology **964–969**, 1416, 1905
 atomization 966
 characters *versus* character states 964–965
 convergent characters 968
 definition 964
 homonomy 967, 970
 levels
  iterative 965
  ontogenetic 965
  polymorphic 965
  supraspecific 965
 mobile introns *1048t*
 molecular genetics
  allozyme electrophoresis 968
  DNA hybridization 968
  positional homology 968
  random amplified polymorphic DNA (RAPD) 968
  restriction enzyme analysis 968
 ontology 965–966
 parallel characters 968
 phylogeny
  homoplasy 966, 968, 969–970
  independence 966–967
  orthologous genes 966, *967f*, 968
  paralogous genes 966, *967f*, 968, 970
  plerologous genes 967
  xenologous genes 967, 970
 synapsis (DNA transactions) 1906
 taxonomy
  apomorphy 965–967, 968
  autapomorphy 965
  symplesiomorphy 965
  synapomorphy 965–967, 968, 969–970
 testing
  congruence test 968, 970
  conjunction test 967, 970
  similarity test 967, 970
homology-dependent gene silencing (HDGS) 1473
homonomy 967, 970
homoplasmy 1150
homoplasy **969–970**
 analogous similarities 969
 congruence test 970
 convergent similarities 969–970
 homonomy 970
 parallel similarities 969–970
Homo sapiens 859, 1045–1046
 BRG1 chromatin remodeling complex *342t*
 CHD4 chromatin remodeling complex *342t*
 hBRM chromatin remodeling complex *342t*
 hISWI chromatin remodeling complex *342t*
 h SWI/SNF chromatin remodeling complex *342t*
 retrotransposons *1699f*

Homo sapiens (*continued*)
RSF chromatin remodeling complex *342t*
homozygosity **970**
Hooker, Joseph 514
Hoover T R **163–165**
HOP1 gene product 1912
Hoppe-Seyler, Felix 553
hordein (*Hordeum vulgare*) 1784
Hordeum bulbosum 971–972
Hordeum chilense 972
Hordeum sp. **971–972**
cytogenetics 972
diseases
barley mild mosaic virus (BaMMV) 973
barley yellow dwarf virus (BYDV) 973
barley yellow mosaic virus (BaYMV) 973
Cochliobolus sativa 973
Erysiphe graminis 973
Puccinia graminis 973
Puccinia hordei 973
Puccinia striformis 973
Pyrenophora graminea 973
Pyrenophora teres 973
Rhynchosporium secalis 973
gene resistance genes 973
genetic engineering
Agrobacterium 973
applications 973
biolistics 973
genetic maps 972
genetic markers
amplified fragment length polymorphism (AFLP) 972
polymerase chain reaction (PCR) 972
random amplified polymorphic DNA (RAPD) 972
restriction fragment length polymorphism (RFLP) 972
simple sequence repeats (SSR) 972
hybridization 971
inbred strain 972
morphology 971
nomenclature 972
origins 971
phylogeny 971
polymorphism 972
systemic acquired resistance (SAR) *1914t*
uses 971
Hordeum spontaneum 971
Hordeum vulgare 971
horizontal resistance in plants 1535
horizontal transfer **973–975**
detection 974
DNA sequences 973–974
endosymbiont 973–974
eukaryotes 973–975
frequency 974
mechanisms 974
plasmids 1489
prokaryotes 974
significance 974
transkingdom 974
transposable elements 975
see also pathogenicity islands
Horvitz, Robert 254
host-lethal gene **975–976**
bacteriophage T4 975–976
cloning process 975–976
host-lethal gene (*continued*)
RNA polymerases 975–976
subtilis phage SPO1 975
host-range (*h*) mutants 1927
host-range mutant **976**
host-restriction modification 1124
Hotchkiss, Rollin 1492
hot spots **977**
Benzer, Seymour 977
characteristics 976
Chi sites 1624
gene conversion 777
genetic markers 976
meiotic hot spots **976–977**
att site 977
bacteria 977
Chi sequence 977
cog 976
cos 976–977
M26 976
phage λ 976–977
phage T4 977
prokaryotes 976–977
RecA protein 977
RecBCD enzyme 977
5-methylcytosine (m^5C) 977
molecular studies 976
mouse **1640**
mutation 1271
mutation rate 1279
pyrimidine–pyrimidine sequences 977
RecBCD enzyme 1624
rII locus 977
tandem repeats 977
see also Chi sequences
housekeeping genes 630, **978**
glucose 6-phosphate dehydrogenase (G6PD) deficiency 881
house of cards model 1801
Howard-Flanders, Paul 565, 1635
Hoxa1 gene 1996
Hoxd4 gene 1996
Hox genes 426–427, **978–979**, 1427
function 979
gene family 784
Hox clusters 978
MLL gene 1228
mutational studies 978–979
sex determination 1814
see also homeobox
hPMS1 gene 423, *2084–2085t*
hPMS2 gene 423, *2084–2085t*
Hprt gene 978, 1087
HPRT (hypoxanthine-guanine phosphoribosyl transferase) 1087
HPS gene 27
hRAD1 protein 1912
H/RS cells 953
3β-HSD *446t*
HsdM protein 1693
HsdR protein 1693
HsdS protein 1693
hSNF5 gene *2084–2085t*
hsp *see* heat shock proteins
Hst (hybrid sterility) 1–7 genes 985–986
h SWI/SNF chromatin remodeling complex *342t*
hth (homothorax) gene 961
HTLV-1 (human T-cell lymphotrophic virus 1) **979**
ATLL (adult T-cell lymphoma/ leukemia) 979
env gene 979
HTLV-1 (human T-cell lymphotrophic virus 1) (*continued*)
Gag gene 979
HAM (HTLV-1-associated myelopathy) 979
non-Hodgkin's lymphoma (NHL) 1347
Pol gene 979
Rex gene 979
Tax gene 979
TSP (tropical spastic paraparesis) 979
Hu J C **1376–1377**
Huang C **1741–1743**
Hubby, John 1516
Huber R E **212–214**
Hudson, Richard 1517
Hughes-Jones N **1713–1715**
Hull, Fred 1401
Hülskamp M **2045–2048**
Hultén M **18–19**, **343–344**, **1064–1065**, **1094–1095**, **1969–1970**
Hultén, Maj 357
humal immunodeficiency virus (HIV) 2111
human chromosomes **980**
Human Gene Mutation Database (HGMD) 1188
human genetics **980**
Human Genome Organizations Mutation Database Initiative 1188
Human Genome Project 91, 355–356, **980–981**, 1067, 2135
BACs (bacterial artificial chromosomes) 135
Brenner, Sydney 245
Celera 356
DNA sequencing 355–356
DOE 355
human linkage maps 1108
National Institutes of Health (NIH) 355
nomenclature 355
nomenclature, genetic 1341
see also chromosome mapping
human herpes virus-8 (HHV-8) 1347
human immunodeficiency virus (HIV) 247, 586
human leucocyte antigen (HLA) 730, 751
Human, Mary L. 1124, 1692
humans
chromosome mapping 354–356, *364f*
chromosome number 1969
cloning 391
codon usage bias *403t*
cranial sutures *478f*
EST number *366t*
genes 364
genetic mapping 2012–2013
linkage mapping *358t*
Mendelian Trait Database 2013
multifactorial genetic disorders *2015t*
Online Mendelian Inheritance in Man (OMIM) 365
origins 1518–1519
pedigrees *2013f*
piebald trait 1468
pigmentation 401
replication errors *1676t*
sex chromosomes 1408
transmission genetics 2012–2015
triploidy 2055
trisomy 2056, 2056–2057
see also Human Genome Project
human T-cell leukemia virus type I (HTLV-1) 271

human T-cell lymphotrophic virus 1 (HTLV-1) *see* HTLV-1
human transmissible spongiform encephalopathies 1881
Humphrey, R.R. 854
Hunt D M **420–422**
Hunter syndrome **981**
Huntington's disease 354, 357, **981–983**, 1421, *2049t*
  clinical features 982
  epidemiology 982
  genetic studies 982–983
  human behavioral genetics 211
  pathological mechanisms 982–983
  pathology 982
  prenatal diagnosis 1540
  single-gene inheritance 1836
HU proteins 542
Hurler, Gertrude 983
Hurler syndrome 756, **983–984**
Huson S M **1307–1309**
Hutchinson–Guilford syndrome 22
hut operon 1079, 1896
Huxley, Thomas Henry **984**
  Darwin, Charles 984
  Evidence on Man's Place in Nature 984
Huynen, M.A. 1323
HXK1 gene 885
HXK2 gene 885
hy4 gene 1456–1457
Hyacinthus 2055–2056
Hybosciara 763
hybrid **984**
hybrid-arrested translation **984**
hybrid dysgenesis **984–985**, 1403
  dysgenic traits 984–985
  molecular drive 1234
  selfish DNA 1806
  sterility 985
  transposable elements 985
hybrid element insertion (HEI) 1407
hybridization **1379–1381**, **989**
  duplex detection 1380
  duplex formation 1380
  evolutionary studies 1381
  gene expression 1380
  mathematical calculations 1380
  organismal **1379–1381**
  organism identification 1381
  phylogenetic relationships 1381
  procedure 1379–1380
  sequence requirements 1380
  in situ hybridization (ISH) 1380–1381
  thermal stability 1380
  wolf–dogs 265
  see also in situ hybridization
hybrid sterility (mouse) **985–986**
  chromosome 17 986
  gene mapping 985–986
  Haldane's rule 985
  Hst (hybrid sterility) 1–7 genes 985–986
  hybrid vigor 987
  inbred strain 985–986
  meiosis 986
  meiotic drive 986
  Mus musculus domesticus 985–986
  Mus musculus musculus 986
  Mus musculus spretus 986
  postzygotic reproductive isolating mechanism (RIM) 985
  t haplotype 986
hybrid vigor **987**
  associative overdominance 987
hybrid vigor (*continued*)
  hybrid sterility (mouse) 987
  inbreeding depression 987
  true overdominance 987
  see also heterosis; overdominance
hybrid zone (mouse) **987–988**
  chromosome 17 988
  Europe 988
  inbred strain 987–988
  Japan 988
  Mus molossinus 988
  Mus musculus bactrianus 988
  Mus musculus castaneus 988
  Mus musculus domesticus 988
  Mus musculus musculus 988
  species versus subspecies debate 987–988
  t haplotype 988
hydatidiform moles **989–991**, 1000
  abnormal placental morphogenesis 989–990
  androgenone 65
  complete hydatiform mole (CHM) 989–991, *990f*
    clinical features 989, *990f*
    diploid androgenesis 989
    dispermy 989
  fetal phenotypes 990–991
  genomic imprinting 989
  parthenogenesis 2098
  partial hydatiform mole (PHM) 989–991, *990f*
    clinical features 989, *990f*
    morphology 989
    triploidy 989–991
  uniparental inheritance 2098
hydrazine 1010
hydrid dysgenesis *1404f*
hydrogenosomes 1379
hydroxyaminopurine (6-hydroxyaminopurine) 191
5-hydroxymethyluracil 557
hy gene 1455
Hymenoptera 730
hyperchromicity **991**
hyperplasia 448
hypertension *2015t*
hyperthermophile 92
hypervariable region **991**
hypochondroplasia 2
hypochromic effect 550
hypogonadism 1066
Hypolymnas misippus 731
hypoplasia 448
hypothyroid (hyt) mutation 592
hypoxanthine-aminopterin-thymidine (HAT) 359
hypoxanthine phosphoribosyl transferase (HPRT) 2145
hypoxia 69

I

I1307K allele 263
iatrogenic CJD 2008
iceA gene 916–917
ICENP protein 1065
ichthyosis 599, **993–995**
  characteristics 993–994
  classification 993
  collodion baby 994
  epidermolytic hyperkeratosis (EH) 993, *994f*
  harlequin fetus 994, *995f*
ichthyosis (*continued*)
  ichthyosiform erythroderma 994
  ichthyosis bullosa of Siemens 993–994
  ichthyosis vulgaris simplex 993
  lamellar ichthyosis 994
  Netherton syndrome 994
  Refsum disease 994
  trichothiodystrophy (TTD) 568, 994
  X-linked ichthyosis 993
IciA protein *1383f*
identity by descent 725, 995
idiogram **996**
idiopathic epididymal obstruction 1025
idiotype 40
IF1 initiation factor 323, 1029, 1366, 2001
IF2 initiation factor 323, 1029, 1366, 2001
IF3 initiation factor 323, 1029, 1366, 2001
IF gene *430t*, *431t*
Igf2 gene 209, 903, **996**
Igf2r gene **996**
IGF (insulin-like growth factors) 592
IHF (integration host factor) protein 121, 544, 800, *1040t*, 1040–1041, *1383f*, 1435
IkappaB 1896
IL3RA gene 2157
illegitimate recombination **996–998**
  DNA ligation 997, *997f*
  DNA synthesis 997, *997f*
  site-specific recombination 998
    Escherichia coli 998
    immunoglobulins 998
    phage λ 998
    recombination signal sequences 998
ILR9 gene 2160
IMAC *see* immobilized metal ion affinity chromatography (IMAC)
imidazole-acetol phosphate (IAP) 944
imidazole-glycerol phosphate (IGP) 943–944
imm gene 1895
immobilized metal ion affinity chromatography (IMAC) 334
immotile cilia syndromes 1023
immune response 998
immunity **998**
immunoglobulin gene superfamily **998–999**, 1057
  gene family 784
  histocompatibility genes 999
  hypervariable region 991
  immune response 998
  immunoglobulin heavy (IGH) chain 953
  immunoglobulins 249, 998
  monoclonal antibodies 1235
  neuronal guidance 1320
imprinting, genomic **999–1001**
  androgenone 65
  chromatin 1001
  developmental consequences
    androgenesis 1000
    embryogenesis 1000
    hydatidiform mole 989, 1000
    parthenogenesis 999–1000
  DNA methylation 1001
  evolution 1001
  function 1001
  genetic studies 1000
  imprinted gene *999f*
  imprinting disorders 1000
  mechanisms 1000–1001
  uniparental disomy (UPD) 1000
inborn errors of metabolism **1014–1015**

129 inbred strain **1016**
inbred strain 250, **1015–1016**
F$_1$ hybrid 1015
filial (F) generation 1015–1016
genetic homeostasis 837–838
heterosis 933
hybrid sterility (mouse) 985–986
hybrid zone (mouse) 987–988
inbreeding 1015
inbreeding depression 987, 1016
incross 1017
Little, Clarence 1117
mouse 1243–1244
Mus musculus musculus 1260
Mus spretus 1262
nomenclature, genetic 1339
overdominance 1400
panmixis 1410
process 1015
rats 1609
recombinant congenic strains 1636
see also coisogenic strain; congenic strain
inbreeding *see* inbred strain
inbreeding depression 987, **1016**, 1805
see also conservation genetics
In chromosome 1166
incompatibility 1152, **1016**
incomplete dominance 682, **1016–1017**
see also codominance
incomplete penetrance *see* penetrance
incontentia pigmentii 600
incross **1017**, 1400
indel **1017**
independent assortment **1017–1018**
Drosophila melanogaster 1018
linkage 1018
meiosis 1017–1018
Mendel's laws 1017
Saccharomyces cerevisiae 1018
independent segregation **1018–1019**, 1019
Indian *Drosophila* Stock Centre 847
Indigofera (indigo dye) 1083
Indigofereae 1083
induced chromosome *363t*
inducer **1019**, 1678
see also induction of transcription
inducer exclusion 281
inducible enzyme *see* induction of transcription
inducible system *see* induction of transcription
induction of prophage **1019**
induction of transcription **1019–1021**
Escherichia coli 1019–1020
inducer 1019–1020
instruction theory 1019
mechanisms 1020
infertility **1021–1026**
azoospermia factor (AZF) *1023f*, 1024
azoospermia gene 1023, *1023f*
female reproductive tract abnormalities 1024
gene deletions *1023f*, 1024
intracytoplasmic sperm injection (ICSI) 1022, *1022f*
Klinefelter syndrome 1066
male reproductive tract abnormalities 1024–1025
meiosis 1021
oocyte production 1021–1022
scrotal anatomy *1025f*
sperm production 1022–1024
infertility (*continued*)
in vitro fertilization (IVF) 1009
infinite allele model 490
influenza virus **1026**
antigenic shift 1026
RNA replication 1026
type A 1026
type B 1026
type C 1026
INFOBIOGEN web site 519
inheritance **1026–1027**
inherited rickets **1027–1030**
autosomal dominant hypophosphatemia 1029
causes 1027
1,25-dihydroxyvitamin D 1027–1028
DNA-binding domain 1028
hypercalciuric 1028
vitamin D hydroxylation 1027
vitamin D-resistant 1027–1028
X-linked dominant hypophosphatemic rickets 1028
X-linked recessive hypophosphatemic rickets 1028–1029
initiation factors **1029**
Ochoa, Severo 1366
transcription 1029
translation 1029
see also translation initiation factors
initiator region (INR) 379, 1550
INK4a gene *2084–2085t*
INK4 family 505
INR *see* initiator region (INR)
insecticide resistance 490
insertion sequence (IS) **1029–1037**
autoregulation strategies 1035–1037, *1036f*
compound transposon 1030
discovery 1029
gene expression 1030
helix–turn–helix motif *1036f*
indel 1017
occurrence 1030
structure *1030f*, 1030–1031
target specificity 1035
transposases (Tpases) 1031–1037
DDE motif *1032t*, 1032–1033
domain structure 1031–1033, *1032t*
retroviral integrases (IN) 1032, *1032t*
transposition strategies 1033–1035, *1034f*
variety 1030, *1031t*
in situ hybridization (ISH) **1002–1004**
applications 1003–1004
chromosome painting 1002–1004
comparative genome hybridization (CGH) 1003–1004
DNA probes 1002–1003
filter hybridization **698**
fluorescence microscopy (FISH) 535, 1002
fluorochromes 1002
function 1002
gene mapping 1003
lampbrush chromosomes 1075
maps 361–362, *363t*
multicolor fluorescent ISH (M-FISH) 980, 1002–1003
radioisotopes 1002
Institute for Laboratory Animal Research (ILAR) 1337
Institute of Physical and Chemical Research (RIKEN) 852
insulinoma **1037**
int22h gene 918
integrase family **1038–1041**
attachment sites 120–121
Cre recombinase 1040, *1040t*
dimeric targets 1040
FIS (factor for inversion stimulation) 121, *1040t*, 1040–1041
Flp recombinase **717–721**, 1040, *1040t*
function 1038
heterobivalent recombinases 1040–1041
heteroduplex DNA 1039
IHF (integration host factor) 121, *1040t*, 1040–1041
integrase **1038**
integrons 1041
Int protein 1038, 1434–1435
lambda (λ) Int recombinase *1040t*, 1040–1041
mechanisms 1038–1039, *1039f*
monomeric targets 1040
overlap region 1039
retrotransposons *1699f*, 1700
site-specific recombination 717, 1843, *1844t*, 1845, *1845f*
structure 1041
target specificity 1039–1040
topoisomerase 1038
Xer recombinase 1040, *1040t*
Xis (excisionase) protein 121, *1040t*, 1041
integration **1041**
integration of chromosome 260–262
integrin adhesion receptors 1884
integrons **1041–1045**
59-be (59-base element) 1041–1044, *1042f*
antibiotic resistance genes 1042–1044, 1687
attI site 1041–1045, *1042f*, *1044f*
chromosomal 1043, 1044
class 1 1042–1045, *1044f*
class 2 1044
class 3 1044
function 1043
gene cassettes 773, 1041–1044, *1042f*
gene expression 1043–1044
insertion sequence (IS) 1045
IntI recombinase 1041–1044
mobile 1043, 1044
open reading frames *1042f*, 1043–1044
Pc promoter 1041–1045, *1042f*, *1044f*
structure 1045
Tn*402* transposon 1045
transposition genes 1045
transposons 1042
Vibrio cholerae 1043–1044
see also gene cassettes
inteins 1052, **1565–1567**
biochemistry 1566
discovery 1565
organization 1565–1566
protein manipulation *1567f*
uses 1566–1567
intelligence **1045–1046**
genetic origin 1045–1046
handedness 1046
Homo sapiens 1045–1046
human cognitive abilities 1045–1046
intelligence quotient 1045
lateralization 1045–1046

intelligence (*continued*)
Stanford–Binet test 1045
Wechsler adult intelligence scale (WAIS) 1045
interchange heterozygote 1787
interchromomeric fiber 1074–1075, *1075f*
intercross **1046**, 1400
interference **1046**
see also chimera; chromatid interference
interferon-β gene 630
International Board for Plant Genetic Resources 852
International Committee on Standardized Genetic Nomenclature for Mice 1337
International Maize and Wheat Improvement Center 1786
International Mycological Institute (IMI) 851
International Potato Center 1850
International Resource Center for Zebrafish 853
International Rice Genetic Cooperative (RGC) 853
interphase 288, **1047**
interpolated Markov models 1197
Interpro database 1201
intersex **1047**, 1809
interspecific cross **1047**
intervening sequence **1047**
see also introns
Intestiphage 176
intI gene 1041–1045, *1042f*, *1044f*
int (integrase) gene 1687, 1859, 2021
IntI recombinase 1041–1044, 1843
intracytoplasmic sperm injection (ICSI) 689, 1009, 1022, *1022f*, 1066
introns **1047–1051**, **1052–1053**
coat color mutations, animals 401–402
consensus sequences *1536f*
distribution 1052
eukaryotic genes 663
exons 1052
gene expression 1052
intron homing **1047–1051**
applications 1051
evolutionary implications 1051
group I homing mechanism 1048, *1049f*
group II homing mechanism 1048–1050, *1050f*
history 1047
mobile introns 1047–1051, *1048t*
mRNA (messenger ribonucleic acid) 1184
mutational studies 1052–1053
selfish DNA 1806–1807
spliceosomal 1052
split genes 1878
T phages 1929–1930
see also coding sequences
invariants, phylogenetic **1053–1054**, *1054f*
inversion **1054**
see also crossover suppressor
invertasome 939–941, *940f*
inverted repeat sequences (IRS) *1030f*, 1030–1031, **1054**
inverted terminal repeats **1054**
in vitro evolution **1004–1008**
applications 1008
aptamers 1006
DNAzymes 1006
evolutionary reconstruction 1005
history 1004–1005
in vitro evolution (*continued*)
mechanisms *1006f*
nucleotides 1008
Qβ replicase 1005, *1005f*
RNA catalytic ability 1006, *1007f*
RNA (ribonucleic acid) 1005–1008
selection techniques 1006
SELEX (systematic evolution of ligands by exponential amplification) 1006
in vitro fertilization (IVF) **1008–1009**
in vitro mutagenesis **1010–1014**
bacteriophage M13 1012
BcgI 1011
BspMI 1011
cassette mutagenesis 1011, *1011f*
chemical mutagenesis 1010
DNA libraries 1013
DNA ligation 1012
DNA polymerases 1010–1012
DNA shuffling 1014
doped mutagenesis 1013–1014
error-prone PCR 1014
nonselective mutagenesis 1010
oligonucleotides 1010–1012
PCR mutagenesis *1011f*, 1012
plasmid vectors 1011
polymerase chain reaction (PCR) 1012
primer-directed mutagenesis *1011f*, 1012
restriction endonucleases 1010–1011
reverse genetics 1014
saturated mutagenesis 1013–1014
site-directed mutagenesis 1010
in vitro packaging **1014**
ion channels 314–318
ionizing radiation 271
iron repressor protein (IRP) 2003
iron responsive element (IRE) 2002–2003
Isl1 gene 1097
Isl2 gene 1097
IS2 insertion sequence 936–937
IS3 insertion sequence 936–937
IS*903* transposase 381–382
Isaacson P G **722**, **1347–1349**
isochromosome 348, **1054**
isoenzymes 625
isolation by distance **1054–1055**
demes 528
gene frequencies 1055
linkage disequilibrium 1055, 1105
population genetics 1055
isoleucine **1055**
isomerization **1055–1057**
crossovers 1055–1056, *1056f*
Holliday junction 955, 1055–1056, *1056f*
process 1056
isopropyl-1-thio-β-D-galactoside (IPTG) 1020
isotype **1057**
isozymes 625
ISWI1 chromatin remodeling complex *342t*
ISWI2 chromatin remodeling complex *342t*
ISWI family *342t*
IT15 gene 982, *2049t*
iterons 1486
ITGAM gene *430t*, *431t*
ITGAX gene *430t*, *431t*
ITGB2 gene *430t*, *431t*

J

JA *see* jasmonic acid (JA)
Jack W E **2104–2106**
jackknifing *see* trees
Jackson I **1336–1340**
Jackson Laboratory 397, 847, 849, 1118
Jackson Laboratory Backcross Mapping Panels 912
Jacob, François **1059**
allostery 39
Brenner, Sydney 244
F (fertility) factor 679
gene therapy (human) 815
lactose metabolism 1070, 1652
Monod, Jacques 1237
mRNA 1059
Nobel Prize 1070
operators 1376
operon theory 655, 1059, 1070, 1071, 1377
see also Brenner, Sydney; Monod, Jacques
Jacob (Genesis 30:27–42) 2009–2010
Jacobsen syndrome *2049t*
Jacobs P A **1810**
Jacobs, Patricia 1086, 2056
Jaenisch, Rudolf 630
Jansen E **1109–1117**
Japan Collection of Microorganisms (JCM) 852
Japanese Rice Genome Research Program 1394
jasmonic acid (JA) 1481
Jaspers N G J **558**, **1586**, **2150**
Jayaram M **717–721**
jellyfish *see* coelenterates (jellyfish)
J gene *see* recombination in the immune system
Johannsen, Wilhelm Ludvig 521, 759, 1298
John Innes Institute 1179
Johnson R **1099–1103**
Johnson R C **938–942**
Jones, D.F. 1400
Jordan, D.S. 1861
Jordan, K. 1861
Jorgensen E M **1309–1314**
Jukes–Cantor correction **1059–1060**
Jukes T H **1342–1345**
Jukes, T.H. 1326, 1800
jumping genes *see* horizontal transfer
Jurata L W **1096–1099**
juxtacrine mode *see* autocrine mode

K

Kaback M M **1941–1943**
Kado C **491–493**
kaiA gene 389–390
kaiB gene 389–390
KaiC gene 389–390
Kaiser, Dale 1131, 1637
Kaiser-Wilhelm-Institut für Züchtungsforschung Müncheberg 202
kala-azar 1736
Kalanchoë daigremontiana 492
Kallioniemi A **424–426**
Kalousek D K **989–991**
KALp gene 2159
kanamycin 611, 1707
Kane C **1973–1983**
Karaolis D K R **1422–1424**
Karn J **1701–1706**
Karposi's sarcoma 1768, *1769t*
karyopherins 1353–1354
karyoplast 1357
karyotype 249–250, 996, **1061**, 1239
see also idiogram
kasugamycin 77

Katso R M T **626–628**
Kaufman, Thomas 847, 958
kb (kilobase) **1061**
Kearns–Sayre disease 1218, 1224
KEGG (Kyoto Encyclopedia of Genes and Genomes) database 519, 1201
Kellogg E A **889–890**
Kelly T **1290–1291**
Kemphues K J **1148–1149**
Kennedy's disease *2049t*
Kettlewell, H.B.D. 1534
keule mutation 1476
Khorana, Har Gobind 555, **1061**
 genetic code 1061
 Nobel Prize 1061, 1329
 rhodopsin 1061
Khrushchev, Nikita 1130
Kidwell M G **939–975, 984–985**
Kimble, Judith 254
Kimmel M **233–235**
Kimura correction **1062**
 see also Jukes–Cantor correction
Kimura, Motoo 490, **1063**
 An Introduction to Population Genetics Theory 490
 neutral mutation 1324
 neutral theory 603, 833, 905, 1063, 1326, 1343, 1516, 1800
 theory of population genetics and evolution 1063
kinases **1063–1064**
 eukaryotes 1063–1064
 histidine 1063
 MAP kinase kinases (MAPKK) 1064
 MAP kinase kinase kinases (MAPKKK) 1064
 microorganisms 282
 mitogen-activated kinases (MAPK) 1064
 mitosis 1226
 oncogenesis 1063
 phosphorylation 1063–1064
 prokaryotes 1063
 protein kinases 1063–1064
 serine/threonine 1063
 transcription 1064
 tyrosine 1063
kinetochore **1064–1065**
 centromere 956–957, 1064
 chromatin 1064
 chromatin-associated proteins 1064–1065
 DNA (deoxyribonucleic acid) 1064
 holocentric chromosomes 956–957
 meiosis 875
 mitosis 956, 1225–1226
 nondisjunction 1346
 nuclear envelope (NE) 1356
 phosphorylation 1065
 protein function 1065
 protein location *1065f*
 spindle microtubules 956–957, 1064–1065
 structure 1064, *1065f*
 see also cell cycle; chromosome banding
Kinetoplastida 1735
King, J.L. 906, 1326, 1342, 1800
Kingman, J.F.C. 392
kin selection *see* Hamilton's theory
Kinsey, J.A. 849
ki (pollen killer) gene 2064
kirromycin 77, 611
Kishino H **1157–1160**
Kit gene 1469
KK-γ gene 600
Klar A J S **1153–1157**
Klar, Amar 636
Klebsiella 1079, 1328, 1922
Klebsiella pneumoniae 707, 1382
 Chi sequences 328
Klein G **599, 641–644, 1289–1290**
KLHL1 (sense) gene *2050t*
Klinefelter syndrome **1065–1066**, 1086, 1314, 1823, 2056
 aneuploidy 66
 chromosome aberrations 346
 clinical features 1066
 Ford, Charles 723
 hermaphrodite 925, 1066
 human chromosomes 980
 hypogonadism 1066
 infertility 1022–1023, 1066
 mosaicism 1066
 sex chromatin 1066, 1809–1810
 sex chromosomes 1811
 X chromosome 2147
 see also chromosome aberrations; Turner syndrome
Klobutcher L A **1133–1135, 1135–1136, 1204**
Kloecker, Albert 1761
Kloepper J **1477–1480**
Klug, Aaron 555
Kluyvera ascorbata 1480
Knight, Andrew 1172
knockout 590, 963, **1066–1067**, 1934
knolle mutation 1476
Knotted 1 gene 962
knotted-1 mutation 1473
Knudson, Alfred 81
Koch M A **86–87**
Kohara Y **286**
Kolmogorov diffusion equations 1062
Kölreuter, J.G. 1172
Konopka, Ronald 386
Koornneef, Martin 88
Kopf G S **3–4, 1875–1876**
Kornberg, Arthur **1067–1068**
 DNA polymerases 558
 DNA structure 555
 For the Love of Enzymes: The Odyssey of a Biochemist 1067
 Nobel Prize 1067, 1366
 Ochoa, Severo 1366
 synthesis of RNA and DNA 1067
Kornberg enzyme *see* DNA polymerases; genetic recombination
Korswagen H C **933–934**
Kosambi's mapping function 357, 1142
Kr1 gene 2064
kr2 gene 2064
kr3 gene 2064
kr4 gene 2064
Kraepelin, E. 1774
Krakauer D C **845–846**
Krebs cycle 1365
Kreitman M **952–953, 1324–1325, 1803–1804**
Kreitman, Martin 1516
Kreuzer K N **168–175**
Krimbas C B **702–706**
Krishnan H B **1780–1787**
Kropinski, Andrew 185
Krumlauf R **978–979**
Kullback–Leibler information content 1158
Kumar A **2034–2040**
Kurland C **649–650**
kuru **1068**, 2008
 bovine spongiform encephalopathy (BSE) 1068, 1881
 cannibalism 1068
 characteristics 1068
 Creutzfeldt–Jakob disease 1068
 see also spongiform encephalopathies (transmissible)
Kutateladze M **175–179**
Kutter E **84–85, 179–186, 244–246, 599, 695–698, 917, 975–976, 1077, 1080–1081, 1130, 1131–1132, 1441–1442, 1895–1896, 1921–1930, 1970, 1971–1973, 2099**
Kutter, Elizabeth *1925f*
Kyoto Encyclopedia of Genes and Genomes (KEGG) database 519
Kyriacou C P **209–212, 386–390, 1314–1318**

L

lab (labial) gene *960f*, 961
Laboratory Registration Code (Lab Code) 1337
Labouesse M **612–621**
lacA gene *155f*, 1070
lac gene 453, 655
lacI gene 977, 1069, 1491, 1652, *1653f*
Lacks S A **1213, 1491–1494, 1649, 1661–1669**
lacL gene 1070
lac mutants **1069–1070**
 beta (β)-galactosidase 1069
 cloning vectors 1069
 Lederberg, Joshua 1069
 permease 1069
 relation to *lac* operon 1069
lacO gene 1069, 1070
lac operon *153f*, 630, 943, 1029, **1070**, 1896
 beta (β)-galactosidase 212, 1070
 CAP (catabolite activator protein) 1070
 catabolite gene activator protein (CAP) 278–279
 catabolite repression 279
 cloning vectors 1070
 CRP (cAMP receptor protein) 1070
 DNA denaturation 551, *552f*
 Escherichia coli 1070
 gene expression 1070
 gene regulation 1070, 1652
 induction of transcription 1020
 lactose metabolism 1070
 mRNA (messenger ribonucleic acid) 1070
 operators 1376
 operon 1377
 permease 1070
 repressor 1678
 Shine–Dalgarno sequence 152
 transacetylase 1070
LaCour, Len 512
lacP gene 1070
lac repressor 165, 1678
Lactococcus lactis 1892
 Chi sequences 328
 histidine operon *946f*
lactose **1070–1071**
lacY gene
 bacterial genes *155f*, 156
 lactose metabolism 1070
 Lederberg, Joshua 1069
 microbial genetics 1193–1194
 plesiomorphy 1491
 reverse mutation 1709–1710

Lacy, Robert 459
lacZ fusion 334, 1195
lacZ gene
  bacterial genes *155f*, 156
  beta (β)-galactosidase 212
  Escherichia coli 656
  gene fusion 739
  lactose metabolism 1070
  Lederberg, Joshua 1069
  microbial genetics 1193–1194
  plesiomorphy 1491
  reverse mutation 1709–1710
  transgenes 1990
  transposons 2039
LacZ protein 1554
La Du B N **35–36**
lagging strand **1071**
Laibach, Friedrich 87
Lamarckism **1072–1073**
  Darwinian theory of evolution 1072
  Lysenko, Trofim 1073
  Soviet genetics 1129
  theory of evolution 1072
Lamarck, Jean Baptiste **1071–1072**
  Dictionnaire de Botanique 1071
  Histoire naturelle des animaux sans vertèbres 1071
  Philosophie zoologique 1072
  Recherches sur l'organization des corps vivants 1072
  theory of evolution 1071
Lamb B C **101–104**
lambda (λ) Int recombinase *1040t*, 1040–1041
lambda (λ) *rex* gene 1895
lamellar ichthyosis 994
Laminaria 748
lampbrush chromosomes **1073–1077**
  characteristics 1074–1075
  chromatids 1073, *1075f*
  chromomere loops 1073–1074, *1074f*
  chromomeres 1073–1074, *1074f*
  DNP (deoxyribonucleoprotein) 1073
  function 1075, 1077
  history 1073
  interchromomeric fiber 1074–1075, *1075f*
  meiosis 1073
  morphology 1076–1077
  oocyte *1073f*
  oogenesis 1074
  read-through hypothesis 1076
    DNase-1 1076
    Notophthalmus viridescens 1076
    restriction enzymes 1076
  RNA polymerases 1074–1075
  RNA transcription 1074–1075, *1075f*
  in situ hybridization (ISH) 1075
  transcription process 1075, *1076f*
    read-through hypothesis 1075
    RNA polymerases 1075
  transcription units 1075, 1077, *1077f*
Lanceolate (La) mutation 1472
Landsteiner, Karl 1713
Landy A **1038–1041, 1434–1437**
Langer–Gideon syndrome 361
Langridge, Bob 555
large granular lymphocytes (LGL) 1094
large T antigen *see* T antigens
LaRossa R A **1362–1363**
Latarjet, Raymond 1124
Latchman D S **1971–1973**
late genes **1077**, 2111
Lathyrus (grass pea) 1083
Lathyrus odoratus 353, 745
Lathyrus sativus L. 1470
Latif F **81–81**
Laupala kohalensis 1683t
Laupala paranigra 1683t
Laurén J **2103–2104**
Lau Y-K **648–649**
Laval J **888–889**
Lawler, Sylvia 356
Lawrence, J. B. 361
Laybourn P **803–813**
Lck protein 1883
LD *see* linkage disequilibrium
LD gene 715
leader peptide **1077–1078**
  histidyl-tRNA 1078
  leader polypeptide 1077
  mRNA (messenger ribonucleic acid) 1078
  open reading frames 1078
  signal peptides 1078
  signal sequences 1078
  translational attenuation 1078
leader sequence **1078–1079**
  bacteria
    ami operon 1079
    antibiotic resistance genes 1079
    Bacillus subtilis 1078
    bgl operon 1079
    Escherichia coli 1079
    glp operon 1079
    hut operon 1079
    Klebsiella 1079
    lic operon 1079
    mRNA (messenger ribonucleic acid) 1078
    nas regulon 1079
    nucleotide sequences 1078
    Pseudomonas 1079
    pur operon 1079
    pyr operon 1079
    RNA polymerases 1078
    sac regulon 1079
    transcription antitermination 1078
    transcription attenuation 1078
    transcription termination 1078
    translational repression 1079
    translation attenuation 1079
    trp operon 1078
  eukaryotes
    internal ribosome entry site (IRES) 1079
    translation 1079
    upstream open reading frames (uORFs) 1079
  transcription antitermination 1078
  transcription attenuation 1078
  transcription regulation 1078
  transcription repression 1079
  transcription termination 1078
leading strand **1080**
leaky mutation **1276**
least squares **1080**
LEC gene 1476
lec (leafy cotyledon) mutation 1476
Lederberg, Esther 654
Lederberg, Joshua 450, **1080–1081**
  Escherichia coli 654
  F (fertility) factor 453, 679
  genetic recombination in bacteria 1080, 1614
  Hfr bacteria strains 936, 1614
Lederberg, Joshua (*continued*)
  lac mutants 1069
  microbial genetics 1080
  Neurospora 1080
  Nobel Prize 1070
  phage transduction 1080
  selection process 1709
Leder P **1328–1329**
Leder, Philip 1328, 1372
LeDouarin, Nicolle 331
Lee J Y **338–339, 340–343, 344–345, 350**
Lefranc G **456–457**
Lefranc M-P **456–457**
Le gene 1179
leghemoglobin 1084, 1716
Legionella pneumophila 452, 1892
Leguminosae **1081–1085**, 1445
  Arabidopsis 1084
  Caesalpinioideae 1081, *1082f*, 1121
    Ceratonia (carob) 1083
    Cercis (judas tree, redbud) 1083
    Gleditsia (honey locust) 1083
  characteristics 1081
  economic importance 1083
  fossil record 1084
  genomes 1084–1085
    chloroplast genome 1084–1085
    mitochondrial genome 1085
    neopolyploidy 1085
    nuclear genome 1085
    paleopolyploidy 1085
    polyploidization 1085
    tetraploidy 1085
  linkage map 1085
  Mimosoideae 1081–1082, *1082f*, 1121
    Acacia 1083
    Leucaena 1083
    Mimosa 1083
    Neptunia 1083
    Parkia (locust bean) 1083
    Prosopis (mesquite) 1083
  nodulation 1081, *1082f*, 1084
    leghemoglobin 1084
    nitrogen-fixing symbioses 1084
    nodulation ability 1084
    rhizobia 1084
  Papilionoideae 1082, *1082f*, 1121
    Aeschynomeneae 1083
    Arachis hypogaea (peanut) 1083, 1121
    Astragalus 1083
    Cajanus (pigeon pea) 1083
    Canavalia (sword or jack bean) 1083
    Cicer (chickpea) 1083
    Cicereae 1083
    Cyamopsis (guar gum) 1083
    Dalbergieae 1083
    Desmodieae 1083
    Galegeae 1083
    Genisteae 1083
    Glycine max (soybean) 1083, 1121
    Indigofera (indigo dye) 1083
    Indigofereae 1083
    Lathyrus (grass pea) 1083
    Lens (lentil) 1083
    Loteae 1083
    Lotus japonicus 1083, **1121–1122**
    Lupinus spp. (lupin) 1083
    Medicago truncatula 1083
    Melilotus (sweet clover) 1083
    Millettieae 1083
    Phaseoleae 1083

Leguminosae (*continued*)
Phaseolus vulgaris (common bean) 1083, 1121
Pisum sativum (pea) 1083, 1121
Podalyrieae 1083
Psoraleeae 1083
Robinia (locust) 1083
Robinieae 1083
Sesbania (sesban) 1083
Trifolieae 1083
Trifolium (clover) 1083
Vicia (vetch, broad bean) 1083
Vicieae 1083
Vigna (cowpea, mung bean) 1083
Wisteria 1083
phylogenetic relationships 1081, *1082f*
scientific importance 1083
synteny conservation 1085
leiomyoma *see* lipoma and uterine leiomyoma
Leishmania 765, 1741
Lejeune, Jérôme **1085–1087**
cri-du-chat syndrome 1086
cytogenetic analysis 1086
Down syndrome 1085
Dutrilleaux, Bernard 1086
trisomy 21 syndrome 1085
trisomy 9p 1086
Leloir pathway 741
Lengeler J W **281–284**
Lens culinaris 1470
Lenski R E **671–672**
Lens (lentil) 1083
Lep^{ob} gene 1244
Lepomis punctatus 1522, *1523f*
Leptosphaeria maculans 1914t
Leptospira interrogans 859
Lesch–Nyhan syndrome (LNS) **1087–1088**
chromosome Xq26-q27 1087
clinical features 1087
classical Lesch-Nyhan syndrome 1087
hyperuricemic LNS variant 1087
neurological LNS variant 1087
gene therapy (human) 816
HPRT (hypoxanthine-guanine phosphoribosyl transferase) 816, 1087
purine metabolism disorder 1087
uric acid production 1087
Lessard P A **54–56**
let-502 gene 620
let-805 gene *620f*
let-99 gene 301
let genes 255
lethal locus **1088**
lethal mutation **1088**
lethal yellow gene 23
Leucaena 1083
leucine **1088**, *1088f*
leucine zipper 458
leukemia, acute **1088–1091**, **1091–1092**, 1760
acute lymphoblastic leukemia (ALL) 1088, 1449, 1506
B-cells 1090
chromosome abnormalities *1089t*, 1090
chromosome translocations 1090
Ets family 662
minimal residual disease 1206
MLL gene 1228
leukemia, acute (*continued*)
T-cells 1090
acute myelodysplastic syndrome (MDS) 1228
acute myelogenous leukemia (AML) 1088
chromosome abnormalities *1090t*, 1091
chromosome translocations 1091
minimal residual disease 1206
MLL gene 1228
acute promyelocytic leukemia (APL) 1506
adult T-cell lymphoma/leukemia (ATLL) 979
B-cell precursor acute lymphoblastic leukemia (BCP-ALL) 1091
chromosome translocations 1092, *1092t*
11q23 1092
14q11 1091
14q32 1091
2p12 1091
2q11 1091
7p13 1091
7q35 1091
8q24 1091
immunoglobulin genes 1091
MLL gene 1092
MYC oncogene 1091
cytogenetic analysis 1092
diagnosis 1091–1092
cytogenetics 1091
fluorescence microscopy (FISH) 1091
reverse transcriptase polymerase chain reaction (RT-PCR) 1091
FAB classification of leukemia 680
fusion genes 1092
MIC and MIC-M classification of leukemia 1190–1191
origination *in utero* 1092
Philadelphia chromosome 1449
T-cell acute lymphoblastic leukemia (T-ALL) 1118, 1932
T-cell receptor 1090, 1092
transcription 1092
WHO classification of leukemia **2135–2136**
see also Philadelphia chromosome
leukemia, chronic 207–208, *1093t*, **1093–1094**
B-cell chronic lymphocytic leukemia (CLL) 1093
B-cell prolymphocytic leukemia (B-PLL) 1093
CD5+ B cells 1093
chromosome abnormalities 1093
IGHV (immunoglobulin heavy chain variable) mutations 1093
p53 gene mutation 1093
splenic lymphoma with villous lymphocytes (SLVL) 1093
chromosome translocations 1093
chronic myelogenous leukemia (CML) 189, 250, 1088, 1093, 1206, 1371, 1449, 1506
FAB classification of leukemia 680
minimal residual disease 1206
Philadelphia chromosome 189, 207–208, 1449
T-cell malignancies
adult T-cell lymphoma/leukemia (ATLL) 1094
leukemia, chronic (*continued*)
large granular lymphocytes (LGL) 1094
Sezary syndrome 1094
T-cell prolymphocytic leukemia (T-PLL) 1093
T-cell receptor 1093
WHO classification of leukemia **2135–2136**
see also Philadelphia chromosome
leukemia inhibitory factor (LIF) 1967
leu marker 1984
Leupold, Urs 1777
Leutwieller, Leslie 88
Levan, Albert **1094–1095**
Allium test 1094
Cancer Chromosome Laboratory 1095
chromosome studies 1094
clinical cytogenetics 1094
contributions 1094–1095
double-minute chromosomes (*dmin*) 1095
Tjio, Joe-Hin 1094, 1969
Levan G **580**, **963**, **1607–1610**
Levene, Phoebus A. 325, 553
Levine, Mike 958
Levine, Paul 1461
Levine, Philip 650, 1713
Levinton J **1869**
Leviviridae 1744–1746
Levy S B **1687**
Lew D **286–296**
Lewis, Edward 847, 963, **1095**
colinearity principle 1095
Drosophila melanogaster 1095
embryonic development 1095
Nobel Prize 1095
Nusslein-Volhard, Christiane 1095
Wieschaus, Eric F. 1095
Lewis, I.M. 1124
Lewis R **1171–1180**
Lewis, R.C. 383
Lewontin R C **577–578**
Lewontin, Richard 836, 1516
lexA gene 1651, 1853–1854
LexA protein 1622, 1853–1854
Leydig cells 1812, 1816, 1817, 1875
LFY gene 715
L' Hernault S W **1871–1872**, **1872–1873**, **1873–1874**, **1876**
Lhx1 gene 1096, 1818
Lhx2 gene 1097
Lhx3 gene 1097
Lhx4 gene 1097
Lhx5 gene 1096
Lhx6 gene 1097
Lhx8 gene 1097
Lhx9 gene 1097
library **1095**
see also gene library; genomic library
lic operon 1079
Lieber M R **1731–1732**
Liebig, Justus 522
LIF *see* leukemia inhibitory factor (LIF)
Li–Fraumeni syndrome (LFS) *569–570t*
breast adenocarcinomas 11
breast cancer 243
p53 gene 571
sarcomas 1770
ligation **1096**
light chains *see* immunoglobulin gene superfamily

light receptor kinases *see* photomorphogenesis in plants
Lilium henryi 860, 2022, 2025
Liljas A **78–79, 125–126, 323–324, 608, 1184–1185, 1719–1723, 1723–1729, 1986–1989, 1999–2002**
Lilley D M J **73–74, 575–577, 915–916**
Lim1 gene 1814, 1818
lim-4 gene 1097
lim-6 gene 1097
lim-7 gene 1097
limb development **1099–1103**
  anterior/posterior (A/P) axis
    mechanisms 1100
    ZPA (zone of polarizing activity) 1100
  apical ectodermal ridge (AER) 1099
  asymmetries 1100
    anterior/posterior (A/P) axis 1100
    dorsal/ventral (D/V) axis 1100
    function 1100
    mechanisms 1100
    proximal/distal axis 1100
  bmps (bone morphogenetic proteins) 1101
  cell death 1101
  characteristics *1099f*, 1100
  dorsal/ventral (D/V) axis
    mechanisms 1100
    wnt-7a glycoprotein 1100
  endochondrial ossification 1101
  fibroblast growth factor (fgf) family 1099
  malformations
    Greig cephalopolysyndactyly (GCPS) 1101
    nail–patella syndrome (nps) 1101
  morphology 1099
  regulatory pathways 1101, *1101f*
  tetrapod limb 1099
limb girdle muscular dystrophies *1263f*, 1263–1264
LIM domain genes **1096–1099**
  cLMO (cytoplasmic LIM domain) proteins 1097
  Enigma family 1098
  Islet subfamily 1097
  Lhx1 subfamily 1096
  Lhx2 subfamily 1097
  Lhx3 subfamily 1097
  Lhx6 subfamily 1097
  LIM-kinase 1097
  Lmx subfamily 1097
  nLMO family 1097
  paxillin family 1098
  structure 1096
  transcription factors 1096
Limon J **1167–1168**
Limulus polyphemus 672
lin-4 gene 928–930, 2003
lin-5 gene 301
lin-6 gene 301
lin-11 gene 1096
lin-12 gene 306, *307f*, 963, 1313
lin-14 gene 306, *307f*, 928–930, *929f*, 2003
lin-26 gene 317
lin-28 gene 928–930
lin-29 gene 928–930
lin-32 gene 1310
lin-42 gene 928, 532
LIN-6 protein 301
LIN-12 protein 306, *307f*
LIN-14 protein 306, *307f*
Linanthus parryae 528
Lindegren, Carl 851, 1322
  linkage maps 1322
  Saccharomyces cerevisiae 1761
LINE **1103–1104**, 1699, 1702, 2022
  process 1104
  repetitive DNA sequences 1672
  retroviruses 1104
  reverse transcriptase 1104
  selfish gene 1104
  see also retroposon; SINE
lin genes 255
linkage 353–354, **1104–1105**
  centromere 1104
  detection 1104
  independent assortment 1018
  LINKAGE (computer program) 357
  linkage map 1105
  LIPED (computer program) 357
  MAP-MAKER (computer program) 357
  prokaryotes 1105
  recombination frequency 1105
  somatic cell hybrid analysis 359–360
  statistical analysis 356–357
  X chromosome 354
  see also concordance
LINKAGE (computer program) 357
linkage disequilibrium 1055, **1105**, 1518
  see also gametic disequilibrium
linkage group **1106**
linkage map **1106–1109**
  Bayesian analysis 204
  composite maps 1107
  construction 1106
  Escherichia coli 657
  haplotype 912
  homology relationships 1109
  human linkage maps
    CEPH families 1107
    Co-operative Human Linkage Centre (CHLC) 1107
    Genethon Human Genome Research Centre 1107
    genetic linkage map markers 1108
    Genome Data Base 1107
    Human Genome Project 1108
    Marshfield Center for Medical Research 1107
    radiation hybrid maps 1108
    Stanford University genome site 1108
    utilization 1108
    Whitehead Institutes genome site 1108
  interference 1046
  Lindegren, Carl 1322
  map distance 1140
  map expansion **1140–1141**
  mapping function 1141
  marker 1146
  meiosis 1106
  mouse linkage maps
    backcross 1107
    crossover events 1106
    Encyclopedia of the Mouse Genome 1108
    haplotype analysis 1106
    intercross 1107
    Mouse Genome Informatics (MGI) website 1107
    Mus musculus domesticus 1107
    Mus spretus 1107
linkage map (*continued*)
    utilization 1108–1109
  purpose 1106
  virtual maps 1109
linker DNA **1109**
Linnaeus 1860, 1864
LIPED (computer program) 357
lipid-containing phages 184
lipoma and uterine leiomyoma **1109–1117**
  lipoma
    chromosome abnormalities 1114
    clinical features 1114
    cytogenetics 1114, *1115f*
    HMGIC/LPP fusion transcripts *1116f*, 1116–1117
    LIM domain genes 1116
    molecular genetics 1116
    multiple tumors 1115
    pathology 1114
    soft tissue tumors 1114
  sarcomas 1768, *1769t*
  tumor-type specific genes 1109–1110
  uterine leiomyoma
    architectural transcription factors 1113
    chromosome abnormalities 1110
    cytogenetics 1110, *1111f*
    fibroids 1110
    gene expression 1113, *1113f*
    HMGI subfamily 1111
    HMG protein family 1111, *1112f*
    pathology 1110
    treatment 1110
    trisomy 21 1110
liposarcoma *see* myxoid liposarcoma
Lippman, Fritz 1290
Lisch nodules 1307
lit-1 gene 617
Lithgow G J **21–23**
Little, Clarence Cook 278, 397, **1117–1118**
  coat color mutations 1118
  inbred strain 1117
  Jackson Laboratory 1118
  mouse genetic nomenclature 1337
  mouse genetics 1117–1118
Littlefield, John 359
little (*lit*) mouse mutation 592
liver glucose-6-phosphatase enzyme (G6PC) 2120–2122, *2121f*
Li W-H **1080, 1212, 1229–1233**
LKB1 gene *2084–2085t*
Lloyd A C **65, 752–753**
LMO family of genes *1118t*, **1118–1119**
  chromosome translocations 1118
  embryonic stem (ES) cells 1118
  LIM domain genes 1118–1119
  LIM-Only genes 1118
  LMO1–4 gene 1097
  LMO1 gene 1118
  LMO2 gene 1118–1119, *1119f*
  LMO3 gene 1118
  LMO4 gene 1118
  protein interaction 1118–1119
  transcriptional regulators 1118
  tumorigenesis 1119
  see also LIM domain genes
LMP1 protein 642–643, 1289
LMP2A protein 642–644, 1289
LMP2B protein 642–644, 1289
Lmp2 gene 1640
Lmx1a gene 1097
Lmx1b gene 1097, 1101
Lobban, Peter 1637

locus **1119**
Locus Specific Databases 1188
Locusta migratoria 1646, *1646f*
locusts 319
logarithm of the odds (LOD) score 204, 356–357, 1106, **1119–1120**
Long A **263–264**
long branch attraction 1417–1418
long interspersed nuclear elements *see* LINE
long-period interspersion **1120**
long-range restriction maps *363t*
long terminal repeats (LTRs) **1120, 2021**
Lönnig W-E **199–203, 2020–2033**
loss of heterozygosity (LOH) **1120–1121**
  adenocarcinomas (ADCs) 10
  allelotyping 1121
  cancer research 1609
  chromosomal deletions 1121
  definition 1120
  detection 1120
  LOH analysis 1121
  lung cancer 1123
  positional cloning 1121
  rats 1609
  tumor suppressor genes 1121
Loteae 1083
Lotus corniculatus (birdsfoot trefoil) 1121
Lotus japonicus 1083, **1121–1122**, *1903f*
  genome resources 1122
  genome sequencing 1122
  Leguminosae 1121
  model plant 1122
  molecular genetic analysis 1122
  mycorrhizal fungi 1122
  nitrogen-fixing symbionts 1122
  nodulation 1122, 1122
  nodulins 1334
  origins 1121
  plant–microbe interaction 1122
  plant mutants 1122
  rhizobia 1122
Louis E J **2160–2161**
Lovett P S **1078–1079, 1612–1614**
Lovett S T **1932–1933**
Low D A **1442–1444**
Low K B **133, 638, 1446, 1447, 1490–1491**
Lox gene 2003
loxP gene 482
loxP site 2039
loxR site 2039
LPP gene 1110, *1113f*, *1116f*
LTRs *see* long terminal repeats (LTRs)
LUG gene 715
lung cancer **1122–1123**
  chromosome abnormalities 1122
  cytogenetic abnormalities 1123
  cytogenetic analysis 1123
Lu P **39–40**
Lupinus angustifolius 201
Lupinus luteus 201
Lupinus spp. (lupin) 1083
Luria–Delbrück experiment 522, 1124, **1124–1125**
  Delbrück, Max 1124
  Luria–Delbrück Fluctuation Test 1125
  Luria, Salvador E. 1124
  mutational studies 1124–1125
  phage-resistant bacteria 1124–1125
Luria–Latarjet (L–L) experiment 1124
Luria, Salvador E. 452, 522, 554, 925, **1123–1124**, 1921
  American Phage Group 1123
  bacteria/bacteriophage studies 169, 1123
  bacteriocines 1124
  Dulbecco, Renato 589
  Escherichia coli 654
  host-restriction modification 1124
  Human, Mary L. 1124
  Latarjet, Raymond 1124
  Luria–Delbrück experiment 522, 1124, **1124–1125**
  Luria–Latarjet (L–L) experiment 1124
  Nobel Prize 523, 925, 1123
  radiobiological target theory 1124
  restriction and modification 1692
  selection process 1709
  Watson, James Dewey 2133
Lusetti S **1618–1623**
Lush, J.L. 1797, 2142
luteinizing hormone (LH) 1875
Lutheran blood group **1125**
Luzula 322
Luzzatto L **683**, 881
LW antigen 1713
Lwoff, André 212, 1130–1131, 1237, 2109
Lwoff–Tournier system 2112
lycopene 1125
Lycopersicon esculentum (tomato) **1125–1127**
  C.M. Ricks Tomato Genetics Resource Center (TGRC) 1126–1127
  gene function 1126
  genetic maps 1126
  hybridization 1126
  introduction to Europe 1125, *1126f*
  lycopene 1125
  physiological studies 1126
  resources
    GRIN database 1127
    SolGenes database 1127
    TGRC database 1127
  restriction fragment length polymorphism (RFLP) 1126
  role in genetic studies 1127
  structure 1126
  systemic acquired resistance (SAR) *1914t*
  tomato genome 1126
  transgenic plants 1126
  use in genetic studies 1125–1126
  wild populations 1127
Lyell, Charles 514
LYL1 gene 1930–1932
Lyme disease 80
lymphoblastoid cell lines (LCLs) 641–642
lymphocryptovirus 641
lymphoepithelioma 1289
Lynch syndrome 10, *569–570t*
Lyn protein 1883
Lyon hypothesis *see* X-chromosome inactivation
Lyon, Mary 629, 1337, 2148
Lyon M F **579, 1367, 1970–1971, 2148–2149**
Lysenko, Trofim **1127–1130**, *1128f*
  agrobiology 1128
  environmental manipulation 1128
  Lysenko Affair 1129
  Lysenkoism 1128
    Michurinist 1129
    Michurin, Ivan V. 1129
    theory of mentoring 1129
  purge of Soviet genetics 1129
Lysenko, Trofim (*continued*)
  Soviet genetics 1129
  study of heredity 1127
    Darwinian pangenesis 1128
    Lamarckism 1073, 1128
    Marxist ideology 1128
  VASKhNIL (Lenin All-Union Academy of Agricultural Sciences) 1129
  vernalization 1128
lys gene 1786
lysine 972, **1130**, *1130f*
lysis **1130**
  see also phage Mu
lysis inhibition 1926
lysogeny **1130–1131**
  Bacillus megaterium 1131
  bacterial DNA 1131
  bacteriophage growth 1130
  Delbrück, Max 1130
  Gutmann, Antoinette 1130–1131
  Kaiser, Dale 1131
  lambda (λ) integration 1131
  Lwoff, André 1130–1131
  lysogenic state 1131
  phage induction 1131
  phage λ 1131
  prophage 1131
  Wollman, Elie 1131
  Wollman, Jacob 1131
  see also phage Mu
lysozyme 1130, **1131–1132**
  chemotherapeutic properties 1131
  endolysins 1131–1132
  Fleming, Alexander 1131
  murein structure 1131
  occurrence 1131
  peptidoglycan 1131
lytic phage **1132**
Lyttle, Terry 490

M

M33 gene 1814, 1818
m^5C *631f*, 631–632
m^5CpG 636
mab-3 gene 532, 1813
mab genes 255
MacCloud, Colin 654
MacDowell, E.C. 1118
MacDowell, L. 1118
Macgregor H C **249–250, 1073–1077**
Machado–Joseph disease 130–131, *2050t*
macronucleus **1135–1136**
  amitosis 1135
  macronuclear development **1133–1135**, 1135
    chromatin structure 1134–1135
    chromosome fragmentation 1133, *1134f*
    ciliates 1133–1135
    conjugation 1133
    DNA rearrangement 1133, *1134f*
    DNA scrambling 1134
    internal eliminated sequences (IES) 1133, *1134f*
    telomeres 1133
  minichromosomes 1135
  size 1135
  telomeres 1136
macrosatellites 1205
MADS-box genes 715
Mager J **1599–1601**
Magnaporthe grisea 1914t

Magnello M E **63–64**
Magnuson T **1599–1601**
Maher E **81–81, 1697–1698, 2122–2123**
Maine E M **1373–1374**
Maitra A **9–12**
maize 464
  Ac/Ds superfamily 2026
  CACTA superfamily 2026–2027
Maize Genetics Cooperation Stock Center 849
Maize Genome Database (Maize DB) 849
Maizels N **1640–1642**
major histocompatibility complex (MHC) 668, 730, **1136**, 1251, 1640, 1847
Makova K **1080, 1212, 1229–1233**
malaria 80, 338
Malcolm S **209, 942**
MALDI *see* matrix-assisted laser desorption/ionization
Malécot, Gustave 995, 1055
malignant fibrous histiocytoma 1768, *1769t*
Maloy S **156–163**
Malthus, Thomas Robert
  Essay on the Principle of Population 515
  fitness 703
maltose-binding protein (MBP) 1553–1554
Malvaceae 238
mammalian (mouse) genetics **1137–1140**
  background 1137
  chromosomal aberrations 1139–1140
  cloning 1138–1139
  cloning vectors 1138
  DNA markers 1139
  genetic databases 1139
  linkage maps 1138
  mammalian genomes 1137–1138
  model organisms 1138
  Mus musculus musculus 1137–1140
  mutant phenotypes 1139–1140
Mange, Arthur 490
Manhattan metric **1189**
Maniatis T **1878**
Maniatis, Tom 630
mannose-binding lectin 429
Mansouri A **1426–1428**
MAOA gene 213
map distance **1140**
map expansion **1140–1141**
MAPK *see* kinases
MAPKK *see* kinases
MAPKKK *see* kinases
MAP-MAKER (computer program) 357
mapping function **1141–1144**
  centimorgan (cM) 1141
  circular functions 1143
  counting model 1142, *1143f*
  definition 1141
  Haldane's interference function 1142, *1142f*
  Haldane's mapping function 357, 360, 1141, *1142f*
  Haldane's no-interference function 1141, *1142f*, *1143f*
  Kosambi's function 1142, *1143f*
  linkage map 1141
  recombination frequency 1141
  Sturtevant's function 1142, *1142f*, *1143f*
  see also centimorgan (cM)
mapping panel **1144**
  mouse linkage maps 1144
  radiation hybrid maps 1144
  usefulness 1144
Marcus R **1649**
Marfan, Antonin 90
Marfan syndrome 90, **1144–1146**
  cause 1145
  clinical features 1144–1145
  Ehlers–Danlos syndrome (EDS) 605
  fibrillin 1145
  management 1145
  microfibrils 1145
  prevalence 1145
marine amphibians 291
marine invertebrates 291
marker **1146**
  linkage maps 1146
  restriction endonucleases 1146
  restriction fragment length polymorphism (RFLP) 1146
  restriction mapping 1146
marker effect **1146–1147**
  mismatch repair 1146
  short patch repair 1146
marker rescue **1147**
Markov chains 392
Marshall-Graves, J. 636
Marshfield Center for Medical Research 1107
Martes americana 840
Martinez Arias A **1415**
Marvin, Don 555
masked mRNA **1147–1148**
MASP1 gene *430t*, *431t*
MASP-1 protein 429
MASP2 gene *430t*, *431t*
MASP-2 protein 429
Masson P H **890–894**
mat1 gene 962, 1154
mat1 locus 1154, *1155f*
Mat2 gene 962
mat2 locus 1154, *1155f*
mat3 locus 1154, *1155f*
*MAT*a gene 276–278, 1152, 1154, 1398
*MAT*α gene 276–278, 1152, 1154, *1155f*, 1398
maternal effect **1148–1149**
  cryptic maternal effect 1149
  embryogenesis 1148
  lethal mutation 1149
  maternal necessity 1149
  maternal sufficiency 1149
  mutational studies 1149
maternal inheritance **1149–1151**
  chloroplasts 338
  diagnosis of mitochondrial DNA (mtDNA) diseases 1151
  homoplasmy 1150
  mitochondrial bottlenecks 1150
  mitochondrial diseases 1149–1150
  mitochondrial DNA (mtDNA) 1149
  parent of origin effects 1150
Mather, Kenneth 512
Mathew C **683**
Mathur J **2045–2048**
mating table 912, *912t*
mating types **1151–1153**
  asymmetric cell division 1156
  Coprinus cinereus 1153
  determination 1152
  genetic mechanisms 1152–1153
  incompatibility systems 1152
  loci
    cell determination 276
    HML locus 1154, *1155f*
    HMR locus 1154, *1155f*
    HO locus 1154
mating types (*continued*)
    mat1 locus 1154, *1155f*
    mat2 locus 1154, *1155f*
    mat3 locus 1154, *1155f*
    MAT locus 276, 1154, *1155f*
    Saccharomyces cerevisiae 1154
    Schizosaccharomyces pombe 1154
    yeast 276
  mat1 imprinting 1156–1157
  Neurospora crassa 1153
  pheromones 1152
  Saccharomyces cerevisiae 1152
  selfing 1152
  sexual cycle 1152
  switching **1153–1157**, *1155f*
    asymmetric cell division 1155
    donor choice 1157
    double-strand break 1156
    epigenetics 636, *637f*
    gene conversion 1154, *1155f*
    loci 1154, *1155f*
    mat1 imprinting 1156
    mechanisms 1156
    regulation 1155
    Saccharomyces cerevisiae 1154–1157
    Schizosaccharomyces pombe 1154–1157
    transposition model 1154, *1155f*
    yeast **1153–1157**
  yeasts 1154
MAT locus 1152, 1154, *1155f*
matR gene 238
matrix-assisted laser desorption/ionization 1576
Matthaei, Heinrich 1328
Matthews, K.A. 847
Maxam, Allan 572, 878
Maxam–Gilbert sequencing *see* DNA sequencing
maximum likelihood **1157–1160**
  Bayesian hierarchical model 1159
  conditional likelihood 1158, *1158f*
  empirical Bayes model 1160
  evolutionary rate 1157–1160
  genome sequencing 1160
  Kullback–Leibler information content 1158
  likelihood of a tree 1158, *1158f*
  Markov process 1159
  model comparisons 1159
  parsimony 1417
  phylogeny 1465
  statistical techniques 1157–1160
maximum parsimony 1465
Maxson L **1203**
Mayer–Rokitansky–Kuster–Hauser syndrome 1024
Mayetiola destructor 410
Mayr E **39, 62, 384, 958, 1238–1239, 1414, 1860–1864, 1864–1869, 1904, 1934–1937**
Mayr, Ernst 1861
  evolution 664
  The Growth of Biological Thought: Diversity, Evolution, and Inheritance 1298
  population thinking 515
  reproductive isolation 1679
May, S. 852
MBL2 gene *430t*, *431t*
MBP *see* maltose-binding protein (MBP)
Mc1r gene 400
MC29 (avian myelocytomatosis virus) **1160**
MCAK protein 1225

McCarty, Maclyn 654, 815
McClintock, Barbara 42, 329, 464, 1029, **1161–1162**, 2016
contributions
Ac (Activator) locus 1161
discovery of transposons 1161
Ds (Dissociation) locus 1161
epigenetic regulation 1162
maize genetics 1161
Suppressor-mutator (Spm) family 1161
epigenetics 629
Greighton, Harriet 1161
Nobel Prize 1162, 2020
McCluskey, K. 849
McGinnis, Bill 958
McKee S **1697–1698**
McKusick, Victor 365, 1180
McLeod, Colin 815
mClk gene 389
MCP gene *430t*, *431t*
MCP protein 433
Mcr1 (melanocortin receptor 1) gene 1470
mCry gene 389
MDM2 proto-oncogene 1772
MDR1 gene 1431
MDR3 gene 1431
mealybugs 927
mec-3 gene 1096, 1311, 1322
mec-4(d) gene 314–316
mec-4 gene 314–316
mec-6 gene 317
mecA gene 75
mec gene 255, 1687
Meckel syndrome 724
MeCP2 633
MEDEA gene 1476
medea mutation 1476
Medicago (alfalfa) 1083, 1330
Medicago littoralis 1162
Medicago sativa 1162
Medicago truncatula 1083, **1162**, 1334
medical genetics *see* clinical genetics
Meinke D W **1780–1782**
Meins F **1471–1474**
meiosis **1163–1165**
adjacent/alternate disjunction 18
alleles 38
aneuploidy 66
Aspergillus nidulans 108
balanced translocation 189, *189f*
benefits 1163
biological costs 1163–1164
bouquet stage 369–370, *371f*
Caenorhabditis elegans 1373
chimera *329f*, **330–333**
chromatid 339–340, 339
chromosome 344
chromosome bridge 350–351
chromosome disjunction 1345–1346, *1347f*
chromosome pairing, synapsis **369–372**, *370f*
conversion gradient **465–468**
crossing-over **488**
Darlington, Cyril Dean 512
definition 1163
diploidy 535
exchange 673
female meiosis 1165
fertilization 689
first and second division segregation 698, *699f*
meiosis (*continued*)
gametes **748–749**
gene conversion 774, *774f*
genetic effects 1163
genetic recombination 841
genetic variation 1163
germ cell 874–875
Hardy–Weinberg law 913
heterochromatin 927
heteroduplex DNA 931
holocentric chromosomes 956–957, *957f*
homologous chromosomes 963–964
hybrid sterility (mouse) 986
independent assortment 1017–1018
infertility 1021
interference 1046
kinetochore 1064
lampbrush chromosomes 1073
linkage 1104
linkage map 1106
male meiosis 1165
marker chromosomes 2004–2005
meiotic drive 986
meiotic mistakes 1164
Mendel's laws 1175, *1175f*
mouse 1374
negative interference 1303
nondisjunction 1345
nonreciprocal exchange **1350**
nuclear envelope (NE) 1355
postmeiotic segregation **1531–1532**
recombination 339, 1163
recombination frequency 1106
recombination models 1643
Saccharomyces cerevisiae 1154
Schizosaccharomyces pombe 1154, 1779
segmental interchange 1787
sex determination 1164–1165
spermatids **1871–1872**
spermatocytes **1872–1873**
spermatogenesis 634, 1875
spermatogonia **1876**
synaptonemal complex 1910
tetrad analysis **1952–1957**
translocation 2004, *2004f*
transmission genetics 2011
unequal crossing-over 2095
uniparental inheritance 2097
zygote **2165–2166**
see also chromatid; chromosome movement
meiotic drive (mouse) **1165–1167**
In chromosome 1166
chromosome 17 1166
deviation from Mendelian inheritance (DMI) 1167
Drosophila melanogaster 1166
Drosophila pseudoobscura 1166
Mus musculus musculus 1166
segregation distortion 1165
t haplotype 1166, 1921
ultraselfish gene 1165
meiotic product **1167**
Mekalanos, John 697
melanin 399–400, 1469
melanocytes 1469
melanoma **1167–1168**
aneuploidy 1168
chromosomal aberrations 1167–1168
cytogenetic studies 1167–1168
metastatic melanoma 1167
murine melanoma 1168
melanoma (*continued*)
transplantable melanomas 1168
melanosomes 1469
Melilotus (sweet clover) 1083, 1330
Melnick M **385**
melting temperature **1168**
MEN1 gene 1257
MEN 2 *see* multiple endocrine neoplasia
Mendel, Gregor Johann 521, 540, 553, **1168–1171**, 1298
Darwin, Charles 516
early life 1170, 1173
experimental process 1169, 1172, 1173–1177
Experiments in Plant Hybridization 1172, 2009
Galton, Francis 1170
genes 759
Lysenko, Trofim 1128
Nägeli, Karl Wilhelm von 1169, 1177
Phaseolus nanus 1177
Phaseolus vulgaris 1177
Pisum sativum (garden pea) 1169, 1470, *1470t*
plant genetics 87
transmission genetics 2009–2011
Mendelian genetics **1180**
Mendelian population **1181**
Mendelian ratio **1181**
Mendelian Trait Database 2013
Mendelian traits **1180**
Mendel's laws 590, **1171–1180**
conclusions 1177, *1177t*
current research 1179–1180
disjunction 538
disruptions *1177t*, 1178–1179
dominance 1178
epistasis 1178
genetic heterogeneity 1178
lethal alleles 1178
linkage 1178
multiple alleles 1178
phenotypic expression 1178
pleiotropy 1179
uniparental disomy (UPD) 1179
early genetic research 1172–1173
Bateson, William 1173
Festetics, Imre (Emerich) 1173
Flemming, Walther 1172
Knight, Andrew 1172
Kölreuter, J.G. 1172
Schleiden, Matthias 1172
Schwann, Theodor 1172
Virchow, Rudolf 1172
experimental process 1173–1177
gametic selection 1296
independent assortment 1017
law of independent assortment 1176, *1176f*
law of segregation 1175, *1175f*
laws of inheritance 1171
meiosis 1175, *1175f*
Mendelian genetics 1180
Mendelian Inheritance in Man 1180
Mendelian traits 1180
Mendel's first law 1175, *1175f*
Mendel's second law 1176, *1176f*
Pisum sativum
dominance 1174, *1174t*
Nordic Gene Bank 1179
phenotypic ratio 1174, *1174t*
Pisum Genetic Stocks Collection 1179

Mendel's laws (*continued*)
recessiveness 1174, *1174t*
traits 1174, *1174t*
Punnett square 1176, *1176f*
reaction to concept 1177
rediscovery 1178
see also Mendel, Gregor Johann
Menke syndrome 498
Merickel S K **938–942**
Merlin protein 1308
merodiploidy 677
merosin *587f*
merP promoter 164
Merriam J **759–761, 839, 1017–1018, 1026–1027, 1180, 1952**
MerR protein 164
Meryon, Edward 586
Meryon's disease *see* Duchenne muscular dystrophy
mes-2 gene 619
mes-6 gene 619
Meselson, Matthew 245, 257, 1059, 1182
Meselson–Radding model *1181f*, **1181–1182**, 1643–1644
Meselson–Stahl experiment **1182–1184**
conservative replication 1182, *1183f*
distribution of DNA molecules 1183, *1183f*
inference reasoning 1183
replication process 1182–1184
semiconservative replication 1182, *1183f*
Mesorhizobium loti 1122, 1902
messenger RNA (mRNA) 249, **1184–1185**, 1358
chloroplasts 338
codons **1184**
DNA cloning 545, 546
DNA replication 541
dynamic mutations 595
genetic code 1184
histone genes 948–949
housekeeping genes 978
introns 1184
Jacob, François 1059
lac operon 1070
leader peptide 1078
masked mRNA 1147
Northern blotting 1350–1352
Rh blood group system 651
rRNA 1184
Shine–Dalgarno sequence 1185
transcription 1184
transcription process 1184
translation 323
translation process 1185
tRNA 1184
Watson–Crick model 1184
see also Cap; poly(A) tail; polycistronic mRNA; transfer RNA (tRNA)
metabolic disorders **1187–1188**
expression studies 1188
galactosemia 1187
Human Gene Mutation Database (HGMD) 1188
Human Genome Organizations Mutation Database Initiative 1188
Locus Specific Databases 1188
missense mutations 1188
mutational studies 1188
Online Mendelian Inheritance in Man (OMIM) 1188
phenylketonuria (PKU) 1187
metacentric chromosome **1189**
metallothionein gene 978
Methanobacterium 2114
Methanobacterium thermoautotrophicum 92, 1566
Methanococcus jannaschii 92, 192, *946f*, 1693
Methanococcus voltae 93, 2114
MET-HGF/SF system **1185–1187**
activation 1185–1186
angiogenesis 1186
cancer, role in 1186
chromosome 17q21 1185
domain structure 1185, *1185f*
hepatocyte growth factor/scatter factor (HGF/SF) 1185
MET locus 1185
pathways 1185
physiological role 1186
receptors 1185, *1185f*
methionine 58, 323, **1189**
Methuselah gene 22, 212
methylated DNA binding domain (MBD) 633
methylation induced premeiotically (MIP) 636
5-methylcytosine (m^5C) 557, *631f*, 631–632, 977
methylmethanesulfonate 557
MET locus 1185
MET proto-oncogene 1660
Métraux J P **1913–1917**
metrics **1190**
see also four-point metric; Manhattan metric; ultrametric
Metz, Charles 629
mex-1 gene 619
mex-3 gene 619
Mexican axolotl 1304
Mexican wolf 461
Meyerhoff, Otto 1365
Meyerowitz, Elliot 88
M-FISH 361–362, *362f*, 368, *368f*, 980, 1002–1003
Mgf (mast cell growth factor) gene 398, 525, 1469
*MIC2*gene 2157
MIC and MIC-M classification of leukemia **1190–1191**
acute lymphoblastic leukemia (ALL) 1190–1191
acute myelogenous leukemia (AML) 1190–1191
cytogenetic analysis 1190–1191
French-American-British (FAB) cooperative group 1190–1191
see also FAB classification of leukemia; WHO classification of leukemia
Michiels J **707–709**
Michurin, Ivan V. 1129
microarray technology **1191, 1200–1201**
microbial genetics **1191–1196**
advantages 1192
cis–trans tests 1194
cistron 1194–1195
epistasis 1195
Escherichia coli 1192
fine-structure genetics 1192
genetic analysis 1192
genetic mapping
gene transfer 1194
homologous recombination 1194
microbial systems 1191
mutagenesis
microbial genetics (*continued*)
lac system 1193
null phenotypes 1193
phenotypes 1192–1193
transposable elements 1193
regulation 1195–1196
gene fusion 1195
lacZ fusion 1195
regulons 1195
reversion 1195
second-site suppression 1195
Microbial Genetics Center (Japan) 850
microbial genomics **1196–1203**
bioinformatic analysis 1197
closure 1197
current applications 1197–1200
future applications 1200–1201
gene prediction programs 1197
Haemophilus influenzae 1196
random shotgun sequencing 1196–1197
reconstruction 1197, *1198f*
Vibrio cholerae 1196, *1198f*
Microbial Germplasm Database (MGD) 852
microchromosomes **1203**
Micrococcus luteus 147f, 192, 1665
microcomplement fixation (MC'F) **1203**
2-μ plasmid 2162
micronucleus **1204**
ciliates 1204
macronuclear development 1204
Oxytricha nova 1204f
sexual reproduction 1204
Stylonychia lemnae 1204
Tetrahymena thermophila 1204
see also macronucleus
microorganisms
alarmones 281–282
cAMP nucleotide 281–283
diauxy 281
kinases (protein kinases) 282
microsatellite 250, **1205**
deletion mapping (mouse) 526
dogs 268
genetic distance 830
linkage maps 232
macrosatellites 1205
midisatellites 1205
minisatellites 1205
repetitive DNA sequences 1672
tandem repeats 1205
unequal crossover 1205
microsatellite instability 1285
microscopy *see* electron microscopy
Microspora 1882
microtubules 287, 288, *290f*, 320, **1205**
Microtus (American voles) 688
midisatellites 1205
Miescher, Johan Friedrich 553
mig genes 255
Miklos G L G **859–863**
Miller H I **224–227**
Miller, Jeffrey 1710
Miller J H **3, 208–209, 210–211, 212, 383, 530, 537, 556–557, 558, 571–572, 572, 698, 722, 792, 932, 976, 977, 1070–1071, 1181, 1237–1238, 1286–1287, 1459, 1567, 1599, 1675, 1677–1678, 1809, 1853, 1999**
Miller O J **914**
Millettieae 1083
Mimosa 1083
Mimulus cardinalis 1683t

Mimulus cupriphilus 1683t
Mimulus guttatus 1683t
Mimulus lewisii 1683t
Mimulus micranthus 1683t
Mindich L **1744–1746**
Miniature Inverted Repeat Transposable Elements (MITEs) 1673
minicells **1205**
minichromosome 1135, **1205**
minimal residual disease **1206–1211**, 1506
  detection techniques
    cytometric immunophenotyping 1206
    cytomorphological 1206, *1206f*
    polymerase chain reaction (PCR) 1206, *1206f*, *1206t*
    RT-PCR analysis 1208, *1208f*
  hemopoietic malignancies 1206–1211
    acute lymphoblastic leukemia (ALL) 1206
    acute myelogenous leukemia (AML) 1206
    chronic myelogenous leukemia (CML) 1206
    follicular cell lymphoma (FCL) 1208
    non-Hodgkin's lymphoma (NHL) 1206
  monitoring methods 1209–1210, *1211f*
  quantification methods 1209
  targets
    chromosomal aberrations 1207–1209
    fusion genes 1207–1209
    Ig/TCR gene rearrangements 1206
    junctional regions 1206, *1207f*
minimum change **1212**
Miniopterus schreibersi 859
minisatellite 1205, **1212**
minute (*mi*) mutants 1927
mioC gene *1383f*
Mis gene 1818
mismatch repair (long/short patch) 955–956, 1182, **1213**, 1267, 1286, 1666–1668, 1842
missense mutations 57, 198, 1188, **1276–1277**
mistranslation **1213–1215**
  alternative readings 1214–1215
  start, stop codon 1214
  translation errors 1213–1214
Mitchell, Mary 1323
Mitchison, Murdoch 1777
MITF gene 1469
mit mutation 1223
mitochondria **1215–1217**
  biogenesis 1216
  DNA 1421–1422, 1466
  evolutionary origins 1216–1217
  function 1216
  genetics **1217–1219**
    editing 1218
    electron transport 1217
    genetic code 1218
    genomes 1217
    horizontal transfer 1217–1218
    Krebs cycle 1217
    maternal inheritance 1218–1219
    mitochondrial diseases 1218
    mitochondrial DNA (mtDNA) 1219
    recombination 1218
    shared coding 1217–1218
  genome 335–336
mitochondria (*continued*)
  organelles 1378–1379
  petite **1430**
  structure 1215–1216
  syncytium 1216
  TIM complex 1216
  TOM complex 1216
  translation system 1216
  universal genetic code 2099, *2099f*
  see also chloroplasts; protein secretion systems
mitochondrial DNA (mtDNA) **1219–1220**
  Acanthamoeba castellanii 1220
  definition 1219
  Dictyostelium discoidium 1220
  dogs 265
  evolutionary origins 1220
  gene organization 1220
  isolation 1219
  Plasmodium falciparum 1219
  Reclinomonas americana 1220
  replication 539, *539f*
  size 1219–1220
  structure 1219
mitochondrial genome **1220–1222**
  definition 1220–1221
  electron transport 1221
  function 1221
  gene expression 1222
  genetic code 1221
  Plasmodium falciparum 1221
  Reclinomonas americana 1221
  translation 1221
mitochondrial inheritance **1222**, *2014t*
mitochondrial mutants **1222–1224**
  angiosperms 1223
  filamentous fungi 1223
  heteroplasmy 1222
  humans 1224
  yeast 1223
mitogen-activated kinases (MAPK) *see* MAPK
mitosis 288, *289f*, **1224–1227**
  aneuploidy 66
  Aspergillus nidulans 108
  cell cycle 288, *289f*
  cell replication 1224
  centromere 321–322, 1225
  chromatid 339, 1225
  chromosome 1225–1226, 344
  chromosome break 350
  chromosome bridge 350–351
  chromosome movement 1226
  chromosome scaffold **372–373**
  chromosome segregation 288, *290f*, 300–301
  condensation 1225
  congression 1225
  contractile ring **464**
  cytokinesis 1226
  Darlington, Cyril Dean 512
  diploidy 535
  exchange 673
  gene conversion 774
  Golgi body 1225
  heteroduplex DNA 931
  holocentric chromosomes 956
  kinetochore 1064, 1225–1226
  mechanisms 1225
  mitotic spindle 289–290, 300–301, 1225
  mosaicism (human) 1240–1241
  motility 1225–1226
mitosis (*continued*)
  nondisjunction 1345
  nuclear envelope (NE) 1355
  nucleolus 1359
  recombination models 1643
  regulation 1226–1227
  research techniques 1224
  spermatogenesis 1875
  spindle microtubules 1225
  telophase 1225
  therapeutic targets 1227
  ubiquitin 2092
  zygote **2165–2166**
  see also cell cycle; chromatid; chromosome movement
Mitton J B **785–790**, **822–823**, **839–840**
MJD1 gene *2050t*
Mla gene 973
MLH1 gene 1285
MLH1 protein 1646, 1667, 1912
MLL gene 1092, **1227–1228**
  acute lymphoblastic leukemia (ALL) 1228
  acute myelodysplastic syndrome (MDS) 1228
  acute myelogenous leukemia (AML) 1228
  chromosomal translocations 1228
  homeotic genes 1228
  Hox genes 1228
  WHO classification of leukemia **2135–2136**
Mlo gene 973
mob gene 454
model organisms
  Arabidopsis thaliana 86, 87
  Bacillus subtilis 136
  Brachydanio rerio 1320
  Caenorhabditis elegans 612, 653, 934, 1320
  Dictyostelium discoidium 934
  Drosophila melanogaster 584, 653, 934, 1320
  Escherichia coli 653–655
  Mus musculus musculus 1138, 1320
  Neurospora crassa 653
  Paramecium aurelia 653
  Rattus norvegicus 1608
  Saccharomyces cerevisiae 934, 1153
  Schizosaccharomyces pombe 1153, 1776
Mod protein 1693
Moens P B **328–330**, **368–372**, **1163–1165**, **1645–1647**, **1910–1912**
Mohr, Jan 356
molecular clock **1229–1233**
  evolutionary rate 672
  fossil record 1229
  hominoid lineages *1231t*, 1231–1232
  hominoid slowdown hypothesis 1231–1232
  local clock *1230t*, 1230–1232
  macromolecules 1229
  molecular clock hypothesis 1229
  nonsynonymous substitutions 1230–1232
  nucleotide substitutions 1229–1232, *1230t*, *1231t*, *1232t*
  relative-rate test 1229–1230, *1230f*
  rodent lineages *1230t*, 1230–1232
  selective neutrality 1801
  synonymous substitutions 1230–1232
molecular drive **1233–1234**
  Drosophila melanogaster 1234

molecular drive (*continued*)
gene conversion 1234
hybrid dysgenesis 1234
repeated sequences 1233
tandem repetition 1233
transposable elements 1234
unequal recombination 1233
molecular genetics **1234–1235**
mom gene 1440–1441
monochromosomal somatic cell hybrids **1235**
monocistronic mRNA **1235**
monoclonal antibodies **1235–1237**
background 1235–1236
antigens 1236
epitopes 1236
heavy chains 1235
immunoglobulin gene superfamily 1235
light chains 1235
somatic rearrangement 1236
CDR grafting 1236
Epstein–Barr virus (EBV) 1236
humanized antibodies 1236
libraries 1236–1237
plasmacytomas 1236
Monod, Jacques 815, 1019, 1020, 1059, **1237–1238**
allostery 39, 1238
Audreau, Alice 1237
Chance and Necessity 1238
diauxic growth 1237
Jacob, François 1237
lactose metabolism 1070, 1237, 1652
Lwoff, André 1237
Nobel Prize 1070, 1238
operators 1376
operon theory 655, 1070, 1237, 1377
Pasteur Institute 1238
monomorphic locus **1238**
mononucleosis 641
monophyly 970, **1238–1239**
MONOPTEROS gene 1475, 1755
monopteros mutation 1475
monosomy 346, **1239**
Morata G **426–427**
Moraxella bovis 43
Moraxella lacunata 43
morgan *see* centimorgan (cM)
Morgan, Thomas Hunt 522, 530, 847, **1239–1240**
centimorgan (cM) 319
chromosome mapping 353–354
chromosome studies 1239–1240
The Development of the Frog's Egg: An Introduction to Experimental Embryology 1239
differential gene activity 1472
Drosophila melanogaster 584, 1239
genetic mapping 1239
linkage 2011
Lysenko, Trofim 1128
Mechanisms of Mendelian Heredity 1240
Mendel's laws 1239
Muller, Hermann Joseph 1256
Nobel Prize 354, 1240
Sturtevant, A.H. 1239
Mori I **1368–1370**
Morse H C **1248–1252**
Mortimer R K **1761–1763**, **1763–1765**
Mortimer, R.K. 851
Morton N E **1054–1055**, **1105**
Morton, Newton 356, 357, 490
morula *see* embryonic development, mouse; mouse
MOS **1240**
mosaicism (human) **1240–1242**
chromosomal aberrations 1241
chromosome specific behavior 1242
confined placental mosaicism (CPM) 1241
Down syndrome 583
hermaphrodite 925
Klinefelter syndrome 1066
mitosis 1240–1241
origins 1240–1241
preimplantation embryos 1242
timing 1241
trisomic zygote rescue 1241
trisomy 1241
uniparental disomy (UPD) 1241
Moschel R C **271–272**
Moscow Regional *Drosophila melanogaster* Stock Center 847
mos oncogene 1240
most recent common ancestor (MRCA) 392
mot genes 711
Mottus R **1523–1530**
mouse **1242–1247**
Abelson leukemia virus 250
background 1242–1243
breeding systems 1246
Cattanach's translocation 284–286
chimera *331f*, 331–332, *332f*
chromosomal aberrations 1244
chromosome 7 284–286
chromosome 17 1919
chromosome mapping 367, *367t*
circadian feedback loop *388f*
classical genetics **1247–1249**
The Biology of the Laboratory Mouse 1247
cloning 1248
DNA markers 1248
early years 1247
F_1 hybrid 1248
genetic mapping 1247
Genetic Variants and Strains of the Laboratory Mouse 1247
Haldane, J.B.S. 1247
interspecific backcross 1248
microsatellites 1248
Molecular Cloning: A Laboratory Manual 1248
Mus musculus musculus 1248
Mus spretus 1248
recombinant inbred strains 1247–1248
clock mutants 389
cloning 391
coat color mutations 397–398
colony management 1246
cryobiology 1246
deletion mapping **524–528**
embryogenesis 1245–1246
embryonic development **621–623**
embryonic differentiation 622–623
preimplantation phase 621–622, *622f*
zona pellucida 623
embryonic stem (ES) cells *331f*, 332
EST number *366t*
F_1 hybrid 1243
gene analogies
aging 1245
mouse (*continued*)
cancer 1245
cystic fibrosis 1245
epilepsy 1245
eye genetics 1245
heart disease 1245
juvenile diabetes 1245
muscular dystrophy 1244
obesity 1244
Parkinson's disease 1244
severe combined immunodeficiency 1244–1245
spinal cord injury 1244
genome size *367t*
germ cells 1245
hot spots of recombination **1640**
inbred strain 1243–1244
informatics 1246
knockout lines 332
nomenclature 1246
oogenesis **1374–1375**
differentiation process 1374–1375
estrus cycle 1374
germ cell production 1374
meiosis 1374
ovulation 1375
osteoporosis 1244
parthenogenesis 1419–1420
Pax genes **1426–1428**
piebald trait *1467f*, *1468f*, 1469
plasmacytomas **1482–1485**
postimplantation 623
proteome 1577
replication errors *1676t*
research strategies 1243
Robertsonian translocation 2007
selection experiments 1244
sex-reversed rearrangement *1253f*, **1253–1254**
spermatogenesis **1875–1876**
follicle stimulating hormone (FSH) 1875
gonadotropin-releasing hormone (GnRH) 1875
Leydig cells 1875
luteinizing hormone (LH) 1875
meiosis 1875
mitosis 1875
Sertoli cells 1875
spermiogenesis 1875
testosterone 1875
strains 250
targeted mutagenesis **1933–1934**
tetraparental **1957**
t haplotype **1919–1921**, *1920f*
transgenic 1933, 1989, 1991, *1991f*
translocations 284–286
Mouse Genome Database (MGD) 1337, 1636
Mouse Genome Informatics (MGI) website 849, 1144
mouse leukemia viruses (MLV) **1249–1252**
classification 1249, *1250f*
genetic vectors 1252
immunization 1252
infection mechanisms 1249–1251
MLV genome 1249, *1250f*
neoplasia *1250f*, 1251–1252
spongiform encephalopathy 1251
virology 1249
Mrak, Emil 1761

MRCA *see* most recent common ancestor (MRCA)
MRC Mammalian Genetics Unit (UK) 849
MRC Radiobiological Research Unit (UK) 1871
MRD *see* minimal residual disease
MRE11 gene 571
MRE11 protein 1645
mRNA *see* messenger RNA
MRP6 gene 1580–1581
Msh2 gene 2082
MSH2 gene 1285
MSH2 protein 1667
MSH3 gene 1285
MSH3 protein 1667
MSH6 gene 1285
MSH6 protein 1667
MSX2 gene *480t*
MTCP1 gene 114, 1094
mtDNA **1254**
  see also mitochondrial DNA (mtDNA)
mTim gene 389
MTS1 gene *2084–2085t*
Mucor 749
Mucorales 749
Mucronate mutant 1786
mudrA gene 2028
mudrB gene 2028
MuDR gene 2028
Mueller L D **1519–1523**
Muenke syndrome 479 *see* craniosynostosis
Muir–Torre syndrome 422
Mukai, Terumi 490
mulberry silkworm *see Bombyx mori*
mule **1255**
  hinny 1255
  sexual reproduction 1255
  uses 1255
  X-chromosome inactivation 1255
Muller, Hermann Joseph 490, 1129, 1240, **1255–1256**, 1409
  chromosome mapping 353–354
  contributions to genetics 1256
  Drosophila melanogaster 1256
  eugenics 1256
  evolution 1256
  genetic load 838
  Haldane–Muller principle 906
  Morgan, Thomas Hunt 1256
  Muller's ratchet 1518
  Nobel Prize 1255
  population genetics 1516
  position effects 1525
  postzygotic isolation 1685
  radiation crusade 1256
  reciprocal translocations 2003–2004
  Watson, James Dewey 2133
Müller-Hill B **1019–1021**
Müller-Hill, Benno 877
Mullerian ducts 925, 1024, 1812, 1816
Mullerian inhibitory substance (MIS) 1816, 1812
Muller's ratchet 640, 1518
Mulligan L M **262–263**, **920–921**, **1257–1258**, **1285–1286**, **1696**
Mullis, Kary 1499
multicolor fluorescent *in situ* hybridization *see* M-FISH
multicopy plasmids **1257**
multidimension chromatography 1577
multifactorial inheritance **1257**
  see also progeny testing
multiple endocrine neoplasia **1257–1258**
  multiple endocrine neoplasia type 1 (MEN-1) 1037, 1257
  multiple endocrine neoplasia type 2 (MEN-2) 262, 1257–1258, 1696
multiplicity of infection **1258**
multisite mutation **1258–1259**
Mu (Mutator) gene 2028
Muntiacus muntjac 859
Muntiacus muntjac reevesi 344
Muntiacus muntjac vaginalis 320, 344
Muntiacus reevesi 860
muntjac deer 320, *321f*, 344
Murgola E J **9**, **25**, **59–60**, **95**, **105–106**, **106**, **122–123**, **198–199**, **900–901**, **1088**, **1130**, **1447**, **1495–1496**, **1549**, **1809**, **1901**, **2002–2003**, **2054–2055**, **2101**, **2103**
Musa acuminata 2055–2056
Musa balbisiana 2055–2056
muscular dystrophies **1262–1264**, *1263f*
  Becker muscular dystrophy 1262, *1263f*
  congenital muscular dystrophy 1264
  distal muscular dystrophy *1263f*, 1264
  Duchenne muscular dystrophy (DMD) 1262, *1263f*
  Emery–Dreifuss muscular dystrophy 1262–1263, *1263f*
  facioscapulohumeral muscular dystrophy *1263f*, 1264
  limb girdle muscular dystrophies *1263f*, 1263–1264
  oculopharyngeal muscular dystrophy *1263f*, 1264
muskmelon *1914t*
Mus macedonicus 1261
Mus molossinus 988
Mus musculus 653
Mus musculus bactrianus 988
Mus musculus castaneus 526, 988, 1259–1260, **1261**
Mus musculus domesticus
  gametic selection 1296
  hybrid sterility (mouse) 985–986
  hybrid zone (mouse) 988
  linkage map 1107
  Mus musculus musculus 1259–1260
  Mus spretus 1262
Mus musculus molossinus 1260
Mus musculus musculus 526, 688, **1259–1261**
  classical genetics 1248
  evolutionary history 1259
  gametic selection 1296
  genetic divergence 1260
  genetic introgression 1259–1260
  geographical distribution 1259, *1260f*
  hybrid sterility (mouse) 986
  hybrid zone (mouse) 988
  inbred strain 1260
  laboratory strains 1260
  mammalian (mouse) genetics **1137–1140**
  Mus musculus castaneus 1259–1260
  Mus musculus domesticus 1259–1260
  Mus musculus molossinus 1260
  neuronal specification 1320
  origins 1259
Mus spicilegus 1261
Mus spretus 526, **1261–1262**
  classical genetics 1248
  geographical distribution *1261f*, 1261–1301
  hybrid sterility (mouse) 986
  inbred strain 1262
Mus spretus (*continued*)
  linkage map 1107
  Mus musculus domesticus 1262
  role in genetic studies 1262
Mustela erminea 840
mutagenic specificity **1264–1268**
  background mutations 1267
  chemical structure 1264
  DNA modification 1265
  DNA repair 1267
  electrophiles 1265–1267
  mismatch repair 1267
  mutational spectra *1266f*
  sequence-specific reactions 1265–1267
  target genes 1267–1268
mutagens **1268**
mutant allele **1268**
mutation **1268–1275**
  achondroplasia 1271
  auxotrophic 443
  back mutation **1275**
  categories 1269–1271
    addition/deletion mutations 1269–1270
    base substitution mutations 1269, *1269f*
    chromosomal 1271
    complex mutations 1270
    duplications 1270–1271
    insertions 1271
    inversions 1270
    multilocus deletions 1270
  cause 1272–1275
    damaged DNA 1272
    deamination 1272
    DNA polymerases 1272–1273
    misalignments 1273–1275, *1274f*, *1274f*
    misreading *1270f*, 1273
    unforced errors 1272
  Charcot–Marie–Tooth disease 1271
  cistron 383
  conditional lethality **441–443**, *443f*
  constitutive mutations **463**
  craniosynostosis 479–481
  definition 1268–1275
  DNA repair 1272
  Down syndrome 1271
  frequency 1271–1272, **1275**
  genotypes 1269
  genetic rescue 1995
  green fluorescent protein (GFP) 312
  hot spots 1271
  ion channels 316
  leaky mutation **1276**
  lethal 441
  missense mutations **1276–1277**
  multisite mutation **1258–1259**
  mutational site **1285**
  mutation rate 1271, **1277–1279**
  nonrandom distribution 1271
  null mutation **1277**
  paternal inheritance 1421–1422
  phenotypes 1269
  point mutation 1269
  polymorphisms 1272
  random insertional 1995
  repair mechanisms 1273, *1273f*
  reverse mutation 1275
  silent mutations **1279**
  spontaneous mutations **1279**
  targeted 1995
  transversion mutation **2042**

mutation (*continued*)
  trisomy 18 1271
  trisomy 21 1271
mutational analysis **1279–1284**
  p53 protein 1280
  positional scanning mutagenesis 1281, *1283f*
  protease enzymes 1280–1282, *1282f*
  protein stability 1284
  scanning mutagenesis 1281–1284, *1283f*
  tumorigenesis 1280
mutational site **1285**
mutation–drift balance 187
mutation frequency **1275**
mutation load **1276**
mutation rate **1277–1279**, *1278f*
  hot spots 1279
  mutation frequency 1277
  population history 1277
  retroelements 1278
  riboviruses 1277
  RNA phages 1744–1746
  spontaneous mutations 1277
mutation–selection balance 835
mutator phenotype **1285–1286**
  chromosomal instability 1285
  hereditary nonpolyposis colorectal cancer (HNPCC) 1285
  microsatellite instability 1285
  nucleotide excision repair (NER) genes 1285
mutators **1286–1287**
  biodiversity 1287
  mismatch repair 1286
  mutator phenotype
    hereditary nonpolyposis colorectal cancer (HNPCC) 1286
    xeroderma pigmentosum (XP) 1286
mu-Tbr1 gene *1944f*
mu-Tbx1 gene *1944f*
mu-Tbx2 gene *1944f*
mu-Tbx3 gene *1944f*
mu-Tbx4 gene *1944f*
mu-Tbx5 gene *1944f*
mu-Tbx6 gene *1944f*
mutD gene 1286
mu-T gene *1944f*
mut genes 1666
mutH gene 1286
mutL gene 1286
mutM gene 1286
mutS gene 1286
Mut system 143, *563f*, 932, 1146, 1667
mutT gene 1286
mutualism 411
mutY gene 1286
myasthenia gravis 1150
Myb oncogene **1287**
mycelium 1323
Myc gene 1470
myc locus *see* oncogenes
Mycobacterium tuberculosis 150t, 192, 586, 1891, 2039
  histidine operon *946f*
MYC oncogene 247, 1091, 1605, 1773
Mycoplasma capricolum 192, *403t*, 405
Mycoplasma genitalium 150t, 859, 1201
Mycoplasma pneumoniae 150t, 1201
myc proto-oncogene 1372
myeloid antigens 953
Myo5a (d) gene 524–525
Myo5a gene 398, 1469
myoclonic epilepsy (*EPM1*) 595, *596f*
myotonic dystrophy 357, 593, *2049t*
Myoviridae 181, 1922
Myrmecia pilosula 860
Mytilus edulis 786, *786f*, 823, *1522t*
Mytilus galloprovincialis 823
Myxococcus xanthus 147, 859, 1892
myxoid liposarcoma **1287**
myxoma virus growth factor (MGF) 627
Myxomycota 1883
Myxospora 1882

N

N_2 generation **1289**
nad1 gene 2041
nad2 gene 2041
nad5 gene 2041
nad gene 1221, 1224
Nagai K **1733–1735**
Ngeli, Karl Wilhelm von 63, 1169, 1177
nag operon 1896
NahG gene 1916
nail–patella syndrome (nps) 1101
Na^+ ion channel 315
NAP-1 343
Nash equilibrium 669
Nasmyth, Kim 636
Nasonia 730, 1685
nasopharyngeal carcinoma (NPC) **1289–1290**
  anaplastic form 1289
  causes 1289
  cytogenetic changes 1290
  desmosomes 1289
  Epstein–Barr virus (EBV) 643, 1289
  geographic distribution 1289
  lymphoepithelioma 1289
  p53 mutations 1290
  radiosensitivity 1289
  role of virus 1289
  tonofilaments 1289
nas regulon 1079
Nathans, Daniel **1290–1291**
  Arber, Werner 1290
  cleavage map 1290–1291
  Haemophilus influenzae 1290
  Howard Hughes Medical Institute 1291
  Johns Hopkins University 1291
  Nobel Prize 90, 1290, 1846
  protein synthesis 1290
  restriction endonucleases 1290
  restriction enzymes 1847
  Selman Waksman Award 1290
  Smith, Hamilton O. 1290, 1847
  SV40 1290
  viral carcinogenesis 1290
National Collection of Type Cultures (NCTC) 851–852
National Culture Collection of Bacteria of the Netherlands 850
National Institute of Allergy and Infectious Diseases (NIAID) 519
National Institute of Genetics (Japan) 847
National Institutes of Health (NIH) 355, 818
National Institutes of Health (NIH)
  National Center for Research Resources 847
National Science Foundation (NSF) Living Stock Collections Program 847
National Seed Storage Laboratory 852
natural selection **1291–1297**
  allele frequency 1292, *1292t*
  Darwin, Charles 514–515, 1291
  directional selection 1293, *1293f*
natural selection (*continued*)
  disruptive selection 1293
  evolutionary biology 516
  fecundity selection 1295, *1296t*
  fundamental theorem of natural selection 1294
  gametic selection 1295
  genetic consequences 1292
  heritability 1294, *1295f*
  heterozygote advantage 1293, *1294f*
  hitchhiking effect 952
  levels of selection 1296–1297
  mean fitness 1292
  non-Darwinian evolution 1343
  process of evolution 1291–1297
  selection differential 1294
  sexual selection 1295
  stabilizing selection 1293
  survival probability 1292, *1292t*
  theory of natural selection 1291–1292
  viability selection 1295, *1295t*
  Wallace, Alfred Russel 514, 1291
nature–nurture controversy 742, **1297–1301**
  adoption studies 1299
  environmentalism 1298–1299
  Galton, Francis 1297–1298
  Mayr, Ernst 1298
  Shakespeare, William 1298
  systems approach 1299–1300, *1300f*
  twin studies 1298, 1299, 1300
  Watson, J.B. 1298–1299
  see also Galton, Francis
NBS1 gene 571, *2084–2085t*
NCBI web site 156, 517
nearly neutral theory *1301f*, **1301–1302**
  history 1301
  implications 1301–1302
  neutral theory 1301, *1301f*
  Ohta, T. 1301
  population genetic studies 1302
  selection theory 1301, *1301f*
  selective neutrality 1800
necrosis
  Caenorhabditis elegans 315f
  chromatin 313
  DNA 313
negative complementation **1302**
negative interference **1302–1304**
  bacteriophage crosses 1303
  heteroalleles 1304
  map expansion 1304
  meiosis 1303
  recombination interactions
    coefficient of coincidence 1302, 1303
    correlation 1302
    null hypothesis 1303
    patches 1303, *1303f*
    recombination frequency 1303
    splices 1303, *1303f*
negative regulators **1304**
negative supercoiling **1304**
Negrin, Juan 1365
Neidhardt F C **653–659**
Nei M **828–832**
Neisseria gonorrhoeae 80, 1443
Neisseria meningitidis 46, 1197, 1423, 1443
Neisseria sicca 1694t
Nelson H C M **1730–1731**
Nelson K E **1196–1203**
Nematoda 251–252
nematology 252
neocentromere 322
neo gene 482

neomycin 1890
neoteny **1304–1305**
developmental shifts during evolution 1304
heterochrony 1304
larval reproduction 1304
paedomorphosis 1304
peramorphosis 1304
role in evolution 1305
Urodela 1304
neovascularization 69
Neptunia 1083
nerve growth factor (NGF) 1967
Netherton syndrome 994
N-ethyl-*N*-nitrosourea 557
netrins 1320
network **1305–1306**
graph theory 1305
molecular evolution
non-tree networks 1305
phylogenetic networks 1305, *1305f*
tree networks 1305
parsimonious trees 1306
reticulations (loop structures) 1305
biological mechanisms 1305
gene conversion 1306
gene exchange 1306
mutually incompatible sites 1306, *1306f*, *1306t*
parallel changes 1305, *1305f*
recombination *1305f*, 1306
Neuhauser C **392–397**
neu oncogene **1306**
dimerization 1306
role in tumorigenesis 1306
transmembrane tyrosine kinase receptor 1306
neural tube defects *2015t*
neuraminidase (N) 1026
neurofibromatosis 357, **1307–1309**
classification 1307
clinical features
café-au-lait spots 1307
neurofibromas 1307
ophthalmological abnormalities 1307
Schwannomas 1307
neurofibromatosis type 1 (Nf1)
chromosome 17 1308
clinical features *1307t*
dermal neurofibromas 1307
diagnostic criteria 1307–1308
epidemiology 1307
freckles 1307
GAPase 1308
genetics 1308
Lisch nodules 1307
management 1308
neurofibromin protein 1308
tumor suppressor genes 1308
neurofibromatosis type 2 (Nf2)
acoustic neuromas 1308
brain tumors 1308
café-au-lait spots 1308
cataracts 1308
clinical features *1307t*
diagnostic criteria 1308
epidemiology 1308
genetics 1308
guanosine triphosphatase (GTPase) 1308
management 1308
Merlin protein 1308
neurofibromatosis (*continued*)
Schwannomas 1308
Schwannomin protein 1308
von Recklinghausen disease 1307
Watson syndrome 1308
neurofibromin protein 1308
neuronal ceroid lipofuscinosis 316
neuronal guidance **1318–1320**
axon growth cones 1318
axon pathfinding 1319
axons 1318
chemoaffinity theory 1319
guidance forces 1319
guidance molecules 1319
guidepost cells 1319
history 1319
neurons 1318
pioneer axon 1319
target recognition 1320
neuronal intranuclear inclusions (NII) 983
neuronal specification **1320–1322**
cell diversity
asymmetric cell division *1322f*
cell-intrinsic mechanisms 1322
extrinsic mechanisms 1321
cell fate 1320
LIM domain 1322
model organisms 1320
neurogenesis 1321, *1321f*
neurogenetics
neurogenic genes 1321, *1321f*
neuron-specific genes 1321, *1321f*
proneural genes *1321f*
POU domain 1322
transcription factors 1322
neurons
Caenorhabditis elegans 314–318
cell death 314–318
degeneration **313–318**
neuropathic amyloidosis 499
Neurospora 636, 738, 749, 795, 1080, 1954
Neurospora crassa 405, 653, 1154, 1223, **1322–1323**
Aspergillus nidulans 106
Beadle, George 1323
centromere 860
clock mutants 389
colinearity 419
gene silencing 813
genetic studies 1322–1323
growth cycle 1323
heterokaryon 1323
homokaryon 1323
Lindegren, Carl 1322
mating types 1153, 1154
meiotic crossing-over 1322
meiotic hot spots 976
Mitchell, Mary 1323
mycelium 1323
poky mutants 1323
replication errors *1676t*
Tatum, Edward 1323
Yanofsky, Charles 2161
neutral drift **1323–1324**
gene regulatory network 1323
genotype to phenotype mapping 1323
neutral mutation 1323
tRNA secondary structure 1323
neutral mutation **1324–1325**
codons 1325
evolutionary rate 671–672
genetic drift 1324
genome structure 1325
neutral mutation (*continued*)
hitchhiking effect 952
Kimura, Motoo 1324
molecular clock 1324
nearly neutral mutations 1325
non-Darwinian evolution 1343
nonfunctional DNA 1325
population genetics 1324
rate of neutral evolution 1324
synonymous mutation 1325
variance in rate of evolution 1325
neutral theory **1326–1327**
behavior of mutant genes 1326, *1326f*
genetic drift 833
history 1326
Jukes, T. 1326
Kimura, Motoo 603, 1063, 1326
King, J.L. 1326
molecular clock 1327
non-Darwinian evolution 1343
random genetic drift 1326
selective constraint 1327
selective neutrality 1800
Newton G **980–981**
Newton, W.C.F. 512
NF1 gene *81t*, 1770, *2084–2085t*
NF2 gene *81t*, *2084–2085t*
NGF *see* nerve growth factor (NGF)
NHEJ *see* nonhomologous end-jointing (NHEJ)
niaA gene 110, *110f*
niacin 2118
NIAID (National Institute of Allergy and Infectious Diseases) 519
nick **1327**
nick translation **1327–1328**
DNA polymerases 1327
exonucleases 1327
radioactivity incorporation 1327–1328
Nicotiana 763, 1422
nicotinamide adenine dinucleotide (NAD) 1365
Niemann–Pick disease 187
Nierman W C **865–872**
nif genes 708, **1328**
bacteria 1328
Azobacter 1328
Azorhizobium 1328
Klebsiella 1328
Rhizobium 1328, 1715
cyanobacteria 1328
gene rearrangement 802
nif gene cluster 1328
nifA gene 708
nifB gene 1328
nifD gene 1328
nifE gene 1328
nifH gene 1328
nifK gene 1328
nifN gene 1328
nitrogenase 1328
nitrogen fixation 1328
see also fix genes
Nigon, Victor 251
niiA gene 110, *110f*
Nijhut, F. 1534
Nijmegen breakage syndrome 114, 351, *569–570t*
Nilsson, Heribert 200
Nirenberg, Marshall Warren 1061, **1328–1329**, 1366
codons 1328
genetic code 1328

Nirenberg, Marshall Warren (*continued*)
Leder, Philip 1328
Matthaei, Heinrich 1328
Nobel Prize 1328
nitrogen fixation 1478, 1479, 1715
nitrogen-fixing symbioses 1084
4-nitroquinoline 1-oxide 557
Nitrospira 149
N-methyladenine 557
N-methyl-*N*'nitro-*N*-nitrosoguanidine 557
N-myc gene 780
Nobel Prize
Altman, Sidney 1730
Arber, Werner 90, 1290, 1846
Baltimore, David 590
Beadle, George 1070
Berg, Paul 877
Cech, Thomas 1730
Crick, Francis 2134
Delbrück, Max 523, 925, 1123
Dulbecco, Renato 590
Gilbert, Walter 877
Hershey, Alfred 523, 925, 1123
Holley, Robert W. 1061, 1328
Jacob, François 1070
Khorana, Har Gobind 1061, 1329
Kornberg, Arthur 1067, 1366
Lederberg, Joshua 1070
Lewis, Edward 1095
Luria, Salvador E. 523, 925, 1123
McClintock, Barbara 1162, 2020
Monod, Jacques 1070, 1238
Morgan, Thomas Hunt 354, 1240
Muller, Hermann Joseph 1255
Mullis, Kary 1499
Nathans, Daniel 90, 1290, 1846
Nirenberg, Marshall Warren 1061, 1328
Noble M E M **500–506**
Nusslein-Volhard, Christiane 1095
Ochoa, Severo 1067, 1365
Sanger, Fred 877, 1768
Smith, Hamilton O. 90, 1290, 1846
Snell, George 1847
Tatum, Edward 1070
Temin, Howard 590
Waksman, Selman A. 1890
Watson, James Dewey 2134
Wieschaus, Eric F. 1095
Wilkins, Maurice 2134
nodABC gene 1330
nod-box **1329–1330**
DNA sequences 1329
NodD binding site 1329
nod gene expression 1329
nod operon 1329
nod promoters 1329
nodD gene 1330
Nod factors **1330–1332**
biological activity 1331–1332
lipochito-oligosaccharides 1330
nod genes 1330
Nod proteins
NodA protein *1332f*
NodB protein *1332f*
NodC protein *1332f*
NodD protein 1333, 1715
NodE protein *1332f*
NodF protein *1332f*
NodH protein *1332f*
NodL protein *1332f*
NodO protein 1332
NodPQ protein *1332f*
Nod factors (*continued*)
NodT protein 1332
NodV protein 1334
NodW protein 1334
rhizobium–legume symbioses 1330
Rhizobium sp. NGR234 1330
Sinorhizobium meliloti 1330
Medicago 1330
Melilotus 1330
Trigonella 1330
structure 1330, *1330f*, *1331t*
nod genes 1715, 1330
nodulation genes **1332–1334**
classification *1333t*
flavonoid inducers
genistein/daidzein 1333
luteolin 1333
naringenin 1333
function *1333t*
gene expression 1333
gene regulation 1334
nodABCMFE genes 1332
nodD gene 1333
nodIJ genes 1332
nod regulon 1333
Nod signal 1332
Rhizobium leguminosarum 1333
role in nitrogen-fixing symbiosis 1332
Sinorhizobium meliloti 1333
nodulins **1334–1335**
classification 1335
endosymbiont 1334
identification techniques 1335
Leguminosae 1334
nodule development 1334, 1334–1335
Rosid clade I 1334
symbiotic nitrogen fixation 1334
noek gene 2048
noggin gene 2150
Noll K M **1961–1963**
Nomarski optics 531, 614
Nomenclature Committee 1341
nomenclature, genetic **1335–1336**
alternative names 1336
human **1340–1342**
background 1340
gene names 1341
gene symbol 1340–1341
genetic databases 1341
genetic research 1341
guidelines 1341
Human Genome Project 1341
Nomenclature Committee 1341
Institute for Laboratory Animal Research (ILAR) 1337
International Committee on Standardized Genetic Nomenclature for Mice 1337
Laboratory Registration Code (Lab Code) 1337
mouse **1336–1340**
chromosome anomalies 1339–1340
gene names 1337–1338
guidelines 1337
history 1337
Institute for Laboratory Animal Research (ILAR) 1337
International Committee on Standardized Genetic Nomenclature for Mice 1337
Laboratory Registration Code (Lab Code) 1337
nomenclature, genetic (*continued*)
Mouse Genome Database (MGD) 1337
naming rules 1337
strains 1338–1339
symbol names 1337–1338
transgenes 1338
multiple objects 1336
naming rules 1335–1336
nomenclature systems 1335, *1336t*
Nomura, M. 468
nonA gene 1317
nonautonomous controlling elements **1342**
nonchromosomal striped (NCS) mutant 1223
non-Darwinian evolution 1342–1345, *1343t*
nondisjunction **1345–1347**
aneuploidy 66
chromosome disjunction 1345–1346, *1347f*
effect on human reproduction
Down syndrome 1346
spontaneous abortions 1346
trisomies 1346
heterodisomy 1345
isodisomy 1345
meiosis 875, 1345, *1346f*
mitosis 1345
molecular biology 1346
monosomy 1345
semisterility 1346
spontaneous abortions 1346
translocation heterozygotes 1345–1346, *1347f*
trisomy 1345, *1346f*
uniparental disomy (UPD) 1345, *1346f*
see also Cattanach's translocation
nonfluorescent pseudomonads 1478
non-Hodgkin's lymphoma (NHL) **1347–1349**
B-cells 247, 1347
Burkitt's lymphoma (BL) 247
classification 1347–1348
clinical features 1348
etiology
Epstein–Barr virus (EBV) 1347
human herpes virus-8 (HHV-8) 1347
human T-cell lymphotrophic virus 1 (HTLV-1) 1347
FAB classification of leukemia 680
histology 1348
immunophenotype 1348
incidence 1347
International Prognostic Index (IPI) 1348
lymphocyte tumors 1347
minimal residual disease 1206
molecular genetics 1348
normal cell counterpart 1348
prognosis 1348
Revised European and American Lymphoma (REAL) classification 1348
T-cells 1347
treatment 1349
WHO classification of leukemia **2135–2136**
nonhomologous end-jointing (NHEJ) 351
non-Mendelian inheritance **1349**
complex genetic transmission 1349
epigenetics 636
gene function 1349
linkage 1349

non-Mendelian inheritance (*continued*)
simple genetic transmission 1349
nonreciprocal exchange **1350**
nonrepetitive DNA **1350**
nonsense codon 49, **1350**, 1366, 1375
nonsense mutation 198, **1350**
nonsense suppressor **1350**
nontranscribed spacer **1350**
Noonan syndrome 1023
NOR *see* nucleolar organizer region (NOR)
Norbury C J **297–298**
Nordic Gene Bank 1179
Northern blotting **1350–1352**
hybridization 1351
mRNA analysis 1350–1352
Northern blot *1352f*
procedures 1350–1352, *1351f*, *1352f*
reverse Northernn blot 1352
no tail gene 1945
Notch gene 1315
Notophthalmus viridescens 1076, *1611f*
Nova proteins 1735
Nowak M A **845–846**
nph1 gene 1457
nph4 gene 1457
nrdB gene 1929
nrdD gene 1929
NRG-1α 627
NRG-1β 627
NRG2-α 627
NRG2-β 627
NRG3 627
NRG4 627
NTRK3 gene 1772
nuclear envelope (NE) **1352–1356**
chromatin organization 1355
function 1352
kinetochore 1356
meiosis 1355
mitosis 1355
nuclear pore complex (NPC) 1353, *1353f*
nucleocytoplasmic transport
karyopherins 1353–1354
mechanisms 1354–1355
nuclear export signals (NESs) 1353–1354
nuclear localization signals (NLSs) 1353–1354
Ran-GDP 1354
Ran-GTPase cycle 1354, *1354f*
ribonucleoprotein particles (RNPs) 1353–1354
RNP export 1355
structure 1352, *1353f*
nuclear matrix **1356**
nuclear pore complex (NPC) *see* nuclear envelope
nuclear pores **1356**
nuclear transfer **1356–1357**
cloning 1356
egg cytoplasm 1357
somatic cell hybridization
chromosomal gene mapping 1357
cybrid 1357
cytoplast 1357
electrofusion 1356
hemagglutinatin virus of Japan (HVJ) 1356
karyoplast 1357
monoclonal antibodies 1357
polyethylene glycon (PEG) 1356
nuclear transfer (*continued*)
Sendai virus 1356
therapeutic cloning 1357
transgenes 1994
nuclease **1357–1358**
degradation of nucleic acids 1357
DNase 1357
endonuclease 1357
exonuclease 1357
homing endonucleases 1358
importance 1358
molecular biological tools 1358
restriction endonucleases 1358
RNase 1357
nucleic acid **1358–1359**
adenine (A) 1358
cytosine (C) 1358
DNA (deoxyribonucleic acid) 1358
genome composition 1358
guanine (G) 1358
mRNA (messenger ribonucleic acid) 1358
polarity 1358
purines 1358
pyrimidines 1358
RNA (ribonucleic acid) 1358
rRNA (ribosomal) 1359
smRNA (small nuclear) 1358
structure 1358
thymine (T) 1358
tRNA (transfer) 1358
uracil (U) 1358
nuclein 553
nucleocytoplasmic transport *see* nuclear envelope
nucleolar organizer **1359**
nucleolar organizer region (NOR) 374
nucleolus **1359–1360**
mechanisms 1359
mitosis 1359
nucleolar organizer region (NOR) 1359
nucleolus organizer region (NOR) 349
ribosomal subunit assembly 1359
rRNA (ribosomal) synthesis 1359
structure 1359
nucleoporins 1353
nucleosome 340, 633, **1360**
nucleotides and nucleosides 540, **1360–1361**
bases 1360
CA microsatellite repeats 250
composition 1360
importance 1360
phosphorylation 1360
structure 1360
sugar 1360
in vitro evolution 1008
nucleotide sequence *see* DNA sequencing
nucleotide sequence relationships 1053, *1054f*
nucleotide substitutions 1059–1060
horizontal transfer 974
Jukes–Cantor correction 1059–1060
Kimura correction 1062
nucleotypes 249–250
nucleus **1361**
perinuclear space **1430**
nude mouse **1361**
null hypothesis **1361–1362**
Bayesian analysis 204
degrees of freedom 729
deviations 1362
negative interference 1303
null hypothesis (*continued*)
population genetics 1361
statistical hypothesis 1361
nullisomy 66, 346, 1362
null mutation **1277**
numb gene 1322
NURF chromatin remodeling complex *342t*
Nurse, Paul 1777
NusG protein 1718
Nusslein-Volhard, Christiane 1095
nutritional mutations **1362–1363**
auxotrophs 1362
importance
biochemistry 1362
biotechnology 1363
genetics 1362
physiology 1362
recombinant DNA technology 1363
regulation 1362
metabolic capacity effect 1362
mutational studies 1362
amino acid biosynthesis 1363
carbon utilization 1362–1363
use of organic compounds 1362
occurrence 1363
Nyhan W L **1087–1088**
nystatin 1890

O

o2 mutant 1786
OA1 gene 27
Oak Ridge National Laboratory 1871
Oak Ridge National Laboratory Mutant Mouse Collection 849
Oberbauer R **82–83**
OB fold 1840
O'Brien S J **684–685**
Occam's razor 1415
ochre codon **1366**, 1887
ochre mutation 245, **1366**
ochre suppressor **1366**
Ochoa, Severo **1365–1366**
Center of Molecular Biology 'Severo Ochoa' 1366
codons 1366
DNA structure 555
genetic code 1366
initiation factors 1366
Kornberg, Arthur 1067, 1366
Krebs cycle 1365
Meyerhoff, Otto 1365
Nirenberg, Marshall Warren 1366
Nobel Prize 1365
oxidative phosphorylation 1365
polynucleotide phosphorylase 1365
RNA synthesis 1366
O'Connell K **298–302**
Oct-1 gene 959, 962
Oct-2 gene 959, 962
ocular albinism (OA) 25
oculocutaneous albinism (OCA) 25
oculopharyngeal muscular dystrophy *1263f*, 1264, *2050t*
odr-10 gene 1313
ODR-10 protein 1313
Oenothera 1788
Oesch F **51–54**
Ogle R F **650–653**
Ogren, Bill 88
ogt gene 36
Ohno's law **1367**
autosomal genes 1367
comparative genetic maps 1367

Ohno's law (*continued*)
  Graves, Jenny 1367
  Ohno, Susumu 1367
  pseudoautosomal genes 1367
  X-chromosome arrangement 1367
  X-chromosome inactivation 1367
  X-chromosome linkage 1367
Ohno, Susumu 629, 1367, 2147
Ohta T **709–710**, **814**, **1301–1302**, **1323–1324**, **1326–1327**, **1800–1803**
Ohta, T. 1325, 1800
oilseed rape *1914t*
O'Kane C J **584–585**
O'Kane S L O Jr **237–238**
Okazaki fragments **1367**
  bacteriophage recombination 172
  discontinuous replication 537
  DNA synthesis 577
  dynamic mutations 595
  epigenetics 636
  mating type switch 1156
  primase 1542
  primer RNA 1546, 1547, *1547f*
  repair mechanisms 1667
  trinucleotide repeats *2052f*, 2052–2053
Olby R **742–748**
olfaction **1368–1370**
  Caenorhabditis elegans 1369
  mammals 1368
  mechanisms 1369
  olfactory adaptation 1369
  olfactory receptor neurons 1368
  olfactory system 1368
  receptors 1368
  signal transduction 1369
Olimarabidopsis 86
Olsen G J **1415–1419**
OMIM *see* Online Mendelian Inheritance in Man (OMIM)
Omnione database 1201
OMPs *see* outer membrane proteins (OMPs)
Onagraceae 238
oncogenes **1370–1372**
  abl oncogene 1063, 1484
  activation *1370t*, 1372
  BCL2 proto-oncogene 1932
  BCR/ABL oncogene 207–208, 1208, *1208f*
  c-ABL oncogene 207–208, 250
  c-abl oncogene 1484
  c-Ha-ras oncogene 978
  C-Ha-ras oncogene 1484
  c-*met*-proto-oncogene 1185
  c-myc oncogene 888, 1160, 1483, 1739
  c-Myc oncogene 2086
  c-onc gene *1250f*, 1252
  c-raf oncogene 1484
  crown gall tumors 492
  definition 1370–1371
  discovery 1701
    retroviruses 1371
  Eμ-myc oncogene 1484
  erbB-1 gene 648–649
  erbB-2 (HER-2/neu) gene 11, 648–649
  erbB-3 (HER-3) gene 648–649
  erbB-4 (HER-4) gene 648–649
  FLI1 (Friend leukemia virus integration 1) oncogene 1772, *1772t*
  FLI1 gene 1772, *1772t*
  FMS oncogene **721–722**
  function *1371t*, 1372
  gene expression 1370
oncogenes (*continued*)
  growth factors 1370
  HRAS1 oncogene 1602
  Hras oncogene 1602
  human cancer 1371
  Ki-ras oncogene 423
  KRAS2 oncogene 1602
  k-*ras* oncogene 9, 310, 1602
  MDM2 proto-oncogene 1772
  MET proto-oncogene 1660
  MOS 1240
  mos oncogene 1063
  Myb oncogene **1287**
  myc oncogene 1372
  MYC oncogene 247, 1091, 1605, 1773, 1932
  Nras oncogene 1602
  NRAS oncogene 1602
  Philadelphia chromosome *1370t*, 1371
  Pim1 oncogene 1470
  Pim2 oncogene 1470
  proto-oncogenes 1370
  PTEN/MMAC oncogene 11
  raf oncogene 1063, 1484
  ras oncogene 69, 1484
  Rel oncogene **1658–1659**
  RET proto-oncogene 942, **1696**
  RET1 proto-oncogene 1257–1258
  Rous sarcoma virus 1371
  src oncogene 1063
  tumor development 1370
  upilson-Ha-ras oncogene 1484
  v-fms oncogene 721
  v-myc oncogene 1484
  v-onc oncogene 1251–1252
  v-raf oncogene 1484
  v-sis oncogene 126
  v-src oncogene 1371
  Wt1 oncogene 1818
  WT1 oncogene *81t*, 1772, *1772t*, 1814, 1818, *2084–2085t*, 2138
  yes oncogene 1063
oncogenesis
  epidermal growth factor (EGF) 648–649
  kinases 1063
  tyrosine kinases 648–649
OncoMouse® **1372–1373**
  biological importance 1372
  E.I. Dupont de Nemours and Company, Inc. 1372
  genetically engineered transgenic mouse 1372
  Harvard University 1372
  model animal 1372
  research issues 1372–1373
  tumorigenesis 1372
Oncorhyncus mykiss 2098
O'Neill O **659–661**
Online Mendelian Inheritance in Man (OMIM) 365, 1188, 1835
On the Origin of Species (Darwin, 1859) 1935
oogenesis 689, 691, 875
Oomycota 1883
opal codon **1375**, 1887
opaque2 mutant 1786
opaque6 mutant 1786
opaque7 mutant 1786
opaque15 mutant 1786
open reading frame (ORF) **1375**
  bacteria *150t*, 155
  bacteriophages 183
open reading frame (ORF) (*continued*)
  definition 1375
  DNA sequences 1375
  epigenetics 634
  erythroblastosis fetalis 651
  eukaryotes 1375
  functionality 1375
  gene cassettes 771
  genes 760
  infertility 1022
  insertion sequence (IS) 1030
  integrons *1042f*, 1043–1044
  introns 1375
  leader peptide 1078
  mobile introns *1048t*, 1048–1051
  prokaryotes 1375
  retrotransposons *1699f*, 1699–1700
  Shine–Dalgarno sequence 1375
  viruses 2114
operational taxonomic units (OTUs) 1939–1940, 2042
operators **1376–1377**
  attenuation 1377
  binding process 1376
  cis configuration 1376
  gene regulation 1376
  Jacob, François 1376
  lac operon 1376
  Monod, Jacques 1376
  mRNA (messenger ribonucleic acid) 1376
  O^c mutation 1376
  operon model 1376
  repressors 1376
  structure 1376
  trans configuration 1376
  translation 1377
  upstream, downstream sites 2100
operon **1377**
  attenuation 1377
  gene regulation 1377
  Jacob, François 1059, 1377
  lac operon 1377
  leader sequence 1377
  Monod, Jacques 1377
  mRNA (messenger ribonucleic acid) 1377
  operon theory 1059, 1070, 1071, 1377
  repressor–operator interaction 1377
  types 1377
  see also pleiotropy
opine genes 492
Orcyteropus afer 859
Oreochromis niloticus 2098
organelles **1377–1379**
  chloroplast 1379
  definition 1377
  endoplasmic reticulum (ER) 1378
  glycosomes 1379
  Golgi body 1378
  hydrogenosomes 1379
  lysosomes 1378
  mitochondrion 1378–1379
  nucleus 1378
  peroxisomes 1379
  vacuoles 1379
organogenesis 1427
Orgel, Leslie 649–650, 1751
oriC 1382
Oriental white storks 540
origin of life **1387–1394**
  alternative models 1389, *1390f*
  autocatalytic subsystems 1387

origin of life (*continued*)
  definition of life 1388
  Eigen's paradox 1393
  error threshold 1393
  evolution criteria 1388
  experimental replication 1393
  genetic code 1393
  hereditary membrane replication 1392
  homochirality 1392
  lipid replication 1392
  metabolic theories 1389–1392, *1391f*
  model of minimal life 1388, *1389f*
  molecular self-replication 1392
  population structure 1393
  ribozymes 1392
  RNA world hypothesis 1392
  unlimited heredity 1392
origin (*ori*) 718, *718f*, **1387**
O'Riordan J L H **1027–1029**
Ori sequences **1381–1387**
  bacteria 1382
  chromatin structure 1385
  ColE1 1381
  CpG islands 1386
  DNA replication process 1381
  eukaryotic replication origins 1384, *1385f*
  mammals 1385, *1386f*
  methylation 1382
  origin features 1386–1387
  plasmids 1381
  prokaryotes 1381
  pseudomonads 1382
  replication initiation 1387
  rolling-circle replication 1382
  Saccharomyces cerevisiae 1384–1385, *1386f*
  Schizosaccharomyces pombe 1385, *1386f*
  sequence context 1385
  structural organization of *oriC* region 1382, *1383f*
  SV40 1384, *1386f*
  SV40 T antigen 1384
  theta-type replication 1381
  Xenopus laevis 1384, *1385f*
oriT 657, 677, 937
oriV 677
orphan receptor **1394**
Orr-Weaver T L **338–339**, **340–343**, **344–345**, **350**
Orth A **1259–1261**
orthologs 436, *437f*
orthology 819, **1394**
Orthoptera 319
Oryzabase web site 853
Oryza sativa (rice) **1394–1395**
  agronomically important genes 1395
  cultivation 1394
  disease resistance genes 1395
  gene analysis 1395
  gene disruption 1395
  gene functions 1395
  genetic map 1394
  genome sequencing 859
    BAC (bacterial artificial chromosomes) 1394–1395
    PAC 1394–1395
    physical map 1394–1395
    YAC (yeast artificial chromosomes) 1394–1395
  indica 1394
  Japanese Rice Genome Research Program 1394
Oryza sativa (rice) (*continued*)
  japonica 1394
  molecular genetic analysis 1394
  molecular genetic map 1394
  Oryza glaberrima 1394
  photoperiod sensitivity genes 1395
  systemic acquired resistance (SAR) *1914t*
Osborne, Thomas 1783
osm genes 255
osmium tetroxide 551
osteogenesis imperfecta (OI) **1395–1398**
  biochemistry 1397
  Bruck syndrome 1397
  classification 1395
  clinical features 1396–1397
  Cole–Carpenter syndrome 1397
  international nomenclature *1396t*
  molecular pathology 1397
osteoporosis 1395
osteosarcoma 1768, *1769t*, 1770, *1770t*
Osterwalder, A. 1777
Ostrander E A **264–270**
Ostrina nubialis 1681
OTUs *see* operational taxonomic units (OTUs)
outbreeding **1398–1400**
  Coprinus 1398
  evolutionary importance 1400
  flowering plants 1399
  homeodomain proteins 1399
  inbreeding depression 1400
  mating incompatibilities 1400
  Saccharomyces cerevisiae 1398
  Schizophyllum 1399
  sexual reproduction 1398
  Ustilago maydis 1398
outcross **1400**
outer membrane proteins (OMPs) 1555
outgroup rooting 1416
ovarian development 1816
overdominance 647, **1400–1401**
  dominance hypothesis 1400–1401
  genetic equilibrium 835
  heterosis 933, 1400
  Hull, Fred 1401
  hybrid vigor 1400
  hybrid yield 1400
  inbred strains 1400
  Jones, D.F. 1400
  overdominance hypothesis 1400
  pseudo-overdominance 1401
  Shull, G.H. 1400
overwinding **1401**
ovulation **1401–1402**
Owen, R. 964
oxidative DNA damage repair 1661, *1662f*, **1669–1671**, *1670f*
8-oxoguanine *1669f*, 1671
Oxytricha 1135
Oxytricha fallax 1134
Oxytricha nova 1204f
Oxytricha similis 860
Oxytricha trifallax 1134

P

p16 gene 10, *81t*, 2082, *2084–2085t*
$p16^{INKA}$ gene 1123
P1 plasmid 1486
P1 vectors 869
p300 gene 2083
P450c1-α 1027
p53 gene 10, 69, 423, 571, 1770, 2082, 2083, *2084–2085t*
  anti-oncogenes *81t*
  leukemia, chronic 1093
  see also oncogenes
p53 protein 310, 2127
  BRCA1/BRCA2 genes 239
  carcinoid tumors 273–274
  mutational analysis 1280
  tumorigenesis 1280
PABP2 gene *2050t*, 2053
Pace, Norman 1751
packing ratio 373
PACs (P1 artificial chromosomes) 869, 1394–1395
Padua R A **721–722**
2D PAGE *see* two-dimensional polyacrylamide gel electrophoresis
Paget's disease 1770, *1770t*
PAH *see* phenylalanine hydroxylase (PAH)
PAI *see* pathogenicity islands
Painter, Robert 565
Painter, Theophilus Schickel 354, **1407–1410**
pairing *see* base pairing/mispairing; chromosome pairing
Palacios R **61–62**
palindrome **1410**
Pamilo P **1059–1060**
PAN gene 716
pangenes 521
pangenesis 521
Panicoideae 890
panmixis **1410–1411**
  see also Hardy–Weinberg law; inbred strain; Wahlund effect
panspermia 1389
pantothenic acid 2118, *2119f*
Papaioannou V E **1943–1945**
pap gene 1443
Papilio glaucus 731
Papilio polyxenes 1535
Papilionoideae 1445
papilloma virus 16 310
Papovavirus family 2079
Paracentrorus 236
paracrine 400
paracrine mode *see* autocrine mode
parallel similarities 969–970
paralogs 436, *437f*, *439f*
paralogy **1411**
paramecia **1411–1414**
  autogamy 1412–1413
  conjugation 1412
  cortical inheritance 1413–1414
  cytogamy 1413
  macronuclear regeneration 1413
  Mendelian genetics 1412–1413
  mitochondrial inheritance 1413
  non-Medelian genetics 1413–1414
  nuclear reorganization 1412
  symbiont organisms 1413
Paramecium 450, 749, 1135, 1411
Paramecium aurelia 653, 1865
Paramecium tetraurelia 1412–1414
paramorphosis *see* neoteny
parapatric **1414**
paraphyly 970, **1414**
parasegment **1415**
  see also cell lineage
paraxial mesoderm 1427
parB gene 1488
Pardue, Mary Lou 348, 361

parental **1415**
  see also cross
par gene 308, 616–618
Parker J **12, 57–58, 60, 76–78, 115–116, 146–151, 191–192, 192, 401–402, 522, 585–586, 601–602, 610–611, 780–782, 797, 1069–1070, 1077–1078, 1213–1215, 1276–1277, 1375, 1429–1430, 1496, 1509, 1686, 1715–1716, 1746–1747, 1882–1883, 1885–1886, 1889, 1890–1891, 1899–1901, 2100**
Parker R **1748–1751**
Parkia (locust bean) 1083
Paro, Renato 636
PAR (proteins) *616f*, 616–618
parsimony **1415–1419**
  algorithms 1416–1417
  long branch attraction 1417–1418
  maximum likehood 1417
  systematic errors 1417–1418, *1418f*
  tree 1415–1418, *1416f*
  see also phylogeny
parthenogenesis 1000
  chimera 2098
  cyclical parthenogenesis 2097
  hydatidiform moles 2098
  invertebrates 2097
  mammalian **1419–1420**
  vertebrates 2097
parvovirus 1830
Passmore H C **1640**
Pastinaca sativa 1535
Paszkowski J **813–814**
pat-2 gene *620f*
pat-3 gene *620f*
Patau, Klaus 1420
Patau syndrome 66, 346, 348, **1420**, 2056
  see also chromosome aberrations
paternal inheritance **1420–1422**
  Caenorhabditis elegans 1422
  cytoplasmic components 1421–1422
  diversity 1423
  Drosophila melanogaster 1422
  evolution 1423–1424
  examples 1423
  mutations 1421–1422
  nuclear factors 1420–1421
pathogenesis-related (PR) proteins 1913
pathogenic agents of plants *1914t*
pathogenicity islands 916, **1422–1424**, 1423
  see also horizontal transfer
Pato M L **1438–1441**
Paton, Noel 1365
pattern formation **1424–1426**
  apoptosis 1426
  asymmetric cell division 1425
  control of cell division axes 1425
  control of cell proliferation 1425
  demarcation of fields of cells 1425
  establishment of boundaries 1425–1426
  external cues 1425
  lateral inhibition 1426
  long-range signaling between cells 1426
  organized cell growth 1425
  self-assembly of large molecular structures 1425
  short-range signaling between cells 1426
  sorting of cells 1426
  stochastic assignment of cell fate 1425
  see also apoptosis; cell division genetics; cell lineage; compartmentalization
Patterson, John T. 1408
Pauling, Linus 193, 554, 1828
Paulsen I T **1196–1203**
Pavan W J **1467–1469**
PAX3 gene 1715, 1772, *1772t*
PAX7–FKHR fusion gene 425, 1715
PAX7 gene 1715, 1772, *1772t*
Pax genes **1426–1428**
  brain development 1427
  cell differentiation and oncogenesis 1427–1428
  central nervous system 1427
  differentiation boundaries 1427
  expression during development 1427
  genes
    Pax1 gene 1427
    Pax2 gene 1427
    Pax3 gene 1427
    PAX3 gene 1469
    Pax5 gene 1427
    Pax6 gene 1427
    Pax7 gene 1427
    Pax9 gene 1427
  mutations, phenotypes, and function 1427
  organogenesis 1427
  paraxial mesoderm 1427
  proteins 1426
pb (proboscopedia) gene *960f*
pBR322 plasmid 656, **1428**
  see also cloning vectors
PCDHY gene 2159
Pcg gene 1814
Pcg proteins 1814
PCR *see* polymerase chain reaction (PCR)
PCR end-point analysis 1503
Pctr1 gene 1483
Pctr2 gene 1483
pdf (pigment dispersing factor) gene 389
PDGF *see* platelet-derived growth factor (PDGF)
Pdha2 retroposon 667
pe (pale ear) mutation 1469
pearly millet *1914t*
Pearson, Karl 64, 742, 1516
Pearson P L **353–367, 1085–1087, 2055–2056, 2056–2058**
Pearson syndrome 1224
pedigree analysis **1428–1429**, 2013
  horizontal pedigree 1428
  knight's move pedigree 1428–1429
  nonspecific pedigree 1429
  vertical pedigree 1428
pedomorphosis **1429**
  see also neoteny
P elements **1403–1407**, 1708, 1806
  cis-acting components 1406
  enhancer traps 1407
  germline transformation 1406
  hybrid dysgenesis *1404f*
  hybrid element insertion 1407
  mutagenesis 1406–1407
  population biology 1405
  structure *1403f*
  structure and function 1403–1405
  transposition *1405f*
Peltonen L **724–726**
penetrance **1429**
penicillin 1687
Penicillium chrysogenum 225
Penny D **473–477**
Penrose, L.S. 701
PEP *see* phosphoenolpyruvate (PEP)
PepA protein 1040, *1040t*
pepsin 1573
peptide bond **1429–1430**, *1568f*
peptidoglycan 1131
peptidyl-tRNA 602, 609
Pera R R **1021–1026**
per gene 210, 387, 1316
perichromosomal material 374, 374–375
perinuclear space **1430**
  see also nucleus
peripatric 39, 1862
permanganate 551
permease 1069, 1070
permissive cells **1430**
  see also virus
PeroBase 849
Peromyscus (American deer mice) 688
Peromyscus californicus 1522t
Peromyscus Genetic Stock Center 849
Peromyscus maniculatus 987
Peromyscus polionotus 635, 987, *1522t*
Perona J J **543–541, 572–574**
Peronospora tabacina 1914t
peroxisomes 1379
PER protein 387
Perret X **214–215**
Perutz, Max 815
Peters, Rudolf 1365
pet gene 1461
petite **1430**
petite mutation 1223
Petit M M R **1109–1117**
Petroselinium (Pts) mutation 1472
Petroselinum 2026
Petrosino J F **1623–1631**
Pettijohn, David 565
Petunia 2026, 2055–2056
Petunia hybrida 716
PEX gene 1028–1029
PFC gene *430t*, *431t*
Pfeiffer syndrome 478–479
PFGE *see* pulsed field gel electrophoresis (PFGE)
PFK1 gene 886
PFK2 gene 886
PFK26 gene 886
PFK27 gene 886
Pfu DNA polymerase 1502
P gene 26
p (pink-eyed dilution) gene 398, 1469
p (pink-eyed dilution) locus 1599, 1871
PGI1 gene 885
PGK1 gene 886
Pgk2 retroposon 667
Pgk gene 477, 978
P-glycoprotein **1430–1433**
  ABC transporters 1431
  cancer and regulation of gene expression 1432
  inhibitors 1433
  mechanism 1432–1433
  structure *1431f*
  substrate profiles and binding 1431–1432
  tissue distribution and function 1432
PGM gene 892
PGM1 gene 886
Pgp *see* P-glycoprotein
PGPR *see* plant growth promoting rhizobacteria (PGPR)
P granules 615–616
Ph *see* Philadelphia chromosome
ph1b gene 2067
ph1c gene 2067
Ph1 gene 2067

pha-4 gene 619
Phabagen 850
phage 21 413
phage 434 413
phage (bacteriophage) **1434**
phage BF23 417
Phage Course, The 522
phage crosses **1437–1438**
phage HK022 85, 1895
phage HP1 413
phage λ 183, 542, 547, 1772
  circular linkage maps 377
  cohesive ends 412, *413f*
  integration and excision **1434–1437**, *1434f*
    att site structure 1435–1436
    accessory proteins 1435
    Int protein 1434–1435
    P and P′ arms 1436–1437
    secondary *att* sites 1436
    strand exchange 1436
  replication errors *1676t*
  specialized transduction 1858–1859, *1859f*
  transduction 1983–1984
  see also site-specific recombination
phage M13 547, 1546, *1676t see* filamentous bacteriophages
phage Mu **1438–1441**
  Escherichia coli 1439
  expression 1439
  genetic map 1439, *1439f*
  genetic tool 1441
  G inversion 1441
  lysis and lysogeny 1439–1440
  mom modification 1440–1441
  transposition 1440, *1440f*
  see also lysis; lysogeny; recombination
phage N4 1975
phage Ox2 1442
phage P1 182, 481, 658, 1442
phages P2 and P4 182–183
phage P22 183, 1766, 1895
phage P4 1542
phage P4 α protein *1547t*
phage ϕ6 184
phage ϕ29 184
phage ϕ80 413, 417
phage ϕ105 413
phage ϕCTX 413
phage receptor **1441–1442**
phage SP01 1975
phage T1 417, 654, 1923
phage T2 654, 926, 977, 1442, 1447, *1676t*
phage T3 1922–1923
phage T4 185–186, 419–420, 655, 977, 1437, 1441–1442, 1447, 1542, *1676t*, 1895, 1923–1927, 1975
  conditional lethality *443f*
  genetic map *1925f*
  host-lethal gene 975–976
  infection 1924–1927, *1926f*
  lysis inhibition 1926
  micrograph *1924f*
  mutants 1927
  phage T4 gene 61 protein *1547t*
phage T5 1923
phage T6 417
phage T7 1542, 1922–1923
phage T7 gene 4 protein *1547t*
"phage truce" 1921
Phagobioderm 177
phagocytosis 313
pharming 661
Phaseoleae 1083
Phaseolus acutifolius A. Gray 1445
Phaseolus coccineus L. 1445
Phaseolus (common bean) 1511
Phaseolus lunatus L. 1444–1445
Phaseolus nanus 1177
Phaseolus polyanthus Greenman 1445
Phaseolus vulgaris 1177, **1444–1445**
phase variation **1442–1444**
  biological significance 1443–1444
  DNA methylation 1443, *1444f*
  general recombination 1443
  site-specific recombination *1442f*, 1442–1443
  slipped-strand mispairing 1443
Phelan J **835–838**, **1239–1240**, **1585**
phenetics 1937, **1446**
phenocopy **1446**
phenogram **1446**
phenotype **1446**
phenotypic lag **1447**
phenotypic mixing **1447**
phenylalanine 1447, **1447**
phenylalanine hydroxylase (PAH) 1448
phenylketonuria 1187, 1363, **1447–1449**, 1490
  biochemistry 1448
  description 1447–1448
  genetics 1448–1449
  screening 1448
  see also genetic diseases
pheromone 277, 1368
Phex gene 1028
Phibbs, P.V. 851
Philadelphia chromosome 189, 207–208, 250, *1370t*, 1371, **1449–1450**, *1450f*
  genetics 1449–1450
  incidence and outcome 1449
  translocation *1450f*
  see also leukemia, acute; leukemia, chronic
phi (ϕ)X174 **1450–1454**
  evolution 1453
  gene expression 1452
  genetic map development 1451
  genome replication and packaging 1451–1452
  particle morphogenesis 1452–1453
phi (ϕ) bond 1569, *1569f*
phi beta (ϕβ) globin gene 2096
Pho85 502
Phoma foveata 1850
phosphatonin 1028
phosphoenolpyruvate (PEP) 260
5-phosphoribosyl1-pyrophosphate (PRPP) 943–945
N′-5′-phosphoribosyl-AMP (PRAMP) 943–944
N′-5′-phosphoribosyl-ATP (PRATP) 943–944
N′-[(5′-phosphoribosyl) formimino]-5-aminoimidazole-4 carboxamide ribonucleotide (5′-Pro-FAR) 943–944
N′-[(5′-phosphoribulosyl) formimino]-5-aminoimidazole-4 carboxamide ribonucleotide (5′-PRFAR) 943–944
phosphorylase 259
phosphorylation
  ataxia telangiectasia (AT) 114
  Caenorhabditis elegans 617
  enzymes 626
  Ets family 662
phosphorylation (*continued*)
  histone genes 950
  kinases 1063–1064
  kinetochore 1065
  LIM domain genes 1097
  masked mRNA 1148
  mitosis 1226
  Ochoa, Severo 1365
  tyrosine kinases 648
phosphoryltransferases 1032
photoautotrophs 147
photoheterotrophs 147
photomorphogenesis in plants **1454–1458**
  Arabidopsis thaliana 1454–1455
  cryptochromes 1456–1457
  phototropin 1457
  phototropism 1457
  phytochromes 1455–1456
  seedling development and light *1455f*
  signaling 1457–1458
photoreactivation **1458–1459**
photorepair **1459**
photosynthesis 1379, **1459–1465**, *1460f*
  Chlamydomonas reinhardtii 335
  chloroplast 1460–1461
  chloroplast genetics 1462–1463
  eukaryotes 337
  evolution 337
  genes 1463–1464
  genetic origin of photosystem I subunits *1461t*
  model organisms 1461–1462
  mutants and their phenotypes 1463
  nuclear participation *1464f*
  regulation 1464–1465
  symbionts 337
  see also Chlamydomonas reinhardtii
phototropin 1457
phototropism 522, 1457
PHYA gene 1455
Phycomyces blakesleeanus 522, 749
phyletic evolution 665
Phylloscopus trochilus 1865
phylogenetic invariants **1053–1054**, *1054f*
phylogenetic tree *see* trees
phylogeny 966–968, **1465–1466**
  see also cladistics; parsimony; trees
phylogeography **1466–1467**
Physarum 763
Physarum polycephalum 1740
Physcomitrella patens 1708
physical mapping **1467**
  chromosome breakpoint maps *363t*
  clone contig maps *363t*
  DNA sequence maps *363t*
  Drosophila melanogaster 354
  fragmentation maps *363t*
  induced chromosome *363t*
  long-range restriction map 363, *363t*
  in situ hybridization (ISH) *363t*
  transcript maps *363t*
  types *363t*
  see also chromosome mapping
Phythium ultimum 1478
phytochromes 1455–1456
Phytomonas tumefaciens 491
Phytophthora cryptogea 1914t
Phytophthora infestans 1850, *1914t*
Phytophthora parasitica 1479, 1480, *1914t*
Pianka E **669–671**
Pib gene 1395
pie-1 gene 619
piebaldism *see* piebald trait

piebald (s) gene 398
piebald (*s*) locus 1599, 1871
piebald trait **1467–1469**
  Hirschsprung's disease 1468–1469
  humans 1468
  mouse *1467f*, *1468f*, 1469
  Waardenburg syndrome 1468–1469
  see also coat color mutations, animals
pied flycatcher 1296
PI gene 716
pigmentation, mouse **1469–1470**
Pilder S H **985–986**
pilE gene 1443
pili 1556
pilS gene 1443
Pim1 oncogene 1470
Pim2 oncogene 1470
Pim oncogenes **1470**
PINCH protein 1098
P Insertion Mutant Stock Centre 847
Pinus flexilis (limber pine) *788f*, 788–789
Piophage 176
Pipas J M **2078–2081**
Pirrotta, Vincenzo 636
Pisum 1422, 2026
Pisum Genetic Stocks Collection 1179
Pisum sativum (garden pea) 87, 1083, 1169, **1470–1471**
  AFRC Institute of Plant Science Research (UK) 1179
  cotyledon (yellow/green) gene *1470t*
  dominance 1174, *1174t*
  flower (violet/white) gene *1470t*
  genome 1471
  John Innes Institute 1179
  Mendel, Gregor 1470
  molecular markers 1471
  Nordic Gene Bank 1179
  phenotypic ratio 1174, *1174t*
  Pisum Genetic Stocks Collection 1179
  pod (green/yellow) gene *1470t*
  pods (dispersed/clustered) gene *1470t*
  pod wall (stiff/collapsing) gene *1470t*
  recessiveness 1174, *1174t*
  seeds (round/wrinkled) gene *1470t*
  stem length (tall/dwarf) gene *1470t*
  traits 1174, *1174t*
  transmission genetics 2009
Pit1 (pituitary specific transcription factor1) gene 592, 959
Pittsburgh Bacteriophage Institute 182
pKM101 plasmid *453f*
PKU *see* phenylketonuria
Planctomyces 149
plant development **1471–1474**
  basic processes 1472–1473
  determination 1472–1473
  differentiational gene activity 1472
  epigenetic regulation 1473–1474
  epimutation 1473
  growth and morphogenesis 1472
  homology-dependent gene silencing 1473–1474
  posttranscriptional gene silencing (PTGS) 1473–1474
  stabilization of developmental states 1473
plant embryogenesis **1474–1477**
  Arabidopsis thaliana 1474–1475
  genetic screeing for embryonic mutants 1475
  mutations in apical–basal axis formation 1475–1476
plant embryogenesis (*continued*)
  mutations in radial axis formation 1476
plant genetics 87
plant growth promoting rhizobacteria (PGPR) **1477–1480**
  control of plant pathogens 1478–1479
  direct plant growth promotion 1479
  microorganisms effects on plants 1477
  nonfluorescent pseudomonads 1478
  plant growth promotion 1478
  rhizosphere effects on microoganisms 1477
  species 1478
plant hormones **1480–1481**
  abscisic acid (ABA) 1481
  Arabidopsis thaliana 1480
  auxin 1480–1481
  brassinosteroids (BRs) 1481
  cytokinins 1481
  ethylene 1481
  gibberellins (GAs) 1481
  jasmonic acid (JA) 1481
  salicylic acid (SA) 1481
plant hybrids 2009
plants
  DNA transposons 2026–2029
  retrotransposons 2021–2026
  transposons **2020–2033**, *2021f*
  trichome development **2045–2048**
  triploidy 2055–2056
plaques **1482**
plasmacytomas (mouse) **1482–1485**
  chromosome translocations 1483–1484
  environmental factors 1484
  genetic changes 1484
  induced plasmacytomagenesis 1482–1483
  microenvironmental factors 1484
  oncogene-driven plasmacytomagenesis 1484
  spontaneous plasmacytomagenesis 1482, 1484
  susceptibiliy and resistance 1483
  see also cancer susceptibility
plasmid DNA (T-DNA) 1781
plasmids **1485–1487**
  Agrobacterium 24
  antibiotic resistance 75
  ColE1 656
  colicins 418
  copy number control 1488
  description 1485
  DNA cloning 546–547
  episome 638
  Flp recombinase 717, *718f*
  F plasmid *603f*, 657
  genetic engineering 1490
  genome databases 517
  history 1485
  horizontal transfer 1489
  immunity 998
  incompatibility 1489, 1016
  integrons 1042
  minichromosome 1205
  pBR322 plasmid 656
  plasmid vectors 1011, 2104–2105
  pneumonia bacteria 1492
  resistance plasmids **1686**
  rolling circle replication *1487f*, 1487–1488
  selfish DNA 1807
  stable maintenance 1488–1489
  Sym plasmids 1715
plasmids (*continued*)
  theta replication 1485–1487, *1486f*
  see also bacterial genetics; yeast plasmids
Plasmodium 80, 338, 1217, 1379
Plasmodium falciparum 405, 519, *520f*, 589, 749, 765, 1219, 1221
Plasmodium vivax 588
Plasterk R H A **933–934, 2016–2020**
plastids 1379
platelet-derived growth factor (PDGF) 1967, 1884
Plato 209
Platt, John R. 1183
PLE gene 716
pleiotropy 824, 846, **1490–1491**, 1826
  see also concordance; operon
plesiomorphy **1491**, 1904, 1905
Plethodon 249, 344
Plethodon cinereus 344, 1522, *1522t*
ploidy **1491**
PLOSL (early adulthood-onset progressive dementia) 724
PMS1 gene 1285
PMS2 gene 1285
PMS2 protein 1667
pneumolysin 1493–1494
pneumonia bacteria **1491–1494**
  capsule synthesis 1494
  cloning 1493
  competence regulation 1492, 1494
  discovery 1491–1492
  DNA binding and entry 1492
  folate biosynthesis 1493
  gene expression 1493
  gene mapping 1493
  genetic chemistry 1492–1493
  genetic transformation 1491–1492
  glycosylases 1494
  insertion sequences 1492
  mismatch repair 1493
  mosaic genes 1494
  phages 1492
  plasmids 1492
  pneumolysin 1493–1494
  population genetics 1494
  recombination modes 1493
  restriction enzymes 1494
  surface proteins 1493
  transposons 1492
  virulence genes 1493–1494
Poaceae 971
Podalyrieae 1083
Podospora 749
Podospora anserina 1154, 1223
Podoviridae 181, 1922
Poeciliopsis 730
point mutations 345, 524, 1188, **1494**
poky mutant 1223
polarity **1495**
polaron **1495**
Pol gene 979
pol gene 1704, 2021
PolI *see* DNA polymerases
PolII *see* DNA polymerases
PolIII *see* DNA polymerases
Polisky, Barry 958
Pollak E **826–827, 1595–1597, 1597–1598, 1791–1792**
Pollard J W **1966–1969**
Pollock, Martin 1019
Polocco, M. 849
poly(A)-binding proteins 1733–1734

polyacrylamide gel electrophoresis (PAGE) 758
polyadenylation **1496**
poly(A) tail **1495–1496**
see also messenger RNA (mRNA)
polycistronic mRNA **1496**
see also messenger RNA (mRNA)
polycystic kidney disease 337, 357
polycystic ovarian syndrome (PCO) 1024, 689
polyethylene glycon (PEG) 1356
polygenes *see* complex traits
polyglutamine diseases 594
dentatorubral pallidoluysian atrophy 983
Huntington's disease 981–983
spinobulbar muscular atrophy (SBMA) 983
spinocerebellar ataxias (SCA) 983
trinucleotide repeat disorders 594
polymerase **1497–1499**
DNA repair 1498
function 1497–1498
HIV-1 reverse transcriptase 1498
RNA polymerases 1499
structure 1497, *1498f*
T7 DNA polymerase *1498f*
polymerase chain reaction (PCR) 651, 652, 972, 1002, **1499–1503**
chromosome mapping 354
contamination 1503
DNA cloning 546
DNA typing 1502
gene construction and mutants 1501
linkage map 1107
linker PCR 1501–1502
long and accurate PCR 1502
nesting 1502
PCR cycle 1499–1501, *1500f*
PCR end-point analysis 1503
proofreading 1502
real-time quantitative PCR **1503–1506**
ASO primer *1505f*, 1506
ASO probe *1505f*, 1506
hybridization probe 1504
medical applications *1504t*
minimal residual disease 1506
principles *1503f*, 1503–1504
SYBR green I dye 1504–1505
TaqMan probe 1504
Taq DNA polymerase 1501
thermophilic bacteria 1961
in vitro mutagenesis 1012
see also DNA polymerases; template
polymerase I 1497
polymorphism **1507–1509**
blood groups polymorphism 1507
DNA polymorphism 1508
heterozygote 935
inversion polymorphism 1507
protein electrophoresis 1516
protein polymorphisms 1507–1508
tree reconstruction **1509**
uses 1508–1509
polynucleotide phosphorylase 1067
polyomavirus 589
polypeptides 998, **1509**
polyploidy **1509–1511**, 1684
agriculture 1510–1511
allotetraploids 1510
Arabidopsis thaliana 1510
autotetraploids 1510
detection 1510
polyploidy (*continued*)
evolution 1510
nomenclature 1509–1510
origin 1510
see also chromosome aberrations; chromosome number; triploidy
polyposis syndromes 422
polysaccharide intracellular adhesin (PIA) 46
polysome (polyribosome) **1511**
Polystichum munitum 840
polytene chromosomes 344, 963, **1511–1513**
Balbiani rings 1513
Drosophila melanogaster 1511, 1512–1513
endoreplication 1511–1512
gene isolation and cloning 1513
gene localization 1512–1513
puffs 1513
radiation genetics 1512–1513
Pomeranz Krummel D **1733–1735**
Pompe disease 756
Pontecorvo, G. 383
Pooideae 890
pop-1 gene 617
POP-1 protein 308
Pope F M **599–601, 603–605, 993–995, 1580–1581, 1581–1582**
population genetics **1513–1519**
allelic frequency 1514–1515
basis of complex traits 1518
bottleneck effect 233
Darlington, Cyril Dean 513
epistasis 640
fundamental theorem of natural selection 734
Galton, Francis 743
genetic conflicts and evolution 1517–1518
genetic load 839
genetic variations 1515–1516
Hardy–Weinberg law 914
hemophilia 918
history 490, 1513–1514
human origins 1518–1519
identity by descent 995
isolation by distance 1055
linkage disequilibrium mapping 1518
mathematical models 1516–1517
migration and population structure 1515
Muller's ratchet 1518
mutant allele **1268**
mutations 1514
natural selection 1515
nucleotide diversity and recombination rate 1517
QTL (quantitative trait locus) 1518
random genetic drift 1514–1515
random mating **1601–1602**
t haplotype 1921
see also conservation genetics
population substructure **1519–1523**
allelic frequency 1520–1522
gene flow *1522t*
genetic consequences 1520
gene tree *1522f*, 1522–1523, *1523f*
Hardy–Weinberg law 1519
migration 1521–1522
origins 1519, *1520f*
single locus 1520–1521
two or more loci 1521
Wahlund effect 1520
Wright's *F* statistic 1521
Populus 1866
pos-1 gene 619
positional homology 968
position effects **1523–1530**, *1525f*
Drosophila melanogaster 1525f, 1525–1526
historical background 1525–1526
normal genetic structure 1524–1525
position effect variegation (PEV) 1527–1529
Saccharomyces cerevisiae 1527, 1529
stable position effects 1526–1527
euchromatic 1526–1527
yeast mating locus 1527
yeast rDNA genes 1527
telemeric position effect 1529
position effect variegation 340, 1852
positive interference *see* interference
positive regulator proteins **1530**
positive supercoiling **1530**
postmeiotic segregation **1531–1532**
definition 1532
heteroduplex DNA 931
Holliday's model 956
origin *1531f*, 1532
postranscriptional modification **1532–1533**
posttranscriptional gene silencing (PTGS) 1473–1474
posttranslational modification **1533**
postzygotic reproductive isolating mechanism (RIM) 985
potato *see Solanum tuberosum* (potato)
Potter M **1482–1485**
Poulton, E. 1862
Poulton J **1149–1151, 1222**
Pour P M **310–311**
Powell W **971–973**
PP2A (putative) gene *2050t*
*PPARγ*gene 2083, *2084–2085t*
(p)ppGpp nucleotide synthesis 1891
PRA *see* progressive retinal atrophy (PRA)
Prader–Willi syndrome 346, 361, 1000, 1179, 1421
epigenetics 634
maternal inheritance 1150
oculocutaneous albinism (OCA) 26
uniparental disomy (UPD) 1241
Prangishvili D **2114–2116**
pRb protein 310
prcd gene 269
predator–prey and parasite–host interactions **1533–1536**
complex defense traits 1535
gene for gene interactions in plants 1535, *1536t*
horizontal resistance in plants 1535
sequential structure 1533–1534
simple defense traits 1534
vertical resistance in plants 1535
premature menopause 1021
premature ovarian failure 1021
premature termination codon (PTC) mutations 1188
pre-mRNA splicing **1536–1539**
commitment complex (CC) 1538
nuclear pre-mRNA introns 1536
proteins used for splicing 1537–1538
spliceosomal smRNAs 1537
spliceosome 1537
spliceosome assembly pathway 1538–1539, *1539f*
splicing reaction 1536–1537, *1537f*

prenatal diagnosis **1539–1541**
  amniocentesis 1540
  chorion villus sampling (CVS) 1540
  diagnostic procedures 1540–1541
  Down syndrome 1540
  Huntington disease 1540
  screening for chromosome aberrations 1541
  see also genetic counseling
presenilin gene family 49
pre-Sertoli cells 1812, 1816
Prezent, Isaac I. 1129
primary ciliary dyskinesia 337
primary transcript **1541**
primase **1542–1546**
  bacteria and bacteriophages 1542
  biochemical properties 1542
  definition 1542
  discovery 1542
  Escherichia coli 1542, *1543f*
  eukaryotic primases *1544f*, 1544–1545
  genetic control 1545
  macromolecular synthesis operon 1543, *1544f*
  physiological properties 1543
primer **1546**
primer RNA **1546–1548**
  definition 1546
  DNA synthesis 1546–1547
  initiation specificity 1547, *1547t*
  Okazaki fragments 1546, *1547f*
primitive character **1548**
primosome **1548**
primosome assembly sequences (*pas*) 1486–1487
Primula 112
printrons 1052
prions 2007
  see also spongiform encephalopathies (transmissible)
prisoner's dilemma 669–670
  see also Hamilton's theory
pristane 1482
PRKY gene 2159
prlD gene 1555
probe **1548**
procentriole *see* centrioles
processed pseudogene **1548**
processive single-strand exonucleases 542
proflavin 2054
progeny testing **1548–1549**
  see also multifactorial inheritance; selective breeding
programmed cell death *see* apoptosis
progressive retinal atrophy (PRA) 269
prokaryotes **1549**
  DNA repair 563
  elongation factors 611
  excision repair 673
  genetic code *822f*
  heat shock proteins (Hsps) 915
  horizontal transfer 974
  hot spots 976–977
  kinases 1063
  linkage 1105
  nonreciprocal exchange **1350**
  open reading frame (ORF) 1375
  protein secretion systems 1553–1558, *1554f*
  regulatory RNA 1658
prolamins 1783–1784
proline **1549**
promoter occlusion 1982
promoters **1549–1551**
  cis-acting locus 379–380
  detection 380
  enhancers 624–625
  eukaryotic promoters *1550f*, 1550–1551
  function 379
  histidine biosynthesis 945
  histone genes 948–949
  hot spots 976, 977
  housekeeping genes 978
  lac mutants 1069
  lac operon 1070
  operators 1376
  prokaryotic promoters 1550, *1550f*
  structure 379
  upstream, downstream sites 2100
  see also cis-acting proteins
pronuclear microinjection 1991–1992, *1993f*
proofreading **1551**
  DNA polymerases 558
  ribosome 601–602
  spontaneous mutations 1279
  see also editing and proofreading
prophage 658, 998, 1131, 1434, **1551**
Prop1 (Prophet of Pit1) gene 592
propionic acidemia (PAA) 1363
proplastids 1379
Prosopis (mesquite) 1083
prospero gene 1322
protease gene 2022
protein interaction domains **1551–1552**
  see also proteins and protein structure
proteins
  Agouti protein 23
  Agr proteins 1886
  alpha (α)-catenin 620
  AP180 protein 1313
  AraC 543
  ArgR protein 1040, *1040t*
  ATM protein 113–114
  ATR protein 1912
  Bcr/Abl fusion protein 208
  beta (β)-catenin 620
  Blk protein 1883
  BLM protein 1646, 1912
  BRCA1 protein 239–240
  BRCA2 protein 240
  c-Abl protein 208, 251
  cadherin 620
  CAP (catabolite activator protein) 1070
  Cdc45 protein 1384
  Cdc6 protein 1384
  CED-3 protein 205
  CED-4 protein 205
  CED-9 protein 205
  CENP-A 1065
  CENP-B 1065
  CENP-C 1065
  CENP-E 1065
  CENP-E protein 1225
  CENP-F 1065
  CHS protein 27
  cis-acting proteins 938
  CisA protein 380
  COR1 protein *1910f*, 1911
  Cro protein 1131
  CRP (cAMP receptor protein) 1070
  CRP (catabolite repressor protein) 544
  CRP (cysteine-rich protein) family 1098
  CTCF 634–635
  Dab1 protein 1650
  DMC1 protein 1645, *1645f*, *1646f*, 1912
proteins (*continued*)
  Dmc1 protein 1906
  DnaA protein *1383f*
  DnaB 916
  Dps 542, 544
  DYN-1 protein 1313
  EBERs protein 643, 1289
  EBNA1 protein 642–644, 1289
  EBNA2 protein 642–643
  EBNA3A protein 642–644
  EBNA3B protein 642–644
  EBNA3C protein 642–644
  EBNA5 protein 642
  EBNA6 protein 642
  EBNA-LP protein 642
  E family proteins 15–17
  EF-Tu 934
  Eg5 protein 1225
  EGL-1protein 205
  eIF4a 916
  en-1 (engrailed-1) protein 1100
  Fgr protein 1883
  FIS (factor for inversion stimulation) 121, 542, 939–941, 1040–1041, *1040t*, *1383f*
  FixJ protein *708t*
  FixK protein *708t*
  FixL protein *708t*
  folding 324–325
  Fpg 566
  Fyn protein 1883
  GATA4 protein 1814, 1818
  gp32 protein 1839
  gp43 protein 1841
  gp59 protein 1841
  gp61 protein 1841
  Hck protein 1883
  HMG1/2 951
  H-NS 542
  HNS protein *1383f*
  HobH protein *1383f*
  hRAD1 protein 1912
  HsdM protein 1693
  HsdR protein 1693
  HsdS protein 1693
  HU 542
  ICENP 1065
  IciA protein *1383f*
  IHF (integration host factor) protein 121, 800, *1040t*, 1040–1041, *1383f*
  LacI protein 165
  Lck protein 1883
  LexA protein 1622, 1853–1854
  LIN-6 protein 301
  LIN-12 protein 306, *307f*
  LIN-14 protein 306, *307f*
  LMP1 protein 642–643, 1289
  LMP2A protein 642–644, 1289
  LMP2B protein 642–644, 1289
  Lyn protein 1883
  MASP-1 protein 429
  MCAK protein 1225
  MCP protein 433
  MeCP2 633
  MerR protein 164
  MLH1 protein 1646, 1667, 1912
  Mod protein 1693
  MRE11 protein 1645
  MSH2 protein 1667
  MSH3 protein 1667
  MSH6 protein 1667
  neurofibromin protein 1308

proteins (*continued*)
NodA protein *1332f*
NodB protein *1332f*
NodC protein *1332f*
NodD protein 1333
NodE protein *1332f*
NodF protein *1332f*
NodH protein *1332f*
NodL protein *1332f*
NodO protein 1332
NodPQ protein *1332f*
NodT protein 1332
NodV protein 1334
NodW protein 1334
ODR-10 protein 1313
P450c1-α 1027
p53 protein 239, 273–274
PAR (proteins) *616f*, 616–618
Pcg proteins 1814
PepA protein 1040, *1040t*
PER protein 387
PINCH protein 1098
PMS2 protein 1667
POP-1 protein 308
PriA 916
RAB proteins 1602
Rad1 protein 1906
Rad4 protein 1666
Rad14 protein 1666
Rad23 protein 1666
Rad51 protein 582, *608f*, 931, 1645, *1645f*, *1646f*, 1668, 1841, 1906, 1912
RAG1 protein 998, 1641
RAG2 protein 998, 1641
RAP1 950
RAS proteins 1602, 1603
RecA protein 163, 175, 582, 931, 938, 977, **1618–1623**, 1839, 1841, 1853–1854, 1906
RecBCD enzyme 916, 938, 977, 1841
RecE protein 1647
RecO protein 1841
RecR protein 1841
RecT protein 1647
RelA protein 1892
Res protein 1693
RET protein 1257–1258
Rho 916
RHO/RAC proteins 1602
ROB protein *1383f*
RPA protein 675
Ruv proteins 1668
SCP2 protein *1910f*, 1911
SEL-12 protein 1313
SF1 protein 1814
SIN1 951
SIN2 951
SIR2 950
SIR3 950
SIR4 950
SmLIM (smooth muscle) protein 1098
Sm proteins 1538
SNAP-25/RIC-4 protein 1312–1313
SNF2 950
SNF5 950
SNF6 950
SOX9 protein 1814, 1817
SPO11 protein 1645
SpoT protein 1892
Src protein 251, 1883
SR proteins 1538
SRY protein 1812, 1816
SV40 T-antigen 916

proteins (*continued*)
SWI1/ADR6 950–951
SWI2/SNF2 950–951
SWI3 950
synaptobrevin/SNB-1 protein 1312–1313
synaptotagmin/SNT-1 1313
synpatojanin/UNC-26 protein 1313
syntaxin/UNC-64 protein 1312–1313
TATA binding protein 544
TFIIH transcription factor 562, 568, 675
TrpA protein *419f*
UNC-4 protein 1312
UNC-5 protein 1311
UNC-6 protein 1311
UNC-11 protein 1313
UNC-13 protein 1312–1313
UNC-17 protein 1312
UNC-18 protein 1312–1313
UNC-25 protein 1312
UNC-30 protein 1312
UNC-34 protein 1311
UNC-40 protein 1311
UNC-47 protein 1312
UNC-49 protein 1312
UNC-73 protein 1311
UvrAB protein 916
UvrA protein 674, *674f*, 1665, *1665f*
UvrB protein 674, *674f*, 1665, *1665f*
UvrC protein 674, *674f*, 1665, *1665f*
UvrD protein *1665f*, 1666
UvsX protein 1841
VirA protein *1985t*
VirB1–11 protien *1985t*
VirC protein *1985t*
VirD1 enzyme *1985t*
VirD2 enzyme *1985t*
VirD4 protein *1985t*
VirE2 protein *1985t*
VirG protein *1985t*
Xis (excisionase) protein 121, *1040t*, 1041
XMAP215 protein 301
XPA protein 675, 1666
XPB protein 675, 1666
XPC-HR23B protein 675, 1666
XPD protein 675, 1666
XPF-ERCC1 protein 675
XPF protein 1666
XPG protein 675, 1666
Yes protein 1883
ZYG-9 protein 301
proteins and protein structure **1568–1573**
alpha (α)-helix 1569, *1569f*
amino acids 1568
beta (β)-strand 1569–1570, *1570f*
covalent modifications of proteins 1572
domains 1570–1572, *1571f*
fibrous proteins 1572
globular proteins 1570–1571
levels of structure 1568–1569
peptide bond *1568f*
phi (ϕ) bond 1569, *1569f*
psi (Ψ) bond 1569, *1569f*
quaternary structure 1572
secondary structure 1569–1570
structural genomics 1572
structural motifs 1570
superfolds 1571, *1571f*
tertiary structure 1570–1572
web sites 1572–1573
see also protein interaction domains; protein synthesis

protein secretion systems **1552–1565**
description 1552
diseases caused by misfolding 1564–1565
eukaryotes *1554f*, 1558–1564
endoplasmic reticulum *1554f*, 1558–1561
Golgi apparatus *1554f*, 1558–1564
transport vesicles 1561–1562
prokaryotes 1553–1558, *1554f*
bacterial pathway 1553–1555
organelle assembly in periplasm 1555–1556
outer membrane proteins (OMPs) 1555
pili 1556
type III secretion systems 1556–1558
signal hypothesis 1552–1553
signal sequence 1552
see also mitochondria; ribosomes
protein splicing *1565f*, **1565–1567**
biochemistry 1566
cis-splicing *1565f*
discovery of inteins 1565
intein organization 1565–1566
protein manipulation *1567f*
trans-splicing *1565f*
uses 1566–1567
protein synthesis **1567**
see also proteins and protein structure
Proteobacteria (Purple Bacteria) 149, 1378
proteolysis **1573–1575**
apoptosis 1574
destruction of damaged proteins 1573–1574
digestion 1573
endocytosis 1573
mitosis 1226–1227
modification of new proteins 1574
regulatory proteins 1574
site-specific 1574
ubiquitin system 1573
proteome **1575–1578**
analysis by physical techniques 1576–1577
databases 1577–1578
Drosophila melanogaster 1577
functional proteomics 1577
mouse 1577
Saccharomyces cerevisiae 1577
significance 1575–1576
web sites 1577
proteomics 1201
Proteus 176
Proteus vulgaris 147f
prot gene 2022
proto-oncogene 1370, **1578**
see also oncogenes
Protopterus aethiopicus 860
Prout T **730–731**
Providencia stuartii 1694t
provirus **1578–1579**
long terminal repeats (LTRs) 1120
proximal **1579**
PRP (precursor RNA processing) genes 1537
Pruitt, Bob 88
Prusiner, Stanley 1879
PRY gene 2159
psaA gene *1461t*, 1464
psaB gene *1461t*
psaC gene *1461t*
PsaD gene *1461t*
PsaE gene *1461t*

PsaF gene *1461t*
psa gene 1461
PsaG gene *1461t*
PsaK gene *1461t*
psal gene *1461t*
PsaL gene *1461t*
PsaM gene *1461t*
PsaN gene *1461t*
psb gene 1461
pSC101 plasmid 1486
pseudoalleles *see* alleles
Pseudoalteromonas atlantica 45
pseudoautosomal linkage **1579**
see also X-chromosome inactivation
pseudogene 667, **1579–1580**
Pseudomonas 176, 1079, 1556, 1903
Pseudomonas aeruginosa 75, 586, 796, 1198, 1382
Pseudomonas alcaligenes 1694t
Pseudomonas fluorescens 1914t
Pseudomonas Genetic Stock Center 850–851
Pseudomonas lachrymans 1914t
Pseudomonas putida 155, 1382, 1892, 1480, *1914t*
Pseudomonas savastanoi 492
Pseudomonas syringae 226, 492, *1914t*
Pseudomonas tabaci 1914t
Pseudoperonospora cubensis 1914t
pseudouridine 1988
pseudovaginal perineoscrotal hypospadias 1823
pseudoxanthoma elasticum (PXE) **1580–1581**, *1580f*
cutis laxa 498, *498f*
ectodermal dysplasias 599
Ehlers–Danlos syndrome (EDS) 605
Psi (Ψ) bond 1569, *1569f*
Psi (Ψ)-loop 1988
Psoraleeae 1083
psoriasis *1581f*, **1581–1582**
pT181 plasmid 1488
PTC gene *81t*, 191, *2084–2085t*
PTEN gene *2084–2085t*
PTEN/MMAC oncogene 11
PTGS *see* posttranscriptional gene silencing (PTGS)
pTiA6 plasmid *453f*
pTiC58 plasmid *453f*
ptl gene *453f*
ptr gene 1624, *1624f*
PTS *see* carbohydrate phosphotransferase system (PTS)
Puccinia graminis 973
Puccinia hordei 973
Puccinia striformis 973
Puck, T.T. 1969
puff 1513, **1582**
Pugh, J. 629
pulse-chase **1582**
pulsed field gel electrophoresis (PFGE) 363, **1582–1583**, *1583f*
applications 1584–1585
history 1583
large DNA fragments 1584
parameters affecting resolution 1584
resolving power 1584
punctuated equilibrium 665–666
Punnett, Reginald Crundall 1178, 2015
Punnett square *1176f*, **1585**, *1585f*
PU (Palindromic Unit) 214
purine **1585**
puromycin 1729
pur operon 1079
put operon 1896
putR repressor 1896
Pvt-1 gene 1483
pVT745 plasmid *453f*
Pwo DNA polymerase 1502
PXE *see* pseudoxanthoma elasticum (PXE)
Pyeritz R E **90, 1144–1146**
PYK1 gene 887
PYK2 gene 887
Pyrenopeziza brassicae 1154
Pyrenophora graminea 973
Pyrenophora teres 973
pyrimidine **1586**
pyrimidine dimers 557, **1586**
pyrimidine–pyrimidine sequences 977
Pyrococcus furiosus 1198
Pyrococcus horikoshii 92
pyrogenic staphylococcal enterotoxins 185
pyrogenic streptococcal enterotoxins 185
pyr operon 1079

Q

Q antitermination protein 380
Q-banding 348, 349, **1587**
chromosome banding 1587
quinacrine 1587
Q gene 1045
Qβ replicase 1005, *1005f*
QTL mapping 28, **1587–1593**
composite interval mapping *1590f*
advantages 1590
interval mapping 1589
likelihood ratio 1589
limitations 1590
multiple regression analysis 1589
nonlinearity 1589
interval mapping 1588, *1590f*
mapping data
genetic linkage map 1587
polymorphic genetic markers 1587
quantitative trait values 1587
trait variations 1587
mathematical calculations 1588
multiple interval mapping 1590–1591, *1592f*
Drosophila mauritiana 1592
Drosophila simulans 1592
epistasis 1591
estimation procedure 1591
evaluation procedure 1591
likelihood ratio 1591
prediction procedure 1591
search strategy 1591
quantitative genetics 1587
quantitative trait locus (QTL) 1587
recombination frequency 1587
relationship between genotype and phenotype 1587
single marker analysis 1588
statistical calculations
chi-square test 1589
critical threshold 1589
ghost gene phenomenon 1589
human linkage analysis 1589
likelihood ratio 1588
LOD score 1589, *1590f*, *1592f*
statistical methods
identity-by-descent mapping 1593
Markov chain Monte Carlo algorithms 1593
sib-pair method 1593
stepwise model selection
epistasis step 1592
QTL mapping (*continued*)
estimation step 1592
evaluation step 1592
optimization step 1592
search step 1592
stopping rule 1592
QTL (quantitative trait locus) 358, 435, 464, 1518, **1593–1595**
comparison techniques 1593
continuous traits 1593
genetic factors 1593
genetic mapping 1593
human trait analysis
computer algorithms 1594
polymorphic molecular markers 1594
mouse trait analysis
advantages 1595
disadvantages 1595
human homologs 1595
inbred strains 1594
polygenic traits 1594
nature–nurture controversy 1299, *1300f*
neoteny 1304
nongenetic factors 1593
numerical expression 1593
phenotypic traits 1593
recombinant congenic strains 1636
selective breeding 1799
Quality Protein Maize (QPM) 1786
quantitative genetics 463
quantitative inheritance **1595–1597**
classical quantitative genetics
covariance 1596
dominance deviation 1595
Fisher, R.A. 1595
genetic variance 1596
Mendelian inheritance 1595
statistical methods 1595
long-term selection experiments 1596
molecular genetic analysis 1596–1597
quantitative trait **1597–1598**
analysis 639
Fisher, R.A. 1597
genetic variability 1597
major genes 1597
measurable traits 1597
QTL (quantitative trait locus) 1597
resemblance of relatives 1597
selection intensity 1793
selection limit 1794
see also character
quantitative trait locus (QTL) *see* QTL (quantitative trait locus)
Quercus 840, 1866
quinolones 77, 1687

R

R1 plasmid 380, 382, 1486
R6K plasmid 1486
Rabbitts P **1120–1121, 1122–1123**
Rabbitts T H **1118–1119**
RAB proteins 1602
Race, R.R. 1713
rachis 1873
RAD10 gene 1838
Rad14 protein 1666
RAD1 gene 1838
Rad23 protein 1666
RAD26 gene 563
rad27 gene 595
RAD30A gene 558
Rad4 protein 1666

RAD51L1 gene 1110, *1112f*
RAD51 protein 1668, 1841, 1912
double-strand break repair model 582
electron microscopy (EM) *608f*
recombination nodules (RNs) 1645, *1645f*, *1646f*
RAD genes **1599**
RADIALIS gene 716
radiation genetics
Drosophila melanogaster 1512–1513
polytene chromosomes 1512–1513
see also translocation
radiation genetics (mouse) **1599–1601**
Delbank 1600
deletion experiments 1600
embryonic stem (ES) cells 1599
mouse genome 1600
mutational studies 1599
screening 1600
radiation hybrid mapping 354, 360
radiation hybrids 270
radish *1914t*
Radman, Miroslav 1651, 1853
raf oncogene 1484
RAG1 protein 998, 1641, *1738f*
RAG2 protein 998, 1641, *1738f*
Raju T N K **90–91**, **1067–1068**, **1846–1847**
Raleigh E A **1–2**, **151–156**
Ralls, Katherine 459
Ralstonia solanacearum 1850
Rana 762
Rana japonica 2097
Rana nigromaculata 2097
Rana pipiens 2097
Randall, John 554
random amplified polymorphic DNA (RAPD) 232, 831, 968, 972
random genetic drift 1514–1515
random mating 913, **1601–1602**
RAP1 gene 1948
Rap enzyme 1760
Raphanus sativus (radish) 237
rapid-lysis (*r*) mutants 1927
RAS1 gene 1603
RAS2 gene 1603, 22
RAS gene family **1602–1607**
crystallographic structure 1607
functional diversity 1605
oncogenes 1602
protein families 1602
RAS proteins
cellular roles 1603
function 1603–1605
guanosine diphosphate (GDP) 1603, *1603f*
guanosine triphosphatase (GTPase) 1603, *1603f*
guanosine triphosphate (GTP) 1603, *1603f*
MAPK (mitogen-activated kinases) 1604–1605
MAPK pathways *1604f*
MAPKK 1604
MAPKKK 1604
molecular switch 1603, *1603f*
phosphatidylinositol-3 kinase 1605
RHO/RAC proteins 1605
signaling pathways 1603–1604, *1604f*
Src family tyrosine kinases 1603–1604
structure *1606f*, 1606–1607
yeast 1603
transcriptional control 1605
RAS gene family (*continued*)
tumorigenesis 1606
ras oncogene 69, 1484
RAS proteins 1602, 1603
RATMAP database 1608
rats **1607–1610**
cancer research 1609, *1609f*
comparative genomic hybridization (CGH) *1609f*, 1609–1610
comparative mapping 1610
cytogenetics 1608
expressed sequence tags (ESTs) *366t*
historical background 1607–1608
inbred strain 1609
loss of heterozygosity (LOH) 1609
model organisms 1608
rat liver S9 53
RATMAP database 1608
Rattus norvegicus 1607
Rattus rattus 1607
Raven, P.H. 411
RB1 gene 263, 1770, *2084–2085t*
R-banding 349
rbcL gene 238
RbcS gene 1463
Rb gene 2081, 2083
RB gene 274, 1123
RBI gene *81t*
RBM gene 2156, 2159
Rbm gene cluster *1253f*, 1254
RBTN1/TTG1 see LMO family of genes
RBTN2/TTG2 see LMO family of genes
RBTN3 see LMO family of genes
rde gene 930
rDNA amplification **1610–1612**
mechanisms 1610–1611
rRNA repeating sequences *1611f*, 1610–1612
Tetrahymena thermophila 1612
transcription process *1611f*, 1611–1612
Xenopus laevis 1611–1612
rDT gene 2026
reading frame **1612**, 2000
Reardon J T **673–675**
rearrangements **1612–1614**
consequences 1613
mechanisms 1613
varieties 1612–1613
recA gene 1651, 1853–1854
RecA protein 161, 175, **1618–1623**, 1839
ATP hydrolysis 1621
DNA repair 1621–1622, *1622f*
DNA strand exchange reaction *1620f*, 1620–1621
extension and product formation 1621
homology alignment 1620
nucleoprotein filament formation 1620
strand switch 1621
double-strand break repair model 582
filament structure *1619f*, 1619–1620, *1620f*
monomer structure *1618f*, 1618–1619
Rad51 homolog 1621
reckless DNA degradation 1635
recombination pathways 1647
repair mechanisms 1668
RuvAB enzyme 1756, *1757f*
single-stranded DNA-binding proteins (SSBs) 1841
SOS repair 1622–1623, 1853–1854
RecA protein (*continued*)
synapsis (DNA transactions) 1906
recA/RAD51 gene family 1110
RecBCD enzyme 175, 325–328, *327f*, *1622f*, **1623–1631**, 1841
ATP hydrolysis 1624
Chi interaction *1625f*, 1625–1626, *1626f*
Chi sequences 1623
Chi sites 1624
DNA replication restart 1628–1629, *1629f*
double-strand break repair 1668
enzymology 1626–1628, *1627f*
exonuclease V 1623
function 1623, 1624
genetic organization 1624, *1624f*
helix–loop–helix motif 1624
heteroduplex DNA 1625
Holliday junction 1625, *1626f*
hot spots 1624
phage λ 1625
prevention of sigma replication 1629
RecBCD recombination pathway 1623
recD gene 1624
reckless DNA degradation 1635
repair process 1626–1628
Ruv proteins 1623
recB gene 1623, *1624f*, 1647
recC gene 1623, *1624f*, 1647
Recchia G D **351–353**
recD gene 1624, *1624f*
recE gene 1647
RecE protein 1647
receptor protein tyrosine kinases (RPTKs) 1884
recessive **1631**
recessive inheritance **1631**
recessive lethal 525, **1631**
RecFOR protein *1622f*
rec genes 938, **1614–1618**, *1615–1617t*
DNA polymerases 1614
exonucleases 1614
historical background 1614
homologs 1614
Rec proteins *1615–1617t*
Ruv proteins 1614
Saccharomyces cerevisiae 1614
RecG protein *1622f*
reciprocal cross **1631–1632**, 2009–2010
reciprocality **1635**
reciprocal recombination **1632–1635**
double-strand break repair model 1632, *1633f*
gene conversion 1632, *1633f*
Holliday junction 1632, *1633f*
meiosis 1632
meiotic tetrads 1632, *1632t*
nonreciprocal recombination 1634
site-specific recombination 1633–1634
reciprocal translocation 347, **1635**
reckless DNA degradation **1635–1636**
Escherichia coli 1635
RecA protein 1635
RecBCD enzyme 1635
recombination process 1635
ultraviolet radiation 1635
Reclinomonas 1217
Reclinomonas americana 1220, 1221
recombinant **1636**
recombinant congenic strains **1636–1637**
backcross 1636
BALB/c mouse 1636

recombinant congenic strains (*continued*)
genetic mapping 1636
inbred strain 1636
Mouse Genome Database (MGD) 1636
mouse strains 1636
QTL (quantitative trait locus) 1636
recombinant DNA **1637–1639**
DNA cloning 1637
historical background 1637–1638
inserts 1637–1639
phage λ 1638
plasmid vectors 1638
polymerase chain reaction (PCR) 1639
Recombinant DNA Advisory Committee 1638
replicon 1637
structure 1637, *1637f*
vectors 1637–1639
yeast vectors 1638
Recombinant DNA Advisory Committee 1638
recombinant DNA guidelines **1639**
recombinant DNA technology *see* DNA cloning
recombinant inbred strains **1639–1640**
see also inbred strain
recombination 261–262
chloroplast DNA 338
chromatid 339
coefficient of coincidence **413–414**
meiosis 339
see also phage Mu
recombinational repair **1649**
recombination frequencies (*R*) 319
recombination frequency
linkage 1105
map expansion **1140–1141**
mapping function 1141
meiosis 1106
recombination hot spots (mouse) *see* hot spots
recombination in the immune system **1640–1642**
allelic exclusion 1641
antigen diversity 1641–1642
B-cells 1640
immunoglobulin heavy (IGH) chain switch recombination *1641f*, 1642
lymphocyte development 1640–1641
T-cells 1640
V(D)J recombination 1640–1642, *1641f*
recombination models **1642–1644**
asymmetrical heteroduplex 1642
chiasma-type hypothesis 1642
copy-choice hypothesis 1642
crossover 1643
double-strand break repair model 1643
gene conversion 1642
heteroduplex DNA 1642
Holliday junction 1642
hybrid DNA 1642
meiotic recombination 1643
Meselson–Radding model 1642
mitosis 1643
replication restart 1643
see also Holliday's model
recombination nodules (RNs) **1645–1647**
Bloom's syndrome (BS) 1646
chiasma 330
chromatin loops *1645f*
definition 1645
early nodules 1645, *1645f*, *1646f*
late nodules *1645f*, 1646, *1646f*
recombination nodules (RNs) (*continued*)
Chloealtis conspersa 1646, *1646f*
chromosomal locations 1646, *1646f*
detection 1647
distribution 1646
function 1646
Locusta migratoria 1646, *1646f*
meiosis 1645, *1645f*
nuclear envelope (NE) *1645f*
synaptonemal complex *1645f*, *1646f*
recombination pathways 1647, **1647–1648**
Escherichia coli 1647
RecBCD pathway 1647
RecE pathway 1647
RecF pathway 1648
Ruv proteins 1647
see also Chi sequences
recombination signal sequences 998
recombination suppression **1648–1649**
RecO protein 1841
recQ gene 568
recQ homologs 571
RecR protein 1841
recT gene 1647
RecT protein 1647
RED *see* repeat expansion detection (RED)
RED1 gene product 1912
red clover *1914t*
Rédei, George P. 87
Redfield R J **165–168, 1984**
red gene 1625
red–green color blindness 354
Red Queen coevolution 410
Red Queen hypothesis 706
Reed, Dorothy 953
Reed–Sternberg cells **1649**
reeler locus **1649–1650**
neuronal development 1649–1650
scrambler mutation 1650
signaling pathways 1650
yotari mutation 1650
Reeves R H **1649–1650**
Refsum disease 994
regulatory genes **1652–1657**
activators 1652
autoregulatory genes 1654, *1654f*
definition 1652
DNA-binding proteins 1654–1655
effector molecules 1652
lac operon 1652, *1653f*
mechanisms 1655–1656, *1656f*
negative control 1652
postive control 1652
regulation strategies *1653t*
regulators
bifunctional 1654
global 1654
regulons 1654
repressors 1652
RNA polymerase interaction 1655
structures
DNA loops *1656f*
enhanceosome 1656, *1656f*
repressosome *1656f*, 1656–1657
superimposed controls
cAMP receptor protein (CRP) 1653
cyclic AMP (cAMP) 1653
independent activators 1654, *1654f*
independent control 1653, *1653f*
mixed control 1653, *1653f*
regulatory RNA **1657–1658**
Caenorhabditis elegans 1658
eukaryotes 1658
regulatory RNA (*continued*)
function 1657–1658
posttranscription regulation 1658
prokaryotes 1658
regulons 655, 1195
Reha-Krantz L J **1711–1713**
Reichheld S **2117–2120**
Reik W **65, 901–902**
Reinhart B J **1657–1658**
relA gene 1891
RelA protein 1892
release (termination) factors **1659**
Rel oncogene **1658–1659**
Reln (reelin) gene 1649
renal cell cancer **1659–1660**
chromosomal aberrations 1660
classification 1659
genetic classification 1659–1660
Grawitz tumor 1659
pathology 1659
von Hippel–Lindau disease 1659–1660
Y chromosome (human) 1660
renaturation **1661**
Renwich, Jim 356, 359
repA gene 382
repair mechanisms **1661–1669**
biological effects of damage 1663
direct reversal of damage
alkyl group removal *1662f*, 1663
ligation 1663
photoreactivation *1662f*, 1663
indirect reversal of damage 1663–1668
base excision repair (BER) 1663–1664, *1664f*
mismatched bases 1664–1665
mismatch repair 1666–1668, *1667f*
oligonucleotide excision *1665f*, 1665–1666
ribonucleotide excision 1665
interaction 1669
Mut system 1667
recombinational repair 1668–1669
double-strand break repair 1668, *1668f*
strand gaps 1669
redundancy 1669
regulation 1669
significance 1669
types of damage 1661–1662, *1662f*
uv radiation 1665
RepA protein 380
repeat expansion detection (RED) 2051
repeat-induced point mutation (RIP) 636
repetitive DNA sequences **1671–1675**
chromosome translocations 1674
definition 1671
Drosophila pattern 1673
ectopic exchange 1675
evolutionary considerations 1673–1674
amplification 1674
concerted evolution 1673
gene conversion 1673
unequal crossover 1673
functionality 1674–1675
heterochromatin 1674
historical background 1671–1672
influence on gene expression 1674–1675
insertional mutagenesis 1674
selfish propagation 1675
triple repeat expansion 1675
types 1672–1673
gene families 1672

repetitive DNA sequences (*continued*)
long-range repeats 1673
miniature inverted repeat transposable elements (MITEs) 1673
retroposons 1673
retroviruses 1673
retrovirus-related sequences 1673
short simple repeats 1673
SINEs 1673
tandem repeats 1672
transposon-derived sequences 1672
transposons 1672
uses
fingerprinting 1675
oncogene isolation 1675
phylogenetic analysis 1675
population studies 1675
recombination mapping 1675
transposons 1675
Xenopus pattern 1673
replication **1675**
see also theta replication; topoisomerases
replication A protein (RPA) 1839
replication errors *1676t*, **1676–1677**
DNA polymerases 1676
Drake, John W. 1676
Drake's rule 1676, *1676t*
genome size 1676
mismatch repair 1676
mutation rates 1676
proofreading 1676
replication eye **1677**
replication fork **1677**
replicon 1485, 1637, **1677–1678**
replisome **1678**
reporter gene **1678**
Rep proteins 1381
REP (repetitive extragenic palindrome) 214
repression **1678**
repressor **1678–1679**
corepressors 1678
effector molecules 1678
function 1678
inducers 1678
lac operon 1678
lac repressor 1678
trp operon 1679
trp repressor 1679
reproductive isolation **1679–1686**
biological species concept 1679
causes 1679
definition 1679
Dobzhansky, Theodosius 1679
ethological isolation 1682–1684
genetic analysis
Drosophila mauritiana 1682f
Drosophila sechellia 1682f
existing studies 1681–1682, *1683t*
hybrid sterility 1681, *1682f*
principles 1679–1681
Haldane's rule 1681
Mayr, Ernst 1679
mechanisms 1679, *1680t*
postzygotic isolation *1680t*, 1684–1685
chromosome rearrangements 1684
Dobzhansky–Muller model 1685, *1685f*
genetic changes 1685
infectious microorganisms 1685
polyploidy 1684
reproductive isolation (*continued*)
transposons 1685
prezygotic isolation *1680t*
speciation 1679–1681
see also speciation
resistance plasmids 586, **1686**
resistance to antibiotics *see* antibiotic resistance
resolvase/invertase family 1038, **1687–1688**
Gin invertase 938–942
Hin invertase 938–942
Holliday junction 1688
res site 1688
site-specific recombination 717, 1687–1688, 1843, *1844t*, *1845f*, 1845–1846
resolvase-mediated deletion **1688–1692**
resolution process 1690–1691
resolvase structure 1688, *1690f*
res site 1688–1689, *1689f*
synaptic complex 1690–1691, *1691f*
Tn*3* transposon 1688, *1689f*
Tn*552* transposon 1688
transposition pathway 1688, *1689f*
$\gamma\delta$ transposon 1688, *1689f*
resolvasome 1757
Res protein 1693
res site 1688
restriction and modification **1692–1693**
restriction endonuclease **1693–1696**
BcgI 1011
BspMI 1011
cleavage ability 1693–1696
DNA regulation 542
enzymes 625
linker DNA **1109**
marker 1146
Nathans, Daniel 1290
restriction and modification 1692
Type I 1693, *1694t*
Type II *1694t*, 1694–1696
Type III 1693, *1694t*
restriction enzymes
Arber, Werner 1847
DNA cloning 546, 548, *548f*
Nathans, Daniel 1847
restriction enzyme analysis 968
Smith, Hamilton 1847
restriction fragment length polymorphism (RFLP) 972, **1696**
chromosome mapping 357
deletion mapping (mouse) 526
genetic distance 831–832
linkage map 232, 1107
marker 1146
minisatellite 1212
Oryza sativa (rice) 1394
Southern blotting 1857
restriction map **1696**
RET allele 262
reticulation **1696–1697**
see also trees
retinitis pigmentosa 269, 337, 1023, **1697**
retinoblastoma 357, **1697–1698**
clinical features 1697
genetics 1697
molecular genetics 1697–1698
treatment 1697
retinoic acid 2119–2120
RetNet website 1697
RET protein 1257–1258
RET proto-oncogene **1696**
RET1 proto-oncogene 1257–1258
retroelements 1806
retroposon **1698–1699**
retroregulation **1699**
retrotransposition 667, 1857
retrotransposons 1051, **1699–1701**
classification 1699
definition 1699
distribution 1701
host relationship 1701
long terminal repeats (LTRs) 1699
Metaviridae 1700
Pseudoviridae 1700
reverse transcription mechansim 1700
structure *1699f*, 1699–1701
retroviruses **1701–1706**
cancer, role in 1705
carcinogens 271–272
genetic information transfer 1702, *1703f*
genetic manipulation 1706
historical background 1701–1702, *1702t*
HTLV-1 (human T-cell lymphotrophic virus 1) 979
human immunodeficiency virus (HIV) 1706
long terminal repeats (LTRs) 1120
mouse leukemia viruses (MLV) 1249
retroposon **1698–1699**
retroviral life cycle 1702–1705
structure 1704, *1704f*
virus 2111
Rett syndrome 636, 2147
reverse genetics 1014, **1706–1709**
Arabidopsis thaliana 1708
background 1706
Caenorhabditis elegans 1707–1708
Cre/*loxP* technology 1707
Dictyostelium 1707
Drosophila 1708
Drosophila melanogaster 1706
mouse 1708
P element 1406, 1708
Saccharomyces cerevisiae 1706
Schizosaccharomyces pombe 1706
yeast 1706–1707
reverse mutation 1275, **1709–1710**
adaptive mutation 1710
Ames test 1710
forward mutations 1709–1710
pseudo-reversions 1709
reversion assays 1710
selection 1709
true reversions 1709
reverse transcriptase **1710–1711**
reverse transcription **1711**
see also cDNA
reverse translation **1711**
reversion **1711**
reversion tests **1711–1713**
applications 1712
base substitution mutations 1712, *1712f*
frameshift mutations 1712, *1712f*
genetic stability 1711
mutation rates 1712
tryptophan (Trp) revertants *1711f*, 1711–1712
see also Ames test; Luria–Delbrück experiment; mutational analysis; mutation rate; reverse mutation
Rex gene 979
RF1 release factor 323–324
RF2 release factor 323–324
RF3 release factor 324

RFLP *see* restriction fragment length polymorphism (RFLP)
R gene 1045
rhabdomyosarcoma **1715**, 1768, *1769t*, *1770t*
Rh blood group system 650–653, **1713–1715**
amino acid sequences *1714t*
C/c antigen 651
chemical structure 1713–1714
cloning 651, 1713
D antigen 650, 1713–1714
E/e antigen 651
Fisher–Race nomenclature system 651
gene identification 1713–1714
molecular studies 651
preimplantation determination 652–653
prenatal determination
amniocentesis 652
noninvasive 652
Rh polypeptides 651–652, *1714t*
function 651–652
structure 1713
C/c polypeptide 651
cDNA 651
E/e polypeptide 651
mRNA (messenger ribonucleic acid) 651
RhCcEe gene 651, 1713–1714
Rh compatibility 1410
RHD gene 1713–1714
RhD gene 1714
RHG gene 892
rhizobia 1084
Rhizobiaceae 24
Rhizobium 24, 1328, 1334, 1478, 1556, **1715–1716**
Rhizobium etli 708, 1479
Rhizobium leguminosarum 708, 1333, 1479
Rhizobium meliloti 707, 859, 1162
Rhizobium sp. NGR234 1122, 1330
Rhizoctonia solani 1850
Rhizopus 1882
rhizosphere 1477
see also plant growth promoting rhizobacteria (PGPR)
rhl1 gene 2045
rhl2 gene 2045
rhl3 gene 2045
Rhoades, M.M. 849
RhoAy1 gene *1253f*, 1254
RhoAy2 gene *1253f*, 1254
Rhodes, Marcus 464
Rhodococcus faciens 859
rhodopsin 1061
Rho factor **1716–1719**
ATP hydrolysis 1716
mechanism 1716–1718, *1717f*
NusG protein 1718
Rho-dependent terminator 1718
rho gene 1716
rut sequences 1718
structure 1716
transcription termination 1716, *1717f*
RHO/RAC proteins 1602
Rh polypeptides 651–652
Rhynchosciara americana 723, 762
Rhynchosporium secalis 973
Ribes bracteosum 840
ribonucleic acid *see* RNA (ribonucleic acid)
ribosomal RNA (rRNA) 1358, **1719–1723**
function 1722
ribosomal RNA (rRNA) (*continued*)
rRNA molecules 1719
structure 1719–1722, *1720–1721f*
transcription factors 1029
see also core particle
ribosome binding site **1723**
ribosome recycling factor (RRF) 324, 2001, 2002
ribosomes **1723–1730**
aminoacyl-tRNA 601–602, 609, 610
antibiotics 1729
composition 1724, *1724t*
crosslinking 1726–1727
DNA replication 541
elongation 601–602, 609
elongation factors 1729
exit channel 1729
history 1724
mRNA binding 1728
peptidyl transfer site 1728
peptidyl-tRNA 602, 609, 610
proofreading 601–602
ribosomal proteins 1724–1725, *1725t*
ribosome recycling 2001
rRNA molecules 1724
structure 1725–1728, *1726–1727f*
structure analysis methods
electron microscopy (EM) 1727
neutron scattering 1727
nuclear magnetic resonance 1727
X-ray diffraction method 1727–1728
translation 2000–2001, 2002
tRNA binding sites 1728
A-site 1728
E-site 1728
P-site 1728
see also elongation; protein secretion systems
ribozymes 1392, **1730**, 1751
ribulokinase 1362
rice *see Oryza sativa*
RiceGenes 853
Rice Genetic Cooperative 852
Rich, Alex 555
Richardson J P **1716–1719**
Richardson N **1740–1741**
Richards R I **593–597**, **726–728**
Richmond, Timothy 555
ricin 611
Rickettsia prowazekii 149, *150t*, 1217, 1892
Ridge R W **1753–1756**
rifampicin 77, 1687
rifamycins **1730**
Riggs, A. 629
Riggs P **739**, **739–740**
right/left handed DNA *see* DNA structure
rII mutants 1927
RIME (*Rhizobium*-specific intergenic mosaic elements) 214
Rimoin D L **2–3**, **236–237**
Rinchik E M **524–528**
Rinkenberger J L **2070–2071**
Ripley L S **1268–1275**
Ris, Hans 412
RI strain *see* recombinant inbred strains
Riva, Geo 1123
rII locus 383
R-loop **1731–1732**, *1732f*
Rmcf gene 1250
RNAases **1732–1733**
RNA-binding protein domains **1733–1735**
double-stranded RNA-binding domain (dsRBD) 1734
K homology (KH) sequence domain 1734–1735
RNP domain 1733–1734
RNA degradation **1748–1751**
3′→5′ exonucleolytic degradation pathway 1749, *1749f*, *1750f*
biological role 1748
deadenylation-dependent decapping pathway 1748, *1749f*
deadenylation-independent decapping pathway *1749f*, 1749–1750
degradation pathways 1748
endonucleolytic cleavage *1749f*, 1750–1751
exosome complex 1749, *1750f*
RNA editing
animals **1735–1740**
A-to-I editing 1739–1740
adenosine diaminases (ADARs) 1739
function 1739–1740
mechanism 1739, *1740f*
origins 1740
C-to-U editing 1737–1739
ACF protein 1737
antibody production 1738, *1738f*
APOBEC-1 1737, *1737f*
apolipoprotein (ApoB) mRNA 1737, *1737f*
mechanisms 1737
tumorigenesis 1738–1739
guide RNA (gRNA) 1741–1742, *1742f*
human immunodeficiency virus (HIV) 1740
insertion/deletion editing 1735–1736
differences 1742
mechanisms 1736, 1742, *1742f*
origins 1736
U deletion editing 1742, *1742f*
U insertion editing *1736f*, 1742, *1742f*
mitochondria 1740
plants **1740–1741**
substitution editing 1736–1740
trypanosomes **1741–1743**
U-to-C editing 1739
RNA interference **1743–1744**
see also antisense RNA; gene silencing
RNA phages **1744–1746**
RNA polymerase 609, **1746–1747**
archaea 1746
ATP hydrolysis *1717f*
attenuation 122, 123
bacteria 146, 1746–1747
sigma factors 1746
transcription initiation 1746–1747
crystal structures *1974f*
DNA-binding domains (DBDs) *808f*, 808–809, *809f*
DNA regulation 542
DNA repair 563
elongation *1979t*
epigenetics 632
eukaryotes 1746, 1747
function 1746
gene regulation 803, *804f*
histone genes 948
Hogness Box 954

RNA polymerase (*continued*)
host-lethal gene 975–976
lampbrush chromosomes 1074–1075
polymerase 1499
Rho factor 1716, *1717f*
RNA phages 1744–1746
RNA polymerase III (RNAP III) 1746–1747
RNA polymerase II (RNAP II) 1746–1747
RNA polymerase I (RNAP I) 1746–1747
RNA synthesis 1746
sigma factors 1831–1832
structure 807, *808f*
substrates 1746
transcription 1746
RNA (ribonucleic acid) **1732**
aptamers 1006
catalytic ability 1006, *1007f*
evolutionary rate 671
helicases 915–916
hemophilia 919
hybridization 1380
mobile introns 1047–1051
synthesis 1067
virus 2109, *2113f*
in vitro evolution 1005–1008
see also messenger RNA (mRNA); ribosomal RNA (rRNA); RNA polymerase; RNA world; snRNA (small nuclear); transfer RNA (tRNA)
RNase 1357
RNaseE 947
RNaseH 1032, 1358, 1382, 1546, *1699f*, 1700
RNaseL 1751
RNaseP 947, 1006, *1007f*
RNA turnover *see* RNA degradation
RNA world hypothesis 1392, **1751–1752**
RNs *see* recombination nodules (RNs)
Robb F T **91–95**
Robertsonian translocation 583, *723*, **1752**, 2006, *2006f*
centric fusion 319–320
chromosome aberrations 347–348
Down syndrome 348
mouse 2007
Patau syndrome 348, 1420
satellited chromosome **1773–1774**
Sorex araneus 2007
see also centric fusion; translocation
Robertson, W. 1752
Robinia (locust) 1083
Robinieae 1083
Robinson, A. 1970
ROB protein *1383f*
Rochaix J-D **334–337**
Rocky Mountain relapsing fever 80
Rodeck C H **650–653**
Roderick T H **1242–1247**
Roe B A **757–759**
Roeder, Robert 804
Rogers' distance 828
Rohlf F J **1189**, **1189–1190**, **1190**
rolling circle replication 877, **1752–1753**, *1753f*
Ronson C **1902–1903**
root development **1753–1756**
Arabidopsis 1755
cell fate 1753
cell lineage 1753–1755
genetic analysis 1755
root development (*continued*)
meristem composition 1753, *1754f*
Rosaceae 112
Rosche W A **20–21**, **197–198**
Rosenberg S M **677–680**, **683**, **936–938**, **1614–1618**, **1623–1631**, **1635–1636**, **1647–1648**, **1709–1710**, **1756–1759**, **1759–1760**, **1823–1824**
rosy gene 1989
Roth J R **1766–1768**
Rothmund–Thompson syndrome (RTS) *569–570t*, 571
Rougier N **1419–1420**
Rougvie A E **927–930**
roundworms 251–252
Rous, Peyton 1371, 1701
Rous sarcoma virus 1371, 2111
Rout M P **1352–1356**
Rowley J D **1088–1091**
Royer C A **1896**
RP4 plasmid 453
R plasmids *see* resistance plasmids
rpl gene 1461
rpoD gene 1543
rpo gene 1461
RPS24Y gene 2159
RPS4Y gene 2159
rps gene 1461
rpsL gene 1891
rpsU gene 1543
RRF *see* ribosome recycling factor (RRF)
Rrn3 transcription factors 1029
rrs gene 1891
RSC chromatin remodeling complex *342t*
RSF chromatin remodeling complex *342t*
RT/RNaseH gene 2021
rts (retsina) gene 2048
Rubinsztein D C **48–49**, **981–983**
Ruddle F **857–859**, **932**
Rus enzyme 1760
Russell, L.B. 527
Russell, W.L. 524, 1599, 1871
Rustgi A K **422–423**
Rutaceae 238
rut sequences 1718
Rutter, William 804
RuvAB enzyme **1756–1758**
biological functions 1757
branch migration 1756–1758, *1758f*
crossed-strand DNA junctions
formation 1756
heteroduplex DNA joints 1756, *1757f*
Holliday junction 1756, *1757f*
D-loop 1756, *1757f*
RecA protein 1756, *1757f*
single-strand DNA 1756, *1757f*
three- and four-strand junctions 1756
see also heteroduplex DNA; Holliday junction; RecA protein; RecBCD enzyme
RuvC enzyme **1759–1760**
crossed-strand DNA junctions 1759
endonucleolytic cleavage 1759, *1760f*
Holliday junction 1759–1760
RuvC protein 1759
RuvC gene 1032
ruv genes 1648
Ruvinsky I **1167**, **1548–1549**, **1790**
Ruvkun G **1657–1658**
Ruv proteins 1614, *1622f*, 1623, 1647, 1668
ry gene 795

S

S1 nuclease **1761**
SA *see* salicylic acid (SA)
SABRE gene 1755
Saccharomyces SIR genes 1948
Saccharomyces 738
Saccharomyces cerevisiae 606f, 636, 748, 964, 1064, 1384, *1385f*, *1386f*, **1761–1763**, 1882
aging 22
Aspergillus nidulans 106
cAMP nucleotide 258–259
catabolite repression 284
cell differentiation 1153–1154
cell division 303
cell division genetics 298
cell division mutants 500
centromere 322, *322f*, 860
chromosome break 330
chromosome mapping *367t*
chromosomes **1763–1765**
centrioles 1764
centromeres 1764
chromatin 1764
crossing-over 1765, *1765f*
gene organization 1765
meiotic division 1765
meiotic recombination 1765
mitotic division 1764–1765
nucleosomes 1764
recombination nodules 1765
replication origins 1764
spindle fibers 1764
synaptonemal complex 1765
telomeres 1764
YACs (yeast artificial chromosomes) 1764
codon usage bias *403t*, 405
conversion gradient 466–467
developmental genetics 531
dispersed transposable sequences 861
gene mapping 793
gene number 796
genetic map 1763
genome organization 860
genome sequence 1763, *1764t*
genome sequencing 859
genome size *367t*
haploid phase 535
heteroduplex DNA 931
heterothallic yeast 1761, *1762f*
heterotrimeric G proteins 934
histidine operon *946f*
histone genes 949
homothallic yeast 1761, *1762f*
illegitimate recombination 997
independent assortment 1018
integrase family 1040
ISWI1 chromatin remodeling complex *342t*
kinases 1063
life cycle 1761, *1762f*
mating types 1152
mating type switch **1153–1157**
asymmetric cell division 1155
double-strand break 1156
regulation 1155
MAT locus 1154, *1155f*
meiosis 1154
mitochondrial mutants 1223
mitochondrial RNA polymerase 1922

Saccharomyces cerevisiae (*continued*)
mobile introns 1047, 1052
model organisms 1153
mutagenesis 1761–1763
mutant varieties 1763
nondisjunction 1346
outbreeding 1398
position effects 1527, 1529
proteome 1577
rearrangements 1613
rec gene homologs 1614
recombination models 1642
repair mechanisms 1666
replication errors *1676t*
retrotransposons *1699f*
reverse genetics 1706
RNA polymerases 1976
RSC chromatin remodeling complex *342t*
single-strand annealing 1838
sth 1 chromatin remodeling complex *342t*
subcellular RNA localization 1893
Swi2/Snf2 chromatin remodeling complex *342t*
SWI/SNF chromatin remodeling complex *342t*
synaptonemal complex 1912
telemeric position effect 1529
telomeres 861
telomeric DNA *1947f*
tetrad analysis 1763, 1954
tetrads 1953–1954
transgenes 1990
uniparental inheritance 2096
yeast plasmids 2162
see also yeast
Saccharomyces Genome Database 851
Sachs, Bernard 1941
Sachs, M.M. 849
sac regulon 1079
Saedler H **199–203, 2020–2033**
Saethre–Chotzen syndrome *see* craniosynostosis
Sager, Ruth 1462
Saitou N **215–216, 752, 819–821, 1062, 1305–1306, 1509, 1869–1870, 2042–2045**
salamanders 344
Salas M **1365–1366**
salicylic acid (SA) 1481, 1913
Salk Institute 590
Salmonella 653, 658, 1080, 1198, 1423, 1556, **1766–1768**, 1922
Escherichia coli 1767
genetic map 1766
genetic system 1766–1767
histidine operon 1766
leucine operon 1766
pathogenicity 1767
phage Mu 1439
phage P22 1766
phase variation 1442
Salmonella dublin 1766
Salmonella enterica 1766–1768
Salmonella typhi 1766
Salmonella typhimurium 1766–1768
taxonomy 1766
Tn*10* transposon 1767
transductional crosses 1766
transposable elements 1767
Salmonella enterica 711
transcription attenuation 122
Salmonella Genetic Stock Center (SGSC) 850
Salmonella paratyphi 176
Salmonella Reference Collection (SARC) 850
Salmonella typhimurium 1–2, 176, 405, 1382, 1423, 1983
alternation of gene expression 43
Chi sequences 328
gene amplification 764
Hin invertase 938
histidine operon 943, *946f*, 1712
microbial genetics 1191
reverse mutation 1710
RNA phages 1744
short bacterial repeated elements 214
transcription attenuation 122
Samson L D **36–37, 1650–1651**
Samson, Leona 257, 1651
Sancar A **673–675**
Sanderson, K. 850
Sanderson M R **553–556**
Sandhoff disease 756
Sandler, Larry 490
Sanger Centre 355, 360, 517
Sanger, Fred 1451, **1768**
DNA sequencing 572
gene therapy (human) 815
genetic code 821
Nobel Prize 877, 1768
Sapindaceae 238
SAR
see scaffold-attached region (SAR)
sarcoglycan *587f*
sarcomas **1768–1773**
Cancer Genome Project 1773
chromosomal translocations 1771–1772, *1772t*
classes 1769, *1769t*
cytogenetic studies 1771–1772
distribution 1768
etiology 1769
genetic diseases 1769, *1770t*
APC (adenomatous polyposis coli) gene 1769
Beckwith–Wiedemann syndrome 1770, *1770t*
familial adenomatous polyposis (FAP) 1769
familial retinoblastoma 1770, *1770t*
Gardner's syndrome 1769, *1770t*
Li–Fraumeni syndrome (LFS) 1770, *1770t*
osteosarcoma 1770
Paget's disease 1770, *1770t*
von Recklinghausen disease 1770, *1770t*
Wilms' tumor 1770
pathology 1770
treatment 1771
Sasaki T **1394–1395**
SAS gene 1773
satellited chromosome **1773–1774**
satellite DNA *see* microsatellite
Sauer B 481–486
Savarirayan R **2–3, 236–237**
sbc genes 1648
SBMA
see spinal and bulbar muscular atrophy (SBMA)
SCA *see* spinocerebellar ataxia type (SCA)
SCA1 gene *2049t*
SCA2 gene *2049t*
SCA3 gene *2050t*
SCA7 gene *2050t*
SCA 8 (antisense) gene *2050t*
scaffold-attached region (SAR) 372–373, 374
scaffold/matrix attachment region (S/MAR)
cis-acting locus 377, 378
detection 378
function 378
structure 378
scanning electron microscopy (SEM) *607f*, 608
Schedl T **873–876**
Scheie's disease 756
Scheller, Richard 210
Schesser K **1552–1565**
Schildkraut I **90, 105, 391–392, 524, 608, 793, 834, 1014, 1109, 1147, 1205, 1327–1328, 1357–1358, 1410, 1430, 1447, 1467, 1639, 1678–1679, 1692–1693, 1779–1780, 1795–1796, 1825, 1895**
Schimmel P **1897–1899, 2140–2141**
schizophrenia 534, **1774–1776**, *2015t*, *2050t*
background 1774
genetic implications 1776
psychopathology 1775
structural changes in the brain 1775
symptoms 1775
Schizophyllum 1399, 1954
Schizophyllum commune 1154
Schizosaccharomyces 738
Schizosaccharomyces japonicus 1776
Schizosaccharomyces octosporus 1776
Schizosaccharomyces pombe 405, 535, *1386f*, **1776–1779**
cell cycle *292f*, 1778
cell differentiation 1153–1154
cell division genetics 298
cell division mutants 500
centromere *322f*, 860
chromosome break 330
chromosome structure 1777–1778
epigenetics 636, *637f*
genetic studies 1777
genome organization 860
genome sequence 1777
history 1776–1777
mat1 locus *1155f*
mat2 locus 1154, *1155f*
mat3 locus 1154, *1155f*
mating type switch **1153–1157**, 1778
asymmetric cell division 1155
mat1 imprinting 1156
regulation 1155
meiosis 1154, 1779
meiotic hot spots 976–977
mitochondrial DNA (mtDNA) 1779
model organisms 1153, 1776
morphology 1778
reverse genetics 1706, 1777
sexual differentiation 1778
tetrad analysis 1777
uniparental inheritance 2096
web site locations 1779
yeast plasmids 2163
Schleiden, Matthias 1172
Scholl, R. 852
Schoner B E **854–855**
Schrödinger, Erwin 554, 2133
Schüler K **1848–1850**
Schultz R M **1568–1572**
Schuster, P. 1323
Schwannomas 1307

Schwannomin protein 1308
Schwann, Theodor 1172
Sciara 629, 763, 1420–1421
ScII protein 372
ScI protein 372
Sclerospora graminicola 1914t
Sclerotinia sclerotiorum 1850
Scocca, J.J. 351
Scott J **1735–1740**
Scott, Matthew 958
SCP2 protein *1910f*, 1911
Sc proteins 372–373
scrapie 2008
screening **1779–1780**
SCR (SCARECROW) gene 1476, 1755
scr gene 892
scr (scarecrow) mutation 1476
scRNA **1780**
scRNP **1780**
Scrophulariaceae 112
Scr (Sex combs reduced) gene *960f*
sdh gene 1221
SD locus (*Drosophila melanogaster*) 1790
SDP (strain distribution pattern) *see* strain distribution pattern (SDP)
se (short ear) gene 398
se (short-ear) locus 1599, 1871
Searle A G **2003–2007**
secA gene 1555
secB gene 1554
second division segregation *see* first and second division segregation
secretion *see* protein secretion systems
seed development **1780–1782**
  angiosperms 1781
  Arabidopsis thaliana 1780–1782
  double fertilization 1780–1781
  embryogenesis 1781
  endosperm 1781
  mutant varieties 1781
  mutational studies 1781
seed storage proteins **1782–1787**
  amino acids 1786
  cereal storage proteins 1784
  characteristics 1783
  classification
    albumins 1783–1784
    globulins 1783–1784
    glutelins 1783–1784
    glycinin (*Glycine max*) 1784
    hordein (*Hordeum vulgare*) 1784
    prolamins 1783–1784
    zeins (*Zea mays*) 1784
  gene regulation 1786
  genomic organization 1785
  history 1783
  International Maize and Wheat Improvement Center 1786
  legume storage proteins 1785
  modification 1786
  mutant varieties 1786
  protein bodies 1783, *1783f*
  Quality Protein Maize (QPM) 1786
  seed constituents 1782–1783
Seemungal D **733, 980–981**
Segall A M **120–122**
Segal S **741–742**
segmental interchange *1787f*, **1787–1790**
  interchange heterozygote 1787
  meiosis 1787
  meiotic products 1787–1789, *1789f*
  Oenothera 1788
segmentation genes **1790**
segregation **1790**
  chromosomal theory of heredity 1790
  fungal genetics 1790
  law of segregation 1790
  Mendel's laws 1790
  segregation distortion **1790–1791**
    SD locus (*Drosophila melanogaster*) 1790
    Spore killer (*Neurospora*) 1790
    t-complex (mouse) 1790, 1791
    t haplotype 1791
seizure disorders *2015t*
sel-12 gene 1313
SEL-12 protein 1313
Seldin M F **1106–1109, 1144**
selection coefficient **1791**, 1801, *1801f*
selection differential **1791–1792**
selection index **1792–1793**
  best linear unbiased prediction (BLUP) method 1792
  breeding of animals 1792
selection intensity **1793–1794**
  see also artificial selection
selection limit **1794–1795**
  artificial selection 1794
  fixation probability 1794
  natural selection 1795
  quantitative trait 1794
  see also artificial selection
selection pressure **1795**
selection techniques **1795–1796**
  see also antibiotic-resistance mutants
selective breeding **1796–1799**
  background 1796
  best linear unbiased prediction (BLUP) method 1799
  genetic changes 1797–1799
    cattle 1798
    chickens 1798, *1798t*
    dogs 1797
    plants 1798
    racehorses 1798
    wheat 1798, *1798t*
  genetic engineering 1799
  genetic principles 1796–1797
  Henderson, C.R. 1797
  Lush, J.L. 1797
  QTL (quantitative trait locus) 1799
  see also progeny testing
selective neutrality **1800–1803**
  bottleneck effect 1802
  evolutionary rate 1801–1802
  fixation probability 1800
  fixed model 1801, *1801f*
  house of cards model 1801
  molecular clock 1801
  nearly neutral theory 1800
  neutral theory 1800
  nonsynonymous substitutions 1802
  polymorphism 1802
  population size 1802
  selection coefficient 1801, *1801f*
  shift model 1801
  synonymous substitutions 1802
selective sweep **1803–1804**
  allele frequency 1804
  background selection 1804
  Drosophila melanogaster 1804
  Drosophila simulans 1804
  fixation 1803–1804
  gene genealogies 1803
  hitchhiking effect 1804
  polymorphism 1803–1804
selective sweep (*continued*)
  process 1803
  recombination rates 1804
  superoxide dismutase (SOD) 1804
  Wolbachia 1804
SELEX (systematic evolution of ligands by exponential amplification) 1006
self-fertilization 1805
  advantages 1805
  Caenorhabditis elegans 1805
  hermaphrodite 1805
  inbreeding depression 1805
  meiotic recombination 1805
  parthenogenesis 1805
selfing 1152, 1867
selfish DNA **1805–1808**
  coevolution 1808
  C value 1807
  C-value paradox 1807
  definition 1805–1806
    junk DNA 1806
    selfish gene 1805–1806
    ultraselfish gene 1805
  Drosophila melanogaster 1808
  hybrid dysgenesis 1806
  intron evolution 1808
  P elements 1806
  theory 1805
  varieties 1806–1807
selfish gene 906–907, 1165, 1805
Seller M J **1876–1877**
semaphorins 1320
Semibalanus balanoides 786
semiconservative replication **1808**
semidiscontinuous replication **1808–1809**
semidominance *see* incomplete dominance
Sendai virus 1356
senescence 310
  cells 311
sense codon 522, 1277, **1809**
SEP1 gene 716
SEP2 gene 716
SEP3 gene 716
sequence alignment *see* alignment problem
sequenced genomes
  see genome databases
serine **1809**, *1809f*
serine protease inhibitor family (serpins) 40
Serratia 1439, 1478
Serratia marcescens 1914t
Sertoli cells 691, 1023, 1066, 1812, 1816, 1875
Sesbania (sesban) 1083
Setlow, Richard 565
sex chromatin 1066, **1809–1810**
  Barr body 1809–1810
  congenital adrenal hyperplasia 1809
  gender verification 1810
  intersex 1809
  isochromosome 1054
  Klinefelter syndrome 1809–1810
  Turner syndrome 1809–1810
  X chromatin 1809
sex chromosomes **1810–1811**
  aneuploidy **1810**
  DMRT gene 1811
  Drosophila melanogaster 1810
  evolution 1811
  history 1408–1409
  Klinefelter syndrome 1811
  sex determination 1810
  sex-determining genes 1811
  SRY gene 1811

sex chromosomes (*continued*)
  Turner syndrome 1811
  XYY males **1810**
  see also W chromosome; X chromosome; Y chromosome; Z chromosome
sex determination
  homeotic genes 1814
  human **1811–1816**, *1815f*
    anti-Mullerian hormone (AMH) 1812, 1814
    campomelic dysplasia 1817, 1812
    Denys–Drash syndrome (DDS) 1818, 1814
    Frasier syndrome 1818, 1814
    gene regulation 1814
    HMG domain proteins 1816, 1812
    Leydig cells 1812
    Mullerian ducts 1812
    Mullerian inhibitory substance (MIS) 1812
    pre-Sertoli cells 1812
    Sertoli cells 1812
    SRY protein 1816
    testis-determining factor (TDF) 1812
    testis suppression 1813
    testosterone 1812
    Wilms' tumor 1818, 1814
    Wolffian ducts 1815
  mouse **1816–1819**
    anti-Mullerian hormone (AMH) 1816
    ELP1 locus 1817
    ELP2 locus 1817
    ELP3 locus 1817
    Ftzf1 locus 1817
    Leydig cells 1816, 1817
    Mullerian ducts 1816
    Mullerian inhibitory substance (MIS) 1816
    ovarian development 1815, 1816
    pre-Sertoli cells 1816
    Sertoli cells 1816
    testis suppression 1815
    testosterone 1816
    Wolffian ducts 1816
    Y chromosome 1816
sexduction **1823–1824**
  see also F-duction
sex-limited character **1824**
sex linkage **1819**
  pseudoautosomal genes 1819
  X-linked genes 1819, *1819t*
sex plasmid **1820**
sex ratios **1820–1821**
  environmental effects 1821
  Fisher's theory of equal investment in the sexes 1820
  fitness 1820
  sex-biased interactions between relatives
    local mate competition (LMC) 1820–1821
    local resource competition (LRC) 1820–1821
    local resource enhancement (LRE) 1820–1821
  see also frequency-dependent selection
sex reversal **1822–1823**
  androgen insensitivity 1822
  campomelic dysplasia 1822
  Denys–Drash syndrome (DDS) 1822
  Klinefelter syndrome 1823
sex reversal (*continued*)
  pseudovaginal perineoscrotal hypospadias 1823
  testis differentiation 1822
  Turner syndrome 1822
  XX male syndrome 1823
  XY female syndrome 1822
Seychelles warbler 1821
sey gene 1427
Sezary syndrome 1094, **1824**
Sf1 gene 1814, 1817
SF1 gene 1813–1814
SF1 protein 1814
SFTPA1 gene *430t*, *431t*
SFTPA2 gene *430t*, *431t*
SFTPD gene *430t*, *431t*
Sgaramella V **544–550**
SGS1 gene 22
SH domains **1824**
SH2 domain 1603–1604, **1824–1825**, 1883
SH3 domain 1604, **1825**, 1883
Shaffer H B **1304–1305**
Shakespeare, William
  nature–nurture controversy 1298
  The Tempest 1298
Sham P **1965**
Sharp P M **402–406**
shearing **1825**
sheep scrapie 1880
Sheppard, P. M. 1534
Sherman paradox 728
Sherratt D J **351–353**
sh gene 1395
shh (sonic hedgehog) gene 1100
shifting balance theory of evolution 706, 832–833, **1825–1828**
  adaptive landscapes 1825–1826, *1826f*
  allele frequency 1825–1826, *1826f*
  demes 1825
  epistasis 1826
  fitness 1825–1826, *1826f*
  genetic drift 603
  importance 1827–1828
  pleiotropy 1826
  process 1826–1827
  Wright, Sewall 1825–1828
shift model 1801
Shiga-like toxins 184
Shigella 653, 1439, 1556
Shigella dysenteriae 1922
Shigella flexneri 176
Shigella sonnei 176
Shine–Dalgarno sequence 152, 855, 1185, 1722, **1828**, 1886
  open reading frame (ORF) 1375
  polycistronic mRNA 1496
shope fibroma growth factor (SFGF) 627
short bacterial repeated elements **214–215**
short interspersed nuclear element (SINE) *see* SINE
shotgun cloning 981, **1828**
SHOX gene 2157
shr gene 892
SHR (SHORT-ROOT) gene 1755
shr (short-root) mutation 1476
Shull, G.H. 933, 1337, 1400
Shulman J **2125–2127**
shuttle vector **1828**
Siamois gene 2150
sickle-cell anemia 187, **1828–1831**
  carrier 275
  cause 1828
  cell distortion 1829, *1830f*
sickle-cell anemia (*continued*)
  clinical features 1829–1830
  geographical distribution 1828, *1829f*
  hemoglobin changes 1828
  mutant allele **1268**
  mutation effects 1829
  single-gene inheritance 1835
  treatment 1830
sicklepod *1914t*
sieA gene 1895
sigma^{32} 1833
sigma^{54}-holoenzyme 164–165
sigma^{70} 1550, 1746, 1831–1833, *1833f*
sigma^{F} 1833
sigma factors **1831–1834**
  future research 1833
  history 1831–1832, *1832t*
  regulation mechanisms 1833
  sigma^{32} 1833
  sigma^{70} 1550, 1746, 1831–1833, *1833f*
  sigma^{F} 1833
  transcription, role in 1831–1832
signal sequence **1834**
signal transduction **1834**
  heterotrimeric G proteins 934
  LIM domain genes 1097
  signal transduction pathways 532, 662
  see also cAMP nucleotide
silencing
  DNA methylation 636
  epigenetic regulation 635
  eukaryotes 636
silent mutations **1279**
SilkBase database 232
Sillence D O **1395–1398**
Silver L **37, 39, 40, 64–65, 74, 105, 144–145, 190, 237, 243–244, 250, 278, 402, 533, 537, 537–538, 538, 557, 621–623, 623, 666–669, 680, 688, 698, 750, 783–785, 819, 840, 873, 970, 984, 987–988, 996, 998–999, 1015–1016, 1016, 1017, 1046, 1047, 1066–1067, 1103–1104, 1117–1118, 1119, 1137–1140, 1205, 1212, 1238, 1247–1249, 1257, 1261, 1268, 1289, 1349, 1361, 1374–1375, 1400, 1401–1402, 1415, 1446, 1579, 1593–1595, 1639–1640, 1696, 1698–1699, 1790–1791, 1834–1835, 1847–1848, 1870–1871, 1889, 1913, 1919, 1933–1934, 1952, 1957, 2138**
Silver–Russell syndrome 1000, 1241
silver staining 348, 349
simian virus 40 (SV40) 310, 547, 590, 1384, *1386f*, **2078–2081**
  alternative splicing 47
  enhancers 624
  episome 638
  infectious cycle 2079
  Nathans, Daniel 1290, 1847
  restriction enzymes 1847
  single-stranded DNA-binding proteins (SSBs) 1841
  viral tumorigenesis 2080–2081
similarity **1834**
similarity test 967, 970
Simmons, Michael 490
Simon E H **274–275**
Simons R W **83–84**
simple sequence length polymorphism (SSLP) 526
  see also microsatellite

simple sequence repeats (SSR) 972
  linkage map 1107
  Oryza sativa (rice) 1394
  see also microsatellite
Simpson, G.G. 665
Simpson, Larry 1741
Simpson's paradox 639
Sinapis alba (white mustard) 237
Sinc gene 1879
Sinclair A **1811–1816**
SINE **1834–1835**, 2022
  function 1835
  repetitive DNA sequences 1672
  reverse transcriptase 1835
  selfish gene 1834
  see also LINE; retroposon
Singh R S **186–188, 1507–1509**
single-copy plasmids **1835**
single-gene inheritance **1835–1838**
  anticipation 1837
  dominance traits 1836
  history 1835
  inheritance patterns 1835–1836
  Mendelian inheritance 1837
  penetrance 1837
  recessive traits 1836
  variable expressivity 1837
single-nucleotide polymorphisms (SNPs) 359, 935, 981, **1838**
single-strand annealing **1838–1839**
  double-strand break repair 1838–1839, *1839f*
  mechanisms 1838, *1839f*
  Saccharomyces cerevisiae 1838
single-strand assimilation **1839**
single-stranded DNA-binding proteins (SSBs) **1839–1842**
  function 1840
  helix destabilization 1840
  roles 1840–1842
  structure 1840, *1841f*
single-stranded DNA replication 1451
single-strand exchange **1839**
single-strand gap repair model *777f*, 778
Sinorhizobium 24
Sinorhizobium fredii 885
Sinorhizobium meliloti 707, 1329, 1330, 1333, 1479
Sinsheimer, Robert 1451
Sinskey A J **54–56**
Siphoviridae 181, 1922
SIR2 gene 22
sister chromatid exchange (SCE) **1774**
sister chromatids **1842**
site-specific recombination 998
  attachment sites **120–122**
  biological consequences 1843, *1844t*
  definition 1842
  Escherichia coli 998
  immunoglobulins 998
  integrase family 717
  mechanisms 1843–1845, *1845f*
  phage λ 998
  recombination signal sequences 998
  resolvase/invertase family 717
    double-strand breaks 1846
    mechanisms *1845f*, 1845–1846
    serines 1845–1846
  specialized recombination 1857–1858
  structural consequences 1842–1843, *1843f*
site-specific recombination (*continued*)
  see also phage λ
SIX5 gene 1838
skeletal disorder *see* achondroplasia
Ski oncoprotein **1846**
skn-1 gene 618
Skopek T R **1264–1268**
SKY *see* spectral karyotyping (SKY)
Sl (*Steel*) locus 398, **1887–1888**, 2125
  Kit ligand (KL) 1887
  mast cell growth factor (MGF) 1887
  membrane-bound SLF 1887–1888
  Sl gene family 1887–1888
  soluble SLF 1887–1888
  Steel factor (SLF) 1887, *1888f*
  stem cell factor (SCF) 1887
  structure *1888f*
  W (*white spotting*) locus 1887
SL1 transcription factors 1029
Slatkin, Montgomery 1517, 1522
sleeping sickness 1736
Slopek, Stefan 177
SMAD4 gene 423
small T antigen *see* T antigens
SMC1 protein 374
SMC2 protein 372, 372–373
SMC4 protein 372
SMC family 374
Smcy gene *1253f*, 1254
SMCY gene 2159
Smith B J **1136**
Smith, Cedric A.B. 356–357
Smith D W **1381–1387**
Smith, Erwin F. 491
Smith, Gerald 1624
Smith G R **325–328**
Smith, Hamilton O. **1846–1847**
  Haemophilus influenzae 1847
  Nobel Prize 90, 1290, 1846
Smith, John Maynard 1862
SmLIM (smooth muscle) protein 1098
smoking 272
Sm proteins 1538
smRNA (small nuclear) 1358
SNAP-25/RIC-4 protein 1312–1313
Snape J W **971–973**
Sneath P H A **1446, 1834, 1937–1941**
Snell dwarf (*dw*) mutation 592
Snell, George 278, 1337, **1847–1848**
  major histocompatibility complex (MHC) 1847
  mouse genetics 1847
  Nobel Prize 1847
SNPs
  see single-nucleotide polymorphisms (SNPs)
snRNA (small nuclear) **1848**
snRNPs **1848**
Snyder M **2034–2040**
SOC1 gene 715
sodium bisulfite 1010
Solanaceae 112
Solanum tuberosum (potato) **1848–1850**
  characteristics 1849
  gene mapping 1850
  history 1849
  importance 1849
  International Potato Center 1850
  pathogens 1850
  potato breeding 1849–1850
  species
    Solanum ajanhuiri 1849
    Solanum bulbocastanum 1850
Solanum tuberosum (potato) (*continued*)
    Solanum chaucha 1849
    Solanum curtilobum 1849
    Solanum juzepczukii 1849
    Solanum phureja 1849
    Solanum stenotomum 1849
    Solanum tuberosum 1849
  systemic acquired resistance (SAR) *1914t*
SolGenes database 853, 1127
Sollner-Webb B **1741–1743**
Solter D **390–391, 1356–1357**
Solter, Davor 629
soluble RNA **1850–1851**
  see also tRNA
somatic mutation **1851**
somatic pairing **1851–1853**
  chromosome arrangement 1851–1852
  Drosophila melanogaster 1851f, 1851–1852
  homolog pairing 1851, *1851f*
  mechanisms 1852–1853
  mitosis 1851
  somatic pairing 1852
  transvection 1852
  see also chromosome pairing, synapsis
somatostatin 545
somatotropes 592
Somerville, Chris 88
Somerville R L **598, 901, 1258–1259, 1328, 1585, 1586, 1889, 1965–1966, 2100–2101**
Sonneborn, Tracy 1411, 2133
Sordaria 749, 1954, *1955f*
Sordaria fimicola 1953f
sordarin 611
Sorex araneus 2007
SOS bypass **1853**
SOS repair 1492, 1651, **1853–1855**
  mutagenesis 1854
  phage λ 1854
  prophage induction 1854
  SOS box 1853
  see also DNA repair; excision repair
Soulé, Michael 459
soup approach 1389
Southern blotting **1855–1857**, 2051
  applications 1857
  hybridization analysis 1856–1857
  methodology 1855–1856, *1856f*
Southern, Edward 629
Soviet genetics
  Lamarckism 1129
  Lysenko, Trofim 1129
  Muller, Hermann J. 1129
  neo-Mendelian 1129
  purge of Soviet genetics 1129
Sowers L C **193–197**
Sox9 gene 1813, 1817
SOX9 gene 1812–1813, 1817
SOX9 protein 1814, 1817
SOX10 gene 1469
spd-2 gene 300–301
Spear B T **42**
specialized recombination **1857–1858**
  DNA transposition 1857
  immunoglobulin heavy chain class switch 1858
  intron (and intein) homing 1858
  mating type switch 1858
  retrohoming 1858
  retrotransposition 1857
  site-specific recombination 1857–1858
  V(D)J joining 1858

specialized transduction **1858–1860**
  applications 1860
  mechanisms 1858–1859, *1859f*
  range extension 1859–1860
speciation 384, **1860–1864**
  agamospecies 1861
  allele frequency 665
  Darwin, Charles 1860–1861
  definition 1860–1861
  evolution 664
  founder effect 665
  genetic drift 665
  geographic speciation 1861
  hybridization 1863
  instantaneous speciation 1861
  introgressive hybridization 665
  isolating mechansims 1862–1863
  rates of speciation 1863
  reproductive isolation 1679
  sympatric speciation 1862
    host shift 1862
    mate preference switch 1862
species **1864–1869**
  agamospecies 1866–1867
  biological species concept 1865–1867
  cohesion species concept 1868
  definition 1679, 1864
  evolutionary species concept 1868
  geographical isolation 1867
  isolating mechansims 1866
  nominalistic species concept 1867
  phylogenetic species concept 1868
  recognition species concept 1868
  typological species concept 1865
species selection **1869**
species trees **1869–1870**, *1870f*
specific locus test 1870, **1870–1871**
  agouti (*a*) locus 1599, 1871
  albino (*c*) locus 1599, 1871
  brown (*b*) locus 1599, 1871
  Carter, T.C. 1871
  dilute (*d*) locus 1599, 1871
  piebald (*s*) locus 1599, 1871
  pink-eyed dilution (*p*) locus 1599, 1871
  Russell, W.L. 1599, 1871
  short-ear (*se*) locus 1599, 1871
spectral karyotyping (SKY) 368
Speed T P **729–730**, **1361–1362**
spe genes 255, 1874, *1874f*
Spemann, Hans 235, 331
Spencer, Herbert 63, 702
Spenser, Mike 555
spermatids **1871–1872**, 1875, 1876
spermatocytes **1872–1873**
spermatogenesis 689, 691, 1421
  Caenorhabditis elegans **1873–1874**
  chromosome imprinting 634
  DNA methylation 635
  meiosis 634
  mouse **1875–1876**
spermatogonia **1876**
sp gene 1427
Sphaerotheca fuliginea 1914t
S phase **1761**
Spiegelman, Sol 1004–1005, 1019
spina bifida **1876–1877**
  cause 1877
  characteristics 1876–1877
  distribution 1877
  folic acid 1877
spinal and bulbar muscular atrophy (SBMA) 1421
spindle **1877**
spindle microtubules *606f*
  double-minute chromosomes (*dmin*) 580
  kinetochore 1064–1065
  meiosis 875
  mitosis 1225
  nondisjunction 1346
spinobulbar muscular atrophy (SBMA) 593, 983, *2049t*
spinocerebellar ataxias (SCA) 983, 1421
  spinocerebellar ataxia type 1 *2049t*
  spinocerebellar ataxia type 2 *2049t*
  spinocerebellar ataxia type 3 *2050t*
  spinocerebellar ataxia type 6 *2050t*
  spinocerebellar ataxia type 7 *2050t*
  spinocerebellar ataxia type 8 *2050t*
  spinocerebellar ataxia type 12 *2050t*
Spirin, A.S. 1147
Spirochetes 149
Spiroplasma citri 1384
Spizellomyces punctatus 1740
Spl7 gene 1395
splenic lymphoma with villous lymphocytes (SLVL) 1093
spliceosomal proteins 1733–1734
spliceosome 1378, 1537
splicing **1877**
splicing junctions **1878**
split genes **1878**
  alternative splicing 1878
  introns and exons 1878
  RNA splicing 1878
Spm (suppressor of mutation) gene 2027
Spm transposon 636
SPO1 183–184
Spo11 enzyme 350
Spo11 gene 350
SPO11 protein 1645
Spodoptera descoinsi 1683t
Spodoptera latifascia 1683t
spongiform encephalopathies (transmissible) **1879–1882**
  bovine spongiform encephalopathy (BSE) 1879, 1880
  cause 1879
  Creutzfeldt–Jakob disease 1879
  familial fatal insomnia (FFI) 1881
  Gerstmann–Straussler disease (GSD) 1881
  host genetic effects 1879
  human transmissible spongiform encephalopathies 1881
  kuru 1881
  prion protein (PrP) 1879
  sheep scrapie 1880
  sporadic Creutzfeldt–Jakob disease 1881
  strain variations 1880
  transmissible spongiform encephalopathy **2007–2008**
Spongospora subterranea 1850
spontaneous mutations **1279**
sporadic Creutzfeldt–Jakob disease 1881
Spore killer (*Neurospora*) 1790
spores **1882–1883**
spoT gene 1892
SpoT protein 1892
Spritz R A **25–27**
Spurr N K **1235**
SQUA gene 715
squamata 2097
Src family tyrosine kinases **1883–1885**
  epidermal growth factor (EGF) 1884
Src family tyrosine kinases (*continued*)
  G-protein-coupled receptors (GPCRs) 1884
  human cancer 1885
  integrin adhesion receptors 1884
  oncogenic transformation 1884–1885
  platelet-derived growth factor (PDGF) 1884
  proteins 251, 1883
  RAS proteins 1603–1604
  receptor protein tyrosine kinases (RPTKs) 1884
  SH2 domain 1883
  SH3 domain 1883
  signaling functions 1884
  structure 1883
src operon 1896
Src protein 251, 1883
SR proteins 1538
sry gene 1812
Sry gene 1816–1817
SRY gene 1811, 1812, *1815f*, 1816, 1822–1823, 2159
SRY protein 1812, 1816
SSX1 gene 1771, *1772t*
SSX2 gene 1771, *1772t*
stable equilibrium *see* equilibrium
Stacey G **1329–1330**, **1332–1334**, **1903–1904**
stachel gene 2048
Stadler D **104–105**, **738**, **1322–1323**
Staehelin C **1162**
staggered cuts **1885**
Stahl, Franklin W. 257, 1182, 1437–1438
Stahl F W **319**, **339–340**, **376–377**, **681–682**, **774–778**, **926**, **976–977**, **1104–1105**, **1140**, **1140–1141**, **1141–1144**, **1285**, **1302–1304**, **1350**, **1531–1532**, **1632–1635**, **1635**, **2116–2117**
Stalin, Joseph 1130
standard genetic distance 829–830
Standart N **1147–1148**
Stanford–Binet test 1045
Stanford Human Genome Center 360
Stanford University genome site 1108
Stanier, Roger 1019
Staphylococcus 176
Staphylococcus aureus 138, 192, 586, 1423, 1687, **1885–1886**
  antibiotic resistance 74
  antibiotic-resistant strains 1886
  pathogenesis 1886
  plasmid 1488
  regulatory system 1886
Staphylococcus epidermidis 46
start, stop codons **1886–1887**
stationarity 647–648
Staufen protein 1734
Stebbins, G.L. 664
Steinberg, Charles 1927
Steinberg, J. 1437–1438
stem cells *see* embryonic stem (ES) cells
Stent G S **521–522**, **1071–1072**, **1072–1073**
stepwise mutation model 830, *830f*
sterilization laws 2015
Stern, Curt 1240
steroids **1889**
Stetson, R.E. 650, 1713
Stevens, Tom 1565
Stewart C L **330–333**
Stewart, Timothy 1372
sth 1 chromatin remodeling complex *342t*
Sticker syndrome 605
sticky ends **1889**

STM (shootmeristemless) gene 962, 1475
stm (shoot meristemless) mutation 1475
Stoltzfus A **1052–1053**
Stone, Wilson 1409
stopper mutant 1223
storage plastids 1379
Stougaard J **1121–1122**
strain **1889**
strain distribution pattern (SDP) **1889**
strand placement **1890**
Strausbaugh L D **436–441**
Strefford J C **368**
Streptococcus 176
Streptococcus pneumoniae 80, 138, 654, 658, 1423, 1491, 1666
  antibiotic resistance 74
  cloning 1493
Streptomyces 1882, **1890**
  antibiotic production 1890
  genetic tools 1890
  structure 1890
Streptomyces ambofaciens 859
Streptomyces aureofaciens 1890
Streptomyces clavuligerus 1890
Streptomyces coelicolor *403t*, 405, *946f*, 1892
Streptomyces erythreus 1890
Streptomyces fimbriatus *1694t*
Streptomyces fradiae 1890
Streptomyces griseus 1890
Streptomyces mediterranei 1730
Streptomyces nouseri 1890
Streptomyces scabies 1850
Streptomyces venezuelae 1890
Streptomycetes 1382
streptomycin 77–78, 602, 650, 1687, 1729, **1890–1891**
  Escherichia coli 654
  Mycobacterium tuberculosis 1891
  resistance plasmids 1686
  Streptomyces griseus 1890
  streptomycin-resistant mutants 1890–1891
  Waksman, Selman A. 1890
Strickberger, Monroe W. 1177
stringent response **1891–1892**
  eubacteria 1891
  (p)ppGpp nucleotide synthesis 1891
stroma 1379
Strong, Leonell 1118
structural gene **1892**
STSp gene 2159
Stuart, Ken 1742
Stubbe, W. 1463
Stubbs L **375–376, 1582–1585**
Studier, F. W. 1922
Sturtevant, Alfred Henry 847, 1337, 2011, 2095
  chromosome mapping 353–354
  genes 760
  Morgan, Thomas Hunt 1239
Sturtevant's function *see* mapping function
Stylonychia 763, 1135
Stylonychia lemnae 1204
Stylosanthes guianensis *1914t*
subcellular RNA localization **1892–1895**
  biological functions 1893
  Drosophila melanogaster 1893
  mechanisms 1893–1894
  Saccharomyces cerevisiae 1893
  Xenopus 1893
subcloning **1895**
  see also DNA cloning
substitution mutations
  see base substitution mutations
subtilis phage SPO1 975
Sugimoto Y **1350–1352**
SuH gene 1317, *1318f*
Sulfolobus 2114
Sulfolobus acidocaldarium 92
Sulfolobus islandicum 94
Sulfolobus solfataricus 94
sulfonylurea receptor (SUR1) 1431
Sullivan J **1902–1903**
Sullivan K F **1224–1227**
Sulston, John 254, 613
Summers W C **511–512, 522–524, 877–878, 925–926, 1123–1124, 1124–1125, 1127–1130**
Sumner A T **348–350, 373–375, 663, 926–927, 932–933**
superinfection exclusion 1895
superinfection immunity **1895–1896**
  bacteriophage T4 1895
  phage HK022 1895
  phage P22 1895
  superinfection exclusion 1895
SUPERMAN gene 1473
superoxide dismutase (SOD) 530, 1804
superrepressor **1896**
  arg repressor 1896
  birA repressor 1896
  Gal80 repressor protein 1896
  IkappaB 1896
  operons
    gal operon 1896
    hut operon 1896
    lac operon 1896
    nag operon 1896
    put operon 1896
    src operon 1896
  putR repressor 1896
  translational superrepressor 1896
  trp repressor 1896
sup genes 255
suppression **1897–1899**
  compensatory changes 1898
  protein synthesis errors 1898
  suppressor tRNAs *1897f*, 1897–1898
  therapeutic applications 1898
  translational errors 1897
suppressor mutations **1899–1901**
  intergenic suppressors 1900
    allele-specific suppressors 1900
    bypass suppressors 1900
    informational suppressors 1900
    mutant tRNAs 1900
    nonsense suppressors 1900
  intragenic suppression 1899
    revertant 1899
    second-site mutation 1899–1900
  Yanofsky, Charles 2161
suppressor tRNA **1901**
  anticodon base changes 1901
  frameshift mutations 1901
  missense mutations 1901
  missense suppression 656
  nonsense mutations 1901
  nonsense suppression 656
  suppression *1897f*, 1897–1898
  suppressor mutations 1900
  translational suppression 1901
Surani, Azim 629
Susman M **441–443**
su-Ta gene *1944f*
Sutherland, Earl 499
Sutherland G R **513–597, 726–728**
Sutton, Walter, S. 1177, 2011
SV40 *see* simian virus 40 (SV40)
SVP gene 715
Sweet syndrome 499
swi1 gene 1156
Swi2/Snf2 chromatin remodeling complex *342t*
swi3 gene 1156
swi7 gene 1156
SWI/SNF chromatin remodeling complex *342t*
SWI/SNF family *342t*
Sxl gene 48, 129
Sxl protein 1734
Sxr gene cluster *1253f*, 1253–1254
SYBL1 gene 2160
symbionts 337, **1902**
  see also chloroplasts
symbiosis islands **1902–1903**
symbiosome *1903f*, **1903–1904**
sympatric 665, 988, 1862, **1904**
Sym plasmids 1715
symplesiomorphy 965, **1904–1905**
sym (symbiosis) gene 1471
synapomorphy 965–967, 968, 969–970, 1904, **1905–1906**
  autapomorphy 1905
  homology 1905
  phylogenetic trees 1905
synapsis (DNA transactions) **1906–1909**
  Dmc1 protein 1906
  DNA-mediated synapsis 1906, *1907f*
  homologous genetic recombination 1906
  homology 1906
  protein-mediated synapsis 1906–1907
    Holliday junction *1908f*
    integrase family 1907, *1908f*
    resolvase/invertase family 1907, *1908f*
    site-specific recombination 1907, *1908f*
    topological filter 1907, *1908f*
    transposition 1909, *1909f*
    transposons 1909
  Rad1 protein 1906
  Rad51 protein 1906
  RecA protein 1906
synaptobrevin/SNB-1 protein 1312–1313
synaptonemal complex **1910–1912**
  chromatin organization 1912
  development *1910f*, 1911, *1911f*, *1912f*
  historical background 1911
  HOP1 gene product 1912
  meiosis 1910
  RED1 gene product 1912
  Saccharomyces cerevisiae 1912
  structure *1910f*, 1911–1912
  ZIP1 gene product 1912
  see also chromosome pairing, synapsis
synaptotagmin/SNT-1 1313
Synchytrium endobioticum 1850
syndactyly **1912**
Synechococcus 389–390, 1462
Synechocystis *150t*, 404, 859
syngamy 689
syngenic **1913**
syn mutation 1223
synovial sarcoma 1769, *1769t*, **1913**
synpatojanin/UNC-26 protein 1313
syntaxin/UNC-64 protein 1312–1313
synteny 86, 269, 359, 1338, 1394, **1913**

systemic acquired resistance (SAR) **1913–1917**
Arabidopsis thaliana 1913
induced resistance 1913
pathogenesis-related (PR) proteins 1913
pathogenic agents of plants *1914t*
reactions after first infection *1914t*, 1915, *1915f*
salicylic acid 1913, 1915–1917
systemic reactions 1916–1917
systemic signal 1916
tobacco mosaic virus (TMV) 1913
SYT gene 1771, *1772t*
Szathmáry E **1387–1394**
Szczelkun M D **1693–1696**
Szczepánski T **1206–1211**, **1503–1506**
Szymanski D **87–90**

T

T2 phage 377
T4 phage 377, 383
T7 DNA polymerase *1498f*
T7 polymerase gene 975–976
Tabby gene 400–401
tag gene 1664
TAL1 gene 1930–1932
TAL2 gene 1930–1932
TAL gene family (SCL) **1930–1932**
Basic helix–loop–helix (bHLH) proteins *1931f*, 1931–1932
DNA-binding transcription factors 1931–1932
T-cell leukemia 1932
T-ALL *see* leukemia, acute
Tam1 gene 2026
tandem repeats 977, 1205, **1932–1933**
see also concerted evolution
T antigens **2078–2081**
large T antigen 2079–2080
small T antigen 2080
Taq DNA polymerase 1501, 1502, 1504
targeted mutagenesis (mouse) **1933–1934**
gene knockouts 1934
transgenic mice 1933
Taricha 409
TATA binding protein 544
TATA box 458, 551, 954, 1550, **1934**
housekeeping genes 978
RNA polymerases 805, *805f*, *806f*
Tata F **1314–1318**
Tatum, Edward L. 936, 1080, 1323
colinearity 419
conjugation 450
F (fertility) factor 679
Nobel Prize 1070
plasmids 657
tau gene 389
Tavaré S **392–397**
Tavaré, Simon 1517
Tavernarakis N **313–318**
taxa 384
Tax gene 979
taxonomy, evolutionary **1934–1937**
cladification 1935–1936
cladistic analysis 1935
classification 1934–1935
Darwinian classication 1935
incompatibilities between Darwian and cladistic classifications 1936–1937
method cladification 1936
objective of classification 1935
phenetics 1937
special purpose classification 1937
taxonomy, evolutionary (*continued*)
see also cladistics
taxonomy, numerical **1937–1941**
bootstrap analysis 1941
character weight 1938–1939
operational taxonomic units (OTUs) 1939–1940
philosophical basis 1938
predictivity 1938
procedures 1939–1941
Q analysis 1941
R analysis 1941
Taylor A M R **113–115**
Tay–Sachs disease 756, **1941–1943**
balanced polymorphism 187
carrier 275
clinical features 1942
diagnosis 1942
genetic counseling 1943
inheritance, distribution, and frequency 1942
molecular genetics 1942–1943
pathogenesis 1942
prevention 1943
treatment 1943
Tay, Warren 1941
TB4Y gene 2159
T-banding 349
T-box genes **1943–1945**
brachyury locus 237, 1943
defining features 1943–1944
history 1943
phylogenetic analysis 1944–1945
phylogenetic tree *1944f*
see also gene family
TBP transcription factors 805, 1029
Tbr1 gene subfamily *1944f*
tbr (trichome birefringence) gene 2048
Tbx1 gene subfamily *1944f*
Tbx2 gene 1945
Tbx2 gene subfamily *1944f*
Tbx3 gene 1945
TBX3 gene 1945
TBX5 gene 1945
Tbx6 gene subfamily *1944f*
Tcd-1 gene 1919, *1920f*
Tcd-2 gene 1919, *1920f*
Tcd-3 gene 1919, *1920f*
Tcd-4 gene 1919, *1920f*
Tcd-5 gene 1919, *1920f*
Tcd gene 1919
T-cell acute lymphoblastic leukemia (T-ALL) *see* leukemia, acute
T-cell prolymphocytic leukemia (T-PLL) *see* leukemia, chronic
T-cell receptor gene family **1919**, 1932
T-cells 953, 998, 1251
TCL1 gene 114, 1094
t-complex (mouse) 1790
Tcp-10b gene 1920
Tcp-10 gene *1920f*
TCP-1 (group II) chaperonins 324
TCR *see* T-cell receptor gene family
Tcr gene 1919, *1920f*
TDF *see* testis determining factor (TDF)
TDF gene 1154, 1816
TDH1 gene 886
TDH2 gene 886
TDH3 gene 886
T-DNA 1984–1986, *1985f*
TDS *see* Tay–Sachs disease
Tdy gene 1154, 1816–1817
Tease C **18–19**, **1064–1065**
Tegeticula 411
Tellima grandiflora 840
telomerase 310, **1945–1946**, 1947
telomeres 310, 519, 538, 1136, **1946–1950**
aging and cancer 1949–1950
binding proteins 1947–1948
carcinoid tumors 274
changes in length *1949f*
DNA probes 1003
Drosophila melanogaster 1948–1949, *1949f*
end-replication problem 1946, *1946f*
gene rearrangement 798
holocentric chromosomes 956
macronuclear development 1133
maintenance *1947f*
Saccharomyces cerevisiae 1529, 1946–1947, *1947f*
subtelomeric structure 1946–1947
telomerase function 1947
terminal repeats 1946
Tetrahymena thermophila 1612
see also aging; chromosome structure
Temin, Howard 257, 590, 1702
Temin, Rayla Greenbert 490
temperate phage **1950**
temperature-sensitive (*ts*) mutant 441–442, 1927, **1950–1951**
see also conditional lethality; T phages
template **1951–1952**
see also DNA repair; DNA replication; polymerase chain reaction (PCR)
Teopod1 gene 930
Teopod2 gene 930
Teopod3 gene 930
TER1 gene 1948
teratogen 449
terminal inverted repeats 1030–1031, **1054**, 2026
terminal redundancy **1952**
termination codon *see* amber codon; nonsense codon; ochre codon; opal codon
termination factors
see release (termination) factors
terminator **1952**
test cross **1952**
testis determining factor (TDF) 1812, 1952
testis determining locus **1952**
testis differentiation 1822
testosterone 1812, 1816, 1875
TEs (transposable elements) *see* transposons
tetracycline 611, 1686, 1687, 1890
tetrad analysis 699, 1763, 1777, **1952–1957**
1:1 segregation 1954
4:4 segregation *1953f*
double crossovers *1957f*
explanation of three tetrad classes *1956f*
half-tetrad analysis 1953
independent and distant linkage 1956–1957
linked segregations 1955
meiosis 1952–1953
ordered tetrads 1954
strand relationships in double crossovers 1955–1956
tetrads and octads in fungi 1953–1954
two-locus segregation 1954–1955
Tetrahymena 749, 763, 1135, 1910, 1947
Tetrahymena pyriformis 860, 1219
Tetrahymena thermophila 1133, 1204, 1612

tetramethyl rhodamine isothiocyanate 1002
tetranucleotide hypothesis 553
tetraparental mouse **1957**
see also chimera
tetraploid 346, 536
tetratype **1958**
T-even phages 1923–1930
TF activating proteins 378, 379, 379–380
TFIIA transcription factors 805, 1029
TFIIB transcription factors 805, 1029, 1972
TFIID transcription factors 805, 949, 1029, 1972
TFIIE transcription factors 805, 1029
TFIIF transcription factors 804, 805, 1029
TFIIH transcription factors 562, 568, 805, 1029
TFIIIA transcription factors 1029
TFIIIB transcription factors 805, 1029
TFIIIC transcription factors 1029
TFL1 gene 715
T gene subfamily *1944f*
TGF-α *see* tumor growth factor-α
TGFβIIR gene 423, 2083
Tgn (thyroglobulin) gene 592–593
thalassemias 1831, **1958–1961**
alpha (α)-thalassemias 1958
beta (β)-thalassemias 1958–1959
clinical description 1959–1960
clinical variability 1960
control and treatment 1961
distribution *1959f*, 1960
hemoglobin variants 1960
molecular pathology 1958–1959
population genetics 1960–1961
types 1958
Thaler, David 1624
Thamnophis sirtalis 1535
thanatophoric dysplasia 2
t haplotype 591, 986, 988, **1919–1921**, *1920f*
evolution 1920–1921
gametic selection 1296
history and genetics 1919
meiotic drive 1921
meiotic drive (mouse) 1166
population genetics 1921
segregation distortion 1791
transmission ratio distortion 1919–1920
The Arabidopsis Information Resource (TAIR) 90
Theissen G **713–717**
Theorell, Hugo 1067
Thermatoga maritima 1198
Thermoanaerobacter 1962
Thermococcus litoralis 1198
thermophilic bacteria **1961–1963**
anaerobic thermophiles 1962
Bacillus stearothermophilus 1962
description 1961
DNA transfer 1962
evolution 1962
genetic engineering 1961–1962
polymerase chain reaction (PCR) 1961
Thermus thermophilus 1963
in vivo vs. in *vitro* genetics 1962
see also Archaea
Thermoproteus 2114
Thermotagales 149
Thermotoga 1962
Thermus aquaticus 1501, 1833, 1961
Thermus thermophilus 1726–1727f, 1962, 1963
theta replication 1381, **1963**, *1963f*
see also replication
thiamine 2118
Thielaviopsis basicola 1914t
Thiomargarita namibiensis 147
thiostrepton 611
Thomas E **1258**, **1482**, **1579–1580**, 1613, **1886–1887**, **1950**, **1963**, **2107–2108**
Thomas, Mary Anna 1408
Thomomys bottae 1522t
Thomson G **750–752**
Thomson, Nichol 254
three-point cross (test-cross) *1964t*, **1964–1965**
threonine **1965**
threshold characters **1965**
Tbx6 gene 1945
thyA gene 1624, *1624f*
thymidine 1360
thymidine kinase (TK) 359
thymidylate synthase (*td*) gene 1929
thymidylic acid 1360
thymine dimer **1966**
thymine (T) 540, 572, 1358, **1965–1966**
Tiarella trifoliata 840
tiger salamander 1304
TIGR web site 517
Tilghman, Shirley 630
Timberlake W E **106–111**
tim (timeless) gene 387
Timoféeff-Ressovsky, Nicolai 522
TIM protein 387, 387–388
Ti plasmids 493, **1966**
Tippett, Patricia 1713
TIRs *see* terminal inverted repeats
Tiselius, Arne 757
tissue culture 310, **1966–1969**
aging 1967
CHO cell line 1968
culture media 1967
culture methods 1968–1969
embryonic stem (ES) cells 1967
genetics 1968
HAT selection technique 1968
Hayflick limit 1967
history 1966–1967
see also cell lines
tit for tat 670
Tjio, Joe-Hin 1094, **1969–1970**
TK *see* thymidine kinase (TK)
Tlsty T D **761–771**
Tm *see* melting temperature
Tn*3* transposon 1688, *1689f*, 1843, *1844t*
Tn*10* transposon 1658, 1767
Tn*402* transposon 1045
Tn*552* transposon 1688
Tn916 plasmid 453–454
Tn*916* transposon 1843, *1844t*
Tn*1545* transposon 1843, *1844t*
Tn*4430* transposon *1844t*
Tn*4451* transposon 1843, *1844t*
TNF *see* tumor necrosis factor (TNF)
tniA gene 1045
tnpA gene 1689
tnpR gene 1688–1689
Tnt1 gene 2029, 2022
tobacco *1914t*
tobacco mosaic virus (TMV) 1913
tobacco necrosis virus *1914t*
tobacco rattle virus *1914t*
Tolman, Edward 209
Tolmiea menziesii 840
Tolmie J L **583–584**
tomato *see Lycopersicon esculentum*
tonB gene 1923
tonofilaments 1289
tonoplast 1379
Topo II topoisomerase 372, 373–374, 378
topoisomerases 576, 1038, 1844, **1970**
see also replication
tortoiseshell coloring 1819, **1970–1971**
see also coat color mutations
totipotency 618, 621
Townsend, C. O. 491
TP53 gene 1123
tpfQ/I gene 43, *45f*
TpG 636
T phages 522, **1921–1930**
complementation 1928
genetic fine-structure analysis 1927–1928
introns 1929–1930
mutants 1927
operational definition of a gene 1928
phage T1 1923
phage T3 1922–1923
phage T4 1923–1927
phage T4 genetic map *1925f*
phage T4 infection 1924–1927, *1926f*
phage T4 micrograph *1924f*
phage T5 1923
phage T7 1922–1923
special properties 1928–1929
T-even phages 1923–1930
see also temperature-sensitive (*ts*) mutant
TPI1 gene 886
Tracey M **834–835**, **1791**, **1795**
tra gene 678, 1489
transacetylase 1070
trans-acting factors **1971–1973**
DNA binding 1971–1972
enhancers 624
Ets family 661–663
homeobox genes 958–962
Rrn3 1029
SL1 1029
structure 807, *808f*
synthesis regulation 1973
TBP 805, 1029
TFIIA 805, 1029, 1499
TFIIB 805, 1029, 1499
TFIID 805, 1029, 1499, 1550
TFIIE 805, 1029, 1499
TFIIF 804, 805, 1029, 1499
TFIIH 805, 1029
TFII-I 1499
TFIIIA 1029
TFIIIB 805, 1029
TFIIIC 1029
transcription activation 1972
transcription factors regulation 1973
transcription repression 1972
UBF 1029
trans, cis configurations
see cis–trans configurations
transcribed spacer **1973**
transcript cleavage reaction *1980f*
transcription **1973–1983**, 2000
activation 1972
chromatid 340
colliding RNA polymerases *1982f*
elongation 609, *1979t*, 1979–1981, *1980f*
elongation regulation 1980–1981
eukaryotes 250

transcription (*continued*)
initiation factors 1029
initiation reaction regulation 1979
kinases 1064
preinitiation complex formation 1977–1978
promoter clearance 1978–1979
promoter escape 1978
promoter occlusion 1982
promoter recognition 1977, 1978
regulation 949–951
repression 1972
RNA polymerases 1974, 1975–1977, *1976t*, *1979t*
RNA synthesis initiation 1978–1979
sigma factors 1029
termination 1981–1982
termination regulation 1981–1982
transcript cleavage reaction *1980f*
transcription cycle 1974, *1975f*, 1977–1982
upstream, downstream sites 2100
see also cis-acting locus; *cis*-acting proteins; topoisomerases
transcriptional start site (TSS) 379
transcription-coupled repair (TCR) 566, 568
transcription factors *see trans*-acting factors
transcript maps *363t*, 364
transduction **1983–1984**
transfection **1984**
see also green fluorescent protein (GFP)
transferred DNA *see* T-DNA
transfer RNA (tRNA) 610, 1358, **1986–1989**
adaptor molecules 1988
Bacillus subtilis 141
codon–anticodon relationships 1988
elongation factors 609
history 1987
mimicry 1988–1989
structure 1987
synthesis 1987–1988
transcription factors 1029
transcription termination 124
translation 323–324
tRNA synthetases 1988
yeast phenylalanine structure *1987f*
yeast structure *1987f*
see also messenger RNA (mRNA)
transformation **1989**
transformer gene 1316
transforming growth factor α (TGF-α) 626
transforming principle 553
transgenes 545, **1989–1990**
Caenorhabditis elegans 1990
Drosophila melanogaster 1989
embryonic stem (ES) cells 1992–1994
expression 1994, 1996–1997, *1997f*
mouse 1989
nuclear transfer 1994
pronuclear microinjection 1991–1992, *1993f*
retroviral vectors 1992
transgenic animals **1990–1998**
agriculture 1997–1998
applications 1994–1998
embryonic stem (ES) cells 1992–1994, *1993f*
gene disruption 1995
gene product overexpression 1996, *1996f*
genetic ablation 1997
genetic markers 1997
transgenic animals (*continued*)
genetic rescue of mutations 1995
gene traps 1995
history 1991
mouse 1933, 1991, *1991f*
nuclear transfer 1994
pharmaceutical industry 1997–1998
production 1991–1994
pronuclear microinjection 1991–1992, *1993f*
random insertional mutations 1995
regulatory DNA sequences 1995–1996
retroviral vectors 1992, *1993f*
targeted mutations 1995
transgene expression 1994, 1996–1997, *1997f*
transgenes 1990
transgenic organisms 545
transgenic plants 1984
transglutaminases 994
trans-heteroduplex 582
transient neonatal diabetes 1000
transient polymorphism 187, **1998–1999**, *1999f*
transition **1999**
translation **1999–2002**
aminoacylation 601
DNA replication 541
editing 601
elongation 323, 2001
inhibition 2001
initiation 323, 2000–2001
mRNA 323
product transport 2001–2002
proofreading 601–602, 609
reading frames 2000
ribosome recycling 2001
ribosomes 609, 2000–2001, 2002
termination 323–324, 2001
transcribed RNA processing 2000
tRNA 323
upstream, downstream sites 2100
translational control **2002–2003**
translation initiation factors
eIF1 1029
eIF2 1029
eIF3 1029
eIF4 1029
eIF5 1029
IF1 1029
IF2 1029
IF3 1029
TFIID 949, 954
translocation 610, **2003–2007**
bacteria 610
Cattanach's translocation *2005f*
chromosome 7 (mouse) 284–286
EF-G-GTP 610
genomic imprinting 2005
insertion 2005
linkage maps 2005
marker chromosomes 2004–2005
meiosis 2004, *2004f*
mouse 284–286
reciprocal translocations 2003–2004, *2005f*
Robertsonian translocation *2006f*
shift (transposition) 2006
X–autosome translocations 2004
X chromosome (mouse) 284–286
see also Cattanach's translocation; radiation genetics; Robertsonian translocation
transmembrane conductance regulator (CFTR) 1565, 1565
transmissible mink encephalopathy 2008
transmission electron microscopy (TEM) 605
transmission genetics **2008–2016**
eugenics 2015
genetic mapping of humans 2012–2013
hereditarianism 2014–2015
humans 2012–2015
Mendel, Gregor Johann 2009–2011
Mendelian inheritance 2009–2010
Mendelian Trait Database 2013
mode of inheritance and pedigree analysis *2014t*
modes of inheritance 2013–2014
multifactorial genetic diseases *2015t*
non-Mendelian inheritance 2011–2012
selective breeding 2008–2010
sterilization laws 2015
transmission ratio distortion (TRD) 1919–1920
transposable elements *see* transposons
transposase 1038, 2018–2019, **2033**
transposons 630, 636, 1042, **2016–2020**
applications *1935f*, 2019–2020, **2034–2040**, *2037f*
DNA sequencing 2038, *2039f*
gene indentification and tagging 2035–2036
gene mapping and cloning 2036
gene mutagenesis 2036–2037
gene transfer 2038
multipurpose transposons *2039f*, 2039–2040
P elements 2035
protein taps 2038
reporter fusions and gene expression 2037–2038
bacterial genetics 162
classification 2017
compound transposon 1030
DNA transposons 2017
donor cleavage *2018f*
donor cleavage and strand transfer *2019f*
drug resistance 586
excision *2030f*, **2033–2034**
immunity 998
intron homing 1051
long terminal repeats (LTRs) 1120
molecular drive 1234
phosphoryltransferases 1032
plants **2020–2033**, *2021f*
Ac/Ds superfamily 2026, *2027f*
CACTA superfamily 2026–2027
class 1: retrotransposons 2021–2026
class 2: DNA transposons 2026–2029
contribution to genetic material 2030
discovery in maize 2020–2021
DNA-diversity 2030–2031
EN/Spm element *2028f*
EN/Spm in maize 2026
evolution 2031–2032
excision *2030f*
genome distribution and chromosomal organization 2025–2026
long interspersed nuclear elements (LINES) 2022
LTR-retrotransposons 2021–2022, *2023f*
mutator elements 2028–2029, *2029f*

transposons (*continued*)
non-LTR-retrotransposons 2022–2025, *2024f*
overview and classification 2021
regulation 2029–2030
short interspersed nuclear elements (SINES) 2022, 2025
tagging and reverse genetics 2030
terminal inverted repeats 2026
retrotransposons 2017
selfish DNA 1806
structure 2016–2017
synapsis (DNA transactions) 1909
target choice 2018
transposase proteins 2018–2019
transposases (Tpases) 1030–1037
transposition mechanism 2017–2018
transposition regulation 2019
transposons
copia family 975
Escherichia coli 659
gamma delta (γδ) transposon 1688, *1689f*
Helena family 985
hobo family 985
I family 985
mariner family 975
Paris family 985
Penelope family 985
P family 585, 975, 985
Spm transposon 636
Telemac family 985
Tn*3* transposon 1688, *1689f*
Tn*552* transposon 1688
Tn*1000* 936–937
Ulysses family 985
V(D)J recombination 2034
see also controlling elements
trans-splicing **2040–2041**
group II introns 2041
other cases 2041
spliced leader addition 2040–2041
transvection **2041**
transversion mutation **2042**
Trask, B. 361
tra/trb gene operon 451
Travers A **542–544**
TRD *see* transmission ratio distortion (TRD)
trees 384, **2042–2045**
algorithms 1416–1417
bifurcating or multifurcating 2043
clock tree 2043
estimated gene tree 2044–2045
operational taxonomic units (OTUs) 2042
parsimony 1415–1418, *1416f*
phylogenetic *2044f*
possible tree topologies 2043
rooted *2044f*
rooted or unrooted 2043
types *2042f*, *2043f*
see also coalescent; phylogeny
Treponema pallidum *150t*, 1892
TRF1 gene 1948
TRF2 gene 1948
Tribolium 730
Trichocomaceae 106
trichome development **2045–2048**
branching mutants 2045–2046
differentiation mutants 2045
endoreduplication 2045
expansion 2048
genetic models *2047f*
trichome development (*continued*)
initiation 2045
maturation 2048
mutants *2046f*
Trichomonas vaginalis 405
trichothiodystrophy (TTD) *569–570t*, 994
clinical features 568
ichthyosis 568
transcription-coupled repair (TCR) 568
transcription deficiency 568
Trifolieae 1083
Trifolium (clover) 1083
Trifoliume 730
Trigona lineata 1170
Trigonella 1330
trinucleotide repeat disorders 594
trinucleotide repeats **2048–2053**
DNA expansion mechanism 2052–2053
gene coding region mutants 2051–2052
gene noncoding region mutants 2052
genetic anticipation 2048
human disorders *2049t*
triplet code **2054–2055**
triploidy 345, 989–991, 2055, **2055–2056**
see also polyploidy
triptychon gene 2045
trisomic zygote rescue 1241
trisomy 346, 1241, 1346, **2056–2058**
trisomy 13 **1420**
trisomy 18 346, 1271, **2058–2060**
see also chromosome aberrations
trisomy 21 346, 583–584, 1085, 2057
amniocentesis 61
aneuploidy 66
meiosis 1164
mutation 1271
nondisjunction 1346
uterine leiomyoma 1110
trisomy 9p 1086
tritanopia 421
Triticea 971
Triticum aestivum 66, 537, 964, 1164
Triticum monococcum 1511
Triticum species (wheat) **2060–2068**
cytogenetic stocks 2064–2066, *2065t*
cytogenetic structure 2061–2064
domestication 2060–2061
evolution 2060, *2060f*
gene mapping with monosomics 2066
genetic mapping 2067–2068
genetic transmission 2067
genome and plasmon constitution *2063t*
idiogram *2062–2063f*
intervarietal substitution lines 2066–2067
mating systems 2067
ploidy and domestication status *2061t*
Triticum turgidum 1511
Triturus 762
Triturus cristatus 840
trivial equilibrium **2068–2070**, *2069f*
tRNA *see* transfer RNA (tRNA)
trophoblast **2070–2071**
tropical spastic paraparesis (TSP) 979
Trp53 gene 1245
trpA gene 57, *218f*, 219, 419, *419f*, 655, 1900, 2078
TrpA protein *419f*
trpB gene *218f*, 2078
trpC gene *218f*
trpD gene 218, *218f*, 2078
trpE gene *218f*, 2078
trp gene 453
trp operon 122, 123, 655, 943, 1078, 1679, **2076–2078**
trp repressor 1679, 1896
Tryon, Robert 209
Trypanosoma 765, 1741
trypanosomes 80, 278, 1735
trypsin 1573, **2071–2075**
altering protein function 2075
catalysis 2072–2073, 2074
history 2071–2072
mutational analysis 2074–2075
structure 2072, *2072f*, *2073f*
substrate recognition 2074, 2074
substrate specificity 2074–2075
zymogen activation 2072
tryptophan operon **2076–2078**
biosynthetic enzymes 2077
Escherichia coli 2076f, 2076–2077, 2077–2078
tryptophan biosynthesis 2076, 2077
tryptophan (Trp) 59, 216, **2075**
TSC1 gene *81t*
TSC1/TSC2 gene *2084–2085t*
TSC2 gene *81t*
Tschermak von Seysenegg, Erich 353, 521, 1137, 1178
TSE
see spongiform encephalopathies (transmissible)
tsf gene 610
ts gene 442
TSH receptor (Tshr) gene 592
TSIX gene 2151
TSP (tropical spastic paraparesis) 979
Tspy gene *1253f*, 1254
TSPY gene 2159
TSS *see* transcriptional start site (TSS)
Tsui L-C **507–509**
TTD-A gene 568
ttg1 gene 2045
Tto1 gene 2022
ttx-3 gene 1097
TTY1 gene 2159
TTY2 gene 2159
tufA gene 1221
tuf gene 610
tumor angiogenesis 68
tumor growth factor-α 310, 1605
tumor growth factor-β 1605
tumorlets 272
tumor necrosis factor (TNF) **2081**
tumor suppressor genes 263, **2081–2088**, *2084–2085t*
biological basis 2082–2083
breast cancer genes 2086–2087
cancer pathogenesis 2087–2088
caretakers 2082
gatekeepers 2081–2082
identification 2083
mismatch repair genes 2087
Rb–p53 pathway 2083–2086
tumorigenesis *2082f*
Wnt signaling 2086
turbid (*tu*) mutants 1927
Turcot syndrome 14, 422
Turek P J **1021–1026**
T-*urf13* gene 1223
Turleau C **486–487**
Turner syndrome 346, 688, 1086, 2056, **2088–2089**
aneuploidy 66
characteristics 1021
Ford, Charles 723

Turner syndrome (*continued*)
human chromosomes 980
isochromosome 1054
monosomy 1239
sex chromatin 1809–1810
sex chromosomes 1811
X chromosome 2147
XY female syndrome 1822
Y chromosome (human) 2156
turnip crinkle virus *1914t*
TWIST gene *480t*, 481
twisting number **2089**
two-dimensional polyacrylamide gel electrophoresis 1576
two-hit model 81, 262, 1697
Twort, Frederick 179
Ty1/copia gene 2022, 2029
Ty1 gag gene 2022
Ty3/copia gene 2026
Ty3/gypsy gene 2022
Ty elements *see* transposons
Tyr (tyrosinase) gene 398
Tyr(c) gene 527
TYR gene 26
tyrosinase 400, 1469
tyrosine 938, 1038, 1469, **2090**
tyrosine kinases 251, **648–649**
Hirschsprung disease 942
inhibitors 649
phosphorylation 648
role in oncogenesis 648–649
see also epidermal growth factor (EGF)
tyrosinemia 36
Tyrp1 (tyrosinase-related protein 1*)* gene 398
Tyrp1(b) gene 527
TYRP1 gene 27

U

Ube1y gene *1253f*, 1254
UBF transcription factors 1029
ubiquitin **2091–2093**
26S proteasome 2091, *2092f*
cell cycle 2092
cyclins 2092
degradation signals 2091
destruction box 2092
eukaryotes 2091
function 2092
mitosis 1226, 2092
pathways 2091, *2092f*
process 2091
proteolysis 1573
Ub-proteasome system 2091
Ub system 2091–2093
Ubx (ultrabithorax) gene 958, *960f*, 961, 962, 1852
udt (underdeveloped trichome) gene 2048
UFO gene 716
Ulrich A B **310–311**
ultrametric **1189–1190**
ultraselfish gene 1165
Ulva 535
umber mutation *see* start, stop codons
umuC gene 1622, 1651, 1854
umuD gene 1622, 1651, 1854
unc-4 gene 961, 1312
unc-6 gene 532
unc-13 gene 1312–1313
unc-18 gene 1312–1313
unc-25 gene 1312
unc-26 gene 1313
unc-30 gene 1312
unc-34 gene 1311
unc-47 gene 1312
unc-52 gene *620f*
unc-54 gene 317, 1750
unc-73 gene 1311
unc-86 gene *307f*, 308, 959, 1310, 1322
unc-104 gene 1312
UNC-4 protein 1312
UNC-5 protein 1311
UNC-6 protein 1311
UNC-11 protein 1313
UNC-13 protein 1312–1313
UNC-17 protein 1312
UNC-18 protein 1312–1313
UNC-25 protein 1312
UNC-30 protein 1312
UNC-34 protein 1311
UNC-40 protein 1311
UNC-47 protein 1312
UNC-49 protein 1312
UNC-73 protein 1311
UNC-86 protein *307f*, 308
uncB gene *1383f*
unc genes 255
uncI gene *1383f*
underdominance **2093–2094**
allele frequency 2093–2094, *2094f*
evolutionary outcome 2093
heterozygote inferiority 2093
natural selection 2093
relative fitness 2093
unstable equilibrium 2094
unstable polymorphism 2094
underwinding **2094**
unequal crossing-over **2095–2096**, *2095f*
chromatid 339
Drosophila melanogaster 2095, *2095f*
evolution 667–668, *668f*
hemoglobin anti-Lepore 2096, *2096f*
hemoglobin Lepore 2096, *2096f*
humans 2096
meiosis 2095
microsatellite 1205
Saccharomyces cerevisiae 2096
X chromosome 339
see also chromatid; concerted evolution; crossing-over
un gene 1427
unidentified reading frame (URF) **2101**
unidirectional replication **2095**
UniGene set 364
unineme 373
uniparental disomy (UPD) 209, 346, 1000, 1179, 1241
uniparental inheritance **2096–2099**
androgenesis 2098
chimera 2098
Chlamydomonas 2097
cloning 2098
genomic imprinting 2097
hydatidiform moles 2098
meiosis 2097
mitochondria 2097
parthenogenesis 2097–2098
Saccharomyces cerevisiae 2096
Schizosaccharomyces pombe 2096
see also parthenogenesis
unique DNA **2099**
see also repetitive DNA
universal genetic code **2099**, *2099f*
unscheduled DNA synthesis **2099**
unstable equilibrium *see* equilibrium
unstable polymorphism 2094
UPD *see* uniparental disomy (UPD)
UPD14 syndrome 1000
upilson-Ha-ras oncogene 1484
upstream, downstream sites **2100**
codons 2100
DNA sequences 2100
enhancers 624, 2100
operator 2100
promoters 2100
transcription 2100
translation 2100
upstream **2100**
URA3 gene 2163
ura4 gene 2163
uracil (U) 541, 566, 1358, **2100–2101**
uridine 1360
uridine triphosphate (UTP) **2101**
uridylic acid 1360
Urnov F D **628–637**
Urodela 1304
Uromyces phaseoli 1914t
Ursus americanus 840
Ursus arctos 840
U.S. Department of Agriculture 847
National Genetic Resources Collection 852
National Plant Germplasm System 852
U.S. Department of Energy (DOE) 519
Usher syndrome 1023
Usp9y gene *1253f*, 1254
USP9Y gene 2159
Ussery D W **517–521**, **550–553**
Ustilago maydis 535, 1154, 1398, 1910
uterine abnormalities 1024
Uty gene *1253f*, 1254
UTY gene 2159
UV radiation 271, 1020, 1458–1459, *1459f*, 1665
UvrAB protein 916
UvrA protein 674, *674f*, 1665, *1665f*
UvrB protein 674, *674f*, 1665, *1665f*
UvrC protein 674, *674f*, 1665, *1665f*
uvrD gene 1286
UvrD protein *1665f*, 1666
Uvr genes *see* excision repair
UvsX protein 1841
Uyenoyama M K **1398–1400**

V

vacA gene 916–917
vaccinia growth factor (VGF) 627
vacuoles 1379
valine **2103**, *2103f*
vanA gene 75
van Boom, Jacques 555
vancomycin 586, 1687, 1886
van den Berg E **1659–1660**
Vanderleyden J **707–709**
van der Marel, Gijs 555
van der Velden V H J **1503–1506**
Van de Ven W J M **1109–1117**
Van Dommelen J **707–709**
van Dongen J J M **1206–1211**, **1503–1506**
Van Etten R A **207–208**, **250–251**
Van Montagu M **23–25**
variable number of tandem repeat (VNTR) *see* minisatellite
variable region **2103**, *2103f*
variegation **2103**
Varshavsky A **1573–1575**, **2091–2093**
vascular endothelial growth factor (VEGF) 68, 127, 1605, **2103–2104**
VASKhNIL (Lenin All-Union Academy of Agricultural Sciences) 1129

Vavilov, Nikolai 201
V(D)J recombination 1640–1642, *1641f*, 2016
VDJ segments 1640–1642, *1738f*
vectors 656, **2104–2106**
characteristics 2104
cloning vectors 2104
DNA cloning 545
Escherichia coli 2104
horizontal transfer 974–975
library construction 2105–2106
multiple cloning sites 2104
plasmid vectors 2104–2105
protein expression 2106
shuttle vector **1828**
viral vectors 2105
velocardiofacial syndrome
see DiGeorge syndrome
velopharyngeal insufficiency 534
Venkitaraman A R **2081–2088**
Vent DNA polymerase 1502
Ventor, Craig 356
Verma I M **1658–1659**
vertebrates 249, 309
vertical resistance in plants 1535
vertical transmission **2106–2107**
Verticillium 1850
v-fms oncogene 721
VHL gene *81t*, *2084–2085t*, 2087
Vibrio cholerae 772, 1043–1044, 1198, 1423, 1903
Vibrio harveyi 1382
Vicia faba 683, 723, 1470, 2022
Vicia (vetch, broad bean) 1083
Vicieae 1083
Vidal M **1551–1552**
Vigna (cowpea, mung bean) 1083
Virchow, Rudolf 1172
VirD1 enzyme 1985
VirD2 enzyme 1985
vir gene 493
viroids **2107**
virology *see* virus
virulence and avirulence 410–411
virulent phage **2107–2108**, *2108f*
virus **2108–2114**
Abelson murine leukemia virus 250
archaea viruses **2114–2116**
background 2108
carcinogens 271–272
characteristics 2109
classification
DNA viruses *2112f*
Lwoff–Tournier system 2112
RNA viruses *2113f*
Epstein–Barr virus (EBV) 271
genomes 2111, *2111t*
hepatitis B virus 271
permissive cells **1430**
plaque method 2108–2109
replication 2111
retroviruses 2111
structure 2109–2110, *2110f*
study techniques 2108–2109
virions 658, 2109
virus–host interactions 2110
virus multiplication 2110–2111
vir (virulence) allele 1535
visceral leishmaniasis 1736
Viscounti–Delbrück hypothesis **2116–2117**
Delbrück, Max 2116
linkage relationships 2116–2117
Viscounti–Delbrück hypothesis (*continued*)
phage crosses 2116–2117
recombinant frequency 2116
Viscounti, N. 2116
vitamins **2117–2120**
antioxidants 2118–2119
calcium and phosphorus metabolism 2120
cell development 2119–2120
cell metabolism 2118, *2119f*
DNA synthesis 2118
enzymes 625
purpose 2117
vitamin A 2119–2120
vitamin B_1 2118
vitamin B_5 2118
vitamin B_6 2118
vitamin B_{12} 2118
vitamin C 2118, 2118–2119
vitamin C synthesis 2120
vitamin E 2118–2119
vitamin K 2118
VMA1 gene 1565
v-myc oncogene 1484
Vogt, Marguerite 589
Vogt T F **1372–1373**
Volvox 309, 748
vomeronasal organ (VNO) 1368
v-onc oncogene 1251–1252
von Gierke disease **2120–2122**
clinical features 2122
Cori, Carl 2120
Cori, Gerty 2120
gluconeogenesis 2121, *2122f*
history 2120
liver glucose-6-phosphatase enzyme (G6PC) 2120–2122, *2121f*
treatment 2122
von Hippel–Lindau disease 1659–1660, **2122–2123**
von Hippel–Lindau gene
see tumor suppressor genes
von Recklinghausen disease 1307, 1770
von Willebrand disease **2123**
von Willebrand factor 919
v-raf oncogene 1484
v-sis oncogene 126
v-src oncogene 1371
V (variable) gene **2103**

W

Waardenburg syndrome 1427, 1468–1469
wac gene *443f*
Wade M J **894–899**
WAF1 gene **2127**
Wagner, Moritz 1861
WAGR syndrome 2138
Wahlund effect 1520, **2127–2132**
allele frequency 2127–2132, *2128t*, *2129t*
chi-square test 2131
Hard–Weinberg law 2127–2132
multiple alleles 2132
numerical examples *2128t*, *2129t*
subpopulations 2127–2132, *2128f*, *2129t*
Wahlund's formula 2130–2131
Wahlund, Sten Gösta William 1520
Wakimoto B T **1420–1422**
Waksman, Selman A. 1890
Waldeyer, W. 344
Waldor, Matthew 697
Wallace, Alfred Russel **2132–2133**
Bates, Henry Walter 2132
Contributions to the Theory of Natural Selection 2133
Darwin, Charles 2132
Geographical Distribution of Animals 2133
Island Life 2133
natural selection 514, 1291
theory of evolution 2132
Wallace effect 112
Wallace, B. 578
Wallace effect 112
Wang, Andrew 555
Warburg, Otto 235
Warramaba virgo 2097
Warren S T **2048–2053**
Wassarman P M **690–695**, **2165–2166**
Waterfield M D **626–628**
watermelon *1914t*
Watson–Crick model *see* DNA structure
Watson, James Dewey 540, 554, 815, 877, 1178, **2133–2135**
Bragg, Lawrence 2134
Cold Spring Harbor 2134, 2135
Crick, Francis 2134
Delbrück, Max 2133
DNA (deoxyribonucleic acid) 2134
The Double Helix 2135
Dulbecco, Renato 2133
Franklin, Rosalind 2134
Honest Jim 2135
Human Genome Project 2135
Luria, Salvador E. 2133
The Molecular Biology of the Gene 2135
Muller, Hermann Joseph 2133
Nobel Prize 2134
Schrödinger, Erwin 2133
Sonneborn, Tracy 2133
Wilkins, Maurice 2134
Wright, Sewall 2133
Watson, J.B. 209, 1298–1299
Watson syndrome 1308
Wayne R K **264–270**
wc1 gene 389
wc2 gene 389
W chromosome **2125**
Weatherall D J **878–881**, **1828–1831**, **1958–1961**
Wechsler adult intelligence scale (WAIS) 1045
Weeden N F **1470–1471**
Weinberg, Wilhelm 912, 1514
Weiner, Amy 958
Weinstock G M **135**, **1191–1196**
Weis A E **1533–1536**
Weissmann, August 63, 199, 516, 1128
Weissmann, Charles 1366
Weiss, Mary 359–360
Weldon, Raphael 742
Wellcome Trust 355
Wells R D **1061**
Werb Z **1419–1420**, **2070–2071**
Werner syndrome (WS) 22, *569–570t*, 571
Werner T **377–380**
Western blotting 334
West S A **1820–1821**
West, Steven 1757
wetA gene 110, *110f*
w^+ gene *1525f*, 1525–1526
wg (wingless) gene 427
Wheat Gene Catalog 2068

white collar genes 389
White, Gilbert 1865
Whitehead Institute 360
  genome site 1108
White J A **1340–1342**
White, John 254, 1309
White, O.E. 1337
WHO classification of leukemia 1770–1771, **2135–2136**
  see also FAB classification of leukemia; MIC and MIC-M classification of leukemia
whole organism cloning **2136–2138**
  benefits 2137
  Dolly, the sheep 2137
  gene reprogramming 2137
  genome constancy 2137
  human reproductive cloning 2137
  mammal 2137
  nuclear transplantation 2136
  procedures 2136–2137
  success rate 2137
  Xenopus 2136
Wichman, H. 1453
Wieczorek D J **412–413**
Wiener, A.S. 1713
Wieschaus, Eric F. 1095
WIG gene 715
Wilcox D E **129–133, 685–688,** 920, **934–936, 1824, 1835–1838, 2106–2107**
wild-type (WT) **2138**
Wiley E O **62–63, 85,** 384, **384–385, 964–969, 969–970, 1491, 1548, 1904–1905, 1905–1906**
Wilkie A O M **478–481**
Wilkins, Marc 1575
Wilkins, Maurice 540, 554, 2134
Willan, Robert 1581
Williams syndrome 911, 1097
Willis L B **54–56**
Wilms' tumor 209, 1427, 1818, **2138–2139**
  Beckwith–Wiedemann syndrome 1770, 2139
  Denys–Drash syndrome (DDS) 2138
  Frasier syndrome 2139
  imprinting, genomic 1000
  role of *WT1* gene 2139
  sarcomas 1770
  sex determination 1814
  U-to-C editing 1739
  WAGR syndrome 2138
  WT1 oncogene 2138
Wilson disease **2139–2140**
Wilson, Edmund Beecher 235, 590, **2140**
  The Cell in Development and Inheritance 2140
  The Physical Basis of Life 2140
  significance of the sex chromosome 2140
Wilson, Herbert 555
Winking H **1752**
Winkler, Hans 200
Winkler M E **218–224**
Wiskott-Aldrich syndrome **2140**
Wisteria 1083
WIT database 1201
Witkin, Evelyn 1854
Wnt4 gene 1815
wnt-7a glycoprotein 1100
Wnt pathway 308
wobble hypothesis 79, **2140–2141**
  base pairs 2141, *2141f*
  Crick, Francis 2140
wobble hypothesis (*continued*)
  tRNA 2140
Woese, Carl 91, 146, 1724, 1751
Wöhler, Friedrich 1067
Wolbachia 952, 1684, 1804
Wolffe, Alan 630
Wolffe A P **628–637, 948–952**
Wolffian ducts 924–925, 1024–1025, 1815, 1816, 1822
Wolf–Hirschorn syndrome 982–983
Wollman, Elie 679, 1123, 1131
Wollman, Elisabeth 1130
Wollman, Eugene 1130
Wollman, Jacob 1131
Wolstenholme J **1240–1242**
wolves 264, 265
Wong H Y **1805–1808**
Woo S L C **1447–1449**
World Conservation Union (IUCN) 458
WormBase 254
Wright–Fisher population model 145
Wright, Margaret 656
Wright, Sewall 278, 397, 490, 516, 995, 1054, **2141–2143**
  adaptive landscapes 4
  effective population number 602
  evolution 664
  F statistic 1521
  fitness 703
  fitness landscape 706
  fundamental theorem of natural selection 734
  genetic drift 832
  heritability 921
  inbreeding coefficient 2142
  mathematical theory of evolution 2141
  mouse genetic nomenclature 1337
  path analysis 2142
  population genetics 905, 1514, 1515, 2141
  shifting balance theory of evolution 5, 603, 640, 706, 832–833, 1825–1828, 2142
  Watson, James Dewey 2133
Wright's *F* statistic 1521
Wt1 gene 1814, 1818
WT1 oncogene *81t*, 1772, *1772t*, 1814, 1818, *2084–2085t*, 2138
wunschel mutation 1476
W (*white spotting*) locus 398, **2125–2127**
  chromosome 10 2125
  gain-of-function *2126f*, 2127
  gastrointestinal stromal tumors (GIST) 2127
  Kit receptor 2125, *2126f*
  loss of function 2126, *2126f*
  phenotypic abnormalities 2125
  piebald trait *2126f*, 2127
  specific locus test 1871
  splice variants 2126
  Sl (Steel) locus 1887, 2125
  tyrosine kinases 2125, *2126f*
wx (waxy) gene 2020
Wyllie A H **12–14**

X

X25 gene *2049t*
Xa1 gene 1395
Xa21 gene 1395
Xanthomonas 1556
Xanthomonas campestris 1480
X–autosome translocation 688
Xbra gene *1944f*
X chromosome **2145–2147**
  Duchenne muscular dystrophy gene *362f*
  evolutionary origins 2145–2147
  Genome Data Base 2145
  human X chromosome *2145f*, *2146f*
  human X-linked diseases
    Duchenne muscular dystrophy (DMD) 2147
    fragile X syndrome (FRAXA) 2147
    Klinefelter syndrome 2147
    Rett syndrome 2147
    Turner syndrome 2147
  isochromosome 1054
  linkage 354
  mouse X chromosome 284–286, *2146f*
  role in genetic studies 2145
  unequal crossing-over 339
X-chromosome inactivation **2148–2149**
  chromatin 343
  dosage compensation 2148
  epigenetics 634
  heterochromatin 927
  mechanisms 2148
  mule 1255
  tortoiseshell coloring **1970–1971**
  X-inactivation center (*XIC*) 2148, 2150
  see also Cattanach's translocation
X chromosome (mouse) 2005
XE7 gene 2157
xenology **2149**
  see also horizontal transfer
Xenopus laevis *2149f*, **2149–2150**
  cell degeneration 315
  chromosome structure 372
  cyclin-dependent kinases 500
  DNA cloning 545
  embryological research 2149–2150
  gene amplification 762
  histone genes 948
  model organisms 2149
  Ori sequences 1384, *1385f*
  rDNA amplification 1611–1612
  single-strand annealing 1838
  subcellular RNA localization 1893
  whole organism cloning 2136
Xenopus tropicalis 2150
xenotransplantation 661
x-Eomes gene *1944f*
XerC recombinases 351–353, *352f*, 1040
XerD recombinases 351–353, *352f*, 1040
xeroderma pigmentosum (XP) **558, 2150**
  Cleaver, James 566
  clinical features 567
  defective DNA repair 563–564
  DNA repair *569–570t*
  excision repair 566, 567, 674–675
  mutators 1286
  nucleotide excision repair (NER) genes 1285
  repair mechanisms 1666
  sunlight sensitivity 567
Xer recombinase 1040, *1040t*
Xer recombination 352–353, 353
x-ET gene *1944f*
XG gene 2157
XGPY gene 2159
X-inactivation center (*XIC*) 634, 2148, 2150
Xiphophorus helleri 1683t
Xiphophorus maculatus 1683t
xisA gene 802

Xis (excisionase) protein 121, *1040t*, 1041, 1435, 1843, *1844t*
xis gene 1859
XIST/xist gene **2150–2154**
aneuploidy 2152
discovery 2150–2151
expression 2151–2152
male meiosis 2153
mechanisms 2152–2153
TSIX gene 2151
X-inactivation 2151
Xist gene 579, 634, 2148
XIST RNA 2152–2153
XKRY gene 2160
X-linked dominant *2014t*
X-linked ichthyosis 993
X-linked recessive *2014t*
X-linked recessive hypophosphatemic rickets 1028–1029
X-linked recessive nephrocalcinosis 1028
XMAP215 protein 301
XPA gene 2150
XP(A-H) gene *2084–2085t*
XPA protein 1666
XPB gene 568, 2150
XPB protein 1666
XPC gene 568, 2150
XPC-HR23B protein 1666
XPD gene 568, 2150
XPD protein 1666
XPE gene 568, 2150
XPF gene 2150
XPF protein 1666
XP genes 674–675
XPG gene 568, 2150
XPG protein 1666
X-ray crystallography **2154**
XRN1 gene 1748
Xu M-Q **1565–1567**
x-VegT gene *1944f*
XX male syndrome 1023, 1823
XXY syndrome 1023
XY female syndrome 1822
xylB gene 977

Y

YACs (yeast artificial chromosomes) **95–96, 2160–2161**
cloning systems 95–96, 2161
disadvantages 2161
DNA cloning 547
Oryza sativa (rice) 1394–1395
Saccharomyces cerevisiae 1764
structure 95, *95f*, *2155f*, 2160–2161
uses 96, 2161
vectors 870
YAC libraries 96, 870, 2161
Yamamoto Y **539–540**
Yanagida, Mitsuhiro 1778
Yanofsky C **86, 530, 761, 2076–2078**
Yanofsky, Charles 57, 419, **2161–2162**
ycf gene 1461
Y chromosome 1952
Y chromosome (human) **2155–2160**
amplification 2156
centromere *322f*
deletion mapping 2157, *2158f*
evolution 2155
genetic mapping 2157–2160
heterochromatin 2156
isochromosome 1054
model 2156–2157, *2157f*
noncombining regions (NRY) 2155
pairing region 2155–2156, *2157f*
PAR1 2157
pseudoautosomal region (PAR) 2155–2156, *2157f*
renal cell cancer 1660
sequence content 2155–2156, *2157f*
sex determination 1816
Turner syndrome 2156
Yp NRY euchromatin 2157–2159
Yq NRY euchromatin 2159–2160
Y chromosome (mouse) *1253f*, 1253–1254
Yd2 gene 973
yeast 298
ant^{R} mutation 1223
cassette model 275–278
cell cycle *292f*
cell determination 275–277, *277f*
cell differentiation 276–277, *277f*, 1153–1154
cell growth 1153
cell signaling 276–277
cyclins 291–292
mating type locus 276, 1154
mating type switch 277–278, *278f*, **1153–1157**
mit mutation 1223
petite **1430**
petite mutation 1223
phenylalanine tRNA structure *1987f*
reverse genetics 1706–1707
syn mutation 1223
tRNA structure *1987f*
see also Saccharomyces cerevisiae
yeast artificial chromosome (YAC)
see YACs (yeast artificial chromosomes)
Yeast Genetic Stock Center 851
yeast plasmids **2162–2164**
cellular components 2162–2163
genetic markers 2163
2-μ plasmid 2162
recombinant plasmids 2162–2163
Saccharomyces cerevisiae 2162
uses 2163–2164
see also plasmids
yeast two-hybrid system **2164**
Yersinia 1556
Yersinia pestis 1558
Yes protein 1883
Y-linkage *2014t*
Yook K J **1309–1314**
York J Lyndal **625–626**
Young syndrome 1025

Z

Z chromosome **2165**
Z DNA **2165**
Zea mays 671, 1029
heterochronic mutation 930
homologous chromosomes 963
photosynthesis 1461–1462
zeins (*Zea mays*) 1784
Zea parviglumis 671
Z (Zebra) gene 643
zeins (*Zea mays*) 1784
Zeitschrift für induktive Abstammungs- und Vererbungslehre (Molecular and General Genetics) 201
zen2 (zerknüllt-related) gene *960f*
Zeng Z-B **1587–1593**
zen (zerknüllt) gene *960f*
Zepp 2025
zeta (ζ) globin gene 878
ZFIN (Zebrafish Information Network) 853
zf-tbx6 gene *1944f*
zf-T gene *1944f*
Zfy1 gene 1253, *1253f*
Zfy2 gene *1253f*, 1254
ZFY gene 2159
Zharkov D O **1669–1671**
Zhelonkina A **1741–1743**
Ziegler, D.R. 850
Zillig W **2114–2116**
Zillig, Wolfram 92
Zimmer, Elizabeth 436
Zimmer, Karl 522
Zimmermann F K **885–888**
zinc finger proteins
see DNA-binding proteins
Zinder, Norton 658, 1080, 1290, 1983
ZIP1 gene product 1912
ZNF217 gene 425
zona pellucida 623, 689
zoo blot **2165**
zoos 460
Zubay G **278–280**
zwichel gene 2048
ZWILLE gene 1475–1476
zwille/pinhead mutation 1475
zyg-1 gene 300
zyg-9 gene 301
ZYG-9 protein 301
Zygomycetes 1882
zygote 689, **2165–2166**
zygotic lethal gene **2166–2167**
Zyskind J W **1381–1387**